General, Organic, and Biochemistry

Sixth Edition

Katherine J. Denniston
Towson University

Joseph J. Topping
Towson University

Robert L. Caret
Towson University

McGraw-Hill
Higher Education

Boston   Burr Ridge, IL   Dubuque, IA   New York   San Francisco   St. Louis
Bangkok   Bogotá   Caracas   Kuala Lumpur   Lisbon   London   Madrid   Mexico City
Milan   Montreal   New Delhi   Santiago   Seoul   Singapore   Sydney   Taipei   Toronto

The McGraw-Hill Companies

McGraw-Hill
Higher Education

GENERAL, ORGANIC, AND BIOCHEMISTRY, SIXTH EDITION

Published by McGraw-Hill, a business unit of The McGraw-Hill Companies, Inc., 1221 Avenue of the Americas, New York, NY 10020. Copyright © 2008 by The McGraw-Hill Companies, Inc. All rights reserved. Previous editions © 2007, 2004, 2001, 1997 and 1993. No part of this publication may be reproduced or distributed in any form or by any means, or stored in a database or retrieval system, without the prior written consent of The McGraw-Hill Companies, Inc., including, but not limited to, in any network or other electronic storage or transmission, or broadcast for distance learning.

Some ancillaries, including electronic and print components, may not be available to customers outside the United States.

This book is printed on recycled, acid-free paper containing 10% postconsumer waste.

2 3 4 5 6 7 8 9 0 DOW/DOW 0 9 8

ISBN 978–0–07–351110–8
MHID 0–07–351110–2

Publisher: *Thomas D. Timp*
Senior Sponsoring Editor: *Tamara Hodge*
Developmental Editor: *Jodi Rhomberg*
Marketing Manager: *Todd Turner*
Senior Project Manager: *Gloria G. Schiesl*
Lead Production Supervisor: *Sandy Ludovissy*
Lead Media Project Manager: *Judi David*
Executive Media Producer: *Linda Meehan Avenarius*
Lead Producer: *Daryl Bruflodt*
Senior Coordinator of Freelance Design: *Michelle D. Whitaker*
Cover/Interior Designer: *Elise Landson*
Senior Photo Research Coordinator: *Lori Hancock*
Photo Research: *Connie Mueller*
Supplement Producer: *Melissa Leick*
Compositor: *Laserwords Private Limited*
Typeface: *10/12 Palatino*
Printer: *R. R. Donnelley Willard, OH*
Cover photo: © *age fotostock / SuperStock; Cover art: Imagineering Media Services, Inc.*

The credits section for this book begins on page C-1 and is considered an extension of the copyright page.

Library of Congress Cataloging-in-Publication Data

Denniston, K. J. (Katherine J.)
    General, organic, and biochemistry/Katherine J. Denniston, Joseph J. Topping, Robert L. Caret.–6th ed.
        p. cm.
    Includes index.
    ISBN 978-0-07-351110-8 — ISBN 0-07-351110-2 (hard copy : alk. paper)
    1. Chemistry, Organic–Textbooks. 2. Biochemistry–Textbooks. I. Topping, Joseph J. II. Caret, Robert L., 1947- III. Title.
QD253.2.D46 2008
547—dc22   2007030779

www.mhhe.com

List of the Elements with their Symbols and Atomic Masses*

Element	Symbol	Atomic Number	Atomic Mass†	Element	Symbol	Atomic Number	Atomic Mass†
Actinium	Ac	89	(227)	Mendelevium	Md	101	(256)
Aluminum	Al	13	26.98	Mercury	Hg	80	200.6
Americium	Am	95	(243)	Molybdenum	Mo	42	95.94
Antimony	Sb	51	121.8	Neodymium	Nd	60	144.2
Argon	Ar	18	39.95	Neon	Ne	10	20.18
Arsenic	As	33	74.92	Neptumium	Np	93	(237)
Astatine	At	85	(210)	Nickel	Ni	28	58.69
Barium	Ba	56	137.3	Niobium	Nb	41	92.91
Berkelium	Bk	97	(247)	Nitrogen	N	7	14.01
Beryllium	Be	4	9.012	Nobelium	No	102	(253)
Bismuth	Bi	83	209.0	Osmium	Os	76	190.2
Bohrium	Bh	107	(262)	Oxygen	O	8	16.00
Boron	B	5	10.81	Palladium	Pd	46	106.4
Bromine	Br	35	79.90	Phosphorus	P	15	30.97
Cadmium	Cd	48	112.4	Platinum	Pt	78	195.1
Calcium	Ca	20	40.08	Plutonium	Pu	94	(242)
Californium	Cf	98	(249)	Polonium	Po	84	(210)
Carbon	C	6	12.01	Potassium	K	19	39.10
Cerium	Ce	58	140.1	Praseodymium	Pr	59	140.9
Cesium	Cs	55	132.9	Promethium	Pm	61	(147)
Chlorine	Cl	17	35.45	Protactinium	Pa	91	(231)
Chromium	Cr	24	52.00	Radium	Ra	88	(226)
Cobalt	Co	27	58.93	Radon	Rn	86	(222)
Copper	Cu	29	63.55	Rhenium	Re	75	186.2
Curium	Cm	96	(247)	Rhodium	Rh	45	102.9
Darmstadtium	Ds	110	(281)	Roentgenium	Rg	111	272
Dysprosium	Dy	66	162.5	Rubidium	Rb	37	85.47
Einsteinium	Es	99	(254)	Ruthenium	Ru	44	101.1
Erbium	Er	68	167.3	Rutherfordium	Rf	104	(257)
Europium	Eu	63	152.0	Samarium	Sm	62	150.4
Fermium	Fm	100	(253)	Scandium	Sc	21	44.96
Fluorine	F	9	19.00	Seaborgium	Sg	106	(263)
Francium	Fr	87	(223)	Selenium	Se	34	78.96
Gadolinium	Gd	64	157.3	Silicon	Si	14	28.09
Gallium	Ga	31	69.72	Silver	Ag	47	107.9
Germanium	Ge	32	72.59	Sodium	Na	11	22.99
Gold	Au	79	197.0	Strontium	Sr	38	87.62
Hafnium	Hf	72	178.5	Sulfur	S	16	32.07
Hahnium	Ha	105	(260)	Tantalum	Ta	73	180.9
Hassium	Hs	108	(265)	Technetium	Tc	43	(99)
Helium	He	2	4.003	Tellurium	Te	52	127.6
Holmium	Ho	67	164.9	Terbium	Tb	65	158.9
Hydrogen	H	1	1.008	Thallium	Tl	81	204.4
Indium	In	49	114.8	Thorium	Th	90	232.0
Iodine	I	53	126.9	Thulium	Tm	69	168.9
Iridium	Ir	77	192.2	Tin	Sn	50	118.7
Iron	Fe	26	55.85	Titanium	Ti	22	47.88
Krypton	Kr	36	83.80	Tungsten	W	74	183.9
Lanthanum	La	57	138.9	Uranium	U	92	238.0
Lawrencium	Lr	103	(257)	Vanadium	V	23	50.94
Lead	Pb	82	207.2	Xenon	Xe	54	131.3
Lithium	Li	3	6.941	Ytterbium	Yb	70	173.0
Lutetium	Lu	71	175.0	Yttrium	Y	39	88.91
Magnesium	Mg	12	24.31	Zinc	Zn	30	65.39
Manganese	Mn	25	54.94	Zirconium	Zr	40	91.22
Meitnerium	Mt	109	(266)				

*All atomic masses have four significant figures. These values are recommended by the Committee on Teaching of Chemistry, International Union of Pure and Applied Chemistry.

†Masses of the longest-lived isotope for radioactive elements are given in parentheses.

Brief Contents

iii

Contents

ORGANIC CHEMISTRY

Chemistry Connections and Perspectives

A Medical *Perspective*

An Environmental *Perspective*

Preface

The sixth edition of *General, Organic, and Biochemistry*, like our earlier editions, has been designed to help undergraduate majors in health-related fields understand key concepts and appreciate the significant connections between chemistry, health, and the treatment of disease. We have tried to strike a balance between theoretical and practical chemistry, while emphasizing material that is unique to health-related studies. We have written at a level intended for students whose professional goals do not include a mastery of chemistry, but for whom an understanding of the principles and practice of chemistry is a necessity.

While we have stressed the importance of chemistry to the health-related professions, this book was written for all students that need a one- or two-semester introduction to chemistry. Our focus on the relationship between chemistry, the environment, medicine, and the function of the human body is an approach that can engage students in a variety of majors. We have integrated the individual disciplines of inorganic, organic, and biochemistry to emphasize their interrelatedness rather than their differences. This approach provides a sound foundation in chemistry and teaches students that life is not a magical property, but rather depends on a complex sequence of chemical reactions that obey the scientific laws.

Key Features of the Sixth Edition

In preparing the sixth edition, we have been guided by the collective wisdom of over thirty reviewers who are experts in one or more of the three subdisciplines covered in the book and who represent a diversity of experience, including community colleges and four-year colleges and universities. We have retained the core approach of our successful earlier editions, updated material where necessary, and expanded or removed material consistent with retention of the original focus and mission of the book. Throughout the project, we have been careful to ensure that the final product is as student-oriented and readable as its predecessors.

New Features

- All examples have been enhanced to include a relevant practice problem within the context of the example. Further practice problems are also noted within the example so that students can test their mastery of information and build self-confidence.
- Boxed topics have been enhanced with photos and figures intended to motivate the student to go beyond what is written and/or solidify the relationship between the boxed topic and the chapter material.
- Icons that identify animations and appendices have been added to the margins. These easily recognizable icons provide quick reference to supplemental online materials.
- A new Scientific Calculator Appendix has been added to the ARIS site that accompanies this text. Students often struggle with how to properly use the scientific calculator. This step-by-step appendix walks students through how to use the calculator and points out common mistakes. This appendix offers students yet another tool to help clarify chemistry.
- Nearly 100 new photos have been added to this edition, many of which have been positioned in the margin with inquiry-based captions to help relate chemistry to everyday life.
- The ARIS website and other media supplements, as described later in this Preface, have been enhanced.

We designed the sixth edition to promote student learning and facilitate teaching. It is important to engage students, to appeal to visual learners, and to provide a variety of pedagogical tools to help them organize and summarize information. We have utilized a variety of strategies to accomplish our goals.

Engaging Students

Students learn better when they can see a clear relationship between the subject material they are studying and real life. We wrote the text to help students make connections between the principles of chemistry and their previous life experiences and/or their future professional experiences. Our strategy to accomplish this integration includes the following:

- **Boxed Readings—"Chemistry Connection":** Introductory vignettes allow the student to see the significance of chemistry in their daily lives and in their future professions.
- **Boxed Perspectives:** These short essays present real-world situations that involve one or more topics students will encounter in the chapter. The "Medical Perspective" boxes relate chemistry to a health concern or a diagnostic application. The "Environmental Perspective" boxes deal with environmental issues, including the impact of chemistry on the ecosystem and how environmental changes affect human health. "Human Perspective" boxes delve into chemistry and society and include such topics as gender issues in science and historical viewpoints. Each box covers a pertinent topic of interest to students and society. Self-tanning lotions, sugar substitutes, as well as the most recent strategies for the treatment of HIV/AIDS, are just a few examples of boxed perspective topics.

Learning Tools

In designing the original learning system we asked ourselves the question: "If we were students, what would help us organize and understand the material covered in this chapter?" With valuable suggestions from our reviewers, we have made some modifications to improve the learning system. However, with the blessings of those reviewers, we have retained all of the elements of the system which have been shown to support student learning:

- **Learning Goals:** A set of chapter objectives at the beginning of each chapter previews concepts that will be covered in the chapter. Icons locate text material that supports the learning goals.

- **Detailed Chapter Outline:** A listing of topic headings is provided for each chapter. Topics are arranged in outline form to help students organize the material in their own minds.
- **Chapter Cross-References:** To help students locate the pertinent background material, references to previous chapters, sections, and perspectives are noted in the margins of the text. These marginal cross references also alert students to upcoming topics related to the information currently being studied.

A Medical Perspective

Enzymes and Acute Myocardial Infarction

A patient is brought into the emergency room with acute, squeezing chest pains; shallow, irregular breathing; and pale, clammy skin. The immediate diagnosis is myocardial infarction, a heart attack. The first thoughts of the attending nurses and physicians concern the series of treatments and procedures that will save the patient's life. It is a short time later, when the patient's condition has stabilized, that the doctor begins to consider the battery of enzyme assays that will confirm the diagnosis.

Acute myocardial infarction (AMI) occurs when the blood supply to the heart muscle is blocked for an extended time. If this lack of blood supply, called *ischemia*, is prolonged, the myocardium suffers irreversible cell damage and muscle death, or infarction. When this happens, the concentration of cardiac enzymes in the blood rises dramatically as the dead cells release their contents into the bloodstream.

Three cardiac biomarkers have become the primary tools used to assess myocardial disease and suspected AMI. These are myoglobin, creatine kinase-MB (CK-MB), and cardiac troponin I. Of these three, only troponin is cardiac specific. In fact, it is so reliable that the American College of Cardiology has stated that any elevation of troponin is "abnormal and represents cardiac injury."

Myoglobin is the smallest of these three proteins and diffuses most rapidly through the vascular system. Thus, it is the first cardiac biomarker to appear, becoming elevated as early as 30 minutes after onset of chest pain. Myoglobin has another benefit in following a myocardial infarction. It is rapidly cleared from the body by the kidneys, returning to normal levels within 16 to 36 hours after a heart attack. If the physician sees this decline in myoglobin levels, followed by a subsequent rise, it is an indication that the patient has had a second myocardial infarction.

Creatine kinase-MB is one of the most important cardiac biomarkers, even though it is found primarily in muscle and brain. Levels typically rise 3 to 8 hours after chest pains begin. Within another 48 to 72 hours, the CK-MB levels return to normal.

As a result, like myoglobin, CK-MB can also be used to diagnose a second AMI.

The physician also has enzymes available to treat a heart attack patient. Most AMIs are the result of a *thrombus*, or clot, within a coronary blood vessel. The clot restricts blood flow to the heart muscle. One technique that shows promise for treatment following a coronary thrombosis, a heart attack caused by the formation of a clot, is destruction of the clot by intravenous or intracoronary injection of an enzyme called *streptokinase*. This enzyme, formerly purified from the pathogenic bacterium *Streptococcus pyogenes* but now available through recombinant DNA techniques, catalyzes the production of the proteolytic enzyme plasmin from its proenzyme, plasminogen. Plasmin can degrade a fibrin clot into subunits. This has the effect of dissolving the clot that is responsible for restricted blood flow to the heart, but there is an additional protective function as well. The subunits produced by plasmin degradation of fibrin clots are able to inhibit further clot formation by inhibiting thrombin.

Recombinant DNA technology has provided medical science with yet another, perhaps more promising, clot-dissolving enzyme. *Tissue-type plasminogen activator (TPA)* is a proteolytic enzyme that occurs naturally in the body as a part of the anticlotting mechanisms. TPA converts the proenzyme, plasminogen, into the active enzyme, plasmin. Injection of TPA within two hours of the initial chest pain can significantly improve the circulation to the heart and greatly improve the patient's chances of survival.

For Further Understanding

Why is myoglobin an effective biomarker to follow the status of the patient when using a thrombolytic agent such as TPA or streptokinase?

Aspirin has also been suggested as a treatment to enhance a patient's probability of surviving a heart attack. What mechanism can you devise to explain this suggestion?

19.12 Uses of Enzymes in Medicine

Analysis of blood serum for levels (concentrations) of certain enzymes can provide a wealth of information about a patient's medical condition. Often, such tests are used to confirm a preliminary diagnosis based on the disease symptoms or clinical picture.

For example, when a heart attack occurs, a lack of blood supplied to the heart muscle causes some of the heart muscle cells to die. These cells release their contents, including their enzymes, into the bloodstream. Simple tests can be done to

12 LEARNING GOAL
Provide examples of medical uses of enzymes.

---

242   **Chapter 7** Energy, Rate, and Equilibrium

Section 6.3 describes molar concentration.

of seconds, minutes, hours, or even months or years, depending on the rate of the reaction. The reaction mixture is then analyzed to determine the molar concentration of each of the products and reactants. These concentrations are substituted in the equilibrium-constant expression and the equilibrium constant is calculated. The following example illustrates this process.

EXAMPLE 7.8   **Calculating an Equilibrium Constant**

9 LEARNING GOAL
Write equilibrium-constant expressions and use these expressions to calculate equilibrium constants.

Hydrogen iodide is placed in a sealed container and allowed to come to equilibrium. The equilibrium reaction is:

$$2HI(g) \rightleftharpoons H_2(g) + I_2(g)$$

and the equilibrium concentrations are:

- **Summary of Reactions:** In the organic chemistry chapters, each major reaction type is highlighted on a green background. Major chemical equations are summarized at the end of the chapter, facilitating review.
- **Chapter Summary:** Each major topic of the chapter is briefly reviewed in paragraph form in the end of chapter summary. These summaries serve as a mini-study guide, covering the major concepts in the chapter.
- **Key Terms:** Key terms are printed in boldface in the text, listed at the end of the chapter, and defined immediately. Each end-of-chapter key term is accompanied by a section number for rapid reference.
- **Glossary of Key Terms:** In addition to being listed at the end of the chapter, each key term from the text is also defined in the alphabetical glossary at the end of the book.

Detailed List of Changes

Changes and updates are evident in every chapter of this sixth edition. Major changes to individual chapters include:

- **Chapter 1:** This chapter has been reorganized; significant figures and scientific notation are now discussed before units and unit conversion. A new perspective on body-mass index has been added, and the specific gravity perspective has been broadened to include both medical and industrial applications (winemaking is discussed).
- **Chapter 3:** A new section has been added to introduce the concept of isomers as a part of a larger discussion of covalent bonding. The existence of isomers is used to help explain why there are so many hydrocarbons, partly explaining the complexity of petroleum.
- **Chapter 6:** The section about osmosis and osmotic pressure has been rewritten to include the discussions in Chapters 6 and 17 of the fifth edition. This has the effect of improving the information flow in both chapters and provides the student full coverage of the topic in one place.
- **Chapter 7:** A section describing the use of equilibrium constants, along with an example and its practice problems, has been added to the discussion of equilibrium in this chapter.
- **Chapter 8:** The section "Reactions Between Acids and Bases" has been reorganized and expanded. Steps in a titration are now summarized in a table. The worked example is more detailed, various acid-base reactions are supported with animations, and ten new questions have been added to the end-of-chapter problems. For convenience, they now have their own section in the end-of-chapter problems.

- **Chapter 9:** The section on radiocarbon dating has been integrated into Section 9.3, immediately following the discussion of half-life. This is a nice transition, owing to the fact that the concept of half-life is essential to understanding radiocarbon dating, and the dating material is a good, practical example of an application of natural radioactivity.
- **Chapter 10:** Two new examples have been added to Chapter 10. Example 10.1, "Using Different Types of Formulas to Represent Organic Compounds," now uses the example of molecules found in gasoline and highlighted in "An Environmental Perspective: The Petroleum Industry and Gasoline Production," which is also found in this chapter. Example 10.2, "Naming Substituted Alkanes Using the I.U.P.A.C. System" uses examples of molecules relevant to gasoline, refrigerants, and the tsetse fly pheromone. Chapter 10 also contains a new table that shows students how to name alkanes longer than 10 carbons.

Summary of Reactions   349

Summary of Reactions

Reactions of Alkanes

Combustion:

$$C_nH_{2n+2} + O_2 \longrightarrow CO_2 + H_2O + \text{heat energy}$$
Alkane   Oxygen   Carbon   Water
Dioxide

Halogenation:

$$R-\overset{\overset{\textstyle H}{|}}{\underset{\underset{\textstyle H}{|}}{C}}-H + X_2 \xrightarrow{\text{light or heat}} R-\overset{\overset{\textstyle H}{|}}{\underset{\underset{\textstyle H}{|}}{C}}-X + H-X$$
Alkane   Halogen   Alkyl   Hydrogen
halide   halide

SUMMARY

10.1 The Chemistry of Carbon
The modern science of organic chemistry began with Wöhler's synthesis of urea in 1828. At that time, people believed that it was impossible to synthesize an organic molecule outside of a living system. We now define organic chemistry as the study of carbon-containing compounds. The differences between the ionic bond, which is characteristic of many inorganic substances, and the covalent bond in organic compounds are responsible for the great contrast in properties and reactivity between organic and inorganic compounds. All organic compounds are classified as either *hydrocarbons* or *substituted hydrocarbons*. In substituted hydrocarbons a hydrogen atom is replaced by a functional group. A *functional group* is an atom or group of atoms arranged in a particular way that imparts specific chemical or physical properties to a molecule. The major families of organic molecules are defined by the specific functional groups that they contain.

10.2 Alkanes
The *alkanes* are *saturated hydrocarbons*, that is, hydrocarbons that have only carbon and hydrogen atoms that are bonded together by carbon-carbon and carbon-hydrogen single bonds. They have the general molecular formula C_nH_{2n+2} and are nonpolar, water-insoluble compounds with low melting and boiling points. In the *I.U.P.A.C. Nomenclature System* the alkanes are named by determining the number of carbon atoms in the parent compound and numbering the carbon chain to provide the lowest possible number for all substituents. The substituent names and numbers are used as prefixes before the name of the parent compound.

Constitutional or structural isomers are molecules that have the same molecular formula but different structures. They have different physical and chemical properties because the atoms are bonded to one another in different patterns.

10.3 Cycloalkanes
Cycloalkanes are a family of organic molecules having C—C single bonds in a ring structure. They are named by adding the prefix *cyclo-* to the name of the alkane parent compound.

A *cis-trans isomer* is a type of stereoisomer. *Stereoisomers* are molecules that have the same structural formula and bonding pattern but different arrangements of atoms in space. A cycloalkane is in the *cis* configuration if two substituents are on the same side of the ring (either both above or both below). A cycloalkane is in the *trans* configuration when one substituent is above the ring and the other is below the ring. The *cis-trans* isomers are not interconvertible.

10.4 Conformations of Alkanes and Cycloalkanes
As a result of *free rotation* around carbon-carbon single bonds, infinitely many *conformations* or *conformers* exist for any alkane.

Limited rotation around the carbon-carbon single bonds of cycloalkanes also results in a variety of conformations of cycloalkanes. In cyclohexane the *chair conformation* is the most energetically favored. Another conformation is the *boat conformation*.

10.5 Reactions of Alkanes and Cycloalkanes
Alkanes can participate in *combustion* reactions. In complete combustion reactions they are oxidized to produce carbon dioxide, water, and heat energy. They can also undergo *halogenation* reactions to produce *alkyl halides*.

KEY TERMS

aliphatic hydrocarbon (10.1)
alkane (10.2)
alkyl group (10.2)
alkyl halide (10.5)
aromatic hydrocarbon (10.1)
axial atom (10.4)
boat conformation (10.4)
chair conformation (10.4)
cis-trans isomers (10.3)
combustion (10.5)
condensed formula (10.2)
conformations (10.4)
conformers (10.4)
constitutional isomers (10.2)
cycloalkane (10.3)
equatorial atom (10.4)
functional group (10.1)
geometric isomers (10.3)

halogenation (10.5)
hydrocarbon (10.1)
I.U.P.A.C. Nomenclature System (10.2)
line formula (10.2)
molecular formula (10.2)
parent compound (10.2)
primary (1°) carbon (10.2)
quaternary (4°) carbon (10.2)
saturated hydrocarbon (10.1)
secondary (2°) carbon (10.2)
stereoisomers (10.3)
structural formula (10.2)
structural isomer (10.2)
substituted hydrocarbon (10.1)
substitution reaction (10.5)
tertiary (3°) carbon (10.2)
unsaturated hydrocarbon (10.1)

10-31

- **Chapter 11:** Updates have been made to two examples in Chapter 11. Example 11.1, "Naming Alkenes and Alkynes Using I.U.P.A.C. Nomenclature," has been made more relevant by including the biological molecule isoprene. Example 11.6, "Writing Equations for the Hydrogenation of Alkenes," now includes the biological molecule linoleic acid and emphasizes the change from a liquid oil to a solid fat in the process. This is further applied to the production of margarine.
- **Chapter 12:** Example 12.1, "Using I.U.P.A.C. Nomenclature to Name an Alcohol," has been modified to use the example of a pheromone molecule.
- **Chapter 13:** Molecules involved in the taste of cheddar cheese and blue cheese have been added to Example 13.1, "Using the I.U.P.A.C. Nomenclature System to Name Aldehydes" and Example 13.3, "Using the I.U.P.A.C. Nomenclature System to Name Ketones."
- **Chapter 14:** The section "Some Important Carboxylic Acids" has been substantially enhanced. In addition to this new material, two examples have been updated. Example 14.7, "Naming Esters Using the I.U.P.A.C. and Common Nomenclature Systems," now uses an ester responsible for the flavor of pineapple; and Example 14.1, "Using the I.U.P.A.C. Nomenclature System to Name a Carboxylic Acid," now uses an example related to the synthesis of a biodegradable plastic called Biopol, which is featured in "An Environmental Perspective: Garbage Bags from Potato Peels" in this chapter.
- **Chapter 15:** A completely new example has been added to this chapter, Example 15.4, "Naming Amides Using Common and I.U.P.A.C. Nomenclature Systems." Example 15.3, "Writing the Systematic Name for an Amine," has been modified to use the cockroach pheromone, *N*-methylmethanamine, as the featured molecule.
- **Chapter 17:** A new section on omega-3 fatty acids has been added. In addition, a forensic science angle has been added to the Human Perspective: Mummies Made of Soap." A number of in-chapter and end-of-chapter problems that focus on reactions involving fatty acids have been added. The information on diffusion, osmosis, and active transport have been removed from this chapter. Diffusion and osmosis were moved to an earlier chapter where their coverage is more appropriate. Active transport, considered by many reviewers to be a biology topic rather than a chemistry topic, was removed in response to their recommendations.
- **Chapter 18:** The new information addressing the use of collagen in cosmetic procedures and clinical applications involving collagen has been added to the Human Perspective box on collagen.

- **Chapter 19:** This chapter has a completely updated Medical Perspectives box that focuses on enzymes and acute myocardial infarction. Additionally, two sections of this chapter ("The Transition State and Product Formation" and "Environmental Effects") have been significantly reworked to facilitate better student understanding.
- **Chapter 20:** Information on abnormalities of chromosome numbers, such as Down Syndrome, has been added to this chapter.
- **Chapter 23:** Two boxes in this chapter have been significantly updated. "Chemistry Connection: Obesity: A Genetic Disorder?" now includes information on the hormones ghrelin and obestatin, and "A Human Perspective: Losing Those Unwanted Pounds of Adipose Tissue" includes information about the development of an antiobesity drug and current surgical procedures for weight loss.

The Art Program

Today's students are much more visually oriented than any previous generation. Television and the computer represent alternate modes of learning. We have built upon this observation through the use of color, figures, and three-dimensional computer-generated models. This art program enhances the readability of the text and provides alternative pathways to learning.

Dynamic Illustrations

Each chapter is amply illustrated using figures, tables, and chemical formulas. All of these illustrations are carefully annotated for clarity. To help students better understand difficult concepts, there are approximately 350 illustrations and 250 photos in the sixth edition.

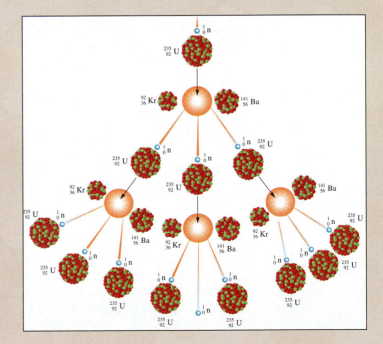

Color-Coding Scheme

We have color-coded reactions so that chemical groups being added or removed in a reaction can be quickly recognized.

- Red print is used in chemical equations or formulas to draw the reader's eye to key elements or properties in a reaction or structure.

Aldehydes and ketones can be distinguished on the basis of differences in their reactivity. The most common laboratory test for aldehydes is the **Tollens' test.** When exposed to the Tollens' reagent, a basic solution of $Ag(NH_3)_2^+$, an aldehyde undergoes oxidation. The silver ion (Ag^+) is reduced to silver metal (Ag^0) as the aldehyde is oxidized to a carboxylic acid anion.

Aldehyde	Silver ammonia complex— Tollens' reagent		Carboxylate anion	Silver metal mirror

Silver metal precipitates from solution and coats the flask, producing a smooth silver mirror, as seen in Figure 13.4. The test is therefore often called the Tollens' silver mirror test. The commercial manufacture of silver mirrors uses a similar process. Ketones cannot be oxidized to carboxylic acids and do not react with the Tollens' reagent.

- Blue print is used when additional features must be highlighted.
- Green background screens denote generalized chemical and mathematical equations.

Neutralization

The reaction of an acid with a base to produce a salt and water is referred to as **neutralization.** In the strictest sense, neutralization requires equal numbers of moles of H_3O^+ and OH^- to produce a neutral solution (no excess acid or base). Consider the reaction of a solution of hydrochloric acid and sodium hydroxide:

$$HCl(aq) + NaOH(aq) \longrightarrow NaCl(aq) + H_2O(l)$$
$$\text{Acid} \qquad \text{Base} \qquad\qquad \text{Salt} \qquad \text{Water}$$

In the organic chemistry chapters, the Summary of Reactions at the end of the chapter is also highlighted with a green background screen for ease of recognition.

- Yellow background illustrates energy, either as energy stored in electrons or groups of atoms, in the general and biochemistry sections of the text. In the organic chemistry section of the text, yellow background screens also reveal the parent chain of an organic compound.

Ionization Energy

The energy required to remove an electron from an isolated atom is the **ionization energy.** The process for sodium is represented as follows:

$$\text{ionization energy} + Na \longrightarrow Na^+ + e^-$$

- Certain situations make it necessary to adopt a unique color convention tailored to the material in a particular chapter. For example, in Chapter 18, the structures of amino acids require three colors to draw attention to key features of these molecules. For consistency, red is used to denote the acid portion of an amino acid and blue is used to denote the basic portion of an amino acid. Green print is used to denote the R groups, and a yellow background screen directs the eye to the α-carbon.

Figure 18.1

General structure of an α-amino acid. All amino acids isolated from proteins, with the exception of proline, have this general structure.

Computer-Generated Models

The students' ability to understand the geometry and three-dimensional structure of molecules is essential to the understanding of organic and biochemical reactions. Computer-generated models are used throughout the text because they are both accurate and easily visualized.

Problem Solving and Critical Thinking

Perhaps the best preparation for a successful and productive career is the development of problem-solving and critical thinking skills. To this end, we created a variety of problems that require recall, fundamental calculations, and complex reasoning. In this edition, we have used suggestions from our reviewers, as well as our own experience, to enhance the problem sets to include more practice problems for difficult concepts and further integration of the subject areas.

- **In-Chapter Examples, Solutions, and Problems:** Each chapter includes a number of examples that show the student, step-by-step, how to properly reach the correct solution to model problems. Each example contains a practice problem question as well as a referral to further practice questions. These questions allow students to test their mastery of information and to build self-confidence.

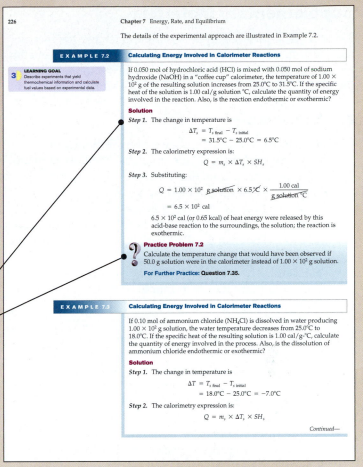

- **In-Chapter and End-of-Chapter Problems:** We have created a wide variety of paired concept problems. The answers to the odd-numbered questions are found in the back of the book as reinforcement for students as they develop problem-solving skills. However, the students must then be able to apply the same principles to the related even-numbered problems.
- **Critical Thinking Problems:** Each chapter includes a set of critical thinking problems. These problems are intended to challenge students to integrate concepts to solve more complex problems. They make a perfect complement to the classroom lecture because they provide an opportunity for in-class discussion of complex problems dealing with daily life and the health care sciences.

Over the course of the last five editions, hundreds of reviewers have shared their knowledge and wisdom with us, as well as the reaction of their students to elements of this book. Their contributions, as well as our own continuing experience in the area of teaching and learning science, have resulted in a text that we are confident will provide a strong foundation in chemistry, while enhancing the learning experience of the students.

Supplementary Materials

This text is supported by a complete package for instructors and students. Several print and media supplements have been prepared to accompany the text and make learning as meaningful and up-to-date as possible.

For the Instructor

- **Instructor's Manual:** Written by the authors and developed by Danáe Quirk Dorr, this ancillary contains suggestions for organizing lectures, instructional objectives, perspectives on boxed readings from the text, answers to the even-numbered problems from the text, a list of each chapter's key problems and concepts, and more. The Instructor's Manual is available through the ARIS website for this text.
- **Test Bank:** The test bank offers questions that can be used for homework assignments or the preparation of exams.
- **EZ Test:** Computerized Classroom Management System. This program can be utilized to quickly create customized exams. It allows instructors to sort questions by format or level of difficulty, edit existing questions or add new ones, and scramble questions and answer keys for multiple versions of the same test.
- *A Laboratory Manual for General, Organic, and Biochemistry, Sixth* Edition, by Charles H. Henrickson, Larry C. Byrd, and Norman W. Hunter of Western Kentucky University, offers clear and concise laboratory experiments that reinforce students' understanding of concepts. Prelaboratory exercises, questions, and report sheets are coordinated with each experiment to ensure students' active involvement and comprehension. A new student tutorial on graphing with Excel® has been added to this edition.
- **Laboratory Instructor's Manual:** Written by Charles H. Henrickson, Larry C. Byrd, and Norman W. Hunter of Western Kentucky University, this helpful guide contains hints that the authors have learned over the years to ensure students' success in the laboratory. This Resource Guide is available through the ARIS course website for this text.
- **ARIS:** McGraw-Hill's *General, Organic, and Biochemistry* ARIS website (Assessment, Review, and Instruction System) makes homework meaningful—and manageable—for instructors and students. Instructors can assign and grade chapter-specific homework within the industries most robust and versatile homework management system. They can also create and share course materials and assignments with colleagues with a few clicks of the mouse. ARIS

allows instructors to edit questions, import their own content, and create announcements and due dates for assignments. Homework questions can be imported into a variety of course management systems such as WebCT, Blackboard, and WebAssign. These course cartridges also provide online testing and powerful student tracking features. Chapter-specific study tools are available on the ARIS website including:

—Self-quizzes
—Animations
—PowerPoint® presentations
—Key terms
—Appendix materials

Go to www.mhhe.com/aris to learn more about ARIS.
Instructors: To access ARIS, request registration information from your McGraw-Hill sales representative.

- **McGraw-Hill Presentation Center:** Build instructional material wherever, whenever, and however you want! McGraw-Hill Presentation Center is an online digital library containing assets such as photos, artwork, and other media types that can be used to create customized lectures, visually enhanced tests and quizzes, compelling course websites, or attractive printed support materials. The McGraw-Hill Presentation Center Library includes thousands of assets from many McGraw-Hill titles. This ever-growing resource gives instructors the power to utilize assets specific to an adopted textbook as well as content from all other books in the library. The Presentation Center can be accessed from the instructor side of your textbook's ARIS website, and the Presentation Center's dynamic search engine allows you to explore by discipline, course, textbook chapter, asset type, or keyword. Simply browse, select, and download the files you need to build engaging course materials. All assets are copyrighted by McGraw-Hill Higher Education but can be used by instructors for classroom purposes.

- **Over 300 animations are available through the ARIS site:** Many animations are linked to appropriate

  sections of the textbook using the [globe icon] icon. The animations supplement the textbook material in much the same way as do instructor demonstrations. However, for the students, they are only a few mouse-clicks away, anytime, day or night. Realizing that students are visual learners and quite computer literate, the animations add another dimension to learning; they bring a greater degree of reality to the written word.

For the Student

- **Student Study Guide/Solutions Manual:** A separate Student Study Guide/Solutions Manual, prepared by Danáe Quirk Dorr and the authors of this text, is available. It contains the answers and complete solutions for the odd-numbered problems. It also offers students a variety of exercises and keys for testing their comprehension of basic, as well as difficult, concepts.
- **Schaum's Outline of General, Organic, and Biological Chemistry:** Written by George Odian and Ira Blei, this supplement provides students with over 1400 solved problems with complete solutions. It also teaches effective problem-solving techniques.
- **ARIS:** McGraw-Hill's Assessment, Review, and Instruction System for *General, Organic, and Biochemistry* is available to students and instructors using this text. The website offers quizzes, key definitions, a review of mathematics applied to problem solving, important tables, definitions, and more. This website can be found at www.mhhe.com/aris.

- **Quantum Intelligent Tutors:** Personal tutoring and homework help. It's just like working with a human tutor!

- Real-time personal tutoring help
- Step-by-step feedback and detailed instruction based on your *own* work
- Immediate answers to your questions over the internet, day or night
- Scientifically proven to increase conceptual understanding and problem-solving skills
- Designed by an award-winning master instructor with over 35 years of teaching experience
- Select from McGraw-Hill end-of-chapter book problems or enter your own problems

Acknowledgments

We are thankful to our families, whose patience and support made it possible for us to undertake this project. We are also grateful to our many colleagues at McGraw-Hill for their support, guidance, and assistance. In particular, we would like to thank Gloria Schiesl, Senior Project Manager, Jodi Rhomberg, Developmental Editor, Tami Hodge, Senior Sponsoring Editor, and Thomas Timp, Publisher. We also wish to acknowledge the assistance of Danáe Quirk Dorr, who in conjunction with the authors, carefully prepared the Instructor's Solutions Manual and Student Solutions Manual to accompany this text.

A revision cannot move forward without the feedback of professors teaching the course. The reviewers have our gratitude and assurance that their comments received serious consideration. The following professors provided reviews, participated in a focus group, or gave valuable advice for the preparation of the sixth edition:

Stanley Bajue   *Medgar Evers College*
Brenda Broers   *Clark College*
Kent Chambers   *Hardin Simmons University*
Mark Champagne   *Macomb Community College–Warren*
Pamela Doyle   *Essex County College*
Karen Duda   *Kennesaw State University*
Cory Emal   *Eastern Michigan University*
Kevin Gratton   *Johnson County Community College*
Balazs Hargittai   *Saint Francis University*
Veronica Jaramillo   *East Los Angeles College*
Booker Juma   *Fayetteville State University*
Ira Krull   *Northeastern Michigan University*
Li-June Ming   *University of South Florida–Tampa*
Kimberly Meyers   *Saint Francis University*
Elva Mae Nicholson   *Eastern Michigan University*
Michael Ogawa   *Bowling Green State University*
Beng Guat Ooi   *Middle Tennessee State University*
Charles Osborne   *Northeast State University*
Chasta Parker   *Winthrop University*
Manoj Patil   *Western Iowa Technical Community College*
Leslie Putman   *Northern Michigan University*
Susan Reid   *North Hennepin Community College*
Kimberly Royal   *Cuyahoga Community College–Eastern Campus*
Kim Salt   *Crafton Hill College*
John Singer   *Jackson Community College*
Jason Tasch   *Cuyahoga Community College–Metro Campus*
Steven Trail   *Elgin Community College*
Philip Verhalen   *Panola College*
Thomas Wilson   *University of Massachusetts–Lowell*
Kim Woodrum   *University of Kentucky–Lexington*

1 Chemistry
Methods and Measurement

Learning Goals

1 ▸ Describe the interrelationship of chemistry with other fields of science and medicine.

2 ▸ Discuss the approach to science, the scientific method, and distinguish among the terms *hypothesis, theory,* and *scientific law.*

3 ▸ Distinguish between data and results.

4 ▸ Describe the properties of the solid, liquid, and gaseous states.

5 ▸ Provide specific examples of physical and chemical properties and physical and chemical change.

6 ▸ Distinguish between intensive and extensive properties.

7 ▸ Classify matter as element, compound, or mixture.

8 ▸ Report data and results using scientific notation and the proper number of significant figures.

9 ▸ Learn the major units of measure in the English and metric systems, and be able to convert from one system to another.

10 ▸ Know the three common temperature scales and be able to convert from one to another.

11 ▸ Use density, mass, and volume in problem solving, and calculate the specific gravity of a substance from its density.

Outline

Introduction

Chemistry Connection:
Chance Favors the Prepared Mind

1.1 The Discovery Process

A Human Perspective:
The Scientific Method

A Medical Perspective:
Curiosity, Science, and Medicine

1.2 Matter and Properties

1.3 Significant Figures and Scientific Notation

1.4 Units and Unit Conversion

1.5 Experimental Quantities

A Human Perspective:
Food Calories

A Human Perspective:
Assessing Obesity: The Body-Mass Index

A Human Perspective:
Specific Gravity: Quick and Useful Analysis

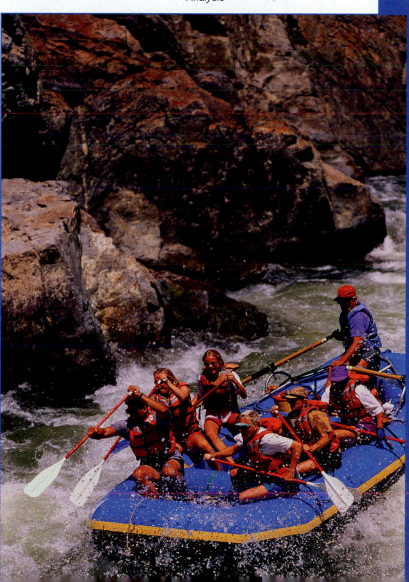

Name the types of measurement associated with this activity.

Introduction

When you awoke this morning, a flood of chemicals called neurotransmitters was sent from cell to cell in your nervous system. As these chemical signals accumulated, you gradually became aware of your surroundings. Chemical signals from your nerves to your muscles propelled you out of your warm bed to prepare for your day.

For breakfast you had a glass of milk, two eggs, and buttered toast, thus providing your body with needed molecules in the form of carbohydrates, proteins, lipids, vitamins, and minerals. As you ran out the door, enzymes of your digestive tract were dismantling the macromolecules of your breakfast. Other enzymes in your cells were busy converting the chemical energy of food molecules into adenosine triphosphate (ATP), the universal energy currency of all cells.

As you continue through your day, thousands of biochemical reactions will keep your cells functioning optimally. Hormones and other chemical signals will regulate

Chemistry Connection

Chance Favors the Prepared Mind

Most of you have chosen a career in medicine because you want to help others. In medicine, helping others means easing pain and suffering by treating or curing diseases. One important part of the practice of medicine involves observation. The physician must carefully observe the patient and listen to his or her description of symptoms to arrive at a preliminary diagnosis. Then appropriate tests must be done to determine whether the diagnosis is correct. During recovery the patient must be carefully observed for changes in behavior or symptoms. These changes are clues that the treatment or medication needs to be modified.

These practices are also important in science. The scientist makes an observation and develops a preliminary hypothesis or explanation for the observed phenomenon. Experiments are then carried out to determine whether the hypothesis is reasonable. When performing the experiment and analyzing the data, the scientist must look for any unexpected results that indicate that the original hypothesis must be modified.

Several important discoveries in medicine and the sciences have arisen from accidental observations. A health care worker or scientist may see something quite unexpected. Whether this results in an important discovery or is simply ignored depends on the training and preparedness of the observer.

It was Louis Pasteur, a chemist and microbiologist, who said, "Chance favors the prepared mind." In the history of science and medicine there are many examples of individuals who have made important discoveries because they recognized the value of an unexpected observation.

One such example is the use of ultraviolet (UV) light to treat infant jaundice. Infant jaundice is a condition in which the skin and the whites of the eyes appear yellow because of high levels of the bile pigment bilirubin in the blood. Bilirubin is a breakdown product of the oxygen-carrying blood protein hemoglobin. If bilirubin accumulates in the body, it can cause brain damage and death. The immature liver of the baby cannot remove the bilirubin.

An observant nurse in England noticed that when jaundiced babies were exposed to sunlight, the jaundice faded. Research based on her observation showed that the UV light changes the bilirubin into another substance that can be excreted. To this day, jaundiced newborns are treated with UV light.

The Pap smear test for the early detection of cervical and uterine cancer was also developed because of an accidental observation. Dr. George Papanicolaou, affectionately called Dr. Pap, was studying changes in the cells of the vagina during the stages of the menstrual cycle. In one sample he recognized cells that looked like cancer cells. Within five years, Dr. Pap had perfected a technique for staining cells from vaginal fluid and observing them microscopically for the presence of any abnormal cells. The lives of countless women have been saved because a routine Pap smear showed early stages of cancer.

In this first chapter of your study of chemistry you will learn more about the importance of observation and accurate, precise measurement in medical practice and scientific study. You will also study the scientific method, the process of developing hypotheses to explain observations, and the design of experiments to test those hypotheses.

the conditions within your body. They will let you know if you are hungry or thirsty. If you injure yourself or come into contact with a disease-causing microorganism, chemicals in your body will signal cells to begin the necessary repair or defense processes.

Life is an organized array of large, carbon-based molecules maintained by biochemical reactions. To understand and appreciate the nature of a living being, we must understand the principles of science and chemistry as they apply to biological molecules.

1.1 The Discovery Process

Chemistry

Chemistry is the study of matter, its chemical and physical properties, the chemical and physical changes it undergoes, and the energy changes that accompany those processes. **Matter** is anything that has mass and occupies space. The changes that matter undergoes always involve either gain or loss of energy. **Energy** is the ability to do work to accomplish some change. The study of chemistry involves matter, energy, and their interrelationship. Matter and energy are at the heart of chemistry.

Major Areas of Chemistry

LEARNING GOAL
1 Describe the interrelationship of chemistry with other fields of science and medicine.

Chemistry is a broad area of study covering everything from the basic parts of an atom to interactions between huge biological molecules. Because of this, chemistry encompasses the following specialties.

Biochemistry is the study of life at the molecular level and the processes associated with life, such as reproduction, growth, and respiration. *Organic chemistry* is the study of matter that is composed principally of carbon and hydrogen. Organic chemists study methods of preparing such diverse substances as plastics, drugs, solvents, and a host of industrial chemicals. *Inorganic chemistry* is the study of matter that consists of all of the elements other than carbon and hydrogen and their combinations. Inorganic chemists have been responsible for the development of unique substances such as semiconductors and high-temperature ceramics for industrial use. *Analytical chemistry* involves the analysis of matter to determine its composition and the quantity of each kind of matter that is present. Analytical chemists detect traces of toxic chemicals in water and air. They also develop methods to analyze human body fluids for drugs, poisons, and levels of medication. *Physical chemistry* is a discipline that attempts to explain the way in which matter behaves. Physical chemists develop theoretical concepts and try to prove them experimentally. This helps us understand how chemical systems behave.

Over the last thirty years, the boundaries between the traditional sciences of chemistry, physics, and biology, as well as mathematics and computer science have gradually faded. Medical practitioners, physicians, nurses, and medical technologists use therapies that contain elements of all these disciplines. The rapid expansion of the pharmaceutical industry is based on a recognition of the relationship between the function of an organism and its basic chemical makeup. Function is a consequence of changes that chemical substances undergo.

For these reasons, an understanding of basic chemical principles is essential for anyone considering a medically related career; indeed, a worker in any science-related field will benefit from an understanding of the principles and applications of chemistry.

The study of the causes of rapid melting of glaciers is a global application of chemistry. How does this illustrate the interaction of matter and energy?

A Human *Perspective*

The Scientific Method

The discovery of penicillin by Alexander Fleming is an example of the scientific method at work. Fleming was studying the growth of bacteria. One day, his experiment was ruined because colonies of mold were growing on his plates. From this failed experiment, Fleming made an observation that would change the practice of medicine: Bacterial colonies could not grow in the area around the mold colonies. Fleming hypothesized that the mold was making a chemical compound that inhibited the growth of the bacteria. He performed a series of experiments designed to test this hypothesis.

The key to the scientific method is the design of carefully controlled experiments that will either support or disprove the hypothesis. This is exactly what Fleming did.

In one experiment he used two sets of tubes containing sterile nutrient broth. To one set he added mold cells. The second set (the control tubes) remained sterile. The mold was allowed to grow for several days. Then the broth from each of the tubes (experimental and control) was passed through a filter to remove any mold cells. Next, bacteria were placed in each tube. If Fleming's hypothesis was correct, the tubes in which the mold had grown would contain the chemical that inhibits growth, and the bacteria would not grow. On the other hand, the control tubes (which were never used to grow mold) would allow bacterial growth. This is exactly what Fleming observed.

Within a few years this *antibiotic,* penicillin, was being used to treat bacterial infections in patients.

A nurse administers an injection of penicillin to a young patient.

For Further Understanding

What is the purpose of the control tubes used in this experiment?

What common characteristics do you find in this story and the Medical Perspective on page 5?

---

2 **LEARNING GOAL**
Discuss the approach to science, the scientific method, and distinguish among the terms *hypothesis, theory,* and *scientific law.*

The Scientific Method

The **scientific method** is a systematic approach to the discovery of new information. How do we learn about the properties of matter, the way it behaves in nature, and how it can be modified to make useful products? Chemists do this by using the scientific method to study the way in which matter changes under carefully controlled conditions.

The scientific method is not a "cookbook recipe" that, if followed faithfully, will yield new discoveries; rather, it is an organized approach to solving scientific problems. Every scientist brings his or her own curiosity, creativity, and imagination to scientific study. But scientific inquiry still involves some of the "cookbook approach."

Characteristics of the scientific process include the following:

- *Observation.* The description of, for example, the color, taste, or odor of a substance is a result of observation. The measurement of the temperature of a liquid or the size or mass of a solid results from observation.
- *Formulation of a question.* Humankind's fundamental curiosity motivates questions of why and how things work.
- *Pattern recognition.* If a scientist finds a cause-and-effect relationship, it may be the basis of a generalized explanation of substances and their behavior.

A Medical Perspective

Curiosity, Science, and Medicine

Curiosity is one of the most important human traits. Small children constantly ask "why?". As we get older, our questions become more complex, but the curiosity remains.

Curiosity is also the basis of the scientific method. A scientist observes an event, wonders why it happens, and sets out to answer the question. Dr. Michael Zasloff's curiosity may lead to the development of an entirely new class of antibiotics. When he was a geneticist at the National Institutes of Health, his experiments involved the surgical removal of the ovaries of African clawed frogs. After surgery he sutured (sewed up) the incision and put the frogs back in their tanks. These water-filled tanks were teeming with bacteria, but the frogs healed quickly, and the incisions did not become infected!

Of all the scientists to observe this remarkable healing, only Zasloff was curious enough to ask whether there were chemicals in the frogs' skin that defended the frogs against bacterial infections—a new type of antibiotic. All currently used antibiotics are produced by fungi or are synthesized in the laboratory. One big problem in medicine today is more and more pathogenic (disease-causing) bacteria are becoming resistant to these antibiotics. Zasloff hoped to find an antibiotic that worked in an entirely new way so the current problems with antibiotic resistance might be overcome.

Dr. Zasloff found two molecules in frog skin that can kill bacteria. Both are small proteins. Zasloff named them *magainins*, from the Hebrew word for shield. Most of the antibiotics that we now use enter bacteria and kill them by stopping some biochemical process inside the cell. Magainins are more direct; they simply punch holes in the bacterial membrane, and the bacteria explode.

One of the magainins, now chemically synthesized in the laboratory so that no frogs are harmed, may be available to the public in the near future. This magainin can kill a wide variety of bacteria (broad-spectrum antibiotic), and it has passed the Phase I human trials. If this compound passes all the remaining tests, it will be used in treating deep infected wounds and ulcers, providing an alternative to traditional therapy.

The curiosity that enabled Zasloff to advance the field of medicine also catalyzed the development of chemistry. We will see the product of this fundamental human characteristic as we study the work of many extraordinary chemists throughout this chapter.

Dr. Zasloff's uninfected patient.

For Further Understanding

Why is it important for researchers to continually design and develop new antibacterial substances?

What common characteristics do you find in this work and the discovery discussed in the Chemistry Connection on page 2?

- *Developing theories.* When scientists observe a phenomenon, they want to explain it. The process of explaining observed behavior begins with a hypothesis. A **hypothesis** is simply an attempt to explain an observation, or series of observations, in a commonsense way. If many experiments support a hypothesis, it may attain the status of a theory. A **theory** is a hypothesis supported by extensive testing (experimentation) that explains scientific facts and can predict new facts.
- *Experimentation.* Demonstrating the correctness of hypotheses and theories is at the heart of the scientific method. This is done by carrying out carefully designed experiments that will either support or disprove the theory or hypothesis.

Figure 1.1

The scientific method, an organized way of doing science. A degree of trial and error is apparent here. If experimentation does not support the hypothesis, one must begin the cycle again.

- *Summarizing information.* A scientific **law** is nothing more than the summary of a large quantity of information. For example, the law of conservation of matter states that matter cannot be created or destroyed, only converted from one form to another. This statement represents a massive body of chemical information gathered from experiments.

The scientific method involves the interactive use of hypotheses, development of theories, and thorough testing of theories using well-designed experiments and is summarized in Figure 1.1.

Models in Chemistry

Hypotheses, theories, and laws are frequently expressed using mathematical equations. These equations may confuse all but the best of mathematicians. For this reason a *model* of a chemical unit or system is often used to make ideas more clear. A good model based on everyday experience, although imperfect, gives a great deal of information in a simple fashion. Consider the fundamental unit of methane, the major component of natural gas, which is composed of one carbon atom (symbolized by C) and four hydrogen atoms (symbolized by H).

A geometrically correct model of methane can be constructed from balls and sticks. The balls represent the individual units (atoms) of hydrogen and carbon, and the sticks correspond to the attractive forces that hold the hydrogen and carbon together. The model consists of four balls representing hydrogen symmetrically arranged around a center ball representing carbon. The "carbon" ball is attached to each "hydrogen" ball by sticks, as shown:

Color-coding the balls distinguishes one type of matter from another; the geometrical form of the model, all of the angles and dimensions of a tetrahedron, are the same for each methane unit found in nature. Methane is certainly not a collection of balls and sticks; but such models are valuable because they help us understand the chemical behavior of methane and other, more complex substances.

Chemists and physicists have used the observed properties of matter to develop models of the individual units of matter. These models collectively make up what we now know as the atomic theory of matter.

These models have developed from experimental observations over the past two hundred years. Theory and experiment are mutually reinforcing. We must gain some insight into atomic structure to appreciate the behavior of the atoms themselves as well as larger aggregates of atoms: compounds.

The structure-properties concept has advanced so far that compounds are designed and synthesized in the laboratory with the hope that they will perform very specific functions, such as curing diseases that have been resistant to other forms of treatment. Figure 1.2 shows some of the variety of modern technology that has its roots in the understanding of the atom.

Molecules that maximize therapeutic properties and minimize undesirable side effects can be designed on the computer and synthesized and purified in the laboratory.

(a) (b)

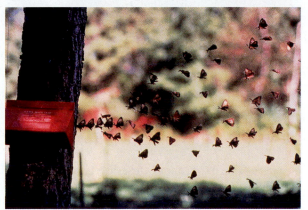

(c) (d)

Figure 1.2
Examples of technology originating from scientific inquiry: (a) synthesis of a new drug, (b) solar energy cells, (c) preparation of solid-state electronics, (d) use of a gypsy moth sex attractant for insect control.

1.2 Matter and Properties

Properties are characteristics of matter and are classified as either physical or chemical. In this section we will learn the meaning of physical and chemical properties and how they are used to characterize matter.

Data and Results

A scientific experiment produces **data.** Each piece of data is the individual result of a single measurement or observation. Examples include the *mass* of a sample and the *time* required for a chemical reaction to occur. Mass, length, volume, time, temperature, and energy are common types of data obtained from chemical experiments.

A **result** is the outcome of an experiment. Data and results may be identical, but more often several related pieces of data are combined, and logic is used to produce a result.

LEARNING GOAL
3 Distinguish between data and results.

| Distinguishing Between Data and Results | EXAMPLE 1.1 |

In many cases, a drug is less stable if moisture is present, and excess moisture can hasten the breakdown of the active ingredient, leading to loss of potency. Therefore we may wish to know how much water a certain quantity of a drug gains when exposed to air. To do this experiment, we

Continued—

EXAMPLE 1.1 —*Continued*

(a)

(b)

(c)

Figure 1.3
The three states of matter exhibited by water: (a) solid, as ice; (b) liquid, as ocean water; (c) gas, as humidity in the air.

Animation
The Three States of Matter

LEARNING GOAL
4 Describe the properties of the solid, liquid, and gaseous states.

LEARNING GOAL
5 Provide specific examples of physical and chemical properties and physical and chemical change.

must first weigh the drug sample, then expose it to the air for a period of time and reweigh it. The change in weight,

$$[\text{weight}_{\text{final}} - \text{weight}_{\text{initial}}] = \text{weight difference}$$

indicates the weight of water taken up by the drug formulation. The initial and final weights are individual bits of *data;* by themselves they do not answer the question, but they do provide the information necessary to calculate the answer: the results. The difference in weight and the conclusions based on the observed change in weight are the *results* of the experiment.

 Practice Problem 1.1
Describe an experiment demonstrating that the boiling point of water changes when salt (sodium chloride) is added to the water.

For Further Practice: Questions 1.25 and 1.26.

The experiment described in Example 1.1 was really not a very good experiment because many other environmental conditions were not measured. Measurement of the temperature and humidity of the atmosphere and the length of time that the drug was exposed to the air (the creation of a more complete set of data) would make the results less ambiguous.

States of Matter

There are three *states of matter:* the **gaseous state,** the **liquid state,** and the **solid state.** A gas is made up of particles that are widely separated. In fact, a gas will expand to fill any container; it has no definite shape or volume. In contrast, particles of a liquid are closer together; a liquid has a definite volume but no definite shape; it takes on the shape of its container. A solid consists of particles that are close together and that often have a regular and predictable pattern of particle arrangement (crystalline). A solid has both fixed volume and fixed shape. Attractive forces, which exist between all particles, are very pronounced in solids and much less so in gases.

Matter and Physical Properties

Water is the most common example of a substance that can exist in all three states over a reasonable temperature range (Figure 1.3). Conversion of water from one state to another constitutes a *physical change*. A **physical change** produces a recognizable difference in the appearance of a substance without causing any change in its composition or identity. For example, we can warm an ice cube and it will melt, forming liquid water. Clearly its appearance has changed; it has been transformed from the solid to the liquid state. It is, however, still water; its composition and identity remain unchanged. A physical change has occurred. We could in fact demonstrate the constancy of composition and identity by refreezing the liquid water, re-forming the ice cube. This melting and freezing cycle could be repeated over and over. This very process is a hallmark of our global weather changes. The continual interconversion of the three states of water in the environment (snow, rain, and humidity) clearly demonstrates the retention of the identity of water particles or *molecules.*

Figure 1.4
An example of separation based on differences in physical properties. Magnetic iron is separated from other nonmagnetic substances. A large-scale version of this process is important in the recycling industry.

A **physical property** can be observed or measured without changing the composition or identity of a substance. As we have seen, melting ice is a physical change. We can measure the temperature when melting occurs; this is the *melting point* of water. We can also measure the *boiling point* of water, when liquid water becomes a gas. Both the melting and boiling points of water, and of any other substance, are physical properties.

A practical application of separation of materials based upon their differences in physical properties is shown in Figure 1.4.

Matter and Chemical Properties

We have noted that physical properties can be exhibited, measured, or observed without any change in identity or composition. In contrast, **chemical properties** do result in a change in composition and can be observed only through chemical reactions. A **chemical reaction** is a process of rearranging, removing, replacing, or adding atoms to produce new substances. For example, the process of photosynthesis can be shown as

$$\text{carbon dioxide} + \text{water} \xrightarrow[\text{Chlorophyll}]{\text{Light}} \text{sugar} + \text{oxygen}$$

This chemical reaction involves the conversion of carbon dioxide and water (the *reactants*) to a sugar and oxygen (the *products*). The products and reactants are clearly different. We know that carbon dioxide and oxygen are gases at room temperature and water is a liquid at this temperature; the sugar is a solid white powder. A chemical property of carbon dioxide is its ability to form sugar under certain conditions. The process of formation of this sugar is the *chemical change*.

Light is the energy needed to make the reaction happen. Chlorophyll is the energy absorber, converting light energy to chemical energy.

Identifying Properties

EXAMPLE 1.2

Can the process that takes place when an egg is fried be described as a physical or chemical change?

5 **LEARNING GOAL**
Provide specific examples of physical and chemical properties and physical and chemical change.

Continued—

EXAMPLE 1.2 —*Continued*

Solution

Examine the characteristics of the egg before and after frying. Clearly, some significant change has occurred. Furthermore, the change appears irreversible. More than a simple physical change has taken place. A chemical reaction (actually, several) must be responsible; hence chemical change.

Practice Problem 1.2

Classify each of the following as either a chemical change or a physical change:
a. water boiling to become steam
b. butter becoming rancid
c. combustion of wood
d. melting of ice in spring
e. decay of leaves in winter

For Further Practice: Questions 1.37 and 1.38.

Question 1.1

Classify each of the following as either a chemical property or a physical property:

a. color
b. flammability
c. hardness

Question 1.2

Classify each of the following as either a chemical property or a physical property:

a. odor
b. taste
c. temperature

Intensive and Extensive Properties

See page 31 for a discussion of density and specific gravity.

It is important to recognize that properties can also be classified according to whether they depend on the size of the sample. Consequently, there is a fundamental difference between properties such as density and specific gravity and properties such as mass and volume.

An **intensive property** is a property of matter that is *independent* of the *quantity* of the substance. Density, boiling and melting points, and specific gravity are intensive properties. For example, the boiling point of one single drop of water is exactly the same as the boiling point of a liter of water.

An **extensive property** *depends* on the *quantity* of a substance. Mass and volume are extensive properties. There is an obvious difference between 1 g of silver and 1 kg of silver; the quantities and, incidentally, the value, differ substantially.

LEARNING GOAL

6   Distinguish between intensive and extensive properties.

Is temperature an extensive or intensive property?

Solution

Imagine two glasses each containing 100 g of water, and each at 25°C. Now pour the contents of the two glasses into a larger glass. You would predict that the mass of the water in the larger glass would be 200 g (100 g + 100 g) because mass is an *extensive property*, dependent on quantity. However, we would expect the temperature of the water to remain the same (not 25°C + 25°C); hence temperature is an *intensive property* . . . independent of quantity.

 Practice Problem 1.3

Is the boiling point of water an intensive or extensive property?

For Further Practice: Questions 1.45 and 1.46.

Classification of Matter

Chemists look for similarities in properties among various types of materials. Recognizing these likenesses simplifies learning the subject and allows us to predict the behavior of new substances on the basis of their relationship to substances already known and characterized.

LEARNING GOAL

7 Classify matter as element, compound, or mixture.

Many classification systems exist. The most useful system, based on composition, is described in the following paragraphs (see also Figure 1.5).

All matter is either a *pure substance* or a *mixture*. A **pure substance** has only one component. Pure water is a pure substance. It is made up only of particles containing two hydrogen atoms and one oxygen atom, that is, water molecules (H_2O).

There are different types of pure substances. Elements and compounds are both pure substances. An **element** is a pure substance that cannot be changed into a simpler form of matter by any chemical reaction. Hydrogen and oxygen, for example, are elements. Alternatively, a **compound** is a substance resulting from the combination of two or more elements in a definite, reproducible way. The elements hydrogen and oxygen, as noted earlier, may combine to form the compound water, H_2O.

A **mixture** is a combination of two or more pure substances in which each substance retains its own identity. Alcohol and water can be combined in a mixture. They coexist as pure substances because they do not undergo a chemical reaction; they exist as thoroughly mixed discrete molecules. This collection of dissimilar particles is the mixture. A mixture has variable composition; there are an infinite number of combinations of quantities of alcohol and water that can be mixed. For example, the mixture may contain a small amount of alcohol and a large amount of water or vice versa. Each is, however, an alcohol–water mixture.

At present, more than one hundred elements have been characterized. A complete listing of the elements and their symbols is found on the inside front cover of this textbook.

Figure 1.5

Classification of matter. All matter is either a pure substance or a mixture of pure substances. Pure substances are either elements or compounds, and mixtures may be either homogeneous (uniform composition) or heterogeneous (nonuniform composition).

Figure 1.6
Schematic representation of some classes of matter. (a) A pure substance, water, consists of a single component.(b) A homogeneous mixture, ethanol and water, has a uniform distribution of components. (c) A heterogeneous mixture, marble, has a nonuniform distribution of components. The lack of homogeneity is readily apparent.

(a)   (b)   (c)

A detailed discussion of solutions (homogeneous mixtures) and their properties is presented in Chapter 6.

A mixture may be either *homogeneous* or *heterogeneous* (Figure 1.6). A **homogeneous mixture** has uniform composition. Its particles are well mixed, or thoroughly intermingled. A homogeneous mixture, such as alcohol and water, is described as a *solution*. Air, a mixture of gases, is an example of a gaseous solution. A **heterogeneous mixture** has a nonuniform composition. A mixture of salt and pepper is a good example of a heterogeneous mixture. Concrete is also composed of a heterogeneous mixture of materials (various types and sizes of stone and sand present with cement in a nonuniform mixture).

EXAMPLE 1.4   **Categorizing Matter**

LEARNING GOAL
7 Classify matter as element, compound, or mixture.

Is seawater a pure substance, a homogeneous mixture, or a heterogeneous mixture?

Solution

Imagine yourself at the beach, filling a container with a sample of water from the ocean. Examine it. You would see a variety of solid particles suspended in the water: sand, green vegetation, perhaps even a small fish! Clearly, it is a mixture, and one in which the particles are not uniformly distributed throughout the water; hence a heterogeneous mixture.

  Practice Problem 1.4

Is each of the following materials a pure substance, a homogeneous mixture, or a heterogeneous mixture?
  a. ethyl alcohol
  b. blood
  c. Alka-Seltzer dissolved in water
  d. oxygen in a hospital oxygen tank

For Further Practice: Questions 1.49 and 1.50.

1.3 Significant Figures and Scientific Notation

Information-bearing figures in a number are termed *significant figures.* Data and results arising from a scientific experiment convey information about the way in which the experiment was conducted. The degree of uncertainty or doubt associated with a measurement or series of measurements is indicated by the number of figures used to represent the information.

Significant Figures

Consider the following situation: A student was asked to obtain the length of a section of wire. In the chemistry laboratory, several different types of measuring devices are usually available. Not knowing which was most appropriate, the student decided to measure the object using each device that was available in the laboratory. The following data were obtained:

5.4 cm

(a)

5.36 cm

(b)

Two questions should immediately come to mind:

Are the two answers equivalent?

If not, which answer is correct?

In fact, the two answers are *not* equivalent, but *both* are correct. How do we explain this apparent contradiction?

The data are not equivalent because each is known to a different degree of certainty. The answer 5.36 cm, containing three significant figures, specifies the length of the object more exactly than 5.4 cm, which contains only two significant figures. The term **significant figures** is defined to be all digits in a number representing data or results that are known with certainty *plus one uncertain digit.*

In case (a), we are certain that the object is at least 5 cm long and equally certain that it is *not* 6 cm long because the end of the object falls between the calibration lines 5 and 6. We can only estimate between 5 and 6, because there are no calibration indicators between 5 and 6. The end of the wire appears to be approximately four-tenths of the way between 5 and 6, hence 5.4 cm. The 5 is known with certainty, and 4 is estimated; there are two significant figures.

In case (b), the ruler is calibrated in tenths of centimeters. The end of the wire is at least 5.3 cm and not 5.4 cm. Estimation of the second decimal place between the two closest calibration marks leads to 5.36 cm. In this case, 5.3 is certain, and the 6 is estimated (or uncertain), leading to three significant digits.

Both answers are correct because each is consistent with the measuring device used to generate the data. An answer of 5.36 cm obtained from a measurement using ruler (a) would be *incorrect* because the measuring device is not capable of that exact specification. On the other hand, a value of 5.4 cm obtained from ruler (b) would be erroneous as well; in that case the measuring device is capable of generating a higher level of certainty (more significant digits) than is actually reported.

In summary, the number of significant figures associated with a measurement is determined by the measuring device. Conversely, the number of significant figures reported is an indication of the sophistication of the measurement itself.

The uncertain digit represents the degree of doubt in a single measurement.

The uncertain digit results from an estimation.

Recognition of Significant Figures

Only *significant* digits should be reported as data or results. However, are all digits, as written, significant digits? Let's look at a few examples illustrating the rules that are used to represent data and results with the proper number of significant digits.

- All nonzero digits are significant.
  7.314 has *four* significant digits.
- The number of significant digits is independent of the position of the decimal point.
  73.14 has *four* significant digits, as does 7.314.
- Zeros located between nonzero digits are significant.
  60.052 has *five* significant figures.
- Zeros at the end of a number (often referred to as trailing zeros) are significant if the number contains a decimal point.
  4.70 has *three* significant figures.
  Helpful Hint: Trailing zeros are ambiguous; the next section offers a solution for this ambiguity.
- Trailing zeros are insignificant if the number does not contain a decimal point and are significant if a decimal point is indicated.
  100 has *one* significant figure; 100. has three significant figures.
- Zeros to the left of the first nonzero integer are not significant; they serve only to locate the position of the decimal point.
  0.0032 has *two* significant figures.

Question 1.3

How many significant figures are contained in each of the following numbers?

a. 7.26   d. 7.0
b. 726   e. 0.0720
c. 700.2

Question 1.4

How many significant figures are contained in each of the following numbers?

a. 0.042   d. 240
b. 4.20   e. 204
c. 24.0

Scientific Notation

LEARNING GOAL

8 Report data and results using scientific notation and the proper number of significant figures.

It is often difficult to express very large numbers to the proper number of significant figures using conventional notation. The solution to this problem lies in the use of **scientific notation,** also referred to as *exponential notation*, which involves the representation of a number as a power of ten.

The speed of light is 299,792,458 m/s. For many calculations, two or three significant figures are sufficient. Using scientific notation, two significant figures, and rounding (p. 18), the speed of light is 3.0×10^8 m/s. The conversion is illustrated using simpler numbers:

$$6200 = 6.2 \times 1000 = 6.2 \times 10^3$$

or

$$5340 = 5.34 \times 1000 = 5.34 \times 10^3$$

 A Review of Mathematics

RULE: To convert a number greater than 1 to scientific notation, the original decimal point is moved x places to the left, and the resulting number is multiplied by 10^x. The exponent (x) is a *positive* number equal to the number of places the original decimal point was moved. ■

Scientific notation is also useful in representing numbers less than 1. For example, the mass of a single helium atom is

$$0.0000000000000000000000006692 \text{ gram}$$

a rather cumbersome number as written. Scientific notation would represent the mass of a single helium atom as 6.692×10^{-24} gram. The conversion is illustrated by using simpler numbers:

$$0.0062 = 6.2 \times \frac{1}{1000} = 6.2 \times \frac{1}{10^3} = 6.2 \times 10^{-3}$$

or

$$0.0534 = 5.34 \times \frac{1}{100} = 5.34 \times \frac{1}{10^2} = 5.34 \times 10^{-2}$$

RULE: To convert a number less than 1 to scientific notation, the original decimal point is moved x places to the right, and the resulting number is multiplied by 10^{-x}. The exponent ($-x$) is a *negative* number equal to the number of places the original decimal point was moved. ■

Question 1.5

Represent each of the following numbers in scientific notation, showing only significant digits:

a. 0.0024   b. 0.0180   c. 224

Question 1.6

Represent each of the following numbers in scientific notation, showing only significant digits:

a. 48.20   b. 480.0   c. 0.126

Error, Accuracy, Precision, and Uncertainty

Error is the difference between the true value and our estimation, or measurement, of the value. Some degree of error is associated with any measurement. Two types of error exist: random error and systematic error. *Random error* causes data from multiple measurements of the same quantity to be scattered in a more or less uniform way around some average value. *Systematic error* causes data to be either smaller or larger than the accepted value. Random error is inherent in the experimental approach to the study of matter and its behavior; systematic error can be found and, in many cases, removed or corrected.

Examples of systematic error include such situations as:

- Dust on the balance pan, which causes all objects weighed to appear heavier than they really are.
- Impurities in chemicals used for the analysis of materials, which may interfere with (or block) the desired process.

Accuracy is the degree of agreement between the true value and the measured value. **Uncertainty** is the degree of doubt in a single measurement.

When measuring quantities that show continuous variation, for example, the weight of this page or the volume of one of your quarters, some doubt or uncertainty is present because the answer cannot be expressed with an infinite number of meaningful digits. The number of meaningful digits is determined by the measuring device. The presence of some error is a natural consequence of any measurement.

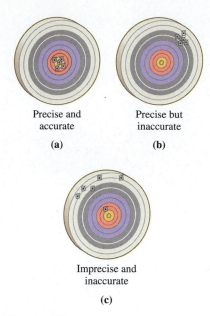

Figure 1.7

An illustration of precision and accuracy in replicate experiments.

(a) Precise and accurate

(b) Precise but inaccurate

(c) Imprecise and inaccurate

The simple process of converting the fraction $^2/_3$ to its decimal equivalent can produce a variety of answers that depend on the device used to perform the calculation: pencil and paper, calculator, computer. The answer might be

$$0.67$$
$$0.667$$
$$0.6667$$

and so forth. All are correct, but each value has a different level of uncertainty. The first number listed, 0.67, has the greatest uncertainty.

It is always best to measure a quantity several times. Modern scientific instruments are designed to perform measurements rapidly; this allows many more measurements to be completed in a reasonable period. Replicate measurements of the same quantity minimize the uncertainty of the result. **Precision** is a measure of the agreement of replicate measurements.

It is important to recognize that accuracy and precision are not the same thing. It is possible to have one without the other. However, when scientific measurements are carefully made, the two most often go hand in hand; high-quality data are characterized by high levels of precision and accuracy.

In Figure 1.7, bull's-eye (a) shows the goal of all experimentation: accuracy *and* precision. Bull's-eye (b) shows the results to be repeatable (good precision); however, some error in the experimental procedure has caused the results to center on an incorrect value. This error is systematic, occurring in each replicate measurement. Occasionally, an experiment may show "accidental" accuracy. The precision is poor, but the average of these replicate measurements leads to a correct value. We don't want to rely on accidental success; the experiment should be repeated until the precision inspires faith in the accuracy of the method. Modern measuring devices in chemistry, equipped with powerful computers with immense storage capacity, are capable of making literally thousands of individual replicate measurements to enhance the quality of the result. Bull's-eye (c) describes the most common situation. A low level of precision is all too often associated with poor accuracy.

Significant Figures in Calculation of Results

Addition and Subtraction

If we combine the following numbers:

$$37.68 \quad \text{liters}$$
$$108.428 \quad \text{liters}$$
$$6.71862 \quad \text{liters}$$

our calculator will show a final result of

$$152.82662 \quad \text{liters}$$

Clearly, the answer, with eight digits, defines the volume of total material much more accurately than *any* of the individual quantities being combined. This cannot be correct; *the answer cannot have greater significance than any of the quantities that produced the answer.* We rewrite the problem:

$$37.68xxx \quad \text{liters}$$
$$108.428xx \quad \text{liters}$$
$$+ \quad 6.71862 \quad \text{liters}$$
$$\overline{152.82662} \quad \text{(should be 152.83) liters}$$

LEARNING GOAL

8 Report data and results using scientific notation and the proper number of significant figures.

Remember the distinction between the words *zero* and *nothing*. Zero is one of the ten digits and conveys as much information as 1, 2, and so forth. *Nothing* implies no information; the digits in the positions indicated by x could be 0, 1, 2, or any other.

where x = no information; x may be any integer from 0 to 9. Adding 2 to two unknown numbers (in the right column) produces no information. Similar logic prevails for the next two columns. Thus, five digits remain, all of which are significant. Conventional rules for rounding off would dictate a final answer of 152.83.

See rules for rounding off discussed on page 18.

Question 1.7

Report the result of each of the following to the proper number of significant figures:

   a. $4.26 + 3.831 =$
   b. $8.321 - 2.4 =$
   c. $16.262 + 4.33 - 0.40 =$

Question 1.8

Report the result of each of the following to the proper number of significant figures:

   a. $7.939 + 6.26 =$
   b. $2.4 - 8.321 =$
   c. $2.333 + 1.56 - 0.29 =$

Multiplication and Division

In the preceding discussion of addition and subtraction, the position of the decimal point in the quantities being combined has a bearing on the number of significant figures in the answer. In multiplication and division this is not the case. The decimal point position is irrelevant when determining the number of significant figures in the answer. It is the number of significant figures in the data that is important. Consider

LEARNING GOAL

8 Report data and results using scientific notation and the proper number of significant figures.

$$\frac{4.237 \times 1.21 \times 10^{-3} \times 0.00273}{11.125} = 1.26 \times 10^{-6}$$

The answer is limited to three significant figures; the answer can have *only* three significant figures because two numbers in the calculation, 1.21×10^{-3} and 0.00273, have three significant figures and "limit" the answer. Remember, *the answer can be no more precise than the least precise number from which the answer is derived.* The *least precise number* is the number with the fewest significant figures.

Question 1.9

Report the results of each of the following operations using the proper number of significant figures:

   a. $63.8 \times 0.80 =$

   b. $\dfrac{63.8}{0.80} =$

   c. $\dfrac{53.8 \times 0.90}{0.3025} =$

Question *1.10*

Report the results of each of the following operations using the proper number of significant figures:

a. $\dfrac{27.2 \times 15.63}{1.84} =$

b. $\dfrac{13.6}{18.02 \times 1.6} =$

c. $\dfrac{12.24 \times 6.2}{18.02 \times 1.6} =$

 A Review of Mathematics

Exponents

Now consider the determination of the proper number of significant digits in the results when a value is multiplied by any power of ten. In each case the number of significant figures in the answer is identical to the number contained in the original term. Therefore

$$(8.314 \times 10^2)^3 = 574.7 \times 10^6 = 5.747 \times 10^8$$

and

$$(8.314 \times 10^2)^{1/2} = 2.883 \times 10^1$$

Each answer contains four significant figures.

Exact (Counted) and Inexact Numbers

Inexact numbers, by definition, have uncertainty (the degree of doubt in the final significant digit). *Exact numbers*, on the other hand, have no uncertainty. Exact numbers may arise from a definition; there are *exactly* 60 minutes in 1 hour or there are exactly 1000 mL in 1 liter.

Exact numbers are a consequence of counting. Counting the number of dimes in your pocket or the number of letters in the alphabet are common examples. The fact that exact numbers have no uncertainty means that they do not limit the number of significant figures in the result of a calculation.

For example,

$$4.00 \text{ oz} \times 1 \text{ lb}/16 \text{ oz} = 0.250 \text{ lb (3 significant figures)}$$

or

$$2568 \text{ oz} \times 1 \text{ lb}/16 \text{ oz} = 160.5 \text{ lb (4 significant figures)}$$

In both examples, the number of significant figures in the result is governed by the data (the number of ounces) not the conversion factor, which is *exact*, because it is defined.

A good rule of thumb to follow is: In the metric system the quantity being converted, not the conversion factor, generally determines the number of significant figures.

Rounding Off Numbers

The use of an electronic calculator generally produces more digits for a result than are justified by the rules of significant figures on the basis of the data input. For example, on your calculator,

$$3.84 \times 6.72 = 25.8048$$

The most correct answer would be 25.8, dropping 048.

A number of acceptable conventions for rounding exist. Throughout this book we will use the following:

RULE: When the number to be dropped is less than 5, the preceding number is not changed. When the number to be dropped is 5 or larger, the preceding number is increased by one unit. ∎

Rounding Numbers   EXAMPLE 1.5

Round off each of the following to three significant figures.

Solution

 a. 63.6<u>6</u>9 becomes 63.7. *Rationale:* 6 > 5.
 b. 8.77<u>1</u>5 becomes 8.77. *Rationale:* 1 < 5.
 c. 2.22<u>4</u>5 becomes 2.22. *Rationale:* 4 < 5.
 d. 0.0004<u>0</u>9 becomes 0.000411. *Rationale:* 9 > 5.

Helpful Hint: Symbol $x > y$ implies "x greater than y." Symbol $x < y$ implies "x less than y."

Practice Problem 1.5

Round off each of the following numbers to three significant figures.
 a. 61.40
 b. 6.171
 c. 0.066494
 d. 6.2262
 e. 3895
 f. 6.885

For Further Practice: Questions 1.53 and 1.54.

1.4 Units and Unit Conversion

Any measurement made in the experiment must specify the units of that measurement. An initial weight of three *ounces* is clearly quite different than three *pounds*. A **unit** defines the basic quantity of mass, volume, time, or whatever quantity is being measured. A number that is not followed by the correct unit usually conveys no useful information.

Proper use of units is central to all aspects of science. The following sections are designed to develop a fundamental understanding of this vital topic.

English and Metric Units

The *English system* is a collection of functionally unrelated units. In the *English system of measurement,* the standard *pound* (lb) is the basic unit of weight. The fundamental unit of *length* is the standard *yard* (yd), and the basic unit of *volume* is the standard *gallon* (gal). The English system is used in the United States in business and industry. However, it is not used in scientific work, primarily because it is difficult to convert from one unit to another. For example,

LEARNING GOAL

9 Learn the major units of measure in the English and metric systems, and be able to convert from one system to another.

$$1 \text{ foot} = 12 \text{ inches} = 0.33 \text{ yard} = \frac{1}{5280} \text{ mile} = \frac{1}{6} \text{ fathom}$$

Clearly, operations such as the conversion of 1.62 yards to units of miles are not straightforward. In fact, the English "system" is not really a system at all. It is simply a collection of measures accumulated throughout English history. Because they have no common origin, it is not surprising that conversion from one unit to another is not straightforward.

The United States, the last major industrial country to retain the English system, has begun efforts to convert to the metric system. The *metric system* is truly "systematic." It is composed of a set of units that are related to each other decimally, in other words, as powers of ten. Because the *metric system* is a decimal-based system, it is inherently simpler to use and less ambiguous. For example, the length of an object may be represented as

$$1 \text{ meter} = 10 \text{ decimeters} = 100 \text{ centimeters} = 1000 \text{ millimeters}$$

The metric system was originally developed in France just before the French Revolution in 1789. The more extensive version of this system is the *Système International*, or *S.I. system*. Although the S.I. system has been in existence for over forty years, it has yet to gain widespread acceptance. To make the S.I. system truly systematic, it utilizes certain units, especially those for pressure, that many find unwieldy.

In this text we will use the metric system, not the S.I. system, and we will use the English system only to the extent of converting *from* it to the more scientifically useful metric system.

In the metric system, there are three basic units. Mass is represented as the *gram*, length as the *meter*, and volume as the *liter*. Any subunit or multiple unit contains one of these units preceded by a prefix indicating the power of ten by which the base unit is to be multiplied to form the subunit or multiple unit. The most common metric prefixes are shown in Table 1.1.

Other metric units, for time, temperature, and energy, will be treated in Section 1.5.

The same prefix may be used for volume, mass, length, time, and so forth. Consider the following examples:

$$1 \text{ milliliter (mL)} = \frac{1}{1000} \text{ liter} = 0.001 \text{ liter} = 10^{-3} \text{ liter}$$

A volume unit is indicated by the base unit, liter, and the prefix *milli-*, which indicates that the unit is one thousandth of the base unit. In the same way,

A Review of Mathematics

$$1 \text{ milligram (mg)} = \frac{1}{1000} \text{ gram} = 0.001 \text{ gram} = 10^{-3} \text{ gram}$$

The representation of numbers as powers of ten was discussed in Section 1.3.

and

$$1 \text{ millimeter (mm)} = \frac{1}{1000} \text{ meter} = 0.001 \text{ meter} = 10^{-3} \text{ meter}$$

TABLE 1.1	Some Common Prefixes Used in the Metric System	
Prefix	**Multiple**	**Decimal Equivalent**
mega (M)	10^6	1,000,000.
kilo (k)	10^3	1,000.
deka (da)	10^1	10.
deci (d)	10^{-1}	0.1
centi (c)	10^{-2}	0.01
milli (m)	10^{-3}	0.001
micro (μ)	10^{-6}	0.000001
nano (n)	10^{-9}	0.000000001

Unit Conversion: English and Metric Systems

To convert from one unit to another, we must have a *conversion factor* or series of conversion factors that relate two units. The proper use of these conversion factors is called the *factor-label method*. This method is also termed *dimensional analysis*.

This method is used for two kinds of conversions: to convert from one unit to another within the *same system* or to convert units from *one system to another*.

Conversion of Units Within the Same System

We know, for example, that in the English system,

$$1 \text{ gallon} = 4 \text{ quarts}$$

Because dividing both sides of the equation by the same term does not change its identity,

$$\frac{1 \text{ gallon}}{1 \text{ gallon}} = \frac{4 \text{ quarts}}{1 \text{ gallon}}$$

The expression on the left is equal to unity (1); therefore

$$1 = \frac{4 \text{ quarts}}{1 \text{ gallon}} \quad \text{or} \quad 1 = \frac{1 \text{ gallon}}{4 \text{ quarts}}$$

Now, multiplying any other expression by the ratio 4 quarts/1 gallon or 1 gallon/4 quarts will not change the value of the term, because multiplication of any number by 1 produces the original value. However, there is one important difference: The units will have changed.

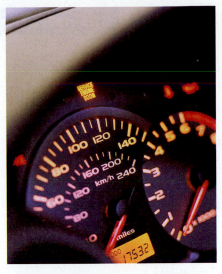

The speed of an automobile is indicated in both English (miles per hour) and metric (kilometers per hour) units.

9 **LEARNING GOAL**
Learn the major units of measure in the English and metric systems, and be able to convert from one system to another.

Using Conversion Factors	EXAMPLE 1.6

Convert 12 gallons to units of quarts.

Solution

$$12 \ \cancel{\text{gal}} \times \frac{4 \text{ qt}}{1 \ \cancel{\text{gal}}} = 48 \text{ qt}$$

The conversion factor, 4 qt/1 gal, serves as a bridge, or linkage, between the unit that was given (gallons) and the unit that was sought (quarts).

The conversion factor may be written as 4 qt/1 gal or 1 gal/4 qt, because both are equal to 1. However, only the first factor, 4 qt/1 gal, will give us the units we need to solve the problem. If we had set up the problem incorrectly, we would obtain

$$12 \text{ gal} \times \frac{1 \text{ gal}}{4 \text{ qt}} = 3 \frac{\text{gal}^2}{\text{qt}}$$

Incorrect units

Clearly, units of gal²/qt are not those asked for in the problem, nor are they reasonable units. The factor-label method is therefore a self-indicating system; the correct units (those required by the problem) will result only if the factor is set up properly.

Continued—

EXAMPLE 1.6 —*Continued*

Practice Problem 1.6

Convert 360 feet to miles.

For Further Practice: Questions 1.63a, b and 1.64a, b.

TABLE 1.2	Some Common Relationships Used in the English System
A. Weight	1 pound = 16 ounces
	1 ton = 2000 pounds
B. Length	1 foot = 12 inches
	1 yard = 3 feet
	1 mile = 5280 feet
C. Volume	1 gallon = 4 quarts
	1 quart = 2 pints
	1 quart = 32 fluid ounces

Table 1.2 lists a variety of commonly used English system relationships that may serve as the basis for useful conversion factors.

Conversion of units within the metric system may be accomplished by using the factor-label method as well. Unit prefixes that dictate the conversion factor facilitate unit conversion (refer to Table 1.1).

EXAMPLE 1.7   **Using Conversion Factors**

Convert 10.0 centimeters to meters.

Solution

LEARNING GOAL
9 Learn the major units of measure in the English and metric systems, and be able to convert from one system to another.

First, recognize that the prefix *centi-* means $\frac{1}{100}$ of the base unit, the meter (m), just as one cent is $\frac{1}{100}$ of a dollar. There are 100 cents in a dollar and there are 100 cm in one meter. Thus, our conversion factor is either

$$\frac{1\ m}{100\ cm} \quad \text{or} \quad \frac{100\ cm}{1\ m}$$

each being equal to 1. Only one, however, will result in proper cancellation of units, producing the correct answer to the problem. If we proceed as follows:

$$10.0\ \cancel{cm} \times \frac{1\ m}{100\ \cancel{cm}} = 0.100\ m$$

Data given	Conversion factor	Desired result

Continued—

EXAMPLE 1.7 —*Continued*

we obtain the desired units, meters. If we had used the conversion factor 100 cm/1 m, the resulting units would be meaningless and the answer would have been incorrect:

$$10.0 \text{ cm} \times \frac{100 \text{ cm}}{1 \text{ m}} = 1000 \frac{\text{cm}^2}{\text{m}}$$

Incorrect units

Practice Problem 1.7

a. Convert 1.0 liter to each of the following units, using the factor-label method:

    ____ milliliters   ____ centiliters

    ____ microliters   ____ dekaliters

    ____ kiloliters

b. Convert 1.0 gram to each of the following units:

    ____ micrograms   ____ centigrams

    ____ milligrams   ____ decigrams

    ____ kilograms

For Further Practice: Questions 1.65c, d, e and 1.66d, e.

Conversion of Units from One System to Another

The conversion of a quantity expressed in units of one system to an equivalent quantity in the other system (English to metric or metric to English) requires a *bridging* conversion unit. Examples are shown in Table 1.3.

English and metric conversions are shown in Tables 1.1 and 1.2.

The conversion may be represented as a three-step process:

Step 1. Conversion from the units given in the problem to a bridging unit.

Step 2. Conversion to the other system using the bridge.

Step 3. Conversion within the desired system to units required by the problem.

TABLE 1.3	Commonly Used "Bridging" Units for Intersystem Conversions		
Quantity	**English**		**Metric**
Mass	1 pound	=	454 grams
	2.2 pounds	=	1 kilogram
Length	1 inch	=	2.54 centimeters
	1 yard	=	0.91 meter
Volume	1 quart	=	0.946 liter
	1 gallon	=	3.78 liters

| EXAMPLE 1.8 | **Using Conversion Factors Between Systems** |

Convert 4.00 ounces to kilograms.

Solution

Step 1. A convenient bridging unit for mass is 1 pound = 454 grams. To use this conversion factor, we relate ounces (given in the problem) to pounds:

$$4.00 \text{ ounces} \times \frac{1 \text{ pound}}{16 \text{ ounces}} = 0.250 \text{ pound}$$

Step 2. Using the bridging unit conversion, we get

$$0.250 \text{ pound} \times \frac{454 \text{ grams}}{1 \text{ pound}} = 114 \text{ grams}$$

Step 3. Grams may then be directly converted to kilograms, the desired unit:

$$114 \text{ grams} \times \frac{1 \text{ kilogram}}{1000 \text{ grams}} = 0.114 \text{ kilogram}$$

The calculation may also be done in a single step by arranging the factors in a chain:

$$4.00 \text{ oz} \times \frac{1 \text{ lb}}{16 \text{ oz}} \times \frac{454 \text{ g}}{1 \text{ lb}} \times \frac{1 \text{ kg}}{1000 \text{ g}} = 0.114 \text{ kg}$$

Helpful Hint: Refer to the discussion of rounding off numbers on page 18.

Practice Problem 1.8

Convert:
a. 0.50 inch to meters
b. 0.75 quart to liters
c. 56.8 grams to ounces
d. 0.50 inch to centimeters
e. 0.75 quart to milliliters
f. 56.8 milligrams to ounces

For Further Practice: Questions 1.63c, d, e and 1.64c, d, e.

| EXAMPLE 1.9 | **Using Conversion Factors Involving Exponents** |

Convert 1.5 meters² to centimeters².

Solution

The problem is similar to the conversion performed in previous examples. However, we must remember to include the exponent in the units. Thus

$$1.5 \text{ m}^2 \times \left(\frac{10^2 \text{ cm}}{1 \text{ m}}\right)^2 = 1.5 \text{ m}^2 \times \frac{10^4 \text{ cm}^2}{1 \text{ m}^2} = 1.5 \times 10^4 \text{ cm}^2$$

Continued—

EXAMPLE 1.9 —*Continued*

Note: The exponent affects both the number *and* unit within the parentheses.

Practice Problem 1.9

Convert:
a. 1.5 cm² to m²
b. 3.6 m² to cm²

For Further Practice: Questions 1.71 and 1.72.

1.5 Experimental Quantities

Thus far we have discussed the scientific method and its role in acquiring data and converting the data to obtain the results of the experiment. We have seen that such data must be reported in the proper units with the appropriate number of significant figures. The quantities that are most often determined include mass, length, volume, time, temperature, and energy. Now let's look at each of these quantities in more detail.

Mass

Mass describes the quantity of matter in an object. The terms *weight* and *mass,* in common usage, are often considered synonymous. They are not, in fact. **Weight** is the force of gravity on an object:

Weight = mass × acceleration due to gravity

When gravity is constant, mass and weight are directly proportional. But gravity is not constant; it varies as a function of the distance from the center of the earth. Therefore weight cannot be used for scientific measurement because the weight of an object may vary from one place on the earth to the next.

Mass, on the other hand, is independent of gravity; it is a result of a comparison of an unknown mass with a known mass called a *standard mass.* Balances are instruments used to measure the mass of materials.

Examples of common balances used for the determination of mass are shown in Figure 1.8.

The common conversion units for mass are as follows:

$$1 \text{ gram (g)} = 10^{-3} \text{ kilogram (kg)} = \frac{1}{454} \text{ pound (lb)}$$

In chemistry, when we talk about incredibly small bits of matter such as individual atoms or molecules, units such as grams and even micrograms are much too large. We don't say that a 100-pound individual weighs 0.0500 ton; the unit does not fit the quantity being described. Similarly, an atom of a substance such as hydrogen is very tiny. Its mass is only 1.661×10^{-24} gram.

One *atomic mass unit* (amu) is a more convenient way to represent the mass of one hydrogen atom, rather than 1.661×10^{-24} gram:

$$1 \text{ amu} = 1.661 \times 10^{-24} \text{ g}$$

Figure 1.8

Three common balances that are useful for the measurement of mass. (a) A two-pan comparison balance for approximate mass measurement suitable for routine work requiring accuracy to 0.1 g (or perhaps 0.01 g). (b) A top-loading single-pan electronic balance that is similar in accuracy to (a) but has the advantages of speed and ease of operation. The revolution in electronics over the past twenty years has resulted in electronic balances largely supplanting the two-pan comparison balance in routine laboratory usage. (c) An analytical balance that is capable of precise mass measurement (three to five significant figures beyond the decimal point). A balance of this type is used when the highest level of precision and accuracy is required.

(a)

(b)

(c)

Units should be chosen to suit the quantity being described. This can easily be done by choosing a unit that gives an exponential term closest to 10^0.

Length

The standard metric unit of *length,* the distance between two points, is the meter. Large distances are measured in kilometers; smaller distances are measured in millimeters or centimeters. Very small distances such as the distances between atoms on a surface are measured in *nanometers* (nm):

$$1 \text{ nm} = 10^{-7} \text{ cm} = 10^{-9} \text{ m}$$

Common conversions for length are as follows:

$$1 \text{ meter (m)} = 10^2 \text{ centimeters (cm)} = 3.94 \times 10^1 \text{ inch (in.)}$$

Volume

The standard metric unit of *volume,* the space occupied by an object, is the liter. A liter is the volume occupied by 1000 grams of water at 4 degrees Celsius (°C). The volume, 1 liter, also corresponds to:

$$1 \text{ liter (L)} = 10^3 \text{ milliliters (mL)} = 1.06 \text{ quarts (qt)}$$

The relationship between the liter and the milliliter is shown in Figure 1.9.

Typical laboratory devices used for volume measurement are shown in Figure 1.10. These devices are calibrated in units of milliliters (mL) or microliters (μL); one mL is, by definition, equal to one cm³. The volumetric flask is designed to *contain* a specified volume, and the graduated cylinder, pipet, and buret *dispense* a desired volume of liquid.

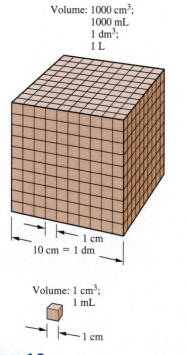

Volume: 1000 cm³; 1000 mL 1 dm³; 1 L

10 cm = 1 dm   1 cm

Volume: 1 cm³; 1 mL

1 cm

Figure 1.9

The relationships among various volume units.

(a)

(b)

(c)

(d)

Figure 1.10
Common laboratory equipment used for the measurement of volume. Graduated (a) cylinders, (b) pipets, and (c) burets are used for the delivery of liquids. (d) Volumetric flasks are used to contain a specific volume. A graduated cylinder is usually used for measurement of approximate volume; it is less accurate and precise than either pipets or burets.

Time

The standard metric unit of time is the second. The need for accurate measurement of time by chemists may not be as apparent as that associated with mass, length, and volume. It is necessary, however, in many applications. In fact, matter may be characterized by measuring the time required for a certain process to occur. The rate of a chemical reaction is a measure of change as a function of time.

LEARNING GOAL

10 Know the three common temperature scales and be able to convert from one to another.

Temperature

Temperature is the degree of "hotness" of an object. This may not sound like a very "scientific" definition, and, in a sense, it is not. We know intuitively the difference between a "hot" and a "cold" object, but developing a precise definition to explain this is not easy. We may think of the temperature of an object as a measure of the amount of heat in the object. However, this is not strictly true. An object increases in temperature because its heat content has increased and vice versa; however, the relationship between heat content and temperature depends on the quantity and composition of the material.

Many substances, such as mercury, expand as their temperature increases, and this expansion provides us with a way to measure temperature and temperature changes. If the mercury is contained within a sealed tube, as it is in a thermometer, the height of the mercury is proportional to the temperature. A mercury thermometer may be calibrated, or scaled, in different units, just as a ruler can be. Three common temperature scales are *Fahrenheit (°F)*, *Celsius (°C)*, and *Kelvin (K)*. Two convenient reference temperatures that are used to calibrate a thermometer are the freezing and boiling temperatures of water. Figure 1.11 shows the relationship between the scales and these reference temperatures.

Figure 1.11
The freezing point and boiling point of water, body temperature, and room temperature expressed in the three common units of temperature.

The Kelvin scale is of particular importance because it is directly related to molecular motion. As molecular speed increases, the Kelvin temperature proportionately increases.

The Kelvin symbol does not have a degree sign. The degree sign implies a value that is *relative* to some standard. Kelvin is an *absolute* scale.

Although Fahrenheit temperature is most familiar to us, Celsius and Kelvin temperatures are used exclusively in scientific measurements. It is often necessary to convert a temperature reading from one scale to another. To convert from Fahrenheit to Celsius, we use the following formula:

$$^\circ C = \frac{^\circ F - 32}{1.8}$$

To convert from Celsius to Fahrenheit, we solve this formula for °F, resulting in

$$^\circ F = 1.8\,^\circ C + 32$$

To convert from Celsius to Kelvin, we use the formula

$$K = {}^\circ C + 273.15$$

EXAMPLE 1.10   Converting from Fahrenheit to Celsius and Kelvin

Normal body temperature is 98.6°F. Calculate the corresponding temperature in degrees Celsius:

Solution

Using the expression relating °C and °F,

$$^\circ C = \frac{^\circ F - 32}{1.8}$$

Substituting the information provided,

$$= \frac{98.6 - 32}{1.8} = \frac{66.6}{1.8}$$

results in:

$$= 37.0\,^\circ C$$

Calculate the corresponding temperature in Kelvin units:

Solution

Using the expression relating K and °C,

$$K = {}^\circ C + 273.15$$

substituting the value obtained in the first part,

$$= 37.0 + 273.15$$

results in:

$$= 310.2\ K$$

LEARNING GOAL

10 Know the three common temperature scales and be able to convert from one to another.

Practice Problem 1.10

a. The freezing temperature of water is 32°F. Calculate the freezing temperature of water in Celsius units and Kelvin units.
b. When a patient is ill, his or her temperature may increase to 104°F. Calculate the temperature of this patient in Celsius units and Kelvin units.

For Further Practice: Questions 1.67 and 1.68.

Energy

Energy, the ability to do work, may be categorized as either **kinetic energy,** the energy of motion, or **potential energy,** the energy of position. Kinetic energy may be considered as energy in process; potential energy is stored energy. All energy is either kinetic or potential.

Another useful way of classifying energy is by form. The principal forms of energy include light, heat, electrical, mechanical, and chemical energy. All of these forms of energy share the following set of characteristics:

- In chemical reactions, energy cannot be created or destroyed.
- Energy may be converted from one form to another.
- Conversion of energy from one form to another always occurs with less than 100% efficiency. Energy is not lost (remember, energy cannot be destroyed) but rather, is not useful. We buy gasoline to move our car from place to place; however, much of the energy stored in the gasoline is released as heat.
- All chemical reactions involve either a "gain" or a "loss" of energy.

Energy absorbed or liberated in chemical reactions is usually in the form of heat energy. Heat energy may be represented in units of *calories* or *joules,* their relationship being

$$1 \text{ calorie (cal)} = 4.18 \text{ joules (J)}$$

One calorie is defined as the amount of heat energy required to increase the temperature of 1 gram of water 1°C.

Heat energy measurement is a quantitative measure of heat content. It is an extensive property, dependent upon the quantity of material. Temperature, as we have mentioned, is an intensive property, independent of quantity.

Not all substances have the same capacity for holding heat; 1 gram of iron and 1 gram of water, even if they are at the same temperature, do *not* contain the same amount of heat energy. One gram of iron will absorb and store 0.108 calorie of heat energy when the temperature is raised 1°C. In contrast, 1 gram of water will absorb almost ten times as much energy, 1.00 calorie, when the temperature is increased an equivalent amount.

Units for other forms of energy will be introduced in later chapters.

The *kilocalorie* (kcal) is the familiar nutritional calorie. It is also known as the large Calorie; note that in this term the *C* is uppercase to distinguish it from the normal calorie. The large calorie is 1000 small calories. Refer to Section 7.2 and A Human Perspective: Food Calories for more information.

Water in the environment (lakes, oceans, and streams) has a powerful effect on the climate because of its ability to store large quantities of energy. In summer, water stores heat energy, moderating temperatures of the surrounding area. In winter, some of this stored energy is released to the air as the water temperature falls; this prevents the surroundings from experiencing extreme changes in temperature.

Concentration

Concentration is a measure of the number of particles of a substance, or the mass of those particles, that are contained in a specified volume. Concentration is a widely used way of representing mixtures of different substances. Examples include:

- The concentration of oxygen in the air
- Pollen counts, given during the hay fever seasons, which are simply the number of grains of pollen contained in a measured volume of air
- The amount of an illegal drug in a certain volume of blood, indicating the extent of drug abuse
- The proper dose of an antibiotic, based on a patient's weight.

We will describe many situations in which concentration is used to predict useful information about chemical reactions (Sections 6.4 and 8.2, for example). In Chapter 6 we calculate a numerical value for concentration from experimental data.

A Human Perspective

Food Calories

The body gets its energy through the processes known collectively as metabolism, which will be discussed in detail in subsequent chapters on biochemistry and nutrition. The primary energy sources for the body are carbohydrates, fats, and proteins, which we obtain from the foods we eat. The amount of energy available from a given foodstuff is related to the Calories (C) available in the food. Calories are a measure of the energy and heat content that can be derived from the food. One (food) Calorie (symbolized by C) equals 1000 (metric) calories (symbolized by c):

$$1 \text{ Calorie} = 1000 \text{ calories} = 1 \text{ kilocalorie}$$

The energy available in food can be measured by totally burning the food; in other words, we are using the food as a fuel. The energy given off in the form of heat is directly related to the amount of chemical energy, energy stored in chemical bonds, that is available in the food and that the food could provide to the body through the various metabolic pathways.

The classes of food molecules are not equally energy rich. For instance, when oxidized via metabolic pathways, carbohydrates and proteins provide the cell with 4 Calories per gram, whereas fats generate approximately 9 Calories per gram.

In addition, as with all processes, not all the available energy can be efficiently extracted from the food; a certain percentage is always lost. The average person requires between 2000 and 3000 Calories per day to maintain normal body functions such as the regulation of body temperature, muscle movement, and so on. If a person takes in more Calories than the body uses, the Calorie-containing substances will be stored as fat, and the person will gain weight. Conversely, if a person uses more Calories than are ingested, the individual will lose weight.

Excess Calories are stored in the form of fat, the form that provides the greatest amount of energy per gram. Too many Calories lead to too much fat. Similarly, a lack of Calories (in the form of food) forces the body to raid its storehouse, the fat. Weight is lost in this process as the fat is consumed. Unfortunately, it always seems easier to add fat to the storehouse than to remove it.

The "rule of thumb" is that 3500 Calories are equivalent to approximately 1 pound of body weight. You have to take in 3500 Calories more than you use to gain a pound, and you have to expend 3500 Calories more than you normally use to

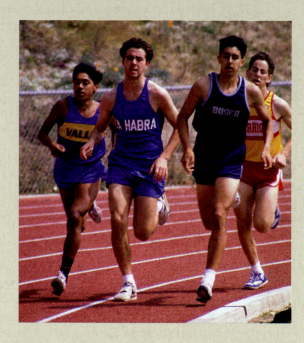

lose a pound. If you eat as little as 100 Calories a day above your body's needs, you could gain about 10–11 pounds per year:

$$\frac{100 \text{ } \cancel{C}}{\text{day}} \times \frac{365 \text{ } \cancel{day}}{1 \text{ year}} \times \frac{1 \text{ lb}}{3500 \text{ } \cancel{C}} = \frac{10.4 \text{ lb}}{\text{year}}$$

A frequently recommended procedure for increasing the rate of weight loss involves a combination of dieting (taking in fewer Calories) and exercise. Running, swimming, jogging, and cycling are particularly efficient forms of exercise. Running burns 0.11 Calories per minute per pound of body weight; swimming burns approximately 0.05 Calories per minutes per pound of body weight.

For Further Understanding

Sarah runs 1 hour each day, and Nancy swims 2 hours each day. Assuming that Sarah and Nancy are the same weight, which girl burns more calories in 1 week?

Would you expect a runner to burn more calories in summer or winter? Why?

TABLE 1.4	Densities of Some Common Materials		
Substance	**Density (g/mL)**	**Substance**	**Density (g/mL)**
Air	0.00129 (at 0°C)	Methyl alcohol	0.792
Ammonia	0.000771 (at 0°C)	Milk	1.028–1.035
Benzene	0.879	Oxygen	0.00143 (at 0°C)
Bone	1.7–2.0	Rubber	0.9–1.1
Carbon dioxide	0.001963 (at 0°C)	Turpentine	0.87
Ethyl alcohol	0.789	Urine	1.010–1.030
Gasoline	0.66–0.69	Water	1.000 (at 4°C)
Gold	19.3	Water	0.998 (at 20°C)
Hydrogen	0.000090 (at 0°C)	Wood	0.3–0.98
Kerosene	0.82	(balsa, least dense; ebony	
Lead	11.3	and teak, most dense)	
Mercury	13.6		

Figure 1.12
Density (mass/volume) is a unique property of a material. A mixture of wood, water, brass, and mercury is shown, with the cork—the least dense—floating on water. Additionally, brass, with a density greater than water but less than liquid mercury, floats on the interface between these two liquids.

Density and Specific Gravity

Both mass and volume are a function of the *amount* of material present (extensive property). **Density,** the ratio of mass to volume,

$$d = \frac{\text{mass}}{\text{volume}} = \frac{m}{V}$$

is *independent* of the amount of material (intensive property). Density is a useful way to characterize or identify a substance because each substance has a unique density (Figure 1.12).

In density calculations, the mass is usually represented in grams, and volume is given in either milliliters (mL) or cubic centimeters (cm^3 or cc):

$$1 \text{ mL} = 1 \text{ cm}^3 = 1 \text{ cc}$$

The unit of density would therefore be g/mL, g/cm^3, or g/cc.

One milliliter of air and 1 milliliter of iron do not weigh the same amount. There is much more mass in 1 milliliter of iron; its density is greater. Density measurements were used to discriminate between real gold and "fool's gold" during the gold rush era. Today the measurement of the density of a substance is still a valuable analytical technique. The densities of a number of common substances are shown in Table 1.4.

LEARNING GOAL

11 Use density, mass, and volume in problem solving, and calculate the specific gravity of a substance from its density.

Animation
Solid-Liquid Density

Intensive and extensive properties were described on page 10.

Calculating the Density of a Solid	EXAMPLE 1.11

2.00 cm^3 of aluminum are found to weigh 5.40 g. Calculate the density of aluminum in units of g/cm^3 and g/mL.

Continued—

LEARNING GOAL

11 Use density, mass, and volume in problem solving, and calculate the specific gravity of a substance from its density.

EXAMPLE 1.11 —*Continued*

Solution

The density expression is:

$$d = \frac{m}{V} = \frac{g}{cm^3}$$

Substituting the information given in the problem,

$$= \frac{5.40 \ g}{2.00 \ cm^3}$$

results in:

$$= 2.70 \ g/cm^3$$

Since 1 mL = 1 cm³ (p. 26), we can use this identity as a conversion factor

$$2.70 \ \frac{g}{cm^3} \times \frac{1 \ cm^3}{1 \ mL} = 2.70 \ g/mL$$

Practice Problem 1.11

0.500 mL of mercury has a mass of 6.80 grams. Calculate the density of mercury in units of g/mL and g/cm³.

For Further Practice: Questions 1.81 and 1.82.

EXAMPLE 1.12   **Calculating the Mass of a Gas from Its Density**

LEARNING GOAL

11 Use density, mass, and volume in problem solving, and calculate the specific gravity of a substance from its density.

Air has a density of 0.0013 g/mL. What is the mass of a 6.0-L sample of air?

Solution

$$0.0013 \ g/mL = 1.3 \times 10^{-3} \ g/mL$$

(The decimal point is moved three positions to the right.) This problem can be solved by using conversion factors:

$$6.0 \ L \ air \times \frac{10^3 \ mL \ air}{1 \ L \ air} \times \frac{1.3 \times 10^{-3} \ g \ air}{mL \ air} = 7.8 \ g \ air$$

Practice Problem 1.12

What mass of air (grams) would be found in a 2.0 L party balloon?

For Further Practice: Questions 1.83 and 1.84.

EXAMPLE 1.13   **Using the Density to Calculate the Mass of a Liquid**

Calculate the mass, in grams, of 10.0 mL of mercury (symbolized Hg) if the density of mercury is 13.6 g/mL.

Continued—

A Human *Perspective*

Assessing Obesity: The Body-Mass Index

Density, the ratio of two extensive properties, mass and volume, is an intensive property that can provide useful information about the identity and properties of a substance. The Body-Mass Index (BMI) is also a ratio of two extensive properties, the weight and height (actually the square of the height) of an individual. As a result, the BMI is also an intensive property. It is widely used by physical trainers, medical professionals, and life insurance companies to quantify obesity, which is a predictor of a variety of potential medical problems.

In metric units, the Body-Mass Index is expressed:

$$BMI = \frac{Weight\ (kg)}{Height\ (m^2)}$$

This can be converted to the English system:

$$BMI = \frac{Weight\ (lb)}{Height\ (in^2)} \times 703$$

The number 703 is the commonly-used conversion factor to convert from English units (inches and pounds) to metric units (meters and kilograms) that are the units in the definition of BMI.

The conversion is accomplished in the following way:

$$BMI = \frac{kg}{m^2} = \frac{lb}{in^2} \times \frac{1kg}{2.205\ lb} \times \left(\frac{39.37\ in}{1m}\right)^2$$

Weight and height (metric)   Weight and height (English)

$$= \frac{lb}{in^2} \times 703\ \frac{kg\text{-}in^2}{lb\text{-}m^2}$$

The units of the conversion factor are generally not shown and the BMI in English units is reduced to:

$$BMI = \frac{Weight\ in\ pounds}{(Height\ in\ inches)^2} \times 703$$

Computer website BMI calculators generally use this form of the equation.

An individual with a BMI of 25 or greater is considered overweight; if the BMI is 30 or greater, the individual is described as obese. One's BMI, once known, can be used as a guideline in the design of suitable diet and exercise programs.

BMI values for a variety of weights and heights are shown as a function of individuals height and weight:

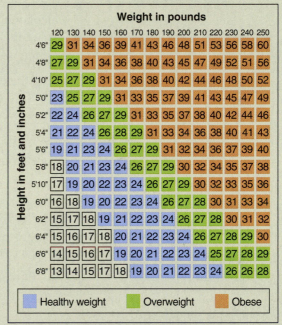

Developed by the National Center for Health Statistics in collaboration with the National Center for Chronic Disease Prevention and Health Promotion

For Further Understanding

Refer to A Human Perspective: Food Calories (p. 30) and describe connections between these two perspectives.

Calculate your BMI in both metric and English units. Should they agree? Do they?

EXAMPLE 1.13 —*Continued*

Solution

Using the density as a conversion factor from volume to mass, we have

$$m = (10.0\ \text{mL Hg})\left(13.6\ \frac{g\ Hg}{mL\ Hg}\right)$$

Continued—

EXAMPLE 1.13 —*Continued*

Cancellation of units results in:

$$= 136 \text{ g Hg}$$

Practice Problem 1.13

The density of ethyl alcohol (200 proof, or pure alcohol) is 0.789 g/mL at 20°C. Calculate the mass of a 30.0-mL sample.

For Further Practice: Questions 1.85 and 1.86.

EXAMPLE 1.14   **Using the Density to Calculate the Volume of a Liquid**

Calculate the volume, in milliliters, of a liquid that has a density of 1.20 g/ mL and a mass of 5.00 grams.

Solution

Using the density as a conversion factor from mass to volume, we have

$$V = (5.00 \text{ g liquid})\left(\frac{1 \text{ mL liquid}}{1.20 \text{ g liquid}}\right)$$

Cancellation of units results in:

$$= 4.17 \text{ mL liquid}$$

Practice Problem 1.14

Calculate the volume, in milliliters, of 10.0 g of a saline solution that has a density of 1.05 g/mL.

For Further Practice: Questions 1.90 and 1.93.

Specific gravity is frequently referenced to water at 4°C, its temperature of maximum density (1.000 g/mL). Other reference temperatures may be used. However, the temperature must be specified.

For convenience, values of density are often related to a standard, well-known reference, the density of pure water at 4°C. This "referenced" density is called the **specific gravity,** the ratio of the density of the object in question to the density of pure water at 4°C.

$$\text{specific gravity} = \frac{\text{density of object (g/mL)}}{\text{density of water (g/mL)}}$$

Specific gravity is a *unitless* term. Because the density of water at 4.0°C is 1.00 g/mL, the numerical values for the density and specific gravity of a substance are equal. That is, an object with a density of 2.00 g/mL has a specific gravity of 2.00 at 4°C.

Routine hospital tests involving the measurement of the specific gravity of urine and blood samples are frequently used as diagnostic tools. For example, diseases such as kidney disorders and diabetes change the composition of urine. This compositional change results in a corresponding change in the specific gravity. This change is easily measured and provides the basis for a quick preliminary diagnosis. This topic is discussed in greater detail in A Medical Perspective: Diagnosis Based on Waste.

A Human Perspective

Quick and Useful Analysis

Measurement of the specific gravity of a liquid is fast, easy, and nondestructive of the sample. Changes in specific gravity over time can provide a wealth of information. Two examples follow:

Urine, a waste product consisting of a wide variety of metabolites, may be analyzed to indicate abnormalities in various metabolic processes or even unacceptable behavior (recall the steroid tests in Olympic competition).

Many of these tests must be performed by using sophisticated and sensitive instrumentation. However, a very simple test, the measurement of the specific gravity of urine, can be an indicator of diabetes mellitus or Bright's disease. The normal range for human urine specific gravity is 1.010–1.030.

A hydrometer, a weighted glass bulb inserted in a liquid, may be used to determine specific gravity. The higher it floats in the liquid, the more dense the liquid. A hydrometer that is calibrated to indicate the specific gravity of urine is called a urinometer.

Winemaking is a fermentation process (chapter 12). The flavor, aroma, and composition of wine depend upon the extent of fermentation. As fermentation proceeds, the specific gravity of the wine gradually changes. Periodic measurement of the specific gravity during fermentation enables the winemaker to determine when the wine has reached its optimal composition.

For Further Understanding

Give reasons that may account for such a broad range of "normal" values for urine specific gravity.

Could results for a diabetes test depend on food or medicine consumed prior to the test?

Monitoring the winemaking process.

A hydrometer, used in the measurement of the specific gravity of urine.

SUMMARY

1.1 The Discovery Process

Chemistry is the study of matter and the changes that matter undergoes. *Matter* is anything that has mass and occupies space. The changes that matter undergoes always involve either gain or loss of energy. *Energy* is the ability to do work (to accomplish some change). Thus a study of chemistry involves matter, energy, and their interrelationship.

The major areas of chemistry include biochemistry, organic chemistry, inorganic chemistry, analytical chemistry, and physical chemistry.

The *scientific method* consists of six interrelated processes: observation, questioning, pattern recognition, development of *theories* from *hypotheses*, experimentation, and summarizing information. A *law* summarizes a large quantity of information.

The development of the scientific method has played a major role in civilization's rapid growth during the past two centuries.

1.2 Matter and Properties

A scientific experiment produces *data*. Each piece of data arises from a single measurement. Mass, length, volume, time, temperature, and energy are the most common types of data obtained from chemical experiments.

Results are the outcome of an experiment. Usually, several pieces of data are combined, often using a mathematical equation, to produce a result.

Properties (characteristics) of matter may be classified as either physical or chemical. *Physical properties* can be observed without changing the chemical composition of the sample. *Chemical properties* result in a change in composition and can be observed only through *chemical reactions*. *Intensive properties* are independent of the quantity of the substance. *Extensive properties* depend on the quantity of a substance.

Three states of matter exist (*solid, liquid,* and *gas*); these states of matter are distinguishable by differences in physical properties.

All matter is classified as either a *pure substance* or a *mixture*. A pure substance is a substance that has only one component. A mixture is a combination of two or more pure substances in which the combined substances retain their identity.

A *homogeneous mixture* has uniform composition. Its particles are well mixed. A *heterogeneous mixture* has a non-uniform composition.

An *element* is a pure substance that cannot be converted into a simpler form of matter by any chemical reaction. A *compound* is a substance produced from the combination of two or more elements in a definite, reproducible fashion.

1.3 Significant Figures and Scientific Notation

Significant figures are all digits in a number representing data or results that are known with certainty plus the first uncertain digit. The number of significant figures associated with a measurement is determined by the measuring device. Results should be rounded off to the proper number of significant figures.

Error is defined as the difference between the true value and our estimation, or measurement, of the value. *Accuracy* is the degree of agreement between the true and measured values. *Uncertainty* is the degree of doubt in a single measurement. The number of meaningful digits in a measurement is determined by the measuring device. *Precision* is a measure of the agreement of replicate measurements.

Very large and very small numbers may be represented with the proper number of significant figures by using *scientific notation*.

1.4 Units and Unit Conversion

Science is the study of humans and their environment. Its tool is experimentation.

A *unit* defines the basic quantity of mass, volume, time, and so on. A number that is not followed by the correct unit usually conveys no useful information.

The metric system is a decimal-based system in contrast to the English system. In the metric system, mass is represented as the gram, length as the meter, and volume as the liter. Any subunit or multiple unit contains one of these units preceded by a prefix indicating the power of ten by which the base unit is to be multiplied to form the subunit or multiple unit. Scientists favor this system over the not-so-systematic English units of measurement.

To convert one unit to another, we must set up a *conversion factor* or series of conversion factors that relate two units. The proper use of these conversion factors is referred to as the factor-label method. This method is used either to convert from one unit to another within the same system or to convert units from one system to another. It is a very useful problem-solving tool.

1.5 Experimental Quantities

Mass describes the quantity of matter in an object. The terms weight and mass are often used interchangeably, but they are not equivalent. *Weight* is the force of gravity on an object. The fundamental unit of mass in the metric system is the gram. One atomic mass unit (amu) is equal to 1.661×10^{-24} g.

The standard metric unit of length is the meter. Large distances are measured in kilometers; smaller distances are measured in millimeters or centimeters. Very small distances (on the atomic scale) are measured in nanometers (nm). The standard metric unit of volume is the liter. A liter is the volume occupied by 1000 grams of water at 4 degrees Celsius. The standard metric unit of time is the second, a unit that is used in the English system as well.

Temperature is the degree of "hotness" of an object. Many substances, such as liquid mercury, expand as their temperature increases, and this expansion provides us with a way to measure temperature and temperature changes. Three common temperature scales are Fahrenheit (°F), Celsius (°C), and Kelvin (K).

Energy, the ability to do work, may be categorized as either *kinetic energy,* the energy of motion, or *potential energy,* the energy of position. The principal forms of energy are light, heat, mechanical, electrical, nuclear, and chemical energy.

Energy absorbed or liberated in chemical reactions is most often in the form of heat energy. Heat energy may be represented in units of calories or joules: 1 calorie (cal) = 4.18 joules (J). One calorie is defined as the amount of heat energy required to change the temperature of 1 gram of water 1°C.

Concentration is a measure of the number of particles of a substance, or the mass of those particles, that are contained in a specified volume. Concentration is a widely used way of representing relative quantities of different substances in a mixture of those substances.

Density is the ratio of mass to volume and is a useful way of characterizing a substance. Values of density are often related to a standard reference, the density of pure water at 4°C. This "referenced" density is the *specific gravity,* the ratio of the density of the object in question to the density of pure water at 4°C.

KEY TERMS

accuracy (1.3)	mass (1.5)
chemical property (1.2)	matter (1.1)
chemical reaction (1.2)	mixture (1.2)
chemistry (1.1)	physical change (1.2)
compound (1.2)	physical property (1.2)
concentration (1.5)	potential energy (1.5)
data (1.2)	precision (1.3)
density (1.5)	properties (1.2)
element (1.2)	pure substance (1.2)
energy (1.1)	result (1.2)
error (1.3)	scientific method (1.1)
extensive property (1.2)	scientific notation (1.3)
gaseous state (1.2)	significant figures (1.3)
heterogeneous mixture (1.2)	solid state (1.2)
homogeneous mixture (1.2)	specific gravity (1.5)
hypothesis (1.1)	temperature (1.5)
intensive property (1.2)	theory (1.1)
kinetic energy (1.5)	uncertainty (1.3)
law (1.1)	unit (1.4)
liquid state (1.2)	weight (1.5)

QUESTIONS AND PROBLEMS

The Discovery Process

Foundations

1.11 Define each of the following terms:
   a. chemistry
   b. matter
   c. energy

1.12 Define each of the following terms:
   a. hypothesis
   b. theory
   c. law

1.13 Define each of the following terms:
   a. potential energy
   b. kinetic energy
   c. data

1.14 Define each of the following terms:
   a. results
   b. mass
   c. weight

1.15 Give the base unit for each of the following in the metric system:
   a. mass
   b. volume
   c. length

1.16 Give the base unit for each of the following in the metric system:
   a. time
   b. temperature
   c. energy

Applications

1.17 Discuss the difference between the terms *mass* and *weight*.

1.18 Discuss the difference between the terms *data* and *results*.

1.19 Distinguish between specific gravity and density.

1.20 Distinguish between kinetic energy and potential energy.

1.21 Discuss the meaning of the term *scientific method*.

1.22 Describe an application of reasoning involving the scientific method that has occurred in your day-to-day life.

1.23 Stem-cell research has the potential to provide replacement "parts" for the human body. Is this statement a hypothesis, theory, or law? Explain your reasoning.

1.24 Observed increases in global temperatures are caused by elevated levels of carbon dioxide. Is this statement a hypothesis, theory, or law? Explain your reasoning.

1.25 Describe an experiment demonstrating that the freezing point of water changes when salt (sodium chloride) is added to water.

1.26 Describe an experiment that would enable you to determine the amount (grams) of solids suspended in a 1-L sample of seawater.

Matter and Properties

Foundations

1.27 Describe what is meant by a physical property.

1.28 Describe what is meant by a physical change.

1.29 Describe several chemical properties of matter.

1.30 Describe what is meant by a chemical reaction.

1.31 Distinguish between a pure substance and a mixture.

1.32 Give examples of pure substances and mixtures.

1.33 Distinguish between a homogeneous mixture and a heterogeneous mixture.

1.34 Distinguish between an intensive property and an extensive property.

1.35 Describe the general properties of the gaseous state.

1.36 Contrast the physical properties of the gaseous and solid states.

Applications

1.37 Label each of the following as either a physical change or a chemical reaction:
   a. An iron nail rusts.
   b. An ice cube melts.
   c. A limb falls from a tree.

1.38 Label each of the following as either a physical change or a chemical reaction:
   a. A puddle of water evaporates.
   b. Food is digested.
   c. Wood is burned.

1.39 Label each of the following properties of sodium as either a physical property or a chemical property:
   a. Sodium is a soft metal (can be cut with a knife).
   b. Sodium reacts violently with water to produce hydrogen gas and sodium hydroxide.

1.40 Label each of the following properties of sodium as either a physical property or a chemical property:
   a. When exposed to air, sodium forms a white oxide.
   b. Sodium melts at 98°C.
   c. The density of sodium metal at 25°C is 0.97 g/cm³.

1.41 Label each of the following as either a pure substance or a mixture:
   a. water
   b. table salt (sodium chloride)
   c. blood

1.42 Label each of the following as either a pure substance or a mixture:
   a. sucrose (table sugar)
   b. orange juice
   c. urine

1.43 Label each of the following as either a homogeneous mixture or a heterogeneous mixture:
   a. a soft drink
   b. a saline solution
   c. gelatin

1.44 Label each of the following as either a homogeneous mixture or a heterogeneous mixture:
   a. gasoline
   b. vegetable soup
   c. concrete

1.45 Label each of the following as either an intensive property or an extensive property:
   a. mass
   b. volume
   c. density

1.46 Label each of the following as either an intensive property or an extensive property:
   a. specific gravity
   b. temperature
   c. heat content

1.47 Describe the difference between the terms *atom* and *element*.

1.48 Describe the difference between the terms *atom* and *compound*.

1.49 Give at least one example of each of the following:
   a. an element
   b. a pure substance

1.50 Give at least one example of each of the following:
   a. a homogeneous mixture
   b. a heterogeneous mixture

Significant Figures and Scientific Notation

Foundations

1.51 How many significant figures are contained in each of the following numbers?
   a. 10.0   **d.** 2.062
   b. 0.214   **e.** 10.50
   c. 0.120   **f.** 1050

1.52 How many significant figures are contained in each of the following numbers?
   a. 3.8×10^{-3} **d.** 24
   b. 5.20×10^2 **e.** 240
   c. 0.00261   **f.** 2.40

1.53 Round the following numbers to three significant figures:
   a. 3.873×10^{-3} **d.** 24.3387
   b. 5.202×10^{-2} **e.** 240.1
   c. 0.002616   **f.** 2.407

1.54 Round the following numbers to three significant figures:
   a. 123700   **d.** 53.2995
   b. 0.00285792   **e.** 16.96
   c. 1.421×10^{-3} **f.** 507.5

1.55 Define each of the following terms:
   a. precision
   b. accuracy

1.56 Define each of the following terms:
   a. error
   b. uncertainty

Applications

1.57 Perform each of the following arithmetic operations, reporting the answer with the proper number of significant figures:
   a. (23)(657)   **d.** $1157.23 - 17.812$
   b. $0.00521 + 0.236$ **e.** $\dfrac{(1.987)(298)}{0.0821}$
   c. $\dfrac{18.3}{3.0576}$

1.58 Perform each of the following arithmetic operations, reporting the answer with the proper number of significant figures:
   a. $\dfrac{(16.0)(0.1879)}{45.3}$ **d.** $18 + 52.1$
   b. $\dfrac{(76.32)(1.53)}{0.052}$ **e.** $58.17 - 57.79$
   c. $(0.0063)(57.8)$

1.59 Express the following numbers in scientific notation (use the proper number of significant figures):
   a. 12.3   **e.** 92,000,000
   b. 0.0569   **f.** 0.005280
   c. -1527 **g.** 1.279
   d. 0.000000789   **h.** -531.77

1.60 Using scientific notation, express the number two thousand in terms of:
   a. one significant figure   **d.** four significant figures
   b. two significant figures   **e.** five significant figures
   c. three significant figures

1.61 Express each of the following numbers in decimal notation:
   a. 3.24×10^3 **e.** -8.21×10^{-2}
   b. 1.50×10^{-4} **f.** 2.9979×10^8
   c. 4.579×10^{-1} **g.** 1.50×10^0
   d. -6.83×10^5 **h.** 6.02×10^{23}

1.62 Which of the following numbers have two significant figures? Three significant figures? Four significant figures?
   a. 327   **e.** 7.8×10^3
   b. 1.049×10^4 **f.** 1507
   c. 1.70   **g.** 4.8×10^2
   d. 0.000570   **h.** 7.389×10^{15}

Units and Unit Conversion

Foundations

1.63 Convert 2.0 pounds to:
   a. ounces   **d.** milligrams
   b. tons   **e.** dekagrams
   c. grams

1.64 Convert 5.0 quarts to:
   a. gallons   **d.** milliliters
   b. pints   **e.** microliters
   c. liters

1.65 Convert 3.0 grams to:
   a. pounds   **d.** centigrams
   b. ounces   **e.** milligrams
   c. kilograms

1.66 Convert 3.0 meters to:
   a. yards   **d.** centimeters
   b. inches   **e.** millimeters
   c. feet

1.67 Convert 50.0°F to:
   a. °C
   b. K

1.68 Convert -10.0°F to:
   a. °C
   b. K

1.69 Convert 20.0°C to:
   a. K
   b. °F

1.70 Convert 300.0 K to:
   a. °C
   b. °F

1.71 A typical office has 144 ft² of floor space. Calculate the floor space in m².

1.72 Tire pressure is measured in units of lb/in². Convert 32 lb/in² to g/cm².

Applications

1.73 A 150-lb adult has approximately 9 pints of blood. How many liters of blood does the individual have?

1.74 If a drop of blood has a volume of 0.05 mL, how many drops of blood are in the adult described in Problem 1.73?

1.75 A patient's temperature is found to be 38.5°C. To what Fahrenheit temperature does this correspond?

1.76 A newborn is 21 inches in length and weighs 6 lb 9 oz. Describe the baby in metric units.

1.77 Which distance is shorter: 5.0 cm or 5.0 in.?

1.78 Which volume is smaller: 50.0 mL or 0.500 L?

1.79 Which mass is smaller: 5.0 mg or 5.0 μg?

1.80 Which volume is smaller: 1.0 L or 1.0 qt?

Experimental Quantities

Foundations

1.81 Calculate the density of a 3.00×10^2-g object that has a volume of 50.0 mL.

1.82 Calculate the density of 50.0 g of an isopropyl alcohol–water mixture (commercial rubbing alcohol) that has a volume of 63.6 mL.

1.83 What volume, in liters, will 8.00×10^2 g of air occupy if the density of air is 1.29 g/L?

1.84 In Question 1.83, you calculated the volume of 8.00×10^2 g of air with a density of 1.29 g/L. The temperature of the air sample was lowered and the density increased to 1.50 g/L. Calculate the new volume of the air sample.

1.85 What is the mass, in grams, of a piece of iron that has a volume of 1.50×10^2 mL and a density of 7.20 g/mL?

1.86 What is the mass of a femur (leg bone) having a volume of 118 cm³? The density of bone is 1.8 g/cm³.

Applications

1.87 You are given a piece of wood that is maple, teak, or oak. The piece of wood has a volume of 1.00×10^2 cm³ and a mass of 98 g. The densities of maple, teak, and oak are as follows:

Wood	Density (g/cm³)
Maple	0.70
Teak	0.98
Oak	0.85

What is the identity of the piece of wood?

1.88 The specific gravity of a patient's urine sample was measured to be 1.008. Given that the density of water is 1.000 g/mL at 4°C, what is the density of the urine sample?

1.89 The density of grain alcohol is 0.789 g/mL. Given that the density of water at 4°C is 1.00 g/mL, what is the specific gravity of grain alcohol?

1.90 The density of mercury is 13.6 g/mL. If a sample of mercury weighs 272 g, what is the volume of the sample in milliliters?

1.91 You are given three bars of metal. Each is labeled with its identity (lead, uranium, platinum). The lead bar has a mass of 5.0×10^1 g and a volume of 6.36 cm³. The uranium bar has a mass of 75 g and a volume of 3.97 cm³. The platinum bar has a mass of 2140 g and a volume of 1.00×10^2 cm³. Which of these metals has the lowest density? Which has the greatest density?

1.92 Refer to Problem 1.91. Suppose that each of the bars had the same mass. How could you determine which bar had the lowest density or highest density?

1.93 The density of methanol at 20°C is 0.791 g/mL. What is the volume of a 10.0-g sample of methanol?

1.94 The density of methanol at 20°C is 0.791 g/mL. What is the mass of a 50.0-mL sample of methanol?

CRITICAL THINKING PROBLEMS

1. An instrument used to detect metals in drinking water can detect as little as one microgram of mercury in one liter of water. Mercury is a toxic metal; it accumulates in the body and is responsible for the deterioration of brain cells. Calculate the number of mercury atoms you would consume if you drank one liter of water that contained only one microgram of mercury. (The mass of one mercury atom is 3.3×10^{-22} grams.)

2. Yesterday's temperature was 40°F. Today it is 80°F. Bill tells Sue that it is twice as hot today. Sue disagrees. Do you think Sue is correct or incorrect? Why or why not?

3. Aspirin has been recommended to minimize the chance of heart attacks in persons who have already had one or more occurrences. If a patient takes one aspirin tablet per day for ten years, how many pounds of aspirin will the patient consume? (Assume that each tablet is approximately 325 mg.)

4. Design an experiment that will allow you to measure the density of your favorite piece of jewelry.

5. The diameter of an aluminum atom is 250 picometers (1 picometer = 10^{-12} meters). How many aluminum atoms must be placed end to end to make a "chain" of aluminum atoms one foot long?

2

The Structure of the Atom and the Periodic Table

Learning Goals

1 Describe the important properties of protons, neutrons, and electrons.

2 Calculate the number of protons, neutrons, and electrons in any atom.

3 Distinguish among atoms, ions, and isotopes and calculate atomic masses from isotopic abundance.

4 Trace the history of the development of atomic theory, beginning with Dalton.

5 Explain the critical role of spectroscopy in the development of atomic theory and in our everyday lives.

6 State the basic postulates of Bohr's theory, its utility, and its limitations.

7 Recognize the important subdivisions of the periodic table: periods, groups (families), metals, and nonmetals.

8 Use the periodic table to obtain information about an element.

9 Describe the relationship between the electronic structure of an element and its position in the periodic table.

10 Write electron configurations for atoms of the most commonly occurring elements.

11 Use the octet rule to predict the charge of common cations and anions.

12 Utilize the periodic table and its predictive power to estimate the relative sizes of atoms and ions, as well as relative magnitudes of ionization energy and electron affinity.

Organization and understanding go hand-in-hand.

Introduction

Why does ice float on water? Why don't oil and water mix? Why does blood transport oxygen to our cells, whereas carbon monoxide inhibits this process? Questions such as these are best explained by understanding the behavior of substances at the atomic level.

In this chapter we will learn some of the properties of the major particles that make up the atom and look at early experiments that enabled us to develop theories of atomic structure. These theories, in turn, help us to explain the behavior of atoms themselves, as well as the compounds that result from their combination.

The structure of atoms of each element is unique, so it is useful to consider relationships and differences among the elements themselves. The unifying concept is called the periodic law, and it gives rise to an organized "map" of the elements that relates their structure to their chemical and physical properties. This "map" is the periodic table.

As we study the periodic law and periodic table, we shall see that the chemical and physical properties of elements follow directly from the electronic structure of the atoms that make up these elements. A thorough familiarity with the arrangement of the periodic table is vital to the study of chemistry. It not only allows us to predict the structure and properties of the various elements, but it also serves as the basis for developing an understanding of chemical bonding, or the process of forming molecules. Additionally, the properties and behavior of these larger units (bulk properties) are fundamentally related to the properties of the atoms that compose them.

Chemistry Connection

Managing Mountains of Information

Recall for a moment the first time that you sat down in front of a computer. Perhaps it was connected to the Internet; somewhere in its memory was a word processor program, a spreadsheet, a few games, and many other features with strange-sounding names. Your challenge, very simply, was to use this device to access and organize information. Several manuals, all containing hundreds of pages of bewilderment, were your only help. How did you overcome this seemingly impossible task?

We are quite sure that you did not succeed without doing some reading and talking to people who had experience with computers. Also, you did not attempt to memorize every single word in each manual.

Success with a computer or any other storehouse of information results from developing an overall understanding of the way in which the system is organized. Certain facts must be memorized, but seeing patterns and using these relationships allows us to accomplish a wide variety of tasks that involve similar logic.

The study of chemistry is much like "real life." Just as it is impossible to memorize every single fact that will allow you to run a computer or drive an automobile in traffic, it is equally impossible to learn every fact in chemistry. Knowing the organization and logic of a process, along with a few key facts, makes a task manageable.

One powerful organizational device in chemistry is the periodic table. Its use in organizing and predicting the behavior of all of the known elements (and many of the compounds formed from these elements) is the subject of this chapter.

2.1 Composition of the Atom

The basic structural unit of an element is the **atom,** which is the smallest unit of an element that retains the chemical properties of that element. A tiny sample of the element copper, too small to be seen by the naked eye, is composed of billions of copper atoms arranged in some orderly fashion. Each atom is incredibly small. Only recently have we been able to "see" atoms using modern instruments such as the scanning tunneling microscope (Figure 2.1).

Electrons, Protons, and Neutrons

We know from experience that certain kinds of atoms can "split" into smaller particles and release large amounts of energy; this process is *radioactive decay.* We also know that the atom is composed of three primary particles: the *electron,* the *proton,* and the *neutron.* Although other subatomic fragments with unusual names (neutrinos, gluons, quarks, and so forth) have also been discovered, we shall concern ourselves only with the primary particles: the protons, neutrons, and electrons.

We can consider the atom to be composed of two distinct regions:

1. The **nucleus** is a small, dense, positively charged region in the center of the atom. The nucleus is composed of positively charged **protons** and uncharged **neutrons.**
2. Surrounding the nucleus is a diffuse region of negative charge populated by **electrons,** the source of the negative charge. Electrons are very low in mass in contrast to the protons and neutrons.

The properties of these particles are summarized in Table 2.1.

Atoms of various types differ in their number of protons, neutrons, and electrons. The number of protons determines the identity of the atom. As such, the number of protons is *characteristic* of the element. When the number of protons is equal to the number of electrons, the atom is neutral because the charges are balanced and effectively cancel one another.

We may represent an element symbolically as follows:

The **atomic number** (Z) is equal to the number of protons in the atom, and the **mass number** (A) is equal to the *sum* of the number of protons and neutrons (the mass of the electrons is so small as to be insignificant in comparison to that of the nucleus).

Figure 2.1
Sophisticated techniques, such as scanning tunneling electron microscopy, provide visual evidence for the structure of atoms and molecules. Each dot represents the image of a single iron atom. Even more amazing, the iron atoms have been arranged on a copper surface in the form of the Chinese characters representing the word *atom.*

LEARNING GOAL
1 Describe the important properties of protons, neutrons, and electrons.

Radioactivity and radioactive decay are discussed in Chapter 9.

LEARNING GOAL
2 Calculate the number of protons, neutrons, and electrons in any atom.

Recall from Chapter 1 (p. 25) that 1 atomic mass unit (amu) is equivalent to 1.661×10^{-24} g.

TABLE 2.1	Selected Properties of the Three Basic Subatomic Particles		
Name	**Charge**	**Mass (amu)**	**Mass (grams)**
Electron (e)	−1	5.4×10^{-4}	9.1095×10^{-28}
Proton (p)	+1	1.00	1.6725×10^{-24}
Neutron (n)	0	1.00	1.6750×10^{-24}

If

$$\text{number of protons} + \text{number of neutrons} = \text{mass number}$$

then, if the number of protons is subtracted from each side,

$$\text{number of neutrons} = \text{mass number} - \text{number of protons}$$

or, because the number of protons equals the atomic number,

$$\text{number of neutrons} = \text{mass number} - \text{atomic number}$$

For an atom, in which positive and negative charges cancel, the number of protons and electrons must be equal and identical to the atomic number.

EXAMPLE 2.1   Determining the Composition of an Atom

LEARNING GOAL

2 Calculate the number of protons, neutrons, and electrons in any atom.

Calculate the numbers of protons, neutrons, and electrons in an atom of fluorine. The atomic symbol for the fluorine atom is $^{19}_{9}\text{F}$.

Solution

Step 1. The mass number 19 tells us that the total number of protons + neutrons is 19.

Step 2. The atomic number, 9, represents the number of protons.

Step 3. The difference, $19 - 9$, or 10, is the number of neutrons.

Step 4. The number of electrons must be the same as the number of protons, hence 9, for a neutral fluorine atom.

Practice Problem 2.1

Calculate the number of protons, neutrons, and electrons in each of the following atoms:

a. $^{32}_{16}\text{S}$

b. $^{23}_{11}\text{Na}$

c. $^{1}_{1}\text{H}$

d. $^{244}_{94}\text{Pu}$

For Further Practice: Questions 2.21 and 2.22.

Isotopes

LEARNING GOAL

3 Distinguish among atoms, ions, and isotopes and calculate atomic masses from isotopic abundance.

Isotopes are atoms of the same element having different masses *because they contain different numbers of neutrons.* In other words, isotopes have different mass numbers. For example, all of the following are isotopes of hydrogen:

$^{1}_{1}\text{H}$ $\qquad\qquad$ $^{2}_{1}\text{H}$ $\qquad\qquad$ $^{3}_{1}\text{H}$

Hydrogen $\qquad\qquad$ Deuterium $\qquad\qquad$ Tritium

(Hydrogen-1) $\qquad\qquad$ (Hydrogen-2) $\qquad\qquad$ (Hydrogen-3)

Isotopes are often written with the name of the element followed by the mass number. For example, the isotopes $^{12}_{6}\text{C}$ and $^{14}_{6}\text{C}$ may be written as carbon-12 (or C-12) and carbon-14 (or C-14), respectively.

Certain isotopes (radioactive isotopes) of elements emit particles and energy that can be used to trace the behavior of biochemical systems. These isotopes otherwise behave identically to any other isotope of the same element. Their

A detailed discussion of the use of radioactive isotopes in the diagnosis and treatment of diseases is found in Chapter 9.

chemical behavior is identical; it is their nuclear behavior that is unique. As a result, a radioactive isotope can be substituted for the "nonradioactive" isotope, and its biochemical activity can be followed by monitoring the particles or energy emitted by the isotope as it passes through the body.

The existence of isotopes explains why the average masses, measured in atomic mass units (amu), of the various elements are not whole numbers. This is contrary to what we would expect from proton and neutron masses, which are very close to unity.

Consider, for example, the mass of one chlorine atom, containing 17 protons (atomic number) and 18 neutrons:

$$17 \text{ protons} \times \frac{1.00 \text{ amu}}{\text{proton}} = 17.00 \text{ amu}$$

$$18 \text{ neutrons} \times \frac{1.00 \text{ amu}}{\text{neutron}} = 18.00 \text{ amu}$$

$$17.00 \text{ amu} + 18.00 \text{ amu} = 35.00 \text{ amu} \text{ (mass of chlorine atom)}$$

Inspection of the periodic table reveals that the mass of chlorine is actually 35.45 amu, *not* 35.00 amu. The existence of isotopes accounts for this difference. A natural sample of chlorine is composed principally of two isotopes, chlorine-35 and chlorine-37, in approximately a 3:1 ratio, and the tabulated mass is the *weighted average* of the two isotopes. In our calculation the chlorine atom referred to was the isotope that has a mass number of 35 amu.

The weighted average of the masses of all of the isotopes of an element is the **atomic mass** and should be distinguished from the mass number, which is the sum of the number of protons and neutrons in a single isotope of the element.

Example 2.2 demonstrates the calculation of the atomic mass of chlorine.

Atomic mass units are convenient for representing the mass of very small particles, such as individual atoms. Refer to the discussion of units in Chapter 1.

The weighted average is not a true average but is corrected by the relative amounts (the weighting factor) of each isotope present in nature.

Determining Atomic Mass

EXAMPLE 2.2

Calculate the atomic mass of naturally occurring chlorine if 75.77% of chlorine atoms are $^{35}_{17}\text{Cl}$ (chlorine-35) and 24.23% of chlorine atoms are $^{35}_{17}\text{Cl}$ (chlorine-37).

3 **LEARNING GOAL**
Distinguish among atoms, ions, and isotopes and calculate atomic masses from isotopic abundance.

Solution

Step 1. Convert each percentage to a decimal fraction.

$$75.77\% \text{ chlorine-35} \times \frac{1}{100\%} = 0.7577 \text{ chlorine-35}$$

$$24.23\% \text{ chlorine-37} \times \frac{1}{100\%} = 0.2423 \text{ chlorine-37}$$

Step 2. Multiply the decimal fraction of each isotope by the mass of that isotope to determine the isotopic contribution to the average atomic mass.

$$\begin{array}{ll} \text{contribution to} \\ \text{atomic mass} & = \begin{array}{l}\text{(fraction of all} \\ \text{Cl atoms that} \\ \text{are chlorine-35)}\end{array} \times \begin{array}{l}\text{(mass of a} \\ \text{chlorine-35} \\ \text{atom)}\end{array} \\ \text{by chlorine-35} \end{array}$$

$$= 0.7577 \times 35.00 \text{ amu}$$

$$= 26.52 \text{ amu}$$

Continued—

EXAMPLE 2.2 —*Continued*

$$
\begin{matrix}
\text{contribution to} & & \text{(fraction of all} & & \text{(mass of a} \\
\text{atomic mass} & = & \text{Cl atoms that} & \times & \text{chlorine-37} \\
\text{by chlorine-37} & & \text{are chlorine-37)} & & \text{atom)}
\end{matrix}
$$

$$= 0.2423 \qquad \times \quad 37.00 \text{ amu}$$

$$= 8.965 \text{ amu}$$

Step 3. The weighted average is the sum of the isotopic contributions:

$$
\begin{matrix}
\text{atomic mass} & & \text{(contribution} & & \text{(contribution} \\
\text{of naturally} & = & \text{of} & + & \text{of} \\
\text{occurring Cl} & & \text{chlorine-35)} & & \text{chlorine-37)}
\end{matrix}
$$

$$= 26.52 \text{ amu} \quad + \quad 8.965 \text{ amu}$$

$$= 35.49 \text{ amu}$$

which is very close to the tabulated value of 35.45 amu. An even more exact value would be obtained by using a more exact value of the mass of the proton and neutron (experimentally known to a greater number of significant figures).

Practice Problem 2.2

The element nitrogen has two naturally occurring isotopes. One of these has a mass of 14.003 amu and a natural abundance of 99.63%; the other isotope has a mass of 15.000 amu and a natural abundance of 0.37%. Calculate the atomic mass of nitrogen.

For Further Practice: Question 2.29.

Whenever you do calculations such as those in Example 2.2, before even beginning the calculation you should look for an approximation of the value sought. Then do the calculation and see whether you obtain a reasonable number (similar to your anticipated value). In the preceding problem, if the two isotopes have masses of 35 and 37, the atomic mass must lie somewhere between the two extremes. Furthermore, because the majority of a naturally occurring sample is chlorine-35 (about 75%), the value should be closer to 35 than to 37. An analysis of the results often avoids problems stemming from untimely events such as pushing the wrong button on a calculator.

A hint for numerical problem solving: Estimate (at least to an order of magnitude) your answer before beginning the calculation using your calculator.

EXAMPLE 2.3 **Determining Atomic Mass**

LEARNING GOAL

3 Distinguish among atoms, ions, and isotopes and calculate atomic masses from isotopic abundance.

Calculate the atomic mass of naturally occurring carbon if 98.90% of carbon atoms are $^{12}_{6}\text{C}$ (carbon-12) with a mass of 12.00 amu and 1.11% are $^{13}_{6}\text{C}$ (carbon-13) with a mass of 13.00 amu. (Note that a small amount of $^{14}_{6}\text{C}$ is also present but is small enough to ignore in a calculation involving three or four significant figures.)

Continued—

EXAMPLE 2.3 —Continued

Solution

Step 1. Convert each percentage to a decimal fraction.

$$98.90\% \text{ carbon-12} \times \frac{1}{100\%} = 0.9890 \text{ carbon-12}$$

$$1.11\% \text{ carbon-13} \times \frac{1}{100\%} = 0.0111 \text{ carbon-13}$$

Step 2.

$$
\begin{aligned}
\text{contribution to} & \quad \text{(fraction of all} & \text{(mass of a} \\
\text{atomic mass} & = \text{C atoms that} \times \text{carbon-12} \\
\text{by carbon-12} & \quad \text{are carbon-12)} & \text{atom)} \\
& = 0.9890 & \times \; 12.00 \text{ amu} \\
& = 11.87 \text{ amu}
\end{aligned}
$$

$$
\begin{aligned}
\text{contribution to} & \quad \text{(fraction of all} & \text{(mass of a} \\
\text{atomic mass} & = \text{C atoms that} \times \text{carbon-13} \\
\text{by carbon-13} & \quad \text{are carbon-13)} & \text{atom)} \\
& = 0.0111 & \times \; 13.00 \text{ amu} \\
& = 0.144 \text{ amu}
\end{aligned}
$$

Step 3. The weighted average is:

$$
\begin{aligned}
\text{atomic mass} & \quad \text{(contribution} & \text{(contribution} \\
\text{of naturally} & = \text{of} & + \text{of} \\
\text{occurring carbon} & \quad \text{carbon-12)} & \text{carbon-13)} \\
& = 11.87 \text{ amu} & + \; 0.144 \text{ amu} \\
& = 12.01 \text{ amu}
\end{aligned}
$$

Helpful Hint: Because most of the carbon is carbon-12, with very little carbon-13 present, the atomic mass should be very close to that of carbon-12. Approximations, before performing the calculation, provide another check on the accuracy of the final result.

Practice Problem 2.3

The element neon has three naturally occurring isotopes. One of these has a mass of 19.99 amu and a natural abundance of 90.48%. A second isotope has a mass of 20.99 amu and a natural abundance of 0.27%. A third has a mass of 21.99 amu and a natural abundance of 9.25%. Calculate the atomic mass of neon.

For Further Practice: Question 2.30.

Ions

Ions are electrically charged particles that result from a gain of one or more electrons by the parent atom (forming negative ions, or **anions**) or a loss of one or more electrons from the parent atom (forming positive ions, or **cations**).

LEARNING GOAL

3 Distinguish among atoms, ions, and isotopes and calculate atomic masses from isotopic abundance.

Formation of an anion may occur as follows:

9 protons, 9 electrons   9 protons, 10 electrons

$$^{19}_{9}\text{F} + 1e^- \longrightarrow {}^{19}_{9}\text{F}^-$$

The neutral atom   The fluorine anion
gains an electron   is formed

Alternatively, formation of a cation of sodium may proceed as follows:

11 protons, 11 electrons   11 protons, 10 electrons

$$^{23}_{11}\text{Na} \longrightarrow 1e^- + {}^{23}_{11}\text{Na}^+$$

The neutral atom   The sodium cation
loses an electron   is formed

Note that the electrons gained are written to the left of the reaction arrow (they are reactants), whereas the electrons lost are written as products to the right of the reaction arrow. For simplification, the atomic and mass numbers are often omitted, because they do not change during ion formation. For example, the sodium cation would be written as Na^+ and the anion of fluorine as F^-.

2.2 Development of Atomic Theory

With this overview of our current understanding of the structure of the atom, we now look at a few of the most important scientific discoveries that led to modern atomic theory.

Dalton's Theory

The first experimentally based theory of atomic structure was proposed by John Dalton, an English schoolteacher, in the early 1800s. Dalton proposed the following description of atoms:

1. All matter consists of tiny particles called atoms.
2. An atom cannot be created, divided, destroyed, or converted to any other type of atom.
3. Atoms of a particular element have identical properties.
4. Atoms of different elements have different properties.
5. Atoms of different elements combine in simple whole-number ratios to produce compounds (stable aggregates of atoms).
6. Chemical change involves joining, separating, or rearranging atoms.

Although Dalton's theory was founded on meager and primitive experimental information, we regard much of it as correct today. Postulates 1, 4, 5, and 6 are currently regarded as true. The discovery of the processes of nuclear fusion, fission ("splitting" of atoms), and radioactivity has disproved the postulate that atoms cannot be created or destroyed. Postulate 3, that all the atoms of a particular element are identical, was disproved by the discovery of isotopes.

Fusion, fission, radioactivity, and isotopes are discussed in some detail in Chapter 9. Figure 2.2 uses a simple model to illustrate Dalton's theory.

Evidence for Subatomic Particles: Electrons, Protons, and Neutrons

The next major discoveries occurred almost a century later (1879–1897). Although Dalton pictured atoms as indivisible, various experiments, particularly those of

LEARNING GOAL

4 Trace the history of the development of atomic theory, beginning with Dalton.

Atoms of element X   Atoms of element Y

(a)

Compound formed from elements X and Y

(b)

Figure 2.2
An illustration of John Dalton's atomic theory. (a) Atoms of the same element are identical but different from atoms of any other element. (b) Atoms combine in whole-number ratios to form compounds.

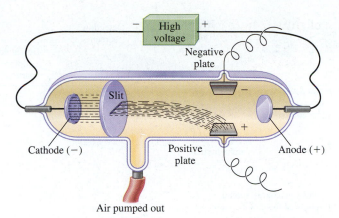

Figure 2.3
Illustration of an experiment demonstrating the charge of cathode rays. The application of an external electric field causes the electron beam to deflect toward a positive charge, implying that the cathode ray is negative.

William Crookes and Eugene Goldstein, indicated that the atom is composed of charged (+ and −) particles.

Crookes connected two metal electrodes (metal discs connected to a source of electricity) at opposite ends of a sealed glass vacuum tube. When the electricity was turned on, rays of light were observed to travel between the two electrodes. They were called **cathode rays** because they traveled from the *cathode* (the negative electrode) to the *anode* (the positive electrode).

Later experiments by J. J. Thomson, an English scientist, demonstrated the electrical and magnetic properties of cathode rays (Figure 2.3). The rays were deflected toward the positive electrode of an external electric field. Because opposite charges attract, this indicates the negative character of the rays. Similar experiments with an external magnetic field showed a deflection as well; hence these cathode rays also have magnetic properties.

A change in the material used to fabricate the electrode discs brought about no change in the experimental results. This suggested that the ability to produce cathode rays is a characteristic of all materials.

In 1897, Thomson announced that cathode rays are streams of negative particles of energy. These particles are *electrons.* Similar experiments, conducted by Goldstein, led to the discovery of particles that are equal in charge to the electron but opposite in sign. These particles, much heavier than electrons (actually 1837 times as heavy), are called *protons.*

As we have seen, the third fundamental atomic particle is the *neutron.* It has a mass virtually identical (it is less than 1% heavier) to that of the proton and has zero charge. The neutron was first postulated in the early 1920s, but it was not until 1932 that James Chadwick demonstrated its existence with a series of experiments involving the use of small particle bombardment of nuclei.

Crookes's cathode ray tube was the forerunner of the computer screen (often called CRT) and the television.

 Animations
Thomson's Cathode Ray Tube
The Rutherford Experiment

Evidence for the Nucleus

In the early 1900s it was believed that protons and electrons were uniformly distributed throughout the atom. However, an experiment by Hans Geiger led Ernest Rutherford (in 1911) to propose that the majority of the mass and positive charge of the atom was actually located in a small, dense region, the *nucleus*, with small, negatively charged electrons occupying a much larger volume outside of the nucleus.

To understand how Rutherford's theory resulted from the experimental observations of Geiger, let us examine this experiment in greater detail. Rutherford and others had earlier demonstrated that some atoms spontaneously "decay" to produce three types of radiation: alpha (α), beta (β), and gamma (γ) radiation. This process is known as *natural radioactivity.* Geiger used radioactive materials, such as *radium,* as projectile sources, "firing" alpha particles at a thin metal foil target

LEARNING GOAL

4 Trace the history of the development of atomic theory, beginning with Dalton.

Figure 2.4

The alpha particle scattering experiment. Most alpha particles passed through the foil without being deflected; a few were deflected from their path by nuclei in the gold atoms.

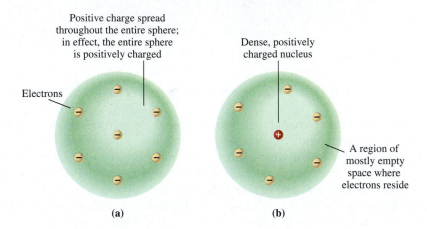

Figure 2.5

(a) A model of the atom (credited to Thomson) prior to the work of Geiger and Rutherford. (b) A model of the atom supported by the alpha-particle scattering experiments of Geiger and Rutherford.

(gold leaf). He then observed the interaction of the metal and alpha particles with a detection screen (Figure 2.4) and found that:

a. Most alpha particles pass through the foil without being deflected.
b. A small fraction of the particles were deflected, some even *directly back to the source.*

Rutherford interpreted this to mean that most of the atom is empty space, because most alpha particles were not deflected. Further, most of the mass and positive charge must be located in a small, dense region; collision of the heavy and positively charged alpha particle with this small dense and positive region (the nucleus) caused the great deflections. Rutherford summarized his astonishment at observing the deflected particles: "It was almost as incredible as if you fired a 15-inch shell at a piece of tissue and it came back and hit you."

The significance of Rutherford's contribution cannot be overstated. It caused a revolutionary change in the way that scientists pictured the atom (Figure 2.5). His discovery of the nucleus is fundamental to our understanding of chemistry. Chapter 9 will provide much more information about the nucleus and its unique properties.

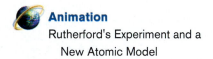

Animation

Rutherford's Experiment and a New Atomic Model

2.3 Light, Atomic Structure, and the Bohr Atom

Light and Atomic Structure

The Rutherford atom leaves us with a picture of a tiny, dense, positively charged nucleus containing protons and surrounded by electrons. The electron arrangement, or configuration, is not clearly detailed. More information is needed regarding the relationship of the electrons to each other and to the nucleus. In dealing with dimensions on the order of 10^{-9} m (the atomic level), conventional methods for measurement of location and distance of separation become impossible. An

LEARNING GOAL

5   Explain the critical role of spectroscopy in the development of atomic theory and in our everyday lives.

alternative approach involves the measurement of *energy* rather than the *position* of the atomic particles to determine structure. For example, information obtained from the absorption or emission of *light* by atoms (energy changes) can yield valuable insight into structure. Such studies are referred to as **spectroscopy.**

In a general sense we refer to light as **electromagnetic radiation.** Electromagnetic radiation travels in *waves* from a source. The most recognizable source of this radiation is the sun. We are aware of a rainbow, in which visible white light from the sun is broken up into several characteristic bands of different colors. Similarly, visible white light, when passed through a glass prism, is separated into its various component colors (Figure 2.6). These various colors are simply light (electromagnetic radiation) of differing *wavelengths*. Light is propagated as a collection of sine waves, and the wavelength is the distance between identical points on successive waves:

Figure 2.6
The visible spectrum of light. Light passes through a prism, producing a continuous spectrum. Color results from the way in which our eyes interpret the various wavelengths.

All electromagnetic radiation travels at a speed of 3.0×10^8 m/s, the **speed of light.** However, each wavelength of light, although traveling with identical velocity, has its own characteristic energy. A collection of all electromagnetic radiation, including each of these wavelengths, is referred to as the **electromagnetic spectrum.** For convenience in discussing this type of radiation we subdivide electromagnetic radiation into various spectral regions, which are characterized by physical properties of the radiation, such as its *wavelength* or its *energy* (Figure 2.7). Some of these regions are quite familiar to us from our everyday experiences; the visible and microwave regions are two common examples.

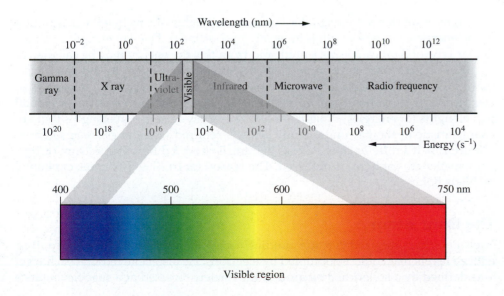

Figure 2.7
The electromagnetic spectrum. Note that the visible spectrum is only a small part of the total electromagnetic spectrum.

An Environmental *Perspective*

Electromagnetic Radiation and Its Effects on Our Everyday Lives

From the preceding discussion of the interaction of electromagnetic radiation with matter—spectroscopy—you might be left with the impression that the utility of such radiation is limited to theoretical studies of atomic structure. Although this is a useful application that has enabled us to learn a great deal about the structure and properties of matter, it is by no means the only application. Useful, everyday applications of the theories of light energy and transmission are all around us. Let's look at just a few examples.

Transmission of sound and pictures is conducted at radio frequencies or radio wavelengths. We are immersed in radio waves from the day we are born. A radio or television is our "detector" of these waves. Radio waves are believed to cause no physical harm because of their very low energy, although some concern for people who live very close to transmission towers has resulted from recent research.

X-rays are electromagnetic radiation, and they travel at the speed of light just like radio waves. However, because of their higher energy, they can pass through the human body and leave an image of the body's interior on a photographic film. X-ray photographs are invaluable for medical diagnosis. However, caution is advised in exposing oneself to X-rays, because the high energy can remove electrons from biological molecules, causing subtle and potentially harmful changes in their chemistry.

The sunlight that passes through our atmosphere provides the basis for a potentially useful technology for providing heat and electricity: *solar energy.* Light is captured by absorbers, referred to as *solar collectors,* which convert the light energy into heat energy. This heat can be transferred to water circulating beneath the collectors to provide heat and hot water for homes or industry. Wafers of a silicon-based material can convert light energy to electrical energy; many believe that if the

The intensity of infrared radiation from a solid or liquid is an indicator of relative temperature. This has been used to advantage in the design of infrared cameras, which can obtain images without the benefit of the visible light that is necessary for conventional cameras. The infrared photograph shows the coastline surrounding the city of San Francisco.

efficiency of these processes can be improved, such approaches may provide at least a partial solution to the problems of rising energy costs and pollution associated with our fossil fuel–based energy economy.

A spectrophotometer, an instrument that utilizes a prism (or similar device) and a light-sensitive detector, is capable of very accurate and precise wavelength measurement.

Light of shorter wavelength has higher energy; this means that the magnitude of the energy and wavelength is inversely proportional. The wavelength of a particular type of light can be measured, and from this the energy may be calculated.

If we take a sample of some element, such as hydrogen, in the gas phase, place it in an evacuated glass tube containing a pair of electrodes, and pass an electrical charge (cathode ray) through the hydrogen gas, light is emitted. Not all wavelengths (or energies) of light are emitted—only certain wavelengths that are characteristic of the gas under study. This is referred to as an *emission spectrum* (Figure 2.8). If a different gas, such as helium, is used, a different spectrum (different wavelengths of light) is observed. The reason for this behavior was explained by Niels Bohr.

The Bohr Atom

Niels Bohr hypothesized that surrounding each atomic nucleus were certain fixed **energy levels** that could be occupied by electrons. He also believed that each level was defined by a spherical *orbit* around the nucleus, located at a specific distance

Animation

Atom Structure

An image of a tumor detected by a CT scanner.

The radiation therapy used in the treatment of many types of cancer has been responsible for saving many lives and extending the span of many others. When radiation is used as a treatment, it destroys cancer cells. This topic will be discussed in detail in Chapter 9.

As a diagnostic tool, spectroscopy has the benefit of providing data quickly and reliably; it can also provide information that might not be available through any other means. Additionally, spectroscopic procedures are often nonsurgical, outpatient procedures. Such procedures involve less risk, can be more routinely performed, and are more acceptable to the general public than surgical procedures. The potential cost savings because of the elimination of many unnecessary surgical procedures is an added benefit.

The most commonly practiced technique uses the CT scanner, an acronym for *computer-accentuated tomography*. In this technique, X-rays are directed at the tissue of interest. As the X-rays pass through the tissue, detectors surrounding the tissue gather the signal, compare it to the original X-ray beam, and, using the computer, produce a three-dimensional image of the tissue.

Microwave radiation for cooking, *infrared* lamps for heating and remote sensing, *ultraviolet* lamps used to kill microorganisms on environmental surfaces, *gamma radiation* from nuclear waste, the *visible* light from the lamp you are using to read this chapter—all are forms of the same type of energy that, for better or worse, plays such a large part in our twenty-first century technological society.

Electromagnetic radiation and spectroscopy also play a vital role in the field of diagnostic medicine. They are routinely used as diagnostic and therapeutic tools in the detection and treatment of disease.

For Further Understanding

Diane says that a medical X-ray is risky, but a CT scan is risk free. Is Diane correct? Explain your answer.

Why would the sensor (detector) for a conventional camera and an IR camera have to be designed differently?

---

from the nucleus. The concept of certain fixed energy levels is referred to as the **quantization** of energy. The implication is that only these orbits, or *quantum levels*, are allowed locations for electrons. If an atom *absorbs* energy, an electron undergoes *promotion* from an orbit closer to the nucleus (lower energy) to one farther from the nucleus (higher energy), creating an *excited state*. Similarly, the release of energy by an atom, or *relaxation*, results from an electron falling into an orbit closer to the nucleus (lower energy level).

Promotion and relaxation processes are referred to as *electronic transitions*. The amount of energy absorbed in jumping from one energy level to a higher energy level is a precise quantity (hence, quantum), and that energy corresponds exactly to the energy differences between the orbits involved. Electron promotion resulting from absorption of energy results in an *excited state atom*; the process of relaxation allows the atom to return to the ground state (Figure 2.9) with the simultaneous release of light energy. The *ground state* is the lowest possible energy state. This emission process, such as the release of energy after excitation of hydrogen atoms by an electric arc, produces the series of emission lines (emission spectrum). Measurement of the wavelengths of these lines enables the

  LEARNING GOAL

6 State the basic postulates of Bohr's theory, its utility, and its limitations.

The line spectrum of each known element is unique. Consequently, spectroscopy is a very useful tool for identifying elements.

Figure 2.8

(a) The emission spectrum of hydrogen. Certain wavelengths of light, characteristic of the atom, are emitted upon electrical excitation. (b) The line spectrum of hydrogen is compared with (c and d) the line spectrum of helium and sodium and the spectrum of visible light (e).

Each electronic transition produces "bundles," or quanta, of energy. These quanta are termed "photons." Photons resulting from a certain electronic transition have their own unique wavelength, frequency, and energy.

calculation of energy levels in the atom. These energy levels represent the location of the atom's electrons.

We may picture the Bohr atom as a series of concentric orbits surrounding the nucleus. The orbits are identified using numbers ($n = 1, 2, 3, \ldots$, etc.). The number n is referred to as a *quantum number.*

The hydrogen spectrum consists of four lines in the visible region of the spectrum. Electronic transitions, calculated from the Bohr theory, account for each of these lines.

Figure 2.9

The Bohr representation of atoms. Excitation involves promotion of an electron to a higher energy level when energy is absorbed. Relaxation is the reverse process; atoms return to the ground state when the electron relaxes to a lower energy level, releasing energy.

A Human *Perspective*

Atomic Spectra and the Fourth of July

At one time or another we have all marveled at the bright, multicolored display of light and sound that is a fireworks display. These sights and sounds are produced by a chemical reaction that generates the energy necessary to excite a variety of elements to their higher-energy electronic states. Light emission results from relaxation of the excited atoms to the ground state. Each atom releases light of specific wavelengths. The visible wavelengths are seen as colored light.

Fireworks need a chemical reaction to produce energy. We know from common experience that oxygen and a fuel will release energy. The fuel in most fireworks preparations is sulfur or aluminum. Each reacts slowly with oxygen; a more potent solid-state source of oxygen is potassium perchlorate ($KClO_4$). The potassium perchlorate reacts with the fuel (an oxidation-reduction reaction, Chapter 8), producing a bright white flash of light. The heat produced excites the various elements packaged with the fuel and oxidant.

Sodium salts, such as sodium chloride, furnish sodium ions, which, when excited, produce yellow light. Red colors arise from salts of strontium, which emit several shades of red corresponding to wavelengths in the 600- to 700-nm region of the visible spectrum. Copper salts produce blue radiation, because copper emits in the 400- to 500-nm spectral region.

The beauty of fireworks is a direct result of the skill of the manufacturer. Selection of the proper oxidant, fuel, and color-producing elements is critical to the production of a spectacular display. Packaging these chemicals in proper quantities so

A fireworks display is a dramatic illustration of light emission by excited atoms.

that they can be stored and used safely is an equally important consideration.

For Further Understanding

Explain why excited sodium emits a yellow color. (Refer to Figure 2.8.)

How does this story illustrate the interconversion of potential and kinetic energy?

A summary of the major features of the Bohr theory is as follows:

- Atoms can absorb and emit energy via *promotion* of electrons to higher energy levels and *relaxation* to lower levels.
- Energy that is emitted upon relaxation is observed as a single wavelength of light, a collection of photons.
- These *spectral lines* are a result of electron transitions between *allowed levels* in the atom.
- The allowed levels are quantized energy levels, or orbits.
- Electrons are found only in these energy levels.
- The highest-energy orbits are located farthest from the nucleus.
- Atoms absorb energy by excitation of electrons to higher energy levels.
- Atoms release energy by relaxation of electrons to lower energy levels.
- Energy differences may be calculated from the wavelengths of light emitted.

 Animation

Emission Spectrum of Hydrogen and Electron Energy Transitions

Modern Atomic Theory

The Bohr model was an immensely important contribution to the understanding of atomic structure. The idea that electrons exist in specific energy states and that transitions between states involve quanta of energy provided the linkage between atomic structure and atomic spectra. However, some limitations of this model quickly became apparent. Although it explained the hydrogen spectrum, it provided only a crude approximation of the spectra for more complex atoms. Subsequent development of more sophisticated experimental techniques demonstrated that there are problems with the Bohr theory even in the case of hydrogen.

Although Bohr's concept of principal energy levels is still valid, restriction of electrons to fixed orbits is too rigorous. All current evidence shows that electrons do *not*, in fact, orbit the nucleus. We now speak of the *probability* of finding an electron in a *region* of space within the principal energy level, referred to as an **atomic orbital.** The rapid movement of the electron spreads the charge into a *cloud* of charge. This cloud is more dense in certain regions, the **electron density** being proportional to the probability of finding the electron at any point in time. Insofar as these atomic orbitals are part of the principal energy levels, they are referred to as sublevels. In Chapter 3 we will see that the orbital model of the atom can be used to predict how atoms can bond together to form compounds. Furthermore, electron arrangement in orbitals enables us to predict various chemical and physical properties of these compounds.

Question *2.1*   What is meant by the term *electron density*?

Question *2.2*   How do *orbits* and *orbitals* differ?

The theory of atomic structure has progressed rapidly, from a very primitive level to its present point of sophistication, in a relatively short time. Before we proceed, let us insert a note of caution. We must not think of the present picture of the atom as final. Scientific inquiry continues, and we should view the present theory as a step in an evolutionary process. *Theories are subject to constant refinement*, as was noted in our discussion of the scientific method.

2.4 The Periodic Law and the Periodic Table

LEARNING GOAL

7 Recognize the important subdivisions of the periodic table: periods, groups (families), metals, and nonmetals.

In 1869, Dmitri Mendeleev, a Russian, and Lothar Meyer, a German, working independently, found ways of arranging elements in order of increasing atomic mass such that elements with similar properties were grouped together in a *table of elements.* The **periodic law** is embodied by Mendeleev's statement, "the elements if arranged according to their atomic weights (masses), show a distinct *periodicity* (regular variation) of their properties." The *periodic table* (Figure 2.10) is a visual representation of the periodic law.

Chemical and physical properties of elements correlate with the electronic structure of the atoms that make up these elements. In turn, the electronic structure correlates with position on the periodic table.

A thorough familiarity with the arrangement of the periodic table allows us to predict electronic structure and physical and chemical properties of the various elements. It also serves as the basis for understanding chemical bonding.

The concept of "periodicity" may be illustrated by examining a portion of the modern periodic table. The elements in the second row (beginning with lithium, Li, and proceeding to the right) show a marked difference in properties. However, sodium (Na) has properties similar to those of lithium, and sodium is therefore

Figure 2.10
Classification of the elements: the periodic table.

placed below lithium; once sodium is fixed in this position, the elements Mg through Ar have properties remarkably similar (though not identical) to those of the elements just above them. The same is true throughout the complete periodic table.

Mendeleev arranged the elements in his original periodic table in order of increasing atomic mass. However, as our knowledge of atomic structure increased, atomic numbers became the basis for the organization of the table. Remarkably, his table was able to predict the existence of elements not known at the time.

The modern periodic law states that *the physical and chemical properties of the elements are periodic functions of their atomic numbers*. If we arrange the elements in order of increasing number of protons, the properties of the elements repeat at regular intervals.

Not all of the elements are of equal importance to an introductory study of chemistry. Table 2.2 lists twenty of the elements that are most important to biological systems, along with their symbols and a brief description of their functions.

We will use the periodic table as our "map," just as a traveler would use a road map. A short time spent learning how to read the map (and remembering to carry it along on your trip!) is much easier than memorizing every highway and

TABLE 2.2	Summary of the Most Important Elements in Biological Systems	
Element	**Symbol**	**Significance**
Hydrogen	H	Components of major biological molecules
Carbon	C	
Oxygen	O	
Nitrogen	N	
Phosphorus	P	
Sulfur	S	
Potassium	K	Produce electrolytes responsible for fluid balance and nerve transmission
Sodium	Na	
Chlorine	Cl	
Calcium	Ca	Bones, nerve function
Magnesium	Mg	
Zinc	Zn	Essential trace metals in human metabolism
Strontium	Sr	
Iron	Fe	
Copper	Cu	
Cobalt	Co	
Manganese	Mn	
Cadmium	Cd	"Heavy metals" toxic to living systems
Mercury	Hg	
Lead	Pb	

intersection. The information learned about one element relates to an entire family of elements grouped as a recognizable unit within the table.

Numbering Groups in the Periodic Table

The periodic table created by Mendeleev has undergone numerous changes over the years. These modifications occurred as more was learned about the chemical and physical properties of the elements. The labeling of groups with Roman numerals followed by the letter *A* (representative elements) or *B* (transition elements) was standard, until 1983, in North America and Russia. However, in other parts of the world, the letters *A* and *B* were used in a different way. Consequently, two different periodic tables were in widespread use. This certainly created some confusion.

The International Union of Pure and Applied Chemistry (IUPAC), in 1983, recommended that a third system, using numbers 1–18 to label the groups, replace both of the older systems. Unfortunately, multiple systems now exist and this can cause confusion for both students and experienced chemists.

The periodic tables in this textbook are "double labeled." Both the old (Roman numeral) and new (1–18) systems are used to label the groups. The label that you use is simply a guide to reading the table; the real source of information is in the structure of the table itself. The following sections will show you how to extract useful information from this structure.

Periods and Groups

A **period** is a horizontal row of elements in the periodic table. The periodic table consists of six periods containing 2, 8, 8, 18, 18, and 32 elements. The seventh period is still incomplete but potentially holds 32 elements. Note that the *lanthanide series*,

Mendeleev's original periodic table included only the elements known at the time, less than half of the current total.

7   LEARNING GOAL
Recognize the important subdivisions of the periodic table: periods, groups (families), metals, and nonmetals.

a collection of 14 elements that are chemically and physically similar to the element lanthanum, is a part of period six. It is written separately for convenience of presentation and is inserted between lanthanum (La), atomic number 57, and hafnium (Hf), atomic number 72. Similarly, the *actinide series*, consisting of 14 elements similar to the element actinium, is inserted between actinium, atomic number 89, and rutherfordium, atomic number 104.

Groups or *families* are columns of elements in the periodic table. The elements of a particular group or family share many similarities, as in a human family. The similarities extend to physical and chemical properties that are related to similarities in electronic structure (that is, the way in which electrons are arranged in an atom).

Group A elements are called **representative elements,** and Group B elements are **transition elements.** Certain families also have common names. For example, Group IA (or 1) elements are also known as the **alkali metals;** Group IIA (or 2), the **alkaline earth metals;** Group VIIA (or 17), the **halogens;** and Group VIIIA (or 18), the **noble gases.**

Copper is a metal that has many uses. Can you provide other uses of copper?

Representative elements **are also known as** *main-group elements.* **These terms are synonymous.**

Metals and Nonmetals

A **metal** is a substance whose atoms tend to lose electrons during chemical change, forming positive ions. A **nonmetal,** on the other hand, is a substance whose atoms may gain electrons, forming negative ions.

A closer inspection of the periodic table reveals a bold zigzag line running from top to bottom, beginning to the left of boron (B) and ending between polonium (Po) and astatine (At). This line acts as the boundary between *metals,* to the left, and *nonmetals,* to the right. Elements straddling the boundary have properties intermediate between those of metals and nonmetals. These elements are referred to as **metalloids.** The metalloids include boron (B), silicon (Si), germanium (Ge), arsenic (As), antimony (Sb), tellurium (Te), polonium (Po), and astatine (At).

Metals and nonmetals may be distinguished by differences in their physical properties in addition to their chemical tendency to lose or gain electrons. Metals have a characteristic luster and generally conduct heat and electricity well. Most (except mercury) are solids at room temperature. Nonmetals, on the other hand, are poor conductors, and several are gases at room temperature.

Note that aluminum (Al) is classified as a metal, not a metalloid.

> **LEARNING GOAL**
>
> **7** Recognize the important subdivisions of the periodic table: periods, groups (families), metals, and nonmetals.

Atomic Number and Atomic Mass

The atomic number is the number of protons in the nucleus of an atom of an element. It also corresponds to the nuclear charge, the positive charge from the nucleus. Both the atomic number and the average atomic mass of each element are readily available from the periodic table. For example,

> **LEARNING GOAL**
>
> **8** Use the periodic table to obtain information about an element.

$$20 \longleftarrow \text{atomic number}$$
$$\text{Ca} \longleftarrow \text{symbol}$$
$$\text{calcium} \longleftarrow \text{name}$$
$$40.08 \longleftarrow \text{atomic mass}$$

More detailed periodic tables may also include such information as the electron arrangement, relative sizes of atoms and ions, and most probable ion charges.

Question 2.3

Refer to the periodic table (Figure 2.10) and find the following information:

a. the symbol of the element with an atomic number of 40
b. the mass of the element sodium (Na)
c. the element whose atoms contain 24 protons
d. the known element that should most resemble the as-yet undiscovered element with an atomic number of 117

Question 2.4

Refer to the periodic table (Figure 2.10) and find the following information:

   a. the symbol of the noble gas in period 3
   b. the element in Group IVA with the smallest mass
   c. the only metalloid in Group IIIA
   d. the element whose atoms contain 18 protons

Question 2.5

For each of the following element symbols, give the name of the element, its atomic number, and its atomic mass.

   a. He
   b. F
   c. Mn

Question 2.6

For each of the following element symbols, give the name of the element, its atomic number, and its atomic mass:

   a. Mg
   b. Ne
   c. Se

2.5 Electron Arrangement and the Periodic Table

LEARNING GOAL

9 Describe the relationship between the electronic structure of an element and its position in the periodic table.

A primary objective of studying chemistry is to understand the way in which atoms join together to form chemical compounds. The most important factor in this *bonding process* is the arrangement of the electrons in the atoms that are combining. The **electron configuration** describes the arrangement of electrons in atoms. The periodic table is helpful because it provides us with a great deal of information about the electron arrangement or electronic configuration of atoms.

We have seen (p. 59) that elements in the periodic table are classified as either representative or transition. Representative elements consist of all group 1, 2, and 13–18 elements (IA–VIIIA). All others are transition elements. The guidelines that we will develop for writing electron configurations are intended for representative elements. Electron configurations for transition elements include several exceptions to the rule.

Valence Electrons

If we picture two spherical objects that we wish to join together, perhaps with glue, the glue can be applied to the surface and the two objects can then be brought into contact. We can extend this analogy to two atoms that are modeled as spherical objects. Although this is not a perfect analogy, it is apparent that the surface interaction is of primary importance. Although the positively charged nucleus and "interior" electrons certainly play a role in bonding, we can most easily understand the process by considering only the outermost electrons. We refer to these as *valence electrons*. **Valence electrons** are the outermost electrons in an atom, which are involved, or have the potential to become involved, in the bonding process.

Metals tend to have fewer valence electrons, and nonmetals tend to have more valence electrons.

For representative elements the number of valence electrons in an atom corresponds to the number of the *group* or *family* in which the atom is found. For example, elements such as hydrogen and sodium (in fact, all alkali metals, Group

A Medical Perspective

Copper Deficiency and Wilson's Disease

An old adage tells us that we should consume all things in moderation. This is very true of many of the trace minerals, such as copper. Too much copper in the diet causes toxicity and too little copper results in a serious deficiency disease.

Copper is extremely important for the proper functioning of the body. It aids in the absorption of iron from the intestine and facilitates iron metabolism. It is critical for the formation of hemoglobin and red blood cells in the bone marrow. Copper is also necessary for the synthesis of collagen, a protein that is a major component of the connective tissue. It is essential to the central nervous system in two important ways. First, copper is needed for the synthesis of norepinephrine and dopamine, two chemicals that are necessary for the transmission of nerve signals. Second, it is required for the deposition of the myelin sheath (a layer of insulation) around nerve cells. Release of cholesterol from the liver depends on copper, as does bone development and proper function of the immune and blood clotting systems.

The estimated safe and adequate daily dietary intake (ESADDI) for adults is 1.5–3.0 mg. Meats, cocoa, nuts, legumes, and whole grains provide significant amounts of copper.

Although getting enough copper in the diet would appear to be relatively simple, it is estimated that Americans often ingest only marginal levels of copper, and we absorb only 25–40% of that dietary copper. Despite these facts, it appears that copper deficiency is not a serious problem in the United States.

Individuals who are at risk for copper deficiency include people who are recovering from abdominal surgery, which causes decreased absorption of copper from the intestine. Others at risk are premature babies and people who are sustained solely by intravenous feedings that are deficient in copper. In addition, people who ingest high doses of antacids or take excessive supplements of zinc, iron, or vitamin C can develop copper deficiency because of reduced copper absorption. Because copper is involved in so many processes in the body, it is not surprising that the symptoms of copper deficiency are many and diverse. They include anemia; decreased red and white blood cell counts; heart disease; increased levels of serum cholesterol; loss of bone; defects in the nervous system, immune system, and connective tissue; and abnormal hair.

Just as too little copper causes serious problems, so does an excess of copper. At doses greater than about 15 mg, copper causes toxicity that results in vomiting. The effects of extended exposure to excess copper are apparent when we look at Wilson's disease. This is a genetic disorder in which excess copper cannot be removed from the body and accumulates in the cornea of the eye, liver, kidneys, and brain. The symptoms include a greenish ring around the cornea, cirrhosis of the liver, copper in the urine, dementia and paranoia, drooling,

Examples of foods rich in copper.

and progressive tremors. As a result of the condition, the victim generally dies in early adolescence. Wilson's disease can be treated with medication and diet modification, with moderate success, if it is recognized early, before permanent damage has occurred to any tissues.

For Further Understanding

Why is there an upper limit on the recommended daily amount of copper?

Iron is another essential trace metal in our diet. Go to the Web and find out if upper limits exist for daily iron consumption.

TABLE 2.3   The Electron Distribution for the First Twenty Elements of the Periodic Table

Element Symbol and Name	Total Number of Electrons	Total Number of Valence Electrons	Electrons in $n = 1$	Electrons in $n = 2$	Electrons in $n = 3$	Electrons in $n = 4$
H, hydrogen	1	1	1	0	0	0
He, helium	2	2	2	0	0	0
Li, lithium	3	1	2	1	0	0
Be, beryllium	4	2	2	2	0	0
B, boron	5	3	2	3	0	0
C, carbon	6	4	2	4	0	0
N, nitrogen	7	5	2	5	0	0
O, oxygen	8	6	2	6	0	0
F, fluorine	9	7	2	7	0	0
Ne, neon	10	8	2	8	0	0
Na, sodium	11	1	2	8	1	0
Mg, magnesium	12	2	2	8	2	0
Al, aluminum	13	3	2	8	3	0
Si, silicon	14	4	2	8	4	0
P, phosphorus	15	5	2	8	5	0
S, sulfur	16	6	2	8	6	0
Cl, chlorine	17	7	2	8	7	0
Ar, argon	18	8	2	8	8	0
K, potassium	19	1	2	8	8	1
Ca, calcium	20	2	2	8	8	2

IA or 1) have one valence electron. From left to right in period 2, beryllium, Be (Group IIA or 2), has two valence electrons; boron, B (Group IIIA or 3), has three; carbon, C (Group IVA or 4), has four; and so forth.

We have seen that an atom may have electrons in several different energy levels. These energy levels are symbolized by n, the lowest energy level being assigned a value of $n = 1$. Each energy level may contain up to a fixed maximum number of electrons. For example, the $n = 1$ energy level may contain a maximum of two electrons. Thus hydrogen (atomic number = 1) has one electron and helium (atomic number = 2) has two electrons in the $n = 1$ level. Only these elements have electrons *exclusively* in the first energy level:

$n = 1$

Hydrogen: one-electron atom

$n = 1$

Helium: two-electron atom

These two elements make up the first period of the periodic table. Period 1 contains all elements whose *maximum* energy level is $n = 1$. In other words, the $n = 1$ level is the *outermost* electron region for hydrogen and helium. Hydrogen has one electron and helium has two electrons in the $n = 1$ level.

The valence electrons of elements in the second period are in the $n = 2$ energy level. (Remember that you must fill the $n = 1$ level with two electrons before adding electrons to the next level.) The third electron of lithium (Li) and the remaining electrons of the second period elements must be in the $n = 2$ level and are considered the valence electrons for lithium and remaining second period elements.

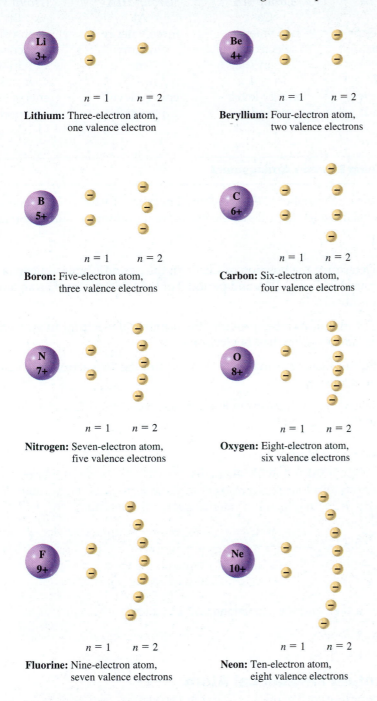

Lithium: Three-electron atom,
one valence electron

Beryllium: Four-electron atom,
two valence electrons

Boron: Five-electron atom,
three valence electrons

Carbon: Six-electron atom,
four valence electrons

Nitrogen: Seven-electron atom,
five valence electrons

Oxygen: Eight-electron atom,
six valence electrons

Fluorine: Nine-electron atom,
seven valence electrons

Neon: Ten-electron atom,
eight valence electrons

The electron distribution (arrangement) of the first twenty elements of the periodic table is given in Table 2.3.

Two general rules of electron distribution are based on the periodic law:

RULE 1: The number of valence electrons in an atom equals the *group* number for all representative (A group) elements. ■

RULE 2: The energy level ($n = 1, 2$, etc.) in which the valence electrons are located corresponds to the *period* in which the element may be found. ■

For example,

Group IA	Group IIA	Group IIIA	Group VIIA
Li	Ca	Al	Br
one valence electron	two valence electrons	three valence electrons	seven valence electrons
in	in	in	in
$n = 2$	$n = 4$	$n = 3$	$n = 4$
energy level; period 2	energy level; period 4	energy level; period 3	energy level; period 4

EXAMPLE 2.4   **Determining Electron Arrangement**

Provide the total number of electrons, total number of valence electrons, and energy level in which the valence electrons are found for the silicon (Si) atom.

Solution

Step 1. Determine the position of silicon in the periodic table. Silicon is found in Group IVA and period 3 of the table. Silicon has an atomic number of 14.

Step 2. The atomic number provides the number of electrons in an atom. Silicon therefore has 14 electrons.

Step 3. Because silicon is in Group IV, only 4 of the 14 electrons are valence electrons.

Step 4. Silicon has 2 electrons in $n = 1$, 8 electrons in $n = 2$, and 4 electrons in the $n = 3$ level.

Practice Problem 2.4

For each of the following elements, provide the *total* number of electrons and *valence* electrons in its atom as well as the number of the energy level in which the valence electrons are found:
a. Na
b. Mg
c. S
d. Cl
e. Ar

For Further Practice: Questions 2.67 and 2.68.

LEARNING GOAL

9 Describe the relationship between the electronic structure of an element and its position in the periodic table.

The Quantum Mechanical Atom

As we noted in Section 2.3, the success of Bohr's theory was short-lived. Emission spectra of multi-electron atoms (recall that the hydrogen atom has only one electron) could not be explained by Bohr's theory. Evidence that electrons have wave properties served to intensify the problem. Bohr stated that electrons in atoms had very specific locations, now termed *principal energy levels.* The very nature of waves, spread out in space, defies such an exact model of electrons in atoms. Furthermore, the exact model is contradictory to theory and subsequent experiments.

The basic concept of the Bohr theory, that the energy of an electron in an atom is quantized, was refined and expanded by an Austrian physicist,

Erwin Schröedinger. He described electrons in atoms in probability terms, developing equations that emphasize the wavelike character of electrons. Although Schröedinger's approach was founded on complex mathematics, we can readily use models of electron probability regions to enable us to gain a reasonable insight into atomic structure without the need to understand the underlying mathematics.

Schröedinger's theory, often described as quantum mechanics, incorporates Bohr's principal energy levels ($n = 1, 2$, and so forth); however, it is proposed that each of these levels is made up of one or more sublevels. Each sublevel, in turn, contains one or more atomic orbitals. In the following section we shall look at each of these regions in more detail and learn how to predict the way that electrons are arranged in stable atoms.

Energy Levels and Sublevels

Principal Energy Levels

The principal energy levels are designated $n = 1, 2, 3$, and so forth. The number of possible sublevels in a principal energy level is also equal to n. When $n = 1$, there can be only one sublevel; $n = 2$ allows two sublevels, and so forth.

The total electron capacity of a principal level is $2(n)^2$. For example:

$n = 1$	$2(1)^2$	Capacity $= 2e^-$
$n = 2$	$2(2)^2$	Capacity $= 8e^-$
$n = 3$	$2(3)^2$	Capacity $= 18e^-$

Sublevels

A **sublevel** is a set of equal-energy orbitals within a principal energy level. The sublevels, or subshells, are symbolized as s, p, d, f, and so forth; they increase in energy in the following order:

$$s < p < d < f$$

We specify both the principal energy level and type of sublevel when describing the location of an electron—for example, 1s, 2s, 2p. Energy level designations for the first four principal energy levels follow:

- The first principal energy level ($n = 1$) has one possible sublevel: 1s.
- The second principal energy level ($n = 2$) has two possible sublevels: 2s and 2p.
- The third principal energy level ($n = 3$) has three possible sublevels: 3s, 3p, and 3d.
- The fourth principal energy level ($n = 4$) has four possible sublevels: 4s, 4p, 4d, and 4f.

Orbitals

An **atomic orbital** is a specific region of a sublevel containing a maximum of two electrons.

Figure 2.11 depicts a model of an s orbital. It is spherically symmetrical, much like a Ping-Pong ball. Its volume represents a region where there is a high probability of finding electrons of similar energy. This probability decreases as we approach the outer region of the atom. The nucleus is at the center of the s orbital. At that point the probability of finding the electron is zero; electrons cannot reside in the nucleus. Only one s orbital can be found in any n level. Atoms with many electrons, occupying a number of n levels, have an s orbital in each n level. Consequently 1s, 2s, 3s, and so forth are possible orbitals.

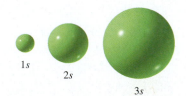

Figure 2.11
Representation of s orbitals.

$$p_x \qquad\qquad p_y \qquad\qquad p_z$$

Figure 2.12 describes the shapes of the three possible *p* orbitals within a given level. Each has the same shape, and that shape appears much like a dumbbell; these three orbitals differ only in the direction they extend into space. Imaginary coordinates *x*, *y*, and *z* are superimposed on these models to emphasize this fact. These three orbitals, termed p_x, p_y, and p_z, may coexist in a single atom.

In a similar fashion, five possible *d* orbitals and seven possible *f* orbitals exist. The *d* orbitals exist only in $n = 3$ and higher principal energy levels; *f* orbitals exist only in $n = 4$ and higher principal energy levels. Because of their complexity, we will not consider the shapes of *d* and *f* orbitals.

Electrons in Sublevels

We can deduce the maximum electron capacity of each sublevel based on the information just given.

For the *s* sublevel:

$$1 \text{ orbital} \times \frac{2e^- \text{ capacity}}{\text{orbital}} = 2e^- \text{ capacity}$$

For the *p* sublevel:

$$3 \text{ orbitals} \times \frac{2e^- \text{ capacity}}{\text{orbital}} = 6e^- \text{ capacity}$$

For the *d* sublevel:

$$5 \text{ orbitals} \times \frac{2e^- \text{ capacity}}{\text{orbital}} = 10e^- \text{ capacity}$$

For the *f* sublevel:

$$7 \text{ orbitals} \times \frac{2e^- \text{ capacity}}{\text{orbital}} = 14e^- \text{ capacity}$$

Electron Spin

Section 2.2 discusses the properties of electrons demonstrated by Thomson.

As we have noted, each atomic orbital has a maximum capacity of two electrons. The electrons are perceived to *spin* on an imaginary axis, and the two electrons in the same orbital must have opposite spins: clockwise and counterclockwise. Their behavior is analogous to two ends of a magnet. Remember, electrons have magnetic properties. The electrons exhibit sufficient magnetic attraction to hold themselves together despite the natural repulsion that they "feel" for each other, owing to their similar charge (remember, like charges repel). Electrons must therefore have opposite spins to coexist in an orbital. A pair of electrons in one orbital that possess opposite spins are referred to as *paired* electrons.

Electron Configuration and the Aufbau Principle

The arrangement of electrons in atomic orbitals is referred to as the atom's electron configuration. The *aufbau*, or building up, *principle* helps us to represent the electron configuration of atoms of various elements. According to this principle, electrons fill the lowest-energy orbital that is available first. We should also recall that the maximum capacity of an *s* level is two, that of a *p* level is six, that of a *d* level is ten, and that of an *f* level is fourteen electrons. Consider the following guidelines for writing electron configurations:

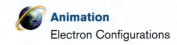

LEARNING GOAL

10 Write electron configurations for atoms of the most commonly occurring elements.

Animation
Electron Configurations

Guidelines for Writing Electron Configurations

- Obtain the total number of electrons in the atoms from the atomic number found on the periodic table.
- Electrons in atoms occupy the lowest energy orbitals that are available, beginning with 1s.
- Each principal energy level, *n*, can contain only *n* subshells.
- Each sublevel is composed of one (*s*) or more (three *p*, five *d*, seven *f*) orbitals.
- No more than two electrons can be placed in any orbital.
- The maximum number of electrons in any principal energy level is $2(n)^2$.
- The theoretical order of orbital filling is depicted in Figure 2.13.

Now let us look at several elements:

Hydrogen

Hydrogen is the simplest atom; it has only one electron. That electron must be in the lowest principal energy level ($n = 1$) and the lowest orbital (*s*). We indicate the number of electrons in a region with a *superscript*, so we write $1s^1$.

Helium

Helium has two electrons, which will fill the lowest energy level. The ground state (lowest energy) electron configuration for helium is $1s^2$.

Lithium

Lithium has three electrons. The first two are configured as helium. The third must go into the orbital of the lowest energy in the second principal energy level; therefore the configuration is $1s^2\,2s^1$.

Beryllium Through Neon

The second principal energy level can contain eight electrons [$2(2)^2$], two in the *s* level and six in the *p* level. The "building up" process results in

Be	$1s^2\,2s^2$	
B	$1s^2\,2s^2\,2p^1$	Three 2*p* orbitals can
C	$1s^2\,2s^2\,2p^2$	hold a maximum of 6 e⁻.
N	$1s^2\,2s^2\,2p^3$	Neon has a complete
O	$1s^2\,2s^2\,2p^4$	$n = 2$ level; that
F	$1s^2\,2s^2\,2p^5$	explains its unusual
Ne	$1s^2\,2s^2\,2p^6$	stability.

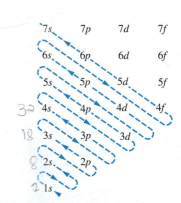

Figure 2.13
A useful way to remember the filling order for electrons in atoms. Begin adding electrons at the bottom (lowest energy) and follow the arrows. Remember: no more than two electrons in each orbital.

Sodium Through Argon

Electrons in these elements retain the basic $1s^2\,2s^2\,2p^6$ arrangement of the preceding element, neon; new electrons enter the third principal energy level:

Na	$1s^2\,2s^2\,2p^6\,3s^1$
Mg	$1s^2\,2s^2\,2p^6\,3s^2$
Al	$1s^2\,2s^2\,2p^6\,3s^2\,3p^1$
Si	$1s^2\,2s^2\,2p^6\,3s^2\,3p^2$
P	$1s^2\,2s^2\,2p^6\,3s^2\,3p^3$
S	$1s^2\,2s^2\,2p^6\,3s^2\,3p^4$
Cl	$1s^2\,2s^2\,2p^6\,3s^2\,3p^5$
Ar	$1s^2\,2s^2\,2p^6\,3s^2\,3p^6$

Three $3p$ orbitals can hold a maximum of 6 e$^-$. Argon has a complete $n = 3$ level; that explains its unusual stability.

By knowing the order of filling of atomic orbitals, lowest to highest energy, you may write the electron configuration for any element. The order of orbital filling can be represented by the diagram in Figure 2.13. Such a diagram provides an easy way of predicting the electron configuration of the elements. Remember that the diagram is based on an energy scale, with the lowest energy orbital at the beginning of the "path" and the highest energy orbital at the end of the "path." An alternative way of representing orbital energies is through the use of an energy level diagram, such as the one in Figure 2.14.

E X A M P L E   2.5   **Writing the Electron Configuration of Tin**

LEARNING GOAL
10 ▸ Write electron configurations for atoms of the most commonly occurring elements.

Write the electron configuration for tin.

Solution

Step 1. Tin, Sn, has an atomic number of 50; thus we must place fifty electrons in atomic orbitals.

Step 2. We must also remember the total electron capacities of orbital types: s, 2; p, 6; d, 10; and f, 14. The first principal energy level has one sublevel, the second has two sublevels, and so on.

Step 3. The order of filling (Figure 2.13) is $1s$, $2s$, $2p$, $3s$, $3p$, $4s$, $3d$, $4p$.

Step 4. The electron configuration is as follows:

$$1s^2\,2s^2\,2p^6\,3s^2\,3p^6\,4s^2\,3d^{10}\,4p^6\,5s^2\,4d^{10}\,5p^2$$

Helpful Hint: As a check, count electrons in the electron configuration (add all of the superscripted numbers) to see that we have accounted for all fifty electrons of the Sn atom.

Practice Problem 2.5

Give the electron configuration for an atom of:
a. sulfur
b. calcium
c. potassium
d. phosphorus

For Further Practice: Questions 2.79 and 2.80.

Shorthand Electron Configurations

As we noted earlier the electron configuration for the sodium atom (Na, atomic number 11) is

$$1s^2 \, 2s^2 \, 2p^6 \, 3s^1$$

The electron configuration for the preceding noble gas, neon (Ne, atomic number 10), is

$$1s^2 \, 2s^2 \, 2p^6$$

The electron configuration for sodium is really the electron configuration of Ne, with $3s^1$ added to represent one additional electron. So it is permissible to write

$$[Ne] \, 3s^1 \text{ as equivalent to } 1s^2 \, 2s^2 \, 2p^6 \, 3s^1$$

$[Ne] \, 3s^1$ is the *shorthand electron configuration* for sodium.

Similarly,

$$[Ne] \, 3s^2 \text{ representing Mg}$$
$$[Ne] \, 3s^2 \, 3p^5 \text{ representing Cl}$$
$$[Ar] \, 4s^1 \text{ representing K}$$

are valid electron configurations.

The use of abbreviated electron configurations, in addition to being faster and easier to write, serves to highlight the valence electrons, those electrons involved in bonding. The symbol of the noble gas represents the *core*, nonvalence electrons and the valence electron configuration follows the noble gas symbol.

Figure 2.14
An orbital energy-level diagram. Electrons fill orbitals in the order of increasing energy.

Give the *shorthand* electron configuration for:

a. sulfur
b. calcium

Question *2.7*

Give the *shorthand* electron configuration for:

a. potassium
b. phosphorus

Question *2.8*

2.6 The Octet Rule

Elements in the last family, the noble gases, have either two valence electrons (helium) or eight valence electrons (neon, argon, krypton, xenon, and radon). These elements are extremely stable and were often termed *inert gases* because they do not readily bond to other elements, although they can be made to do so under extreme experimental conditions. A full $n = 1$ energy level (as in helium) or an outer *octet* of electrons (eight valence electrons, as in all of the other noble gases) is responsible for this unique stability.

Atoms of elements in other groups are more reactive than the noble gases because in the process of chemical reaction they are trying to achieve a more stable "noble gas" configuration by gaining or losing electrons. This is the basis of

We may think of stability as a type of contentment; a noble gas atom does not need to rearrange its electrons or lose or gain any electrons to get to a more stable, lower energy, or more "contented" configuration.

the **octet rule** which states that elements usually react in such a way as to attain the electron configuration of the noble gas closest to them in the periodic table (a stable octet of electrons). In chemical reactions they will gain, lose, or share the minimum number of electrons necessary to attain this more stable energy state. The octet rule, although simple in concept, is a remarkably reliable predictor of chemical change, especially for representative elements.

Ion Formation and the Octet Rule

Metals and nonmetals differ in the way in which they form ions. Metallic elements (located at the left of the periodic table) tend to form positively charged ions called *cations.* Positive ions are formed when an atom loses one or more electrons, for example,

$$Na \longrightarrow Na^+ + e^-$$
Sodium atom   Sodium ion
(11e$^-$, 1 valence e$^-$)   (10e$^-$)

$$Mg \longrightarrow Mg^{2+} + 2e^-$$
Magnesium atom   Magnesium ion
(12e$^-$, 2 valence e$^-$)   (10e$^-$)

$$Al \longrightarrow Al^{3+} + 3e^-$$
Aluminum atom   Aluminum ion
(13e$^-$, 3 valence e$^-$)   (10e$^-$)

In each of these cases the atom has lost *all* of its valence electrons. The resulting ion has the same number of electrons as the nearest noble gas atom:

$Na^+(10e^-)$ and $Mg^{2+}(10e^-)$ and $Al^{3+}(10e^-)$ are all isoelectronic with Ne (10e$^-$).

Recall that the prefix *iso* (Greek *isos*) means equal.

These ions are particularly stable. Each ion is **isoelectronic** (that is, it has the same number of electrons) with its nearest noble gas neighbor and has an octet of electrons in its outermost energy level.

Sodium is typical of each element in its group. Knowing that sodium forms a 1+ ion leads to the prediction that H, Li, K, Rb, Cs, and Fr also will form 1+ ions. Furthermore, magnesium, which forms a 2+ ion, is typical of each element in its group; Be^{2+}, Ca^{2+}, Sr^{2+}, and so forth are the resulting ions.

Nonmetallic elements, located at the right of the periodic table, tend to gain electrons to become isoelectronic with the nearest noble gas element, forming negative ions called *anions.*

Section 3.2 discusses the naming of ions.

Consider:

The ion of fluorine is the *fluoride ion;* the ion of oxygen is the *oxide ion;* and the ion of nitrogen is the *nitride ion.*

$$F + 1e^- \longrightarrow F^-$$ (isoelectronic with Ne, 10e$^-$)
Fluorine atom   Fluoride ion
(9e$^-$, 7 valence e$^-$)   (10e$^-$)

$$O + 2e^- \longrightarrow O^{2-}$$ (isoelectronic with Ne, 10e$^-$)
Oxygen atom   Oxide ion
(8e$^-$, 6 valence e$^-$)   (10e$^-$)

$$N + 3e^- \longrightarrow N^{3-}$$ (isoelectronic with Ne, 10e$^-$)
Nitrogen atom   Nitride ion
(7e$^-$, 5 valence e$^-$)   (10e$^-$)

As in the case of positive ion formation, each of these negative ions has an octet of electrons in its outermost energy level.

A Medical Perspective

Dietary Calcium

"Drink your milk!" "Eat all of your vegetables!" These imperatives are almost universal memories from our childhood. Our parents knew that calcium, present in abundance in these foods, was an essential element for the development of strong bones and healthy teeth.

Recent studies, spanning the fields of biology, chemistry, and nutrition science indicate that the benefits of calcium go far beyond bones and teeth. This element has been found to play a role in the prevention of disease throughout our bodies.

Calcium is the most abundant mineral (metal) in the body. It is ingested as the calcium ion (Ca^{2+}) either in its "free" state or "combined," as a part of a larger compound; calcium dietary supplements often contain ions in the form of calcium carbonate. The acid naturally present in the stomach produces the calcium ion:

$$CaCO_3 + 2H^+ \longrightarrow Ca^{2+} + H_2O + CO_2$$

calcium carbonate	stomach acid	calcium ion	water	carbon dioxide

Calcium is responsible for a variety of body functions including:

- transmission of nerve impulses
- release of "messenger compounds" that enable communication among nerves
- blood clotting
- hormone secretion
- growth of living cells throughout the body

The body's storehouse of calcium is bone tissue. When the supply of calcium from external sources, the diet, is insufficient, the body uses a mechanism to compensate for this shortage. With vitamin D in a critical role, this mechanism removes calcium from bone to enable other functions to continue to take place. It is evident then that prolonged dietary calcium deficiency can weaken the bone structure. Unfortunately, current studies show that as much as 75% of the American population may not be consuming sufficient amounts of calcium. Developing an understanding of the role of calcium in premenstrual syndrome, cancer, and blood pressure regulation is the goal of three current research areas.

Calcium and premenstrual syndrome (PMS). Dr. Susan Thys-Jacobs, a gynecologist at St. Luke's-Roosevelt Hospital Center in New York City, and colleagues at eleven other medical centers are conducting a study of calcium's ability to relieve the discomfort of PMS. They believe that women with chronic PMS have calcium blood levels that are normal only because calcium is continually being removed from the bone to maintain an adequate supply in the blood. To complicate the situation, vitamin D levels in many young women are very low (as much as 80% of a person's vitamin D is made in the skin, upon exposure to sunlight; many of us now minimize our exposure to the sun because of concerns about ultraviolet radiation and skin cancer). Because vitamin D plays an essential role in calcium metabolism, even if sufficient calcium is consumed, it may not be used efficiently in the body.

Colon cancer. The colon is lined with a type of cell (epithelial cell) that is similar to those that form the outer layers of skin. Various studies have indicated that by-products of a high-fat diet are irritants to these epithelial cells and produce abnormal cell growth in the colon. Dr. Martin Lipkin, Rockefeller University in New York, and his colleagues have shown that calcium ions may bind with these irritants, reducing their undesirable effects. It is believed that a calcium-rich diet, low in fat, and perhaps use of a calcium supplement can prevent or reverse this abnormal colon cell growth, delaying or preventing the onset of colon cancer.

Blood pressure regulation. Dr. David McCarron, a blood pressure specialist at the Oregon Health Sciences University, believes that dietary calcium levels may have a significant influence on hypertension (high blood pressure). Preliminary studies show that a diet rich in low-fat dairy products, fruits, and vegetables, all high in calcium, may produce a significant lowering of blood pressure in adults with mild hypertension.

The take-home lesson appears clear: a high calcium, low fat diet promotes good health in many ways. Once again, our parents were right!

For Further Understanding

Distinguish between "free" and "combined" calcium in the diet.

Why might calcium supplements be ineffective in treating all cases of calcium deficiency?

The element fluorine, forming F^-, indicates that the other halogens, Cl, Br, and I, behave as a true family and form Cl^-, Br^-, and I^- ions. Also, oxygen and the other nonmetals in its group form 2^- ions; nitrogen and phosphorus form 3^- ions. It is important to recognize that ions are formed by gain or loss of *electrons*. No change occurs in the nucleus; the number of protons remains the same.

Question *2.9*

Give the charge of the most probable ion resulting from each of the following elements. With what element is the ion isoelectronic?

a. Ca   d. Mg
b. Sr   e. P
c. S

Question *2.10*

Which of the following pairs of atoms and ions are isoelectronic?

a. Cl^-, Ar   d. Li^+, Ne
b. Na^+, Ne   e. O^{2-}, F^-
c. Mg^{2+}, Na^+

The transition metals tend to form positive ions by losing electrons, just like the representative metals. Metals, whether representative or transition, share this characteristic. However, the transition elements are characterized as "variable valence" elements; depending on the type of substance with which they react, they may form more than one stable ion. For example, iron has two stable ionic forms:

$$Fe^{2+} \text{ and } Fe^{3+}$$

Copper can exist as

$$Cu^+ \text{ and } Cu^{2+}$$

and elements such as vanadium, V, and manganese, Mn, each can form four different stable ions.

Predicting the charge of an ion or the various possible ions for a given transition metal is not an easy task. Energy differences between valence electrons of transition metals are small and not easily predicted from the position of the element in the periodic table. In fact, in contrast to representative metals, the transition metals show great similarities within a *period* as well as within a *group*.

2.7 Trends in the Periodic Table

Atomic Size

Many atomic properties correlate with electronic structure, hence, with their position in the periodic table. Given the fact that interactions among multiple charged particles are very complex, we would not expect the correlation to be perfect. Nonetheless, the periodic table remains an excellent guide to the prediction of properties.

If our model of the atom is a tiny sphere whose radius is determined by the distance between the center of the nucleus and the boundary of the region where the valence electrons have a probability of being located, the size of the atom will be determined principally by two factors.

1. The energy level (*n* level) in which the outermost electron(s) is (are) found increases as we go *down* a group. (Recall that the outermost *n* level correlates with period number.) Thus the size of atoms should increase from top to bottom of the periodic table as we fill successive energy levels of the atoms with electrons (Figure 2.15).
2. As the magnitude of the positive charge of the nucleus increases, its "pull" on all of the electrons increases, and the electrons are drawn closer to the nucleus. This results in a contraction of the atomic radius and therefore a decrease in atomic size. This effect is apparent as we go *across* the periodic table within a period. Atomic size decreases from left to right in the periodic table. See how many exceptions you can find in Figure 2.15.

12 | **LEARNING GOAL**
Utilize the periodic table and its predictive power to estimate the relative sizes of atoms and ions, as well as relative magnitudes of ionization energy and electron affinity.

The radius of an atom is traditionally defined as one-half of the distance between atoms in a covalent bond. The covalent bond is discussed in Section 3.1.

Animation
Atomic Radius

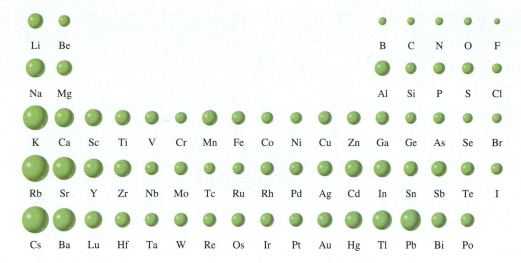

Figure 2.15
Variation in the size of atoms as a function of their position in the periodic table. Note particularly the decrease in size from left to right in the periodic table and the increase in size as we proceed down the table, although some exceptions do exist. (Lanthanide and actinide elements are not included here.)

Ion Size

Positive ions (cations) are smaller than the parent atom. The cation has more protons than electrons (an increased nuclear charge). The excess nuclear charge pulls the remaining electrons closer to the nucleus. Also, cation formation often results in the loss of all outer-shell electrons, resulting in a significant decrease in radius.

Negative ions (anions) are larger than the parent atom. The anion has more electrons than protons. Owing to the excess negative charge, the nuclear "pull" on each individual electron is reduced. The electrons are held less tightly, resulting in a larger anion radius in contrast to the neutral atom.

Ions with multiple positive charge (such as Cu^{2+}) are even *smaller* than their corresponding monopositive ion (Cu^+); ions with multiple negative charge (such as O^{2-}) are *larger* than their corresponding less negative ion.

Figure 2.16 depicts the relative sizes of several atoms and their corresponding ions.

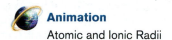

Animation
Atomic and Ionic Radii

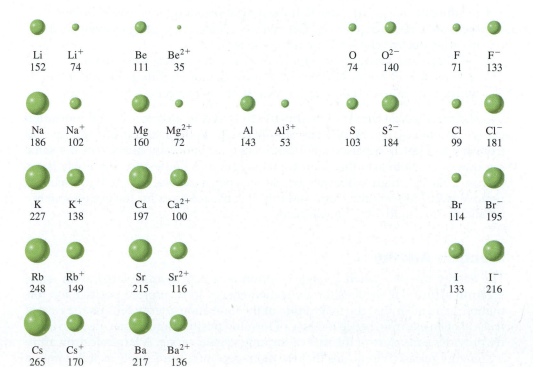

Figure 2.16
Relative size of ions and their parent atoms. Atomic radii are provided in units of picometers.

(a)

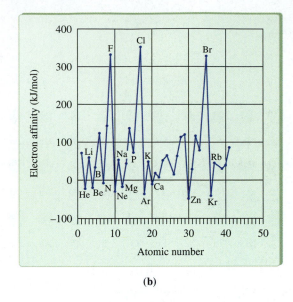

(b)

Figure 2.17

(a) The ionization energies of the first forty elements versus their atomic numbers. Note the very high values for elements located on the right in the periodic table, and low values for those on the left. Some exceptions to the trends are evident. (b) The periodic variation of electron affinity. Note the very low values for the noble gases and the elements on the far left of the periodic table. These elements do not form negative ions. In contrast, F, Cl, and Br readily form negative ions.

Remember: ionization energy and electron affinity are predictable from *trends* in the periodic table. As with most trends, exceptions occur.

Ionization Energy

The energy required to remove an electron from an isolated atom is the **ionization energy.** The process for sodium is represented as follows:

$$\text{ionization energy} + Na \longrightarrow Na^+ + e^-$$

The magnitude of the ionization energy should correlate with the strength of the attractive force between the nucleus and the outermost electron.

- Reading *down* a group, note that the ionization energy decreases, because the atom's size is increasing. The outermost electron is progressively farther from the nuclear charge, hence easier to remove.
- Reading *across* a period, note that atomic size decreases, because the outermost electrons are closer to the nucleus, more tightly held, and more difficult to remove. Therefore the ionization energy generally increases.

A correlation does indeed exist between trends in atomic size and ionization energy. Atomic size generally *decreases* from the bottom to top of a group and from left to right in a period. Ionization energies generally *increase* in the same periodic way. Note also that ionization energies are highest for the noble gases (Figure 2.17a). A high value for ionization energy means that it is difficult to remove electrons from the atom, and this, in part, accounts for the extreme stability and nonreactivity of the noble gases.

Electron Affinity

The energy released when a single electron is added to an isolated atom is the **electron affinity.** If we consider ionization energy in relation to positive ion formation (remember that the magnitude of the ionization energy tells us the ease of *removal* of an electron, hence the ease of forming positive ions), then electron affinity provides a measure of the ease of forming negative ions. A large electron affinity (energy released) indicates that the atom becomes more stable as it becomes a negative ion (through gaining an electron). Consider the gain of an electron by a bromine atom:

$$Br + e^- \longrightarrow Br^- + energy$$

Electron affinity

Periodic trends for electron affinity are as follows:

- Electron affinities generally decrease down a group.
- Electron affinities generally increase across a period.

Remember these trends are not absolute. Exceptions exist, as seen in the irregularities in Figure 2.17b.

Question 2.11

Rank Be, N, and F in order of increasing

a. atomic size
b. ionization energy
c. electron affinity

Question 2.12

Rank Cl, Br, I, and F in order of increasing

a. atomic size
b. ionization energy
c. electron affinity

SUMMARY

2.1 Composition of the Atom

The basic structural unit of an element is the *atom*, which is the smallest unit of an element that retains the chemical properties of that element.

The atom has two distinct regions. The *nucleus* is a small, dense, positively charged region in the center of the atom composed of positively charged *protons* and uncharged *neutrons*. Surrounding the nucleus is a diffuse region of negative charge occupied by *electrons*, the source of the negative charge. Electrons are very low in mass in comparison to protons and neutrons.

The *atomic number* (Z) is equal to the number of protons in the atom. The *mass number* (A) is equal to the sum of the protons and neutrons (the mass of the electrons is insignificant).

Isotopes are atoms of the same element that have different masses because they have different numbers of neutrons (different mass numbers). Isotopes have chemical behavior identical to that of any other isotope of the same element.

Ions are electrically charged particles that result from a gain or loss of one or more electrons by the parent atom. *Anions,* negative ions, are formed by a gain of one or more electrons by the parent atom. *Cations,* positive ions, are formed by a loss of one or more electrons from the parent atom.

2.2 Development of Atomic Theory

The first experimentally based theory of atomic structure was proposed by John Dalton. Although Dalton pictured atoms as indivisible, the experiments of William Crookes, Eugene Goldstein, and J. J. Thomson indicated that the atom is composed of charged particles: protons and electrons. The third fundamental atomic particle is the neutron. An experiment conducted by Hans Geiger led Ernest Rutherford to propose that the majority of the mass and positive charge of the atom is located in a small, dense region, the *nucleus,* with small, negatively charged electrons occupying a much larger, diffuse space outside of the nucleus.

2.3 Light, Atomic Structure, and the Bohr Atom

The study of the interaction of *light* and *matter* is termed *spectroscopy.* Light, *electromagnetic radiation,* travels at a speed of 3.0×10^8 m/s, the *speed of light.* Light is made up of many wavelengths. Collectively, they make up the *electromagnetic spectrum.* Samples of elements emit certain wavelengths of light when an electrical current is passed through the sample. Different elements emit a different pattern (different wavelengths) of light.

Niels Bohr proposed an atomic model that described the atom as a nucleus surrounded by fixed *energy levels* (or *quantum levels*) that can be occupied by electrons. He believed that each level was defined by a spherical *orbit* located at a specific distance from the nucleus.

Promotion and relaxation processes are referred to as *electronic transitions.* Electron promotion resulting from absorption of energy results in an excited state atom; the process of relaxation allows the atom to return to the ground state by emitting a certain wavelength of light.

The modern view of the atom describes the probability of finding an electron in a region of space within the principal energy level, referred to as an *atomic orbital.* The rapid movement of the electrons spreads them into a cloud of charge. This cloud is more dense in certain regions, the density being proportional to the probability of finding the electron at any point in time. The orbital is strikingly different from Bohr's orbit. The electron does not orbit the nucleus; rather, its behavior is best described as that of a wave.

2.4 The Periodic Law and the Periodic Table

The *periodic law* is an organized "map" of the elements that relates their structure to their chemical and physical properties. It states that the elements, when arranged according to their atomic numbers, show a distinct periodicity (regular variation) of their properties. The periodic table is the result of the periodic law.

The modern periodic table exists in several forms. The most important variation is in group numbering. The tables in this text use the two most commonly accepted numbering systems.

A horizontal row of elements in the periodic table is referred to as a *period.* The periodic table consists of seven periods. The lanthanide series is a part of period 6; the actinide series is a part of period 7.

The columns of elements in the periodic table are called *groups* or *families.* The elements of a particular family share many similarities in physical and chemical properties because of the similarities in electronic structure. Some of the most important groups are named; for example, the *alkali metals* (IA or 1), *alkaline earth metals* (IIA or 2), the *halogens* (VIIA or 17), and the *noble gases* (VIII or 18).

Group A elements are called *representative elements;* Group B elements are *transition elements.* A bold zigzag line runs from top to bottom of the table, beginning to the left of boron (B) and ending between polonium (Po) and astatine (At). This line acts as the boundary between *metals* to the left and *nonmetals* to the right. Elements straddling the boundary, *metalloids,* have properties intermediate between those of metals and nonmetals.

2.5 Electron Arrangement and the Periodic Table

The outermost electrons in an atom are *valence electrons.* For representative elements the number of valence electrons in an atom corresponds to the group or family number (old numbering system using Roman numerals). Metals tend to have fewer valence electrons than nonmetals.

Electron configuration of the elements is predictable, using the aufbau principle. Knowing the electron configuration, we can identify valence electrons and begin to predict the kinds of reactions that the elements will undergo.

Elements in the last family, the noble gases, have either two valence electrons (helium) or eight valence electrons (neon, argon, krypton, xenon, and radon). Their most important properties are their extreme stability and lack of reactivity. A full valence level is responsible for this unique stability.

2.6 The Octet Rule

The *octet rule* tells us that in chemical reactions, elements will gain, lose, or share the minimum number of electrons necessary to achieve the electron configuration of the nearest noble gas.

Metallic elements tend to form cations. The ion is *isoelectronic* with its nearest noble gas neighbor and has a stable octet of electrons in its outermost energy level. Nonmetallic elements tend to gain electrons to become isoelectronic with the nearest noble gas element, forming anions.

2.7 Trends in the Periodic Table

Atomic size decreases from left to right and from bottom to top in the periodic table. *Cations* are smaller than the parent atom. *Anions* are larger than the parent atom. Ions with multiple positive charge are even smaller than their corresponding monopositive ion; ions with multiple negative charge are larger than their corresponding less negative ion.

The energy required to remove an electron from the atom is the *ionization energy.* Down a group, the ionization energy generally decreases. Across a period, the ionization energy generally increases.

The energy released when a single electron is added to a neutral atom in the gaseous state is known as the *electron affinity.* Electron affinities generally decrease proceeding down a group and increase proceeding across a period.

KEY TERMS

alkali metal (2.4)	electron density (2.3)
alkaline earth metal (2.4)	energy level (2.3)
anion (2.1)	group (2.4)
atom (2.1)	halogen (2.4)
atomic mass (2.1)	ion (2.1)
atomic number (2.1)	ionization energy (2.7)
atomic orbital (2.3 and 2.5)	isoelectronic (2.6)
cathode rays (2.2)	isotope (2.1)
cation (2.1)	mass number (2.1)
electromagnetic radiation (2.3)	metal (2.4)
electromagnetic spectrum (2.3)	metalloid (2.4)
electron (2.1)	neutron (2.1)
electron affinity (2.7)	noble gas (2.4)
electron configuration (2.5)	nonmetal (2.4)

Atomic Symbol	No. of Protons	No. of Neutrons	No. of Electrons	Charge
Example:				
$^{27}_{13}$Al	13	14	13	0
$^{39}_{19}$K	19	___	19	0
$^{31}_{15}$P^{3-}	15	16	___	___
___	29	34	27	2+
$^{55}_{26}$Fe^{2+}	___	29	___	2+
___	8	8	10	___

2.26 Fill in the blanks:

2.27 Fill in the blanks:
  a. An isotope of an element differs in mass because the atom has a different number of ____.
  b. The atomic number gives the number of ____ in the nucleus.
  c. The mass number of an atom is due to the number of ____ and ____ in the nucleus.
  d. A charged atom is called a(n) ____.
  e. Electrons surround the ____ and have a ____ charge.

2.28 Label each of the following statements as true or false:
  a. An atom with an atomic number of 7 and a mass of 14 is identical to an atom with an atomic number of 6 and a mass of 14.
  b. Neutral atoms have the same number of electrons as protons.
  c. The mass of an atom is due to the sum of the number of protons, neutrons, and electrons.

2.29 The element copper has two naturally occurring isotopes. One of these has a mass of 62.93 amu and a natural abundance of 69.09%. A second isotope has a mass of 64.9278 amu and a natural abundance of 30.91%. Calculate the atomic mass of copper.

2.30 The element lithium has two naturally occurring isotopes. One of these has a mass of 6.0151 amu and a natural abundance of 7.49%. A second isotope has a mass of 7.0160 amu and a natural abundance of 92.51%. Calculate the atomic mass of lithium.

QUESTIONS AND PROBLEMS

Composition of the Atom

Foundations

2.13 Calculate the number of protons, neutrons, and electrons in:
  a. $^{16}_{8}$O
  b. $^{31}_{15}$P

2.14 Calculate the number of protons, neutrons, and electrons in:
  a. $^{136}_{56}$Ba
  b. $^{209}_{84}$Po

2.15 State the mass and charge of the:
  a. electron
  b. proton
  c. neutron

2.16 Calculate the number of protons, neutrons, and electrons in:
  a. $^{37}_{17}$Cl
  b. $^{23}_{11}$Na
  c. $^{84}_{36}$Kr

2.17 a. What is an ion?
  b. What process results in the formation of a cation?
  c. What process results in the formation of an anion?

2.18 a. What are isotopes?
  b. What is the major difference among isotopes of an element?
  c. What is the major similarity among isotopes of an element?

Applications

2.19 How many protons are in the nucleus of the isotope Rn-220?

2.20 How many neutrons are in the nucleus of the isotope Rn-220?

2.21 Selenium-80 is a naturally occurring isotope. It is found in over-the-counter supplements.
  a. How many protons are found in one atom of selenium-80?
  b. How many neutrons are found in one atom of selenium-80?

2.22 Iodine-131 is an isotope used in thyroid therapy.
  a. How many protons are found in one atom of iodine-131?
  b. How many neutrons are found in one atom of iodine-131?

2.23 Write symbols for each isotope:
  a. Each atom contains 1 proton and 0 neutrons.
  b. Each atom contains 6 protons and 8 neutrons.

2.24 Write symbols for each isotope:
  a. Each atom contains 1 proton and 2 neutrons.
  b. Each atom contains 92 protons and 146 neutrons.

2.25 Fill in the blanks:

Symbol	No. of Protons	No. of Neutrons	No. of Electrons	Charge
Example:				
$^{40}_{20}$Ca	20	20	20	0
$^{23}_{11}$Na	11	12	11	0
$^{32}_{16}$S^{2-}	16	16	18	2−
O	8	8	8	0
$^{24}_{12}$Mg^{2+}	12	12	10	2+
K	19	20	18	1+

Development of Atomic Theory

Foundations

2.31 What are the major postulates of Dalton's atomic theory?

2.32 What points of Dalton's theory are no longer current?

2.33 Note the major accomplishment of each of the following:
  a. Chadwick
  b. Goldstein

2.34 Note the major accomplishment of each of the following:
  a. Geiger
  b. Bohr

2.35 Note the major accomplishment of each of the following:
  a. Dalton
  b. Crookes

2.36 Note the major accomplishment of each of the following:
  a. Thomson
  b. Rutherford

Applications

2.37 Describe the experiment that provided the basis for our understanding of the nucleus.

2.38 Describe the series of experiments that characterized the electron.

2.39 List at least three properties of the electron.

2.40 Describe the process that occurs when electrical energy is applied to a sample of hydrogen gas.

2.41 What is a cathode ray? Which subatomic particle is detected?

2.42   Pictured is a cathode ray tube. Show the path that an electron would follow in the tube.

Light, Atomic Structure, and the Bohr Atom

Foundations

2.43   Rank the various regions of the electromagnetic spectrum in order of increasing wavelength.

2.44   Rank the various regions of the electromagnetic spectrum in order of increasing energy.

2.45   Which form of radiation has greater energy, microwave or infrared?

2.46   Which form of radiation has the longer wavelength, ultraviolet or infrared?

2.47   What is meant by the term *spectroscopy?*

2.48   What is meant by the term *electromagnetic spectrum?*

Applications

2.49   Critique this statement: Electrons can exist in any position outside of the nucleus.

2.50   Critique this statement: Promotion of electrons is accompanied by a release of energy.

2.51   What are the most important points of the Bohr theory?

2.52   Give two reasons why the Bohr theory did not stand the test of time.

2.53   What was the major contribution of Bohr's atomic model?

2.54   What was the major deficiency of Bohr's atomic model?

The Periodic Law and the Periodic Table

Foundations

2.55   Provide the name of the element represented by each of the following symbols:
   a. Na
   b. K
   c. Mg
   d. B

2.56   Provide the name of the element represented by each of the following symbols:
   a. Ca
   b. Cu
   c. Co
   d. Si

2.57   Which group of the periodic table is known as the alkali metals? List them.

2.58   Which group of the periodic table is known as the alkaline earth metals? List them.

2.59   Which group of the periodic table is known as the halogens? List them.

2.60   Which group of the periodic table is known as the noble gases? List them.

Applications

2.61   Label each of the following statements as true or false:
   a. Elements of the same group have similar properties.
   b. Atomic size decreases from left to right across a period.

2.62   Label each of the following statements as true or false:
   a. Ionization energy increases from top to bottom within a group.
   b. Representative metals are located on the left in the periodic table.

2.63   For each of the elements Na, Ni, Al, P, Cl, and Ar provide the following information:
   a. Which are metals?
   b. Which are representative metals?
   c. Which tend to form positive ions?
   d. Which are inert or noble gases?

2.64   For each of the elements Ca, K, Cu, Zn, Br, and Kr provide the following information:
   a. Which are metals?
   b. Which are representative metals?
   c. Which tend to form positive ions?
   d. Which are inert or noble gases?

2.65   Using the information below, for Group I elements:

Element	Atomic Number	Melting Point(°C)
Li	3	180.5
Na	11	97.8
K	19	63.3
Rb	37	38.9
Cs	55	28.4

Prepare a graph relating melting point and atomic number. How does this demonstrate the periodic law?

2.66   Use the graph prepared in Question 2.65 to predict the melting point of francium.

Electron Arrangement and the Periodic Table

Foundations

2.67   How many total electrons and valence electrons are found in an atom of each of the following elements? What is the number of the principal energy level in which the valence electrons are found?
   a. H
   b. Na
   c. B
   d. F
   e. Ne
   f. He

2.68   How many total electrons and valence electrons are found in an atom of each of the following elements? What is the number of the principal energy level in which the valence electrons are found?
   a. Mg
   b. K
   c. C
   d. Br
   e. Ar
   f. Xe

2.69   Distinguish between a principal energy level and a sublevel.

2.70   Distinguish between a sublevel and an orbital.

2.71   Sketch a diagram and describe our current model of an *s* orbital.

2.72   How is a 2*s* orbital different from a 1*s* orbital?

2.73   How many *p* orbitals can exist in a given principal energy level?

2.74   Sketch diagrams of a set of *p* orbitals. How does a p_x orbital differ from a p_y orbital? From a p_z orbital?

2.75   How does a 3*p* orbital differ from a 2*p* orbital?

2.76   What is the maximum number of electrons that an orbital can hold?

Applications

2.77 What is the maximum number of electrons in each of the following energy levels?
  a. $n = 1$
  b. $n = 2$
  c. $n = 3$

2.78 **a.** What is the maximum number of s electrons that can exist in any one principal energy level?
  b. How many p electrons?
  c. How many d electrons?
  d. How many f electrons?

2.79 Using the periodic table, write the electron configuration of each of the following atoms:
  a. Al
  b. Na
  c. Sc
  d. Ca
  e. Fe
  f. Cl

2.80 Using only the periodic table or list of elements, write the electron configuration of each of the following atoms:
  a. B
  b. S
  c. Ar
  d. V
  e. Cd
  f. Te

2.81 Which of the following electron configurations are not possible? Why?
  a. $1s^2 1p^2$
  b. $1s^2 2s^2 2p^2$
  c. $2s^2, 2s^2, 2p^6, 2d^1$
  d. $1s^2, 2s^3$

2.82 For each incorrect electron configuration in Question 2.81, assume that the number of electrons is correct, identify the element, and write the correct electron configuration.

The Octet Rule

Foundations

2.83 Give the most probable ion formed from each of the following elements:
  a. Li
  b. O
  c. Ca
  d. Br
  e. S
  f. Al

2.84 Using only the periodic table or list of elements, write the electron configuration of each of the following ions:
  a. I^-
  b. Ba^{2+}
  c. Se^{2-}
  d. Al^{3+}

2.85 Which of the following pairs of atoms and/or ions are isoelectronic with one another?
  a. O^{2-}, Ne
  b. S^{2-}, Cl^-

2.86 Which of the following pairs of atoms and/or ions are isoelectronic with one another?
  a. F^-, Cl^-
  b. K^+, Ar

Applications

2.87 Which species in each of the following groups would you expect to find in nature?
  a. Na, Na^+, Na^-
  b. S^{2-}, S^-, S^+
  c. Cl, Cl^-, Cl^+

2.88 Which atom or ion in each of the following groups would you expect to find in nature?
  a. K, K^+, K^-
  b. O^{2-}, O, O^{2+}
  c. Br, Br^-, Br^+

2.89 Write the electron configuration of each of the following biologically important ions:
  a. Ca^{2+}
  b. Mg^{2+}

2.90 Write the electron configuration of each of the following biologically important ions:
  a. K^+
  b. Cl^-

Trends in the Periodic Table

Foundations

2.91 Arrange each of the following lists of elements in order of increasing atomic size:
  a. N, O, F
  b. Li, K, Cs
  c. Cl, Br, I

2.92 Arrange each of the following lists of elements in order of increasing atomic size:
  a. Al, Si, P, Cl, S
  b. In, Ga, Al, B, Tl
  c. Sr, Ca, Ba, Mg, Be
  d. P, N, Sb, Bi, As

2.93 Arrange each of the following lists of elements in order of increasing ionization energy:
  a. N, O, F
  b. Li, K, Cs
  c. Cl, Br, I

2.94 Arrange each of the following lists of elements in order of decreasing electron affinity:
  a. Na, Li, K
  b. Br, F, Cl
  c. S, O, Se

Applications

2.95 Explain why a positive ion is always smaller than its parent atom.

2.96 Explain why a negative ion is always larger than its parent atom.

2.97 Explain why a fluoride ion is commonly found in nature but a fluorine atom is not.

2.98 Explain why a sodium ion is commonly found in nature but a sodium atom is not.

2.99 Cl^- and Ar are isoelectronic. Which is larger? Why?

2.100 K^+ and Ar are isoelectronic. Which is larger? Why?

CRITICAL THINKING PROBLEMS

1. A natural sample of chromium, taken from the ground, will contain four isotopes: Cr-50, Cr-52, Cr-53, and Cr-54. Predict which isotope is in greatest abundance. Explain your reasoning.

2. Crookes's cathode ray tube experiment inadvertently supplied the basic science for a number of modern high-tech devices. List a few of these devices and describe how they involve one or more aspects of this historic experiment.

3. Name five elements that you came in contact with today. Were they in combined form or did they exist in the form of atoms? Were they present in pure form or in mixtures? If mixtures, were they heterogeneous or homogeneous? Locate each in the periodic table by providing the group and period designation, for example: Group IIA (2), period 3.

4. The periodic table is incomplete. It is possible that new elements will be discovered from experiments using high-energy particle accelerators. Predict as many properties as you can that might characterize the element that would have an atomic number of 118. Can you suggest an appropriate name for this element?

5. The element titanium is now being used as a structural material for bone and socket replacement (shoulders, knees). Predict properties that you would expect for such applications; go to the library or Internet and look up the properties of titanium and evaluate your answer.

6. Imagine that you have undertaken a voyage to an alternate universe. Using your chemical skills, you find a collection of elements quite different than those found here on Earth. After measuring their properties and assigning symbols for each, you wish to organize them as Mendeleev did for our elements. Design a periodic table using the information you have gathered:

Symbol	Mass (amu)	Reactivity	Electrical Conductivity
A	2.0	High	High
B	4.0	High	High
C	6.0	Moderate	Trace
D	8.0	Low	0
E	10.0	Low	0
F	12.0	High	High
G	14.0	High	High
H	16.0	Moderate	Trace
I	18.0	Low	0
J	20.0	None	0
K	22.0	High	High
L	24.0	High	High

Predict the reactivity and conductivity of an element with a mass of 30.0 amu. What element in our universe does this element most closely resemble?

3 Structure and Properties of Ionic and Covalent Compounds

Learning Goals

1 Classify compounds as having ionic, covalent, or polar covalent bonds.

2 Write the formula of a compound when provided with the name of the compound.

3 Name common inorganic compounds using standard conventions and recognize the common names of frequently used substances.

4 Predict differences in physical state, melting and boiling points, solid-state structure, and solution chemistry that result from differences in bonding.

5 Draw Lewis structures for covalent compounds and polyatomic ions.

6 Describe the relationship between stability and bond energy.

7 Predict the geometry of molecules and ions using the octet rule and Lewis structures.

8 Understand the role that molecular geometry plays in determining the solubility and melting and boiling points of compounds.

9 Use the principles of VSEPR theory and molecular geometry to predict relative melting points, boiling points, and solubilities of compounds.

Outline

Structure determines properties. Provide other examples of objects that illustrate the structure-property relationship.

Introduction

A chemical compound is formed when two or more atoms of different elements are joined by attractive forces called chemical bonds. These bonds result from either a transfer of electrons from one atom to another (the ionic bond) or a sharing of electrons between two atoms (the covalent bond). The elements, once converted to a compound, cannot be recovered by any physical process. A chemical reaction must take place to regenerate the individual elements. The chemical and physical properties of a compound are related to the structure of the compound, and this structure is, in turn, determined by the arrangement of electrons in the atoms that produced the compounds. Properties such as solubility, boiling point, and melting point correlate well with the shape and charge distribution in the individual units of the compound.

We need to learn how to properly name and write formulas for ionic and covalent compounds. We should become familiar with some of their properties and be able to relate these properties to the structure and bonding of the compounds.

3.1 Chemical Bonding

When two or more atoms form a chemical compound, the atoms are held together in a characteristic arrangement by attractive forces. The **chemical bond** is the force of attraction between any two atoms in a compound. The attraction is the force that overcomes the repulsion of the positively charged nuclei of the two atoms.

Chemistry Connection

Magnets and Migration

All of us, at one time or another, have wondered at the magnificent sight of thousands of migrating birds, flying in formation, heading south for the winter and returning each spring.

Less visible, but no less impressive, are the schools of fish that travel thousands of miles, returning to the same location year after year. Almost instantly, when faced with some external stimulus such as a predator, they snap into a formation that rivals an army drill team for precision.

The questions of how these life-forms know when and where they are going and how they establish their formations have perplexed scientists for many years. The explanations so far are really just hypotheses.

Some clues to the mystery may be hidden in very tiny particles of magnetite, Fe_3O_4. Magnetite contains iron that is naturally magnetic, and collections of these particles behave like a compass needle; they line up in formation aligned with the earth's magnetic field.

Magnetotactic bacteria contain magnetite in the form of magnetosomes, small particles of Fe_3O_4. Fe_3O_4 is a compound whose atoms are joined by chemical bonds. Electrons in the iron atoms have an electron configuration that results in single electrons (not pairs of electrons) occupying orbitals. These unpaired electrons impart magnetic properties to the compound.

The normal habitat of magnetotactic bacteria is either fresh water or the ocean; the bacteria orient themselves to the earth's magnetic field and swim to the nearest pole (north or south). This causes them to swim into regions of nutrient-rich sediment.

Could the directional device, the simple F_3O_4 unit, also be responsible for direction finding in higher organisms in much the same way that an explorer uses a compass? Perhaps so! Recent studies have shown evidence of magnetosomes in the brains of birds, tuna, green turtles, and dolphins.

Most remarkably, at least one study has shown evidence that magnetite is present in the human brain.

These preliminary studies offer hope of unraveling some of the myth and mystery of guidance and communication in living systems. The answers may involve a very basic compound that is like those we will study in this chapter.

Interactions involving valence electrons are responsible for the chemical bond. We shall focus our attention on these electrons and the electron arrangement of atoms both before and after bond formation.

Lewis Symbols

The **Lewis symbol,** or Lewis structure, developed by G. N. Lewis early in the twentieth century, is a convenient way of representing atoms singly or in combination. Its principal advantage is that *only* valence electrons (those that may participate in bonding) are shown. Lewis symbolism is based on the octet rule that was described in Chapter 2.

To draw Lewis structures, we first write the chemical symbol of the atom; this symbol represents the nucleus and all of the lower energy nonvalence electrons. The valence electrons are indicated by dots arranged around the atomic symbol. For example:

Recall that the number of valence electrons can be determined from the position of the element in the periodic table (see Figure 2.10).

$$H\cdot \qquad He:$$
$$\text{Hydrogen} \qquad \text{Helium}$$

$$Li\cdot \qquad \cdot Be\cdot$$
$$\text{Lithium} \qquad \text{Beryllium}$$

$$\cdot B\cdot \qquad \cdot \overset{\cdot}{C}\cdot$$
$$\text{Boron} \qquad \text{Carbon}$$

$$\cdot \overset{\cdot\cdot}{N}\cdot \qquad \cdot \overset{\cdot\cdot}{O}\cdot$$
$$\text{Nitrogen} \qquad \text{Oxygen}$$

$$:\overset{\cdot\cdot}{F}\cdot \qquad :\overset{\cdot\cdot}{Ne}:$$
$$\text{Fluorine} \qquad \text{Neon}$$

Note particularly that the number of dots corresponds to the number of valence electrons in the outermost shell of the atoms of the element. The four "sides" of the chemical symbol represent an atomic orbital capable of holding one or two valence electrons. Because each atomic orbital can hold no more than two electrons, we can show a maximum of two dots on each side of the element's symbol. Using the same logic employed in writing electron configurations in Chapter 2, we place one dot on each side then sequentially add a second dot, filling each side in turn. This process is limited by the total number of available valence electrons. Each unpaired dot (representing an unpaired electron) is available to form a chemical bond with another element, producing a compound. Figure 3.1 depicts the Lewis dot structures for the representative elements.

Principal Types of Chemical Bonds: Ionic and Covalent

Two principal classes of chemical bonds exist: ionic and covalent. Both involve valence electrons.

Ionic bonding involves a transfer of one or more electrons from one atom to another, leading to the formation of an ionic bond. **Covalent bonding** involves a sharing of electrons resulting in the covalent bond.

Before discussing each type, we should recognize that the distinction between ionic and covalent bonding is not always clear-cut. Some compounds are clearly ionic, and some are clearly covalent, but many others possess both ionic and covalent characteristics.

Ionic Bonding

Representative elements form ions that obey the octet rule. Ions of opposite charge attract each other and this attraction is the essence of the ionic bond. Consider the reaction of a sodium atom and a chlorine atom to produce sodium chloride:

$$Na + Cl \longrightarrow NaCl$$

Animation
Ionic, Covalent, and Polar Covalent Bonds

1 **LEARNING GOAL**
Classify compounds as having ionic, covalent, or polar covalent bonds.

IA (1)																	VIIIA (18)
·H	IIA (2)											IIIA (13)	IVA (14)	VA (15)	VIA (16)	VIIA (17)	He:
·Li	·Be·											·B·	·C·	·N·	·O·	:F·	:Ne:
·Na	·Mg·	IIIB (3)	IVB (4)	VB (5)	VIB (6)	VIIB (7)	(8)	VIIIB (9)	(10)	IB (11)	IIB (12)	·Al·	·Si·	·P·	·S·	:Cl·	:Ar:
·K	·Ca·											·Ga·	·Ge·	·As·	·Se·	:Br·	:Kr:
·Rb	·Sr·											·In·	·Sn·	·Sb·	·Te·	:I·	:Xe:
·Cs	·Ba·											·Tl·	·Pb·	·Bi·	·Po·	:At·	:Rn:
·Fr	·Ra·																

Figure 3.1
Lewis dot symbols for representative elements. Each unpaired electron is a potential bond.

Refer to Section 2.7 for a discussion of ionization energy and electron affinity.

Recall that the sodium atom has

- a low ionization energy (it readily loses an electron) and
- a low electron affinity (it does not want more electrons).

If sodium loses its valence electron, it will become isoelectronic (same number of electrons) with neon, a very stable noble gas atom. This tells us that the sodium atom would be a good electron donor, forming the sodium ion:

$$\text{Na} \cdot \longrightarrow \text{Na}^+ + \boxed{e^-}$$

Recall that the chlorine atom has

- a high ionization energy (it will not easily give up an electron) and
- a high electron affinity (it readily accepts another electron).

Chlorine will gain one more electron. By doing so, it will complete an octet (eight outermost electrons) and be isoelectronic with argon, a stable noble gas. Therefore, chlorine behaves as a willing electron acceptor, forming a chloride ion:

$$:\ddot{\text{Cl}}\cdot + \boxed{e^-} \longrightarrow [:\ddot{\text{Cl}}:]^-$$

The electron released by sodium (*electron donor*) is the electron received by chlorine (*electron acceptor*):

$$\text{Na} \cdot \longrightarrow \text{Na}^+ + \boxed{e^-}$$

$$\boxed{e^-} + \cdot\ddot{\text{Cl}}: \longrightarrow [:\ddot{\text{Cl}}:]^-$$

The resulting ions of opposite charge, Na^+ and Cl^-, are attracted to each other (opposite charges attract) and held together by this *electrostatic force* as an **ion pair:** Na^+Cl^-.

This electrostatic force, the attraction of opposite charges, is quite strong and holds the ions together. It is the ionic bond.

The essential features of ionic bonding are the following:

- Atoms of elements with low ionization energy and low electron affinity tend to form positive ions.
- Atoms of elements with high ionization energy and high electron affinity tend to form negative ions.

Figure 3.2
The arrangement of ions in a crystal of NaCl (sodium chloride, table salt). (a) A sodium atom loses one electron to become a smaller sodium ion, and a chlorine atom gains that electron, becoming a larger chloride ion. (b) Attraction of Na^+ and Cl^- forms NaCl ion pairs that aggregate in a three-dimensional crystal lattice structure. (c) A microscopic view of NaCl crystals shows their cubic geometry. Each tiny crystal contains billions of sodium and chloride ions.

- Ion formation takes place by an electron transfer process.
- The positive and negative ions are held together by the electrostatic force between ions of opposite charge in an ionic bond.
- Reactions between representative metals and nonmetals (elements far to the left and right, respectively, in the periodic table) tend to result in ionic bonds.

Although ionic compounds are sometimes referred to as ion pairs, in the solid state these ion pairs do not actually exist as individual units. The positive ions exert attractive forces on several negative ions, and the negative ions are attracted to several positive centers. Positive and negative ions arrange themselves in a regular three-dimensional repeating array to produce a stable arrangement known as a **crystal lattice.** The lattice structure for sodium chloride is shown in Figure 3.2.

Covalent Bonding

The octet rule is not just for ionic compounds. Covalently bonded compounds share electrons to complete the octet of electrons for each of the atoms participating in the bond. Consider the bond formed between two hydrogen atoms, producing the diatomic form of hydrogen: H_2. Individual hydrogen atoms are not stable, and two hydrogen atoms readily combine to produce diatomic hydrogen:

$$H + H \longrightarrow H_2$$

LEARNING GOAL

1 Classify compounds as having ionic, covalent, or polar covalent bonds.

Animation
Covalent Bonding: Sharing Electrons

A diatomic compound is one that is composed of two atoms joined by a covalent bond.

If a hydrogen atom were to gain a second electron, it would be isoelectronic with the stable electron configuration of helium. However, because two identical hydrogen atoms have an equal tendency to gain or lose electrons, an electron transfer from one atom to the other is unlikely to occur under normal conditions. Each atom may attain a noble gas structure only by *sharing* its electron with the other, as shown with Lewis symbols:

$$H\cdot + \cdot H \longrightarrow H:H$$

When electrons are shared rather than transferred, the *shared electron pair* is referred to as a *covalent bond* (Figure 3.3). Compounds characterized by covalent bonding are called *covalent compounds.* Covalent bonds tend to form between atoms with similar tendencies to gain or lose electrons. The most obvious examples are the diatomic molecules H_2, N_2, O_2, F_2, Cl_2, Br_2, and I_2. Bonding in these molecules is *totally covalent* because there can be no net tendency for electron transfer between identical atoms. The formation of F_2, for example, may be represented as

$$:\ddot{F}\cdot + \cdot\ddot{F}: \longrightarrow :\ddot{F}:\ddot{F}:$$

Fourteen valence electrons are arranged in such a way that each fluorine atom is surrounded by eight electrons. The octet rule is satisfied for each fluorine atom.

As in H_2, a single covalent bond is formed. The bonding electron pair is said to be *localized,* or largely confined to the region between the two fluorine nuclei.

Two atoms do not have to be identical to form a covalent bond. Consider compounds such as the following:

		H	H
$H:\ddot{F}:$	$H:\ddot{O}:H$	$H:C:H$	$H:\ddot{N}:H$
		H	

Hydrogen fluoride	Water	Methane	Ammonia
$7e^-$ from F	$6e^-$ from O	$4e^-$ from C	$5e^-$ from N
$1e^-$ from H	$2e^-$ from 2H	$4e^-$ from 4H	$3e^-$ from 3H
$8e^-$ for F	$8e^-$ for O	$8e^-$ for C	$8e^-$ for N
$2e^-$ for H	$2e^-$ for H	$2e^-$ for H	$2e^-$ for H

In each of these cases, bond formation satisfies the octet rule. A total of eight electrons surround each atom other than hydrogen. Hydrogen has only two electrons (corresponding to the electronic structure of helium).

Hydrogen atoms approach at high velocity.

Polar Covalent Bonding and Electronegativity

The Polar Covalent Bond

Covalent bonding is the sharing of an electron pair by two atoms. However, just as we may observe in our day-to-day activities, sharing is not always equal. In a molecule like H_2 (or N_2, or any other diatomic molecule composed of only one element), the electrons, on average, spend the same amount of time in the vicinity of each atom; the electrons have no preference because both atoms are identical.

Hydrogen nuclei begin to attract each other's electrons.

Now consider a diatomic molecule composed of two different elements; HF is a common example. It has been experimentally shown that the electrons in the H—F bond are not equally shared; the electrons spend more time in the vicinity of the fluorine atom. This unequal sharing can be described in various ways:

Partial electron transfer: This describes the bond as having both covalent and ionic properties.

Hydrogen atoms form the hydrogen molecule; atoms are held together by the shared electrons, the covalent bond.

Unequal electron density: The density of electrons around F is greater than the density of electrons around H.

Figure 3.3
Covalent bonding in hydrogen.

Polar covalent bond is the preferred term for a bond made up of unequally shared electron pairs. One end of the bond (in this case, the F atom) is more electron rich (higher electron density), hence, more negative. The other end of the bond (in this case, the H atom) is less electron rich (lower electron density), hence, more positive. These two ends, one somewhat positive (δ^+) and the other somewhat negative (δ^-) may be described as electronic poles, hence the term polar covalent bonds.

The water molecule is perhaps the best-known example of a molecule that exhibits polar covalent bonding (Figure 3.4). In Section 3.4 we will see that the water molecule itself is polar and this fact is the basis for many of water's unique properties.

Once again, we can use the predictive power of the periodic table to help us determine whether a particular bond is polar or nonpolar covalent. We already know that elements that tend to form negative ions (by gaining electrons) are found to the right of the table whereas positive ion formers (that may lose electrons) are located on the left side of the table. Elements whose atoms strongly attract electrons are described as electronegative elements. Linus Pauling, a chemist noted for his theories on chemical bonding, developed a scale of relative electronegativities that correlates reasonably well with the positions of the elements in the periodic table.

Electronegativity

Electronegativity (E_n) is a measure of the ability of an atom to attract electrons in a chemical bond. Elements with high electronegativity have a greater ability to attract electrons than do elements with low electronegativity. Pauling developed a method to assign values of electronegativity to many of the elements in the periodic table. These values range from a low of 0.7 to a high of 4.0, 4.0 being the most electronegative element.

Figure 3.5 shows that the most electronegative elements (excluding the nonreactive noble gas elements) are located in the upper right corner of the periodic table, whereas the least electronegative elements are found in the lower left corner of the table. In general, electronegativity values increase as we proceed left to right and bottom to top of the table. Like other periodic trends, numerous exceptions occur.

If we picture the covalent bond as a competition for electrons between two positive centers, it is the difference in electronegativity, ΔE_n, that determines the extent of polarity. Consider: H_2 or H—H

$$\Delta E_n = \begin{bmatrix} \text{Electronegativity} \\ \text{of hydrogen} \end{bmatrix} - \begin{bmatrix} \text{Electronegativity} \\ \text{of hydrogen} \end{bmatrix}$$

$$\Delta E_n = 2.1 - 2.1 = 0$$

The bond in H_2 is nonpolar covalent. Bonds between *identical* atoms are *always* nonpolar covalent. Also, Cl_2 or Cl—Cl

$$\Delta E_n = \begin{bmatrix} \text{Electronegativity} \\ \text{of chlorine} \end{bmatrix} - \begin{bmatrix} \text{Electronegativity} \\ \text{of chlorine} \end{bmatrix}$$

$$\Delta E_n = 3.0 - 3.0 = 0$$

The bond in Cl_2 is nonpolar covalent. Now consider HCl or H—Cl

$$\Delta E_n = \begin{bmatrix} \text{Electronegativity} \\ \text{of chlorine} \end{bmatrix} - \begin{bmatrix} \text{Electronegativity} \\ \text{of hydrogen} \end{bmatrix}$$

$$\Delta E_n = 3.0 - 2.1 = 0.9$$

The bond in HCl is polar covalent.

(a)

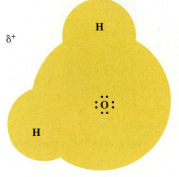

(b)

Figure 3.4
Polar covalent bonding in water. Oxygen is electron rich (δ^-) and hydrogen is electron deficient (δ^+) due to unequal electron sharing. Water has two polar covalent bonds.

 Animation
Electronegativity

Linus Pauling is the only person to receive two Nobel Prizes in very unrelated fields; the chemistry award in 1954 and eight years later, the Nobel Peace Prize. His career is a model of interdisciplinary science, with important contributions ranging from chemical physics to molecular biology.

By convention, the electronegativity difference is calculated by subtracting the less electronegative element's value from the value for the more electronegative element. In this way, negative numbers are avoided.

IA (1)	IIA (2)	IIIB (3)	IVB (4)	VB (5)	VIB (6)	VIIB (7)	(8)	VIIIB (9)	(10)	IB (11)	IIA (12)	IIIA (13)	IVA (14)	VA (15)	VIA (16)	VIIA (17)
H 2.1																
Li 1.0	Be 1.5											B 2.0	C 2.5	N 3.0	O 3.5	F 4.0
Na 0.9	Mg 1.2											Al 1.5	Si 1.8	P 2.1	S 2.5	Cl 3.0
K 0.8	Ca 1.0	Sc 1.3	Ti 1.5	V 1.6	Cr 1.6	Mn 1.5	Fe 1.8	Co 1.9	Ni 1.9	Cu 1.9	Zn 1.6	Ga 1.6	Ge 1.8	As 2.0	Se 2.4	Br 2.8
Rb 0.8	Sr 1.0	Y 1.2	Zr 1.4	Nb 1.6	Mo 1.8	Tc 1.9	Ru 2.2	Rh 2.2	Pd 2.2	Ag 1.9	Cd 1.7	In 1.7	Sn 1.8	Sb 1.9	Te 2.1	I 2.5
Cs 0.7	Ba 0.9	La* 1.1	Hf 1.3	Ta 1.5	W 1.7	Re 1.9	Os 2.2	Ir 2.2	Pt 2.2	Au 2.4	Hg 1.9	Tl 1.8	Pb 1.9	Bi 1.9	Po 2.0	At 2.2
Fr 0.7	Ra 0.9	Ac† 1.1														

Less than 1.0

Between 1.0–3.0

Greater than or equal to 3.0

*Lathanides: 1.1 – 1.3
†Actinides: 1.1 – 1.5

Figure 3.5
Electronegativities of the elements.

3.2 Naming Compounds and Writing Formulas of Compounds

Nomenclature is the assignment of a correct and unambiguous name to each and every chemical compound. Assignment of a name to a structure or deducing the structure from a name is a necessary first step in any discussion of these compounds.

Ionic Compounds

The **formula** is the representation of the fundamental compound using chemical symbols and numerical subscripts. It is the "shorthand" symbol for a compound—for example,

$$NaCl \quad \text{and} \quad MgBr_2$$

The formula identifies the number and type of the various atoms that make up the compound unit. The number of like atoms in the unit is shown by the use of a subscript. The presence of only one atom is understood when no subscript is present.

The formula $NaCl$ indicates that each ion pair consists of one sodium cation (Na^+) and one chloride anion (Cl^-). Similarly, the formula $MgBr_2$ indicates that one magnesium ion and two bromide ions combine to form the compound.

In Chapter 2 we learned that positive ions are formed from elements that

- are located at the left of the periodic table,
- are referred to as *metals*, and
- have low ionization energies, low electron affinities, and hence easily *lose* electrons.

Elements that form negative ions, on the other hand,

- are located at the right of the periodic table (but exclude the noble gases),
- are referred to as *nonmetals,* and
- have high ionization energies, high electron affinities, and hence easily *gain* electrons.

In short, metals and nonmetals usually react to produce ionic compounds resulting from the transfer of one or more electrons from the metal to the nonmetal.

An electronegativity difference of 1.9 is generally accepted as the boundary between polar covalent and ionic bonds. Although, strictly speaking, any electronegativity difference, no matter how small, produces a polar bond, the degree of polarity for bonds with electronegativity differences less than 0.5 is minimal. Consequently, we shall classify these bonds as nonpolar. The *formula* of an ionic compound is the smallest whole-number ratio of ions in the substance.

Writing Formulas of Ionic Compounds from the Identities of the Component Ions

It is important to be able to write the formula of an ionic compound when provided with the identities of the ions that make up the compound. The charge of each ion can usually be determined from the group (family) of the periodic table in which the parent element is found. The cations and anions must combine in such a way that the resulting formula unit has a net charge of zero.

Consider the following examples.

LEARNING GOAL

2 Write the formula of a compound when provided with the name of the compound.

Predicting the Formula of an Ionic Compound   **EXAMPLE 3.1**

Predict the formula of the ionic compound formed from the reaction of sodium and oxygen atoms.

Solution

Step 1. Sodium is in group IA (or 1); it has *one* valence electron. Loss of this electron produces Na^+.

Step 2. Oxygen is in group VIA (or 16); it has *six* valence electrons. A gain of two electrons (to create a stable octet) produces O^{2-}.

Step 3. Two positive charges are necessary to counterbalance two negative charges on the oxygen anion. Because each sodium ion carries a 1+ charge, two sodium ions are needed for each O^{2-}.

Step 4. The subscript 2 is used to indicate that the formula unit contains two sodium ions. Thus the formula of the compound is Na_2O.

Practice Problem 3.1

Predict the formulas of the compounds formed from the combination of ions of the following elements:
a. lithium and bromine
b. calcium and bromine
c. potassium and chlorine

For Further Practice: Questions 3.17 and 3.18.

EXAMPLE 3.2   **Predicting the Formula of an Ionic Compound**

Predict the formula of the compound formed by the reaction of aluminum and oxygen atoms.

Solution

Step 1. Aluminum is in group IIIA (or 13) of the periodic table; we predict that it has three valence electrons. Loss of these electrons produces Al^{3+}.

Step 2. Oxygen is in group VIA (or 16) of the periodic table and has six valence electrons. A gain of two electrons (to create a stable octet) produces O^{2-}.

Step 3. How can we combine Al^{3+} and O^{2-} to yield a unit of zero charge? It is necessary that *both* the cation and anion be multiplied by factors that will result in a zero net charge:

Step 4. $2 \times (+3) = +6$ and $3 \times (-2) = -6$
$2 \times Al^{3+} = +6$ and $3 \times O^{2-} = -6$

Hence the formula is Al_2O_3.

Practice Problem 3.2

Predict the formulas of the compounds formed from the combination of ions of the following elements:
a. calcium and nitrogen
b. magnesium and bromine
c. magnesium and nitrogen

For Further Practice: Questions 3.19 and 3.20.

Writing Names of Ionic Compounds from the Formula of the Compound

Nomenclature, the way in which compounds are named, is based on their formulas. The name of the cation appears first, followed by the name of the anion. The positive ion has the name of the element; the negative ion is named by using the *stem* of the name of the element joined to the suffix *-ide*. Some examples follow.

Formula	cation	and	anion stem	+	ide	=	Compound name
NaCl	sodium		chlor	+	ide		sodium chloride
Na_2O	sodium		ox	+	ide		sodium oxide
Li_2S	lithium		sulf	+	ide		lithium sulfide
$AlBr_3$	aluminum		brom	+	ide		aluminum bromide
CaO	calcium		ox	+	ide		calcium oxide

If the cation and anion exist in only one common charged form, there is no ambiguity between formula and name. Sodium chloride *must be* NaCl, and lithium sulfide *must be* Li_2S, so that the sum of positive and negative charges is zero. With many elements, such as the transition metals, several ions of different charge may exist. Fe^{2+}, Fe^{3+} and Cu^+, Cu^{2+} are two common examples. Clearly, an ambiguity exists if we use the name iron for both Fe^{2+} and Fe^{3+} or copper for both Cu^+ and Cu^{2+}. Two systems have been developed to avoid this problem: the *Stock system* and the *common nomenclature system*.

In the Stock system (systematic name), a Roman numeral placed immediately after the name of the ion indicates the magnitude of the cation's charge. In the older common nomenclature system, the suffix -*ous* indicates the lower ionic charge, and the suffix -*ic* indicates the higher ionic charge. Consider the examples in Table 3.1.

Systematic names are easier and less ambiguous than common names. Whenever possible, we will use this system of nomenclature. The older, common names (-ous, -ic) are less specific; furthermore, they often use the Latin names of the elements (for example, iron compounds use *ferr-*, from *ferrum*, the Latin word for iron).

Monatomic ions are ions consisting of a single atom. Common monatomic ions are listed in Table 3.2. The ions that are particularly important in biological systems are highlighted in red.

Polyatomic ions, such as the hydroxide ion, OH^-, are composed of two or more atoms bonded together. These ions, although bonded to other ions with ionic bonds, are themselves held together by covalent bonds.

The polyatomic ion has an *overall* positive or negative charge. Some common polyatomic ions are listed in Table 3.3. The formulas, charges, and names of these polyatomic ions, especially those highlighted in red, should be memorized.

TABLE 3.1	Systemic (Stock) and Common Names for Iron and Copper Ions		
For systematic name:			
Formula	*+ Ion Charge*	*Cation Name*	*Compound Name*
$FeCl_2$	2+	Iron(II)	Iron(II) chloride
$FeCl_3$	3+	Iron(III)	Iron(III) chloride
Cu_2O	1+	Copper(I)	Copper(I) oxide
CuO	2+	Copper(II)	Copper(II) oxide
For common nomenclature:			
Formula	*+ Ion Charge*	*Cation Name*	*Common -ous/ic Name*
$FeCl_2$	2+	Ferr*ous*	Ferrous chloride
$FeCl_3$	3+	Ferr*ic*	Ferric chloride
Cu_2O	1+	Cupr*ous*	Cuprous oxide
CuO	2+	Cupr*ic*	Cupric oxide

TABLE 3.2	Common Monatomic Cations and Anions		
Cation	**Name**	**Anion**	**Name**
H^+	Hydrogen ion	H^-	Hydride ion
Li^+	Lithium ion	F^-	Fluoride ion
Na^+	Sodium ion	Cl^-	Chloride ion
K^+	Potassium ion	Br^-	Bromide ion
Cs^+	Cesium ion	I^-	Iodide ion
Be^{2+}	Beryllium ion	O^{2-}	Oxide ion
Mg^{2+}	Magnesium ion	S^{2-}	Sulfide ion
Ca^{2+}	Calcium ion	N^{3-}	Nitride ion
Ba^{2+}	Barium ion	P^{3-}	Phosphide ion
Al^{3+}	Aluminum ion		
Ag^+	Silver ion		

Note: The ions of principal importance are highlighted in magenta.

TABLE 3.3	**Common Polyatomic Cations and Anions**
Ion	**Name**
H_3O^+	Hydronium
NH_4^+	Ammonium
NO_2^-	Nitrite
NO_3^-	Nitrate
SO_3^{2-}	Sulfite
SO_4^{2-}	Sulfate
HSO_4^-	Hydrogen sulfate
OH^-	Hydroxide
CN^-	Cyanide
PO_4^{3-}	Phosphate
HPO_4^{2-}	Hydrogen phosphate
$H_2PO_4^-$	Dihydrogen phosphate
CO_3^{2-}	Carbonate
HCO_3^-	Bicarbonate
ClO^-	Hypochlorite
ClO_2^-	Chlorite
ClO_3^-	Chlorate
ClO_4^-	Perchlorate
CH_3COO^- (or $C_2H_3O_2^-$)	Acetate
MnO_4^-	Permanganate
$Cr_2O_7^{2-}$	Dichromate
CrO_4^{2-}	Chromate
O_2^{2-}	Peroxide

Note: The most commonly encountered ions are highlighted in magenta.

The following examples are formulas of several compounds containing polyatomic ions.

Formula	Cation	Anion	Name
NH_4Cl	NH_4^+	Cl^-	ammonium chloride
$Ca(OH)_2$	Ca^{2+}	OH^-	calcium hydroxide
Na_2SO_4	Na^+	SO_4^{2-}	sodium sulfate
$NaHCO_3$	Na^+	HCO_3^-	sodium bicarbonate

Sodium bicarbonate may also be named sodium hydrogen carbonate, a preferred and less ambiguous name. Likewise, Na_2HPO_4 is named sodium hydrogen phosphate, and other ionic compounds are named similarly.

Question 3.1

Name each of the following compounds:

a. KCN
b. MgS
c. $Mg(CH_3COO)_2$

Question 3.2

Name each of the following compounds:

a. Li_2CO_3
b. $FeBr_2$
c. $CuSO_4$

Writing Formulas of Ionic Compounds from the Name of the Compound

It is also important to be able to write the correct formula when given the compound name. To do this, we must be able to predict the charge of monatomic ions and remember the charge and formula of polyatomic ions. Equally important, the relative number of positive and negative ions in the unit must result in a net (compound) charge of zero. The compounds are electrically neutral. Two examples follow.

LEARNING GOAL

2 Write the formula of a compound when provided with the name of the compound.

Writing a Formula When Given the Name of the Compound　　　E X A M P L E　3.3

Write the formula of sodium sulfate.

Solution

Step 1. The sodium ion is Na^+, a group I (or 1) element. The sulfate ion is SO_4^{2-} (from Table 3.3).

Step 2. Two positive charges, two sodium ions, are needed to cancel the charge on one sulfate ion (two negative charges).

Hence the formula is Na_2SO_4.

Practice Problem 3.3

Write the formula for each of the following compounds:
a. calcium carbonate
b. sodium bicarbonate
c. copper(I) sulfate

For Further Practice: Questions 3.37 and 3.38.

Writing a Formula When Given the Name of the Compound　　　E X A M P L E　3.4

Write the formula of ammonium sulfide.

Solution

Step 1. The ammonium ion is NH_4^+ (from Table 3.3). The sulfide ion is S^{2-} (from its position on the periodic table).

Step 2. Two positive charges are necessary to cancel the charge on one sulfide ion (two negative charges).

Hence the formula is $(NH_4)_2S$.
    Note that parentheses must be used whenever a subscript accompanies a polyatomic ion.

Practice Problem 3.4

Write the formula for each of the following compounds:
a. sodium phosphate
b. potassium bromide
c. iron(II) nitrate

For Further Practice: Questions 3.39 and 3.40.

TABLE **3.4**	Prefixes Used to Denote Numbers of Atoms in a Compound	
Prefix	**Number of Atoms**	
Mono-	1	
Di-	2	
Tri-	3	
Tetra-	4	
Penta-	5	
Hexa-	6	
Hepta-	7	
Octa-	8	
Nona-	9	
Deca-	10	

Covalent Compounds

Naming Covalent Compounds

Most covalent compounds are formed by the reaction of nonmetals. **Molecules** are compounds characterized by covalent bonding. We saw earlier that ionic compounds are not composed of single units but are a part of a massive three-dimensional crystal structure in the solid state. Covalent compounds exist as discrete molecules in the solid, liquid, and gas states. This is a distinctive feature of covalently bonded substances.

The conventions for naming covalent compounds follow:

By convention the prefix *mono-* is often omitted from the second element as well (dinitrogen oxide, not dinitrogen monoxide). In other cases, common usage retains the prefix (carbon monoxide, not carbon oxide).

1. The names of the elements are written in the order in which they appear in the formula.
2. A prefix (Table 3.4) indicating the number of each kind of atom found in the unit is placed before the name of the element.
3. If only one atom of a particular kind is present in the molecule, the prefix mono- is usually omitted from the first element.
4. The stem of the name of the last element is used with the suffix -ide.
5. The final vowel in a prefix is often dropped before a vowel in the stem name.

EXAMPLE 3.5   **Naming a Covalent Compound**

Name the covalent compound N_2O_4.

Solution

Step 1. two nitrogen atoms   four oxygen atoms

Step 2. di-   tetra-

Step 3. dinitrogen   tetr(a)oxide

The name is dinitrogen tetroxide.

 Practice Problem 3.5

Name each of the following compounds:
a. B_2O_3 c. ICl   e. PCl_5
b. NO   d. PCl_3 f. P_2O_5

For Further Practice: Questions 3.31 and 3.32.

The following are examples of other covalent compounds.

Formula	Name
N_2O	dinitrogen monoxide
NO_2	nitrogen dioxide
SiO_2	silicon dioxide
CO_2	carbon dioxide
CO	carbon monoxide

Writing Formulas of Covalent Compounds

Many compounds are so familiar to us that their *common names* are generally used. For example, H_2O is water, NH_3 is ammonia, C_2H_5OH (ethanol) is ethyl alcohol, and $C_6H_{12}O_6$ is glucose. It is useful to be able to correlate both systematic and common names with the corresponding molecular formula and vice versa.

> **LEARNING GOAL**
>
> **2** Write the formula of a compound when provided with the name of the compound.

When common names are used, formulas of covalent compounds can be written *only* from memory. You *must* remember that water is H_2O, ammonia is NH_3, and so forth. This is the major disadvantage of common names. Because of their widespread use, however, they cannot be avoided and must be memorized.

Compounds named by using Greek prefixes are easily converted to formulas. Consider the following examples.

Writing the Formula of a Covalent Compound   **E X A M P L E 3.6**

Write the formula of nitrogen monoxide.

Solution

Step 1. Nitrogen has no prefix; one is understood.

Step 2. Oxide has the prefix mono—one oxygen.

Step 3. Hence the formula is NO.

Practice Problem 3.6

Write the formula of each of the following compounds:
a. nitrogen trifluoride
b. carbon monoxide

For Further Practice: Question 3.45.

Writing the Formula of a Covalent Compound   **E X A M P L E 3.7**

Write the formula of dinitrogen tetroxide.

Solution

Step 1. Nitrogen has the prefix *di*—two nitrogen atoms.

Step 2. Oxygen has the prefix *tetr(a)*—four oxygen atoms.

Step 3. Hence the formula is N_2O_4.

Practice Problem 3.7

Write the formula of each of the following compounds:
a. diphosphorus pentoxide
b. silicon dioxide

For Further Practice: Question 3.46.

A Human *Perspective*

Origin of the Elements

The current, most widely held theory of the origin of the universe is the "big bang" theory. An explosion of very dense matter was followed by expansion into space of the fragments resulting from this explosion. This is one of the scenarios that have been created by scientists fascinated by the origins of matter, the stars and planets, and life as we know it today.

The first fragments, or particles, were protons and neutrons moving with tremendous velocity and possessing large amounts of energy. Collisions involving these high-energy protons and neutrons formed deuterium atoms (^2H), which are isotopes of hydrogen. As the universe expanded and cooled, tritium (^3H), another hydrogen isotope, formed as a result of collisions of neutrons with deuterium atoms. Subsequent capture of a proton produced helium (He). Scientists theorize that a universe that was principally composed of hydrogen and helium persisted for perhaps 100,000 years until the temperature decreased sufficiently to allow the formation of a simple molecule, hydrogen, two atoms of hydrogen bonded together (H_2).

Many millions of years later, the effect of gravity caused these small units to coalesce, first into clouds and eventually into stars, with temperatures of millions of degrees. In this setting, these small collections of protons and neutrons combined to form larger atoms such as carbon (C) and oxygen (O), then sodium (Na), neon (Ne), magnesium (Mg), silicon (Si), and so forth. Subsequent explosions of stars provided the conditions that formed many larger atoms. These fragments, gathered together by the force of gravity, are the most probable origin of the planets in our own solar system.

The reactions that formed the elements as we know them today were a result of a series of *fusion reactions,* the joining of nuclei to produce larger atoms at very high temperatures (millions of degrees Celsius). These fusion reactions are similar to processes that are currently being studied as a possible alternative source of nuclear power. We shall study such nuclear processes in more detail in Chapter 9.

Nuclear reactions of this type do not naturally occur on the earth today. The temperature is simply too low. As a result we have, for the most part, a collection of stable elements existing as chemical compounds, atoms joined together by chemical bonds while retaining their identity even in the combined state. Silicon exists all around us as sand and soil in a combined form, silicon dioxide; most metals exist as a part of a chemical compound, such as iron ore. We are learning more about the structure and properties of these compounds in this chapter.

For Further Understanding

How does tritium differ from "normal" hydrogen?

Would you expect to find similar atoms on other planets?

3.3 Properties of Ionic and Covalent Compounds

LEARNING GOAL

4 Predict differences in physical state, melting and boiling points, solid-state structure, and solution chemistry that result from differences in bonding.

The differences in ionic and covalent bonding result in markedly different properties for ionic and covalent compounds. Because covalent molecules are distinct units, they have less tendency to form an extended structure in the solid state. Ionic compounds, with ions joined by electrostatic attraction, do not have definable units but form a crystal lattice composed of enormous numbers of positive and negative ions in an extended three-dimensional network. The effects of this basic structural difference are summarized in this section.

Physical State

All ionic compounds (for example, NaCl, KCl, and $NaNO_3$) are solids at room temperature; covalent compounds may be solids (sugar), liquids (H_2O, ethanol), or gases (carbon monoxide, carbon dioxide). The three-dimensional crystal structure that is characteristic of ionic compounds holds them in a rigid, solid arrangement, whereas molecules of covalent compounds may be fixed, as in a solid, or more mobile, a characteristic of liquids and gases.

Melting and Boiling Points

The **melting point** is the temperature at which a solid is converted to a liquid, and the **boiling point** is the temperature at which a liquid is converted to a gas at a specified pressure. Considerable energy is required to break apart an ionic crystal lattice with uncountable numbers of ionic interactions and convert the ionic substance to a liquid or a gas. As a result, the melting and boiling temperatures for ionic compounds are generally higher than those of covalent compounds, whose molecules interact less strongly in the solid state. A typical ionic compound, sodium chloride, has a melting point of 801°C; methane, a covalent compound, melts at −182°C. Exceptions to this general rule do exist; diamond, a covalent solid with an extremely high melting point, is a well-known example.

Structure of Compounds in the Solid State

Ionic solids are *crystalline,* characterized by a regular structure, whereas covalent solids may either be crystalline or have no regular structure. In the latter case they are said to be *amorphous.*

Solutions of Ionic and Covalent Compounds

In Chapter 1 we saw that mixtures are either heterogeneous or homogeneous. A homogeneous mixture is a solution. Many ionic solids dissolve in solvents, such as water. An ionic solid, if soluble, will form positive and negative ions in solution by **dissociation.**

Because ions in water are capable of carrying (conducting) a current of electricity, we refer to these compounds as **electrolytes,** and the solution is termed an **electrolytic solution.** Covalent solids dissolved in solution usually retain their neutral (molecular) character and are **nonelectrolytes.** The solution is not an electrical conductor.

The role of the solvent in the dissolution of solids is discussed in Section 3.5.

3.4 Drawing Lewis Structures of Molecules and Polyatomic Ions

Lewis Structures of Molecules

In Section 3.1, we used Lewis structures of individual atoms to help us understand the bonding process. To begin to explain the relationship between molecular structure and molecular properties, we will first need a set of guidelines to help us write Lewis structures for more complex molecules.

LEARNING GOAL

5 Draw Lewis structures for covalent compounds and polyatomic ions.

1. *Use chemical symbols for the various elements to write the skeletal structure of the compound.* To accomplish this, place the bonded atoms next to one another. This is relatively easy for simple compounds; however, as the number of atoms in the compound increases, the possible number of arrangements increases dramatically. We may be told the pattern of arrangement of the atoms in advance; if not, we can make an intelligent guess and see if a reasonable Lewis structure can be constructed. Three considerations are very important here:
   • the least electronegative atom will be placed in the central position (the central atom),
   • hydrogen and fluorine (and the other halogens) often occupy terminal positions,
   • carbon often forms chains of carbon-carbon covalent bonds.
2. *Determine the number of valence electrons associated with each atom; combine them to determine the total number of valence electrons in the compound.* However, if we are representing polyatomic cations or anions, we must account for the charge on the ion. Specifically:
   • for polyatomic cations, subtract one electron for each unit of positive charge. This accounts for the fact that the positive charge arises from electron loss.
   • for polyatomic anions, add one electron for each unit of negative charge. This accounts for excess negative charge resulting from electron gain.

The skeletal structure indicates only the relative positions of atoms in the molecule or ion. Bonding information results from the Lewis structure.

The central atom is often the element farthest to the left and/or lowest in the periodic table.
The central atom is often the element in the compound for which there is only one atom.
Hydrogen is *never* the central atom.

A Medical Perspective

Blood Pressure and the Sodium Ion/Potassium Ion Ratio

When you have a physical exam, the physician measures your blood pressure. This indicates the pressure of blood against the walls of the blood vessels each time the heart pumps. A blood pressure reading is always characterized by two numbers. With every heartbeat there is an increase in pressure; this is the systolic blood pressure. When the heart relaxes between contractions, the pressure drops; this is the diastolic pressure. Thus the blood pressure is expressed as two values—for instance, 117/72—measured in millimeters of mercury. Hypertension is simply defined as high blood pressure. To the body it means that the heart must work too hard to pump blood, and this can lead to heart failure or heart disease.

Heart disease accounts for 50% of all deaths in the United States. Epidemiological studies correlate the following major risk factors with heart disease: heredity, sex, race, age, diabetes, cigarette smoking, high blood cholesterol, and hypertension. Obviously, we can do little about our age, sex, and genetic heritage, but we can stop smoking, limit dietary cholesterol, and maintain a normal blood pressure.

The number of Americans with hypertension is alarmingly high: 60 million adults and children. More than 10 million of these individuals take medication to control blood pressure, at a cost of nearly $2.5 billion each year. In many cases, blood pressure can be controlled without medication by increasing physical activity, losing weight, decreasing consumption of alcohol, and limiting intake of sodium.

It has been estimated that the average American ingests 7.5–10 g of salt (NaCl) each day. Because NaCl is about 40% (by mass) sodium ions, this amounts to 3–4 g of sodium daily. Until 1989 the Food and Nutrition Board of the National Academy of Sciences National Research Council's defined *e*stimated *s*afe and *a*dequate *d*aily *d*ietary *i*ntake (ESADDI) of sodium ion was 1.1–3.3 g. Clearly, Americans exceed this recommendation.

Recently, studies have shown that excess sodium is not the sole consideration in the control of blood pressure. More important is the sodium ion/potassium ion (Na^+/K^+) ratio. That ratio should be about 0.6; in other words, our diet should contain about 67% more potassium than sodium. Does the typical American diet fall within this limit? Definitely not! Young American males (25–30 years old) consume a diet with a $Na^+/K^+ = 1.07$, and the diet of females of the same age range has a $Na^+/K^+ = 1.04$. It is little wonder that so many Americans suffer from hypertension.

How can we restrict sodium in the diet, while increasing the potassium? The top photo depicts a variety of foods that are low in sodium and high in potassium. These include fresh fruits and vegetables and fruit juices, a variety of cereals, unsalted nuts, and cooked dried beans (legumes). The lower photo shows the unfortunate popularity of some high-sodium, low-potassium foods. Notice that most of these are processed or prepared foods. This points out how difficult it can be to control sodium in the diet. The majority of the sodium that we ingest comes from commercially prepared foods. The consumer must read the nutritional information printed on cans and packages to determine whether the sodium levels are within acceptable limits.

Low Sodium Ion, High Potassium Ion Foods

High Sodium Ion, Low Potassium Ion Foods

For Further Understanding

Find several commercial food products on the shelves of your local grocery store; read the label and calculate the sodium ion–potassium ion ratio.

Describe each product that you have chosen in terms of its suitability for inclusion in the diet of a person with moderately elevated blood pressure.

3. **Connect the central atom to each of the surrounding atoms using electron pairs.** Then complete the octets of all of the atoms bonded to the central atom. Recall that hydrogen needs only two electrons to complete its valence shell. Electrons not involved in bonding must be represented as lone pairs and the total number of electrons in the structure must equal the number of valence electrons computed in our second step.

4. **If the octet rule is not satisfied for the central atom, move one or more electron pairs from the surrounding atoms.** Use these electrons to create double or triple bonds until all atoms have an octet.

5. **After you are satisfied with the Lewis structure that you have constructed, perform a final electron count.** This allows you to verify that the total number of electrons and the number around each atom are correct.

Now, let us see how these guidelines are applied in the examples that follow.

Drawing Lewis Structures of Covalent Compounds

EXAMPLE 3.8

Draw the Lewis structure of carbon dioxide, CO_2.

LEARNING GOAL

5 Draw Lewis structures for covalent compounds and polyatomic ions.

Solution

Step 1. Draw a skeletal structure of the molecule, arranging the atoms in their most probable order.

For CO_2, two possibilities exist:

$$C—O—O \quad \text{and} \quad O—C—O$$

Referring to Figure 3.5, we find that the electronegativity of oxygen is 3.5 whereas that of carbon is 2.5. Our strategy dictates that the least electronegative atom, in this case carbon, is the central atom. Hence the skeletal structure O—C—O may be presumed correct.

Step 2. Next, we want to determine the number of valence electrons on each atom and add them to arrive at the total for the compound.

For CO_2,

$$1 \text{ C atom} \times 4 \text{ valence electrons} = 4 \text{ e}^-$$
$$\underline{2 \text{ O atoms} \times 6 \text{ valence electrons} = 12 \text{ e}^-}$$
$$16 \text{ e}^- \text{ total}$$

Step 3. Now, use electron pairs to connect the central atom, C, to each oxygen with a single bond.

$$O:C:O$$

Distribute the electrons around the atoms (in pairs if possible) in an attempt to satisfy the octet rule, eight electrons around each element.

$$:\overset{..}{O}:C:\overset{..}{O}:$$

This structure satisfies the octet rule for each oxygen atom, but not the carbon atom (only four electrons surround the carbon).

Continued—

EXAMPLE 3.8 —*Continued*

Step 4. However, when this structure is modified by moving two electrons from each oxygen atom to a position between C and O, each oxygen and carbon atom is surrounded by eight electrons. The octet rule is satisfied, and the structure below is the most probable Lewis structure for CO_2.

$$\ddot{\text{O}}::\text{C}::\ddot{\text{O}}$$

In this structure, four electrons (two electron pairs) are located between C and each O, and these electrons are shared in covalent bonds. Because a **single bond** is composed of two electrons (one electron pair) and because four electrons "bond" the carbon atom to each oxygen atom in this structure, there must be two bonds between each oxygen atom and the carbon atom, a **double bond**:

The notation for a single bond : is equivalent to — (one pair of electrons).

The notation for a double bond : : is equivalent to ═ (two pairs of electrons).

We may write CO_2 as shown above or, replacing dots with dashes to indicate bonding electron pairs,

$$\ddot{\text{O}}{=}\text{C}{=}\ddot{\text{O}}$$

Step 5. As a final step, let us do some "electron accounting." There are eight electron pairs, and they correspond to sixteen valence electrons (8 pair × 2e⁻/pair). Furthermore, there are eight electrons around each atom and the octet rule is satisfied. Therefore

$$\ddot{\text{O}}{=}\text{C}{=}\ddot{\text{O}}$$

is a satisfactory way to depict the structure of CO_2.

Practice Problem 3.8

Draw a Lewis structure for each of the following covalent compounds:
a. H_2O (water)
b. CH_4 (methane)

For Further Practice: Questions 3.63 and 3.64.

EXAMPLE 3.9   **Drawing Lewis Structures of Covalent Compounds**

Draw the Lewis structure of ammonia, NH_3.

Solution

Step 1. When trying to implement the first step in our strategy we may be tempted to make H our central atom because it is less electronegative than N. But, remember the margin note in this section:

"Hydrogen is *never* the central atom."

Continued—

EXAMPLE 3.9 —Continued

Hence:

$$\begin{array}{c} H \\ | \\ H-N-H \end{array}$$

is our skeletal structure.

Step 2. Applying our strategy to determine the total valence electrons for the molecule, we find that there are five valence electrons in nitrogen and one in each of the three hydrogens, for a total of eight valence electrons.

Step 3. Applying our strategy for distribution of valence electrons results in the following Lewis diagram:

$$\begin{array}{c} H \\ H:\overset{..}{N}:H \end{array}$$

Step 4. This satisfies the octet rule for nitrogen (eight electrons around N) and hydrogen (two electrons around each H) and is an acceptable structure for ammonia. Ammonia may also be written:

$$\begin{array}{c} H \\ | \\ H-\underset{..}{N}-H \end{array}$$

Step 5. There are eight valence electrons in this structure; this agrees with the electron count in Step 2. Note the pair of nonbonding electrons on the nitrogen atom. These are often called a **lone pair,** or *unshared* pair, of electrons. As we will see later in this section, lone pair electrons have a profound effect on molecular geometry. The geometry, in turn, affects the reactivity of the molecule.

Practice Problem 3.9

Draw a Lewis structure for each of the following covalent compounds:
a. C_2H_6 (ethane)
b. N_2 (nitrogen gas)

For Further Practice: Questions 3.71 and 3.72.

Lewis Structures of Polyatomic Ions

The strategies for writing the Lewis structures of polyatomic ions are similar to those for neutral compounds. There is, however, one major difference: the charge on the ion must be accounted for when computing the total number of valence electrons.

LEARNING GOAL

5 Draw Lewis structures for covalent compounds and polyatomic ions.

EXAMPLE 3.10 | **Drawing Lewis Structures of Polyatomic Cations**

Draw the Lewis structure of the ammonium ion, NH_4^+.

LEARNING GOAL

5 Draw Lewis structures for covalent compounds and polyatomic ions.

Solution

Step 1. The ammonium ion has the following skeletal structure and charge:

$$\begin{bmatrix} & H & \\ & | & \\ H\!-\!&N&\!-\!H \\ & | & \\ & H & \end{bmatrix}^{+}$$

Step 2. The total number of valence electrons is determined by subtracting one electron for each unit of positive charge.

$$
\begin{array}{ll}
1 \text{ N atom} \times 5 \text{ valence electrons} = & 5 \text{ e}^- \\
4 \text{ H atoms} \times 1 \text{ valence electron} = & 4 \text{ e}^- \\
-1 \text{ electron for} +1 \text{ charge} = & -1 \text{ e}^- \\
\hline
& 8 \text{ e}^- \text{ total}
\end{array}
$$

Steps 3 and 4. Distribute these eight electrons around our skeletal structure:

$$\begin{bmatrix} & H & \\ & \cdot\cdot & \\ H\!:\!&N&\!:\!H \\ & & \\ & H & \end{bmatrix}^{+} \quad \text{or} \quad \begin{bmatrix} & H & \\ & | & \\ H\!-\!&N&\!-\!H \\ & | & \\ & H & \end{bmatrix}^{+}$$

Step 5. A final check shows eight total electrons, eight around the central atom, nitrogen, and two electrons associated with each hydrogen. Hence the structure is satisfactory.

Practice Problem 3.10

Draw the Lewis structure of H_3O^+ (the hydronium ion)

For Further Practice: Question 3.67.

EXAMPLE 3.11 | **Drawing Lewis Structures of Polyatomic Anions**

Draw the Lewis structure of the carbonate ion, CO_3^{2-}.

LEARNING GOAL

5 Draw Lewis structures for covalent compounds and polyatomic ions.

Solution

Step 1. Carbon is less electronegative than oxygen. Therefore carbon is the central atom. The carbonate ion has the following skeletal structure and charge:

$$\begin{bmatrix} & O & \\ & | & \\ O\!-\!&C&\!-\!O \end{bmatrix}^{2-}$$

Continued—

EXAMPLE 3.11 —*Continued*

Step 2. The total number of valence electrons is determined by adding one electron for each unit of negative charge:

$$1 \text{ C atom} \times 4 \text{ valence electrons} = 4 \text{ e}^-$$
$$3 \text{ O atoms} \times 6 \text{ valence electrons} = 18 \text{ e}^-$$
$$+ 2 \text{ negative charges} = 2 \text{ e}^-$$

$$24 \text{ e}^- \text{ total}$$

Step 3. Distributing the electron dots around the central carbon atom (forming four bonds) and around the surrounding oxygen atoms in an attempt to satisfy the octet rule results in the structure:

$$\left[\ddot{\underset{..}{:\ddot{O}:}} \atop {:\ddot{O}:C:\ddot{O}:}\right]^{2-} \quad \text{or} \quad \left[{:\ddot{O}: \atop {| \atop :\ddot{O}-C-\ddot{O}:}}\right]^{2-}$$

Step 4. Although the octet rule is satisfied for the three O atoms, it is not for the C atom. We must move a lone pair from one of the O atoms to form another bond with C:

$$\left[{:O: \atop {\| \atop :\ddot{O}:C:\ddot{O}:}}\right]^{2-} \quad \text{or} \quad \left[{|O| \atop {\| \atop |\overline{O}-C-\overline{O}|}}\right]^{2-}$$

Now the octet rule is also satisfied for the C atom.

Step 5. A final check shows twenty-four electrons and eight electrons around each atom. Hence the structure is a satisfactory representation of the carbonate ion.

Practice Problem 3.11

Draw the Lewis structure of O_2^{2-}.

For Further Practice: Question 3.68.

| **Drawing Lewis Structures of Polyatomic Anions** | **EXAMPLE 3.12** |

Draw the Lewis structure of the acetate ion, CH_3COO^-.

Solution

Step 1. A commonly encountered anion, the acetate ion has a skeletal structure that is more complex than any of the examples that we have studied thus far. Which element should we choose as the central atom? We have three choices: H, O, and C. H is eliminated because hydrogen can never be the central atom. Oxygen is more electronegative than carbon, so carbon must be the central atom. There are two carbon atoms; often they are joined. Further clues are obtained from the formula itself; CH_3COO^- implies three hydrogen

Continued—

EXAMPLE 3.12 —*Continued*

atoms attached to the first carbon atom and two oxygen atoms joined to the second carbon. A plausible skeletal structure is:

$$
\begin{bmatrix}
& \text{H} & & \text{O} & \\
& | & & | & \\
\text{H} & - \text{C} & - \text{C} & - \text{O} \\
& | & & & \\
& \text{H} & & &
\end{bmatrix}^{-}
$$

Step 2. The pool of valence electrons for anions is determined by adding one electron for each unit of negative charge:

$$
\begin{aligned}
2 \text{ C atoms} \times 4 \text{ valence electrons} &= 8 \text{ e}^- \\
3 \text{ H atoms} \times 1 \text{ valence electron} &= 3 \text{ e}^- \\
2 \text{ O atoms} \times 6 \text{ valence electrons} &= 12 \text{ e}^- \\
+ \; 1 \text{ negative charge} &= 1 \text{ e}^- \\
\hline
& 24 \text{ e}^- \text{ total}
\end{aligned}
$$

Step 3. Distributing these twenty-four electrons around our skeletal structure gives

$$
\begin{bmatrix}
& \text{H} \!:\! \ddot{\text{O}} \!: & \\
\text{H} \!:\! \ddot{\text{C}} \!:\! \ddot{\text{C}} \!:\! \ddot{\text{O}} \!: \\
& \ddot{\text{H}} &
\end{bmatrix}^{-}
\quad \text{or} \quad
\begin{bmatrix}
\text{H} & & |\text{O}| \\
| & & \| \\
\text{H} - \text{C} - \text{C} - \overline{\text{O}}| \\
| & & \\
\text{H} & &
\end{bmatrix}^{-}
$$

Step 4. This Lewis structure satisfies the octet rule for carbon and oxygen and surrounds each hydrogen with two electrons.

Step 5. All twenty-four electrons are used in this process.

Practice Problem 3.12

Write a Lewis structure describing the bonding in each of the following polyatomic ions:
a. the bicarbonate ion, HCO_3^-
b. the phosphate ion, PO_4^{3-}

For Further Practice: Question 3.69.

Lewis Structure, Stability, Multiple Bonds, and Bond Energies

LEARNING GOAL

6 Describe the relationship between stability and bond energy.

Hydrogen, oxygen, and nitrogen are present in the atmosphere as diatomic gases, H_2, O_2, and N_2. All are covalent molecules. Their stability and reactivity, however, are quite different. Hydrogen is an explosive material, sometimes used as a fuel. Oxygen, although more stable than hydrogen, reacts with fuels in combustion. The explosion of the space shuttle *Challenger* resulted from the reaction of massive amounts of hydrogen and oxygen. Nitrogen, on the other hand, is extremely non-reactive. Because nitrogen makes up about 80% of the atmosphere, it dilutes the oxygen, which accounts for only about 20% of the atmosphere.

Breathing pure oxygen for long periods, although necessary in some medical situations, causes the breakdown of nasal and lung tissue over time. Oxygen diluted with nonreactive nitrogen is an ideal mixture for humans and animals to breathe.

Why is there such a great difference in reactivity among these three gases? We can explain this, in part, in terms of their bonding characteristics. The Lewis structure for H_2 (two valence electrons) is

$$H:H \quad \text{or} \quad H—H$$

For oxygen (twelve valence electrons, six on each atom), the only Lewis structure that satisfies the octet rule is

$$\ddot{O}::\ddot{O} \quad \text{or} \quad \ddot{O}=\ddot{O}$$

The Lewis structure of N_2 (ten total valence electrons) must be

$$:N:::N: \quad \text{or} \quad :N\equiv N:$$

Therefore

N_2 has a *triple bond* (six bonding electrons).

O_2 has a *double bond* (four bonding electrons).

H_2 has a *single bond* (two bonding electrons).

A **triple bond,** in which three pairs of electrons are shared by two atoms, is very stable. More energy is required to break a triple bond than a double bond, and a double bond is stronger than a single bond. Stability is related to the bond energy. The **bond energy** is the amount of energy, in units of kilocalories or kilojoules, required to break a bond holding two atoms together. Bond energy is therefore a *measure* of stability. The values of bond energies decrease in the order *triple bond > double bond > single bond.* The H—H bond energy in 436 kJ/mole. This amount of energy is necessary to break a H—H bond. In contrast, the O=O bond energy is 499 kJ/mole and the N≡N bond energy is 941 kJ/mole.

The bond length is related to the presence or absence of multiple bonding. The distance of separation of two nuclei is greatest for a single bond, less for a double bond, and still less for a triple bond. The *bond length* decreases in the order *single bond > double bond > triple bond.*

> The term *bond order* is sometimes used to distinguish among single, double, and triple bonds. A bond order of 1 corresponds to a single bond, 2 corresponds to a double bond, and 3 corresponds to a triple bond.

> The mole is simply a unit denoting quantity, just as a *dozen* eggs represents 12 eggs. A mole of bonds is 6.022×10^{23} bonds (see chapter 4).

Question 3.3

Contrast a single and double bond with regard to:

a. distance of separation of the bonded nuclei
b. strength of the bond

How are these two properties related?

Question 3.4

Two nitrogen atoms in a nitrogen molecule are held together more strongly than the two chlorine atoms in a chlorine molecule. Explain this fact by comparing their respective Lewis structures.

Isomers

Isomers are compounds that share the same molecular formula but have different structures. Hydrocarbons (compounds that contain only hydrogen and carbon atoms) frequently exhibit this property. For example, butane, C_4H_{10}, has two isomeric forms. One isomer, termed normal butane, *n*-butane, has a structure characterized by the four carbon atoms being linked in a chain; the other, termed isobutane, has a three-carbon chain, with the fourth carbon attached to the middle carbon.

n-pentane

2-methylbutane

2,2-dimethylpropane

$$H-\overset{\overset{\displaystyle H}{|}}{C}-\overset{\overset{\displaystyle H}{|}}{C}-\overset{\overset{\displaystyle H}{|}}{C}-\overset{\overset{\displaystyle H}{|}}{C}-H$$

n-Butane (C$_4$H$_{10}$)

$$H-\overset{\overset{\displaystyle H}{|}}{C}-\overset{\overset{\displaystyle \overset{\displaystyle H}{|}}{\overset{\displaystyle C}{|}}}{\underset{\underset{\displaystyle H}{|}}{C}}-\overset{\overset{\displaystyle H}{|}}{C}-H$$

Isobutane (C$_4$H$_{10}$)

Notice that both structures, although clearly different, contain four carbon atoms and ten hydrogen atoms. Owing to their structural differences, each has a different melting point and boiling point. In fact, all of their physical properties differ at least slightly. These differences in properties clearly show that, in fact, *n*-butane and isobutane are different compounds.

As the size of the hydrocarbon increases, the number of possible isomers increases dramatically. A 5-carbon hydrocarbon has three isomers, but a 30-carbon hydrocarbon has over 400 million possible isomers. Petroleum is largely a complex mixture of hydrocarbons and the principal reason for this complexity lies in the tremendous variety of possible isomers present.

The three isomers of pentane (a 5-carbon hydrocarbon) are depicted in the figure in the margin.

Lewis Structures and Resonance

In some cases we find that it is possible to write more than one Lewis structure that satisfies the octet rule for a particular compound. Consider sulfur dioxide, SO$_2$. Its skeletal structure is

$$O-S-O$$

Total valence electrons may be calculated as follows:

$$
\begin{aligned}
1 \text{ sulfur atom} \times 6 \text{ valence e}^- / \text{atom} &= 6 \text{ e}^- \\
+ \ 2 \text{ oxygen atoms} \times 6 \text{ valence e}^- / \text{atom} &= 12 \text{ e}^- \\
\hline
&\quad 18 \text{ e}^- \text{ total}
\end{aligned}
$$

The resulting Lewis structures are

$$\ddot{\overset{..}{O}}::\overset{..}{S}:\ddot{\overset{..}{O}}: \qquad \text{and} \qquad :\ddot{\overset{..}{O}}:\overset{..}{S}::\ddot{O}$$

Both satisfy the octet rule. However, experimental evidence shows no double bond in SO$_2$. The two sulfur-oxygen bonds are equivalent. Apparently, neither structure accurately represents the structure of SO$_2$, and neither actually exists. The actual structure is said to be an average or *hybrid* of these two Lewis structures. When a compound has two or more Lewis structures that contribute to the real structure, we say that the compound displays the property of **resonance.** The contributing Lewis structures are **resonance forms.** The true structure, a hybrid of the resonance forms, is known as a **resonance hybrid** and may be represented as:

$$\ddot{\overset{..}{O}}::\overset{..}{S}:\ddot{\overset{..}{O}}: \qquad \leftrightarrow \qquad :\ddot{\overset{..}{O}}:\overset{..}{S}::\ddot{O}$$

A common analogy might help to clarify this concept. A horse and a donkey may be crossbred to produce a hybrid, the mule. The mule doesn't look or behave exactly like either parent, yet it has attributes of both. The resonance hybrid of a molecule has properties of each resonance form but is not identical to any one form. Unlike the mule, resonance hybrids *do not actually exist.* Rather, they comprise a model that results from the failure of any one Lewis structure to agree with experimentally obtained structural information.

The presence of resonance enhances molecular stability. The more resonance forms that exist, the greater is the stability of the molecule they represent. This concept is important in understanding the chemical reactions of many complex organic molecules and is used extensively in organic chemistry.

> It is a misconception to picture the real molecule as oscillating back and forth among the various resonance structures. Resonance is a modeling strategy designed to help us visualize electron arrangements too complex to be adequately explained by the simple Lewis dot structure.

Drawing Resonance Hybrids of Covalently Bonded Compounds　　　　EXAMPLE 3.13

Draw the possible resonance structures of the nitrate ion, NO_3^-, and represent them as a resonance hybrid.

Solution

Step 1. Nitrogen is less electronegative than oxygen; therefore, nitrogen is the central atom and the skeletal structure is:

$$\left[\begin{array}{c} O \\ | \\ O-N-O \end{array}\right]^-$$

Step 2. The pool of valence electrons for anions is determined by adding one electron for each unit of negative charge:

$$
\begin{array}{llr}
1\ N\ atom \times 5\ valence\ electrons & = & 5\ e^- \\
3\ O\ atoms \times 6\ valence\ electrons & = & 18\ e^- \\
+\ 1\ negative\ charge & = & 1\ e^- \\
\hline
& & 24\ e^-\ total
\end{array}
$$

Step 3. Distributing the electrons throughout the structure results in three legitimate Lewis structures:

$$\left[\begin{array}{c} :\ddot{O}: \\ \ddot{O}::\ddot{N}:\ddot{O}: \end{array}\right]^- \quad \text{and} \quad \left[\begin{array}{c} :\ddot{O}: \\ :\ddot{O}:\ddot{N}:\ddot{O}: \end{array}\right]^- \quad \text{and} \quad \left[\begin{array}{c} :\ddot{O}: \\ :\ddot{O}:\ddot{N}::\ddot{O} \end{array}\right]^-$$

Step 4. All contribute to the true structure of the nitrate ion, represented as a resonance hybrid.

$$\left[\begin{array}{c} :\ddot{O}: \\ \ddot{O}::\ddot{N}:\ddot{O}: \end{array}\right]^- \leftrightarrow \left[\begin{array}{c} :\ddot{O}: \\ :\ddot{O}:\ddot{N}:\ddot{O}: \end{array}\right]^- \leftrightarrow \left[\begin{array}{c} :\ddot{O}: \\ :\ddot{O}:\ddot{N}::\ddot{O} \end{array}\right]^-$$

Practice Problem 3.13

a. SeO_2, like SO_2, has two resonance forms. Draw their Lewis structures.
b. Explain any similarities between the structures for SeO_2 and SO_2 in light of periodic relationships.

For Further Practice: Questions 3.73 and 3.74.

Lewis Structures and Exceptions to the Octet Rule

The octet rule is remarkable in its ability to realistically model bonding and structure in covalent compounds. But, like any model, it does not adequately describe all systems. Beryllium, boron, and aluminum, in particular, tend to form compounds in which they are surrounded by fewer than eight electrons. This situation is termed an *incomplete octet.* Other molecules, such as nitric oxide:

$$\overset{..}{\underset{.}{N}} = \overset{..}{\underset{..}{O}}$$

are termed *odd electron* molecules. Note that it is impossible to pair all electrons to achieve an octet simply because the compound contains an odd number of valence electrons. Elements in the third period and beyond may involve *d* orbitals and form an *expanded octet,* with ten or even twelve electrons surrounding the central atom. Examples 3.14 and 3.15 illustrate common exceptions to the octet rule.

EXAMPLE 3.14

Drawing Lewis Structures of Covalently Bonded Compounds That Are Exceptions to the Octet Rule

Draw the Lewis structure of beryllium hydride, BeH_2.

Solution

Step 1. A reasonable skeletal structure of BeH_2 is:

$$H-Be-H$$

Step 2. The total number of valence electrons in BeH_2 is:

$$1 \text{ beryllium atom } \times 2 \text{ valence e}^-/\text{atom} = 2 \text{ e}^-$$
$$\underline{2 \text{ hydrogen atoms} \times 1 \text{ valence e}^-/\text{atom} = 2 \text{ e}^-}$$
$$4 \text{ e}^- \text{ total}$$

Step 3. The resulting Lewis structure must be:

$$H:Be:H \qquad \text{or} \qquad H-Be-H$$

Step 4. It is apparent that there is no way to satisfy the octet rule for Be in this compound. Consequently, BeH_2 is an exception to the octet rule. It contains an incomplete octet.

Practice Problem 3.14

The BCl_3 molecule has an incomplete octet around B. Draw the Lewis structure of BCl_3.

For Further Practice: Question 3.83.

EXAMPLE 3.15

Drawing Lewis Structures of Covalently Bonded Compounds That Are Exceptions to the Octet Rule

Draw the Lewis structure of phosphorus pentafluoride.

Solution

Step 1. A reasonable skeletal structure of PF_5 is:

$$F-P\begin{matrix} F \\ | \\ \\ | \\ F \end{matrix}\begin{matrix} F \\ \\ \\ F \end{matrix}$$

Continued—

EXAMPLE 3.15 —*Continued*

Step 2. Phosphorus is a third-period element; it may have an expanded octet.

The total number of valence electrons is:

$$1 \text{ phosphorus atom} \times 5 \text{ valence } e^-/\text{atom} = 5 e^-$$
$$\underline{5 \text{ fluorine atoms} \times 7 \text{ valence } e^-/\text{atom} = 35 e^-}$$
$$40 e^- \text{ total}$$

Step 3. Distributing the electrons around each F in the skeletal structure results in the Lewis structure:

PF$_5$ is an example of a compound with an expanded octet.

Practice Problem 3.15

The SF$_6$ molecule has an expanded octet around S. Draw the Lewis structure of SF$_6$.

For Further Practice: Question 3.84.

Lewis Structures and Molecular Geometry; VSEPR Theory

The shape of a molecule plays a large part in determining its properties and reactivity. We may predict the shapes of various molecules by inspecting their Lewis structures for the orientation of their electron pairs. The covalent bond, for instance, in which bonding electrons are localized between the nuclear centers of the atoms, is *directional*; the bond has a specific orientation in space between the bonded atoms. Electrostatic forces in ionic bonds, in contrast, are *nondirectional*; they have no specific orientation in space. The specific orientation of electron pairs in covalent molecules imparts a characteristic shape to the molecules. Consider the following series of molecules whose Lewis structures are shown.

LEARNING GOAL

7 Predict the geometry of molecules and ions using the octet rule and Lewis structures.

 Animations

Valence Shell Electron Pair Repulsion Theory

VSEPR and Molecular Geometry

BeH$_2$ H : Be : H

BF$_3$:F:B:F:

CH$_4$ H : C : H

NH$_3$ H : N : H

H$_2$O   H : O : H

The electron pairs around the central atom of the molecule arrange themselves to minimize electronic repulsion. This means that the electron pairs arrange themselves so that they can be as far as possible from each other. We may use this fact to predict molecular shape. This approach is termed the *valence shell electron pair repulsion* (VSEPR) **theory.**

Let's see how the VSEPR theory can be used to describe the bonding and structure of each of the preceding molecules.

BeH₂

> Only four electrons surround the beryllium atom in BeH₂. Consequently, BeH₂ is a stable exception to the octet rule.

As we saw in Example 3.14, beryllium hydride has two shared electron pairs around the beryllium atom. These electron pairs have minimum repulsion if they are located as far apart as possible while still bonding the hydrogen to the central atom. This condition is met if the electron pairs are located on opposite sides of the molecule, resulting in a **linear structure,** 180° apart:

$$H : Be : H \qquad or \qquad H - Be - H$$

The *bond angle,* the angle between H—Be and Be—H bonds, formed by the two bonding pairs is 180° (Figure 3.6).

Figure 3.6

Bonding and geometry in beryllium hydride, BeH₂. (a) Linear geometry in BeH₂. (b) Computer-generated model of linear BeH₂.

Animations

The Geometry of BeF₂

The Geometry of BF₃

(a)   (b)

BeF₂, CS₂, and HCN are other examples of molecules that exhibit linear geometry.

BF₃

> BF₃ has only six electrons around the central atom, B. It is one of a number of stable compounds that are exceptions to the octet rule.

Boron trifluoride has three shared electron pairs around the central atom. Placing the electron pairs in a plane, forming a triangle, minimizes the electron pair repulsion in this molecule, as depicted in Figure 3.7 and the following sketches:

Such a structure is *trigonal planar,* and each F—B—F bond angle is 120°. We also find that compounds with central atoms in the same group of the periodic table have similar geometry. Aluminum, in the same group as boron, produces compounds such as AlH₃, which is also trigonal planar.

(a)   (b)

Figure 3.7

Bonding and geometry in boron trifluoride, BF₃: (a) trigonal planar geometry in BF₃; (b) computer-generated model of trigonal planar BF₃.

CH₄

Methane has four shared pairs of electrons. Here, minimum electron repulsion is achieved by arranging the electrons at the corners of a tetrahedron (Figure 3.8). Each H—C—H bond angle is 109.5°. Methane has a three-dimensional **tetrahedral structure.** Silicon, in the same group as carbon, forms compounds such as SiCl₄ and SiH₄ that also have tetrahedral structures.

(a)   (b)   (c)   (d)

Figure 3.8

Representations of the three-dimensional structure of methane, CH_4. (a) Tetrahedral methane structure. (b) Computer-generated model of tetrahedral methane. (c) Three-dimensional representation of structure (b).

NH₃

Ammonia also has four electron pairs about the central atom. In contrast to methane, in which all four pairs are bonding, ammonia has three pairs of bonding electrons and one nonbonding lone pair of electrons. We might expect CH_4 and NH_3 to have electron pair arrangements that are similar but not identical. The lone pair in ammonia is more negative than the bonding pairs; some of the negative charge on the bonding pairs is offset by the presence of the hydrogen atoms with their positive nuclei. Thus the arrangement of electron pairs in ammonia is distorted.

The hydrogen atoms in ammonia are pushed closer together than in methane (Figure 3.9). The bond angle is 107° because lone pair–bond pair repulsions are greater than bond pair–bond pair repulsions. The structure or shape is termed *trigonal pyramidal*, and the molecule is termed a **trigonal pyramidal molecule.**

H₂O

Water also has four electron pairs around the central atom; two pairs are bonding, and two pairs are nonbonding. These four electron pairs are approximately tetrahedral to each other; however, because of the difference between bonding and nonbonding electrons, noted earlier, the tetrahedral relationship is only approximate.

CH_4, NH_3, and H_2O all have eight electrons around their central atoms; all obey the octet rule.

 Animations

The Geometry of CH_4

The Geometry of NH_3

The Geometry of H_2O

NH_3

(a)   (b)   (c)   (d)

Figure 3.9

The structure of the ammonia molecule. (a) Pyramidal ammonia structure. (b) Computer-generated model of pyramidal ammonia. (c) A three-dimensional sketch. (d) The H—N—H bond angle in ammonia.

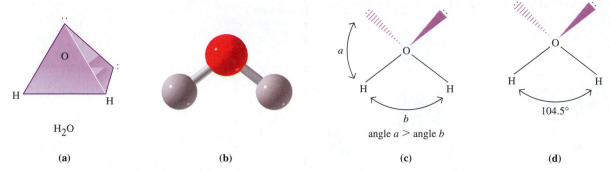

Figure 3.10
The structure of the water molecule. (a) Angular water structure. (b) Computer-generated model of angular water. (c) A three-dimensional sketch. (d) The H—O—H bond angle in water.

T A B L E **3.5**	Molecular Structure: The Geometry of a Molecule Is Affected by the Number of Nonbonded Electron Pairs Around the Central Atom and the Number of Bonded Atoms					
Bonded Atoms	**Nonbonding Electron Pairs**	**Bond Angle**	**Molecular Structure**	**Example**	**Structure**	
2	0	180°	Linear	CO_2	
3	0	120°	Trigonal planar	SO_3	
2	1	<120°	Angular	SO_2	
4	0	~109°	Tetrahedral	CH_4	
3	1	~107°	Trigonal pyramidal	NH_3	
2	2	~104.5°	Angular	H_2O	

Molecules with five and six electron pairs also exist. They may have structures that are *trigonal bipyramidal* (forming a six-sided figure) or *octahedral* (forming an eight-sided figure).

The **angular** (or *bent*) **structure** has a bond angle of 104.5°, which is 5° smaller than the tetrahedral angle, because of the repulsive effects of the lone pairs of electrons (as shown in Figure 3.10).

The characteristics of linear, trigonal planar, angular, and tetrahedral structures are summarized in Table 3.5.

Periodic Structural Relationships

The molecules considered previously contain the central atoms Be (Group IIA), B (Group IIIA), C (Group IVA), N (Group VA), and O (Group VIA). We may expect that a number of other compounds, containing the same central atom, will have structures with similar geometries. This is an approximation, not always true, but still useful in expanding our ability to write reasonable, geometrically accurate structures for a large number of compounds.

The periodic similarity of group members is also useful in predictions involving bonding. Consider Group VI, oxygen, sulfur, and selenium (Se). Each has six valence electrons. Each needs two more electrons to complete its octet. Each should react with hydrogen, forming H_2O, H_2S, and H_2Se.

If we recall that H_2O is an angular molecule with the following Lewis structure,

$$H : \overset{..}{\underset{H}{O}} :$$

it follows that H_2S and H_2Se would also be angular molecules with similar Lewis structures, or

$$H : \overset{..}{\underset{H}{S}} : \quad \text{and} \quad H : \overset{..}{\underset{H}{Se}} :$$

This logic applies equally well to the other representative elements.

More Complex Molecules

A molecule such as dimethyl ether, CH_3—O—CH_3, has two different central atoms: oxygen and carbon. We could picture the parts of the molecule containing the CH_3 group (commonly referred to as the *methyl group*) as exhibiting tetrahedral geometry (analogous to methane):

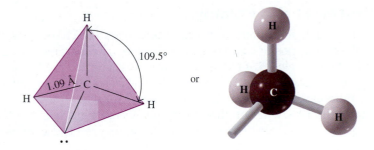

The part of the molecule connecting these two methyl groups (the oxygen) would have a bond angle similar to that of water (in which oxygen is also the central atom), approximately 104° , as seen in Figure 3.11. This is a reasonable way to represent the molecule dimethyl ether.

Trimethylamine, $(CH_3)_3N$, is a member of the amine family. As in the case of ether, two different central atoms are present. Carbon and nitrogen determine the geometry of amines. In this case the methyl group should assume the tetrahedral geometry of methane, and the nitrogen atom should have the methyl groups in a pyramidal arrangement, similar to the hydrogen atoms in ammonia, as seen in Figure 3.12. This creates a pyramidal geometry around nitrogen. H—N—H bond angles in ammonia are 107° ; experimental information shows a very similar C—N—C bond angle in trimethylamine.

Known bond angle, 104.5°

Predicted bond angle is approximately 104°

Known bond angle, 107°

Predicted bond angle is approximately 107°

Figure 3.11
A comparison of the bonding in water and dimethyl ether.

Figure 3.12
A comparison of the bonding in ammonia and trimethylamine.

Question *3.5*

Sketch the geometry of each of the following molecules (basing your structure on the Lewis electron dot representation of the molecule):

a. PH_3
b. SiH_4

Question *3.6*

Sketch the geometry of each of the following molecules (basing your structure on the Lewis electron dot representation of the molecule):

a. C_2H_4
b. C_2H_2

It is essential to represent the molecule in its correct geometric form, using the Lewis and VSEPR theories, to understand its physical and chemical behavior. In the rest of this section we use these models to predict molecular behavior.

Lewis Structures and Polarity

 Animation

Molecular Polarity

A molecule is *polar* if its centers of positive and negative charges do not coincide. Molecules whose positive and negative charges are separated when the molecules are placed in an electric field align themselves with the field. The molecule behaves as a *dipole* (having two "poles" or ends, one pole is more negative and the other pole is more positive) and is said to be polar.

Nonpolar molecules will not align with the electric field because their positive and negative centers are the same; no dipole exists. These molecules are nonpolar.

The hydrogen molecule is the simplest nonpolar molecule:

$$H : H \quad \text{or} \quad H - H$$

Both electrons, on average, are located at the center of the molecule and positively charged nuclei are on either side. The center of both positive and negative charge is at the center of the molecule; therefore the bond is nonpolar.

We may arrive at the same conclusion by considering the equality of electron sharing between the atoms being bonded. Electron sharing is related to the concept of electronegativity introduced in Section 3.1.

Remember: Electronegativity deals with atoms in molecules, whereas electron affinity and ionization energy deal with isolated atoms.

The atoms that comprise H_2 are identical; their electronegativity (electron attracting power) is the same. Thus the electrons remain at the center of the molecule, and the molecule is nonpolar.

Similarly, O_2, N_2, Cl_2, and F_2 are nonpolar molecules with nonpolar bonds. Arguments analogous to those made for hydrogen explain these observations as well.

Let's next consider hydrogen fluoride, HF. Fluorine is more electronegative than hydrogen. This indicates that the electrons are more strongly attracted to a fluorine atom than they are to a hydrogen atom. This results in a bond and molecule that are polar. The symbol

Less electronegative part of bond ⟋⟶ More electronegative part of bond

placed below a bond indicates the direction of polarity. The more negative end of the bond is near the head of the arrow, and the less negative end of the bond is next to the tail of the arrow. Symbols using the Greek letter *delta* may also be used to designate polarity. In this system the more negative end of the bond is designated δ^- (partial negative), and the less negative end is designated δ^+ (partial positive). The symbols are applied to the hydrogen fluoride molecule as follows:

$$\delta^+ \; H\!\!-\!\!F \; \delta^-$$

Less electronegative end of bond $\nearrow$ $\longleftarrow$ $\nwarrow$ More electronegative end of bond

HF is a **polar covalent molecule** characterized by **polar covalent bonding.** This implies that the electrons are shared unequally.

A molecule containing all nonpolar bonds must also be nonpolar. In contrast, a molecule containing polar bonds may be either polar or nonpolar depending on the relative arrangement of the bonds and any lone pairs of electrons.

Let's now examine the bonding in carbon tetrachloride. All four bonds of CCl_4 are polar because of the electronegativity difference between C and Cl. However, because of the symmetrical arrangement of the four C—Cl bonds, their polarities cancel, and the molecule is nonpolar covalent:

Now look at H_2O. Because of its angular (bent) structure, the polar bonds do not cancel, and the molecule is polar covalent:

The electron density is shifted away from the hydrogens toward oxygen in the water molecule. In carbon tetrachloride, equal electron "pull" in all directions results in a nonpolar covalent molecule.

> **The word *partial*** implies less than a unit charge. Thus δ^+ and δ^- do not imply overall charge on a unit (such as the + or − sign on an ion); they are meant to show only the relative distribution of charge within a unit. The HF molecule shown is *neutral* but has an unequal charge distribution *within* the molecule.

 Animations

> The Structure and Dipole
> Moment of HF
>
> The Structure and Dipole
> Moment of CCl_4
>
> The Structure and Dipole
> Moment of H_2O

Question 3.7

Predict which of the following bonds are polar, and, if polar, in which direction the electrons are pulled:

a. O—S   c. Cl—Cl
b. C≡N   d. I—Cl

Question 3.8

Predict which of the following bonds are polar, and, if polar, in which direction the electrons are pulled:

a. Si—Cl   c. H—C
b. S—Cl   d. C—C

Question 3.9

Predict whether each of the following molecules is polar:

a. BCl_3 c. HCl
b. NH_3 d. $SiCl_4$

Question 3.10

Predict whether each of the following molecules is polar:

a. CO_2 c. BrCl
b. SCl_2 d. CS_2

3.5 Properties Based on Electronic Structure and Molecular Geometry

LEARNING GOAL

8 Understand the role that molecular geometry plays in determining the solubility and melting and boiling points of compounds.

LEARNING GOAL

9 Use the principles of VSEPR theory and molecular geometry to predict relative melting points, boiling points, and solubilities of compounds.

The solute is the substance that is present in lesser quantity, and the solvent is the substance that is present in the greater amount (see Section 6.3).

The interaction of water and ammonia is an example of a particularly strong intermolecular force, the hydrogen bond; this phenomenon is discussed in Chapter 5.

Intramolecular forces are attractive forces *within* molecules. They are the chemical bonds that determine the shape and polarity of individual molecules. **Intermolecular forces,** on the other hand, are forces *between* molecules.

It is important to distinguish between these two kinds of forces. It is the *intermolecular* forces that determine such properties as the solubility of one substance in another and the freezing and boiling points of liquids. But, at the same time we must realize that these forces are a direct consequence of the *intramolecular* forces in the individual units, the molecules.

In the following section we will see some of the consequences of bonding that are directly attributable to differences in intermolecular forces (solubility, boiling and melting points). In Section 5.2 we will investigate, in some detail, the nature of the intermolecular forces themselves.

Solubility

Solubility is defined as the maximum amount of solute that dissolves in a given amount of solvent at a specified temperature. Polar molecules are most soluble in polar solvents, whereas nonpolar molecules are most soluble in nonpolar solvents. This is the rule of *"like dissolves like."* Substances of similar polarity are mutually soluble, and large differences in polarity lead to insolubility.

Case I: Ammonia and Water

Ammonia is soluble in water because both ammonia and water are polar molecules:

Dissolution of ammonia in water is a consequence of the intermolecular forces present among the ammonia and water molecules. The δ^- end (a nitrogen) of the ammonia molecule is attracted to the δ^+ end (a hydrogen) of the water molecule; at the same time the δ^+ end (a hydrogen) of the ammonia molecule is attracted to the δ^- end (an oxygen) of the water molecule. These attractive forces thus "pull" ammonia into water (and water into ammonia), and the ammonia molecules are randomly distributed throughout the solvent, forming a homogeneous solution (Figure 3.13).

Case II: Oil and Water

Oil and water do not mix; oil is a nonpolar substance composed primarily of molecules containing carbon and hydrogen. Water molecules, on the other hand, are quite polar. The potential solvent, water molecules, have partially charged ends, whereas the molecules of oil do not. As a result, water molecules exert their

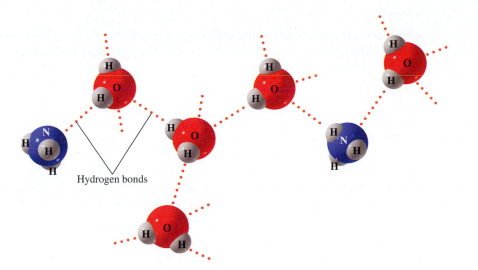

Figure 3.13
The interaction of polar covalent water molecules (the solvent) with polar covalent solute molecules such as ammonia, resulting in the formation of a solution.

attractive forces on other water molecules, not on the molecules of oil; the oil remains insoluble, and because it is less dense than water, the oil simply floats on the surface of the water. This is illustrated in Figure 3.14.

Boiling Points of Liquids and Melting Points of Solids

Boiling a liquid requires energy. The energy is used to overcome the intermolecular attractive forces in the liquid, driving the molecules into the less associated gas phase. The amount of energy required to accomplish this is related to the boiling temperature. This, in turn, depends on the strength of the intermolecular attractive forces in the liquid, which parallel the polarity. This is not the only determinant of boiling point. Molecular mass is also an important consideration. The larger the mass of the molecule, the more difficult it becomes to convert the collection of molecules to the gas phase.

A similar argument can be made for the melting points of solids. The ease of conversion of a solid to a liquid also depends on the magnitude of the attractive forces in the solid. The situation actually becomes very complex for ionic solids because of the complexity of the crystal lattice.

As a general rule, polar compounds have strong attractive (intermolecular) forces, and their boiling and melting points tend to be higher than those of nonpolar substances of similar molecular mass.

Melting and boiling points of a variety of substances are included in Table 3.6.

Figure 3.14
The interaction of polar water molecules and nonpolar oil molecules. The familiar salad dressing, oil and vinegar, forms two layers. The oil does not dissolve in vinegar, an aqueous solution of acetic acid.

T A B L E **3.6**	Melting and Boiling Points of Selected Compounds in Relation to Their Bonding Type		
Formula (Name)	**Bonding Type**	**M.P. (°C)**	**B.P. (°C)**
N_2 (nitrogen)	Nonpolar covalent	−210	−196
O_2 (oxygen)	Nonpolar covalent	−219	−183
NH_3 (ammonia)	Polar covalent	−78	−33
H_2O (water)	Polar covalent	0	100
NaCl (sodium chloride)	Ionic	801	1413
KBr (potassium bromide)	Ionic	730	1435

Question *3.11*

Predict which compound in each of the following groups should have the higher melting and boiling points (Hint: Write the Lewis dot structure and determine whether the molecule is polar or nonpolar.):

a. H_2O and C_2H_4
b. CO and CH_4
c. NH_3 and N_2
d. Cl_2 and ICl

Question *3.12*

Predict which compound in each of the following groups should have the higher melting and boiling points (Hint: Write the Lewis dot structure and determine whether the molecule is polar or nonpolar.):

a. C_2H_6 and CH_4
b. CO and NO
c. F_2 and Br_2
d. $CHCl_3$ and CCl_4

SUMMARY

3.1 Chemical Bonding

When two atoms are joined to make a chemical compound, the force of attraction between the two atoms is the *chemical bond. Ionic bonding* is characterized by an electron transfer process occurring before bond formation, forming an *ion pair.* In *covalent bonding,* electrons are shared between atoms in the bonding process. *Polar covalent bonding,* like covalent bonding, is based on the concept of electron sharing; however, the sharing is unequal and based on the *electronegativity* difference between joined atoms. The *Lewis symbol,* showing only valence electrons, is a convenient way of representing atoms singly or in combination.

3.2 Naming Compounds and Writing Formulas of Compounds

The "shorthand" symbol for a compound is its *formula.* The formula identifies the number and type of atoms in the compound.

An ion that consists of only a single atom is said to be *monatomic. Polyatomic ions,* such as the hydroxide ion, OH^-, are composed of two or more atoms bonded together.

Names of ionic compounds are derived from the names of their ions. The name of the cation appears first, followed by the name of the anion. In the Stock system for naming an ion (the systematic name), a Roman numeral indicates the charge of the cation. In the older common nomenclature system, the suffix *-ous* indicates the lower of the ionic charges, and the suffix *-ic* indicates the higher ionic charge.

Most covalent compounds are formed by the reaction of nonmetals. Covalent compounds exist as *molecules.*

The convention used for naming covalent compounds is as follows:

- The names of the elements are written in the order in which they appear in the formula.
- A prefix indicating the number of each kind of atom found in the unit is placed before the name of the element.
- The stem of the name of the last element is used with the suffix *-ide.*

Many compounds are so familiar to us that their common names are used. It is useful to be able to correlate both systematic and common names with the corresponding molecular formula.

3.3 Properties of Ionic and Covalent Compounds

Covalently bonded molecules are discrete units, and they have less tendency to form an extended structure in the solid state. Ionic compounds, with ions joined by electrostatic attraction, do not have definable units but form a *crystal lattice* composed of positive and negative ions in an extended three-dimensional network.

The *melting point* is the temperature at which a solid is converted to a liquid; the *boiling point* is the temperature at which a liquid is converted to a gas. Melting and boiling temperatures for ionic compounds are generally higher than those of covalent compounds.

Ionic solids are crystalline, whereas covalent solids may be either crystalline or amorphous.

Many ionic solids dissolve in water, *dissociating* into positive and negative ions (an *electrolytic solution*). Because these ions can carry (conduct) a current of electricity, they are called *electrolytes*. Covalent solids in solution usually retain their neutral character and are *nonelectrolytes*.

3.4 Drawing Lewis Structures of Molecules and Polyatomic Ions

The procedure for drawing Lewis structures of molecules involves writing a skeletal structure of the molecule, arranging the atoms in their most probable order, determining the number of valence electrons on each atom, and combining them to get the total for the compound. The electrons are then distributed around the atoms (in pairs if possible) to satisfy the octet rule. At this point electron pairs may be moved, creating double or triple bonds, in an effort to satisfy the octet rule for all atoms and produce the final structure.

Stability of a covalent compound is related to the *bond energy*. The magnitude of the bond energy decreases in the order *triple bond > double bond > single bond*. The bond length decreases in the order *single bond > double bond > triple bond*.

The valence shell electron pair repulsion theory states that electron pairs around the central atom of the molecule arrange themselves to minimize electronic repulsion; the electrons orient themselves as far as possible from each other. Two electron pairs around the central atom lead to a *linear* arrangement of the attached atoms; three indicate a *trigonal planar* arrangement, and four result in a *tetrahedral* geometry. Both lone pair and bonding pair electrons must be taken into account when predicting structure. Molecules with fewer than four and as many as five or six electron pairs around the central atom also exist. They are exceptions to the octet rule.

A molecule is polar if its centers of positive and negative charges do not coincide. A *polar covalent* molecule has at least one *polar covalent bond*. An understanding of the concept of *electronegativity*, the relative electron-attracting power of atoms in molecules, helps us to assess the polarity of a bond.

A molecule containing all nonpolar bonds must be nonpolar. A molecule containing polar bonds may be either polar or nonpolar, depending on the relative position of the bonds.

3.5 Properties Based on Electronic Structure and Molecular Geometry

Attractions between molecules are called *intermolecular forces*. *Intramolecular forces*, on the other hand, are the attractive forces within molecules. It is the intermolecular forces that determine such properties as the solubility of one substance in another and the freezing and boiling points of liquids.

Solubility is the maximum amount of solute that dissolves in a given amount of solvent at a specified temperature. Polar molecules are most soluble in polar solvents; nonpolar molecules are most soluble in nonpolar solvents. This is the rule of "like dissolves like."

As a general rule, polar compounds have strong intermolecular forces, and their boiling and melting points tend to be higher than nonpolar compounds of similar molecular mass.

KEY TERMS

angular structure (3.4)	melting point (3.3)
boiling point (3.3)	molecule (3.2)
bond energy (3.4)	monatomic ion (3.2)
chemical bond (3.1)	nomenclature (3.2)
covalent bond (3.1)	nonelectrolyte (3.3)
crystal lattice (3.1)	polar covalent bonding (3.4)
dissociation (3.3)	polar covalent molecule (3.4)
double bond (3.4)	polyatomic ion (3.2)
electrolyte (3.3)	resonance (3.4)
electrolytic solution (3.3)	resonance form (3.4)
electronegativity (3.1)	resonance hybrid (3.4)
formula (3.2)	single bond (3.4)
intermolecular force (3.5)	solubility (3.5)
intramolecular force (3.5)	tetrahedral structure (3.4)
ionic bonding (3.1)	trigonal pyramidal
ion pair (3.1)	molecule (3.4)
isomers (3.4)	triple bond (3.4)
Lewis symbol (3.1)	*valence shell electron*
linear structure (3.4)	*pair repulsion* (VSEPR)
lone pair (3.4)	theory (3.4)

QUESTIONS AND PROBLEMS

Chemical Bonding

Foundations

3.13 Classify each of the following compounds as ionic or covalent:
   a. $MgCl_2$
   b. CO_2
   c. H_2S
   d. NO_2

3.14 Classify each of the following compounds as ionic or covalent:
   a. NaCl
   b. CO
   c. ICl
   d. H_2

3.15 Classify each of the following compounds as ionic or covalent:
   a. SiO_2
   b. SO_2
   c. SO_3
   d. $CaCl_2$

3.16 Classify each of the following compounds as ionic or covalent:
   a. NF_3
   b. NaF
   c. CsF
   d. $SiCl_4$

Applications

3.17 Using Lewis symbols, write an equation predicting the product of the reaction of:
   a. Li + Br
   b. Mg + Cl

3.18 Using Lewis symbols, write an equation predicting the product of the reaction of:
   a. Na + O
   b. Na + S

3.19 Using Lewis symbols, write an equation predicting the product of the reaction of:
   a. S + H
   b. P + H

3.20 Using Lewis symbols, write an equation predicting the product of the reaction of:
   a. Si + H
   b. Ca + F

3.21 Explain, using Lewis symbols and the octet rule, why helium is so nonreactive.

3.22 Explain, using Lewis symbols and the octet rule, why neon is so nonreactive.

Naming Compounds and Writing Formulas of Compounds

Foundations

3.23 Name each of the following ions:
   a. Na^+
   b. Cu^+
   c. Mg^{2+}

3.24 Name each of the following ions:
   a. Cu^{2+}
   b. Fe^{2+}
   c. Fe^{3+}

3.25 Name each of the following ions:
   a. S^{2-}
   b. Cl^-
   c. CO_3^{2-}

3.26 Name each of the following ions:
   a. ClO^-
   b. NH_4^+
   c. CH_3COO^-

3.27 Name each of the following compounds:
   a. $MgCl_2$
   b. $AlCl_3$

3.28 Name each of the following compounds:
   a. Na_2O
   b. $Fe(OH)_3$

3.29 Name each of the following covalent compounds:
   a. NO_2
   b. SO_3

3.30 Name each of the following covalent compounds:
   a. N_2O_4
   b. CCl_4

3.31 Name each covalent compound:
   a. SO_2
   b. SO_3

3.32 Name each covalent compound:
   a. N_2O_5
   b. CO

3.33 Write the formula for each of the following monatomic ions:
   a. the potassium ion
   b. the bromide ion

3.34 Write the formula for each of the following monatomic ions:
   a. the calcium ion
   b. the chromium(VI) ion

3.35 Write the formula for each of the following complex ions:
   a. the sulfate ion
   b. the nitrate ion

3.36 Write the formula for each of the following complex ions:
   a. the phosphate ion
   b. the bicarbonate ion

3.37 Write the correct formula for each of the following:
   a. sodium chloride
   b. magnesium bromide

3.38 Write the correct formula for each of the following:
   a. potassium oxide
   b. potassium nitride

3.39 Write the correct formula for each of the following:
   a. silver cyanide
   b. ammonium chloride

3.40 Write the correct formula for each of the following:
   a. magnesium carbonate
   b. magnesium bicarbonate

3.41 Write the correct formula for each of the following:
   a. copper(II) oxide
   b. iron(III) oxide

3.42 Write the correct formula for each of the following:
   a. manganese(II) oxide
   b. manganese(III) oxide

Applications

3.43 Predict the formula of a compound formed from:
   a. aluminum and oxygen
   b. lithium and sulfur

3.44 Predict the formula of a compound formed from:
   a. boron and hydrogen
   b. magnesium and phosphorus

3.45 Write a suitable formula for:
   a. silicon dioxide
   b. sulfur dioxide

3.46 Write a suitable formula for:
   a. vanadium pentoxide
   b. vanadium trioxide

3.47 Write a suitable formula for:
   a. sodium nitrate
   b. magnesium nitrate

3.48 Write a suitable formula for:
   a. aluminum nitrate
   b. ammonium nitrate

3.49 Write a suitable formula for:
   a. ammonium iodide
   b. ammonium sulfate

3.50 Write a suitable formula for:
   a. ammonium acetate
   b. ammonium cyanide

Properties of Ionic and Covalent Compounds

Foundations

3.51 Contrast ionic and covalent compounds with respect to their solid state structure.

3.52 Contrast ionic and covalent compounds with respect to their behavior in solution.

3.53 Contrast ionic and covalent compounds with respect to their relative boiling points.

3.54 Contrast ionic and covalent compounds with respect to their relative melting points.

Applications

3.55 Would KCl be expected to be a solid at room temperature? Why?

3.56 Would CCl_4 be expected to be a solid at room temperature? Why?

3.57 Would H_2O or CCl_4 be expected to have a higher boiling point? Why?

3.58 Would H_2O or CCl_4 be expected to have a higher melting point? Why?

Drawing Lewis Structures of Molecules and Polyatomic Ions

Foundations

3.59 Draw the appropriate Lewis structure for each of the following atoms:
   a. H
   b. He
   c. C
   d. N

3.60 Draw the appropriate Lewis structure for each of the following atoms:
   a. Be
   b. B
   c. F
   d. S

3.61 Draw the appropriate Lewis structure for each of the following ions:
   a. Li^+
   b. Mg^{2+}
   c. Cl^-
   d. P^{3-}

3.62 Draw the appropriate Lewis structure for each of the following ions:
   a. Be^{2+}
   b. Al^{3+}
   c. O^{2-}
   d. S^{2-}

Applications

3.63 Give the Lewis structure for each of the following compounds:
   a. NCl_3
   b. CH_3OH
   c. CS_2

3.64 Give the Lewis structure for each of the following compounds:
   a. HNO_3
   b. CCl_4
   c. PBr_3

3.65 Using the VSEPR theory, predict the geometry, polarity, and water solubility of each compound in Question 3.63.

3.66 Using the VSEPR theory, predict the geometry, polarity, and water solubility of each compound in Question 3.64.

3.67 Draw the Lewis structure of NO^+.

3.68 Draw the Lewis structure of NO_2^-.

3.69 Draw the Lewis structure of OH^-.

3.70 Draw the Lewis structure of HS^-.

3.71 Discuss the concept of resonance, being certain to define the terms *resonance, resonance form,* and *resonance hybrid.*

3.72 Why is resonance an important concept in bonding?

3.73 The acetate ion (Example 3.12) exhibits resonance. Draw two resonance forms of the acetate ion.

3.74 The nitrate ion (Table 3.3) has three resonance forms. Draw each form.

3.75 Ethanol (ethyl alcohol or grain alcohol) has a molecular formula of C_2H_5OH. Represent the structure of ethanol using the Lewis electron dot approach.

3.76 Formaldehyde, H_2CO, in water solution has been used as a preservative for biological specimens. Represent the Lewis structure of formaldehyde.

3.77 Acetone, C_3H_6O, is a common solvent. It is found in such diverse materials as nail polish remover and industrial solvents. Draw its Lewis structure if its skeletal structure is

3.78 Ethylamine is an example of an important class of organic compounds. The molecular formula of ethylamine is $CH_3CH_2NH_2$. Draw its Lewis structure.

3.79 Predict whether the bond formed between each of the following pairs of atoms would be ionic, nonpolar, or polar covalent:
   a. S and O
   b. Si and P

c. Na and Cl
d. Na and O
e. Ca and Br

3.80 Predict whether the bond formed between each of the following pairs of atoms would be ionic, nonpolar, or polar covalent:
   a. Cl and Cl
   b. H and H
   c. C and H
   d. Li and F
   e. O and O

3.81 Draw an appropriate covalent Lewis structure formed by the simplest combination of atoms in Problem 3.79 for each solution that involves a nonpolar or polar covalent bond.

3.82 Draw an appropriate covalent Lewis structure formed by the simplest combination of atoms in Problem 3.80 for each solution that involves a nonpolar or polar covalent bond.

3.83 $BeCl_2$ has an incomplete octet around Be. Draw the Lewis structure of $BeCl_2$.

3.84 SeF_6 has an expanded octet around Se. Draw the Lewis structure of SeF_6.

Properties Based on Electronic Structure and Molecular Geometry

Foundations

3.85 What is the relationship between the polarity of a bond and the polarity of the molecule?

3.86 What effect does polarity have on the solubility of a compound in water?

3.87 What effect does polarity have on the melting point of a pure compound?

3.88 What effect does polarity have on the boiling point of a pure compound?

Applications

3.89 Would you expect KCl to dissolve in water?

3.90 Would you expect ethylamine (Question 3.78) to dissolve in water?

CRITICAL THINKING PROBLEMS

1. Predict differences in our global environment that may have arisen if the freezing point and boiling point of water were 20° C higher than they are.

2. Would you expect the compound $C_2S_2H_4$ to exist? Why or why not?

3. Draw the resonance forms of the carbonate ion. What conclusions, based on this exercise, can you draw about the stability of the carbonate ion?

4. Which of the following compounds would be predicted to have the higher boiling point? Explain your reasoning.

5. Why does the octet rule not work well for compounds of lanthanide and actinide elements? Suggest a number other than eight that may be more suitable.

4

Calculations and the Chemical Equation

Learning Goals

1 Know the relationship between the mole and Avogadro's number, and the usefulness of these quantities.

2 Perform calculations using Avogadro's number and the mole.

3 Write chemical formulas for common inorganic substances.

4 Calculate the formula weight and molar mass of a compound.

5 Know the major function served by the chemical equation, the basis for chemical calculations.

6 Classify chemical reactions by type: combination, decomposition, or replacement.

7 Recognize the various classes of chemical reactions: precipitation, reactions with oxygen, acid-base, and oxidation-reduction.

8 Balance chemical equations given the identity of products and reactants.

9 Calculate the number of moles or grams of product resulting from a given number of moles or grams of reactants or the number of moles or grams of reactant needed to produce a certain number of moles or grams of product.

10 Calculate theoretical and percent yield.

Outline

Introduction

Chemistry Connection:
The Chemistry of Automobile Air Bags

4.1 The Mole Concept and Atoms

4.2 The Chemical Formula, Formula Weight, and Molar Mass

4.3 The Chemical Equation and the Information It Conveys

4.4 Balancing Chemical Equations

A Medical Perspective:
Carbon Monoxide Poisoning: A Case of Combining Ratios

4.5 Calculations Using the Chemical Equation

A Medical Perspective:
Pharmaceutical Chemistry: The Practical Significance of Percent Yield

Careful measurements validate chemical equations.

Introduction

The calculation of chemical quantities based on chemical equations, termed stoichiometry, is the application of logic and arithmetic to chemical systems to answer questions such as the following:

A pharmaceutical company wishes to manufacture 1000 kg of a product next year. How much of each of the starting materials must be ordered? If the starting materials cost $20/g, how much money must be budgeted for the project?

We often need to predict the quantity of a product produced from the reaction of a given amount of material. This calculation is possible. It is equally possible to calculate how much of a material would be necessary to produce a desired amount of product. One of many examples is shown in the following Chemistry Connection: the need to solve a very practical problem.

What is required is a recipe: a procedure to follow. The basis for our recipe is the chemical equation. A properly written chemical equation provides all of the necessary information for the chemical calculation. That critical information is the combining ratio of elements or compounds that must occur to produce a certain amount of product or products.

In this chapter we define the mole, the fundamental unit of measure of chemical arithmetic, learn to write and balance chemical equations, and use these tools to perform calculations of chemical quantities.

Chemistry Connection

The Chemistry of Automobile Air Bags

Each year, thousands of individuals are killed or seriously injured in automobile accidents. Perhaps most serious is the front-end collision. The car decelerates or stops virtually on impact; the momentum of the passengers, however, does not stop, and the driver and passengers are thrown forward toward the dashboard and the front window. Suddenly, passive parts of the automobile, such as control knobs, the rearview mirror, the steering wheel, the dashboard, and the windshield, become lethal weapons.

Automobile engineers have been aware of these problems for a long time. They have made a series of design improvements to lessen the potential problems associated with front-end impact. Smooth switches rather than knobs, recessed hardware, and padded dashboards are examples. These changes, coupled with the use of lap and shoulder belts, which help to immobilize occupants of the car, have decreased the frequency and severity of the impact and lowered the death rate for this type of accident.

An almost ideal protection would be a soft, fluffy pillow, providing a cushion against impact. Such a device, an air bag inflated only on impact, is now standard equipment for the protection of the driver and front-seat passenger.

How does it work? Ideally, it inflates only when severe front-end impact occurs; it inflates very rapidly (in approximately 40 milliseconds), then deflates to provide a steady deceleration, cushioning the occupants from impact. A remarkably simple chemical reaction makes this a reality.

When solid sodium azide (NaN_3) is detonated by mechanical energy produced by an electric current, it decomposes to form solid sodium and nitrogen gas:

$$2NaN_3(s) \longrightarrow 2Na(s) + 3N_2(g)$$

The nitrogen gas inflates the air bag, cushioning the driver and front-seat passenger.

The solid sodium azide has a high density (characteristic of solids) and thus occupies a small volume. It can easily be stored in the center of a steering wheel or in the dashboard. The rate of the detonation is very rapid. In milliseconds it produces three moles of N_2 gas for every two moles of NaN_3. The N_2 gas occupies a relatively large volume because its density is low. This is a general property of gases.

Figuring out how much sodium azide is needed to produce enough nitrogen to properly inflate the bag is an example of a practical application of the chemical arithmetic that we are learning in this chapter.

4.1 The Mole Concept and Atoms

Atoms are exceedingly small, yet their masses have been experimentally determined for each of the elements. The unit of measurement for these determinations is the **atomic mass unit,** abbreviated amu:

$$1 \text{ amu} = 1.661 \times 10^{-24} \text{ g}$$

The Mole and Avogadro's Number

The exact value of the atomic mass unit is defined in relation to a standard, just as the units of the metric system represent defined quantities. The carbon-12 isotope has been chosen and is assigned a mass of exactly 12 atomic mass units. Hence this standard reference point defines an atomic mass unit as exactly one-twelfth the mass of a carbon-12 atom.

The periodic table provides atomic weights in atomic mass units. These atomic weights are average values, based on the contribution of all naturally occurring isotopes of the particular element. For example, the average mass of a carbon atom is 12.01 amu and

$$\frac{12.01 \text{ amu C}}{\text{C atom}} \times \frac{1.661 \times 10^{-24} \text{ g C}}{1 \text{ amu C}} = 1.995 \times \frac{10^{-23} \text{ g C}}{\text{C atom}}$$

The average mass of a helium atom is 4.003 amu and

$$\frac{4.003 \text{ amu He}}{\text{He atom}} \times \frac{1.661 \times 10^{-24} \text{ g He}}{1 \text{ amu He}} = 6.649 \times \frac{10^{-24} \text{ g He}}{\text{He atom}}$$

In everyday work, chemists use much larger quantities of matter (typically, grams or kilograms). A more practical unit for defining a "collection" of atoms is the **mole:**

$$1 \text{ mol of atoms} = 6.022 \times 10^{23} \text{ atoms of an element}$$

This number is **Avogadro's number.** Amedeo Avogadro, a nineteenth-century scientist, conducted a series of experiments that provided the basis for the mole concept. This quantity is based on the number of carbon-12 atoms in one mole of carbon-12.

The practice of defining a unit for a quantity of small objects is common; a *dozen* eggs, a *ream* of paper, and a *gross* of pencils are well-known examples. Similarly, a mole is 6.022×10^{23} individual units of anything. We could, if we desired, speak of a mole of eggs or a mole of pencils. However, in chemistry we use the mole to represent a specific quantity of atoms, ions, or molecules.

The mole (mol) and the atomic mass unit (amu) are related. The atomic mass of an element corresponds to the average mass of a single atom in amu *and* the mass of a mole of atoms in grams.

The mass of 1 mol of atoms, in grams, is defined as the **molar mass.** Consider this relationship for sodium in Example 4.1.

LEARNING GOAL

1 Know the relationship between the mole and Avogadro's number and the usefulness of these quantities.

The term *atomic weight* is not correct but is a fixture in common usage. Just remember that atomic weight is really "average atomic mass."

LEARNING GOAL

2 Perform calculations using Avogadro's number and the mole.

Relating Avogadro's Number to Molar Mass	EXAMPLE 4.1

Calculate the mass, in grams, of Avogadro's number of sodium atoms.

Continued—

EXAMPLE 4.1 —*Continued*

Solution

Step 1. The periodic table indicates that the average mass of one sodium atom is 22.99 amu. This may be represented as:

$$\frac{22.99 \text{ amu Na}}{1 \text{ atom Na}}$$

Step 2. In order to answer the question, we must calculate the molar mass in units of g/mol. We need two conversion factors; one to convert amu $\longrightarrow$ grams and another to convert atoms $\longrightarrow$ mol.

Step 3. As previously noted, 1 amu is 1.661×10^{-24} g, and 6.022×10^{23} atoms of sodium is Avogadro's number. These relationships may be written as:

$$1.661 \times 10^{-24} \frac{\text{g Na}}{\text{amu}} \quad \text{and} \quad 6.022 \times 10^{23} \frac{\text{atoms Na}}{\text{mol Na}}$$

Step 4. Representing this information as a series of conversion factors, using the factor-label method, we have

$$22.99 \frac{\cancel{\text{amu}} \text{ Na}}{\cancel{\text{atom}} \text{ Na}} \times 1.661 \times 10^{-24} \frac{\text{g Na}}{\cancel{\text{amu}} \text{ Na}} \times 6.022 \times 10^{23} \frac{\cancel{\text{atoms}} \text{ Na}}{\text{mol Na}} = 22.99 \frac{\text{g Na}}{\text{mol Na}}$$

Step 5. The average mass of one *atom* of sodium, in units of amu, is *numerically idential* to the mass of *Avogadro's number of atoms*, expressed in units of grams. Hence the molar mass of sodium is 22.99 g Na/mol.

Helpful Hint: Section 1.4 discusses the use of conversion factors.

 Practice Problem 4.1

Calculate the mass, in grams, of Avogadro's number of:
a. aluminum atoms
b. mercury atoms

For Further Practice: Questions 4.7 and 4.8.

Figure 4.1

The comparison of approximately one mole each of silver (as Morgan and Peace dollars), gold (as Canadian Maple Leaf coins), and copper (as pennies) shows the considerable difference in mass (as well as economic value) of equivalent moles of different substances.

The sodium example is not unique. The relationship holds for every element in the periodic table.

Because Avogadro's number of particles (atoms) is 1 mol, it follows that the average mass of one atom of hydrogen is 1.008 amu and the mass of 1 mol of hydrogen atoms is 1.008 g or the average mass of one atom of carbon is 12.01 amu and the mass of 1 mol of carbon atoms is 12.01 g.

In fact, one mole of atoms of *any element* contains the same number, Avogadro's number, of atoms, 6.022×10^{23} atoms.

The difference in mass of a mole of two different elements can be quite striking (Figure 4.1). For example, a mole of hydrogen atoms is 1.008 g, and a mole of lead atoms is 207.19 g.

Calculating Atoms, Moles, and Mass

Performing calculations based on a chemical equation requires a facility for relating the number of atoms of an element to a corresponding number of moles of that element and ultimately to their mass in grams. Such calculations involve the use of

LEARNING GOAL

2 Perform calculations using Avogadro's number and the mole.

conversion factors. The use of conversion factors was first described in Chapter 1. Some examples follow.

Converting Moles to Atoms   **E X A M P L E 4.2**

How many iron atoms are present in 3.0 mol of iron metal?

 LEARNING GOAL
2   Perform calculations using Avogadro's number and the mole.

Solution

Step 1. The calculation is based on the choice of the appropriate conversion factor. The relationship

$$\frac{6.022 \times 10^{23} \text{ atoms Fe}}{1 \text{ mol Fe}}$$

follows directly from

$$1 \text{ mol Fe} = 6.022 \times 10^{23} \text{ atoms Fe}$$

Step 2. Using this conversion factor, we have

$$\text{number of atoms of Fe} = 3.0 \text{ mol Fe} \times \frac{6.022 \times 10^{23} \text{ atoms Fe}}{1 \text{ mol Fe}}$$

$$= 18 \times 10^{23} \text{ atoms of Fe, or}$$

$$= 1.8 \times 10^{24} \text{ atoms of Fe}$$

Practice Problem 4.2

How many oxygen atoms are present in 2.50 mol of:
a. oxygen atoms
b. oxygen molecules

For Further Practice: Questions 4.9 and 4.10.

Converting Atoms to Moles   **E X A M P L E 4.3**

Calculate the number of moles of sulfur represented by 1.81×10^{24} atoms of sulfur.

Solution

Step 1. Just as in the previous example, the calculation is based on the choice of the appropriate conversion factor. The relationship

$$\frac{1 \text{ mol S}}{6.022 \times 10^{23} \text{ atoms S}}$$

follows directly from

$$1 \text{ mol S} = 6.022 \times 10^{23} \text{ atoms S}$$

Step 2. $1.81 \times 10^{24} \text{ atoms S} \times \dfrac{1 \text{ mol S}}{6.022 \times 10^{23} \text{ atoms S}} = 3.01 \text{ mol S}$

Note that this conversion factor is the inverse of that used in Example 4.2. Remember, the conversion factor must cancel units that should not appear in the final answer.

Practice Problem 4.3

How many moles of sodium are represented by 9.03×10^{23} atoms of sodium?

For Further Practice: Questions 4.11 and 4.12.

| EXAMPLE 4.4 | Converting Moles of a Substance to Mass in Grams |

LEARNING GOAL

2 Perform calculations using Avogadro's number and the mole.

What is the mass, in grams, of 3.01 mol of sulfur?

Solution

Step 1. We know from the periodic table that 1 mol of sulfur has a mass of 32.06 g. Setting up a suitable conversion factor between grams and moles results in

Step 2.
$$\frac{32.06 \text{ g S}}{1 \text{ mol S}}$$

which is based on:
$$1 \text{ mol S} = 32.06 \text{ g S}$$

Step 3. Using this conversion factor (ensuring that the units mol S cancel):

$$3.01 \text{ mol S} \times \frac{32.06 \text{ g S}}{1 \text{ mol S}} = 96.5 \text{ g S}$$

Practice Problem 4.4

What is the mass, in grams, of 3.50 mol of the element helium?

For Further Practice: Questions 4.13 and 4.14.

The following examples demonstrate the use of a sequence of conversion factors to proceed from the information *provided* in the problem to the information *requested* by the problem.

| EXAMPLE 4.5 | Converting Grams to Number of Atoms |

LEARNING GOAL

2 Perform calculations using Avogadro's number and the mole.

Calculate the number of atoms of sulfur in 1.00 g of sulfur.

Solution

It is generally useful to map out a pattern for the required conversion. We are given the number of grams and need the number of atoms that correspond to that mass.

Begin by "tracing a path" to the answer:

Two transformations, or conversions, are required:

Step 1. Convert grams to moles.

Step 2. Convert moles to atoms.

To perform Step 1, we could consider either

$$\frac{1 \text{ mol S}}{32.06 \text{ g S}} \quad \text{or} \quad \frac{32.06 \text{ g S}}{1 \text{ mol S}}$$

If we want grams to cancel, $\dfrac{1 \text{ mol S}}{32.06 \text{ g S}}$ is the correct choice, resulting in

$$g S \times \frac{1 \text{ mol S}}{32.06 \text{ g S}} = \text{value in mol S}$$

Continued—

E X A M P L E 4.5 —*Continued*

To perform Step 2, the conversion of moles to atoms, the moles of S must cancel; therefore

$$\text{mol S} \times \frac{6.022 \times 10^{23} \text{ atoms S}}{1 \text{ mol S}} = \text{number of atoms S}$$

which are the units necessary for the solution. Combining Step 1 and Step 2 we have

$$1.00 \text{ g S} \times \frac{1 \text{ mol S}}{32.06 \text{ g S}} \times \frac{6.022 \times 10^{23} \text{ atoms S}}{1 \text{ mol S}} = 1.88 \times 10^{22} \text{ atoms S}$$

Practice Problem 4.5

How many oxygen atoms are present in 40.0 g of oxygen molecules?

For Further Practice: Questions 4.21 and 4.22.

Converting Kilograms to Moles   **E X A M P L E 4.6**

Calculate the number of moles of sulfur in 1.00 kg of sulfur.

Solution

Using the strategy developed in Example 4.5

$$1.00 \text{ kg S} \times \frac{10^3 \text{ g S}}{1 \text{ kg S}} \times \frac{1 \text{ mol S}}{32.06 \text{ g S}} = 31.2 \text{ mol S}$$

Practice Problem 4.6

Calculate the number of moles of silver in a silver ring that has a mass of 3.42 g.

For Further Practice: Questions 4.17 and 4.18.

The conversion between the three principal measures of quantity of matter—the number of grams (mass), the number of moles, and the number of individual particles (atoms, ions, or molecules)—is essential to the art of problem solving in chemistry. Their interrelationship is depicted in Figure 4.2.

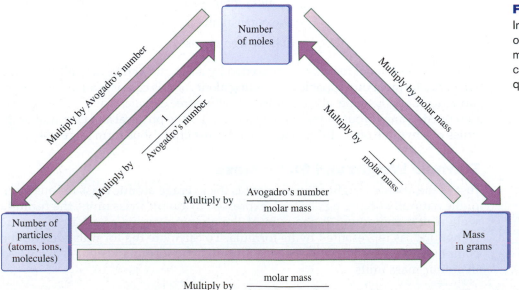

Figure 4.2

Interconversion between numbers of moles, particles, and grams. The mole concept is central to chemical calculations involving measured quantities of matter.

(a)   **(b)**

Figure 4.3

The marked difference in color of (a) hydrated and (b) anhydrous copper sulfate is clear evidence that they are, in fact, different compounds.

3 ▸ **LEARNING GOAL**
Write chemical formulas for common inorganic substances.

Determination of Composition and Formulas of Compounds

It is possible to determine the correct molecular formula of a compound from experimental data.

4 ▸ **LEARNING GOAL**
Calculate the formula weight and molar mass of a compound.

4.2 The Chemical Formula, Formula Weight, and Molar Mass

The Chemical Formula

Compounds are pure substances. They are composed of two or more elements that are chemically combined. A **chemical formula** is a combination of symbols of the various elements that make up the compound. It serves as a convenient way to represent a compound. The chemical formula is based on the formula unit. The **formula unit** is the smallest collection of atoms that provides two important pieces of information:

- the identity of the atoms present in the compound and
- the relative numbers of each type of atom.

Let's look at the following formulas:

- *Hydrogen gas*, H_2. This indicates that two atoms of hydrogen are chemically bonded forming diatomic hydrogen, hence the subscript 2.
- *Water*, H_2O. Water is composed of molecules that contain two atoms of hydrogen (subscript 2) and one atom of oxygen (lack of a subscript means *one* atom).
- *Sodium chloride*, NaCl. One atom of sodium and one atom of chlorine combine to make sodium chloride.
- *Calcium hydroxide*, $Ca(OH)_2$. Calcium hydroxide contains one atom of calcium and two atoms each of oxygen and hydrogen. The subscript outside the parentheses applies to *all* atoms inside the parentheses.
- *Ammonium sulfate*, $(NH_4)_2SO_4$. Ammonium sulfate contains two ammonium ions (NH_4^+) and one sulfate ion (SO_4^{2-}). Each ammonium ion contains one nitrogen and four hydrogen atoms. The formula shows that ammonium sulfate contains two nitrogen atoms, eight hydrogen atoms, one sulfur atom, and four oxygen atoms.
- *Copper(II) sulfate pentahydrate*, $CuSO_4 \cdot 5H_2O$. This is an example of a compound that has water in its structure. Compounds containing one or more water molecules as an integral part of their structure are termed **hydrates.** Copper sulfate pentahydrate has five units of water (or ten H atoms and five O atoms) in addition to one copper atom, one sulfur atom, and four oxygen atoms for a total atomic composition of:

  1 copper atom

  1 sulfur atom

  9 oxygen atoms

  10 hydrogen atoms

Note that the symbol for water is preceded by a dot, indicating that, although the water is a formula unit capable of standing alone, in this case it is a part of a larger structure. Copper sulfate also exists as a structure free of water, $CuSO_4$. This form is described as anhydrous (no water) copper sulfate. The physical and chemical properties of a hydrate often differ markedly from the anhydrous form (Figure 4.3).

Formula Weight and Molar Mass

Just as the atomic weight of an element is the average atomic mass for one atom of the naturally occurring element, expressed in atomic mass units, the **formula weight** of a compound is the sum of the atomic weights of all atoms in the compound, as represented by its formula. To calculate the formula weight of a compound we *must* know the correct formula. The formula weight is expressed in atomic mass units.

When working in the laboratory, we do not deal with individual molecules; instead, we use units of moles or grams. Eighteen grams of water (less than one ounce) contain approximately Avogadro's number of molecules (6.022×10^{23} molecules). Defining our working units as moles and grams makes good chemical sense.

We earlier concluded that the atomic mass of an element in amu from the periodic table corresponds to the mass of a mole of atoms of that element in units of grams/mol. It follows that *molar mass,* the mass of a mole of compound, is numerically equal to the formula weight in atomic mass units.

Calculating Formula Weight and Molar Mass   **E X A M P L E 4.7**

Calculate the formula weight and molar mass of water, H_2O.

LEARNING GOAL
4 Calculate the formula weight and molar mass of a compound.

Solution

Step 1. Each water molecule contains two hydrogen atoms and one oxygen atom.

Step 2. The formula weight is

$$2 \text{ atoms of hydrogen} \times 1.008 \text{ amu/atom} = 2.016 \text{ amu}$$
$$1 \text{ atom of oxygen} \quad \times 16.00 \text{ amu/atom} = 16.00 \text{ amu}$$
$$\overline{ 18.02 \text{ amu}}$$

Step 3. The average mass of a single molecule of H_2O is 18.02 amu and is the formula weight. Therefore the mass of a mole of H_2O is 18.02 g or 18.02 g/mol.

Helpful Hint: Adding 2.016 and 16.00 shows a result of 18.016 on your calculator. Proper use of significant figures (Chapter 1) dictates rounding that result to 18.02.

Practice Problem 4.7

Calculate the formula weight and molar mass of NH_3 (ammonia).

For Further Practice: Questions 4.25 and 4.26.

Calculating Formula Weight and Molar Mass   **E X A M P L E 4.8**

Calculate the formula weight and molar mass of sodium sulfate.

Solution

Step 1. The sodium ion is Na^+, and the sulfate ion is SO_4^{2-}. Two sodium ions must be present to neutralize the negative charges on the sulfate ion. The formula is Na_2SO_4. Sodium sulfate contains two sodium atoms, one sulfur atom, and four oxygen atoms.

Step 2. The formula weight is

$$2 \text{ atoms of sodium} \times 22.99 \text{ amu/atom} = 45.98 \text{ amu}$$
$$1 \text{ atom of sulfur} \quad \times 32.06 \text{ amu/atom} = 32.06 \text{ amu}$$
$$4 \text{ atoms of oxygen} \times 16.00 \text{ amu/atom} = 64.00 \text{ amu}$$
$$\overline{ 142.04 \text{ amu}}$$

Continued—

EXAMPLE 4.8 —*Continued*

Step 3. The average mass of a single unit of Na_2SO_4 is 142.04 amu and is the formula weight. Therefore the mass of a mole of Na_2SO_4 is 142.04 g, or 142.04 g/mol.

Practice Problem 4.8

Calculate the formula weight and molar mass of $C_6H_{12}O_6$ (a sugar, glucose).

For Further Practice: Questions 4.27 and 4.28.

A formula unit   A formula unit (molecule)

In Example 4.8, Na_2SO_4 is an ionic compound. As we have seen, it is not technically correct to describe ionic compounds as molecules; similarly, the term *molecular weight* is not appropriate for Na_2SO_4. The term *formula weight* may be used to describe the formula unit of a substance, whether it is made up of ions, ion pairs, or molecules. We shall use the term *formula weight* in a general way to represent each of these species.

Figure 4.4 illustrates the difference between molecules and ion pairs.

Figure 4.4

Formula units of (a) sodium chloride, an ionic compound, and (b) methane, a covalent compound.

(a)   (b)

EXAMPLE 4.9   **Calculating Formula Weight and Molar Mass**

Calculate the formula weight and molar mass of calcium phosphate.

Solution

Step 1. The calcium ion is Ca^{2+}, and the phosphate ion is PO_4^{3-}. To form a neutral unit, three Ca^{2+} must combine with two PO_4^{3-}; $[3 \times (+2)]$ calcium ion charges are balanced by $[2 \times (-3)]$, the phosphate ion charge.

Step 2. Thus, for calcium phosphate, $Ca_3(PO_4)_2$, the subscript 2 for phosphate dictates that there are two phosphorus atoms and eight oxygen atoms (2×4) in the formula unit.

Step 3. Therefore

$$
\begin{aligned}
3 \text{ atoms of Ca} &\times 40.08 \text{ amu/atom} = 120.24 \text{ amu} \\
2 \text{ atoms of P} &\times 30.97 \text{ amu/atom} = 61.94 \text{ amu} \\
8 \text{ atoms of O} &\times 16.00 \text{ amu/atom} = 128.00 \text{ amu} \\
\hline
&\phantom{\times 16.00 \text{ amu/atom} = } 310.18 \text{ amu}
\end{aligned}
$$

Step 4. The formula weight of calcium phosphate is 310.18 amu, and the molar mass is 310.18 g/mol.

Continued—

Practice Problem 4.9

Calculate the formula weight and molar mass of $CoCl_2 \cdot 6H_2O$ (cobalt chloride hexahydrate).

For Further Practice: Questions 4.29 and 4.30.

4.3 The Chemical Equation and the Information It Conveys

A Recipe for Chemical Change

The **chemical equation** is the shorthand notation for a chemical reaction. It describes all of the substances that react and all the products that form. **Reactants,** or starting materials, are all substances that undergo change in a chemical reaction; **products** are substances produced by a chemical reaction.

The chemical equation also describes the physical state of the reactants and products as solid, liquid, or gas. It tells us whether the reaction occurs and identifies the solvent and experimental conditions employed, such as heat, light, or electrical energy added to the system.

Most important, the relative number of moles of reactants and products appears in the equation. According to the **law of conservation of mass,** matter cannot be either gained or lost in the process of a chemical reaction. The total mass of the products must be equal to the total mass of the reactants. In other words, the law of conservation of mass tells us that we must have a balanced chemical equation.

LEARNING GOAL

5 Know the major function served by the chemical equation, the basis for chemical calculations.

Animation

Conservation of Mass

Features of a Chemical Equation

Consider the decomposition of calcium carbonate:

$$CaCO_3(s) \xrightarrow{\Delta} CaO(s) + CO_2(g)$$

Calcium carbonate   Calcium oxide   Carbon dioxide

The factors involved in writing equations of this type are described as follows:

1. *The identity of products and reactants must be specified using chemical symbols.* In some cases it is possible to predict the products of a reaction. More often, the reactants and products must be verified by chemical analysis. (Generally, you will be given information regarding the identity of the reactants and products.)
2. *Reactants are written to the left of the reaction arrow ($\rightarrow$), and products are written to the right.* The direction in which the arrow points indicates the direction in which the reaction proceeds. In the decomposition of calcium carbonate, the reactant on the left ($CaCO_3$) is converted to products on the right ($CaO + CO_2$) during the course of the reaction.
3. *The physical states of reactants and products may be shown in parentheses.* For example:
   - $Cl_2(g)$ means that chlorine is in the gaseous state.
   - $Mg(s)$ indicates that magnesium is a solid.
   - $Br_2(l)$ indicates that bromine is present as a liquid.
   - $NH_3(aq)$ tells us that ammonia is present as an aqueous solution (dissolved in water).

This equation reads: One mole of solid calcium carbonate decomposes upon heating to produce one mole of solid calcium oxide and one mole of gaseous carbon dioxide.

4. *The symbol Δ over the reaction arrow means that energy is necessary for the reaction to occur.* Often, this and other special conditions are noted above or below the reaction arrow. For example, "light" means that a light source provides energy necessary for the reaction. Such reactions are termed photochemical reactions.

5. *The equation must be balanced.* All of the atoms of every reactant must also appear in the products, although in different compounds. We will treat this topic in detail later in this chapter.

According to the factors outlined, the equation for the decomposition of calcium carbonate may now be written as

$$CaCO_3(s) \xrightarrow{\Delta} CaO(s) + CO_2(g)$$

The Experimental Basis of a Chemical Equation

The chemical equation must represent a chemical change: One or more substances are changed into new substances, with different chemical and physical properties. Evidence for the reaction may be based on observations such as

See discussion of acid-base reactions in Chapter 8.

- the release of carbon dioxide gas when an acid is added to a carbonate,
- the formation of a solid (or precipitate) when solutions of iron ions and hydroxide ions are mixed,

See A Medical Perspective: Hot and Cold Packs in Chapter 7.

- the production of heat when using hot packs for treatment of injury, and
- the change in color of a solution upon addition of a second substance.

Many reactions are not so obvious. Sophisticated instruments are now available to the chemist. These instruments allow the detection of subtle changes in chemical systems that would otherwise go unnoticed. Such instruments may measure

- heat or light absorbed or emitted as the result of a reaction,
- changes in the way the sample behaves in an electric or magnetic field before and after a reaction, and
- changes in electrical properties before and after a reaction.

Whether we use our senses or a million dollar computerized instrument, the "bottom line" is the same: We are measuring a change in one or more chemical or physical properties in an effort to understand the changes taking place in a chemical system.

Disease can be described as a chemical system (actually a biochemical system) gone awry. Here, too, the underlying changes may not be obvious. Just as technology has helped chemists see subtle chemical changes in the laboratory, medical diagnosis has been revolutionized in our lifetimes using very similar technology. Some of these techniques are described in the Medical Perspective: Magnetic Resonance Imaging, in Chapter 9.

Writing Chemical Reactions

LEARNING GOAL

6 Classify chemical reactions by type: combination, decomposition, or replacement.

Chemical reactions, whether they involve the formation of precipitate, reaction with oxygen, acids and bases, or oxidation-reduction, generally follow one of a few simple patterns: combination, decomposition, and single- or double-replacement. Recognizing the underlying pattern will improve your ability to write and understand chemical reactions.

Combination Reactions

Combination reactions involve the joining of two or more elements or compounds, producing a product of different composition. The general form of a combination reaction is

$$A + B \longrightarrow AB$$

in which *A* and *B* represent reactant elements or compounds and *AB* is the product.

Examples include

1. combination of a metal and a nonmetal to form a salt,

$$Ca(s) + Cl_2(g) \longrightarrow CaCl_2(s)$$

2. combination of hydrogen and chlorine molecules to produce hydrogen chloride,

$$H_2(g) + Cl_2(g) \longrightarrow 2HCl(g)$$

3. formation of water from hydrogen and oxygen molecules,

$$2H_2(g) + O_2(g) \longrightarrow 2H_2O(g)$$

4. reaction of magnesium oxide and carbon dioxide to produce magnesium carbonate,

$$MgO(s) + CO_2(g) \longrightarrow MgCO_3(s)$$

Decomposition Reactions

Decomposition reactions produce two or more products from a single reactant. The general form of these reactions is the reverse of a combination reaction:

$$AB \longrightarrow A + B$$

Some examples are

1. the heating of calcium carbonate to produce calcium oxide and carbon dioxide,

$$CaCO_3(s) \longrightarrow CaO(s) + CO_2(g)$$

2. the removal of water from a hydrated material (a *hydrate* is a substance that has water molecules incorporated in its structure),

$$CuSO_4 \cdot 5H_2O(s) \longrightarrow CuSO_4(s) + 5H_2O(g)$$

Hydrated compounds are described on page 130.

Replacement Reactions

Replacement reactions include both *single-replacement* and *double-replacement*. In a **single-replacement reaction,** one atom replaces another in the compound, producing a new compound:

$$A + BC \longrightarrow AC + B$$

Examples include

1. the replacement of copper by zinc in copper sulfate,

$$Zn(s) + CuSO_4(aq) \longrightarrow ZnSO_4(aq) + Cu(s)$$

2. the replacement of aluminum by sodium in aluminum nitrate,

$$3Na(s) + Al(NO_3)_3(aq) \longrightarrow 3NaNO_3(aq) + Al(s)$$

A **double-replacement reaction,** on the other hand, involves *two compounds* undergoing a "change of partners." Two compounds react by exchanging atoms to produce two new compounds:

$$AB + CD \longrightarrow AD + CB$$

Examples include

1. the reaction of an acid (hydrochloric acid) and a base (sodium hydroxide) to produce water and salt, sodium chloride,

$$HCl(aq) + NaOH(aq) \longrightarrow H_2O(l) + NaCl(aq)$$

2. the formation of solid barium sulfate from barium chloride and potassium sulfate,

$$BaCl_2(aq) + K_2SO_4(aq) \longrightarrow BaSO_4(s) + 2KCl(aq)$$

Question 4.1

Classify each of the following reactions as decomposition (D), combination (C), single-replacement (SR), or double-replacement (DR):

a. $HNO_3(aq) + KOH(aq) \longrightarrow KNO_3(aq) + H_2O(aq)$
b. $Al(s) + 3NiNO_3(aq) \longrightarrow Al(NO_3)_3(aq) + 3Ni(s)$
c. $KCN(aq) + HCl(aq) \longrightarrow HCN(aq) + KCl(aq)$
d. $MgCO_3(s) \longrightarrow MgO(s) + CO_2(g)$

Question 4.2

Classify each of the following reactions as decomposition (D), combination (C), single-replacement (SR), or double-replacement (DR):

a. $2Al(OH)_3(s) \xrightarrow{\Delta} Al_2O_3(s) + 3H_2O(g)$
b. $Fe_2S_3(s) \xrightarrow{\Delta} 2Fe(s) + 3S(s)$
c. $Na_2CO_3(aq) + BaCl_2(aq) \longrightarrow BaCO_3(s) + 2NaCl(aq)$
d. $C(s) + O_2(g) \xrightarrow{\Delta} CO_2(g)$

Types of Chemical Reactions

The most commonly encountered chemical reactions involve:

- the combination of soluble ions to produce an insoluble solid, a *precipitate*
- the reaction of a substance with oxygen, *oxidation,* to produce a new substance
- the reaction of acids and bases involving the transfer of hydrogen ions
- the transfer of one or more electrons from one reactant to another, *oxidation-reduction*

Animations

Types of Reactions

Predicting Precipitation Reactions

LEARNING GOAL

7 Recognize the various classes of chemical reactions: precipitation, reactions with oxygen, acid-base, and oxidation-reduction.

Writing Net Ionic Equations

Precipitation reactions may be written as net ionic equations.

Precipitation Reactions

Precipitation reactions include any chemical change in solution that results in one or more insoluble product(s). For aqueous solution reactions the product is insoluble in water.

An understanding of precipitation reactions is useful in many ways. They may explain natural phenomena, such as the formation of stalagmites and stalactites in caves; they are simply precipitates in rocklike form. Kidney stones may result from the precipitation of calcium oxalate (CaC_2O_4). The routine act of preparing a solution requires that none of the solutes will react to form a precipitate.

How do you know whether a precipitate will form? Readily available solubility tables, such as Table 4.1, make prediction rather easy.

TABLE 4.1	Solubilities of Some Common Ionic Compounds

Solubility Predictions

Sodium, potassium, and ammonium compounds are generally *soluble.*

Nitrates and acetates are generally *soluble.*

Chlorides, bromides, and iodides (halides) are generally *soluble.* However, halide compounds containing lead(II), silver(I), and mercury(I) are *insoluble.*

Carbonates and phosphates are generally *insoluble.* Sodium, potassium, and ammonium carbonates and phosphates are, however, *soluble.*

Hydroxides and sulfides are generally *insoluble.* Sodium, potassium, calcium, and ammonium compounds are, however, *soluble.*

The following example illustrates the process.

Predicting Whether Precipitation Will Occur   EXAMPLE 4.10

Will a precipitate form if two solutions of the soluble salts NaCl and AgNO$_3$ are mixed?

Solution

Step 1. Two soluble salts, if they react to form a precipitate, will probably "exchange partners":

$$NaCl(aq) + AgNO_3(aq) \longrightarrow AgCl(?) + NaNO_3(?)$$

Step 2. Refer to Table 4.1 to determine the solubility of AgCl and NaNO$_3$. We predict that NaNO$_3$ is soluble and AgCl is not:

$$NaCl(aq) + AgNO_3(aq) \longrightarrow AgCl(s) + NaNO_3(aq)$$

Step 3. The fact that the solid AgCl is predicted to form classifies this double-replacement reaction as a precipitation reaction.

Helpful Hints: (aq) indicates a soluble species; (s) indicates a solid, an insoluble species.

Practice Problem 4.10

Predict whether the following reactants, when mixed in aqueous solution, undergo a precipitation reaction. Write a balanced equation for each precipitation reaction.
a. potassium chloride and silver nitrate
b. potassium acetate and silver nitrate
c. sodium hydroxide and ammonium chloride
d. sodium hydroxide and iron(II) chloride

For Further Practice: Questions 4.51 and 4.52.

LEARNING GOAL

7 Recognize the various classes of chemical reactions: precipitation, reactions with oxygen, acid-base, and oxidation-reduction.

Animations

Precipitation of Barium Sulfate

Precipitation of Lead Iodide

Reactions with Oxygen

Many substances react with oxygen. These reactions generally release energy. The combustion of gasoline is used for transportation. Fossil fuel combustion is used to heat homes and provide energy for industry. Reactions involving oxygen provide energy for all sorts of biochemical processes.

When organic (carbon-containing) compounds react with the oxygen in air (burning), carbon dioxide is usually produced. If the compound contains hydrogen, water is the other product.

The reaction between oxygen and methane, CH$_4$, the major component of natural gas, is

$$CH_4(g) + 2O_2(g) \longrightarrow CO_2(g) + 2H_2O(g)$$

CO$_2$ and H$_2$O are waste products, and CO$_2$ may contribute to the greenhouse effect and global warming. The really important, unseen product is heat energy. That is why we use this reaction in our furnaces!

Inorganic substances also react with oxygen and produce heat, but these reactions usually proceed more slowly. *Corrosion* (rusting iron) is a familiar example:

$$4Fe(s) + 3O_2(g) \longrightarrow 2Fe_2O_3(s)$$
$$\text{Rust}$$

Some reactions of metals with oxygen are very rapid. A dramatic example is the reaction of magnesium with oxygen.

$$2Mg(s) + O_2(g) \longrightarrow 2MgO(s)$$

4-15

See discussion of acid-base reactions in Chapter 8.

Acid-Base Reactions

Another approach to the classification of chemical reactions is based on the gain or loss of hydrogen ions. **Acid-base reactions** involve the transfer of a *hydrogen ion*, H^+, from one reactant (the acid) to another (the base).

A common example of an acid-base reaction involves hydrochloric acid and sodium hydroxide:

$$HCl(aq) + NaOH(aq) \longrightarrow H_2O(l) + Na^+(aq) + Cl^-(aq)$$
$$\text{Acid} \qquad \text{Base} \qquad \text{Water} \qquad \text{Salt}$$

Acid-base reactions may also be written as net ionic equations.

A hydrogen ion is transferred from the acid to the base, producing water and a salt in solution.

Oxidation-Reduction Reactions

Another important reaction type, **oxidation-reduction,** takes place because of the transfer of negative charge (one or more *electrons*) from one reactant to another.

The reaction of zinc metal with copper(II) ions is one example of oxidation-reduction:

 Writing Net Ionic Equations

$$Zn(s) \quad + \quad Cu^{2+}(aq) \quad \longrightarrow \quad Zn^{2+}(aq) \quad + \quad Cu(s)$$
$$\uparrow \qquad\qquad \uparrow \qquad\qquad\qquad \uparrow \qquad\qquad \uparrow$$

| Substance to be oxidized | Substance to be reduced | Oxidized product | Reduced product |

Zinc metal atoms each donate two electrons to copper(II) ions; consequently zinc atoms become zinc(II) ions and copper(II) ions become copper atoms. Zinc is oxidized (increased positive charge) and copper is reduced (decreased positive charge) as a result of electron transfer.

All of the reactions with oxygen (discussed earlier) are oxidation-reduction reactions as well.

The principles and applications of acid-base reactions will be discussed in Sections 8.1 through 8.4, and oxidation-reduction processes will be discussed in Section 8.5.

Passing an electrical current through water causes an oxidation-reduction reaction. The products are H_2 and O_2 in a 2:1 ratio. Is this ratio of products predicted by the equation for the decomposition of water, $2H_2O(l) \longrightarrow 2H_2(g) + O_2(g)$? Explain.

4.4 Balancing Chemical Equations

The chemical equation shows the *molar quantity* of reactants needed to produce a certain *molar quantity* of products.

The relative number of moles of each product and reactant is indicated by placing a whole-number *coefficient* before the formula of each substance in the chemical equation. A coefficient of 2 (for example, 2NaCl) indicates that 2 mol of sodium chloride are involved in the reaction. Also, $3NH_3$ signifies 3 mol of ammonia; it means that 3 mol of nitrogen atoms and 3×3, or 9, mol of hydrogen atoms are involved in the reaction. The coefficient 1 is understood, not written. Therefore H_2SO_4 would be interpreted as 1 mol of sulfuric acid, or 2 mol of hydrogen atoms, 1 mol of sulfur atoms, and 4 mol of oxygen atoms.

The equation

$$CaCO_3(s) \xrightarrow{\Delta} CaO(s) + CO_2(g)$$

is balanced as written. On the reactant side we have

1 mol Ca

1 mol C

3 mol O

On the product side there are

1 mol Ca

1 mol C

3 mol O

Therefore the law of conservation of mass is obeyed, and the equation is balanced as written.

Now consider the reaction of aqueous hydrogen chloride with solid calcium metal in aqueous solution:

$$HCl(aq) + Ca(s) \longrightarrow CaCl_2(aq) + H_2(g)$$

The equation, as written, is not balanced.

Reactants	Products
1 mol H atoms	2 mol H atoms
1 mol Cl atoms	2 mol Cl atoms
1 mol Ca atoms	1 mol Ca atoms

We need 2 mol of both H and Cl on the left, or reactant, side. An *incorrect* way of balancing the equation is as follows:

$$H_2Cl_2(aq) + Ca(s) \longrightarrow CaCl_2(aq) + H_2(g)$$

NOT a correct equation

The equation satisfies the law of conservation of mass; however, we have altered one of the reacting species. Hydrogen chloride is HCl, not H_2Cl_2. We must remember that *we cannot alter any chemical substance in the process of balancing the equation.* We can *only* introduce coefficients into the equation. Changing subscripts changes the identity of the chemicals involved, and that is not permitted. The equation must represent the reaction accurately. The correct equation is

$$2HCl(aq) + Ca(s) \longrightarrow CaCl_2(aq) + H_2(g)$$

Correct equation

LEARNING GOAL

8 Balance chemical equations given the identity of products and reactants.

The coefficients indicate *relative* numbers of moles: 10 mol of $CaCO_3$ produce 10 mol of CaO; 0.5 mol of $CaCO_3$ produce 0.5 mol of CaO; and so forth.

Animation
Conservation of Mass

Coefficients placed in front of the formula indicate the relative numbers of moles of compound (represented by the formula) that are involved in the reaction. Subscripts placed to the lower right of the atomic symbol indicate the relative number of atoms in the compound.

Water (H_2O) and hydrogen peroxide (H_2O_2) illustrate the effect a subscript can have. The two compounds show marked differences in physical and chemical properties.

Figure 4.5
Balancing the equation HCl + Ca →
CaCl$_2$ + H$_2$. (a) Neither product is
the correct chemical species. (b) The
reactant, HCl, is incorrectly represented
as H$_2$Cl$_2$. (c) This equation is correct;
all species are correct, and the law of
conservation of mass is obeyed.

(a) Incorrect equation

(b) Incorrect equation

(c) Correct equation

This process is illustrated in Figure 4.5.

Many equations are balanced by trial and error. After the identity of the products and reactants, the physical state, and the reaction conditions are known, the following steps provide a method for correctly balancing a chemical equation:

Step 1.   Count the number of moles of atoms of each element on both product and reactant side.

Step 2.   Determine which elements are not balanced.

Step 3.   Balance one element at a time using coefficients.

Step 4.   After you believe that you have successfully balanced the equation, check, as in Step 1, to be certain that mass conservation has been achieved.

Let us apply these steps to the reaction of calcium with aqueous hydrogen chloride:

$$HCl(aq) + Ca(s) \longrightarrow CaCl_2(aq) + H_2(g)$$

Step 1.	**Reactants**	**Products**
	1 mol H atoms	2 mol H atoms
	1 mol Cl atoms	2 mol Cl atoms
	1 mol Ca atoms	1 mol Ca atoms

Step 2.   The numbers of moles of H and Cl are not balanced.

Step 3.   Insertion of a 2 before HCl on the reactant side should balance the equation:

$$2HCl(aq) + Ca(s) \longrightarrow CaCl_2(aq) + H_2(g)$$

Step 4. Check for mass balance:

Reactants	Products
2 mol H atoms	2 mol H atoms
2 mol Cl atoms	2 mol Cl atoms
1 mol Ca atoms	1 mol Ca atoms

Hence the equation is balanced.

Balancing Equations   EXAMPLE 4.11

Balance the following equation: Hydrogen gas and oxygen gas react explosively to produce gaseous water.

Solution

Recall that hydrogen and oxygen are diatomic gases; therefore

$$H_2(g) + O_2(g) \longrightarrow H_2O(g)$$

Step 1. Count the number of moles of atoms of each element on both product and reactant side.

Reactants	Products
2 mol H atoms	2 mol H atoms
2 mol O atoms	1 mol O atoms

Step 2. Note that the moles of hydrogen atoms are balanced but that the moles of oxygen atoms are not.

Step 3. We must first balance the moles of oxygen atoms. Insert 2 before H_2O.

$$H_2(g) + O_2(g) \longrightarrow 2H_2O(g)$$

Balancing moles of oxygen atoms creates an imbalance in the number of moles of hydrogen atoms, so

$$2H_2(g) + O_2(g) \longrightarrow 2H_2O(g)$$

Step 4. The equation is balanced, with 4 mol of hydrogen atoms and 2 mol of oxygen atoms on each side of the reaction arrow.

> **LEARNING GOAL**
> **8** Balance chemical equations given the identity of products and reactants.

Practice Problem 4.11

Balance the chemical equation: $Fe(s) + O_2(g) \longrightarrow Fe_2O_3(s)$

For Further Practice: Questions 4.59 and 4.60.

Balancing Equations   EXAMPLE 4.12

Balance the following equation: Propane gas, C_3H_8, a fuel, reacts with oxygen gas to produce carbon dioxide and water vapor. The reaction is

$$C_3H_8(g) + O_2(g) \longrightarrow CO_2(g) + H_2O(g)$$

Continued—

EXAMPLE 4.12 —*Continued*

Solution

Step 1. Count the number of moles of atoms of each element on both product and reactant side.

Reactants	Products
3 mol C atoms	1 mol C atoms
8 mol H atoms	2 mol H atoms
2 mol O atoms	3 mol O atoms

Step 2. Note that C, H, and O atoms are not balanced.

Step 3. First, balance the carbon atoms; there are 3 mol of carbon atoms on the left and only 1 mol of carbon atoms on the right. We need $3CO_2$ on the right side of the equation:

$$C_3H_8(g) + O_2(g) \longrightarrow 3CO_2(g) + H_2O(g)$$

Next, balance the hydrogen atoms; there are 2 mol of hydrogen atoms on the right and 8 mol of hydrogen atoms on the left. We need $4H_2O$ on the right:

$$C_3H_8(g) + O_2(g) \longrightarrow 3CO_2(g) + 4H_2O(g)$$

There are now 10 mol of oxygen atoms on the right and 2 mol of oxygen atoms on the left. To balance, we must have $5O_2$ on the left side of the equation:

$$C_3H_8(g) + 5O_2(g) \longrightarrow 3CO_2(g) + 4H_2O(g)$$

Step 4. *Remember:* In every case, be sure to check the final equation for mass balance.

Practice Problem 4.12

Balance the chemical equation:
$$C_2H_5OH(l) + O_2(g) \longrightarrow CO_2(g) + H_2O(g)$$

For Further Practice: Questions 4.61 and 4.62.

EXAMPLE 4.13   **Balancing Equations**

LEARNING GOAL

8 Balance chemical equations given the identity of products and reactants.

Balance the following equation: Butane gas, C_4H_{10}, a fuel used in pocket lighters, reacts with oxygen gas to produce carbon dioxide and water vapor. The reaction is

$$C_4H_{10}(g) + O_2(g) \longrightarrow CO_2(g) + H_2O(g)$$

Solution

Step 1. Count the number of moles of atoms of each element on both product and reactant side.

Continued—

EXAMPLE 4.13 —*Continued*

Reactants	Products
4 mol C atoms	1 mol C atoms
10 mol H atoms	2 mol H atoms
2 mol O atoms	3 mol O atoms

Step 2. Note that C, H, and O atoms are not balanced.

Step 3. First, balance the carbon atoms; there are 4 mol of carbon atoms on the left and only 1 mol of carbon atoms on the right:

$$C_4H_{10}(g) + O_2(g) \longrightarrow 4CO_2(g) + H_2O(g)$$

Next, balance hydrogen atoms; there are 10 mol of hydrogen atoms on the left and only 2 mol of hydrogen atoms on the right:

$$C_4H_{10}(g) + O_2(g) \longrightarrow 4CO_2(g) + 5H_2O(g)$$

There are now 13 mol of oxygen atoms on the right and only 2 mol of oxygen atoms on the left. Therefore a coefficient of 6.5 is necessary for O_2.

$$C_4H_{10}(g) + 6.5O_2(g) \longrightarrow 4CO_2(g) + 5H_2O(g)$$

Fractional or decimal coefficients are often needed and used. However, the preferred form requires all integer coefficients. Multiplying each term in the equation by a suitable integer (2, in this case) satisfies this requirement. Hence

$$2C_4H_{10}(g) + 13O_2(g) \longrightarrow 8CO_2(g) + 10H_2O(g)$$

Step 4. The equation is balanced, with 8 mol of carbon atoms, 20 mol of hydrogen atoms, and 26 mol of oxygen atoms on each side of the reaction arrow.

Helpful Hint: When balancing equations, we find that it is often most efficient to begin by balancing the atoms in the most complicated formulas.

Practice Problem 4.13

Balance the chemical equation: $C_6H_6(l) + O_2(g) \longrightarrow CO_2(g) + H_2O(g)$

For Further Practice: Questions 4.63 and 4.64.

Balancing Equations

EXAMPLE 4.14

Balance the following equation: Aqueous ammonium sulfate reacts with aqueous lead nitrate to produce aqueous ammonium nitrate and solid lead sulfate. The reaction is

LEARNING GOAL

8 Balance chemical equations given the identity of products and reactants.

$$(NH_4)_2SO_4(aq) + Pb(NO_3)_2(aq) \longrightarrow NH_4NO_3(aq) + PbSO_4(s)$$

Solution

In this case the polyatomic ions remain as intact units. Therefore we can balance them as we would balance molecules rather than as atoms.

Continued—

A Medical Perspective

Carbon Monoxide Poisoning: A Case of Combining Ratios

A fuel, such as methane, CH_4, burned in an excess of oxygen produces carbon dioxide and water:

$$CH_4(g) + 2O_2(g) \longrightarrow CO_2(g) + 2H_2O(g)$$

The same combustion in the presence of insufficient oxygen produces carbon monoxide and water:

$$2CH_4(g) + 3O_2(g) \longrightarrow 2CO(g) + 4H_2O(g)$$

The combustion of methane, repeated over and over in millions of gas furnaces, is responsible for heating many of our homes in the winter. The furnace is designed to operate under conditions that favor the first reaction and minimize the second; excess oxygen is available from the surrounding atmosphere. Furthermore, the vast majority of exhaust gases (containing principally CO, CO_2, H_2O, and unburned fuel) are removed from the home through the chimney. However, if the chimney becomes obstructed, or the burner malfunctions, carbon monoxide levels within the home can rapidly reach hazardous levels.

Why is exposure to carbon monoxide hazardous? Hemoglobin, an iron-containing compound, binds with O_2 and transports it throughout the body. Carbon monoxide also combines with hemoglobin, thereby blocking oxygen transport. The binding affinity of hemoglobin for carbon monoxide is about two hundred times as great as for O_2. Therefore, to maintain O_2 binding and transport capability, our exposure to carbon monoxide must be minimal. Proper ventilation and suitable oxygen-to-fuel ratio are essential for any combustion process in the home, automobile, or workplace. In recent years carbon monoxide sensors have been developed. These sensors sound an alarm when toxic levels of CO are reached. These warning devices have helped to create a safer indoor environment.

The example we have chosen is an illustration of what is termed the *law of multiple proportions*. This law states that

identical reactants may produce different products, depending on their combining ratio. The experimental conditions (in this case, the quantity of available oxygen) determine the preferred path of the chemical reaction. In Section 4.5 we will learn how to use a properly balanced equation, representing the chemical change occurring, to calculate quantities of reactants consumed or products produced.

For Further Understanding

Why may new, more strict insulation standards for homes and businesses inadvertently increase the risk of carbon monoxide poisoning?

Explain the link between smoking and carbon monoxide that has motivated many states and municipalities to ban smoking in restaurants, offices, and other indoor spaces.

EXAMPLE 4.14 —*Continued*

There are two ammonium ions on the left and only one ammonium ion on the right. Hence

$$(NH_4)_2SO_4(aq) + Pb(NO_3)_2(aq) \longrightarrow 2NH_4NO_3(aq) + PbSO_4(s)$$

No further steps are necessary. The equation is now balanced. There are two ammonium ions, two nitrate ions, one lead ion, and one sulfate ion on each side of the reaction arrow.

Practice Problem 4.14

Balance the chemical equation:
$$S_2Cl_2(s) + NH_3(g) \longrightarrow N_4S_4(s) + NH_4Cl(s) + S_8(s)$$

For Further Practice: Questions 4.65 and 4.66.

4.5 Calculations Using the Chemical Equation

General Principles

The calculation of quantities of products and reactants based on a balanced chemical equation is termed *stoichiometry*, and is important in many fields. The synthesis of drugs and other complex molecules on a large scale is conducted on the basis of a balanced equation. This minimizes the waste of expensive chemical compounds used in these reactions. Similarly, the ratio of fuel and air in a home furnace or automobile must be adjusted carefully, according to their combining ratio, to maximize energy conversion, minimize fuel consumption, and minimize pollution.

In carrying out chemical calculations we apply the following guidelines.

1. The chemical formulas of all reactants and products must be known.
2. The basis for the calculations is a balanced equation because the conservation of mass must be obeyed. If the equation is not properly balanced, the calculation is meaningless.
3. The calculations are performed in terms of moles. The coefficients in the balanced equation represent the relative number of moles of products and reactants.

We have seen that the number of moles of products and reactants often differs in a balanced equation. For example,

$$C(s) + O_2(g) \longrightarrow CO_2(g)$$

is a balanced equation. Two moles of reactants combine to produce one mole of product:

$$1 \text{ mol C} + 1 \text{ mol O}_2 \longrightarrow 1 \text{ mol CO}_2$$

However, 1 mol of C *atoms* and 2 mol of O *atoms* produce 1 mol of C *atoms* and 2 mol of O *atoms*. In other words, the number of moles of reactants and products may differ, but the number of moles of atoms cannot. The formation of CO_2 from C and O_2 may be described as follows:

$$C(s) + O_2(g) \longrightarrow CO_2(g)$$

$$1 \text{ mol C} + 1 \text{ mol O}_2 \longrightarrow 1 \text{ mol CO}_2$$

$$12.0 \text{ g C} + 32.0 \text{ g O}_2 \longrightarrow 44.0 \text{ g CO}_2$$

The mole is the basis of our calculations. However, moles are generally measured in grams (or kilograms). A facility for interconversion of moles and grams is fundamental to chemical arithmetic (see Figure 4.2). These calculations, discussed earlier in this chapter, are reviewed in Example 4.15.

Use of Conversion Factors

Conversion Between Moles and Grams

Conversion from moles to grams, and vice versa, requires only the formula weight of the compound of interest. Consider the following examples.

LEARNING GOAL

9 Calculate the number of moles or grams of product resulting from a given number of moles or grams of reactants or the number of moles or grams of reactant needed to produce a certain number of moles or grams of product.

 Animation
Stoichiometry

Converting Between Moles and Grams   **E X A M P L E 4.15**

a. Convert 1.00 mol of oxygen gas, O_2, to grams.

Continued—

EXAMPLE 4.15 —*Continued*

Solution

Step 1. Use the following path:

$$\boxed{\text{moles of oxygen}} \longrightarrow \boxed{\text{grams of oxygen}}$$

Step 2. The molar mass of oxygen (O_2) is 32.0 g O_2 and the conversion factor becomes:

$$\frac{32.0 \text{ g } O_2}{1 \text{ mol } O_2}$$

Step 3. Using the conversion factor (ensure that moles cancel):

$$1.00 \ \cancel{\text{mol } O_2} \times \frac{32.0 \text{ g } O_2}{1 \ \cancel{\text{mol } O_2}} = 32.0 \text{ g } O_2$$

b. How many grams of carbon dioxide are contained in 10.0 mol of carbon dioxide?

Solution

Step 1. Use the following path:

$$\boxed{\text{moles of carbon dioxide}} \longrightarrow \boxed{\text{grams of carbon dioxide}}$$

Step 2. The molar mass of CO_2 is 44.0 g CO_2 and the conversion factor becomes:

$$\frac{44.0 \text{ g } CO_2}{1 \text{ mol } CO_2}$$

Step 3. Using the conversion factor (ensure that moles cancel):

$$10.0 \ \cancel{\text{mol } CO_2} \times \frac{44.0 \text{ g } CO_2}{1 \ \cancel{\text{mol } CO_2}} = 4.40 \times 10^2 \text{ g } CO_2$$

c. How many moles of sodium are contained in 1 lb (454 g) of sodium metal?

Solution

Step 1. Use the following path:

$$\boxed{\text{grams of sodium}} \longrightarrow \boxed{\text{moles of sodium}}$$

Step 2. The molar mass of Na is 22.99 g Na and the conversion factor becomes:

$$\frac{1 \text{ mol Na}}{22.99 \text{ g Na}}$$

Step 3. Using the conversion factor (ensure that g Na cancel):

$$454 \ \cancel{\text{g Na}} \times \frac{1 \text{ mol Na}}{22.99 \ \cancel{\text{g Na}}} = 19.7 \text{ mol Na}$$

Continued—

EXAMPLE 4.15 —*Continued*

Helpful Hint: Note that each factor can be inverted producing a second possible factor. Only one will allow the appropriate unit cancellation.

Practice Problem 4.15

Perform each of the following conversions:
a. 5.00 mol of water to grams of water
b. 25.0 g of LiCl to moles of LiCl
c. 1.00×10^{-5} mol of $C_6H_{12}O_6$ to micrograms of $C_6H_{12}O_6$
d. 35.0 g of $MgCl_2$ to moles of $MgCl_2$

For Further Practice: Questions 4.19 and 4.20.

Conversion of Moles of Reactants to Moles of Products

In Example 4.12 we balanced the equation for the reaction of propane and oxygen as follows:

$$C_3H_8(g) + 5O_2(g) \rightarrow 3CO_2(g) + 4H_2O(g)$$

In this reaction, 1 mol of C_3H_8 corresponds to, or results in,

5 mol of O_2 being consumed and

3 mol of CO_2 being formed and

4 mol of H_2O being formed.

This information may be written in the form of a conversion factor or ratio:

$$1 \text{ mol } C_3H_8 / 5 \text{ mol } O_2$$

Translated: One mole of C_3H_8 reacts with five moles of O_2.

$$1 \text{ mol } C_3H_8 / 3 \text{ mol } CO_2$$

Translated: One mole of C_3H_8 produces three moles of CO_2.

$$1 \text{ mol } C_3H_8 / 4 \text{ mol } H_2O$$

Translated: One mole of C_3H_8 produces four moles of H_2O.

Conversion factors, based on the chemical equation, permit us to perform a variety of calculations.

Let us look at a few examples, based on the combustion of propane and the equation that we balanced in Example 4.12.

Calculating Reacting Quantities   EXAMPLE 4.16

Calculate the number of grams of O_2 that will react with 1.00 mol of C_3H_8.

Solution

Step 1. Two conversion factors are necessary to solve this problem:

- conversion from moles of C_3H_8 to moles of O_2 and
- conversion of moles of O_2 to grams of O_2.

Continued—

LEARNING GOAL
9 Calculate the number of moles or grams of product resulting from a given number of moles or grams of reactants or the number of moles or grams of reactant needed to produce a certain number of moles or grams of product.

EXAMPLE 4.16 —*Continued*

Step 2. Therefore our path is:

$$\boxed{\begin{array}{c}\text{moles} \\ C_3H_8\end{array}} \longrightarrow \boxed{\begin{array}{c}\text{moles} \\ O_2\end{array}} \longrightarrow \boxed{\begin{array}{c}\text{grams} \\ O_2\end{array}}$$

Step 3. Set up conversion factors to cancel mol C_3H_8 and mol O_2:

$$1.00 \ \text{mol } C_3H_8 \times \frac{5 \ \text{mol } O_2}{1 \ \text{mol } C_3H_8} \times \frac{32.0 \ \text{g } O_2}{1 \ \text{mol } O_2} = 1.60 \times 10^2 \ \text{g } O_2$$

Practice Problem 4.16

When potassium cyanide (KCN) reacts with acids, a poisonous gas, hydrogen cyanide (HCN), is produced. The equation is

$$KCN(aq) + HCl(aq) \longrightarrow KCl(aq) + HCN(g)$$

Calculate the number of grams of KCN that will react with 1.00 mol of HCl.

For Further Practice: Questions 4.73 and 4.74.

EXAMPLE 4.17   **Calculating Grams of Product from Moles of Reactant**

Calculate the number of grams of CO_2 produced from the combustion of 1.00 mol of C_3H_8.

Solution

Steps 1. and *2.* Employ logic similar to that used in Example 4.16 and use the following path:

$$\boxed{\begin{array}{c}\text{moles} \\ C_3H_8\end{array}} \longrightarrow \boxed{\begin{array}{c}\text{moles} \\ CO_2\end{array}} \longrightarrow \boxed{\begin{array}{c}\text{grams} \\ CO_2\end{array}}$$

Step 3. Set up conversion factors to cancel mol C_3H_8 and mol CO_2:

$$1.00 \ \text{mol } C_3H_8 \times \frac{3 \ \text{mol } CO_2}{1 \ \text{mol } C_3H_8} \times \frac{44.0 \ \text{g } CO_2}{1 \ \text{mol } CO_2} = 132 \ \text{g } CO_2$$

Practice Problem 4.17

Fermentation is a critical step in the process of wine making. The reaction is

$$\underset{\text{glucose}}{C_6H_{12}O_6(aq)} \longrightarrow \underset{\text{ethanol}}{2CH_3CH_2OH(aq)} + 2CO_2(g)$$

Calculate the number of grams of ethanol produced from the fermentation of 5.00 mol of glucose.

For Further Practice: Questions 4.75 and 4.76.

Relating Masses of Reactants and Products   E X A M P L E 4.18

Calculate the number of grams of C_3H_8 required to produce 36.0 g of H_2O using the balanced equation from Example 4.12.

LEARNING GOAL

9 Calculate the number of moles or grams of product resulting from a given number of moles or grams of reactants or the number of moles or grams of reactant needed to produce a certain number of moles or grams of product.

Solution

Steps 1. and *2.* It is necessary to convert

grams of H_2O to moles of H_2O,
moles of H_2O to moles of C_3H_8, and
moles of C_3H_8 to grams of C_3H_8.

Use the following path:

$$\boxed{\text{grams } H_2O} \longrightarrow \boxed{\text{moles } H_2O} \longrightarrow \boxed{\text{moles } C_3H_8} \longrightarrow \boxed{\text{grams } C_3H_8}$$

Step 3. Setup conversion factors to cancel g H_2O, mol H_2O, and mol C_3H_8:

$$36.0 \ \cancel{\text{g } H_2O} \times \frac{1 \ \cancel{\text{mol } H_2O}}{18.0 \ \cancel{\text{g } H_2O}} \times \frac{1 \ \cancel{\text{mol } C_3H_8}}{4 \ \cancel{\text{mol } H_2O}} \times \frac{44.0 \ \text{g } C_3H_8}{1 \ \cancel{\text{mol } C_3H_8}} = 22.0 \ \text{g } C_3H_8$$

Practice Problem 4.18

The balanced equation for the combustion of ethanol (ethyl alcohol) is:

$$C_2H_5OH(l) + 3O_2(g) \longrightarrow 2CO_2(g) + 3H_2O(g)$$

a. How many moles of O_2 will react with 1 mol of ethanol?
b. How many grams of O_2 will react with 1 mol of ethanol?
c. How many grams of CO_2 will be produced by the combustion of 1 mol of ethanol?

For Further Practice: Questions 4.77 and 4.78.

Let's consider an example that requires us to write and balance the chemical equation, use conversion factors, and calculate the amount of a reactant consumed in the chemical reaction.

Calculating a Quantity of Reactant   E X A M P L E 4.19

Calcium hydroxide may be used to neutralize (completely react with) aqueous hydrochloric acid. Calculate the number of grams of hydrochloric acid that would be neutralized by 0.500 mol of solid calcium hydroxide.

Solution

Step 1. The formula for calcium hydroxide is $Ca(OH)_2$ and that for hydrochloric acid is HCl. The unbalanced equation produces calcium chloride and water as products:

$$Ca(OH)_2(s) + HCl(aq) \longrightarrow CaCl_2(aq) + H_2O(l)$$

Continued—

Step 2. Balance the equation:

$$Ca(OH)_2(s) + 2HCl(aq) \longrightarrow CaCl_2(aq) + 2H_2O(l)$$

Step 3. Determine the necessary conversion:

* moles of $Ca(OH)_2$ to moles of HCl and
* moles of HCl to grams of HCl.

Step 4. Use the following path:

$$\boxed{\text{moles } Ca(OH)_2} \longrightarrow \boxed{\text{moles HCl}} \longrightarrow \boxed{\text{grams HCl}}$$

Step 5. $0.500 \ \text{mol } Ca(OH)_2 \times \dfrac{2 \ \text{mol HCl}}{1 \ \text{mol } Ca(OH)_2} \times \dfrac{36.5 \ \text{g HCl}}{1 \ \text{mol HCl}} = 36.5 \ \text{g HCl}$

This reaction is illustrated in Figure 4.6.

Helpful Hints:
1. The reaction between an acid and a base produces a salt and water (Chapter 8).
2. Remember to balance the chemical equation; the proper coefficients are essential parts of the subsequent calculations.

Practice Problem 4.19

Metallic iron reacts with O_2 gas to produce iron(III) oxide
a. Write and balance the equation.
b. Calculate the number of grams of iron needed to produce 5.00 g of product.

For Further Practice: Questions 4.81 and 4.82.

Figure 4.6

An illustration of the law of conservation of mass. In this example, 1 mol of calcium hydroxide and 2 mol of hydrogen chloride react to produce 3 mol of product (2 mol of water and 1 mol of calcium chloride). The total mass, in grams, of reactant(s) consumed is equal to the total mass, in grams, of product(s) formed. *Note:* In reality, HCl does not exist as discrete molecules in water. The HCl separates to form H^+ and Cl^-. Ionization in water will be discussed with the chemistry of acids and bases in Chapter 8.

$Ca(OH)_2$	+	2 HCl	$\longrightarrow$	$CaCl_2$	+	2 H_2O
1 mol		2 mol		1 mol		2 mol
74 g/mol		36.5 g/mol		111 g/mol		18 g/mol
74 g		73 g		111 g		36 g

147 g of reactants   147 g of products

Calculating Reactant Quantities

EXAMPLE 4.20

What mass of sodium hydroxide, NaOH, would be required to produce 8.00 g of the antacid milk of magnesia, $Mg(OH)_2$, by the reaction of $MgCl_2$ with NaOH?

Solution

Step 1. Write and balance the equation:

$$MgCl_2(aq) + 2NaOH(aq) \longrightarrow Mg(OH)_2(s) + 2NaCl(aq)$$

Step 2. Determine the strategy:

The equation tells us that 2 mol of NaOH form 1 mol of $Mg(OH)_2$. If we calculate the number of moles of $Mg(OH)_2$ in 8.00 g of $Mg(OH)_2$, we can determine the number of moles of NaOH necessary and then the mass of NaOH required:

$$\boxed{\text{mass } Mg(OH)_2} \longrightarrow \boxed{\text{moles } Mg(OH)_2} \longrightarrow \boxed{\text{moles } NaOH} \longrightarrow \boxed{\text{mass } NaOH}$$

Step 3. Determine the conversion factor to convert mass to mol $Mg(OH)_2$:

$$\text{Since } 58.3 \text{ g } Mg(OH)_2 = 1 \text{ mol } Mg(OH)_2$$

The conversion factor is:

$$\frac{1 \text{ mol } Mg(OH)_2}{58.3 \text{ g } Mg(OH)_2}$$

Step 4. $8.00 \text{ g } Mg(OH)_2 \times \dfrac{1 \text{ mol } Mg(OH)_2}{58.3 \text{ g } Mg(OH)_2} = 0.137 \text{ mol } Mg(OH)_2$

Step 5. Two moles of NaOH react to give one mole of $Mg(OH)_2$. Therefore

$$0.137 \text{ mol } Mg(OH)_2 \times \frac{2 \text{ mol } NaOH}{1 \text{ mol } Mg(OH)_2} = 0.274 \text{ mol } NaOH$$

Step 6. 40.0 g of NaOH = 1 mol of NaOH. Therefore

$$0.274 \text{ mol } NaOH \times \frac{40.0 \text{ g } NaOH}{1 \text{ mol } NaOH} = 11.0 \text{ g } NaOH$$

However, the calculation may be done in a single step:

$$8.00 \text{ g } Mg(OH)_2 \times \frac{1 \text{ mol } Mg(OH)_2}{58.3 \text{ g } Mg(OH)_2} \times \frac{2 \text{ mol } NaOH}{1 \text{ mol } Mg(OH)_2}$$

$$\times \frac{40.0 \text{ g } NaOH}{1 \text{ mol } NaOH} = 11.0 \text{ g } NaOH$$

Note once again that we have followed a logical and predictable path to the solution:

$$\boxed{\text{grams } Mg(OH)_2} \longrightarrow \boxed{\text{moles } Mg(OH)_2} \longrightarrow \boxed{\text{moles } NaOH} \longrightarrow \boxed{\text{grams } NaOH}$$

Helpful Hint: Mass is a laboratory unit, whereas moles is a calculation unit. The laboratory balance is calibrated in units of mass (grams). Although moles are essential for calculation, often the starting point and objective are in mass units. As a result, our path is often grams → moles → grams.

Continued—

EXAMPLE 4.20 —*Continued*

Practice Problem 4.20

Barium carbonate decomposes upon heating to barium oxide and carbon dioxide.
a. Write and balance the equation.
b. Calculate the number of grams of carbon dioxide produced by heating 50.0 g of barium carbonate.

For Further Practice: Questions 4.83 and 4.84.

A general problem-solving strategy is summarized in Figure 4.7. By systematically applying this strategy, you will be able to solve virtually any problem requiring calculations based on the chemical equation.

Theoretical and Percent Yield

LEARNING GOAL

10 Calculate theoretical and percent yield.

The **theoretical yield** is the *maximum* amount of product that can be produced (in an ideal world). In the "real" world it is difficult to produce the amount calculated as the theoretical yield. This is true for a variety of reasons. Some experimental error is unavoidable. Moreover, many reactions simply are not complete; some amount of reactant remains at the end of the reaction. We will study these processes, termed *equilibrium reactions* in Chapter 7.

A **percent yield,** the ratio of the actual and theoretical yields multiplied by 100%, is often used to show the relationship between predicted and experimental quantities. Thus

$$\% \text{ yield} = \frac{\text{actual yield}}{\text{theoretical yield}} \times 100\%$$

Figure 4.7

A general problem-solving strategy, using molar quantities.

For a reaction of the general type:
$$A + B \longrightarrow C$$

(a) Given a specified number of grams of A, calculate moles of C.

(b) Given a specified number of grams of A, calculate grams of C.

(c) Given a volume of A in milliliters, calculate grams of C.

A Medical *Perspective*

Pharmaceutical Chemistry: The Practical Significance of Percent Yield

In recent years the major pharmaceutical industries have introduced a wide variety of new drugs targeted to cure or alleviate the symptoms of a host of diseases that afflict humanity.

The vast majority of these drugs are synthetic; they are made in a laboratory or by an industrial process. These substances are complex molecules that are patiently designed and constructed from relatively simple molecules in a series of chemical reactions. A series of ten to twenty "steps," or sequential reactions, is not unusual to put together a final product that has the proper structure, geometry, and reactivity for efficacy against a particular disease.

Although a great deal of research occurs to ensure that each of these steps in the overall process is efficient (having a large percent yield), the overall process is still very inefficient (low percent yield). This inefficiency, and the research needed to minimize it, at least in part determines the cost and availability of both prescription and over-the-counter preparations.

Consider a hypothetical five-step sequential synthesis. If each step has a percent yield of 80% our initial impression might be that this synthesis is quite efficient. However, on closer inspection we find quite the contrary to be true.

The overall yield of the five-step reaction is the product of the decimal fraction of the percent yield of each of the sequential reactions. So, if the decimal fraction corresponding to 80% is 0.80:

$$0.80 \times 0.80 \times 0.80 \times 0.80 \times 0.80 = 0.33$$

Converting the decimal fraction to percentage:

$$0.33 \times 100\% = 33\% \text{ yield}$$

Many reactions are considerably less than 80% efficient, especially those that are used to prepare large molecules with complex arrangements of atoms. Imagine a more realistic scenario in which one step is only 20% efficient (a 20% yield) and the other four steps are 50%, 60%, 70%, and 80% efficient. Repeating the calculation with these numbers (after conversion to a decimal fraction):

$$0.20 \times 0.50 \times 0.60 \times 0.70 \times 0.80 = 0.0336$$

Converting the decimal fraction to a percentage:

$$0.0336 \times 100\% = 3.36\% \text{ yield}$$

a very inefficient process.

If we apply this logic to a fifteen- or twenty-step synthesis we gain some appreciation of the difficulty of producing modern pharmaceutical products. Add to this the challenge of predicting the most appropriate molecular structure that will have the desired biological effect and be relatively free of side effects. All these considerations give new meaning to the term *wonder drug* that has been attached to some of the more successful synthetic products.

We will study some of the elementary steps essential to the synthesis of a wide range of pharmaceutical compounds in later chapters, beginning with Chapter 10.

For Further Understanding

Explain the possible connection of this perspective to escalating costs of pharmaceutical products.

Can you describe other situations, not necessarily in the field of chemistry, where multiple-step processes contribute to inefficiency?

In Example 4.17, the theoretical yield of CO_2 is 132 g. For this reaction let's assume that a chemist actually obtained 125 g CO_2. This is the actual yield and would normally be provided as a part of the data in the problem.

Calculate the percent yield as follows:

$$\% \text{ yield} = \frac{\text{actual yield}}{\text{theoretical yield}} \times 100\%$$

$$= \frac{125 \text{ g } CO_2 \text{ actual}}{132 \text{ g } CO_2 \text{ theoretical}} \times 100\% = 94.7\%$$

E X A M P L E 4.21   **Calculation of Percent Yield**

Assume that the theoretical yield of iron in the process

$$2Al(s) + Fe_2O_3(s) \longrightarrow Al_2O_3(l) + 2Fe(l)$$

was 30.0 g.

If the actual yield of iron were 25.0 g in the process, calculate the percent yield.

Solution

$$\% \text{ yield} = \frac{\text{actual yield}}{\text{theoretical yield}} \times 100\%$$

$$= \frac{25.0 \text{ g}}{30.0 \text{ g}} \times 100\%$$

$$= 83.3\%$$

Practice Problem 4.21

Given the reaction represented by the balanced equation

$$CH_4(g) + 3Cl_2(g) \longrightarrow 3HCl(g) + CHCl_3(g)$$

a. Calculate the number of grams of $CHCl_3$ produced by mixing 105 g Cl_2 with excess CH_4.
b. If 10.0 g $CHCl_3$ were produced, calculate the % yield.

For Further Practice: Questions 4.89 and 4.90.

SUMMARY

4.1 The Mole Concept and Atoms

Atoms are exceedingly small, yet their masses have been experimentally determined for each of the elements. The unit of measurement for these determinations is the *atomic mass unit*, abbreviated amu:

$$1 \text{ amu} = 1.661 \times 10^{-24} \text{ g}$$

The periodic table provides atomic masses in atomic mass units.

A more practical unit for defining a "collection" of atoms is the *mole:*

$$1 \text{ mol of atoms} = 6.022 \times 10^{23} \text{ atoms of an element}$$

This number is referred to as *Avogadro's number.*

The mole and the atomic mass unit are related. The atomic mass of a given element corresponds to the average mass of a single atom in atomic mass units and the mass of a mole of atoms in grams. The mass of one mole of atoms is termed the *molar mass* of the element. One mole of atoms of any element contains the same number, Avogadro's number, of atoms.

4.2 The Chemical Formula, Formula Weight, and Molar Mass

Compounds are pure substances that are composed of two or more elements that are chemically combined. They are represented by their *chemical formula*, a combination of symbols of the various elements that make up the compounds. The chemical formula is based on the *formula unit*. This is the smallest collection of atoms that provides the identity of the atoms present in the compound and the relative numbers of each type of atom.

Just as a mole of atoms is based on the atomic mass, a mole of a compound is based on the formula mass or *formula weight*. The formula weight is calculated by addition of the masses of all the atoms or ions of which the unit is composed. To calculate the formula weight, the formula unit must be known. The formula weight of one mole of a compound is its *molar mass* in units of g/mol.

4.3 The Chemical Equation and the Information It Conveys

The *chemical equation* is the shorthand notation for a chemical reaction. It describes all of the substances that react to produce the product(s). *Reactants*, or starting materials, are all substances that undergo change in a chemical reaction; *products* are substances produced by a chemical reaction.

According to the *law of conservation of mass,* matter can neither be gained nor lost in the process of a chemical reaction. The law of conservation of mass states that we must have a balanced chemical equation.

Features of a suitable equation include the following:

- The identity of products and reactants must be specified.
- Reactants are written to the left of the reaction arrow ($\rightarrow$) and products to the right.
- The physical states of reactants and products are shown in parentheses.
- The symbol Δ over the reaction arrow means that heat energy is necessary for the reaction to occur.
- The equation must be balanced.

Chemical reactions involve the *combination* of reactants to produce products, the *decomposition* of reactant(s) into products, or the *replacement* of one or more elements in a compound to yield products. Replacement reactions are subclassified as either *single-* or *double*-replacement.

Reactions that produce products with similar characteristics are often classified as a single group. The formation of an insoluble solid, a *precipitate,* is very common. Such reactions are precipitation reactions.

Chemical reactions that have a common reactant may be grouped together. Reactions involving oxygen, *combustion reactions,* are such a class.

Another approach to the classification of chemical reactions is based on transfer of hydrogen ions (+ charge) or electrons (– charge). *Acid-base reactions* involve the transfer of a hydrogen ion, H^+, from one reactant to another. Another important reaction type, *oxidation-reduction,* takes place because of the transfer of negative charge, one or more electrons, from one reactant to another.

4.4 Balancing Chemical Equations

The chemical equation enables us to determine the quantity of reactants needed to produce a certain molar quantity of products. The chemical equation expresses these quantities in terms of moles.

The relative number of moles of each product and reactant is indicated by placing a whole-number coefficient before the formula of each substance in the chemical equation.

Many equations are balanced by trial and error. If the identity of the products and reactants, the physical state, and the reaction conditions are known, the following steps provide a method for correctly balancing a chemical equation:

- Count the number of atoms of each element on both product and reactant sides.
- Determine which atoms are not balanced.

- Balance one element at a time using coefficients.
- After you believe that you have successfully balanced the equation, check to be certain that mass conservation has been achieved.

4.5 Calculations Using the Chemical Equation

Calculations involving chemical quantities are based on the following requirements:

- The basis for the calculations is a balanced equation.
- The calculations are performed in terms of moles.
- The conservation of mass must be obeyed.

The mole is the basis for calculations. However, masses are generally measured in grams (or kilograms). Therefore you must be able to interconvert moles and grams to perform chemical arithmetic.

KEY TERMS

acid-base reaction (4.3)	law of conservation of mass (4.3)
atomic mass unit (4.1)	molar mass (4.1)
Avogadro's number (4.1)	mole (4.1)
chemical equation (4.3)	oxidation-reduction reaction (4.3)
chemical formula (4.2)	percent yield (4.5)
combination reaction (4.3)	product (4.3)
decomposition reaction (4.3)	reactant (4.3)
double-replacement reaction (4.3)	single-replacement reaction (4.3)
formula unit (4.2)	theoretical yield (4.5)
formula weight (4.2)	
hydrate (4.2)	

QUESTIONS AND PROBLEMS

The Mole Concept and Atoms

Foundations

4.3 We purchase eggs by the dozen. Name several other familiar packaging units.

4.4 One dozen eggs is a convenient consumer unit. Explain why the mole is a convenient chemist's unit.

4.5 What is the average molar mass of:
   a. Si
   b. Ag

4.6 What is the average molar mass of:
   a. S
   b. Na

4.7 What is the mass of Avogadro's number of argon atoms?

4.8 What is the mass of Avogadro's number of iron atoms?

Applications

4.9 How many carbon atoms are present in 1.0×10^{-4} moles of carbon?

4.10 How many mercury atoms are present in 1.0×10^{-10} moles of mercury?

4.11 How many moles of arsenic correspond to 1.0×10^2 atoms of arsenic?

4.12 How many moles of sodium correspond to 1.0×10^{15} atoms of sodium?

4.13 How many grams are contained in 2.00 mol of neon atoms?

4.14 How many grams are contained in 3.00 mol of carbon atoms?

4.15 What is the mass in grams of 1.00 mol of helium atoms?

4.16 What is the mass in grams of 1.00 mol of nitrogen atoms?

4.17 Calculate the number of moles corresponding to:
   a. 20.0 g He
   b. 0.040 kg Na
   c. 3.0 g Cl_2

4.18 Calculate the number of moles corresponding to:
   a. 0.10 g Ca
   b. 4.00 g Fe
   c. 2.00 kg N_2

4.19 What is the mass, in grams, of 15.0 mol of silver?

4.20 What is the mass, in grams, of 15.0 mol of carbon?

4.21 Calculate the number of atoms in 15.0 g of silver.

4.22 Calculate the number of atoms in 15.0 g of carbon.

The Chemical Formula, Formula Weight, and Molar Mass

Foundations

4.23 Distinguish between the terms *molecule* and *ion pair*.

4.24 Distinguish between the terms *formula weight* and *molecular weight*.

4.25 Calculate the molar mass, in grams per mole, of each of the following formula units:
   a. NaCl
   b. Na_2SO_4
   c. $Fe_3(PO_4)_2$

4.26 Calculate the molar mass, in grams per mole, of each of the following formula units:
   a. S_8
   b. $(NH_4)_2SO_4$
   c. CO_2

4.27 Calculate the molar mass, in grams per mole, of oxygen gas, O_2.

4.28 Calculate the molar mass, in grams per mole, of ozone, O_3.

4.29 Calculate the molar mass of $CuSO_4 \cdot 5H_2O$.

4.30 Calculate the molar mass of $CaCl_2 \cdot 2H_2O$.

Applications

4.31 Calculate the number of moles corresponding to:
   a. 15.0 g NaCl
   b. 15.0 g Na_2SO_4

4.32 Calculate the number of moles corresponding to:
   a. 15.0 g NH_3
   b. 16.0 g O_2

4.33 Calculate the mass in grams corresponding to:
   a. 1.000 mol H_2O
   b. 2.000 mol NaCl

4.34 Calculate the mass in grams corresponding to:
   a. 0.400 mol NH_3
   b. 0.800 mol $BaCO_3$

4.35 Calculate the mass in grams corresponding to:
   a. 10.0 mol He
   b. 1.00×10^2 mol H_2

4.36 Calculate the mass in grams corresponding to:
   a. 2.00 mol CH_4
   b. 0.400 mol $Ca(NO_3)_2$

4.37 How many grams are required to have 0.100 mol of each of the following?
   a. Mg
   b. $CaCO_3$

4.38 How many grams are required to have 0.100 mol of each of the following?
   a. $C_6H_{12}O_6$ (glucose)
   b. NaCl

4.39 How many grams are required to have 0.100 mol of each of the following compounds?
   a. NaOH
   b. H_2SO_4

4.40 How many grams are required to have 0.100 mol of each of the following compounds?
   a. C_2H_5OH (ethanol)
   b. $Ca_3(PO_4)_2$

4.41 How many moles are in 50.0 g of each of the following substances?
   a. KBr
   b. $MgSO_4$

4.42 How many moles are in 50.0 g of each of the following substances?
   a. Br_2
   b. NH_4Cl

4.43 How many moles are in 50.0 g of each of the following substances?
   a. CS_2
   b. $Al_2(CO_3)_3$

4.44 How many moles are in 50.0 g of each of the following substances?
   a. $Sr(OH)_2$
   b. $LiNO_3$

The Chemical Equation and the Information It Conveys

Foundations

4.45 What law is the ultimate basis for a correct chemical equation?

4.46 List the general types of information that a chemical equation provides.

4.47 What is the meaning of the subscript in a chemical formula?

4.48 What is the meaning of the coefficient in a chemical equation?

4.49 What is the meaning of Δ over the reaction arrow?

4.50 What is the meaning of (s), (l), or (g) immediately following the symbol for a chemical substance?

Applications

4.51 Will a precipitate form if solutions of the soluble salts $Pb(NO_3)_2$ and KI are mixed?

4.52 Will a precipitate form if solutions of the soluble salts $AgNO_3$ and NaOH are mixed?

Balancing Chemical Equations

Foundations

4.53 When you are balancing an equation, why must the subscripts in the chemical formula remain unchanged?

4.54 Describe the process of checking to ensure that an equation is properly balanced.

4.55 What is a reactant?

4.56 On which side of the reaction arrow are reactants found?

4.57 What is a product?

4.58 On which side of the reaction arrow are products found?

Applications

4.59 Balance each of the following equations:
   a. $C_2H_6(g) + O_2(g) \longrightarrow CO_2(g) + H_2O(g)$
   b. $K_2O(s) + P_4O_{10}(s) \longrightarrow K_3PO_4(s)$
   c. $MgBr_2(aq) + H_2SO_4(aq) \longrightarrow HBr(g) + MgSO_4(aq)$

4.60 Balance each of the following equations:
   a. $C_6H_{12}O_6(s) + O_2(g) \longrightarrow CO_2(g) + H_2O(g)$
   b. $H_2O(l) + P_4O_{10}(s) \longrightarrow H_3PO_4(aq)$
   c. $PCl_5(g) + H_2O(l) \longrightarrow HCl(aq) + H_3PO_4(aq)$

4.61 Complete, then balance, each of the following equations:
 a. $Ca(s) + F_2(g) \longrightarrow$
 b. $Mg(s) + O_2(g) \longrightarrow$
 c. $H_2(g) + N_2(g) \longrightarrow$

4.62 Complete, then balance, each of the following equations:
 a. $Li(s) + O_2(g) \longrightarrow$
 b. $Ca(s) + N_2(g) \longrightarrow$
 c. $Al(s) + S(s) \longrightarrow$

4.63 Balance each of the following equations:
 a. $C_4H_{10}(g) + O_2(g) \longrightarrow H_2O(g) + CO_2(g)$
 b. $Au_2S_3(s) + H_2(g) \longrightarrow Au(s) + H_2S(g)$
 c. $Al(OH)_3(s) + HCl(aq) \longrightarrow AlCl_3(aq) + H_2O(l)$
 d. $(NH_4)_2Cr_2O_7(s) \longrightarrow Cr_2O_3(s) + N_2(g) + H_2O(g)$
 e. $C_2H_5OH(l) + O_2(g) \longrightarrow CO_2(g) + H_2O(g)$

4.64 Balance each of the following equations:
 a. $Fe_2O_3(s) + CO(g) \longrightarrow Fe_3O_4(s) + CO_2(g)$
 b. $C_6H_6(l) + O_2(g) \longrightarrow CO_2(g) + H_2O(g)$
 c. $I_4O_9(s) + I_2O_6(s) \longrightarrow I_2(s) + O_2(g)$
 d. $KClO_3(s) \longrightarrow KCl(s) + O_2(g)$
 e. $C_6H_{12}O_6(s) \longrightarrow C_2H_6O(l) + CO_2(g)$

4.65 Write a balanced equation for each of the following reactions:
 a. Ammonia is formed by the reaction of nitrogen and hydrogen.
 b. Hydrochloric acid reacts with sodium hydroxide to produce water and sodium chloride.

4.66 Write a balanced equation for each of the following reactions:
 a. Nitric acid reacts with calcium hydroxide to produce water and calcium nitrate.
 b. Butane (C_4H_{10}) reacts with oxygen to produce water and carbon dioxide.

4.67 Write a balanced equation for each of the following reactions:
 a. Glucose, a sugar, $C_6H_{12}O_6$, is oxidized in the body to produce water and carbon dioxide.
 b. Sodium carbonate, upon heating, produces sodium oxide and carbon dioxide.

4.68 Write a balanced equation for each of the following reactions:
 a. Sulfur, present as an impurity in coal, is burned in oxygen to produce sulfur dioxide.
 b. Hydrofluoric acid (HF) reacts with glass (SiO_2) in the process of etching to produce silicon tetrafluoride and water.

Calculations Using the Chemical Equation

Foundations

4.69 What is the law that forms the basis of chemical calculations?

4.70 Why is it essential to use balanced equations to solve mole problems?

4.71 Balancing equations involves changing coefficients, not subscripts. Why?

4.72 Describe the steps used in the calculation of grams of product resulting from the reaction of a specified number of grams of reactant.

Applications

4.73 How many grams of B_2H_6 will react with 3.00 moles of O_2?

$$B_2H_6(l) + 3O_2(g) \longrightarrow B_2O_3(s) + 3H_2O(l)$$

4.74 How many grams of Al will react with 3.00 moles of O_2?

$$4Al(s) + 3O_2(g) \longrightarrow 2Al_2O_3(s)$$

4.75 Calculate the amount of $CrCl_3$ that could be produced from 50.0 g Cr_2O_3 according to the equation

$$Cr_2O_3(s) + 3CCl_4(l) \longrightarrow 2CrCl_3(s) + 3COCl_2(aq)$$

4.76 A 3.5-g sample of water reacts with PCl_3 according to the following equation:

$$3H_2O(l) + PCl_3(g) \longrightarrow H_3PO_3(aq) + 3HCl(aq)$$

How many grams of H_3PO_3 are produced?

4.77 For the reaction

$$N_2(g) + H_2(g) \longrightarrow NH_3(g)$$

 a. Balance the equation.
 b. How many moles of H_2 would react with 1 mol of N_2?
 c. How many moles of product would form from 1 mol of N_2?
 d. If 14.0 g of N_2 were initially present, calculate the number of moles of H_2 required to react with all of the N_2.
 e. For conditions outlined in part (d), how many grams of product would form?

4.78 Aspirin (acetylsalicylic acid) may be formed from salicylic acid and acetic acid as follows:

$$C_7H_6O_3(aq) + CH_3COOH(aq) \longrightarrow C_9H_8O_4(s) + H_2O(l)$$
Salicylic   Acetic   Aspirin
acid   acid

 a. Is this equation balanced? If not, complete the balancing.
 b. How many moles of aspirin may be produced from 1.00×10^2 mol salicylic acid?
 c. How many grams of aspirin may be produced from 1.00×10^2 mol salicylic acid?
 d. How many grams of acetic acid would be required to react completely with the 1.00×10^2 mol salicylic acid?

4.79 The proteins in our bodies are composed of molecules called amino acids. One amino acid is methionine; its molecular formula is $C_5H_{11}NO_2S$. Calculate:
 a. the formula weight of methionine
 b. the number of oxygen atoms in a mole of this compound
 c. the mass of oxygen in a mole of the compound
 d. the mass of oxygen in 50.0 g of the compound

4.80 Triglycerides (Chapters 17 and 23) are used in biochemical systems to store energy; they can be formed from glycerol and fatty acids. The molecular formula of glycerol is $C_3H_8O_3$. Calculate:
 a. the formula weight of glycerol
 b. the number of oxygen atoms in a mole of this compound
 c. the mass of oxygen in a mole of the compound
 d. the mass of oxygen in 50.0 g of the compound

4.81 Joseph Priestley discovered oxygen in the eighteenth century by using heat to decompose mercury(II) oxide:

$$2HgO(s) \xrightarrow{\Delta} 2Hg(l) + O_2(g)$$

How much oxygen is produced from 1.00×10^2 g HgO?

4.82 Dinitrogen monoxide (also known as nitrous oxide and used as an anesthetic) can be made by heating ammonium nitrate:

$$NH_4NO_3(s) \longrightarrow N_2O(g) + 2H_2O(g)$$

How much dinitrogen monoxide can be made from 1.00×10^2 g of ammonium nitrate?

4.83 The burning of acetylene (C_2H_2) in oxygen is the reaction in the oxyacetylene torch. How much CO_2 is produced by burning 20.0 kg of acetylene in an excess of O_2? The unbalanced equation is

$$C_2H_2(g) + O_2(g) \longrightarrow CO_2(g) + H_2O(g)$$

4.84 The reaction of calcium hydride with water can be used to prepare hydrogen gas:

$$CaH_2(s) + 2H_2O(l) \longrightarrow Ca(OH)_2(aq) + 2H_2(g)$$

How many grams of hydrogen gas are produced in the reaction of 1.00×10^2 g calcium hydride with water?

4.85 Various members of a class of compounds, alkenes (Chapter 11), react with hydrogen to produce a corresponding alkane (Chapter 10). Termed hydrogenation, this type of reaction is used to produce products such as margarine. A typical hydrogenation reaction is

$$C_{10}H_{20}(l) + H_2(g) \longrightarrow C_{10}H_{22}(s)$$

Decene   Decane

How much decane can be produced in a reaction of excess decene with 1.00 g hydrogen?

4.86 A Human Perspective: Alcohol Consumption and the Breathalyzer Test (Chapter 12), describes the reaction between the dichromate ion and ethanol to produce acetic acid. How much acetic acid can be produced from a mixture containing excess of dichromate ion and 1.00×10^{-1} g of ethanol?

4.87 A rocket can be powered by the reaction between dinitrogen tetroxide and hydrazine:

$$N_2O_4(l) + 2N_2H_4(l) \longrightarrow 3N_2(g) + 4H_2O(g)$$

An engineer designed the rocket to hold 1.00 kg N_2O_4 and excess N_2H_4. How much N_2 would be produced according to the engineer's design?

4.88 A 4.00-g sample of Fe_3O_4 reacts with O_2 to produce Fe_2O_3:

$$4Fe_3O_4(s) + O_2(g) \longrightarrow 6Fe_2O_3(s)$$

Determine the number of grams of Fe_2O_3 produced.

4.89 If the actual yield of decane in Problem 4.85 is 65.4 g, what is the % yield?

4.90 If the actual yield of acetic acid in Problem 4.86 is 0.110 g, what is the % yield?

4.91 If the % yield of nitrogen gas in Problem 4.87 is 75.0%, what is the actual yield of nitrogen?

4.92 If the % yield of Fe_2O_3 in Problem 4.88 is 90.0%, what is the actual yield of Fe_2O_3?

CRITICAL THINKING PROBLEMS

1. Which of the following has fewer moles of carbon: 100 g of $CaCO_3$ or 0.5 mol of CCl_4?
2. Which of the following has fewer moles of carbon: 6.02×10^{22} molecules of C_2H_6 or 88 g of CO_2?
3. How many molecules are found in each of the following?
    a. 1.0 lb of sucrose, $C_{12}H_{22}O_{11}$ (table sugar)
    b. 1.57 kg of N_2O (anesthetic)
4. How many molecules are found in each of the following?
    a. 4×10^5 tons of SO_2 (produced by the 1980 eruption of the Mount St. Helens volcano)
    b. 25.0 lb of SiO_2 (major constituent of sand)

5 States of Matter
Gases, Liquids, and Solids

Learning Goals

1 Describe the major points of the kinetic molecular theory of gases.

2 Explain the relationship between the kinetic molecular theory and the physical properties of measurable quantities of gases.

3 Describe the behavior of gases expressed by the gas laws: Boyle's law, Charles's law, combined gas law, Avogadro's law, the ideal gas law, and Dalton's law.

4 Use gas law equations to calculate conditions and changes in conditions of gases.

5 Describe properties of the liquid state in terms of the properties of the individual molecules that comprise the liquid.

6 Describe the processes of melting, boiling, evaporation, and condensation.

7 Describe the dipolar attractions known collectively as van der Waals forces.

8 Describe hydrogen bonding and its relationship to boiling and melting temperatures.

9 Relate the properties of the various classes of solids (ionic, covalent, molecular, and metallic) to the structure of these solids.

Outline

Introduction
Chemistry Connection:
The Demise of the Hindenburg

5.1 The Gaseous State
An Environmental Perspective:
The Greenhouse Effect and Global Climate Change

5.2 The Liquid State
A Medical Perspective:
Blood Gases and Respiration

5.3 The Solid State

Volcanic activity is a dramatic example of interconversion among the states of matter.

Introduction

We have learned that the major differences between solids, liquids, and gases are due to the relationships among particles. These relationships include:

- the average distance of separation of particles in each state,
- the kinds of interactions between the particles, and
- the degree of organization of particles.

We have already discovered that the solid state is the most organized, with particles close together, allowing significant interactions among the particles. This results in high melting and boiling points for solid substances. Large amounts of energy are needed to overcome the attractive forces and disrupt the orderly structure.

Section 1.2 introduces the properties of the three states of matter.

Substances that are gases, on the other hand, are disordered, with particles widely separated and weak interactions between particles. Their melting and boiling points are relatively low. Gases at room temperature must be cooled a great deal for them to liquefy or solidify. For example, the melting and boiling points of N_2 are $-210°C$ and $-196°C$, respectively.

Chemistry Connection

The Demise of the Hindenburg

One of the largest and most luxurious airships of the 1930s, the Hindenburg, completed thirty-six transatlantic flights within a year after its construction. It was the flagship of a new era of air travel. But, on May 6, 1937, while making a landing approach near Lakehurst, New Jersey, the hydrogen-filled airship exploded and burst into flames. In this tragedy, thirty-seven of the ninety-six passengers were killed and many others were injured.

We may never know the exact cause. Many believe that the massive ship (it was more than 800 feet long) struck an overhead power line. Others speculate that lightning ignited the hydrogen and some believe that sabotage may have been involved.

In retrospect, such an accident was inevitable. Hydrogen gas is very reactive, it combines with oxygen readily and rapidly, and this reaction liberates a large amount of energy. An explosion is the result of rapid, energy-releasing reactions.

Why was hydrogen chosen? Hydrogen is the lightest element. One mole of hydrogen has a mass of 2 grams. Hydrogen can be easily prepared in pure form, an essential requirement; more than seven million cubic feet of hydrogen were needed for each airship. Hydrogen has a low density; hence it provides great lift. The lifting power of a gas is based on the difference in density of the gas and the surrounding air (air is composed of gases with much greater molar masses; N_2 is 28 g and O_2 is 32 g). Engineers believed that the hydrogen would be safe when enclosed by the hull of the airship.

Hindenburg

Today, airships are filled with helium (its molar mass is 4 g) and are used principally for advertising and television. A Goodyear blimp can be seen hovering over almost every significant outdoor sporting event.

In this chapter we will study the relationships that predict the behavior of gases in a wide variety of applications from airships to pressurized oxygen for respiration therapy.

TABLE 5.1	A Comparison of Physical Properties of Gases, Liquids, and Solids		
	Gas	**Liquid**	**Solid**
Volume and Shape	Expands to fill the volume of its container; consequently, it takes the shape of the container	Has a fixed volume at a given mass and temperature; volume principally dependent on its mass and secondarily on temperature; it assumes the shape of its container	Has a fixed volume; volume principally dependent on its mass and secondarily on temperature; it has a definite shape
Density	Low (typically ~10^{-3} g/mL)	High (typically ~1g/mL)	High (typically 1–10 g/mL)
Compressibility	High	Very low	Virtually incompressible
Particle Motion	Virtually free	Molecules or atoms "slide" past each other	Vibrate about a fixed position
Intermolecular Distance	Very large	Molecules or atoms are close to each other	Molecules, ions, or atoms are close to each other

Liquids are intermediate in character. The molecules of a liquid are close together, like those of solids. However, the molecules of a liquid are disordered, like those of a gas.

Changes in state are described as physical changes. When a substance undergoes a change in state, many of its physical properties change. For example, when ice forms from liquid water, changes occur in density and hardness, but it is still water. Table 5.1 summarizes the important differences in physical properties among gases, liquids, and solids.

5.1 The Gaseous State

Ideal Gas Concept

An **ideal gas** is simply a model of the way that particles (molecules or atoms) behave at the microscopic level. The behavior of the individual particles can be inferred from the measurable behavior of samples of real gases. We can easily measure temperature, volume, pressure, and quantity (mass) of real gases. Similarly, when we systematically change one of these properties, we can determine the effect on each of the others. For example, putting more molecules in a balloon (the act of blowing up a balloon) causes its volume to increase in a predictable way. In fact, careful measurements show a direct proportionality between the quantity of molecules and the volume of the balloon, an observation made by Amadeo Avogadro more than 200 years ago.

We owe a great deal of credit to the efforts of scientists Boyle, Charles, Avogadro, Dalton, and Gay-Lussac, whose careful work elucidated the relationships among the gas properties. Their efforts are summarized in the ideal gas law and are the subject of the first section of this chapter.

Measurement of Gases

The most important gas laws (Boyle's law, Charles's law, Avogadro's law, Dalton's law, and the ideal gas law) involve the relationships between pressure (P), volume (V), temperature (T), and number of moles (n) of gas. We are already familiar with the measurement of temperature, volume, and mass (allowing the calculation of number of moles) from our laboratory experience. Measurement of pressure is perhaps not as obvious.

76 cm

Atmospheric pressure

Figure 5.1

A mercury barometer of the type invented by Torricelli. The height of the column of mercury (*h*) is a function of the magnitude of the surrounding atmospheric pressure. The mercury in the tube is supported by atmospheric pressure.

Gas pressure is a result of the force exerted by the collision of particles with the walls of the container. **Pressure** is force per unit area. The pressure of a gas may be measured with a **barometer,** invented by Evangelista Torricelli in the mid-1600s. The most common type of barometer is the mercury barometer depicted in Figure 5.1. A tube, sealed at one end, is filled with mercury and inverted in a dish of mercury. The pressure of the atmosphere pushing down on the mercury surface in the dish supports the column of mercury. The height of the column is proportional to the atmospheric pressure. The tube can be calibrated to give a numerical reading in millimeters, centimeters, or inches of mercury. A commonly used unit of measurement is the atmosphere (atm). One standard atmosphere (1 atm) of pressure is equivalent to a height of mercury that is equal to

　　760 mm Hg (millimeters of mercury)

　　76.0 cm Hg (centimeters of mercury)

　　1 mm of Hg is also = 1 torr, in honor of Torricelli.

The English system equivalent is a pressure of 14.7 lb/in.^{2} (pounds per square inch) or 29.9 in. Hg (inches of mercury). A recommended, yet less frequently used, systematic unit is the pascal (or kilopascal), named in honor of Blaise Pascal, a seventeenth-century French mathematician and scientist:

$$1 \text{ atm} = 1.01 \times 10^5 \text{ Pa (pascal)} = 101 \text{ kPa (kilopascal)}$$

Atmospheric pressure is due to the cumulative force of the air molecules (N_2 and O_2, for the most part) that are attracted to the earth's surface by gravity.

Question 5.1

Express each of the following in units of atmospheres:
a. 725 mm Hg
b. 29.0 cm Hg
c. 555 torr

Question 5.2

Express each of the following in units of atmospheres:
a. 10.0 torr
b. 61.0 cm Hg
c. 275 mm Hg

Kinetic Molecular Theory of Gases

LEARNING GOAL

1 Describe the major points of the kinetic molecular theory of gases.

The kinetic molecular theory of gases provides a reasonable explanation of the behavior of gases that we have studied in this chapter. The macroscopic properties result from the action of the individual molecules comprising the gas.

　　The **kinetic molecular theory** can be summarized as follows:

1. Gases are made up of small atoms or molecules that are in constant, random motion.
2. The distance of separation among these atoms or molecules is very large in comparison to the size of the individual atoms or molecules. In other words, a gas is mostly empty space.
3. All of the atoms and molecules behave independently. No attractive or repulsive forces exist between atoms or molecules in a gas.
4. Atoms and molecules collide with each other and with the walls of the container without *losing* energy. The energy is *transferred* from one atom or molecule to another.
5. The average kinetic energy of the atoms or molecules increases or decreases in proportion to absolute temperature.

Properties of Gases and the Kinetic Molecular Theory

We know that gases are easily *compressible*. The reason is that a gas is mostly empty space, providing space for the particles to be pushed closer together.

Gases will *expand* to fill any available volume because they move freely with sufficient energy to overcome their attractive forces.

Gases have a *low density*. Density is defined as mass per volume. Because gases are mostly empty space, they have a low mass per volume.

Gases readily *diffuse* through each other simply because they are in continuous motion and paths are readily available because of the large space between adjacent atoms or molecules. Light molecules diffuse rapidly; heavier molecules diffuse more slowly (Figure 5.2).

Gases exert *pressure* on their containers. Pressure is a force per unit area resulting from collisions of gas particles with the walls of their container.

Gases behave most *ideally at low pressures and high temperatures*. At low pressures, the average distance of separation among atoms or molecules is greatest, minimizing interactive forces. At high temperatures, the atoms and molecules are in rapid motion and are able to overcome interactive forces more easily.

> **2** **LEARNING GOAL**
> Explain the relationship between the kinetic molecular theory and the physical properties of measurable quantities of gases.

Kinetic energy (K.E.) is equal to $1/2\ mv^2$, in which m = mass and v = velocity. Thus increased velocity at higher temperature correlates with an increase in kinetic energy.

Boyle's Law

The Irish scientist Robert Boyle found that the volume of a gas varies *inversely* with the pressure exerted by the gas if the number of moles and temperature of gas are held constant. This relationship is known as **Boyle's law.**

Mathematically, the *product* of pressure (P) and volume (V) is a constant:

$$PV = k_1$$

This relationship is illustrated in Figure 5.3.

Boyle's law is often used to calculate the volume resulting from a pressure change or vice versa. We consider

$$P_iV_i = k_1$$

the *initial* condition and

$$P_fV_f = k_1$$

> **3** **LEARNING GOAL**
> Describe the behavior of gases expressed by the gas laws: Boyle's law, Charles's law, combined gas law, Avogadro's law, the ideal gas law, and Dalton's law.

(a)

(b)

Figure 5.2

Gaseous diffusion. (a) Ammonia (17.0 g/mol) and hydrogen chloride (36.5 g/mol) are introduced into the ends of a glass tube containing indicating paper. Red indicates the presence of hydrogen chloride and blue indicates ammonia. (b) Note that ammonia has diffused much farther than hydrogen chloride in the same amount of time. This is a verification of the kinetic molecular theory. Light molecules move faster than heavier molecules at a specified temperature.

Figure 5.3
An illustration of Boyle's law. Note the inverse relationship of pressure and volume.

A Review of Mathematics

the final condition. Because *PV*, initial or final, is constant and is equal to k_1,

$$P_iV_i = P_fV_f$$

Consider a gas occupying a volume of 10.0 L at 1.00 atm of pressure. The product, $PV = (10.0 \text{ L})(1.00 \text{ atm})$, is a constant, k_1. Doubling the pressure, to 2.0 atm, decreases the volume to 5.0 L:

$$(2.0 \text{ atm})(V_x) = (10.0 \text{ L})(1.00 \text{ atm})$$
$$V_x = 5.0 \text{ L}$$

Tripling the pressure decreases the volume by a factor of 3:

$$(3.0 \text{ atm})(V_x) = (10.0 \text{ L})(1.00 \text{ atm})$$
$$V_x = 3.3 \text{ L}$$

E X A M P L E 5.1   **Calculating a Final Pressure**

4 **LEARNING GOAL**
Use gas law equations to calculate conditions and changes in conditions of gases.

A sample of oxygen, at 25°C, occupies a volume of 5.00×10^2 mL at 1.50 atm pressure. What pressure must be applied to compress the gas to a volume of 1.50×10^2 mL, with no temperature change?

Solution

Step 1. Boyle's law applies directly, because there is no change in temperature or number of moles (no gas enters or leaves the container).

Step 2. Begin by identifying each term in the Boyle's law expression:

$$P_i = 1.50 \text{ atm}$$
$$V_i = 5.00 \times 10^2 \text{ mL}$$
$$V_f = 1.50 \times 10^2 \text{ mL}$$

Continued—

EXAMPLE 5.1 —Continued

Step 3. The Boyle's law expression is:

$$P_iV_i = P_fV_f$$

Step 4. Solving for P_f:

$$P_f = \frac{P_iV_i}{V_f}$$

 A Review of Mathematics

Step 5. Substituting:

$$P_f = \frac{(1.50 \text{ atm})(5.00 \times 10^2 \text{ mL})}{1.50 \times 10^2 \text{ mL}}$$

$$= 5.00 \text{ atm}$$

Helpful Hint:

The calculation can be done with any volume units. It is important only that the units be the *same* on both sides of the equation.

Practice Problem 5.1

Complete the following table:

Sample Number	Initial Pressure (atm)	Final Pressure (atm)	Initial Volume (L)	Final Volume (L)
1	X	5.0	1.0	7.5
2	5.0	X	1.0	0.20
3	1.0	0.50	X	0.30
4	1.0	2.0	0.75	X

For Further Practice: Questions 5.31 and 5.32.

Charles's Law

Jacques Charles, a French scientist, studied the relationship between gas volume and temperature. This relationship, **Charles's law,** states that the volume of a gas varies *directly* with the absolute temperature (K) if pressure and number of moles of gas are constant.

Mathematically, the *ratio* of volume (V) and temperature (T) is a constant:

$$\frac{V}{T} = k_2$$

In a way analogous to Boyle's law, we may establish a set of initial conditions,

$$\frac{V_i}{T_i} = k_2$$

and final conditions,

$$\frac{V_f}{T_f} = k_2$$

Because k_2 is a constant, we may equate them, resulting in

$$\frac{V_i}{T_i} = \frac{V_f}{T_f}$$

and use this expression to solve some practical problems.

LEARNING GOAL

3 Describe the behavior of gases expressed by the gas laws: Boyle's law, Charles's law, combined gas law, Avogadro's law, the ideal gas law, and Dalton's law.

Temperature is a measure of the energy of molecular motion. The Kelvin scale is *absolute*, that is, directly proportional to molecular motion. Celsius and Fahrenheit are simply numerical scales based on the melting and boiling points of water. It is for this reason that Kelvin is used for energy-dependent relationships such as the gas laws.

Figure 5.4
An illustration of Charles's law. Note the direct relationship between volume and temperature.

A Review of Mathematics

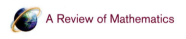

Animation
Charles's Law

Consider a gas occupying a volume of 10.0 L at 273 K. The ratio V/T is a constant, k_2. Doubling the temperature, to 546 K, increases the volume to 20.0 L as shown here:

$$\frac{10.0 \text{ L}}{273 \text{ K}} = \frac{V_f}{546 \text{ K}}$$

$$V_f = 20.0 \text{ L}$$

Tripling the temperature, to 819 K, increases the volume by a factor of 3:

$$\frac{10.0 \text{ L}}{273 \text{ K}} = \frac{V_f}{819 \text{ K}}$$

$$V_f = 30.0 \text{ L}$$

These relationships are illustrated in Figure 5.4.

EXAMPLE 5.2   **Calculating a Final Volume**

4

LEARNING GOAL
Use gas law equations to calculate conditions and changes in conditions of gases.

A balloon filled with helium has a volume of 4.0×10^3 L at 25°C. What volume will the balloon occupy at 50°C if the pressure surrounding the balloon remains constant?

Solution

Step 1. Remember, the temperature must be converted to Kelvin before Charles's law is applied:

$$T_i = 25°C + 273 = 298 \text{ K}$$
$$T_f = 50°C + 273 = 323 \text{ K}$$
$$V_i = 4.0 \times 10^3 \text{ L}$$
$$V_f = ?$$

Continued—

EXAMPLE 5.2 —*Continued*

Step 2. Using the Charles's law expression relating initial and final conditions:

$$\frac{V_i}{T_i} = \frac{V_f}{T_f}$$

Step 3. Rearrange and solve for V_f

$$V_f = \frac{(V_i)(T_f)}{T_i}$$

Step 4. Substituting our data, we get

$$V_f = \frac{(V_i)(T_f)}{T_i} = \frac{(4.0 \times 10^3\ L)(323\ K)}{298\ K} = 4.3 \times 10^3\ L$$

Practice Problem 5.2

A sample of nitrogen gas has a volume of 3.00 L at 25°C. What volume will it occupy at each of the following temperatures if the pressure and number of moles are constant?
a. 100°C
b. 150°F
c. 273 K
d. 546 K
e. 0°C
f. 373 K

For Further Practice: Questions 5.39 and 5.40.

Figure 5.5
Charles's law predicts that the volume of air in the balloon will increase when heated. We assume that the volume of the balloon is fixed; consequently, some air will be pushed out. The air remaining in the balloon is less dense (same volume, less mass) and the balloon will rise. When the heater is turned off the air cools, the density increases, and the balloon returns to earth.

The behavior of a hot-air balloon is a commonplace consequence of Charles's law. The balloon rises because air expands when heated (Figure 5.5). The volume of the balloon is fixed because the balloon is made of an inelastic material; as a result, when the air expands some of the air must be forced out. Hence the density of the remaining air is less (less mass contained in the same volume), and the balloon rises. Turning down the heat reverses the process, and the balloon descends.

Combined Gas Law

Boyle's law describes the inverse proportional relationship between volume and pressure; Charles's law shows the direct proportional relationship between volume and temperature. Often, a sample of gas (a fixed number of moles of gas) undergoes change involving volume, pressure, and temperature simultaneously. It would be useful to have one equation that describes such processes.

The **combined gas law** is such an equation. It can be derived from Boyle's law and Charles's law and takes the form:

$$\frac{P_i V_i}{T_i} = \frac{P_f V_f}{T_f}$$

Let's look at two examples that use this expression.

Animation
Interactive Gas Law

LEARNING GOAL
3 Describe the behavior of gases expressed by the gas laws: Boyle's law, Charles's law, combined gas law, Avogadro's law, the ideal gas law, and Dalton's law.

EXAMPLE 5.3   **Using the Combined Gas Law**

LEARNING GOAL
4 Use gas law equations to calculate conditions and changes in conditions of gases.

A Review of Mathematics

Calculate the volume of N_2 that results when 0.100 L of the gas is heated from 300.0 K to 350.0 K at 1.00 atm.

Solution

Step 1. Summarize the data:

$$P_i = 1.00 \text{ atm} \qquad P_f = 1.00 \text{ atm}$$
$$V_i = 0.100 \text{ L} \qquad V_f = ? \text{ L}$$
$$T_i = 300.0 \text{ K} \qquad T_f = 350.0 \text{ K}$$

Step 2. The combined gas law expression is:

$$\frac{P_i V_i}{T_i} = \frac{P_f V_f}{T_f}$$

Step 3. Rearrange:

$$P_f V_f T_i = P_i V_i T_f$$

and

$$V_f = \frac{P_i V_i T_f}{P_f T_i}$$

Step 4. Because $P_i = P_f$

$$V_f = \frac{V_i T_f}{T_i}$$

Step 5. Substituting gives

$$V_f = \frac{(0.100 \text{ L})(350.0 \text{ K})}{300.0 \text{ K}}$$
$$= 0.117 \text{ L}$$

Helpful Hint:

In this case, because the pressure is constant, the combined gas law reduces to Charles's law.

 Practice Problem 5.3

Hydrogen sulfide, H_2S, has the characteristic odor of rotten eggs. If a sample of H_2S gas at 760.0 torr and 25.0°C in a 2.00-L container is allowed to expand into a 10.0-L container at 25.0°C, what is the pressure in the 10.0-L container?

For Further Practice: Questions 5.45 and 5.46.

EXAMPLE 5.4   **Using the Combined Gas Law**

A sample of helium gas has a volume of 1.27 L at 149 K and 5.00 atm. When the gas is compressed to 0.320 L at 50.0 atm, the temperature increases markedly. What is the final temperature?

Continued—

EXAMPLE 5.4 —*Continued*

Solution

Step 1. Summarize the data:

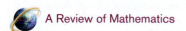
A Review of Mathematics

$$P_i = 5.00 \text{ atm} \qquad P_f = 50.0 \text{ atm}$$
$$V_i = 1.27 \text{ L} \qquad V_f = 0.320 \text{ L}$$
$$T_i = 149 \text{ K} \qquad T_f = ? \text{ K}$$

Step 2. The combined gas law expression is

$$\frac{P_i V_i}{T_i} = \frac{P_f V_f}{T_f}$$

Step 3. Rearrange:

$$P_f V_f T_i = P_i V_i T_f$$

and

$$T_f = \frac{P_f V_f T_i}{P_i V_i}$$

Step 4. Substituting yields

$$T_f = \frac{(50.0 \text{ atm})(0.320 \text{ L})(149 \text{ K})}{(5.00 \text{ atm})(1.27 \text{ L})}$$
$$= 375 \text{ K}$$

Practice Problem 5.4

Cyclopropane, C_3H_6, is used as a general anesthetic. If a sample of cyclopropane stored in a 2.00-L container at 10.0 atm and 25.0°C is transferred to a 5.00-L container at 5.00 atm, what is the resulting temperature?

For Further Practice: Questions 5.47 and 5.48.

Avogadro's Law

The relationship between the volume and number of moles of a gas at constant temperature and pressure is known as **Avogadro's law.** It states that equal volumes of any ideal gas contain the same number of moles if measured under the same conditions of temperature and pressure.

Mathematically, the *ratio* of volume (V) to number of moles (n) is a constant:

$$\frac{V}{n} = k_3$$

Consider 1 mol of gas occupying a volume of 10.0 L; using logic similar to the application of Boyle's and Charles's laws, 2 mol of the gas would occupy 20.0 L, 3 mol would occupy 30.0 L, and so forth. As we have done with the previous laws, we can formulate a useful expression relating initial and final conditions:

$$\frac{V_i}{n_i} = \frac{V_f}{n_f}$$

LEARNING GOAL

3 Describe the behavior of gases expressed by the gas laws: Boyle's law, Charles's law, combined gas law, Avogadro's law, the ideal gas law, and Dalton's law.

EXAMPLE 5.5	Using Avogadro's Law

LEARNING GOAL

4 Use gas law equations to calculate conditions and changes in conditions of gases.

If 5.50 mol of CO occupy 20.6 L, how many liters will 16.5 mol of CO occupy at the same temperature and pressure?

Solution

Step 1. The quantities moles and volume are related through Avogadro's law. Summarizing the data:

$$V_i = 20.6 \text{ L} \qquad V_f = ? \text{ L}$$
$$n_i = 5.50 \text{ mol} \qquad n_f = 16.5 \text{ mol}$$

Step 2. Using the mathematical expression for Avogadro's law:

$$\frac{V_i}{n_i} = \frac{V_f}{n_f}$$

Step 3. Rearranging:

$$V_f = \frac{V_i n_f}{n_i}$$

Step 4. Substitution yields:

$$V_f = \frac{(20.6 \text{ L})(16.5 \text{ mol})}{(5.50 \text{ mol})}$$
$$= 61.8 \text{ L of CO}$$

Practice Problem 5.5

a. 1.00 mole of hydrogen gas occupies 22.4 L. How many moles of hydrogen are needed to fill a 100.0 L container at the same pressure and temperature?

b. How many moles of hydrogen are needed to triple the volume occupied by 0.25 mol of hydrogen, assuming no changes in pressure or temperature?

For Further Practice: Questions 5.51 and 5.52.

Molar Volume of a Gas

The volume occupied by *1 mol* of any gas is referred to as its **molar volume.** At **standard temperature and pressure (STP)** the molar volume of any gas is 22.4 L. STP conditions are defined as follows:

$$T = 273 \text{ K (or } 0°\text{C)}$$
$$P = 1 \text{ atm}$$

Thus, 1 mol of N_2, O_2, H_2, or He all occupy the *same volume, 22.4 L,* at STP.

Gas Densities

It is also possible to compute the density of various gases at STP. If we recall that density is the mass/unit volume,

$$d = \frac{m}{V}$$

and that 1 mol of helium weighs 4.00 g,

$$d_{He} = \frac{4.00 \text{ g}}{22.4 \text{ L}} = 0.178 \text{ g/L at STP}$$

or, because 1 mol of nitrogen weighs 28.0 g, then

$$d_{N_2} = \frac{28.0 \text{ g}}{22.4 \text{ L}} = 1.25 \text{ g/L at STP}$$

The large difference in gas densities of helium and nitrogen (which makes up about 80% of the air) accounts for the lifting power of helium. A balloon filled with helium will rise through a predominantly nitrogen atmosphere because its gas density is less than 15% of the density of the surrounding atmosphere:

$$\frac{d_{He}}{d_{N_2}} \times 100\% = \% \text{ density}$$

$$\frac{0.178 \text{ g/L}}{1.25 \text{ g/L}} \times 100\% = 14.2\%$$

> Heating a gas, such as air, will decrease its density and have a lifting effect as well.

The Ideal Gas Law

Boyle's law (relating volume and pressure), Charles's law (relating volume and temperature), and Avogadro's law (relating volume to the number of moles) may be combined into a single expression relating all four terms. This expression is the **ideal gas law:**

$$PV = nRT$$

in which R, based on k_1, k_2, and k_3 (Boyle's, Charles's, and Avogadro's law constants), is a constant and is referred to as the *ideal gas constant:*

$$R = 0.0821 \text{ L-atm K}^{-1} \text{ mol}^{-1}$$

if the units

atmospheres, for P

liters, for V

number of moles, for n

and

Kelvin for T

are used.

Consider some examples of the application of the ideal gas equation.

 LEARNING GOAL
3 Describe the behavior of gases expressed by the gas laws: Boyle's law, Charles's law, combined gas law, Avogadro's law, the ideal gas law, and Dalton's law.

> Remember that 0.0821 L-atm/K-mol is identical to 0.0821 L-atm K^{-1} mol^{-1}.

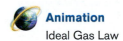 **Animation**
Ideal Gas Law

Calculating a Molar Volume   E X A M P L E 5.6

Demonstrate that the molar volume of oxygen gas at STP is 22.4 L.

 LEARNING GOAL
4 Use gas law equations to calculate conditions and changes in conditions of gases.

Solution

Step 1. The ideal gas expression is:

$$PV = nRT$$

Continued—

EXAMPLE 5.6 —*Continued*

Step 2. Rearrange and solve for V:

$$V = \frac{nRT}{P}$$

Step 3. At standard temperature and pressure,

$$T = 273 \text{ K}$$
$$P = 1.00 \text{ atm}$$

and the other terms are

$$n = 1.00 \text{ mol}$$
$$R = 0.0821 \text{ L-atm K}^{-1} \text{ mol}^{-1}$$

Step 4. Substitute and solve:

$$V = \frac{(1.00 \text{ mol})(0.0821 \text{ L-atm K}^{-1} \text{ mol}^{-1})(273 \text{ K})}{(1.00 \text{ atm})}$$

$$= 22.4 \text{ L}$$

Practice Problem 5.6

Demonstrate that the molar volume of helium (or any other ideal gas) is also 22.4 L.

For Further Practice: Questions 5.55 and 5.56.

EXAMPLE 5.7 | **Calculating the Number of Moles of a Gas**

LEARNING GOAL

4 Use gas law equations to calculate conditions and changes in conditions of gases.

Calculate the number of moles of helium in a 1.00-L balloon at 27°C and 1.00 atm of pressure.

Solution

Step 1. The ideal gas expression is:

$$PV = nRT$$

Step 2. Rearrange and solve for n:

$$n = \frac{PV}{RT}$$

Step 3. The data are:

$$P = 1.00 \text{ atm}$$
$$V = 1.00 \text{ L}$$
$$T = 27°C + 273 = 3.00 \times 10^2 \text{ K}$$
$$R = 0.0821 \text{ L-atm K}^{-1} \text{ mol}^{-1}$$

Step 4. Substitute and solve:

$$n = \frac{(1.00 \text{ atm})(1.00 \text{ L})}{(0.0821 \text{ L-atm K}^{-1} \text{ mol}^{-1})(3.00 \times 10^2 \text{ K})}$$

$$n = 0.0406 \text{ or } 4.06 \times 10^{-2} \text{ mol}$$

Continued—

EXAMPLE 5.7   —Continued

Practice Problem 5.7

How many moles of N_2 gas will occupy a 5.00-L container at standard temperature and pressure?

For Further Practice: Questions 5.57 and 5.61.

Converting Mass to Volume

EXAMPLE 5.8

Oxygen used in hospitals and laboratories is often obtained from cylinders containing liquefied oxygen. If a cylinder contains 1.00×10^2 kg of liquid oxygen, how many liters of oxygen can be produced at 1.00 atm of pressure at room temperature (20.0°C)?

4 LEARNING GOAL
Use gas law equations to calculate conditions and changes in conditions of gases.

Solution

Step 1. The ideal gas expression is:

$$PV = nRT$$

Step 2. Rearrange and solve for V:

$$V = \frac{nRT}{P}$$

Step 3. Using conversion factors, we obtain

$$n_{O_2} = 1.00 \times 10^2 \ \cancel{kg \ O_2} \times \frac{10^3 \ \cancel{g \ O_2}}{1 \ \cancel{kg \ O_2}} \times \frac{1 \ mol \ O_2}{32.0 \ \cancel{g \ O_2}}$$

and

$$n_{O_2} = 3.13 \times 10^3 \ mol \ O_2$$

Step 4. Convert °C to K:

$$T = 20.0°C + 273 = 293 \ K$$

and

$$P = 1.00 \ atm$$

Step 5. Substitute and solve:

$$V = \frac{(3.13 \times 10^3 \ mol)(0.0821 \ L\text{-atm} \ K^{-1} \ mol^{-1})(293 \ K)}{1.00 \ atm}$$

$$= 7.53 \times 10^4 \ L$$

Practice Problem 5.8

What volume is occupied by 10.0 g N_2 at 30.0°C and a pressure of 750 torr?

For Further Practice: Questions 5.59 and 5.60.

An Environmental *Perspective*

The Greenhouse Effect and Global Climate Change

A greenhouse is a bright, warm, and humid environment for growing plants, vegetables, and flowers even during the cold winter months. It functions as a closed system in which the concentration of water vapor is elevated and visible light streams through the windows; this creates an ideal climate for plant growth.

Some of the visible light is absorbed by plants and soil in the greenhouse and radiated as infrared radiation. This radiated energy is blocked by the glass or absorbed by water vapor and carbon dioxide (CO_2). This trapped energy warms the greenhouse and is a form of solar heating: light energy is converted to heat energy.

On a global scale, the same process takes place. Although more than half of the sunlight that strikes the earth's surface is reflected back into space, the fraction of light that is absorbed produces sufficient heat to sustain life. How does this happen? Greenhouse gases, such as CO_2, trap energy radiated from the earth's surface and store it in the atmosphere. This moderates our climate. The earth's surface would be much colder and more inhospitable if the atmosphere was not able to capture some reasonable amount of solar energy.

Can we have too much of a good thing? It appears so. Since 1900 the atmospheric concentration of CO_2 has increased from 296 parts per million (ppm) to over 350 ppm (approximately 17% increase). The energy demands of technological and population growth have caused massive increases in the combustion of organic matter and carbon-based fuels (coal, oil, and natural gas), adding over 50 billion tons of CO_2 to that already present in the atmosphere. Photosynthesis naturally removes CO_2 from the atmosphere. However, the removal of forestland to create living space and cropland has decreased the amount of vegetation available to consume atmospheric CO_2 through photosynthesis. The rapid destruction of the Amazon rain forest is just the latest of many examples.

If our greenhouse model is correct, an increase in CO_2 levels should produce global warming, perhaps changing our climate in unforeseen and undesirable ways.

For Further Understanding

What steps might be taken to decrease levels of CO_2 in the atmosphere over time?

In what ways might our climate and our lives change as a consequence of significant global warming?

(a) A greenhouse traps solar radiation as heat. (b) Our atmosphere also acts as a solar collector. Carbon dioxide, like the windows of a greenhouse, allows the visible light to enter and traps the heat.

Question 5.3

A 20.0-L gas cylinder contains 4.80 g H_2 at 25°C. What is the pressure of this gas?

Question 5.4

At what temperature will 2.00 moles of He fill a 2.00-L container at standard pressure?

LEARNING GOAL

3 Describe the behavior of gases expressed by the gas laws: Boyle's law, Charles's law, combined gas law, Avogadro's law, the ideal gas law, and Dalton's law.

Dalton's Law of Partial Pressures

Our discussion of gases so far has presumed that we are working with a single pure gas. A *mixture* of gases exerts a pressure that is the *sum* of the pressures that each gas would exert if it were present alone under the same conditions. This is known as **Dalton's law** of partial pressures.

Stated another way, the total pressure of a mixture of gases is the sum of the **partial pressures.** That is,

$$P_t = p_1 + p_2 + p_3 + \cdots$$

in which P_t = total pressure and $p_1, p_2, p_3, \ldots$, are the partial pressures of the component gases. For example, the total pressure of our atmosphere is equal to the sum of the pressures of N_2 and O_2 (the principal components of air):

$$P_{air} = p_{N_2} + p_{O_2}$$

Other gases, such as argon (Ar), carbon dioxide (CO_2), carbon monoxide (CO), and methane (CH_4) are present in the atmosphere at very low partial pressures. However, their presence may result in dramatic consequences; one such gas is carbon dioxide. Classified as a "greenhouse gas," it exerts a significant effect on our climate. Its role is described in An Environmental Perspective: The Greenhouse Effect and Global Climate Change.

> The ideal gas law applies to mixtures of gases as well as pure gases.

Ideal Gases Versus Real Gases

To this point we have assumed, in both theory and calculations, that all gases behave as ideal gases. However, in reality there is no such thing as an ideal gas. As we noted at the beginning of this section, the ideal gas is a model (a very useful one) that describes the behavior of individual atoms and molecules; this behavior translates to the collective properties of measurable quantities of these atoms and molecules. Limitations of the model arise from the fact that interactive forces, even between the widely spaced particles of gas, are not totally absent in any sample of gas.

Attractive forces are present in gases composed of polar molecules. Nonuniform charge distribution on polar molecules creates positive and negative regions, resulting in electrostatic attraction and deviation from ideality.

Calculations involving polar gases such as HF, NO, and SO_2 based on ideal gas equations (which presume no such interactions) are approximations. However, at low pressures, such approximations certainly provide useful information. Nonpolar molecules, on the other hand, are only weakly attracted to each other and behave much more ideally in the gas phase.

> *See Sections 3.5 and 5.2 for a discussion of interactions of polar molecules.*

5.2 The Liquid State

Molecules in the liquid state are close to one another. Attractive forces are large enough to keep the molecules together in contrast to gases, whose cohesive forces are so low that a gas expands to fill any volume. However, these attractive forces in a liquid are not large enough to restrict movement, as in solids. Let's look at the various properties of liquids in more detail.

LEARNING GOAL
5 Describe properties of the liquid state in terms of the properties of the individual molecules that comprise the liquid.

Compressibility

Liquids are practically incompressible. In fact, the molecules are so close to one an other that even the application of many atmospheres of pressure does not significantly decrease the volume. This makes liquids ideal for the transmission of force, as in the brake lines of an automobile. The force applied by the driver's foot on the brake pedal does not compress the brake fluid in the lines; rather, it transmits the force directly to the brake pads, and the friction between the brake pads and rotors (that are attached to the wheel) stops the car.

Viscosity

The **viscosity** of a liquid is a measure of its resistance to flow. Viscosity is a function of both the attractive forces between molecules and molecular geometry.

A Medical Perspective

Blood Gases and Respiration

Respiration must deliver oxygen to cells and the waste product, carbon dioxide, to the lungs to be exhaled. Dalton's law of partial pressures helps to explain the way in which this process occurs.

Gases (such as O_2 and CO_2) move from a region of higher partial pressure to one of lower partial pressure in an effort to establish an equilibrium. At the interface of the lung, the membrane barrier between the blood and the surrounding atmosphere, the following situation exists: Atmospheric O_2 partial pressure is high, and atmospheric CO_2 partial pressure is low. The reverse is true on the other side of the membrane (blood). Thus CO_2 is efficiently removed from the blood, and O_2 is efficiently moved into the bloodstream.

At the other end of the line, capillaries are distributed in close proximity to the cells that need to expel CO_2 and gain O_2. The partial pressure of CO_2 is high in these cells, and the partial pressure of O_2 is low, having been used up by the energy-harvesting reaction, the oxidation of glucose:

$$C_6H_{12}O_6 + 6O_2 \longrightarrow 6CO_2 + 6H_2O + \text{energy}$$

The O_2 diffuses into the cells (from a region of high to low partial pressure), and the CO_2 diffuses from the cells to the blood (again from a region of high to low partial pressure).

The net result is a continuous process proceeding according to Dalton's law. With each breath we take, oxygen is distributed to the cells and used to generate energy, and the waste product, CO_2, is expelled by the lungs.

For Further Understanding

Carbon dioxide and carbon monoxide can be toxic, but for different reasons. Use the Internet to research this topic and:

Explain why carbon dioxide is toxic.

Explain why carbon monoxide is toxic.

Molecules with complex structures, which do not "slide" smoothly past each other, and polar molecules, tend to have higher viscosity than less structurally complex, less polar liquids. Glycerol, which is used in a variety of skin treatments, has the structural formula:

It is quite viscous, owing to its polar nature and its significant intermolecular attractive forces. This is certainly desirable in a skin treatment because its viscosity keeps it on the area being treated. Gasoline, on the other hand, is much less viscous and readily flows through the gas lines of your auto; it is composed of nonpolar molecules.

Viscosity generally decreases with increasing temperature. The increased kinetic energy at higher temperatures overcomes some of the intermolecular attractive forces. The temperature effect is an important consideration in the design of products that must remain fluid at low temperatures, such as motor oils and transmission fluids found in automobiles.

Surface Tension

The **surface tension** of a liquid is a measure of the attractive forces exerted among molecules at the surface of a liquid. It is only the surface molecules that are not totally surrounded by other liquid molecules (the top of the molecule faces the

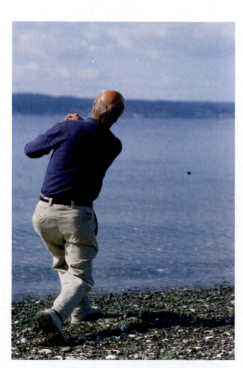

Skimming stones is possible due to the surface tension of water. Explain.

atmosphere). These surface molecules are surrounded and attracted by fewer liquid molecules than those below and to each side. Hence the net attractive forces on surface molecules pull them downward, into the body of the liquid. As a result, the surface molecules behave as a "skin" that covers the interior.

This increased surface force is responsible for the spherical shape of drops of liquid. Drops of water "beading" on a polished surface, such as a waxed automobile, illustrate this effect.

Because surface tension is related to the attractive forces exerted among molecules, surface tension generally decreases with an increase in temperature or a decrease in the polarity of molecules that make up the liquid.

Substances known as **surfactants** can be added to a liquid to decrease surface tension. Common surfactants include soaps and detergents that reduce water's surface tension; this promotes the interaction of water with grease and dirt, making it easier to remove.

Vapor Pressure of a Liquid

Evaporation, condensation, and the meaning of the term *boiling point* are all related to the concept of liquid vapor pressure. Consider the following example. A liquid, such as water, is placed in a sealed container. After a time the contents of the container are analyzed. Both liquid water and water vapor are found at room temperature, when we might expect water to be found only as a liquid. In this closed system, some of the liquid water was converted to a gas:

$$\text{energy} + H_2O(l) \longrightarrow H_2O(g)$$

How did this happen? The temperature is too low for conversion of a liquid to a gas by boiling. According to the kinetic theory, liquid molecules are in continuous motion, with their *average* kinetic energy directly proportional to the Kelvin temperature. The word *average* is the key. Although the average kinetic energy is too low to allow "average" molecules to escape from the liquid phase to the gas phase, there exists a range of molecules with different energies, some low and some high, that make up the "average" (Figure 5.6). Thus some of these high-energy molecules possess sufficient energy to escape from the bulk liquid.

At the same time a fraction of these gaseous molecules lose energy (perhaps by collision with the walls of the container) and return to the liquid state:

$$H_2O(g) \longrightarrow H_2O(l) + \text{energy}$$

The process of conversion of liquid to gas, at a temperature too low to boil, is **evaporation.** The reverse process, conversion of the gas to the liquid state, is **condensation.** After some time the rates of evaporation and condensation become *equal,* and this sets up a dynamic equilibrium between liquid and vapor states. The **vapor pressure of a liquid** is defined as the pressure exerted by the vapor *at equilibrium.*

$$H_2O(g) \rightleftharpoons H_2O(l)$$

The equilibrium process of evaporation and condensation of water is depicted in Figure 5.7.

The boiling point of a liquid is defined as the temperature at which the vapor pressure of the liquid becomes equal to the atmospheric pressure. The "normal" atmospheric pressure is 760 torr, or 1 atm, and the **normal boiling point** is the temperature at which the vapor pressure of the liquid is equal to 1 atm.

It follows from the definition that the boiling point of a liquid is not constant. It depends on the atmospheric pressure. At high altitudes, where the atmospheric pressure is low, the boiling point of a liquid, such as water, is lower than the normal boiling point (for water, 100°C). High atmospheric pressure increases the boiling point.

Animation
Surface Tension of Water

LEARNING GOAL
6 Describe the processes of melting, boiling, evaporation, and condensation.

Animation
Vapor Pressure

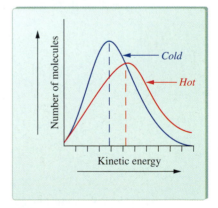

Figure 5.6
The temperature dependence of liquid vapor pressure is illustrated. The average molecular kinetic energy increases with temperature. Note that the average values are indicated by dashed lines. The small number of high-energy molecules may evaporate.

The process of evaporation of perspiration from the skin produces a cooling effect, because heat is stored in the evaporating molecules.

Figure 5.7
Liquid water in equilibrium with
water vapor. (a) Initiation: process of
evaporation exclusively. (b, c) After a
time, both evaporation and condensation
occur, but evaporation predominates.
(d) Dynamic equilibrium established.
Rates of evaporation and condensation
are equal.

Apart from its dependence on the surrounding atmospheric pressure, the boiling point depends on the nature of the attractive forces between the liquid molecules. Polar liquids, such as water, with large intermolecular attractive forces have *higher* boiling points than nonpolar liquids, such as gasoline, which exhibit weak attractive forces.

Question 5.5

Distinguish between the terms *evaporation* and *condensation*.

Question 5.6

Distinguish between the terms *evaporation* and *boiling*.

Van der Waals Forces

LEARNING GOAL

7 Describe the dipolar attractions known collectively as van der Waals forces.

Physical properties of liquids, such as those discussed in the previous section, can be explained in terms of their intermolecular forces. We have seen (see Section 3.5) that attractive forces between polar molecules, **dipole-dipole interactions,** significantly decrease vapor pressure and increase the boiling point. However, nonpolar substances can exist as liquids as well; many are liquids and even solids at room temperature. What is the nature of the attractive forces in these nonpolar compounds?

In 1930 Fritz London demonstrated that he could account for a weak attractive force between any two molecules, whether polar or nonpolar. He postulated that the electron distribution in molecules is not fixed; electrons are in continuous motion, relative to the nucleus. So, for a short time a nonpolar molecule could experience an *instantaneous dipole,* a short-lived polarity caused by a temporary dislocation of the electron cloud. These temporary dipoles could interact with other temporary dipoles, just as permanent dipoles interact in polar molecules. We now call these intermolecular forces **London forces.**

London forces and dipole-dipole interactions are collectively known as **van der Waals forces.** London forces exist among polar and nonpolar molecules because electrons are in constant motion in all molecules. Dipole-dipole attractions occur only among polar molecules. In addition to van der Waals forces, a special type of dipole-dipole force, the *hydrogen bond,* has a very significant effect on molecular properties, particularly in biological systems.

Hydrogen Bonding

LEARNING GOAL

8 Describe hydrogen bonding and its relationship to boiling and melting temperatures.

Typical forces in polar liquids, discussed above, are only about 1–2% as strong as ionic and covalent bonds. However, certain liquids have boiling points that are much higher than we would predict from these dipolar interactions alone. This indicates the presence of some strong intermolecular force. This attractive force is due to **hydrogen bonding.** Molecules in which a hydrogen atom is bonded to a small, highly electronegative atom such as nitrogen, oxygen, or fluorine exhibit this effect. The presence of a highly electronegative atom bonded to a hydrogen atom creates a large dipole:

This arrangement of atoms produces a very polar bond, often resulting in a polar molecule with strong intermolecular attractive forces. Although the hydrogen bond is weaker than bonds formed *within* molecules (covalent and polar covalent *intra*molecular forces), it is the strongest attractive force *between* molecules (intermolecular force).

Consider the boiling points of four small molecules:

$$CH_4 \qquad NH_3 \qquad H_2O \qquad HF$$
$$-161°C \quad -33°C \quad +100°C \quad +19.5°C$$

Clearly, ammonia, water, and hydrogen fluoride boil at significantly higher temperatures than methane. The N—H, O—H, and F—H bonds are far more polar than the C—H bond, owing to the high electronegativity of N, O, and F.

It is interesting to note that the boiling points increase as the electronegativity of the element bonded to hydrogen increases, with one exception: Fluorine, with the highest electronegativity should cause HF to have the highest boiling point. This is not the case. The order of boiling points is

water > hydrogen fluoride > ammonia > methane

not

hydrogen fluoride > water > ammonia > methane

Why? To answer this question we must look at the *number of potential bonding sites* in each molecule. Water has two partial positive sites (located at each hydrogen atom) and two partial negative sites (two lone pairs of electrons on the oxygen atom); it can form hydrogen bonds at each site. This results in a complex network of attractive forces among water molecules in the liquid state and the strength of the forces holding this network together accounts for water's unusually high boiling point. This network is depicted in Figure 5.8.

Ammonia and hydrogen fluoride can form only one hydrogen bond per molecule. Ammonia has three partial positive sites (three hydrogen atoms bonded to nitrogen) but only one partial negative site (the lone pair); the single lone pair is the limiting factor. One positive site and one negative site are needed for each hydrogen bond. Hydrogen fluoride has only one partial positive site and one partial negative site. It too can form only one hydrogen bond per molecule. Consequently, the network of attractive forces in ammonia and hydrogen fluoride is much less extensive than that found in water, and their boiling points are considerably lower than that of water.

Hydrogen bonding has an extremely important influence on the behavior of many biological systems. Molecules such as proteins and DNA require extensive hydrogen bonding to maintain their structures and hence functions. DNA (deoxyribonucleic acid, Section 20.2) is a giant among molecules with intertwined chains of atoms held together by thousands of hydrogen bonds.

Recall that the most electronegative elements are in the upper right corner of the periodic table, and these elements exert strong electron attraction in molecules as described in Chapter 3.

Figure 5.8
Hydrogen bonding in water. Note that the central water molecule is hydrogen bonded to four other water molecules. The attractive force between the hydrogen (δ^+) part of one water molecule and the oxygen (δ^-) part of another water molecule constitutes the hydrogen bond.

Intramolecular hydrogen bonding between polar regions helps keep proteins folded in their proper three-dimensional structure. See Chapter 18.

Question 5.7

Arrange the following compounds in order of increasing boiling point:

$$CO_2 \qquad CH_3OH \qquad CH_3Cl$$

Explain your logic.

Question *5.8*

Butanol and diethyl ether are isomers:

$$
\begin{array}{cccc}
\text{H} & \text{H} & \text{H} & \text{H} \\
| & | & | & | \\
\text{H}-\text{C}-\text{C}-\text{C}-\text{C}-\text{O}-\text{H} \\
| & | & | & | \\
\text{H} & \text{H} & \text{H} & \text{H}
\end{array}
\qquad
\begin{array}{cccc}
\text{H} & \text{H} & \text{H} & \text{H} \\
| & | & | & | \\
\text{H}-\text{C}-\text{C}-\text{O}-\text{C}-\text{C}-\text{H} \\
| & | & | & | \\
\text{H} & \text{H} & \text{H} & \text{H}
\end{array}
$$

butanol
b.p. = 117°C

diethyl ether
b.p. = 34.5°C

Explain the large difference in boiling point for these isomers.

5.3 The Solid State

The close packing of the particles of a solid results from attractive forces that are strong enough to restrict motion. This occurs because the kinetic energy of the particles is insufficient to overcome the attractive forces among particles. The particles are "locked" together in a defined and highly organized fashion. This results in fixed shape and volume, although, at the atomic level, vibrational motion is observed.

Properties of Solids

LEARNING GOAL

9 Relate the properties of the various classes of solids (ionic, covalent, molecular, and metallic) to the structure of these solids.

Solids are virtually incompressible, owing to the small distance between particles. Most will convert to liquids at a higher temperature, when the increased heat energy overcomes some of the attractive forces within the solid. The temperature at which a solid is converted to the liquid phase is its **melting point.** The melting point depends on the strength of the attractive forces in the solid, hence its structure. As we might expect, polar solids have higher melting points than nonpolar solids of the same molecular weight.

A solid may be a **crystalline solid,** having a regular repeating structure, or an **amorphous solid,** having no organized structure. Diamond and sodium chloride (Figure 5.9) are examples of crystalline substances; glass, plastic, and concrete are examples of amorphous solids.

Types of Crystalline Solids

Intermolecular forces are also discussed in Sections 3.5 and 5.2.

Crystalline solids may exist in one of four general groups:
1. *Ionic solids.* The units that comprise an **ionic solid** are positive and negative ions. Electrostatic forces hold the crystal together. They generally have high melting points, and are hard and brittle. A common example of an ionic solid is sodium chloride.
2. *Covalent solids.* The units that comprise a **covalent solid** are atoms held together by covalent bonds. They have very high melting points (1200°C to 2000°C or more is not unusual) and are extremely hard. They are insoluble in most solvents. Diamond is a covalent solid composed of covalently bonded carbon atoms. Diamonds are used for industrial cutting because they are so hard and as gemstones because of their crystalline beauty.
3. *Molecular solids.* The units that make up a **molecular solid,** molecules, are held together by intermolecular attractive forces (London forces, dipole-dipole interactions, and hydrogen bonding). Molecular solids are usually soft and have low melting points. They are frequently volatile and are poor electrical conductors. A common example is ice (solid water; Figure 5.10).
4. *Metallic solids.* The units that comprise a **metallic solid** are metal atoms held together by metallic bonds. **Metallic bonds** are formed by the overlap of orbitals of metal atoms, resulting in regions of high electron density

Figure 5.9
Crystalline solids.

(a) The crystal structure of diamond

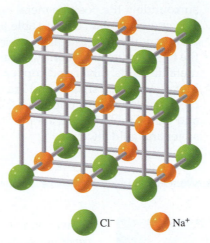

(b) The crystal structure of sodium chloride

(c) The crystal structure of methane, a frozen molecular solid. Only one methane molecule is shown in detail.

(d) The crystal structure of a metallic solid. The gray area represents mobile electrons around fixed metal cations.

Figure 5.10

The structure of ice, a molecular solid. Hydrogen bonding among water molecules produces a regular open structure that is less dense than liquid water.

= O

= H

surrounding the positive metal nuclei. Electrons in these regions are extremely mobile. They are able to move freely from atom to atom through pathways that are, in reality, overlapping atomic orbitals. This results in the high *conductivity* (ability to carry electrical current) exhibited by many metallic solids. Silver and copper are common examples of metallic solids. Metals are easily shaped and are used for a variety of purposes. Most of these are practical applications such as hardware, cookware, and surgical and dental tools. Others are purely for enjoyment and decoration, such as silver and gold jewelry.

SUMMARY

5.1 The Gaseous State

The *kinetic molecular theory* describes an *ideal gas* in which gas particles exhibit no interactive or repulsive forces and the volumes of the individual gas particles are assumed to be negligible.

Boyle's law states that the volume of a gas varies inversely with the pressure exerted by the gas if the number of moles and temperature of gas are held constant ($PV = k_1$).

Charles's law states that the volume of a gas varies directly with the absolute temperature (K) if pressure and number of moles of gas are constant ($V/T = k_2$).

Avogadro's law states that equal volumes of any gas contain the same number of moles if measured at constant temperature and pressure ($V/n = k_3$).

The volume occupied by 1 mol of any gas is its *molar volume*. At *standard temperature and pressure* (STP) the molar volume of any ideal gas is 22.4 L. STP conditions are defined as 273 K (or 0°C) and 1 atm pressure.

Boyle's law, Charles's law, and Avogadro's law may be combined into a single expression relating all four terms, the *ideal gas law: PV = nRT*. R is the ideal gas constant (0.0821 L-atm K^{-1}mol^{-1}) if the units P (atmospheres), V (liters), n (number of moles), and T (Kelvin) are used.

The *combined gas law* provides a convenient expression for performing gas law calculations involving the most common variables: pressure, volume, and temperature.

Dalton's law of partial pressures states that a mixture of gases exerts a pressure that is the sum of the pressures that each gas would exert if it were present alone under similar conditions ($P_t = p_1 + p_2 + p_3 + \ldots$).

5.2 The Liquid State

Liquids are practically incompressible because of the closeness of the molecules. The *viscosity* of a liquid is a measure of its resistance to flow. Viscosity generally decreases with increasing temperature. The *surface tension* of a liquid is a measure of the attractive forces at the surface of a liquid. *Surfactants* decrease surface tension.

The conversion of liquid to vapor at a temperature below the boiling point of the liquid is *evaporation*.

Conversion of the gas to the liquid state is *condensation*. The *vapor pressure of the liquid* is defined as the pressure exerted by the vapor at equilibrium at a specified temperature. The *normal boiling point* of a liquid is the temperature at which the vapor pressure of the liquid is equal to 1 atm.

Molecules in which a hydrogen atom is bonded to a small, highly electronegative atom such as nitrogen, oxygen, or fluorine exhibit *hydrogen bonding*. Hydrogen bonding in liquids is responsible for lower than expected vapor pressures and higher than expected boiling points. The presence of *van der Waals forces* and hydrogen bonds significantly affects the boiling points of liquids as well as the melting points of solids.

5.3 The Solid State

Solids have fixed shapes and volumes. They are *incompressible*, owing to the closeness of the particles. Solids may be *crystalline*, having a regular, repeating structure, or *amorphous*, having no organized structure.

Crystalline solids may exist as *ionic solids, covalent solids, molecular solids*, or *metallic solids*. Electrons in metallic solids are extremely mobile, resulting in the high *conductivity* (ability to carry electrical current) exhibited by many metallic solids.

KEY TERMS

amorphous solid (5.3)	London forces (5.2)
Avogadro's law (5.1)	melting point (5.3)
barometer (5.1)	metallic bond (5.3)
Boyle's law (5.1)	metallic solid (5.3)
Charles's law (5.1)	molar volume (5.1)
combined gas law (5.1)	molecular solid (5.3)
condensation (5.2)	normal boiling point (5.2)
covalent solid (5.3)	partial pressure (5.1)
crystalline solid (5.3)	pressure (5.1)
Dalton's law (5.1)	standard temperature and
dipole-dipole interactions (5.2)	pressure (STP) (5.1)
evaporation (5.2)	surface tension (5.2)
hydrogen bonding (5.2)	surfactant (5.2)
ideal gas (5.1)	van der Waals forces (5.2)
ideal gas law (5.1)	vapor pressure of a liquid (5.2)
ionic solid (5.3)	viscosity (5.2)
kinetic molecular theory (5.1)	

QUESTIONS AND PROBLEMS

Kinetic Molecular Theory

Foundations

5.9 Compare and contrast the gas, liquid, and solid states with regard to the average distance of particle separation.

5.10 Compare and contrast the gas, liquid, and solid states with regard to the nature of the interactions among the particles.

5.11 Describe the molecular/atomic basis of gas pressure.

5.12 Describe the measurement of gas pressure.

Applications

5.13 Why are gases easily compressible?

5.14 Why are gas densities much lower than those of liquids or solids?

5.15 Why do gases expand to fill any available volume?

5.16 Why do gases with lower molar masses diffuse more rapidly than gases with higher molar masses?

5.17 Do gases exhibit more ideal behavior at low or high pressures? Why?

5.18 Do gases exhibit more ideal behavior at low or high temperatures? Why?

5.19 Use the kinetic molecular theory to explain why dissimilar gases mix more rapidly at high temperatures than at low temperatures.

5.20 Use the kinetic molecular theory to explain why aerosol cans carry instructions warning against heating or disposing of the container in a fire.

5.21 Predict and explain any observed changes taking place when an inflated balloon is cooled (perhaps refrigerated).

5.22 Predict and explain any observed changes taking place when an inflated balloon is heated (perhaps microwaved).

Boyle's Law

Foundations

5.23 State Boyle's law in words.

5.24 State Boyle's law in equation form.

5.25 The pressure on a fixed mass of a gas is tripled at constant temperature. Will the volume increase, decrease, or remain the same?

5.26 By what factor will the volume of the gas in Question 5.25 change?

Applications

A sample of helium gas was placed in a cylinder and the volume of the gas was measured as the pressure was slowly increased. The results of this experiment are shown graphically.

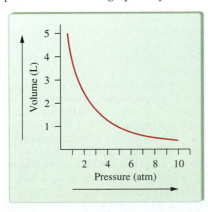

Boyle's Law

Questions 5.27–5.30 are based on this experiment.

5.27 At what pressure does the gas occupy a volume of 5 L?

5.28 What is the volume of the gas at a pressure of 5 atm?

5.29 Calculate the Boyle's law constant at a volume of 2 L.

5.30 Calculate the Boyle's law constant at a pressure of 2 atm.

5.31 Calculate the pressure, in atmospheres, required to compress a sample of helium gas from 20.9 L (at 1.00 atm) to 4.00 L.

5.32 A balloon filled with helium gas at 1.00 atm occupies 15.6 L. What volume would the balloon occupy in the upper atmosphere, at a pressure of 0.150 atm?

Charles's Law

Foundations

5.33 State Charles's law in words.

5.34 State Charles's law in equation form.

5.35 Explain why the Kelvin scale is used for gas law calculations.

5.36 The temperature on a summer day may be 90°F. Convert this value to Kelvin units.

Applications

5.37 The temperature of a gas is raised from 25°C to 50°C. Will the volume double if mass and pressure do not change? Why or why not?

5.38 Verify your answer to Question 5.37 by calculating the temperature needed to double the volume of the gas.

5.39 Determine the change in volume that takes place when a 2.00-L sample of $N_2(g)$ is heated from 250°C to 500°C.

5.40 Determine the change in volume that takes place when a 2.00-L sample of $N_2(g)$ is heated from 250 K to 500 K.

5.41 A balloon containing a sample of helium gas is warmed in an oven. If the balloon measures 1.25 L at room temperature (20°C), what is its volume at 80°C?

5.42 The balloon described in Problem 5.41 was then placed in a refrigerator at 39°F. Calculate its new volume.

Combined Gas Law

Foundations

5.43 Will the volume of gas increase, decrease, or remain the same if the temperature is increased and the pressure is decreased? Explain.

5.44 Will the volume of gas increase, decrease, or remain the same if the temperature is decreased and the pressure is increased? Explain.

Applications

Use the combined gas law,

$$\frac{P_i V_i}{T_i} = \frac{P_f V_f}{T_f}$$

to answer Questions 5.45 and 5.46.

5.45 Solve the combined gas law expression for the final volume.

5.46 Solve the combined gas law expression for the final temperature.

5.47 If 2.25 L of a gas at 16°C and 1.00 atm is compressed at a pressure of 125 atm at 20°C, calculate the new volume of the gas.

5.48 A balloon filled with helium gas occupies 2.50 L at 25°C and 1.00 atm. When released, it rises to an altitude where the temperature is 20°C and the pressure is only 0.800 atm. Calculate the new volume of the balloon.

Avogadro's Law

Foundations

5.49 State Avogadro's law in words.

5.50 State Avogadro's law in equation form.

Applications

5.51 If 5.00 g helium gas is added to a 1.00 L balloon containing 1.00 g of helium gas, what is the new volume of the balloon? Assume no change in temperature or pressure.

5.52 How many grams of helium must be added to a balloon containing 8.00 g helium gas to double its volume? Assume no change in temperature or pressure.

Molar Volume and the Ideal Gas Law

Foundations

5.53 Will 1.00 mol of a gas always occupy 22.4 L?

5.54 H_2O and CH_4 are gases at 150°C. Which exhibits more ideal behavior? Why?

5.55 What are the units and numerical value of standard temperature?

5.56 What are the units and numerical value of standard pressure?

Applications

5.57 A sample of nitrogen gas, stored in a 4.0-L container at 32°C, exerts a pressure of 5.0 atm. Calculate the number of moles of nitrogen gas in the container.

5.58 Seven moles of carbon monoxide are stored in a 30.0-L container at 65°C. What is the pressure of the carbon monoxide in the container?

5.59 Calculate the volume of 44.0 g of carbon monoxide at STP.

5.60 Calculate the volume of 44.0 g of carbon dioxide at STP.

5.61 Calculate the number of moles of a gas that is present in a 7.55-L container at 45°C, if the gas exerts a pressure of 725 mm Hg.

5.62 Calculate the pressure exerted by 1.00 mol of gas, contained in a 7.55-L cylinder at 45°C.

5.63 A sample of argon (Ar) gas occupies 65.0 mL at 22°C and 750 torr. What is the volume of this Ar gas sample at STP?

5.64 A sample of O_2 gas occupies 257 mL at 20°C and 1.20 atm. What is the volume of this O_2 gas sample at STP?

5.65 Calculate the molar volume of Ar gas at STP.

5.66 Calculate the molar volume of O_2 gas at STP.

5.67 Calculate the volume of 4.00 mol Ar gas at 8.25 torr and 27°C.

5.68 Calculate the volume of 6.00 mol O_2 gas at 30 cm Hg and 72°F.

5.69 What is the temperature (°C) of 1.75 g of O_2 gas occupying 2.00 L at 1.00 atm?

5.70 How many grams of O_2 gas occupy 10.0 L at STP?

Dalton's Law

Foundations

5.71 State Dalton's law in words.

5.72 State Dalton's law in equation form.

Applications

5.73 A gas mixture has three components: N_2, F_2, and He. Their partial pressures are 0.40 atm, 0.16 atm, and 0.18 atm, respectively. What is the pressure of the gas mixture?

5.74 A gas mixture has a total pressure of 0.56 atm and consists of He and Ne. If the partial pressure of the He in the mixture is 0.27 atm, what is the partial pressure of the Ne in the mixture?

The Liquid State

Foundations

5.75 Compare the strength of intermolecular forces in liquids with those in gases.

5.76 Compare the strength of intermolecular forces in liquids with those in solids.

5.77 What is the relationship between the temperature of a liquid and the vapor pressure of that liquid?

5.78 What is the relationship between the strength of the attractive forces in a liquid and its vapor pressure?

5.79 Describe the process occurring at the molecular level that accounts for the property of viscosity.

5.80 Describe the process occurring at the molecular level that accounts for the property of surface tension.

Applications

Questions 5.81–5.84 are based on the following:

methane   chloromethane   methanol

5.81 Which of these molecules exhibit London forces? Why?

5.82 Which of these molecules exhibit dipole-dipole forces? Why?

5.83 Which of these molecules exhibit hydrogen bonding? Why?

5.84 Which of these molecules would you expect to have the highest boiling point? Why?

The Solid State

5.85 Explain why solids are essentially incompressible.

5.86 Distinguish between amorphous and crystalline solids.

5.87 Describe one property that is characteristic of:
   a. ionic solids
   b. covalent solids

5.88 Describe one property that is characteristic of:
   a. molecular solids
   b. metallic solids

5.89 Predict whether beryllium or carbon would be a better conductor of electricity in the solid state. Why?

5.90 Why is diamond used as an industrial cutting tool?

5.91 Mercury and chromium are toxic substances. Which element is more likely to be an air pollutant? Why?

5.92 Why is the melting point of silicon much higher than that of argon, even though argon has a greater molar mass?

CRITICAL THINKING PROBLEMS

1. An elodea plant, commonly found in tropical fish aquaria, was found to produce 5.0×10^{22} molecules of oxygen per hour. What volume of oxygen (STP) would be produced in an eight-hour period?

2. A chemist measures the volume of 1.00 mol of helium gas at STP and obtains a value of 22.4 L. After changing the temperature to 137 K, the experimental value was found to be 11.05 L. Verify the chemist's results using the ideal gas law and explain any apparent discrepancies.

3. A chemist measures the volumes of 1.00 mol of H_2 and 1.00 mol of CO and finds that they differ by 0.10 L. Which gas produced the larger volume? Do the results contradict the ideal gas law? Why or why not?

4. A 100.0-g sample of water was decomposed using an electric current (electrolysis) producing hydrogen gas and oxygen gas. Write the balanced equation for the process and calculate the volume of each gas produced (STP). Explain any relationship you may observe between the volumes obtained and the balanced equation for the process.

5. An autoclave is used to sterilize surgical equipment. It is far more effective than steam produced from boiling water in the open atmosphere because it generates steam at a pressure of 2 atm. Explain why an autoclave is such an efficient sterilization device.

6 Solutions

Learning Goals

1 Distinguish among the terms *solution, solute,* and *solvent.*

2 Describe various kinds of solutions, and give examples of each.

3 Describe the relationship between solubility and equilibrium.

4 Calculate solution concentration in weight/volume percent, weight/weight percent, parts per thousand, and parts per million.

5 Calculate solution concentration using molarity.

6 Perform dilution calculations.

7 Interconvert molar concentration of ions and milliequivalents/liter.

8 Describe and explain concentration-dependent solution properties.

9 Describe why the chemical and physical properties of water make it a truly unique solvent.

10 Explain the role of electrolytes in blood and their relationship to the process of dialysis.

Outline

Introduction

Chemistry Connection:
Seeing a Thought

6.1 Properties of Solutions

6.2 Concentration Based on Mass

A Human Perspective: *Scuba Diving: Nitrogen and the Bends*

6.3 Concentration of Solutions: Moles and Equivalents

6.4 Concentration-Dependent Solution Properties

A Medical Perspective:
Oral Rehydration Therapy

6.5 Water as a Solvent

A Human Perspective:
An Extraordinary Molecule

6.6 Electrolytes in Body Fluids

A Medical Perspective:
Hemodialysis

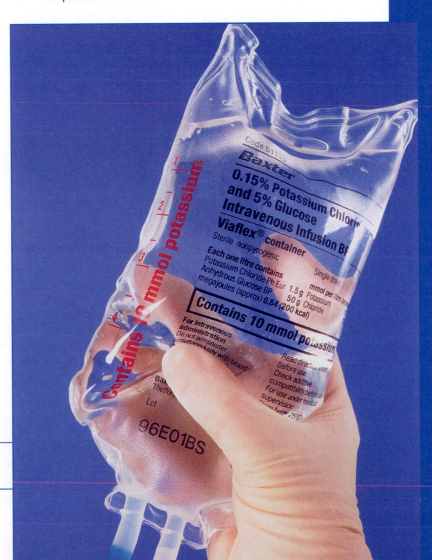

Composition and concentration are critically important in medical intervention.

Introduction

Many chemical reactions, and virtually all important organic and biochemical reactions, take place as reactants dissolved in solution. For this reason the major emphasis of this chapter will be on aqueous solution reactions.

We will see that the properties of solutions depend not only on the types of substances that make up the solution but also on the amount of each substance that is contained in a certain volume of the solution. The latter is termed the concentration of the solution.

6.1 Properties of Solutions

A **solution** is a homogeneous (or uniform) mixture of two or more substances. A solution is composed of one or more *solutes,* dissolved in a *solvent.* The **solute** is a compound of a solution that is present in lesser quantity than the solvent. The **solvent** is the solution component present in the largest quantity. For example, when sugar (the solute) is added to water (the solvent), the sugar dissolves in the water to produce a solution. In those instances in which the solvent is water, we refer to the homogeneous mixture as an **aqueous solution,** from the Latin *aqua,* meaning water.

The dissolution of a solid in a liquid is perhaps the most common example of solution formation. However, it is also possible to form solutions in gases and solids as well as in liquids. For example:

- Air is a gaseous mixture, but it is also a solution; oxygen and a number of trace gases are dissolved in the gaseous solvent, nitrogen.
- Alloys, such as brass and silver and the gold used to make jewelry, are also homogeneous mixtures of two or more kinds of metal atoms in the solid state.

Chemistry Connection

Seeing a Thought

At one time, not very long ago, mental illness was believed to be caused by some failing of the human spirit. Thoughts are nonmaterial (you can't hold a thought in your hand), and the body is quite material. No clear relationship, other than the fact that thoughts somehow come from the brain, could be shown to link the body and the spirit.

A major revolution in the diagnosis and treatment of mental illness has taken place in the last three decades. Several forms of depression, paranoia, and schizophrenia have been shown to have chemical and genetic bases. Remarkable improvement in behavior often results from altering the chemistry of the brain by using chemical therapy. Similar progress may result from the use of gene therapy (discussed in Chapter 20).

Although a treatment of mental illness, as well as of memory and logic failures, may occasionally arise by chance, a cause-and-effect relationship, based on the use of scientific methodology, certainly increases the chances of developing successful treatment. If we understand the chemical reactions involved in the thought process, we can perhaps learn to "repair" them when, for whatever reason, they go astray.

Recently, scientists at Massachusetts General Hospital in Boston have developed sophisticated versions of magnetic resonance imaging devices (MRI, discussed in Medical Perspective in Chapter 9). MRI is normally used to locate brain tumors and cerebral damage in patients. The new generation of instruments is so sensitive that it is able to detect chemical change in the brain resulting from an external stimulus. A response to a question or the observation of a flash of light produces a measurable signal. This signal is enhanced with the aid of a powerful computer that enables the location of the signal to be determined with pinpoint accuracy. So there is evidence not only for the chemical basis of thought, but for its location in the brain as well.

In this chapter and throughout your study of chemistry, you will be introduced to a wide variety of chemical reactions, some rather ordinary, some quite interesting. All are founded on the same principles that power our thoughts and actions.

Although solid and gaseous solutions are important in many applications, our emphasis will be on *liquid solutions* because so many important chemical reactions take place in liquid solutions.

General Properties of Liquid Solutions

Liquid solutions are clear and transparent with no visible particles of solute. They may be colored or colorless, depending on the properties of the solute and solvent. Note that the terms *clear* and *colorless* do not mean the same thing; a clear solution has only one state of matter that can be detected; *colorless* simply means the absence of color.

Recall that solutions of **electrolytes** are formed from solutes that are soluble ionic compounds. These compounds dissociate in solution to produce ions that behave as charge carriers. Solutions of electrolytes are good conductors of electricity. For example, sodium chloride dissolving in water:

$$NaCl(s) \xrightarrow{H_2O} Na^+(aq) + Cl^-(aq)$$

Solid sodium   Dissolved sodium
chloride   chloride

In contrast, solutions of **nonelectrolytes** are formed from nondissociating *molecular* solutes (nonelectrolytes), and these solutions are nonconducting. For example, dissolving sugar in water:

$$C_6H_{12}O_6(s) \xrightarrow{H_2O} C_6H_{12}O_6(aq)$$

Solid glucose   Dissolved glucose

A *true solution* is a homogeneous mixture with uniform properties throughout. In a true solution the solute cannot be isolated from the solution by filtration. The particle size of the solute is about the same as that of the solvent, and solvent and solute pass directly through the filter paper. Furthermore, solute particles will not "settle out" after a time. All of the molecules of solute and solvent are intimately mixed. The continuous particle motion in solution maintains the homogeneous, random distribution of solute and solvent particles.

Volumes of solute and solvent are not additive; 1L of alcohol mixed with 1L of water does not result in exactly 2L of solution. The volume of pure liquid is determined by the way in which the individual molecules "fit together." When two or more kinds of molecules are mixed, the interactions become more complex. Solvent interacts with solvent, solute interacts with solvent, and solute may interact with other solute. This will be important to remember when we solve concentration problems later.

Solutions and Colloids

How can you recognize a solution? A beaker containing a clear liquid may be a pure substance, a true solution, or a colloid. Only chemical analysis, determining the identity of all substances in the liquid, can distinguish between a pure substance and a solution. A pure substance has *one* component, pure water being an example. A true solution will contain more than one substance, with the tiny particles homogeneously intermingled.

A **colloidal suspension** also consists of solute particles distributed throughout a solvent. However, the distribution is not completely homogeneous, owing to the size of the colloidal particles. Particles with diameters of 1×10^{-9} m (1 nm) to 2×10^{-7}m (200 nm) are colloids. Particles smaller than 1 nm are solution particles; those larger than 200 nm are precipitates (solid in contact with solvent).

To the naked eye, a colloidal suspension and a true solution appear identical; neither solute nor colloid can be seen by the naked eye. However, a simple experiment, using only a bright light source, can readily make the distinction based

LEARNING GOAL
2 Describe various kinds of solutions, and give examples of each.

Section 3.3 discusses properties of compounds.

Animations
Dissolution of Compounds
Salt Dissolving in Water
Strong, Weak, and Nonelectrolytes

Particles in electrolyte solutions are ions, making the solution an electrical conductor.

Particles in solution are individual molecules. No ions are formed in the dissolution process.

Section 3.5 relates properties and molecular geometry.

See Section 4.3 for more information on precipitates.

The Tyndall effect. The beaker on the left contains a colloidal suspension, which scatters the light. This scattered light is visible as a haze. The beaker on the right contains a true solution; no scattered light is observed.

upon differences in their interaction with light. Colloid particles are large enough to scatter light; solute particles are not. When a beam of light passes through a colloidal suspension, the large particles scatter light, and the liquid appears hazy. We see this effect in sunlight passing through fog. Fog is a colloidal suspension of tiny particles of liquid water dispersed throughout a gas, air. The haze is light scattered by droplets of water. You may have noticed that your automobile headlights are not very helpful in foggy weather. Visibility becomes worse rather than better because light scattering increases.

The light-scattering ability of colloidal suspensions is termed the *Tyndall effect*. True solutions, with very tiny particles, do not scatter light—no haze is observed—and true solutions are easily distinguished from colloidal suspensions by observing their light-scattering properties (Figure 6.1).

A **suspension** is a heterogeneous mixture that contains particles much larger than a colloidal suspension; over time, these particles may settle, forming a second phase. A suspension is not a true solution, nor is it a precipitate.

Question 6.1

Describe how you would distinguish experimentally between a pure substance and a true solution.

Question 6.2

Describe how you would distinguish experimentally between a true solution and a colloidal suspension.

Degree of Solubility

Section 3.5 describes solute-solvent interactions in detail.

The term *qualitative* **implies identity, and the term** *quantitative* **relates to quantity.**

In our discussion of the relationship of polarity and solubility, the rule *"like dissolves like"* was described as the fundamental condition for solubility. Polar solutes are soluble in polar solvents, and nonpolar solutes are soluble in nonpolar solvents. Thus, knowing a little bit about the structure of the molecule enables us to predict qualitatively the solubility of the compound.

The *degree* of **solubility,** *how much* solute can dissolve in a given volume of solvent, is a quantitative measure of solubility. It is difficult to predict the solubility of each and every compound. However, general solubility trends are based on the following considerations:

- *The magnitude of difference between polarity of solute and solvent.* The greater the difference, the less soluble is the solute.
- *Temperature.* An increase in temperature usually, but not always, increases solubility. Often, the effect is dramatic. For example, an increase in temperature from 0°C to 100°C increases the water solubility of KCl from 28 g/100 mL to 58 g/100 mL.
- *Pressure.* Pressure has little effect on the solubility of solids and liquids in liquids. However, the solubility of a gas in liquid is directly proportional to the applied pressure. Carbonated beverages, for example, are made by dissolving carbon dioxide in the beverage under high pressure (hence the term *carbonated*).

When a solution contains all the solute that can be dissolved at a particular temperature, it is a **saturated solution.** When solubility values are given—for example, 13.3 g of potassium nitrate in 100 mL of water at 24°C—they refer to the concentration of a saturated solution.

As we have already noted, *increasing* the temperature generally increases the amount of solute a given solution may hold. Conversely, *cooling* a saturated solution often results in a decrease in the amount of solute in solution. The excess solute falls to the bottom of the container as a **precipitate** (a solid in contact with the solution). Occasionally, on cooling, the excess solute may remain in solution for a time. Such a solution is described as a **supersaturated solution.** This type of solution is inherently unstable. With time, excess solute will precipitate, and the solution will revert to a saturated solution, which is stable.

Solubility and Equilibrium

When an excess of solute is added to a solvent, it begins to dissolve and continues until it establishes a *dynamic equilibrium* between dissolved and undissolved solute.

Initially, the rate of dissolution is large. After a time the rate of the reverse process, precipitation, increases. The rates of dissolution and precipitation eventually become equal, and there is no further change in the composition of the solution. There is, however, a continual exchange of solute particles between solid and liquid phases because particles are in constant motion. The solution is saturated. The most precise definition of a saturated solution is a solution that is in equilibrium with undissolved solute.

3   **LEARNING GOAL**
Describe the relationship between solubility and equilibrium.

The concept of equilibrium was introduced in Section 5.2 and will be discussed in detail in Section 7.4.

Solubility of Gases: Henry's Law

When a liquid and a gas are allowed to come to equilibrium, the amount of gas dissolved in the liquid reaches some maximum level. This quantity can be predicted from a very simple relationship. **Henry's law** states that the number of moles of a gas dissolved in a liquid at a given temperature is proportional to the partial pressure of the gas. In other words, the gas solubility is directly proportional to the pressure of that gas in the atmosphere that is in contact with the liquid.

Carbonated beverages are bottled at high pressures of carbon dioxide. When the cap is removed, the fizzing results from the fact that the partial pressure of carbon dioxide in the atmosphere is much less than that used in the bottling process. As a result, the equilibrium quickly shifts to one of lower gas solubility.

Gases are most soluble at low temperatures, and the gas solubility decreases markedly at higher temperatures. This explains many common observations. For example, a chilled container of carbonated beverage that is opened quickly goes

The concept of partial pressure is a consequence of Dalton's law, discussed in Section 5.1.

flat as it warms to room temperature. As the beverage warms up, the solubility of the carbon dioxide decreases.

Henry's law helps to explain the process of respiration. Respiration depends on a rapid and efficient exchange of oxygen and carbon dioxide between the atmosphere and the blood. This transfer occurs through the lungs. The process, oxygen entering the blood and carbon dioxide released to the atmosphere, is accomplished in air sacs called *alveoli*, which are surrounded by an extensive capillary system. Equilibrium is quickly established between alveolar air and the capillary blood. The temperature of the blood is effectively constant. Therefore the equilibrium concentration of both oxygen and carbon dioxide are determined by the partial pressures of the gases (Henry's law). The oxygen is transported to cells, a variety of reactions takes place, and the waste product of respiration, carbon dioxide, is brought back to the lungs to be expelled into the atmosphere.

The exchange of O_2 and CO_2 in the lungs and other tissues is a complex series of events described in greater detail in Section 18.9.

See A Medical Perspective: Blood Gases and Respiration, Chapter 5.

Question 6.3

Explain why, over time, a bottle of soft drink goes "flat" after it is opened.

Question 6.4

Would the soft drink in Question 6.3 go "flat" faster if the bottle warmed to room temperature? Why?

6.2 Concentration Based on Mass

LEARNING GOAL

4 Calculate solution concentration in weight/volume percent, weight/weight percent, parts per thousand, and parts per million.

Solution **concentration** is defined as the amount of solute dissolved in a given amount of solution. The concentration of a solution has a profound effect on the properties of a solution, both *physical* (melting and boiling points) and *chemical* (solution reactivity). Solution concentration may be expressed in many different units. Here we consider concentration units based on percentage.

Weight/Volume Percent

The concentration of a solution is defined as the amount of solute dissolved in a specified amount of solution,

$$\text{concentration} = \frac{\text{amount of solute}}{\text{amount of solution}}$$

If we define the amount of solute as the *mass* of solute (in grams) and the amount of solution in *volume* units (milliliters), concentration is expressed as the ratio

$$\text{concentration} = \frac{\text{grams of solute}}{\text{milliliters of solution}}$$

This concentration can then be expressed as a percentage by multiplying the ratio by the factor 100%. This results in

$$\% \text{ concentration} = \frac{\text{grams of solute}}{\text{milliliters of solution}} \times 100\%$$

The percent concentration expressed in this way is called **weight/volume percent,** or **% (W/V).** Thus

$$\% \left(\frac{W}{V} \right) = \frac{\text{grams of solute}}{\text{milliliters of solution}} \times 100\%$$

A Human *Perspective*

Scuba Diving: Nitrogen and the Bends

A deep-water diver's worst fear is the interruption of the oxygen supply through equipment malfunction, forcing his or her rapid rise to the surface in search of air. If a diver must ascend too rapidly, he or she may suffer a condition known as "the bends."

Key to understanding this problem is recognition of the tremendous increase in pressure that divers withstand as they descend, because of the weight of the water above them. At the surface the pressure is approximately 1 atm. At a depth of 200 feet the pressure is approximately six times as great.

At these pressures the solubility of nitrogen in the blood increases dramatically. Oxygen solubility increases as well, although its effect is less serious (O_2 is 20% of air, N_2 is 80%). As the diver quickly rises, the pressure decreases rapidly, and the nitrogen "boils" out of the blood, stopping blood flow and impairing nerve transmission. The joints of the body lock in a bent position, hence the name of the condition: the bends.

To minimize the problem, scuba tanks are often filled with mixtures of <u>helium</u> and <u>oxygen</u> rather than <u>nitrogen</u> and <u>oxygen</u>. Helium has a much lower solubility in blood and, like nitrogen, is inert.

Scuba diving.

For Further Understanding

Why are divers who slowly rise to the surface less likely to be adversely affected?

What design features would be essential in deep-water manned exploration vessels?

Consider the following examples.

Calculating Weight/Volume Percent

EXAMPLE 6.1

Calculate the weight/volume percent composition, or % (W/V), of 3.00×10^2 mL of solution containing 15.0 g of glucose.

LEARNING GOAL

4 ▸ Calculate solution concentration in weight/volume percent, weight/weight percent, parts per thousand, and parts per million.

Solution

Step 1. The expression for weight/volume percent is:

$$\% \left(\frac{W}{V} \right) = \frac{\text{grams of solute}}{\text{milliliters of solution}} \times 100\%$$

Continued—

EXAMPLE 6.1 —*Continued*

Step 2. There are 15.0 g of glucose, the solute, and 3.00×10^2 mL of total solution. Therefore, substituting in our expression for weight/volume percent:

$$\% \left(\frac{W}{V} \right) = \frac{15.0 \text{ g glucose}}{3.00 \times 10^2 \text{ mL solution}} \times 100\%$$

$$= 5.00\% \left(\frac{W}{V} \right) \text{ glucose}$$

Practice Problem 6.1

a. Calculate the % (W/V) of 0.0600 L of solution containing 10.0 g NaCl.
b. Calculate the % (W/V) of 0.200 L of solution containing 15.0 g KCl.
c. 20.0 g of oxygen gas are diluted with 80.0 g of nitrogen gas in a 78.0-L container at standard temperature and pressure. Calculate the % (W/V) of oxygen gas.
d. 50.0 g of argon gas are diluted with 80.0 g of helium gas in a 476-L container at standard temperature and pressure. Calculate the % (W/V) of argon gas.

For Further Practice: Questions 6.9 and 6.10.

EXAMPLE 6.2   **Calculating the Weight or Volume of Solute from a Weight/Volume Percent**

LEARNING GOAL

4 Calculate solution concentration in weight/volume percent, weight/weight percent, parts per thousand, and parts per million.

Calculate the number of grams of NaCl in 5.00×10^2 mL of a 10.0% solution.

Solution

Step 1. The expression for weight/volume percent is:

$$\% \left(\frac{W}{V} \right) = \frac{\text{grams of solute}}{\text{milliliters of solution}} \times 100\%$$

Step 2. Substitute the data from the problem:

$$10.0\% \left(\frac{W}{V} \right) = \frac{X \text{ g NaCl}}{5.00 \times 10^2 \text{ mL solution}} \times 100\%$$

Step 3. Cross-multiplying to simplify:

$$X \text{ g NaCl} \times 100\% = \left(10.0\% \frac{W}{V} \right) (5.00 \times 10^2 \text{ mL solution})$$

Step 4. Dividing both sides by 100% to isolate grams NaCl on the left side of the equation:

$$X = 50.0 \text{ g NaCl}$$

Continued—

E X A M P L E 6.2 —*Continued*

Practice Problem 6.2

a. Calculate the mass (in grams) of sodium hydroxide required to make 2.00 L of a 1.00% (W/V) solution.
b. Calculate the volume (in milliliters) of a 25.0% (W/V) solution containing 10.0 g NaCl.

For Further Practice: Questions 6.19 and 6.20.

If the units of mass are other than grams, or if the solution volume is in units other than milliliters, the proper conversion factor must be used to arrive at the units used in the equation.

Section 1.4 discusses units and unit conversion.

Weight/Weight Percent

The **weight/weight percent,** or **% (W/W),** is most useful for mixtures of solids, whose weights (masses) are easily obtained. The expression used to calculate weight/weight percentage is analogous in form to % (W/V):

$$\% \left(\frac{W}{W}\right) = \frac{\text{grams solute}}{\text{grams solution}} \times 100\%$$

Calculating Weight/Weight Percent

E X A M P L E 6.3

Calculate the % (W/W) of platinum in a gold ring that contains 14.00 g gold and 4.500 g platinum.

LEARNING GOAL

4 Calculate solution concentration in weight/ volume percent, weight/weight percent, parts per thousand, and parts per million.

Solution

Step 1. Using our definition of weight/weight percent

$$\% \left(\frac{W}{W}\right) = \frac{\text{grams solute}}{\text{grams solution}} \times 100\%$$

Step 2. Substituting,

$$= \frac{4.500 \text{ g platinum}}{4.500 \text{ g platinum} + 14.00 \text{ g gold}} \times 100\%$$

$$= \frac{4.500 \text{ g}}{18.50 \text{ g}} \times 100\%$$

$$= 24.32\% \text{ platinum}$$

Practice Problem 6.3

a. Calculate the % (W/W) of oxygen gas in a mixture containing 20.0 g of oxygen gas and 80.0 g of nitrogen gas.
b. Calculate the % (W/W) of argon gas in a mixture containing 50.0 g of argon gas and 80.0 g of helium gas.

For Further Practice: Question 6.13 and 6.14.

Parts Per Thousand (ppt) and Parts Per Million (ppm)

The calculation of concentration in parts per thousand or parts per million is based on the same logic as weight/weight percent. Percentage is actually the number of parts of solute in 100 parts of solution. For example, a 5.00% (W/W) is made up of 5.00 g solute in 100 g solution.

$$5.00\% \ (\text{W/W}) = \frac{5.00 \text{ g solute}}{100 \text{ g solution}} \times 100\%$$

It follows that a 5.00 parts per thousand (ppt) solution is made up of 5.00 g solute in 1000 g solution.

$$5.00 \text{ ppt} = \frac{5.00 \text{ g solute}}{1000 \text{ g solution}} \times 10^3 \text{ ppt}$$

Using similar logic, a 5.00 parts per million solution (ppm) is made up of 5.00 g solute in 1,000,000 g solution.

$$5.00 \text{ ppm} = \frac{5.00 \text{ g solute}}{1,000,000 \text{ g solution}} \times 10^6 \text{ ppm}$$

The general expressions are:

$$\text{ppt} = \frac{\text{grams solute}}{\text{grams solution}} \times 10^3 \text{ ppt}$$

and

$$\text{ppm} = \frac{\text{grams solute}}{\text{grams solution}} \times 10^6 \text{ ppm}$$

Ppt and ppm are most often used for expressing the concentrations of very dilute solutions.

EXAMPLE 6.4   **Calculating ppt and ppm**

LEARNING GOAL

4 Calculate solution concentration in weight/volume percent, weight/weight percent, parts per thousand, and parts per million.

A 1.00 g sample of stream water was found to contain 1.0×10^{-6} g lead. Calculate the concentration of lead in the stream water in units of % (W/W), ppt, and ppm. Which is the most suitable unit?

Solution

weight percent: $\% \ (\text{W/W}) = \dfrac{\text{grams solute}}{\text{grams solution}} \times 100\%$

$$\% \ (\text{W/W}) = \frac{1.0 \times 10^{-6} \text{ g Pb}}{1.0 \text{ g solution}} \times 100\%$$

$$\% \ (\text{W/W}) = 1.0 \times 10^{-4} \ \%$$

parts per thousand: $\text{ppt} = \dfrac{\text{grams solute}}{\text{grams solution}} \times 10^3 \text{ ppt}$

$$\text{ppt} = \frac{1.0 \times 10^{-6} \text{ g Pb}}{1.0 \text{ g solution}} \times 10^3 \text{ ppt}$$

$$\text{ppt} = 1.0 \times 10^{-3} \text{ ppt}$$

Continued—

E X A M P L E 6.4 —*Continued*

parts per million: $\text{ppm} = \dfrac{\text{grams solute}}{\text{grams solution}} \times 10^6 \text{ ppm}$

$\text{ppm} = \dfrac{1.0 \times 10^{-6} \text{ g Pb}}{1.0 \text{ g solution}} \times 10^6 \text{ ppm}$

$\text{ppm} = 1.0 \text{ ppm}$

Parts per million is the most reasonable unit.

Practice Problem 6.4

Calculate the ppt and ppm of oxygen gas in a mixture containing

a. 20.0 g of oxygen gas and 80.0 g of nitrogen gas.
b. 50.0 g of argon gas and 80.0 g of helium gas.

For Further Practice: Questions 6.23 and 6.24.

6.3 Concentration of Solutions: Moles and Equivalents

In our discussion of the chemical arithmetic of reactions in Chapter 4, we saw that the chemical equation represents the relative number of *moles* of reactants producing products. When chemical reactions occur in solution, it is most useful to represent their concentrations on a *molar* basis.

Molarity

The most common mole-based concentration unit is molarity. **Molarity,** symbolized M, is defined as the number of moles of solute per liter of solution, or

$$M = \frac{\text{moles solute}}{\text{L solution}}$$

Calculating Molarity from Moles   **E X A M P L E 6.5**

Calculate the molarity of 2.0 L of solution containing 5.0 mol NaOH.

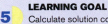

5 **LEARNING GOAL**
Calculate solution concentration using molarity.

Solution

Using our expression for molarity

$$M = \frac{\text{moles solute}}{\text{L solution}}$$

Substituting,

$$M_{\text{NaOH}} = \frac{5.0 \text{ mol solute}}{2.0 \text{ L solution}}$$

$$= 2.5 \ M$$

Practice Problem 6.5

Calculate the molarity of 2.5 L of solution containing 0.75 mol $MgCl_2$.

For Further Practice: Questions 6.29 and 6.30.

Section 1.4 discussed units and unit conversion.

Remember the need for conversion factors to convert from mass to number of moles. Consider the following example:

EXAMPLE 6.6 **Calculating Molarity from Mass**

LEARNING GOAL

5 Calculate solution concentration using molarity.

If 5.00 g glucose are dissolved in 1.00×10^2 mL of solution, calculate the molarity, M, of the glucose solution.

Solution

Step 1. To use our expression for molarity it is necessary to convert from units of grams of glucose to moles of glucose. The molar mass of glucose is 1.80×10^2 g/mol. Therefore

$$5.00 \ \cancel{g} \ \times \ \frac{1 \ \text{mol}}{1.80 \times 10^2 \ \cancel{g}} = 2.78 \times 10^{-2} \ \text{mol glucose}$$

Step 2. We must convert mL to L:

$$1.00 \times 10^2 \ \cancel{mL} \ \times \ \frac{1 \ \text{L}}{10^3 \ \cancel{mL}} = 1.00 \times 10^{-1} \ \text{L}$$

Step 3. Substituting these quantities:

$$M_{\text{glucose}} = \frac{2.78 \times 10^{-2} \ \text{mol}}{1.00 \times 10^{-1} \ \text{L}}$$
$$= 2.78 \times 10^{-1} \ M$$

 Practice Problem 6.6

Calculate the molarity, M, of KCl when 2.33 g KCl are dissolved in 2.50×10^3 mL of solution.

For Further Practice: Questions 6.31 and 6.32.

EXAMPLE 6.7 **Calculating Volume from Molarity**

LEARNING GOAL

5 Calculate solution concentration using molarity.

Calculate the volume of a 0.750 M sulfuric acid (H_2SO_4) solution containing 0.120 mol of solute.

Solution

Substituting in our basic expression for molarity, we obtain

$$0.750 \ M \ H_2SO_4 = \frac{0.120 \ \text{mol} \ H_2SO_4}{X \ \text{L}}$$
$$X \ \text{L} = 0.160 \ \text{L}$$

 Practice Problem 6.7

Calculate the volume of a 0.200 M KCl solution containing 5.00×10^{-2} mol of solute.

For Further Practice: Questions 6.35 and 6.36.

Question 6.5

Calculate the number of moles of solute in 5.00×10^2 mL of 0.250 M HCl.

Question 6.6

Calculate the number of grams of silver nitrate required to prepare 2.00 L of 0.500 M $AgNO_3$.

Dilution

Laboratory reagents are often purchased as concentrated solutions (for example, 12 M HCl or 6 M NaOH) for reasons of safety, economy, and space limitations. We must often *dilute* such a solution to a larger volume to prepare a less concentrated solution for the experiment at hand. The approach to such a calculation is as follows.

We define

LEARNING GOAL

6 Perform dilution calculations.

Animation

Dilution

$$M_1 = \text{molarity of solution } before \text{ dilution}$$
$$M_2 = \text{molarity of solution } after \text{ dilution}$$
$$V_1 = \text{volume of solution } before \text{ dilution}$$
$$V_2 = \text{volume of solution } after \text{ dilution}$$

and

$$M = \frac{\text{moles solute}}{\text{L solution}}$$

This equation can be rearranged:

$$\text{moles solute} = (M)\,(\text{L solution})$$

The number of moles of solute *before* and *after* dilution is unchanged, because dilution involves only addition of extra solvent:

$$\text{moles}_1 \text{ solute} = \text{moles}_2 \text{ solute}$$

| Initial | Final |
| condition | condition |

or

$$(M_1)(\text{L}_1 \text{ solution}) = (M_2)(\text{L}_2 \text{ solution})$$
$$(M_1)(V_1) = (M_2)(V_2)$$

Knowing any three of these terms enables us to calculate the fourth.

| **Calculating Molarity After Dilution** | **EXAMPLE 6.8** |

Calculate the molarity of a solution made by diluting 0.050 L of 0.10 M HCl solution to a volume of 1.0 L.

LEARNING GOAL

6 Perform dilution calculations.

Continued—

E X A M P L E 6.8 —*Continued*

Solution

Step 1. Summarize the information provided in the problem:

$$M_1 = 0.10\ M$$
$$M_2 = X\ M$$
$$V_1 = 0.050\ L$$
$$V_2 = 1.0\ L$$

Step 2. Use the dilution expression:

$$(M_1)\,(V_1) = (M_2)\,(V_2)$$

Step 3. Solve for M_2, the final solution concentration:

$$M_2 = \frac{(M_1)\,(V_1)}{V_2}$$

Step 4. Substituting,

$$X\ M = \frac{(0.10\ M)\,(0.050\ L)}{(1.0\ L)}$$
$$= 0.0050\ M \text{ or } 5.0 \times 10^{-3}\ M\ \text{HCl}$$

Practice Problem 6.8

What volume of 0.200 M sugar solution can be prepared from 50.0 mL of 0.400 M solution?

For Further Practice: Questions 6.38 and 6.39.

E X A M P L E 6.9 **Calculating a Dilution Volume**

Calculate the volume, in liters, of water that must be added to dilute 20.0 mL of 12.0 M HCl to 0.100 M HCl.

Solution

Step 1. Summarize the information provided in the problem:

$$M_1 = 12.0\ M$$
$$M_2 = 0.100\ M$$
$$V_1 = 20.0\ \text{mL (0.0200 L)}$$
$$V_2 = V_{\text{final}}$$

Step 2. Then, using the dilution expression:

$$(M_1)\,(V_1) = (M_2)\,(V_2)$$

Step 3. Solve for V_2, the final volume:

$$V_2 = \frac{(M_1)\,(V_1)}{(M_2)}$$

Continued—

EXAMPLE 6.9 —*Continued*

Step 4. Substituting,

$$V_{final} = \frac{(12.0\ M)\ (0.0200\ L)}{0.100\ M}$$
$$= 2.40\ L\ \text{solution}$$

Note that this is the *total final volume*. The amount of water added equals this volume *minus* the original solution volume, or

$$2.40\ L - 0.0200\ L = 2.38\ L\ \text{water}$$

Practice Problem 6.9

How would you prepare 1.0×10^2 mL of 2.0 M HCl, starting with concentrated (12.0 M) HCl?

For Further Practice: Questions 6.37 and 6.40.

The dilution equation is valid with any concentration units, such as % (W/V) *as well as* molarity, which was used in Examples 6.8 and 6.9. However, you must use the same units for both initial *and* final concentration values. Only in this way can you cancel units properly.

Representation of Concentration of Ions in Solution

The concentration of ions in solution may be represented in a variety of ways. The most common include moles per liter (molarity) and equivalents per liter.

When discussing solutions of ionic compounds, molarity emphasizes the number of individual ions. A one molar solution of Na^+ contains Avogadro's number, 6.022×10^{23}, of Na^+ per liter. In contrast, equivalents per liter emphasize charge; one equivalent of Na^+ contains Avogadro's number of positive charge.

We defined 1 mol as the number of grams of an atom, molecule, or ion corresponding to Avogadro's number of particles. One **equivalent** of an ion is the number of grams of the ion corresponding to Avogadro's number of electrical charges. Some examples follow:

1 mol Na^+ = 1 equivalent Na^+	(one Na^+ = 1 unit of charge/ion)
1 mol Cl^- = 1 equivalent Cl^-	(one Cl^- = 1 unit of charge/ion)
1 mol Ca^{2+} = 2 equivalents Ca^{2+}	(one Ca^{2+} = 2 units of charge/ion)
1 mol CO_3^{2-} = 2 equivalents CO_3^{2-}	(one CO_3^{2-} = 2 units of charge/ion)
1 mol PO_4^{3-} = 3 equivalents PO_4^{3-}	(one PO_4^{3-} = 3 units of charge/ion)

Changing from moles per liter to equivalents per liter (or the reverse) can be accomplished by using conversion factors.

Milliequivalents (meq) or milliequivalents/liter (meq/L) are often used when describing small amounts or low concentration of ions. These units are routinely used when describing ions in blood, urine, and blood plasma.

LEARNING GOAL

7 Interconvert molar concentration of ions and milliequivalents/liter.

Calculating Ion Concentration

EXAMPLE 6.10

Calculate the number of equivalents per liter (eq/L) of phosphate ion, PO_4^{3-}, in a solution that is $5.0 \times 10^{-3}\ M$ phosphate.

LEARNING GOAL

7 Interconvert molar concentration of ions and milliequivalents/liter.

Continued—

EXAMPLE 6.10 —*Continued*

Solution

Step 1. It is necessary to use two conversion factors:

$$mol\ PO_4^{3-} \longrightarrow mol\ charge$$

and

$$mol\ charge \longrightarrow eq\ PO_4^{3-}$$

Step 2. Arranging these factors in sequence yields:

$$\frac{5.0 \times 10^{-3}\ mol\ PO_4^{3-}}{1\ L} \times \frac{3\ mol\ charge}{1\ mol\ PO_4^{3-}} \times \frac{1\ eq\ PO_4^{3-}}{1\ mol\ charge} = \frac{1.5 \times 10^{-2}\ eq\ PO_4^{3-}}{L}$$

Practice Problem 6.10

Calculate the number of equivalents per liter (eq/L) of carbonate ion, CO_3^{2-}, in a solution that is $6.4 \times 10^{-4}\ M$ carbonate ion.

For Further Practice: Questions 6.47 and 6.48.

6.4 Concentration-Dependent Solution Properties

Colligative properties are solution properties that depend on the *concentration of the solute particles*, rather than the *identity of the solute*.

There are four colligative properties of solutions:

- vapor pressure lowering
- freezing point depression
- boiling point elevation
- osmotic pressure

Each of these properties has widespread practical application. We look at each in some detail in the following sections.

Vapor Pressure Lowering

Raoult's law states that, when a nonvolatile solute is added to a solvent, the vapor pressure of the solvent decreases in proportion to the concentration of the solute.

Perhaps the most important consequence of Raoult's law is the effect of the solute on the freezing and boiling points of a solution.

When a nonvolatile solute is added to a solvent, the freezing point of the resulting solution decreases (a lower temperature is required to convert the liquid to a solid). The boiling point of the solution is found to increase (it requires a higher temperature to form the gaseous state).

Raoult's law may be explained in molecular terms by using the following logic: Vapor pressure of a solution results from the escape of solvent molecules from the liquid to the gas phase, thus increasing the partial pressure of the gas phase solvent molecules until the equilibrium vapor pressure is reached. Presence of solute molecules hinders the escape of solvent molecules, thus lowering the equilibrium vapor pressure (Figure 6.2).

Figure 6.2

An illustration of Raoult's law: lowering of vapor pressure by addition of solute molecules. White units represent solvent molecules, and red units are solute molecules. Solute molecules present a barrier to escape of solvent molecules, thus decreasing the vapor pressure.

Freezing Point Depression and Boiling Point Elevation

The freezing point depression may be explained by examining the equilibrium between solid and liquid states. At the freezing point, ice is in equilibrium with liquid water:

Recall that the concept of liquid vapor pressure was discussed in Section 5.2.

$$H_2O \ (l) \underset{(r)}{\overset{(f)}{\rightleftharpoons}} H_2O \ (s)$$

The solute molecules interfere with the rate at which liquid water molecules associate to form the solid state, decreasing the rate of the forward reaction. For a true equilibrium, the rate of the forward (f) and reverse (r) processes must be equal. Lowering the temperature eventually slows the rate of the reverse (r) process sufficiently to match the rate of the forward reaction. At the lower temperature, equilibrium is established, and the solution freezes.

The boiling point elevation can be explained by considering the definition of the boiling point, that is, the temperature at which the vapor pressure of the liquid equals the atmospheric pressure. Raoult's law states that the vapor pressure of a solution is decreased by the presence of a solute. Therefore a higher temperature is necessary to raise the vapor pressure to the atmospheric pressure, hence the boiling point elevation.

Section 7.4 discusses equilibrium.

The extent of the freezing point depression (ΔT_f) is proportional to the solute concentration over a limited range of concentration:

$$\Delta T_f = k_f \times (\text{solute concentration})$$

The boiling point elevation (ΔT_b) is also proportional to the solute concentration:

$$\Delta T_b = k_b \times (\text{solute concentration})$$

If the value of the proportionality factor (k_f or k_b) is known for the solvent of interest, the magnitude of the freezing point depression or boiling point elevation can be calculated for a solution of known concentration.

Solute concentration must be in *mole*-based units. The number of particles (molecules or ions) is critical here, not the mass of solute. One *heavy* molecule will have exactly the same effect on the freezing or boiling point as one *light* molecule. A mole-based unit, because it is related directly to Avogadro's number, will correctly represent the number of particles in solution.

We have already worked with one mole-based unit, *molarity*, and this concentration unit can be used to calculate either the freezing point depression or the boiling point elevation.

A second mole-based concentration unit, molality, is more commonly used in these types of situations. **Molality** (symbolized m) is defined as the number of moles of solute per kilogram of solvent in a solution:

$$m = \frac{\text{moles solute}}{\text{kg solvent}}$$

Molality does not vary with temperature, whereas molarity is temperature dependent. For this reason, molality is the preferred concentration unit for studies such as freezing point depression and boiling point elevation, in which measurement of *change* in temperature is critical.

Practical applications that take advantage of freezing point depression of solutions by solutes include the following:

Molarity is temperature dependent simply because it is expressed as mole/volume. Volume is temperature dependent—most liquids expand measurably when heated and contract when cooled. Molality is moles/mass; both moles and mass are temperature independent.

- Salt is spread on roads to melt ice in winter. The salt lowers the freezing point of the water, so it exists in the liquid phase below its normal freezing point, 0°C or 32°F.
- Solutes such as ethylene glycol, "antifreeze," are added to car radiators in the winter to prevent freezing by lowering the freezing point of the coolant.

We have referred to the concentration of *particles* in our discussion of colligative properties. Why did we stress this term? The reason is that there is a very important difference between electrolytes and nonelectrolytes. That difference is the way in which they behave when they dissolve. For example, if we dissolve 1 mol of glucose ($C_6H_{12}O_6$) in 1L of water,

$$1\ C_6H_{12}O_6(s) \xrightarrow{\ H_2O\ } 1\ C_6H_{12}O_6(aq)$$

1 mol (Avogadro's number, 6.022×10^{23} particles) of glucose is present in solution. *Glucose is a covalently bonded nonelectrolyte.* Dissolving 1 mol of sodium chloride in 1L of water,

$$1\ NaCl(s) \xrightarrow{\ H_2O\ } 1\ Na^+(aq) + 1\ Cl^-(aq)$$

produces 2 mol of particles (1 mol of sodium ions and 1 mol of chloride ions). *Sodium chloride is an ionic electrolyte.*

$$1\ \text{mol glucose} \longrightarrow 1\ \text{mol of particles in solution}$$
$$1\ \text{mol sodium chloride} \longrightarrow 2\ \text{mol of particles in solution}$$

It follows that 1 mol of sodium chloride will decrease the vapor pressure, increase the boiling point, or depress the freezing point of 1L of water *twice as much* as 1 mol of glucose in the same quantity of water.

Question 6.7

Comparing pure water and a 10% (W/V) glucose solution, which has the higher freezing point?

Question 6.8

Comparing pure water and a 10% (W/V) glucose solution, which has the higher boiling point?

Osmotic Pressure, Osmosis, and Osmolarity

Animation
Osmosis

The cell membrane mediates the interaction of the cell with its environment and is responsible for the controlled passage of material into and out of the cell. One of the principle means of transport is termed *diffusion*. **Diffusion** is the net movement of solute or solvent molecules from an area of high concentration to an area of low concentration. This region where the concentration decreases over a distance is termed the **concentration gradient.** Because of the structure of the cell membrane, only small molecules are able to diffuse freely across this barrier. Large molecules and highly charged ions are restricted by that barrier. In other words, the cell membrane is behaving in a selective fashion. Such membranes are termed **selectively permeable membranes.**

The term *selectively permeable* or *differentially permeable* is used to describe biological membranes because they restrict passage of particles based both on size and charge. Even small ions, such as H^+, cannot pass freely across a cell membrane.

Because a cell membrane is selectively permeable, it is not always possible for solutes to pass through it in response to a concentration gradient. In such cases, the solvent diffuses through the membrane. Such membranes, permeable to solvent but not to solute, are specifically called **semipermeable membranes. Osmosis** is the diffusion of a solvent (water in biological systems) through a semipermeable membrane in response to a water concentration gradient.

Suppose that we place a 0.5 M glucose solution in a dialysis bag that is composed of a membrane with pores that allow the passage of water molecules but not glucose molecules. Consider what will happen when we place this bag into a beaker of pure water. We have created a gradient in which there is a higher concentration of glucose inside the bag than outside, but the glucose cannot diffuse through the bag to achieve equal concentration on both sides of the membrane.

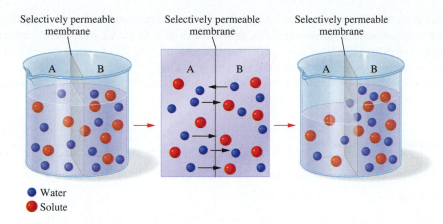

Water
Solute

Figure 6.3
Osmosis across a membrane. The solvent, water, diffuses from an area of lower solute concentration (side A) to an area of higher solute concentration (side B).

Now let's think about this situation in another way. We have a higher concentration of water molecules outside the bag (where there is only pure water) than inside the bag (where some of the water molecules are occupied in dipole-dipole interactions with solute particles and are consequently unable to move freely in the system). Because water can diffuse through the membrane, a net diffusion of water will occur through the membrane into the bag. This is the process of osmosis (Figure 6.3).

As you have probably already guessed, this system can never reach equilibrium (equal concentrations inside and outside the bag). Regardless of how much water diffuses into the bag, diluting the glucose solution, the concentration of glucose will always be higher inside the bag (and the accompanying free water concentration will always be lower).

What happens when the bag has taken in as much water as it can, when it has expanded as much as possible? Now the walls of the bag exert a force that will stop the *net* flow of water into the bag. **Osmotic pressure** is the pressure that must be exerted to stop the flow of water across a selectively permeable membrane by osmosis. Stated more precisely, the osmotic pressure of a solution is the net pressure with which water enters it by osmosis from a pure water compartment when the two compartments are separated by a semipermeable membrane.

The osmotic pressure, like the pressure exerted by a gas, may be treated quantitatively. Osmotic pressure, symbolized by π, follows the same form as the ideal gas equation:

Ideal Gas	**Osmotic Pressure**
$PV = nRT$	$\pi V = nRT$
or	or
$P = \dfrac{n}{V}RT$	$\pi = \dfrac{n}{V}RT$
and since	and since
$M = \dfrac{n}{V}$	$M = \dfrac{n}{V}$
then	then
$P = MRT$	$\pi = MRT$

The osmotic pressure can be calculated from the solution concentration at any temperature. How do we determine "solution concentration"? Recall that osmosis is a colligative property, dependent on the concentration of solute particles. Again, it becomes necessary to distinguish between solutions of electrolytes and

Chapter 6 Solutions

nonelectrolytes. For example, a 1 M glucose solution consists of 1 mol of particles per liter; glucose is a nonelectrolyte. A solution of 1 M NaCl produces 2 mol of particles per liter (1 mol of Na$^+$ and 1 mol of Cl$^-$). A 1 M CaCl$_2$ solution is 3 M in particles (1 mol of Ca^{2+} and 2 mol of Cl$^-$ per liter).

Osmolarity, the molarity of particles in solution, and abbreviated osmol, is used for osmotic pressure calculations.

EXAMPLE 6.11 **Calculating Osmolarity**

Determine the osmolarity of $5.0 \times 10^{-3}\,M$ Na$_3$PO$_4$.

Solution

Step 1. Na$_3$PO$_4$ is an ionic compound and produces an electrolytic solution:

$$\text{Na}_3\text{PO}_4 \xrightarrow{\text{H}_2\text{O}} 3\text{Na}^+ + \text{PO}_4^{3-}$$

Step 2. 1 mol of Na$_3$PO$_4$ yields four product ions; consequently

$$5.0 \times 10^{-3}\,\frac{\text{mol Na}_3\text{PO}_4}{\text{L}} \times \frac{4\text{ mol particles}}{1\text{ mol Na}_3\text{PO}_4} = 2.0 \times 10^{-2}\,\frac{\text{mol particles}}{\text{L}}$$

Step 3. Using our expression for osmolarity,

$$2.0 \times 10^{-2}\,\frac{\text{mol particles}}{\text{L}} = 2.0 \times 10^{-2}\text{ osmol}$$

Practice Problem 6.11

Determine the osmolarity of the following solutions:
a. $5.0 \times 10^{-3}\,M$ NH$_4$NO$_3$ (electrolyte)
b. $5.0 \times 10^{-3}\,M$ C$_6$H$_{12}$O$_6$ (nonelectrolyte)

For Further Practice: Questions 6.77 and 6.78.

EXAMPLE 6.12 **Calculating Osmotic Pressure**

Calculate the osmotic pressure of a $5.0 \times 10^{-2}\,M$ solution of NaCl at 25°C (298 K).

Solution

Step 1. Using our definition of osmotic pressure, π:

$$\pi = MRT$$

Step 2. M should be represented as osmolarity as we have shown in Example 6.11

$$M = 5.0 \times 10^{-2}\,\frac{\text{mol NaCl}}{\text{L}} \times \frac{2\text{ mol particles}}{1\text{ mol NaCl}} = 1.0 \times 10^{-1}\,\frac{\text{mol particles}}{\text{L}}$$

Step 3. Substituting in our osmotic pressure expression:

$$\pi = 1.0 \times 10^{-1}\,\frac{\text{mol particles}}{\text{L}} \times 0.0821\,\frac{\text{L-atm}}{\text{K-mol}} \times 298\text{ K}$$

$$= 2.4\text{ atm}$$

Continued—

EXAMPLE 6.12 —Continued

Practice Problem 6.12

Calculate the osmotic pressure of each solution described in Practice Problem 6.11. Assume that the solutions are at 298 K.

For Further Practice: Questions 6.63 and 6.64.

Blood plasma has an osmolarity equivalent to a 0.30 *M* glucose solution or a 0.15 *M* NaCl solution. The latter is true because NaCl in solution dissociates into Na^+ and Cl^- and thus contributes twice the number of solute particles as a molecule that does not ionize. If red blood cells, which have an osmolarity equal to blood plasma, are placed in a 0.30 *M* glucose solution, no net osmosis will occur because the osmolarity and water concentration inside the red blood cell are equal to those of the 0.30 *M* glucose solution. The solutions inside and outside the red blood cell are said to be **isotonic** (*iso* means "same," and *tonic* means "strength") **solutions.** Because the osmolarity is the same inside and outside, the red blood cell will remain the same size (Figure 6.4b).

What happens if we now place the red blood cells into a **hypotonic solution,** in other words, a solution having a lower osmolarity than the cytoplasm of the cell? In this situation there will be a net movement of water into the cell as water diffuses down its concentration gradient. The membrane of the red blood cell does not have the strength to exert a sufficient pressure to stop this flow of water, and the cell will swell and burst (Figure 6.4c). Alternatively, if we place the red blood cells into a **hypertonic solution** (one with a greater osmolarity than the cell), water will pass out of the cells, and they will shrink dramatically in size (Figure 6.4a).

These principles have important applications in the delivery of intravenous (IV) solutions into an individual. Normally, any fluids infused intravenously must have the correct osmolarity; they must be isotonic with the blood cells and the blood plasma. Such infusions are frequently either 5.5% dextrose (glucose) or "normal saline." The first solution is composed of 5.5 g of glucose per 100 mL of solution (0.30 *M*), and the latter of 9.0 g of NaCl per 100 mL of solution (0.15 *M*). In either case, they have the same osmotic pressure and osmolarity as the plasma and blood cells and can therefore be safely administered without upsetting the osmotic balance between the blood and the blood cells.

Practical examples of osmosis abound, including the following:

- A sailor, lost at sea in a lifeboat, dies of dehydration while surrounded by water. Seawater, because of its high salt concentration, dehydrates the cells of the body as a result of the large osmotic pressure difference between itself and intracellular fluids.

Figure 6.4

Scanning electron micrographs of red blood cells exposed to (a) hypertonic, (b) isotonic, and (c) hypotonic solutions.

(a)   (b)   (c)

A Medical Perspective

Oral Rehydration Therapy

Diarrhea kills millions of children before they reach the age of five years. This is particularly true in third world countries where sanitation, water supplies, and medical care are poor. In the case of diarrhea, death results from fluid loss, electrolyte imbalance, and hypovolemic shock (multiple organ failure due to insufficient perfusion). Cholera is one of the best-understood bacterial diarrheas. The organism *Vibrio cholera*, seen in the micrograph below, survives passage through the stomach and reproduces in the intestine, where it produces a toxin called choleragen. The toxin causes the excessive excretion of Na^+, Cl^-, and HCO_3^- from epithelial cells lining the intestine. The increased ion concentration (hypertonic solution) outside the cell results in movement of massive quantities of water into the intestinal lumen. This causes the severe, abundant, clear vomit and diarrhea that can result in the loss of 10–15 L of fluid per day. Over the four- to six-day progress of the disease, a patient may lose from one to two times his or her body mass!

The need for fluid replacement is obvious. Oral rehydration is preferred over intravenous administration of fluids and electrolytes since it is noninvasive. In many third world countries, it is the only therapy available in remote areas. The rehydration formula includes 50–80 g/L rice (or other starch), 3.5 g/L sodium chloride, 2.5 g/L sodium bicarbonate, and 1.5 g/L potassium chloride. Oral rehydration takes advantage of the cotransport of Na^+ and glucose across the cells lining the intestine. Thus, the channel protein brings glucose into the cells, and Na^+ is carried along. Movement of these materials into the cells will help alleviate the osmotic imbalance, reduce the diarrhea, and correct the fluid and electrolyte imbalance.

The disease runs its course in less than a week. In fact, antibiotics are not used to combat cholera. The only effective therapy is oral rehydration, which reduces mortality to less than 1%. A much better option is prevention. In the photo below, a woman is shown filtering water through sari cloth. This simple practice has been shown to reduce the incidence of cholera significantly.

Vibrio cholera.

A woman is shown filtering water through sari cloth.

For Further Understanding

Explain dehydration in terms of osmosis.

Explain why even severely dehydrated individuals continue to experience further fluid loss.

- A cucumber, soaked in brine, shrivels into a pickle. The water in the cucumber is drawn into the brine (salt) solution because of a difference in osmotic pressure (Figure 6.5).
- A Medical Perspective: Oral Rehydration Therapy describes one of the most lethal and pervasive examples of cellular fluid imbalance.

(a)

(b)

Figure 6.5
A cucumber (a) in an acidic salt solution undergoes considerable shrinkage on its way to becoming a pickle (b) because of osmosis.

6.5 Water as a Solvent

Water is by far the most abundant substance on earth. It is an excellent solvent for most inorganic substances. In fact, it is often referred to as the "universal solvent" and is the principal biological solvent. Approximately 60% of the adult human body is water, and maintenance of this level is essential for survival. These characteristics are a direct consequence of the molecular structure of water.

As we saw in Chapter 3, water is a bent molecule with a 104.5° bond angle. This angular structure, resulting from the effect of the two lone pairs of electrons around the oxygen atom, is responsible for the polar nature of water. The polarity, in turn, gives water its unique properties.

Because water molecules are polar, water is an excellent solvent for other polar substances ("like dissolves like"). Because much of the matter on earth is polar, hence at least somewhat water soluble, water has been described as the universal solvent. It is readily accessible and easily purified. It is nontoxic and quite nonreactive. The high boiling point of water, 100°C, compared with molecules of similar size such as N_2 (b.p. = −196°C), is also explained by water's polar character. Strong dipole-dipole interactions between a δ^+ hydrogen of one molecule and δ^- oxygen of a second, referred to as hydrogen bonding, create an interactive molecular network in the liquid phase (see Figure 5.8a). The strength of these interactions requires more energy (higher temperature) to cause water to boil. The higher than expected boiling point enhances water's value as a solvent; often, reactions are carried out at higher temperatures to increase their rate. Other solvents, with lower boiling points, would simply boil away, and the reaction would stop.

This idea is easily extended to our own chemistry—because 60% of our bodies is water, we should appreciate the polarity of water on a hot day. As a biological solvent in the human body, water is involved in the transport of ions, nutrients, and waste into and out of cells. Water is also the solvent for biochemical reactions in cells and the digestive tract. Water is a reactant or product in some biochemical processes.

 LEARNING GOAL
9 Describe why the chemical and physical properties of water make it a truly unique solvent.

 Animation
The Properties of Water

See A Human Perspective: An Extraordinary Molecule, in this chapter.

Refer to Sections 3.4 and 3.5 for a more complete description of the bonding, structure, and polarity of water.

Recall the discussion of intermolecular forces in Chapters 3 and 5.

A Human Perspective

An Extraordinary Molecule

Think for a moment. What is the only common molecule that exists in all three physical states of matter (solid, liquid, and gas) under natural conditions on earth? This molecule is absolutely essential for life; in fact, life probably arose in this substance. It is the most abundant molecule in the cells of living organisms (70–95%) and covers 75% of the earth's surface. Without it, cells quickly die, and without it the earth would not be a fit environment in which to live. By now you have guessed that we are talking about the water molecule. It is so abundant on earth that we take this deceptively simple molecule for granted.

What are some of the properties of water that cause it to be essential to life as we know it? Water has the ability to stabilize temperatures on the earth and in the body. This ability is due in part to the energy changes that occur when water changes physical state; but ultimately, this ability is due to the polar nature of the water molecule.

Life can exist only within a fairly narrow range of temperatures. Above or below that range, the chemical reactions necessary for life, and thus life itself, will cease. Water can moderate temperature fluctuation and maintain the range necessary for life, and one property that allows it to do so is its unusually high specific heat, 1 cal/g °C. This means that water can absorb or lose more heat energy than many other substances without a significant temperature change.

This is because in the liquid state, every water molecule is hydrogen bonded to other water molecules. Because a temperature increase is really just a measure of increased (more rapid) molecular movement, we must get the water molecules moving more rapidly, independent of one another, to register a temperature increase. Before we can achieve this independent, increased activity, the hydrogen bonds between molecules must be broken. Much of the heat energy that water absorbs is involved in breaking hydrogen bonds and is *not* used to increase molecular movement. Thus a great deal of heat is needed to raise the temperature of water even a little bit.

Water also has a very high heat of vaporization. It takes 540 calories to change 1 g of liquid water at 100°C to a gas and even more, 603 cal/g, when the water is at 37°C, human body temperature. That is about twice the heat of vaporization of alcohol. As water molecules evaporate, the surface of the liquid cools because only the highest-energy (or "hottest") molecules leave as a gas. Only the "hottest" molecules have enough energy to break the hydrogen bonds that bind them to other water molecules. Indeed, evaporation of water molecules from the surfaces of lakes and oceans helps to maintain stable temperatures in those bodies of water. Similarly, evaporation of perspiration from body surfaces helps to prevent overheating on a hot day or during strenuous exercise.

EXAMPLE 6.13   **Predicting Structure from Observable Properties**

LEARNING GOAL

9 Describe why the chemical and physical properties of water make it a truly unique solvent.

Sucrose is a common sugar and we know that it is used as a sweetener when dissolved in many beverages. What does this allow us to predict about the structure of sucrose?

Solution

Sucrose is used as a sweetener in teas, coffee, and a host of soft drinks. The solvent in all of these beverages is water, a polar molecule. The rule "like dissolves like" implies that sucrose must also be a polar molecule. Without even knowing the formula or structure of sucrose, we can infer this important information from a simple experiment—dissolving sugar in our morning cup of coffee.

Practice Problem 6.13

a. Predict whether carbon monoxide or carbon dioxide would be more soluble in water. Explain your answer. (Hint: Refer to Section 5.2, the discussion of interactions in the liquid state.)

b. Predict whether ammonia or methane would be more soluble in water. Explain your answer. (Hint: Refer to Section 5.2, the discussion of interactions in the liquid state.)

For Further Practice: Questions 6.83 and 6.84.

Even the process of freezing helps stabilize and moderate temperatures. This is especially true in the fall. Water releases heat when hydrogen bonds are formed. This is an example of an exothermic process. Thus, when water freezes, solidifying into ice, additional hydrogen bonds are formed, and heat is released into the environment. As a result, the temperature change between summer and winter is more gradual, allowing organisms to adjust to the change.

One last feature that we take for granted is the fact that when we put ice in our iced tea on a hot summer day, the ice floats. This means that the solid state of water is actually *less* dense than the liquid state! In fact, it is about 10% less dense, having an open lattice structure with each molecule hydrogen bonded to the maximum of four other water molecules. What would happen if ice did sink? All bodies of water, including the mighty oceans would eventually freeze solid, killing all aquatic and marine plant and animal life. Even in the heat of summer, only a few inches of ice at the surface would thaw. Instead, the ice forms at the surface and provides a layer of insulation that prevents the water below from freezing.

As we continue our study of chemistry, we will refer again and again to this amazing molecule. In other Human Perspective features we will examine other properties of water that make it essential to life.

Beauty is also a property of water.

For Further Understanding

Why is the high heat of vaporization of water important to our bodies?

Why is it cooler at the ocean shore than in the desert during summer?

6.6 Electrolytes in Body Fluids

The concentrations of cations, anions, and other substances in biological fluids are critical to health. Consequently, the osmolarity of body fluids is carefully regulated by the kidney.

The two most important cations in body fluids are Na^+ and K^+. Sodium ion is the most abundant cation in the blood and intercellular fluids whereas potassium ion is the most abundant intracellular cation. In blood and intercellular fluid, the Na^+ concentration is 135 milliequivalents/L and the K^+ concentration is 3.5–5.0 meq/L. Inside the cell, the situation is reversed. The K^+ concentration is 125 meq/L and the Na^+ concentration is 10 meq/L.

If osmosis and simple diffusion were the only mechanisms for transporting water and ions across cell membranes, these concentration differences would not occur. One positive ion would be just as good as any other. However, the situation is more complex than this. Large protein molecules embedded in cell membranes actively pump sodium ions to the outside of the cell and potassium ions into the cell. This is termed *active transport* because cellular energy must be expended to transport those ions. Proper cell function in the regulation of muscles and the nervous system depends on the sodium ion/potassium ion ratio inside and outside of the cell.

If the Na^+ concentration in the blood becomes too low, urine output decreases, the mouth feels dry, the skin becomes flushed, and a fever may develop. The blood

LEARNING GOAL

10 Explain the role of electrolytes in blood and their relationship to the process of dialysis.

A Medical Perspective

Hemodialysis

As we have seen in Section 6.6, blood is the medium for exchange of both nutrients and waste products. The membranes of the kidneys remove waste materials such as urea and uric acid (Chapter 22) and excess salts and large quantities of water. This process of waste removal is termed **dialysis,** a process similar in function to osmosis (Section 6.4). Semipermeable membranes in the kidneys, dialyzing membranes, allow small molecules (principally water and urea) and ions in solution to pass through and ultimately collect in the bladder. From there they can be eliminated from the body.

Unfortunately, a variety of diseases can cause partial or complete kidney failure. Should the kidneys fail to perform their primary function, dialysis of waste products, urea and other waste products rapidly increase in concentration in the blood. This can become a life-threatening situation in a very short time.

The most effective treatment of kidney failure is the use of a machine, an artificial kidney, that mimics the function of the kidney. The artificial kidney removes waste from the blood using the process of hemodialysis (blood dialysis). The blood is pumped through a long semipermeable membrane, the dialysis membrane. The dialysis process is similar to osmosis. However, in addition to water molecules, larger molecules (including the waste products in the blood) and ions can pass across the membrane from the blood into a dialyzing fluid. The dialyzing fluid is isotonic with normal blood; it also is similar in its concentration of all other essential blood components.

The waste materials move across the dialysis membrane (from a higher to a lower concentration, as in diffusion). A successful dialysis procedure selectively removes the waste from the body without upsetting the critical electrolyte balance in the blood.

Hemodialysis, although lifesaving, is not by any means a pleasant experience. The patient's water intake must be severely limited to minimize the number of times each week that treatment must be used. Many dialysis patients require two or three treatments per week and each session may require one-half (or more) day of hospitalization, especially when the patient suffers from complicating conditions such as diabetes.

Improvements in technology, as well as the growth and sophistication of our health care delivery systems over the past several years, have made dialysis treatment much more patient friendly. Dialysis centers, specializing in the treatment of kidney patients, are now found in most major population centers. Smaller, more automated dialysis units are available for home use, under the supervision of a nursing practitioner. With the remarkable progress in kidney transplant success, dialysis is becoming, more and more, a temporary solution, sustaining life until a suitable kidney donor match can be found.

For Further Understanding

In what way is dialysis similar to osmosis?

In what way does dialysis differ from osmosis?

Dialysis patient.

level of Na^+ may be elevated when large amounts of water are lost. Diabetes, certain high-protein diets, and diarrhea may cause elevated blood Na^+ level. In extreme cases, elevated Na^+ levels may cause confusion, stupor, or coma.

Concentrations of K^+ in the blood may rise to dangerously high levels following any injury that causes large numbers of cells to rupture, releasing their intracellular K^+. This may lead to death by heart failure. Similarly, very low levels of K^+ in the blood may also cause death from heart failure. This may occur following prolonged exercise that results in excessive sweating. When this happens, both body fluids and electrolytes must be replaced. Salt tablets containing both NaCl and KCl taken with water and drinks such as Gatorade effectively provide water and electrolytes and prevent serious symptoms.

The cationic charge in blood is neutralized by two major anions, Cl^- and HCO_3^-. The chloride ion plays a role in acid-base balance, maintenance of osmotic pressure within an acceptable range, and oxygen transport by hemoglobin. The bicarbonate anion is the form in which most waste CO_2 is carried in the blood.

A variety of proteins is also found in the blood. Because of their larger size, they exist in colloidal suspension. These proteins include blood clotting factors, immunoglobulins (antibodies) that help us fight infection, and albumins that act as carriers of nonpolar, hydrophobic substances (fatty acids and steroid hormones) that cannot dissolve in water.

Additionally, blood is the medium for exchange of nutrients and waste products. Nutrients, such as the polar sugar glucose, enter the blood from the intestine or the liver. Because glucose molecules are polar, they dissolve in body fluids and are circulated to tissues throughout the body. As noted above, nonpolar nutrients are transported with the help of carrier proteins. Similarly, nitrogen-containing waste products, such as urea, are passed from cells to the blood. They are continuously and efficiently removed from the blood by the kidneys.

In cases of loss of kidney function, mechanical devices—dialysis machines— mimic the action of the kidney. The process of blood dialysis—hemodialysis—is discussed in A Medical Perspective: Hemodialysis on page 210.

Calculating Electrolyte Concentrations   E X A M P L E **6.14**

A typical concentration of calcium ion in blood plasma is 4 meq/L. Represent this concentration in moles/L.

10 | **LEARNING GOAL**
Explain the role of electrolytes in blood and their relationship to the process of dialysis.

Solution

Step 1. The calcium ion has a 2+ charge (recall that calcium is in Group IIA of the periodic table; hence, a 2+ charge on the calcium ion).

Step 2. We will need three conversion factors:

$$\text{meq (milliequivalents)} \longrightarrow \text{eq (equivalents)}$$

$$\text{eq (equivalents)} \longrightarrow \text{moles of charge}$$

$$\text{moles of charge} \longrightarrow \text{moles of calcium ion}$$

Step 3. Using dimensional analysis as in Example 6.10,

$$\frac{4 \text{ meq Ca}^{2+}}{1L} \times \frac{1 \text{ eq Ca}^{2+}}{10^3 \text{ meq Ca}^{2+}} \times \frac{1 \text{ mol charge}}{1 \text{ eq Ca}^{2+}} \times \frac{1 \text{ mol Ca}^{2+}}{2 \text{ mol charge}} = \frac{2 \times 10^{-3} \text{ mol Ca}^{2+}}{L}$$

Continued—

EXAMPLE 6.14 —*Continued*

Practice Problem 6.14

Sodium chloride [0.9% (W/V)] is a solution administered intravenously to replace fluid loss. It is frequently used to avoid dehydration. The sodium ion concentration is 15.4 meq/L. Calculate the sodium ion concentration in moles/L.

For Further Practice: Questions 6.93 and 6.94.

SUMMARY

6.1 Properties of Solutions

A majority of chemical reactions, and virtually all important organic and biochemical reactions, take place not as a combination of two or more pure substances, but rather as reactants dissolved in solution, *solution reactions*.

A *solution* is a homogeneous (or uniform) mixture of two or more substances. A solution is composed of one or more *solutes*, dissolved in a *solvent*. When the solvent is water, the solution is called an *aqueous solution*.

Liquid solutions are clear and transparent with no visible particles of solute. They may be colored or colorless, depending on the properties of the solute and solvent.

In solutions of *electrolytes* the solutes are ionic compounds that dissociate in solution to produce ions. They are good conductors of electricity. Solutions of *nonelectrolytes* are formed from nondissociating molecular solutes (non-electrolytes), and their solutions are nonconducting.

The rule "like dissolves like" is the fundamental condition for solubility. Polar solutes are soluble in polar solvents, and nonpolar solutes are soluble in nonpolar solvents.

The degree of solubility depends on the difference between the polarity of solute and solvent, the temperature, and the pressure. Pressure considerations are significant only for solutions of gases.

When a solution contains all the solute that can be dissolved at a particular temperature, it is *saturated*. Excess solute falls to the bottom of the container as a *precipitate*. Occasionally, on cooling, the excess solute may remain in solution for a time before precipitation. Such a solution is a *supersaturated solution*. When excess solute, the precipitate, contacts solvent, the dissolution process reaches a state of dynamic equilibrium. *Colloidal suspensions* have particle sizes between those of true solutions and precipitates. A *suspension* is a heterogeneous mixture that contains particles much larger than a colloidal suspension. Over time, these particles may settle, forming a second phase.

Henry's law describes the solubility of gases in liquids. At a given temperature the solubility of a gas is proportional to the partial pressure of the gas.

6.2 Concentration Based on Mass

The amount of solute dissolved in a given amount of solution is the solution *concentration*. The more widely used percentage-based concentration units are *weight/volume percent* and *weight/weight percent*. Parts per thousand (ppt) and parts per million (ppm) are used with very dilute solutions.

6.3 Concentration of Solutions: Moles and Equivalents

Molarity, symbolized *M*, is defined as the number of moles of solute per liter of solution.

Dilution is often used to prepare less concentrated solutions. The expression for this calculation is $(M_1)(V_1) = (M_2)(V_2)$. Knowing any three of these terms enables one to calculate the fourth. The concentration of solute may be represented as moles per liter (molarity) or any other suitable concentration units. However, both concentrations must be in the same units when using the dilution equation.

When discussing solutions of ionic compounds, molarity emphasizes the number of individual ions. A 1 *M* solution of Na^+ contains Avogadro's number of sodium ions. In contrast, equivalents per liter emphasizes charge; a solution containing one equivalent of Na^+ per liter contains Avogadro's number of positive charge.

One *equivalent* of an ion is the number of grams of the ion corresponding to Avogadro's number of electrical charges. Changing from moles per liter to equivalents per liter (or the reverse) is done using conversion factors.

6.4 Concentration-Dependent Solution Properties

Solution properties that depend on the concentration of solute particles, rather than the identity of the solute, are *colligative properties*.

There are four colligative properties of solutions, all of which depend on the concentration of *particles* in solution.

1. *Vapor pressure lowering. Raoult's law* states that when a solute is added to a solvent, the vapor pressure of the solvent decreases in proportion to the concentration of the solute.

2. and 3. *Freezing point depression and boiling point elevation.* When a nonvolatile solid is added to a solvent, the freezing point of the resulting solution decreases, and the boiling point increases. The magnitudes of both the freezing point depression (ΔT_f) and the boiling point elevation (ΔT_b) are proportional to the solute concentration over a limited range of concentrations. The mole-based concentration unit, molality, is more commonly used in calculations involving colligative properties. This is due to the fact that molality is temperature independent. *Molality* (symbolized m) is defined as the number of moles of solute per kilogram of solvent in a solution.

4. *Osmosis and osmotic pressure. Osmosis* is the movement of solvent from a dilute solution to a more concentrated solution through a *semipermeable membrane*. The pressure that must be applied to the more concentrated solution to stop this flow is the *osmotic pressure*. The osmotic pressure, like the pressure exerted by a gas, may be treated quantitatively by using an equation similar in form to the ideal gas equation: $\pi = MRT$. By convention the molarity of particles that is used for osmotic pressure calculations is termed *osmolarity (osmol)*.

In biological systems, if the concentration of the fluid surrounding red blood cells is higher than that inside the cell (a *hypertonic* solution), water flows from the cell, causing it to collapse. Too low a concentration of this fluid relative to the solution within the cell (a *hypotonic* solution) will cause cell rupture.

Two solutions are *isotonic* if they have identical osmotic pressures. In that way the osmotic pressure differential across the cell is zero, and no cell disruption occurs.

6.5 Water as a Solvent

The role of water in the solution process deserves special attention. It is often referred to as the "universal solvent" because of the large number of ionic and polar covalent compounds that are at least partially soluble in water. It is the principal biological solvent. These characteristics are a direct consequence of the molecular geometry and structure of water and its ability to undergo hydrogen bonding.

6.6 Electrolytes in Body Fluids

The concentrations of cations, anions, and other substances in biological fluids are critical to health. As a result, the osmolarity of body fluids is carefully regulated by the kidney using the process of *dialysis*.

KEY TERMS

aqueous solution (6.1)
colligative property (6.4)
colloidal suspension (6.1)
concentration (6.2)
concentration gradient (6.4)
dialysis (6.6)
diffusion (6.4)
electrolyte (6.1)
equivalent (6.3)
Henry's law (6.1)
hypertonic solution (6.4)
hypotonic solution (6.4)
isotonic solution (6.4)
molality (6.4)
molarity (6.3)
nonelectrolyte (6.1)
osmolarity (6.4)
osmosis (6.4)

osmotic pressure (6.4)
precipitate (6.1)
Raoult's law (6.4)
saturated solution (6.1)
selectively permeable
  membrane (6.4)
semipermeable
  membrane (6.4)
solubility (6.1)
solute (6.1)
solution (6.1)
solvent (6.1)
supersaturated solution (6.1)
suspension (6.1)
weight/volume percent
  (% [W/V]) (6.2)
weight/weight percent
  (% [W/W]) (6.2)

QUESTIONS AND PROBLEMS

Concentration Based on Mass

Foundations

6.9 Calculate the composition of each of the following solutions in weight/volume %:
  a. 20.0 g NaCl in 1.00 L solution
  b. 33.0 g sugar, $C_6H_{12}O_6$, in 5.00×10^2 mL solution

6.10 Calculate the composition of each of the following solutions in weight/volume %:
  a. 0.700 g KCl per 1.00 mL
  b. 1.00 mol $MgCl_2$ in 2.50×10^2 mL solution

6.11 Calculate the composition of each of the following solutions in weight/volume %:
  a. 50.0 g ethanol dissolved in 1.00 L solution
  b. 50.0 g ethanol dissolved in 5.00×10^2 mL solution

6.12 Calculate the composition of each of the following solutions in weight/volume %:
  a. 20.0 g acetic acid dissolved in 2.50 L solution
  b. 20.0 g benzene dissolved in 1.00×10^2 mL solution

6.13 Calculate the composition of each of the following solutions in weight/weight %:
  a. 21.0 g NaCl in 1.00×10^2 g solution
  b. 21.0 g NaCl in 5.00×10^2 mL solution ($d = 1.12$ g/mL)

6.14 Calculate the composition of each of the following solutions in weight/weight %:
  a. 1.00 g KCl in 1.00×10^2 g solution
  b. 50.0 g KCl in 5.00×10^2 mL solution ($d = 1.14$ g/mL)

Applications

6.15 How many grams of solute are needed to prepare each of the following solutions?
  a. 2.50×10^2 g of 0.900% (W/W) NaCl
  b. 2.50×10^2 g of 1.25% (W/W) NaC$_2$H$_3$O$_2$ (sodium acetate)

6.16 How many grams of solute are needed to prepare each of the following solutions?
  a. 2.50×10^2 g of 5.00% (W/W) NH$_4$Cl (ammonium chloride)
  b. 2.50×10^2 g of 3.50% (W/W) Na$_2$CO$_3$

6.17 A solution was prepared by dissolving 14.6 g of KNO$_3$ in sufficient water to produce 75.0 mL of solution. What is the weight/volume % of this solution?

6.18 A solution was prepared by dissolving 12.4 g of NaNO$_3$ in sufficient water to produce 95.0 mL of solution. What is the weight/volume % of this solution?

6.19 How many grams of sugar would you use to prepare 100 mL of a 1.00 weight/volume % solution?

6.20 How many mL of 4.0 weight/volume % Mg(NO$_3$)$_2$ solution would contain 1.2 g of magnesium nitrate?

6.21 Which solution is more concentrated: a 0.04% (W/W) solution or a 50 ppm solution?

6.22 Which solution is more concentrated: a 20 ppt solution or a 200 ppm solution?

6.23 A solution contains 1.0 mg of Cu^{2+} per 0.50 kg solution. Calculate the concentration in ppt.

6.24 A solution contains 1.0 mg of Cu^{2+} per 0.50 kg solution. Calculate the concentration in ppm.

Concentration of Solutions: Moles and Equivalents

Foundations

6.25 Why is molarity often preferred over mass-based concentration units?

6.26 Write the expression for molarity.

6.27 Why is it often necessary to dilute solutions in laboratory?

6.28 Write the dilution expression and define each term.

Applications

6.29 Calculate the molarity of 5.0 L of solution containing 2.5 mol HNO$_3$.

6.30 Calculate the molarity of 2.75 L of solution containing 1.35×10^{-2} mol HCl.

6.31 Calculate the molarity of each solution in Problem 6.9.

6.32 Calculate the molarity of each solution in Problem 6.10.

6.33 Calculate the number of grams of solute that would be needed to make each of the following solutions:
  a. 2.50×10^2 mL of 0.100 M NaCl
  b. 2.50×10^2 mL of 0.200 M C$_6$H$_{12}$O$_6$ (glucose)

6.34 Calculate the number of grams of solute that would be needed to make each of the following solutions:
  a. 2.50×10^2 mL of 0.100 M NaBr
  b. 2.50×10^2 mL of 0.200 M KOH

6.35 Calculate the volume of a 0.500 M sucrose solution (table sugar, C$_{12}$H$_{22}$O$_{11}$) containing 0.133 mol of solute.

6.36 Calculate the volume of a $1.00 \times 10^{-2} M$ KOH solution containing 3.00×10^{-1} mol of solute.

6.37 It is desired to prepare 0.500 L of a 0.100 M solution of NaCl from a 1.00 M stock solution. How many milliliters of the stock solution must be taken for the dilution?

6.38 50.0 mL of a 0.250 M sucrose solution was diluted to 5.00×10^2 mL. What is the molar concentration of the resulting solution?

6.39 A 50.0-mL portion of a stock solution was diluted to 500.0 mL. If the resulting solution was 2.00 M, what was the molarity of the original stock solution?

6.40 A 6.00-mL portion of an 8.00 M stock solution is to be diluted to 0.400 M. What will be the final volume after dilution?

6.41 Calculate the molarity of a solution that contains 2.25 mol of NaNO$_3$ dissolved in 2.50 L.

6.42 Calculate the molarity of a solution that contains 1.75 mol of KNO$_3$ dissolved in 3.00 L.

6.43 How many grams of glucose (C$_6$H$_{12}$O$_6$) are present in 1.75 L of a 0.500 M solution?

6.44 How many grams of sodium hydroxide are present in 675 mL of a 0.500 M solution?

6.45 50.0 mL of 0.500 M NaOH were diluted to 500.0 mL. What is the new molarity?

6.46 300.0 mL of H$_2$O are added to 300.0 mL of 0.250 M H$_2$SO$_4$. What is the new molarity?

6.47 Calculate the number of equivalents/L of Ca^{2+} in a solution that is $5.0 \times 10^{-2} M$ in Ca^{2+}.

6.48 Calculate the number of equivalents/L of SO$_4{}^{2-}$ in a solution that is $2.5 \times 10^{-3} M$ in SO$_4{}^{2-}$.

Concentration-Dependent Solution Properties

Foundations

6.49 What is meant by the term *colligative property*?

6.50 Name and describe four colligative solution properties.

6.51 Explain, in terms of solution properties, why salt is used to melt ice in the winter.

6.52 Explain, in terms of solution properties, why a wilted plant regains its "health" when watered.

Applications

6.53 In what way do colligative properties and chemical properties differ?

6.54 Look up the meaning of the term "colligative." Why is it an appropriate title for these properties?

6.55 State Raoult's law.

6.56 What is the major importance of Raoult's law?

6.57 Why does one mole of CaCl$_2$ lower the freezing point of water more than one mole of NaCl?

6.58 Using salt to try to melt ice on a day when the temperature is –20 °C will be unsuccessful. Why?

Answer questions 6.59–6.62 by comparing two solutions: 0.50 M sodium chloride (an ionic compound) and 0.50 M sucrose (a covalent compound).

6.59 Which solution has the higher melting point?

6.60 Which solution has the higher boiling point?

6.61 Which solution has the higher vapor pressure?

6.62 Each solution is separated from water by a semipermeable membrane. Which solution has the higher osmotic pressure?

6.63 Calculate the osmotic pressure of 0.50 M sodium chloride.

6.64 Calculate the osmotic pressure of 0.50 M sucrose.

Answer questions 6.65–6.68 based on the following scenario: Two solutions, A and B, are separated by a semipermeable membrane. For each case, predict whether there will be a net flow of water in one direction and, if so, which direction.

6.65 A is pure water and B is 5% glucose.

6.66 A is 0.10 M glucose and B is 0.10 M KCl.

6.67 A is 0.10 M NaCl and B is 0.10 M KCl.

6.68 A is 0.10 M NaCl and B is 0.20 M glucose.

In questions 6.69–6.72, label each solution as isotonic, hypotonic, or hypertonic in comparison to 0.9% NaCl (0.15 M NaCl).

6.69 0.15 M CaCl$_2$

6.70 0.35 M glucose

6.71 0.15 M glucose

6.72 3% NaCl

Water as a Solvent

Foundations

6.73 What properties make water such a useful solvent?

6.74 Sketch the "interactive network" of water molecules in the liquid state.

6.75 Sketch the interaction of water with an ammonia molecule.

6.76 Sketch the interaction of water with an ethanol molecule.

Applications

6.77 Determine the osmolarity of 5.0×10^{-4} M KNO$_3$ (electrolyte).

6.78 Determine the osmolarity of 2.5×10^{-4} M C$_6$H$_{12}$O$_6$ (nonelectrolyte).

6.79 Solutions of ammonia in water are sold as window cleaner. Why do these solutions have a long "shelf life"?

6.80 Why does water's abnormally high boiling point help to make it a desirable solvent?

6.81 Sketch the interaction of a water molecule with a sodium ion.

6.82 Sketch the interaction of a water molecule with a chloride ion.

6.83 What type of solute dissolves readily in water?

6.84 What type of solute dissolves readily in gasoline?

Electrolytes in Body Fluids

Foundations

6.85 Why is it important to distinguish between electrolytes and nonelectrolytes when discussing colligative properties?

6.86 Name the two most important cations in biological fluids.

6.87 Explain why a dialysis solution must have a low sodium ion concentration if it is designed to remove excess sodium ion from the blood.

6.88 Explain why a dialysis solution must have an elevated potassium ion concentration when loss of potassium ion from the blood is a concern.

Applications

6.89 Describe the clinical effects of elevated concentrations of sodium ion in the blood.

6.90 Describe the clinical effects of depressed concentrations of potassium ion in the blood.

6.91 Describe conditions that can lead to elevated concentrations of sodium in the blood.

6.92 Describe conditions that can lead to dangerously low concentrations of potassium in the blood.

6.93 A potassium chloride solution that also contains 5% (W/V) dextrose is administered intravenously to treat some forms of malnutrition. The potassium ion concentration in this solution is 40 meq/L. Calculate the potassium ion concentration in moles/L.

6.94 If the potassium ion concentration in the solution described in Question 6.93 was only 35 meq/L, calculate the potassium ion concentration in units of mol/L.

CRITICAL THINKING PROBLEMS

1. Which of the following compounds would cause the greater freezing point depression, per mole, in H$_2$O: C$_6$H$_{12}$O$_6$ (glucose) or NaCl?

2. Which of the following compounds would cause the greater boiling point elevation, per mole, in H$_2$O: MgCl$_2$ or HOCH$_2$CH$_2$OH (ethylene glycol, antifreeze)? (Hint: HOCH$_2$CH$_2$OH is covalent.)

3. Analytical chemists often take advantage of differences in solubility to separate ions. For example, adding Cl$^-$ to a solution of Cu^{2+} and Ag$^+$ causes AgCl to precipitate; Cu^{2+} remains in solution. Filtering the solution results in a separation. Design a scheme to separate the cations Ca^{2+} and Pb^{2+}.

4. Using the strategy outlined in the above problem, design a scheme to separate the anions S^{2-} and CO$_3{}^{2-}$.

5. Design an experiment that would enable you to measure the degree of solubility of a salt such as KI in water.

6. How could you experimentally distinguish between a saturated solution and a supersaturated solution?

7. Blood is essentially an aqueous solution, but it must transport a variety of nonpolar substances (hormones, for example). Colloidal proteins, termed *albumins*, facilitate this transport. Must these albumins be polar or nonpolar? Why?

7 Energy, Rate, and Equilibrium

Learning Goals

1 Correlate the terms *endothermic* and *exothermic* with heat flow between a *system* and its *surroundings*.

2 State the meaning of the terms *enthalpy, entropy,* and *free energy* and know their implications.

3 Describe experiments that yield thermochemical information and calculate fuel values based on experimental data.

4 Describe the concept of reaction rate and the role of kinetics in chemical and physical change.

5 Describe the importance of *activation energy* and the *activated complex* in determining reaction rate.

6 Predict the way reactant structure, concentration, temperature, and catalysis affect the rate of a chemical reaction.

7 Write rate equations for elementary processes.

8 Recognize and describe equilibrium situations.

9 Write equilibrium-constant expressions and use these expressions to calculate equilibrium constants.

10 Use LeChatelier's principle to predict changes in equilibrium position.

Outline

Describe the relationship of the modern automobile engine to the topics discussed in this chapter.

Introduction

In Chapter 4 we calculated quantities of matter involved in chemical change, assuming that all of the reacting material was consumed and that only products of the reaction remain at the end of the reaction. Often this is not true. Furthermore, not all chemical reactions take place at the same speed; some occur almost instantaneously (explosions), whereas others may proceed for many years (corrosion).

Two concepts play important roles in determining the extent and speed of a chemical reaction: thermodynamics, which deals with energy changes in chemical reactions, and kinetics, which describes the rate or speed of a chemical reaction.

Although both thermodynamics and kinetics involve energy, they are two separate considerations. A reaction may be thermodynamically favored but very slow; conversely, a reaction may be very fast because it is kinetically favorable yet produce very little (or no) product because it is thermodynamically unfavorable.

In this chapter we investigate the fundamentals of thermodynamics and kinetics, with an emphasis on the critical role that energy changes play in chemical reactions. We consider physical change and chemical change, including the conversions that take place among the states of matter (solid, liquid, and gas). We use these concepts to explain the behavior of reactions that do not go to completion, equilibrium reactions. We develop the equilibrium-constant expression and demonstrate how equilibrium composition can be altered using LeChatelier's principle.

7.1 Thermodynamics

Thermodynamics is the study of energy, work, and heat. It may be applied to chemical change, such as the calculation of the quantity of heat obtainable from the combustion of one gallon of fuel oil. Similarly, energy released or

Chemistry Connection

The Cost of Energy? More Than You Imagine

When we purchase gasoline for our automobiles or oil for the furnace, we are certainly buying matter. That matter is only a storage device; we are really purchasing the energy stored in the chemical bonds. Combustion, burning in oxygen, releases the stored potential energy in a form suited to its function: mechanical energy to power a vehicle or heat energy to warm a home.

Energy release is a consequence of change. In fuel combustion, this change results in the production of waste products that may be detrimental to our environment. This necessitates the expenditure of time, money, and *more* energy to clean up our surroundings.

If we are paying a considerable price for our energy supply, it would be nice to believe that we are at least getting full value for our expenditure. Even that is not the case. Removal of energy from molecules also extracts a price. For example, a properly tuned automobile engine is perhaps 30% efficient. That means that less than one-third of the available energy actually moves the car. The other two-thirds is released into the atmosphere as wasted energy, mostly heat energy. The law of conservation of energy tells us that the energy is not destroyed, but it is certainly not available to us in a useful form.

Can we build a 100% efficient energy transfer system? Is there such a thing as cost-free energy? No, on both counts. It is theoretically impossible, and the laws of thermodynamics, which we discuss in this chapter, tell us why this is so.

consumed in physical change, such as the boiling or freezing of water, may be determined.

There are three basic laws of thermodynamics, but only the first two will be of concern here. They help us to understand why some chemical reactions may occur spontaneously while others do not.

The Chemical Reaction and Energy

John Dalton believed that chemical change involved joining, separating, or re-arranging atoms. Two centuries later, this statement stands as an accurate description of chemical reactions. However, we now know much more about the nonmaterial energy changes that are an essential part of every reaction.

Throughout the discussion of thermodynamics and kinetics it will be useful to remember the basic ideas of the kinetic molecular theory (Section 5.1):

- molecules and atoms in a reaction mixture are in constant, random motion;
- these molecules and atoms frequently collide with each other;
- only some collisions, those with sufficient energy, will break bonds in molecules; and
- when reactant bonds are broken, new bonds may be formed and products result.

It is worth noting that we cannot measure an absolute value for energy stored in a chemical system. We can only measure the *change* in energy (energy absorbed or released) as a chemical reaction occurs. Also, it is often both convenient and necessary to establish a boundary between the *system* and its *surroundings*.

The **system** contains the process under study. The **surroundings** encompass the rest of the universe. Energy is lost from the system to the surroundings or energy may be gained by the system at the expense of the surroundings. This energy change, most often in the form of heat, may be determined because the temperature of the system or surroundings will change, and this change can be measured. This process is illustrated in Figure 7.1.

Consider the combustion of methane in a Bunsen burner, the system. The temperature of the air surrounding the burner increases, indicating that some of the potential energy of the system has been converted to heat energy. The heat energy of the system (methane, oxygen, and the Bunsen burner) is being lost to the surroundings.

Now, an exact temperature measurement of the air before and after the reaction is difficult. However, if we could insulate a portion of the surroundings, to isolate and trap the heat, we could calculate a useful quantity, the heat of the reaction. Experimental strategies for measuring temperature change and calculating heats of reactions, termed *calorimetry*, are discussed in Section 7.2.

The concept of energy, its various forms, and commonly used energy units were introduced in Chapter 1. It will be useful to reread this section before proceeding with this chapter.

(a)   (b)

Figure 7.1
Illustration of heat flow in (a) exothermic and (b) endothermic reactions.

Exothermic and Endothermic Reactions

The first law of thermodynamics states that the energy of the universe is constant. It is the law of conservation of energy. The study of energy changes that occur in chemical reactions is a very practical application of the first law. Consider, for example, the generalized reaction:

$$A—B + C—D \longrightarrow A—D + C—B$$

An **exothermic reaction** releases energy to the surroundings. The surroundings become warmer.

Each chemical bond is stored chemical energy (potential energy). For the reaction to take place, bond A—B and bond C—D must break; this process *always* requires energy. At the same time, bonds A—D and C—B must form; this process always releases energy.

If the energy required to break the A—B and C—D bonds is *less* than the energy given off when the A—D and C—B bonds form, the reaction will release the excess energy. The energy is a *product*, and the reaction is called an exothermic (*Gr. exo,* out, and *Gr. therm,* heat) reaction. This conversion of chemical energy to heat is represented in Figure 7.2a.

An example of an exothermic reaction is the combustion of methane, represented by a *thermochemical equation:*

$$CH_4(g) + 2O_2(g) \longrightarrow CO_2(g) + 2H_2O(g) + \boxed{211 \text{ kcal}}$$

Exothermic reaction

This thermochemical equation reads: the combustion of one mole of methane releases 211 kcal of heat.

An **endothermic reaction** absorbs energy from the surroundings. The surroundings become colder.

If the energy required to break the A—B and C—D bonds is *greater* than the energy released when the A—D and C—B bonds form, the reaction will need an external supply of energy (perhaps from a Bunsen burner). Insufficient energy is available in the system to initiate the bond-breaking process. Such a reaction is called an endothermic (*Gr. endo,* to take on, and *Gr. therm,* heat) reaction, and energy is a *reactant.* The conversion of heat energy into chemical energy is represented in Figure 7.2b.

The decomposition of ammonia into nitrogen and hydrogen is one example of an endothermic reaction:

$$\boxed{22 \text{ kcal}} + 2NH_3(g) \longrightarrow N_2(g) + 3H_2(g)$$

Endothermic reaction

> In an exothermic reaction, heat is released *from the system* to the surroundings. In an endothermic reaction, heat is absorbed *by the system* from the surroundings.

Figure 7.2

(a) An exothermic reaction. ΔE represents the energy released during the progress of the exothermic reaction: $A + B \longrightarrow C + D + \Delta E$. (b) An endothermic reaction. ΔE represents the energy absorbed during the progress of the endothermic reaction: $\Delta E + A + B \longrightarrow C + D$.

(a)

(b)

This thermochemical equation reads: the decomposition of two moles of ammonia requires 22 kcal of heat.

The examples used here show the energy absorbed or released as heat energy. Depending on the reaction and the conditions under which the reaction is run, the energy may take the form of light energy or electrical energy. A firefly releases energy as a soft glow of light on a summer evening. An electrical current results from a chemical reaction in a battery, enabling your car to start.

Determining Whether a Process Is Exothermic or Endothermic

EXAMPLE 7.1

An ice cube is dropped into a glass of water at room temperature. The ice cube melts. Is the melting of the ice exothermic or endothermic?

Solution

Step 1. Consider the ice cube to be the system and the water, the surroundings.

Step 2. Recognize that for the cube to melt, it must gain energy and its energy source must be the water.

Step 3. The heat flow is from surroundings to system.

Step 4. The system gains energy (+energy); hence, the melting process (physical change) is endothermic.

LEARNING GOAL

1 ▶ Correlate the terms *endothermic* and *exothermic* with heat flow between a *system* and its *surroundings*.

Practice Problem 7.1

Are the following processes exothermic or endothermic?

a. Fuel oil is burned in a furnace.
b. When solid NaOH is dissolved in water, the solution becomes hotter.

For Further Practice: Questions 7.23 and 7.24.

Enthalpy

Enthalpy is the term used to represent heat. The *change in enthalpy* is the energy difference between the products and reactants of a chemical reaction and is symbolized as ΔH. By convention, energy released is represented with a negative sign (indicating an exothermic reaction), and energy absorbed is shown with a positive sign (indicating an endothermic reaction).

For the combustion of methane, an exothermic process, energy is a *product* in the thermochemical equation, and

$$\Delta H = -211 \text{ kcal}$$

For the decomposition of ammonia, an endothermic process, energy is a *reactant* in the thermochemical equation, and

$$\Delta H = +22 \text{ kcal}$$

LEARNING GOAL

2 ▶ State the meaning of the terms *enthalpy, entropy,* and *free energy* and know their implications.

 Animation

Heat Flow in Endothermic and Exothermic Reactions

In these discussions, we consider the enthalpy change and energy change to be identical. This is true for most common reactions carried out in lab, with minimal volume change.

Spontaneous and Nonspontaneous Reactions

It seems that all exothermic reactions should be spontaneous. After all, an external supply of energy does not appear to be necessary; in fact, energy is a product of the reaction. It also seems that all endothermic reactions should be nonspontaneous: energy is a reactant that we must provide. However, these hypotheses are not supported by experimentation.

Experimental measurement has shown that most *but not all* exothermic reactions are spontaneous. Likewise, most *but not all*, endothermic reactions are not spontaneous. There must be some factor in addition to enthalpy that will help us to explain the less obvious cases of nonspontaneous exothermic reactions and spontaneous endothermic reactions. This other factor is entropy.

Entropy

The first law of thermodynamics considers the enthalpy of chemical reactions. The second law states that the universe spontaneously tends toward increasing disorder or randomness.

A measure of the randomness of a chemical system is its **entropy.** The entropy of a substance is represented by the symbol S. A random, or disordered, system is characterized by *high entropy*; a well-organized system has *low entropy*.

What do we mean by disorder in chemical systems? Disorder is simply the absence of a regular repeating pattern. Disorder or randomness increases as we convert from the solid to the liquid to the gaseous state. As we have seen, solids often have an ordered crystalline structure, liquids have, at best, a loose arrangement, and gas particles are virtually random in their distribution. Therefore gases have high entropy, and crystalline solids have very low entropy. Figures 7.3 and 7.4 illustrate properties of entropy in systems.

> **2 LEARNING GOAL**
> State the meaning of the terms *enthalpy*, *entropy*, and *free energy* and know their implications.

A *system* is a part of the universe upon which we wish to focus our attention. For example, it may be a beaker containing reactants and products.

Chapter 5 compares the physical properties of solids, liquids, and gases.

Animation

Entropy and Rubber Bands

Question 7.1

Are the following processes exothermic or endothermic?

a. $C_6H_{12}O_6(s) \longrightarrow 2C_2H_5OH(l) + 2CO_2(g)$, $\Delta H = -16$ kcal
b. $N_2O_5(g) + H_2O(l) \longrightarrow 2HNO_3(l) + 18.3$ kcal

Question 7.2

Are the following processes exothermic or endothermic?

a. $S(s) + O_2(g) \longrightarrow SO_2(g)$, $\Delta H = -71$ kcal
b. $N_2(g) + 2O_2(g) + 16.2$ kcal $\longrightarrow 2NO_2(g)$

The second law describes the entire universe or any isolated system within the universe. On a more personal level, we all fall victim to the law of increasing

Figure 7.3

(a) Gas particles, trapped in the left chamber, spontaneously diffuse into the right chamber, initially under vacuum, when the valve is opened. (b) It is unimaginable that the gas particles will rearrange themselves and reverse the process to create a vacuum. This can only be accomplished using a pump, that is, by doing work on the system.

Spontaneous Process
(a)

Nonspontaneous Process
(b)

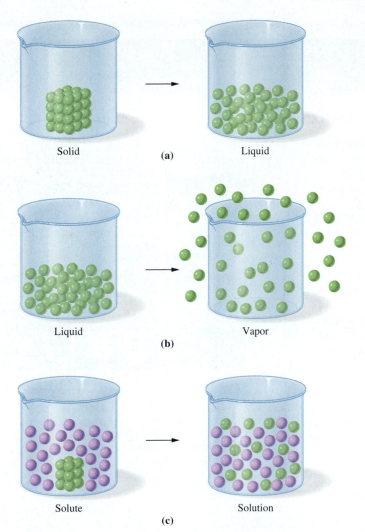

Figure 7.4
Processes such as (a) melting, (b) vaporization, and (c) dissolution increase entropy, or randomness, of the particles.

disorder. Chaos in our room or workplace is certainly not our intent! It happens almost effortlessly. However, reversal of this process requires work and energy. The same is true at the molecular level. The gradual deterioration of our cities' infrastructure (roads, bridges, water mains, and so forth) is an all-too-familiar example. Millions of dollars (translated into energy and work) are needed annually just to try to maintain the status quo.

The entropy of a reaction is measured as a difference, ΔS, between the entropies, S, of products and reactants.

The drive toward increased entropy, along with a tendency to achieve a lower potential energy, is responsible for spontaneous chemical reactions. Reactions that are exothermic and whose products are more disordered (higher in entropy) will occur spontaneously, whereas endothermic reactions producing products of lower entropy will not be spontaneous. If they are to take place at all, they will need some energy input.

Question 7.3

Which substance has the greatest entropy, He(g) or Na(s)? Explain your reasoning.

Question 7.4

Which substance has the greatest entropy, $H_2O(l)$ or $H_2O(g)$? Explain your reasoning.

A Human Perspective

Triboluminescence: Sparks in the Dark with Candy

Generations of children have inadvertently discovered the phenomenon of triboluminescence. Crushing a wintergreen candy (Lifesavers) with the teeth in a dark room (in front of a few friends or a mirror) or simply rubbing two pieces of candy together may produce the effect—transient sparks of light!

Triboluminescence is simply the production of light upon fracturing a solid. It is easily observed and straightforward to describe but difficult to explain. It is believed to result from charge separation produced by the disruption of a crystal lattice. The charge separation has a very short lifetime. When the charge distribution returns to equilibrium, energy is released, and that energy is the light that is observed.

Dr. Linda M. Sweeting and several other groups of scientists have tried to reproduce these events under controlled circumstances. Crystals similar to the sugars in wintergreen candy are prepared with a very high level of purity. Some theories attribute the light emission to impurities in a crystal rather than to the crystal itself. Devices have been constructed that will crush the crystal with a uniform and reproducible force. Light-measuring devices, spectrophotometers, accurately measure the various wavelengths of light and the intensity of the light at each wavelength.

Through the application of careful experimentation and measurement of light-emitting properties of a variety of related compounds, these scientists hope to develop a theory of light emission from fractured solids.

This is one more example of the scientific method improving our understanding of everyday occurrences.

Charles Schulz's "Peanuts" vision of triboluminescence.

Reprinted by permission of UFS, Inc.

For Further Understanding

Can you suggest an experiment that would support or refute the hypothesis that impurities in crystals are responsible for triboluminescence?

Would you expect that amorphous substances would exhibit triboluminescence? Why or why not?

Free Energy

The two situations described above are clear-cut and unambiguous. In any other situation the reaction may or may not be spontaneous. It depends on the relative size of the enthalpy and entropy values.

Free energy, symbolized by ΔG, represents the combined contribution of the enthalpy *and* entropy values for a chemical reaction. Thus free energy is the ultimate predictor of reaction spontaneity and is expressed as

$$\Delta G = \Delta H - T\Delta S$$

ΔH represents the change in enthalpy between products and reactants, ΔS represents the change in entropy between products and reactants, and T is the Kelvin temperature of the reaction. A reaction with a negative value of ΔG will *always* be spontaneous. Reactions with a positive ΔG will *always* be nonspontaneous.

We need to know both ΔH and ΔS in order to predict the sign of ΔG and make a definitive statement regarding the spontaneity of the reaction. Additionally, the temperature may determine the direction of spontaneity. Consider the four possible situations:

- ΔH positive and ΔS negative: ΔG is always positive, regardless of the temperature. The reaction is always nonspontaneous.
- ΔH negative and ΔS positive: ΔG is always negative, regardless of the temperature. The reaction is always spontaneous.
- Both ΔH and ΔS positive: The sign of ΔG depends on the temperature.
- Both ΔH and ΔS negative: The sign of ΔG depends on the temperature.

Question 7.5

Predict whether a reaction with positive ΔH and negative ΔS will be spontaneous, nonspontaneous, or temperature dependent. Explain your reasoning.

Question 7.6

Predict whether a reaction with positive ΔH and positive ΔS will be spontaneous, nonspontaneous, or temperature dependent. Explain your reasoning.

7.2 Experimental Determination of Energy Change in Reactions

The measurement of heat energy changes in a chemical reaction is **calorimetry**. This technique involves the measurement of the change in the temperature of a quantity of water or solution that is in contact with the reaction of interest and isolated from the surroundings. A device used for these measurements is a *calorimeter*, which measures heat changes in calories.

A Styrofoam coffee cup is a simple design for a calorimeter, and it produces surprisingly accurate results. It is a good insulator, and, when filled with solution, it can be used to measure temperature changes taking place as the result of a chemical reaction occurring in that solution (Figure 7.5). The change in the temperature of the solution, caused by the reaction, can be used to calculate the gain or loss of heat energy for the reaction.

For an exothermic reaction, heat released by the reaction is absorbed by the surrounding solution. For an endothermic reaction, the reactants absorb heat from the solution.

The **specific heat** of a substance is defined as the number of calories of heat needed to raise the temperature of 1 g of the substance 1 degree Celsius. Knowing the specific heat of the water or the aqueous solution along with the total number of grams of solution and the temperature increase (measured as the difference between the final and initial temperatures of the solution), enables the experimenter to calculate the heat released during the reaction.

The solution behaves as a "trap" or "sink" for energy released in the exothermic process. The temperature increase indicates a gain in heat energy. Endothermic reactions, on the other hand, take heat energy away from the solution, lowering its temperature.

The quantity of heat absorbed or released by the reaction (Q) is the product of the mass of solution in the calorimeter (m_s), the specific heat of the solution (SH_s), and the change in temperature (ΔT_s) of the solution as the reaction proceeds from the initial to final state.

The heat is calculated by using the following equation:

$$Q = m_s \times \Delta T_s \times SH_s$$

with units

$$\text{calories} = \cancel{\text{gram}} \times \cancel{\degree C} \times \frac{\text{calories}}{\cancel{\text{gram-}\degree C}}$$

LEARNING GOAL

3 Describe experiments that yield thermochemical information and calculate fuel values based on experimental data.

Figure 7.5
A "coffee cup" calorimeter used for the measurement of heat change in chemical reactions. The concentric Styrofoam cups insulate the system from its surroundings. Heat released by the chemical reaction enters the solution, raising its temperature, which is measured by using a thermometer.

The details of the experimental approach are illustrated in Example 7.2.

EXAMPLE 7.2   **Calculating Energy Involved in Calorimeter Reactions**

LEARNING GOAL

3   Describe experiments that yield thermochemical information and calculate fuel values based on experimental data.

If 0.050 mol of hydrochloric acid (HCl) is mixed with 0.050 mol of sodium hydroxide (NaOH) in a "coffee cup" calorimeter, the temperature of 1.00×10^2 g of the resulting solution increases from 25.0°C to 31.5°C. If the specific heat of the solution is 1.00 cal/g solution °C, calculate the quantity of energy involved in the reaction. Also, is the reaction endothermic or exothermic?

Solution

Step 1. The change in temperature is

$$\Delta T_s = T_{s\,final} - T_{s\,initial}$$
$$= 31.5°C - 25.0°C = 6.5°C$$

Step 2. The calorimetry expression is:

$$Q = m_s \times \Delta T_s \times SH_s$$

Step 3. Substituting:

$$Q = 1.00 \times 10^2 \text{ g solution} \times 6.5°C \times \frac{1.00 \text{ cal}}{\text{g solution °C}}$$

$$= 6.5 \times 10^2 \text{ cal}$$

6.5×10^2 cal (or 0.65 kcal) of heat energy were released by this acid-base reaction to the surroundings, the solution; the reaction is exothermic.

 Practice Problem 7.2

Calculate the temperature change that would have been observed if 50.0 g solution were in the calorimeter instead of 1.00×10^2 g solution.

For Further Practice: Question 7.35.

EXAMPLE 7.3   **Calculating Energy Involved in Calorimeter Reactions**

If 0.10 mol of ammonium chloride (NH_4Cl) is dissolved in water producing 1.00×10^2 g solution, the water temperature decreases from 25.0°C to 18.0°C. If the specific heat of the resulting solution is 1.00 cal/g-°C, calculate the quantity of energy involved in the process. Also, is the dissolution of ammonium chloride endothermic or exothermic?

Solution

Step 1. The change in temperature is

$$\Delta T = T_{s\,final} - T_{s\,initial}$$
$$= 18.0°C - 25.0°C = -7.0°C$$

Step 2. The calorimetry expression is:

$$Q = m_s \times \Delta T_s \times SH_s$$

Continued—

EXAMPLE 7.3 —*Continued*

Step 3. Substituting:

$$Q = 1.00 \times 10^2 \text{ g solution} \times (-7.0°C) \times \frac{1.00 \text{ cal}}{\text{g solution °C}}$$

$$= -7.0 \times 10^2 \text{ cal}$$

7.0×10^2 cal (or 0.70 kcal) of heat energy were absorbed by the dissolution process because the solution lost (− sign) 7.0×10^2 cal of heat energy to the system. The reaction is endothermic.

Practice Problem 7.3

Calculate the temperature change that would have been observed if 1.00×10^2 g of another liquid, producing a solution with a specific heat capacity of 0.800 cal/g-°C, were substituted for the water in the calorimeter.

For Further Practice: Question 7.37.

Convert the energy released in Example 7.2 to joules.

Question *7.7*

Convert the energy absorbed in Example 7.3 to joules.

Question *7.8*

Many chemical reactions that produce heat are combustion reactions. In our bodies many food substances (carbohydrates, proteins, and fats, Chapters 21 and 22) are oxidized to release energy. **Fuel value** is the amount of energy per gram of food.

The fuel value of food is an important concept in nutrition science. The fuel value is generally reported in units of *nutritional Calories.* One **nutritional Calorie** is equivalent to one kilocalorie (1000 calories). It is also known as the *large Calorie* (uppercase C).

Energy necessary for our daily activity and bodily function comes largely from the "combustion" of carbohydrates. Chemical energy from foods that is not used to maintain normal body temperature or in muscular activity is stored in the bonds of chemical compounds known collectively as fat. Thus "high-calorie" foods are implicated in obesity.

A special type of calorimeter, a *bomb calorimeter,* is useful for the measurement of the fuel value (Calories) of foods. Such a device is illustrated in Figure 7.6. Its design is similar, in principle, to that of the "coffee cup" calorimeter discussed earlier. It incorporates the insulation from the surroundings, solution pool, reaction chamber, and thermometer. Oxygen gas is added as one of the reactants, and an electrical igniter is inserted to initiate the reaction. However, it is not open to the atmosphere. In the sealed container the reaction continues until the sample is completely oxidized. All of the heat energy released during the reaction is captured in the water.

Note: Refer to A Human Perspective: Food Calories, Section 1.5.

LEARNING GOAL

3 Describe experiments that yield thermochemical information and calculate fuel values based on experimental data.

Figure 7.6

A bomb calorimeter that may be used to measure heat released upon combustion of a sample. This device is commonly used to determine the fuel value of foods. The bomb calorimeter is similar to the "coffee cup" calorimeter. However, note the electrical component necessary to initiate the combustion reaction.

Thermometer — Source of electric current — Stirrer — Insulation — Water — Oxygen inlet — Resistance wire for igniting sample — Reaction chamber — Sample

---

E X A M P L E 7.4 | **Calculating the Fuel Value of Foods**

LEARNING GOAL

3 Describe experiments that yield thermochemical information and calculate fuel values based on experimental data.

One gram of glucose (a common sugar or carbohydrate) was burned in a bomb calorimeter. The temperature of 1.00×10^3 g H_2O was raised from 25.0°C to 28.8°C ($\Delta T_w = 3.8$°C). Calculate the fuel value of glucose.

Solution

Step 1. Recall that the fuel value is the number of nutritional Calories liberated by the combustion of 1 g of material and 1 g of material was burned in the calorimeter.

Step 2. Then

$$\text{Fuel value} = Q = m_w \times \Delta T_w \times SH_w$$

Step 3. Water is the surroundings in the calorimeter; it has a specific heat capacity equal to 1.00 cal/g H_2O °C.

Substituting in our expression for fuel value:

$$\text{Fuel value} = Q = \text{g } H_2O \times \text{°C} \times \frac{1.00 \text{ cal}}{\text{g } H_2O \text{ °C}}$$

$$= 1.00 \times 10^3 \text{ g } H_2O \times 3.8 \text{°C} \times \frac{1.00 \text{ cal}}{\text{g } H_2O \text{ °C}}$$

$$= 3.8 \times 10^3 \text{ cal}$$

Step 4. Converting from calories to nutritional calories:

$$3.8 \times 10^3 \text{ cal} \times \frac{1 \text{ nutritional Calorie}}{10^3 \text{ cal}} = 3.8 \text{ C (nutritional Calories, or kcal)}$$

The fuel value of glucose is 3.8 kcal/g.

Continued—

EXAMPLE 7.4 —*Continued*

Practice Problem 7.4

A 1.0-g sample of a candy bar (which contains lots of sugar!) was burned in a bomb calorimeter. A 3.0°C temperature increase was observed for 1.00×10^3 g of water. The entire candy bar weighed 2.5 ounces. Calculate the fuel value (in nutritional Calories) of the sample and the total caloric content of the candy bar.

For Further Practice: Questions 7.36 and 7.38.

7.3 Kinetics

LEARNING GOAL

4 Describe the concept of reaction rate and the role of kinetics in chemical and physical change.

Thermodynamics help us to decide whether a chemical reaction is spontaneous. Knowing that a reaction can occur spontaneously tells us nothing about the time that it may take.

Chemical **kinetics** is the study of the **rate** (or speed) **of chemical reactions.** Kinetics also gives an indication of the *mechanism* of a reaction, a step-by-step description of how reactants become products. Kinetic information may be represented as the *disappearance* of reactants or *appearance* of product over time. A typical graph of concentration versus time is shown in Figure 7.7.

Information about the rate at which various chemical processes occur is useful. For example, what is the "shelf life" of processed foods? When will slow changes in composition make food unappealing or even unsafe? Many drugs lose their potency with time because the active ingredient decomposes into other substances. The rate of hardening of dental filling material (via a chemical reaction) influences the dentist's technique. Our very lives depend on the efficient transport of oxygen to each of our cells and the rapid use of the oxygen for energy-harvesting reactions.

The diagram in Figure 7.8 is a useful way of depicting the kinetics of a reaction at the molecular level.

Often a color change, over time, can be measured. Such changes are useful in assessing the rate of a chemical reaction (Figure 7.9).

Let's see what actually happens when two chemical compounds react and what experimental conditions affect the rate of a reaction.

 Animations

Kinetics

Rates of Chemical Reactions

Figure 7.7

For a hypothetical reaction $A \longrightarrow B$ the concentration of A molecules (reactant molecules) decreases over time and B molecules (product molecules) increase in concentration over time.

The Chemical Reaction

Consider the exothermic reaction that we discussed in Section 7.1:

$$CH_4(g) + 2O_2(g) \longrightarrow CO_2(g) + 2H_2O(l) + \boxed{211 \text{ kcal}}$$

For the reaction to proceed, C—H and O—O bonds must be broken, and C—O and H—O bonds must be formed. Sufficient energy must be available to cause the bonds to break if the reaction is to take place. This energy is provided by the collision of molecules. If sufficient energy is available at the temperature of the reaction, one or more bonds will break, and the atoms will recombine in a lower energy arrangement, in this case as carbon dioxide and water. A collision producing product molecules is termed an *effective collision*. Only effective collisions lead to chemical reaction.

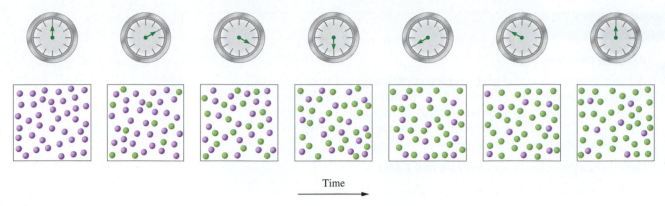

Time

Figure 7.8
An alternate way of representing the information contained in Figure 7.7.

Figure 7.9
The conversion of reddish brown Br_2 in solution to colorless Br^- over time.

Time

Activation Energy and the Activated Complex

LEARNING GOAL

5 Describe the importance of *activation energy* and the *activated complex* in determining reaction rate.

Animation

Activation Energy

This reaction will take place when an electrical current is passed through water. The process is called *electrolysis*.

The minimum amount of energy required to initiate a chemical reaction is called the **activation energy** for the reaction.

We can picture the chemical reaction in terms of the changes in potential energy that occur during the reaction. Figure 7.10a graphically shows these changes for an exothermic reaction. Important characteristics of this graph include the following:

- The reaction proceeds from reactants to products through an extremely unstable state that we call the **activated complex.** The activated complex cannot be isolated from the reaction mixture but may be thought of as a short-lived group of atoms structured in such a way that it quickly and easily breaks apart into the products of the reaction.
- Formation of the activated complex requires energy. The difference between the energy of reactants and that of the activated complex is the activation energy. This energy must be provided by the collision of the reacting molecules or atoms at the temperature of the reaction.
- Because this is an exothermic reaction, the overall energy change must be a *net* release of energy. The *net* release of energy is the difference in energy between products and reactants.

For an endothermic reaction, such as the decomposition of water,

$$\text{energy} + 2H_2O(l) \longrightarrow 2H_2(g) + O_2(g)$$

the change of potential energy with reaction time is shown in Figure 7.10b.

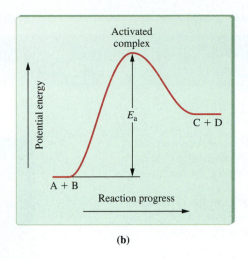

(a) (b)

Figure 7.10
(a) The change in potential energy as a function of reaction time for an exothermic chemical reaction. Note particularly the energy barrier associated with the formation of the activated complex. This energy barrier (E_a) *is* the activation energy. (b) The change in potential energy as a function of reaction time for an endothermic chemical reaction. In contrast to the exothermic reaction in (a), the energy of the products is greater than the energy of the reactants.

The reaction takes place slowly because of the large activation energy required for the conversion of water into the elements hydrogen and oxygen.

Question 7.9

The act of striking a match illustrates the role of activation energy in a chemical reaction. Explain.

Question 7.10

Distinguish between the terms *net energy* and *activation energy*.

Factors That Affect Reaction Rate

Factors influencing reaction rate include:

- structure of the reacting species,
- molecular shape and orientation,
- concentration of reactants,
- temperature of reactants,
- physical state of reactants, and
- presence of a catalyst.

6 **LEARNING GOAL**
Predict the way reactant structure, concentration, temperature, and catalysis affect the rate of a chemical reaction.

Structure of the Reacting Species

Reactions among ions in solution are usually very rapid. Ionic compounds in solution are dissociated; consequently, their bonds are already broken, and the activation energy for their reaction should be very low. On the other hand, reactions involving covalently bonded compounds may proceed more slowly. Covalent bonds must be broken and new bonds formed. The activation energy for this process would be significantly higher than that for the reaction of free ions. Bond strengths certainly play a role in determining reaction rates because the magnitude of the activation energy, or energy barrier, is related to bond strength.

Molecular Shape and Orientation

The size and shape of reactant molecules influence the rate of the reaction. Large molecules, containing bulky groups of atoms, may block the reactive part of the molecule from interacting with another reactive substance, causing the reaction to proceed slowly. Only molecular collisions that have the correct

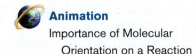
Animation
Importance of Molecular Orientation on a Reaction

A Medical *Perspective*

Hot and Cold Packs

Hot packs provide "instant warmth" for hikers and skiers and are used in treatment of injuries such as pulled muscles. Cold packs are in common use today for the treatment of injuries and the reduction of swelling.

These useful items are an excellent example of basic science producing a technologically useful product. (Recall our discussion in Chapter 1 of the relationship of science and technology.)

Both hot and cold packs depend on large energy changes taking place during a chemical reaction. Cold packs rely on an endothermic reaction, and hot packs generate heat energy from an exothermic reaction.

A cold pack is fabricated as two separate compartments within a single package. One compartment contains NH_4NO_3, and the other contains water. When the package is squeezed, the inner boundary between the two compartments ruptures, allowing the components to mix, and the following reaction occurs:

$$6.7 \text{ kcal/mol} + NH_4NO_3(s) \longrightarrow NH_4^+(aq) + NO_3^-(aq)$$

A cold pack.

Heating pads.

This reaction is endothermic; heat taken from the surroundings produces the cooling effect.

The design of a hot pack is similar. Here, finely divided iron powder is mixed with oxygen. Production of iron oxide results in the evolution of heat:

$$4Fe + 3O_2 \longrightarrow 2Fe_2O_3 + 198 \text{ kcal/mol}$$

This reaction occurs via an oxidation-reduction mechanism (see Chapter 8). The iron atoms are oxidized, O_2 is reduced. Electrons are transferred from the iron atoms to O_2 and Fe_2O_3 forms exothermically. The rate of the reaction is slow; therefore the heat is liberated gradually over a period of several hours.

For Further Understanding

What is the sign of ΔH for each equation in this story?

Would reactions with small rate constants be preferred for applications such as those described here? Why or why not?

Animation

Effect of Molecular Orientation
on Collision Effectiveness

Concentration is introduced in Section 1.5, and units and calculations are discussed in Sections 6.2 and 6.3.

collision orientation lead to product formation. These collisions are termed *effective collisions.*

The Concentration of Reactants

The rate of a chemical reaction is often a complex function of the concentration of one or more of the reacting substances. The rate will generally *increase* as concentration *increases* simply because a higher concentration means more reactant molecules in a given volume and therefore a greater number of collisions per unit time. If we assume that other variables are held constant, a larger number of collisions leads to a larger number of effective collisions. For example, the rate at which a fire burns depends on the concentration of oxygen in the atmosphere surrounding the fire, as well as the concentration of the fuel (perhaps methane or propane). A common fire-fighting strategy is the use of fire extinguishers filled with carbon dioxide. The carbon dioxide dilutes the oxygen to a level where the combustion process can no longer be sustained.

The Temperature of Reactants

The rate of a reaction *increases* as the temperature increases, because the average kinetic energy of the reacting particles is directly proportional to the Kelvin temperature. Increasing the speed of particles increases the likelihood of collision, and the higher kinetic energy means that a higher percentage of these collisions will result in product formation (effective collisions). A 10°C rise in temperature has often been found to double the reaction rate.

The Physical State of Reactants

The rate of a reaction depends on the physical state of the reactants: solid, liquid, or gas. For a reaction to occur the reactants must collide frequently and have sufficient energy to react. In the solid state, the atoms, ions, or molecules are restricted in their motion. In the gaseous and liquid states the particles have both free motion and proximity to each other. Hence reactions tend to be fastest in the gaseous and liquid states and slowest in the solid state.

The Presence of a Catalyst

A **catalyst** is a substance that *increases* the reaction rate. If added to a reaction mixture, the catalytic substance undergoes no net change, nor does it alter the outcome of the reaction. However, the catalyst interacts with the reactants to create an alternative pathway for production of products. This alternative path has a lower activation energy. This makes it easier for the reaction to take place and thus increases the rate. This effect is illustrated in Figure 7.11.

Catalysis is important industrially; it may often make the difference between profit and loss in the sale of a product. For example, catalysis is useful in converting double bonds to single bonds. An important application of this principle involves the process of hydrogenation. Hydrogenation converts one or more of the carbon-carbon double bonds of unsaturated fats (e.g., corn oil, olive oil) to single bonds characteristic of saturated fats (such as margarine). The use of a metal catalyst, such as nickel, in contact with the reaction mixture dramatically increases the rate of the reaction.

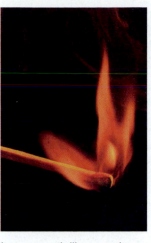

How does a match illustrate the concept of activation energy?

These factors were considered in our discussion of the states of matter (Chapter 5).

Section 11.5 describes the role of catalysis in organic reactions.

Sections 19.1 through 19.6 describe enzyme catalysis.

(a) Uncatalyzed reaction

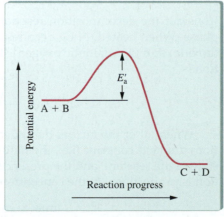

(b) Catalyzed reaction

Figure 7.11
The effect of a catalyst on the magnitude of the activation energy of a chemical reaction: (a) uncatalyzed reaction, (b) catalyzed reaction. Note that the presence of a catalyst decreases the activation energy ($E'_a < E_a$), thus increasing the rate of the reaction.

Figure 7.12

The synthesis of ammonia, an important industrial product, is facilitated by a solid phase catalyst (the Haber process). H_2 and N_2 bind to the surface, their bonds are weakened, dissociation and reformation as ammonia occur, and the newly formed ammonia molecules leave the surface. This process is repeated over and over, with no change in the catalyst.

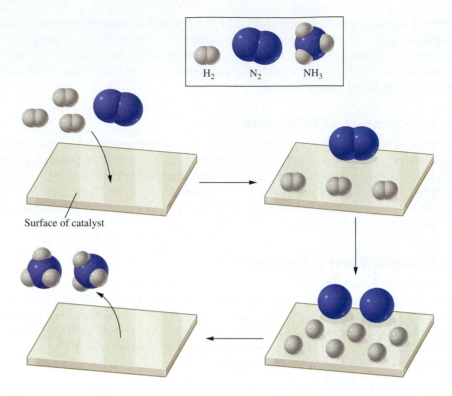

H_2 N_2 NH_3

Surface of catalyst

Thousands of essential biochemical reactions in our bodies are controlled and speeded up by biological catalysts called *enzymes*.

A molecular level view of the action of a solid catalyst widely used in industrial synthesis of ammonia is presented in Figure 7.12.

Question 7.11

Would you imagine that a substance might act as a poison if it interfered with the function of an enzyme? Why?

Question 7.12

Bacterial growth decreases markedly in a refrigerator. Why?

LEARNING GOAL

7 ▷ Write rate equations for elementary processes.

Mathematical Representation of Reaction Rate

Consider the decomposition reaction of N_2O_5 (dinitrogen pentoxide) in the gas phase. When heated, N_2O_5 decomposes and forms two products: NO_2 (nitrogen dioxide) and O_2 (diatomic oxygen). The balanced chemical equation for the reaction is

$$2N_2O_5(g) \xrightarrow{\Delta} 4NO_2(g) + O_2(g)$$

When all of the factors that affect the rate of the reaction (except concentration) are held constant (i.e., the nature of the reactant, temperature and physical state of the reactant, and the presence or absence of a catalyst) the rate of the reaction is proportional to the concentration of N_2O_5.

$$\text{rate} \propto \text{concentration } N_2O_5$$

We will represent the concentration of N_2O_5 in units of molarity and represent molar concentration using brackets.

$$\text{concentration } N_2O_5 = [N_2O_5]$$

Then,

$$\text{rate} \propto [N_2O_5]$$

Laboratory measurement shows that the rate of the reaction depends on the molar concentration raised to an experimentally determined exponent that we will symbolize as n

$$\text{rate} \propto [N_2O_5]^n$$

In expressions such as the one shown, the proportionality symbol, $\propto$, may be replaced by an equality sign and a proportionality constant that we represent as k, the **rate constant.**

$$\text{rate} = k[N_2O_5]^n$$

The exponent, n, is the **order of the reaction.** For the reaction described here, which has been studied in great detail, n is numerically equal to 1, hence the reaction is first order in N_2O_5 and the rate equation for the reaction is:

$$\text{rate} = k[N_2O_5]$$

Equations that follow this format, the rate being equal to the rate constant multiplied by the reactant concentration raised to an exponent that is the order, are termed **rate equations.**

Note that the exponent, n, in the rate equation is not the same as the coefficient of N_2O_5 in the balanced equation. However, in reactions that occur in a single step, the coefficient in the balanced equation and the exponent n (the order of the reaction) are numerically the same.

In general, the rate of reaction for an equation of the general form:

$$A \longrightarrow product$$

is

$$rate = k[A]^n$$

in which

$$n = \text{order of the reaction}$$
$$k = \text{the rate constant of the reaction}$$

An equation of the form

$$A + B \longrightarrow products$$

has a rate expression

$$rate = k[A]^n[B]^{n'}$$

Both the value of the rate constant and the order of the reaction are deduced from a series of experiments. We cannot predict them by simply looking at the chemical equation. Only the *form* of the rate expression can be found by inspection of the chemical equation, and, even then, only for reactions that occur in a single step.

EXAMPLE 7.5 | **Writing Rate Equations**

LEARNING GOAL

7 Write rate equations for elementary processes.

Write the form of the rate equation for the oxidation of ethanol (C_2H_5OH). The reaction has been experimentally determined to be first order in ethanol and third order in oxygen (O_2).

Solution

Step 1. The rate expression involves only the reactants, C_2H_5OH and O_2. Depict their concentrations as

$$[C_2H_5OH][O_2]$$

Step 2. Now raise each to an exponent corresponding to its experimentally determined order

$$[C_2H_5OH][O_2]^3$$

Step 3. This is proportional to the rate:

$$\text{rate} \propto [C_2H_5OH][O_2]^3$$

Step 4. Proportionality ($\propto$) is incorporated into an equation using a proportionality constant, k.

$$\text{rate} = k[C_2H_5OH][O_2]^3$$

is the rate expression. (Remember that 1 is understood as an exponent; $[C_2H_5OH]$ is correct and $[C_2H_5OH]^1$ is not.)

Practice Problem 7.5

Write the general form of the rate equation for each of the following processes.
a. $N_2(g) + O_2(g) \longrightarrow 2NO(g)$
b. $2C_4H_6(g) \longrightarrow C_8H_{12}(g)$
c. $CH_4(g) + 2O_2(g) \longrightarrow CO_2(g) + 2H_2O(g)$
d. $2NO_2(g) \longrightarrow 2NO(g) + O_2(g)$

For Further Practice: Questions 7.49 and 7.50.

Knowledge of the form of the rate equation, coupled with the experimental determination of the value of the rate constant, k, and the order, n, are valuable in a number of ways. Industrial chemists use this information to establish optimum conditions for preparing a product in the shortest practical time. The design of an entire manufacturing facility may, in part, depend on the rates of the critical reactions.

In Section 7.4 we will see how the rate equation forms the basis for describing equilibrium reactions.

7.4 Equilibrium

Rate and Reversibility of Reactions

LEARNING GOAL

8 Recognize and describe equilibrium situations.

We have assumed that most chemical and physical changes considered thus far proceed to completion. A complete reaction is one in which all reactants have been converted to products. However, many important chemical reactions do not go to completion. As a result, after no further obvious change is taking place, measurable quantities of reactants and products remain. Reactions of this type (incomplete reactions) are called **equilibrium reactions**.

Examples of physical and chemical equilibria abound in nature. Many environmental systems depend on fragile equilibria. The amount of oxygen dissolved in a certain volume of lake water (the oxygen concentration) is governed by the principles of equilibrium. The lives of plants and animals within this system are critically related to the levels of dissolved oxygen in the water.

The very form and function of the earth is a consequence of a variety of complex equilibria. Stalactite and stalagmite formations in caves are made up of solid calcium carbonate ($CaCO_3$). They owe their existence to an equilibrium process described by the following equation:

$$Ca^{2+}(aq) + 2HCO_3^-(aq) \rightleftharpoons CaCO_3(s) + CO_2(aq) + H_2O(l)$$

Animations

Equilibrium

Dynamic Equilibrium

Physical Equilibrium

A physical equilibrium, such as sugar dissolving in water, is a reversible reaction. A **reversible reaction** is a process that can occur in both directions. It is indicated by using a double arrow ($\rightleftharpoons$) symbol.

Dissolution of sugar in water, producing a saturated solution, is a convenient illustration of a state of *dynamic equilibrium.*

A **dynamic equilibrium** is a situation in which the rate of the forward process in a reversible reaction is exactly balanced by the rate of the reverse process.

Let's now look at the sugar and water equilibrium in more detail.

Sugar in Water

Imagine that you mix a small amount of sugar (2 or 3 g) in 100 mL of water. After you have stirred it for a short time, all of the sugar dissolves; there is no residual solid sugar because the sugar has dissolved *completely.* The reaction clearly has converted all solid sugar to its dissolved state, an aqueous solution of sugar, or

$$sugar(s) \longrightarrow sugar(aq)$$

Now, suppose that you add a very large amount of sugar (100 g), more than can possibly dissolve, to the same volume of water. As you stir the mixture you observe more and more sugar dissolving. After some time the amount of solid sugar remaining in contact with the solution appears constant. Over time, you observe no further decrease in the amount of undissolved sugar. Although nothing further appears to be happening, in reality a great deal of activity is taking place!

An equilibrium situation has been established. Over time the amount of sugar dissolved in the measured volume of water (the concentration of sugar in water) does not change. Hence the amount of undissolved sugar remains the same. However, if you could look at the individual sugar molecules, you would see something quite amazing. Rather than sugar molecules in the solid simply staying in place, you would see them continuing to leave the solid state and go into solution. At the same time, a like number of dissolved sugar molecules would leave the water and form more solid. This active process is described as a *dynamic equilibrium.* The reaction is proceeding in a forward (left to right) and a reverse (right to left) direction at the same time and is a reversible reaction:

$$sugar(s) \rightleftharpoons sugar(aq)$$

The double arrow serves as

- an indicator of a reversible process,
- an indicator of an equilibrium process, and
- a reminder of the dynamic nature of the process.

An excess of sugar in water produces a saturated solution, discussed in Section 6.1.

Dynamic equilibrium can be particularly dangerous for living cells because it represents a situation in which nothing is getting done. There is no gain. Let's consider an exothermic reaction designed to produce a net gain of energy for the cell. In a dynamic equilibrium the rate of the forward (energy-releasing) reaction is equal to the rate of the backward (energy-requiring) reaction. Thus there is no net gain of energy to fuel cellular activity, and the cell will die.

How can we rationalize the apparent contradiction: continuous change is taking place yet no observable change in the amount of sugar in either the solid or dissolved form is observed.

The only possible explanation is that the rate of the forward process

$$\text{sugar}(s) \longrightarrow \text{sugar}(aq)$$

must be equal to the rate of the reverse process

$$\text{sugar}(s) \longleftarrow \text{sugar}(aq)$$

Under this condition, the number of sugar molecules leaving the solid in a given time interval is identical to the number of sugar molecules returning to the solid state.

Question 7.13

Construct an example of a dynamic equilibrium using a subway car at rush hour.

Question 7.14

A certain change in reaction conditions for a process was found to increase the rate of the forward reaction much more than that of the reverse reaction. Did the amount of product increase, decrease, or remain the same? Why?

Chemical Equilibrium

The Reaction of N_2 and H_2

When we mix nitrogen gas (N_2) and hydrogen gas (H_2) at an elevated temperature (perhaps 500°C), some of the molecules will collide with sufficient energy to break N—N and H—H bonds. Rearrangement of the atoms will produce the product (NH_3):

$$N_2(g) + 3H_2(g) \rightleftharpoons 2NH_3(g)$$

Beginning with a mixture of hydrogen and nitrogen, the rate of the reaction is initially rapid, because the reactant concentration is high; as the reaction proceeds, the concentration of reactants decreases. At the same time the concentration of the product, ammonia, is increasing. At equilibrium the *rate of depletion* of hydrogen and nitrogen *is equal to* the *rate of depletion* of ammonia. In other words, *the rates of the forward and reverse reactions are equal.*

The concentration of the various species is fixed at equilibrium because product is being *consumed and formed at the same rate.* In other words, the reaction continues indefinitely (dynamic), but the concentration of products and reactants is fixed (equilibrium). This is a *dynamic equilibrium.* The composition of this reaction mixture as a function of time is depicted in Figure 7.13.

For systems such as the ammonia/hydrogen/nitrogen equilibrium, an **equilibrium constant** expression can be written; it summarizes the relationship between the concentration of reactants and products in an equilibrium reaction.

The Generalized Equilibrium-Constant Expression for a Chemical Reaction

We write the general form of an equilibrium chemical reaction as

$$aA + bB \rightleftharpoons cC + dD$$

in which A and B represent reactants, C and D represent products, and a, b, c, and d are the coefficients of the balanced equation. The equilibrium constant expression for this general case is

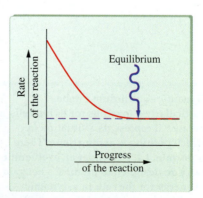

Figure 7.13
The change of the rate of reaction as a function of time. The rate of reaction, initially rapid, decreases as the concentration of reactant decreases and approaches a limiting value at equilibrium.

$$K_{eq} = \frac{[C]^c[D]^d}{[A]^a[B]^b}$$

Products of the overall equilibrium reaction are in the numerator, and *reactants* are in the denominator.

For the ammonia system, it follows that the appropriate equilibrium expression is:

$$K_{eq} = \frac{[NH_3]^2}{[N_2][H_2]^3}$$

[ ] represents molar concentration, *M*.

It does not matter what initial amounts (concentrations) of reactants or products we choose. When the system reaches equilibrium, the calculated value of K_{eq} will not change. The magnitude of K_{eq} can be altered only by changing the temperature. Thus K_{eq} is temperature dependent. The chemical industry uses this fact to advantage by choosing a reaction temperature that will maximize the yield of a desired product.

How could one determine when a reaction has reached equilibrium?

Question 7.15

Does the attainment of equilibrium imply that no further change is taking place in the system?

Question 7.16

Writing Equilibrium-Constant Expressions

An equilibrium-constant expression can be written only after a correct, balanced chemical equation that describes the equilibrium system has been developed. A balanced equation is essential because the *coefficients* in the equation become the *exponents* in the equilibrium-constant expression.

LEARNING GOAL

9 Write equilibrium-constant expressions and use these expressions to calculate equilibrium constants.

Each chemical reaction has a unique equilibrium constant value at a specified temperature. Equilibrium constants listed in the chemical literature are often reported at 25°C, to allow comparison of one system with any other. For any equilibrium reaction, the value of the equilibrium constant changes with temperature.

The brackets represent molar concentration or molarity; recall that molarity has units of mol/L. Although the equilibrium constant may have units (owing to the units on each concentration term), by convention units are usually not used. In our discussion of equilibrium, all equilibrium constants are shown as *unitless*.

The exponents correspond to the *coefficients* of the balanced equation.

A properly written equilibrium-constant expression may not include all of the terms in the chemical equation upon which it is based. Only the concentration of gases and substances in solution are shown, because their concentrations can change. Concentration terms for liquids and solids are *not* shown. The concentration of a liquid is constant. Most often, the liquid is the solvent for the reaction under consideration. A solid also has a fixed concentration and, for solution reactions, is not really a part of the solution. When a solid is formed it exists as a solid phase in contact with a liquid phase (the solution).

Writing an Equilibrium-Constant Expression   EXAMPLE 7.6

Write an equilibrium-constant expression for the reversible reaction:

$$H_2(g) + F_2(g) \rightleftharpoons 2HF(g)$$

LEARNING GOAL

9 Write equilibrium-constant expressions and use these expressions to calculate equilibrium constants.

Continued—

EXAMPLE 7.6 —*Continued*

Solution

Step 1. Inspection of the chemical equation reveals that no solids or liquids are present. Hence all reactants and products appear in the equilibrium-constant expression.

Step 2. The numerator term is the product term $[HF]^2$.

Step 3. The denominator terms are the reactants $[H_2]$ and $[F_2]$. Note that each term contains an exponent identical to the corresponding coefficient in the balanced equation.

Step 4. Arranging the numerator and denominator terms as a fraction and setting the fraction equal to K_{eq} yields

$$K_{eq} = \frac{[HF]^2}{[H_2][F_2]}$$

Practice Problem 7.6

Write an equilibrium-constant expression for each of the following reversible reactions.

a. $2NO_2(g) \rightleftharpoons N_2(g) + 2O_2(g)$

b. $2H_2O(l) \rightleftharpoons 2H_2(g) + O_2(g)$

For Further Practice: Questions 7.73 and 7.74.

EXAMPLE 7.7   **Writing an Equilibrium-Constant Expression**

LEARNING GOAL

9   Write equilibrium-constant expressions and use these expressions to calculate equilibrium constants.

Write an equilibrium-constant expression for the reversible reaction:

$$MnO_2(s) + 4H^+(aq) + 2Cl^-(aq) \rightleftharpoons Mn^{2+}(aq) + Cl_2(g) + 2H_2O(l)$$

Solution

Step 1. MnO_2 is a solid and H_2O, although a product, is negligible compared with the water solvent. Thus they are not written in the equilibrium-constant expression.

$$\boxed{MnO_2(s)} + 4H^+(aq) + 2Cl^-(aq) \rightleftharpoons Mn^{2+}(aq) + Cl_2(g) + \boxed{2H_2O(l)}$$

Not a part of the K_{eq} expression

Step 2. The numerator term includes the remaining products:

$$[Mn^{2+}] \quad \text{and} \quad [Cl_2]$$

Step 3. The denominator term includes the remaining reactants:

$$[H^+]^4 \quad \text{and} \quad [Cl^-]^2$$

Note that each exponent is identical to the corresponding coefficient in the chemical equation.

Continued—

EXAMPLE 7.7 —*Continued*

Step 4. Arranging the numerator and denominator terms as a fraction and setting the fraction equal to K_{eq} yields

$$K_{eq} = \frac{[Mn^{2+}][Cl_2]}{[H^+]^4[Cl^-]^2}$$

Practice Problem 7.7

Write an equilibrium-constant expression for each of the following reversible reactions.

a. $AgCl(s) \rightleftharpoons Ag^+(aq) + Cl^-(aq)$

b. $PCl_5(s) \rightleftharpoons PCl_3(g) + Cl_2(g)$

For Further Practice: Questions 7.77 and 7.79.

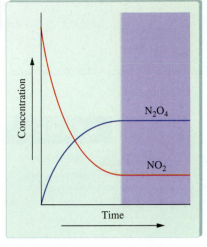

Figure 7.14
The combination reaction of NO_2 molecules produces N_2O_4. Initially, the concentration of reactant (NO_2) diminishes rapidly while the N_2O_4 concentration builds. Eventually, the concentrations of both reactant and product become constant over time (blue area). The equilibrium condition has been attained.

Animation
NO_2/N_2O_4 Equilibrium

Interpreting Equilibrium Constants

What utility does the equilibrium constant have? The reversible arrow in the chemical equation alerts us to the fact that an equilibrium exists. Some measurable quantity of the product and reactant remain. However, there is no indication whether products predominate, reactants predominate, or significant concentrations of both products and reactants are present at equilibrium.

The numerical value of the equilibrium constant provides this additional information. It tells us the extent to which reactants have converted to products. This is important information for anyone who wants to manufacture and sell the product. It also is important to anyone who studies the effect of equilibrium reactions on environmental systems and living organisms.

Although an absolute interpretation of the numerical value of the equilibrium constant depends on the form of the equilibrium-constant expression, the following generalizations are useful:

- K_{eq} greater than 1×10^2. A large numerical value of K_{eq} indicates that the numerator (product term) is much larger than the denominator (reactant term) and that at equilibrium mostly product is present.
- K_{eq} less than 1×10^{-2}. A small numerical value of K_{eq} indicates that the numerator (product term) is much smaller than the denominator (reactant term) and that at equilibrium mostly reactant is present.
- K_{eq} between 1×10^{-2} and 1×10^2. In this case the equilibrium mixture contains significant concentrations of both reactants and products.

Question 7.17

At a given temperature, the equilibrium constant for a certain reaction is 1×10^{20}. Does this equilibrium favor products or reactants? Why?

Question 7.18

At a given temperature, the equilibrium constant for a certain reaction is 1×10^{-18}. Does this equilibrium favor products or reactants? Why?

Calculating Equilibrium Constants

The magnitude of the equilibrium constant for a chemical reaction is determined experimentally. The reaction under study is allowed to proceed until the composition of products and reactants no longer changes (Figure 7.14). This may be a matter

LEARNING GOAL

9 Write equilibrium-constant expressions and use these expressions to calculate equilibrium constants.

of seconds, minutes, hours, or even months or years, depending on the rate of the reaction. The reaction mixture is then analyzed to determine the molar concentration of each of the products and reactants. These concentrations are substituted in the equilibrium-constant expression and the equilibrium constant is calculated. The following example illustrates this process.

Section 6.3 describes molar concentration.

| EXAMPLE 7.8 | Calculating an Equilibrium Constant |

 LEARNING GOAL

9 Write equilibrium-constant expressions and use these expressions to calculate equilibrium constants.

Hydrogen iodide is placed in a sealed container and allowed to come to equilibrium. The equilibrium reaction is:

$$2HI(g) \rightleftharpoons H_2(g) + I_2(g)$$

and the equilibrium concentrations are:

$$[HI] = 0.54 \ M$$
$$[H_2] = 1.72 \ M$$
$$[I_2] = 1.72 \ M$$

Calculate the equilibrium constant.

Solution

Step 1. Write the equilibrium-constant expression:

$$K_{eq} = \frac{[H_2][I_2]}{[HI]^2}$$

Step 2. Substitute the equilibrium concentrations of products and reactants to obtain

$$K_{eq} = \frac{[1.72][1.72]}{[0.54]^2} = \frac{2.96}{0.29}$$

$$= 10.1 \text{ or } 1.0 \times 10^1 \text{ (two significant figures)}$$

 Practice Problem 7.8

A container holds the following mixture at equilibrium:

$$[NH_3] = 0.25 \ M$$
$$[N_2] = 0.11 \ M$$
$$[H_2] = 1.91 \ M$$

If the reaction is:

$$N_2(g) + 3H_2(g) \rightleftharpoons 2NH_3(g)$$

Calculate the equilibrium constant.

For Further Practice: Questions 7.78 and 7.80.

Using Equilibrium Constants

We have seen that the equilibrium constant for a reaction can be calculated if we know the equilibrium concentrations of all of the reactants and products. Once known, the equilibrium constant can be used to obtain equilibrium concentrations of one or more reactants or products for a variety of different situations. These calculations can be quite complex. Let's look at one straightforward but useful case, one where the equilibrium concentration of reactants is known and we wish to calculate the product concentration.

Using an Equilibrium Constant

EXAMPLE 7.9

LEARNING GOAL

9 ▶ Write equilibrium-constant expressions and use these expressions to calculate equilibrium constants.

Given the equilibrium reaction studied in Example 7.8:

$$2HI(g) \rightleftharpoons H_2(g) + I_2(g)$$

A sample mixture of HI, H_2, and I_2, at equilibrium, was found to have $[H_2] = 1.0 \times 10^{-2} M$ and $[HI] = 4.0 \times 10^{-2} M$. Calculate the molar concentration of I_2 in the equilibrium mixture.

Solution

Step 1. From Example 7.8, the equilibrium expression and equilibrium constant are:

$$K_{eq} = \frac{[H_2][I_2]}{[HI]^2}; \quad K_{eq} = 1.0 \times 10^1$$

Step 2. Solve the equilibrium expression for $[I_2]$
Cross-multiply:

$$[H_2][I_2] = K_{eq}[HI]^2$$

Divide both sides by $[H_2]$

$$[I_2] = \frac{K_{eq}[HI]^2}{[H_2]}$$

Step 3. Substitute the values:

$$K_{eq} = 1.0 \times 10^1$$
$$[H_2] = 1.0 \times 10^{-2} M$$
$$[HI] = 4.0 \times 10^{-2} M$$

Step 4. Solve:

$$[I_2] = \frac{[1.0 \times 10^1][4.0 \times 10^{-2}]^2}{[1.0 \times 10^{-2}]}$$
$$[I_2] = 1.6 M$$

Practice Problem 7.9

Using the reaction (above), calculate the $[I_2]$ if both $[H_2]$ and $[HI]$ were $1.0 \times 10^{-4} M$.

For Further Practice: Questions 7.81 and 7.82.

LeChatelier's Principle

In the nineteenth century the French chemist LeChatelier discovered that changes in equilibrium depend on the amount of "stress" applied to the system. The stress may take the form of an increase or decrease of the temperature of the system at

LEARNING GOAL

10 ▶ Use LeChatelier's principle to predict changes in equilibrium position.

Animation
LeChatelier's Principle

equilibrium or perhaps a change in the amount of reactant or product present in a fixed volume (the concentration of reactant or product).

LeChatelier's principle states that if a stress is placed on a system at equilibrium, the system will respond by altering the equilibrium composition in such a way as to minimize the stress.

Consider the equilibrium situation discussed earlier:

$$N_2(g) + 3H_2(g) \rightleftharpoons 2NH_3(g)$$

If the reactants and products are present in a fixed volume (such as 1L) and more NH_3 (the *product*) is introduced into the container, the system will be stressed—the equilibrium will be disturbed. The system will try to alleviate the stress (as we all do) by *removing* as much of the added material as possible. How can it accomplish this? By converting some NH_3 to H_2 and N_2. The equilibrium shifts to the left, and the dynamic equilibrium is soon reestablished.

Adding extra H_2 or N_2 would apply the stress to the other side of the equilibrium. To minimize the stress, the system would "use up" some of the excess H_2 or N_2 to make product, NH_3. The equilibrium would shift to the right.

In summary,

$$N_2(g) + 3H_2(g) \rightleftharpoons 2NH_3(g)$$

Product introduced: Equilibrium shifted ←

Reactant introduced: Equilibrium shifted →

Addition of products or reactants may have a profound effect on the composition of a reaction mixture but *does not* affect the value of the equilibrium constant.

What would happen if some of the ammonia molecules were *removed* from the system? The loss of ammonia represents a stress on the system; to relieve that stress, the ammonia would be replenished by the reaction of hydrogen and nitrogen. The equilibrium would shift to the right.

Effect of Concentration

Addition of extra product or reactant to a fixed reaction volume is just another way of saying that we have increased the concentration of product or reactant. Removal of material from a fixed volume decreases the concentration. Therefore changing the concentration of one or more components of a reaction mixture is a way to alter the equilibrium composition of an equilibrium mixture (Figure 7.15). Let's look at some additional experimental variables that may change equilibrium composition.

Figure 7.15
The effect of concentration on equilibrium position of the reaction:

$FeSCN^{2+}(aq) \rightleftharpoons Fe^{3+}(aq) + SCN^-(aq)$
(red)   (yellow) (colorless)

Solution (a) represents this reaction at equilibrium; addition of SCN^- shifts the equilibrium to the left (b) intensifying the red color. Removal of SCN^- shifts the equilibrium to the right (c) shown by the disappearance of the red color.

(a)   (b)   (c)

Effect of Heat

The change in equilibrium composition caused by the addition or removal of heat from an equilibrium mixture can be explained by treating heat as a product or reactant. The reaction of nitrogen and hydrogen is an exothermic reaction:

$$N_2(g) + 3H_2(g) \rightleftharpoons 2NH_3(g) + 22 \text{ kcal}$$

Adding heat to the reaction is similar to increasing the amount of product. The equilibrium will shift to the left, increasing the amounts of N_2 and H_2 and decreasing the amount of NH_3. If the reaction takes place in a fixed volume, the concentrations of N_2 and H_2 increase and the NH_3 concentration decreases.

Removal of heat produces the reverse effect. More ammonia is produced from N_2 and H_2, and the concentrations of these reactants must decrease.

In the case of an endothermic reaction such as

$$39 \text{ kcal} + 2N_2(g) + O_2(g) \rightleftharpoons 2N_2O(g)$$

addition of heat is analogous to the addition of reactant, and the equilibrium shifts to the right. Removal of heat would shift the reaction to the left, favoring the formation of reactants.

The dramatic effect of heat on the position of equilibrium is shown in Figure 7.16.

Effect of Pressure

Only gases are affected significantly by changes in pressure because gases are free to expand and compress in accordance with Boyle's law. However, liquids and solids are not compressible, so their volumes are unaffected by pressure.

Expansion and compression of gases and Boyle's law are discussed in Section 5.1.

Therefore pressure changes will alter equilibrium composition only in reactions that involve a gas or variety of gases as products and/or reactants. Again, consider the ammonia example,

$$N_2(g) + 3H_2(g) \rightleftharpoons 2NH_3(g)$$

One mole of N_2 and three moles of H_2 (total of four moles of reactants) convert to two moles of NH_3 (two moles of product). An increase in pressure favors a decrease in volume and formation of product. This decrease in volume is made possible by a shift to the right in equilibrium composition. Two moles of ammonia require less volume than four moles of reactant.

The industrial process for preparing ammonia, the Haber process, uses pressures of several hundred atmospheres to increase the yield.

Figure 7.16

The effect of heat on equilibrium position. For the reaction:

$$CoCl_4^{2-}(aq) + 6H_2O(l) \rightleftharpoons$$
(blue)

$$Co(H_2O)_6^{2+}(aq) + 4Cl^-(aq)$$
(pink)

Heating the solution favors the blue $CoCl_4^{2-}$ species; cooling favors the pink $Co(H_2O)_6^{2+}$ species.

A decrease in pressure allows the volume to expand. The equilibrium composition shifts to the left and ammonia decomposes to form more nitrogen and hydrogen.

In contrast, the decomposition of hydrogen iodide,

$$2HI(g) \rightleftharpoons H_2(g) + I_2(g)$$

is unaffected by pressure. The number of moles of gaseous product and reactant are identical. No volume advantage is gained by a shift in equilibrium composition. In summary:

- Pressure affects the equilibrium composition only of reactions that involve at least one gaseous substance.
- Additionally, the relative number of moles of gaseous products and reactants must differ.
- The equilibrium composition will shift to increase the number of moles of gas when the pressure decreases; it will shift to decrease the number of moles of gas when the pressure increases.

Effect of a Catalyst

A catalyst has no effect on the equilibrium composition. A catalyst increases the rates of both forward and reverse reactions to the same extent. The equilibrium composition *and* equilibrium concentration do not change when a catalyst is used, but the equilibrium composition is achieved in a shorter time. The role of a solid-phase catalyst in the synthesis of ammonia is shown in Figure 7.12.

EXAMPLE 7.10   **Predicting Changes in Equilibrium Composition**

Earlier in this section we considered the geologically important reaction that occurs in rock and soil.

$$Ca^{2+}(aq) + 2HCO_3^-(aq) \rightleftharpoons CaCO_3(s) + CO_2(aq) + H_2O(l)$$

Predict the effect on the equilibrium composition for each of the following changes.

a. The $[Ca^{2+}]$ is increased.
b. The amount of $CaCO_3$ is increased.
c. The $[HCO_3^-]$ is decreased.
d. A catalyst is added.

Solution

a. The concentration of reactant increases; the equilibrium shifts to the right, and more products are formed.
b. $CaCO_3$ is a solid; solids are not written in the equilibrium-constant expression, so there is no effect on the equilibrium composition.
c. The concentration of reactant decreases; the equilibrium shifts to the left, and more reactants are formed.
d. A catalyst has no effect on the equilibrium composition.

Continued—

EXAMPLE 7.10 —Continued

Practice Problem 7.10

For the hypothetical equilibrium reaction

$$A(g) + B(g) \rightleftharpoons C(g) + D(g)$$

predict whether the amount of A in a 5.0-L container would increase, decrease, or remain the same for each of the following changes.

a. Addition of excess B
b. Addition of excess C
c. Removal of some D
d. Addition of a catalyst

For Further Practice: Questions 7.83 and 7.84.

SUMMARY

7.1 Thermodynamics

Thermodynamics is the study of energy, work, and heat. Thermodynamics can be applied to the study of chemical reactions because we can determine the quantity of heat flow (by measuring the temperature change) between the *system* and the *surroundings. Exothermic reactions* release energy and products that are lower in energy than the reactants. *Endothermic reactions* require energy input. Heat energy is represented as *enthalpy, H.* The energy gain or loss is the change in enthalpy, ΔH, and is one factor that is useful in predicting whether a reaction is spontaneous or non-spontaneous.

Entropy, S, is a measure of the randomness of a system. A random, or disordered system has high entropy; a well-ordered system has low entropy. The change in entropy in a chemical reaction, ΔS, is also a factor in predicting reaction spontaneity.

Free energy, ΔG, incorporates both factors, enthalpy and entropy; as such, it is an absolute predictor of the spontaneity of a chemical reaction.

7.2 Experimental Determination of Energy Change in Reactions

A *calorimeter* measures heat changes (in calories or joules) that occur in chemical reactions.

The *specific heat* of a substance is the number of calories of heat needed to raise the temperature of 1 g of the substance 1 degree Celsius.

The amount of energy per gram of food is referred to as its *fuel value.* Fuel values are commonly reported in units of *nutritional Calories* (1 nutritional Calorie = 1 kcal). A bomb

calorimeter is useful for measurement of the fuel value of foods.

7.3 Kinetics

Chemical *kinetics* is the study of the *rate* or speed of a chemical reaction. Energy for reactions is provided by molecular collisions. If this energy is sufficient, bonds may break, and atoms may recombine in a different arrangement, producing product. A collision producing one or more product molecules is termed an effective collision.

The minimum amount of energy needed for a reaction is the *activation energy.* The reaction proceeds from reactants to products through an intermediate state, the *activated complex.*

Experimental conditions influencing the reaction rate include the structure of the reacting species, the concentration of reactants, the temperature of reactants, the physical state of reactants, and the presence or absence of a catalyst.

A *catalyst* increases the rate of a reaction. The catalytic substance undergoes no net change in the reaction, nor does it alter the outcome of the reaction.

7.4 Equilibrium

Many chemical reactions do not completely convert reactants to products. A mixture of products and reactants exists, and its composition will remain constant until the experimental conditions are changed. This mixture is in a state of *chemical equilibrium.* The reaction continues indefinitely (dynamic), but the concentrations of products and reactants are fixed (equilibrium) because the rates of the forward and reverse reactions are equal. This is a *dynamic equilibrium.*

LeChatelier's principle states that if a stress is placed on an equilibrium system, the system will respond by altering the equilibrium in such a way as to minimize the stress.

KEY TERMS

activated complex (7.3)
activation energy (7.3)
calorimetry (7.2)
catalyst (7.3)
dynamic equilibrium (7.4)
endothermic reaction (7.1)
enthalpy (7.1)
entropy (7.1)
equilibrium constant (7.4)
equilibrium reaction (7.4)
exothermic reaction (7.1)
free energy (7.1)
fuel value (7.2)

kinetics (7.3)
LeChatelier's principle (7.4)
nutritional Calorie (7.2)
order of the reaction (7.3)
rate constant (7.3)
rate equation (7.3)
rate of chemical reaction (7.3)
reversible reaction (7.4)
specific heat (7.2)
surroundings (7.1)
system (7.1)
thermodynamics (7.1)

QUESTIONS AND PROBLEMS

Energy and Thermodynamics

Foundations

7.19 What is the energy unit most commonly employed in chemistry?

7.20 What energy unit is commonly employed in nutrition science?

7.21 Describe what is meant by an exothermic reaction.

7.22 Describe what is meant by an endothermic reaction.

7.23 The oxidation of fuels (coal, oil, gasoline) are exothermic reactions. Why?

7.24 Provide an explanation for the fact that most decomposition reactions are endothermic but most combination reactions are exothermic.

7.25 Describe how a calorimeter is used to distinguish between exothermic and endothermic reactions.

7.26 Construct a diagram of a coffee-cup calorimeter.

7.27 Why does a calorimeter have a "double-walled" container?

7.28 Explain why the fuel value of foods is an important factor in nutrition science.

7.29 Explain what is meant by the term *free energy*.

7.30 Explain what is meant by the term *specific heat*.

7.31 State the first law of thermodynamics.

7.32 State the second law of thermodynamics.

7.33 Explain what is meant by the term *enthalpy*.

7.34 Explain what is meant by the term *entropy*.

Applications

7.35 5.00 g of octane are burned in a bomb calorimeter containing 2.00×10^2 g H_2O. How much energy, in calories, is released if the water temperature increases 6.00°C?

7.36 0.0500 mol of a nutrient substance is burned in a bomb calorimeter containing 2.00×10^2 g H_2O. If the formula weight of this nutrient substance is 114 g/mol, what is the fuel value (in nutritional Calories) if the temperature of the water increased 5.70°C.

7.37 Calculate the energy released, in joules, in Question 7.35 (recall conversion factors, Chapter 1).

7.38 Calculate the fuel value, in kilojoules, in Question 7.36 (recall conversion factors, Chapter 1).

7.39 Predict whether each of the following processes increases or decreases entropy, and explain your reasoning.
   a. melting of a solid metal
   b. boiling of water

7.40 Predict whether each of the following processes increases or decreases entropy, and explain your reasoning.
   a. burning a log in a fireplace
   b. condensation of water vapor on a cold surface

7.41 Predict whether a reaction with a negative ΔH and a positive ΔS will be spontaneous, nonspontaneous, or temperature dependent. Explain your reasoning.

7.42 Predict whether a reaction with a negative ΔH and a negative ΔS will be spontaneous, nonspontaneous, or temperature dependent. Explain your reasoning.

7.43 Isopropyl alcohol, commonly known as rubbing alcohol, feels cool when applied to the skin. Explain why.

7.44 Energy is required to break chemical bonds during the course of a reaction. When is energy released?

Kinetics

Foundations

7.45 Provide an example of a reaction that is extremely slow, taking days, weeks, or years to complete.

7.46 Provide an example of a reaction that is extremely fast, perhaps quicker than the eye can perceive.

7.47 Define the term *activated complex* and explain its significance in a chemical reaction.

7.48 Define and explain the term *activation energy* as it applies to chemical reactions.

Applications

7.49 Distinguish among the terms *rate*, *rate constant*, and *order*.

7.50 Write the rate equation for:

$$CH_4(g) + 2O_2(g) \longrightarrow 2H_2O(l) + CO_2(g)$$

if the order of all reactants is one.

7.51 Will the rate of the reaction in Question 7.50 increase, decrease, or remain the same if the rate constant doubles?

7.52 Will the rate of the reaction in Question 7.50 increase, decrease, or remain the same if the concentration of methane increases?

7.53 Describe the general characteristics of a catalyst.

7.54 Select one enzyme from a later chapter in this book and describe its biochemical importance.

7.55 Sketch a potential energy diagram for a reaction that shows the effect of a catalyst on an exothermic reaction.

7.56 Sketch a potential energy diagram for a reaction that shows the effect of a catalyst on an endothermic reaction.

7.57 Give at least two examples from the life sciences in which the rate of a reaction is critically important.

7.58 Give at least two examples from everyday life in which the rate of a reaction is an important consideration.

7.59 Describe how an increase in the concentration of reactants increases the rate of a reaction.

7.60 Describe how an increase in the temperature of reactants increases the rate of a reaction.

7.61 Write the rate expression for the single-step reaction:

$$N_2O_4(g) \rightleftharpoons 2NO_2(g)$$

7.62 Write the rate expression for the single-step reaction:

$$H_2S(aq) + Cl_2(aq) \rightleftharpoons S(s) + 2HCl(aq)$$

7.63 Describe how a catalyst speeds up a chemical reaction.

7.64 Explain how a catalyst can be involved in a chemical reaction without being consumed in the process.

Equilibrium

Foundations

7.65 Does a large equilibrium constant mean that products or reactants are favored?

7.66 Does a large equilibrium constant mean that the reaction must be rapid?

7.67 Provide an example of a physical equilibrium.

7.68 Provide an example of a chemical equilibrium.

7.69 Explain LeChatelier's principle.

7.70 How can LeChatelier's principle help us to increase yields of chemical reactions?

7.71 Describe the meaning of the term *dynamic equilibrium.*

7.72 What is the relationship between the forward and reverse rates for a reaction at equilibrium?

Applications

7.73 Write a valid equilibrium constant for the reaction shown in Question 7.61.

7.74 Write a valid equilibrium constant for the reaction shown in Question 7.62.

7.75 Distinguish between a physical equilibrium and a chemical equilibrium.

7.76 Distinguish between the rate constant and the equilibrium constant for a reaction.

7.77 Write the equilibrium constant expression for the reaction:

$$N_2(g) + 3H_2(g) \rightleftharpoons 2NH_3(g)$$

7.78 Using the equilibrium constant expression in Question 7.77, calculate the equilibrium constant if:

$$[N_2] = 0.071 \, M$$
$$[H_2] = 9.2 \times 10^{-3} \, M$$
$$[NH_3] = 1.8 \times 10^{-4} \, M$$

7.79 Write the equilibrium constant expression for the reaction:

$$2H_2(g) + S_2(g) \rightleftharpoons 2H_2S(g)$$

7.80 Using the equilibrium constant expression in Question 7.79, calculate the equilibrium constant if:

$$[H_2] = 2.1 \times 10^{-1} \, M$$
$$[S_2] = 1.1 \times 10^{-6} \, M$$
$$[H_2S] = 7.3 \times 10^{-1} \, M$$

7.81 Use the equilibrium constant expression you wrote in Question 7.77 and the equilibrium constant you calculated in Question 7.78 to determine the equilibrium concentration of NH_3 if:

$$[N_2] = 8.0 \times 10^{-2} \, M$$
$$[H_2] = 5.0 \times 10^{-3} \, M$$

7.82 Use the equilibrium constant expression you wrote in Question 7.79 and the equilibrium constant you calculated in Question 7.80 to determine the equilibrium concentration of H_2S if:

$$[H_2] = 1.0 \times 10^{-1} \, M$$
$$[S_2] = 1.0 \times 10^{-5} \, M$$

7.83 For the reaction

$$CH_4(g) + Cl_2(g) \rightleftharpoons CH_3Cl(g) + HCl(g) + 26.4 \, kcal$$

predict the effect on the equilibrium (will it shift to the left or to the right, or will there be no change?) for each of the following changes.
  a. The temperature is increased.
  b. The pressure is increased by decreasing the volume of the container.
  c. A catalyst is added.

7.84 For the reaction

$$47 \, kcal + 2SO_3(g) \rightleftharpoons 2SO_2(g) + O_2(g)$$

predict the effect on the equilibrium (will it shift to the left or to the right, or will there be no change?) for each of the following changes.
  a. The temperature is increased.
  b. The pressure is increased by decreasing the volume of the container.
  c. A catalyst is added.

7.85 Label each of the following statements as true or false and explain why.
  a. A slow reaction is an incomplete reaction.
  b. The rates of forward and reverse reactions are never the same.

7.86 Label each of the following statements as true or false and explain why.
  a. A reaction is at equilibrium when no reactants remain.
  b. A reaction at equilibrium is undergoing continual change.

7.87 Use LeChatelier's principle to predict whether the amount of PCl_3 in a 1.00-L container is increased, is decreased, or remains the same for the equilibrium

$$PCl_3(g) + Cl_2(g) \rightleftharpoons PCl_5(g) + heat$$

when each of the following changes is made.
  a. PCl_5 is added.   d. The temperature is decreased.
  b. Cl_2 is added.   e. A catalyst is added.
  c. PCl_5 is removed.

7.88 Use LeChatelier's principle to predict the effects, if any, of each of the following changes on the equilibrium system, described below, in a closed container.

$$C(s) + 2H_2(g) \rightleftharpoons CH_4(g) + 18 \, kcal$$

  a. C is added.   d. The temperature is increased.
  b. H_2 is added.   e. A catalyst is added.
  c. CH_4 is removed.

7.89 Will an increase in pressure increase, decrease, or have no effect on the concentration of $H_2(g)$ in the reaction:

$$C(s) + H_2O(g) \rightleftharpoons CO(g) + H_2(g)$$

7.90 Will an increase in pressure increase, decrease, or have no effect on the concentration of $NO(g)$ in the reaction:

$$N_2(g) + O_2(g) \rightleftharpoons 2NO(g)$$

7.91 Write the equilibrium-constant expression for the reaction described in Question 7.89.

7.92 Write the equilibrium-constant expression for the reaction described in Question 7.90.

7.93 True or false: The equilibrium will shift to the right when a catalyst is added to the mixture described in Question 7.89. Explain your reasoning.

7.94 True or false: The equilibrium for an endothermic reaction will shift to the right when the reaction mixture is heated. Explain your reasoning.

7.95 A bottle of carbonated beverage slowly goes "flat" (loses CO_2) after it is opened. Explain, using LeChatelier's principle.

7.96 Carbonated beverages quickly go flat (lose CO_2) when heated. Explain, using LeChatelier's principle.

CRITICAL THINKING PROBLEMS

1. Predict the sign of ΔG for perspiration evaporating. Would you expect the ΔH term or the ΔS term to be more dominant? Explain your reasoning.

2. Can the following statement ever be true? "Heating a reaction mixture increases the rate of a certain reaction but decreases the yield of product from the reaction." Explain why or why not.

3. Molecules must collide for a reaction to take place. Sketch a model of the orientation and interaction of HI and Cl that is most favorable for the reaction:

$$HI(g) + Cl(g) \longrightarrow HCl(g) + I(g)$$

4. Silver ion reacts with chloride ion to form the precipitate, silver chloride:

$$Ag^+(aq) + Cl^-(aq) \rightleftharpoons AgCl(s)$$

After the reaction reached equilibrium, the chemist filtered 99% of the solid silver chloride from the solution, hoping to shift the equilibrium to the right, to form more product. Critique the chemist's experiment.

5. Human behavior often follows LeChatelier's principle. Provide one example and explain in terms of LeChatelier's principle.

6. A clever device found in some homes is a figurine that is blue on dry, sunny days and pink on damp, rainy days. These figurines are coated with substances containing chemical species that undergo the following equilibrium reaction:

$$Co(H_2O)_6^{2+}(aq) + 4Cl^-(aq) \rightleftharpoons CoCl_4^{2-}(aq) + 6H_2O(l)$$

   a. Which substance is blue?
   b. Which substance is pink?
   c. How is LeChatelier's principle applied here?

7. You have spent the entire morning in a 20°C classroom. As you ride the elevator to the cafeteria, six persons enter the elevator after being outside on a subfreezing day. You suddenly feel chilled. Explain the heat flow situation in the elevator in thermodynamic terms.

8 Acids and Bases and Oxidation-Reduction

Learning Goals

1 Identify acids and bases and acid-base reactions.

2 Describe the role of the solvent in acid-base reactions.

3 Write equations describing acid-base dissociation and label the conjugate acid-base pairs.

4 Calculate pH from concentration data.

5 Calculate hydronium and/or hydroxide ion concentration from pH data.

6 Provide examples of the importance of pH in chemical and biochemical systems.

7 Describe the meaning and utility of neutralization reactions.

8 Describe the applications of buffers to chemical and biochemical systems, particularly blood chemistry.

9 Explain the meaning of the terms *oxidation* and *reduction,* and describe some practical examples of redox processes.

10 Diagram a voltaic cell and describe its function.

11 Compare and contrast voltaic and electrolytic cells.

Outline

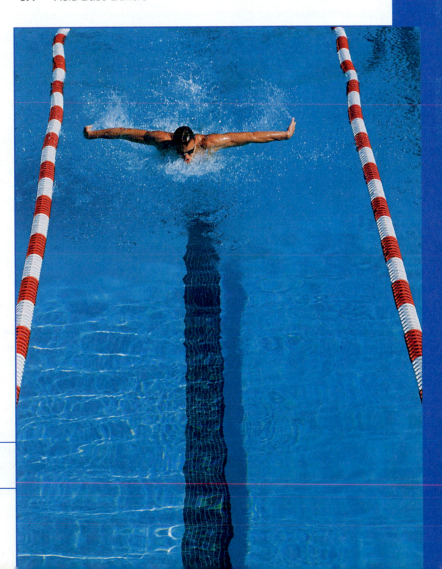

Solution properties, including clarity and bacteria levels, are often pH dependent.

Introduction

In this chapter we will learn about two general classes of chemical change: acid-base reactions and oxidation-reduction reactions. Although superficially quite different, their underlying similarity is that both are essentially charge-transfer processes. An acid-base reaction involves the transfer of one or more positively charged units, protons or hydrogen ions; an oxidation-reduction reaction involves the transfer of one or more negatively charged particles, electrons.

Acids and bases include some of the most important compounds in nature. Historically, it was recognized that certain compounds, acids, had a sour taste, were able to dissolve some metals, and caused vegetable dyes to change color. Bases have long been recognized by their bitter taste, slippery feel, and corrosive nature. Bases react strongly with acids and cause many metal ions in solution to form a solid precipitate.

Digestion of proteins is aided by stomach acid (hydrochloric acid) and many biochemical processes such as enzyme catalysis depend on the proper level of acidity. Indeed, a wide variety of chemical reactions critically depend on the acid-base composition of the solution (Figure 8.1). This is especially true of the biochemical reactions occurring in the cells of our bodies. For this reason the level of acidity must be very carefully regulated. This is done with substances called buffers.

Chemistry Connection

Drug Delivery

When a doctor prescribes medicine to treat a disease or relieve its symptoms, the medication may be administered in a variety of ways. Drugs may be taken orally, injected into a muscle or a vein, or absorbed through the skin. Specific instructions are often provided to regulate the particular combination of drugs that can or cannot be taken. The diet, both before and during the drug therapy, may be of special importance.

To appreciate why drugs are administered in a specific way, it is necessary to understand a few basic facts about medications and how they interact with the body.

Drugs function by undergoing one or more chemical reactions in the body. Few compounds react in only one way, to produce a limited set of products, even in the simple environment of a beaker or flask. Imagine the number of possible reactions that a drug can undergo in a complex chemical factory like the human body. In many cases a drug can react in a variety of ways other than its intended path. These alternative paths are side reactions, sometimes producing *side effects* such as nausea, vomiting, insomnia, or drowsiness. Side effects may be unpleasant and may actually interfere with the primary function of the drug.

The development of safe, effective medication, with minimal side effects, is a slow and painstaking process and deter-

mining the best drug delivery system is a critical step. For example, a drug that undergoes an unwanted side reaction in an acidic solution would not be very effective if administered orally. The acidic digestive fluids in the stomach could prevent the drug from even reaching the intended organ, let alone retaining its potency. The drug could be administered through a vein into the blood; blood is not acidic, in contrast to digestive fluids. In this way the drug may be delivered intact to the intended site in the body, where it is free to undergo its primary reaction.

Drug delivery has become a science in its own right. Pharmacology, the study of drugs and their uses in the treatment of disease, has a goal of creating drugs that are highly selective. In other words, they will undergo only one reaction, the intended reaction. Encapsulation of drugs, enclosing them within larger molecules or collections of molecules, may protect them from unwanted reactions as they are transported to their intended site.

In this chapter we will explore the fundamentals of solutions and solution reactions, including acid-base and oxidation-reduction reactions. Knowing a few basic concepts that govern reactions in beakers will help us to understand the conditions that affect the reactivity of a host of biochemically interesting molecules that we will encounter in later chapters.

Oxidation-reduction processes are also common in living systems. Respiration is driven by oxidation-reduction reactions. Additionally, oxidation-reduction reactions generate heat that warms our homes and workplaces and fuels our industrial civilization. Moreover, oxidation-reduction is the basis for battery design. Batteries are found in automobiles and electronic devices such as cameras and radios, and are even implanted in the human body to regulate heart rhythm.

8.1 Acids and Bases

LEARNING GOAL

1 Identify acids and bases and acid-base reactions.

The properties of acids and bases are related to their chemical structure. All acids have common characteristics that enable them to increase the hydrogen ion concentration in water. All bases lower the hydrogen ion concentration in water.

Two theories, one developed from the other, help us to understand the unique chemistry of acids and bases.

Arrhenius Theory of Acids and Bases

One of the earliest definitions of acids and bases is the **Arrhenius theory.** According to this theory, an **acid,** dissolved in water, dissociates to form *hydrogen ions or protons* (H^+), and a **base,** dissolved in water, dissociates to form *hydroxide ions* (OH^-). For example, hydrochloric acid dissociates in solution according to the reaction

$$HCl(aq) \longrightarrow H^+(aq) + Cl^-(aq)$$

Sodium hydroxide, a base, produces hydroxide ions in solution:

$$NaOH(aq) \longrightarrow Na^+(aq) + OH^-(aq)$$

The Arrhenius theory satisfactorily explains the behavior of many acids and bases. However, a substance such as ammonia, NH_3, has basic properties but cannot be an Arrhenius base, because it contains no OH^-. The *Brønsted-Lowry theory* explains this mystery and gives us a broader view of acid-base theory by considering the central role of the solvent in the dissociation process.

Brønsted-Lowry Theory of Acids and Bases

The **Brønsted-Lowry theory** defines an **acid** as a proton (H^+) donor and a **base** as a proton acceptor.

Hydrochloric acid in solution *donates* a proton to the solvent water thus behaving as a Brønsted-Lowry acid:

$$HCl(aq) + H_2O(l) \longrightarrow H_3O^+(aq) + Cl^-(aq)$$

H_3O^+ is referred to as the hydrated proton or **hydronium ion.**

The basic properties of ammonia are clearly accounted for by the Brønsted-Lowry theory. Ammonia *accepts* a proton from the solvent water, producing OH^-. An equilibrium mixture of NH_3, H_2O, NH_4^+, and OH^- results.

Figure 8.1
The yellow solution on the left, containing $CrO_4{}^{2-}$ (chromate ion), was made acidic producing the reddish brown solution on the right. The principal component in solution is now $Cr_2O_7{}^{2-}$. Addition of base to this solution removes H^+ ions and regenerates the yellow $CrO_4{}^{2-}$. This is an example of an acid-base dependent chemical equilibrium.

$$NH_3(aq) + H{-}OH(l) \rightleftarrows NH_4^+(aq) + OH^-(aq)$$

For aqueous solutions, the Brønsted-Lowry theory adequately describes the behavior of acids and bases. We shall limit our discussion of acid-base chemistry to aqueous solutions and use the Brønsted-Lowry definition described here.

Acid-Base Properties of Water

LEARNING GOAL

2) Describe the role of the solvent in acid-base reactions.

The role that the solvent, water, plays in acid-base reactions is noteworthy. In the example above, the water molecule accepts a proton from the HCl molecule. The water is behaving as a proton acceptor, a base.

However, when water is a solvent for ammonia (NH_3), a base, the water molecule donates a proton to the ammonia molecule. The water, in this situation, is acting as a proton donor, an acid.

Water, owing to the fact that it possesses *both* acid and base properties, is termed **amphiprotic.** Water is the most commonly used solvent for acids and bases. Solute-solvent interactions between water and acids or bases promote both the solubility and the dissociation of acids and bases.

Acid and Base Strength

Concentration of solutions is discussed in Section 6.3.

The terms *acid or base strength* and *acid or base concentration* are easily confused. *Strength* is a measure of the *degree of dissociation* of an acid or base in solution, independent of its concentration. The degree of dissociation is the fraction of acid or base molecules that produces ions in solution. Concentration, as we have learned, refers to the amount of solute (in this case, the amount of acid or base) per quantity of solution.

The concentration of an acid or base does affect the degree of dissociation. However, the major factor in determining the degree of dissociation is the strength of the acid or base.

The strength of acids and bases in water depends on the extent to which they react with the solvent, water. Acids and bases are classified as *strong* when the reaction with water is virtually 100% complete and as *weak* when the reaction with water is much less than 100% complete.

Important strong acids include:

Animations

The Dissociation of Strong and Weak Acids

Ionization of a Strong Base and a Weak Base

Dissociation of Acetic Acid

Hydrochloric acid	$HCl(aq) + H_2O(l) \longrightarrow H_3O^+(aq) + Cl^-(aq)$
Nitric acid	$HNO_3(aq) + H_2O(l) \longrightarrow H_3O^+(aq) + NO_3^-(aq)$
Sulfuric acid	$H_2SO_4(aq) + H_2O(l) \longrightarrow H_3O^+(aq) + HSO_4^-(aq)$

Reversibility of reactions is discussed in Section 7.4.

Note that the equation for the dissociation of each of these acids is written with a single arrow. This indicates that the reaction has little or no tendency to proceed in the reverse direction to establish equilibrium. Virtually all of the acid molecules are dissociated to form ions.

All common strong bases are *metal hydroxides.* Strong bases completely dissociate in aqueous solution to produce hydroxide ions and metal cations. Of the common metal hydroxides, only NaOH and KOH are soluble in water and are readily usable strong bases:

| Sodium hydroxide | $NaOH(aq) \longrightarrow Na^+(aq) + OH^-(aq)$ |
| Potassium hydroxide | $KOH(aq) \longrightarrow K^+(aq) + OH^-(aq)$ |

Weak acids and weak bases dissolve in water principally in the molecular form. Only a small percentage of the molecules dissociate to form the hydronium or hydroxide ion.

Two important weak acids are:

The double arrow implies an equilibrium between dissociated and undissociated species.

| Acetic acid | $CH_3COOH(aq) + H_2O(l) \rightleftharpoons H_3O^+(aq) + CH_3COO^-(aq)$ |
| Carbonic acid | $H_2CO_3(aq) + H_2O(l) \rightleftharpoons H_3O^+(aq) + HCO_3^-(aq)$ |

We have already mentioned the most common weak base, ammonia. Many organic compounds function as weak bases. Several examples of weak bases follow:

Pyridine $\quad$ $C_5H_5N(aq) + H_2O(l) \rightleftharpoons C_5H_5NH^+(aq) + OH^-(aq)$

Aniline $\quad$ $C_6H_5NH_2(aq) + H_2O(l) \rightleftharpoons C_6H_5NH_3^+(aq) + OH^-(aq)$

Methylamine $\quad$ $CH_3NH_2(aq) + H_2O(l) \rightleftharpoons CH_3NH_3^+(aq) + OH^-(aq)$

Many organic compounds have acid or base properties. The chemistry of organic acids and bases will be discussed in Chapter 14 (Carboxylic Acids and Carboxylic Acid Derivatives) and 15 (Amines and Amides).

The fundamental chemical difference between strong and weak acids or bases is their equilibrium ion concentration. A strong acid, such as HCl, does not, in aqueous solution, exist to any measurable degree in equilibrium with its ions, H_3O^+ and Cl^-. On the other hand, a weak acid, such as acetic acid, establishes a dynamic equilibrium with its ions, H_3O^+ and CH_3COO^-.

Conjugate Acids and Bases

The Brønsted-Lowry theory contributed several fundamental ideas that broadened our understanding of solution chemistry. First of all, an acid-base reaction is a charge-transfer process. Second, the transfer process usually involves the solvent. Water may, in fact, accept or donate a proton. Last, and perhaps most important, the acid-base reaction is seen as a reversible process. This leads to the possibility of a reversible, dynamic equilibrium (see Section 7.4).

Consequently, any acid-base reaction can be represented by the general equation

$$\underset{\text{(acid)}}{HA} + \underset{\text{(base)}}{B} \rightleftharpoons BH^+ + A^-$$

LEARNING GOAL

3 Write equations describing acid-base dissociation and label the conjugate acid-base pairs.

In the forward reaction, the acid (HA) donates a proton (H^+) to the base (B) leading to the formation of BH^+ and A^-. However, in the reverse reaction, it is the BH^+ that behaves as an acid; it donates its "extra" proton to A^-. A^- is therefore a base in its own right because it accepts the proton.

These *product* acids and bases are termed *conjugate acids and bases.*

A **conjugate acid** is the species formed when a base accepts a proton.

A **conjugate base** is the species formed when an acid donates a proton.

The acid and base on the opposite sides of the equation are collectively termed a **conjugate acid-base pair.** In the above equation:

BH^+ is the conjugate acid of the base B.

A^- is the conjugate base of the acid HA.

B and BH^+ constitute a conjugate acid-base pair.

HA and A^- constitute a conjugate acid-base pair.

Rewriting our model equation:

Although we show the forward and reverse arrows to indicate the reversibility of the reaction, seldom are the two processes "equal but opposite." One reaction, either forward or reverse, is usually favored. Consider the reaction of hydrochloric acid in water:

$$\downarrow \text{Forward reaction: significant}$$

$$\underset{\text{(acid)}}{HCl(aq)} + \underset{\text{(base)}}{H_2O(l)} \rightleftharpoons \underset{\text{(acid)}}{H_3O^+(aq)} + \underset{\text{(base)}}{Cl^-(aq)}$$

$$\uparrow \text{Reverse reaction: not significant}$$

Animations

Reaction of a Strong Acid
    with Water

Reaction of a Weak Acid
    with Water

HCl is a much better proton donor than H_3O^+. Consequently the forward reaction predominates, the reverse reaction is inconsequential, and hydrochloric acid is termed a *strong acid*. As we learned in Chapter 7, reactions in which the forward reaction is strongly favored have large equilibrium constants. The dissociation of hydrochloric acid is so favorable that we describe it as 100% dissociated and use only a single forward arrow to represent its behavior in water:

$$HCl(aq) + H_2O(l) \longrightarrow H_3O^+(aq) + Cl^-(aq)$$

The degree of dissociation, or strength, of acids and bases has a profound influence on their aqueous chemistry. For example, vinegar (a 5% [W/V] solution of acetic acid in water) is a consumable product; aqueous hydrochloric acid in water is not. Why? Acetic acid is a *weak* acid and, as a result, a dilute solution does no damage to the mouth and esophagus. The following section looks at the strength of acids and bases in solution in more detail.

Question 8.1

Write an equation for the reaction of each of the following with water:

a. HF (a weak acid)
b. NH_3 (a weak base)

Question 8.2

Write an equation for the reaction of each of the following with water:

a. H_2S (a weak acid)
b. CH_3NH_2 (a weak base)

Question 8.3

Select the conjugate acid-base pairs for each reaction in Question 8.1.

Question 8.4

Select the conjugate acid-base pairs for each reaction in Question 8.2.

The relative strength of an acid or base is determined by the ease with which it donates or accepts a proton. Acids with the greatest proton-donating capability (strongest acids) have the weakest conjugate bases. Good proton acceptors (strong bases) have weak conjugate acids. This relationship is clearly indicated in Figure 8.2. This figure can be used to help us compare and predict relative acid-base strength.

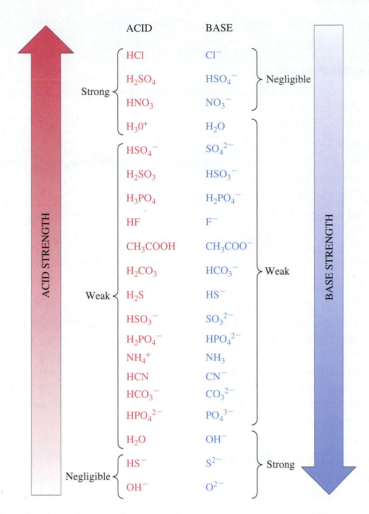

Figure 8.2
Conjugate acid-base pairs. Strong acids have weak conjugate bases; strong bases have weak conjugate acids. Note that, in every case, the conjugate base has one fewer H^+ than the corresponding conjugate acid.

Predicting Relative Acid-Base Strengths

EXAMPLE 8.1

LEARNING GOAL

3 Write equations describing acid-base dissociation and label the conjugate acid-base pairs.

a. Write the conjugate acid of HS^-.

Solution

The conjugate acid may be constructed by adding a proton (H^+) to the base structure, consequently, H_2S.

b. Using Figure 8.2, identify the stronger base, HS^- or F^-.

Solution

HS^- is the stronger base because it is located farther down the right-hand column.

c. Using Figure 8.2 identify the stronger acid, H_2S or HF.

Solution

HF is the stronger acid because its conjugate base is weaker *and* because it is located farther up the left-hand column.

Continued—

EXAMPLE 8.1 —*Continued*

Practice Problem 8.1

a. In each pair, write the conjugate base of each acid and identify the stronger acid.

H_2O or NH_4^+

H_2SO_4 or H_2SO_3

b. In each pair, write the conjugate acid of each base and identify the stronger base.

CO_3^{2-} or PO_4^{3-}

HCO_3^- or HPO_4^{2-}

For Further Practice: Questions 8.23 and 8.24.

Solutions of acids and bases used in the laboratory must be handled with care. Acids burn because of their exothermic reaction with water present on and in the skin. Bases react with proteins, which are principal components of the skin and eyes.

Such solutions are more hazardous if they are strong or concentrated. A strong acid or base produces more H_3O^+ or OH^- than does the corresponding weak acid or base. More-concentrated acids or bases contain more H_3O^+ or OH^- than do less-concentrated solutions of the same strength.

The Dissociation of Water

Solutions of electrolytes are discussed in Section 6.4.

Aqueous solutions of acids and bases are electrolytes. The dissociation of the acid or base produces ions that can conduct an electrical current. As a result of the differences in the degree of dissociation, *strong acids and bases are strong electrolytes; weak acids and bases are weak electrolytes.* The conductivity of these solutions is principally dependent on the solute and not the solvent (water).

Although pure water is virtually 100% molecular, a small number of water molecules do ionize. This process occurs by the transfer of a proton from one water molecule to another, producing a hydronium ion and a hydroxide ion:

$$H_2O(l) + H_2O(l) \rightleftharpoons H_3O^+(aq) + OH^-(aq)$$

This process is the **autoionization,** or self-ionization, of water. Water is therefore a *very* weak electrolyte and a very poor conductor of electricity. Water has *both* acid and base properties; dissociation produces both the hydronium and hydroxide ion.

Pure water at room temperature has a hydronium ion concentration of $1.0 \times 10^{-7} M$. One hydroxide ion is produced for each hydronium ion. Therefore, the hydroxide ion concentration is also $1.0 \times 10^{-7} M$. Molar equilibrium concentration is conveniently indicated by brackets around the species whose concentration is represented:

$$[H_3O^+] = 1.0 \times 10^{-7} \ M$$
$$[OH^-] = 1.0 \times 10^{-7} \ M$$

The product of hydronium and hydroxide ion concentration in pure water is referred to as the **ion product for water,** symbolized by K_w.

$$K_w = \text{ion product} = [H_3O^+][OH^-]$$
$$= [1.0 \times 10^{-7}][1.0 \times 10^{-7}]$$
$$= 1.0 \times 10^{-14}$$

The ion product is constant because its value does not depend on the nature or concentration of the solute, as long as the temperature does not change. The ion product is a temperature-dependent quantity.

The nature and concentration of the solutes added to water do alter the relative concentrations of H_3O^+ and OH^- present, but the product, $[H_3O^+][OH^-]$, always equals 1.0×10^{-14} at 25°C. This relationship is the basis for a scale that is useful in the measurement of the level of acidity or basicity of solutions. This scale, the pH scale, is discussed next.

8.2 pH: A Measurement Scale for Acids and Bases

A Definition of pH

The **pH scale** gauges the hydronium ion concentration and reflects the degree of acidity or basicity of a solution. The pH scale is somewhat analogous to the temperature scale used for assignment of relative levels of hot or cold. The temperature scale was developed to allow us to indicate how cold or how hot an object is. The pH scale specifies how acidic or how basic a solution is. The pH scale has values that range from 0 (very acidic) to 14 (very basic). A pH of 7, the middle of the scale, is neutral, neither acidic nor basic.

To help us to develop a concept of pH, let's consider the following:

- Addition of an acid (proton donor) to water *increases* the $[H_3O^+]$ and decreases the $[OH^-]$.
- Addition of a base (proton acceptor) to water *decreases* the $[H_3O^+]$ by increasing the $[OH^-]$.
- $[H_3O^+] = [OH^-]$ when *equal* amounts of acid and base are present.
- In all three cases, $[H_3O^+][OH^-] = 1.0 \times 10^{-14} =$ the ion product for water at 25°C.

LEARNING GOAL
4 Calculate pH from concentration data.

pH values greater than 14 and less than zero are possible, but largely meaningless, due to ion association characteristics of very concentrated solutions.

Measuring pH

The pH of a solution can be calculated if the concentration of either H_3O^+ or OH^- is known. Alternatively, measurement of pH allows the calculation of H_3O^+ or OH^- concentration. The pH of aqueous solutions may be approximated by using indicating paper (pH paper) that develops a color related to the solution pH. Alternatively, a pH meter can give us a much more exact pH measurement. A sensor measures an electrical property of a solution that is proportional to pH (Figure 8.3).

The clarity and purity of a swimming pool are critically related to the pH of the pool water.

Calculating pH

One of our objectives in this chapter is to calculate the pH of a solution when the hydronium or hydroxide ion concentration is known, and to calculate $[H_3O^+]$ or $[OH^-]$ from the pH.

The pH of a solution is defined as the negative logarithm of the molar concentration of the hydronium ion:

LEARNING GOAL
4 Calculate pH from concentration data.

$$pH = -\log[H_3O^+]$$

Figure 8.3
The measurement of pH. (a) A strip of test paper impregnated with indicator (a material that changes color as the acidity of the surroundings changes) is put in contact with the solution of interest. The resulting color is matched with a standard color chart (colors shown as a function of pH) to obtain the approximate pH. (b) A pH meter uses a sensor (a pH electrode) that develops an electrical potential that is proportional to the pH of the solution.

(a)   (b)

E X A M P L E 8.2   **Calculating pH from Acid Molarity**

LEARNING GOAL

5 Calculate hydronium and/or hydroxide ion concentration from pH data.

Calculate the pH of a $1.0 \times 10^{-3} M$ solution of HCl.

Solution

Step 1. Recognize that HCl is a strong acid.

Step 2. If 1 mol HCl dissolves and dissociates in 1 L of aqueous solution, it produces 1 mol H_3O^+ (a 1 M solution of H_3O^+). Therefore a $1.0 \times 10^{-3} M$ HCl solution has $[H_3O^+] = 1.0 \times 10^{-3} M$, and

Step 3. Using our expression for pH:

$$pH = -\log [H_3O^+]$$

Step 4. Substituting for $[H_3O^+]$:

$$pH = -\log [1.0 \times 10^{-3}]$$
$$= -[-3.00] = 3.00$$

  Practice Problem 8.2

Calculate the pH of a $1.0 \times 10^{-4} M$ solution of HNO_3.

For Further Practice: Questions 8.43 and 8.44.

E X A M P L E 8.3   **Calculating [H₃O⁺] from pH**

LEARNING GOAL

5 Calculate hydronium and/or hydroxide ion concentration from pH data.

Calculate the $[H_3O^+]$ of a solution of hydrochloric acid with pH = 4.00.

Solution

Step 1. We use the pH expression:

$$pH = -\log [H_3O^+]$$
$$4.00 = -\log [H_3O^+]$$

Continued—

E X A M P L E 8.3 —*Continued*

Step 2. Multiplying both sides of the equation by –1, we get

$$-4.00 = \log [H_3O^+]$$

Step 3. Taking the antilogarithm of both sides (the reverse of a logarithm), we have

$$\text{antilog} \; -4.00 = [H_3O^+]$$

Step 4. The antilog is the exponent of 10; therefore

$$1.0 \times 10^{-4} \; M = [H_3O^+]$$

Practice Problem 8.3

Calculate the $[H_3O^+]$ of a solution of HNO_3 that has a pH = 5.00.

For Further Practice: Questions 8.45 and 8.46.

Calculating the pH of a Base

E X A M P L E 8.4

Calculate the pH of a $1.0 \times 10^{-5} \, M$ solution of NaOH.

> **5** **LEARNING GOAL**
> Calculate hydronium and/or hydroxide ion concentration from pH data.

Solution

Step 1. Recognize that NaOH is a strong base.

Step 2. If 1 mol NaOH dissolves and dissociates in 1L of aqueous solution, it produces 1 mol OH^- (a 1 M solution of OH^-). Therefore a $1.0 \times 10^{-5} \, M$ NaOH solution has $[OH^-] = 1.0 \times 10^{-5} \, M$.

Step 3. To calculate pH, we need $[H_3O^+]$. Recall that

$$[H_3O^+][OH^-] = 1.0 \times 10^{-14}$$

Step 4. Solving this equation for $[H_3O^+]$,

$$[H_3O^+] = \frac{1.0 \times 10^{-14}}{[OH^-]}$$

Step 5. Substituting the information provided in the problem,

$$[H_3O^+] = \frac{1.0 \times 10^{-14}}{1.0 \times 10^{-5}}$$
$$= 1.0 \times 10^{-9} \, M$$

Step 6. The solution is now similar to that in Example 8.2:

$$pH = -\log [H_3O^+]$$
$$= -\log [1.0 \times 10^{-9}]$$
$$= 9.00$$

Continued—

EXAMPLE 8.4 *—Continued*

Practice Problem 8.4

a. Calculate the pH corresponding to a $1.0 \times 10^{-2} M$ solution of sodium hydroxide.

b. Calculate the pH corresponding to a $1.0 \times 10^{-6} M$ solution of sodium hydroxide.

For Further Practice: Questions 8.47 and 8.48.

EXAMPLE 8.5   **Calculating Both Hydronium and Hydroxide Ion Concentrations from pH**

LEARNING GOAL

Calculate hydronium and/or hydroxide ion concentration from pH data.

Calculate the $[H_3O^+]$ and $[OH^-]$ of a sodium hydroxide solution with a pH = 10.00.

Solution

Step 1. First, calculate $[H_3O^+]$ using our expression for pH:

$$pH = -\log [H_3O^+]$$
$$10.00 = -\log [H_3O^+]$$
$$-10.00 = \log [H_3O^+]$$
$$\text{antilog } -10.00 = [H_3O^+]$$
$$1.0 \times 10^{-10} M = [H_3O^+]$$

Step 2. To calculate the $[OH^-]$, we need to solve for $[OH^-]$ by using the following expression:

$$K_w = [H_3O^+][OH^-] = 1.0 \times 10^{-14}$$
$$[OH^-] = \frac{1.0 \times 10^{-14}}{[H_3O^+]}$$

Step 3. Substituting the $[H_3O^+]$ from the first step, we have

$$[OH^-] = \frac{1.0 \times 10^{-14}}{[1.0 \times 10^{-10}]}$$
$$= 1.0 \times 10^{-4} M$$

Practice Problem 8.5

Calculate the $[H_3O^+]$ and $[OH^-]$ of a potassium hydroxide solution with a pH = 8.00.

For Further Practice: Questions 8.49 and 8.50.

Often, the pH or $[H_3O^+]$ will not be a whole number (pH = 1.5, pH = 5.3, $[H_3O^+] = 1.5 \times 10^{-3}$ and so forth). With the advent of inexpensive and versatile calculators, calculations with noninteger numbers pose no great problems. Consider Examples 8.6 and 8.7.

Calculating pH with Noninteger Numbers

EXAMPLE 8.6

Calculate the pH of a sample of lake water that has a $[H_3O^+] = 6.5 \times 10^{-5} M$.

LEARNING GOAL
5 Calculate hydronium and/or hydroxide ion concentration from pH data.

Solution

Step 1. Use the expression for pH:

$$pH = -\log[H_3O^+]$$

Step 2. Substituting

$$pH = -\log[6.5 \times 10^{-5}]$$
$$= 4.19$$

Note: The pH, 4.19, is low enough to suspect acid rain. (See An Environmental Perspective: Acid Rain in this chapter.)

Practice Problem 8.6

Calculate the pH of a sample of blood that has a $[H_3O^+] = 3.3 \times 10^{-8} M$.

For Further Practice: Questions 8.61 and 8.62.

Calculating [H₃O⁺] from pH

EXAMPLE 8.7

The measured pH of a sample of lake water is 6.40. Calculate $[H_3O^+]$.

LEARNING GOAL
5 Calculate hydronium and/or hydroxide ion concentration from pH data.

Solution

Step 1. An alternative mathematical form of

$$pH = -\log[H_3O^+]$$

is the expression

$$[H_3O^+] = 10^{-pH}$$

Step 2. We can use this expression to solve for $[H_3O^+]$.

$$[H_3O^+] = 10^{-6.40}$$

Step 3. Performing the calculation on your calculator results in 3.98×10^{-7} or $4.0 \times 10^{-7} M = [H_3O^+]$.

Practice Problem 8.7

a. Calculate the $[H_3O^+]$ corresponding to pH = 8.50.
b. Calculate the $[H_3O^+]$ corresponding to pH = 4.50.

For Further Practice: Questions 8.57 and 8.58.

Examples 8.2–8.7 illustrate the most frequently used pH calculations. It is important to remember that in the case of a base you must convert the $[OH^-]$ to $[H_3O^+]$, using the expression for the ion product for the solvent, water. It is also useful to remember the following points:

- The pH of a 1 M solution of any strong acid is 0.
- The pH of a 1 M solution of any strong base is 14.
- Each tenfold change in concentration changes the pH by one unit. A tenfold change in concentration is equivalent to moving the decimal point one place.
- A *decrease* in acid concentration *increases* the pH.
- A *decrease* in base concentration *decreases* the pH.

Figure 8.4 provides a convenient overview of solution pH.

Question 8.5

Calculate the $[OH^-]$ of the solution in Example 8.2.

Question 8.6

Calculate the $[OH^-]$ of the solution in Example 8.3.

Figure 8.4

The pH scale. A pH of 7 is neutral ($[H_3O^+] = [OH^-]$). Values less than 7 are acidic (H_3O^+ predominates) and values greater than 7 are basic (OH^- predominates).

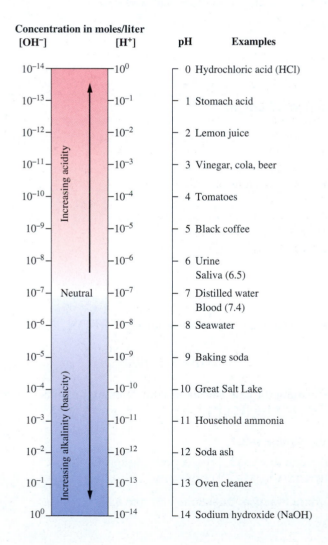

The Importance of pH and pH Control

Solution pH and pH control play a major role in many facets of our lives. Consider a few examples:

LEARNING GOAL

6 Provide examples of the importance of pH in chemical and biochemical systems.

- *Agriculture:* Crops grow best in a soil of proper pH. Proper fertilization involves the maintenance of a suitable pH.
- *Physiology:* If the pH of our blood were to shift by one unit, we would die. Many biochemical reactions in living organisms are extremely pH dependent.
- *Industry:* From manufacture of processed foods to the manufacture of automobiles, industrial processes often require rigorous pH control.
- *Municipal services:* Purification of drinking water and treatment of sewage must be carried out at their optimum pH.
- *Acid rain:* Nitric acid and sulfuric acid, resulting largely from the reaction of components of vehicle emissions and electric power generation (nitrogen and sulfur oxides) with water, are carried down by precipitation and enter aquatic systems (lakes and streams), lowering the pH of the water. A less than optimum pH poses serious problems for native fish populations.

See An Environmental Perspective: Acid Rain in this chapter.

The list could continue on for many pages. However, in summary, any change that takes place in aqueous solution generally has at least some pH dependence.

8.3 Reactions Between Acids and Bases

Neutralization

The reaction of an acid with a base to produce a salt and water is referred to as **neutralization.** In the strictest sense, neutralization requires equal numbers of moles of H_3O^+ and OH^- to produce a neutral solution (no excess acid or base).

Consider the reaction of a solution of hydrochloric acid and sodium hydroxide:

LEARNING GOAL

7 Describe the meaning and utility of neutralization reactions.

$$\underset{\text{Acid}}{\text{HCl}(aq)} + \underset{\text{Base}}{\text{NaOH}(aq)} \longrightarrow \underset{\text{Salt}}{\text{NaCl}(aq)} + \underset{\text{Water}}{\text{H}_2\text{O}(l)}$$

Our objective is to make the balanced equation represent the process actually occurring. We recognize that HCl, NaOH, and NaCl are dissociated in solution:

Equation balancing is discussed in Chapter 4.

$$H^+(aq) + Cl^-(aq) + Na^+(aq) + OH^-(aq) \longrightarrow Na^+(aq) + Cl^-(aq) + H_2O(l)$$

We further know that Na^+ and Cl^- are unchanged in the reaction; they are termed *spectator ions.* If we write only those components that actually change, ignoring the spectator ions, we produce a *net, balanced ionic equation:*

$$H^+(aq) + OH^-(aq) \longrightarrow H_2O(l)$$

If we realize that the H^+ occurs in aqueous solution as the hydronium ion, H_3O^+, the most correct form of the net, balanced ionic equation is

 Writing Net Ionic Equations

$$H_3O^+(aq) + OH^-(aq) \longrightarrow 2H_2O(l)$$

The equation for any strong acid/strong base neutralization reaction is the same as this equation.

TABLE 8.1	**Conducting an Acid-Base Titration**

1. A known volume (perhaps 25.00 mL) of the unknown acid of unknown concentration is measured into a flask using a pipet.

2. An **indicator,** a substance that changes color as the solution reaches a certain pH (Figure 8.5), is added to the unknown solution. We must know, from prior experience, the expected pH at the equivalence point (see Step 4). For this titration, phenolphthalein or phenol red would be a logical choice, since the equivalence-point pH is known to be seven.

3. A solution of sodium hydroxide (perhaps 0.1000 M) is carefully added to the unknown solution using a **buret** (Figure 8.6), which is a long glass tube calibrated in milliliters. A stopcock at the bottom of the buret regulates the amount of liquid dispensed. The standard solution is added until the indicator changes color.

4. At this point, the **equivalence point,** the number of moles of hydroxide ion added is equal to the number of moles of hydronium ion present in the unknown acid.

5. The volume dispensed by the buret (perhaps 35.00 mL) is measured.

6. Using the data from the experiment (volume of the unknown, volume of the titrant, and molarity of the titrant), calculate the molar concentration of the unknown substance.

Animations

The Neutralization Reaction
of NaOH and HCl

Titration of HCl with NaOH

A neutralization reaction may be used to determine the concentration of an unknown acid or base solution. The technique of **titration** involves the addition of measured amounts of a **standard solution** (one whose concentration is known with certainty) to neutralize the second, unknown solution. From the volumes of the two solutions and the concentration of the standard solution the concentration of the unknown solution may be determined.

A strategy for carrying out an acid-base titration is summarized in Table 8.1. The calculations involved in an acid-base titration are illustrated in Example 8.8.

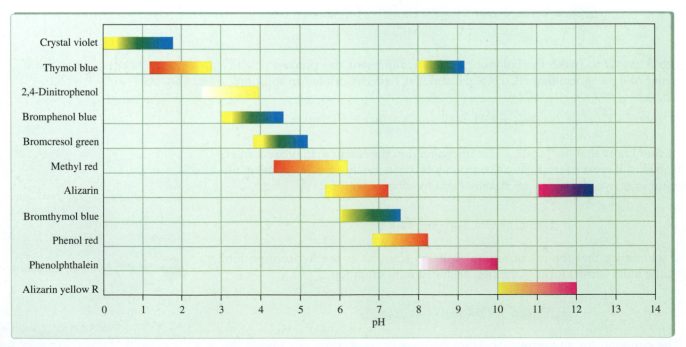

Figure 8.5

The relationship between pH and color of a variety of compounds, some of which are commonly used as acid-base indicators. Many indicators are naturally occurring substances.

Determining the Concentration of a Solution of Hydrochloric Acid EXAMPLE 8.8

A 25.00-mL sample of an acid of unknown concentration is transferred to a flask, a few drops of the indicator phenolphthalein are added, and the resulting solution is titrated with 0.1000 M sodium hydroxide solution. After 35.00 mL of sodium hydroxide solution were added, the indicator turned pink, signaling the chemist that the unknown and titrant had reached their equivalence point. Calculate the M of the acid.

LEARNING GOAL

7 Describe the meaning and utility of neutralization reactions.

Solution

Step 1. Pertinent information for this titration includes:

Volume of the unknown acid solution, 25.00 mL

Volume of sodium hydroxide solution added, 35.00 mL

Concentration of the sodium hydroxide solution, 0.1000 M

Step 2. From the balanced equation, we know that HCl and NaOH react in a 1:1 combining ratio:

$$HCl(aq) + NaOH(aq) \longrightarrow NaCl(aq) + H_2O(l)$$

Note: The net, balanced, ionic equation for this reaction provides the same information; one mole of H_3O^+ reacts with one mole of OH^-.

$$H_3O^+(aq) + OH^-(aq) \longrightarrow 2H_2O(l)$$

Step 3. Using a strategy involving conversion factors

$$35.00 \text{ mL NaOH} \times \frac{1 \text{ L NaOH}}{10^3 \text{ mL NaOH}} \times \frac{0.1000 \text{ mol NaOH}}{\text{L NaOH}} = 3.500 \times 10^{-3} \text{ mol NaOH}$$

(a)

Step 4. Knowing that HCl and NaOH undergo a 1:1 reaction,

$$3.500 \times 10^{-3} \text{ mol NaOH} \times \frac{1 \text{ mol HCl}}{1 \text{ mol NaOH}} = 3.500 \times 10^{-3} \text{ mol HCl}$$

3.500×10^{-3} mol HCl are contained in 25.00 mL of HCl solution.

Step 5. Thus,

$$\frac{3.500 \times 10^{-3} \text{ mol HCl}}{25.00 \text{ mL HCl soln}} \times \frac{10^3 \text{ mL HCl soln}}{1 \text{ L HCl soln}} = 1.400 \times 10^{-1} \text{ mol HCl/L HCl soln}$$

$$= 0.1400 \ M$$

The titration of an acid with a base is depicted in Figure 8.6.

Note: An alternate problem-solving strategy produces the same result:

$$(M_{acid})(V_{acid}) = (M_{base})(V_{base})$$

and

$$M_{acid} = M_{base}\left(\frac{V_{base}}{V_{acid}}\right)$$

$$M_{acid} = (0.1000 \ M)\left(\frac{35.00 \text{ mL}}{25.00 \text{ mL}}\right)$$

$$M_{acid} = 0.1400 \ M$$

Continued—

(b)

Figure 8.6

An acid-base titration. (a) An exact volume of a standard solution (in this example, a base) is added to a solution of unknown concentration (in this example, an acid). (b) From the volume (read from the buret) and concentration of the standard solution, coupled with the mass or volume of the unknown, the concentration of the unknown may be calculated.

An Environmental *Perspective*

Acid Rain

Acid rain is a global environmental problem that has raised public awareness of the chemicals polluting the air through the activities of our industrial society. Normal rain has a pH of about 5.6 as a result of the chemical reaction between carbon dioxide gas and water in the atmosphere. The following equation shows this reaction:

$$CO_2(g) \quad + \quad H_2O(l) \rightleftharpoons H_2CO_3(aq)$$

Carbon dioxide   Water   Carbonic acid

Acid rain refers to conditions that are much more acidic than this. In upstate New York the rain has as much as 25 times the acidity of normal rainfall. One rainstorm, recorded in West Virginia, produced rainfall that measured 1.5 on the pH scale. This is approximately the pH of stomach acid or about ten thousand times more acidic than "normal rain" (remember that the pH scale is logarithmic; a 1 pH unit decrease represents a tenfold increase in hydronium ion concentration).

Acid rain is destroying life in streams and lakes. More than half the highland lakes in the western Adirondack Mountains have no native game fish. In addition to these 300 lakes, 140 lakes in Ontario have suffered a similar fate. It is estimated that 48,000 other lakes in Ontario and countless others in the northeastern and central United States are threatened. Our forests are endangered as well. The acid rain decreases soil pH, which in turn alters the solubility of minerals needed by plants. Studies have shown that about 40% of the red spruce and maple trees in New England have died. Increased acidity of rainfall appears to be the major culprit.

What is the cause of this acid rain? The combustion of fossil fuels (gas, oil, and coal) by power plants produces oxides of sulfur and nitrogen. Nitrogen oxides, in excess of normal levels, arise mainly from conversion of atmospheric nitrogen to nitrogen oxides in the engines of gasoline and diesel powered vehicles. Sulfur oxides result from the oxidation of sulfur in fossil fuels. The sulfur atoms were originally a part of the amino acids and proteins of plants and animals that became, over the millenia, our fuel. These react with water, as does the CO_2 in normal rain, but the products are strong acids: sulfuric and nitric acids. Let's look at the equations for these processes.

pH Values for a Variety of Substances Compared with the pH of Acid Rain

Indicates the range of pH values ascribed to acid rain

EXAMPLE 8.8 —*Continued*

Practice Problem 8.8

a. Calculate the molar concentration of a sodium hydroxide solution if 40.00 mL of this solution were required to neutralize 20.00 mL of a 0.2000 *M* solution of hydrochloric acid.

b. Calculate the molar concentration of a sodium hydroxide solution if 36.00 mL of this solution were required to neutralize 25.00 mL of a 0.2000 *M* solution of hydrochloric acid.

For Further Practice: Questions 8.71 and 8.72.

Damage caused by acid rain.

A similar chemistry is seen with the sulfur oxides. Coal may contain as much as 3% sulfur. When the coal is burned, the sulfur also burns. This produces choking, acrid sulfur dioxide gas:

$$S(s) + O_2(g) \longrightarrow SO_2(g)$$

By itself, sulfur dioxide can cause serious respiratory problems for people with asthma or other lung diseases, but matters are worsened by the reaction of SO_2 with atmospheric oxygen:

$$2SO_2(g) + O_2(g) \longrightarrow 2SO_3(g)$$

Sulfur trioxide will react with water in the atmosphere:

$$SO_3(g) + H_2O(l) \longrightarrow H_2SO_4(aq)$$

The product, sulfuric acid, is even more irritating to the respiratory tract. When the acid rain created by the reactions shown above falls to earth, the impact is significant.

It is easy to balance these chemical equations, but decades could be required to balance the ecological systems that we have disrupted by our massive consumption of fossil fuels. A sudden decrease of even 25% in the use of fossil fuels would lead to worldwide financial chaos. Development of alternative fuel sources, such as solar energy and safe nuclear power, will help to reduce our dependence on fossil fuels and help us to balance the global equation.

In the atmosphere, nitric oxide (NO) can react with oxygen to produce nitrogen dioxide as shown:

$$2NO(g) + O_2(g) \longrightarrow 2NO_2(g)$$

Nitric oxide   Oxygen   Nitrogen dioxide

Nitrogen dioxide (which causes the brown color of smog) then reacts with water to form nitric acid:

$$3NO_2(g) + H_2O(l) \longrightarrow 2HNO_3(aq) + NO(g)$$

For Further Understanding

Criticize this statement: "Passing and enforcing strong legislation against sulfur and nitrogen oxide emission will solve the problem of acid rain in the United States."

Research the literature to determine the percentage of electricity that is produced from coal in your state of residence.

Polyprotic Substances

Not all acid-base reactions occur in a 1:1 combining ratio (as hydrochloric acid and sodium hydroxide in the previous example). Acid-base reactions with other than 1:1 combining ratios occur between what are termed *polyprotic substances*. **Polyprotic substances** donate (as acids) or accept (as bases) more than one proton per formula unit.

Reactions of Polyprotic Substances

HCl dissociates to produce one H^+ ion for each HCl. For this reason, it is termed a *monoprotic acid*. Its reaction with sodium hydroxide is:

$$HCl(aq) + NaOH(aq) \longrightarrow H_2O(l) + Na^+(aq) + Cl^-(aq)$$

Sulfuric acid, in contrast, is a *diprotic acid*. Each unit of H_2SO_4 produces two H^+ ions (the prefix *di-* indicating two). Its reaction with sodium hydroxide is:

$$H_2SO_4(aq) + 2NaOH(aq) \longrightarrow 2H_2O(l) + 2Na^+(aq) + SO_4^{2-}(aq)$$

Phosphoric acid is a *triprotic acid*. Each unit of H_3PO_4 produces three H^+ ions. Its reaction with sodium hydroxide is:

$$H_3PO_4(aq) + 3NaOH(aq) \longrightarrow 3H_2O(l) + 3Na^+(aq) + PO_4^{3-}(aq)$$

Dissociation of Polyprotic Substances

Sulfuric acid, and other diprotic acids, dissociate in two steps:

Step 1. $H_2SO_4(aq) + H_2O(l) \longrightarrow H_3O^+(aq) + HSO_4^-(aq)$

Step 2. $HSO_4^-(aq) + H_2O(l) \rightleftharpoons H_3O^+(aq) + SO_4^{2-}(aq)$

Notice that H_2SO_4 behaves as a strong acid (Step 1) and HSO_4^- behaves as a weak acid, indicated by a double arrow (Step 2).

Phosphoric acid dissociates in three steps, all forms behaving as weak acids.

Step 1. $H_3PO_4(aq) + H_2O(l) \rightleftharpoons H_3O^+(aq) + H_2PO_4^-(aq)$

Step 2. $H_2PO_4^-(aq) + H_2O(l) \rightleftharpoons H_3O^+(aq) + HPO_4^{2-}(aq)$

Step 3. $HPO_4^{2-}(aq) + H_2O(l) \rightleftharpoons H_3O^+(aq) + PO_4^{3-}(aq)$

Bases exhibit this property as well.
NaOH produces one OH^- ion per formula unit:

$$NaOH(aq) \longrightarrow Na^+(aq) + OH^-(aq)$$

$Ba(OH)_2$, barium hydroxide, produces two OH^- ions per formula unit:

$$Ba(OH)_2(aq) \longrightarrow Ba^{2+}(aq) + 2OH^-(aq)$$

8.4 Acid-Base Buffers

A **buffer solution** contains components that enable the solution to resist large changes in pH when either acids or bases are added. Buffer solutions may be prepared in the laboratory to maintain optimum conditions for a chemical reaction. Buffers are routinely used in commercial products to maintain optimum conditions for product behavior.

Buffer solutions also occur naturally. Blood, for example, is a complex natural buffer solution maintaining a pH of approximately 7.4, optimum for oxygen transport. The major buffering agent in blood is the mixture of carbonic acid (H_2CO_3) and bicarbonate ions (HCO_3^-).

The color of the petals of the hydrangea is formed by molecules that behave as acid-base indicators. The color is influenced by the pH of the soil in which the hydrangea is grown. The plant (a) was grown in soil with lower pH (more acidic) than plant (b).

(a)   **(b)**

The Buffer Process

The basis of buffer action is the establishment of an equilibrium between either a weak acid and its conjugate base or a weak base and its conjugate acid. Let's consider the case of a weak acid and its salt.

A common buffer solution may be prepared from acetic acid (CH_3COOH) and sodium acetate (CH_3COONa). Sodium acetate is a salt that is the source of the conjugate base CH_3COO^-. An *equilibrium* is established in solution between the weak acid and the conjugate base.

We ignore Na^+ in the description of the buffer. Na^+ does not actively participate in the reaction.

$$CH_3COOH(aq) \; + \; H_2O(l) \; \rightleftharpoons \; H_3O^+(aq) \; + \; CH_3COO^-(aq)$$

| Acetic acid (weak acid) | Water | Hydronium ion | Acetate ion (conjugate base) |

The acetate ion is the conjugate base of acetic acid.

A buffer solution functions in accordance with LeChatelier's principle, which states that an equilibrium system, when stressed, will shift its equilibrium to relieve that stress. This principle is illustrated by the following examples.

Addition of Base (OH⁻) to a Buffer Solution

Addition of a basic substance to a buffer solution causes the following changes.

- OH^- from the base reacts with H_3O^+ producing water.
- Molecular acetic acid *dissociates* to replace the H_3O^+ consumed by the base, maintaining the pH close to the initial level.

This is an example of LeChatelier's principle, because the loss of H_3O^+ (the *stress*) is compensated by the dissociation of acetic acid to produce more H_3O^+.

Addition of Acid (H₃O⁺) to a Buffer Solution

Addition of an acidic solution to a buffer results in the following changes.

- H_3O^+ from the acid increases the overall $[H_3O^+]$.
- The system reacts to this stress, in accordance with LeChatelier's principle, to form more molecular acetic acid; the acetate ion combines with H_3O^+. Thus, the H_3O^+ concentration and therefore, the pH, remain close to the initial level.

 Animation
Effect of Addition of a Strong Acid and a Strong Base on a Buffer

These effects may be summarized as follows:

$$CH_3COOH(aq) + H_2O(l) \rightleftharpoons H_3O^+(aq) + CH_3COO^-(aq)$$

OH^- added, equilibrium shifts to the right

$\longrightarrow$

H_3O^+ added, equilibrium shifts to the left

$\longleftarrow$

Buffer Capacity

Buffer capacity is a measure of the ability of a solution to resist large changes in pH when a strong acid or strong base is added. More specifically, buffer capacity is described as the amount of strong acid or strong base that a buffer can neutralize without significantly changing its pH. Buffering capacity against base is a function of the concentration of the weak acid (in this case CH_3COOH). Buffering capacity against acid is dependent on the concentration of the anion of the salt, the conjugate base (CH_3COO^- in this example). Buffer solutions are often designed to have identical buffer capacity for both acids and bases. This is achieved when,

Commercial products that claim improved function owing to their ability to control pH. Can you name other products whose performance is pH dependent?

in the above example, $[CH_3COO^-]/[CH_3COOH] = 1$. As an added bonus, making the $[CH_3COO^-]$ and $[CH_3COOH]$ as large as is practical ensures a high buffer capacity for both added acid and added base.

Question 8.7

Explain how the molar concentration of H_2CO_3 in the blood would change if the partial pressure of CO_2 in the lungs were to increase. (Refer to A Medical Perspective: Control of Blood pH on page 276.)

Question 8.8

Explain how the molar concentration of H_2CO_3 in the blood would change if the partial pressure of CO_2 in the lungs were to decrease. (Refer to A Medical Perspective: Control of Blood pH on page 276.)

Question 8.9

Explain how the molar concentration of hydronium ion in the blood would change under each of the conditions described in Questions 8.7 and 8.8.

Question 8.10

Explain how the pH of blood would change under each of the conditions described in Questions 8.7 and 8.8.

Preparation of a Buffer Solution

It is useful to understand how to prepare a buffer solution and how to determine the pH of the resulting solution. Many chemical reactions produce the largest amount of product only when they are run at an optimal, constant pH. The study of biologically important processes in the laboratory often requires conditions that approximate the composition of biological fluids. A constant pH would certainly be essential.

The buffer process is an equilibrium reaction and is described by an equilibrium-constant expression. For acids, the equilibrium constant is represented as K_a, the subscript a implying an acid equilibrium. For example, the acetic acid/sodium acetate system is described by

$$CH_3COOH(aq) + H_2O(l) \rightleftharpoons H_3O^+(aq) + CH_3COO^-(aq)$$

and

$$K_a = \frac{[H_3O^+][CH_3COO^-]}{[CH_3COOH]}$$

Using a few mathematical maneuvers we can turn this equilibrium-constant expression into one that will allow us to calculate the pH of the buffer if we know how much acid (acetic acid) and salt (sodium acetate) are present in a known volume of the solution.

First, multiply both sides of the equation by the concentration of acetic acid, $[CH_3COOH]$. This will eliminate the denominator on the right side of the equation.

$$[CH_3COOH]K_a = \frac{[H_3O^+][CH_3COO^-][CH_3COOH]}{[CH_3COOH]}$$

or

$$[CH_3COOH]K_a = [H_3O^+][CH_3COO^-]$$

Now, dividing both sides of the equation by the acetate ion concentration [CH_3COO^-] will give us an expression for the hydronium ion concentration [H_3O^+]

$$\frac{[CH_3COOH]K_a}{[CH_3COO^-]} = [H_3O^+]$$

The calculation of pH from [H_3O^+] is discussed in Section 8.2.

Once we know the value for $[H_3O^+]$, we can easily find the pH.
To use this equation:

- assume that [CH_3COOH] represents the concentration of the acid component of the buffer.
- assume that [CH_3COO^-] represents the concentration of the conjugate base (principally from the dissociation of the salt, sodium acetate) component of the buffer.

$$\frac{[CH_3COOH]K_a}{[CH_3COO^-]} = [H_3O^+]$$

$$\frac{[acid]K_a}{[conjugate\ base]} = [H_3O^+]$$

Let's look at examples of practical applications of this equation.

Calculating the pH of a Buffer Solution

EXAMPLE 8.9

Calculate the pH of a buffer solution in which both the acetic acid and sodium acetate concentrations are $1.0 \times 10^{-1} M$. The equilibrium constant, K_a, for acetic acid is 1.8×10^{-5}.

LEARNING GOAL

8 Describe the applications of buffers to chemical and biochemical systems, particularly blood chemistry.

Solution

Step 1. Acetic acid is the acid; [acid] = $1.0 \times 10^{-1} M$

Sodium acetate is the salt, furnishing the conjugate base; [conjugate base] = $1.0 \times 10^{-1} M$

Step 2. The equilibrium is

$$CH_3COOH(aq) + H_2O(l) \rightleftharpoons H_3O^+(aq) + CH_3COO^-(aq)$$

acid   conjugate base

Step 3. The hydronium ion concentration is expressed as

$$[H_3O^+] = \frac{[acid]K_a}{[conjugate\ base]}$$

Step 4. Substituting the values given in the problem

$$[H_3O^+] = \frac{[1.0 \times 10^{-1}]1.8 \times 10^{-5}}{[1.0 \times 10^{-1}]}$$

$$[H_3O^+] = 1.8 \times 10^{-5}$$

Continued—

EXAMPLE 8.9 —*Continued*

Step 5. Now we can substitute in our expression for pH:

$$pH = -\log [H_3O^+]$$
$$pH = -\log 1.8 \times 10^{-5}$$
$$= 4.74$$

The pH of the buffer solution is 4.74.

Practice Problem 8.9

A buffer solution is prepared in such a way that the concentrations of propanoic acid and sodium propanoate are each $2.00 \times 10^{-1} M$. If the buffer equilibrium is described by

$$C_2H_5COOH(aq) + H_2O(l) \rightleftharpoons H_3O^+(aq) + C_2H_5COO^-(aq)$$

Propanoic acid   Propanoate anion

with $K_a = 1.34 \times 10^{-5}$, calculate the pH of the solution.

For Further Practice: Questions 8.81 and 8.82.

EXAMPLE 8.10 **Calculating the pH of a Buffer Solution**

LEARNING GOAL

8 Describe the applications of buffers to chemical and biochemical systems, particularly blood chemistry.

Calculate the pH of a buffer solution similar to that described in Example 8.9 except that the acid concentration is doubled, while the salt concentration remains the same.

Solution

Step 1. Acetic acid is the acid; [acid] = $2.0 \times 10^{-1} M$ (remember, the acid concentration is twice that of Example 8.9; $2 \times [1.0 \times 10^{-1}] = 2.0 \times 10^{-1} M$

Sodium acetate is the salt, furnishing the conjugate base; [conjugate base] = $1.0 \times 10^{-1} M$

Step 2. The equilibrium is

$$CH_3COOH(aq) + H_2O(l) \rightleftharpoons H_3O^+(aq) + CH_3COO^-(aq)$$

acid   conjugate base

Step 3. The hydronium ion concentration is expressed as,

$$[H_3O^+] = \frac{[acid]K_a}{[conjugate\ base]}$$

Step 4. Substituting the values given in the problem

$$[H_3O^+] = \frac{[2.0 \times 10^{-1}]1.8 \times 10^{-5}}{[1.00 \times 10^{-1}]}$$
$$[H_3O^+] = 3.60 \times 10^{-5}$$

Continued—

EXAMPLE 8.10 —*Continued*

Step 5. Now we can substitute in our expression for pH:

$$pH = -\log [H_3O^+]$$
$$pH = -\log 3.60 \times 10^{-5}$$
$$= 4.44$$

The pH of the buffer solution is 4.44.

Practice Problem 8.10

Calculate the pH of the buffer solution in Practice Problem 8.9 if the concentration of the sodium propanoate were doubled while the acid concentration remained the same.

For Further Practice: Questions 8.85 and 8.86.

A comparison of the two solutions described in Examples 8.9 and 8.10 demonstrates a buffer solution's most significant attribute: the ability to stabilize pH. Although the acid concentration of these solutions differs by a factor of two, the difference in their pH is only 0.30 units.

The Henderson-Hasselbalch Equation

The solution of the equilibrium-constant expression and the pH are sometimes combined into one operation. The combined expression is termed the **Henderson-Hasselbalch equation.**

For the acetic acid/sodium acetate buffer system,

$$CH_3COOH(aq) + H_2O(l) \rightleftharpoons H_3O^+(aq) + CH_3COO^-(aq)$$

$$K_a = \frac{[H_3O^+][CH_3COO^-]}{[CH_3COOH]}$$

Taking the $-\log$ of both sides of the equation:

$$-\log K_a = -\log [H_3O^+] - \log \frac{[CH_3COO^-]}{[CH_3COOH]}$$

$$pK_a = pH - \log \frac{[CH_3COO^-]}{[CH_3COOH]}$$

pKa = −log Ka, analogous to pH = −log[H₃O⁺].

the Henderson-Hasselbalch expression is:

$$pH = pK_a + \log \frac{[CH_3COO^-]}{[CH_3COOH]}$$

The form of this equation is especially amenable to buffer problem calculations. In this expression, $[CH_3COOH]$ represents the molar concentration of the weak acid and $[CH_3COO^-]$ is the molar concentration of the conjugate base of the weak acid. The generalized expression is:

$$pH = pK_a + \log \frac{[\text{conjugate base}]}{[\text{weak acid}]}$$

A Medical *Perspective*

Control of Blood pH

A pH of 7.4 is maintained in blood partly by a carbonic acid–bicarbonate buffer system based on the following equilibrium:

$$H_2CO_3(aq) + H_2O(l) \rightleftharpoons H_3O^+(aq) + HCO_3^-(aq)$$

Carbonic acid   Bicarbonate ion
(weak acid)   (salt)

The regulation process based on LeChatelier's principle is similar to the acetic acid–sodium acetate buffer, which we have already discussed.

Red blood cells transport O_2, bound to hemoglobin, to the cells of body tissue. The metabolic waste product, CO_2, is picked up by the blood and delivered to the lungs.

The CO_2 in the blood also participates in the carbonic acid–bicarbonate buffer equilibrium. Carbon dioxide reacts with water in the blood to form carbonic acid:

$$CO_2(aq) + H_2O(l) \rightleftharpoons H_2CO_3(aq)$$

As a result the buffer equilibrium becomes more complex:

$$CO_2(aq) + 2H_2O(l) \rightleftharpoons H_2CO_3(aq) + H_2O(l) \rightleftharpoons H_3O^+(aq) + HCO_3^-(aq)$$

Through this sequence of relationships the concentration of CO_2 in the blood affects the blood pH.

Higher than normal CO_2 concentrations shift the above equilibrium to the right (LeChatelier's principle), increasing $[H_3O^+]$ and lowering the pH. The blood becomes too acidic, leading to numerous medical problems. A situation of high blood CO_2 levels and low pH is termed *acidosis*. Respiratory acidosis results from various diseases (emphysema, pneumonia) that restrict the breathing process, causing the buildup of waste CO_2 in the blood.

Lower than normal CO_2 levels, on the other hand, shift the equilibrium to the left, decreasing $[H_3O^+]$ and making the pH more basic. This condition is termed *alkalosis* (from "alkali," implying basic). Hyperventilation, or rapid breathing, is a common cause of respiratory alkalosis.

For Further Understanding

Write the Henderson-Hasselbalch expression for the equilibrium between carbonic acid and the bicarbonate ion.

Calculate the $[HCO_3^-]/[H_2CO_3]$ that corresponds to a pH of 7.4. The K_a for carbonic acid is 4.2×10^{-7}.

Substituting concentrations along with the value for the pK_a of the acid allows the calculation of the pH of the buffer solution in problems such as those shown in Examples 8.9 and 8.10.

Question 8.11

Solve the problem illustrated in Example 8.9 using the Henderson-Hasselbalch equation.

Question 8.12

Solve the problem illustrated in Example 8.10 using the Henderson-Hasselbalch equation.

Question *8.13*

Solve Practice Problem 8.9 using the Henderson-Hasselbalch equation.

Question *8.14*

Solve Practice Problem 8.10 using the Henderson-Hasselbalch equation.

8.5 Oxidation-Reduction Processes

Oxidation-reduction processes are responsible for many types of chemical change. Corrosion, the operation of a battery, and biochemical energy-harvesting reactions are a few examples. In this section we explore the basic concepts underlying this class of chemical reactions.

LEARNING GOAL

9 Explain the meaning of the terms *oxidation* and *reduction*, and describe some practical examples of redox processes.

Oxidation and Reduction

Oxidation is defined as a loss of electrons, loss of hydrogen atoms, or gain of oxygen atoms. *Sodium metal,* is, for example, oxidized to a *sodium ion,* losing one electron when it reacts with a nonmetal such as chlorine:

Animation
Oxidation-Reduction Reactions

$$Na \longrightarrow Na^+ + e^-$$

Reduction is defined as a gain of electrons, gain of hydrogen atoms, or loss of oxygen atoms. A *chlorine atom* is reduced to a *chloride ion* by gaining one electron when it reacts with a metal such as sodium:

$$Cl + e^- \longrightarrow Cl^-$$

Oxidation and reduction are complementary processes. The *oxidation half-reaction* produces an electron that is the reactant for the *reduction half-reaction.* The combination of two half-reactions, one oxidation and one reduction, produces the complete reaction:

Oxidation half-reaction:	$Na \longrightarrow Na^+ + e^-$
Reduction half-reaction:	$Cl + e^- \longrightarrow Cl^-$
Complete reaction:	$Na + Cl \longrightarrow Na^+ + Cl^-$

Oxidation-reduction reactions are often termed *redox reactions.*

Half-reactions, one oxidation and one reduction, are exactly that: one-half of a complete reaction. The two half-reactions combine to produce the complete reaction. Note that the electrons cancel: in the electron transfer process, no free electrons remain.

In the preceding reaction, sodium metal is the **reducing agent.** It releases electrons for the reduction of chlorine. Chlorine is the **oxidizing agent.** It accepts electrons from the sodium, which is oxidized.

The reducing agent becomes oxidized and the oxidizing agent becomes reduced.

The characteristics of oxidizing and reducing agents may be summarized as follows:

Oxidizing Agent
- Is reduced
- Gains electrons
- Causes oxidation

Reducing Agent
- Is oxidized
- Loses electrons
- Causes reduction

Question *8.15*

Write the oxidation half-reaction, the reduction half-reaction, and the complete reaction for the formation of calcium sulfide from the elements Ca and S.

A Medical Perspective

Oxidizing Agents for Chemical Control of Microbes

Before the twentieth century, hospitals were not particularly sanitary establishments. Refuse, including human waste, was disposed of on hospital grounds. Because many hospitals had no running water, physicians often cleaned their hands and instruments by wiping them on their lab coats and then proceeded to treat the next patient! As you can imagine, many patients died of infections in hospitals.

By the late nineteenth century a few physicians and microbiologists had begun to realize that infectious diseases are transmitted by microbes, including bacteria and viruses. To decrease the number of hospital-acquired infections, physicians like Joseph Lister and Ignatz Semmelweis experimented with chemicals and procedures that were designed to eliminate pathogens from environmental surfaces and from wounds.

Many of the common disinfectants and antiseptics are oxidizing agents. A disinfectant is a chemical that is used to kill or inhibit the growth of pathogens, disease-causing microorganisms, on environmental surfaces. An antiseptic is a milder chemical that is used to destroy pathogens associated with living tissue.

Hydrogen peroxide is an effective antiseptic that is commonly used to cleanse cuts and abrasions. We are all familiar with the furious bubbling that occurs as the enzyme catalase from our body cells catalyzes the breakdown of H_2O_2:

$$2H_2O_2(aq) \longrightarrow 2H_2O(l) + O_2(g)$$

A highly reactive and deadly form of oxygen, the superoxide radical (O_2^-), is produced during this reaction. This superoxide inactivates proteins, especially critical enzyme systems.

At higher concentrations (3–6%), H_2O_2 is used as a disinfectant. It is particularly useful for disinfection of soft contact lenses, utensils, and surgical implants because there is no residual toxicity. Concentrations of 6–25% are even used for complete sterilization of environmental surfaces.

Benzoyl peroxide is another powerful oxidizing agent. Ointments containing 5–10% benzoyl peroxide have been used as antibacterial agents to treat acne. The compound is currently found in over-the-counter facial scrubs because it is also an exfoliant, causing sloughing of old skin and replacement with smoother-looking skin. A word of caution is in order: in sensitive individuals, benzoyl peroxide can cause swelling and blistering of tender facial skin.

Chlorine is a very widely used disinfectant and antiseptic. Calcium hypochlorite [$Ca(OCl)_2$] was first used in hospital maternity wards in 1847 by the pioneering Hungarian physician Ignatz Semmelweis. Semmelweis insisted that hospital workers cleanse their hands in a $Ca(OCl)_2$ solution and dramatically reduced the incidence of infection. Today, calcium hypochlorite is more commonly used to disinfect bedding, clothing, restaurant eating utensils, slaughterhouses, barns, and dairies.

Sodium hypochlorite (NaOCl), sold as Clorox, is used as a household disinfectant and deodorant but is also used to disinfect swimming pools, dairies, food-processing equipment, and kidney dialysis units. It can be used to treat drinking water of questionable quality. Addition of 1/2 teaspoon of household bleach (5.25% NaOCl) to 2 gallons of clear water renders it drinkable after 1/2 hour. The Centers for Disease Control even

---

Question 8.16

Write the oxidation half-reaction, the reduction half-reaction, and the complete reaction for the formation of calcium iodide from calcium metal and I_2. Remember, the electron gain *must* equal the electron loss.

LEARNING GOAL

9 Explain the meaning of the terms *oxidation* and *reduction*, and describe some practical examples of redox processes.

At the same time that iron is oxidized, O_2 is being reduced to O^{2-} and is incorporated into the structure of iron(III) oxide. Electrons lost by iron reduce oxygen. This again shows that oxidation and reduction processes go hand in hand.

Applications of Oxidation and Reduction

Oxidation-reduction processes are important in many areas as diverse as industrial manufacturing and biochemical processes.

Corrosion

The deterioration of metals caused by an oxidation-reduction process is termed **corrosion.** Metal atoms are converted to metal ions; the structure, hence the properties, changes dramatically, and usually for the worse.

Millions of dollars are spent annually in an attempt to correct the damage resulting from corrosion. A current area of chemical research is concerned with

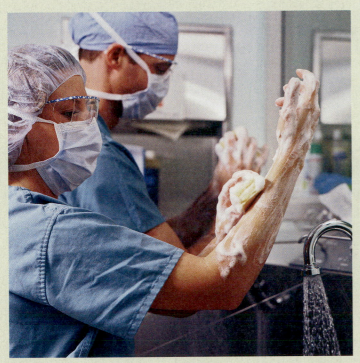

recommend a 1:10 dilution of bleach as an effective disinfectant against human immunodeficiency virus, the virus that causes acquired immune deficiency syndrome (AIDS).

Chlorine gas (Cl_2) is used to disinfect swimming pool water, sewage, and municipal water supplies. This treatment has successfully eliminated epidemics of waterborne diseases. However, chlorine is inactivated in the presence of some organic materials and, in some cases, may form toxic chlorinated organic compounds. For these reasons, many cities are considering the use of ozone (O_3) rather than chlorine.

Ozone is produced from O_2 by high-voltage electrical discharges. (That fresh smell in the air after an electrical storm is ozone.) Several European cities use ozone to disinfect drinking water. It is a more effective killing agent than chlorine, especially with some viruses: less ozone is required for disinfection; there is no unpleasant residual odor or flavor; and there appear to be fewer toxic by-products. However, ozone is more expensive than chlorine, and maintaining the required concentration in the water is more difficult. Nonetheless, the benefits seem to outweigh the drawbacks, and many U.S. cities may soon follow the example of European cities and convert to the use of ozone for water treatment.

For Further Understanding

Describe the difference between the terms *disinfectant* and *antiseptic*.

Explain why hydrogen peroxide, at higher concentration, is used as a disinfectant, whereas lower concentrations are used as antiseptics.

(a)

(b)

(c)

The rust (an oxide of iron) that diminishes structural strength and ruins the appearance of (a) automobiles, (b) bridges, and (c) other iron-based objects is a common example of an oxidation-reduction reaction. Can you provide examples of other electron transfer processes that produce changes in properties?

A Medical Perspective

Electrochemical Reactions in the Statue of Liberty and in Dental Fillings

Throughout history, we have suffered from our ignorance of basic electrochemical principles. For example, during the Middle Ages, our chemistry ancestors (alchemists) placed an iron rod into a blue solution of copper sulfate. They noticed that bright shiny copper plated out onto an iron rod and they thought that they had changed a base metal, iron, into copper. What actually happened was the redox reaction shown in Equation 1.

$$2Fe(s) + 3Cu^{2+}(aq) \longrightarrow 2Fe^{3+}(aq) + 3Cu(s) \qquad (1)$$

This misunderstanding encouraged them to embark on a futile, one-thousand-year attempt to change base metals into gold.

Statue of Liberty redox reaction.

Over one hundred years ago, France presented the United States with the Statue of Liberty. Unfortunately, the French did not anticipate the redox reaction shown in Equation 1 when they mounted the copper skin of the statue on iron support rods. Oxygen in the atmosphere oxidized the copper skin to produce copper ions. Then, because iron is more active than copper, the displacement shown in Equation 1 aided the corrosion of the support bars. As a result of this and other reactions, the statue needed refurbishing before we celebrated its one hundredth anniversary in 1986.

Sometimes dentists also overlook possible redox reactions when placing gold caps over teeth next to teeth with amalgam fillings. The amalgam in tooth fillings is an alloy of mercury, silver, tin, and copper. Because the metals in the amalgam are more active than gold, contact between the amalgam fillings and minute numbers of gold ions results in redox reactions such as the following.*

$$3Sn(s) + 2Au^{3+}(aq) \longrightarrow 3Sn^{2+}(aq) + 2Au(s) \qquad (2)$$

As a result, the dental fillings dissolve and the patients are left with a constant metallic taste in their mouths.

These examples show that like our ancestors, we continue to experience unfortunate results because of a lack of understanding of basic electrochemical principles.

Source: Ronald DeLorenzo, *Journal of Chemical Education*, May 1985, pp. 424–425.

*Equation 2 is oversimplified to illustrate more clearly the basic displacement of gold ions by metallic tin atoms. Actually, only complex ions of gold and tin can exist in aqueous solutions, not the simple cations that are shown.

For Further Understanding

Label the oxidizing agent, reducing agent, substance oxidized, and substance reduced in each equation in this perspective.

For each equation, state the substance that gains electrons and the substance that loses electrons.

the development of corrosion-inhibiting processes. In one type of corrosion, elemental iron is oxidized to iron(III) oxide (rust):

$$4Fe(s) + 3O_2(g) \longrightarrow 2Fe_2O_3(s)$$

Combustion of Fossil Fuels

Burning fossil fuel is an extremely exothermic process. Energy is released to heat our homes, offices, and classrooms. The simplest fossil fuel is methane, CH_4, and its oxidation reaction is written:

$$CH_4(g) + 2O_2(g) \longrightarrow CO_2(g) + 2H_2O(g)$$

Methane is a hydrocarbon. The complete oxidation of any hydrocarbon (including those in gasoline, heating oil, liquid propane, and so forth) produces carbon dioxide and water. The energy released by these reactions is of paramount importance. The water and carbon dioxide are viewed as waste products, and the carbon dioxide contributes to the greenhouse effect (see An Environmental Perspective: The Greenhouse Effect and Global Climate Change on page 174).

Bleaching

Bleaching agents are most often oxidizing agents. Sodium hypochlorite (NaOCl) is the active ingredient in a variety of laundry products. It is an effective oxidizing agent. Products containing NaOCl are advertised for their stain-removing capabilities.

Stains are a result of colored compounds adhering to surfaces. Oxidation of these compounds produces products that are not colored or compounds that are subsequently easily removed from the surface, thus removing the stain.

Biological Processes

Respiration

There are many examples of biological oxidation-reduction reactions. For example, the electron-transport chain of aerobic respiration involves the reversible oxidation and reduction of iron atoms in cytochrome c,

$$\text{cytochrome } c \ (Fe^{3+}) + e^- \longrightarrow \text{cytochrome } c \ (Fe^{2+})$$

The reduced iron ion transfers an electron to an iron ion in another protein, called cytochrome c oxidase, according to the following reaction:

$$\text{cytochrome } c \ (Fe^{2+}) + \text{cytochrome } c \text{ oxidase } (Fe^{3+})$$
$$\text{cytochrome } c \ (Fe^{3+}) + \text{cytochrome } c \text{ oxidase } (Fe^{2+})$$

Cytochrome c oxidase eventually passes four electrons to O_2, the final electron acceptor of the chain:

$$O_2 + 4e^- + 4H^+ \longrightarrow 2H_2O$$

See Chapters 21 and 22 for the details of these energy-harvesting cellular oxidation-reduction reactions.

Metabolism

When ethanol is metabolized in the liver, it is oxidized to acetaldehyde (the molecule partially responsible for hangovers). Continued oxidation of acetaldehyde produces acetic acid, which is eventually oxidized to CO_2 and H_2O. These reactions, summarized as follows, are catalyzed by liver enzymes.

$$\underset{\text{Ethanol}}{CH_3CH_2-OH} \longrightarrow \underset{\text{Acetaldehyde}}{CH_3\overset{\overset{\displaystyle O}{\|}}{C}-H} \longrightarrow \underset{\text{Acetic acid}}{CH_3\overset{\overset{\displaystyle O}{\|}}{C}-OH} \longrightarrow CO_2 + H_2O$$

It is more difficult to recognize these reactions as oxidations because neither the product nor the reactant carries a charge. In previous examples we looked for an increase in positive charge as an indication that an oxidation had occurred. A decrease in positive charge (or increased negative charge) would signify reduction.

Alternative descriptions of oxidation and reduction are useful in identifying these reactions.

Oxidation is the *gain* of oxygen or *loss* of hydrogen.

Reduction is the *loss* of oxygen or *gain* of hydrogen.

In the conversion of ethanol to acetaldehyde, ethanol has six hydrogen atoms per molecule; the product acetaldehyde has four hydrogen atoms per molecule. This represents a loss of two hydrogen atoms per molecule. Therefore, ethanol has been oxidized to acetaldehyde, based on the interpretation of the above-mentioned rules.

This strategy is most useful for recognizing oxidation and reduction of *organic compounds* and organic compounds of biological interest, *biochemical compounds*. Organic compounds and their structures and reactivity are the focus of Chapters 10 through 15 and biochemical compounds are described in Chapters 16 through 23.

Voltaic Cells

When zinc metal is dipped into a copper(II) sulfate solution, zinc atoms are oxidized to zinc ions and copper(II) ions are reduced to copper metal, which deposits on the surface of the zinc metal (Figure 8.7). This reaction is summarized as follows:

$$\overset{\text{Oxidation/e}^- \text{ loss}}{Zn(s) + Cu^{2+}(aq) \longrightarrow Zn^{2+}(aq) + Cu(s)}$$

Reduction/e$^-$ gain

In the reduction of aqueous copper(II) ions by zinc metal, electrons flow from the zinc rod directly to copper(II) ions in the solution. If electron transfer from the zinc rod to the copper ions in solution could be directed through an external electrical circuit, this spontaneous oxidation-reduction reaction could be used to produce an electrical current that could perform some useful function.

However, when zinc metal in one container is connected by a copper wire with a copper(II) sulfate solution in a separate container, no current flows through

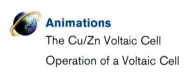

LEARNING GOAL

10 Diagram a voltaic cell and describe its function.

Animations

The Cu/Zn Voltaic Cell

Operation of a Voltaic Cell

Recall that solutions of ionic salts are good conductors of electricity (Chapter 6).

Figure 8.7

The spontaneous reaction of zinc metal and Cu^{2+} ions is the basis of the cell depicted in Figure 8.8.

$$Zn(s) \quad + \quad Cu^{2+}(aq) \quad \longrightarrow \quad Zn^{2+}(aq) \quad + \quad Cu(s)$$

the wire. A complete, or continuous circuit is necessary for current to flow. To complete the circuit, we connect the two containers with a tube filled with a solution of an electrolyte such as potassium chloride. This tube is described as a *salt bridge*.

Current now flows through the external circuit (Figure 8.8). The device shown in Figure 8.8 is an example of a *voltaic cell*. A **voltaic cell** is an *electrochemical* cell that converts stored *chemical* energy into *electrical* energy.

This cell consists of two *half-cells*. The oxidation half-reaction occurs in one half-cell and the reduction half-reaction occurs in the other half-cell. The sum of the two half-cell reactions is the overall oxidation-reduction reaction that describes the cell. The electrode at which oxidation occurs is called the **anode,** and the electrode at which reduction occurs is the **cathode.** In the device shown in Figure 8.8, the zinc metal is the anode. At this electrode the zinc atoms are oxidized to zinc ions:

$$\text{Anode half-reaction: } Zn(s) \longrightarrow Zn^{2+}(aq) + 2e^-$$

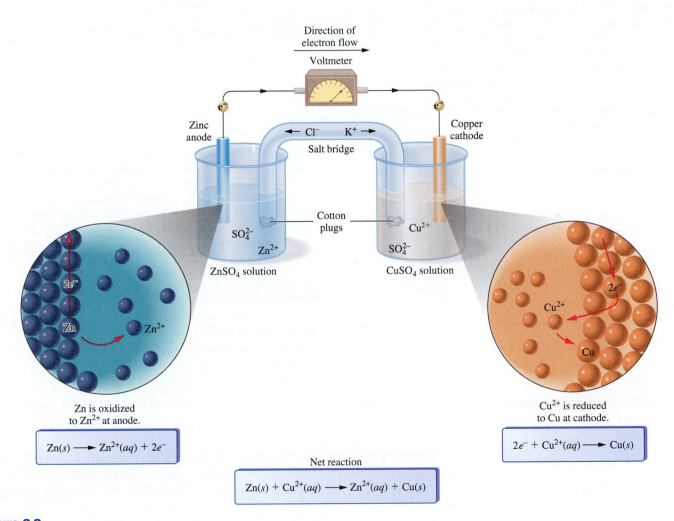

Figure 8.8

A voltaic cell generating electrical current by the reaction:

$$Zn(s) + Cu^{2+}(aq) \longrightarrow Zn^{2+}(aq) + Cu(s)$$

Each electrode consists of the pure metal, zinc or copper. Zinc is oxidized, releasing electrons that flow to the copper, reducing Cu^{2+} to Cu. The salt bridge completes the circuit and the voltmeter displays the voltage (or chemical potential) associated with the reaction.

A Medical *Perspective*

Turning the Human Body into a Battery

The heart has its own natural pacemaker that sends nerve impulses (pulses of electrical current) throughout the heart approximately seventy-two times per minute. These electrical pulses cause your heart muscles to contract (beat), which pumps blood through the body. The fibers that carry the nerve impulses can be damaged by disease, drugs, heart attacks, and surgery. When these heart fibers are damaged, the heart may run too slowly, stop temporarily, or stop altogether. To correct this condition, artificial heart pacemakers (see figure below) are surgically inserted in the human body. A pacemaker (pacer) is a battery-driven device that sends an electrical current (pulse) to the heart about seventy-two times per minute. Over 300,000 Americans are now wearing artificial pacemakers with an additional 30,000 pacemakers installed each year.

Yearly operations used to be necessary to replace the pacemaker's batteries. Today, pacemakers use improved batteries that last much longer, but even these must be replaced eventually.

It would be very desirable to develop a permanent battery to run pacemakers. Some scientists began working on ways of converting the human body itself into a battery (voltaic cell) to power artificial pacemakers.

Several methods for using the human body as a voltaic cell have been suggested. One of these is to insert platinum and

The "body battery."

$$Zn \longrightarrow Zn^{2+} + 2e^-$$

$$2e^- + \tfrac{1}{2}O_2 \longrightarrow O^{2-}$$

$$\text{or } 2e^- + \tfrac{1}{2}O_2 + 2H^+ \longrightarrow H_2O$$

zinc electrodes into the human body as diagrammed in the figure above. The pacemaker and the electrodes would be worn internally. This "body battery" could easily generate the small amount of current (5×10^{-5} ampere) that is required by most pacemakers. This "body battery" has been tested on animals for periods exceeding four months without noticeable problems.

Source: Ronald DeLorenzo, *Problem Solving in General Chemistry*, 2nd ed., Wm. C. Brown, Publishers, Dubuque, Iowa, 1993, pp. 336–338.

Artificial heart pacemaker.

For Further Understanding

What are some criteria that must be considered when choosing the electrode material?

Combine the two half-reactions for the "body battery" to yield a complete oxidation-reduction equation.

Electrons released at the anode travel through the external circuit to the cathode (the copper rod) where they are transferred to copper(II) ions in the solution. Copper(II) ions are reduced to copper atoms that deposit on the copper metal surface, the cathode:

$$\text{Cathode half-reaction: } Cu^{2+}(aq) + 2e^- \longrightarrow Cu(s)$$

The sum of these half-cell reactions is the cell reaction:

$$Zn(s) + Cu^{2+}(aq) \longrightarrow Zn^{2+}(aq) + Cu(s)$$

Voltaic cells are found in many aspects of our life, as convenient and reliable sources of electrical energy, the battery. Batteries convert stored chemical energy to an electrical current to power a wide array of different commercial appliances: radios, portable televisions and computers, flashlights, a host of other useful devices.

Technology has made modern batteries smaller, safer, and more dependable than our crudely constructed copper-zinc voltaic cell. In fact, the silver cell (Figure 8.9) is sufficiently safe and nontoxic that it can be implanted in the human body as a part of a pacemaker circuit that is used to improve heart rhythm. A rather futuristic potential application of voltaic cells is noted in A Medical Perspective: Turning the Human Body into a Battery on page 284.

Electrolysis

Electrolysis reactions use electrical energy to cause nonspontaneous oxidation-reduction reactions to occur. They are the reverse of voltaic cells. One common application is the rechargeable battery. When it is being used to power a device, such as a laptop computer, it behaves as a voltaic cell. After some time, the chemical reaction approaches completion and the voltaic cell "runs down." The cell reaction is reversible and the battery is plugged into a battery charger. The charger is really an external source of electrical energy that reverses the chemical reaction in the battery, bringing it back to its original state. The cell has been operated as an electrolytic cell. Removal of the charging device turns the cell back into a voltaic device, ready to spontaneously react to produce electrical current once again.

The relationship between a voltaic cell and an electrolytic cell is illustrated in Figure 8.10.

Figure 8.9
A silver battery used in cameras, heart pacemakers, and hearing aids. This battery is small, stable, and nontoxic (hence implantable in the human body).

LEARNING GOAL
11 Compare and contrast voltaic and electrolytic cells.

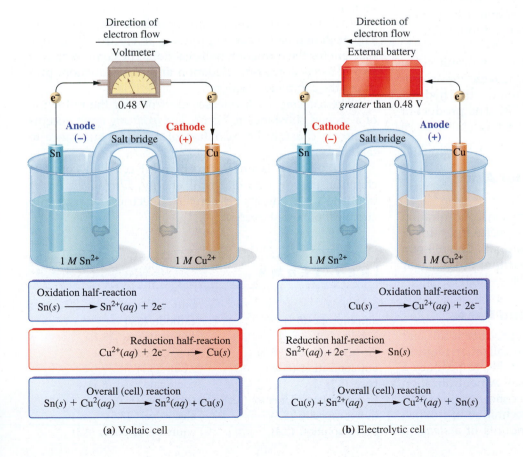

(a) Voltaic cell

(b) Electrolytic cell

Figure 8.10
(a) A voltaic cell is converted to (b) an electrolytic cell by attaching a battery with a voltage sufficiently large to reverse the reaction. This process underlies commercially available rechargeable batteries.

8.1 Acids and Bases

One of the earliest definitions of acids and bases is the *Arrhenius theory*. According to this theory, an *acid* dissociates to form hydrogen ions, H^+, and a *base* dissociates to form hydroxide ions, OH^-. The *Brønsted-Lowry theory* defines an acid as a proton (H^+) donor and a base as a proton acceptor.

Water, the solvent in many acid-base reactions, is *amphiprotic*. It has both acid and base properties.

The strength of acids and bases in water depends on their degree of dissociation, the extent to which they react with the solvent, water. Acids and bases are strong when the reaction with water is virtually 100% complete and weak when the reaction with water is much less than 100% complete.

Weak acids and weak bases dissolve in water principally in the molecular form. Only a small percentage of the molecules dissociate to form the *hydronium* ion or *hydroxide* ion.

Aqueous solutions of acids and bases are electrolytes. The dissociation of the acid or base produces ions, which conduct an electrical current. Strong acids and bases are strong electrolytes. Weak acids and bases are weak electrolytes.

Although pure water is virtually 100% molecular, a small number of water molecules do ionize. This process occurs by the transfer of a proton from one water molecule to another, producing a hydronium ion and a hydroxide ion. This process is the *autoionization,* or self-ionization, of water.

Pure water at room temperature has a hydronium ion concentration of 1.0×10^{-7} *M*. One hydroxide ion is produced for each hydronium ion. Therefore, the hydroxide ion concentration is also 1.0×10^{-7} *M*. The product of hydronium and hydroxide ion concentration (1.0×10^{-14}) is the *ion product for water.*

8.2 pH: A Measurement Scale for Acids and Bases

The *pH scale* correlates the hydronium ion concentration with a number, the pH, that serves as a useful indicator of the degree of acidity or basicity of a solution. The pH of a solution is defined as the negative logarithm of the molar concentration of the hydronium ion ($pH = -\log[H_3O^+]$).

8.3 Reactions Between Acids and Bases

The reaction of an acid with a base to produce a salt and water is referred to as *neutralization*. Neutralization requires equal numbers of moles of H_3O^+ and OH^- to produce a neutral solution (no excess acid or base). A neutralization reaction may be used to determine the concentration of an unknown acid or base solution. The technique of *titration* involves the addition of measured amounts of a *standard solution* (one whose concentration is known) from a *buret* to neutralize the second, unknown solution. The *equivalence point* is signaled by an *indicator.*

8.4 Acid-Base Buffers

A *buffer solution* contains components that enable the solution to resist large changes in pH when acids or bases are added. The basis of buffer action is an equilibrium between either a weak acid and its salt or a weak base and its salt.

A buffer solution follows LeChatelier's principle, which states that an equilibrium system, when stressed, will shift its equilibrium to alleviate that stress.

Buffering against base is a function of the concentration of the weak acid for an acidic buffer. Buffering against acid is dependent on the concentration of the anion of the salt.

A buffer solution can be described by an equilibrium-constant expression. The equilibrium-constant expression for an acidic system can be rearranged and solved for $[H_3O^+]$. In that way, the pH of a buffer solution can be obtained, if the composition of the solution is known. Alternatively, the *Henderson-Hasselbalch equation,* derived from the equilibrium constant expression, may be used to calculate the pH of a buffer solution.

8.5 Oxidation-Reduction Processes

Oxidation is defined as a loss of electrons, loss of hydrogen atoms, or gain of oxygen atoms. *Reduction* is defined as a gain of electrons, gain of hydrogen atoms, or loss of oxygen atoms.

Oxidation and reduction are complementary processes. The oxidation half-reaction produces an electron that is the reactant for the reduction half-reaction. The combination of two half-reactions, one oxidation and one reduction, produces the complete reaction.

The *reducing agent* releases electrons for the reduction of a second substance to occur. The *oxidizing agent* accepts electrons, causing the oxidation of a second substance to take place.

A *voltaic cell* is an electrochemical cell that converts chemical energy into electrical energy. *Electrolysis* is the opposite of a battery. It converts electrical energy into chemical potential energy.

KEY TERMS

acid (8.1)
amphiprotic (8.1)
anode (8.5)
Arrhenius theory (8.1)
autoionization (8.1)
base (8.1)
Brønsted-Lowry theory (8.1)
buffer capacity (8.4)
buffer solution (8.4)

buret (8.3)
cathode (8.5)
conjugate acid (8.1)
conjugate acid-base
   pair (8.1)
conjugate base (8.1)
corrosion (8.5)
electrolysis (8.5)
equivalence point (8.3)

Henderson-Hasselbalch
  equation (8.4)
hydronium ion (8.1)
indicator (8.3)
ion product for water (8.1)
neutralization (8.3)
oxidation (8.5)
oxidizing agent (8.5)

pH scale (8.2)
polyprotic substance (8.3)
reducing agent (8.5)
reduction (8.5)
standard solution (8.3)
titration (8.3)
voltaic cell (8.5)

I   II

III   IV

QUESTIONS AND PROBLEMS

Acids and Bases

Foundations

8.17 **a.** Define an acid according to the Arrhenius theory.
   b. Define an acid according to the Brønsted-Lowry theory.

8.18 **a.** Define a base according to the Arrhenius theory.
   b. Define a base according to the Brønsted-Lowry theory.

8.19 What are the essential differences between the Arrhenius and Brønsted-Lowry theories?

8.20 Why is ammonia described as a Brønsted-Lowry base and not an Arrhenius base?

Applications

8.21 Write an equation for the reaction of each of the following with water:
   a. HNO_2
   b. HCN

8.22 Write an equation for the reaction of each of the following with water:
   a. HNO_3
   b. HCOOH

8.23 **a.** Write the conjugate acid of NO_3^-.
   b. Which is the stronger base, NO_3^- or CN^-?
   c. Which is the stronger acid, HNO_3 or HCN?

8.24 **a.** Write the conjugate acid of F^-.
   b. Which is the stronger base, F^- or CH_3COO^-?
   c. Which is the stronger acid, HF or CH_3COOH?

8.25 Label each of the following as a strong or weak acid (consult Figure 8.2, if necessary):
   a. H_2SO_3
   b. H_2CO_3
   c. H_3PO_4

8.26 Label each of the following as a strong or weak base (consult Figure 8.2, if necessary):
   a. KOH
   b. CN^-
   c. SO_4^{2-}

8.27 Identify the conjugate acid-base pairs in each of the following chemical equations:
   a. $NH_4^+(aq) + CN^-(aq) \rightleftharpoons NH_3(aq) + HCN(aq)$
   b. $CO_3^{2-}(aq) + HCl(aq) \rightleftharpoons HCO_3^-(aq) + Cl^-(aq)$

8.28 Identify the conjugate acid-base pairs in each of the following chemical equations:
   a. $HCOOH(aq) + NH_3(aq) \rightleftharpoons HCOO^-(aq) + NH_4^+(aq)$
   b. $HCl(aq) + OH^-(aq) \rightleftharpoons H_2O(l) + Cl^-(aq)$

8.29 Distinguish between the terms acid-base *strength* and acid-base *concentration*.

8.30 Of the diagrams shown here, which one represents:
   a. a concentrated strong acid
   b. a dilute strong acid
   c. a concentrated weak acid
   d. a dilute weak acid

8.31 Classify each of the following as a Brønsted acid, Brønsted base, or both:
   a. H_3O^+
   b. OH^-
   c. H_2O

8.32 Classify each of the following as a Brønsted acid, Brønsted base, or both:
   a. NH_4^+
   b. NH_3

8.33 Classify each of the following as a Brønsted acid, Brønsted base, or both:
   a. H_2CO_3
   b. HCO_3^-
   c. CO_3^{2-}

8.34 Classify each of the following as a Brønsted acid, Brønsted base, or both:
   a. H_2SO_4
   b. HSO_4^-
   c. SO_4^{2-}

8.35 Write the formula of the conjugate acid of CN^-.

8.36 Write the formula of the conjugate acid of Br^-.

8.37 Write the formula of the conjugate base of HI.

8.38 Write the formula of the conjugate base of HCOOH.

pH of Acid and Base Solutions

Foundations

8.39 Calculate the $[H_3O^+]$ of an aqueous solution that is:
   a. $1.0 \times 10^{-7}\,M$ in OH^-
   b. $1.0 \times 10^{-3}\,M$ in OH^-

8.40 Calculate the $[H_3O^+]$ of an aqueous solution that is:
   a. $1.0 \times 10^{-9}\,M$ in OH^-
   b. $1.0 \times 10^{-5}\,M$ in OH^-

8.41 Label each solution in Problem 8.39 as acidic, basic, or neutral.

8.42 Label each solution in Problem 8.40 as acidic, basic, or neutral.

8.43 Calculate the pH of a solution that is:
   a. $1.0 \times 10^{-2}\,M$ in HCl
   b. $1.0 \times 10^{-4}\,M$ in HNO_3

8.44 Calculate the pH of a solution that is:
   a. $1.0 \times 10^{-1}\,M$ in HCl
   b. $1.0 \times 10^{-5}\,M$ in HNO_3

8.45 Calculate $[H_3O^+]$ for a solution of nitric acid that is:
   a. pH = 1.00
   b. pH = 5.00

8.46   Calculate $[H_3O^+]$ for a solution of hydrochloric acid that is:
    a.   pH = 2.00
    b.   pH = 3.00

8.47   Calculate the pH of a $1.0 \times 10^{-3}\,M$ solution of KOH.

8.48   Calculate the pH of a $1.0 \times 10^{-5}\,M$ solution of NaOH.

8.49   Calculate *both* $[H_3O^+]$ and $[OH^-]$ for a solution that is:
    a.   pH = 1.30
    b.   pH = 9.70

8.50   Calculate *both* $[H_3O^+]$ and $[OH^-]$ for a solution that is:
    a.   pH = 5.50
    b.   pH = 7.00

8.51   What is a neutralization reaction?

8.52   Describe the purpose of a titration.

Applications

8.53   The pH of urine may vary between 4.5 and 8.2. Determine the H_3O^+ concentration and OH^- concentration if the measured pH is:
    a.   6.00
    b.   5.20
    c.   7.80

8.54   The hydronium ion concentration in blood of three different patients was:

Patient	$[H_3O^+]$
A	5.0×10^{-8}
B	3.1×10^{-8}
C	3.2×10^{-8}

What is the pH of each patient's blood? If the normal range is 7.30–7.50, which, if any, of these patients have an abnormal blood pH?

8.55   Determine how many times more acidic a solution is at:
    a.   pH 2 relative to pH 4
    b.   pH 7 relative to pH 11
    c.   pH 2 relative to pH 12

8.56   Determine how many times more basic a solution is at:
    a.   pH 6 relative to pH 4
    b.   pH 10 relative to pH 9
    c.   pH 11 relative to pH 6

8.57   What is the H_3O^+ concentration of a solution with a pH of:
    a.   5.00
    b.   12.00
    c.   5.50

8.58   What is the H_3O^+ concentration of a solution with a pH of:
    a.   6.80
    b.   4.60
    c.   2.70

8.59   Calculate the pH of a solution with a H_3O^+ concentration of:
    a. $1.0 \times 10^{-6}\,M$
    b. $1.0 \times 10^{-8}\,M$
    c. $5.6 \times 10^{-4}\,M$

8.60   What is the OH^- concentration of each solution in Question 8.59?

8.61   Calculate the pH of a solution that has $[H_3O^+] = 7.5 \times 10^{-4}\,M$.

8.62   Calculate the pH of a solution that has $[H_3O^+] = 6.6 \times 10^{-5}\,M$.

8.63   Calculate the pH of a solution that has $[OH^-] = 5.5 \times 10^{-4}\,M$.

8.64   Calculate the pH of a solution that has $[OH^-] = 6.7 \times 10^{-9}\,M$.

Reactions Between Acids and Bases

Foundations

8.65   Write an equation to represent the neutralization of an aqueous solution of HNO_3 with an aqueous solution of NaOH.

8.66   Write an equation to represent the neutralization of an aqueous solution of HCl with an aqueous solution of KOH.

8.67   Rewrite the equation in Question 8.65 as a net, balanced, ionic equation.

8.68   Rewrite the equation in Question 8.66 as a net, balanced, ionic equation.

8.69   What function does an indicator perform?

8.70   Choose an indicator from Figure 8.5 that would appear yellow in acid solution and blue in basic solution.

Applications

8.71   Titration of 15.00 mL of HCl solution requires 22.50 mL of $0.1200\,M$ NaOH solution. What is the molarity of the HCl solution?

8.72   Titration of 17.85 mL of HNO_3 solution requires 16.00 mL of $0.1600\,M$ KOH solution. What is the molarity of the HNO_3 solution?

8.73   What volume of $0.1500\,M$ NaOH is required to titrate 20.00 mL of $0.1000\,M$ HCl?

8.74   What volume of $0.2000\,M$ KOH is required to titrate 25.00 mL of $0.1500\,M\ HNO_3$?

Buffer Solutions

Foundations

8.75   Which of the following are capable of forming a buffer solution?
    a. NH_3 and NH_4Cl
    b. HNO_3 and KNO_3

8.76   Which of the following are capable of forming a buffer solution?
    a.   HBr and $MgCl_2$
    b. H_2CO_3 and $NaHCO_3$

8.77   Define:
    a.   buffer solution
    b.   acidosis (refer to A Medical Perspective: Control of Blood pH on page 276)

8.78   Define:
    a.   alkalosis (refer to A Medical Perspective: Control of Blood pH on page 276)
    b.   standard solution

Applications

8.79   For the equilibrium situation involving acetic acid,

$$CH_3COOH(aq) + H_2O(l) \rightleftharpoons CH_3COO^-(aq) + H_3O^+(aq)$$

explain the equilibrium shift occurring for the following changes:
    a.   A strong acid is added to the solution.
    b.   The solution is diluted with water.

8.80   For the equilibrium situation involving acetic acid,

$$CH_3COOH(aq) + H_2O(l) \rightleftharpoons CH_3COO^-(aq) + H_3O^+(aq)$$

explain the equilibrium shift occurring for the following changes:
    a.   A strong base is added to the solution.
    b.   More acetic acid is added to the solution.

8.81   What is $[H_3O^+]$ for a buffer solution that is $0.200\,M$ in acid and $0.500\,M$ in the corresponding salt if the weak acid $K_a = 5.80 \times 10^{-7}$?

8.82   What is the pH of the solution described in Question 8.81?

8.83   What does K_a tell us about acid strength?

8.84   What does K_b tell us about base strength?

8.85   Calculate the pH of a buffer system containing $1.0\,M$ CH_3COOH and $1.0\,M$ CH_3COONa. (K_a of acetic acid, CH_3COOH, is 1.8×10^{-5})

8.86   Calculate the pH of a buffer system containing $1.0\,M\ NH_3$ and $1.0\,M\ NH_4Cl$. (K_a of NH_4^+, the acid in this system, is 5.6×10^{-10})

8.87 The pH of blood plasma is 7.40. The principal buffer system is HCO_3^-/H_2CO_3. Calculate the ratio $[HCO_3^-]/[H_2CO_3]$ in blood plasma. (K_a of H_2CO_3, carbonic acid, is 4.5×10^{-7})

8.88 The pH of blood plasma from a patient was found to be 7.6, a life-threatening situation. Calculate the ratio $[HCO_3^-]/[H_2CO_3]$ in this sample of blood plasma. (K_a of H_2CO_3, carbonic acid, is 4.5×10^{-7})

Oxidation-Reduction Reactions

Foundations

8.89 During an oxidation process in an oxidation-reduction reaction, does the species oxidized gain or lose electrons?

8.90 During an oxidation-reduction reaction, is the oxidizing agent oxidized or reduced?

8.91 During an oxidation-reduction reaction, is the reducing agent oxidized or reduced?

8.92 Do metals tend to be good oxidizing agents or good reducing agents?

Applications

8.93 In the following reaction, identify the oxidized species, reduced species, oxidizing agent, and reducing agent:

$$Cl_2(aq) + 2KI(aq) \longrightarrow 2KCl(aq) + I_2(aq)$$

8.94 In the following reaction, identify the oxidized species, reduced species, oxidizing agent, and reducing agent:

$$Zn(s) + Cu^{2+}(aq) \longrightarrow Zn^{2+}(aq) + Cu(s)$$

8.95 Write the oxidation and reduction half-reactions for the equation in Question 8.93.

8.96 Write the oxidation and reduction half-reactions for the equation in Question 8.94.

8.97 Explain the relationship between oxidation-reduction and voltaic cells.

8.98 Compare and contrast a battery and electrolysis.

8.99 Describe one application of voltaic cells.

8.100 Describe one application of electrolytic cells.

CRITICAL THINKING PROBLEMS

1. Acid rain is a threat to our environment because it can increase the concentration of toxic metal ions, such as Cd^{2+} and Cr^{3+}, in rivers and streams. If cadmium and chromium are present in sediment as $Cd(OH)_2$ and $Cr(OH)_3$, write reactions that demonstrate the effect of acid rain. Use the library or internet to find the properties of cadmium and chromium responsible for their environmental impact.

2. Aluminum carbonate is more soluble in acidic solution, forming aluminum cations. Write a reaction (or series of reactions) that explains this observation.

3. Carbon dioxide reacts with the hydroxide ion to produce the bicarbonate anion. Write the Lewis dot structures for each reactant and product. Label each as a Brønsted acid or base. Explain the reaction using the Brønsted theory. Why would the Arrhenius theory provide an inadequate description of this reaction?

4. Maalox is an antacid composed of $Mg(OH)_2$ and $Al(OH)_3$. Explain the origin of the trade name Maalox. Write chemical reactions that demonstrate the antacid activity of Maalox.

5. Acid rain has been described as a regional problem, whereas the greenhouse effect is a global problem. Do you agree with this statement? Why or why not?

9 The Nucleus, Radioactivity, and Nuclear Medicine

Learning Goals

1 Enumerate the characteristics of alpha, beta, positron, and gamma radiation.

2 Write balanced equations for nuclear processes.

3 Calculate the amount of radioactive substance remaining after a specified period of time has elapsed.

4 Explain the process of radiocarbon dating.

5 Describe how nuclear energy can generate electricity: fission, fusion, and the breeder reactor.

6 Cite examples of the use of radioactive isotopes in medicine.

7 Describe the use of ionizing radiation in cancer therapy.

8 Discuss the preparation and use of radioisotopes in diagnostic imaging studies.

9 Explain the difference between natural and artificial radioactivity.

10 Describe the characteristics of radioactive materials that relate to radiation exposure and safety.

11 Be familiar with common techniques for the detection of radioactivity.

12 Know the common units of radiation intensity: the curie, roentgen, rad, and rem.

Outline

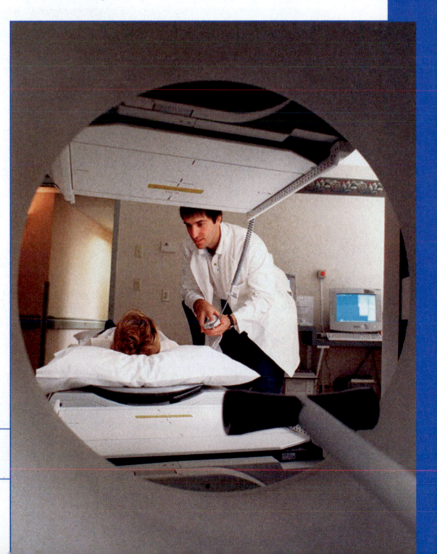

Nuclear technology has revolutionized the practice of medicine.

Introduction

Our discussion of the atom and atomic structure in Chapter 2 revealed a nucleus containing protons and neutrons surrounded by electrons. Until now, we have treated the nucleus as simply a region of positive charge in the center of the atom. The focus of our interest has been the electrons and their arrangement around the nucleus. Electron arrangement is an essential part of a discussion of bonding or chemical change.

In this chapter we consider the nucleus and nuclear properties. The behavior of nuclei may have as great an effect on our everyday lives as any of the thousands of synthetic compounds developed over the past several decades. Examples of nuclear technology range from everyday items (smoke detectors) to sophisticated instruments for medical diagnosis and treatment and electrical power generation (nuclear power plants).

Beginning in 1896 with Becquerel's discovery of radiation emitted from uranium ore, the technology arising from this and related findings has produced both risks and benefits. Although early discoveries of radioactivity and its properties expanded our fundamental knowledge and brought fame to the investigators, it was not accomplished without a price. Several early investigators died prematurely of cancer and other diseases caused by the radiation they studied.

Chemistry Connection

An Extraordinary Woman in Science

The path to a successful career in science, or any other field for that matter, is seldom smooth or straight. That was certainly true for Madame Marie Sklodowska Curie. Her lifelong ambition was to raise a family and do something interesting for a career. This was a lofty goal for a nineteenth-century woman.

The political climate in Poland, coupled with the prevailing attitudes toward women and careers, especially careers in science, certainly did not make it any easier for Mme. Curie. To support herself and her sister, she toiled at menial jobs until moving to Paris to resume her studies.

It was in Paris that she met her future husband and fellow researcher, Pierre Curie. Working with crude equipment in a laboratory that was primitive, even by the standards of the time, she and Pierre made a most revolutionary discovery only two years after Henri Becquerel discovered radioactivity. Radioactivity, the emission of energy from certain substances, was released from *inside* the atom and was independent of the molecular form of the substance. The absolute proof of this assertion came only after the Curies processed over one *ton* of a material (pitchblende) to isolate less than a gram of pure radium. The difficult conditions under which this feat was accomplished are perhaps best stated by Sharon Bertsch McGrayne in her book *Nobel Prize Women in Science* (Birch Lane Press, New York, p. 23):

The only space large enough at the school was an abandoned dissection shed. The shack was stifling hot in summer and freezing cold in winter. It had no ventilation system for removing poisonous fumes, and its roof leaked. A chemist accustomed to Germany's modern laboratories called it "a cross between a stable and a potato cellar and, if I had not seen the work table with the chemical apparatus, I would have thought it a practical joke." This ramshackle shed became the symbol of the Marie Curie legend.

The pale green glow emanating from the radium was beautiful to behold. Mme. Curie would go to the shed in the middle of the night to bask in the light of her accomplishment. She did not realize that this wonderful accomplishment would, in time, be responsible for her death.

Mme. Curie received not one, but two Nobel Prizes, one in physics and one in chemistry. She was the first woman in France to earn the rank of professor.

As you study this chapter, the contributions of Mme. Curie, Pierre Curie, and the others of that time will become even more clear. Ironically, the field of medicine has been a major beneficiary of advances in nuclear and radiochemistry, despite the toxic properties of those same radioactive materials.

Even today, the existence of nuclear energy and its associated technology is a mixed blessing. On one side, the horrors of Nagasaki and Hiroshima, the fear of nuclear war, and potential contamination of populated areas resulting from the peaceful application of nuclear energy are critical problems facing society. Conversely, hundreds of thousands of lives have been saved because of the early detection of disease such as cancer by diagnosis based on the interaction of radiation and the body and the cure of cancer using techniques such as cobalt-60 treatment. Furthermore, nuclear energy is an alternative energy source, providing an opportunity for us to compensate for the depletion of oil reserves.

9.1 Natural Radioactivity

Radioactivity is the process by which some atoms emit energy and particles. The energy and particles are termed *radiation*. Nuclear radiation occurs as a result of an alteration in nuclear composition or structure. This process occurs in a nucleus that is unstable and hence radioactive. Radioactivity is a nuclear event: *matter and energy released during this process come from the nucleus.*

We shall designate the nucleus using *nuclear symbols,* analogous to the *atomic symbols* that were discussed in Chapter 2. The nuclear symbols consist of the *symbol* of the element, the *atomic number* (the number of protons in the nucleus), and the *mass number,* which is defined as the sum of neutrons and protons in the nucleus.

With the use of nuclear symbols, the fluorine nucleus is represented as

$$\text{Mass number} \longrightarrow {}^{19}_{9}\text{F} \longleftarrow \text{Symbol of the element}$$
$$\text{Atomic number} \longrightarrow$$
$$\text{(or nuclear charge)}$$

This symbol is equivalent to writing *fluorine-19.* This alternative representation is frequently used to denote specific isotopes of elements.

Not all nuclei are unstable. Only unstable nuclei undergo change and produce radioactivity, the process of radioactive decay. Recall that different atoms of the same element having different masses exist as *isotopes.* One isotope of an element may be radioactive, whereas others of the same element may be quite stable. It is important to distinguish between the terms *isotope* and **nuclide.** The term *isotope* refers to any atoms that have the same atomic number but different mass number. The term **nuclide** refers to any atom characterized by an atomic number and a mass number.

Many elements in the periodic table occur in nature as mixtures of isotopes. Two common examples include carbon (Figure 9.1),

| ${}^{12}_{6}\text{C}$ | ${}^{13}_{6}\text{C}$ | ${}^{14}_{6}\text{C}$ |
| Carbon-12 | Carbon-13 | Carbon-14 |

and hydrogen,

${}^{1}_{1}\text{H}$	${}^{2}_{1}\text{H}$	${}^{3}_{1}\text{H}$
Hydrogen-1	Hydrogen-2	Hydrogen-3
Protium	Deuterium	Tritium
	(symbol D)	(symbol T)

Protium is a stable isotope and makes up more than 99.9% of naturally occurring hydrogen. Deuterium (D) can be isolated from hydrogen; it can form compounds such as "heavy water," D_2O. Heavy water is a potential source of deuterium for fusion processes. Tritium (T) is rare and unstable, hence radioactive.

 Animation
Radioactive Decay

Be careful not to confuse the mass number (a simple count of the neutrons and protons) with the atomic mass, which includes the contribution of electrons and is a true *mass* figure.

Isotopes are introduced in Section 2.1.

Figure 9.1

Three isotopes of carbon. Each nucleus contains the same number of protons. Only the number of neutrons is different; hence, each isotope has a different mass.

Carbon-12 has
6 protons and
6 neutrons

Carbon-13 has
6 protons and
7 neutrons

Carbon-14 has
6 protons and
8 neutrons

Alpha, beta, positron, and gamma radiation have widespread use in the field of medicine.

 Animation

The Types of Radioactive Decay

 LEARNING GOAL

Enumerate the characteristics of alpha, beta, positron, and gamma radiation.

Other radiation particles, such as neutrinos and deuterons, will not be discussed here.

In writing the symbols for a nuclear process, it is essential to indicate the particular isotope involved. This is why the mass number and atomic number are used. These values tell us the number of neutrons in the species, hence the isotope's identity.

Natural radiation emitted by unstable nuclei include *alpha particles*, *beta particles*, *positrons*, and *gamma rays*.

Alpha Particles

An **alpha particle** (α) contains two protons and two neutrons. An alpha particle is identical to the nucleus of the helium atom (He) or a *helium ion* (He^{2+}), which also contains two protons (atomic number = 2) and two neutrons (mass number − atomic number = 2). Having no electrons to counterbalance the nuclear charge, the alpha particle may be symbolized as

$$^{4}_{2}He^{2+} \quad \text{or} \quad ^{4}_{2}He \quad \text{or} \quad \alpha$$

Alpha particles have a relatively large mass compared to other nuclear particles. Consequently, alpha particles emitted by radioisotopes are relatively slow-moving particles (approximately 10% of the speed of light), and they are stopped by barriers as thin as a few pages of this book.

Beta Particles and Positrons

The **beta particle** (β), in contrast, is a fast-moving electron traveling at approximately 90% of the speed of light as it leaves the nucleus. It is formed in the nucleus by the conversion of a neutron into a proton. The beta particle is represented as

$$^{0}_{-1}e \quad \text{or} \quad ^{0}_{-1}\beta \quad \text{or} \quad \beta$$

The subscript −1 is written in the same position as the atomic number and, like the atomic number (number of protons), indicates the charge of the particle.

Beta particles are smaller and faster than alpha particles. They are more penetrating and are stopped only by more dense materials such as wood, metal, or several layers of clothing.

A *positron* has the same mass as a beta particle but carries a positive charge and is symbolized as $^{0}_{+1}e$ or $^{0}_{+1}\beta$. Positrons are produced by the conversion of a proton to a neutron in the nucleus of the isotope. The proton, in effect, loses its positive charge as well as a tiny bit of mass. This positively charged mass that is released is the positron.

Gamma Rays

Gamma rays (γ) are the most energetic part of the electromagnetic spectrum (see Section 2.2), and result from nuclear processes; in contrast, alpha radiation and

TABLE 9.1	A Summary of the Major Properties of Alpha, Beta, Positron, and Gamma Radiation					
Name and Symbol	Identity	Charge	Mass (amu)	Velocity		Penetration
Alpha (α)	Helium nucleus	+2	4.0026	5–10% of the speed of light		Low
Beta $_{-1}^{0}\beta$	Electron	−1	0.000549	Up to 90% of the speed of light		Medium
Positron $_{+1}^{0}\beta$	Electron	+1	0.000549	Up to 90% of the speed of light		Medium
Gamma (γ)	Radiant energy	0	0	Speed of light		High

beta radiation are matter. Because electromagnetic radiation has no protons, neutrons, or electrons, the symbol for a gamma ray is simply

$$\gamma$$

Gamma radiation is highly energetic and is the most penetrating form of nuclear radiation. Barriers of lead, concrete, or, more often, a combination of the two are required for protection from this type of radiation.

Properties of Alpha, Beta, Positron, and Gamma Radiation

Important properties of alpha, beta, positron, and gamma radiation are summarized in Table 9.1.

Alpha, beta, positron, and gamma radiation are collectively termed *ionizing radiation*. **Ionizing radiation** produces a trail of ions throughout the material that it penetrates. The ionization process changes the chemical composition of the material. When the material is living tissue, radiation-induced illness may result (Section 9.6).

The penetrating power of alpha radiation is very low. Damage to internal organs from this form of radiation is negligible except when an alpha particle emitter is actually ingested. Beta particles and positrons have much higher velocities than alpha particles; still, they have limited penetrating power. They cause skin and eye damage and, to a lesser extent, damage to internal organs. Shielding is required when working with beta emitters. Pregnant women must take special precautions. The great penetrating power and high energy of gamma radiation make it particularly difficult to shield. Hence, it can damage internal organs.

Anyone working with any type of radiation must take precautions. Radiation safety is required, monitored, and enforced in the United States under provisions of the Occupational Safety and Health Act (OSHA).

 Animations
Alpha, Beta, and Gamma Radiation
Alpha, Beta, and Gamma Rays

LEARNING GOAL
1 Enumerate the characteristics of alpha, beta, positron, and gamma radiation.

Question 9.1

Gamma radiation is a form of *electromagnetic radiation*. Provide examples of other forms of electromagnetic radiation.

Question 9.2

How does the energy of gamma radiation compare with that of other regions of the electromagnetic spectrum?

9.2 Writing a Balanced Nuclear Equation

Nuclear equations represent nuclear change in much the same way as chemical equations represent chemical change.

A **nuclear equation** can be used to represent the process of radioactive decay. In radioactive decay a *nuclide* breaks down, producing a *new nuclide, smaller particles, and/or energy*. The concept of mass balance, required when writing

LEARNING GOAL
2 Write balanced equations for nuclear processes.

chemical equations, is also essential for nuclear equations. When writing a balanced equation, remember that:

- the total mass on each side of the reaction arrow must be identical, and
- the sum of the atomic numbers on each side of the reaction arrow must be identical.

Alpha Decay

Consider the decay of one isotope of uranium, $^{238}_{92}U$, into thorium and an alpha particle. Because an alpha particle is lost in this process, this decay is called *alpha decay*. Examine the balanced equation for this nuclear reaction:

$$^{238}_{92}U \longrightarrow \,^{234}_{90}Th + \,^{4}_{2}He$$

Uranium-238   Thorium-234   Helium-4

The sum of the mass numbers on the right ($234 + 4 = 238$) is equal to the mass number on the left. The sum of atomic numbers on the right ($90 + 2 = 92$) is equal to the atomic number on the left.

Beta Decay

Beta decay is illustrated by the decay of one of the less-abundant nitrogen isotopes, $^{16}_{7}N$. Upon decomposition, nitrogen-16 produces oxygen-16 and a beta particle. Conceptually, a neutron = proton + electron. In beta decay, one neutron in nitrogen-16 is converted to a proton and the electron, the beta particle, is released. The reaction is represented as

$$^{16}_{7}N \longrightarrow \,^{16}_{8}O + \,^{0}_{-1}e$$

or

$$^{16}_{7}N \longrightarrow \,^{16}_{8}O + \beta$$

Note that the mass number of the beta particle is zero, because the electron includes no protons or neutrons. Sixteen nuclear particles are accounted for on both sides of the reaction arrow. Note also that the product nuclide has the same mass number as the parent nuclide but the atomic number has *increased* by one unit.

The atomic number on the left ($+7$) is counterbalanced by $[8 + (-1)]$ or ($+7$) on the right. Therefore the equation is correctly balanced.

Positron Emission

The decay of carbon-11 to a stable isotope, boron-11, is one example of positron emission.

$$^{11}_{6}C \longrightarrow \,^{11}_{5}B + \,^{0}_{1}e$$

or

$$^{11}_{6}C \longrightarrow \,^{11}_{5}B + \,^{0}_{+1}\beta$$

A **positron** has the same mass as an electron, or beta particle, but opposite ($+$) charge. In contrast to beta emission, the product nuclide has the same mass number as the parent nuclide, but the atomic number has *decreased* by one unit.

The atomic number on the left ($+6$) is counterbalanced by $[5 + (+1)]$ or ($+6$) on the right. Therefore, the equation is correctly balanced.

Gamma Production

If *gamma radiation* were the only product of nuclear decay, there would be no measurable change in the mass or identity of the radioactive nuclei. This is so because the gamma emitter has simply gone to a lower energy state. An example of an isotope that decays in this way is technetium-99m. It is described as a **metastable isotope,** meaning that it is unstable and increases its stability through gamma decay without change in the mass or charge of the isotope. The letter m is used to denote a metastable isotope. The decay equation for $^{99m}_{43}\text{Tc}$ is

$$^{99m}_{43}\text{Tc} \longrightarrow {}^{99}_{43}\text{Tc} + \gamma$$

More often, gamma radiation is produced along with other products. For example, iodine-131 decays as follows:

$$^{131}_{53}\text{I} \longrightarrow {}^{131}_{54}\text{Xe} + {}^{0}_{-1}\beta + \gamma$$

| Iodine-131 | Xenon-131 | Beta particle | Gamma ray |

This reaction may also be represented as

$$^{131}_{53}\text{I} \longrightarrow {}^{131}_{54}\text{Xe} + {}^{0}_{-1}\text{e} + \gamma$$

An isotope of xenon, a beta particle, and gamma radiation are produced.

Predicting Products of Nuclear Decay

It is possible to use a nuclear equation to predict one of the products of a nuclear reaction if the others are known. Consider the following example, in which we represent the unknown product as ?:

$$^{40}_{19}\text{K} \longrightarrow ? + {}^{0}_{-1}\text{e}$$

Step 1. The mass number of this isotope of potassium is 40. Therefore the sum of the mass number of the products must also be 40, and ? must have a mass number of 40.

Step 2. Likewise, the atomic number on the left is 19, and the sum of the unknown atomic number plus the charge of the beta particle (-1) must equal 19.

Step 3. The unknown atomic number must be 20, because [20 + (-1) = 19]. The unknown is

$$^{40}_{20}?$$

Step 4. If we consult the periodic table, the element that has atomic number 20 is calcium; therefore ? = $^{40}_{20}\text{Ca}$.

Predicting the Products of Radioactive Decay EXAMPLE 9.1

Determine the identity of the unknown product of the alpha decay of curium-245:

LEARNING GOAL
2 Write balanced equations for nuclear processes.

$$^{245}_{96}\text{Cm} \longrightarrow {}^{4}_{2}\text{He} + ?$$

Solution

Step 1. The mass number of the curium isotope is 245. Therefore the sum of the mass numbers of the products must also be 245, and ? must have a mass number of 241.

Continued—

EXAMPLE 9.1 —*Continued*

Step 2. Likewise, the atomic number on the left is 96, and the sum of the unknown atomic number plus the atomic number of the alpha particle (2) must equal 96.

Step 3. The unknown atomic number must be 94, because [94 + 2 = 96]. The unknown is

$$^{241}_{94}?$$

Step 4. Referring to the periodic table, we find that the element that has atomic number 94 is plutonium; therefore ? = $^{241}_{94}$Pu.

Practice Problem 9.1

Complete each of the following nuclear equations:

a. $^{85}_{36}$Kr $\longrightarrow$? + $^{0}_{-1}$e
b. ? $\longrightarrow$ $^{4}_{2}$He + $^{222}_{86}$Rn
c. $^{239}_{92}$U $\longrightarrow$? + $^{0}_{-1}$e
d. $^{11}_{5}$B $\longrightarrow$ $^{7}_{3}$Li + ?

For Further Practice: Questions 9.33 and 9.34.

9.3 Properties of Radioisotopes

Why are some isotopes radioactive but others are not? Do all radioactive isotopes decay at the same rate? Are all radioactive materials equally hazardous? We address these and other questions in this section.

Nuclear Structure and Stability

A measure of nuclear stability is the **binding energy** of the nucleus. The binding energy of the nucleus is the energy required to break up a nucleus into its component protons and neutrons. This must be very large, because identically charged protons in the nucleus exert extreme repulsive forces on one another. These forces must be overcome if the nucleus is to be stable. When a nuclide decays, some energy is released because the products are more stable than the parent nuclide. This released energy is the source of the high-energy radiation emitted and the basis for all nuclear technology.

Why are some isotopes more stable than others? The answer to this question is not completely clear. Evidence obtained so far points to several important factors that describe stable nuclei:

- Nuclear stability correlates with the ratio of neutrons to protons in the isotope. For example, for light atoms a neutron:proton ratio of 1 characterizes a stable atom.
- Nuclei with large numbers of protons (84 or more) tend to be unstable.
- Naturally occurring isotopes containing 2, 8, 20, 50, 82, or 126 protons or neutrons are stable. These *magic numbers* seem to indicate the presence of energy levels in the nucleus, analogous to electronic energy levels in the atom.
- Isotopes with even numbers of protons or neutrons are generally more stable than those with odd numbers of protons or neutrons.
- All isotopes (except hydrogen-1) with more protons than neutrons are unstable. However, the reverse is not true.

TABLE 9.2	Half-Lives of Selected Radioisotopes	
Name	**Symbol**	**Half-Life**
Carbon-14	$^{14}_{6}C$	5730 years
Cobalt-60	$^{60}_{27}Co$	5.3 years
Hydrogen-3	$^{3}_{1}H$	12.3 years
Iodine-131	$^{131}_{53}I$	8.1 days
Iron-59	$^{59}_{26}Fe$	45 days
Molybdenum-99	$^{99}_{42}Mo$	67 hours
Sodium-24	$^{24}_{11}Na$	15 hours
Strontium-90	$^{90}_{38}Sr$	28 years
Technetium-99m	$^{99m}_{43}Tc$	6 hours
Uranium-235	$^{235}_{92}U$	710 million years

Half-Life

The **half-life** ($t_{1/2}$) is the time required for one-half of a given quantity of a substance to undergo change. Not all radioactive isotopes decay at the same rate. The rate of nuclear decay is generally represented in terms of the half-life of the isotope. Each isotope has its own characteristic half-life that may be as short as a few millionths of a second or as long as billions of years. Half-lives of some naturally occurring and synthetic isotopes are given in Table 9.2.

The stability of an isotope is indicated by the isotope's half-life. Isotopes with short half-lives decay rapidly; they are very unstable. This is not meant to imply that substances with long half-lives are less hazardous. Often, just the reverse is true.

Imagine that we begin with 100 mg of a radioactive isotope that has a half-life of 24 hours. After one half-life, or 24 hours, 1/2 of 100 mg will have decayed to other products, and 50 mg remain. After two half-lives (48 hours), 1/2 of the remaining material has decayed, leaving 25 mg, and so forth:

$$100 \text{ mg} \xrightarrow[\substack{\text{Half-life} \\ (24 \text{ h})}]{\text{One}} 50 \text{ mg} \xrightarrow[\substack{\text{Half-life} \\ (48 \text{ h total})}]{\text{A Second}} 25 \text{ mg} \xrightarrow{\hspace{1cm}} \text{etc.}$$

Decay of a radioisotope that has a reasonably short $t_{1/2}$ is experimentally determined by following its activity as a function of time. Graphing the results produces a radioactive decay curve as shown in Figure 9.2.

The mass of any radioactive substance remaining after a period may be calculated with a knowledge of the initial mass and the half-life of the isotope, following the scheme just outlined. The general equation for this process is:

$$m_f = m_i (.5)^n$$

where m_f = final or remaining mass
  m_i = initial mass
   n = number of half-lives

LEARNING GOAL

3 Calculate the amount of radioactive substance remaining after a specified period of time has elapsed.

 Animation

Radioactive Half-Life

Refer to the discussion of radiation exposure and safety in Sections 9.6 and 9.7.

Predicting the Extent of Radioactive Decay   EXAMPLE 9.2

A 50.0-mg supply of iodine-131, used in hospitals in the treatment of hyperthyroidism, was stored for 32.4 days. If the half-life of iodine-131 is 8.1 days, how many milligrams remain?

Continued—

EXAMPLE 9.2 —*Continued*

Solution

First calculate *n*, the number of half-lives elapsed using the half-life as a conversion factor:

$$n = 32.4 \ \text{days} \times \frac{1 \ \text{half-life}}{8.1 \ \text{days}} = 4.0 \ \text{half-lives}$$

Then calculate the amount remaining:

$$50.0 \ \text{mg} \xrightarrow[\text{half-life}]{\text{first}} 25.0 \ \text{mg} \xrightarrow[\text{half-life}]{\text{second}} 12.5 \ \text{mg} \xrightarrow[\text{half-life}]{\text{third}} 6.25 \ \text{mg} \xrightarrow[\text{half-life}]{\text{fourth}} 3.13 \ \text{mg}$$

Hence, 3.13 mg of iodine-131 remain after 32.4 days.

An Alternate Strategy

Use the equation:

$$m_f = m_i(.5)^n$$

Where
m_i is the initial mass
m_f is the final mass
and
n is the half-life
Substituting:

$$m_f = 50.0 \ \text{mg}(.5)^4$$

$$m_f = 3.13 \ \text{mg of iodine-131 remain after 32.4 days.}$$

Note that both strategies produce the same answer.

Practice Problem 9.2

a. A 100.0-ng sample of sodium-24 was stored in a lead-lined cabinet for 2.5 days. How much sodium-24 remained? See Table 9.2 for the half-life of sodium-24.
b. If a patient is administered 10 ng of technetium-99m, how much will remain one day later, assuming that no technetium has been eliminated by any other process? See Table 9.2 for the half-life of technetium-99m.

For Further Practice: Questions 9.55 and 9.56.

Figure 9.2

The decay curve for the medically useful radioisotope technetium-99m. Note that the number of radioactive atoms remaining—hence the radioactivity—approaches zero.

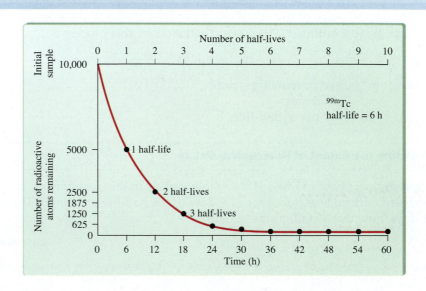

TABLE **9.3**	Isotopes Useful in Radioactive Dating		
Isotope	**Half-Life (years)**	**Upper Limit (years)**	**Dating Applications**
Carbon-14	5730	5×10^4	Charcoal, organic material, artwork
Tritium (^3_1H)	12.3	1×10^2	Aged wines, artwork
Potassium-40	1.3×10^9	Age of earth (4×10^9)	Rocks, planetary material
Rhenium-187	4.3×10^{10}	Age of earth (4×10^9)	Meteorites
Uranium-238	4.5×10^9	Age of earth (4×10^9)	Rocks, earth's crust

Radiocarbon Dating

Natural radioactivity is useful in establishing the approximate age of objects of archaeological, anthropological, or historical interest. **Radiocarbon dating** is the estimation of the age of objects through measurement of isotopic ratios of carbon.

Radiocarbon dating is based on the measurement of the relative amounts (or ratio) of $^{14}_6C$ and $^{12}_6C$ present in an object. The $^{14}_6C$ is formed in the upper atmosphere by the bombardment of $^{14}_7N$ by high-speed neutrons (cosmic radiation):

$$^{14}_7N + ^1_0n \longrightarrow ^{14}_6C + ^1_1H$$

The carbon-14, along with the more abundant carbon-12, is converted into living plant material by the process of photosynthesis. Carbon proceeds up the food chain as the plants are consumed by animals, including humans. When a plant or animal dies, the uptake of both carbon-14 and carbon-12 ceases. However, the amount of carbon-14 slowly decreases because carbon-14 is radioactive ($t_{1/2} = 5730$ years). Carbon-14 decay produces nitrogen:

$$^{14}_6C \longrightarrow ^{14}_7N + ^0_{-1}e$$

When an artifact is found and studied, the relative amounts of carbon-14 and carbon-12 are determined. By using suitable equations involving the $t_{1/2}$ of carbon-14, it is possible to approximate the age of the artifact.

This technique has been widely used to increase our knowledge about the history of the earth, to establish the age of objects (Figure 9.3), and even to detect art forgeries. Early paintings were made with inks fabricated from vegetable dyes (plant material that, while alive, metabolized carbon).

The carbon-14 dating technique is limited to objects that are less than fifty thousand years old, or approximately nine half-lives, which is a practical upper limit. Older objects that have geological or archaeological significance may be dated using naturally occurring isotopes having much longer half-lives.

Examples of useful dating isotopes are listed in Table 9.3.

Figure 9.3
Radiocarbon dating was used in the authentication study of the Shroud of Turin. It is a minimally destructive technique and is valuable in estimating the age of historical artifacts.

> **4** **LEARNING GOAL**
> Explain the process of radiocarbon dating.

9.4 Nuclear Power

Energy Production

Einstein predicted that a small amount of nuclear mass corresponds to a very large amount of energy that is released when the nucleus breaks apart. Einstein's equation is

$$E = mc^2$$

> **5** **LEARNING GOAL**
> Describe how nuclear energy can generate electricity: fission, fusion, and the breeder reactor.

An Environmental *Perspective*

Nuclear Waste Disposal

Nuclear waste arises from a variety of sources. A major source is the spent fuel from nuclear power plants. Medical laboratories generate significant amounts of low-level waste from tracers and therapy. Even household items with limited lifetimes, such as certain types of smoke detectors, use a tiny amount of radioactive material.

Virtually everyone is aware, through television and newspapers, of the problems of solid waste (nonnuclear) disposal that our society faces. For the most part, this material will degrade in some reasonable amount of time. Still, we are disposing of trash and garbage at a rate that far exceeds nature's ability to recycle it.

Now imagine the problem with nuclear waste. We cannot alter the rate at which it decays. This is defined by the half-life. We can't heat it, stir it, or add a catalyst to speed up the process as we can with chemical reactions. Furthermore, the half-lives of many nuclear waste products are very long: plutonium, for example, has a half-life in excess of 24,000 years. Ten half-lives represents the approximate time required for the radioactivity of a substance to reach background levels. So we are talking about a *very* long storage time.

Where on earth can something so very hazardous be contained and stored with reasonable assurance that it will lie undisturbed for a quarter of a million years? Perhaps this is a rhetorical question. Scientists, engineers, and politicians have debated this question for almost fifty years. As yet, no permanent disposal site has been agreed upon. Most agree that the best solution is burial in a stable rock formation, but there is no firm agreement on the location. Fear of earthquakes, which may release large quantities of radioactive materials into the underground water system, is the most serious consideration. Such a disaster could render large sections of the country unfit for habitation.

Many argue for the continuation of temporary storage sites with the hope that the progress of science and technology will, in the years ahead, provide a safer and more satisfactory long-term solution.

A photograph of the earth, taken from the moon, clearly illustrates the limits of resources and the limits to waste disposal.

The nuclear waste problem, important for its own sake, also affects the development of future societal uses of nuclear chemistry. Before we can fully enjoy its benefits, we must learn to use and dispose of it safely.

For Further Understanding

Summarize the major arguments supporting expanded use of nuclear power for electrical energy.

Enumerate the characteristics of an "ideal" solution to the nuclear waste problem.

in which

$$E = \text{energy}$$
$$m = \text{mass}$$
$$c = \text{speed of light}$$

This kinetic energy, when rapidly released, is the basis for the greatest instruments of destruction developed by humankind, nuclear bombs. However, when heat energy is released in a controlled fashion, as in a nuclear power plant, the heat energy converts liquid water into steam. The steam, in turn, drives an electrical generator, producing electricity.

Nuclear Fission

Fission (splitting) occurs when a heavy nuclear particle is split into smaller nuclei by a smaller nuclear particle (such as a neutron). This splitting process is accompanied by the release of large amounts of energy.

A nuclear power plant uses a fissionable material (capable of undergoing fission), such as uranium-235, as fuel. The energy released by the fission process in the nuclear core heats water in an adjoining chamber, producing steam. The high pressure of the steam drives a turbine, which converts this heat energy into electricity using an electric power generator. The energy transformation may be summarized as follows:

$$\text{nuclear energy} \longrightarrow \text{heat energy} \longrightarrow \text{mechanical energy} \longrightarrow \text{electrical energy}$$

| Nuclear reactor | Steam | Turbine | Electricity |

The fission reaction, once initiated, is self-perpetuating. For example, neutrons are used to initiate the reaction:

$$^{1}_{0}\text{n} + {}^{235}_{92}\text{U} \longrightarrow {}^{236}_{92}\text{U} \longrightarrow {}^{92}_{36}\text{Kr} + {}^{141}_{56}\text{Ba} + 3{}^{1}_{0}\text{n} + \text{energy}$$

| Fuel | Unstable | | Products of reaction |

Note that three neutrons are released as product for each single reacting neutron. Each of the three neutrons produced is available to initiate another fission process. Nine neutrons are released from this process. These, in turn, react with other nuclei. The fission process continues and intensifies, producing very large amounts of energy (Figure 9.4). This process of intensification is referred to as a **chain reaction.**

Animations

Nuclear Fission

Nuclear Chain Reaction

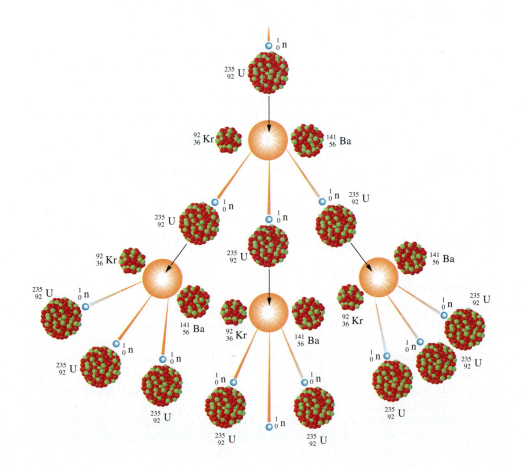

Figure 9.4

The fission of uranium-235 producing a chain reaction. Note that the number of available neutrons, which "trigger" the decomposition of the fissionable nuclei to release energy, increases at each step in the "chain." In this way the reaction builds in intensity. Control rods stabilize (or limit) the extent of the chain reaction to a safe level.

Figure 9.5

A representation of the "energy zones" of a nuclear reactor. Heat produced by the reactor core is carried by water in a second zone to a boiler. Water in the boiler (third zone) is converted to steam, which drives a turbine to convert heat energy to electrical energy. The isolation of these zones from each other allows heat energy transfer without actual physical mixing. This minimizes the transport of radioactive material into the environment.

Figure 9.6

The core of a nuclear reactor located at Oak Ridge National Laboratories in Tennessee.

 Animation

Operation of a Nuclear
        Power Plant

5 **LEARNING GOAL**

Describe how nuclear energy can generate electricity: fission, fusion, and the breeder reactor.

To maintain control over the process and to prevent dangerous overheating, rods fabricated from cadmium or boron are inserted into the core. These rods, which are controlled by the reactor's main operating system, absorb free neutrons as needed, thereby moderating the reaction.

A nuclear fission reactor may be represented as a series of energy transfer zones, as depicted in Figure 9.5. A view of the core of a fission reactor is shown in Figure 9.6.

Nuclear Fusion

Fusion (meaning *to join together*) results from the combination of two small nuclei to form a larger nucleus with the concurrent release of large amounts of energy. The best example of a fusion reactor is the sun. Continuous fusion processes furnish our solar system with light and heat.

An example of a fusion reaction is the combination of two isotopes of hydrogen, deuterium (^2_1H) and tritium (^3_1H), to produce helium, a neutron, and energy:

$$^2_1H + ^3_1H \longrightarrow ^4_2He + ^1_0n + energy$$

Although fusion is capable of producing tremendous amounts of energy, no commercially successful fusion plant exists in the United States. Safety concerns relating to problems of containment of the reaction, resulting directly from the technological problems associated with containing high temperatures (millions of degrees) and pressures required to sustain a fusion process, have slowed the development of fusion reactors.

Breeder Reactors

A **breeder reactor** is a variation of a fission reactor that literally manufactures its own fuel. A perceived shortage of fissionable isotopes makes the breeder an attractive alternative to conventional fission reactors. A breeder reactor uses $^{238}_{92}U$, which is abundant but nonfissionable. In a series of steps, the uranium-238 is converted to plutonium-239, which *is* fissionable and undergoes a fission chain reaction, producing energy. The attractiveness of a reactor that makes its own fuel from abundant starting materials is offset by the high cost of the system, potential environmental damage, and fear of plutonium proliferation. Plutonium can be readily used to manufacture nuclear bombs. Currently only France and Japan operate breeder reactors for electrical power generation.

9.5 Medical Applications of Radioactivity

The use of radiation in the treatment of various forms of cancer, as well as the newer area of **nuclear medicine,** the use of radioisotopes in diagnosis, has become widespread in the past quarter century. Let's look at the properties of radiation that make it an indispensable tool in modern medical care.

LEARNING GOAL
6 ▸ Cite examples of the use of radioactive isotopes in medicine.

Cancer Therapy Using Radiation

When high-energy radiation, such as gamma radiation, passes through a cell, it may collide with one of the molecules in the cell and cause it to lose one or more electrons, causing a series of events that result in the production of ion pairs. For this reason, such radiation is termed *ionizing radiation* (Section 9.1).

LEARNING GOAL
7 ▸ Describe the use of ionizing radiation in cancer therapy.

Ions produced in this way are highly energetic. Consequently they may damage biological molecules and cause changes in cellular biochemical processes. Interaction of ionizing radiation with intracellular water produces free electrons and other particles that can damage DNA. This may result in diminished or altered cell function or, in extreme cases, the death of the cell.

An organ that is cancerous is composed of both healthy cells and malignant cells. Tumor cells are more susceptible to the effects of gamma radiation than normal cells because they are undergoing cell division more frequently. Therefore exposure of the tumor area to carefully targeted and controlled dosages of high-energy gamma radiation from cobalt-60 (a high-energy gamma ray source) will kill a higher percentage of abnormal cells than normal cells. If the dosage is administered correctly, a sufficient number of malignant cells will die, destroying the tumor, and enough normal cells will survive to maintain the function of the affected organ.

Gamma radiation can cure cancer. Paradoxically, the exposure of healthy cells to gamma radiation can actually cause cancer. For this reason, radiation therapy for cancer is a treatment that requires unusual care and sophistication.

Nuclear Medicine

The diagnosis of a host of biochemical irregularities or diseases of the human body has been made routine through the use of radioactive tracers. Medical **tracers** are small amounts of radioactive substances used as probes to study internal organs. Medical techniques involving tracers are **nuclear imaging** procedures.

LEARNING GOAL
8 ▸ Discuss the preparation and use of radioisotopes in diagnostic imaging studies.

Animation
Nuclear Medical Techniques

A small amount of the tracer, an isotope of an element that is known to be attracted to the organ of interest, is administered to the patient. For a variety of reasons, such as ease of administration of the isotope to the patient and targeting the organ of interest, the isotope is often a part of a larger molecule or ion. Because the isotope is radioactive, its path may be followed by using suitable detection devices. A "picture" of the organ is obtained, often far more detailed than is possible with conventional X-rays. Such techniques are noninvasive; that is, surgery is not required to investigate the condition of the internal organ, eliminating the risk associated with an operation.

The radioactive isotope of an element chosen for tracer studies has chemical behavior similar to any other isotope of the same element. For example, iodine-127, the most abundant nonradioactive isotope of iodine, is used by the body in the synthesis of thyroid hormones and tends to concentrate in the thyroid gland. Both radioactive iodine-131 and iodine-127 behave in the same way, making it possible to use iodine-131 to study the thyroid. The rate of uptake of the radioactive isotope gives valuable information regarding underactivity or overactivity (hypoactive or hyperactive thyroid).

Isotopes with short half-lives are preferred for tracer studies. These isotopes emit their radiation in a more concentrated burst (short half-life materials have

A close up of a scanned image of human intestines.

TABLE 9.4	Isotopes Commonly Used in Nuclear Medicine	
Area of Body	**Isotope**	**Use**
Blood	Red blood cells tagged with chromium-51	Determine blood volume in body
Bone	*Technetium-99m, barium-131	Allow early detection of the extent of bone tumors and active sites of rheumatoid arthritis
Brain	*Technetium-99m	Detect and locate brain tumors and stroke
Coronary artery	Thallium-201	Determine the presence and location of obstructions in coronary arteries
Heart	*Technetium-99m	Determine cardiac output, size, and shape
Kidney	*Technetium-99m	Determine renal function and location of cysts; a common follow-up procedure for kidney transplant patients
Liver-spleen	*Technetium-99m	Determine size and shape of liver and spleen; location of tumors
Lung	Xenon-133	Determine whether lung fills properly; locate region of reduced ventilation and tumors
Thyroid	Iodine-131	Determine rate of iodine uptake by thyroid

*The destination of this isotope is determined by the identity of the compound in which it is incorporated.

greater activity), facilitating their detection. If the radioactive decay is easily detected, the method is more sensitive and thus capable of providing more information. Furthermore, an isotope with a short half-life decays to background more rapidly. This is a mechanism for removal of the radioactivity from the body. If the radioactive element is also rapidly metabolized and excreted, this is obviously beneficial as well.

The following examples illustrate the use of imaging procedures for diagnosis of disease.

- *Bone disease and injury.* The most widely used isotope for bone studies is technetium-99m, which is incorporated into a variety of ions and molecules that direct the isotope to the tissue being investigated. Technetium compounds containing phosphate are preferentially adsorbed on the surface of bone. New bone formation (common to virtually all bone injuries) increases the incorporation of the technetium compound. As a result, an enhanced image appears at the site of the injury. Bone tumors behave in a similar fashion.
- *Cardiovascular diseases.* Thallium-201 is used in the diagnosis of coronary artery disease. The isotope is administered intravenously and delivered to the heart muscle in proportion to the blood flow. Areas of restricted flow are observed as having lower levels of radioactivity, indicating some type of blockage.
- *Pulmonary disease.* Xenon is one of the noble gases. Radioactive xenon-133 may be inhaled by the patient. The radioactive isotope will be transported from the lungs and distributed through the circulatory system. Monitoring the distribution, as well as the reverse process, the removal of the isotope from the body (exhalation), can provide evidence of obstructive pulmonary disease, such as cancer or emphysema.

Examples of useful isotopes and the organ(s) in which they tend to concentrate are summarized in Table 9.4.

For many years, imaging with radioactive tracers was used exclusively for diagnosis. Recent applications have expanded to other areas of medicine. Imaging is now used extensively to guide surgery, assist in planning radiation therapy, and support the technique of angioplasty.

Question 9.3

Technetium-99m is used in diagnostic imaging studies involving the brain. What fraction of the radioisotope remains after 12 hours have elapsed? See Table 9.2 for the half-life of technetium-99m.

Question 9.4

Barium-131 is a radioisotope used to study bone formation. A patient ingested barium-131. How much time will elapse until only one-fourth of the barium-131 remains, assuming that none of the isotope is eliminated from the body through normal processes? The half-life of barium-131 is 11.6 minutes.

Making Isotopes for Medical Applications

In early experiments with radioactivity, the radioactive isotopes were naturally occurring. For this reason the radioactivity produced by these unstable isotopes is described as **natural radioactivity.** If, on the other hand, a normally stable, non-radioactive nucleus is made radioactive, the resulting radioactivity is termed **artificial radioactivity.** The stable nucleus is made unstable by the introduction of "extra" protons, neutrons, or both.

The process of forming radioactive substances is often accomplished in the core of a **nuclear reactor,** in which an abundance of small nuclear particles, particularly neutrons, is available. Alternatively, extremely high-velocity charged particles (such as alpha and beta particles) may be produced in **particle accelerators,** such as a cyclotron. Accelerators are extremely large and use magnetic and electric fields to "push and pull" charged particles toward their target at very high speeds. A portion of the accelerator at the Brookhaven National Laboratory is shown in Figure 9.7.

8   **LEARNING GOAL**
Discuss the preparation and use of radioisotopes in diagnostic imaging studies.

9   **LEARNING GOAL**
Explain the difference between natural and artificial radioactivity.

Figure 9.7
A portion of a linear accelerator located at Brookhaven National Laboratory in New York. Particles can be accelerated at velocities close to the speed of light and accurately strike small "target" nuclei. At such facilities, rare isotopes can be synthesized and their properties studied.

A Medical Perspective

Magnetic Resonance Imaging

The Nobel prize in physics was awarded to Otto Stern in 1943 and to Isidor Rabi in 1944. They discovered that certain atomic nuclei have a property known as spin, analogous to the spin associated with electrons which we discussed in Chapter 2. The spin of electrons is responsible for the magnetic properties of atoms. Spinning nuclei behave as tiny magnets, producing magnetic fields as well.

One very important aspect of this phenomenon is the fact that the atoms in close proximity to the spinning nuclei (its chemical environment) exert an effect on the nuclear spin. In effect, measurable differences in spin are indicators of their surroundings. This relationship has been exhaustively studied for one atom in particular, hydrogen, and magnetic resonance techniques have become useful tools for the study of molecules containing hydrogen.

Human organs and tissue are made up of compounds containing hydrogen atoms. In the 1970s and 1980s the experimental technique was extended beyond tiny laboratory samples of pure compounds to the most complex sample possible—the human body. The result of these experiments is termed *magnetic resonance imaging (MRI)*.

Dr. Paul Barnett of the Greater Baltimore Medical Center studies images obtained using MRI.

MRI is noninvasive to the body, requires no use of radioactive substances, and is quick, safe, and painless. A person is placed in a cavity surrounded by a magnetic field, and an image (based on the extent of radio frequency energy absorption) is generated, stored, and sorted in a computer. Differences between normal and malignant tissue, atherosclerotic thickening of an aortal wall, and a host of other problems may be seen clearly in the final image.

Advances in MRI technology have provided medical practitioners with a powerful tool in diagnostic medicine. This is but one more example of basic science leading to technological advancement.

A patient entering an MRI scanner.

For Further Understanding

Why is hydrogen a useful atom to study in biological systems?

Why would MRI provide minimal information about bone tissue?

Many isotopes that are useful in medicine are produced by particle bombardment. A few examples include the following:

- Gold-198, used as a tracer in the liver, is prepared by neutron bombardment.

$$^{197}_{79}\text{Au} + ^{1}_{0}\text{n} \longrightarrow ^{198}_{79}\text{Au}$$

- Gallium-67, used in the diagnosis of Hodgkin's disease, is prepared by proton bombardment.

$$^{66}_{30}\text{Zn} + ^{1}_{1}\text{p} \longrightarrow ^{67}_{31}\text{Ga}$$

Figure 9.8
Preparation of technetium-99m.
(a) A diagram depicting the conversion of $^{99}MoO_4^{2-}$ to $^{99m}TcO_4^-$ through radioactive decay. The radioactive pertechnetate ion is periodically removed from the generator in saline solution and used in tracer studies.
(b) A photograph of a commercially available technetium-99m generator suitable for use in a hospital laboratory.

Some medically useful isotopes, with short half-lives, must be prepared near the site of the clinical test. Preparation and shipment from a reactor site would waste time and result in an isotopic solution that had already undergone significant decay, resulting in diminished activity.

A common example is technetium-99m. It has a half-life of only six hours. It is prepared in a small generator, often housed in a hospital's radiology laboratory (Figure 9.8). The generator contains radioactive molybdate ion (MoO_4^{2-}). Molybdenum-99 is more stable than technetium-99m; it has a half-life of 67 hours.

The molybdenum in molybdate ion decays according to the following nuclear equation:

$$^{99}_{42}Mo \longrightarrow\ ^{99m}_{43}Tc +\ ^{0}_{-1}e$$

Chemically, radioactive molybdate MoO_4^{2-} converts to radioactive pertechnetate ion (TcO_4^-). The radioactive TcO_4^- is removed from the generator when needed. It is administered to the patient as an aqueous salt solution that has an osmotic pressure identical to that of human blood.

9.6 Biological Effects of Radiation

It is necessary to use suitable precautions in working with radioactive substances. The chosen protocol is based on an understanding of the effects of radiation, dosage levels and "tolerable levels," the way in which radiation is detected and measured, and the basic precepts of radiation safety.

Radiation Exposure and Safety

In working with radioactive materials, the following factors must be considered.

The Magnitude of the Half-Life

In considering safety, isotopes with short half-lives have, at the same time, one major disadvantage and one major advantage.

On one hand, short-half-life radioisotopes produce a larger amount of radioactivity per unit time than a long-half-life substance. For example, consider equal amounts of hypothetical isotopes that produce alpha particles. One has a half-life of ten days; the other has a half-life of one hundred days. After one half-life, each substance will produce exactly the same number of alpha particles. However, the

LEARNING GOAL

10 Describe the characteristics of radioactive materials that relate to radiation exposure and safety.

Higher levels of exposure in a short time produce clearer images.

first substance generates the alpha particles in only one-tenth of the time, hence emits ten times as much radiation per unit time. Equal exposure times will result in a higher level of radiation exposure for substances with short half-lives, and lower levels for substances with long half-lives.

On the other hand, materials with short half-lives (weeks, days, or less) may be safer to work with, especially if an accident occurs. Over time (depending on the magnitude of the half-life) radioactive isotopes will decay to **background radiation** levels. This is the level of radiation attributable to our surroundings on a day-to-day basis.

Virtually all matter is composed of both radioactive and nonradioactive isotopes. Small amounts of radioactive material in the air, water, soil, and so forth make up a part of the background levels. Cosmic rays from outer space continually bombard us with radiation, contributing to the total background. Owing to the inevitability of background radiation, there can be no situation on earth where we observe zero radiation levels.

An isotope with a short half-life, for example 5.0 min, may decay to background in as few as ten half-lives

$$10 \text{ half-lives} \times \frac{5.0 \text{ min}}{1 \text{ half-life}} = 50 \text{ min}$$

See An Environmental Perspective: Nuclear Waste Disposal on page 302.

A spill of such material could be treated by waiting ten half-lives, perhaps by going to lunch. When you return to the laboratory, the material that was spilled will be no more radioactive than the floor itself. An accident with plutonium-239, which has a half-life of 24,000 years, would be quite a different matter! After fifty minutes, virtually all of the plutonium-239 would still remain. Long-half-life isotopes, by-products of nuclear technology, pose the greatest problems for safe disposal. Finding a site that will remain undisturbed "forever" is quite a formidable task.

Question 9.5

Describe the advantage of using isotopes with short half-lives for tracer applications in a medical laboratory.

Question 9.6

Can you think of any disadvantage associated with the use of isotopes described in Question 9.5? Explain.

Shielding

Alpha and beta particles, being relatively low in penetrating power, require low-level **shielding.** A lab coat and gloves are generally sufficient protection from this low-penetration radiation. On the other hand, shielding made of lead, concrete, or both is required for gamma rays (and X-rays, which are also high-energy radiation). Extensive manipulation of gamma emitters is often accomplished in laboratory and industrial settings by using robotic control: computer-controlled mechanical devices that can be programmed to perform virtually all manipulations normally carried out by humans.

Distance from the Radioactive Source

Radiation intensity varies *inversely* with the *square* of the distance from the source. Doubling the distance from the source *decreases* the intensity by a factor of four (2^2). Again, the use of robot manipulators is advantageous, allowing a greater distance between the operator and the radioactive source.

An Environmental *Perspective*

Radon and Indoor Air Pollution

Marie and Pierre Curie first discovered that air in contact with radium compounds became radioactive. Later experiments by Ernest Rutherford and others isolated the radioactive substance from the air. This substance was an isotope of the noble gas radon (Rn).

We now know that radium (Ra) produces radon by spontaneous decay:

$$^{226}_{88}Ra \longrightarrow {}^{4}_{2}He + {}^{222}_{86}Rn$$

Radium in trace quantities is found in the soil and rock and is unequally distributed in the soil. The decay product, radon, is emitted from the soil to the surrounding atmosphere. Radon is also found in higher concentrations when uranium is found in the soil. This is not surprising, because radium is formed as a part of the stepwise decay of uranium.

If someone constructs a building over soil or rock that has a high radium content (or uses stone with a high radium content to build the foundation!), the radon gas can percolate through the basement and accumulate in the house. Couple this with the need to build more energy-efficient, well-insulated dwellings, and the radon levels in buildings in some regions of the country can become quite high.

Radon itself is radioactive; however, its radiation is not the major problem. Because it is a gas and chemically inert, it is rapidly exhaled after breathing. However, radon decays to polonium:

$$^{222}_{86}Rn \longrightarrow {}^{4}_{2}He + {}^{218}_{84}Po$$

This polonium isotope is radioactive and is a nonvolatile heavy metal that can attach itself to bronchial or lung tissue, emitting hazardous radiation and producing other isotopes that are also radioactive.

In the United States, homes are now being tested and monitored for radon. In many states, proof of acceptable levels of radon is a condition of sale of the property. Studies continue to attempt to find reasonable solutions to the radon problem. Current recommendations include sealing cracks and openings in basements, increasing ventilation, and evaluating sites

A radon sensor is often used to monitor radon levels in basements, especially in regions known to have high levels of radium in the ground.

before construction of buildings. Debate continues within the scientific community regarding a safe and attainable indoor air quality standard for radon.

For Further Understanding

Why is indoor radon more hazardous than outdoor radon?

Polonium-218 has a very long half-life. Explain why this constitutes a potential health problem.

Time of Exposure

The effects of radiation are cumulative. Generally, potential damage is directly proportional to the time of exposure. Workers exposed to moderately high levels of radiation on the job may be limited in the time that they can perform that task. For example, workers involved in the cleanup of the Three Mile Island nuclear plant, incapacitated in 1979, observed strict limits on the amount of time that they could be involved in the cleanup activities.

Types of Radiation Emitted

Alpha and beta emitters are generally less hazardous than gamma emitters, owing to differences in energy and penetrating power that require less shielding.

However, ingestion or inhalation of an alpha emitter or beta emitter can, over time, cause serious tissue damage; the radioactive substance is in direct contact with sensitive tissue. An Environmental Perspective: Radon and Indoor Air Pollution (p. 311) expands on this problem.

Waste Disposal

Virtually all applications of nuclear chemistry create radioactive waste and, along with it, the problems of safe handling and disposal. Most disposal sites, at present, are considered temporary, until a long-term safe solution can be found. Figure 9.9 conveys a sense of the enormity of the problem. Also, An Environmental Perspective: Nuclear Waste Disposal on page 302, examines this problem in more detail.

9.7 Measurement of Radiation

The changes that take place when radiation interacts with matter (such as photographic film) provide the basis of operation for various radiation detection devices.

The principal detection methods involve the use of either photographic film to create an image of the location of the radioactive substance or a counter that allows the measurement of intensity of radiation emitted from some source by converting the radiation energy to an electrical signal.

Nuclear Imaging

This approach is often used in nuclear medicine. An isotope is administered to a patient, perhaps iodine-131, which is used to study the thyroid gland, and the isotope begins to concentrate in the organ of interest. Nuclear images (photographs) of that region of the body are taken at periodic intervals using a special type of film. The emission of radiation from the radioactive substance creates the image, in much the same way as light causes the formation of images on conventional film in a camera. Upon development of the series of photographs, a record of the organ's uptake of the isotope over time enables the radiologist to assess the condition of the organ.

Computer Imaging

The coupling of rapid developments in the technology of television and computers, resulting in the marriage of these two devices, has brought about a versatile alternative to photographic imaging.

A specialized television camera, sensitive to emitted radiation from a radioactive substance administered to a patient, develops a continuous and instantaneous record of the voyage of the isotope throughout the body. The signal, transmitted to the computer, is stored, sorted, and portrayed on a monitor. Advantages include increased sensitivity, allowing a lower dose of the isotope, speed through elimination of the developing step, and versatility of application, limited perhaps only by the creativity of the medical practitioners.

A particular type of computer imaging, useful in diagnostic medicine, is the CT scanner. The CT scanner measures the interaction of X-rays with biological tissue, gathering huge amounts of data and processing the data to produce detailed information, all in a relatively short time. Such a device may be less hazardous than conventional X-ray techniques because it generates more useful information per unit of radiation. It often produces a superior image. A photograph of a CT scanner is shown in Figure 9.10, and an image of a damaged spinal bone, taken by a CT scanner, is shown in Figure 9.11.

Figure 9.9
Photograph of the construction of one-million-gallon capacity storage tanks for radioactive waste. Located in Hanford, Washington, they are now covered with 6–8 ft of earth.

11 LEARNING GOAL
Be familiar with common techniques for the detection of radioactivity.

CT represents *Computer-aided Tomography*: the computer reconstructs a series of measured images of tissue density (tomography). Small differences in tissue density may indicate the presence of a tumor.

The Geiger Counter

A Geiger counter is an instrument that detects ionizing radiation. Ions, produced by radiation passing through a tube filled with an ionizable gas, can conduct an electrical current between two electrodes. This current flow can be measured and is proportional to the level of radiation (Figure 9.12). Such devices, which were routinely used in laboratory and industrial monitoring, have been largely replaced by more sophisticated devices, often used in conjunction with a computer.

Film Badges

A common sight in any hospital or medical laboratory or any laboratory that routinely uses radioisotopes is the film badge worn by all staff members exposed in any way to low-level radioactivity.

A film badge is merely a piece of photographic film that is sensitive to energies corresponding to radioactive emissions. It is shielded from light, which would interfere, and mounted in a clip-on plastic holder that can be worn throughout the workday. The badges are periodically collected and developed. The degree of darkening is proportional to the amount of radiation to which the worker has been exposed, just as a conventional camera produces images on film in proportion to the amount of light that it "sees."

Proper record keeping thus allows the laboratory using radioactive substances to maintain an ongoing history of each individual's exposure and, at the same time, promptly pinpoint any hazards that might otherwise go unnoticed.

Units of Radiation Measurement

The amount of radiation emitted by a source or received by an individual is reported in a variety of ways, using units that describe different aspects of radiation. The *curie* and the *roentgen* describe the intensity of the emitted radiation, whereas the *rad* and the *rem* describe the biological effects of radiation.

The Curie

The **curie** is a measure of the amount of radioactivity in a radioactive source. The curie is independent of the nature of the radiation (alpha, beta, or gamma) and its effect on biological tissue. A curie is defined as the amount of radioactive material that produces 3.7×10^{10} atomic disintegrations per second.

The Roentgen

The **roentgen** is a measure of very high energy ionizing radiation (X-ray and gamma ray) only. The roentgen is defined as the amount of radiation needed to produce 2×10^9 ion pairs when passing through one cm^3 of air at 0°C. The roentgen is a measure of radiation's interaction with air and gives no information about the effect on biological tissue.

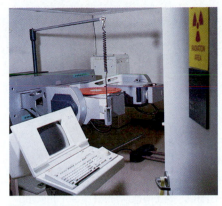

Figure 9.10
An imaging laboratory at the Greater Baltimore Medical Center.

Figure 9.11
Damage observed in a spinal bone on a CT scan image.

LEARNING GOAL

12 Know the common units of radiation intensity: the curie, roentgen, rad, and rem.

Argon gas

Thin window penetrated by radiation

Anode (+)

Cathode (−)

Amplifier and counter

High voltage

Figure 9.12
The design of a Geiger counter used for the measurement of radioactivity.

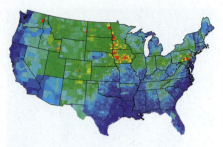

Figure 9.13
Relative yearly radiation dosages for individuals in the continental United States. Red, yellow, and green shading indicates higher levels of background radiation. Blue shading indicates regions of lower background exposure.

The Rad

The **rad,** or *radiation absorbed dosage*, provides more meaningful information than either of the previous units of measure. It takes into account the nature of the absorbing material. It is defined as the dosage of radiation able to transfer 2.4×10^{-3} cal of energy to one kg of matter.

The Rem

The **rem,** or *roentgen equivalent for man*, describes the biological damage caused by the absorption of different kinds of radiation by the human body. The *rem* is obtained by multiplication of the *rad* by a factor called the *relative biological effect (RBE)*. The RBE is a function of the type of radiation (alpha, beta, or gamma). Although a beta particle is more penetrating than an alpha particle, an alpha particle is approximately ten times more damaging to biological tissue. As a result, the RBE is ten for alpha particles and one for beta particles. Relative yearly radiation dosages received by Americans are shown in Figure 9.13.

The **lethal dose (LD$_{50}$)** of radiation is defined as the acute dosage of radiation that would be fatal for 50% of the exposed population within 30 days. An estimated lethal dose is 500 rems. Some biological effects, however, may be detectable at a level as low as 25 rem.

Question 9.7

From a clinical standpoint, what advantages does expressing radiation in rems have over the use of other radiation units?

Question 9.8

Is the roentgen unit used in the measurement of alpha particle radiation? Why or why not?

SUMMARY

9.1 Natural Radioactivity

Radioactivity is the process by which atoms emit energetic, ionizing particles or rays. These particles or rays are termed radiation. Nuclear radiation occurs because the nucleus is unstable, hence radioactive. Nuclear symbols consist of the elemental symbol, the atomic number, and the mass number.

Not all *nuclides* are unstable. Only unstable nuclides undergo change and produce radioactivity in the process of radioactive decay. Natural radiation emitted by unstable nuclei includes *alpha particles, beta particles, positrons,* and *gamma rays*. This radiation is collectively termed *ionizing radiation*.

9.2 Writing a Balanced Nuclear Equation

A *nuclear equation* represents a nuclear process such as radioactive decay. The total of the mass numbers on each side of the reaction arrow must be identical, and the sum of the atomic numbers of the reactants must equal the sum of the

atomic numbers of the products. Nuclear equations can be used to predict products of nuclear reactions.

9.3 Properties of Radioisotopes

The *binding energy* of the nucleus is a measure of nuclear stability. When an isotope decays, energy is released. Nuclear stability correlates with the ratio of neutrons to protons in the isotope. Nuclei with large numbers of protons tend to be unstable, and isotopes containing 2, 8, 20, 50, 82, or 126 protons or neutrons (magic numbers) are stable. Also, isotopes with even numbers of protons or neutrons are generally more stable than those with odd numbers of protons or neutrons.

The *half-life*, $t_{1/2}$, is the time required for one-half of a given quantity of a substance to undergo change. Each isotope has its own characteristic half-life. The degree of stability of an isotope is indicated by the isotope's half-life. Isotopes with short half-lives decay rapidly; they are very unstable.

Radiocarbon dating is based on the measurement of the relative amounts of carbon-12 and carbon-14 present in an object. The ratio of the masses of these isotopes changes slowly over time, making it useful in determining the age of objects containing carbon.

9.4 Nuclear Power

Einstein predicted that a small amount of nuclear mass would convert to a very large amount of energy when the nucleus breaks apart. *Fission* reactors are used to generate electrical power. Technological problems with *fusion* and *breeder* reactors have prevented their commercialization in the United States.

9.5 Medical Applications of Radioactivity

The use of radiation in the treatment of various forms of cancer, and in the newer area of *nuclear medicine,* has become widespread in the past quarter century.

Ionizing radiation causes changes in cellular biochemical processes that may damage or kill the cell. A cancerous organ is composed of both healthy and malignant cells. Exposure of the tumor area to controlled dosages of high-energy gamma radiation from cobalt-60 will kill a higher percentage of abnormal cells than normal cells and is a valuable cancer therapy.

The diagnosis of a host of biochemical irregularities or diseases of the human body has been made routine through the use of radioactive tracers. *Tracers* are small amounts of radioactive substances used as probes to study internal organs. Because the isotope is radioactive, its path may be followed by using suitable detection devices. A "picture" of the organ is obtained, far more detailed than is possible with conventional X-rays.

The radioactivity produced by unstable isotopes is described as *natural radioactivity*. A normally stable, non-radioactive nucleus can be made radioactive, and this is termed *artificial radioactivity* (the process produces synthetic isotopes). Synthetic isotopes are often used in clinical situations. Isotopic synthesis may be carried out in the core of a *nuclear reactor* or in a *particle accelerator*. Short-lived isotopes, such as technetium-99m, are often produced directly at the site of the clinical testing.

9.6 Biological Effects of Radiation

Safety considerations are based on the magnitude of the *half-life, shielding,* distance from the radioactive source, time of exposure, and type of radiation emitted. We are never entirely free of the effects of radioactivity. *Background radiation* is normal radiation attributable to our surroundings.

Virtually all applications of nuclear chemistry create radioactive waste and, along with it, the problems of safe handling and disposal. Most disposal sites are considered temporary, until a long-term safe solution can be found.

9.7 Measurement of Radiation

The changes that take place when radiation interacts with matter provide the basis for various radiation detection devices. Photographic imaging, computer imaging, the Geiger counter, and film badges represent the most frequently used devices for detecting and measuring radiation.

Commonly used radiation units include the *curie,* a measure of the amount of radioactivity in a radioactive source; the *roentgen,* a measure of high-energy radiation (X-ray and gamma ray); the *rad* (radiation absorbed dosage), which takes into account the nature of the absorbing material; and the *rem* (roentgen equivalent for man), which describes the biological damage caused by the absorption of different kinds of radiation by the human body. The *lethal dose* of radiation, LD_{50}, is defined as the dose that would be fatal for 50% of the exposed population within thirty days.

KEY TERMS

alpha particle (9.1)	natural radioactivity (9.5)
artificial radioactivity (9.5)	nuclear equation (9.2)
background radiation (9.6)	nuclear imaging (9.5)
beta particle (9.1)	nuclear medicine (9.5)
binding energy (9.3)	nuclear reactor (9.5)
breeder reactor (9.4)	nuclide (9.1)
chain reaction (9.4)	particle accelerator (9.5)
curie (9.7)	positron (9.2)
fission (9.4)	rad (9.7)
fusion (9.4)	radioactivity (9.1)
gamma ray (9.1)	radiocarbon dating (9.3)
half-life ($t_{1/2}$) (9.3)	rem (9.7)
ionizing radiation (9.1)	roentgen (9.7)
lethal dose (LD_{50}) (9.7)	shielding (9.6)
metastable isotope (9.2)	tracer (9.5)

QUESTIONS AND PROBLEMS

Natural Radioactivity

Foundations

9.9   Describe the meaning of the term *natural radioactivity*.
9.10 What is background radiation?
9.11 What is the composition of an alpha particle?
9.12 What is alpha decay?
9.13 What is the composition of a beta particle?
9.14 What is the composition of a positron?
9.15 In what way do beta particles and positrons differ?
9.16 In what way are beta particles and positrons similar?
9.17 What are the major differences between alpha and beta particles?
9.18 What are the major differences between alpha particles and gamma radiation?
9.19 How do nuclear reactions and chemical reactions differ?
9.20 We can control the rate of chemical reactions. Can we control the rate of natural radiation?

Applications

9.21 Write the nuclear symbol for an alpha particle.
9.22 Write the nuclear symbol for a beta particle.
9.23 Write the nuclear symbol for uranium-235.
9.24 How many protons and neutrons are contained in the nucleus of uranium-235?

9.25 How many protons and neutrons are contained in each of the three isotopes of hydrogen?

9.26 How many protons and neutrons are contained in each of the three isotopes of carbon?

9.27 Write the nuclear symbol for nitrogen-15.

9.28 Write the nuclear symbol for carbon-14.

9.29 Compare and contrast the three major types of radiation produced by nuclear decay.

9.30 Rank the three major types of radiation in order of size, speed, and penetrating power.

9.31 How does an α particle differ from a helium atom?

9.32 What is the major difference between β and δ radiation?

Writing a Balanced Nuclear Equation

Foundations

9.33 Write a nuclear reaction to represent cobalt-60 decaying to nickel-60 plus a beta particle plus a gamma ray.

9.34 Write a nuclear reaction to represent radium-226 decaying to radon-222 plus an alpha particle.

9.35 Complete the following nuclear reaction:

$$^{23}_{11}Na + {}^{2}_{1}H \longrightarrow ? + {}^{1}_{1}H$$

9.36 Complete the following nuclear reaction:

$$^{238}_{92}U + {}^{14}_{7}N \longrightarrow ? + 6{}^{1}_{0}n$$

9.37 Complete the following nuclear reaction:

$$^{24}_{10}Ne \longrightarrow \beta + ?$$

9.38 Complete the following nuclear reaction:

$$^{190}_{78}Pt \longrightarrow \alpha + ?$$

9.39 Complete the following nuclear reaction:

$$? \longrightarrow {}^{140}_{56}Ba + {}^{0}_{-1}e$$

9.40 Complete the following nuclear reaction:

$$? \longrightarrow {}^{214}_{90}Th + {}^{4}_{2}He$$

Applications

9.41 Element 107 was synthesized by bombarding bismuth-209 with chromium-54. Write the equation for this process if one product is a neutron.

9.42 Element 109 was synthesized by bombarding bismuth-209 with iron-58. Write the equation for this process if one product is a neutron.

9.43 Write a balanced nuclear equation for beta emission by magnesium-27.

9.44 Write a balanced nuclear equation for alpha decay of bismuth-212.

9.45 Write a balanced nuclear equation for position emission by nitrogen-12.

9.46 Write a balanced nuclear equation for the formation of polonium-206 through alpha decay.

Properties of Radioisotopes

Foundations

9.47 What is the difference between natural radioactivity and artificial radioactivity?

9.48 Is the fission of uranium-235 an example of natural or artificial radioactivity?

9.49 Summarize the major characteristics of nuclei for which we predict a high degree of stability.

9.50 Explain why the binding energy of a nucleus is expected to be large.

Applications

9.51 Would you predict oxygen-20 to be stable? Explain your reasoning.

9.52 Would you predict cobalt-59 to be stable? Explain your reasoning.

9.53 Would you predict chromium-48 to be stable? Explain your reasoning.

9.54 Would you predict lithium-9 to be stable? Explain your reasoning.

9.55 If 3.2 mg of the radioisotope iodine-131 is administered to a patient, how much will remain in the body after 24 days, assuming that no iodine has been eliminated from the body by any other process? (See Table 9.2 for the half-life of iodine-131.)

9.56 A patient receives 9.0 ng of a radioisotope with a half-life of 12 hours. How much will remain in the body after 2.0 days, assuming that radioactive decay is the only path for removal of the isotope from the body?

9.57 A sample containing 1.00×10^2 mg of iron-59 is stored for 135 days. What mass of iron-59 will remain at the end of the storage period? (See Table 9.2 for the half-life of iron-59.)

9.58 An instrument for cancer treatment containing a cobalt-60 source was manufactured in 1978. In 1995 it was removed from service and, in error, was buried in a landfill with the source still in place. What percentage of its initial radioactivity will remain in the year 2010? (See Table 9.2 for the half-life of cobalt-60.)

9.59 The half-life of molybdenum-99 is 67 hr. A 200 μg quantity decays, over time, to 25 μg. How much time has elapsed?

9.60 The half-life of strontium-87 is 2.8 hr. What percentage of this isotope will remain after 8 hours and 24 minutes?

9.61 Describe the process used to determine the age of the wooden coffin of King Tut.

9.62 What property of carbon enables us to assess the age of a painting?

Nuclear Power

Foundations

9.63 Which type of nuclear process splits nuclei to release energy?

9.64 Which type of nuclear process combines small nuclei to release energy?

9.65 a. Describe the process of fission.
b. How is this reaction useful as the basis for the production of electrical energy?

9.66 a. Describe the process of fusion.
b. How could this process be used for the production of electrical energy?

Applications

9.67 Write a balanced nuclear equation for a fusion reaction.

9.68 What are the major disadvantages of a fission reactor for electrical energy production?

9.69 What is meant by the term *breeder reactor?*

9.70 What are the potential advantages and disadvantages of breeder reactors?

9.71 Describe what is meant by the term *chain reaction.*

9.72 Why are cadmium rods used in a fission reaction?

9.73 What is the greatest barrier to development of fusion reactors?

9.74 What type of nuclear reaction fuels our solar system?

Medical Applications of Radioactivity

Foundations

9.75 Why is radiation therapy an effective treatment for certain types of cancer?

9.76 Describe how medically useful isotopes may be prepared.

9.77 What is the source of background radiation?

9.78 Why do high-altitude jet flights increase a person's exposure to background radiation?

Applications

9.79 The isotope indium-111 is used in medical laboratories as a label for blood platelets. To prepare indium-111, silver-108 is bombarded with an alpha particle, forming an intermediate isotope of indium. Write a nuclear equation for the process, and identify the intermediate isotope of indium.

9.80 Radioactive molybdenum-99 is used to produce the tracer isotope, technetium-99m. Write a nuclear equation for the formation of molybdenum-99 from stable molybdenum-98 bombarded with neutrons.

9.81 Describe an application of each of the following isotopes:
   a. technetium-99m
   b. xenon-133

9.82 Describe an application of each of the following isotopes:
   a. iodine-131
   b. thallium-201

Answer Questions 9.83 through 9.90 based on the assumption that you are employed by a clinical laboratory that prepares radioactive isotopes for medical diagnostic tests. Consider α, β, positron, and γ emission.

9.83 What would be the effect on your level of radiation exposure if you increase your average distance from a radioactive source?

9.84 Would wearing gloves have any significant effect? Why?

9.85 Would limiting your time of exposure have a positive effect? Why?

9.86 Would wearing a lab apron lined with thin sheets of lead have a positive effect? Why?

9.87 Would the use of robotic manipulation of samples have an effect? Why?

9.88 Would the use of concrete rather than wood paneling help to protect workers in other parts of the clinic? Why?

9.89 Would the thickness of the concrete in Question 9.88 be an important consideration? Why?

9.90 Suggest a protocol for radioactive waste disposal.

Measurement of Radiation

Foundations

9.91 What is meant by the term *relative biological effect*?

9.92 What is meant by the term *lethal dose* of radiation?

9.93 Define each of the following units:
   a. curie
   b. roentgen

9.94 Define each of the following radiation units:
   a. rad
   b. rem

Applications

9.95 X-ray technicians often wear badges containing photographic film. How is this film used to indicate exposure to X-rays?

9.96 Why would a Geiger counter be preferred to film for assessing the immediate danger resulting from a spill of some solution containing a radioisotope?

CRITICAL THINKING PROBLEMS

1. Isotopes used as radioactive tracers have chemical properties that are similar to those of a nonradioactive isotope of the same element. Explain why this is a critical consideration in their use.

2. A chemist proposes a research project to discover a catalyst that will speed up the decay of radioactive isotopes that are waste products of a medical laboratory. Such a discovery would be a potential solution to the problem of nuclear waste disposal. Critique this proposal.

3. A controversial solution to the disposal of nuclear waste involves burial in sealed chambers far below the earth's surface. Describe potential pros and cons of this approach.

4. What type of radioactive decay is favored if the number of protons in the nucleus is much greater than the number of neutrons? Explain.

5. If the proton-to-neutron ratio in Question 4 (above) were reversed, what radioactive decay process would be favored? Explain.

6. Radioactive isotopes are often used as "tracers" to follow an atom through a chemical reaction, and the following is an example. Acetic acid reacts with methyl alcohol by eliminating a molecule of water to form methyl acetate. Explain how you would use the radioactive isotope oxygen-18 to show whether the oxygen atom in the water product comes from the —OH of the acid or the —OH of the alcohol.

$$\underset{\text{Acetic acid}}{H_3C-\overset{\overset{\displaystyle O}{\|}}{C}-OH} + \underset{\text{Methyl alcohol}}{HOCH_3} \longrightarrow \underset{\text{Methyl acetate}}{H_3C-\overset{\overset{\displaystyle O}{\|}}{C}-O-CH_3} + H_2O$$

10 An Introduction to Organic Chemistry
The Saturated Hydrocarbons

Learning Goals

1. Compare and contrast organic and inorganic compounds.

2. Draw structures that represent each of the families of organic compounds.

3. Write the names and draw the structures of the common functional groups.

4. Write condensed and structural formulas for saturated hydrocarbons.

5. Describe the relationship between the structure and physical properties of saturated hydrocarbons.

6. Use the basic rules of the I.U.P.A.C. Nomenclature System to name alkanes and substituted alkanes.

7. From the I.U.P.A.C. name of an alkane or substituted alkane, be able to draw the structure.

8. Draw constitutional (structural) isomers of simple organic compounds.

9. Write the names and draw the structures of simple cycloalkanes.

10. Draw *cis* and *trans* isomers of cycloalkanes.

11. Describe conformations of alkanes.

12. Draw the chair and boat conformations of cyclohexane.

13. Write equations for combustion reactions of alkanes.

14. Write equations for halogenation reactions of alkanes.

Outline

Introduction

Chemistry Connection:
The Origin of Organic Compounds

10.1 The Chemistry of Carbon

An Environmental Perspective:
Frozen Methane: Treasure or Threat?

10.2 Alkanes

An Environmental Perspective:
Oil-Eating Microbes

10.3 Cycloalkanes

10.4 Conformations of Alkanes and Cycloalkanes

An Environmental Perspective:
The Petroleum Industry and Gasoline Production

10.5 Reactions of Alkanes and Cycloalkanes

A Medical Perspective:
Polyhalogenated Hydrocarbons Used as Anesthetics

A Medical Perspective:
Chloroform in Your Swimming Pool?

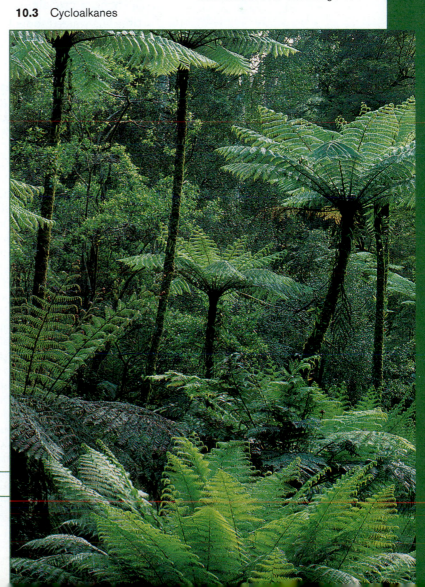

The origins of fossil fuels.

Introduction

Organic chemistry is the study of carbon-containing compounds. The term organic was coined in 1807 by the Swedish chemist Jöns Jakob Berzelius. At that time it was thought that all organic compounds, such as fats, sugars, coal, and petroleum, were formed by living or once living organisms. All early attempts to synthesize these compounds in the laboratory failed, and it was thought that a vital force, available only in living cells, was needed for their formation.

This idea began to change in 1828 when a twenty-seven-year-old German physician, whose first love was chemistry, synthesized the organic molecule urea from inorganic starting materials. This man was Friedrich Wöhler, the "father of organic chemistry."

As a child, Wöhler didn't do particularly well in school because he spent so much time doing chemistry experiments at home. Eventually, he did earn his medical degree, but he decided to study chemistry in the laboratory of Berzelius rather than practice medicine.

After a year he returned to Germany to teach and, as it turned out, to do the experiment that made him famous. The goal of the experiment was to prepare ammonium cyanate from a mixture of potassium cyanate and ammonium sulfate. He heated a solution of the two salts and crystallized the product. But the product didn't look like ammonium cyanate. It was a white crystalline material that looked exactly like urea! Urea is a waste product of protein breakdown in the body and is excreted in the urine.

Chemistry Connection

The Origin of Organic Compounds

About 425 million years ago, mountain ranges rose, and enormous inland seas emptied, producing new and fertile lands. In the next 70 million years the simple aquatic plants evolved into land plants, and huge forests of ferns, trees, and shrubs flourished. Reptiles roamed the forests. During the period between 360 and 280 million years ago the seas rose and fell at least fifty times. During periods of flood, the forests were buried under sediments. When the seas fell again, the forests were reestablished. The cycle was repeated over and over. Each flood period deposited a new layer of peat—partially decayed, sodden, compressed plant matter. These layers of peat were compacted by the pressure of the new sediments forming above them. Much of the sulfur and hydrogen was literally squeezed out of the peat, increasing the percentage of carbon. Slowly, the peat was compacted into seams of coal, which is 55–95% carbon. Oil, consisting of a variety of hydrocarbons, formed on the bottoms of ancient oceans from the remains of marine plants and animals.

Together, coal and oil are the "fossil fuels" that we use to generate energy for transportation, industry, and our homes. In the last two centuries we have extracted many of the known coal reserves from the earth and have become ever more dependent on the world's oil reserves. Coal and oil are products of the chemical reactions of photosynthesis that occurred over millions of years of the earth's history. Our society must recognize that they are nonrenewable resources. We must actively work to conserve the supply that remains and to develop alternative energy sources for the future.

In this chapter we take a closer look at the structure and properties of the hydrocarbons, such as those that make up oil. In this and later chapters we will study the amazing array of organic molecules (molecules made up of carbon, hydrogen, and a few other elements), many of which are essential to life. As we will see, all the structural and functional molecules of the cell, including the phospholipids that make up the cell membrane and the enzymes that speed up biological reactions, are organic molecules. Smaller organic molecules, such as the sugars glucose and fructose, are used as fuel by our cells, whereas others, such as penicillin and aspirin, are useful in the treatment of disease. All these organic compounds, and many more, are the subject of the remaining chapters of this text.

Wöhler recognized urea crystals because he had previously purified them from the urine of dogs and humans. Excited about his accidental discovery, he wrote to his teacher and friend Berzelius, "I can make urea without the necessity of a kidney, or even an animal, whether man or dog."

$$NH_4^+ \ [N{=}C{=}O]^-$$

Ammonium cyanate
(inorganic salt)

$$\underset{H_2N \quad\quad NH_2}{\overset{\overset{\textstyle O}{\|}}{C}}$$

Urea
(organic compound)

Ironically, Wöhler, the first person to synthesize an organic compound from inorganic substances, devoted the rest of his career to inorganic chemistry. However, other chemists continued this work, and as a result, the "vital force theory" was laid to rest, and modern organic chemistry was born.

10.1 The Chemistry of Carbon

The number of possible carbon-containing compounds is almost limitless. The importance of these organic compounds is reflected in the fact that over half of this book is devoted to the study of molecules made with this single element.

Why are there so many organic compounds? There are several reasons. First, carbon can form *stable, covalent* bonds with other carbon atoms. Consider three of the *allotropic forms* of elemental carbon: graphite, diamond, and buckminsterfullerene. Models of these allotropes are shown in Figure 10.1.

Graphite consists of planar layers in which all carbon-to-carbon bonds extend in two dimensions. Because the planar units can slide over one another, graphite is an excellent lubricant. In contrast, diamond consists of a large, three-dimensional network of carbon-to-carbon bonds. As a result, it is an extremely hard substance used in jewelry and cutting tools.

The third allotropic form of carbon is buckminsterfullerene, affectionately called the *buckey ball*. The buckey ball consists of sixty carbon atoms in the shape of a soccer ball. Discovered in the 1980s, buckminsterfullerene was named for Buckminster Fuller, who used such shapes in the design of geodesic domes.

A second reason for the vast number of organic compounds is that carbon atoms can form stable bonds with other elements. Several families of organic compounds (alcohols, aldehydes, ketones, esters, and ethers) contain oxygen atoms

Coal is a rock composed largely of carbon, with some other impurities. Investigate the conditions required to convert coal into a diamond.

Allotropes are forms of an element that have the same physical state but different properties.

(a) Graphite

(b) Diamond

(c) Buckminsterfullerene

Figure 10.1
Three allotropic forms of elemental carbon.

In future chapters we will discuss families of organic molecules containing oxygen atoms (alcohols, aldehydes, ketones, carboxylic acids, ethers, and esters), nitrogen atoms (amides and amines), sulfur atoms, and halogen atoms.

LEARNING GOAL

1 ▷ Compare and contrast organic and inorganic compounds.

Explain why the butane found in the lighter is a liquid only when maintained under pressure.

Polar covalent compounds, such as HCl, dissociate in water and, thus, are electrolytes. Carboxylic acids, the family of organic compounds we will study in Chapter 14, are weak electrolytes when dissolved in water.

bonded to carbon. Others contain nitrogen, sulfur, or halogens. The presence of these elements confers a wide variety of new chemical and physical properties on an organic compound.

Third, carbon can form double or triple bonds with other carbon atoms to produce a variety of organic molecules with very different properties. Finally, the number of ways in which carbon and other atoms can be arranged is nearly limitless. In addition to linear chains of carbon atoms, ring structures and branched chains are common. Two organic compounds may even have the same number and kinds of atoms but completely different structures and thus different properties. Such organic molecules are called *isomers*.

Important Differences Between Organic and Inorganic Compounds

The bonds between carbon and another atom are almost always *covalent bonds*, whereas the bonds in many inorganic compounds are *ionic bonds*. Differences between these two types of bonding are responsible for most of the differences between inorganic and organic substances. Ionic bonds result from the *transfer* of one or more electrons from one atom to another. Thus, ionic bonds are electrostatic, resulting from the attraction between the positive and negative ions formed by the electron transfer. Covalent bonds are formed by sharing one or more pairs of electrons.

Ionic compounds often form three-dimensional crystals made up of many positive and negative ions. Covalent compounds exist as individual units called molecules. Water-soluble ionic compounds often dissociate in water to form ions and are called electrolytes. Most covalent compounds are nonelectrolytes, keeping their identity in solution.

As a result of these differences, ionic substances usually have much higher melting and boiling points than covalent compounds. They are more likely to dissolve in water than in a less-polar solvent, whereas organic compounds, which are typically nonpolar or only moderately polar, are less soluble, or insoluble in water. In Table 10.1, the physical properties of the organic compound butane are compared with those of sodium chloride, an inorganic compound of similar formula weight.

TABLE 10.1	Comparison of the Major Properties of a Typical Organic and an Inorganic Compound: Butane Versus Sodium Chloride	
Property	**Organic Compounds (e.g., Butane)**	**Inorganic Compounds (e.g., Sodium Chloride)**
Formula weight	58	58.5
Bonding	Covalent (C_4H_{10})	Ionic (Na^+ and Cl^- ions)
Physical state at room temperature and atmospheric pressure	Gas	Solid
Boiling point	Low (−0.5°C)	High (1413°C)
Melting point	Low (−139°C)	High (801°C)
Solubility in water	Insoluble	High (36 g/100 mL)
Solubility in organic solvents (e.g., hexane)	High	Insoluble
Flammability	Flammable	Nonflammable
Electrical conductivity	Nonconductor	Conducts electricity in solution and in molten liquid

An Environmental Perspective

Frozen Methane: Treasure or Threat?

Methane is the simplest hydrocarbon, but it has some unusual behaviors. One of these is the ability to form a clathrate, which is an unusual type of matter in which molecules of one substance form a cage around molecules of another substance. For instance, water molecules can form a latticework around methane molecules to form frozen methane hydrate, possibly one of the biggest reservoirs of fossil fuel on earth.

Typically we wouldn't expect a nonpolar molecule, such as methane, to interact with a polar molecule, such as water. So, then, how is this structure formed? As we have studied earlier, water molecules interact with one another by strong hydrogen bonding. In the frozen state, these hydrogen-bonded water molecules form an open latticework. The nonpolar methane molecule is simply trapped inside one of the spaces within the lattice.

Frozen methane is found on the ocean floor. Formed by animals and decaying plant life, there are large pockets of oil and natural gas all over the earth. Methane hydrate forms when methane from one of these pockets under the ocean seeps up through the sea sediments. When the gas reaches the ocean floor, it expands and freezes. Vast regions of the ocean floor are covered by such ice fields.

Fascinating communities of methanogens, organisms that can use methane for a food source, have developed on these ice fields. These creatures live under great pressures, at extremely low temperatures, and with no light. But these unusual communities are not the major focus of interest in the methane ice fields. Scientists would like to "mine" this ice to use the methane as a fuel. In fact, the U.S. Geological Survey estimates that the amount of methane hydrate in the United States is worth over two hundred times the conventional natural gas resources in this country!

But is it safe to harvest the methane from this ice? Caution will certainly be required. Methane is flammable, and, like carbon dioxide, it is a greenhouse gas. In fact, it is about twenty times more efficient at trapping heat than carbon dioxide. (Greenhouse gases are discussed in greater detail in An Environmental Perspective: The Greenhouse Effect and Global Climate Change in Chapter 5.) So the U.S. Department of Energy, which is working with industry to develop ways to harvest the methane, must figure out how to do that without releasing much into the atmosphere where it could intensify global warming.

It may be that a huge release of methane from these frozen reserves was responsible for a major global warming that occurred fifty-five million years ago and lasted for one hundred thousand years. NASA scientists using computer simulations hypothesize that a shift of the continental plates may have released vast amounts of methane gas from the ocean floor. This methane raised the temperature of earth by about 13°F. In fact the persistence of the methane in the atmosphere

Three-dimensional structure of methane hydrate.

A sample of frozen methane hydrate.

warmed earth enough to melt the ice in the oceans and at polar caps and completely change the global climate. This theory, if it turns out to be true, highlights the importance of controlling the amount of methane, as well as carbon dioxide, that we release into the air. Certainly, harvesting the frozen methane of the oceans, if we chose to do it, must be done with great care.

For Further Understanding

News stories alternatively describe frozen methane as a "New Frontier" and "Armageddon." Explain these opposing views.

What are the ethical considerations involved in mining for frozen methane?

Question 10.1

A student is presented with a sample of an unknown substance and asked to classify the substance as organic or inorganic. What tests should the student carry out?

Question 10.2

What results would the student expect if the sample in Question 10.1 were an inorganic compound? What results would the student expect if it were an organic compound?

Families of Organic Compounds

> **LEARNING GOAL**
>
> **2** Draw structures that represent each of the families of organic compounds.

The most general classification of organic compounds divides them into hydrocarbons and substituted hydrocarbons. A **hydrocarbon** molecule contains only carbon and hydrogen. A **substituted hydrocarbon** is one in which one or more hydrogen atoms is replaced by another atom or group of atoms.

The hydrocarbons can be further subdivided into aliphatic and aromatic hydrocarbons (Figure 10.2). The four families of **aliphatic hydrocarbons** are the alkanes, cycloalkanes, alkenes, and alkynes.

Alkanes contain only carbon-to-carbon single bonds (C—C); alkenes have at least one carbon-to-carbon double bond (C=C); and alkynes have at least one carbon-to-carbon triple bond (C≡C).

Alkanes are **saturated hydrocarbons** because they contain only carbon and hydrogen and have only carbon-to-hydrogen and carbon-to-carbon single bonds. The alkenes and alkynes are **unsaturated hydrocarbons** because they contain at least one carbon-to-carbon double or triple bond, respectively.

Recall that each of the lines in these structures represents a shared pair of electrons. See Chapter 3.

Saturated Hydrocarbon   Unsaturated Hydrocarbon

Some hydrocarbons are cyclic. Cycloalkanes consist of carbon atoms bonded to one another to produce a ring. **Aromatic hydrocarbons** contain a benzene ring or a derivative of the benzene ring.

We will learn a more accurate way to represent benzene in Section 11.6.

A Cycloalkane
(Cyclohexane)

Benzene

Figure 10.2

The family of hydrocarbons is divided into two major classes: aliphatic and aromatic. The aliphatic hydrocarbons are further subdivided into four major subclasses: alkanes, cycloalkanes, alkenes, and alkynes.

TABLE 10.2 Common Functional Groups

Type of compound	Structural formula	Condensed formula	Structural formula	Example I.U.P.A.C. name	Common name
Alcohol	R—O—H	ROH	CH_3CH_2—O—H	Ethanol	Ethyl alcohol
Aldehyde	R—C̈—H (=O)	RCHO	CH_3C̈—H (=O)	Ethanal	Acetaldehyde
Amide	R—C̈—N—H (=O, with H)	$RCONH_2$	CH_3C̈—N—H (=O, with H)	Ethanamide	Acetamide
Amine	R—N—H (with H)	RNH_2	CH_3CH_2N—H (with H)	Aminoethane	Ethyl amine
Carboxylic acid	R—C̈—O—H (=O)	RCOOH	CH_3C̈—O—H (=O)	Ethanoic acid	Acetic acid
Ester	R—C̈—O—R' (=O)	RCOOR'	CH_3C̈—OCH_3 (=O)	Methyl ethanoate	Methyl acetate
Ether	R—O—R'	ROR'	CH_3OCH_3	Methoxymethane	Dimethyl ether
Halide	R—Cl (or —Br, —F, —I)	RCl	CH_3CH_2Cl	Chloroethane	Ethyl chloride
Ketone	R—C̈—R' (=O)	RCOR'	CH_3C̈CH_3 (=O)	Propanone	Acetone
Alkene	(R)(R)C=C(H)(H)	RCHCHR	$CH_3CH=CH_2$	Propene	Propylene
Alkyne	R—C≡C—R	RCCR	$CH_3C≡CH$	Propyne	Methyl acetylene

A substituted hydrocarbon is produced by replacing one or more hydrogen atoms with a functional group. A **functional group** is an atom or group of atoms arranged in a particular way that is primarily responsible for the chemical and physical properties of the molecule in which it is found. The importance of functional groups becomes more apparent when we consider that hydrocarbons have little biological activity. However, the addition of a functional group confers unique and interesting properties that give the molecule important biological or medical properties.

All compounds that have a particular functional group are members of the same family. We have just seen that the alkenes are characterized by the presence of carbon-to-carbon double bonds. Similarly, all alcohols contain a hydroxyl group (-OH). Since this group is polar, an alcohol such as ethanol (CH_3CH_2OH) is liquid at room temperature and highly soluble in water, while the alkane of similar molecular weight, propane ($CH_3CH_2CH_3$), is a gas at room temperature and is completely insoluble in water. Other common functional groups are shown in Table 10.2, along with an example of a molecule from each family.

The chemistry of organic and biological molecules is usually controlled by the functional group found in the molecule. Just as members of the same family of the periodic table exhibit similar chemistry, organic molecules with the same functional group exhibit similar chemistry. Although it would be impossible to learn the chemistry of each organic molecule, it is relatively easy to learn the chemistry of each functional group. In this way you can learn the chemistry of all members of a family of organic compounds, or biological molecules, just by learning the chemistry of its characteristic functional group or groups.

LEARNING GOAL

3 Write the names and draw the structures of the common functional groups.

This is analogous to the classification of the elements within the periodic table. See Chapter 2.

10.2 Alkanes

Alkanes are saturated hydrocarbons; that is, alkanes contain only carbon and hydrogen bonded together through carbon-hydrogen and carbon-carbon single bonds. C_nH_{2n+2} is the general formula for alkanes. In this formula, n is the number of carbon atoms in the molecule.

Structure and Physical Properties

Four types of formulas, each providing different information, are used in organic chemistry: the molecular formula, the structural formula, the condensed formula, and the line formula.

The **molecular formula** tells the kind and number of each type of atom in a molecule but does not show the bonding pattern. Consider the molecular formulas for simple alkanes:

CH_4 C_2H_6 C_3H_8 C_4H_{10}
Methane   Ethane   Propane   Butane

For the first three compounds, there is only one possible arrangement of the atoms. However, for C_4H_{10} there are two possible arrangements. How do we know which is correct? The problem is solved by using the **structural formula,** which shows each atom and bond in a molecule. The following are the structural formulas for methane, ethane, propane, and the two isomers of butane:

The advantage of a structural formula is that it shows the complete structure, but for large molecules it is time-consuming to draw and requires too much space. The compromise is the **condensed formula.** It shows all the atoms in a molecule and places them in a sequential order that indicates which atoms are bonded to which. The following are the condensed formulas for the preceding five compounds.

CH_4 CH_3CH_3 $CH_3CH_2CH_3$ $CH_3(CH_2)_2CH_3$ $(CH_3)_3CH$
Methane   Ethane   Propane   Butane   Methylpropane
                                                            (isobutane)

The names and formulas of the first ten straight-chain alkanes are shown in Table 10.3.

The simplest representation of a molecule is the **line formula.** In the line formula we assume that there is a carbon atom at any location where two or more lines intersect. We also assume that there is a carbon at the end of any line and that each carbon in the structure is bonded to the correct number of hydrogen atoms. Compare the structural and line formulas for butane and methylpropane, shown here:

H H H H
| | | |
H—C—C—C—C—H = ∧∨   Butane
| | | |
H H H H

                              H
                              |
                          H—C—H
                     H | H
                     | | |
                  H—C—C—C—H = ⅄   Methylpropane
                     | | |
                     H   H   H

TABLE 10.3   Names and Formulas of the First Ten Straight-Chain Alkanes

Name	Molecular Formula	Condensed Formula	Melting Point, °C*	Boiling Point, °C*
Alkanes	C_nH_{2n+2}			
Methane	CH_4	CH_4	−182.5	−162.2
Ethane	C_2H_6	CH_3CH_3	−183.9	−88.6
Propane	C_3H_8	$CH_3CH_2CH_3$	−187.6	−42.1
Butane	C_4H_{10}	$CH_3CH_2CH_2CH_3$ or $CH_3(CH_2)_2CH_3$	−137.2	−0.5
Pentane	C_5H_{12}	$CH_3CH_2CH_2CH_2CH_3$ or $CH_3(CH_2)_3CH_3$	−129.8	36.1
Hexane	C_6H_{14}	$CH_3CH_2CH_2CH_2CH_2CH_3$ or $CH_3(CH_2)_4CH_3$	−95.2	68.8
Heptane	C_7H_{16}	$CH_3CH_2CH_2CH_2CH_2CH_2CH_3$ or $CH_3(CH_2)_5CH_3$	−90.6	98.4
Octane	C_8H_{18}	$CH_3CH_2CH_2CH_2CH_2CH_2CH_2CH_3$ or $CH_3(CH_2)_6CH_3$	−56.9	125.6
Nonane	C_9H_{20}	$CH_3CH_2CH_2CH_2CH_2CH_2CH_2CH_2CH_3$ or $CH_3(CH_2)_7CH_3$	−53.6	150.7
Decane	$C_{10}H_{22}$	$CH_3CH_2CH_2CH_2CH_2CH_2CH_2CH_2CH_2CH_3$ or $CH_3(CH_2)_8CH_3$	−29.8	174.0

*Melting and boiling points as reported in the National Institute of Standards and Technology Chemistry Webbook, which can be found at http://webbook.nist.gov/.

(a)   (b)   (c)   (d)   (e)

Figure 10.3

The tetrahedral carbon atom: (a) Lewis dot structure; (b) a tetrahedron; (c) the tetrahedral carbon drawn with dashes and wedges; (d) the stick drawing of the tetrahedral carbon atom; (e) ball-and-stick model of methane.

 Animations

The Geometry of CH_4

Valence Shell Electron Pair Repulsion Theory

VSEPR and Molecular Geometry

Each carbon atom forms four single covalent bonds, but each hydrogen atom has only a single covalent bond. Although a carbon atom may be involved in single, double, or triple bonds, it always shares four pairs of electrons. The Lewis Dot structure of the simplest alkane, methane, shows the four shared pairs of electrons (Figure 10.3a). When carbon is involved in four single bonds, the *bond angle*, the angle between two atoms or substituents attached to carbon, is 109.5°, as predicted by the valence shell electron pair repulsion (VSEPR) theory. Thus, alkanes contain carbon atoms that have tetrahedral geometry.

Molecular geometry is described in Section 3.4.

A tetrahedron is a geometric solid having the structure shown in Figure 10.3b. There are many different ways to draw the tetrahedral carbon (Figures 10.3c–10.3e). In Figure 10.3c, solid lines, dashes, and wedges are used to represent the structure of methane. Dashes go back into the page away from you; wedges come out of the page toward you; and solid lines are in the plane of the page. The structure in Figure 10.3d is the same as that in Figure 10.3c; it just leaves a lot more to the imagination. Figure 10.3e is a ball-and-stick model of the methane molecule. Three-dimensional drawings of two other simple alkanes are shown in Figure 10.4.

All hydrocarbons are nonpolar molecules. As a result they are not water soluble but are soluble in nonpolar organic solvents. Furthermore, they have relatively low melting points and boiling points and are generally less dense than water. In general, the longer the hydrocarbon chain (greater the molecular weight), the higher the melting and boiling points and the greater the density (see Table 10.3).

See Section 5.2 for a discussion of the forces responsible for the physical properties of a substance.

Figure 10.4
(a) Drawing and (b) ball-and-stick model of ethane. All the carbon atoms have a tetrahedral arrangement, and all bond angles are approximately 109.5°.
(c) Drawing and (d) ball-and-stick model of a more complex alkane, butane.

(a)   (b)

(c)   (d)

EXAMPLE 10.1   **Using Different Types of Formulas to Represent Organic Compounds**

The following line structure represents 2,2,4-trimethylpentane (also called isooctane), which is the standard of excellence used in determining the octane rating of gasoline. See also An Environmental Perspective: The Petroleum Industry and Gasoline Production later in this chapter.

2,2,4-Trimethylpentane

(isooctane)

Draw the structural and condensed formulas of this molecule.

Solution

Remember that each intersection of lines represents a carbon atom and that each line ends in a carbon atom. This gives us the following carbon skeleton:

$$\begin{array}{ccc} & C & & C \\ & | & & | \\ C-&C&-C-&C&-C \\ & | & & \\ & C & & \end{array}$$

By adding the correct number of hydrogen atoms to the carbon skeleton, we are able to complete the structural formula of this compound.

Continued—

LEARNING GOAL

4 Write condensed and structural formulas for saturated hydrocarbons.

EXAMPLE 10.1 —*Continued*

From the structural formula, we can write the condensed formula as follows:

$$CH_3C(CH_3)_2CH_2CH(CH_3)CH_3$$

Practice Problem 10.1

For each of the molecules in Table 10.4, draw the structural and the line formulas.

For Further Practice: Questions 10.24 and 10.25

Molecules having the same molecular formula but different arrangements of atoms are called structural or constitutional isomers. The carbon chains of these isomers may be straight (Table 10.3) or branched. Compare the following structural isomers of C_6H_{14}. Hexane has a straight chain or carbon backbone:

$$CH_3CH_2CH_2CH_2CH_2CH_3$$

Hexane

All the other isomers have one or more of the carbon atoms branching from the main carbon chain:

CH₃	CH₃	CH₃	CH₃
|	|	|	|
CH₃CHCH₂CH₂CH₃	CH₃CH₂CHCH₂CH₃	CH₃CHCHCH₃	CH₃CCH₂CH₃
		|	|
		CH₃	CH₃
2-Methylpentane	3-Methylpentane	2,3-Dimethylbutane	2,2-Dimethylbutane

These branched-chain forms of the molecule have a much smaller surface area than the straight-chain. As a result, the van der Waals forces attracting the molecules to one another are less strong and these molecules have lower melting and boiling points than the straight-chain isomers. The melting and boiling points of the five structural isomers of C_6H_{14} are found in Table 10.4.

Structural or constitutional isomers are explored in detail in Section 10.2.

Van der Waals forces are discussed in Section 5.2.

Alkyl Groups

Alkyl groups are alkanes with one fewer hydrogen atom. The name of the alkyl group is derived from the name of the alkane containing the same number of carbon atoms. The *-ane* ending of the alkane name is replaced by the *-yl* ending. Thus,

---

TABLE 10.4	Melting and Boiling Points of Five Alkanes of Molecular Formula C_6H_{14}		
Name	**Condensed Formula**	**Boiling Point*** °C	**Melting Point*** °C
Hexane	$CH_3CH_2CH_2CH_2CH_2CH_3$	68.8	−95.2
2-Methylpentane	$CH_3CH(CH_3)CH_2CH_2CH_3$	60.9	−153.2
3-Methylpentane	$CH_3CH_2CH(CH_3)CH_2CH_3$	63.3	−118
2,3-Dimethylbutane	$CH_3CH(CH_3)CH(CH_3)CH_3$	58.1	−130.2
2,2-Dimethylbutane	$CH_3C(CH_3)_2CH_2CH_3$	49.8	−100.2

*Melting and boiling points as reported in the National Institute of Standards and Technology Chemistry Webbook, which can be found at http://webbook.nist.gov/.

TABLE 10.5	Names and Formulas of the First Five Continuous-Chain Alkyl Groups

Alkyl Group Structure	Name
$-CH_3$	Methyl
$-CH_2CH_3$	Ethyl
$-CH_2CH_2CH_3$	Propyl
$-CH_2CH_2CH_2CH_3$	Butyl
$-CH_2CH_2CH_2CH_2CH_3$	Pentyl

$-CH_3$ is a methyl group and $-CH_2CH_3$ is an ethyl group. The dash at the end of these two structures represents the point at which the alkyl group can bond to another atom. The first five continuous-chain alkyl groups are presented in Table 10.5.

Carbon atoms are classified according to the number of other carbon atoms to which they are attached. A **primary carbon (1°)** is directly bonded to one other carbon. A **secondary carbon (2°)** is bonded to two other carbon atoms; a **tertiary carbon (3°)** is bonded to three other carbon atoms, and a **quaternary carbon (4°)** to four.

Alkyl groups are classified according to the number of carbons attached to the carbon atom that joins the alkyl group to a molecule.

Primary alkyl group   Secondary alkyl group   Tertiary alkyl group

All of the continuous-chain alkyl groups are primary alkyl groups (see Table 10.5). Several branched-chain alkyl groups are shown in Table 10.6. Notice that the iso-propyl and *sec*-butyl groups are secondary alkyl groups; the isobutyl group is a primary alkyl group; and the *t*-butyl (*tert*-butyl) is a tertiary alkyl group.

Question 10.3

Classify each of the carbon atoms in the following structures as either primary, secondary, or tertiary.

a.   b.   c.

TABLE 10.6	Structures and Names of Some Branched-Chain Alkyl Groups		
Structure	**Classification**	**Common Name**	**I.U.P.A.C. Name**

CH₃CH—
    |
   CH₃
	2°	Isopropyl*	1-Methylethyl

CH₃ | CH₃CHCH₂—	1°	Isobutyl*	2-Methylpropyl
CH₃ | CH₃CH₂CH—	2°	*sec*-Butyl†	1-Methylpropyl

CH₃
   |
 CH₃C—
   |
   CH₃
	3°	*t*-Butyl or *tert*-Butyl‡	1,1-Dimethylethyl

*The prefix iso(isomeric) is used when there are two methyl groups at the end of the alkyl group.

†The prefix *sec*- (secondary) indicates that there are two carbons bonded to the carbon that attaches the alkyl group to the parent molecule.

‡The prefix *t*- or *tert*- (tertiary) means that three carbons are attached to the carbon that attaches the alkyl group to the parent molecule.

Question 10.4

Classify each of the carbon atoms in the following structures as either primary, secondary, or tertiary.

  a. CH₃CH₂C(CH₃)₂CH₂CH₃
  b. CH₃CH₂CH₂CH₂CH(CH₃)CH(CH₃)CH₃

Nomenclature

Historically, organic compounds were named by the chemist who discovered them. Often the names reflected the source of the compound. For instance, the antibiotic penicillin is named for the mold *Penicillium notatum,* which produces it. The pain reliever aspirin was made by adding an acetate group to a compound first purified from the bark of a willow tree and later from the meadowsweet plant (*Spirea ulmaria*). Thus, the name aspirin comes from *a* (acetate) and *spir* (genus of meadowsweet).

These names are easy for us to remember because we come into contact with these compounds often. However, as the number of compounds increased, organic chemists realized that historical names were not adequate because they revealed nothing about the structure of a compound. Thousands of such compounds and their common names had to be memorized! What was needed was a set of nomenclature (naming) rules that would produce a unique name for every organic compound. Furthermore, the name should be so descriptive that, by knowing the name, a student or scientist could write the structure.

The International Union of Pure and Applied Chemistry (I.U.P.A.C.) is the organization responsible for establishing and maintaining a standard, universal system for naming organic compounds. The system of nomenclature developed by this group is called the **I.U.P.A.C. Nomenclature System.** The following rules are used for naming alkanes by the I.U.P.A.C. system.

1. Determine the name of the **parent compound,** the longest continuous carbon chain in the compound. Refer to Tables 10.3 and 10.7 to determine the parent

LEARNING GOAL

6   Use the basic rules of the I.U.P.A.C. Nomenclature System to name alkanes and substituted alkanes.

TABLE 10.7	Carbon Chain Length and Prefixes Used in the I.U.P.A.C. Nomenclature System

Carbon Chain Length	Prefix	Alkane Name
1	Meth-	Meth*ane*
2	Eth-	Eth*ane*
3	Prop-	Prop*ane*
4	But-	But*ane*
5	Pent-	Pent*ane*
6	Hex-	Hex*ane*
7	Hept-	Hept*ane*
8	Oct-	Oct*ane*
9	Non-	Non*ane*
10	Dec-	Dec*ane*

It is important to learn the prefixes for the carbon chain lengths. We will use them in the nomenclature for all organic molecules.

name. Notice that these names are made up of a prefix related to the number of carbons in the chain and the suffix *-ane,* indicating that the molecule is an alkane (Table 10.7). Write down the name of the parent compound, leaving space before the name to identify the substituents. Parent chains are highlighted in yellow in the following examples:

Parent name: Propane	Pentane	Nonane

2. Number the parent chain to give the lowest number to the carbon bonded to the first group encountered on the parent chain, regardless of the numbers that result for the other substituents.

3. Name and number each atom or group attached to the parent compound. The number tells you the position of the group on the main chain, and the name tells you what type of substituent is present at that position. For example, it may be one of the halogens [F-(fluoro), Cl-(chloro), Br-(bromo), and I-(iodo)] or an alkyl group (Tables 10.5 and 10.6). In the following examples the parent chain is highlighted in yellow:

Substituent:	2-Bromo	3-Methyl	4-Ethyl
I.U.P.A.C. name:	2-Bromopropane	3-Methylpentane	4-Ethyloctane

4. If the same substituent occurs more than once in the compound, a separate position number is given for each, and the prefixes *di-, tri-, tetra-, penta-,* and so forth are used, as shown in the following examples:

Throughout this book we will primarily use the I.U.P.A.C. Nomenclature System. When common names are used, they will be shown in parentheses beneath the I.U.P.A.C. name.

2,5-Dibromo
2,5-Dibromohexane

3,5,7-Trimethyldecane
NOT 4,6,8-Trimethyldecane

An Environmental Perspective

Oil-Eating Microbes

Our highly industrialized society has come to rely more and more on petroleum as a source of energy and a raw material for the manufacture of plastics, drugs, and a host of other consumables. In fact, the United States uses approximately 710 million gallons of oil each day and over half of that is imported from overseas in supertankers.

The EPA reports as many as 14,000 oil spills each year. By some estimates, that amounts to as much as 100 million gallons of this U.S. oil spilled into the environment. The largest spill occurred in 1991 in the Persian Gulf. Two days after the Coalition forces began their air attack, Iraqi forces invading Kuwait released 240 million gallons of oil from onshore terminals and from tankers moored offshore. By comparison, the well-publicized oil spill from the *Exxon Valdez* was only 11 million gallons!

Oil spills can have disastrous effects on the environment. We see film footage of oil spills on the evening news that show us birds and animals such as sea otters whose feathers or fur become saturated with oil and lose their ability to insulate them. These animals may die of exposure if they are not rescued and decontaminated. Other animals are accidentally poisoned because of the compounds they ingest while trying to clean themselves. Underwater vegetation, shellfish, and fish may also be destroyed by an oil spill.

To try to reduce the environmental damage, floating barriers may be used to prevent the spread of the oil. These may be used along with skimmers that separate the oil from the water. In some cases, wood chips, straw, or other absorbent materials may be introduced to "sop up" the oil. Dispersants may be introduced to break the oil slick up into small droplets. The benefit of dispersing the oil into droplets is that they can then be more effectively attacked by microbes in the environment that have enzymes that allow them to use the molecules in oil as a food source.

These oil-eating microbes, or OEMs, require oxygen, water, inorganic nutrients, and a source of organic compounds, such as the hydrocarbons found in oil, in order to survive. Their value in cleaning up oil spills is apparent when you consider

A water bird rescued after an oil spill.

that the products of their metabolic breakdown of oil are harmless carbon compounds, carbon dioxide, and water.

There are at least seventy genera of OEMs in nature. Among these are bacteria and a variety of fungi. They have the ability to survive and break down oil in temperatures ranging from –2 to 60°C, and they are active in environments ranging from polar regions to the equator. Although scientists have created genetically engineered microorganisms that have an enhanced ability to metabolize petroleum compounds, it turns out that those that are present in nature can do the job perfectly well, as long as they have enough oxygen and inorganic nutrients to support their metabolism.

For Further Understanding

List some of the many uses for petroleum products on which the U.S. economy is dependent.

What advantage might there be of relying on the OEMs in the environment rather than introducing genetically modified laboratory strains?

---

5. Place the names of the substituents in alphabetical order before the name of the parent compound, which you wrote down in Step 1. Numbers are separated by commas, and numbers are separated from names by hyphens. By convention, halogen substituents are placed before alkyl substituents.

2-Bromo-3,3-dimethylpentane
NOT 3,3-Dimethyl-2-bromopentane

3-Fluoro-2,4-dimethylhexane
NOT 2,4-Dimethyl-3-fluorohexane

4-Ethyl-3-methyloctane
NOT 3,3-Methyl-4-ehyloctane

TABLE 10.8	I.U.P.A.C. Nomenclature for Alkane Parent Chains Longer Than Ten Carbons

Number of Carbons	I.U.P.A.C. Name
11	*Unde*cane
12	*Dode*cane
13	*Tride*cane
14	*Tetrade*cane
15	*Pentade*cane
16	*Hexade*cane
17	*Heptade*cane
18	*Octade*cane
19	*Nona*decane

EXAMPLE 10.2   **Naming Substituted Alkanes Using the I.U.P.A.C. System**

What is the I.U.P.A.C. name of the molecule below, which is commonly called Freon-12? This compound is a chlorofluorocarbon (CFC) once used as a refrigerant and aerosol propellant. Although it has not been manufactured in the United States since 1995 because of the CFC damage to the ozone layer, it is still used as a propellant in asthma inhalers containing Albuterol.

$$\begin{array}{c} Cl \\ | \\ F - C - Cl \\ | \\ F \end{array}$$

LEARNING GOAL

6 Use the basic rules of the I.U.P.A.C. Nomenclature System to name alkanes and substituted alkanes.

Solution

Helpful Hint: No numbers are necessary if there is only one carbon or if the numbering is clear cut.

Parent chain: methane

Substituents: dichlorodifluoro (no numbers are necessary)

Name: Dichlorodifluoromethane

What is the I.U.P.A.C. name of the following molecule, which is a component of the tsetse fly pheromone? Molecules such as this are used as attractants in tsetse fly control measures.

$$\begin{array}{c} CH_3 \\ | \\ CH_3(CH_2)_{14}CHCH_3 \end{array}$$

Solution

Helpful Hint: For alkanes between 11 and 19 carbons, a prefix is used before the word decane (see Table 10.8).

Parent chain: heptadecane

Substituent: 2-methyl

Name: 2-Methylheptadecane

What is the I.U.P.A.C. name of the following molecule, which is the standard of excellence used in determining the octane rating of gasoline?

Continued—

EXAMPLE 10.2 —*Continued*

$$CH_3CCH_2CHCH_3$$

Solution

Parent chain: pentane

Substituents: 2,2,4-trimethyl

Name: 2,2,4-Trimethylpentane

Practice Problem 10.2

Determine the I.U.P.A.C. name for each of the following molecules.

a.
$$CH_3CHCH_2CHCH_2CH_2CH_2CH_3$$
with CH_3 substituents

b.
$$Cl-C-F$$
with Cl substituents

c.
$$CH_3CH_2CH_2CHCHCHCH_3$$
with F, Cl, CH_3 substituents

For Further Practice: Questions 10.59, 10.60, and 10.63.

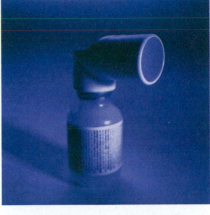

Go online to investigate alternative aerosol propellants that could be used in asthma inhalers.

Having learned to name a compound using the I.U.P.A.C. system, we can easily write the structural formula of a compound, given its name. First, draw and number the parent carbon chain. Add the substituent groups to the correct carbon and finish the structure by adding the correct number of hydrogen atoms.

Drawing the Structure of a Compound Using the I.U.P.A.C. Name

EXAMPLE 10.3

Draw the structural formula for 1-bromo-4-methylhexane.

LEARNING GOAL

7 From the I.U.P.A.C. name of an alkane or substituted alkane, be able to draw the structure.

Solution

Begin by drawing the six-carbon parent chain and indicating the four bonds for each carbon atom.

$$-C-C-C-C-C-C-$$

Next, number each carbon atom:

$$-C-C-C-C-C-C-$$
$$\ \ 1\ \ 2\ \ 3\ \ 4\ \ 5\ \ 6$$

Continued—

EXAMPLE 10.3 —*Continued*

Now add the substituents. In this example a bromine atom is bonded to carbon-1, and a methyl group is bonded to carbon-4:

Finally, add the correct number of hydrogen atoms so that each carbon has four covalent bonds:

As a final check of your accuracy, use the I.U.P.A.C. system to name the compound that you have just drawn, and compare the name with that in the original problem.

The molecular formula and condensed formula can be written from the structural formula shown. The molecular formula is $C_7H_{15}Br$, and the condensed formula is $BrCH_2CH_2CH_2CH(CH_3)CH_2CH_3$.

 Practice Problem 10.3

Draw the structural formula of each of the following compounds:

a. 1-Bromo-2-chlorohexane
b. 2,3-Dimethylpentane
c. 1,3,5-Trichloroheptane
d. 3-Chloro-5-iodo-4-methyloctane
e. 1,2-Dibromo-3-chlorobutane
f. Trifluorochloromethane

For Further Practice: Questions 10.53, 10.54, and 10.55.

Constitutional or Structural Isomers

LEARNING GOAL

8 Draw constitutional (structural) isomers of simple organic compounds.

As we saw earlier, there are two arrangements of the atoms represented by the molecular formula C_4H_{10}: butane and methylpropane. Molecules having the same molecular formula but a different arrangement of atoms are called **constitutional,** or **structural, isomers.** These isomers are unique compounds because of their structural differences, and they have different physical and chemical properties. For instance, the data in Table 10.4 show that the branched-chain isomers of C_6H_{14} have lower boiling and melting points than the straight-chain isomer, hexane. These differences reflect the different shapes of the molecules.

Drawing Constitutional or Structural Isomers of Alkanes

Write all the constitutional isomers having the molecular formula C_6H_{14}.

LEARNING GOAL

8 Draw constitutional (structural) isomers of simple organic compounds.

Solution

1. Begin with the continuous six-carbon chain structure:

$$\underset{1}{CH_3}-\underset{2}{CH_2}-\underset{3}{CH_2}-\underset{4}{CH_2}-\underset{5}{CH_2}-\underset{6}{CH_3}$$

Isomer A

2. Now try five-carbon chain structures with a methyl group attached to one of the internal carbon atoms of the chain:

$$\underset{1}{CH_3}-\underset{2}{CH}-\underset{3}{CH_2}-\underset{4}{CH_2}-\underset{5}{CH_3} \qquad \text{and} \qquad \underset{1}{CH_3}-\underset{2}{CH_2}-\underset{3}{CH}-\underset{4}{CH_2}-\underset{5}{CH_3}$$
$$\qquad\quad |\qquad\qquad\qquad\qquad\qquad\qquad\qquad |$$
$$\qquad\quad CH_3 \qquad\qquad\qquad\qquad\qquad\qquad CH_3$$

Isomer B   Isomer C

3. Next consider the possibilities for a four-carbon structure to which two methyl groups ($-CH_3$) may be attached:

$$\qquad\qquad\qquad\qquad\qquad\qquad\qquad\qquad CH_3$$
$$\underset{1}{CH_3}-\underset{2}{CH}-\underset{3}{CH}-\underset{4}{CH_3} \qquad \text{and} \qquad \underset{1}{CH_3}-\underset{2}{\overset{|}{C}}-\underset{3}{CH_2}-\underset{4}{CH_3}$$
$$\qquad\quad |\quad\;\; | \qquad\qquad\qquad\qquad\qquad\qquad |$$
$$\qquad\quad CH_3\;\; CH_3 \qquad\qquad\qquad\qquad\qquad CH_3$$

Isomer D   Isomer E

These are the five possible constitutional isomers of C_6H_{14}. At first it may seem that other isomers are also possible. But careful comparison will show that they are duplicates of those already constructed. For example, rather than add two methyl groups, a single ethyl group ($-CH_2CH_3$) could be added to the four-carbon chain:

$$CH_3-CH_2-CH-CH_3$$
$$\qquad\qquad\quad |$$
$$\qquad\qquad CH_2CH_3$$

But close examination will show that this is identical to isomer C. Perhaps we could add one ethyl group and one methyl group to a three-carbon parent chain, with the following result:

$$\qquad\qquad CH_2-CH_3$$
$$\qquad\qquad\quad |$$
$$CH_3-C-CH_3$$
$$\qquad\qquad\quad |$$
$$\qquad\qquad CH_3$$

Again we find that this structure is the same as one of the isomers we have already identified, isomer E.

To check whether you have accidentally made duplicate isomers, name them using the I.U.P.A.C. system. All isomers must have different I.U.P.A.C. names. So if two names are identical, the structures are also identical. Use the I.U.P.A.C. system to name the isomers in this example, and prove to yourself that the last two structures are simply duplicates of two of the original five isomers.

Animation

Structural Isomers of Hexane

Practice Problem 10.4

Heptane is a very poor fuel and is given a zero on the octane rating scale. Gasoline octane ratings are determined in test engines by comparison with 2,2,4-trimethylpentane (the standard of high quality) and heptane (standard of poor quality). Draw line formulas and give the I.U.P.A.C. name for each of the nine isomers of heptane.

For Further Practice: Questions 10.64 and 10.65.

10.3 Cycloalkanes

The **cycloalkanes** are a family having C—C single bonds in a ring structure. They have the general molecular formula C_nH_{2n} and thus have two fewer hydrogen atoms than the corresponding alkane (C_nH_{2n+2}). The structures and names of some simple cycloalkanes are shown in Figure 10.5.

In the I.U.P.A.C. system, the cycloalkanes are named by applying the following simple rules.

- Determine the name of the alkane with the same number of carbon atoms as there are within the ring and add the prefix *cyclo*-. For example, cyclopentane is the cycloalkane that has five carbon atoms.
- If the cycloalkane is substituted, place the names of the groups in alphabetical order before the name of the cycloalkane. No number is needed if there is only one substituent.
- If more than one group is present, use numbers that result in the *lowest possible position numbers*.

E X A M P L E 10.5   **Naming a Substituted Cycloalkane Using the I.U.P.A.C. Nomenclature System**

LEARNING GOAL

9 Write the names and draw the structures of simple cycloalkanes.

Name the following cycloalkanes using I.U.P.A.C. nomenclature.

Solution

Parent chain: cyclohexane
Substituent: chloro (no number is required because there is only one substituent)
Name: Chlorocyclohexane

Parent chain: cyclopentane
Substituent: methyl (no number is required because there is only one substituent)
Name: Methylcyclopentane

These cycloalkanes could also be shown as line formulas, as shown below. Each line represents a carbon-carbon bond. A carbon atom and the correct number of hydrogen atoms are assumed to be at the point where the lines meet and at the end of a line.

Chlorocyclohexane

Methylcyclopentane

Continued—

EXAMPLE 10.5 —*Continued*

Practice Problem 10.5

Name each of the following cycloalkanes using I.U.P.A.C. nomenclature:

a.   b. CH_3 c.   d.

For Further Practice: Questions 10.75 and 10.76.

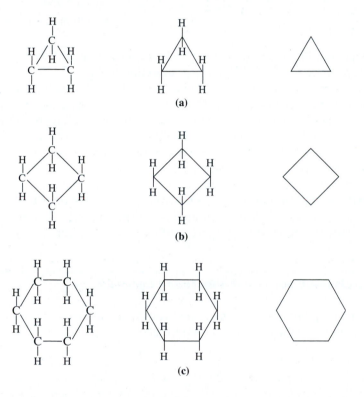

(a)

(b)

(c)

Figure 10.5
Cycloalkanes: (a) cyclopropane; (b) cyclobutane; (c) cyclohexane. All of the cycloalkanes are shown using structural formulas (left column), condensed structural formulas (center column), and line formulas (right column).

cis-trans Isomerism in Cycloalkanes

Atoms of an alkane can rotate freely around the carbon-carbon single bond, resulting in an unlimited number of arrangements. However, rotation around the bonds in a cyclic structure is limited by the fact that the carbons of the ring are all bonded to another carbon within the ring. The formation of ***cis-trans* isomers,** or **geometric isomers,** is a consequence of the absence of free rotation.

Geometric isomers are a type of *stereoisomer*. **Stereoisomers** are molecules that have the same structural formulas and bonding patterns but different arrangements of atoms in space. The *cis-trans* isomers of cycloalkanes are stereoisomers that differ from one another in the arrangement of substituents

LEARNING GOAL

10 Draw *cis* and *trans* isomers of cycloalkanes.

*Stereoisomers are discussed in detail in
Section 16.3.*

Stereochemistry and
Stereoisomers Revisited

in space. Consider the following two views of the *cis*- and *trans*-isomers of
1,2-dichlorocyclohexane:

cis-1,2-Dichlorocyclohexane   *trans*-1,2-Dichlorocyclohexane

cis-1,2-Dichlorocyclohexane   *trans*-1,2-Dichlorocyclohexane

In the wedge and line diagram at the top, it is easy to imagine that you are view-
ing the ring structures as if an edge were projecting toward you. This will help you
understand the more common structural formulas, shown beneath them. In the structure on the left, both Cl atoms are beneath the ring. They are termed *cis* (L., "on the
same side"). The complete name for this compound is *cis*-1,2-dichlorocyclohexane.
In the structure on the right, one Cl is above the ring and the other is below it. They
are said to be *trans* (L., "across from") to one another and the complete name of this
compound is *trans*-1,2-dichlorocyclohexane.

Geometric isomers do not readily interconvert. The cyclic structure prevents
unrestricted free rotation and, thus, prevents interconversion. Only by breaking
carbon-carbon bonds of the ring could interconversion occur. As a result, geometric isomers may be separated from one another in the laboratory.

EXAMPLE 10.6   **Naming *cis-trans* Isomers of Substituted Cycloalkanes**

10 **LEARNING GOAL**
Draw *cis* and *trans* isomers
of cycloalkanes.

Determine whether the following substituted cycloalkanes are *cis* or *trans*
isomers and write the complete name for each.

Solution
Both molecules are cyclopentanes having two methyl group substituents.
Thus both would be named 1,2-dimethylcyclopentane. In the structure on
the left, one methyl group is above the ring and the other is below the ring;
they are in the *trans* configuration, and the structure is named *trans*-1,2-
dimethylcyclopentane. In the structure on the right, both methyl groups are

Continued—

EXAMPLE 10.6 —*Continued*

on the same side of the ring (below it, in this case); they are *cis* to one another, and the complete name of this compound is *cis*-1,2-dimethylcyclopentane.

Practice Problem 10.6

Determine whether each of the following is a *cis*- or a *trans*-isomer.

a.   b.   c.   d.

For Further Practice: Questions 10.87 and 10.88.

Naming a Cycloalkane Having Two Substituents Using the I.U.P.A.C. Nomenclature System

EXAMPLE 10.7

Name the following cycloalkanes using I.U.P.A.C. nomenclature.

Solution

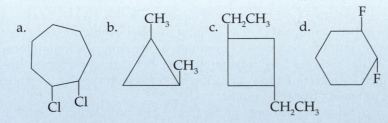

Parent chain: cyclopentane
Substituent: 1,2-dibromo
Isomer: *cis*
Name: *cis*-1,2-Dibromocyclopentane

Parent chain: cyclohexane
Substituent: 1,3-dimethyl
Isomer: *trans*
Name: *trans*-1,3-Dimethylcyclohexane

Practice Problem 10.7

Write the complete I.U.P.A.C. name for each of the following cycloalkanes.

a.   b.   c.   d.

For Further Practice: Questions 10.85 and 10.86.

11 **LEARNING GOAL**
Describe conformations of alkanes.

12 **LEARNING GOAL**
Draw the chair and boat conformations of cyclohexane.

Make a ball-and-stick model of butane and demonstrate these rotational changes for yourself.

The structure of glucose is found in Section 16.4. The physiological roles of glucose are discussed in Chapters 21 and 23.

This conformation gets its name because it resembles a lawn chair.

10.4 Conformations of Alkanes and Cycloalkanes

Because there is *free rotation* around a carbon-carbon single bond, even a very simple alkane, like ethane, can exist in an unlimited number of forms (Figure 10.6a and 10.6b). These different arrangements are called **conformations,** or **conformers.**

Alkanes

Figure 10.6 shows two conformations of a more complex alkane, butane. In addition to these two conformations, an infinite number of intermediate conformers exist. Keep in mind that all these conformations are simply different forms of the same molecule produced by rotation around the carbon-carbon single bonds. Even at room temperature these conformers interconvert rapidly. As a result, they cannot be separated from one another.

Although all conformations can be found in a population of molecules, the *staggered* conformation (see Figure 10.6a and 10.6c) is the most common. One reason for this is that the bonding electrons are farthest from one another in this conformation. This minimizes the repulsion between these bonding electrons.

Cycloalkanes

Cycloalkanes also exist in different conformations. The only exception to this is cyclopropane. Because it has only three carbon atoms, it is always planar.

The conformations of six-member rings have been the most thoroughly studied. One reason is that many important and abundant biological molecules have six-member ring structures. Among these is the simple sugar glucose, also called *blood sugar.* Glucose is the most important sugar in the human body. It is absorbed by the cells of the body and broken down to provide energy for the cells.

The most energetically favorable conformation for a six-member ring is the **chair conformation.** In this conformation the hydrogen atoms are perfectly staggered; that is, they are as far from one another as possible. In addition, the bond angle between carbons is 109.5°, exactly the angle expected for tetrahedral carbon atoms.

Figure 10.6
Conformational isomers of ethane. The hydrogen atoms are much more crowded in the eclipsed conformation, depicted in (b) compared with the staggered conformation shown in (a). The staggered form is energetically favored. The staggered and eclipsed conformations of butane are shown in (c) and (d).

(a) Staggered conformation of ethane

(b) Eclipsed conformation of ethane

(c) Staggered conformation of butane

(d) Eclipsed conformation of butane

An Environmental Perspective

The Petroleum Industry and Gasoline Production

Petroleum consists primarily of alkanes and small amounts of alkenes and aromatic hydrocarbons. Substituted hydrocarbons, such as phenol, are also present in very small quantities. Although the composition of petroleum varies with the source (United States, Persian Gulf, etc.), the mixture of hydrocarbons can be separated into its component parts on the basis of differences in the boiling points of various hydrocarbons (distillation). Often several successive distillations of various fractions of the original mixture are required to completely purify the desired component. In the first distillation, the petroleum is separated into several fractions, each of which consists of a mix of hydrocarbons. Each fraction can be further purified by successive distillations. On an industrial scale, these distillations are carried out in columns that may be hundreds of feet in height.

The gasoline fraction of petroleum, called straight-run gasoline, consists primarily of alkanes and cycloalkanes with six to twelve carbon atoms in the skeleton. This fraction has very poor fuel performance. In fact, branched-chain alkanes are superior to straight-chain alkanes as fuels because they are more volatile, burn less rapidly in the cylinder, and thus reduce "knocking." Alkenes and aromatic hydrocarbons are also good fuels. Methods have been developed to convert hydrocarbons of higher and lower molecular weights than gasoline to the appropriate molecular weight range and to convert straight-chain hydrocarbons into branched ones. *Catalytic cracking* fragments a large hydrocarbon into smaller ones. *Catalytic reforming* results in the rearrangement of a hydrocarbon into a more useful form.

The antiknock quality of a fuel is measured as its octane rating. Heptane is a very poor fuel and is given an octane rating of zero. 2,2,4-Trimethylpentane (commonly called isooctane) is an excellent fuel and is given an octane rating of one hundred. Gasoline octane ratings are experimentally determined by comparison with these two compounds in test engines.

Mining the sea for hydrocarbons.

 Animation
Oil Refining Processes

For Further Understanding

Explain why the mixture of hydrocarbons in crude oil can be separated by distillation.

Draw the structures of heptane and 2,2,4-trimethylpentane (isooctane).

Chair conformation

Six-member rings can also exist in a **boat conformation,** so-called because it resembles a rowboat. This form is much less stable than the chair conformation because the hydrogen atoms are not perfectly staggered.

Compare the structure of this deck chair to the conformation of cyclohexane shown to the left. Explain why this conformation is called the chair conformation.

Boat conformation

Compare the structure of these boats to the conformation of cyclohexane shown to the right. Explain why this conformation is called the boat conformation.

The hydrogen atoms of cyclohexane are described according to their position relative to the ring. Those that lie above or below the ring are said to be **axial atoms.** Those that lie roughly in the plane of the ring are called **equatorial atoms.**

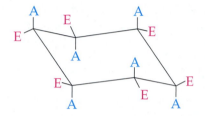

🌐 **Animation**
Conformations of Cyclohexane

Question 10.5

Describe the positions of the six axial hydrogens of the chair conformation of cyclohexane.

Question 10.6

Describe the positions of the six equatorial hydrogens of the chair conformation of cyclohexane.

10.5 Reactions of Alkanes and Cycloalkanes

Combustion

LEARNING GOAL

13 ▷ Write equations for combustion reactions of alkanes.

Alkanes, cycloalkanes, and other hydrocarbons can be oxidized (by burning) in the presence of excess molecular oxygen. In this reaction, called **combustion,** they burn at high temperatures, producing carbon dioxide and water and releasing large amounts of energy as heat.

$$C_nH_{2n+2} \quad + \quad O_2 \quad \longrightarrow \quad CO_2 \quad + \quad H_2O \quad + \quad \text{heat energy}$$

Alkane   Oxygen $\longrightarrow$ Carbon dioxide   Water

Combustion reactions are discussed in Section 4.3.

The following examples show a combustion reaction for a simple alkane and a simple cycloalkane:

$$CH_4 + 2O_2 \longrightarrow CO_2 + 2H_2O + \text{heat energy}$$
Methane

(or C_6H_{12}) + $9O_2 \longrightarrow 6CO_2 + 6H_2O$ + heat energy

Cyclohexane

A Medical Perspective

Polyhalogenated Hydrocarbons Used as Anesthetics

Polyhalogenated hydrocarbons are hydrocarbons containing two or more halogen atoms. Some polyhalogenated compounds are notorious for the problems they have caused humankind. For instance, some insecticides such as DDT, chlordane, kepone, and lindane do not break down rapidly in the environment. As a result, these toxic compounds accumulate in biological tissue of a variety of animals, including humans, and may cause neurological damage, birth defects, or even death.

Other halogenated hydrocarbons are very useful in medicine. They were among the first anesthetics (pain relievers) used routinely in medical practice. These chemicals played a central role as the studies of medicine and dentistry advanced into modern times.

$$CH_3CH_2{-}Cl \qquad CH_3{-}Cl$$

Chloroethane   Chloromethane
(ethyl chloride)   (methyl chloride)

Chloroethane and chloromethane are local anesthetics. A local anesthetic deadens the feeling in a portion of the body. Applied topically (on the skin), chloroethane and chloromethane numb the area. Rapid evaporation of these anesthetics lowers the skin temperature, deadening the local nerve endings. They act rapidly, but the effect is brief, and feeling is restored quickly.

$$CHCl_3$$

Trichloromethane
(chloroform)

In the past, chloroform was used as both a general and a local anesthetic. When administered through inhalation, it rapidly causes loss of consciousness. However, the effects of this powerful anesthetic are of short duration. Chloroform is no longer used because it was shown to be carcinogenic.

Halothane is administered to a patient.

F   H
| |
F—C—C—Br
| |
F   Cl

2-Bromo-2-chloro-1,1,1-trifluoroethane
(Halothane)

Halothane is a general anesthetic that is administered by inhalation. It is considered to be a very safe anesthetic and is widely used.

For Further Understanding

In the first 24 hours following administration, 70% of the halothane is eliminated from the body in exhaled gases. Explain why halothane is so readily eliminated in exhaled gases.

Rapid evaporation of chloroethane from the skin surface causes cooling that causes local deadening of nerve endings. Explain why the skin surface cools dramatically as a result of evaporation of chloroethane.

Balancing Equations for the Combustion of Alkanes

EXAMPLE 10.8

Balance the following equation for the combustion of hexane:

$$C_6H_{14} + O_2 \longrightarrow CO_2 + H_2O$$

Solution

First, balance the carbon atoms; there are 6 mol of carbon atoms on the left and only 1 mol of carbon atoms on the right:

$$C_6H_{14} + O_2 \longrightarrow 6CO_2 + H_2C$$

LEARNING GOAL

13 Write equations for combustion reactions of alkanes.

Continued—

E X A M P L E   10.8 —*Continued*

Next, balance hydrogen atoms; there are 14 mol of hydrogen atoms on the left and only 2 mol of hydrogen atoms on the right:

$$C_6H_{14} + O_2 \longrightarrow 6CO_2 + 7H_2O$$

Now there are 19 mol of oxygen atoms on the right and only 2 mol of oxygen atoms on the left. Therefore, a coefficient of 9.5 is needed for O_2.

$$C_6H_{14} + 9.5O_2 \longrightarrow 6CO_2 + 7H_2O$$

Although decimal coefficients are sometimes used, it is preferable to have all integer coefficients. Multiplying each term in the equation by 2 will satisfy this requirement, giving us the following balanced equation:

$$2C_6H_{14} + 19O_2 \longrightarrow 12CO_2 + 14H_2O$$

The equation is now balanced with 12 mol of carbon atoms, 28 mol of hydrogen atoms and 38 mol of oxygen atoms on each side of the equation.

 Practice Problem 10.8

Write a balanced equation for the complete combustion of each of the following hydrocarbons:
a. cyclobutane
b. ethane
c. decane
d. hexane

For Further Practice: Questions 10.99 and 10.100.

See An Environmental Perspective: The Greenhouse Effect and Global Warming in Chapter 5.

The energy released, along with their availability and relatively low cost, makes hydrocarbons very useful as fuels. In fact, combustion is essential to our very existence. It is the process by which we heat our homes, run our cars, and generate electricity. Although combustion of fossil fuels is vital to industry and society, it also represents a threat to the environment. The buildup of CO_2 may contribute to global warming and change the face of the earth in future generations.

Halogenation

14 | **LEARNING GOAL**
Write equations for halogenation reactions of alkanes.

Alkanes and cycloalkanes can also react with a halogen (usually chlorine or bromine) in a reaction called **halogenation.** Halogenation is a **substitution reaction,** that is, a reaction that results in the replacement of one group for another. In this reaction a halogen atom is substituted for one of the hydrogen atoms in the alkane. The products of this reaction are an **alkyl halide** or *haloalkane* and a hydrogen halide. Alkanes are not very reactive molecules. However, alkyl halides are very useful reactants for the synthesis of other organic compounds. Thus, the halogenation reaction is of great value because it converts unreactive alkanes into versatile starting materials for the synthesis of desired compounds. This is important in the pharmaceutical industry for the synthesis of some drugs. In addition, alkyl halides having two or more halogen atoms are useful solvents, refrigerants, insecticides, and herbicides.

Halogenation can occur only in the presence of heat and/or light, as indicated by the reaction conditions noted over the reaction arrows. The general equation for the halogenation of an alkane follows. The R in the general structure for the alkane may be either a hydrogen atom or an alkyl group.

$$R-\underset{\underset{H}{|}}{\overset{\overset{H}{|}}{C}}-H \;+\; X_2 \quad\xrightarrow{\text{Light or heat}}\quad R-\underset{\underset{H}{|}}{\overset{\overset{H}{|}}{C}}-X \;+\; H-X$$

Alkane   Halogen   Alkyl halide   Hydrogen halide

$$H-\underset{\underset{H}{|}}{\overset{\overset{H}{|}}{C}}-H \;+\; Br_2 \quad\xrightarrow{\text{Light or heat}}\quad H-\underset{\underset{H}{|}}{\overset{\overset{H}{|}}{C}}-Br \;+\; H-Br$$

Methane   Bromine   Bromomethane   Hydrogen bromide

$$CH_3CH_3 \;+\; Cl_2 \quad\xrightarrow{\text{Light}}\quad CH_3CH_2-Cl \;+\; H-Cl$$

Ethane   Chlorine   Chloroethane   Hydrogen chloride

Cyclohexane   Chlorine   Chlorocyclohexane   Hydrogen chloride

The alkyl halide may continue to react forming a mixture of products substituted at multiple sites or substituted multiple times at the same site.

If the halogenation reaction is allowed to continue, the alkyl halide formed may react with other halogen atoms. When this happens, a mixture of products may be formed. For instance, bromination of methane will produce bromomethane (CH_3Br), dibromomethane (CH_2Br_2), tribromomethane ($CHBr_3$), and tetrabromomethane (CBr_4).

In more complex alkanes, halogenation can occur to some extent at all positions to give a mixture of monosubstituted products. For example, bromination of propane produces a mixture of 1-bromopropane and 2-bromopropane.

Question 10.7

Write a balanced equation for each of the following reactions. Show all possible products.

a. the monobromination of propane
b. the monochlorination of butane
c. the monochlorination of cyclobutane
d. the monobromination of pentane

Question 10.8

Provide the I.U.P.A.C. names for the products of the reactions in Question 10.7.

A Medical *Perspective*

Chloroform in Your Swimming Pool?

Has anyone ever asked you to take a shower before swimming in an indoor pool? Perhaps that sounds a bit silly, but there may be a very good reason for doing just that. A team of British researchers has identified the suspected carcinogen chloroform and some other potentially hazardous compounds called trihalomethanes (THMs) in indoor swimming pools in London.

Several questions arise. How did these trihalomethanes get into the pool and why are they a problem in indoor pools? THMs are products of chemical reactions between the chlorine used to disinfect the pool and organic substances from the swimmers themselves. Skin cells are shed into the pool, along with lotions and other body care products. Organic molecules from these substances are chlorinated to produce the trihalomethanes. THMs are volatile compounds. In an outdoor pool, they would evaporate and be blown away by the breeze. In an indoor pool, where there is less air circulation, THMs tend to build to higher concentrations in the air above the pool. These fumes are inhaled by the swimmers and the THMs diffuse into the blood.

It seems logical that some of the factors favoring the production of these compounds include warmer water temperatures and larger numbers of swimmers, which means more organic material in the water. In some of the public pools the level of THMs was as high as 132 µg/L. Compare this to the levels the researchers found in drinking water, only about 3.5 µg/L.

In the past, chloroform was used as an anesthetic. This is no longer the case, since many safer alternatives are available. One reason for discontinuing the use of chloroform as an anesthetic is the determination by the U.S. Department of Health and Human Services that chloroform is a potential carcinogen. Rats and mice exposed to chloroform in their food or water developed liver and kidney cancers.

How concerned should we be about these levels of THMs? Studies have shown that a one-hour swim can increase the blood concentrations of chloroform as much as tenfold. Animal studies have shown that miscarriages occurred in rats and mice that breathed air containing 30–300 parts per million of chloroform during pregnancy. Some studies have suggested that miscarriages, birth defects, and low birthrate might be

Swimmers enjoying an indoor pool.

associated with human exposure to chloroform, as well. But at the current time, researchers feel that these results are inconsistent and that further research needs to be done before any conclusions can be reached. They do recommend, however, that the amount of THMs be reduced as much as possible, while maintaining a high enough level of chlorine to control waterborne infectious diseases. Keeping the water at cooler temperatures and asking swimmers to shower before entering the pool are effective measures to help reduce the production of toxic THMs.

For Further Understanding

Explain in chemical terms why keeping swimming pool water at lower temperatures and asking swimmers to shower before entering the pool help reduce the levels of THMs.

Four trihalomethanes have been identified by the Environmental Protection Agency in water disinfected with chlorine and other disinfectants. These are chloroform, bromodichloromethane, dibromochloromethane, and bromoform. Draw the structures of each of these compounds.

Summary of Reactions

Reactions of Alkanes

Combustion:

$$C_nH_{2n+2} + O_2 \longrightarrow CO_2 + H_2O + \text{heat energy}$$

Alkane   Oxygen   Carbon   Water
dioxide

Halogenation:

$$\underset{\text{Alkane}}{\overset{\displaystyle H \atop |}{R-\underset{|}{\underset{H}{C}}-H}} + \underset{\text{Halogen}}{X_2} \xrightarrow{\text{light or heat}} \underset{\text{Alkyl}\atop\text{halide}}{\overset{\displaystyle H \atop |}{R-\underset{|}{\underset{H}{C}}-X}} + \underset{\text{Hydrogen}\atop\text{halide}}{H-X}$$

SUMMARY

10.1 The Chemistry of Carbon

The modern science of organic chemistry began with Wöhler's synthesis of urea in 1828. At that time, people believed that it was impossible to synthesize an organic molecule outside of a living system. We now define organic chemistry as the study of carbon-containing compounds. The differences between the ionic bond, which is characteristic of many inorganic substances, and the covalent bond in organic compounds are responsible for the great contrast in properties and reactivity between organic and inorganic compounds. All organic compounds are classified as either *hydrocarbons* or *substituted hydrocarbons*. In substituted hydrocarbons a hydrogen atom is replaced by a functional group. A *functional group* is an atom or group of atoms arranged in a particular way that imparts specific chemical or physical properties to a molecule. The major families of organic molecules are defined by the specific functional groups that they contain.

10.2 Alkanes

The *alkanes* are *saturated hydrocarbons*, that is, hydrocarbons that have only carbon and hydrogen atoms that are bonded together by carbon-carbon and carbon-hydrogen single bonds. They have the general molecular formula C_nH_{2n+2} and are nonpolar, water-insoluble compounds with low melting and boiling points. In the *I.U.P.A.C. Nomenclature System* the alkanes are named by determining the number of carbon atoms in the parent compound and numbering the carbon chain to provide the lowest possible number for all substituents. The substituent names and numbers are used as prefixes before the name of the parent compound.

Constitutional or *structural isomers* are molecules that have the same molecular formula but different structures. They have different physical and chemical properties because the atoms are bonded to one another in different patterns.

10.3 Cycloalkanes

Cycloalkanes are a family of organic molecules having C—C single bonds in a ring structure. They are named by adding the prefix *cyclo-* to the name of the alkane parent compound.

A *cis-trans isomer* is a type of stereoisomer. *Stereoisomers* are molecules that have the same structural formula and bonding pattern but different arrangements of atoms in space. A cycloalkane is in the *cis* configuration if two substituents are on the same side of the ring (either both above or both below). A cycloalkane is in the *trans* configuration when one substituent is above the ring and the other is below the ring. The *cis-trans* isomers are not interconvertible.

10.4 Conformations of Alkanes and Cycloalkanes

As a result of *free rotation* around carbon-carbon single bonds, infinitely many *conformations* or *conformers* exist for any alkane.

Limited rotation around the carbon-carbon single bonds of cycloalkanes also results in a variety of conformations of cycloalkanes. In cyclohexane the *chair conformation* is the most energetically favored. Another conformation is the *boat conformation*.

10.5 Reactions of Alkanes and Cycloalkanes

Alkanes can participate in *combustion* reactions. In complete combustion reactions they are oxidized to produce carbon dioxide, water, and heat energy. They can also undergo *halogenation* reactions to produce *alkyl halides*.

KEY TERMS

aliphatic hydrocarbon (10.1)	halogenation (10.5)
alkane (10.2)	hydrocarbon (10.1)
alkyl group (10.2)	I.U.P.A.C. Nomenclature
alkyl halide (10.5)	System (10.2)
aromatic hydrocarbon (10.1)	line formula (10.2)
axial atom (10.4)	molecular formula (10.2)
boat conformation (10.4)	parent compound (10.2)
chair conformation (10.4)	primary (1°) carbon (10.2)
cis-trans isomers (10.3)	quaternary (4°) carbon (10.2)
combustion (10.5)	saturated hydrocarbon (10.1)
condensed formula (10.2)	secondary (2°) carbon (10.2)
conformations (10.4)	stereoisomers (10.3)
conformers (10.4)	structural formula (10.2)
constitutional isomers (10.2)	structural isomer (10.2)
cycloalkane (10.3)	substituted hydrocarbon (10.1)
equatorial atom (10.4)	substitution reaction (10.5)
functional group (10.1)	tertiary (3°) carbon (10.2)
geometric isomers (10.3)	unsaturated hydrocarbon (10.1)

QUESTIONS AND PROBLEMS

The Chemistry of Carbon

Foundations

10.9 Why is the number of organic compounds nearly limitless?

10.10 What are allotropes?

10.11 What are the three allotropic forms of carbon?

10.12 Describe the three allotropes of carbon.

10.13 Why do ionic substances generally have higher melting and boiling points than covalent substances?

10.14 Why are ionic substances more likely to be water-soluble?

Applications

10.15 Rank the following compounds from highest to lowest boiling points:
- a. H_2O CH_4 LiCl
- b. C_2H_6 C_3H_8 NaCl

10.16 Rank the following compounds from highest to lowest melting points:
- a. H_2O CH_4 KCl
- b. C_6H_{14} $C_{16}H_{38}$ NaCl

10.17 What would the physical state of each of the compounds in Question 10.15 be at room temperature?

10.18 Which of the compounds in Question 10.16 would be soluble in water?

10.19 Consider the differences between organic and inorganic compounds as you answer each of the following questions.
- a. Which compounds make good electrolytes?
- b. Which compounds exhibit ionic bonding?
- c. Which compounds have lower melting points?
- d. Which compounds are more likely to be soluble in water?
- e. Which compounds are flammable?

10.20 Describe the major differences between ionic and covalent bonds.

10.21 Give the structural formula for each of the following:

a. $CH_3CHCH_2CHCH_3$

b. $CH_3CHCHCH_3$

10.22 Give the structural formula for each of the following:

a. $CH_3CH_2CHCH_2CHCH_2CH_3$

b. $CH_3CH_2CH_2CH_2CH_2CH$

10.23 Condense each of the following structural formulas:

a.

10.24 Condense each of the following structural formulas:

a.

b.

10.25 Convert the following structural formulas into line formulas:

a.

b.

c.

10.26 Convert the structural formulas in question 10.25 into condensed structural formulas.

10.27 Convert the following structural formulas into line formulas:

a.

b.

c.

10.28 Convert the following line formulas into condensed structural formulas:

a.

b.

c.

d.

10.29 Which of the following structures are not possible? State your reasons.

a. CH₃CHCH₂CH₃ with CH₃

d. CH₃CH₂CH₂CH₂CH₃ with CH₃

b. CH₃CHCH₂CH₃ with CH₃ and CH₃

e. CH₂CH₃CH₂CH₃ with CH₃

c. CH₃CHCH₂CHCH₃ with CH₃ and CH₃

f. CH₃CH₂CH₂CH₃ with CH₂CH₃

a. $CH_3CHCH_2CH_3$ (CH₃)

b. $CH_3CHCH_2CH_3$ (CH₃, CH₃)

c. $CH_3CHCH_2CHCH_3$ (CH₃, CH₃)

d. $CH_3CH_2CH_2CH_2CH_3$ (CH₃)

e. $CH_2CH_3CH_2CH_3$ (CH₃)

f. $CH_3CH_2CH_2CH_3$ (CH₂CH₃)

10.30 Using the octet rule, explain why carbon forms four bonds in a stable compound.

10.31 Convert the following condensed structural formulas into structural formulas:
a. $CH_3CH(CH_3)CH(CH_3)CH_2CH_3$
b. $CH_3CH_2CH_2CH_2CH_3$

10.32 Convert the following condensed structural formulas into structural formulas:
a. $CH_3CH_2CH(CH_2CH_3)CH_2CH_2CH_3$
b. $CH_3CH(CH_3)CH(CH_3)CH_2CH_2CH_2CH_3$

10.33 Convert the following condensed structural formulas into structural formulas:
a. $CH_3CH_2CH(CH_3)CH_2CH_2CH_2CH(CH_2CH_3)CH_3$
b. $CH_3C(CH_3)_2CH_2CH_3$

10.34 Convert the following condensed structural formulas into structural formulas:
a. $CH_3CH(CH_3)CH(CH_3)CH(CH_3)CH_2CH_3$
b. $CH_3C(CH_3)_2CH(CH_2CH_3)CH_2CH_2CH_3$

10.35 Using structural formulas, draw a typical alcohol, aldehyde, ketone, carboxylic acid, and amine. (*Hint:* Refer to Table 10.2.)

10.36 Name the functional group in each of the following molecules:
a. $CH_3CH_2CH_2{-}OH$
b. $CH_3CH_2CH_2{-}NH_2$
c. $CH_3CH_2CH_2{-}C{=}O$ with H
d. $CH_3CH_2CH_2{-}C{=}O$ with OH
e. $CH_3CH_2CH_2{-}C{=}O$ with OCH_2CH_3
f. $CH_3CH_2{-}O{-}CH_2CH_3$
g. $CH_3CH_2CH_2{-}I$

10.37 Give the general formula for each of the following:
a. An alkane
b. An alkyne
c. An alkene
d. A cycloalkane
e. A cycloalkene

10.38 Of the classes of compounds listed in Problem 10.37, which are saturated? Which are unsaturated?

10.39 What major structural feature distinguishes the alkanes, alkenes, and alkynes? Give examples.

10.40 What is the major structural feature that distinguishes between saturated and unsaturated hydrocarbons?

10.41 Give an example, using structural formulas, of each of the following families of organic compounds. Each of your examples should contain a minimum of three carbons. (*Hint:* Refer to Table 10.2.)
a. A carboxylic acid
b. An amine
c. An alcohol
d. An ether

10.42 Folic acid is a vitamin required by the body for nucleic acid synthesis. The structure of folic acid is given below. Circle and identify as many functional groups as possible.

Folic acid

10.43 Aspirin is a pain reliever that has been used for over a hundred years. Today, small doses of aspirin are recommended to reduce the risk of heart attack. The structure of aspirin is given below. Circle and identify the functional groups found in the molecule:

Aspirin
Acetylsalicylic acid

10.44 Vitamin C is one of the water-soluble vitamins. Although most animals can produce vitamin C, humans are not able to do so. We must get our supply of the vitamin from our diet. The structure of vitamin C is shown below. Circle and identify the functional groups found in the molecule:

Vitamin C
Ascorbic acid

Alkanes

Foundations

10.45 Why are hydrocarbons not water soluble?

10.46 Describe the relationship between the length of hydrocarbon chains and the melting points of the compounds.

Applications

10.47 Rank the following compounds from highest to lowest boiling points:
a. heptane   butane   hexane   ethane
b. $CH_3CH_2CH_2CH_2CH_3$ $CH_3CH_2CH_3$
   $CH_3CH_2CH_2CH_2CH_2CH_2CH_2CH_2CH_3$

10.48 Rank the following compounds from highest to lowest melting points:
a. decane   propane   methane   ethane
b. $CH_3CH_2CH_2CH_2CH_3$ $CH_3(CH_2)_8CH_3$
   $CH_3(CH_2)_6CH_3$

10.49 What would the physical state of each of the compounds in Problem 10.47 be at room temperature?

10.50 What would the physical state of each of the compounds in Problem 10.48 be at room temperature?

10.51 Name each of the compounds in Problem 10.47b.

10.52 Name each of the compounds in Problem 10.48b.

10.53 Draw each of the following:
   a. 2-Bromobutane
   b. 2-Chloro-2-methylpropane
   c. 2,2-Dimethylhexane

10.54 Draw each of the following:
   a. Dichlorodiiodomethane
   b. 1,4-Diethylcyclohexane
   c. 2-Iodo-2,4,4-trimethylpentane

10.55 Draw each of the following compounds using structural formulas:
   a. 2,2-Dibromobutane
   b. 2-Iododecane
   c. 1,2-Dichloropentane
   d. 1-Bromo-2-methylpentane

10.56 Draw each of the following compounds using structural formulas:
   a. 1,1,1-Trichlorodecane
   b. 1,2-Dibromo-1,1,2-trifluoroethane
   c. 3,3,5-Trimethylheptane
   d. 1,3,5-Trifluoropentane

10.57 Name each of the following using the I.U.P.A.C. Nomenclature System:

   a. $CH_3CH_2CHCH_2CH_3$
              |
             CH_3

   b. $CH_3CHCH_2CH_2CHCH_3$
              | |
             CH_3 CH_3

   c. $CH_2CH_2CH_2CH_2$—Br
              |
             $CH_2CH_2CH_3$

   d. Cl—$CH_2CH_2CHCH_3$
                        |
                       CH_3

10.58 Provide the I.U.P.A.C. name for each of the following compounds:

   a. $CH_3CH_2CHCH_2CHCH_2CH_3$
                  | |
                 CH_3 CH_2CH_3

   b. $CH_3CHCH_2CH_2CH_2$—Cl
                 |
                Cl

   c. CH_3
            |
           CH_3—C—Br
            |
           CH_3

10.59 Give the I.U.P.A.C. name for each of the following:

   a. CH_3
           |
          CH_3CHCl

   b. I
           |
          $CH_3CHCH_2CH_3$

   c. Br
           |
          CH_3—C—Br
           |
          CH_3

   d. CH_3
           |
          CH_3CHCH_2—Cl

   e. CH_3
           |
          CH_3—C—CH_3
           |
           I

10.60 Name the following using the I.U.P.A.C. Nomenclature System:

   a. Cl
           |
          $CH_3CHCHCH_2CH_3$
                 |
                Cl

   b. CH_3
           |
          $CH_3CH_2CCH_2CHCH_3$
           | |
          CH_3 CH_3

   c. CH_3 CH_3
           | |
          $CH_3CH_2CHCHCHCH_3$
                        |
                       CH_3

   d. $CHCH_2CH_2CH_3$
           |
          Br

10.61 Draw a complete structural formula for each of the straight-chain isomers of the following alkanes:
   a. C_4H_9Br
   b. $C_4H_8Br_2$

10.62 Name all of the isomers that you obtained in Question 10.61.

10.63 Name the following using the I.U.P.A.C. Nomenclature System:
   a. $CH_3(CH_2)_3CH(Cl)CH_3$
   b. $CH_2(Br)(CH_2)_2CH_2Br$
   c. $CH_3CH_2CH(Cl)CH_2CH_3$
   d. $CH_3CH(CH_3)(CH_2)_4CH_3$

10.64 Which of the following pairs of compounds are identical? Which are constitutional isomers? Which are completely unrelated?

   a. Br Br
                 | |
          $CH_3CH_2CHCH_3$ and $CH_3CHCH_2CH_3$

   b. Br CH_3 CH_3
                 | | |
          $CH_3CH_2CHCH_2CHCH_3$ and $CH_3CHCH_2CHCH_2CH_3$
                                                       |
                                                      Br

   c. Br Br
                | |
          $CH_3CCH_2CH_3$ and Br—CCH_2CH_3
                | |
               Br CH_3

   d. CH_3 Br CH_2Br
                     | | |
          $BrCH_2CH_2CCH_2CH_3$ and $CH_2CH_2CHCH_2CH_3$
                     |
                    Br

10.65 Which of the following pairs of molecules are identical compounds? Which are constitutional isomers?

   a. $CH_3CH_2CH_2$ $CH_3CHCH_2CH_2CH_3$
                 | |
          $CH_3CH_2CH_2$ CH_3

   b. $CH_3CH_2CH_2CH_2CH_2CH_2CH_3$ $CH_3CH_2CH_2CH_2CH_2$
                                                              |
                                                          CH_3CH_2

10.66 Which of the following structures are incorrect?

   a. CH_3 Br
               | |
          CH_3—C—CH_2CH_2

   b. H
                    |
          CH_3CH_2—C—CH_3
                    |
                   Br

   c. CH_3
                    |
          $CH_3CH_2CH_2CH_3$
                    |
                   Br

   d. Br
               |
          $CH_3CHCH_2CHCH_3$
                      |
                     Br

10.67 Are the following names correct or incorrect? If they are incorrect, give the correct name.
   a. 1,3-Dimethylpentane
   b. 2-Ethylpropane
   c. 3-Butylbutane
   d. 3-Ethyl-4-methyloctane

10.68 In your own words, describe the steps used to name a compound, using I.U.P.A.C. nomenclature.

10.69 Draw the structures of the following compounds. Are the names provided correct or incorrect? If they are incorrect, give the correct name.
   a. 2,4-Dimethylpentane
   b. 1,3-Dimethylpentane
   c. 1,5-Diiodopentane
   d. 1,4-Diethylheptane
   e. 1,6-Dibromo-6-methyloctane

10.70 Draw the structures of the following compounds. Are the names provided correct or incorrect? If they are incorrect, give the correct name.
  a. 1,4-Dimethylbutane
  b. 1,2-Dichlorohexane
  c. 2,3-Dimethylbutane
  d. 1,2-Diethylethane

Cycloalkanes

Foundations

10.71 Describe the structure of a cycloalkane.
10.72 Describe the I.U.P.A.C. rules for naming cycloalkanes.
10.73 What is the general formula for a cycloalkane?
10.74 How does the general formula of a cycloalkane compare with that of an alkane?

Applications

10.75 Name each of the following cycloalkanes, using the I.U.P.A.C. system:
  a.
  c.
  b.
  d.

10.76 Name each of the following cycloalkanes using the I.U.P.A.C. system:
  a.
  c.
  b.
  d.

10.77 Draw the structure of each of the following cycloalkanes:
  a. 1-Bromo-2-methylcyclobutane
  b. Iodocyclopropane

10.78 Draw the structure of each of the following cycloalkanes:
  a. 1-Bromo-3-chlorocyclopentane
  b. 1,2-Dibromo-3-methylcyclohexane

10.79 Name each of the following substituted cycloalkanes using the I.U.P.A.C. Nomenclature System:

a.
b.

c.

10.80 Name each of the following substituted cycloalkanes using the I.U.P.A.C. Nomenclature System:

a.
b.

c.

10.81 How many geometric and structural isomers of dichloro cyclopropane can you construct? Use a set of molecular models to construct the isomers and to contrast their differences. Draw all these isomers.

10.82 How many isomers of dibromocyclobutane can you construct? As in Question 10.81, use a set of molecular models to construct the isomers and then draw them.

10.83 Which of the following names are correct and which are incorrect? If incorrect, write the correct name.
  a. 2,3-Dibromocyclobutane
  b. 1,4-Diethylcyclobutane
  c. 1,2-Dimethylcyclopropane
  d. 4,5,6-Trichlorocyclohexane

10.84 Which of the following names are correct and which are incorrect? If incorrect, write the correct name.
  a. 1,4,5-Tetrabromocyclohexane
  b. 1,3-Dimethylcyclobutane
  c. 1,2-Dichlorocyclopentane
  d. 3-Bromocyclopentane

10.85 Draw the structures of each of the following compounds:
  a. *cis*-1,3-Dibromocyclopentane
  b. *trans*-1,2-Dimethylcyclobutane
  c. *cis*-1,2-Dichlorocyclopropane
  d. *trans*-1,4-Diethylcyclohexane

10.86 Draw the structures of each of the following compounds:
  a. *trans*-1,4-Dimethylcyclooctane
  b. *cis*-1,3-Dichlorocyclohexane
  c. *cis*-1,3-Dibromocyclobutane

10.87 Name each of the following compounds:

a.

Br   Br

c.

CH₃   CH₃

b. Br

Br

d.

CH₃   CH₃

10.88 Name each of the following compounds:

a.

b.

CH₃

H   C   H
C   H   C
H\Br   H/H
C — C
H   H

c.

d.

CH₃ H

H   H
H   H
H   H
H   H
H   CH₃

Conformations of Alkanes and Cycloalkanes

Foundations

10.89 What are conformational isomers?

10.90 Why is the staggered conformation of ethane more stable than the eclipsed conformation?

Applications

10.91 Make a model of cyclohexane and compare the boat and chair conformations. Use your model to explain why the chair conformation is more energetically favored.

10.92 Why would the ethyl group of ethylcyclohexane generally be found in the equatorial position?

10.93 Why can't conformations be separated from one another?

10.94 What is meant by free rotation around a carbon-carbon single bond?

10.95 Explain why one conformation is more stable than another. (*Hint:* refer to Figure 10.6.)

10.96 Explain why a substituent on a cyclohexane ring would tend to be located in the equatorial position.

Reactions of Alkanes and Cycloalkanes

Foundations

10.97 Define the term *combustion*.

10.98 Explain why halogenation of an alkane is a substitution reaction.

Applications

10.99 Write a balanced equation for the complete combustion of each of the following:
a. propane
b. heptane
c. nonane
d. decane

10.100 Write a balanced equation for the complete combustion of each of the following:
a. pentane
b. hexane
c. octane
d. ethane

10.101 Complete each of the following reactions by supplying the missing reactant or product as indicated by a question mark:

a. $2CH_3CH_2CH_2CH_3 + 13O_2 \xrightarrow{\text{Heat}}$? (Complete combustion)

b. $CH_3-\overset{\overset{\displaystyle CH_3}{|}}{\underset{\underset{\displaystyle CH_3}{|}}{C}}-H + Br_2 \xrightarrow{\text{Light}}$? (Give all possible monobrominated products)

c. ⬡ + ? $\xrightarrow{?}$ Cl—⬡ + HCl

10.102 Give all the possible monochlorinated products for the following reaction:

$$CH_3\overset{\overset{\displaystyle CH_3}{|}}{C}HCH_2CH_3 + Cl_2 \xrightarrow{\text{Light}} ?$$

Name the products, using I.U.P.A.C. nomenclature.

10.103 Draw the constitutional isomers of molecular formula C_6H_{14} and name each using the I.U.P.A.C. system:
a. Which one gives two and only two monobromo derivatives when it reacts with Br_2 and light? Name the products, using the I.U.P.A.C. system.
b. Which give three and only three monobromo products? Name the products, using the I.U.P.A.C. system.
c. Which give four and only four monobromo products? Name the products, using the I.U.P.A.C. system.

10.104 a. Draw and name all of the isomeric products obtained from the monobromination of propane with Br_2/light. If halogenation were a completely random reaction and had an equal probability of occurring at any of the C—H bonds in a molecule, what percentage of each of these monobromo products would be expected?
b. Answer part (a) using 2-methylpropane as the starting material.

10.105 A mole of hydrocarbon formed eight moles of CO_2 and eight moles of H_2O upon combustion. Determine the molecular formula of the hydrocarbon and give the balanced combustion reaction.

10.106 Highly substituted alkyl fluorides, called perfluoroalkanes, are often used as artificial blood substitutes. These perfluoroalkanes have the ability to transport O_2 through the bloodstream as blood does. Some even have twice the O_2 transport capability and are used to treat gangrenous tissue. The structure of perfluorodecalin is shown below. How many moles of fluorine must be reacted with one mole of decalin to produce perfluorodecalin?

⬡⬡ + ? F₂ ⟶ (perfluorodecalin structure)

Decalin   Perfluorodecalin

CRITICAL THINKING PROBLEMS

1. You are given two unlabeled bottles, each of which contains a colorless liquid. One contains hexane and the other contains water. What physical properties could you use to identify the two liquids? What chemical property could you use to identify them?

2. You are given two beakers, each of which contains a white crystalline solid. Both are soluble in water. How would you determine which of the two solids is an ionic compound and which is a covalent compound?

3. Chlorofluorocarbons (CFCs) are man-made compounds made up of carbon and the halogens fluorine and chlorine. One of the most widely used is Freon-12 (CCl_2F_2). It was introduced as a refrigerant in the 1930s. This was an important advance because Freon-12 replaced ammonia and sulfur dioxide, two toxic chemicals that were previously used in refrigeration systems. Freon-12 was hailed as a perfect replacement because it has a boiling point of $-30°C$ and is almost completely inert. To what family of organic molecules do CFCs belong? Design a strategy for the synthesis of Freon-12.

4. Over time, CFC production increased dramatically as their uses increased. They were used as propellants in spray cans, as gases to expand plastic foam, and in many other applications. By 1985 production of CFCs reached 850,000 tons. Much of this leaked into the atmosphere and in that year the concentration of CFCs reached 0.6 parts per billion. Another observation was made by groups of concerned scientists: as the level of CFCs rose, the ozone level in the upper atmosphere declined. Does this correlation between CFC levels and ozone levels prove a relationship between these two phenomena? Explain your reasoning.

5. Although manufacture of CFCs was banned on December 31, 1995, the C—F and C—Cl bonds of CFCs are so strong that the molecules may remain in the atmosphere for 120 years. Within 5 years they diffuse into the upper stratosphere where ultraviolet photons can break the C—Cl bonds. This process releases chlorine atoms, as shown here for Freon-12:

$$CCl_2F_2 + photon \longrightarrow CClF_2 + Cl$$

The chlorine atoms are extremely reactive because of their strong tendency to acquire a stable octet of electrons. The following reactions occur when a chlorine atom reacts with an ozone molecule (O_3). First, chlorine pulls an oxygen atom away from ozone:

$$Cl + O_3 \longrightarrow ClO + O_2$$

Then ClO, a highly reactive molecule, reacts with an oxygen atom:

$$ClO + O \longrightarrow Cl + O_2$$

Write an equation representing the overall reaction (sum of the two reactions). How would you describe the role of Cl in these reactions?

11

The Unsaturated Hydrocarbons
Alkenes, Alkynes, and Aromatics

Learning Goals

1 Describe the physical properties of alkenes and alkynes.

2 Draw the structures and write the I.U.P.A.C. names for simple alkenes and alkynes.

3 Write the names and draw the structures of simple geometric isomers of alkenes.

4 Write equations predicting the products of addition reactions of alkenes and alkynes: hydrogenation, halogenation, hydration, and hydrohalogenation.

5 Apply Markovnikov's rule to predict the major and minor products of the hydration and hydrohalogenation reactions of unsymmetrical alkenes.

6 Write equations representing the formation of addition polymers of alkenes.

7 Draw the structures and write the names of common aromatic hydrocarbons.

8 Write equations for substitution reactions involving benzene.

9 Describe heterocyclic aromatic compounds and list several biological molecules in which they are found.

A crop duster spraying fields with insecticides.

Fatty acids are long hydrocarbon chains having a carboxyl group at the end. Thus by definition they are carboxylic acids. See Chapters 14 and 17.

Lipid-Soluble Vitamins

Oleic acid, the major fatty acid in olive oil, is a monounsaturated fatty acid. What does the term "monounsaturated" mean?

LEARNING GOAL

1 Describe the physical properties of alkenes and alkynes.

Dark green and yellow vegetables, such as carrots, tomatoes, and broccoli, are excellent sources of the precursor of vitamin A, β-carotene. Look up the structure of β-carotene. Develop a hypothesis to explain how the body converts β-carotene into vitamin A.

Introduction

Unsaturated hydrocarbons are those that contain at least one carbon-carbon double or triple bond. They include the alkenes, alkynes, and aromatic compounds. All alkenes have at least one carbon-carbon double bond; all alkynes have at least one carbon-carbon triple bond. Aromatic compounds are particularly stable cyclic compounds and sometimes are depicted as having alternating single and double carbon-carbon bonds. This arrangement of alternating single and double bonds is called a conjugated system of double bonds.

Many important biological molecules are characterized by the presence of double bonds or a linear or cyclic conjugated system of double bonds (Figure 11.1). For instance, we classify fatty acids as either monounsaturated (having one double bond), polyunsaturated (having two or more double bonds), or saturated (having single bonds only). Vitamin A (retinol), a vitamin required for vision, contains a nine-carbon conjugated hydrocarbon chain. Vitamin K, a vitamin required for blood clotting, contains an aromatic ring.

11.1 Alkenes and Alkynes: Structure and Physical Properties

Alkenes and **alkynes** are unsaturated hydrocarbons. The characteristic functional group of an alkene is the carbon-carbon double bond. The functional group that characterizes the alkynes is the carbon-carbon triple bond. The general formulas shown on page 360 compare the structures of alkanes, alkenes, and alkynes.

Oleic acid

(a)

Vitamin A

(b)

Vitamin K

(c)

Figure 11.1

(a) Structural formula of the eighteen-carbon monounsaturated fatty acid oleic acid. (b) Line formula of vitamin A, which is required for vision. Notice that the carbon chain of vitamin A is a conjugated system of double bonds. (c) Line formula of vitamin K, a lipid-soluble vitamin required for blood clotting. The six-member ring with the circle represents a benzene ring. See Figure 11.6 for other representations of the benzene ring.

Chemistry Connection

A Cautionary Tale: DDT and Biological Magnification

We have heard the warnings for years: Stop using non-biodegradable insecticides because they are killing many animals other than their intended victims! Are these chemicals not specifically targeted to poison insects? How then can they be considered a threat to humans and other animals?

DDT, a polyhalogenated hydrocarbon, was discovered in the early 1940s by Paul Müller, a Swiss chemist. Müller showed that DDT is a nerve poison that causes convulsions, paralysis, and eventually death in insects. From the 1940s until 1972, when it was banned in the United States, DDT was sprayed on crops to kill insect pests, sprayed on people as a delousing agent, and sprayed in and on homes to destroy mosquitoes carrying malaria. At first, DDT appeared to be a miraculous chemical, saving literally millions of lives and billions of dollars in crops. However, as time went by, more and more evidence of a dark side of DDT use accumulated. Over time, the chemical had to be sprayed in greater doses as the insect populations evolved to become more and more resistant to it. In 1962, Rachel Carson published her classic work, *Silent Spring*, which revealed that DDT was accumulating in the environment. In particular, high levels of DDT in birds interfered with their calcium metabolism. As a result, the eggshells produced by the birds were too thin to support development of the chick within. It was feared that in spring, when the air should have been filled with bird song, there would be silence. This is the "silent spring" referred to in the title of Carson's book.

DDT is not biodegradable; furthermore, it is not water-soluble, but it is soluble in nonpolar solvents. Thus if DDT is ingested by an animal, it will dissolve in fat tissue and accumulate there, rather than being excreted in the urine. When DDT is introduced into the food chain, which is inevitable when it is sprayed over vast areas of the country, the result is *biological magnification*. This stepwise process begins when DDT applied to crops is ingested by insects. The insects, in turn, are eaten by birds, and the birds are eaten by a hawk. We can imagine another food chain: Perhaps the insects are eaten by mice, which are in turn eaten by a fox, which is then eaten by an owl. Or to make it more personal, perhaps the grass is eaten by a steer, which then becomes your dinner. With each step up one of these food chains, the concentration of DDT in the tissues becomes higher and higher because it is not degraded, it is simply stored. Eventually, the concentration may reach toxic levels in some of the animals in the food chain.

DDT: Dichlorodiphenyltrichloroethane

Consider for a moment the series of events that occurred in Borneo in 1955. The World Health Organization elected to spray DDT in Borneo because 90% of the inhabitants were infected with malaria. As a result of massive spraying, the mosquitoes bearing the malaria parasite were killed. If this sounds like the proverbial happy ending, read on. This is just the beginning of the story. In addition to the mosquitoes, millions of other household insects were killed. In tropical areas it is common for small lizards to live in homes, eating insects found there. The lizards ate the dead and dying DDT-contaminated insects and were killed by the neurotoxic effects of DDT. The house cats ate the lizards, and they, too, died. The number of rats increased dramatically because there were no cats to control the population. The rats and their fleas carried sylvatic plague, a form of bubonic plague. With more rats in contact with humans came the threat of a bubonic plague epidemic. Happily, cats were parachuted into the affected areas of Borneo, and the epidemic was avoided.

The story has one further twist. Many of the islanders lived in homes with thatched roofs. The vegetation used to make these roofs was the preferred food source for a caterpillar that was not affected by DDT. Normally, the wasp population preyed on these caterpillars and kept the population under control. Unfortunately, the wasps were killed by the DDT. The caterpillars prospered, devouring the thatched roofs, which collapsed on the inhabitants.

Every good story has a moral, and this one is not difficult to decipher. The introduction of large amounts of any chemical into the environment, even to eradicate disease, has the potential for long-term and far-reaching effects that may be very difficult to predict. We must be cautious with our fragile environment. Our well-intentioned intervention all too often upsets the critical balance of nature, and in the end we inadvertently do more harm than good.

General formulas:	Alkane C_nH_{2n+2}	Alkene C_nH_{2n}	Alkyne C_nH_{2n-2}

Structural formulas:

	Ethane (ethane)	Ethene (ethylene)	Ethyne (acetylene)
Molecular formulas:	C_2H_6	C_2H_4	C_2H_2
Condensed formulas:	CH_3CH_3	$H_2C=CH_2$	$HC\equiv CH$

The VSEPR theory and molecular geometry are covered in more detail in Section 3.4.

Bananas are picked and shipped while still green. They are treated with the fruit-ripening agent, ethene, once they reach the grocery store. Describe the structure of ethene. What is the common name of ethene?

These compounds have the same number of carbon atoms but differ in the number of hydrogen atoms, a feature of all alkanes, alkenes, and alkynes that contain the same number of carbon atoms. Alkenes contain two fewer hydrogens than the corresponding alkanes, and alkynes contain two fewer hydrogens than the corresponding alkenes.

In alkanes, the four bonds to the central carbon have tetrahedral geometry. When carbon is bonded by one double bond and two single bonds, as in ethene (an alkene), the molecule is *planar*, because all atoms lie in a single plane. Each bond angle is approximately 120°. When two carbon atoms are bonded by a triple bond, as in ethyne (an alkyne), each bond angle is 180°. Thus, the molecule is linear, and all atoms are positioned in a straight line. For comparison, examples of a five-carbon alkane, alkene, and alkyne are shown in Figure 11.2.

All bond angles approximately 109.5°

Ethane

All bond angles approximately 120°

Ethene

Bond angles 180°

Ethyne

Like alkanes, alkenes, alkynes, and aromatic compounds are nonpolar and have properties similar to alkanes of the same carbon chain length. For comparison, the melting points and boiling points of several alkenes and alkynes are presented in Table 11.1. Since they are nonpolar, the "like dissolves like" rule tells us that they are not soluble in water, but they are very soluble in nonpolar solvents such as other hydrocarbons.

A long-chain alkane
(pentane)

A long-chain alkene
(1-pentene)

A long-chain alkyne
(1-pentyne)

Figure 11.2
Three-dimensional drawings and ball-and-stick models of typical alkanes, alkenes, and alkynes.

TABLE 11.1	Physical Properties of Selected Alkenes and Alkynes			
Name	Molecular Formula	Structural Formula	Melting Point (°C)	Boiling Point (°C)
Ethene	C_2H_4	$CH_2\!=\!CH_2$	−169.1	−103.7
Propene	C_3H_6	$CH_2\!=\!CHCH_3$	−185.0	−47.6
1-Butene	C_4H_8	$CH_2\!=\!CHCH_2CH_3$	−185.0	−6.1
Methylpropene	C_4H_8	$CH_2\!=\!C(CH_3)_2$	−140.0	−6.6
Ethyne	C_2H_2	$HC\!\equiv\!CH$	−81.8	−84.0
Propyne	C_3H_4	$HC\!\equiv\!CCH_3$	−101.5	−23.2
1-Butyne	C_4H_6	$HC\!\equiv\!CCH_2CH_3$	−125.9	8.1
2-Butyne	C_4H_6	$CH_3C\!\equiv\!CCH_3$	−32.3	27.0

11.2 Alkenes and Alkynes: Nomenclature

To determine the name of an alkene or alkyne using the I.U.P.A.C. Nomenclature System, use the following simple rules:

- Name the parent compound using the longest continuous carbon chain containing the double bond (alkenes) or triple bond (alkynes).
- Replace the *-ane* ending of the alkane with the *-ene* ending for an alkene or the *-yne* ending for an alkyne. For example:

$$CH_3\!-\!CH_3 \qquad CH_2\!=\!CH_2 \qquad CH\!\equiv\!CH$$
Ethane   Ethene   Ethyne

$$CH_3\!-\!CH_2\!-\!CH_3 \qquad CH_2\!=\!CH\!-\!CH_3 \qquad CH\!\equiv\!C\!-\!CH_3$$
Propane   Propene   Propyne

LEARNING GOAL

2 Draw the structures and write the I.U.P.A.C. names for simple alkenes and alkynes.

• Number the chain to give the lowest number for the first of the two carbons containing the double bond or triple bond. For example:

4 3 2 1
CH₃CH₂CH=CH₂

1 2 3 4 5
CH≡C—CH₂CH₂CH₃

1-Butene
(*NOT* 3-Butene)

1-Pentyne
(*NOT* 4-Pentyne)

Remember, it is the position of the double bond, not the substituent, that determines the numbering of the carbon chain.

• Determine the name and carbon number of each group bonded to the parent alkene or alkyne, and place the name and number in front of the name of the parent compound. Remember that with alkenes and alkynes the double or triple bond takes precedence over a halogen or alkyl group, as shown in the following examples:

4 3 2 1
CH₃—CH=C—CH₃
　　　　|
　　　　Cl

2-Chloro-2-butene

1 2 3 4 5 6
CH₃CH—C≡C—CH₂CH₃
　|
　Br

2-Bromo-3-hexyne

1 2 3 4 5 6 7 8
CH₃CHCH=CHCH₂CHCH₂CH₃
　|　　　　　　　|
　CH₃　　　　　CH₃

2,6-Dimethyl-3-octene

1 2 3 45 6
CH₃CH₂C=CCH₂CH₃
　　　　|　|
　　　　Cl CH₃

3-Chloro-4-methyl-3-hexene

Alkenes with many double bonds are often referred to as polyenes (*poly*— many *enes*—double bonds).

• Alkenes having more than one double bond are called alkadienes (two double bonds) or alkatrienes (three double bonds), as seen in these examples:

1 2 3 4 5 6
CH₃CH=CH—CH=CHCH₃

2,4-Hexadiene

1 2 3 4 5
CH₂=CHCH₂CH=CH₂

1,4-Pentadiene

3-Methyl-1, 4-cyclohexadiene

1 2 3 4 5 6 7 8 9
CH₃CH=CHCH₂CH= CHCH₂CH₂CH₃

2,5-Nonadiene

1 2 3 4 5 6 7
CH₃CH=CHCH=CHCH₂CH₃

2,4-Heptadiene

EXAMPLE 11.1   **Naming Alkenes and Alkynes Using I.U.P.A.C. Nomenclature**

2 LEARNING GOAL
Draw the structures and write the I.U.P.A.C. names for simple alkenes and alkynes.

Name the following alkenes and alkyne using I.U.P.A.C. nomenclature.

Solution

To determine the I.U.P.A.C. name, identify the parent chain, number it to give the lowest possible positions for the carbon or carbons with the double bonds, and finally, identify and number each functional group.

Continued—

EXAMPLE 11.1 —*Continued*

The common name of this molecule is isoprene. It is the major building block for many important biological molecules, including cholesterol and other steroids, β-carotene, and vitamins A, D, E, and K.

$$CH_2{=}CCH{=}CH_2$$
1   2 3   4

Longest chain containing the double bond: butene

Position of the double bond: 1,3-butadiene

Substituents: 2-methyl

Name: 2-Methyl-1,3-butadiene

$$CH_2CH_2CH_3$$
8 7 6 5 4 3 2 1
$$CH_3CH_2CH_2CH_2C{=}CCH_2CH_3$$
$$CH_3$$

Longest chain containing the double bond: octene

Position of double bond: 3-octene (*not* 5-octene)

Substituents: 3-methyl and 4-propyl

Name: 3-Methyl-4-propyl-3-octene

6 5   4 3 2 CH_3 1
$$CH_3CH_2{-}C{\equiv}C{-}C{-}CH_3$$
$$CH_3$$

Longest chain containing the triple bond: hexyne

Position of triple bond: 3-hexyne (must be!)

Substituents: 2,2-dimethyl

Name: 2,2-Dimethyl-3-hexyne

 Practice Problem 11.1

Name each of the following alkenes and alkynes using I.U.P.A.C. nomenclature.

$$\overset{CH_3}{|}\quad\overset{Cl}{|}$$
a. $CH_3CH{=}CHCHCH_2CHCH_3$ b. $CH{\equiv}CCH_2CH{=}CH_2$

$$\overset{Br}{|}$$
c. $CH_3CH_2C{\equiv}CCHCH_2CH_3$ d. $CH_3CH{=}CHCH_2CH_2CH{=}CH_2$

For Further Practice: Questions 11.41 and 11.42.

Naming Cycloalkenes Using I.U.P.A.C. Nomenclature

EXAMPLE 11.2

Name the following cycloalkenes using I.U.P.A.C. nomenclature.

Solution

Parent chain: cyclohexene

Position of double bond: carbon-1 (carbons of the double bond are numbered 1 and 2)

Substituents: 4-chloro

Name: 4-Chlorocyclohexene

Continued—

LEARNING GOAL

2 Draw the structures and write the I.U.P.A.C. names for simple alkenes and alkynes.

A Medical Perspective

Killer Alkynes in Nature

There are many examples of alkynes that are beneficial to humans. Among these are *parsalmide*, a pain reliever, *pargyline*, an antihypertensive, and *17-ethynylestradiol*, a synthetic estrogen that is used as an oral contraceptive.

But in addition to these medically useful alkynes, there are in nature a number that are toxic. Some are extremely toxic to mammals, including humans; others are toxic to fungi, fish, or insects. All of these compounds are plant products that may help protect the plant from destruction by predators.

Capillin is produced by the oriental wormwood plant. Research has shown that a dilute solution of capillin inhibits the growth of certain fungi. Since fungal growth can damage or destroy a plant, the ability to make capillin may provide a survival advantage to the plants. Perhaps it may one day be developed to combat fungal infections in humans.

Ichthyothereol is a fast-acting poison commonly found in plants referred to as fish-poison plants. Ichthyothereol is a very toxic polyacetylenic alcohol that inhibits energy production in the mitochondria. Latin American native tribes use these plants to coat the tips of the arrows used to catch fish. Although ichthyothereol is poisonous to the fish, fish caught by this method pose no risk to the people who eat them!

An extract of the leaves of English ivy has been reported to have antibacterial, analgesic, and sedative effects. The compound thought to be responsible for these characteristics, as well as antifungal activity, is *falcarinol*. Falcarinol, isolated from

Falcarinol is extracted from English ivy, like that covering this stone house.

a tree in Panama, also has been reported by the Molecular Targets Drug Discovery Program, to have antitumor activity. Perhaps one day this compound, or a derivative of it, will be useful in treating cancer in humans.

Cicutoxin has been described as the most lethal toxin native to North America. It is a neurotoxin that is produced by the

Parsalmide

Pargyline

17-Ethynylestradiol

Alkynes used for medicinal purposes.

Capillin

Ichthyothereol

Falcarinol

Cicutoxin

Alkynes that exhibit toxic activity.

water hemlock (*Cicuta maculata*), which is in the same family of plants as parsley, celery, and carrots. Cicutoxin is present in all parts of the plants, but is most concentrated in the root. Eating a portion as small as 2–3 cm² can be fatal to adults. Cicutoxin acts directly on the nervous system. Signs and symptoms of cicutoxin poisoning include dilation of pupils, muscle twitching, rapid pulse and breathing, violent convulsions, coma, and death. Onset of symptoms is rapid and death may occur within two to three hours. No antidote exists for cicutoxin poisoning. The only treatment involves controlling convulsions and seizures in order to preserve normal heart and lung function. Fortunately, cicutoxin poisoning is a very rare occurrence. Occasionally animals may graze on the plants in the spring, resulting in death within fifteen minutes. Humans seldom come into contact with the water hemlock. The most recent cases have involved individuals foraging for wild ginseng, or other wild roots, and mistaking the water hemlock root for an edible plant.

Cicuta maculata, or water hemlock, produces the most deadly toxin indigenous to North America.

For Further Understanding

Circle and name the functional groups in parsalmide and pargyline.

The fungus *Tinea pedis* causes athlete's foot. Describe an experiment you might carry out to determine whether capillin might be effective against athlete's foot.

EXAMPLE 11.2 —*Continued*

Parent chain: cyclopentene

Position of double bond: carbon-1

Substituent: 3-methyl

Name: 3-Methylcyclopentene

Practice Problem 11.2

Name each of the following cycloalkenes using I.U.P.A.C. nomenclature.

For Further Practice: **Questions 11.43d and 11.44d.**

Question *11.1*

Draw a complete structural formula for each of the following compounds:

a. 1-Bromo-3-hexyne
b. 2-Butyne
c. Dichloroethyne
d. 9-Iodo-1-nonyne

Question *11.2*

Name the following compounds using the I.U.P.A.C. Nomenclature System:

a. $CH_3C{\equiv}CCH_2CH_3$

b. $CH_3CH_2CHCHCH_2C{\equiv}CH$
$|\ \ |$
$Br\ Br$

c. $CH_3CHC{=}CCHCH_3$
$\overset{\displaystyle Cl}{|}\ \ \overset{\displaystyle CH_3}{|}$
$\phantom{c.\ CH_3CHC{=}C}|$
$CH_3\ \ \ \ CH_3$

d. $CH_3CHC{\equiv}CCHCH_3$
$\overset{\displaystyle CH_2CH_3}{|}$
$\phantom{d.\ CH_3CHC{\equiv}CC}|$
$\phantom{d.\ CH_3CHC{\equiv}CC}Cl$

11.3 Geometric Isomers: A Consequence of Unsaturation

The carbon-carbon double bond is rigid because of the shapes of the orbitals involved in its formation. As a result, rotation around the carbon-carbon double bond is restricted. In Section 10.3, we saw that the rotation around the carbon-carbon bonds of cycloalkanes was also restricted. As a consequence, these molecules form geometric or *cis-trans* isomers. The *cis* isomers of cycloalkanes have substituent groups on the same side of the ring (L., *cis*, "on the same side"). The *trans* isomers of cycloalkanes have substituent groups located on opposite sides of the ring (L., *trans*, "across from").

In alkenes, **geometric isomers** occur when there are two different groups on each of the carbon atoms attached by the double bond. If both groups are on the same side of the double bond, the molecule is a *cis* isomer. If the groups are on opposite sides of the double bond, the molecule is a *trans* isomer.

Consider the two isomers of 1,2-dichloroethene:

cis-1, 2-Dichloroethene   trans-1, 2-Dichloroethene

In these molecules, each carbon atom of the double bond is also bonded to two different atoms: a hydrogen atom and a chlorine atom. In the molecule on the left, both chlorine atoms are on the same side of the double bond; this is the *cis* isomer and the complete name for this molecule is *cis*-1,2-dichloroethene. In the molecule on the right, the chlorine atoms are on opposite sides of the double bond; this is the *trans* isomer and the complete name of this molecule is *trans*-1,2-dichloroethene.

If one of the two carbon atoms of the double bond has two identical substituents, there are no *cis-trans* isomers for that molecule. Consider the example of 1,1-dichloroethene:

1, 1-Dichloroethene

Restricted rotation around double bonds is partially responsible for the conformation and hence the activity of many biological molecules that we will study later.

Identifying *cis* and *trans* Isomers of Alkenes

E X A M P L E   11.3

Two isomers of 2-butene are shown below. Which is the *cis* isomer and which is the *trans* isomer?

LEARNING GOAL

3 Write the names and draw the structures of simple geometric isomers of alkenes.

Solution

As we saw with cycloalkanes, the prefixes *cis* and *trans* refer to the placement of the substituents attached to a bond that cannot undergo free rotation. In the case of alkenes, it is the groups attached to the carbon-carbon double bond (in this example, the H and CH$_3$ groups). When the groups are on the same side of the double bond, as in the structure on the left, the prefix *cis* is used. When the groups are on the opposite sides of the double bond, as in the structure on the right, *trans* is the appropriate prefix.

cis-2-Butene   trans-2-Butene

Continued—

EXAMPLE 11.3 —*Continued*

Practice Problem 11.3

Which of the following alkenes are *cis*-isomers and which are *trans*-isomers?

a.

$$H_2C=C\quad\begin{matrix}H\\[1ex]\end{matrix}$$

H₃C ... CHBrCHBrCH₃ with H, H on top

b. CH₃CH₂ and Br top; Br and CH₂CH₃ bottom across C=C

c. H and CH₃ top; H₃C and CH₂CH₂CH₃ bottom across C=C

For Further Practice: Questions 11.47 and 11.48.

EXAMPLE 11.4   Naming *cis* and *trans* Isomers of Alkenes

LEARNING GOAL

3 Write the names and draw the structures of simple geometric isomers of alkenes.

Name the following geometric isomers.

Solution

The longest chain of carbon atoms in each of the following molecules is highlighted in yellow. *The chain must also contain the carbon-carbon double bond.* The location of functional groups relative to the double bond is used in determining the appropriate prefix, *cis* or *trans*, to be used in naming each of the molecules.

Parent chain: heptene

Position of double bond: 3-

Substituents: 3,4-dichloro

Configuration: *trans*

Name: *trans*-3,4-Dichloro-3-heptene

Parent chain: octene

Position of double bond: 3-

Substituents: 3,4-dimethyl

Configuration: *cis*

Name: *cis*-3,4-Dimethyl-3-octene

Continued—

EXAMPLE 11.4 —*Continued*

Practice Problem 11.4

Name each of the following geometric isomers:

a.

$$
\underset{H_3C}{\overset{H}{\diagdown}}C=C\underset{CHBrCHBrCH_3}{\overset{H}{\diagup}}
$$

b.

$$
\underset{Br}{\overset{CH_3CH_2}{\diagdown}}C=C\underset{CH_2CH_3}{\overset{Br}{\diagup}}
$$

c.

$$
\underset{H_3C}{\overset{H}{\diagdown}}C=C\underset{CH_2CH_2CH_3}{\overset{CH_3}{\diagup}}
$$

For Further Practice: Questions 11.50 and 11.51.

Identifying Geometric Isomers

EXAMPLE 11.5

Determine whether each of the following molecules can exist as *cis-trans* isomers: (a) 1-pentene, (b) 3-ethyl-3-hexene, and (c) 3-methyl-2-pentene.

LEARNING GOAL

3 Write the names and draw the structures of simple geometric isomers of alkenes.

Solution

a. Examine the structure of 1-pentene,

$$
\underset{H}{\overset{H}{\diagdown}}C=C\underset{H}{\overset{CH_2CH_2CH_3}{\diagup}}
$$

We see that carbon-1 is bonded to two hydrogen atoms, rather than to two different substituents. In this case there can be no *cis-trans* isomers.

b. Examine the structure of 3-ethyl-3-hexene:

$$
\underset{CH_3CH_2}{\overset{CH_3CH_2}{\diagdown}}C=C\underset{H}{\overset{CH_2CH_3}{\diagup}}
$$

We see that one of the carbons of the carbon-carbon double bond is bonded to two ethyl groups. As in example (a), because this carbon is bonded to two identical groups, there can be no *cis* or *trans* isomers of this compound.

c. Finally, examination of the structure of 3-methyl-2-pentene reveals that both a *cis* and *trans* isomer can be drawn.

$$
\underset{CH_3CH_2}{\overset{H_3C}{\diagdown}}C=C\underset{H}{\overset{CH_3}{\diagup}}
\qquad
\underset{CH_3CH_2}{\overset{H_3C}{\diagdown}}C=C\underset{CH_3}{\overset{H}{\diagup}}
$$

cis-3-Methyl-2-pentene   *trans*-3-Methyl-2-pentene

Continued—

EXAMPLE 11.5 —*Continued*

Each of the carbon atoms involved in the double bond is attached to two different groups. As a result, we can determine which is the *cis* isomer and which is the *trans* isomer based on the positions of the methyl groups relative to the double bond.

Practice Problem 11.5

Which of the following molecules can exist as *cis*- and *trans*-isomers? Draw the *cis*- and *trans*-isomers, where possible. For those molecules that cannot exist as *cis*- and *trans*-isomers, explain why.

a. $CH_3CH=CHCH_2CH_3$
b. $CBr_2=CBrCH_2CH_2CH_3$
c. $CH_2=CHCH_3$
d. $CH_3CBr=CBrCH_2CH_2CH_2CH_3$

For Further Practice: Questions 11.48 and 11.49.

The recent debate over the presence of *cis* and *trans* isomers of fatty acids in our diet points out the relevance of geometric isomers to our lives. Fatty acids are long-chain carboxylic acids found in vegetable oils (unsaturated fats) and animal fats (saturated fats). Oleic acid,

$$CH_3(CH_2)_7CH=CH(CH_2)_7\overset{\overset{\displaystyle O}{\|}}{C}-OH$$

is the naturally occurring fatty acid in olive oil. Its I.U.P.A.C. name, *cis*-9-octadecenoic acid, reveals that this is a *cis*-fatty acid.

cis-9-Octadecenoic acid
(oleic acid)

cis-configuration

While oleic acid is V-shaped as a result of the configuration of the double bond, its geometric isomer, *trans*-9-octadecenoic acid, is a rigid linear molecule.

trans-9-Octadecenoic acid
(elaidic acid)

trans-configuration

The majority of *trans*-fatty acids found in the diet result from hydrogenation, which is a reaction used to convert oils into solid fats, such as margarine. It has recently been reported that *trans*-fatty acids in the diet elevate levels of "bad" or LDL cholesterol and lower the levels of "good" or HDL cholesterol, thereby increasing the risk of heart disease. Other studies suggest that *trans*-fatty acids may also increase the risk of type 2 diabetes.

cis- and trans-Fatty acids will be discussed in greater detail in Chapter 17. Hydrogenation is described in Section 11.5.

Question 11.3

In each of the following pairs of molecules, identify the *cis* isomer and the *trans* isomer.

a.

b.

Question 11.4

Provide the complete I.U.P.A.C. name for each of the compounds in Question 11.3.

Question 11.5

Which of the following molecules can exist as both *cis* and *trans* isomers? Explain your reasoning.

a. $CH_3CH_2C{=}CCH_2CH_3$ with Cl and CH_2CH_3

b. $CH_3CH_2C{=}CBr$ with Br and Br

c. $CH_3C{=}CCH_3$ with Cl and Cl

Question 11.6

Draw each of the *cis-trans* isomers in Question 11.5 and provide the complete names using the I.U.P.A.C. Nomenclature System.

Question *11.7*

Draw condensed formulas for each of the following compounds:

   a. *cis*-3-Octene
   b. *trans*-5-Chloro-2-hexene
   c. *trans*-2,3-Dichloro-2-butene

Question *11.8*

Name each of the following compounds, using the I.U.P.A.C. system. Be sure to indicate *cis* or *trans* where applicable.

a.
$$CH_3 \quad \quad CH_3$$
$$\diagdown C=C \diagup$$
$$H \quad \quad CH_3$$

c.
$$CH_3 \quad \quad H$$
$$\diagdown C=C \diagup$$
$$H \quad \quad CH_2\overset{\displaystyle CH_3}{\underset{\displaystyle CH_3}{C}}CH_3$$

b.
$$CH_3CH_2 \quad \quad CH_2CH_3$$
$$\diagdown C=C \diagup$$
$$CH_3 \quad \quad H$$

11.4 Alkenes in Nature

Folklore tells us that placing a ripe banana among green tomatoes will speed up the ripening process. In fact, this phenomenon has been demonstrated experimentally. The key to the reaction is *ethene,* the simplest alkene. Ethene, produced by ripening fruit, is a plant growth substance. It is produced in the greatest abundance in areas of the plant where cell division is occurring. It is produced during fruit ripening, during leaf fall and flower senescence, as well as under conditions of stress, including wounding, heat, cold, or water stress, and disease.

There are surprising numbers of polyenes, alkenes with several double bonds, found in nature. These molecules, which have wildly different properties and functions, are built from one or more five-carbon units called *isoprene.*

$$\overset{\displaystyle CH_3}{\underset{\displaystyle |}{}}$$
$$CH_2=CCH=CH_2$$
Isoprene

The molecules that are produced are called *isoprenoids,* or *terpenes* (Figure 11.3). Terpenes include the steroids; chlorophyll and carotenoid pigments that function in photosynthesis; and the lipid-soluble vitamins A, D, E, and K (Figure 11.1).

Many other terpenes are plant products familiar to us because of their distinctive aromas. *Geraniol,* the familiar scent of geraniums, is a molecule made up of two isoprene units. Purified from plant sources, geraniol is the active ingredient in several natural insect repellants. These can be applied directly to the skin to provide four hours of protection against a variety of insects, including mosquitoes, ticks, and fire ants.

D-*Limonene* is the most abundant component of the oil extracted from the rind of citrus fruits. Because of its pleasing orange aroma, D-limonene is used as a flavor and fragrance additive in foods. However, the most rapidly expanding use of the compound is as a solvent. In this role, D-limonene can be used in place of more toxic solvents, such as mineral spirits, methyl ethyl ketone, acetone, toluene, and fluorinated and chlorinated organic solvents. It can also be formulated

Geraniol
(Roses and geraniums)

Limonene
(Oil of lemon and orange)

Myrcene
(Oil of bayberry)

Farnesol
(Lily of the valley)

Figure 11.3
Many plant products, familiar to us because of their distinctive aromas, are isoprenoids, which are alkenes having several double bonds.

as a water-based cleaning product, such as Orange Glo, that can be used in place of more caustic cleaning solutions. There is a form of limonene that is a molecular mirror image of D-limonene. It is called L-limonene and has a pine or turpentine aroma.

The terpene *myrcene* is found in bayberry. It is used in perfumes and scented candles because it adds a refreshing, spicy aroma to them. Trace amounts of myrcene may be used as a flavor component in root beer.

Farnesol is a terpene found in roses, orange blossom, wild cyclamen, and lily of the valley. Cosmetics companies began to use farnesol in skin care products in the early 1990s. It is claimed that farnesol smoothes wrinkles and increases skin elasticity. It is also thought to reduce skin aging by promoting regeneration of cells and activation of the synthesis of molecules, such as collagen, that are required for healthy skin.

Another terpene, *retinol,* is a form of vitamin A (Figure 11.1). It is able to penetrate the outer layers of skin and stimulate the formation of collagen and elastin. This reduces wrinkles by creating skin that is firmer and smoother.

11.5 Reactions Involving Alkenes and Alkynes

LEARNING GOAL

4 Write equations predicting the products of addition reactions of alkenes and alkynes: hydrogenation, halogenation, hydration, and hydrohalogenation.

Reactions of alkenes involve the carbon-carbon double bond. The key reaction of the double bond is the **addition reaction.** This involves the addition of two atoms or groups of atoms to a double bond. The major alkene addition reactions include addition of hydrogen (H_2), halogens (Cl_2 or Br_2), water (HOH), or hydrogen halides (HBr or HCl). A generalized addition reaction is shown here. The R in these structures represents any alkyl group.

Note that the double bond is replaced by a single bond. The former double bond carbons receive a new single bond to a new atom, producing either an alkane or a substituted alkane.

Hydrogenation: Addition of H_2

Recall that a catalyst itself undergoes no net change in the course of a chemical reaction (see Section 7.3).

Hydrogenation is the addition of a molecule of hydrogen (H_2) to a carbon-carbon double bond to give an alkane. In this reaction the double bond is broken, and two new C—H single bonds result. Platinum, palladium, or nickel is needed as a catalyst to speed up the reaction. Heat and/or pressure may also be required.

Note that the alkene is gaining two hydrogens. Thus, hydrogenation is a reduction reaction (see Sections 8.5 and 12.6).

Writing Equations for the Hydrogenation of Alkenes   EXAMPLE 11.6

Linoleic acid (*cis, cis*-9,12-octadecadienoic acid) is an essential fatty acid. This means that we must obtain it in the diet, because we cannot make it. Fortunately, it is found in abundance in sunflower, safflower, and corn oil, which are often hydrogenated to produce margarine. Write a balanced equation showing the hydrogenation of linoleic acid.

 LEARNING GOAL
4 Write equations predicting the products of addition reactions of alkenes and alkynes: hydrogenation, halogenation, hydration, and hydrohalogenation.

First, notice that there are two double bonds. This means that we will need two molecules of H_2 for every molecule of linoleic acid that is hydrogenated.

Linoleic acid

+ 2H_2

Ni

Stearic acid

The liquid linoleic acid is converted to stearic acid, which is a rather hard solid. In margarine production today, partial hydrogenation is generally used because it produces a fat that can be more easily spread and has a better "mouth feel." They may then add butter flavoring, milk solids, salt, an emulsifying agent, preservatives, vitamin A for nutritional value, and a bit of β-carotene for color.

Margarine and solid shortening are made by partial hydrogenation of vegetable oils. While natural oils contain only *cis*-fatty acids, the hydrogenation reaction produces *trans*-fatty acids. What are the health risks associated with *trans*-fats?

Practice Problem 11.6

Write a balanced equation for the hydrogenation of each of the following alkenes.

a. *cis*-2-Pentene
b. *trans*-2-Pentene
c. 3-Hexene
d. Propene

For Further Practice: Questions 11.55, 11.67a, and 11.83.

The conditions for the hydrogenation of alkynes are very similar to those for the hydrogenation of alkenes. Two moles of hydrogen add to the triple bond of the alkyne to produce an alkane, as seen in the following general reaction:

$$R-C\equiv C-R + 2H_2 \xrightarrow[\text{Heat}]{\text{Pt, Pd, or Ni}}$$

Alkyne   Hydrogen
(2 mol)

Alkane

Figure 11.4
Conversion of a typical oil to a fat involves hydrogenation. In this example, triolein (an oil) is converted to tristearin (a fat).

$$CH_3-(CH_2)_7-CH=CH-(CH_2)_7-\overset{\overset{O}{\|}}{C}-O-CH_2$$

$$CH_3-(CH_2)_7-CH=CH-(CH_2)_7-\overset{\overset{O}{\|}}{C}-O-CH \xrightarrow[\substack{25 \text{ psi,} \\ \text{metal catalyst}}]{3H_2, 200°C,}$$

$$CH_3-(CH_2)_7-CH=CH-(CH_2)_7-\overset{\overset{O}{\|}}{C}-O-CH_2$$

An *oil*

$$CH_3-(CH_2)_7-CH_2-CH_2-(CH_2)_7-\overset{\overset{O}{\|}}{C}-O-CH_2$$

$$CH_3-(CH_2)_7-CH_2-CH_2-(CH_2)_7-\overset{\overset{O}{\|}}{C}-O-CH$$

$$CH_3-(CH_2)_7-CH_2-CH_2-(CH_2)_7-\overset{\overset{O}{\|}}{C}-O-CH_2$$

A *fat*

Question 11.9

The *trans* isomer of 2-pentene was used in Practice Problem 11.6b. Would the result be any different if the *cis* isomer had been used?

Question 11.10

Write balanced equations for the hydrogenation of 1-butene and *cis*-2-butene.

Question 11.11

Write balanced equations for the complete hydrogenation of each of the following alkynes:

   a. $H_3CC\equiv CCH_3$
   b. $H_3CC\equiv CCH_2CH_3$

Question 11.12

Using the I.U.P.A.C. Nomenclature System, name each of the products and reactants in the reactions described in Question 11.11.

Saturated and unsaturated dietary fats are discussed in Section 17.2.

Hydrogenation is used in the food industry to produce margarine, which is a mixture of hydrogenated vegetable oils (Figure 11.4). Vegetable oils are unsaturated, that is, they contain many double bonds and as a result have low melting points and are liquid at room temperature. The hydrogenation of these double bonds to single bonds increases the melting point of these oils and results in a fat, such as Crisco, that remains solid at room temperature. A similar process with highly refined corn oil and added milk solids produces corn oil margarine. As we saw in the previous section, such margarine may contain *trans*-fatty acids as a result of hydrogenation.

Halogenation: Addition of X₂

Chlorine (Cl_2) or bromine (Br_2) can be added to a double bond. This reaction, called **halogenation**, proceeds readily and does not require a catalyst:

Alkene   Halogen   Alkyl dihalide

Writing Equations for the Halogenation of Alkenes

EXAMPLE 11.7

Write a balanced equation showing (a) the chlorination of 1-pentene and (b) the bromination of *trans*-2-butene.

LEARNING GOAL

4 ▶ Write equations predicting the products of addition reactions of alkenes and alkynes: hydrogenation, halogenation, hydration, and hydrohalogenation.

Solution

(a) Begin by drawing the structure of 1-pentene and of diatomic chlorine (Cl_2).

1-Pentene   Chlorine

Knowing that one chlorine atom will form a covalent bond with each of the carbon atoms of the carbon-carbon double bond, we can write the product and complete the equation.

1-Pentene   Chlorine   1,2-Dichloropentane

(b) Begin by drawing the structure of *trans*-2-butene and of diatomic bromine (Br_2).

trans-2-Butene   Bromine

Knowing that one bromine atom will form a covalent bond with each of the carbon atoms of the carbon-carbon double bond, we can write the product and complete the equation.

trans-2-Butene   Bromine   2,3-Dibromobutane

Practice Problem 11.7

Write a balanced equation for the chlorination of each of the following alkenes.

a. 2,4-Hexadiene
b. 1,5-Dibromo-2-pentene
c. 2,3-Dimethyl-1-butene
d. 4,6-Dimethyl-2-heptene

For Further Practice: Questions 11.57, 11.67b, and 11.84.

Alkynes also react with the halogens bromine or chlorine. Two moles of halogen add to the triple bond to produce a tetrahaloalkane:

$$
R-C{\equiv}C-R + 2X_2 \longrightarrow
\begin{array}{c}
X \quad X \\
| \quad | \\
R-C-C-R \\
| \quad | \\
X \quad X
\end{array}
$$

Alkyne   Halogen   Tetrahaloalkane
            (2 mol)

Question 11.13

Write a balanced equation for the addition of bromine to each of the following alkenes. Draw and name the products and reactants for each reaction.

a. $CH_3CH{=}CH_2$
b. $CH_3CH{=}CHCH_3$

Question 11.14

Using the I.U.P.A.C. Nomenclature System, name each of the products and reactants in the reactions described in Question 11.13.

Question 11.15

Write balanced equations for the complete chlorination of each of the following alkynes:

a. $H_3CC{\equiv}CCH_3$
b. $H_3CC{\equiv}CCH_2CH_3$

Question 11.16

Using the I.U.P.A.C. Nomenclature System, name each of the products and reactants in the reactions described in Question 11.15.

Below we see an equation representing the bromination of 1-pentene. Notice that the solution of reactants is red because of the presence of bromine. However, the product is colorless (Figure 11.5).

$$
CH_3CH_2CH_2CH{=}CH_2 + Br_2 \longrightarrow CH_3CH_2CH_2\underset{\underset{Br}{|}}{C}H\underset{\underset{Br}{|}}{C}H_2
$$

1-Pentene   Bromine   1,2-Dibromopentane
(colorless)   (red)   (colorless)

This bromination reaction can be used to show the presence of double or triple bonds in an organic compound. The reaction mixture is red because of the presence of dissolved bromine. If the red color is lost, the bromine has been consumed. Thus bromination has occurred, and the compound must have had a carbon-carbon double or triple bond. The greater the amount of bromine that must be added to the reaction, the more unsaturated the compound is. For instance, a diene or an alkyne would consume twice as much bromine as an alkene with a single double bond.

Hydration: Addition of H₂O

A water molecule can be added to an alkene. This reaction, termed **hydration**, requires a trace of strong acid (H^+) as a catalyst. The product is an alcohol, as shown in the following reaction:

Figure 11.5
Bromination of an alkene. The solution on the left is red because of the presence of bromine and absence of an alkene or alkyne. In the presence of an unsaturated hydrocarbon, the bromine is used in the reaction and the solution becomes colorless.

The following equation shows the hydration of ethene to produce ethanol.

Ethene   Water   Ethanol
(ethyl alcohol)

With alkenes in which the groups attached to the two carbons of the double bond are different (unsymmetrical alkenes), two products are possible. For example:

Propene
(propylene)

Major product
2-Propanol
(isopropyl alcohol)

Minor product
1-Propanol
(propyl alcohol)

LEARNING GOAL

5 Apply Markovnikov's rule to predict the major and minor products of the hydration and hydrohalogenation reactions of unsymmetrical alkenes.

When hydration of an unsymmetrical alkene, such as propene, is carried out in the laboratory, one product (2-propanol) is favored over the other. The Russian chemist Vladimir Markovnikov studied many such reactions and came up with a rule that can be used to predict the major product of such a reaction. **Markovnikov's rule** tells us that the carbon of the carbon-carbon double bond that originally has more hydrogen atoms receives the hydrogen atom being added to the double bond. The remaining carbon forms a bond with the —OH. Simply stated, "the rich get richer"—the carbon with the greater number of hydrogens gets the new one as well. In the preceding example, carbon-1 has two C—H bonds originally, and carbon-2 has only one. The major product, 2-propanol, results from the new C—H bond forming on carbon-1 and the new C—OH bond on carbon-2.

Addition of water to a double bond is a reaction that we find in several biochemical pathways. For instance, the citric acid cycle is a key metabolic pathway in the complete oxidation of the sugar glucose and the release of the majority of the energy used by the body. The citric acid cycle is also the source of starting materials for the synthesis of the biological molecules needed for life. The next-to-last reaction in the citric acid cycle is the hydration of a molecule of fumarate to produce a molecule called malate.

> We have seen that hydration of a double bond requires a trace of acid as a catalyst. In the cell, this reaction is catalyzed by an enzyme, or biological catalyst, called fumarase.

$$
\begin{array}{ccc}
\underset{\text{Fumarate}}{\begin{array}{c} COO^- \\ | \\ C-H \\ \| \\ H-C \\ | \\ COO^- \end{array}} + H_2O & \xrightarrow{\text{Fumarase}} & \underset{\text{Malate}}{\begin{array}{c} COO^- \\ | \\ HO-C-H \\ | \\ H-C-H \\ | \\ COO^- \end{array}}
\end{array}
$$

Hydration of a double bond also occurs in the β-oxidation pathway (see Section 23.2). This pathway carries out the oxidation of dietary fatty acids. Like the citric acid cycle, β-oxidation harvests the energy of the food molecules to use for body functions.

E X A M P L E 11.8   **Writing Equations for the Hydration of Symmetrical and Unsymmetrical Alkenes**

More and more frequently we see ethanol used as an additive in the gasoline we buy. A mixture of 10% ethanol and 90% gasoline significantly raises the octane rating of the fuel. In fact, some cities where auto emissions may reach harmful levels require the use of 10% ethanol gasoline. Bioethanol is produced by fermentation of crops such as corn. However, it can also be manufactured by the hydration of ethene. Write a balanced equation showing the hydration of ethene.

Solution

> **LEARNING GOAL**
> **4** Write equations predicting the products of addition reactions of alkenes and alkynes: hydrogenation, halogenation, hydration, and hydrohalogenation.

$$
\underset{\text{Ethene}}{\begin{array}{c} H \\ \diagdown \\ \\ H \diagup \end{array} C=C \begin{array}{c} H \\ \diagup \\ \\ \diagdown H \end{array}} + HOH \xrightarrow{H^+} \underset{\text{Ethanol}}{\begin{array}{c} H\ \ H \\ | \ \ | \\ H-C-C-H \\ | \ \ | \\ H\ \ OH \end{array}}
$$

Write an equation showing all the products of the hydration of the unsymmetrical alkene 1-pentene.

Solution

Begin by drawing the structure of 1-pentene and of water and indicating the catalyst.

> **LEARNING GOAL**
> **5** Apply Markovnikov's rule to predict the major and minor products of the hydration and hydrohalogenation reactions of unsymmetrical alkenes.

$$
\underset{\text{1-Pentene}}{\begin{array}{c} H \\ \diagdown \\ \\ CH_3CH_2CH_2 \diagup \end{array} C=C \begin{array}{c} H \\ \diagup \\ \\ \diagdown H \end{array}} + \underset{\text{Water}}{H-OH} \xrightarrow{H^+}
$$

Continued—

EXAMPLE 11.8 —*Continued*

Markovnikov's rule tells us that the carbon atom that is already bonded to the greater number of hydrogen atoms is more likely to receive the hydrogen atom from the water molecule. The other carbon atom is more likely to become bonded to the hydroxyl group. Thus we can predict that the major product of this reaction will be 2-pentanol and that the minor product will be 1-pentanol. Now we can complete the equation by showing the products:

$$
\underset{\substack{\text{2-Pentanol}\\\text{(major product)}}}{CH_3CH_2CH_2-\overset{\overset{\displaystyle H}{|}}{\underset{\underset{\displaystyle OH}{|}}{C}}-\overset{\overset{\displaystyle H}{|}}{\underset{\underset{\displaystyle H}{|}}{C}}-H} + \underset{\substack{\text{1-Pentanol}\\\text{(minor product)}}}{CH_3CH_2CH_2-\overset{\overset{\displaystyle H}{|}}{\underset{\underset{\displaystyle H}{|}}{C}}-\overset{\overset{\displaystyle H}{|}}{\underset{\underset{\displaystyle OH}{|}}{C}}-H}
$$

Practice Problem 11.8

Write a balanced equation for the hydration of each of the following alkenes.

a. 1-Butene
b. 2-Methyl-3-hexene
c. Propene
d. 1,4-Dichloro-2-butene

For Further Practice: Questions 11.59, 11.68, and 11.79.

A mixture of 10% ethanol and 90% gasoline has an octane rating of 93–94. One way to produce ethanol for fuel is the fermentation of the glucose made from corn, which we will study in Section 21.4. To some this represents a renewable energy resource; but others argue that too much energy is required to produce it. Go online to read more about the controversy surrounding biofuels.

Hydration of an alkyne is a more complex process because the initial product is not stable and is rapidly isomerized. As you would expect, the product is an alcohol but, in this case, one in which the hydroxyl group is bonded to one of the carbons of a carbon-carbon double bond. This type of molecule is called an *enol* because it is both an alkene *(ene)* and an alcohol *(ol)*. The enol cannot be isolated from the reaction mixture because it is so quickly isomerized into either an aldehyde or ketone, as shown in the following general reaction:

The chemistry of enols, aldehydes, and ketones is found in Chapter 13. In Chapter 21, we will study metabolic reactions involving the enol, phospho-enolpyruvate, which is an essential intermediate in the energy-harvesting pathway called glycolysis.

$$
\underset{\text{Alkyne}}{R-C\equiv C-R'} + \underset{\text{Water}}{H_2O} \longrightarrow \underset{\text{Enol}}{R-\overset{\overset{\displaystyle H}{|}}{C}=\underset{\underset{\displaystyle OH}{|}}{C}-R'} \longrightarrow \underset{\substack{\text{Aldehyde if R}'=\text{H}\\\text{Ketone if R}'=\text{alkyl group}}}{R-\overset{\overset{\displaystyle H}{|}}{\underset{\underset{\displaystyle H}{|}}{C}}-\overset{\overset{\displaystyle }{\|}}{\underset{\underset{\displaystyle O}{}}{C}}-R'}
$$

Question 11.17

Write a balanced equation for the hydration of each of the following alkenes. Predict the major product of each of the reactions.

a. $CH_3CH=CHCH_3$

b. $CH_2=CHCH_2CH_2\underset{\underset{\displaystyle CH_3}{|}}{C}HCH_3$

c. $CH_3CH_2CH_2CH=CHCH_2CH_3$

d. $CH_3CHClCH=CHCHClCH_3$

Question 11.18

Write a balanced equation for the hydration of each of the following alkenes. Predict the major product of each of the reactions.

a. $CH_2{=}CHCH_2CH_2CH_3$
b. $CH_3CH_2CH_2CH{=}CHCH_3$
c. $CH_3CHBrCH_2CH{=}CHCH_2Cl$
d. $CH_3CH_2CH_2CH_2CH_2CH{=}CHCH_3$

Question 11.19

Write balanced equations for the complete hydration of each of the following alkynes:

a. $H_3CC{\equiv}CH$
b. $H_3CC{\equiv}CCH_2CH_3$

Question 11.20

Is the final product in each of the reactions in Question 11.19 an aldehyde or a ketone?

Hydrohalogenation: Addition of HX

A hydrogen halide (HBr, HCl, or HI) also can be added to an alkene. The product of this reaction, called **hydrohalogenation,** is an alkyl halide:

Alkene   Hydrogen halide   Alkyl halide

Ethene   Hydrogen bromide   Bromoethane

This reaction also follows Markovnikov's rule. That is, if HX is added to an unsymmetrical alkene, the hydrogen atom will be added preferentially to the carbon atom that originally had the most hydrogen atoms. Consider the following example:

Propene   Major product   Minor product
                           2-Bromopropane   1-Bromopropane

EXAMPLE 11.9   **Writing Equations for the Hydrohalogenation of Alkenes**

LEARNING GOAL

4 Write equations predicting the products of addition reactions of alkenes and alkynes: hydrogenation, halogenation, hydration, and hydrohalogenation.

Write an equation showing all the products of the hydrohalogenation of 1-pentene with HCl.

Continued—

A Human *Perspective*

Folklore, Science, and Technology

For many years it was suspected that there existed a gas that stimulated fruit ripening and had other effects on plants. The ancient Chinese observed that their fruit ripened more quickly if incense was burned in the room. Early in this century, shippers realized that they could not store oranges and bananas on the same ships because some "emanation" given off by the oranges caused the bananas to ripen too early.

Puerto Rican pineapple growers and Philippine mango growers independently developed a traditional practice of building bonfires near their crops. They believed that the smoke caused the plants to bloom synchronously.

In the mid–nineteenth century, streetlights were fueled with natural gas. Occasionally the pipes leaked, releasing gas into the atmosphere. On some of these occasions, the leaves fell from all the shade trees in the region surrounding the gas leak.

What is the gas responsible for these diverse effects on plants? In 1934, R. Gane demonstrated that the simple alkene ethylene was the "emanation" responsible for fruit ripening. More recently, it has been shown that ethylene induces and synchronizes flowering in pineapples and mangos, induces senescence (aging) and loss of leaves in trees, and effects a wide variety of other responses in various plants.

We can be grateful to ethylene for the fresh, unbruised fruits that we can purchase at the grocery store. These fruits are picked when they are not yet ripe, while they are still firm. They then can be shipped great distances and gassed with ethylene when they reach their destination. Under the influence of the ethylene, the fruit ripens and is displayed in the store.

The history of the use of ethylene to bring fresh ripe fruits and vegetables to markets thousands of miles from the farms is an interesting example of the scientific process and its application for the benefit of society. Scientists began with the curious observations of Chinese, Puerto Rican, and Filipino farmers. Through experimentation they came to understand the phenomenon that caused the observations. Finally, through

Ethylene, the simplest alkene, is used to ripen fruits after shipment to market. Many vegetables must be protected from ethylene because it causes them to yellow or wilt.

technology, scientists have made it possible to harness the power of ethylene so that grocers can "artificially" ripen the fruits and vegetables they sell to us.

For Further Understanding

Try the following experiment: place a ripe banana in a paper bag with an unripe tomato. Place a second unripe tomato on the countertop. Check daily to see which ripens more quickly.

Investigate the development of the painkiller aspirin to learn about another molecule whose biological activity was discovered accidentally.

EXAMPLE 11.9 —*Continued*

Solution

Begin by drawing the structure of 1-pentene and of hydrochloric acid.

$$
\begin{array}{c}
\underset{\displaystyle \text{CH}_3\text{CH}_2\text{CH}_2}{\overset{\displaystyle \text{H}}{\diagdown}} \text{C} = \text{C} \overset{\displaystyle \text{H}}{\underset{\displaystyle \text{H}}{\diagup}} \quad + \quad \text{H—Cl} \longrightarrow
\end{array}
$$

1-Pentene   Hydrochloric
                 acid

LEARNING GOAL

5 Apply Markovnikov's rule to predict the major and minor products of the hydration and hydrohalogenation reactions of unsymmetrical alkenes.

Continued—

EXAMPLE 11.9 —*Continued*

Markovnikov's rule tells us that the carbon atom that is already bonded to the greater number of hydrogen atoms is more likely to receive the hydrogen atom of the hydrochloric acid molecule. The other carbon atom is more likely to become bonded to the chlorine atom. Thus, we can complete the equation by writing the major and minor products.

$$\underset{\substack{\text{2-Chloropentane}\\ \text{(major product)}}}{\text{CH}_3\text{CH}_2\text{CH}_2-\overset{\displaystyle \text{H}}{\underset{\displaystyle \text{Cl}}{\text{C}}}-\overset{\displaystyle \text{H}}{\underset{\displaystyle \text{H}}{\text{C}}}-\text{H}} + \underset{\substack{\text{1-Chloropentane}\\ \text{(minor product)}}}{\text{CH}_3\text{CH}_2\text{CH}_2-\overset{\displaystyle \text{H}}{\underset{\displaystyle \text{H}}{\text{C}}}-\overset{\displaystyle \text{H}}{\underset{\displaystyle \text{Cl}}{\text{C}}}-\text{H}}$$

As predicted by Markovnikov's rule, the major product is 2-chloropentane and the minor product is 1-chloropentane.

 Practice Problem 11.9

Write a balanced equation for the hydrobromination of each of the following alkenes.

a. 2-Pentene
b. Propene
c. 2-Methyl-3-heptene
d. Ethene

For Further Practice: Questions 11.69 and 11.70.

Addition Polymers of Alkenes

 Animation

Natural and Synthetic Polymers

Polymers are macromolecules composed of repeating structural units called **monomers.** A polymer may be made up of several thousand monomers. Many commercially important plastics and fibers are addition polymers made from alkenes or substituted alkenes. They are called **addition polymers** because they are made by the sequential addition of the alkene monomer. The general formula for this addition reaction follows:

The product of the reaction is generally represented in a simplified manner:

$$\left[\begin{array}{c} \overset{\displaystyle \text{R}}{\underset{\displaystyle \text{R}}{\text{C}}}-\overset{\displaystyle \text{R}}{\underset{\displaystyle \text{R}}{\text{C}}} \end{array} \right]_n$$

Polyethylene is a polymer made from the monomer ethylene (ethene):

A Human Perspective

Life Without Polymers?

What do Nike Air-Sole shoes, Saturn automobiles, disposable diapers, tires, shampoo, and artificial joints and skin share in common? These products and a great many other items we use every day are composed of synthetic or natural polymers. Indeed, the field of polymer chemistry has come a long way since the 1920s and 1930s when DuPont chemists invented nylon and Teflon.

Consider the disposable diaper. The outer, waterproof layer is composed of polyethylene. The polymerization reaction that produces polyethylene is shown in Section 11.5. The diapers have elastic to prevent leaking. The elastic is made of a natural polymer, rubber. The monomer from which natural rubber is formed is 2-methyl-1,3-butadiene. The common name of this monomer is *isoprene*. As we will see in coming chapters, isoprene is an important monomer in the synthesis of many natural polymers.

$$nCH_2=\overset{\overset{\displaystyle CH_3}{|}}{C}-CH=CH_2 \longrightarrow \left[CH_2-\overset{\overset{\displaystyle CH_3}{|}}{C}=CH-CH_2\right]_n$$

2-Methyl-1, 3-butadiene   Rubber polymer
(isoprene)

The diaper is filled with a synthetic polymer called poly(acrylic acid). This polymer has the remarkable ability to absorb many times its own weight in liquid. Polymers that have this ability are called superabsorbers, but polymer chemists have no idea why they have this property! The acrylic acid monomer and resulting poly(acrylic acid) polymer are shown here:

$$n\,\overset{\overset{\displaystyle H}{\diagdown}}{\underset{\underset{\displaystyle H}{\diagup}}{C}}=\overset{\overset{\displaystyle H}{\diagup}}{\underset{\underset{\displaystyle \underset{O}{\underset{|}{\overset{|}{C}=O}}}{|}}{C}} \longrightarrow \left[CH_2-\overset{CH}{\underset{\underset{\overset{|}{O}}{\underset{|}{\overset{|}{C}=O}}}{|}}\right]_n$$

Acrylic acid monomer   Poly(acrylic acid)

Another example of a useful polymer is Gore-Tex. This amazing polymer is made by stretching Teflon. Teflon is produced from the monomer tetrafluoroethene, as seen in the following reaction:

$$n\,\overset{\overset{\displaystyle F}{\diagup}}{\underset{\underset{\displaystyle F}{\diagdown}}{C}}=\overset{\overset{\displaystyle F}{\diagdown}}{\underset{\underset{\displaystyle F}{\diagup}}{C} \longrightarrow \left[\begin{matrix}F & F\\ | & |\\ C-C\\ | & |\\ F & F\end{matrix}\right]_n}$$

Tetrafluoroethene   Teflon

Clothing made from this fabric is used to protect firefighters because of its fire resistance. Because it also insulates, Gore-Tex clothing is used by military forces and by many amateur athletes, for protection during strenuous activity in the cold. In addition to its use in protective clothing, Gore-Tex has been used in millions of medical procedures for sutures, synthetic blood vessels, and tissue reconstruction.

For Further Understanding

Visit The Macrogalleria, www.psrc.usm.edu/macrog/index.htm, an Internet site maintained by the Department of Polymer Science of the University of Southern Mississippi, to help you answer these questions:

Why does shrink wrap shrink?

What are optical fibers made of and how do they transmit light?

$$nCH_2=CH_2 \longrightarrow \left[CH_2-CH_2\right]_n$$

Ethene   Polyethylene
(ethylene)

It is used to make bottles, injection-molded toys and housewares, and wire coverings.

Polypropylene is a plastic made from propylene (propene). It is used to make indoor-outdoor carpeting, packaging materials, toys, and housewares. When propylene polymerizes, a methyl group is located on every other carbon of the main chain:

$$nCH_2=\overset{\overset{\displaystyle CH_3}{|}}{CH} \longrightarrow \left[CH_2-\overset{\overset{\displaystyle CH_3}{|}}{CH}\right]_n$$

An Environmental *Perspective*

Plastic Recycling

Plastics, first developed by British inventor Alexander Parkes in 1862, are amazing substances. Some serve as containers for many of our foods and drinks, keeping them fresh for long periods of time. Other plastics serve as containers for detergents and cleansers or are formed into pipes for our plumbing systems. We have learned to make strong, clear sheets of plastic that can be used as windows, and feather-light plastics that can be used as packaging materials. In the United States alone, seventy-five billion pounds of plastics are produced each year.

But plastics, amazing in their versatility, are a mixed blessing. One characteristic that makes them so useful, their stability, has created an environmental problem. It may take forty to fifty years for plastics discarded into landfill sites to degrade. Concern that we could soon be knee-deep in plastic worldwide has resulted in a creative new industry: plastic recycling.

Since there are so many types of plastics, it is necessary to identify, sort, and recycle them separately. To help with this sorting process, manufacturers place recycling symbols on their plastic wares. As you can see in the accompanying table, each symbol corresponds to a different type of plastic.

Polyethylene terephthalate, also known as PETE or simply #1, is a form of polyester often used to make bottles and jars to contain food. When collected, it is ground up into flakes and formed into pellets. The most common use for recycled PETE is the manufacture of polyester carpets. But it may also be spun into a cotton-candy-like form that can be used as a fiber filling for pillows or sleeping bags. It may also be rolled into thin sheets or ribbons and used as tapes for VCRs or tape decks. Reuse to produce bottles and jars is also common.

HDPE, or #2, is high-density polyethylene. Originally used for milk and detergent bottles, recycled HDPE is used to produce pipes, plastic lumber, trash cans, or bottles for storage of materials other than food. LDPE, #4, is very similar to HDPE chemically. Because it is a more highly branched polymer, it is less-dense and more flexible. Originally used to produce plastic bags, recycled LDPE is also used to make trash bags, grocery bags, and plastic tubing and lumber.

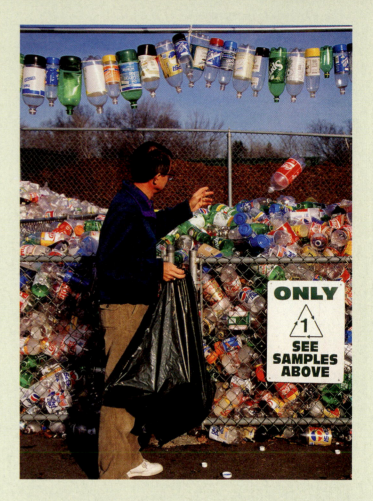

PVC, or #3, is one of the less commonly recycled plastics in the United States, although it is actively recycled in Europe. The recycled material is used to make non-food-bearing containers, shoe soles, flooring, sweaters, and pipes. Polypropylene,

or

$$\sim\!\!\sim\!\!\sim CH_2-\!\!\underset{\underset{CH_3}{|}}{CH}-CH_2-\!\!\underset{\underset{CH_3}{|}}{CH}-CH_2-\!\!\underset{\underset{CH_3}{|}}{CH}\sim\!\!\sim\!\!\sim$$

Polymers made from alkenes or substituted alkenes are simply very large alkanes or substituted alkanes. Like the alkanes, they are typically inert. This chemical inertness makes these polymers ideal for making containers to hold juices, chemicals, and fluids used medically. They are also used to make sutures, catheters, and other indwelling devices. A variety of polymers made from substituted alkenes are listed in Table 11.2.

Code	Type	Name	Formula	Description	Examples
PETE	△1	Polyethylene terephthalate	$-CH_2-CH_2-O-\overset{\overset{O}{\|}}{C}-\hspace{-0.2em}\bigcirc\hspace{-0.2em}-\overset{\overset{\hspace{1em}}{C}}{\underset{O}{\|}}-O-$	Usually clear or green, rigid	Soda bottles, peanut butter jars, vegetable oil bottles
HDPE	△2	High-density polyethylene	$-CH_2-CH_2-$	Semirigid	Milk and water jugs, juice and bleach bottles
PVC	△3	Polyvinyl chloride	$-\overset{\overset{}{\underset{Cl}{\|}}}{CH}-CH_2-$	Semirigid	Detergent and cleanser bottles, pipes
LDPE	△4	Low-density polyethylene	$-CH_2-CH_2-$	Flexible, not crinkly	Six-pack rings, bread bags, sandwich bags
PP	△5	Polypropylene	$-\overset{\overset{}{\underset{CH_3}{\|}}}{CH}-CH_2-$	Semirigid	Margarine tubs, straws, screw-on lids
PS	△6	Polystyrene	$-CH-CH_2-$ with phenyl group	Often brittle	Styrofoam, packing peanuts, egg cartons, foam cups
Other	△7	Multilayer plastics	N/A	Squeezable	Ketchup and syrup bottles

PP or #5, is found in margarine tubs, fabrics, and carpets. Recycled polypropylene has many uses, including fabrication of gardening implements.

You probably come into contact with polystyrene, PS or #6, almost every day. It is used to make foam egg cartons and meat trays, serving containers for fast food chains, CD "jewel boxes," and "peanuts" used as packing material. At the current time, polystyrene food containers are not recycled. PS from nonfood products can be melted down and converted into pellets that are used to manufacture office desktop accessories, hangers, video and audio cassette housings, and plastic trays used to hold plants.

For Further Understanding

Use the Internet or other resources to investigate recycling efforts in your area.

In some areas, efforts to recycle plastics and paper have been abandoned. What factors contributed to this?

TABLE 11.2 Some Important Addition Polymers of Alkenes

Monomer Name	Formula	Polymer	Uses
Styrene	$CH_2=CH-\bigcirc$	Polystyrene	Styrofoam containers
Acrylonitrile	$CH_2=CHCN$	Polyacrylonitrile (Orlon)	Clothing
Methyl methacrylate	$CH_2=C(CH_3)-\overset{\overset{O}{\|}}{C}OCH_3$	Polymethyl methacrylate (Plexiglas, Lucite)	Basketball backboards
Vinyl chloride	$CH_2=CHCl$	Polyvinyl chloride (PVC)	Plastic pipe, credit cards
Tetrafluoroethene	$CF_2=CF_2$	Polytetrafluoroethylene (Teflon)	Nonstick surfaces

11.6 Aromatic Hydrocarbons

In the early part of the nineteenth century, chemists began to discover organic compounds with chemical properties quite distinct from the alkanes, alkenes, and alkynes. They called these substances *aromatic compounds* because many of the first examples were isolated from the pleasant-smelling resins of tropical trees. The carbon:hydrogen ratio of these compounds suggested a very high degree of unsaturation, similar to the alkenes and alkynes. Imagine, then, how puzzled these early organic chemists must have been when they discovered that these compounds do not undergo the kinds of addition reactions common for the alkenes and alkynes.

$$CH_2{=}CH_2 + Br_2 \longrightarrow \underset{\underset{Br}{|}}{CH_2}{-}\underset{\underset{Br}{|}}{CH_2}$$

benzene $+ Br_2 \longrightarrow$ No reaction

We no longer define aromatic compounds as those having a pleasant aroma; in fact, many do not. We now recognize **aromatic hydrocarbons** as those that exhibit a much higher degree of chemical stability than their chemical composition would predict. The most common group of aromatic compounds is based on the six-member aromatic ring, the benzene ring. The structure of the benzene ring is represented in various ways in Figure 11.6.

Structure and Properties

The benzene ring consists of six carbon atoms joined in a planar hexagonal arrangement. Each carbon atom is bonded to one hydrogen atom. Friedrich Kekulé proposed a model for the structure of benzene in 1865. He proposed that single and double bonds alternated around the ring (a conjugated system of double bonds). To explain why benzene did not decolorize bromine—in other words, didn't react like an unsaturated compound—he suggested that the double and single bonds shift positions rapidly.

Resonance models are described in Section 3.4.

Actually, the most accurate way to represent the benzene molecules is as a resonance hybrid of two Keluké structures:

Benzene as a resonance hybrid

The current model of the structure of benzene is based on the idea of overlapping orbitals. Each carbon is bonded to two others by sharing a pair of electrons. Each carbon atom also shares a pair of electrons with a hydrogen atom. The

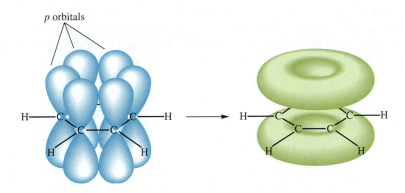

Figure 11.6
Four ways to represent the benzene molecule. Structure (b) is a simplified diagram of structure (a). Structure (d), a simplified diagram of structure (c), is the most commonly used representation.

Figure 11.7
The current model of the bonding in benzene.

remaining six electrons are located in *p* orbitals that are perpendicular to the plane of the ring. These *p* orbitals overlap laterally to form a cloud of electrons above and below the ring (Figure 11.7).

Two symbols are commonly used to represent the benzene ring. The representation in Figure 11.6b is the structure proposed by Kekulé. The structure in Figure 11.6d uses a circle to represent the electron cloud.

Nomenclature

Most simple aromatic compounds are named as derivatives of benzene. Thus benzene is the parent compound, and the name of any atom or group bonded to benzene is used as a prefix, as in these examples:

LEARNING GOAL

7 Draw the structures and write the names of common aromatic hydrocarbons.

Nitrobenzene   Ethylbenzene   Bromobenzene   Benzoic acid   Benzaldehyde

Other members of this family have unique names based on history rather than logic:

| Toluene | Phenol | Aniline | Anisole |

Other common names include xylene, a benzene ring bonded to two methyl groups, and cresol, a benzene ring bonded to a methyl group and a hydroxyl group.

When two groups are present on the ring, three possible orientations exist, and they may be named by either the I.U.P.A.C. Nomenclature System or the common system of nomenclature. If the groups or atoms are located on two adjacent carbons, they are referred to as *ortho (o)* in the common system or with the prefix 1,2- in the I.U.P.A.C. system. If they are on carbons separated by one carbon atom, they are termed *meta (m)* in the common system or 1,3- in the I.U.P.A.C. system. Finally, if the substituents are on carbons separated by two carbon atoms, they are said to be *para (p)* in the common system or 1,4- in the I.U.P.A.C. system. The following examples demonstrate both of these systems:

Two groups 1,2 or *ortho*   Two groups 1,3 or *meta*   Two groups 1,4 or *para*
G = Any group

The three orientations of xylene and cresol are shown here:

| *ortho*-Xylene | *meta*-Xylene | *para*-Xylene |

| *ortho*-Cresol | *meta*-Cresol | *para*-Cresol |

If three or more groups are attached to the benzene ring, numbers must be used to describe their location. The names of the substituents are given in alphabetical order.

Naming Derivatives of Benzene

EXAMPLE 11.10

Name the following compounds using the I.U.P.A.C. and common systems of nomenclature.

a.

b.

c.

Solution

I.U.P.A.C names:

Parent compound:	toluene	phenol	aniline
Substituents:	2-chloro	4-nitro	3-ethyl
Name:	2-Chlorotoluene	4-Nitrophenol	3-Ethylaniline

Common names:

Parent compound:	toluene	phenol	aniline
Substituents:	*ortho*-chloro	*para*-nitro	*meta*-ethyl
Name:	*ortho*-Chlorotoluene	*para*-Nitrophenol	*meta*-Ethylaniline
Abbreviated Name:	*o*-Chlorotoluene	*p*-Nitrophenol	*m*-Ethylaniline

? Practice Problem 11.10

Name the following compounds using the I.U.P.A.C. and common nomenclature systems.

a.

b.

c.

d. (benzene ring with Br and Br substituents)

For Further Practice: Questions 11.90, 11.91, and 11.92.

In I.U.P.A.C. nomenclature, the group derived by removing one hydrogen from benzene (—C$_6$H$_5$), is called the **phenyl group.** An aromatic hydrocarbon with an aliphatic side chain is named as a phenyl substituted hydrocarbon. For example:

$$\overset{1}{C}H_3\overset{2}{C}H\overset{3}{C}H_2\overset{4}{C}H_3$$

2-Phenylbutane

$$\overset{4}{C}H_3\overset{3}{C}H\overset{2}{C}H=\overset{1}{C}H_2$$

3-Phenyl-1-butene

One final special name that occurs frequently in aromatic compounds is the benzyl group:

$$C_6H_5CH_2— \quad \text{or} \quad \text{⬡}—CH_2—$$

The use of this group name is illustrated by:

⬡—CH₂Cl   ⬡—CH₂OH

Benzyl chloride   Benzyl alcohol

Question 11.21

Draw each of the following compounds:

   a. 1,3,5-Trichlorobenzene
   b. *ortho*-Cresol
   c. 2,5-Dibromophenol
   d. *para*-Dinitrobenzene
   e. 2-Nitroaniline
   f. *meta*-Nitrotoluene

Question 11.22

Draw each of the following compounds:

   a. 2,3-Dichlorotoluene   d. *o*-Nitrotoluene
   b. 3-Bromoaniline   e. *p*-Xylene
   c. 1-Bromo-3-ethylbenzene   f. *o*-Dibromobenzene

Polynuclear Aromatic Hydrocarbons

The polynuclear aromatic hydrocarbons (PAH) are composed of two or more aromatic rings joined together. Many of them have been shown to be carcinogenic, that is, they cause cancer.

Naphthalene

Anthracene

Phenanthrene

Benzopyrene

Naphthalene has a distinctive aroma. It has been frequently used as mothballs and may cause hemolytic anemia (a condition causing breakdown of red blood cells) in humans, but has not been associated with human or animal cancers. Anthracene, derived from coal tar, is the parent compound of many dyes and pigments. Phenanthrene is common in the environment, although there is no industrial use of the compound. It is a product of incomplete combustion of fossil fuels and wood. Although anthracene is a suspected carcinogen, phenanthrene has not been shown to be one. Benzopyrene is found in tobacco smoke, smokestack effluents, charcoal-grilled meat, and automobile exhaust. It is one of the most potent carcinogens known.

Reactions Involving Benzene

As we have noted, benzene does not readily undergo addition reactions. The typical reactions of benzene are **substitution reactions,** in which a hydrogen atom is replaced by another atom or group of atoms.

LEARNING GOAL

8 Write equations for substitution reactions involving benzene.

Benzene can react (by substitution) with Cl_2 or Br_2. These reactions require either iron or an iron halide as a catalyst. For example:

Benzene   Chlorine   Chlorobenzene

Benzene   Bromine   Bromobenzene

When a second equivalent of the halogen is added, three isomers—*para, ortho,* and *meta*—are formed.

Benzene also reacts with sulfur trioxide by substitution. Concentrated sulfuric acid is required as the catalyst. Benzenesulfonic acid, a strong acid, is the product:

Benzene   Sulfur trioxide   Benzenesulfonic acid

Benzene can also undergo nitration with concentrated nitric acid dissolved in concentrated sulfuric acid. This reaction requires temperatures in the range of 50–55°C.

Benzene   Nitric acid   Nitrobenzene

11.7 Heterocyclic Aromatic Compounds

Heterocyclic aromatic compounds are those having at least one atom other than carbon as part of the structure of the aromatic ring. The structures and common names of several heterocyclic aromatic compounds are shown:

Pyridine   Pyrimidine   Purine

Imidazole   Furan   Pyrrole

All these compounds are more similar to benzene in stability and chemical behavior than they are to the alkenes. Many of these compounds are components of molecules that have significant effects on biological systems. For instance, the purines and pyrimidines are components of DNA (deoxyribonucleic acid) and RNA (ribonucleic acid). DNA and RNA are the molecules responsible for storing and expressing the genetic information of an organism. The pyridine ring is found in nicotine, the addictive compound in tobacco. The pyrrole ring is a component of the porphyrin ring found in hemoglobin and chlorophyll.

See the Chemistry Connection: The Nicotine Patch in Chapter 15.

Porphyrin

The imidazole ring is a component of cimetidine, a drug used in the treatment of stomach ulcers. The structure of cimetidine is shown below:

Cimetidine

We will discuss a subset of the heterocyclic aromatic compounds, the heterocyclic amines, in Chapter 15.

Summary of Reactions

Addition Reactions of Alkenes

Hydrogenation:

Alkene + Hydrogen $\xrightarrow[\text{Heat or pressure}]{\text{Pt, Pd, or Ni}}$ Alkane

Hydration:

Alkene + Water $\xrightarrow{\text{H}^+}$ Alcohol

Halogenation:

Alkene + Halogen $\longrightarrow$ Alkyl dihalide

Hydrohalogenation:

Alkene + Hydrogen halide $\longrightarrow$ Alkyl halide

Addition Polymers of Alkenes

Alkene monomer   Addition polymer

Reactions of Benzene

Halogenation:

Benzene + Halogen $\xrightarrow{\text{FeX}_3}$ Halobenzene

X + HX

Sulfonation:

Benzene + Sulfur trioxide $\xrightarrow{\text{Concentrated H}_2\text{SO}_4}$

Benzenesulfonic acid

Nitration:

Benzene + Nitric acid $\xrightarrow[50\text{–}55°\text{C}]{\text{Concentrated H}_2\text{SO}_4}$

$-\text{NO}_2 + \text{H}_2\text{O}$

Nitrobenzene

SUMMARY

11.1 Alkenes and Alkynes: Structure and Physical Properties

Alkenes and *alkynes* are *unsaturated hydrocarbons*. Alkenes are characterized by the presence of at least one carbon-carbon double bond and have the general molecular formula C_nH_{2n}. Alkynes are characterized by the presence of at least one carbon-carbon triple bond and have the general molecular formula C_nH_{2n-2}. The physical properties of the alkenes and alkynes are similar to those of alkanes, but their chemical properties are quite different.

11.2 Alkenes and Alkynes: Nomenclature

Alkenes and alkynes are named by identifying the parent compound and replacing the *-ane* ending of the alkane with *-ene* (for an alkene) or *-yne* (for an alkyne). The parent chain is numbered to give the lowest number to the first of the two carbons involved in the double bond (or triple bond). Finally, all groups are named and numbered.

11.3 Geometric Isomers: A Consequence of Unsaturation

The carbon-carbon double bond is rigid. This allows the formation of *geometric isomers*, or isomers that occur when two different groups are bonded to each carbon of the double bond. When groups are on the same side of a double bond, the prefix *cis* is used to describe the compound. When groups are on opposite sides of a double bond, the prefix *trans* is used.

11.4 Alkenes in Nature

Alkenes and polyenes (alkenes with several carbon-carbon double bonds) are common in nature. Ethene, the simplest alkene, is a plant growth substance involved in fruit ripening, senescence and leaf fall, and responses to environmental stresses. Isoprenoids, or terpenes, are polyenes built from one or more isoprene units. Isoprenoids include steroids, chlorophyll and other photosynthetic pigments, and vitamins A, D, E, and K.

11.5 Reactions Involving Alkenes and Alkynes

Whereas alkanes undergo *substitution reactions*, alkenes and alkynes undergo *addition reactions*. The principal addition reactions of the unsaturated hydrocarbons are *halogenation, hydration, hydrohalogenation,* and *hydrogenation. Polymers* can be made from alkenes or substituted alkenes.

11.6 Aromatic Hydrocarbons

Aromatic hydrocarbons contain benzene rings. The rings can be represented as having alternating double and single bonds. However, the most accurate representation is as a resonance hybrid between two Kekulé structures. Simple aromatic compounds are named as derivatives of benzene.

Several members of this family have historical common names, such as aniline, phenol, and toluene. Polynuclear aromatic hydrocarbons consist of two or more aromatic molecules bonded together. Aromatic compounds do not undergo addition reactions. The typical reactions of benzene are *substitution reactions:* halogenation, nitration, and sulfonation.

11.7 Heterocyclic Aromatic Compounds

Heterocyclic aromatic compounds are those having at least one atom other than carbon as part of the structure of the aromatic ring. They are more similar to benzene in stability and chemical behavior than they are to the alkenes. Many of these compounds are components of molecules that have significant effects on biological systems, including DNA, RNA, hemoglobin, and nicotine.

KEY TERMS

addition polymer (11.5)
addition reaction (11.5)
alkene (11.1)
alkyne (11.1)
aromatic hydrocarbon (11.6)
geometric isomers (11.3)
halogenation (11.5)
heterocyclic aromatic compound (11.7)
hydration (11.5)
hydrogenation (11.5)
hydrohalogenation (11.5)
Markovnikov's rule (11.5)
monomer (11.5)
phenyl group (11.6)
polymer (11.5)
substitution reaction (11.6)
unsaturated hydrocarbon (Intro)

QUESTIONS AND PROBLEMS

Alkenes and Alkynes: Structure and Physical Properties
Foundations
11.23 What is the general relationship between the carbon chain length of an alkene and its boiling point?
11.24 Describe and explain the reason for the water solubility of an alkyne.
11.25 Write the general formulas for alkanes, alkenes, and alkynes.
11.26 What are the characteristic functional groups of alkenes and alkynes?

Applications
11.27 Describe the geometry of ethene.
11.28 What are the bond angles in ethene?
11.29 Compare the bond angles in ethane with those in ethene.
11.30 Explain the bond angles of ethene in terms of the valence shell electron pair repulsion (VSEPR) theory.
11.31 Describe the geometry of ethyne.
11.32 What are the bond angles in ethyne?
11.33 Compare the bond angles in ethane, ethene, and ethyne.
11.34 Explain the bond angles of ethyne in terms of the valence shell electron pair repulsion (VSEPR) theory.
11.35 Arrange the following groups of molecules from the highest to lowest boiling points:
   a. ethyne   propyne   2-pentyne
   b. 2-butene   3-decene   ethene

11.36 Arrange the following groups of molecules from the highest to the lowest melting points.

a.

b.

Alkenes and Alkynes: Nomenclature

Foundations

11.37 Briefly describe the rules for naming alkenes and alkynes.

11.38 What is meant by a geometric isomer?

11.39 Describe what is meant by a *cis* isomer of an alkene.

11.40 Describe what is meant by a *trans* isomer of an alkene.

Applications

11.41 Draw a condensed formula for each of the following compounds:
   a. 2-Methyl-2-hexene
   b. *trans*-3-Heptene
   c. *cis*-1-Chloro-2-pentene
   d. *cis*-2-Chloro-2-methyl-3-heptene
   e. *trans*-5-Bromo-2,6-dimethyl-3-octene

11.42 Draw a condensed formula for each of the following compounds:
   a. 2-Hexyne
   b. 4-Methyl-1-pentyne
   c. 1-Chloro-4,4,5-trimethyl-2-heptyne
   d. 2-Bromo-3-chloro-7,8-dimethyl-4-decyne

11.43 Name each of the following using the I.U.P.A.C. Nomenclature System:
   a. $CH_3CH_2CHCH{=}CH_2$
       $\quad\quad\;\; |$
       $\quad\quad\;\; CH_3$
   b. $CH_2CH_2CH_2CH_2{-}Br$
       $\;|$
       $CH_2CH{=}CH_2$
   c. $CH_3CH_2CH{=}CHCHCH_2CH_3$
       $\quad\quad\quad\quad\quad\; |$
       $\quad\quad\quad\quad\quad\; Br$
   d. $CH_3{-}\overset{\displaystyle CH_3}{\underset{\displaystyle CH_3}{C}}{-}$⟨ring⟩$CH_3$

11.44 Name each of the following using the I.U.P.A.C. Nomenclature System:
   a. $CH_3CHCH_2CH{=}CCH_3$
       $\quad\; |\quad\quad\quad\quad |$
       $\quad\; CH_3\quad\quad\; CH_3$
   b. $ClCH_2CHC{\equiv}CH$
       $\quad\quad |$
       $\quad\quad CH_3$
   c. $CH_3CHCH_2CH_2CH_2C{\equiv}CH$
       $\quad\; |$
       $\quad\; Cl$
   d. ⟨cyclopentene ring with Br and Cl⟩

11.45 Draw each of the following compounds using condensed formulas:
   a. 1,3,5-Trifluoropentane
   b. *cis*-2-Octene
   c. Dipropylacetylene

11.46 Draw each of the following compounds using condensed formulas:
   a. 3,3,5-Trimethyl-1-hexene
   b. 1-Bromo-3-chloro-1-heptyne
   c. 3-Heptyne

11.47 Of the following compounds, which can exist as *cis-trans* geometric isomers? Draw the two geometric isomers.
   a. 2,3-Dibromobutane
   b. 2-Heptene
   c. 2,3-Dibromo-2-butene
   d. Propene

11.48 Of the following compounds, which can exist as geometric isomers?
   a. 1-Bromo-1-chloro-2-methylpropene
   b. 1,1-Dichloroethene
   c. 1,2-Dibromoethene
   d. 3-Ethyl-2-methyl-2-hexene

11.49 Which of the following alkenes would not exhibit *cis-trans* geometric isomerism?
   a. $\underset{\displaystyle H}{\overset{\displaystyle CH_3}{}}C{=}C\underset{\displaystyle CH_2CH_3}{\overset{\displaystyle CH_3}{}}$
   b. $\underset{\displaystyle CH_3}{\overset{\displaystyle CH_3}{}}C{=}C\underset{\displaystyle H}{\overset{\displaystyle CH_3}{}}$
   c. $CH_3CH_2{\atop}C{=}C{\atop CH_3}\;\;$ with $CHCH_3$ (CH_2CH_3) and $CHCH_2CH_3$ (CH_3)
   d. $\underset{\displaystyle H}{\overset{\displaystyle CH_3CH_2}{}}C{=}C\underset{\displaystyle CH_2CH_3}{\overset{\displaystyle CH_3}{}}$

11.50 Which of the following structures have incorrect I.U.P.A.C. names? If incorrect, give the correct I.U.P.A.C. name.
   a. $CH_3C{\equiv}CCH_2CHCH_3$
       $\quad\quad\quad\quad\quad\;\; |$
       $\quad\quad\quad\quad\quad\;\; CH_3$

   2-Methyl-4-hexyne
   b. $\underset{\displaystyle CH_3CH_2}{\overset{\displaystyle CH_3CH_2}{}}C{=}C\underset{\displaystyle H}{\overset{\displaystyle CH_2CH_3}{}}$

   3-Ethyl-3-hexyne
   c. $CH_3CHCH_2C{\equiv}CCH_2CHCH_3$
       $\quad\; |\quad\quad\quad\quad\quad\quad\;\; |$
       $\quad\; CH_3\quad\quad\quad\quad\;\; CH_2CH_3$

   2-Ethyl-7-methyl-4-octyne

d.
$$CH_3CH_2\overset{\displaystyle\underset{|}{Cl}}{\underset{|}{C}}=C\overset{CH_2CHCH_3}{\underset{H}{\diagup}}$$

trans-6-Chloro-3-heptene

e.
$$ClCH_2\diagdown_{H}C=C\overset{H}{\underset{CHCH_2CH_3}{\diagup}}CH_3$$

1-Chloro-5-methyl-2-hexene

11.51 Which of the following can exist as *cis* and *trans* isomers?
a. $H_2C{=}CH_2$ d. $ClBrC{=}CClBr$
b. $CH_3CH{=}CHCH_3$ e. $(CH_3)_2C{=}C(CH_3)_2$
c. $Cl_2C{=}CBr_2$

11.52 Draw and name all the *cis* and *trans* isomers in Question 11.51.

11.53 Provide the I.U.P.A.C. name for each of the following molecules:
a. $CH_2{=}CHCH_2CH_2CH{=}CHCH_2CH_2CH_3$
b. $CH_2{=}CHCH_2CH{=}CHCH_2CH{=}CHCH_3$
c. $CH_3CH{=}CHCH_2CH{=}CHCH_2CH_3$
d. $CH_3CH{=}CHCHCH{=}CHCH_3$
   $\quad\quad\quad\quad\quad\underset{|}{\;}$
   $\quad\quad\quad\quad\quad CH_3$

11.54 Provide the I.U.P.A.C. name for each of the following molecules:
a. $CH_3C{=}CHCHCH{=}CCH_3$
   $\quad\;\;\underset{|}{Br}\;\;\;\;\;\underset{|}{CH_3}\;\;\;\underset{|}{Br}$
b. $CH_2{=}CHCHCH{=}CHCHCH_2CH_3$
   $\quad\quad\quad\;\underset{|}{CH_3}\quad\quad\quad\;\;\underset{|}{CH_2CH_3}$
c. $CH_2{=}CHCCH{=}CHCH_2CH{=}CHCHCH_3$
   $\quad\quad\;\;\underset{|}{CH_3}\quad\quad\quad\quad\quad\quad\quad\underset{|}{CH_3}$
d. $CH_3CHCHCH{=}CHCH_2CH{=}CHCHCH_3$
   $\quad\;\underset{|}{CH_2CH_3}\quad\quad\quad\quad\quad\quad\underset{|}{CH_2CH_3}$
   $\quad\;\;\underset{|}{CH_3}$

Reactions Involving Alkenes and Alkynes

Foundations

11.55 Write a general equation representing the hydrogenation of an alkene.

11.56 Write a general equation representing the hydrogenation of an alkyne.

11.57 Write a general equation representing the halogenation of an alkene.

11.58 Write a general equation representing the halogenation of an alkyne.

11.59 Write a general equation representing the hydration of an alkene.

11.60 Write a general equation representing the hydration of an alkyne.

11.61 What is the principal difference between the hydrogenation of an alkene and hydrogenation of an alkyne?

11.62 What is the major difference between the hydration of an alkene and hydration of an alkyne?

Applications

11.63 How could you distinguish between a sample of cyclohexane and a sample of hexene (both C_6H_{12}) using a simple chemical test? (*Hint:* Refer to the subsection entitled "Halogenation: Addition of X_2.")

11.64 Quantitatively, 1 mol of Br_2 is consumed per mole of alkene, and 2 mol of Br_2 are consumed per mole of alkyne. How many moles of Br_2 would be consumed for 1 mol of each of the following:
a. 2-Hexyne
b. Cyclohexene
c.

d.

11.65 Complete each of the following reactions by supplying the missing reactant or product(s) as indicated by question marks:
a. $CH_3CH_2CH{=}CHCH_2CH_3 + ? \longrightarrow$
   $\quad\quad\quad\quad\quad\quad\quad\quad\quad CH_3CH_2CH_2CH_2CH_2CH_3$
b. $CH_3{-}\overset{\displaystyle}{\underset{\displaystyle\underset{|}{CH_2}}{C}}{-}CH_3 + ? \longrightarrow CH_3\overset{\underset{|}{CH_3}}{\underset{\underset{|}{CH_3}}{C}}{-}OH$

c.

$? + $

d. $2CH_3CH_2CH_2CH_2CH_2CH_3 + ?O_2 \xrightarrow{Heat} ? + ?$
   (complete combustion)

e.

$? + $

f.

$? \xrightarrow{H_2O,\ H^+}$

11.66 Draw and name the product in each of the following reactions:
a. Cyclopentene + $H_2O(H^+)$
b. Cyclopentene + HCl
c. Cyclopentene + H_2
d. Cyclopentene + HI

11.67 Write a balanced equation for each of the following reactions:
a. Hydrogenation of 2-butyne
b. Halogenation of 3-pentyne

11.68 Write a balanced equation for each of the following reactions:
a. Hydration of 1-butyne
b. Hydration of 2-butyne

11.69 Predict the major product in each of the following reactions. Name the alkene reactant and the product, using I.U.P.A.C. nomenclature.
a.
$$CH_3\diagdown_{H}C=C\overset{CH_3}{\underset{H}{\diagup}} + H_2 \xrightarrow{Pd} ?$$
b. $CH_3CH_2CH{=}CH_2 + H_2O \xrightarrow{H^+} ?$
c. $CH_3CH{=}CHCH_3 + Cl_2 \longrightarrow ?$
d. $CH_3CH_2CH_2CH{=}CH_2 + HBr \longrightarrow ?$

11.70 Predict the major product in each of the following reactions. Name the alkene reactant and the product, using I.U.P.A.C. nomenclature.

a.

$$CH_3 \quad H$$
$$\diagdown C=C\diagup \quad + H_2 \xrightarrow{Ni} ?$$
$$H \diagup \diagdown CH_3$$

b. $CH_3C{=}CHCH_2CH_2CH_3 + H_2O \xrightarrow{H^+} ?$
  with CH_3 substituent

c. $CH_3C{=}CHCH_3 + Br_2 \longrightarrow ?$
  with CH_3 substituent

d.
$$CH_3$$
$$CH_3CCH{=}CH_2 + HCl \longrightarrow ?$$
$$CH_3$$

11.71 A hydrocarbon with a formula C_5H_{10} decolorized Br_2 and consumed 1 mol of hydrogen upon hydrogenation. Draw all the isomers of C_5H_{10} that are possible based on the above information.

11.72 Triple bonds react in a manner analogous to that of double bonds. The extra pair of electrons in the triple bond, however, generally allows 2 mol of a given reactant to *add* to the triple bond in contrast to 1 mol with the double bond. The "rich get richer" rule holds. Predict the major product in each of the following reactions:
a. Acetylene with 2 mol HCl
b. Propyne with 2 mol HBr
c. 2-Butyne with 2 mol HI

11.73 Complete each of the following by supplying the missing product indicated by the question mark:

a. 2-Butene $\xrightarrow{HBr}$?
b. 3-Methyl-2-hexene $\xrightarrow{HI}$?
c.

$\xrightarrow{HCl}$?

11.74 Bromine is often used as a laboratory spot test for unsaturation in an aliphatic hydrocarbon. Bromine in CCl_4 is red. When bromine reacts with an alkene or alkyne, the alkyl halide formed is colorless; hence, a disappearance of the red color is a positive test for unsaturation. A student tested the contents of two vials, A and B, both containing compounds with a molecular formula, C_6H_{12}. Vial A decolorized bromine, but vial B did not. How may the results for vial B be explained? What class of compound would account for this?

11.75 What is meant by the term *polymer*?

11.76 What is meant by the term *monomer*?

11.77 Write an equation representing the synthesis of Teflon from tetrafluoroethene. (*Hint:* Refer to Table 11.2.)

11.78 Write an equation representing the synthesis of polystyrene. (*Hint:* Refer to Table 11.2.)

11.79 Provide the I.U.P.A.C. name for each of the following molecules. Write a balanced equation for the hydration of each.
a. $CH_3CH{=}CHCH_2CH_3$
b. $CH_2CH{=}CH_2$
  with Br substituent
c.

11.80 Provide the I.U.P.A.C. name for each of the following molecules. Write a balanced equation for the hydration of each.
a.

b. $CH_3CH{=}CHCH_2CH{=}CHCH_2CH{=}CHCH_3$
c. $CH_3CH{=}CHCCH_3$
  with CH_2CH_3 and CH_3 substituents

11.81 Write an equation for the addition reaction that produced each of the following molecules:

a.
$$CH_3$$
$$CH_2CH_2CH_2CHCH_3$$
$$OH$$

b. $CH_3CH_2CHCH_2CH_2CH_3$
  with Br substituent

c.

d.

11.82 Write an equation for the addition reaction that produced each of the following molecules:

a.
$$OH$$
$$CH_3CH_2CHCHCH_3$$
$$CH_2CH_3$$

b. $CH_3CHCH_2CH_3$
  with OH substituent

c.

11.83 Draw the structure of each of the following compounds and write a balanced equation for the complete hydrogenation of each:
a. 1,4-Hexadiene
b. 2,4,6-Octatriene
c. 1,3 Cyclohexadiene
d. 1,3,5-Cyclooctatriene

11.84 Draw the structure of each of the following compounds and write a balanced equation for the bromination of each:
a. 3-Methyl-1,4-hexadiene
b. 4-Bromo-1,3-pentadiene
c. 3-Chloro-2,4-hexadiene
d. 3-Bromo-1,3-Cyclohexadiene

Aromatic Hydrocarbons

Foundations

11.85 Where did the term *aromatic hydrocarbon* first originate?

11.86 What chemical characteristic of the aromatic hydrocarbons is most distinctive?

11.87 What is meant by the term *resonance hybrid?*

11.88 Draw a pair of structures to represent the benzene resonance hybrid.

Applications

11.89 Draw the structure for each of the following compounds:
   a. 2,4-Dibromotoluene
   b. 1,2,4-Triethylbenzene
   c. Isopropylbenzene
   d. 2-Bromo-5-chlorotoluene

11.90 Name each of the following compounds, using the I.U.P.A.C. system.

11.91 Draw each of the following compounds, using condensed formulas:
   a. *meta*-Cresol
   b. Propylbenzene
   c. 1,3,5-Trinitrobenzene
   d. *m*-Chlorotoluene

11.92 Draw each of the following compounds, using condensed formulas:
   a. *p*-Xylene
   b. Isopropylbenzene
   c. *m*-Nitroanisole
   d. *p*-Methylbenzaldehyde

11.93 Describe the Kekulé model for the structure of benzene.

11.94 Describe the current model for the structure of benzene.

11.95 How does a substitution reaction differ from an addition reaction?

11.96 Give an example of a substitution reaction and of an addition reaction.

11.97 Write an equation showing the reaction of benzene with Cl_2 and $FeCl_3$.

11.98 Write an equation showing the reaction of benzene with SO_3. Be sure to note the catalyst required.

Heterocyclic Aromatic Compounds

11.99 Draw the general structure of a pyrimidine.

11.100 What biological molecules contain pyrimidine rings?

11.101 Draw the general structure of a purine.

11.102 What biological molecules contain purine rings?

CRITICAL THINKING PROBLEMS

1. There is a plastic polymer called polyvinylidene difluoride (PVDF) that can be used to sense a baby's breath and thus be used to prevent sudden infant death syndrome (SIDS). The secret is that this polymer can be specially processed so that it becomes piezoelectric (produces an electrical current when it is physically deformed) and pyroelectric (develops an electrical potential when its temperature changes). When a PVDF film is placed beside a sleeping baby, it will set off an alarm if the baby stops breathing. The structure of this polymer is shown here:

$$\left[\begin{array}{cccc} F & H & F & H \\ | & | & | & | \\ -C & -C & -C & -C- \\ | & | & | & | \\ F & H & F & H \end{array} \right]$$

Go to the library and investigate some of the other amazing uses of PVDF. Draw the structure of the alkene from which this compound is produced.

2. Isoprene is the repeating unit of the natural polymer rubber. It is also the starting material for the synthesis of cholesterol and several of the lipid-soluble vitamins, including vitamin A and vitamin K. The structure of isoprene is seen below.

$$CH_2{=}CCH{=}CH_2$$
$$\overset{\displaystyle CH_3}{|}$$

What is the I.U.P.A.C. name for isoprene?

3. When polyacrylonitrile is burned, toxic gases are released. In fact, in airplane fires, more passengers die from inhalation of toxic fumes than from burns. Refer to Table 11.2 for the structure of acrylonitrile. What toxic gas would you predict to be the product of the combustion of these polymers?

4. If a molecule of polystyrene consists of 25,000 monomers, what is the molar mass of the molecule?

5. A factory produces one million tons of polypropylene. How many moles of propene would be required to produce this amount? What is the volume of this amount of propene at 25°C and 1 atm?

12 Alcohols, Phenols, Thiols, and Ethers

Learning Goals

1 Rank selected alcohols by relative water solubility, boiling points, or melting points.

2 Write the names and draw the structures for common alcohols.

3 Discuss the biological, medical, or environmental significance of several alcohols.

4 Classify alcohols as primary, secondary, or tertiary.

5 Write equations representing the preparation of alcohols by the hydration of an alkene.

6 Write equations representing the preparation of alcohols by hydrogenation (reduction) of aldehydes or ketones.

7 Write equations showing the dehydration of an alcohol.

8 Write equations representing the oxidation of alcohols.

9 Discuss the role of oxidation and reduction reactions in the chemistry of living systems.

10 Discuss the use of phenols as germicides.

11 Write names and draw structures for common ethers and discuss their use in medicine.

12 Write equations representing the dehydration reaction between two alcohol molecules.

13 Write names and draw structures for simple thiols and discuss their biological significance.

Outline

Introduction

Chemistry Connection:
Polyols for the Sweet Tooth

12.1 Alcohols: Structure and Physical Properties

12.2 Alcohols: Nomenclature

12.3 Medically Important Alcohols

A Medical Perspective:
Fetal Alcohol Syndrome

12.4 Classification of Alcohols

12.5 Reactions Involving Alcohols

12.6 Oxidation and Reduction in Living Systems

A Human Perspective:
Alcohol Consumption and the Breathalyzer Test

12.7 Phenols

12.8 Ethers

12.9 Thiols

Sugar-free chocolates are not calorie free!

Introduction

The characteristic functional group of the alcohols and phenols is the hydroxyl group (—OH). Alcohols have the general structure R—OH, in which R is any alkyl group. Phenols are similar in structure but contain an aryl group in place of the alkyl group. Both can be viewed as substituted water molecules in which one of the hydrogen atoms has been replaced by an alkyl or aryl group.

An aryl group is an aromatic ring with one hydrogen atom removed.

General formulas: Example:

Alcohol   Phenol   Methanol
                          (methyl alcohol)

Ethers have two alkyl or aryl groups attached to the oxygen atom and may be thought of as substituted alcohols. The functional group characteristic of an ether is R—O—R. Thiols are a family of compounds that contain the sulfhydryl group (—SH). They, too, have a structure similar to that of alcohols.

R and R¹ = alkyl or aryl group.

Ethers   Methoxymethane   Thiol   Methanethiol
              (dimethyl ether)

Many important biological molecules, including sugars (carbohydrates), fats (lipids), and proteins, contain hydroxyl and/or thiol groups.

List the hydroxyl group containing types of molecules that are important elements in the meal shown in this photograph.

D-Glucose, *a sugar*   Lysine vasopressin   Monolaurin, *a lipid*
                         (partial structure), *a protein*

Glycolysis and fermentation are discussed in Chapter 21.

In biological systems, the hydroxyl group is often involved in a variety of reactions such as oxidation, reduction, hydration, and dehydration. In glycolysis (a metabolic pathway by which glucose is degraded and energy is harvested in the form of ATP),

Chemistry Connection

Polyols for the Sweet Tooth

Do you crave sweets, but worry about the empty calories in sugary treats? If so, you are not alone. Research tells us that, even as babies, we demonstrate preference for sweet tastes over all others. But there are many reasons to reduce our intake of refined sugars, in particular sucrose or table sugar. Too many people eat high-calorie, low-nutrition snacks rather than more nutritious foods. This can lead to obesity, a problem that is very common in our society. In addition, sucrose is responsible for tooth decay. Lactic acid, one of the products of the metabolism of sucrose by bacteria on our teeth, dissolves the tooth enamel, which results in a cavity. For those with diabetes, glucose intolerance, or hypoglycemia, sucrose in the diet makes it difficult to maintain a constant blood sugar level.

The food chemistry industry has invested billions of dollars in the synthesis of sugar substitutes. We recognize names such as aspartame (Equal or Nutrasweet), saccharin (Sweet & Low), and SPLENDA®, because they are the most common non-nutritive sweeteners worldwide. We also buy products, including candies, soft drinks, and gums, which are advertised to be "sugar-free." But you might be surprised to find that many of these products are not free of calories. A check of the nutritional label may reveal that these products contain sorbitol, mannitol, or one or more other members of a class of compounds called sugar alcohols, or *polyols* (*poly*—many; *ols*—alcohol or hydroxyl groups).

Sugar alcohols are found in many foods, including fruits, vegetables, and mushrooms. Others are made by hydrogenation or fermentation of carbohydrates from wheat or corn. But all are natural products. Compared to sucrose, they range in sweetness from about half to nearly the same; they also have fewer calories per gram than sucrose (about one-third to one-half the calories). Polyols also cause a cooling sensation in the mouth. This cooling is caused by a negative heat of solution (they must absorb heat from the surroundings in order to dissolve) and is used to advantage in breath freshening mints and gums.

Sorbitol, the most commonly used sugar alcohol, is about 0.6 times the sweetness of sucrose. While sucrose contains four calories per gram, sorbitol is only about 2.6 C/g. Discovered in 1872 in the berries of Mountain Ash trees, sorbitol has a smooth mouthfeel, which makes it ideal as a texturizing agent in foods. It also has a pleasant, cool, sweet flavor and acts as a humectant, keeping foods from losing moisture. No acceptable daily intake (ADI) has been specified for sorbitol, which is an indication that it is considered to be a very safe food additive. However, it has been observed that ingestion of more than 50–80 g/day may have a laxative effect.

Mannitol, a structural isomer of sorbitol, is found naturally in asparagus, olives, pineapple, and carrots. For use in the food industry, it is extracted from seaweed. Mannitol has about 0.7 times the sweetness of sucrose and only about 1.6 C/g. As for sorbitol, no ADI has been specified; but ingestion of more than 20 g/day may cause diarrhea and bloating.

Xylitol was discovered in 1891 and has been used as a sweetener since the 1960s. Found in many fruits and vegetables, it has about the same sweetness as sucrose, but only one-third of the caloric value. Its high cooling effect, as well as sweetness, contribute to its popularity as a sweetening agent in hard candies and gums, as well as in oral health products. In fact, extensive studies suggest that use of xylitol-sweetened gum (7–10 g of xylitol per day) between meals results in a 30–60% decrease in dental cavities.

While polyols give us the sweetness that we enjoy without all of the calories, cavities, and blood sugar peaks of sucrose, they are not without a negative side. Some studies have reported weight gain by individuals who overeat these "sugar-free" foods. The American Diabetes Association has reported that these foods are "acceptable in moderate amounts but should not be eaten in excess." In fact, some diabetics have suffered elevated blood sugar after overeating foods containing polyols. Finally, as we noted above, when ingested in excess, sugar alcohols may cause bloating and diarrhea.

The use of these natural products continues to be investigated by the food and pharmaceutical industries. As we learn more about them, sugar alcohols continue to be versatile food additives. Using them in moderation, we can enjoy the benefits that they confer, without suffering uncomfortable side effects.

$$\begin{array}{ccc}
\text{CH}_2\text{OH} & \text{CH}_2\text{OH} & \text{CH}_2\text{OH} \\
\text{H—C—OH} & \text{HO—C—H} & \text{H—C—OH} \\
\text{HO—C—H} & \text{HO—C—H} & \text{HO—C—H} \\
\text{H—C—OH} & \text{H—C—OH} & \text{H—C—OH} \\
\text{H—C—OH} & \text{H—C—OH} & \text{CH}_2\text{OH} \\
\text{CH}_2\text{OH} & \text{CH}_2\text{OH} & \\
\text{Sorbitol} & \text{Mannitol} & \text{Xylitol}
\end{array}$$

Structures of three of the sugar alcohols used in the food industry.

several steps center on the reactivity of the hydroxyl group. The majority of the consumable alcohol in the world (ethanol) is produced by fermentation reactions carried out by yeasts.

The thiol group is found in the structure of some amino acids and is essential for keeping proteins in the proper three-dimensional shape required for their biological function. Thus, these functional groups play a central role in the structure and chemical properties of biological molecules. The thiol group of the amino acid cysteine is highlighted in blue in the structure of lysine vasopressin shown on page 402.

12.1 Alcohols: Structure and Physical Properties

LEARNING GOAL

1 Rank selected alcohols by relative water solubility, boiling points, or melting points.

Figure 12.1

Ball-and-stick model of the simple alcohol ethanol.

Electronegativity is discussed in Section 3.1. Hydrogen bonding is described in detail in Section 5.2.

An **alcohol** is an organic compound that contains a **hydroxyl group** (—OH) attached to an alkyl group (Figure 12.1). The R—O—H portion of an alcohol is similar to the structure of water. The oxygen and the two atoms bonded to it lie in the same plane, and the R—O—H bond angle is approximately 104°, which is very similar to the H—O—H bond angle of water.

The hydroxyl groups of alcohols are very polar because the oxygen and hydrogen atoms have significantly different electronegativities. Because the two atoms involved in this polar bond are oxygen and hydrogen, hydrogen bonds can form between alcohol molecules (Figure 12.2).

As a result of this intermolecular hydrogen bonding, alcohols boil at much higher temperatures than hydrocarbons of similar molecular weight. These higher boiling points are caused by the large amount of heat needed to break the hydrogen bonds that attract the alcohol molecules to one another. Compare the boiling points of butane and propanol, which have similar molecular weights:

$CH_3CH_2CH_2CH_3$	$CH_3CH_2CH_2OH$
Butane	1-Propanol
M.W. = 58	M.W. = 60
b.p. = −0.5°C	b.p. = 97.2°C

Alcohols of one to four carbon atoms are very soluble in water, and those with five or six carbons are moderately soluble in water. This is due to the ability of the alcohol to form intermolecular hydrogen bonds with water molecules (see Figure 12.2b). As the nonpolar, or hydrophobic, portion of an alcohol (the

Figure 12.2

(a) Hydrogen bonding between alcohol molecules. (b) Hydrogen bonding between alcohol molecules and water molecules.

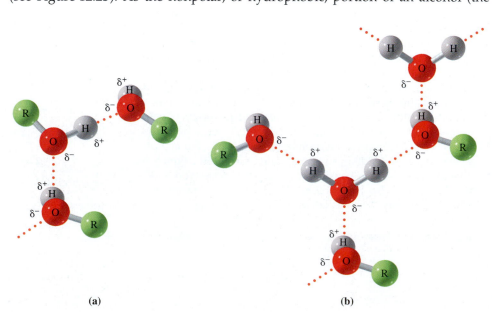

(a)   (b)

carbon chain) becomes larger relative to the polar, hydrophilic, region (the hydroxyl group), the water solubility of an alcohol decreases. As a result alcohols of seven carbon atoms or more are nearly insoluble in water. The term *hydrophobic,* which literally means "water fearing," is used to describe a molecule or a region of a molecule that is nonpolar and, thus, more soluble in nonpolar solvents than in water. Similarly, the term *hydrophilic,* meaning water loving, is used to describe a polar molecule or region of a molecule that is more soluble in the polar solvent water than in a nonpolar solvent.

An increase in the number of hydroxyl groups along a carbon chain will increase the influence of the polar hydroxyl group. It follows, then that diols and triols are more water soluble than alcohols with only a single hydroxyl group.

The presence of polar hydroxyl groups in large biological molecules—for instance, proteins and nucleic acids—allows intramolecular hydrogen bonding that keeps these molecules in the shapes needed for biological function.

Intermolecular hydrogen bonds are attractive forces between two molecules. *Intramolecular* hydrogen bonds are attractive forces between polar groups within the same molecule.

12.2 Alcohols: Nomenclature

I.U.P.A.C. Names

In the I.U.P.A.C. Nomenclature System, alcohols are named according to the following steps:

LEARNING GOAL
2 Write the names and draw the structures for common alcohols.

- Determine the name of the *parent compound,* the longest continuous carbon chain containing the —OH group.
- Replace the *-e* ending of the alkane chain with the *-ol* ending of the alcohol. Following this pattern, an alkane becomes an alkanol. For instance, ethan*e* becomes ethan*ol,* and propan*e* becomes propan*ol.*
- Number the parent chain to give the carbon bearing the hydroxyl group the lowest possible number.
- Name and number all substituents, and add them as prefixes to the "alkanol" name.
- Alcohols containing two hydroxyl groups are named *-diols.* Those bearing three hydroxyl groups are called *-triols.* A number giving the position of each of the hydroxyl groups is needed in these cases.

The way to determine the parent compound was described in Section 10.2.

| **Using I.U.P.A.C. Nomenclature to Name Alcohols** | **EXAMPLE 12.1** |

The molecule shown here is an alarm-defense pheromone in certain populations of leaf-cutter ants. Provide the I.U.P.A.C. name for this molecule.

Solution

    Parent compound: heptane (becomes heptanol)

    Position of —OH: carbon-3 (*not* carbon 5)

    Substituents: 4-methyl

    Name: 4-Methyl-3-heptanol

Continued—

LEARNING GOAL
2 Write the names and draw the structures for common alcohols.

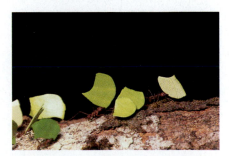

A leaf-cutter ant uses chemicals produced in its mandibles to signal alarm to others. One such pheromone is 4-methyl-3-heptanol. Draw a line structure for this molecule.

E X A M P L E 12.1 —*Continued*

Name the following cyclic alcohol using I.U.P.A.C. nomenclature.

Solution

Remember that this line structure represents a cyclic molecule composed of six carbon atoms and associated hydrogen atoms, as follows:

Parent compound: cyclohexane (becomes cyclohexanol)
Position of —OH: carbon-1 (*not* carbon-3)
Substituents: 3-bromo (*not* 5-bromo)
Name: 3-Bromocyclohexanol (it is assumed that the —OH is on carbon-1 in cyclic structures)

Practice Problem 12.1

Use the I.U.P.A.C. Nomenclature System to name each of the following compounds.

a. $CH_3CHCH_2CH_2CH_2OH$
    |
    CH_3

c. CH_2CHCH_2
    | | |
    OH OH OH
    (Common name: Glycerol)

b. $CH_3CHCH_2CHCH_3$
    | |
    OH CH_2CH_3

d. $CH_3CH_2CHCHCH_2CH_2OH$
          | |
          Cl CH_3

For Further Practice: Questions 12.21 and 12.22.

Common Names

See Section 10.2 for the names of the common alkyl groups.

The common names for alcohols are derived from the alkyl group corresponding to the parent compound. The name of the alkyl group is followed by the word *alcohol*. For some alcohols, such as ethylene glycol and glycerol, historical names are used. The following examples provide the I.U.P.A.C. and common names of several alcohols:

CH_3CHCH_3
|
OH

2-Propanol
(isopropyl
alcohol)

$HOCH_2CH_2OH$

1,2-Ethanediol
(ethylene glycol)

CH_3CH_2OH

Ethanol
(ethyl alcohol)
(grain alcohol)

Question *12.1*

Draw structures for each of the following alcohols.

  a. 2-Methyl-1-propanol
  b. 2-Chlorocyclopentanol
  c. 2,4-Dimethylcyclohexanol
  d. 2,3-Dichloro-3-hexanol

Question *12.2*

Give the common name and the I.U.P.A.C. name for each of the following compounds.

  a. $CH_3CH_2CH_2CH_2CH_2CH_2CH_2OH$

  b. CH_3CHCH_3
        |
       OH

  c.

  d.

12.3 Medically Important Alcohols

Methanol

LEARNING GOAL
3 Discuss the biological, medical, or environmental significance of several alcohols.

Methanol (methyl alcohol), CH_3OH, is a colorless and odorless liquid that is used as a solvent and as the starting material for the synthesis of methanal (formaldehyde). Methanol is often called *wood alcohol* because it can be made by heating wood in the absence of air. In fact, ancient Egyptians produced methanol by this process and, mixed with other substances, used it for embalming. It was not until 1661 that Robert Boyle first isolated pure methanol, which he called *spirit of box*, because he purified it by distillation from boxwood. Methanol is toxic and can cause blindness and perhaps death if ingested. Methanol may also be used as fuel, especially for "formula" racing cars.

Ethanol

Ethanol (ethyl alcohol), CH_3CH_2OH, is a colorless and odorless liquid and is the alcohol in alcoholic beverages. It is also widely used as a solvent and as a raw material for the preparation of other organic chemicals.

    The ethanol used in alcoholic beverages comes from the **fermentation** of carbohydrates (sugars and starches). The beverage produced depends on the starting material and the fermentation process: scotch (grain), bourbon (corn), burgundy wine (grapes and grape skins), and chablis wine (grapes without red skins) (Figure 12.3). The following equation summarizes the fermentation process:

$$C_6H_{12}O_6 \xrightarrow[\text{enzyme action}]{\text{Several steps involving}} 2CH_3CH_2OH + 2CO_2$$

Sugar   Ethanol
(glucose)   (ethyl alcohol)

    The alcoholic beverages listed have quite different alcohol concentrations. Wines are generally 12–13% alcohol because the yeasts that produce the ethanol are killed by ethanol concentrations of 12–13%. To produce bourbon or scotch

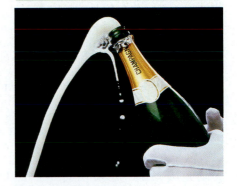

Figure 12.3

Champagne, a sparkling wine, results when fermentation is carried out in a sealed bottle. Under these conditions, the CO_2 produced during fermentation is trapped in the wine.

Fermentation reactions are described in detail in Section 21.4 and in A Human Perspective: Fermentations: The Good, the Bad, and the Ugly.

Distillation is the separation of compounds in a mixture based on differences in boiling points.

A Medical *Perspective*

Fetal Alcohol Syndrome

The first months of pregnancy are a time of great joy and anticipation but are not without moments of anxiety. On her first visit to the obstetrician, the mother-to-be is tested for previous exposure to a number of infectious diseases that could damage the fetus. She is provided with information about diet, weight gain, and drugs that could harm the baby. Among the drugs that should be avoided are alcoholic beverages.

The use of alcoholic beverages by a pregnant woman can cause *fetal alcohol syndrome (FAS)*. A *syndrome* is a set of symptoms that occur together and are characteristic of a particular disease. In this case, physicians have observed that infants born to women with chronic alcoholism showed a reproducible set of abnormalities including mental retardation, poor growth before and after birth, and facial malformations.

Mothers who report only social drinking may have children with *fetal alcohol effects*, a less severe form of fetal alcohol syndrome. This milder form is characterized by a reduced birth weight, some learning disabilities, and behavioral problems.

How does alcohol consumption cause these varied symptoms? No one is exactly sure, but it is well known that the alcohol consumed by the mother crosses the placenta and enters the bloodstream of the fetus. Within about fifteen minutes, the concentration of alcohol in the blood of the fetus is as high as that of the mother! However, the mother has enzymes to detoxify the alcohol in her blood; the fetus does not. Now consider that alcohol can cause cell division to stop or be radically altered. It is thought that even a single night on the town could be enough to cause FAS by blocking cell division during a critical developmental period.

This raises the question "How much alcohol can a pregnant woman safely drink?" As we have seen, the severity of the symptoms seems to increase with the amount of alcohol consumed by the mother. However, it is virtually impossible to do the scientific studies that would conclusively determine the risk to the fetus caused by different amounts of alcohol. There is some evidence that suggests that there is a risk associated with drinking even one ounce of absolute (100%) alcohol each

The American Medical Association recommends abstaining from alcohol during pregnancy.

day. Because of these facts and uncertainties, the American Medical Association and the U.S. Surgeon General recommend that pregnant women completely abstain from alcohol.

For Further Understanding

In July 2004, the U.S. Centers for Disease Control issued a new document entitled *Fetal Alcohol Syndrome: Guidelines for Referral and Diagnosis*, which can be found at the following Web address: http://www.cdc.gov/ncbddd/fas/documents/FAS_guidelines_accessible.pdf. Refer to this document when answering the following questions.

What is the estimate of the number of babies born each year with fetal alcohol syndrome and why is a more accurate number so difficult to determine?

What are some of the issues, practical and ethical, involved in intervention efforts to prevent fetal alcohol syndrome?

with an alcohol concentration of 40–45% ethanol (80 or 90 proof), the original fermentation products must be distilled.

The sale and use of pure ethanol (100% ethanol) are regulated by the federal government. To prevent illegal use of pure ethanol, it is *denatured* by the addition of a denaturing agent, which makes it unfit to drink but suitable for many laboratory applications.

2-Propanol

2-Propanol (isopropyl alcohol),

$$CH_3CHCH_3$$
$$|$$
$$OH$$

was commonly called *rubbing alcohol* because patients with high fevers were often given alcohol baths to reduce body temperature. Rapid evaporation of the alcohol results in skin cooling. This practice is no longer commonly used.

It is also used as a disinfectant (Figure 12.4), an astringent (skin-drying agent), an industrial solvent, and a raw material in the synthesis of organic chemicals. It is colorless, has a very slight odor, and is toxic when ingested.

1,2-Ethanediol

1,2-Ethanediol (ethylene glycol),

$$CH_2\!-\!CH_2$$
$$\ \ |\qquad\ |$$
$$OH\quad OH$$

is used as automobile antifreeze. When added to water in the radiator, the ethylene glycol solute lowers the freezing point and raises the boiling point of the water. Ethylene glycol has a sweet taste but is extremely poisonous. For this reason, color additives are used in antifreeze to ensure that it is properly identified.

1,2,3-Propanetriol

1,2,3-Propanetriol (glycerol),

$$CH_2\!-\!CH\!-\!CH_2$$
$$\ \ |\qquad\ |\qquad\ |$$
$$OH\quad OH\quad OH$$

is a viscous, sweet-tasting, nontoxic liquid. It is very soluble in water and is used in cosmetics, pharmaceuticals, and lubricants. Glycerol is obtained as a by-product of the hydrolysis of fats.

12.4 Classification of Alcohols

Alcohols are classified as **primary (1°), secondary (2°),** or **tertiary (3°),** depending on the number of alkyl groups attached to the **carbinol carbon,** the carbon bearing the hydroxyl (—OH) group. If no alkyl groups are attached, the alcohol is methyl alcohol; if there is a single alkyl group, the alcohol is a primary alcohol; an alcohol with two alkyl groups bonded to the carbon bearing the hydroxyl group is a secondary alcohol, and if three alkyl groups are attached, the alcohol is a tertiary alcohol.

Figure 12.4
Isopropyl alcohol, or rubbing alcohol, is used as a disinfectant before and after an injection or blood test.

LEARNING GOAL
4 Classify alcohols as primary, secondary, or tertiary.

Methyl alcohol   1° Alcohol   2° Alcohol   3° Alcohol

Methanol (methyl alcohol)   Ethanol (1° alcohol)   2-Propanol (2° alcohol)   2-Methyl-2-propanol (3° alcohol)

Two alcohols contribute to the distinctive flavor of mushrooms. These are 1-octanol and 3-octanol. Draw the line structures for these two molecules and provide their common names. Classify these alcohols as primary, secondary, or tertiary.

EXAMPLE 12.2   **Classifying Alcohols**

LEARNING GOAL

4 Classify alcohols as primary, secondary, or tertiary.

Classify each of the following alcohols as primary, secondary, or tertiary.

Solution

In each of the structures shown below, the carbinol carbon is shown in red:

$$CH_3CHCH_3$$
$$|$$
$$OH$$

This alcohol, 2-propanol, is a secondary alcohol because there are two alkyl groups attached to the carbinol carbon.

$$CH_3$$
$$|$$
$$CH_3CCH_3$$
$$|$$
$$OH$$

This alcohol, 2-methyl-2-propanol, is a tertiary alcohol because there are three alkyl groups attached to the carbinol carbon.

$$OH$$
$$|$$
$$CH_3CH_2CHCH_2$$
$$|$$
$$CH_2CH_3$$

This alcohol, 2-ethyl-1-butanol, is a primary alcohol because there is only one alkyl group attached to the carbinol carbon.

   Practice Problem 12.2

Classify each of the following alcohols as 1°, 2°, 3°, or aromatic (phenol).

a. $CH_3CH_2CH_2CH_2OH$

b. $CH_3CH_2CHCH_2CH_3$
   $\quad\quad\quad\; |$
   $\quad\quad\quad\; OH$

c.

d.

e.

For Further Practice: Questions 12.39 and 12.41.

12.5 Reactions Involving Alcohols

Preparation of Alcohols

LEARNING GOAL

5 Write equations representing the preparation of alcohols by the hydration of an alkene.

As we saw in the last chapter, the most important reactions of alkenes are *addition reactions*. Addition of a water molecule to the carbon-carbon double bond of an alkene produces an alcohol. This reaction, called **hydration,** requires a trace of acid (H^+) as a catalyst, as shown in the following equation:

Hydration of alkenes is described in Section 11.5.

Writing an Equation Representing the Preparation of an Alcohol by the Hydration of an Alkene	**E X A M P L E 12.3**

Write an equation representing the preparation of cyclohexanol from cyclohexene.

LEARNING GOAL

5 Write equations representing the preparation of alcohols by the hydration of an alkene.

Solution

Begin by writing the structure of cyclohexene. Recall that cyclohexene is a six-carbon cyclic alkene. Now add the water molecule to the equation.

Cyclohexene   Water

You will recognize that the hydration reaction involves the addition of a water molecule to the carbon-carbon double bond. Recall that the reaction requires a trace of acid as a catalyst. Complete the equation by adding the catalyst and product, cyclohexanol.

Cyclohexene   Water   Cyclohexanol

Practice Problem 12.3

Write a balanced equation showing the hydration of each of the following alkenes. Remember that Markovnikov's rule must be applied in the case of unsymmetrical alkenes.

a. $CH_3CH{=}CH_2$
b. $CH_2{=}CH_2$
c. $CH_3CH_2CH{=}CHCH_2CH_3$

For Further Practice: Questions 12.51 and 12.53.

Alcohols may also be prepared via the hydrogenation (reduction) of aldehydes and ketones. This reaction, summarized as follows, is discussed in Section 13.4, and is similar to the hydrogenation of alkenes.

In an aldehyde, R^1 and R^2 may be either alkyl groups or H. In ketones, R^1 and R^2 are both alkyl groups.

$$
\underset{\substack{\text{Aldehyde} \\ \text{or} \\ \text{Ketone}}}{\overset{\text{O}}{\underset{R^1 \quad R^2}{\parallel}}\!\!\!C} \;+\; \underset{\text{Hydrogen}}{\overset{H}{\underset{H}{\mid}}} \;\xrightarrow{\;\text{Catalyst}\;}\; \underset{\text{Alcohol}}{R^1\!-\!\overset{\text{OH}}{\underset{H}{C}}\!-\!R^2}
$$

E X A M P L E   12.4

Writing an Equation Representing the Preparation of an Alcohol by the Hydrogenation (Reduction) of an Aldehyde

6 **LEARNING GOAL**
Write equations representing the preparation of alcohols by hydrogenation (reduction) of aldehydes or ketones.

Write an equation representing the preparation of 1-propanol from propanal.

Solution

Begin by writing the structure of propanal. Propanal is a three-carbon aldehyde. Aldehydes are characterized by the presence of a carbonyl group (—C=O) attached to the end of the carbon chain of the molecule. After you have drawn the structure of propanal, add diatomic hydrogen to the equation.

$$
\underset{\text{Propanal}}{H-\overset{H}{\underset{H}{C}}-\overset{H}{\underset{H}{C}}-C\!\!\begin{array}{c}\nearrow O \\ \searrow H\end{array}} \;+\; \underset{\text{Hydrogen}}{H\!-\!H} \;\xrightarrow{\;\text{catalyst}\;}
$$

Notice that the general equation reveals this reaction to be another example of a hydrogenation reaction. As the hydrogens are added to the carbon-oxygen double bond, it is converted to a carbon-oxygen single bond, as the carbonyl oxygen becomes a hydroxyl group.

$$
\underset{\text{Propanal}}{H-\overset{H}{\underset{H}{C}}-\overset{H}{\underset{H}{C}}-C\!\!\begin{array}{c}\nearrow O \\ \searrow H\end{array}} \;+\; \underset{\text{Hydrogen}}{H\!-\!H} \;\xrightarrow{\;\text{catalyst}\;}\; \underset{\text{1-Propanol}}{H-\overset{H}{\underset{H}{C}}-\overset{H}{\underset{H}{C}}-\overset{H}{\underset{H}{C}}-OH}
$$

Practice Problem 12.4

Write an equation representing the reduction of butanal. Provide the structures and names for the reactants and products. *Hint:* Butanal is a four-carbon aldehyde.

For Further Practice: Questions 12.65a and 12.66a.

EXAMPLE 12.5

Writing an Equation Representing the Preparation of an Alcohol by the Hydrogenation (Reduction) of a Ketone

Write an equation representing the preparation of 2-propanol from propanone.

LEARNING GOAL

6 Write equations representing the preparation of alcohols by hydrogenation (reduction) of aldehydes or ketones.

Solution

Begin by writing the structure of propanone. Propanone is a three-carbon ketone. Ketones are characterized by the presence of a carbonyl group (—C═O) located anywhere within the carbon chain of the molecule. In the structure of propanone, the carbonyl group must be associated with the center carbon. After you have drawn the structure of propanone, add diatomic hydrogen to the equation.

Notice that this reaction is another example of a hydrogenation reaction. As the hydrogens are added to the carbon-oxygen double bond, it is converted to a carbon-oxygen single bond, as the carbonyl oxygen becomes a hydroxyl group.

Practice Problem 12.5

Write an equation representing the reduction of butanone.
Hint: Butanone is a four-carbon ketone.

For Further Practice: Questions 12.65b and 12.66b.

Dehydration of Alcohols

Alcohols undergo **dehydration** (lose water) when heated with concentrated sulfuric acid (H_2SO_4) or phosphoric acid (H_3PO_4). Dehydration is an example of an **elimination reaction,** that is, a reaction in which a molecule loses atoms or ions from its structure. In this case, the —OH and —H are "eliminated" from adjacent carbons in the alcohol to produce an alkene and water. We have just seen that alkenes can be hydrated to give alcohols. Dehydration is simply the reverse process: the conversion of an alcohol back to an alkene. This is seen in the following general reaction and the examples that follow:

LEARNING GOAL

7 Write equations showing the dehydration of an alcohol.

$$H—\overset{\displaystyle H}{\underset{\displaystyle H}{C}}—\overset{\displaystyle H}{\underset{\displaystyle OH}{C}}—H \xrightarrow{H^+,\ heat} CH_2{=}CH_2 + H—OH$$

Ethanol   Ethene
(ethyl alcohol)   (ethylene)

$$CH_3CH_2CH_2OH \xrightarrow{H^+,\ heat} CH_3CH{=}CH_2 + H_2O$$

1-Propanol   Propene
(propyl alcohol)   (propylene)

In some cases, dehydration of alcohols produces a mixture of products, as seen in the following example:

$$CH_3CH_2\overset{\displaystyle }{\underset{\displaystyle OH}{C}}HCH_3 \xrightarrow[heat]{H^+} CH_3CH_2CH{=}CH_2 + CH_3CH{=}CHCH_3 + H_2O$$

2-Butanol   1-Butene   2-Butene
                       (minor product)   (major product)

Notice in the equation shown above and in Example 12.6, the major product is the more highly substituted alkene. In 1875 the Russian chemist Alexander Zaitsev developed a rule to describe such reactions. **Zaitsev's rule** states that in an elimination reaction, the alkene with the greatest number of alkyl groups on the double bonded carbons (the more highly substituted alkene) is the major product of the reaction.

EXAMPLE 12.6   **Predicting the Products of Alcohol Dehydration**

7 **LEARNING GOAL**
Write equations showing the dehydration of an alcohol.

Predict the products of the dehydration of 3-methyl-2-butanol.

Solution

Assuming that no rearrangement occurs, the product(s) of a dehydration of an alcohol will contain a double bond in which one of the carbons was the original carbinol carbon—the carbon to which the hydroxyl group is attached. Consider the following reaction:

$$2CH_3—\overset{\displaystyle CH_3}{\underset{\displaystyle OH}{\underset{|}{C}H}}—CH—CH_3 \xrightarrow{H^+,\ heat}$$

$$CH_3—\overset{\displaystyle CH_3}{\underset{|}{C}}{=}CH—CH_3 + H_2O$$

2-Methyl-2-butene
(major product)

$$CH_3—\overset{\displaystyle CH_3}{\underset{|}{C}H}—CH{=}CH_2 + H_2O$$

3-Methyl-1-butene
(minor product)

3-Methyl-2-butanol

It is clear that both the major and minor products have a double bond to carbon number 2 in the original alcohol (show in red). Zaitsev's rule tells

Continued—

E X A M P L E   12.6 —*Continued*

us that in dehydration reactions with more than one product possible, the more highly branched alkene predominates. In the reaction shown, 2-methyl-2-butene has three alkyl groups at the double bond, whereas 3-methyl-1-butene has only two alkyl groups at the double bond. The more highly branched alkene is more stable and thus is the major product.

Practice Problem 12.6

Write an equation showing the dehydration of each of the following alcohols. If there are two possible alkene products, indicate which is the major product and which is the minor product.

a. CH_3CH_2OH

$$\begin{array}{c} CH_3 \\ | \\ c.\ CH_3CH_2CHCHCH_2CH_3 \\ | \\ OH \end{array}$$

$$\begin{array}{c} OH \\ | \\ b.\ CH_3CHCH_3 \end{array}$$

$$\begin{array}{c} OH \\ | \\ d.\ CH_3CCH_3 \\ | \\ CH_3 \end{array}$$

For Further Practice: Questions 12.52 and 12.54.

The dehydration of 2-phosphoglycerate to phosphoenolpyruvate is a critical step in the metabolism of the sugar glucose. In the following structures, the circled P represents a phosphoryl group (PO_4^{2-}).

This reaction, and the other reactions of glycolysis, are considered in Section 21.3.

$$\begin{array}{ccc} O^- & & O^- \\ | & & | \\ C{=}O & & C{=}O \\ | & & | \\ CH{-}\text{(P)} & \longrightarrow & C{\sim}\text{(P)} + H_2O \\ | & & \| \\ CH_2{-}OH & & CH_2 \end{array}$$

2-Phosphoglycerate   Phosphoenolpyruvate

The squiggle (~) represents a high energy bond.

Oxidation Reactions

Alcohols may be oxidized with a variety of oxidizing agents to aldehydes, ketones, and carboxylic acids. The most commonly used oxidizing agents are solutions of basic potassium permanganate ($KMnO_4/OH^-$) and chromic acid (H_2CrO_4). The symbol [O] over the reaction arrow is used throughout this book to designate any general oxidizing agent, as in the following reactions:

Oxidation of methanol produces the aldehyde methanal:

LEARNING GOAL

8 Write equations representing the oxidation of alcohols.

An oxidation reaction involves a gain of oxygen or the loss of hydrogen. A reduction reaction involves the loss of oxygen or gain of hydrogen. If two hydrogens are gained or lost for every oxygen gained or lost, the reaction is neither an oxidation nor a reduction.

Methanol
(methyl alcohol)
An alcohol

Methanal
(formaldehyde)
An aldehyde

Note that the symbol [O] is used throughout this book to designate any oxidizing agent.

As we will see in Section 13.4, aldehydes can undergo further oxidation to produce carboxylic acids.

Oxidation of a primary alcohol produces an aldehyde:

$$\underset{\text{1° Alcohol}}{R^1\!\!-\!\!\overset{\displaystyle OH}{\underset{\displaystyle H}{C}}\!\!-\!\!H} \ \xrightarrow{\ [O]\ }\ \underset{\text{An aldehyde}}{\overset{\displaystyle O}{\underset{R^1 \quad H}{C}}}$$

EXAMPLE 12.7 **Writing an Equation Representing the Oxidation of a Primary Alcohol**

LEARNING GOAL

8 Write equations representing the oxidation of alcohols.

Write an equation showing the oxidation of 2,2-dimethylpropanol to produce 2,2-dimethylpropanal.

Solution

Begin by writing the structure of the reactant, 2,2-dimethylpropanol and indicate the need for an oxidizing agent by placing the designation [O] over the reaction arrow:

$$\underset{\text{2,2-Dimethylpropanol}}{CH_3\!\!-\!\!\overset{\displaystyle CH_3}{\underset{\displaystyle CH_3}{C}}\!\!-\!\!\overset{\displaystyle H}{\underset{\displaystyle H}{C}}\!\!-\!\!OH}\ \xrightarrow{\ [O]\ }$$

Now show the oxidation of the hydroxyl group to the aldehyde carbonyl group.

$$\underset{\text{2,2-Dimethylpropanol}}{CH_3\!\!-\!\!\overset{\displaystyle CH_3}{\underset{\displaystyle CH_3}{C}}\!\!-\!\!\overset{\displaystyle H}{\underset{\displaystyle H}{C}}\!\!-\!\!OH}\ \xrightarrow{\ [O]\ }\ \underset{\text{2,2-Dimethylpropanal}}{CH_3\!\!-\!\!\overset{\displaystyle CH_3}{\underset{\displaystyle CH_3}{C}}\!\!-\!\!\overset{\displaystyle O}{\underset{H}{C}}}$$

  Practice Problem 12.7

Write an equation showing the oxidation of the following primary alcohols:

a.
$$CH_3\overset{\displaystyle CH_3}{\underset{\displaystyle CH_3}{C}}CH_2CH_2OH$$

b. CH_3CH_2OH

For Further Practice: Questions 12.48 and 12.56.

Oxidation of a secondary alcohol produces a ketone:

$$\underset{\text{2° Alcohol}}{R^1\!\!-\!\!\overset{\displaystyle OH}{\underset{\displaystyle H}{C}}\!\!-\!\!R^2} \ \xrightarrow{\ [O]\ }\ \underset{\text{A ketone}}{\overset{\displaystyle O}{\underset{R^1 \quad R^2}{C}}}$$

Writing an Equation Representing the Oxidation of a Secondary Alcohol   E X A M P L E 12.8

Write an equation showing the oxidation of 2-propanol to produce propanone.

LEARNING GOAL

8 Write equations representing the oxidation of alcohols.

Solution

Begin by writing the structure of the reactant, 2-propanol, and indicate the need for an oxidizing agent by placing the designation [O] over the reaction arrow:

$$
\begin{array}{c}
\text{OH} \\
| \\
\text{CH}_3\text{—C—CH}_3 \xrightarrow{[O]} \\
| \\
\text{H}
\end{array}
$$

2-Propanol

Now show the oxidation of the hydroxyl group to the ketone carbonyl group.

$$
\begin{array}{c}
\text{OH} \\
| \\
\text{CH}_3\text{—C—CH}_3 \\
| \\
\text{H}
\end{array}
\xrightarrow{[O]}
\begin{array}{c}
\text{O} \\
\| \\
\text{CH}_3\text{—C—CH}_3 \\

\end{array}
$$

2-Propanol   Propanone

Practice Problem 12.8

Write an equation showing the oxidation of the following secondary alcohols:

a.
$$
\begin{array}{c}
\text{OH} \\
| \\
\text{CH}_3\text{CCH}_2\text{CH}_3 \\
| \\
\text{H}
\end{array}
$$

b.
$$
\begin{array}{c}
\text{OH} \\
| \\
\text{CH}_3\text{CHCH}_2\text{CH}_2\text{CH}_3
\end{array}
$$

For Further Practice: Questions 12.49 and 12.55a, b, and c.

Tertiary alcohols cannot be oxidized:

$$
\begin{array}{c}
\text{OH} \\
| \\
\text{R}^1\text{—C—R}^2 \xrightarrow{[O]} \text{No reaction} \\
| \\
\text{R}^3
\end{array}
$$

3° Alcohol

For the oxidation reaction to occur, the carbon bearing the hydroxyl group must contain at least one C—H bond. Because tertiary alcohols contain three C—C bonds to the carbinol carbon, they cannot undergo oxidation.

Question 12.3

Using the I.U.P.A.C Nomenclature System, name the reactants and products for each of the reactions in Practice Problem 12.3 at the end of Example 12.3. Classify each of these alcohols as primary (1°), secondary (2°), or tertiary (3°). Which of the alkenes in Practice Problem 12.3 can be found as both *cis* and *trans* isomers?

Question 12.4

Classify the alcohol products in Practice Problems 12.4 and 12.5, at the end of Examples 12.4 and 12.5, as a primary (1°), secondary (2°), or tertiary (3°) alcohol, and provide the I.U.P.A.C. name of each.

Question 12.5

Name the alcohol reactants in each of the reactions in Practice Problem 12.6, at the end of Example 12.6, using the I.U.P.A.C. Nomenclature System, and classify each of these alcohols as primary (1°), secondary (2°), or tertiary (3°).

Question 12.6

Name each of the reactant alcohols and product aldehydes in Practice Problems 12.7, at the end of Example 12.7, using the I.U.P.A.C. Nomenclature System. *Hint:* Refer to Example 12.7, as well as to Section 13.2 to name the aldehyde products.

Question 12.7

Name each of the reactant alcohols and product ketones in Practice Problem 12.8, at the end of Example 12.8, using the I.U.P.A.C. Nomenclature System. *Hint:* Refer to Example 12.8, as well as to Section 13.2, to name the ketone products.

Question 12.8

Explain why a tertiary alcohol cannot undergo oxidation.

When ethanol is metabolized in the liver, it is oxidized to ethanal (acetaldehyde). If too much ethanol is present in the body, an overabundance of ethanal is formed, which causes many of the adverse effects of the "morning-after hangover." Continued oxidation of ethanal produces ethanoic acid (acetic acid), which is used as an energy source by the cell and eventually oxidized to CO_2 and H_2O. These reactions, summarized as follows, are catalyzed by liver enzymes.

$$CH_3CH_2\!-\!OH \longrightarrow CH_3\overset{\displaystyle O}{\overset{\displaystyle \|}{C}}\!-\!H \longrightarrow CH_3\overset{\displaystyle O}{\overset{\displaystyle \|}{C}}\!-\!OH \longrightarrow CO_2 + H_2O$$

Ethanol	Ethanal	Ethanoic acid
(ethyl alcohol)	(acetaldehyde)	(acetic acid)

12.6 Oxidation and Reduction in Living Systems

LEARNING GOAL

9 Discuss the role of oxidation and reduction reactions in the chemistry of living systems.

Before beginning a discussion of oxidation and reduction in living systems, we must understand how to recognize **oxidation** (loss of electrons) and **reduction** (gain of electrons) in organic compounds. It is easy to determine when an oxidation or a reduction occurs in inorganic compounds because the process is accompanied by a change in charge. For example,

$$Ag^0 \longrightarrow Ag^+ + 1e^-$$

With the loss of an electron, the neutral atom is converted to a positive ion, which is oxidation. In contrast,

$$:\!\overset{..}{\underset{..}{Br}}\!\cdot\; +\, e^- \longrightarrow\; :\!\overset{..}{\underset{..}{Br}}\!:^-$$

With the gain of one electron, the bromine atom is converted to a negative ion, which is reduction.

When organic compounds are involved, however, there may be no change in charge, and it is often difficult to determine whether oxidation or reduction has occurred. The following simplified view may help.

In organic systems, *oxidation* may be recognized as a gain of oxygen or a loss of hydrogen. A *reduction* reaction may involve a loss of oxygen or gain of hydrogen.

Consider the following compounds. An alkane may be oxidized to an alcohol by gaining an oxygen. A primary or secondary alcohol may be oxidized to an aldehyde or ketone, respectively, by the loss of hydrogen. Finally, an aldehyde may be oxidized to a carboxylic acid by gaining an oxygen.

More oxidized form →

$$
\underset{\text{Alkane}}{R-\overset{\displaystyle H}{\underset{\displaystyle H}{C}}-H} \qquad
\underset{\text{Alcohol}}{R-\overset{\displaystyle H}{\underset{\displaystyle H}{C}}-OH} \qquad
\underset{\text{Aldehyde}}{R-\overset{\displaystyle O}{C}-H} \qquad
\underset{\substack{\text{Carboxylic}\\ \text{acid}}}{R-\overset{\displaystyle O}{C}-OH}
$$

← More reduced form

Thus, the conversion of an alkane to an alcohol, an alcohol to a carbonyl compound, and a carbonyl compound (aldehyde) to a carboxylic acid are all examples of oxidations. Conversions in the opposite direction are reductions.

Oxidation and reduction reactions also play an important role in the chemistry of living systems. In living systems these reactions are catalyzed by the action of various enzymes called *oxidoreductases*. These enzymes require compounds called *coenzymes* to accept or donate hydrogen in the reactions that they catalyze.

Nicotinamide adenine dinucleotide, NAD^+, is a coenzyme commonly involved in biological oxidation–reduction reactions (Figure 12.5). We see NAD^+ in action in the final reaction of the citric acid cycle, an energy-harvesting pathway essential to life. In this reaction, catalyzed by the enzyme malate dehydrogenase, malate is oxidized to produce oxaloacetate:

$$
\underset{\text{Malate}}{\overset{\displaystyle COO^-}{\underset{\displaystyle COO^-}{\overset{\displaystyle |}{\underset{\displaystyle |}{HO-\underset{\displaystyle CH_2}{\overset{\displaystyle |}{C}-H}}}}}} + NAD^+
\xrightarrow{\substack{\text{Malate} \\ \text{dehydrogenase}}}
\underset{\text{Oxaloacetate}}{\overset{\displaystyle COO^-}{\underset{\displaystyle COO^-}{\overset{\displaystyle |}{\underset{\displaystyle |}{\underset{\displaystyle CH_2}{\overset{\displaystyle |}{C}=O}}}}}} + NADH
$$

Figure 12.5
Nicotinamide adenine dinucleotide.

Adenine nucleotide   Nicotinamide nucleotide

Nicotinamide adenine dinucleotide (NAD^+)

A Human Perspective

Alcohol Consumption and the Breathalyzer Test

Ethanol has been used widely as a beverage, a medicinal, and a solvent in numerous pharmaceutical preparations. Such common usage often overshadows the fact that ethanol is a toxic substance. Ethanol consumption is associated with a variety of long-term effects, including cirrhosis of the liver, death of brain cells, and alcoholism. Alcohol consumed by the mother can even affect the normal development of her unborn child and result in fetal alcohol syndrome. For these reasons, over-the-counter cough and cold medications that were once prepared in ethanol are now manufactured in alcohol-free form.

Short-term effects, linked to the social use of ethanol, center on its effects on behavior, reflexes, and coordination. Blood alcohol levels of 0.05–0.15% seriously inhibit coordination. Blood levels in excess of 0.10% are considered evidence of intoxication in most states. Blood alcohol levels in the range of 0.30–0.50% produce unconsciousness and the risk of death.

The loss of some coordination and reflex action is particularly serious when the affected individual attempts to drive a car. Law enforcement has come to rely on the "breathalyzer" test to screen for individuals suspected of driving while intoxicated. Those with a positive breathalyzer test are then given a more accurate blood test to establish their guilt or innocence.

The suspect is required to exhale into a solution that will react with the unmetabolized alcohol in the breath. The partial pressure of the alcohol in the exhaled air has been demonstrated to be proportional to the blood alcohol level. The solution is an acidic solution of dichromate ion, which is yellow-orange. The alcohol reduces the chromium in the dichromate ion from +6 to +3, the Cr^{3+} ion, which is green. The intensity of the green color is measured, and it is proportional to the amount of ethanol that was oxidized. The reaction is:

$$16H^+ + 2Cr_2O_7^{2-} + 3CH_3CH_2OH \longrightarrow$$
$$\text{Yellow-orange}$$

$$3CH_3COOH + 4Cr^{3+} + 11H_2O$$
$$\text{Green}$$

The breathalyzer test is a technological development based on a scientific understanding of the chemical reactions that ethanol may undergo—a further example of the dependence of technology on science.

Driving under the influence of alcohol impairs coordination and reflexes.

For Further Understanding

Explain why the intensity of the green color in the reaction solution is proportional to the level of alcohol in the breath.

Draw a graph that represents this relationship.

NAD$^+$ actually accepts a hydride anion, H$^-$, hydrogen with two electrons.

NAD$^+$ participates by accepting hydrogen from the malate. As malate is oxidized, NAD$^+$ is reduced to NADH.

We will study many other biologically important oxidation-reduction reactions in upcoming chapters.

12.7 Phenols

Phenols are compounds in which the hydroxyl group is attached to a benzene ring (Figure 12.6). Like alcohols, they are polar compounds because of the polar hydroxyl group. Thus, the simpler phenols are somewhat soluble in water. They are found in flavorings and fragrances (mint and savory) and are used as preservatives (butylated hydroxytoluene, BHT). Examples include:

Thymol (mint)   Carvacrol (savory)   Butylated hydroxytoluene, BHT (food preservative)

Phenols are also widely used in health care as germicides. In fact, carbolic acid, a dilute solution of phenol, was used as an antiseptic and disinfectant by Joseph Lister in his early work to decrease postsurgical infections. He used carbolic acid to bathe surgical wounds and to "sterilize" his instruments. Other derivatives of phenol that are used as antiseptics and disinfectants include hexachlorophene, hexylresorcinol, and *o*-phenylphenol. The structures of these compounds are shown below:

Phenol (carbolic acid; phenol dissolved in water; antiseptic)

Hexachlorophene (antiseptic)

Hexylresorcinol (antiseptic)

o-Phenylphenol (antiseptic)

Question 12.9

Why are simple phenols somewhat soluble in water?

Question 12.10

What is carbolic acid? How did Joseph Lister use carbolic acid?

12.8 Ethers

Ethers have the general formula R—O—R, and thus they are structurally related to alcohols (R—O—H). The C—O bonds of ethers are polar, so ether molecules are polar (Figure 12.7). However, ethers do not form hydrogen bonds to one another because there is no —OH group. Therefore, they have

LEARNING GOAL

10 Discuss the use of phenols as germicides.

Figure 12.6
Ball-and-stick model of phenol. Keep in mind that this model is not completely accurate because it cannot show the cloud of shared electrons above and below the benzene ring. Review Section 11.6 for a more accurate description of the benzene ring.

A dilute solution of phenol must be used because concentrated phenol causes severe burns and because phenol is not highly soluble in water.

LEARNING GOAL

11 Write names and draw structures for common ethers and discuss their use in medicine.

Figure 12.7
Ball-and-stick model of the ether, methoxymethane (dimethyl ether).

An alkoxy group is an alkyl group bonded to an oxygen atom (—OR).

much lower boiling points than alcohols of similar molecular weight but higher boiling points than alkanes of similar molecular weight. Compare the following examples:

$CH_3CH_2CH_2CH_3$ $CH_3—O—CH_2CH_3$ $CH_3CH_2CH_2OH$

Butane   Methoxyethane   1-Propanol
(butane)   (ethyl methyl ether)   (propyl alcohol)
M.W. = 58   M.W. = 60   M.W. = 60
b.p. = −0.5°C b.p. = 7.9°C   b.p. = 97.2°C

In the I.U.P.A.C. system of naming ethers, the —OR substituent is named as an alkoxy group. This is analogous to the name *hydroxy* for the —OH group. Thus, $CH_3—O—$ is methoxy, $CH_3CH_2—O—$ is ethoxy, and so on.

EXAMPLE 12.9   **Using I.U.P.A.C. Nomenclature to Name an Ether**

11 LEARNING GOAL
Write names and draw structures for common ethers and discuss their use in medicine.

Name the following ether using I.U.P.A.C. nomenclature.

Solution

$$\overset{\overset{\displaystyle O—CH_3}{|}}{CH_3CH_2CHCH_2CH_2CH_2CH_2CH_3}$$
$\quad\quad\ \ 1\ \ \ 2\ \ \ 3\ \ \ 4\ \ \ 5\ \ \ 6\ \ \ 7\ \ \ 8\ \ \ 9$

Parent compound: nonane
Position of alkoxy group: carbon-3 (*not* carbon-7)
Substituents: 3-methoxy
Name: 3-Methoxynonane

Practice Problem 12.9

Name the following ethers using I.U.P.A.C. nomenclature:

a. $CH_3CH_2—O—CH_2CH_2CH_3$
b. $CH_3—O—CH_2CH_2CH_3$
c. $CH_3CH_2—O—CH_2CH_2CH_2CH_2CH_3$
d. $CH_3CH_2CH_2CH_2—O—CH_2CH_2CH_3$

For Further Practice: Questions 12.85 and 12.88.

In the common system of nomenclature, ethers are named by placing the names of the two alkyl groups attached to the ether oxygen as prefixes in front of the word *ether*. The names of the two groups can be placed either alphabetically or by size (smaller to larger), as seen in the following examples:

$CH_3—O—CH_3$ $CH_3—O—CH_2CH_3$ $CH_3CH_2—O—CH(CH_3)_2$

Dimethyl ether   Ethyl methyl ether   Ethyl isopropyl ether
or   or
methyl ether   methyl ethyl ether

Naming Ethers Using the Common Nomenclature System   EXAMPLE 12.10

Write the common name for each of the following ethers.

Solution

$$CH_3CH_2{-}O{-}CH_2CH_3 \qquad CH_3{-}O{-}CH_2CH_2CH_3$$

Alkyl Groups:	two ethyl groups	methyl and propyl
Name:	Diethyl ether	Methyl propyl ether

> LEARNING GOAL
> 11 Write names and draw structures for common ethers and discuss their use in medicine.

Notice that there is only one correct name for methyl propyl ether because the methyl group is smaller than the propyl group and it would be first in an alphabetical listing also.

Practice Problem 12.10

Write the common name for each of the following ethers:

a. $CH_3CH_2{-}O{-}CH_2CH_2CH_3$
b. $CH_3{-}O{-}CH_2CH_2CH_3$
c. $CH_3CH_2{-}O{-}CH_2CH_2CH_2CH_2CH_3$
d. $CH_3CH_2CH_2CH_2{-}O{-}CH_2CH_2CH_3$

For Further Practice: Questions 12.86 and 12.88.

Chemically, ethers are moderately inert. They do not react with reducing agents or bases under normal conditions. However, they are extremely volatile and highly flammable (easily oxidized in air) and hence must always be treated with great care.

Ethers may be prepared by a dehydration reaction (removal of water) between two alcohol molecules, as shown in the following general reaction. The reaction requires heat and acid.

$$R^1{-}OH + R^2{-}OH \xrightarrow[\text{heat}]{H^+} R^1{-}O{-}R^2 + H_2O$$

Alcohol	Alcohol		Ether	Water

Writing an Equation Representing the Synthesis of an Ether via a Dehydration Reaction   EXAMPLE 12.11

Write an equation showing the synthesis of dimethyl ether.

Solution

The alkyl substituents of this ether are two methyl groups. Thus, the alcohol that must undergo dehydration to produce dimethyl ether is methanol.

> LEARNING GOAL
> 12 Write equations representing the dehydration reaction between two alcohol molecules.

$$CH_3OH + CH_3OH \xrightarrow{H^+} CH_3{-}O{-}CH_3 + H_2O$$

Methanol Methanol   Dimethyl ether   Water

Practice Problem 12.11

a. Write an equation showing the dehydration reaction that would produce diethyl ether. Provide structures and names for all reactants and products.
b. Write an equation showing the dehydration reaction between two molecules of 2-propanol. Provide structures and names for all reactants and products.

For Further Practice: Questions 12.83 and 12.84.

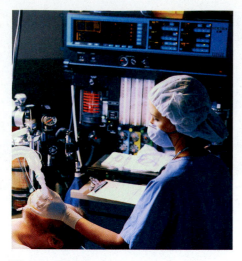

Figure 12.8
An anesthesiologist administers Penthrane to a surgical patient.

Diethyl ether was the first general anesthetic used. The dentist Dr. William Morton is credited with its introduction in the 1800s. Diethyl ether functions as an anesthetic by interacting with the central nervous system. It appears that diethyl ether (and many other general anesthetics) functions by accumulating in the lipid material of the nerve cells, thereby interfering with nerve impulse transmission. This results in analgesia, a lessened perception of pain.

Halogenated ethers are also routinely used as general anesthetics (Figure 12.8). They are less flammable than diethyl ether and are therefore safer to store and work with. *Penthrane* and *Enthrane* (trade names) are two of the more commonly used members of this family:

$$CH_3-O-\overset{\overset{\displaystyle F}{|}}{\underset{\underset{\displaystyle F}{|}}{C}}-\overset{\overset{\displaystyle Cl}{|}}{\underset{\underset{\displaystyle Cl}{|}}{C}}-H \qquad\qquad H-\overset{\overset{\displaystyle F}{|}}{\underset{\underset{\displaystyle F}{|}}{C}}-O-\overset{\overset{\displaystyle F}{|}}{\underset{\underset{\displaystyle F}{|}}{C}}-\overset{\overset{\displaystyle F}{|}}{\underset{\underset{\displaystyle Cl}{|}}{C}}-H$$

Penthrane   Enthrane

Question 12.11

Why do ethers have much lower boiling points than alcohols?

Question 12.12

Describe the structure of ether molecules.

12.9 Thiols

13 | **LEARNING GOAL**
Write names and draw structures for simple thiols and discuss their biological significance.

Compounds that contain the sulfhydryl group (—SH) are called **thiols.** They are similar to alcohols in structure, but the sulfur atom replaces the oxygen atom.

Thiols and many other sulfur compounds have nauseating aromas. They are found in substances as different as the defensive spray of the North American striped skunk, onions, and garlic. The structures of the two most common compounds in the defense spray of the striped skunk, *trans*-2-butene-1-thiol and 3-methyl-1-butanethiol, are seen in Figure 12.9. These structures are contrasted with the structures of the two molecules that make up the far more pleasant scent of roses: geraniol, an unsaturated alcohol, and 2-phenylethanol, an aromatic alcohol.

The I.U.P.A.C. rules for naming thiols are similar to those for naming alcohols, except that the full name of the alkane is retained. The suffix *-thiol* follows the name of the parent compound.

EXAMPLE 12.12   **Naming Thiols Using the I.U.P.A.C. Nomenclature System**

13 | **LEARNING GOAL**
Write names and draw structures for simple thiols and discuss their biological significance.

Write the I.U.P.A.C. name for the thiols shown below.

Solution

Retain the full name of the parent compound and add the suffix *-thiol.*

Continued—

EXAMPLE 12.12 —*Continued*

$$CH_3CH_2—SH \qquad HS—CH_2CH_2—SH$$

Parent compound:	ethane	ethane
Position of —SH:	carbon-1 (must be)	carbon-1 and carbon-2
Name:	Ethanethiol	1,2-Ethanedithiol

$$\overset{\displaystyle CH_3}{\underset{\displaystyle SH}{CH_3\overset{|}{C}HCH_2CH_2}} \qquad \overset{\displaystyle SH}{\underset{\displaystyle SH}{CH_3\overset{|}{C}HCH_2CH_2CH_2}}$$

Parent Compound:	butane	pentane
Position of —SH:	carbon-1	carbon-1 and carbon-4
Substituent:	3 -methyl	
Name:	3-Methyl-1-butanethiol	1,4-Pentanedithiol

Practice Problem 12.12

Draw the structures of each of the following thiols.

a. 1,3-Butanedithiol
b. 2-Methyl-2-pentanethiol
c. 2-Chloro-2-propanethiol
d. Cyclopentanethiol

For Further Practice: Questions 12.93 and 12.94.

trans-2-Butene-1-thiol

3-Methyl-1-butanethiol

3,7-dimethyl-2,6-octadien-1-ol (geraniol)

2-Phenylethanol

Figure 12.9

This skunk on a bed of roses is surrounded by scent molecules. The two most common compounds in the defense spray of the striped skunk are the thiols *trans*-2-butene-1-thiol and 3-methyl-1-butanethiol. Two alcohols, 2-phenylethanol and geraniol, are the major components of the scent of roses.

The amino acid cysteine is a thiol that has an important role to play in the structure and shape of many proteins. Two cysteine molecules can undergo oxidation to form cystine. The new bond formed is called a **disulfide** bond (—S—S—) bond.

2 Cysteine   Cystine

If the two cysteines are in different protein chains, the disulfide bond between them forms a bridge joining them together (Figure 12.10). If the two cysteines are in the same protein chain, a loop is formed. An example of the importance of disulfide bonds is seen in the production and structure of the protein hormone insulin, which controls blood sugar levels in the body. Insulin is initially produced as a protein called preproinsulin. Enzymatic removal of 24 amino acids and formation of disulfide bonds between cysteine amino acids produces proinsulin (Figure 12.10). Regions of the protein are identified as the A, B, and C chains. Notice that the A and B chains are covalently bonded to one another by two disulfide bonds. A third disulfide bond produces a hairpin loop in the A chain of the molecule. The molecule is now ready for the final stage of insulin synthesis in which the C chain is removed by the action of protein degrading enzymes (proteases). The active hormone, shown at the bottom of Figure 12.10, consists of the 21 amino acid A chain bonded to the 30 amino acid B chain by two disulfide bonds. Without these disulfide bonds, functional insulin molecules could not exist because there would be no way to keep the two chains together in the proper shape.

British Anti-Lewisite (BAL) is a dithiol used as an antidote in mercury poisoning. It was originally developed as an antidote to a mustard-gas-like chemical warfare agent called Lewisite, which was developed near the end of World War I and never used. By the onset of World War II, Lewisite was considered to be obsolete because of the discovery of BAL, an effective, inexpensive antidote. The two thiol groups of BAL form a water-soluble complex with mercury (or with the arsenic in Lewisite) that is excreted from the body in the urine.

$$CH_2—CH—CH_2$$
$$OH \quad SH \quad SH$$

BAL

Coenzyme A is a thiol that serves as a "carrier" of acetyl groups (CH_3CO—) in biochemical reactions. It plays a central role in metabolism by shuttling acetyl groups from one reaction to another. When the two-carbon acetate group is attached to coenzyme A, the product is acetyl coenzyme A (acetyl CoA). The bond between coenzyme A and the acetyl group is a high-energy *thioester bond*. In effect, formation of the high-energy thioester bond "energizes" the acetyl group so that it can participate in other biochemical reactions.

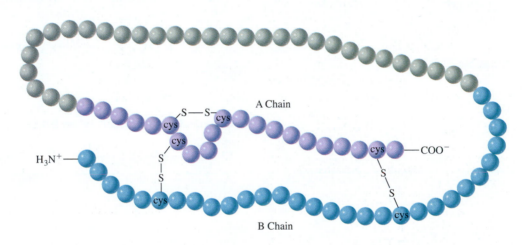

Acetyl group — β-Mercapto-ethylamine group — Pantothenate unit — Phosphorylated ADP

Acetyl coenzyme A
(acetyl CoA)

Acetyl CoA is made and used in the energy-producing reactions that provide most of the energy for life processes. It is also required for the biosynthesis of many biological molecules.

Figure 12.10
Two steps in the synthesis of insulin. Disulfide bonds hold the A and B chains together and form a hairpin loop in the A chain.

Summary of Reactions

Preparation of Alcohols

Hydration of alkenes:

Alkene   Water   Alcohol

Reduction of an aldehyde or ketone:

Aldehyde   Hydrogen   Alcohol
or
ketone

Dehydration of Alcohols

Alcohol   Alkene   Water

Oxidation Reactions

Oxidation of a primary alcohol:

1° Alcohol   An aldehyde

Oxidation of a secondary alcohol:

2° Alcohol   A ketone

Oxidation of a tertiary alcohol:

3° Alcohol

Dehydration Synthesis of an Ether

$$R^1—OH + R^2—OH \xrightarrow[\text{heat}]{H^+} R^1—O—R^2 + H_2O$$

Alcohol   Alcohol   Ether   Water

SUMMARY

12.1 Alcohols: Structure and Physical Properties

Alcohols are characterized by the *hydroxyl group (—OH)* and have the general formula R—OH. They are very polar, owing to the polar hydroxyl group, and are able to form intermolecular hydrogen bonds. Because of hydrogen bonding between alcohol molecules, they have higher boiling points than hydrocarbons of comparable molecular weight. The smaller alcohols are very water soluble.

12.2 Alcohols: Nomenclature

In the I.U.P.A.C. system, alcohols are named by determining the parent compound and replacing the -*e* ending with -*ol*. The chain is numbered to give the hydroxyl group the lowest possible number. Common names are derived from the alkyl group corresponding to the parent compound.

12.3 Medically Important Alcohols

Methanol is a toxic alcohol that is used as a solvent. Ethanol is the alcohol consumed in beer, wine, and distilled liquors. Isopropanol is used as a disinfectant. Ethylene glycol (1,2-ethanediol) is used as antifreeze, and glycerol (1,2,3-propanetriol) is used in cosmetics and pharmaceuticals.

12.4 Classification of Alcohols

Alcohols may be classified as *primary, secondary,* or *tertiary,* depending on the number of alkyl groups attached to the *carbinol carbon,* the carbon bearing the hydroxyl group. A primary alcohol has a single alkyl group bonded to the *carbinol carbon.* Secondary and tertiary alcohols have two and three alkyl groups, respectively.

12.5 Reactions Involving Alcohols

Alcohols can be prepared by the *hydration* of alkenes or reduction of aldehydes and ketones. Alcohols can undergo *dehydration* to yield alkenes. Primary and secondary alcohols undergo *oxidation* reactions to yield aldehydes and ketones, respectively. Tertiary alcohols do not undergo oxidation.

12.6 Oxidation and Reduction in Living Systems

In organic and biological systems, *oxidation* involves the gain of oxygen or loss of hydrogen. *Reduction* involves the loss of oxygen or gain of hydrogen. Nicotinamide adenine dinucleotide, NAD^+, is a coenzyme involved in many biological oxidation and reduction reactions.

12.7 Phenols

Phenols are compounds in which the hydroxyl group is attached to a benzene ring; they have the general formula Ar—OH. Many phenols are important as antiseptics and disinfectants.

12.8 Ethers

Ethers are characterized by the R—O—R functional group. Ethers are generally nonreactive but are extremely flammable. Diethyl ether was the first general anesthetic used in medical practice. It has since been replaced by Penthrane and Enthrane, which are less flammable.

12.9 Thiols

Thiols are characterized by the sulfhydryl group (—SH). The amino acid cysteine is a thiol that is extremely important for maintaining the correct shapes of proteins. Coenzyme A is a thiol that serves as a "carrier" of acetyl groups in biochemical reactions.

KEY TERMS

alcohol (12.1)	oxidation (12.6)
carbinol carbon (12.4)	phenol (12.7)
dehydration (12.5)	primary (1°) alcohol (12.4)
disulfide (12.9)	reduction (12.6)
elimination reaction (12.5)	secondary (2°) alcohol (12.4)
ether (12.8)	tertiary (3°) alcohol (12.4)
fermentation (12.3)	thiol (12.9)
hydration (12.5)	Zaitsev's rule (12.5)
hydroxyl group (12.1)	

QUESTIONS AND PROBLEMS

Alcohols: Structure and Physical Properties

Foundations

12.13 Describe the relationship between the water solubility of alcohols and their hydrocarbon chain length.

12.14 Explain the relationship between the water solubility of alcohols and the number of hydroxyl groups in the molecule.

Applications

12.15 Arrange the following compounds in order of increasing boiling point, beginning with the lowest:
- **a.** $CH_3CH_2CH_2CH_2CH_3$
- **b.** $CH_3CHCH_2CHCH_3$ with OH OH
- **c.** $CH_3CHCH_2CH_2CH_3$ with OH
- **d.** $CH_3CH_2CH_2—O—CH_2CH_3$

12.16 Why do alcohols have higher boiling points than alkanes?

12.17 Which member of each of the following pairs is more soluble in water?
- **a.** CH_3CH_2OH or $CH_3CH_2CH_2CH_2OH$
- **b.** $CH_3CH_2CH_2CH_2CH_3$ or $CH_3CH_2CH_2CH_2OH$
- **c.** OH (cyclopentane ring) or CH_3CHCH_3 with OH

12.18 Arrange the three alcohols in each of the following sets in order of increasing solubility in water:
- **a.** $CH_3CH_2CH_2CH_2CH_2OH$ $CH_3CHCH_2CHCH_2CH_3$ with OH OH
  $CH_3CHCH_2CHCH_2CH_2OH$ with OH OH
- **b.** Pentyl alcohol   1-Hexanol   Ethylene glycol

Alcohols: Nomenclature

Foundations

12.19 Briefly describe the I.U.P.A.C. rules for naming alcohols.

12.20 Briefly describe the rules for determining the common names for alcohols.

Applications

12.21 Give the I.U.P.A.C. name for each of the following compounds:
- **a.** $CH_3CH_2CH_2CH_2CH_2CH_2CH_2OH$
- **b.** CH_3CHCH_3 with OH then CH_3
- **c.** CH_3CCH_3 with CH_2OH

12.22 Give the I.U.P.A.C. name for each of the following compounds:
- **a.** $CH_3CH_2CHCH_2CH_2CH_2OH$ with Br above, CH_3 below
- **b.** $CH_3CH—CCH_2CH_2CH_3$ with OH CH_3 and $CH_2CH_2CH_2CH_3$
- **c.** $CH_3CH_2CCH_2CH_3$ with OH

12.23 Draw each of the following, using complete structural formulas:
- **a.** 3-Hexanol
- **b.** 1,2,3-Pentanetriol
- **c.** 2-Methyl-2-pentanol

12.24 Draw each of the following using condensed structural formulas:
   a. Cyclohexanol
   b. 3,4-Dimethyl-3-heptanol

12.25 Give the I.U.P.A.C. name for each of the following compounds:

a.   **b.**   **c.**

12.26 Draw each of the following alcohols:
   a. 1-Iodo-2-butanol
   b. 1,2-Butanediol
   c. Cyclobutanol

12.27 Give the common name for each of the following compounds:
   a. CH_3OH
   b. CH_3CH_2OH
   c. $CH_2\!-\!CH_2$
         $\;\;|\quad\;\;|$
         $\;\;OH\;\;\;OH$
   d. $CH_3CH_2CH_2OH$

12.28 Draw the structure of each of the following compounds:
   a. Pentyl alcohol
   b. Isopropyl alcohol
   c. Octyl alcohol
   d. Propyl alcohol

12.29 Draw a complete structural formula for each of the following compounds:
   a. 4-Methyl-2-hexanol
   b. Isobutyl alcohol
   c. 1,5-Pentanediol
   d. 2-Nonanol
   e. 1,3,5-Cyclohexanetriol

12.30 Name each of the following alcohols using the I.U.P.A.C. Nomenclature System:

   a.

   b.

   c. $CH_3CHCH_2CHCH_2CHCH_3$
          $\quad\;|\qquad\;\;|\qquad\;\;|$
          $\quad\;OH\quad\;OH\quad\;OH$
          $\qquad\qquad\;\;OH$

   d. $CH_3CH_2CHCHCHCH_3$
          $\qquad\quad\;|\quad\;|$
          $\qquad\quad\;OH\;\;OH$

Medically Important Alcohols

12.31 What is denatured alcohol? Why is alcohol denatured?
12.32 What are the principal uses of methanol, ethanol, and isopropyl alcohol?
12.33 What is fermentation?
12.34 Why do wines typically have an alcohol concentration of 12–13%?
12.35 Why must fermentation products be distilled to produce liquors such as scotch?
12.36 If a bottle of distilled alcoholic spirits—for example, scotch whiskey—is labeled as 80 proof, what is the percentage of alcohol in the scotch?

Classification of Alcohols

Foundations

12.37 Define the term *carbinol carbon*.
12.38 Define the terms *primary, secondary,* and *tertiary alcohol,* and draw a general structure for each.

Applications

12.39 Classify each of the following as a 1°, 2°, or 3° alcohol:
   a. 3-Methyl-1-butanol
   b. 2-Methylcyclopentanol
   c. *t*-Butyl alcohol
   d. 1-Methylcyclopentanol
   e. 2-Methyl-2-pentanol

12.40 1-Heptanol has a pleasant aroma and is sometimes used in cosmetics to enhance the fragrance. Draw the condensed structural formula for 1-heptanol. Is this a primary, secondary, or tertiary alcohol?

12.41 Classify each of the following as a 1°, 2°, or 3° alcohol:
   a. $CH_3CH_2CH_2CH_2CH_2CH_2CH_2OH$
   b. CH_3CHCH_3
         $\quad\;|$
         $\quad\;OH$
         $\quad\;CH_3$
         $\quad\;|$
   c. CH_3CCH_3
         $\quad\;|$
         $\quad\;CH_2OH$
                  $\;\;Br$
                  $\;\;|$
   d. $CH_3CH_2CHCH_2CH_2CH_2OH$
                      CH_3
                      $|$
   e. $CH_3CH\!-\!CCH_2CH_2CH_3$
         $\qquad\;|\quad\;|$
         $\qquad\;OH\;\;CH_3$

12.42 Classify each of the following as a primary, secondary, or tertiary alcohol:
                  CH_2CH_3
                  $|$
   a. CH_3CH_2COH
                  $|$
                  CH_3
   b. $CH_3CHCH_2CHCH_3$
         $\qquad\;|\qquad\;\;|$
         $\qquad\;OH\qquad\;Br$
   c. $CH_3CH_2CH_2OH$
                  CH_3
                  $|$
   d. $CH_3CCH_2CH_2CH_3$
                  $|$
                  OH

12.43 Stingless bees use complex systems to communicate. One aspect of this communication is chemical: the bees produce 2-nonanol, 2-heptanol, and 2-undecanol in their mouth parts (mandibles) to direct other bees to the pollen sources. Draw the condensed structure of each of these alcohols. Classify each as a primary, secondary, or tertiary alcohol.

12.44 Classify each of the following as a primary, secondary, or tertiary alcohol:
a. 2-Methyl-2-butanol
b. 1,2-Dimethylcyclohexanol
c. 2,3,4-Trimethylcyclopentanol
d. 3,3-Dimethyl-2-pentanol

Reactions Involving Alcohols

Foundations

12.45 Write a general equation representing the preparation of an alcohol by hydration of an alkene.
12.46 Write a general equation representing the preparation of an alcohol by hydrogenation of an aldehyde or a ketone.
12.47 Write a general equation representing the dehydration of an alcohol.
12.48 Write a general equation representing the oxidation of a 1° alcohol.
12.49 Write a general equation representing the oxidation of a 2° alcohol.
12.50 Write a general equation representing the oxidation of a 3° alcohol.

Applications

12.51 Predict the products formed by the hydration of the following alkenes:
a. 1-Pentene
b. 2-Pentene
c. 3-Methyl-1-butene
d. 3,3-Dimethyl-1-butene
12.52 Draw the alkene products of the dehydration of the following alcohols:
a. 2-Pentanol
b. 3-Methyl-1-pentanol
c. 2-Butanol
d. 4-Chloro-2-pentanol
e. 1-Propanol
12.53 Write an equation showing the hydration of each of the following alkenes. Name each of the products using the I.U.P.A.C. Nomenclature System.
a. 2-Hexene
b. Cyclopentene
c. 1-Octene
d. 1-Methylcyclohexene
12.54 Write an equation showing the dehydration of each of the following alcohols. Name each of the reactants and products using the I.U.P.A.C. Nomenclature System.

a. $CH_3CHCH_2CH_3$
　　　|
　　　OH

b. (cyclopentane ring with OH and CH₃ substituents)

12.55 What product(s) would result from the oxidation of each of the following alcohols with, for example, potassium permanganate? If no reaction occurs, write N.R.
a. 2-Butanol
b. 2-Methyl-2-hexanol
c. Cyclohexanol
d. 1-Methyl-1-cyclopentanol
12.56 We have seen that ethanol is metabolized to ethanal (acetaldehyde) in the liver. What would be the product formed, under the same conditions, from each of the following alcohols?
a. CH_3OH
b. $CH_3CH_2CH_2OH$
c. $CH_3CH_2CH_2CH_2OH$
12.57 Give the oxidation products of the following alcohols. If no reaction occurs, write N.R.

　　　　　　OH
　　　　　　|
a. $CH_3CH_2CHCH_2CH_3$
b. $CH_3CH_2CH_2OH$
　　　OH　CH₃
　　　|　　|
c. $CH_3CHCH_2CHCH_3$
　　　OH
　　　|
d. $CH_3CCH_2CH_3$
　　　|
　　　CH₃

e. (benzene ring)—$CH_2CH_2CH_2OH$

12.58 Write an equation, using complete structural formulas, demonstrating each of the following chemical transformations:
a. Oxidation of an alcohol to an aldehyde
b. Oxidation of an alcohol to a ketone
c. Dehydration of a cyclic alcohol to a cycloalkene
d. Hydrogenation of an alkene to an alkane
12.59 Write the reaction, occurring in the liver, that causes the oxidation of ethanol. What is the product of this reaction and what symptoms are caused by the product?
12.60 Write the reaction, occurring in the liver, that causes the oxidation of methanol. What is the product of this reaction and what is the possible result of the accumulation of the product in the body?
12.61 Write an equation for the preparation of 2-butanol from 1-butene. What type of reaction is involved?
12.62 Write a general equation for the preparation of an alcohol from an aldehyde or ketone. What type of reaction is involved?
12.63 Show how acetone can be prepared from propene.

　　　　O
　　　　||
　　CH_3CCH_3
　　　Acetone

12.64 Give the oxidation product for cholesterol.

(structure of Cholesterol)

Cholesterol

12.65 Write a balanced equation for the hydrogenation of each of the following:
a. Pentanal (a five-carbon aldehyde)
b. 3-Pentanone (a five-carbon ketone)
12.66 Write a balanced equation for the hydrogenation of each of the following:
a. Ethanal (a two-carbon aldehyde)
b. 2-Hexanone (a six-carbon ketone)

Oxidation and Reduction in Living Systems

Foundations

12.67 Define the terms *oxidation* and *reduction*.

12.68 How do we recognize oxidation and reduction in organic compounds?

Applications

12.69 Arrange the following compounds from the most reduced to the most oxidized:

$$CH_3CH_2\overset{\overset{O}{\|}}{C}-OH \qquad CH_3CH_2CH_3$$

$$CH_3CH_2\overset{\overset{O}{\|}}{C}-H \qquad CH_3CH_2CH_2OH$$

12.70 What is the role of the coenzyme nicotinamide adenine dinucleotide (NAD$^+$) in enzyme-catalyzed oxidation–reduction reactions?

Phenols

Foundations

12.71 What are phenols?

12.72 Describe the water solubility of phenols.

Applications

12.73 2,4,6-Trinitrophenol is known by the common name *picric acid*. Picric acid is a solid but is readily soluble in water. In solution it is used as a biological tissue stain. As a solid, it is also known to be unstable and may explode. In this way it is similar to 2,4,6-trinitrotoluene (TNT). Draw the structures of picric acid and TNT. Why is picric acid readily soluble in water whereas TNT is not?

12.74 Name the following aromatic compounds using the I.U.P.A.C. system:

a.

b.

c.

d.

12.75 List some phenol compounds that are commonly used as antiseptics or disinfectants.

12.76 Why must a dilute solution of phenol be used for disinfecting environmental surfaces?

Ethers

Foundations

12.77 Describe the physical properties of ethers.

12.78 Compare the water solubility of ethers and alcohols.

Applications

12.79 Draw all of the alcohols and ethers of molecular formula $C_4H_{10}O$.

12.80 Name each of the isomers drawn for Problem 12.79.

12.81 Give the I.U.P.A.C. names for Penthrane and Enthrane (see Section 12 .8.)

12.82 Why have Penthrane and Enthrane replaced diethyl ether as a general anesthetic?

12.83 Ethers may be prepared by the removal of water (dehydration) between two alcohols, as shown. Give the structure(s) of the ethers formed by the reaction of the following alcohol(s) under acidic conditions with heat.

Example: $CH_3OH + HOCH_3 \xrightarrow[\text{Heat}]{\text{H}^+} CH_3OCH_3 \quad H_2O$

a. $2CH_3CH_2OH \longrightarrow$?

b. $CH_3OH + CH_3CH_2OH \longrightarrow$?

c. $(CH_3)_2CHOH + CH_3OH \longrightarrow$?

d. $2 \langle \text{cyclopentane} \rangle -CH_2OH \longrightarrow$?

12.84 Write an equation showing the dehydration reaction that would produce each of the following ethers:

a. Methyl propyl ether

b. Dimethyl ether

c. Ethyl pentyl ether

12.85 Name each of the following ethers using the I.U.P.A.C. Nomenclature System:

a. $CH_3\underset{\underset{OCH_2CH_3}{|}}{CH}CH_2CH_2CH_3$

b. $CH_3CH_2\underset{\underset{OCH_3}{|}}{CH}CH_3$

c. $CH_3CH_2CH_2\underset{\underset{OCH_2CH_3}{|}}{CH}$

d. $\langle \text{cyclopentane} \rangle -OCH_3$

12.86 Provide the common names for each of the following ethers

a. $CH_3CH_2-O-CH_2CH_2CH_3$

b. $CH_3-O-CH_2CH_2CH_2CH_2CH_2CH_3$

c. $CH_3CH_2CH_2CH_2-O-CH_2CH_2CH_2CH_3$

d. $CH_3CH_2-O-CH_2CH_2CH_2CH_2CH_2CH_3$

12.87 Draw the structural formula for each of the following ethers:

a. Methyl propyl ether

b. Methyl octyl ether

c. Diisopropyl ether

d. Ethyl pentyl ether

12.88 Write the I.U.P.A.C. and common name for each of the following ethers:

a. $CH_3CH_2-O-CH_2CH_2CH_2CH_3$

b. $CH_3-O-CH_2CH_2CH_2CH_2CH_3$

c. $CH_3CH_2CH_2-O-CH_2CH_2CH_2CH_3$

d. $CH_3CH_2-O-CH_2CH_2CH_2CH_2CH_3$

Thiols

Foundations

12.89 Compare the structure of thiols and alcohols.

12.90 Describe the I.U.P.A.C. rules for naming thiols.

Applications

12.91 Cystine is an amino acid formed from the oxidation of two cysteine molecules to form a disulfide bond. The molecular formula of cystine is $C_6H_{12}O_4N_2S_2$. Draw the structural formula of cystine. (*Hint:* For the structure of cysteine, see page 426.)

12.92 Explain the way in which British Anti-Lewisite acts as an antidote for mercury poisoning.

12.93 Give the I.U.P.A.C. name for each of the following thiols.

  a. $CH_3CH_2CH_2$—SH

  b. $CH_3CHCH_2CH_3$
         |
        SH
        CH_2CH_3

  c. CH_3CCH_3
         |
        SH

  d. HS—⬡—SH

12.94 Give the I.U.P.A.C. name for each of the following thiols.

  a. CH_2CHCH_3
        | |
      SH SH

  b. ⬡—SH

  c. $CH_3CHCH_2CH_2CH_3$
         |
        SH

  d. $CH_3CH_2CH_2CH_2CH_2CH_2CH_2SH$

CRITICAL THINKING PROBLEMS

1. You are provided with two solvents: water (H_2O) and hexane ($CH_3CH_2CH_2CH_2CH_2CH_3$). You are also provided with two biological molecules whose structures are shown here:

$$
\begin{array}{c}
O\\
\parallel\\
C-H\\
H-C-OH\\
HO-C-H\\
H-C-OH\\
H-C-OH\\
CH_2OH
\end{array}
$$

H   O
| ∥
H—C—O—C—$CH_2CH_2CH_2CH_2CH_2CH_2CH_2CH_2CH_2CH_2CH_2CH_2CH_2CH_3$
| O
| ∥
H—C—O—C—$CH_2CH_2CH_2CH_2CH_2CH_2CH_2CH_2CH_2CH_2CH_2CH_2CH_2CH_3$
| O
| ∥
H—C—O—C—$CH_2CH_2CH_2CH_2CH_2CH_2CH_2CH_2CH_2CH_2CH_2CH_2CH_2CH_3$
|
H

Predict which biological molecule would be more soluble in water and which would be more soluble in hexane. Defend your prediction. Design a careful experiment to test your hypothesis.

    Consider the digestion of dietary molecules in the digestive tract. Which of the two biological molecules shown in this problem would be more easily digested under the conditions present in the digestive tract?

2. Cholesterol is an alcohol and a steroid (Chapter 17). Diets that contain large amounts of cholesterol have been linked to heart disease and atherosclerosis, hardening of the arteries. The narrowing of the artery, caused by plaque buildup, is very apparent. Cholesterol is directly involved in this buildup. Describe the various functional groups and principal structural features of the cholesterol molecule. Would you use a polar or nonpolar solvent to dissolve cholesterol? Explain your reasoning.

Cholesterol

3. An unknown compound A is known to be an alcohol with the molecular formula $C_4H_{10}O$. When dehydrated, compound A gave only one alkene product, C_4H_8, compound B. Compound A could not be oxidized. What are the identities of compound A and compound B?

4. Sulfides are the sulfur analogs of ethers, that is, ethers in which oxygen has been substituted by a sulfur atom. They are named in an analogous manner to the ethers with the term *sulfide* replacing *ether*. For example, CH_3—S—CH_3 is dimethyl sulfide. Draw the sulfides that correspond to the following ethers and name them:

  a. diethyl ether

  b. methyl propyl ether

  c. dibutyl ether

  d. ethyl phenyl ether

5. Dimethyl sulfoxide (DMSO) has been used by many sports enthusiasts as a linament for sore joints; it acts as an anti-inflammatory agent and a mild analgesic (pain killer). However, it is no longer recommended for this purpose because it carries toxic impurities into the blood. DMSO is a sulfoxide—it contains the S=O functional group. DMSO is prepared from dimethyl sulfide by mild oxidation, and it has the molecular formula C_2H_6SO. Draw the structure of DMSO.

13 Aldehydes and Ketones

Learning Goals

1 ▸ Draw the structures and discuss the physical properties of aldehydes and ketones.

2 ▸ From the structures, write the common and I.U.P.A.C. names of aldehydes and ketones.

3 ▸ List several aldehydes and ketones that are of natural, commercial, health, and environmental interest and describe their significance.

4 ▸ Write equations for the preparation of aldehydes and ketones by the oxidation of alcohols.

5 ▸ Write equations representing the oxidation of carbonyl compounds.

6 ▸ Write equations representing the reduction of carbonyl compounds.

7 ▸ Write equations for the preparation of hemiacetals, hemiketals, acetals, and ketals.

8 ▸ Draw the keto and enol forms of aldehydes and ketones.

9 ▸ Write equations showing the aldol condensation.

Cinnamon tree blossoms.

Introduction

The **carbonyl group** consists of a carbon atom bonded to an oxygen atom by a double bond.

Carbonyl group

Compounds containing a carbonyl group are called carbonyl compounds. This group includes the aldehydes and ketones covered in this chapter, as well as the carboxylic acids and amides discussed in Chapters 14 and 15.

Chemistry Connection

Genetic Complexity from Simple Molecules

Examine any cellular life-form and you will find the same basic genetic system. One or more deoxyribonucleic acid (DNA) molecules carry the genetic code for all the proteins needed by the cell as enzymes—biological catalysts—as essential structural elements, and much more. But DNA cannot be read directly to produce these critical proteins. Instead, the genetic information carried by the DNA is copied to produce a variety of ribonucleic acid (RNA) molecules. These RNA molecules work together, along with other molecules, to produce the proteins.

For decades scientists have been trying to figure out how this amazing genetic system could have evolved from the small, simple molecules that were found in the shallow seas and atmosphere of earth perhaps four billion years ago. Which molecule might have formed first? After all, for the system to work all three molecules are needed: DNA to carry the information, RNA to carry and interpret it, and proteins to do all the cellular chores.

A startling discovery in the 1980s suggested an answer. It was discovered that some RNA molecules could act as biological catalysts. In other words, these RNA molecules could do two jobs: carry genetic information and catalyze chemical reactions. A hypothesis was developed: that RNA was the first biological molecule and that our genetic system evolved from it.

Could RNA have evolved from simple molecules in the "primordial soup"? In the 1960s two of the components of RNA, adenine and guanine, were synthesized in the laboratory from simple molecules and energy sources thought to be present on early earth. In 1995 researchers discovered that, by adding the carbonyl-group-containing molecule urea to their mixture, they could make large amounts of two other components of RNA, uracil and cytosine.

The remaining requirements for making an RNA molecule are phosphate groups and the sugar ribose. Phosphate would have been readily available, but what about ribose? Ribose, it turned out, could easily be produced from the simplest *aldehyde*, formaldehyde.

Formaldehyde would have been found in the shallow seas. We can find it today in comets, and many comets struck the early earth. In the laboratory, under conditions designed to imitate those on earth four billion years ago, formaldehyde molecules form chains. These chains twist into cyclic ring structures, including the sugar ribose.

These experiments suggest that all the precursors needed to make RNA could have formed spontaneously and thus that RNA may have been the information-carrying molecule from which our genetic system evolved. But other researchers argue that RNA is too fragile to have survived the conditions on early earth.

We may never know exactly how the first self-replicating genetic system formed. But it is intriguing to speculate about the origin of life and to consider the organic molecules and reactions that may have been involved. In this chapter we consider the aldehydes and ketones, two families of organic molecules containing the carbonyl group. As we will see in this and later chapters, the carbonyl group is a functional group that characterizes many biological molecules and affects their properties and reactivity.

Aldehyde
R = H, R, or Ar

An aldehyde
Propanal

Ketone
R = R or Ar

A ketone
Propanone

(a)

(b)

Figure 13.1
The structures of aldehydes and ketones. (a) The general structure of an aldehyde and a ball-and-stick model of the aldehyde propanal. (b) The general structure of a ketone and a ball-and-stick model of the ketone propanone.

Aldehyde

Ketone

Carboxylic Acid

Amide

We will study the aldehydes and ketones together because of their similar chemical and physical properties. They are distinguished by the location of the carbonyl group within the carbon chain. In aldehydes the carbonyl group is always located at the end of the carbon chain (carbon-1). In ketones the carbonyl group is located within the carbon chain of the molecule. Thus, in ketones the carbonyl carbon is attached to two other carbon atoms. However, in aldehydes the carbonyl carbon is attached to at least one hydrogen atom; the second atom attached to the carbonyl carbon of an aldehyde may be another hydrogen or a carbon atom (Figure 13.1).

13.1 Structure and Physical Properties

Aldehydes and **ketones** are polar compounds because of the polar carbonyl group.

Because of the dipole-dipole attractions between molecules, they boil at higher temperatures than hydrocarbons or ethers that have the same number of carbon atoms or are of equivalent molecular weight. Because they cannot form inter-molecular hydrogen bonds, their boiling points are lower than those of alcohols of comparable molecular weight. These trends are clearly demonstrated in the following examples:

LEARNING GOAL

1 Draw the structures and discuss the physical properties of aldehydes and ketones.

$$\delta^- O \cdots\cdots\cdots C \;^{\delta^+}$$
$$\delta^+ \overset{\|}{C} \cdots\cdots\cdots O \;^{\delta^-}$$

Dipole-dipole attraction

$CH_3CH_2CH_2CH_3$	$CH_3-O-CH_2CH_3$	$CH_3CH_2CH_2-OH$	$CH_3CH_2-\overset{O}{\overset{\|}{C}}-H$	$CH_3-\overset{O}{\overset{\|}{C}}-CH_3$
Butane (butane)	Methoxyethane (ethyl methyl ether)	1-Propanol (propyl alcohol)	Propanal (propionaldehyde)	Propanone (acetone)
M.W. = 58	M.W. = 60	M.W. = 60	M.W. = 58	M.W. = 58
b.p. −0.5°C	b.p. 7.0°C	b.p. 97.2°C	b.p. 49°C	b.p. 56°C

Figure 13.2
(a) Hydrogen bonding between the carbonyl group of an aldehyde or ketone and water. (b) Polar interactions between carbonyl groups of aldehydes or ketones.

(a)   (b)

Aldehydes and ketones can form intermolecular hydrogen bonds with water (Figure 13.2). As a result, the smaller members of the two families (five or fewer carbon atoms) are reasonably soluble in water. However, as the carbon chain length increases, the compounds become less polar and more hydrocarbonlike. These larger compounds are soluble in nonpolar organic solvents.

Question 13.1

Which member in each of the following pairs will be more water soluble?

a. $CH_3(CH_2)_2CH_3$ or $CH_3\overset{\overset{\displaystyle O}{\|}}{C}CH_3$

b. $CH_3\overset{\overset{\displaystyle O}{\|}}{C}CH_2CH_2CH_3$ or $CH_3\underset{\underset{\displaystyle OH}{|}}{CH}CH_2CH_2CH_3$

Question 13.2

Which member in each of the following pairs will be more water soluble?

a. CH_3 or $H\!-\!C\!=\!O$

b. $\underset{\underset{\displaystyle OH}{|}}{CH_2}\!-\!\underset{\underset{\displaystyle OH}{|}}{CH_2}$ or $H\!-\!\overset{\overset{\displaystyle O}{\|}}{C}\!-\!\overset{\overset{\displaystyle O}{\|}}{C}\!-\!H$

Question 13.3

Which member in each of the following pairs would have a higher boiling point?

a. $CH_3CH_2\overset{\overset{\displaystyle O}{\|}}{C}\!-\!OH$ or $CH_3CH_2\overset{\overset{\displaystyle O}{\|}}{C}\!-\!H$

b. $CH_3\overset{\overset{\displaystyle O}{\|}}{C}\!-\!OH$ or $CH_3\overset{\overset{\displaystyle O}{\|}}{C}\!-\!CH_3$

Question 13.4

Which member in each of the following pairs would have a higher boiling point?

a. CH_3CH_2OH or $CH_3\overset{\overset{\displaystyle O}{\|}}{C}\!-\!H$

b. $CH_3(CH_2)_6CH_3$ or $CH_3(CH_2)_5\overset{\underset{\displaystyle O}{\|}}{C}\!-\!H$

13.2 I.U.P.A.C. Nomenclature and Common Names

Naming Aldehydes

In the I.U.P.A.C. system, aldehydes are named according to the following set of rules:

- Determine the parent compound, that is, the longest continuous carbon chain containing the carbonyl group.
- Replace the final *-e* of the parent alkane with *-al*.
- Number the chain beginning with the carbonyl carbon (or aldehyde group) as carbon-1.
- Number and name all substituents as usual. No number is used for the position of the carbonyl group because it is always at the end of the parent chain. Therefore, it must be carbon-1.

> **LEARNING GOAL**
> **2** From the structures, write the common and I.U.P.A.C. names of aldehydes and ketones.

Several examples are provided here with common names given in parentheses:

Methanal
(formaldehyde)

Ethanal
(acetaldehyde)

Propanal
(propionaldehyde)

2-Methylpentanal

Using the I.U.P.A.C. Nomenclature System to Name Aldehydes   E X A M P L E **13.1**

For many years, scientists were puzzled about what compounds give the nutty flavor to expensive, aged cheddar cheeses. In 2004, Dr. Mary Anne Drake of North Carolina State University solved this puzzle. She identified a group of aldehydes that impart this desirable flavor. Use the I.U.P.A.C. Nomenclature System to name two of these aldehydes shown below.

> **LEARNING GOAL**
> **2** From the structures, write the common and I.U.P.A.C. names of aldehydes and ketones.

Solution

Parent compound:	propane (becomes propanal)	butane (becomes butanal)
Position of carbonyl group:	carbon-1 (must be!)	carbon-1 (must be!)
Substituents:	2-methyl	3-methyl
Name:	2-Methylpropanal	3-Methylbutanal

Notice that the position of the carbonyl group is not indicated by a number. By definition, the carbonyl group is located at the end of the carbon chain

Continued—

EXAMPLE 13.1 —*Continued*

Another aldehyde associated with the nutty flavor of cheddar cheese is 2-methylbutanal. Draw the condensed structural formula of this molecule.

Carboxylic acid nomenclature is described in Section 14.1.

of an aldehyde. The carbonyl carbon is defined to be carbon-1; thus, it is not necessary to include the position of the carbonyl group in the name of the compound.

Practice Problem 13.1

Many molecules contribute to the complex flavors of olive oils. Among these are hexanal and *trans*-2-hexenal, which have flavors described as "green, grassy" and "green, bitter," respectively. Draw the structures of these two compounds.

For Further Practice: Questions 13.5, 13.29 and 13.36.

The common names of the aldehydes are derived from the same Latin roots as the corresponding carboxylic acids. The common names of the first five aldehydes are presented in Table 13.1.

In the common system of nomenclature, substituted aldehydes are named as derivatives of the straight-chain parent compound (see Table 13.1). Greek letters are used to indicate the position of the substituents. The carbon atom bonded to the carbonyl group is the α-carbon, the next is the β-carbon, and so on.

$$\overset{\delta}{-}C-\overset{\gamma}{C}-\overset{\beta}{C}-\overset{\alpha}{C}-\overset{O}{\overset{\|}{C}}-H$$

Consider the following examples:

$$\underset{\text{2-Methylpentanal}}{\overset{\delta\quad\gamma\quad\beta\quad\alpha\qquad O}{\underset{\overset{|}{CH_3}}{CH_3CH_2CH_2CH}-\overset{\|}{C}-H}}$$

2-Methylpentanal
(α-methylvaleraldehyde)

$$\underset{\text{3-Methylpentanal}}{\overset{\delta\quad\gamma\quad\beta\quad\alpha\qquad O}{\underset{\overset{|}{CH_3}}{CH_3CH_2CHCH_2}-\overset{\|}{C}-H}}$$

3-Methylpentanal
(β-methylvaleraldehyde)

TABLE 13.1	I.U.P.A.C. and Common Names and Formulas for Several Aldehydes	
I.U.P.A.C. Name	**Common Name**	**Formula**
Methanal	Formaldehyde	$H-\overset{O}{\overset{\|}{C}}-H$
Ethanal	Acetaldehyde	$CH_3\overset{O}{\overset{\|}{C}}-H$
Propanal	Propionaldehyde	$CH_3CH_2\overset{O}{\overset{\|}{C}}-H$
Butanal	Butyraldehyde	$CH_3CH_2CH_2\overset{O}{\overset{\|}{C}}-H$
Pentanal	Valeraldehyde	$CH_3CH_2CH_2CH_2\overset{O}{\overset{\|}{C}}-H$

Using the Common Nomenclature System to Name Aldehydes   E X A M P L E 13.2

Name the aldehydes represented by the following condensed formulas.

Solution

Parent compound:	pentane	butane
	(becomes valeraldehyde)	(becomes butyraldehyde)
Position of carbonyl group:	carbon-1 (must be!)	carbon-1 (must be!)
Substituents:	γ-bromo	β-methyl
Name:	γ-Bromovaleraldehyde	β-Methylbutyraldehyde

Notice that the substituents are designated by Greek letters, rather than by Arabic numbers. In the common system of nomenclature for aldehydes, the carbon atom bonded to the carbonyl group is called the α-carbon, the next is the β-carbon, etc. Remember to use these Greek letters to indicate the position of the substituents when naming aldehydes using the common system of nomenclature.

Also remember that by definition, the carbonyl group is located at the beginning of the carbon chain of an aldehyde. Thus, it is not necessary to include the position of the carbonyl group in the name of the compound.

Practice Problem 13.2

Use the common nomenclature system to name each of the following compounds.

a. CH₃CHCHCH₂C—H (with CH₃ substituents and O)

b. CH₃CH₂CH₂CHC—H (with CH₂CH₃ substituent and O)

c. CH₃CHC—H (with Cl substituent and O)

d. CH₃CHCH₂C—H (with OH substituent and O)

For Further Practice: Questions 13.6, 13.41c and d, 13.42b and d.

Naming Ketones

The rules for naming ketones in the I.U.P.A.C. Nomenclature System are directly analogous to those for naming aldehydes. In ketones, however, the *-e* ending of the parent alkane is replaced with the *-one* suffix of the ketone family, and the location of the carbonyl carbon is indicated with a number. The longest carbon chain is numbered to give the carbonyl carbon the lowest possible number. For example,

Propanone	Butanone	4-Octanone
(no number necessary)	(no number necessary)	(*not* 5-octanone)
(acetone)	(methyl ethyl ketone)	(butyl propyl ketone)

LEARNING GOAL

2   From the structures, write the common and I.U.P.A.C. names of aldehydes and ketones.

LEARNING GOAL

2   From the structures, write the common and I.U.P.A.C. names of aldehydes and ketones.

EXAMPLE 13.3   **Using the I.U.P.A.C Nomenclature System to Name Ketones**

> **2** | **LEARNING GOAL**
> From the structures, write the common and I.U.P.A.C. names of aldehydes and ketones.

Two molecules associated with the aroma of blue cheese are the ketones shown below. Name these ketones using the I.U.P.A.C. nomenclature system.

Solution

$$\underset{7\quad6\quad5\quad4\quad3\quad2\,1}{CH_3CH_2CH_2CH_2CH_2\overset{\displaystyle O}{\overset{\|}{C}}CH_3} \qquad \underset{1\quad2\;3\quad4\quad5\quad6\quad7\quad8\quad9}{CH_3\overset{\displaystyle O}{\overset{\|}{C}}CH_2CH_2CH_2CH_2CH_2CH_2CH_3}$$

Parent compound:	heptane	nonane
	(becomes heptanone)	(becomes nonanone)
Position of carbonyl group:	carbon-2 (not carbon-6)	carbon-2 (not carbon-8)
Name:	2-Heptanone	2-Nonanone

 Practice Problem 13.3

Provide the I.U.P.A.C. name for each of the following ketones.

a. $$\underset{\quad\quad\quad\quad|}{CH_3CH_2CHCH_2CH_2CH_2\overset{\displaystyle O}{\overset{\|}{C}}CH_3}$$
       $$CH_2CH_3$$

b. $$CH_3CH_2CH_2\overset{\displaystyle O}{\overset{\|}{C}}\underset{\;\;|}{CHCH_3}$$
       $$CH_3$$

c. $$CH_3\underset{\;\;|}{CH}\overset{\displaystyle O}{\overset{\|}{C}}\underset{\;\;|}{CH}CH_3$$
       $$CH_3\;\;CH_3$$

For Further Practice: Questions 13.7, 13.47, and 13.48.

Two other molecules that contribute to the aroma of blue cheese are hexanoic acid and butanoic acid, two carboxylic acids we will study in Chapter 14. Refer to Table 10.2 for the structure of carboxylic acids and draw the condensed structural formulas for each of these molecules.

The common names of ketones are derived by naming the alkyl groups that are bonded to the carbonyl carbon. These are used as prefixes followed by the word *ketone*. The alkyl groups may be arranged alphabetically or by size (smaller to larger).

EXAMPLE 13.4   **Using the Common Nomenclature System to Name Ketones**

> **2** | **LEARNING GOAL**
> From the structures, write the common and I.U.P.A.C. names of aldehydes and ketones.

Name the ketones represented by the following condensed formulas.

Solution

Identify the alkyl groups that are bonded to the carbonyl carbon.

Alkyl groups:	propyl and methyl	ethyl and pentyl
Name:	Methyl propyl ketone	Ethyl pentyl ketone

Continued—

EXAMPLE 13.4 —*Continued*

Practice Problem 13.4

Provide the common names for each of the following ketones:

$$
\begin{array}{cc}
\text{a. } CH_3CH_2\overset{\displaystyle O}{\overset{\displaystyle \|}{C}}CH_2CH_3 &
\text{b. } CH_3CH_2CH_2CH_2\overset{\displaystyle O}{\overset{\displaystyle \|}{C}}CH_2CH_3
\end{array}
$$

$$
\text{c. } CH_3CH_2CH_2CH_2CH_2CH_2\overset{\displaystyle O}{\overset{\displaystyle \|}{C}}CH_3 \qquad
\text{d. } CH_3\overset{}{\underset{\underset{CH_3}{|}}{C}}H\overset{\displaystyle O}{\overset{\displaystyle \|}{C}}\overset{}{\underset{\underset{CH_3}{|}}{C}}HCH_3
$$

For Further Practice: Questions 13.41a, b, and c and 13.44.

Because the two *groups* bonded to the carbonyl carbon are named, a ketone is actually one carbon longer than an aldehyde with a *similar common name*. For example, methyl butyl ketone has six carbons, but β-methylbutyraldehyde has only five.

$$
\begin{array}{cc}
\overset{\displaystyle O}{\overset{\displaystyle \|}{CH_3-C-CH_2CH_2CH_2CH_3}} &
\underset{}{CH_3\underset{\underset{CH_3}{|}}{C}HCH_2-\overset{\displaystyle O}{\overset{\displaystyle \|}{C}}-H}\\
\;\;1 \quad\; 2 \;\;\; 3 \;\;\; 4 \;\;\; 5 \;\;\; 6 & 4 \;\;\; 3 \;\; 2 \qquad\qquad 1 \\[4pt]
\text{Methyl butyl ketone} & \text{β-Methylbutyraldehyde}
\end{array}
$$

This is because the aldehyde carbonyl carbon is included in the name of the parent chain, butyraldehyde. The carbonyl carbon of the ketone is not included in the common name. It is treated only as the carbon to which the two alkyl or aryl groups are attached.

Question 13.5

From the I.U.P.A.C. names, draw the structural formula for each of the following aldehydes.

a. 2,3-Dichloropentanal   b. 2-Bromobutanal   c. 4-Methylhexanal
d. Butanal   e. 2,4-Dimethylpentanal

Question 13.6

Write the condensed structural formula for each of the following compounds.

a. 3-Methylnonanal   b. β-Bromovaleraldehyde
c. 4-Fluorohexanal   d. α,β-Dimethylbutyraldehyde

Question 13.7

Use the I.U.P.A.C. Nomenclature System to name each of the following compounds.

                   O   O
                   ‖   ‖

a. CH_3CHCCH_3

    |
    I

b. $CH_3CHCH_2CCH_3$

         |
         $CH_2CH_2CH_2CH$

c. CH_3CHCCH_3

    |
    CH_3

d. $CH_3CHCCH_2CH_3$

    |
    CH_3

e. $CH_3CHCCH_2CH_3$

    |
    F

Question 13.8

Write the condensed formula for each of the following compounds.

   a. Methyl isopropyl ketone (What is the I.U.P.A.C. name for this compound?)
   b. 4-Heptanone
   c. 2-Fluorocyclohexanone
   d. Hexachloroacetone (What is the I.U.P.A.C. name of this compound?)

13.3 Important Aldehydes and Ketones

LEARNING GOAL

3 List several aldehydes and ketones that are of natural, commercial, health, and environmental interest and describe their significance.

Methanal   Ethanal

Propanone   Butanone

Methanal (formaldehyde) is a gas (b.p. −21°C). It is available commercially as an aqueous solution called *formalin*. Formalin has been used as a preservative for tissues and as an embalming fluid. See A Medical Perspective: Formaldehyde and Methanol Poisoning for more information on methanal.

    Ethanal (acetaldehyde) is produced from ethanol in the liver. Ethanol is oxidized in this reaction, which is catalyzed by the liver enzyme alcohol dehydrogenase. The ethanal that is produced in this reaction is responsible for the symptoms of a hangover.

    Propanone (acetone), the simplest ketone, is an important and versatile solvent for organic compounds. It has the ability to dissolve organic compounds and is also miscible with water. As a result, it has a number of industrial applications and is used as a solvent in adhesives, paints, cleaning solvents, nail polish, and nail polish remover. Propanone is flammable and should therefore be treated with appropriate care. *Butanone*, a four-carbon ketone, is also an important industrial solvent.

    Many aldehydes and ketones are produced industrially as food and fragrance chemicals, medicinals, and agricultural chemicals. They are particularly important to the food industry, in which they are used as artificial and/or natural additives to food. Vanillin, a principal component of natural vanilla, is shown in Figure 13.3. Artificial vanilla flavoring is a dilute solution of synthetic vanillin dissolved in ethanol. Figure 13.3 also shows other examples of important aldehydes and ketones.

Question 13.9

Draw the structure of the aldehyde synthesized from ethanol in the liver.

Question 13.10

Draw the structure of a ketone that is an important, versatile solvent for organic compounds.

Figure 13.3

Important aldehydes and ketones.

13.4 Reactions Involving Aldehydes and Ketones

Preparation of Aldehydes and Ketones

Aldehydes and ketones are prepared primarily by the **oxidation** of the corresponding alcohol. As we saw in Chapter 12, the oxidation of methyl alcohol gives methanal (formaldehyde). The oxidation of a primary alcohol produces an aldehyde, and the oxidation of a secondary alcohol yields a ketone. Tertiary alcohols do not undergo oxidation under the conditions normally used.

E X A M P L E 13.5   Differentiating the Oxidation of Primary, Secondary, and Tertiary Alcohols

Use specific examples to show the oxidation of a primary, a secondary, and a tertiary alcohol.

Solution

The oxidation of a primary alcohol to an aldehyde:

$$CH_3CH_2CH_2-\overset{\overset{\displaystyle H}{|}}{\underset{\underset{\displaystyle H}{|}}{C}}-OH \xrightarrow[\text{dichromate}]{\text{Pyridinium}} CH_3CH_2CH_2-\overset{\overset{\displaystyle O}{\|}}{C}-H$$

1-Butanol   Butanal
(butyl alcohol)   (butyraldehyde)

A mild oxidizing agent must be used in the oxidation of a primary alcohol to an aldehyde. Otherwise, the aldehyde will be further oxidized to a carboxylic acid.

The oxidation of a secondary alcohol to a ketone:

$$CH_3CH_2CH_2CH_2-\overset{\overset{\displaystyle CH_3}{|}}{\underset{\underset{\displaystyle H}{|}}{C}}-OH \xrightarrow[H_2O]{KMnO_4, OH^-,} CH_3CH_2CH_2CH_2-\overset{\overset{\displaystyle O}{\|}}{C}-CH_3$$

2-Hexanol   2-Hexanone

Tertiary alcohols cannot undergo oxidation:

$$CH_3CH_2CH_2-\overset{\overset{\displaystyle CH_3}{|}}{\underset{\underset{\displaystyle CH_3}{|}}{C}}-OH \xrightarrow{H_2CrO_4} \text{No reaction}$$

2-Methyl-2-pentanol

Practice Problem 13.5

Write equations showing the oxidation of (a) 1-propanol and (b) 2-butanol.

For Further Practice: Questions 13.61 and 13.62.

Oxidation Reactions

Aldehydes are easily oxidized further to carboxylic acids, whereas ketones do not generally undergo further oxidation. The reason is that a carbon-hydrogen bond, present in the aldehyde but not in the ketone, is needed for the reaction to occur. In fact, aldehydes are so easily oxidized that it is often very difficult to

A Medical Perspective

Formaldehyde and Methanol Poisoning

Most aldehydes have irritating, unpleasant odors, and formaldehyde is no exception. Formaldehyde is also an extremely toxic substance. As an aqueous solution, called formalin, it has been used to preserve biological tissues and for embalming. It has also been used to disinfect environmental surfaces, body fluids, and feces. Under no circumstances is it used as an antiseptic on human tissue because of its toxic fumes and the skin irritations that it causes.

Formaldehyde is used in the production of some killed-virus vaccines. When a potentially deadly virus, such as polio virus, is treated with heat and formaldehyde, the genetic information (RNA) is damaged beyond repair. The proteins of the virus also react with formaldehyde. However, the shape of the proteins, which is critical for a protective immune response against the virus, is not changed. Thus, when a child is injected with the Salk killed-virus polio vaccine, the immune system recognizes viral proteins and produces antibodies that protect against polio virus infection.

Formaldehyde can also be produced in the body! As we saw in Chapter 12, methanol can be oxidized to produce formaldehyde. In the body, the liver enzyme alcohol dehydrogenase, whose function it is to detoxify alcohols by oxidizing them, catalyzes the conversion of methanol to formaldehyde (methanal). The formaldehyde then reacts with cellular macromolecules, including proteins, causing severe damage (remember, it is used as an embalming agent!). As a result, methanol poisoning can cause blindness, respiratory failure, convulsions, and death.

Clever physicians have devised a treatment for methanol poisoning that is effective if administered soon enough after ingestion. Since the same enzyme that oxidizes methanol to formaldehyde (methanal) also oxidizes ethanol to acetaldehyde (ethanal), doctors reasoned that administering an intravenous solution of ethanol to the patient could protect against the methanol poisoning. If the ethanol concentration in the body is higher than the methanol concentration, most of the alcohol dehydrogenase enzymes of the liver will be carrying out the oxidation of the ethanol. This is called *competitive inhibition* because the methanol and ethanol molecules are competing for binding to the enzymes. The molecule that is in the higher concentration will more frequently bind to the enzyme and undergo reaction. In this case, the result is that the alcohol dehydrogenase enzymes are kept busy oxidizing ethanol and producing the less-toxic (not nontoxic) product, acetaldehyde. This gives the body time to excrete the methanol before it is oxidized to the potentially deadly formaldehyde.

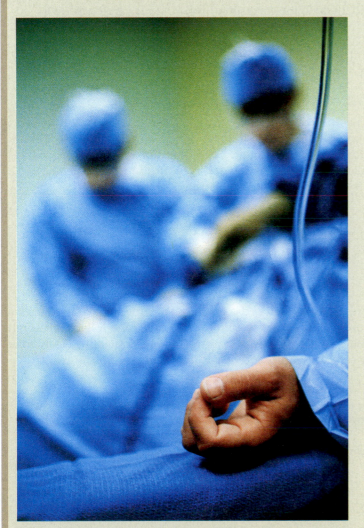

Patient receiving an intravenous solution of ethanol to treat methanol poisoning.

For Further Understanding

Acetaldehyde is described as less toxic than formaldehyde. Do some background research on the effects of these two aldehydes on biological systems.

You have studied enzymes in previous biology courses. Using what you learned in those classes with information from Sections 19.4 and 19.10 in this text, put together an explanation of the way in which competitive inhibition works. Can you think of other types of poisoning for which competitive inhibition might be used to develop an effective treatment?

prepare them because they continue to react to give the carboxylic acid rather than the desired aldehyde. Aldehydes are susceptible to air oxidation, even at room temperature, and cannot be stored for long periods. The following example shows a general equation for the oxidation of an aldehyde to a carboxylic acid:

$$\underset{\text{Aldehyde}}{R-\overset{\displaystyle O}{\overset{\|}{C}}-H} \xrightarrow{[O]} \underset{\text{Carboxylic acid}}{R-\overset{\displaystyle O}{\overset{\|}{C}}-OH}$$

Many oxidizing agents can be used. Both basic potassium permanganate and chromic acid are good oxidizing agents, as the following specific example shows:

$$\underset{\substack{\text{Ethanal}\\ \text{(acetaldehyde)}}}{CH_3-\overset{\displaystyle O}{\overset{\|}{C}}-H} \xrightarrow[\text{H}_2\text{O,OH}^-]{\text{KMnO}_4,} \underset{\substack{\text{Ethanoate anion}\\ \text{(acetate anion)}}}{CH_3-\overset{\displaystyle O}{\overset{\|}{C}}-O^-}$$

The oxidation of benzaldehyde to benzoic acid is an example of the conversion of an *aromatic* aldehyde to the corresponding aromatic carboxylic acid:

Benzaldehyde   Benzoic acid

Aldehydes and ketones can be distinguished on the basis of differences in their reactivity. The most common laboratory test for aldehydes is the **Tollens' test.** When exposed to the Tollens' reagent, a basic solution of $Ag(NH_3)_2{}^+$, an aldehyde undergoes oxidation. The silver ion (Ag^+) is reduced to silver metal (Ag^0) as the aldehyde is oxidized to a carboxylic acid anion.

$$\underset{\text{Aldehyde}}{R-\overset{\displaystyle O}{\overset{\|}{C}}-H} + \underset{\substack{\text{Silver}\\ \text{ammonia complex—}\\ \text{Tollens' reagent}}}{Ag(NH_3)_2{}^+} \longrightarrow \underset{\substack{\text{Carboxylate}\\ \text{anion}}}{R-\overset{\displaystyle O}{\overset{\|}{C}}-O^-} + \underset{\substack{\text{Silver}\\ \text{metal}\\ \text{mirror}}}{Ag^0}$$

Silver metal precipitates from solution and coats the flask, producing a smooth silver mirror, as seen in Figure 13.4. The test is therefore often called the Tollens' silver mirror test. The commercial manufacture of silver mirrors uses a similar process. Ketones cannot be oxidized to carboxylic acids and do not react with the Tollens' reagent.

In basic solution, the product is the carboxylic acid anion:

$$CH_3-\overset{\displaystyle O}{\overset{\|}{C}}-O^-$$

The rules for naming carboxylic acid anions are described in Section 14.1.

Silver ions are very mild oxidizing agents. They will oxidize aldehydes but not alcohols.

EXAMPLE 13.6   **Writing Equations for the Reaction of an Aldehyde and of a Ketone with Tollens' Reagent**

LEARNING GOAL

5 Write equations representing the oxidation of carbonyl compounds.

Write equations for the reaction of propanal and 2-pentanone with Tollens' reagent.

Solution

$$\underset{\text{Propanal}}{CH_3CH_2\overset{\displaystyle O}{\overset{\|}{C}}-H} + Ag(NH_3)_2{}^+ \longrightarrow \underset{\text{Propanoate anion}}{CH_3CH_2\overset{\displaystyle O}{\overset{\|}{C}}-O^-} + Ag^0$$

Continued—

EXAMPLE 13.6 —*Continued*

$$CH_3CH_2CH_2\overset{\overset{\displaystyle O}{\|}}{C}CH_3 + Ag(NH_3)_2{}^+ \longrightarrow \text{No reaction}$$

2-Pentanone

Practice Problem 13.6

Write equations for the reaction of (a) ethanal and (b) propanone with Tollens' reagent.

For Further Practice: Questions 13.66 and 13.67.

Another test that is used to distinguish between aldehydes and ketones is **Benedict's test.** Here, a buffered aqueous solution of copper(II) hydroxide and sodium citrate reacts to oxidize aldehydes but does not generally react with ketones. Cu^{2+} is reduced to Cu^+ in the process. Cu^{2+} is soluble and gives a blue solution, whereas the Cu^+ precipitates as the red solid copper(I) oxide, Cu_2O.

All simple sugars (monosaccharides) are either aldehydes or ketones. Glucose is an aldehyde sugar that is commonly called *blood sugar* because it is the sugar found transported in the blood and used for energy by many cells. In uncontrolled diabetes, glucose may be found in the urine. One early method used to determine the amount of glucose in the urine was to observe the color change of the Benedict's test. The amount of precipitate formed is

Cu(II) is an even milder oxidizing agent than silver ion.

(a)

(b)

(c)

(d)

Figure 13.4

The silver precipitate produced by the Tollens' reaction is deposited on glass. The progress of the reaction is visualized in panels (a) through (d). Silver mirrors are made in a similar process.

A Human *Perspective*

Alcohol Abuse and Antabuse

According to a recent study carried out by the Centers for Disease Control and Prevention,[1] 75,000 Americans die each year as a result of alcohol abuse. Of these, 34,833 people died of cirrhosis of the liver, cancer, or other drinking-related diseases. The remaining 40,933 died in alcohol-related automobile accidents. Of those who died, 72% were men and 6% were under the age of 21. In fact, a separate study has estimated that 1400 college-age students die each year of alcohol-related causes.

These numbers are striking. Alcohol abuse is now the third leading cause of preventable death in the United States, outranked only by tobacco use and poor diet and exercise habits. As the study concluded, "These results emphasize the importance of adopting effective strategies to reduce excessive drinking, including increasing alcohol excise taxes and screening for alcohol misuse in clinical settings."

Tetraethylthiuram disulfide
(disulfiram)
Antabuse

One approach to treatment of alcohol abuse, the drug tetraethylthiuram disulfide or disulfiram, has been used since 1951. The activity of this drug generally known by the trade name Antabuse was discovered accidentally by a group of Danish researchers who were testing it for antiparasitic properties. They made the observation that those who had taken disulfiram became violently ill after consuming any alcoholic beverage. Further research revealed that this compound inhibits one of the liver enzymes in the pathway for the oxidation of alcohols.

In Chapter 12 we saw that ethanol is oxidized to ethanal (acetaldehyde) in the liver. This reaction is catalyzed by the enzyme alcohol dehydrogenase. Acetaldehyde, which is more toxic than ethanol, is responsible for many of the symptoms of a hangover. The enzyme acetaldehyde dehydrogenase oxidizes acetaldehyde into ethanoic acid (acetic acid), which then is used in biochemical pathways that harvest energy for cellular work or that synthesize fats.

Antabuse inhibits acetaldehyde dehydrogenase. This inhibition occurs within one to two hours of taking the drug and continues up to fourteen days. When a person who has taken Antabuse drinks an alcoholic beverage, the level of acetaldehyde quickly reaches levels that are five to ten times higher than would normally occur after a drink. Within just a few minutes, the symptoms of a severe hangover are experienced and may continue for several hours.

The drug Antabuse may be useful in treating alcohol abuse.

Experts in drug and alcohol abuse have learned that drugs such as Antabuse are generally not effective on their own. However, when used in combination with support groups and/or psychotherapy to solve underlying behavioral or psychological problems, Antabuse is an effective deterrent to alcohol abuse.

1. Alcohol-Attributable Deaths and Years of Potential Life Lost—United States, 2001, Morbidity and Mortality Weekly Report, 53 (37): 866–870, September 24, 2004, also available at http://www.cdc.gov/mmwr/preview/mmwrhtml/mm5337a2.htm.

For Further Understanding

Antabuse alone is not a cure for alcoholism. Consider some of the reasons why this is so.

Write equations showing the oxidation of ethanal to ethanoic acid as a pathway with the product of the first reaction serving as the reactant for the second. Explain the physiological effects of Antabuse in terms of these chemical reactions.

directly proportional to the amount of glucose in the urine (Figure 13.5). The reaction of glucose with the Benedict's reagent is represented in the following equation:

Figure 13.5
The amount of precipitate formed and thus the color change observed in the Benedict's test are directly proportional to the amount of reducing sugar in the sample.

We should also note that when the carbonyl group of a ketone is bonded to a —CH_2OH group, the molecule will give a positive Benedict's test. This occurs because such ketones are converted to aldehydes under basic conditions. In Chapter 16 we will see that this applies to the ketone sugars, as well. They are converted to aldehyde sugars and react with Benedict's reagent.

Reduction Reactions

Aldehydes and ketones are both readily reduced to the corresponding alcohol by a variety of reducing agents. Throughout the text the symbol [H] over the reaction arrow represents a reducing agent.

The classical method of aldehyde or ketone reduction is **hydrogenation.** The carbonyl compound is reacted with hydrogen gas and a catalyst (nickel, platinum, or palladium metal) in a pressurized reaction vessel. Heating may also be necessary. The carbon-oxygen double bond (the carbonyl group) is reduced to a carbon-oxygen single bond. This is similar to the reduction of an alkene to an alkane (the reduction of a carbon-carbon double bond to a carbon-carbon single bond). The addition of hydrogen to a carbon-oxygen double bond is shown in the following general equation:

LEARNING GOAL

6 Write equations representing the reduction of carbonyl compounds.

One way to recognize reduction, particularly in organic chemistry, is the gain of hydrogen. Oxidation and reduction are discussed in Section 12.6.

Hydrogenation was first discussed in Section 11.5 for the hydrogenation of alkenes.

$$
\underset{\substack{\text{Aldehyde}\\\text{or ketone}}}{\overset{O}{\underset{R^1}{\overset{\|}{C}}{\underset{R^2}{}}}} + \underset{\text{Hydrogen}}{\overset{H}{\underset{H}{|}}} \xrightarrow{\text{Pt}} \underset{\text{Alcohol}}{\overset{OH}{\underset{R^2}{\overset{|}{R^1-\underset{|}{C}-H}}}}
$$

The hydrogenation (reduction) of a ketone produces a secondary alcohol, as seen in the following equation showing the reduction of the ketone, 3-octanone:

$$
\underset{\substack{\text{3-Octanone}\\\text{(A ketone)}}}{CH_3CH_2-\overset{O}{\overset{\|}{C}}-CH_2CH_2CH_2CH_2CH_3} + \underset{\text{Hydrogen}}{H_2} \xrightarrow{Ni} \underset{\substack{\text{3-Octanol}\\\text{(A secondary alcohol)}}}{CH_3CH_2-\overset{OH}{\underset{H}{\overset{|}{\underset{|}{C}}}}-CH_2CH_2CH_2CH_2CH_3}
$$

A Medical Perspective

That Golden Tan Without the Fear of Skin Cancer

Self-tanning lotions have become very popular in recent years. This seems to be the result of our growing understanding of the link between exposure to the sun and skin cancer and to improvements in the quality of the tan produced by these self-tanners.

The active ingredient in most self-tanners is dihydroxyacetone (DHA).

$$CH_2OH$$
$$|$$
$$C{=}O$$
$$|$$
$$CH_2OH$$

Dihydroxyacetone

This simple, three-carbon molecule has two hydroxyl groups (—OH) and is, thus, an alcohol. It also has a carbonyl group at the center carbon:

$$\backslash$$
$$C{=}O$$
$$/$$

Carbonyl group

This makes the compound a ketone. In fact, because it has both hydroxyl groups and a carbonyl group, DHA is a sugar, more precisely, a keto sugar. This sugar is actually a by-product of our own metabolism; we produce it in the metabolic pathway called glycolysis (Chapter 22).

A researcher by the name of Eva Wittgenstein discovered the tanning reaction while she was studying a human genetic disorder in children. These children were unable to store glycogen, a polysaccharide, or sugar polymer, which is our major energy

storage molecule in the liver. She was trying to treat the disease by feeding large doses of DHA to the children. Sometimes, however, the children spit up some of the sickeningly sweet solution, which ended up on their clothes and skin. Dr. Wittgenstein noticed that the skin darkened at the site of these spills and decided to investigate the observation.

DHA works because of a reaction between its carbonyl group and a free amino group (—NH_3^+) of several amino acids in the skin protein keratin. Amino acids are the building blocks of the biological polymers called proteins (Chapter 18); keratin is just one such protein. The DHA produces brown-colored compounds called melanoids when it bonds to the keratins. These polymeric melanoids are chemically linked to cells of the stratum corneum, the dead, outermost layer of the skin. DHA does not penetrate this outer layer; so the chemical reaction that causes tanning only affects the stratum corneum. As this dead skin sloughs off, so does your tan!

Over the years research has improved the quality of the tan that is produced. Early self-tanning lotions produced an orange tan; the tans from today's lotions are much more natural. The DHA used today is in a much purer form and the other components of the lotion have been redesigned to promote greater penetration. Research has also taught us that the tanning reaction works best at acid pH; so newer formulations are buffered to pH 5. All of these changes have resulted in self-tanners that produce a longer lasting tan with a more natural, golden color.

We have also learned that it is important to exfoliate before using a self-tanner. Anywhere that the dead skin layer is thicker, there will be more keratin. From our study of chemistry, we have learned that when we begin with more reactant, we often get more product. The greater the amount of product

---

EXAMPLE 13.7   **Writing an Equation Representing the Hydrogenation of a Ketone**

Write an equation showing the hydrogenation of 3-pentanone.

Solution

The product of the reduction of a ketone is a secondary alcohol, in this case, 3-pentanol.

$$CH_3CH_2{-}\overset{\overset{\displaystyle O}{\|}}{C}{-}CH_2CH_3 + H_2 \xrightarrow{\;\;Pt\;\;} CH_3CH_2{-}\overset{\overset{\displaystyle OH}{|}}{\underset{\underset{\displaystyle H}{|}}{C}}{-}CH_2CH_3$$

3-Pentanone   3-Pentanol

Practice Problem 13.7

Write an equation for the hydrogenation of (a) propanone and (b) butanone.

For Further Practice: Questions 13.67 and 13.68.

in this case, the darker the color! The resulting tan often looked splotchy or streaky. By gently removing some of the stratum corneum by a gentle exfoliation process, the surface of the skin, and hence the tanning reaction, becomes more uniform.

More recently some companies have added a new sugar, erythrulose, to the self-tanning lotions.

$$\begin{array}{c} CH_2OH \\ | \\ C{=}O \\ | \\ HO{-}CH \\ | \\ CH_2OH \end{array}$$

Erythrulose

Erythrulose is a four-carbon keto sugar that reacts in exactly the same way as DHA. However, since they are different compounds, they do produce different melanoids with slightly different properties, including color. The tan produced by erythrulose is less reddish in tone than that produced by DHA. However, while a DHA tan develops in two to six hours, an erythrulose tan requires two days. For this reason, erythrulose is usually not used alone, but only in combination with DHA.

People often ask whether the tan from a bottle can protect against burn, much as a natural tan does. The melanoids do absorb light of the same wavelengths absorbed by melanin (the substance formed by suntanning), so you might expect some protection against sunburn. However, the protection is minimal, rated at a sun protection factor (SPF) of only 2 or 3. Selftanners offer an excellent substitute to a suntan. They produce the same golden tan without the danger of overexposure to the sun's harmful ultraviolet rays.

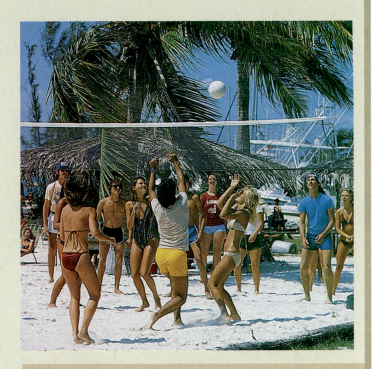

For Further Understanding

Explain in terms of a chemical reaction why using a self-tanner daily results in an increasingly darker tan.

The incidence of skin cancer in men and women has risen dramatically in recent years. Using the Internet and Chapter 20 in this book, develop a hypothesis to explain this observation.

The hydrogenation of an aldehyde results in the production of a primary alcohol, as seen in the following equation showing the reduction of the aldehyde, butanal:

$$\underset{\substack{\text{Butanal} \\ \text{(An aldehyde)}}}{CH_3CH_2CH_2-\overset{\overset{\displaystyle O}{\|}}{C}-H} + \underset{\text{Hydrogen}}{H_2} \xrightarrow{\text{Pt}} \underset{\substack{\text{1-Butanol} \\ \text{(A primary alcohol)}}}{CH_3CH_2CH_2-\overset{\overset{\displaystyle OH}{|}}{\underset{\underset{\displaystyle H}{|}}{C}}-H}$$

Writing an Equation Representing the Hydrogenation of an Aldehyde **E X A M P L E 13.8**

Write an equation showing the hydrogenation of 3-methylbutanal.

Solution

Recall that the reduction of an aldehyde results in the production of a primary alcohol, in this case, 3-methyl-1-butanol.

LEARNING GOAL

6 Write equations representing the reduction of carbonyl compounds.

Continued—

EXAMPLE 13.8 —*Continued*

3-Methylbutanal   3-Methyl-1-butanol

Practice Problem 13.8

Write equations for the hydrogenation of 3,4-dimethylhexanal and 2-chloropentanal.

For Further Practice: Questions 13.69 and 13.70.

The role of the lactate fermentation in exercise is discussed in greater detail in Section 21.4.

A biological example of the reduction of a ketone occurs in the body, particularly during strenuous exercise when the lungs and circulatory system may not be able to provide enough oxygen to the muscles. Under these circumstances, the lactate fermentation begins. In this reaction, the enzyme *lactate dehydrogenase* reduces pyruvate, the product of glycolysis, a pathway for the breakdown of glucose, into lactate. The source of hydrogen ions for this reaction is nicotinamide adenine dinucleotide (NADH), which is oxidized in the course of the reaction.

Pyruvate   Lactate

Question 13.11

Label each of the following as an oxidation or a reduction reaction.
  a. Ethanal to ethanol
  b. Benzoic acid to benzaldehyde
  c. Cyclohexanone to cyclohexanol
  d. 2-Propanol to propanone
  e. 2,3-Butanedione (found in butter) to 2,3-butanediol

Question 13.12

Write an equation for each of the reactions in Question 13.11.

Addition Reactions

7 | LEARNING GOAL
Write equations for the preparation of hemiacetals, hemiketals, acetals, and ketals.

Addition reactions of alkenes are described in detail in Section 11.5.

The principal reaction of the carbonyl group is the **addition reaction** across the polar carbon-oxygen double bond. This reaction is very similar to some that we have already studied, addition across the carbon-carbon double bond of alkenes. Such reactions require that a catalytic amount of acid be present in solution, as shown by the H$^+$ over the arrow for the reactions shown in the following examples.

An example of an addition reaction is the reaction of aldehydes with alcohols in the presence of catalytic amounts of acid. In this reaction, the hydrogen of the alcohol adds to the carbonyl oxygen. The alkoxyl group of the alcohol (—OR) adds to the carbonyl carbon. The predicted product is a **hemiacetal.**

$$\begin{array}{c} OH \\ | \\ R-C-OR \\ | \\ H \end{array}$$

General structure of a hemiacetal

However, this is not the product typically isolated from this reaction. Hemiacetals are quite reactive. In the presence of acid and excess alcohol, they undergo a substitution reaction in which the —OH group of the hemiacetal is exchanged for another —OR group from the alcohol. The product of this reaction is an **acetal.** Acetal formation is a reversible reaction, as the general equation shows:

$$\begin{array}{ccc} O & OH & OR^2 \\ \| & | & | \\ R^1{-}C{-}H + H{-}OR^2 \underset{H^+}{\overset{H^+}{\rightleftharpoons}} R^1{-}C{-}OR^2 + H{-}OR^2 \underset{H^+}{\overset{H^+}{\rightleftharpoons}} R^1{-}C{-}OR^2 + H_2O \\ & | & | \\ & H & H \end{array}$$

Aldehyde   Alcohol   Hemiacetal   Acetal

Consider the acid-catalyzed reaction between propanal and methanol:

$$\begin{array}{ccc} O & OH & OCH_3 \\ \| & | & | \\ CH_3CH_2{-}C{-}H + CH_3OH \overset{H^+}{\rightleftharpoons} CH_3CH_2{-}C{-}OCH_3 + CH_3OH \overset{H^+}{\rightleftharpoons} CH_3CH_2{-}C{-}OCH_3 + H_2O \\ & | & | \\ & H & H \end{array}$$

Propanal   Methanol   Hemiacetal   Propanal dimethyl acetal

Addition reactions will also occur between a ketone and an alcohol. In this case the more reactive intermediate is called a **hemiketal** and the product is called a **ketal.** The general equation for ketal formation is shown here:

$$\begin{array}{ccc} O & OH & OR^3 \\ \| & | & | \\ R^1{-}C{-}R^2 + H{-}OR^3 \underset{H^+}{\overset{H^+}{\rightleftharpoons}} R^1{-}C{-}OR^3 + H{-}OR^3 \underset{H^+}{\overset{H^+}{\rightleftharpoons}} R^1{-}C{-}OR^3 + H_2O \\ & | & | \\ & R^2 & R^2 \end{array}$$

Ketone   Alcohol   Hemiketal   Ketal

A simple scheme is helpful in recognizing these four types of compounds. Begin by drawing a carbon atom with four bonds and follow the flow chart as additional groups are added that will identify the molecules.

A ketal is the final product in the reaction between propanone and ethanol, seen in the following equation:

$$CH_3-\overset{\overset{\displaystyle O}{\|}}{C}-CH_3 + CH_3CH_2OH \overset{H^+}{\rightleftharpoons} CH_3-\overset{\overset{\displaystyle OH}{|}}{\underset{\underset{\displaystyle CH_3}{|}}{C}}-OCH_2CH_3 + CH_3CH_2OH \overset{H^+}{\rightleftharpoons} CH_3-\overset{\overset{\displaystyle OCH_2CH_3}{|}}{\underset{\underset{\displaystyle CH_3}{|}}{C}}-OCH_2CH_3 + H_2O$$

Propanone   Ethanol   Hemiketal   Ketal

Question 13.13

Identify each of the following structures as a hemiacetal, acetal, hemiketal, or ketal.

a. $H-\overset{\overset{\displaystyle CH_3}{|}}{\underset{\underset{\displaystyle OCH_3}{|}}{C}}-OH$ b. $H_3C-\overset{\overset{\displaystyle CH_3}{|}}{\underset{\underset{\displaystyle OCH_3}{|}}{C}}-OCH_3$ c. $H-\overset{\overset{\displaystyle CH_3}{|}}{\underset{\underset{\displaystyle OCH_3}{|}}{C}}-OCH_3$ d. $H_3C-\overset{\overset{\displaystyle CH_3}{|}}{\underset{\underset{\displaystyle OCH_3}{|}}{C}}-OH$

Question 13.14

Identify each of the following structures as a hemiacetal, acetal, hemiketal, or ketal.

a. $H-\overset{\overset{\displaystyle CH_2CH_3}{|}}{\underset{\underset{\displaystyle OCH_3}{|}}{C}}-OH$ b. $CH_3CH_2-\overset{\overset{\displaystyle CH_3}{|}}{\underset{\underset{\displaystyle OCH_3}{|}}{C}}-OH$

c. $H-\overset{\overset{\displaystyle CH_3}{|}}{\underset{\underset{\displaystyle OCH_3}{|}}{C}}-OCH_2CH_3$ d. $H_3C-\overset{\overset{\displaystyle CH_3}{|}}{\underset{\underset{\displaystyle OCH_2CH_3}{|}}{C}}-OCH_3$

Hemiacetals and hemiketals are readily formed in carbohydrates. Monosaccharides contain several hydroxyl groups and one carbonyl group. The linear form of a monosaccharide quickly undergoes an intramolecular reaction in solution to give a cyclic hemiacetal or hemiketal.

Earlier we noted that hemiacetals and hemiketals formed in *intermolecular* reactions were unstable and continued to react, forming acetals and ketals. This is not the case with the *intramolecular* reactions involving five- or six-carbon sugars. In these reactions the cyclic or ring form of the molecule is more stable than the linear form. This reaction is shown for the sugar glucose (blood sugar) in Figure 13.6 and is discussed in detail in Section 16.2.

When the hemiacetal or hemiketal of one monosaccharide reacts with a hydroxyl group of another monosaccharide, the product is an acetal or a ketal. A sugar molecule made up of two monosaccharides is called a *disaccharide.* The C—O—C bond between the two monosaccharides is called a *glycosidic bond* (Figure 13.7).

Keto-Enol Tautomers

LEARNING GOAL

8 Draw the keto and enol forms of aldehydes and ketones.

Many aldehydes and ketones may exist in an equilibrium mixture of two constitutional or structural isomers called *tautomers*. Tautomers differ from one another in the placement of a hydrogen atom and a double bond. One tautomer is the *keto form* (on the left in the following equation). The keto form has the structure typical of an aldehyde or ketone. The other form is called the *enol form* (on the right in the following equation). The enol form has a structure containing a carbon-carbon double bond (*en*) and a hydroxyl group, the functional group characteristic of alcohols (*ol*).

$$R^1-\underset{\underset{R^2}{|}}{\overset{\overset{H}{|}}{C}}-\overset{\overset{O}{||}}{C}-R^3 \rightleftharpoons \underset{R^2}{\overset{R^1}{}}C=C\underset{R^3}{\overset{OH}{}} \quad (R^1, R^2, \text{ and } R^3 = \text{H or alkyl group})$$

Keto form   Enol form

Because the keto form of most simple aldehydes and ketones is more stable, they exist mainly in that form.

Figure 13.6
Hemiacetal formation in sugars, shown for the intramolecular reaction of D-glucose.

α-D-Glucose

D-Glucose
(open-chain form)

β-D-Glucose

Figure 13.7
Acetal formation, demonstrated in the formation of the disaccharide sucrose, common table sugar. The reaction between the hydroxyl groups of the monosaccharides glucose and fructose produces the acetal sucrose. The bond between the two sugars is a glycosidic bond.

Glycosidic bond

α-Glucose   β-Fructose   Sucrose   + H₂O

Writing an Equation Representing the Equilibrium Between the Keto and Enol Forms of a Simple Aldehyde

EXAMPLE 13.9

Draw the keto form of ethanal and write an equation representing the equilibrium between the keto and enol forms of this molecule.

Solution

Ethanal
Keto form
More stable

Enol form
Less stable

LEARNING GOAL

8 Draw the keto and enol forms of aldehydes and ketones.

Continued—

EXAMPLE 13.9 —*Continued*

Practice Problem 13.9

Draw the keto and enol forms of (a) propanal and (b) 3-pentanone.

For Further Practice: Questions 13.91 and 13.92.

Phosphoenolpyruvate is a biologically important enol. In fact, it is the highest energy phosphorylated compound in living systems.

Phosphoenolpyruvate

Phosphoenolpyruvate is produced in the next-to-last step in the metabolic pathway called *glycolysis,* which is the first stage of carbohydrate breakdown. In the final reaction of glycolysis, the phosphoryl group from phosphoenolpyruvate is transferred to adenosine diphosphate (ADP). The reaction produces ATP, the major energy currency of the cell.

The glycolysis pathway is discussed in detail in Chapter 21.

Aldol Condensation

LEARNING GOAL
9 Write equations showing the aldol condensation.

The **aldol condensation** is a reaction in which aldehydes or ketones react to form larger molecules. A new carbon-carbon bond is formed in the process:

$$
\underset{\substack{\text{Aldehyde}\\\text{or}\\\text{Ketone}}}{R^1-CH_2-\overset{\displaystyle O}{\overset{\|}{C}}-R} + \underset{\substack{\text{Aldehyde}\\\text{or}\\\text{Ketone}}}{R^2-CH_2-\overset{\displaystyle O}{\overset{\|}{C}}-R} \underset{\text{enzyme}}{\overset{OH^- \text{ or}}{\rightleftarrows}} \underset{\text{Aldol}}{R^1-CH_2-\overset{\displaystyle OH}{\underset{\displaystyle R}{\overset{\displaystyle |}{\underset{\displaystyle |}{C}}}}-\overset{}{\underset{\displaystyle R^2}{\overset{}{\underset{\displaystyle |}{CH}}}}-\overset{\displaystyle O}{\overset{\|}{C}}-R}
$$

R = H, alkyl, or aryl group

This is actually a very complex reaction that occurs in multiple steps. Here we focus on the end results of the reaction, using the example of the reaction between two molecules of ethanal. As shown in the equation below, the α-carbon (carbon-2) of one aldehyde forms a bond with the carbonyl carbon of a second aldehyde (shown in blue). A bond also forms between a hydrogen atom on that same α-carbon and the carbonyl oxygen (shown in red).

$$
\underset{\text{Ethanal}}{H-\overset{\displaystyle H}{\underset{\displaystyle H}{\overset{\displaystyle |}{\underset{\displaystyle |}{C}}}}-\overset{\displaystyle O}{\overset{\|}{C}}-H} + \underset{\text{Ethanal}}{H-\overset{\displaystyle H}{\underset{\displaystyle H}{\overset{\displaystyle |}{\underset{\displaystyle |}{C}}}}-\overset{\displaystyle O}{\overset{\|}{C}}-H} \overset{OH^-}{\rightleftarrows} \underset{\substack{\text{3-Hydroxybutanal}\\(\beta\text{-hydroxybutyraldehyde})}}{H-\overset{\displaystyle H}{\underset{\displaystyle H}{\overset{\displaystyle |}{\underset{\displaystyle |}{C}}}}-\overset{\displaystyle OH}{\underset{\displaystyle H}{\overset{\displaystyle |}{\underset{\displaystyle |}{C}}}}-CH_2-\overset{\displaystyle O}{\overset{\|}{C}}-H}
$$

The result is similar when two ketones react:

$$CH_3-\overset{\overset{\displaystyle O}{\|}}{C}-CH_3 \;+\; H-\overset{\overset{\displaystyle H}{|}}{\underset{\underset{\displaystyle H}{|}}{C}}-\overset{\overset{\displaystyle O}{\|}}{C}-CH_3 \;\underset{\longleftarrow}{\overset{OH^-}{\longrightarrow}}\; CH_3-\overset{\overset{\displaystyle OH}{|}}{\underset{\underset{\displaystyle CH_3}{|}}{C}}-CH_2-\overset{\overset{\displaystyle O}{\|}}{C}-CH_3$$

Propanone   Propanone   4-Hydroxy-4-methyl-2-pentanone

In the laboratory, the aldol condensation is catalyzed by dilute base. But the same reaction occurs in our cells, where it is catalyzed by an enzyme. This reaction is one of many in a pathway that makes the sugar glucose from smaller molecules. This pathway is called gluconeogenesis (*gluco-* [sugar], *neo-* [new], *genesis* [beginnings]), which simply means origin of new sugar. This pathway is critical during starvation or following strenuous exercise. Under those conditions, blood glucose concentrations may fall dangerously low. Because the brain can use only glucose as an energy source, it is essential that the body be able to produce it quickly.

Gluconeogenesis is described in Chapter 21.

One of the steps in the pathway is an aldol condensation between the ketone dihydroxyacetone phosphate and the aldehyde glyceraldehyde-3-phosphate.

$$\underset{\text{Dihydroxyacetone phosphate}}{\overset{\displaystyle CH_2OPO_3^{2-}}{\underset{\displaystyle H}{\overset{|}{\underset{|}{C}=O}}\,\underset{|}{\overset{|}{H-COH}}}} \;+\; \underset{\text{Glyceraldehyde-3-phosphate}}{\overset{\displaystyle H\diagdown_{C}\diagup^{O}}{H-\overset{|}{\underset{|}{C}}-OH \atop CH_2OPO_3^{2-}}} \;\underset{}{\overset{Aldolase}{\rightleftharpoons}}\; \underset{\text{Fructose-1,6-bisphosphate}}{\overset{\displaystyle CH_2OPO_3^{2-}}{\underset{}{}}}$$

Dihydroxyacetonephosphate is a phosphorylated form of dihydroxyacetone (DHA), the active ingredient in self-tanning lotions. See *A Medical Perspective: That Golden Tan Without the Fear of Skin Cancer* on page 452.

The speed and specificity of this reaction are ensured by the enzyme *aldolase.* The product is the sugar fructose-1,6-bisphosphate, which is converted by another enzyme into glucose-1,6-bisphosphate. Removal of the two phosphoryl groups results in a new molecule of glucose for use by the body as an energy source.

Gluconeogenesis occurs under starvation conditions to provide a supply of blood glucose to nourish the brain. However, when glucose is plentiful, it is broken down to provide ATP energy for the cell. The pathway for glucose degradation is called *glycolysis.* In glycolysis, the reaction just shown is reversed. In general, aldol condensation reactions are reversible. These reactions are called *reverse aldols.*

ATP, the universal energy currency, is discussed in Section 21.1.

Write an equation for the aldol condensation of two molecules of propanal.

Question *13.15*

Write an equation for the aldol condensation of two molecules of butanal.

Question *13.16*

A Human *Perspective*

The Chemistry of Vision

Sight is a complex physiological process that is dependent on two types of cells in the retina of the eye: rods and cones. Rods are primarily responsible for vision in dim light. Cones are responsible for vision in bright light.

Vision in rod cells requires an unsaturated aldehyde, 11-*cis*-retinal (see figure below), which is produced from vitamin A. Vitamin A may be obtained directly in the diet, but it is also produced by the cleavage of β-carotene obtained in the diet.

In the retina, a protein called opsin combines with 11-*cis*-retinal to form a complex protein called rhodopsin. When light strikes the rods, the light energy is absorbed by 11-*cis*-retinal, which is then photochemically converted to 11-*trans*-retinal. This causes a change in the shape in the rhodopsin complex and results in the dissociation of 11-*trans*-retinal from the protein. This, in turn, causes ions to flow more freely into the rod cells. The influx of ions stimulates nerve cells that send signals to the brain. Interpretation of those signals produces the visual image.

Following the initial light stimulus, retinal returns to the *cis*-isomer and reassociates with opsin. The system is then

(a) Rod cell with detail of the membrane discs that carry the visual pigment, rhodopsin. (b) Detail of rhodopsin, a complex of 11-*cis*-retinal and the protein opsin, embedded in the disc membrane. (c) 11-*cis*-retinal and (d) 11-*trans*-retinal.

Summary of Reactions

Aldehydes and Ketones

Oxidation of an Aldehyde

$$R{-}\overset{\displaystyle O}{\overset{\displaystyle \|}{C}}{-}H \xrightarrow{\text{[O]}} R{-}\overset{\displaystyle O}{\overset{\displaystyle \|}{C}}{-}OH$$

Aldehyde   Carboxylic acid

Reduction of Aldehydes and Ketones

$$\overset{\displaystyle O}{\overset{\displaystyle \|}{\underset{R^1 \quad R^2}{C}}} + \overset{H}{\underset{H}{|}} \xrightarrow{\text{Pt}} R^1{-}\overset{\displaystyle OH}{\overset{|}{\underset{R^2}{C}}}{-}H$$

Aldehyde   Hydrogen   Alcohol
or Ketone

Rhodopsin

Rhodopsin absorbs photon of light

Opsin and *cis*-retinal enzymatically combined to regenerate rhodopsin

cis-retinal

trans-Retinal enzymatically converted back to *cis*-retinal

cis-Retinal isomerizes to *trans*-retinal

Opsin

trans-Retinal separates from opsin, which triggers a nerve impulse

Light is absorbed by rhodopsin, converting 11-*cis*-retinal to 11-*trans*-retinal. Loss of retinal from the protein initiates a nerve impulse that is sent to the brain where it is interpreted into the visual image.

ready for the next impulse of light. However, some retinal is lost in the process and must be replaced by conversion of dietary vitamin A or β-carotene to retinal. As you might expect, a deficiency of vitamin A can have terrible consequences. In children, lack of vitamin A causes xerophthalmia, an eye disease that results first in night blindness and eventually in total blindness. This can easily be prevented by a diet rich in vitamin A or β-carotene.

For Further Understanding

What are some sources of vitamin A or β-carotene in the diet?

Xerophthalmia causes blindness in approximately 200,000 children in third world countries each year. Investigate the measures that agencies such as the World Health Organization are using to save the sight of these children.

Addition Reactions

Addition of an alcohol to a ketone—ketal formation:

$$
\underset{\substack{\text{Ketone}}}{\overset{\displaystyle O}{\underset{R^1 \quad R^2}{\overset{\|}{C}}}}
+
\underset{\substack{\text{Alcohol}}}{\overset{H}{\underset{OR^3}{|}}}
\;\overset{H^+}{\rightleftharpoons}\;
\underset{\substack{\text{Hemiketal}}}{\overset{\displaystyle OH}{\underset{R^2}{\overset{|}{\underset{|}{R^1-C-OR^3}}}}}
+
\overset{H}{\underset{OR^3}{|}}
\;\overset{H^+}{\rightleftharpoons}\;
\underset{\substack{\text{Ketal}}}{\overset{\displaystyle OR^3}{\underset{R^2}{\overset{|}{\underset{|}{R^1-C-OR^3}}}}}
+ H_2O
$$

Addition of an alcohol to an aldehyde—acetal formation:

$$\underset{\substack{R^1 \quad H \\ \text{Aldehyde}}}{\overset{O}{\underset{\|}{C}}} + \underset{\substack{OR^2 \\ \text{Alcohol}}}{\overset{H}{\underset{|}{O}}} \xrightleftharpoons{H^+} \underset{\substack{H \\ \text{Hemiacetal}}}{\overset{OH}{\underset{|}{R^1-C-OR^2}}} + \underset{OR^2}{\overset{H}{\underset{|}{O}}} \xrightleftharpoons{H^+} \underset{\substack{H \\ \text{Acetal}}}{\overset{OR^2}{\underset{|}{R^1-C-OR^2}}} + H_2O$$

Keto-enol Tautomerization

$$\underset{\substack{R^2 \\ \text{Keto form}}}{\overset{H \quad O}{\underset{|}{R^1-C-C-R^3}}} \rightleftharpoons \underset{\substack{R^2 \quad R^3 \\ \text{Enol form}}}{\overset{R^1 \quad OH}{C=C}}$$

Aldol Condensation

$$\underset{\text{Aldehyde}}{\overset{O}{\underset{\|}{R^1-CH_2-C-R}}} + \underset{\text{Aldehyde}}{\overset{O}{\underset{\|}{R^2-CH_2-C-R}}} \xrightleftharpoons[\text{enzyme}]{OH^- \text{ or}} \underset{\substack{R \quad R^2 \\ \text{Aldol}}}{\overset{OH \quad O}{\underset{|\quad\|}{R^1-CH_2-C-CH-C-R}}}$$

R = H, alkyl, or aryl group

SUMMARY

13.1 Structure and Physical Properties

The *carbonyl group* ($>C=O$) is characteristic of the *aldehydes* and *ketones*. The carbonyl group and the two groups attached to it are coplanar. In ketones the carbonyl carbon is attached to two carbon-containing groups, whereas in aldehydes the carbonyl carbon is attached to at least one hydrogen; the second group attached to the carbonyl carbon in aldehydes may be another hydrogen or a carbon atom. Owing to the polar carbonyl group, aldehydes and ketones are polar compounds. Their boiling points are higher than those of comparable hydrocarbons but lower than those of comparable alcohols. Small aldehydes and ketones are reasonably soluble in water because of the hydrogen bonding between the carbonyl group and water molecules. Larger carbonyl-containing compounds are less polar and thus are more soluble in nonpolar organic solvents.

13.2 I.U.P.A.C. Nomenclature and Common Names

In the I.U.P.A.C. Nomenclature System, aldehydes are named by determining the parent compound and replacing the final -e of the parent alkane with -al. The chain is numbered beginning with the carbonyl carbon as carbon-1. Ketones are named by determining the parent compound and replacing the -e ending of the parent alkane with the -one suffix of the ketone family. The longest carbon chain is numbered to give the carbonyl carbon the lowest possible number. In the common system of nomenclature, substituted aldehydes are named as derivatives of the parent compound. Greek letters indicate the position of substituents. Common names of ketones are derived by naming the alkyl groups bonded to the carbonyl carbon. These names are followed by the word *ketone*.

13.3 Important Aldehydes and Ketones

Many members of the aldehyde and ketone families are important as food and fragrance chemicals, medicinals, and agricultural chemicals. Methanal (formaldehyde) is used to preserve tissue. Ethanal causes the symptoms of a hangover and is oxidized to produce acetic acid commercially. Propanone is a useful and versatile solvent for organic compounds.

13.4 Reactions Involving Aldehydes and Ketones

In the laboratory, aldehydes and ketones are prepared by the oxidation of alcohols. Oxidation of a primary alcohol produces an aldehyde; oxidation of a secondary alcohol yields a ketone. Tertiary alcohols do not react under these conditions. Aldehydes and ketones can be distinguished from one another on the basis of their ability to undergo oxidation reactions. The *Tollens' test* and *Benedict's test* are the most common such tests. Aldehydes are easily oxidized to carboxylic acids. Ketones do not undergo further oxidation reactions. Aldehydes and ketones are readily reduced to alcohols by *hydrogenation*. The most common reaction of the carbonyl group is *addition* across the highly polar carbon-oxygen double bond. The addition of an alcohol to an aldehyde produces a *hemiacetal*. The hemiacetal reacts with a second alcohol molecule to form an *acetal*. The reaction of a ketone with an alcohol produces a *hemiketal*. A hemiketal reacts with a second alcohol molecule to form a *ketal*. Hemiacetals and hemiketals are readily formed in carbohydrates. Aldol condensation is a reaction in which aldehydes and ketones form larger molecules. Aldehydes and ketones may exist as an equilibrium mixture of keto and enol tautomers.

KEY TERMS

acetal (13.4)
addition reaction (13.4)
aldehyde (13.1)
aldol condensation (13.4)
Benedict's test (13.4)
carbonyl group (Intro)
hemiacetal (13.4)

hemiketal (13.4)
hydrogenation (13.4)
ketal (13.4)
ketone (13.1)
oxidation (13.4)
Tollens' test (13.4)

QUESTIONS AND PROBLEMS

Structure and Physical Properties

Foundations

13.17 Explain the relationship between carbon chain length and water solubility of aldehydes or ketones.

13.18 Explain the dipole-dipole interactions that occur between molecules containing carbonyl groups.

Applications

13.19 Simple ketones (for example, acetone) are often used as industrial solvents for many organically based products such as adhesives and paints. They are often considered "universal solvents," because they dissolve so many diverse materials. Why are these chemicals such good solvents?

13.20 Explain briefly why simple (containing fewer than five carbon atoms) aldehydes and ketones exhibit appreciable solubility in water.

13.21 Draw intermolecular hydrogen bonding between ethanal and water.

13.22 Draw the polar interactions that occur between acetone molecules.

13.23 Why do alcohols have higher boiling points than aldehydes or ketones of comparable molecular weight?

13.24 Why do hydrocarbons have lower boiling points than aldehydes or ketones of comparable molecular weight?

Nomenclature

Foundations

13.25 Briefly describe the rules of the I.U.P.A.C. Nomenclature System for naming aldehydes.

13.26 Briefly describe the rules of the I.U.P.A.C. Nomenclature System for naming ketones.

13.27 Briefly describe how to determine the common name of an aldehyde.

13.28 Briefly describe how to determine the common name of a ketone.

Applications

13.29 Draw each of the following using complete structural formulas:
  a. Methanal
  b. 7,8-Dibromooctanal

13.30 Draw each of the following using condensed structural formulas:
  a. Acetone
  b. Hydroxyethanal

13.31 Draw the structure of each of the following compounds:
  a. 3-Chloro-2-pentanone
  b. Benzaldehyde

13.32 Draw the structure of each of the following compounds:
  a. 4-Bromo-3-hexanone
  b. 2-Chlorocyclohexanone

13.33 Use the I.U.P.A.C. Nomenclature System to name each of the following compounds:

  a. $CH_3CCH_2CH_3$ with O double bonded to second C

  b. $HCCHCH_2CH_3$ with O double bonded to first C and $CH_2CH_2CH_2CH_3$ branch on second C

13.34 Name each of the following using the I.U.P.A.C. Nomenclature System:

  a. $Cl-C-C-CH_3$ with Cl above and Cl below first C, and O double bonded to second C

  b.

13.35 Name each of the following using the I.U.P.A.C. Nomenclature System:

  a. benzaldehyde with NO₂ (NO_2) substituent and H−C=O group

  b.

13.36 Name each of the following using the I.U.P.A.C. Nomenclature System:

  a. $CH_3CH_2CH_2CH$ with O double bonded to last C

  b. $CH_3CCH_2CH_2CH$ with Br on second C and O double bonded to last C

  c. CH_3CHCH_2CH with Br on second C, O double bonded to last C

  d. $CH_3CCH_2CCH_2CH_2CH_3$ with CH_3 and Cl on second C and O double bonded to fourth C

13.37 The molecule shown below has a lovely aroma of lily-of-the-valley. Discovered in 1908, it has been used in hundreds of perfumes. What is the I.U.P.A.C. name of this molecule?

$CH_3CCH_2CH_2CH_2CHCH_2C-H$ with OH and CH_3 on second C, CH_3 on sixth C, and O double bonded to last C

13.38 Give the I.U.P.A.C. name for each of the following compounds:

a. $CH_3CCH_2CCH_2CH_3$ with O and CH_2CH_3, CH_2CH_3 substituents

b. $CH_3CCH_2CHCH_2CH_3$ with O and Cl substituents

13.39 Give the I.U.P.A.C. name for each of the following compounds:

a. $CH_3CHCH_2CHCCH_2CH_3$ with O, CH_3, CH_3 substituents

b. cyclopentanone ring with two CH_3 groups and O

13.40 Give the I.U.P.A.C. name for each of the following compounds:

a. $CH_3CH_2CHCH_2CH$ with CH_3 and O substituents

b. cyclohexanone ring with two Cl groups and O

13.41 Give the common name for each of the following compounds:

a. CH_3CCH_3 with O

b. $CH_3CH_2CCH_3$ with O

c. CH_3CH with O

d. CH_3CH_2CH with O

e. CH_3CHCCH_3 with O and CH_3

13.42 Give the common name for each of the following compounds:

a. $CH_3CH_2CCH_2CH_3$ with O

b. $CH_3CH_2CH_2CHCH$ with O and CH_3

c. $CH_3CCH_2CH_2CH_3$ with O

d. $CH_3CH_2CH_2CH_2CH_2CH$ with O

13.43 Draw the structure of each of the following compounds:
   a. 3-Hydroxybutanal
   b. 2-Methylpentanal
   c. 4-Bromohexanal
   d. 3-Iodopentanal
   e. 2-Hydroxy-3-methylheptanal

13.44 Draw the structure of each of the following compounds:
   a. Dimethyl ketone
   b. Methyl propyl ketone
   c. Ethyl butyl ketone
   d. Diisopropyl ketone

Important Aldehydes and Ketones

13.45 Why is acetone a good solvent for many organic compounds?

13.46 List several uses for formaldehyde.

13.47 Ethanal is produced by the oxidation of ethanol. Where does this reaction occur in the body?

13.48 List several aldehydes and ketones that are used as food or fragrance chemicals.

Reactions Involving Aldehydes and Ketones

Foundations

13.49 Explain what is meant by oxidation in organic molecules and provide an example of an oxidation reaction involving an aldehyde.

13.50 Explain what is meant by reduction in organic reactions and provide an example of a reduction reaction involving an aldehyde or ketone.

13.51 Define the term *addition reaction*. Provide an example of an addition reaction involving an aldehyde or ketone.

13.52 Define the term *aldol condensation*. Provide an example of an aldol condensation using an aldehyde or ketone.

13.53 Write a general equation representing the oxidation of an aldehyde. What is the product of this reaction?

13.54 Write a general equation representing the reduction of an aldehyde. What is the product of this reaction?

13.55 Write a general equation representing the oxidation of a ketone. What is the product of this reaction?

13.56 Write a general equation representing the addition of one alcohol molecule to an aldehyde.

13.57 Write a general equation representing the addition of two alcohol molecules to an aldehyde.

13.58 Write a general equation representing the addition of one alcohol molecule to a ketone.

13.59 Write a general equation representing the addition of two alcohol molecules to a ketone.

13.60 Write a general equation for an aldol condensation.

Applications

13.61 Draw the structures of each of the following compounds. Then draw and name the product that you would expect to produce by oxidizing each of these alcohols:
   a. 2-Butanol
   b. 2-Methyl-1-propanol
   c. Cyclopentanol

13.62 Draw the structures of each of the following compounds. Then draw and name the product that you would expect to produce by oxidizing each of these alcohols:
   a. 2-Methyl-2-propanol
   b. 2-Nonanol
   c. 1-Decanol

13.63 Draw the generalized equation for the oxidation of a primary alcohol.

13.64 Draw the generalized equation for the oxidation of a secondary alcohol.

13.65 Draw the structures of the reactants and products for each of the following reactions. Label each as an oxidation or a reduction reaction:
   a. Ethanal to ethanol
   b. Cyclohexanone to cyclohexanol
   c. 2-Propanol to propanone

13.66 An unknown has been determined to be one of the following three compounds:

$CH_3CH_2CCH_2CH_3$ with O $CH_3CH_2CH_2CH_2CH$ with O

3-Pentanone   Pentanal

$CH_3CH_2CH_2CH_2CH_3$

Pentane

The unknown is fairly soluble in water and produces a silver mirror when treated with the silver ammonia complex. A red precipitate appears when it is treated with the Benedict's reagent. Which of the compounds is the correct structure for the unknown? Explain your reasoning.

13.67 Write a balanced equation for the hydrogenation of each of the following ketones:
 a. 3-Methylbutanone
 b. 2-Pentanone
 c. 1-Chloropropanone

13.68 Write a balanced equation for the hydrogenation of each of the following ketones:
 a. Methyl butyl ketone
 b. Ethyl pentyl ketone
 c. Methyl propyl ketone

13.69 Write a balanced equation for the hydrogenation of each of the following aldehydes:
 a. Butanal
 b. 3-Methylpentanal
 c. 2-Methylpropanal

13.70 Write a balanced equation for the hydrogenation of each of the following aldehydes:
 a. β-Methylbutyraldehyde
 b. γ-Bromovaleraldehyde
 c. Propionaldehyde

13.71 Which of the following compounds would be expected to give a positive Tollens' test?
 a. 3-Pentanone   d. Cyclopentanol
 b. Cyclohexanone   e. 2,2-Dimethyl-1-pentanol
 c. 3-Methylbutanal   f. Acetaldehyde

13.72 Write an equation representing the reaction of glucose with the Benedict's reagent. How was this test used in medicine?

13.73 Write an equation for the addition of one ethanol molecule to each of the following aldehydes and ketones:

$$\underset{\text{a.}}{CH_3\overset{\displaystyle O}{\overset{\|}{C}}CH_3} \qquad \underset{\text{b.}}{CH_3\overset{\displaystyle O}{\overset{\|}{C}}H}$$

13.74 Write an equation for the addition of one ethanol molecule to each of the following aldehydes and ketones:

$$\underset{\text{a.}}{CH_3CH_2\overset{\displaystyle O}{\overset{\|}{C}}H} \qquad \underset{\text{b.}}{CH_3\overset{\displaystyle O}{\overset{\|}{C}}CH_2CH_2CH_3}$$

13.75 What is the general name for the product that is formed when an aldehyde reacts with one molecule of alcohol?

13.76 What is the general name of the product that is formed when a ketone reacts with one molecule of alcohol?

13.77 What is the general name for the product that is formed when an aldehyde reacts with two molecules of alcohol?

13.78 What is the general name of the product that is formed when a ketone reacts with two molecules of alcohol?

13.79 Write an equation for the addition of two methanol molecules to each of the following aldehydes and ketones:

$$\underset{\text{a.}}{CH_3\overset{\displaystyle O}{\overset{\|}{C}}CH_3} \qquad \underset{\text{b.}}{CH_3\overset{\displaystyle O}{\overset{\|}{C}}H}$$

13.80 Write an equation for the addition of two methanol molecules to each of the following aldehydes and ketones:

$$\underset{\text{a.}}{CH_3CH_2\overset{\displaystyle O}{\overset{\|}{C}}H} \qquad \underset{\text{b.}}{CH_3\overset{\displaystyle O}{\overset{\|}{C}}CH_2CH_2CH_3}$$

13.81 An aldehyde can be oxidized to produce a carboxylic acid. Draw the carboxylic acid that would be produced by the oxidation of each of the following aldehydes:
 a. Methanal
 b. Ethanal

13.82 An aldehyde can be oxidized to produce a carboxylic acid. Draw the carboxylic acid that would be produced by the oxidation of each of the following aldehydes:
 a. Propanal
 b. Butanal

13.83 An alcohol can be oxidized to produce an aldehyde or a ketone. What aldehyde or ketone is produced by the oxidation of each of the following alcohols?
 a. Methanol
 b. 1-Propanol

13.84 An alcohol can be oxidized to produce an aldehyde or a ketone. What aldehyde or ketone is produced by the oxidation of each of the following alcohols?
 a. 3-Pentanol
 b. 2-Methyl-2-butanol

13.85 Indicate whether each of the following statements is true or false.
 a. Aldehydes and ketones can be oxidized to produce carboxylic acids.
 b. Oxidation of a primary alcohol produces an aldehyde.
 c. Oxidation of a tertiary alcohol produces a ketone.
 d. Alcohols can be produced by the oxidation of an aldehyde or ketone.

13.86 Indicate whether each of the following statements is true or false.
 a. Ketones, but not aldehydes, react in the Tollens' silver mirror test.
 b. Addition of one alcohol molecule to an aldehyde results in formation of a hemiacetal.
 c. The cyclic forms of monosaccharides are intramolecular hemiacetals or intramolecular hemiketals.
 d. Disaccharides (sugars composed of two covalently joined monosaccharides) are acetals, ketals, or both.

13.87 Write an equation for the aldol condensation of two molecules of ethanal.

13.88 Write an equation for the aldol condensation of two molecules of hexanal.

13.89 Draw the keto and enol forms of propanone.

13.90 Draw the keto and enol forms of 2-butanone.

13.91 Draw the hemiacetal or hemiketal that results from the reaction of each of the following aldehydes or ketones with ethanol:

$$\underset{\text{a.}}{CH_3CH_2CH_2\overset{\displaystyle O}{\overset{\|}{C}}CH_3}$$

b.

c. (cyclopentanone)

13.92 Identify each of the following compounds as a hemiacetal, hemiketal, acetal, or ketal:

a. (ring structure with —OCH₃ and CH₃)

b. (ring structure with OH and OCH₂CH₃)

$$\text{c.}\quad CH_3\overset{\displaystyle OH}{\underset{\displaystyle OCH_2CH_3}{\overset{\|}{C}}}CH_3$$

d. (ring structure with O, OH and CH₃)

$$\text{e.}\quad CH_3\overset{\displaystyle OCH_3}{\underset{\displaystyle OCH_2CH_3}{\overset{\|}{C}}}CH_3$$

$$\text{f.}\quad CH_3CH=CH\overset{\displaystyle OCH_3}{\underset{\displaystyle OH}{\overset{\|}{C}}}CH_3$$

13.93 Complete the following synthesis by supplying the missing reactant(s), reagent(s), or product(s) indicated by the question marks:

$$CH_3\overset{\overset{\displaystyle O}{\|}}{C}CH_3 \xrightarrow{?(1)} CH_3\overset{\overset{\displaystyle OCH_2CH_3}{|}}{\underset{\underset{\displaystyle OCH_2CH_3}{|}}{C}}CH_3$$

$$\Big\uparrow ?(2)$$

$$CH_3\overset{}{\underset{\underset{\displaystyle OH}{|}}{C}}HCH_3 \xrightarrow[\text{Heat}]{H_2SO_4} ?(3)$$

13.94 Which alcohol would you oxidize to produce each of the following compounds?

a. $CH_3\overset{\overset{\displaystyle CH_3}{|}}{C}HCH_2\overset{\overset{\displaystyle O}{\|}}{C}CH_3$

b. $H\overset{\overset{\displaystyle O}{\|}}{C}CH_2CH_2\overset{\overset{\displaystyle O}{\|}}{C}H$

c. —$CH_2\overset{\overset{\displaystyle O}{\|}}{C}H$

d. $H\overset{\overset{\displaystyle O}{\|}}{C}CH_2\overset{\overset{\displaystyle O}{\|}}{C}CH_3$

e. $CH_3\overset{\overset{\displaystyle CH_3}{|}}{\underset{\underset{\displaystyle CH_3}{|}}{C}}CH_2CH_2\overset{\overset{\displaystyle O}{\|}}{C}H$

f. $O{=}$${=}O$

CRITICAL THINKING PROBLEMS

1. Review the material on the chemistry of vision and, with respect to the isomers of retinal, discuss the changes in structure that occur as the nerve impulses (that result in vision) are produced. Provide complete structural formulas of the retinal isomers that you discuss.

2. Classify the structure of β-D-fructose as a hemiacetal, hemiketal, acetal, or ketal. Explain your choice.

3. Design a synthesis for each of the following compounds, using any inorganic reagent of your choice and any hydrocarbon or alkyl halide of your choice:
    a. Octanal
    b. Cyclohexanone
    c. 2-Phenylethanoic acid

4. When alkenes react with ozone, O_3, the double bond is cleaved, and an aldehyde and/or a ketone is produced. The reaction, called *ozonolysis*, is shown in general as:

Predict the ozonolysis products that are formed when each of the following alkenes is reacted with ozone:
    a. 1-Butene
    b. 2-Hexene
    c. *cis*-3,6-Dimethyl-3-heptene

5. Lactose is the major sugar found in mammalian milk. It is a disaccharide composed of the monosaccharides glucose and galactose:

Is lactose a hemiacetal, hemiketal, acetal, or ketal? Explain your choice or choices.

6. The following are the keto and enol tautomers of phenol:

Enol form   Keto form
of phenol   of phenol

We have seen that most simple aldehydes and ketones exist mainly in the keto form because it is more stable. Phenol is an exception, existing primarily in the enol form. Propose a hypothesis to explain this.

14

Carboxylic Acids and Carboxylic Acid Derivatives

Learning Goals

1. Write structures and describe the physical properties of carboxylic acids.

2. Determine the common and I.U.P.A.C. names of carboxylic acids.

3. Describe the biological, medical, or environmental significance of several carboxylic acids.

4. Write equations that show the synthesis of a carboxylic acid.

5. Write equations representing acid-base reactions of carboxylic acids.

6. Write equations representing the preparation of an ester.

7. Write structures and describe the physical properties of esters.

8. Determine the common and I.U.P.A.C. names of esters.

9. Write equations representing the hydrolysis of an ester.

10. Define the term *saponification* and describe how soap works in the emulsification of grease and oil.

11. Determine the common and I.U.P.A.C. names of acid chlorides.

12. Write equations representing the synthesis of acid chlorides.

13. Determine the common and I.U.P.A.C. names of acid anhydrides.

14. Write equations representing the synthesis of acid anhydrides.

15. Discuss the significance of thioesters and phosphoesters in biological systems.

Outline

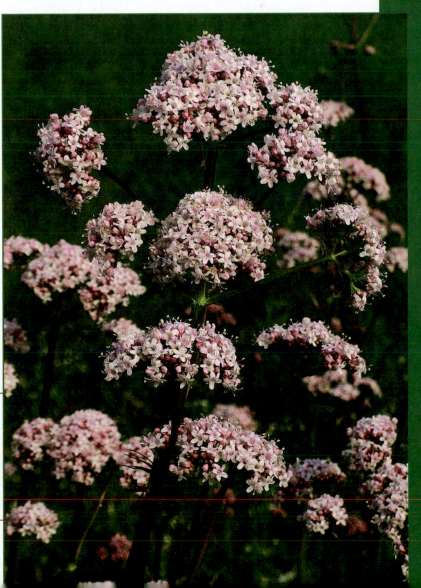

Valerian *(Valerian officinalis)* was used as a sedative by the ancient Greeks and Romans. Even during World War II, it was used by the English to relieve stress. One feature of the plant is the extremely strong odor that is often compared to the smell of a wet dog! The cause of this aroma is valeric acid.

Introduction

Carboxylic acids (Figure 14.1a) have the following general structure:

Aromatic carboxylic acid   Aliphatic carboxylic acid

They are characterized by the carboxyl group, shown in red, which may also be written in condensed form as —COOH or —CO_2H. The name carboxylic acid describes this family of compounds quite well. The term carboxylic is taken from the terms carbonyl and hydroxyl, the two structural units that make up the carboxyl group. The word acid in the name tells us one of the more important properties of these molecules: they dissociate in water to release protons. Thus they are acids.

In this chapter we will also study the esters (Figure 14.1b), which have the following general structures:

Examples of aliphatic and aromatic esters

> In fact, carboxylic acids are weak acids because they partially dissociate in water.

Chemistry Connection

Wake Up, Sleeping Gene

A common carboxylic acid, butyric acid, holds the promise of being an effective treatment for two age-old human genetic diseases. Sickle cell anemia and β-thalassemia are human genetic diseases of the β-globin portion of hemoglobin, the protein that carries oxygen from the lungs to tissues throughout the body. Normal hemoglobin consists of two α-globin proteins, two β-globin proteins, and four heme groups. In sickle cell anemia, the faulty β-globin gene calls for the synthesis of a sticky form of hemoglobin that forms long polymers. This distorts the red blood cells into elongated, sickled shapes that get stuck in capillaries and cannot provide the oxygen needed by the tissues. In β-thalassemia, there may be no β-globin produced at all. This results in short-lived red blood cells and severe anemia.

These two genetic diseases do not affect the fetus because before birth and for several weeks after birth, a fetal globin is made, rather than the adult β-globin. Fetal hemoglobin has a stronger affinity for oxygen than the adult form, ensuring that the fetus gets enough oxygen from the mother's blood through the placenta.

Two observations have led to a possible treatment of these diseases. First, physicians found some sickle cell anemia patients who suffered only mild symptoms because they continued to make high levels of fetal hemoglobin. Second was the observation that some babies born to diabetic mothers continued to produce fetal hemoglobin for an unusually long time after birth. Coincidentally, there was an unexpectedly high concentration of aminobutyric acid, a modified carboxylic acid, in the blood of these infants.

Susan Perrine of the Children's Hospital Oakland Research Center decided to try to reawaken the dormant fetal globin gene. She and her colleagues injected a sodium butyrate solution (the sodium salt of butyric acid) into three sickle cell patients and three β-thalassemia patients. As a result of the two- to three-week treatment, fetal hemoglobin production was boosted as much as 45% in these individuals. One β-thalassemia patient even experienced a complete reversal of the symptoms. Moreover, this treatment had few adverse side effects.

Longer studies with larger numbers of patients will be needed before this treatment can be declared a total success. However, Perrine's results hold the promise of a full and active life for individuals who were previously limited in activity and expected a short life span.

In this chapter we study the properties and reactions of the carboxylic acids; their salts, such as the sodium butyrate used to treat hemoglobin disorders; and their derivatives, the esters. We will focus on the importance of these molecules in biological systems, medicine, and the food industry.

(a)   (b)

Figure 14.1
Ball-and-stick models of (a) a carboxylic acid, propanoic acid, and (b) an ester, methyl ethanoate.

The group shown in red is called the **acyl group.** The acyl group is part of the functional group of the carboxylic acid derivatives, including the esters, acid chlorides, acid anhydrides, and amides.

Amides are discussed in Chapter 15.

14.1 Carboxylic Acids

Structure and Physical Properties

The **carboxyl group** consists of two very polar functional groups, the carbonyl group and the hydroxyl group. Thus **carboxylic acids** are very polar compounds. In addition, they can hydrogen bond to one another and to molecules of a polar solvent such as water. As a result of intermolecular hydrogen bonding, they boil at higher temperatures than aldehydes, ketones, or alcohols of comparable molecular weight. A comparison of the boiling points of an alkane, alcohol, ether, aldehyde, ketone, and carboxylic acid of comparable molecular weight is shown below:

LEARNING GOAL

1 ▶ Write structures and describe the physical properties of carboxylic acids.

$CH_3CH_2CH_2CH_3$

Butane
(butane)
M.W. = 58
b.p. −0.5°C

$CH_3—O—CH_2CH_3$

Methoxyethane
(ethyl methyl ether)
M.W. = 60
b.p. 7.0°C

$CH_3CH_2CH_2—OH$

1-Propanol
(propyl alcohol)
M.W. = 60
b.p. 97.2°C

$$CH_3CH_2\overset{\overset{\displaystyle O}{\|}}{C}—H$$

Propanal
(propionaldehyde)
M.W. = 58
b.p. 49°C

$$CH_3\overset{\overset{\displaystyle O}{\|}}{C}—CH_3$$

Propanone
(acetone)
M.W. = 58
b.p. 56°C

$$CH_3\overset{\overset{\displaystyle O}{\|}}{C}—OH$$

Ethanoic acid
(acetic acid)
M.W. = 60
b.p. 118°C

Formic acid causes the burning sensation at the site of an ant bite. What is the I.U.P.A.C. name for formic acid? Why does treating the bite with baking soda reduce the burning sensation?

As with alcohols, the smaller carboxylic acids are soluble in water (Figure 14.2). However, solubility falls off dramatically as the carbon content of the carboxylic acid increases because the molecules become more hydrocarbonlike and less polar. For example, acetic acid (the two-carbon carboxylic acid found in vinegar) is completely soluble in water, but hexadecanoic acid (a sixteen-carbon carboxylic acid found in palm oil) is insoluble in water.

The lower-molecular-weight carboxylic acids have sharp, sour tastes and unpleasant aromas. Formic acid, HCOOH, is used as a chemical defense by ants and causes the burning sensation of the ant bite. Acetic acid, CH_3COOH, is found in vinegar; propionic acid, CH_3CH_2COOH, is responsible for the tangy flavor of Swiss cheese; and butyric acid, $CH_3CH_2CH_2COOH$, causes the stench associated with rancid butter and gas gangrene.

The longer-chain carboxylic acids are generally called **fatty acids** and are important components of biological membranes and triglycerides, the major lipid storage form in the body.

(a)   (b)

Figure 14.2

Hydrogen bonding (a) among carboxylic acid molecules and (b) between carboxylic acid molecules and water molecules.

Question 14.1

Assuming that each of the following pairs of molecules has the same carbon chain length, which member of each of the following pairs has the lower boiling point?

   a. a carboxylic acid or a ketone
   b. a ketone or an alcohol
   c. an alcohol or an alkane

Question 14.2

Assuming that each of the following pairs of molecules has the same carbon chain length, which member of each of the following pairs has the lower boiling point?

   a. an ether or an aldehyde
   b. an aldehyde or a carboxylic acid
   c. an ether or an alcohol

Question 14.3

Why would you predict that a carboxylic acid would be more polar and have a higher boiling point than an aldehyde of comparable molecular weight?

Question 14.4

Why would you predict that a carboxylic acid would be more polar and have a higher boiling point than an alcohol of comparable molecular weight?

Nomenclature

In the I.U.P.A.C. Nomenclature System, carboxylic acids are named according to the following set of rules:

LEARNING GOAL

2 Determine the common and I.U.P.A.C. names of carboxylic acids.

   • Determine the parent compound, the longest continuous carbon chain bearing the carboxyl group.

- Number the chain so that the carboxyl carbon is carbon-1.
- Replace the *-e* ending of the parent alkane with the suffix *-oic acid*. If there are two carboxyl groups, the suffix *-dioic acid* is used.
- Name and number substituents in the usual way.

The following examples illustrate the naming of carboxylic acids with one carboxyl group:

Ethanoic acid
(acetic acid)

Propanoic acid
(propionic acid)

3-Methylbutanoic acid
(β-methylbutyric acid)

The following examples illustrate naming carboxylic acids with two carboxyl groups:

Hexanedioic acid
(adipic acid)

Propanedioic acid
(malonic acid)

Use the I.U.P.A.C. Nomenclature System to Name a Carboxylic Acid   E X A M P L E 14.1

The following compound is one of the monomers from which biodegradable plastic called Biopol is made (See An Environmental Perspective: Garbage Bags from Potato Peels later in this chapter). Name this carboxylic acid using the I.U.P.A.C. nomenclature system.

Solution

Parent compound: pentane (becomes pentanoic acid)

Position of —COOH: carbon-1 (Must be!)

Substituent: 3-hydroxy

Name: 3-Hydroxypentanoic acid

Name the following carboxylic acid:

Solution

Parent compound: octane (becomes octanoic acid)

Position of —COOH: carbon-1 (Must be!)

Continued—

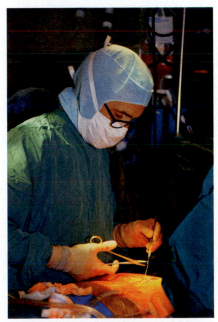

LEARNING GOAL

2 Determine the common and I.U.P.A.C. names of carboxylic acids.

Polymers of lactic acid are used as biodegradable sutures. What is the I.U.P.A.C. name of lactic acid, shown below?

CH₃CHCOOH
|
OH

EXAMPLE 14.1 —*Continued*

Substituents: 3,5,7-bromo

Name: 3,5,7-Tribromooctanoic acid

Practice Problem 14.1

Determine the I.U.P.A.C. name for each of the following structures. Remember that —COOH is an alternative way to represent the carboxyl group.

$$\begin{array}{cc} & \quad\;\; CH_3 \quad\; CH_3 \\ a.\;\; & CH_3CHCH_2CHCOOH \end{array}$$

$$\begin{array}{cc} & \quad\;\; OH \\ c.\;\; & CH_3CHCHCOOH \\ & \qquad\qquad\; OH \end{array}$$

$$\begin{array}{cc} b.\;\; & CH_2CH_2CHCOOH \\ & \;\; Cl \quad\; Cl \end{array}$$

$$\begin{array}{cc} & \qquad\qquad\quad Cl \\ d.\;\; & CH_3CH_2CHCHCHCOOH \\ & \qquad\quad CH_3 \;\; Br \end{array}$$

For Further Practice: Questions 14.34, 14.37, and 14.38.

The carboxylic acid derivatives of cycloalkanes are named by adding the suffix *carboxylic acid* to the name of the cycloalkane or substituted cycloalkane. The carboxyl group is always on carbon-1 and other substituents are named and numbered as usual.

Cyclohexanecarboxylic acid

Question 14.5

Determine the I.U.P.A.C. name for each of the following structures.

a.

b.

Question 14.6

Write the structure for each of the following carboxylic acids.

a. 1,4-Cyclohexanedicarboxylic acid
b. 4-Hydroxycyclohexanecarboxylic acid

LEARNING GOAL

2 Determine the common and I.U.P.A.C. names of carboxylic acids.

As we have seen so often, the use of common names, rather than systematic names, still persists. Often these names have evolved from the source of a given compound. This is certainly true of the carboxylic acids. Table 14.1 shows the

TABLE 14.1   Names and Sources of Some Common Carboxylic Acids

Name	Structure	Source	Root
Formic acid (methanoic acid)	HCOOH	Ants	L: *formica*, ant
Acetic acid (ethanoic acid)	CH$_3$COOH	Vinegar	L: *acetum*, vinegar
Propionic acid (propanoic acid)	CH$_3$CH$_2$COOH	Swiss cheese	Gk: *protos*, first; *pion*, fat
Butyric acid (butanoic acid)	CH$_3$(CH$_2$)$_2$COOH	Rancid butter	L: *butyrum*, butter
Valeric acid (pentanoic acid)	CH$_3$(CH$_2$)$_3$COOH	Valerian root	
Caproic acid (hexanoic acid)	CH$_3$(CH$_2$)$_4$COOH	Goat fat	L: *caper*, goat
Caprylic acid (octanoic acid)	CH$_3$(CH$_2$)$_6$COOH	Goat fat	L: *caper*, goat
Capric acid (decanoic acid)	CH$_3$(CH$_2$)$_8$COOH	Goat fat	L: *caper*, goat
Palmitic acid (hexadecanoic acid)	CH$_3$(CH$_2$)$_{14}$COOH	Palm oil	
Stearic acid (octadecanoic acid)	CH$_3$(CH$_2$)$_{16}$COOH	Tallow (beef fat)	Gk: *stear*, tallow

Note: I.U.P.A.C. names are shown in parentheses.

I.U.P.A.C. and common names of several carboxylic acids, as well as their sources and the Latin or Greek words that gave rise to the common names. Not only are the prefixes different than those used in the I.U.P.A.C. system, the suffix is different as well. Common names end in *-ic acid* rather than *-oic acid.*

In the common system of nomenclature, substituted carboxylic acids are named as derivatives of the parent compound (see Table 14.1). Greek letters are used to indicate the position of the substituent. The carbon atom bonded to the carboxyl group is the α-carbon, the next is the β-carbon, and so on.

Some examples of common names are

β-Hydroxybutyric acid   α-Hydroxypropionic acid

> β-Hydroxybutyric acid is the other monomer used to make the biodegradable plastic Biopol (see Example 14.1).

Naming Carboxylic Acids Using the Common System of Nomenclature   EXAMPLE 14.2

Write the common name for each of the following carboxylic acids.

> **LEARNING GOAL**
> **2** Determine the common and I.U.P.A.C. names of carboxylic acids.

Solution

Parent compound:	caproic acid	valeric acid
Substituents:	β-bromo	γ-chloro
Name:	β-Bromocaproic acid	γ-Chlorovaleric acid

Continued—

EXAMPLE 14.2   —Continued

Practice Problem 14.2

Provide the common name for each of the following molecules. Keep in mind that the carboxyl group can be represented as —COOH.

a. CH₃CHCH₂CHCOOH
       | |
       CH₃ CH₃

b. CH₂CH₂CHCOOH
    | |
    Cl   Cl

c. CH₃CHCHCH₂COOH
      | |
      Br Br

d. CH₃CH₂CHCH₂CH₂COOH
         |
         OH

For Further Practice: Questions 14.33, 14.39, and 14.40.

Benzoic acid is the simplest aromatic carboxylic acid.

Benzoic acid

Nomenclature of aromatic compounds is described in Section 11.6.

In many cases the aromatic carboxylic acids are named, in either system, as derivatives of benzoic acid. Generally, the *-oic acid* or *-ic acid* suffix is attached to the appropriate prefix. However, "common names" of substituted benzoic acids (for example, toluic acid and phthalic acid) are frequently used.

m-Toluic acid   *o*-Bromobenzoic acid   *m*-Iodobenzoic acid   Phthalic acid

The phenyl group is benzene with one hydrogen removed.

Often the phenyl group is treated as a substituent, and the name is derived from the appropriate alkanoic acid parent chain. For example:

2-Phenylethanoic acid
(α-phenylacetic acid)

3-Phenylpropanoic acid
(β-phenylpropionic acid)

Naming Aromatic Carboxylic Acids

Name the following aromatic carboxylic acids.

> **LEARNING GOAL**
> **2** Determine the common and I.U.P.A.C. names of carboxylic acids.

Solution

It is simplest to name the compound as a derivative of benzoic acid. The substituent, Cl, is attached to carbon-4 of the benzene ring. This compound is 4-chlorobenzoic acid or *p*-chlorobenzoic acid.

Solution

This compound is most easily named by treating the phenyl group as a substituent. The phenyl group is bonded to carbon-4 (or the γ-carbon, in the common system of nomenclature). The parent compound is pentanoic acid (valeric acid in the common system). Hence the name of this compound is 4-phenylpentanoic acid or γ-phenylvaleric acid.

Practice Problem 14.3

Draw structures for each of the following compounds.
a. *o*-Toluic acid
b. 2,4,6-Tribromobenzoic acid
c. 2,2,2-Triphenylethanoic acid
d. *p*-Toluic acid
e. 3-Phenylhexanoic acid
f. 3-Phenylcyclohexanecarboxylic acid

For Further Practice: Questions 14.38, 14.41, and 14.42.

Some Important Carboxylic Acids

As Table 14.1 shows, many carboxylic acids occur in nature. The stinging sensation of an ant bite is caused by methanoic (formic) acid, and ethanoic (acetic) acid provides the acidic zip to vinegars. Propanoic (propionic) acid is the product of bacterial fermentation of milk products and is most notable as a tangy component in the characteristic flavor of Swiss cheese.

    Several of the larger carboxylic acids have foul odors. For instance, butanoic (butyric) acid is the odor associated with rancid butter and is produced by the bacteria that cause gas gangrene, contributing to the characteristic smell of the necrotic tissue. Pentanoic (valeric) acid is associated with the valerian plant, which has long been known to have an aroma alternately described as over-ripe cheese

> **LEARNING GOAL**
> **3** Describe the biological, medical, or environmental significance of several carboxylic acids.

An Environmental Perspective

Garbage Bags from Potato Peels?

One problem facing society is its enormous accumulation of trash. To try to control the mountains of garbage that we produce, many institutions and towns practice recycling of aluminum, paper, and plastics. One problem that remains, however, is the plastic trash bag. We stuff the trash bag full of biodegradable garbage and bury it in a landfill; but soil bacteria can't break down the plastic to get to the biodegradable materials inside. Imagine a twenty-fourth century archeologist excavating one of these monuments to our society!

The good news is that laboratory research and products of bacterial metabolism are providing new materials that have the properties of plastics, but are readily biodegradable. For instance, sheets of plastic can be made by making polymers of lactic acid, which is a natural carboxylic acid produced by fermentation of sugars, particularly in milk and working muscle. Because many common soil bacteria can break down poly-lactic acid (PLA), trash bags made from this polymer would be quickly broken down in landfill soil.

Making plastic from lactic acid requires a huge supply of this carboxylic acid. As it turns out, we can produce an enormous quantity of lactic acid from garbage. When french fries are produced, nearly half of the potato is wasted. That amounts to about ten billion pounds of potato waste each year. When cheese is made, the curds are separated from the whey, and several billion gallons of whey are poured down the drain each year. Potato waste and whey can easily be broken down to produce glucose, which, in turn, can be converted into lactic acid used to make biodegradable plastics. PLA plastics have been available since the early 1990's and have been used successfully for sutures, medical implants, and drug delivery systems.

Biodegradable plastic can be made from garbage, such as potato peels left over from making french fries.

$$n\ H_3C-\overset{\overset{\displaystyle H}{|}}{\underset{\underset{\displaystyle OH}{|}}{C}}-\overset{\overset{\displaystyle O}{\|}}{C}-OH \longrightarrow \left[-O-\overset{\overset{\displaystyle CH_3}{|}}{\underset{\underset{\displaystyle H}{|}}{C}}-\overset{\overset{\displaystyle O}{\|}}{C}-O-\overset{\overset{\displaystyle CH_3}{|}}{\underset{\underset{\displaystyle H}{|}}{C}}-\overset{\overset{\displaystyle O}{\|}}{C}-O-\right]_n$$

Lactic acid   Polylactic acid (PLA)

Nature has provided other biodegradable plastics that are produced by bacteria. The most common of these in nature is polyhydroxybutyrate (PHB) made by the bacterium *Alcaligenes eutrophus*. PHB is a homopolymer (made up of a single monomer) of 3-hydroxybutanoic acid (β-hydroxybutyric acid).

To learn more about the fragrances and flavors of esters, see A Human Perspective: The Chemistry of Flavor and Fragrance later in this chapter.

or a wet dog. Nonetheless, extracts of valerian have been used for thousands of years as a natural sedative. Hexanoic (caproic) acid was first isolated from goats and, fittingly, is described as smelling like a goat. Heptanoic (enanthic) acid is also foul smelling and is associated with the odor of rancid oil.

These foul-smelling carboxylic acids have a far more pleasant potential, however. When carboxylic acids react with alcohols, the products are esters, which contribute to the lovely fragrance and flavor of many fruits.

Octanoic (caprylic) acid has an interesting function in the chemistry of human appetite. The hormone ghrelin, produced in the stomach, is sometimes called the "hunger hormone" because it stimulates the hypothalamus of the brain to signal that the body is hungry. However, the hormone alone does not have this affect. Ghrelin must be covalently bonded to a molecule of octanoic acid in order to have the hunger-stimulating effect on the hypothalamus.

β-Hydroxybutyric acid   Polyhydroxybutyric acid (PHB)

β-Hydroxybutyric acid   β-Hydroxyvaleric acid

Biopol—a heteropolymer

Unfortunately, this natural polymer does not have physical properties that would allow it to be a commercially successful plastic. This led to interest in making heteropolymers, polymers composed of two or more monomers. One such polymer with useful properties is a heteropolymer of β-hydroxybutyric acid and β-hydroxyvaleric acid, which has been given the name Biopol.

Biopol has properties that make it commercially useful, and it is completely broken down into carbon dioxide and water by microorganisms in the soil. Thus, it is completely biodegradable. Because the bacteria produced a low yield of Biopol, scientists produced transgenic plants in hopes that they would produce high yields of the polymer. This approach also encountered problems.

For the time being, biodegradable plastics cannot outcompete their nonbiodegradable counterparts. Future research and development will be required to reduce the cost of commercial production and fulfill the promise of an "environmentally friendly" garbage bag.

Computer colorized inclusions of polyhydroxybutyrate in *Alcaligenes eutrophus*.

For Further Understanding

Why are biodegradable plastics useful as sutures?

Two hydrogen atoms are lost in each reaction that adds a carboxylic acid to the polymers described here. What type of chemical reaction is this?

Fatty acids are long-chain monocarboxylic acids and can be isolated from a variety of sources including palm oil, coconut oil, butter, milk, lard, and animal fat, including tallow (beef fat). These fatty acids, in the form of triglycerides, are the major energy storage form in mammals and many plants. When the fatty acids in the triglyceride are saturated, the result is solid fat, as in animal fats. When the fatty acids are unsaturated, the result is a liquid, as in olive oil and canola oil.

More complex carboxylic acids are found in a variety of foods. For example, citric acid is found in citrus fruits and is often used to give the sharp taste to sour candies. It is also added to foods as a preservative and antioxidant. Adipic (hexanedioic) acid gives tartness to soft drinks and helps retard spoilage.

Bacteria in milk produce lactic acid as a product of fermentation of sugars. Lactic acid contributes a tangy flavor to yogurt and buttermilk. It is also used as

A triglyceride is a molecule of glycerol (Section 12.3) bonded to three fatty acid molecules by esterification (Section 14.2 and 17.3).

a food preservative to lower the pH to a level that retards microbial growth that causes food spoilage. Lactic acid is produced in muscle cells when an individual is exercising strenuously. If the level of lactic acid in the muscle and bloodstream becomes high enough, the muscle can't continue to work.

Remember, the carboxyl group can be represented as —COOH, CO$_2$H, or

$$-\overset{\overset{\displaystyle O}{\|}}{C}-OH$$

$$
\begin{array}{ccc}
\text{COOH} & \text{COOH} & \text{COOH} \\
| & | & | \\
\text{H}-\text{C}-\text{H} & \text{H}-\text{C}-\text{OH} & \text{H}-\text{C}-\text{H} \\
| & | & | \\
\text{HO}-\text{C}-\text{COOH} & \text{CH}_3 & \text{H}-\text{C}-\text{H} \\
| & & | \\
\text{H}-\text{C}-\text{H} & & \text{H}-\text{C}-\text{H} \\
| & & | \\
\text{COOH} & & \text{H}-\text{C}-\text{H} \\
& & | \\
& & \text{COOH}
\end{array}
$$

Citric acid	Lactic acid	Adipic acid
(citrus fruit)	*(yogurt)*	*(beet juice)*

Oxalic acid is a dicarboxylic acid found in spinach and rhubarb. Human kidney stones are often formed from the calcium salt of oxalic acid. In fact, it is toxic in high concentrations and foods with high levels of oxalic acid must be boiled before being eaten. While oxalic acid is used in industry as a bleaching agent and spot remover, the potassium salt is used in clinical laboratories to prevent blood samples from coagulating.

Tartaric acid is used in baking powder because it will undergo a reaction with carbonates in the dough, producing CO_2 that will cause the bread or cake to rise. It has also been used as a laxative. Malic acid gives the sour taste to green apples. Since the amount of malic acid decreases as a fruit ripens, the fruit becomes sweeter and less tart as it ripens.

$$
\begin{array}{ccc}
\text{OH} & \text{COOH} & \text{COOH} \\
| & | & | \\
\text{C}=\text{O} & \text{HO}-\text{C}-\text{H} & \text{H}-\text{C}-\text{H} \\
| & | & | \\
\text{C}=\text{O} & \text{HO}-\text{C}-\text{H} & \text{HO}-\text{C}-\text{H} \\
| & | & | \\
\text{OH} & \text{COOH} & \text{COOH}
\end{array}
$$

Oxalic acid	Tartaric acid	Malic acid

Several aromatic carboxylic acids are also of medical interest. The sodium salt of benzoic acid is used as a preservative in soft drinks, pickles, jellies, and many other foods and some cosmetics. It is of value as a preservative because it is colorless, odorless, and tasteless, and will kill bacteria at a concentration of only 0.1%.

Salicylic acid is used as a disinfectant and, in fact, is superior to phenol. It is also used in ointments to remove corns or warts because it causes the top layer of the skin to flake off, leaving the underlying living skin undamaged.

Acetylsalicylic acid is aspirin. As early as the fifth century B.C., the revered physician Hippocrates described a bitter extract from willow bark that could reduce fevers and relieve pain. Ancient texts from the Middle East reveal that Egyptian and Sumerian physicians appreciated the medicinal value of willow bark and Native Americans used it to treat headache, fever, and chills, as well as sore muscles. In 1828, Henri Leroux isolated crystals of the compound that came to be called salicin. Nearly seventy years later in 1897, chemists at Bayer and Company, chemically added an acetyl group to salicin, synthesizing acetylsalicylic acid, a derivative that did not produce the severe gastrointestinal side effects caused by salicin. Thus, aspirin became the first synthetic drug, launching the pharmaceutical industry. Today, aspirin is recommended in low daily doses by the American Heart Association as a preventative measure against heart attacks and strokes caused by blood clots.

Benzoic acid   Salicylic acid   Acetylsalicylic acid

Reactions Involving Carboxylic Acids

Preparation of Carboxylic Acids

Many of the small carboxylic acids are prepared on a commercial scale. For example, ethanoic (acetic) acid, found in vinegar, is produced commercially by the **oxidation** of either ethanol or ethanal as shown here:

Ethanol   Ethanal   Ethanoic acid

A variety of oxidizing agents, including oxygen, can be used, and catalysts are often required to provide acceptable yields. Other simple carboxylic acids can be made by oxidation of the appropriate primary alcohol or aldehyde.

In the laboratory, carboxylic acids are prepared by the oxidation of aldehydes or primary alcohols. Most common oxidizing agents, such as chromic acid, can be used. The general reaction is

Primary alcohol   Aldehyde   Carboxylic acid

LEARNING GOAL

4 Write equations that show the synthesis of a carboxylic acid.

Oxidation reactions involving aldehydes and primary alcohols were discussed in Sections 12.5 and 13.4.

Writing Equations for the Oxidation of a Primary Alcohol to a Carboxylic Acid

EXAMPLE 14.4

Write an equation showing the oxidation of 1-propanol to propanoic acid.

Solution

$$CH_3CH_2\overset{\displaystyle H}{\underset{\displaystyle H}{C}}\text{—OH} \xrightarrow{H_2CrO_4} CH_3CH_2\overset{O}{C}\text{—H} \xrightarrow[\text{oxidation}]{\text{Continued}} CH_3CH_2\text{—}\overset{O}{C}\text{—OH}$$

1-Propanol   Propanal   Propanoic acid
(propyl alcohol)   (propionaldehyde) (propionic acid)

LEARNING GOAL

4 Write equations that show the synthesis of a carboxylic acid.

Practice Problem 14.4

Write equations showing synthesis of (a) ethanoic acid, (b) butanoic acid, and (c) octanoic acid by oxidation of the corresponding primary alcohol.

For Further Practice: Questions 14.51 and 14.52.

Acid-Base Reactions

The carboxylic acids behave as acids because they are proton donors. They are weak acids that dissociate to form a carboxylate ion and a hydrogen ion, as shown in the following equation:

$$\underset{\substack{\text{Carboxylic}\\\text{acid}}}{R-\overset{\overset{\textstyle O}{\|}}{C}-OH} \rightleftharpoons \underset{\substack{\text{Carboxylate}\\\text{anion}}}{R-\overset{\overset{\textstyle O}{\|}}{C}-O^-} + \underset{\substack{\text{Hydrogen}\\\text{ion}}}{H^+}$$

The properties of weak acids are described in Sections 8.1 and 8.2.

Carboxylic acids are weak acids because they dissociate only slightly in solution. The majority of the acid remains in solution in the undissociated form. Typically, less than 5% of the acid is ionized (approximately five carboxylate ions to every ninety-five carboxylic acid molecules).

LeChatelier's principle is described in Section 7.4.

When strong bases are added to a carboxylic acid, neutralization occurs. The acid protons are removed by the OH^- to form water and the carboxylate ion. The equilibrium shown in the reaction above is shifted to the right, owing to removal of H^+. This is an illustration of LeChatelier's principle.

$$\underset{\substack{\text{Carboxylic}\\\text{acid}}}{R-\overset{\overset{\textstyle O}{\|}}{C}-OH} + \underset{\substack{\text{Strong}\\\text{base}}}{NaOH} \longrightarrow \underset{\substack{\text{Carboxylic}\\\text{acid salt}}}{R-\overset{\overset{\textstyle O}{\|}}{C}-O^-Na^+} + \underset{\text{Water}}{H_2O}$$

The carboxylate anion and the cation of the base form the carboxylic acid salt.

The following examples show the neutralization of acetic acid and benzoic acid in solutions of the strong base NaOH.

$$\underset{\text{Acetic acid}}{CH_3\overset{\overset{\textstyle O}{\|}}{C}-OH} + \underset{\substack{\text{Sodium}\\\text{hydroxide}\\\text{(strong base)}}}{NaOH} \longrightarrow \underset{\text{Sodium acetate}}{CH_3\overset{\overset{\textstyle O}{\|}}{C}-O^-Na^+} + \underset{\text{Water}}{H-O-H}$$

Sodium benzoate is commonly used as a food preservative.

Benzoic acid + Sodium hydroxide (strong base) → Sodium benzoate + H—O—H

Note that the salt of a carboxylic acid is named by replacing the *-ic acid* suffix with *-ate*. Thus acetic acid becomes acetate, and benzoic acid becomes benzoate. This name is preceded by the name of the appropriate cation, sodium in the examples above.

E X A M P L E 14.5 **Writing an Equation to Show the Neutralization of a Carboxylic Acid by a Strong Base**

Write an equation showing the neutralization of propanoic acid by potassium hydroxide.

Continued—

EXAMPLE 14.5 —*Continued*

Solution

The protons of the acid are removed by the OH⁻ of the base. This produces water. The cation of the base, in this case potassium ion, forms the salt of the carboxylic acid.

$$CH_3CH_2-\overset{\overset{\displaystyle O}{\|}}{C}-OH \ + \ KOH \ \longrightarrow \ CH_3CH_2-\overset{\overset{\displaystyle O}{\|}}{C}-O^-K^+ \ + \ H_2O$$

| Propanoic acid | Potassium hydroxide | Potassium salt of propanoic acid | Water |

Practice Problem 14.5

Write the formula of the organic product obtained through each of the following reactions.

a. $CH_3CH_2COOH + KOH \longrightarrow ?$
b. $CH_3CH_2CH_2COOH + Ba(OH)_2 \longrightarrow ?$
c. $CH_3CH_2CH_2CH_2CH_2COOH + KOH \longrightarrow ?$
d. Benzoic acid + sodium hydroxide $\longrightarrow ?$

For Further Practice: Questions 14.54, 14.57, and 14.58.

Naming the Salt of a Carboxylic Acid

EXAMPLE 14.6

Write the common and I.U.P.A.C. names of the salt produced in the reaction shown in Example 14.5.

Solution

$$CH_3CH_2-\overset{\overset{\displaystyle O}{\|}}{C}-O^-K^+$$

I.U.P.A.C. name of the parent carboxylic acid:	Propanoic acid
Replace the *-ic acid* ending with *-ate:*	Propanoate
Name of the cation of the base:	Potassium
Name of the carboxylic acid salt:	Potassium propanoate
Common name of the parent carboxylic acid:	Propionic acid
Replace the *-ic acid* ending with *-ate:*	Propionate
Name of the cation of the base:	Potassium
Name of the carboxylic acid salt:	Potassium propionate

Practice Problem 14.6

Name the products of the reactions in Practice Problem 14.5 at the end of Example 14.5.

For Further Practice: Questions 14.59 and 14.60.

Soaps are made by a process called saponification, which is the base-catalyzed hydrolysis of an ester. This is described in detail in Section 14.2.

Carboxylic acid salts are ionic substances. As a result, they are very soluble in water. The long-chain carboxylic acid salts (fatty acid salts) are called *soaps*.

Esterification

Carboxylic acids react with alcohols to form esters and water according to the general reaction:

$$R^1-\overset{\overset{\displaystyle O}{\|}}{C}-OH + R^2OH \xrightarrow{\text{Acid}} R^1-\overset{\overset{\displaystyle O}{\|}}{C}-OR^2 + H_2O$$

Carboxylic   Alcohol   Ester   Water
acid

The details of these reactions will be examined in Section 14.2.

> **LEARNING GOAL**
> **6** Write equations representing the preparation of an ester.

14.2 Esters

Structure and Physical Properties

> **LEARNING GOAL**
> **7** Write structures and describe the physical properties of esters.

Esters are mildly polar and have pleasant aromas. Many esters are found in natural foodstuffs; banana oil (3-methylbutyl ethanoate; common name, isoamyl acetate), pineapples (ethyl butanoate; common name, ethyl butyrate), and raspberries (isobutyl methanoate; common name, isobutyl formate) are but a few examples.

Esters boil at approximately the same temperature as aldehydes or ketones of comparable molecular weight. The simpler ones are somewhat soluble in water.

See A Human Perspective: The Chemistry of Flavor and Fragrance later in this chapter.

Nomenclature

Esters are **carboxylic acid derivatives,** organic compounds derived from carboxylic acids. They are formed from the reaction of a carboxylic acid with an alcohol, and both of these reactants are reflected in the naming of the ester. They are named according to the following set of rules:

- Use the *alkyl* or *aryl* portion of the alcohol name as the first name.
- The *-ic acid* ending of the name of the carboxylic acid is replaced with *-ate* and follows the first name.

> **LEARNING GOAL**
> **8** Determine the common and I.U.P.A.C. names of esters.

For example, in the following reaction, ethanoic acid reacts with methanol to produce methyl ethanoate:

$$CH_3\overset{\overset{\displaystyle O}{\|}}{C}-OH + CH_3OH \underset{\xrightarrow{\hspace{1cm}}}{\overset{H^+,\ heat}{\xrightarrow{\hspace{1cm}}}} CH_3\overset{\overset{\displaystyle O}{\|}}{C}-OCH_3 + H_2O$$

Ethano*ic acid*   *Methanol*   *Methyl* ethano*ate*
(acetic acid)   (methyl alcohol)   (methyl acetate)

Similarly, ace*tic acid* and *ethanol* react to produce *ethyl* ace*tate,* and the product of the reaction between benzo*ic acid* and *isopropyl* alcohol is *isopropyl* benzo*ate.*

Naming Esters Using the I.U.P.A.C. and Common Nomenclature Systems

The molecule shown below contributes to the flavor of pineapple. Write the I.U.P.A.C. and common names for this ester.

> **LEARNING GOAL**
> **8** Determine the common and I.U.P.A.C. names of esters.

$$CH_3CH_2CH_2\overset{\overset{\displaystyle O}{\|}}{C}-OCH_2CH_3$$

Continued—

EXAMPLE 14.7 *—Continued*

Solution

I.U.P.A.C. and common names of parent carboxylic acid:	butanoic acid	butyric acid
Replace the *ic acid* ending of the carboxylic acid with *-ate:*	butanoate	butyrate
Name of the alkyl portion of the alcohol:	ethyl	ethyl
I.U.P.A.C. and common names of the ester:	Ethyl butanoate	Ethyl butyrate

The molecule shown below is associated with the characteristic flavor of apricots. Write the common and I.U.P.A.C. names of this ester.

$$CH_3CH_2CH_2C\overset{O}{\overset{\|}{}}-OCH_2CH_2CH_2CH_2CH_3$$

Solution

I.U.P.A.C. and common names of parent carboxylic acid:	butanoic acid	butyric acid
Replace the *ic acid* ending of the carboxylic acid with *-ate:*	butanoate	butyrate
Name of the alkyl portion of the alcohol:	pentyl	pentyl
I.U.P.A.C. and common names of the ester:	Pentyl butanoate	Pentyl butyrate

 Practice Problem 14.7

Name each of the following esters using both the I.U.P.A.C. and common nomenclature systems.

a.
$$CH_3CH_2CH_2C\overset{O}{\overset{\|}{}}-OCH_2CH_2CH_3$$

b.
$$CH_3CH_2CH_2C\overset{O}{\overset{\|}{}}-OCH_2CH_3$$

c.
$$CH_3C\overset{O}{\overset{\|}{}}-OCH_2CH_2CH_3$$

d.
$$CH_3CH_2C\overset{O}{\overset{\|}{}}-OCH_2CH_2CH_2CH_3$$

For Further Practice: Questions 14.65, 14.66, and 14.68.

Naming esters is analogous to naming the salts of carboxylic acids. Consider the following comparison:

Sodium ethanoate   Ethyl ethanoate
(sodium acetate)   (ethyl acetate)

As shown in this example, the alkyl group of the alcohol, rather than Na^+, has displaced the acidic hydrogen of the carboxylic acid.

Reactions Involving Esters

Preparation of Esters

LEARNING GOAL

6 Write equations representing the preparation of an ester.

The conversion of a carboxylic acid to an ester requires heat and is catalyzed by a trace of acid (H^+). When esters are prepared directly from a carboxylic acid and an alcohol, a water molecule is lost, as in the reaction:

Esterification is reversible. The direction of the reaction is determined by the conditions chosen. Excess alcohol favors ester formation. The carboxylic acid is favored when excess water is present.

$$CH_3CH_2\overset{\displaystyle O}{\overset{\|}{C}}{-}OH \ + \ CH_3OH \ \underset{}{\overset{H^+,\ heat}{\rightleftharpoons}} \ CH_3CH_2\overset{\displaystyle O}{\overset{\|}{C}}{-}OCH_3 \ + \ H{-}O{-}H$$

Propanoic acid   Methanol   Methyl propanoate
(propionic acid) (methyl alcohol) (methyl propionate)

Esterification is a *dehydration* reaction, so called because a water molecule is eliminated during the reaction.

E X A M P L E 14.8   **Writing Equations Representing Esterification Reactions**

LEARNING GOAL

6 Write equations representing the preparation of an ester.

Write an equation showing the esterification reactions that would produce ethyl butanoate and propyl ethanoate.

Solution

The name, ethyl butanoate, tells us that the alcohol used in the reaction is ethanol and the carboxylic acid is butanoic acid. We must remember that a trace of acid and heat are required for the reaction and that the reaction is reversible. With this information, we can write the following equation representing the reaction:

This reaction could be used to produce artificial pineapple flavoring.

$$CH_3CH_2CH_2\overset{\displaystyle O}{\overset{\|}{C}}{-}OH + CH_3CH_2OH \ \underset{}{\overset{H^+,\ heat}{\rightleftharpoons}} \ CH_3CH_2CH_2\overset{\displaystyle O}{\overset{\|}{C}}{-}OCH_2CH_3 + H_2O$$

Butanoic acid   Ethanol   Ethyl butanoate
(butyric acid)   (ethyl butyrate)

Similarly, the name propyl ethanoate reveals that the alcohol used in this reaction is 1-propanol and the carboxylic acid must be ethanoic acid.

Continued—

EXAMPLE 14.8 —Continued

Knowing that we must indicate that the reaction is reversible and that heat and a trace of acid are required, we can write the following equation:

$$CH_3\overset{\overset{\displaystyle O}{\|}}{C}-OH + CH_3CH_2CH_2OH \underset{}{\overset{H^+,\ heat}{\rightleftarrows}} CH_3\overset{\overset{\displaystyle O}{\|}}{C}-OCH_2CH_2CH_3 + H_2O$$

Ethanoic acid   1-Propanol   Propyl ethanoate
(acetic acid)   (propyl acetate)

Practice Problem 14.8

Write an equation showing the esterification reactions that would produce (a) butyl ethanoate and (b) ethyl propanoate.

For Further Practice: Questions 14.75 and 14.76.

Designing the Synthesis of an Ester

EXAMPLE 14.9

Design the synthesis of ethyl propanoate from organic alcohols.

Solution

The ease with which alcohols are oxidized to aldehydes, ketones, or carboxylic acids (depending on the alcohol that you start with and the conditions that you employ), coupled with the ready availability of alcohols, provides the pathway necessary to many successful synthetic transformations. For example, let's develop a method for synthesizing ethyl propanoate, using any inorganic reagent you wish but limiting yourself to organic alcohols that contain three or fewer carbon atoms:

$$CH_3CH_2\overset{\overset{\displaystyle O}{\|}}{C}-O-CH_2CH_3$$

Ethyl propanoate
(ethyl propionate)

Ethyl propanoate can be made from propanoic acid and ethanol:

$$CH_3CH_2\overset{\overset{\displaystyle O}{\|}}{C}-OH + CH_3CH_2OH \underset{}{\overset{H^+,\ heat}{\rightleftarrows}} CH_3CH_2\overset{\overset{\displaystyle O}{\|}}{C}-O-CH_2CH_3 + H_2O$$

Propanoic acid   Ethanol   Ethyl propanoate
(propionic acid)   (ethyl alcohol)   (ethyl propionate)

Ethanol is a two-carbon alcohol that is an allowed starting material, but propanoic acid is not. Can we now make propanoic acid from an alcohol of three or fewer carbons? Yes!

$$CH_3CH_2CH_2OH \overset{[O]}{\longrightarrow} CH_3CH_2\overset{\overset{\displaystyle O}{\|}}{C}-H \overset{[O]}{\longrightarrow} CH_3CH_2\overset{\overset{\displaystyle O}{\|}}{C}-OH$$

Propanol   Propanal   Propanoic acid
(propyl alcohol)   (propionaldehyde)   (propionic acid)

Continued—

LEARNING GOAL

6 Write equations representing the preparation of an ester.

A Human *Perspective*

The Chemistry of Flavor and Fragrance

Carboxylic acids are often foul smelling. For instance, butyric acid is one of the worst smelling compounds imaginable—the smell of rancid butter.

$$CH_3CH_2CH_2\overset{\overset{\displaystyle O}{\|}}{C}{-}OH$$

Butanoic acid
(butyric acid)

Butyric acid is also a product of fermentation reactions carried out by *Clostridium perfringens*. This organism is the most common cause of gas gangrene. Butyric acid contributes to the notable foul smell accompanying this infection.

By forming esters of butyric acid, one can generate compounds with pleasant smells. Ethyl butyrate is the essence of pineapple oil.

Volatile esters are often pleasant in both aroma and flavor. Natural fruit flavors are complex mixtures of many esters and other organic compounds. Chemists can isolate these mixtures and identify the chemical components. With this information they are able to synthesize artificial fruit flavors, using just a few of the esters found in the natural fruit. As a result, the artificial flavors rarely have the full-bodied flavor of nature's original blend.

Raspberries

$$H{-}\overset{\overset{\displaystyle O}{\|}}{C}{-}OCH_2\underset{\underset{\displaystyle CH_3}{|}}{C}HCH_3$$

Isobutyl methanoate
(isobutyl formate)

Pineapple

$$CH_3CH_2CH_2\overset{\overset{\displaystyle O}{\|}}{C}{-}OCH_2CH_3$$

Ethyl butanoate
(ethyl butyrate)

Bananas

$$CH_3\overset{\overset{\displaystyle O}{\|}}{C}{-}OCH_2CH_2\underset{\underset{\displaystyle CH_3}{|}}{C}HCH_3$$

3-Methylbutyl ethanoate
(isoamyl acetate)

Oranges

$$CH_3C\!\!-\!\!OCH_2CH_2CH_2CH_2CH_2CH_2CH_2CH_3$$
(with O double-bonded to C)

Octyl ethanoate
(octyl acetate)

Apples

$$CH_3CH_2CH_2C\!\!-\!\!OCH_3$$
(with O double-bonded to C)

Methyl butanoate
(methyl butyrate)

Apricots

$$CH_3CH_2CH_2C\!\!-\!\!OCH_2CH_2CH_2CH_2CH_3$$
(with O double-bonded to C)

Pentyl butanoate
(pentyl butyrate)

Strawberries

$$CH_3CH_2CH_2C\!\!-\!\!SCH_3$$
(with O double-bonded to C)

Methyl thiobutanoate
(methyl thiobutyrate)
(a thioester in which sulfur replaces oxygen)

For Further Understanding

Draw the structure of methyl salicylate, which is found in oil of wintergreen.

Write an equation for the synthesis of each of the esters shown in this Perspective.

EXAMPLE 14.9 —*Continued*

Propanol is a three-carbon alcohol, an allowed starting material. The synthesis is now complete. By beginning with ethanol and propanol, ethyl propanoate can be synthesized easily.

> **Practice Problem 14.9**
>
> Design the synthesis of (a) methyl butanoate and (b) propyl methanoate from organic alcohols.
>
> **For Further Practice:** Questions 14.77 and 14.78.

Hydrolysis of Esters

LEARNING GOAL

9 Write equations representing the hydrolysis of an ester.

Hydrolysis, sometimes also referred to as *hydration,* refers to cleavage of any bond by the addition of a water molecule. Esters undergo hydrolysis reactions in water, as shown in the general reaction:

$$\underset{\text{Ester}}{R^1\!-\!\overset{\overset{\displaystyle O}{\|}}{C}\!-\!OR^2} + \underset{\text{Water}}{H_2O} \underset{}{\overset{H^+,\ heat}{\rightleftharpoons}} \underset{\substack{\text{Carboxylic} \\ \text{acid}}}{R^1\!-\!\overset{\overset{\displaystyle O}{\|}}{C}\!-\!OH} + \underset{\text{Alcohol}}{R^2OH}$$

This reaction requires heat. A small amount of acid (H^+) may be added to catalyze the reaction, as in the following example:

$$\underset{\substack{\text{Propyl propanoate} \\ \text{(propyl propionate)}}}{CH_3CH_2\overset{\overset{\displaystyle O}{\|}}{C}\!-\!OCH_2CH_2CH_3} + H_2O \overset{H^+,\ heat}{\rightleftharpoons} \underset{\substack{\text{Propanoic acid} \\ \text{(propionic acid)}}}{CH_3CH_2\overset{\overset{\displaystyle O}{\|}}{C}\!-\!OH} + \underset{\substack{\text{1-Propanol} \\ \text{(propanol)}}}{CH_3CH_2CH_2OH}$$

LEARNING GOAL

10 Define the term *saponification* and describe how soap works in the emulsification of grease and oil.

The base-catalyzed hydrolysis of an ester is called **saponification.**

$$\underset{\text{Ester}}{R^1\!-\!\overset{\overset{\displaystyle O}{\|}}{C}\!-\!OR^2} + \underset{\text{Water}}{H_2O} \overset{NaOH,\ heat}{\longrightarrow} \underset{\substack{\text{Carboxylic} \\ \text{acid salt}}}{R^1\!-\!\overset{\overset{\displaystyle O}{\|}}{C}\!-\!O^-Na^+} + \underset{\text{Alcohol}}{R^2OH}$$

Under basic conditions the acid cannot exist. Thus, the reaction yields the salt of the carboxylic acid having the cation of the basic catalyst.

$$\underset{\substack{\text{Butyl ethanoate} \\ \text{(butyl acetate)}}}{CH_3\overset{\overset{\displaystyle O}{\|}}{C}\!-\!OCH_2CH_2CH_2CH_3} \overset{NaOH,\ heat}{\longrightarrow} \underset{\substack{\text{Sodium ethanoate} \\ \text{(sodium acetate)}}}{CH_3\overset{\overset{\displaystyle O}{\|}}{C}\!-\!O^-Na^+} + \underset{\substack{\text{1-Butanol} \\ \text{(butyl alcohol)}}}{CH_3CH_2CH_2CH_2OH}$$

The carboxylic acid is formed when the reaction mixture is neutralized with an acid such as HCl.

$$\underset{\substack{\text{Sodium ethanoate} \\ \text{(sodium acetate)}}}{CH_3\overset{\overset{\displaystyle O}{\|}}{C}\!-\!O^-Na^+} + HCl \longrightarrow \underset{\substack{\text{Ethanoic acid} \\ \text{(acetic acid)}}}{CH_3\overset{\overset{\displaystyle O}{\|}}{C}\!-\!OH} + NaCl$$

Complete each of the following reactions by drawing the structure of the missing products.

a. $CH_3\overset{\overset{O}{\|}}{C}-OCH_2CH_2CH_3 + H_2O \overset{H^+, heat}{\rightleftharpoons} ?$

b. $CH_3CH_2CH_2CH_2CH_2\overset{\overset{O}{\|}}{C}-OCH_2CH_2CH_3 + H_2O \overset{KOH, heat}{\longrightarrow} ?$

c. $CH_3CH_2CH_2CH_2\overset{\overset{O}{\|}}{C}-OCH_3 + H_2O \overset{NaOH, heat}{\longrightarrow} ?$

d. $CH_3CH_2CH_2CH_2CH_2\overset{\overset{O}{\|}}{C}-OCHCH_2CH_2CH_3 + H_2O \overset{H^+, heat}{\rightleftharpoons} ?$
    $\qquad\qquad\qquad\qquad\qquad\quad |$
    $\qquad\qquad\qquad\qquad\quad\ CH_3$

Use the I.U.P.A.C. Nomenclature System to name each of the products in Question 14.7.

Fats and oils are triesters of the alcohol glycerol. When they are hydrolyzed by saponification, the products are **soaps,** which are the salts of long-chain carboxylic acids (fatty acid salts). According to Roman legend, soap was discovered by wash-erwomen following a heavy rain on Mons Sapo ("Mount Soap"). An important sacrificial altar was located on the mountain. The rain mixed with the remains of previous animal sacrifices—wood ash and animal fat—at the base of the altar. Thus, the three substances required to make soap accidentally came together— water, fat, and alkali (potassium carbonate and potassium hydroxide, called *pot-ash,* leached from the wood ash). The soap mixture flowed down the mountain and into the Tiber River, where the washerwomen quickly realized its value.

Triesters of glycerol are more commonly referred to as triglycerides. We know them as solid fats, generally from animal sources, and liquid oils, typically from plants. We will study triglycerides in detail in Section 17.3.

We still use the old Roman recipe to make soap from water, a strong base, and natural fats and oils obtained from animals or plants. The carbon chain length of the fatty acid salts governs the solubility of a soap. The lower-molecular-weight carboxylic acid salts (up to twelve carbons) have greater solubility in water and give a lather containing large bubbles. The higher-molecular-weight carboxylic acid salts (fourteen to twenty carbons) are much less soluble in water and produce a lather with fine bubbles. The nature of the cation also affects the solubility of the soap. In general, the potassium salts of carboxylic acids are more soluble in water than the sodium salts. The synthesis of a soap is shown in Figure 14.3.

The role of soap in the removal of soil and grease is best understood by con-sidering the functional groups in soap molecules and studying the way in which they interact with oil and water. The long, continuous hydrocarbon side chains of soap molecules resemble alkanes, and they dissolve other nonpolar compounds such as oils and greases ("like dissolves like"). The large nonpolar hydrocarbon part of the molecule is described as *hydrophobic,* which means "water-fearing." This part of the molecule is repelled by water. The highly polar carboxylate end of the molecule is called *hydrophilic,* which means "water-loving."

When soap is dissolved in water, the carboxylate end actually dissolves. The hydrocarbon part is repelled by the water molecules so that a thin film of soap is formed on the surface of the water with the hydrocarbon chains pro-truding outward. When soap solution comes in contact with oil or grease, the hydrocarbon part dissolves in the oil or grease, but the polar carboxylate group remains dissolved in water. When particles of oil or grease are surrounded by

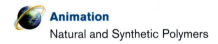

$$\underset{\substack{\text{Fat or oil}\\(\text{triglyceride})}}{\begin{array}{l}CH_2-O-\overset{\overset{\textstyle O}{\|}}{C}-R^1\\[2mm]CH-O-\overset{\overset{\textstyle O}{\|}}{C}-R^2\\[2mm]CH_2-O-\overset{\overset{\textstyle O}{\|}}{C}-R^3\end{array}}\quad\xrightarrow[\substack{\text{heat,}\\H_2O}]{M^+OH^-}\quad\underset{\text{Glycerol}}{\begin{array}{l}CH_2-OH\\[2mm]CH-OH\\[2mm]CH_2-OH\end{array}}\;+\;\underset{\substack{\text{Soap}\\(\text{Mixture of carboxylic}\\\text{acid salts})}}{R^1-\overset{\overset{\textstyle O}{\|}}{C}-O^-M^+ + R^2-\overset{\overset{\textstyle O}{\|}}{C}-O^-M^+ + R^3-\overset{\overset{\textstyle O}{\|}}{C}-O^-M^+}$$

where $M^+ = Na^+$ or K^+

Figure 14.3
Saponification is the base-catalyzed hydrolysis of a glycerol triester.

A more detailed diagram of a micelle is found in Figure 23.1.

An emulsion is a suspension of very fine droplets of one liquid in another. In this case it is oil in water.

🌐 **Animation**
Natural and Synthetic Polymers

soap molecules, the resulting "units" formed are called *micelles*. A simplified view of this phenomenon is shown in Figure 14.4.

Micelles repel one another because they are surrounded on the surface by the negatively charged carboxylate ions. Mechanical action (for example, scrubbing or tumbling in a washing machine) causes oil or grease to be surrounded by soap molecules and broken into small droplets so that relatively small micelles are formed. These small micelles are then washed away. Careful examination of this solution shows that it is an *emulsion* containing suspended micelles.

Condensation Polymers

As we saw in Chapter 11, *polymers* are macromolecules, very large molecules. They result from the combination of many smaller molecules, usually in a repeating pattern, to give molecules whose molar mass may be 10,000 g/mol or greater. The small molecules that make up the polymer are called *monomers*.

A polymer may be made from a single type of monomer. Such a polymer, called a homopolymer, would have the following general structure:

chain continues~A-A-A-A-A-A-A-A-A-A-A-A-A-A-A-A-A-A-A-A-A-A-A-A-A~chain continues

Figure 14.4
Simplified view of the action of a soap. The wiggly lines represent the long, continuous carbon chains of each soap molecule. (a) The thin film of soap molecules that forms at the water surface reduces surface tension. (b) Particles of oil and grease are surrounded by soap molecules to form a micelle.

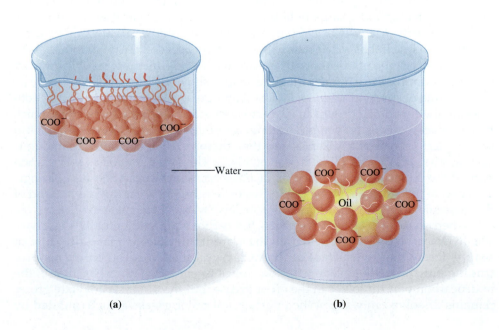

(a)   (b)

The addition polymers of alkenes that we studied in Chapter 11 are examples of this type of polymer. Alternatively, two different monomers may be copolymerized, producing a heteropolymer with the following structure:

chain continues~A-B-A-B-A-B-A-B-A-B-A-B-A-B~chain continues

Polyesters are heteropolymers. They are also known as condensation polymers. **Condensation polymers** are formed by the polymerization of monomers in a reaction that forms a small molecule such as water or an alcohol. Polyesters are synthesized by reacting a dicarboxylic acid and a dialcohol (diol). Notice that each of the combining molecules has two reactive functional groups, highlighted in red here:

Terephthalic acid   1,2-Ethanediol

Another molecule of terephthalic acid can react here.

+ H_2O

Another molecule of 1,2-ethanediol can react here.

Reaction continues

Polyethylene terephthalate
PETE

Each time a pair of molecules reacts using one functional group from each, a new molecule is formed that still has two reactive groups. The product formed in this reaction is polyethylene terephthalate, or PETE.

When formed as fibers, polyesters are used to make fabric for clothing. These polyesters were trendy in the 1970s, during the "disco" period, but lost their popularity soon thereafter. Polyester fabrics, and a number of other synthetic polymers used in clothing, have become even more fashionable since the introduction of microfiber technology. The synthetic polymers are extruded into fibers that are only half the diameter of fine silk fibers. When these fibers are used to create fabrics, the result is a fabric that drapes freely yet retains its shape. These fabrics are generally lightweight, wrinkle resistant, and remarkably strong.

Polyester can be formed into a film called Mylar. These films, coated with aluminum foil, are used to make balloons that remain inflated for long periods. They are also used as the base for recording tapes and photographic film.

PETE can be used to make shatterproof plastic bottles, such as those used for soft drinks. However, these bottles cannot be recycled and reused directly because they cannot withstand the high temperatures required to sterilize them. PETE can't be used for any foods, such as jellies, that must be packaged at high temperatures. For these uses, a new plastic, PEN, or polyethylene naphthalate, is used.

Biodegradable plastics made of both homopolymers and heteropolymers are discussed in An Environmental Perspective: Garbage Bags from Potato Peels? found on page 476.

$$\left[\overset{\overset{\displaystyle O}{\|}}{-C} - \bigcirc\bigcirc - \overset{\overset{\displaystyle O}{\|}}{C} - O - CH_2CH_2 - \right]_n$$

Naphthalate group　　　　　Ethylene group

14.3 Acid Chlorides and Acid Anhydrides

Acid Chlorides

LEARNING GOAL
11 Determine the common and I.U.P.A.C. names of acid chlorides.

Acid chlorides are carboxylic acid derivatives having the general formula

$$R - \overset{\overset{\displaystyle O}{\|}}{C} - Cl$$

They are named by replacing the *-ic acid* ending of the common name with *-yl chloride* or the *-oic acid* ending of the I.U.P.A.C. name of the carboxylic acid with *-oyl chloride*. For example,

$$CH_3CH_2CH_2\overset{\overset{\displaystyle O}{\|}}{C} - Cl$$

Butanoyl chloride
(butyryl chloride)

$$CH_3\overset{\overset{\displaystyle O}{\|}}{C} - Cl$$

Ethanoyl chloride
(acetyl chloride)

$$\underset{\underset{\displaystyle Br}{|}}{CH_2}CH_2\overset{\overset{\displaystyle O}{\|}}{C} - Cl$$

3-Bromopropanoyl chloride
(β-bromopropionyl chloride)

$$Cl - \bigcirc - \overset{\overset{\displaystyle O}{\|}}{C} - Cl$$

4-Chlorobenzoyl chloride
(*p*-chlorobenzoyl chloride)

Acid chlorides are noxious, irritating chemicals and must be handled with great care. They are slightly polar and boil at approximately the same temperature as the corresponding aldehyde or ketone of comparable molecular weight. They react violently with water and therefore cannot be dissolved in that solvent. Acid chlorides have little commercial value other than their utility in the synthesis of esters and amides, two of the other carboxylic acid derivatives.

Acid chlorides are prepared from the corresponding carboxylic acid by reaction with one of several inorganic acid chlorides, including PCl_3, PCl_5, or $SOCl_2$. The general reaction is summarized in the following equation:

LEARNING GOAL
12 Write equations representing the synthesis of acid chlorides.

$$R - \overset{\overset{\displaystyle O}{\|}}{C} - OH \xrightarrow{\underset{\text{chloride}}{\text{inorganic acid}}} R - \overset{\overset{\displaystyle O}{\|}}{C} - Cl + \text{inorganic products}$$

Carboxylic
acid

Acid chloride

The following equations show the synthesis of ethanoyl chloride and benzoyl chloride:

$$CH_3\overset{\overset{\displaystyle O}{\|}}{C} - OH \xrightarrow[\underset{\text{trichloride)}}{\text{(Phosphorus}}]{PCl_3} CH_3\overset{\overset{\displaystyle O}{\|}}{C} - Cl + \text{inorganic products}$$

Ethanoic acid　trichloride)
(acetic acid)

Ethanoyl chloride
(acetyl chloride)

$$\bigcirc - \overset{\overset{\displaystyle O}{\|}}{C} - OH \xrightarrow[\underset{\text{chloride)}}{\text{(Thionyl}}]{SOCl_2} \bigcirc - \overset{\overset{\displaystyle O}{\|}}{C} - Cl + \text{inorganic products}$$

Benzoic acid

Benzoyl chloride

| Writing Equations Representing the Synthesis of Acid Chlorides | E X A M P L E 14.10 |

Write equations representing the following reactions:

a. Butanoic acid with phosphorus trichloride
b. 2-Bromobenzoic acid with thionyl chloride

LEARNING GOAL

12 Write equations representing the synthesis of acid chlorides.

Solution

a. Begin by drawing the structure of the reactant, butanoic acid. The inorganic acid chloride (phosphorus trichloride) can be represented over the reaction arrow. The product is drawn by replacing the —OH of the carboxyl group with —Cl. This gives us the following equation:

$$CH_3CH_2CH_2\overset{\overset{\displaystyle O}{\|}}{C}-OH \xrightarrow{PCl_3} CH_3CH_2CH_2\overset{\overset{\displaystyle O}{\|}}{C}-Cl + \text{inorganic products}$$

b. Begin by drawing the structure of the reactant, 2-bromobenzoic acid. The inorganic acid chloride (thionyl chloride) can be represented over the reaction arrow. The product is drawn by replacing the —OH of the carboxyl group with —Cl. This gives us the following equation:

Practice Problem 14.10

Write equations representing the synthesis of (a) butanoyl chloride and (b) hexanoyl chloride.

For Further Practice: Questions 14.91 and 14.92.

| Naming Acid Chlorides | E X A M P L E 14.11 |

Name the products of the reactions shown in Example 14.10.

LEARNING GOAL

11 Determine the common and I.U.P.A.C. names of acid chlorides.

Solution

$$CH_3CH_2CH_2\overset{\overset{\displaystyle O}{\|}}{C}-OH \xrightarrow{PCl_3} CH_3CH_2CH_2\overset{\overset{\displaystyle O}{\|}}{C}-Cl + \text{inorganic products}$$

Butanoic acid
(butyric acid)

The carboxylic acid that is the reactant in this reaction is butanoic acid (butyric acid). By dropping the *-oic acid* ending of the I.U.P.A.C. name (or the *-ic acid* ending of the common name) and replacing it with *-oyl chloride* (or *-yl chloride*), we can write the I.U.P.A.C. and common names of the product of this reaction. They are butanoyl chloride and butyryl chloride, respectively.

Continued—

EXAMPLE 14.11 —*Continued*

2-Bromobenzoic acid
(*o*-bromobenzoic acid)

The carboxylic acid in this reaction is 2-bromobenzoic acid. By dropping the *-oic acid* ending of the I.U.P.A.C. name and replacing it with *-oyl chloride*, we can name the product of this reaction: 2-bromobenzoyl chloride. It is equally correct to name this compound *o*-bromobenzoyl chloride.

Practice Problem 14.11

Provide the I.U.P.A.C. name for each of the following acid chlorides.

a. CH₃CHCH₂C—Cl
        |
        Br

b. CH₃CH₂CHCH₂C—Cl
            |
            OH

c. CH₃CHCH₂CH₂CH₂CHC—Cl
        | |
        Cl   OH

d. CH₃CHC—Cl
        |
        OH

For Further Practice: Questions 14.97 and 14.98.

The reaction that occurs when acid chlorides react violently with water is hydrolysis. The products are the carboxylic acid and hydrochloric acid.

CH₃C—Cl + H₂O ⟶ CH₃C—OH + HCl

Ethanoyl chloride
(acetyl chloride)

Ethanoic acid
(acetic acid)

Question 14.9

Write an equation showing the synthesis of each of the following acid chlorides. Provide the I.U.P.A.C. names of the carboxylic acid reactants and the acid chloride products.

a. CH₃CHC—Cl
        |
        CH₃

b. CH₃CH₂CH₂CH₂CH₂C—Cl

Question *14.10*

Write an equation showing the synthesis of each of the following acid chlorides. Provide the I.U.P.A.C. names of the carboxylic acid reactants and the acid chloride products.

a. CH₃C—Cl (with =O on carbon)

b. CH₃CH₂CHCH₂CH₂C—Cl (with =O on terminal carbon, and Br substituent)

Question *14.11*

Write an equation showing the synthesis of each of the following acid chlorides. Provide the common names of the carboxylic acid reactants and the acid chloride products.

a. H—C—Cl (with =O on carbon)

b. CH₃CH₂C—Cl (with =O on carbon)

Question *14.12*

Write an equation showing the synthesis of each of the following acid chlorides. Provide the common names of the carboxylic acid reactants and the acid chloride products.

a. CH₃CH₂CHC—Cl (with =O on carbon, CH₃ substituent)

b. CH₃CH₂CH₂CH₂CH₂CH₂C—Cl (with =O on carbon)

Acid Anhydrides

Acid anhydrides are molecules with the following general formula:

$$R^1—C—O—C—R^2$$ (with =O on each carbon)

The name of the family is really quite fitting. The structure above reveals that acid anhydrides are actually two carboxylic acid molecules with a water molecule removed. The word *anhydride* means "without water."

$$R^1—C—O—H + HO—C—R^2$$ (with =O on each carbon)

↓

$$H—OH + R^1—C—O—C—R^2$$ (with =O on each carbon)

Acid anhydrides are classified as *symmetrical* if both acyl groups are the same. Symmetrical acid anhydrides are named by replacing the *acid* ending of the carboxylic acid with the word *anhydride*. For example,

LEARNING GOAL

13 Determine the common and I.U.P.A.C. names of acid anhydrides.

Ethanoic anhydride
(acetic anhydride)

Benzoic anhydride

Unsymmetrical anhydrides are those having two different acyl groups. They are named by arranging the names of the two parent carboxylic acids and following them with the word *anhydride*. The names of the carboxylic acids may be arranged by size or alphabetically. For example:

$$\underset{\substack{\text{Ethanoic propanoic anhydride} \\ \text{(acetic propionic anhydride)}}}{CH_3\overset{\displaystyle O}{\overset{\|}{C}}-O-\overset{\displaystyle O}{\overset{\|}{C}}CH_2CH_3} \qquad \underset{\substack{\text{Ethanoic pentanoic anhydride} \\ \text{(acetic valeric anhydride)}}}{CH_3\overset{\displaystyle O}{\overset{\|}{C}}-O-\overset{\displaystyle O}{\overset{\|}{C}}CH_2CH_2CH_2CH_3}$$

LEARNING GOAL

14 Write equations representing the synthesis of acid anhydrides.

Most acid anhydrides cannot be formed in a reaction between the parent carboxylic acids. One typical pathway for the synthesis of an acid anhydride is the reaction between an acid chloride and a carboxylate anion. This general reaction is seen in the equation below:

$$\underset{\substack{\text{Acid chloride}}}{R^1-\overset{\displaystyle O}{\overset{\|}{C}}-Cl} \quad \xrightarrow[\text{Carboxylate ion}]{R^2-\overset{\displaystyle O}{\overset{\|}{C}}-O^-} \quad \underset{\substack{\text{Acid anhydride}}}{R^1-\overset{\displaystyle O}{\overset{\|}{C}}-O-\overset{\displaystyle O}{\overset{\|}{C}}-R^2} \; + \; \underset{\substack{\text{Chloride} \\ \text{ion}}}{Cl^-}$$

The following equation shows the synthesis of ethanoic anhydride from ethanoic acid:

$$\underset{\substack{\text{Ethanoic acid} \\ \text{(acetic acid)}}}{CH_3\overset{\displaystyle O}{\overset{\|}{C}}-OH} \xrightarrow{SOCl_2} \underset{\substack{\text{Ethanoyl chloride} \\ \text{(acetyl chloride)}}}{CH_3\overset{\displaystyle O}{\overset{\|}{C}}-Cl} \xrightarrow{CH_3\overset{\displaystyle O}{\overset{\|}{C}}-O^-} \underset{\substack{\text{Ethanoic anhydride} \\ \text{(acetic anhydride)}}}{CH_3\overset{\displaystyle O}{\overset{\|}{C}}-O-\overset{\displaystyle O}{\overset{\|}{C}}CH_3}$$

Acid anhydrides readily undergo hydrolysis. The rate of the hydrolysis reaction may be increased by the addition of a trace of acid or hydroxide base to the solution.

$$\underset{\substack{\text{Propanoic anhydride} \\ \text{(propionic anhydride)}}}{CH_3CH_2\overset{\displaystyle O}{\overset{\|}{C}}-O-\overset{\displaystyle O}{\overset{\|}{C}}CH_2CH_3} + H_2O \xrightarrow{\text{Heat}} \underset{\substack{\text{Propanoic acid} \\ \text{(propionic acid)}}}{2CH_3CH_2\overset{\displaystyle O}{\overset{\|}{C}}-OH}$$

E X A M P L E 14.12   **Writing Equations Representing the Synthesis of Acid Anhydrides**

LEARNING GOAL

14 Write equations representing the synthesis of acid anhydrides.

Write an equation representing the synthesis of propanoic anhydride.

Solution

Propanoic anhydride can be synthesized in a reaction between propanoyl chloride and the propanoate anion. This gives us the following equation:

$$\underset{\substack{\text{Propanoyl} \\ \text{chloride}}}{CH_3CH_2\overset{\displaystyle O}{\overset{\|}{C}}-Cl} \xrightarrow[\text{Propanoate ion}]{CH_3CH_2\overset{\displaystyle O}{\overset{\|}{C}}-O^-} \underset{\substack{\text{Propanoic anhydride}}}{CH_3CH_2\overset{\displaystyle O}{\overset{\|}{C}}-O-\overset{\displaystyle O}{\overset{\|}{C}}CH_2CH_3} + \underset{\substack{\text{Chloride} \\ \text{ion}}}{Cl^-}$$

Continued—

EXAMPLE 14.12 —*Continued*

Practice Problem 14.12

Write equations representing the synthesis of (a) butanoic anhydride and (b) hexanoic anhydride.

For Further Practice: Questions 14.99 and 14.100.

Naming Acid Anhydrides   **EXAMPLE 14.13**

Write the I.U.P.A.C. and common names for each of the following acid anhydrides.

13 **LEARNING GOAL**
Determine the common and I.U.P.A.C. names of acid anhydrides.

$$\underset{\|}{\overset{\overset{\displaystyle O}{\|}}{CH_3CH_2CH_2C}}-O-\underset{\|}{\overset{\overset{\displaystyle O}{\|}}{C}}CH_2CH_2CH_3$$

Solution

This is a symmetrical acid anhydride. The I.U.P.A.C. name of the four-carbon parent carboxylic acid is butanoic acid (common name butyric acid). To name the anhydride, simply replace the word *acid* with the word *anhydride.* The I.U.P.A.C. name of this compound is butanoic anhydride (common name butyric anhydride).

$$\underset{\|}{\overset{\overset{\displaystyle O}{\|}}{CH_3C}}-O-\underset{\|}{\overset{\overset{\displaystyle O}{\|}}{C}}CH_2CH_2CH_2CH_2CH_3$$

Solution

This is an unsymmetrical anhydride. The I.U.P.A.C. names of the two parent carboxylic acids are ethanoic acid (two-carbon) and hexanoic acid (six-carbon). To name an unsymmetrical anhydride, the term *anhydride* is preceded by the names of the two parent acids. The I.U.P.A.C. name of this compound is ethanoic hexanoic anhydride. The common names of the two parent carboxylic acids are acetic acid and caproic acid. Thus, the common name of this compound is acetic caproic anhydride.

Practice Problem 14.13

Write the common and I.U.P.A.C. names for each of the following acid anhydrides.

a. $\underset{\|}{\overset{\overset{\displaystyle O}{\|}}{CH_3CH_2CH_2C}}-O-\underset{\|}{\overset{\overset{\displaystyle O}{\|}}{C}}CH_2CH_2CH_2CH_2CH_3$ b. $\underset{\|}{\overset{\overset{\displaystyle O}{\|}}{CH_3C}}-O-\underset{\|}{\overset{\overset{\displaystyle O}{\|}}{C}}CH_2CH_2CH_2CH_3$

c. $\underset{\|}{\overset{\overset{\displaystyle O}{\|}}{CH_3CH_2CH_2CH_2C}}-O-\underset{\|}{\overset{\overset{\displaystyle O}{\|}}{C}}CH_2CH_3$ d. $\underset{\|}{\overset{\overset{\displaystyle O}{\|}}{CH_3CH_2C}}-O-\underset{\|}{\overset{\overset{\displaystyle O}{\|}}{C}}CH_3$

For Further Practice: Questions 14.95 and 14.96.

Acid anhydrides can also react with an alcohol. This reaction produces an ester and a carboxylic acid. This is an example of an acyl group transfer reaction, as shown in the following general reaction:

$$R{-}OH + R{-}\overset{\overset{O}{\|}}{C}{-}O{-}\overset{\overset{O}{\|}}{C}{-}R \longrightarrow R{-}\overset{\overset{O}{\|}}{C}{-}OR + R{-}\overset{\overset{O}{\|}}{C}{-}OH$$

Alcohol   Acid anhydride   Ester   Carboxylic
                                                            acid

The acyl group of the acid anhydride is transferred to the oxygen of the alcohol in this reaction. The alcohol and anhydride reactants and ester product are described below. The carboxylic acid product is omitted.

Carbon-oxygen
bond remains
intact

Acyl group transferred
to oxygen of the alcohol
reactant

$$R{-}O{-}H \qquad + \qquad R'{-}\overset{\overset{O}{\|}}{C}{-}O{-}\overset{\overset{O}{\|}}{C}{-}R''$$

Alcohol reactant   Anhydride reactant

Acyl group from
acid anhydride

Oxygen from
the alcohol
reactant

$$R'{-}\overset{\overset{O}{\|}}{C}{-}O{-}R$$

R group from
alcohol

Ester product

Other acyl group donors include thioesters and esters. As we will see in the final section of this chapter, acyl transfer reactions are very important in nature, particularly in the pathways responsible for breakdown of food molecules and harvesting cellular energy.

Question 14.13

Write an equation showing the synthesis of each of the following acid anhydrides. Provide the I.U.P.A.C. names of the acid chloride and carboxylate anion reactants and the acid anhydride product.

a. $CH_3CHCH_2\overset{\overset{O}{\|}}{C}{-}O{-}\overset{\overset{O}{\|}}{C}CH_2CHCH_3$
       | |
      CH_3 CH_3

b. $H{-}\overset{\overset{O}{\|}}{C}{-}O{-}\overset{\overset{O}{\|}}{C}{-}CH_3$

Question 14.14

Write an equation showing the synthesis of each of the following acid anhydrides. Provide the common names of the acid chloride and carboxylate anion reactants and the acid anhydride product.

a.

b.

14.4 Nature's High-Energy Compounds: Phosphoesters and Thioesters

An alcohol can react with phosphoric acid to produce a phosphate ester, or **phosphoester**, as in

| Alcohol | Phosphoric acid | | Phosphate ester | Water |

LEARNING GOAL

15 Discuss the significance of thioesters and phosphoesters in biological systems.

Phosphoesters of simple sugars or monosaccharides are very important in the energy-harvesting biochemical pathways that provide energy for all life functions. One such pathway is *glycolysis.* This pathway is the first stage in the breakdown of sugars. The first reaction in this pathway is the formation of a phosphoester of the six-carbon sugar, glucose. The phosphorylation of glucose to produce glucose-6-phosphate is represented in the following equation:

The many phosphorylated intermediates in the metabolism of sugars will be discussed in Chapter 21.

β-D-Glucose   β-D-Glucose-6-phosphate

In this reaction the source of the phosphoryl group is **adenosine triphosphate (ATP)**, which is the universal energy currency for all living organisms. As such, ATP is used to store energy released in cellular metabolic reactions and to provide the energy required for most of the reactions that occur in the cell. The transfer of a phosphoryl group from ATP to glucose "energizes" the glucose molecule in preparation for other reactions of the pathway.

ATP consists of a nitrogenous base (adenine) and a phosphate ester of the five-carbon sugar ribose (Figure 14.5). The triphosphate group attached to ribose is made up of three phosphate groups bonded to one another by phosphoanhydride bonds. When two phosphate groups react with one another, a water molecule is lost. Because water is lost, the resulting bond is called a **phosphoanhydride** bond.

In fact the word *glycolysis* comes from two Greek words that mean "splitting sugars" (*glykos*, "sweet," and *lysis*, "to split"). In this pathway, the six-carbon sugar glucose is split, and then oxidized, to produce two three-carbon molecules, called *pyruvate.*

Phosphoryl is the term used to describe the functional group derived from phosphoric acid that is part of another molecule.

A Human Perspective

Carboxylic Acid Derivatives of Special Interest

Analgesics (pain killers) and antipyretics (fever reducers)

Aspirin *(acetylsalicylic acid),* derived from the bark of the willow tree, is the most widely used drug in the world. Hundreds of millions of dollars are spent annually on this compound. It is used primarily as a pain reliever (analgesic) and in the reduction of fever (antipyretic). Aspirin is among the drugs often referred to as NSAIDS, or nonsteroidal anti-inflammatory drugs. These drugs inhibit the inflammatory response by inhibiting an enzyme called cyclooxygenase, which is the first enzyme in the pathway for the synthesis of prostaglandins. Prostaglandins are responsible, in part, for pain and fever. Thus, aspirin and other NSAIDS reduce pain and fever by decreasing prostaglandin synthesis. Aspirin's side effects are a problem for some individuals. Because aspirin inhibits clotting, its use is not recommended during pregnancy, nor should it be used by individuals with ulcers. In those instances, *acetaminophen,* found in the over-the-counter pain-reliever Tylenol, is often prescribed.

The search for NSAIDS that are more effective and yet gentler on the stomach has provided two new analgesics for the over-the-counter market. These are ibuprofen (sold as Motrin, Advil, Nuprin) and naproxen (sold as Naprosyn, Naprelan, Anaprox, and Aleve).

Many types of over-the-counter pain relievers are available.

Some common analgesics.

$$RO-\overset{\overset{O}{\|}}{\underset{\underset{OH}{|}}{P}}-OH + HO-\overset{\overset{O}{\|}}{\underset{\underset{OH}{|}}{P}}-OH \longrightarrow RO-\overset{\overset{O}{\|}}{\underset{\underset{OH}{|}}{P}}-O-\overset{\overset{O}{\|}}{\underset{\underset{OH}{|}}{P}}-OH + H_2O$$

Phosphate ester   Phosphate group   Phosphoanhydride bond

The functions and properties of ATP in energy metabolism are discussed in Section 21.1.

The energy of ATP is made available through hydrolysis of either of the two phosphoanhydride bonds, as shown in Figure 14.5. This is an exothermic process; that is, energy is given off. When the phosphoryl group is transferred to another molecule—for instance, glucose—some of that energy resides in the phosphorylated sugar, thereby "energizing" it. The importance of ATP as an energy source becomes apparent when we realize that we synthesize and break down an amount of ATP equivalent to our body weight each day.

Pheromones

Pheromones, chemicals secreted by animals, influence the behavior of other members of the same species. They often represent the major means of communication among simpler animals. The term *pheromone* literally means "to carry" and "to excite" (Greek, *pherein*, to carry; Greek, *horman*, to excite). They are chemicals carried or shed by one member of the species and used to alert other members of the species.

Pheromones may be involved in sexual attraction, trail marking, aggregation or recruitment, territorial marking, or signaling alarm. Others may be involved in defense or in species socialization—for example, designating various classes within the species as a whole. Among all of the pheromones, insect pheromones have been the most intensely studied. Many of the insect pheromones are carboxylic acids or acid derivatives, as seen here:

9-Keto-*trans*-2-decenoic acid
(queen bee socializing/royalty pheromone)

cis-7-Dodecenyl acetate
(cabbage looper sex pheromone)

$CH_3CH_2CH=CH(CH_2)_9CH_2OCCH_3$

Tetracecenyl acetate
(European corn borer sex pheromone)

For Further Understanding

A nonsteroidal anti-inflammatory drug can cause side effects if used over a long period of time. Do some research on the physiological roles of prostaglandins to develop hypotheses concerning these side effects.

How might you make use of pheromones to control pests such as the corn borer?

Cellular enzymes can carry out a reaction between a thiol and a carboxylic acid to produce a **thioester**:

Thiols are described in Section 12.9.

$$R^1-S-\overset{O}{\overset{\|}{C}}-R^2$$

Thioester

The reactions that produce thioesters are essential in energy-harvesting pathways as a means of "activating" acyl groups for subsequent breakdown reactions. The complex thiol coenzyme A is the most important acyl group activator in the cell. The detailed structure of coenzyme A appears in Section 12.9, but it is generally abbreviated CoA—SH to emphasize the importance of the sulfhydryl group. The most common thioester is the acetyl ester, called **acetyl coenzyme A** (acetyl CoA).

Figure 14.5

The hydrolysis of the phosphoanhydride bond of ATP is accompanied by the release of energy that is used for biochemical reactions in the cell.

β-**Oxidation is the pathway for the breakdown of fatty acids. Like glycolysis, it is an energy-harvesting pathway.**

The acyl group of a carboxylic is named by replacing the -oic acid **or** -ic **suffix with** -yl. **For instance, the acyl group of acetic acid is the acetyl group:**

Acetyl coenzyme A
(acetyl CoA)

Glycolysis, β-oxidation, and the citric acid cycle are cellular energy-harvesting pathways that we will study in Chapters 21, 22, and 23.

Acetyl CoA carries the acetyl group from glycolysis or β-oxidation of a fatty acid to an intermediate of the citric acid cycle. This reaction is an example of an acyl group transfer reaction. In this case, the acyl group donor is a thioester—acetyl CoA. The acyl group being transferred is the acetyl group, which is transferred to the carbonyl carbon of oxaloacetate. This reaction follows:

As we will see in Chapter 22, the citric acid cycle is an energy-harvesting pathway that completely oxidizes the acetyl group to two CO_2 molecules. The electrons that are harvested in the process are used to produce large amounts of ATP. Coenzyme A also serves to activate the acyl group of fatty acids during β-oxidation, the pathway by which fatty acids are oxidized to produce ATP.

Summary of Reactions

Preparation of Carboxylic Acids

Dissociation of Carboxylic Acids

Neutralization of Carboxylic Acids

$$R-\overset{\overset{\displaystyle O}{\|}}{C}-OH + NaOH \longrightarrow R-\overset{\overset{\displaystyle O}{\|}}{C}-O^-Na^+ + H_2O$$

| Carboxylic acid | Strong base | | Carboxylic acid salt | Water |

Esterification

$$R^1-\overset{\overset{\displaystyle O}{\|}}{C}-OH + R^2OH \underset{\Longleftarrow}{\overset{H^+,\,heat}{\Longrightarrow}} R^1-\overset{\overset{\displaystyle O}{\|}}{C}-OR^2 + H_2O$$

| Carboxylic acid | Alcohol | | Ester | Water |

Acid Hydrolysis of Esters

$$R^1-\overset{\overset{\displaystyle O}{\|}}{C}-OR^2 + H_2O \underset{\Longleftarrow}{\overset{H^+,\,heat}{\Longrightarrow}} R^1-\overset{\overset{\displaystyle O}{\|}}{C}-OH + R^2OH$$

| Ester | Water | | Carboxylic acid | Alcohol |

Saponification

$$R^1-\overset{\overset{\displaystyle O}{\|}}{C}-OR^2 + H_2O \xrightarrow{\overset{NaOH,}{heat}} R^1-\overset{\overset{\displaystyle O}{\|}}{C}-O^-Na^+ + R^2OH$$

| Ester | Water | | Carboxylic acid salt | Alcohol |

Synthesis of Acid Chlorides

$$R-\overset{\overset{\displaystyle O}{\|}}{C}-OH \xrightarrow{\overset{inorganic}{\underset{chloride}{acid}}} R-\overset{\overset{\displaystyle O}{\|}}{C}-Cl + \overset{inorganic}{products}$$

| Carboxylic acid | | Acid chloride |

Synthesis of Acid Anhydrides

$$R^1-\overset{\overset{\displaystyle O}{\|}}{C}-Cl \quad R^2-\overset{\overset{\displaystyle O}{\|}}{C}-O^- \quad R^1-\overset{\overset{\displaystyle O}{\|}}{C}-O-\overset{\overset{\displaystyle O}{\|}}{C}-R^2 + Cl^-$$

| Acid chloride | carboxylate ion | Acid anhydride | Chloride ion |

Formation of a Phosphoester

$$ROH + HO-\overset{\overset{\displaystyle O}{\|}}{\underset{\underset{\displaystyle OH}{|}}{P}}-OH \longrightarrow R-O-\overset{\overset{\displaystyle O}{\|}}{\underset{\underset{\displaystyle OH}{|}}{P}}-OH + H_2O$$

| Alcohol | Phosphoric acid | Phosphate ester | Water |

SUMMARY

14.1 Carboxylic Acids

The functional group of the *carboxylic acids* is the *carboxyl group* (—COOH). Because the carboxyl group is extremely polar and carboxylic acids can form intermolecular hydrogen bonds, they have higher boiling points and melting points than alcohols. The lower-molecular-weight carboxylic acids are water soluble and tend to taste sour and have unpleasant aromas. The longer-chain carboxylic acids are called *fatty acids*. Carboxylic acids are named (I.U.P.A.C.) by replacing the *-e* ending of the parent compound with *-oic acid*. Common names are often derived from the source of the carboxylic acid. They are synthesized by the oxidation of primary alcohols or aldehydes. Carboxylic acids are weak acids. They are neutralized by strong bases to form salts. *Soaps* are salts of long-chain carboxylic acids (fatty acids).

14.2 Esters

Esters are mildly polar and have pleasant aromas. The boiling points and melting points of esters are comparable to those of aldehydes and ketones. Esters are formed from the reaction between a carboxylic acid and an alcohol. They can undergo hydrolysis back to the parent carboxylic acid and alcohol. The base-catalyzed hydrolysis of an ester is called *saponification*.

14.3 Acid Chlorides and Acid Anhydrides

Acid chlorides are noxious chemicals formed in the reaction of a carboxylic acid and reagents such as PCl_3 or $SOCl_2$. *Acid anhydrides* are formed by the combination of an acid chloride and a carboxylate anion. Acid anhydrides can react with an alcohol to produce an ester and a carboxylic acid. This is an example of an acyl group transfer reaction.

14.4 Nature's High-Energy Compounds: Phosphoesters and Thioesters

An alcohol can react with phosphoric acid to produce a phosphate ester *(phosphoester)*. When two phosphate groups are joined, the resulting bond is a *phosphoanhydride* bond. These two functional groups are important to the structure and function of *adenosine triphosphate (ATP)*, the universal energy currency of all cells. Cellular enzymes can carry out a reaction between a thiol and a carboxylic acid to produce a *thioester*. This reaction is essential for the activation of acyl groups in carbohydrate and fatty acid metabolism. Coenzyme A is the most important thiol involved in these pathways.

KEY TERMS

acetyl coenzyme A (14.4)

acid anhydride (14.3)

acid chloride (14.3)

acyl group (Intro)

adenosine triphosphate (ATP) (14.4)

carboxyl group (14.1)

carboxylic acid (14.1)

carboxylic acid derivative (14.2)

condensation polymer (14.2)

ester (14.2)

fatty acid (14.1)

hydrolysis (14.2)

oxidation (14.1)

phosphoanhydride (14.4)

phosphoester (14.4)

saponification (14.2)

soap (14.2)

thioester (14.4)

QUESTIONS AND PROBLEMS

Carboxylic Acids: Structure and Properties

Foundations

14.15 The functional group is largely responsible for the physical and chemical properties of the various chemical families. Explain why a carboxylic acid is more polar and has a higher boiling point than an alcohol or an aldehyde of comparable molecular weight.

14.16 Explain why carboxylic acids are weak acids.

Applications

14.17 Which member of each of the following pairs has the lower boiling point?
 a. Hexanoic acid or 3-hexanone
 b. 3-Hexanone or 3-hexanol
 c. 3-Hexanol or hexane

14.18 Which member of each of the following pairs has the lower boiling point?
 a. Dipropyl ether or hexanal
 b. Hexanal or hexanoic acid
 c. Ethanol or ethanoic acid

14.19 Arrange the following from highest to lowest melting points:

14.20 Draw the condensed structural formula for each of the line drawings in Question 14.19 and provide the I.U.P.A.C. name for each.

14.21 Which member in each of the following pairs has the higher boiling point?
 a. Heptanoic acid or 1-heptanol
 b. Propanal or 1-propanol
 c. Methyl pentanoate or pentanoic acid
 d. 1-Butanol or butanoic acid

14.22 Which member in each of the following pairs is more soluble in water?

 a. $CH_3CH_2CH_2CH_2CH_2\overset{\displaystyle O}{\overset{\|}{C}}$—OH or

 $CH_3CH_2CH_2CH_2CH_2\overset{\displaystyle O}{\overset{\|}{C}}$—O⁻Na⁺

 b. $CH_3CH_2CH_2CH_2CH_2CH_2CH_2CH_2CH_3$ or
 $CH_3CH_2CH_2CH_2CH_2CH_2CH_2CH_2OH$

 c. CH_3CH_2—O—CH_2CH_3 or $CH_3CH_2\overset{\displaystyle O}{\overset{\|}{C}}$—$OCH_3$

 d. CH_3CH_2—O—CH_2CH_3 or $CH_3CH_2CH_2CH_2CH_2CH_3$

 e. Decanoic acid or ethanoic acid

14.23 Describe the properties of low-molecular-weight carboxylic acids.

14.24 What are some of the biological functions of the long chain carboxylic acids called fatty acids?

14.25 Why are citric acid and adipic acid added to some food products?

14.26 What is the function of lactic acid in food products? Of what significance is lactic acid in muscle metabolism?

Carboxylic Acids: Structure and Nomenclature

Foundations

14.27 Summarize the I.U.P.A.C. nomenclature rules for naming carboxylic acids.

14.28 Describe the rules for determining the common names of carboxylic acids.

Applications

14.29. Adipic acid occurs naturally in beets and is used as a food additive. What is the I.U.P.A.C. name for adipic acid? Why do you think that adipic acid is used as a food additive?

$HOOCCH_2CH_2CH_2CH_2COOH$

14.30 Propionic acid is a liquid fatty acid found in sweat and milk products. It is a bacterial fermentation product that gives the tangy flavor to Swiss cheese. What is the I.U.P.A.C. name of propionic acid?

$$CH_3CH_2COOH$$

14.31 Write the complete structural formulas for each of the following carboxylic acids:
   a. 2-Bromopentanoic acid
   b. 2-Bromo-3-methylbutanoic acid
   c. 2-Bromocyclohexanecarboxylic acid

14.32 Write the complete structural formulas for each of the following carboxylic acids:
   a. 2,6-Dichlorocyclohexanecarboxylic acid
   b. 2,4,6-Trimethylstearic acid
   c. Propenoic acid

14.33 Name each of the following carboxylic acids, using both the common and the I.U.P.A.C. Nomenclature Systems:

   a. H—C—OH (with =O above the C)

   b. CH₃CHCH₂C—OH (with CH₃ above first C, =O above last C)

   c.   C—OH (with =O above, attached to cyclopentane ring)

14.34 Name each of the following carboxylic acids, using both the common and I.U.P.A.C. Nomenclature Systems:

   a. CH₃CH₂CHCHCH₂C—OH (Br above 3rd C, CH₃ below 4th C, =O above last C)

   b. CH₃CH₂CHCH₂CH₂C—OH (CH₂CH₃ above 3rd C, =O above last C)

   c.   C—OH (with =O above, attached to cyclohexane ring with CH₃)

14.35 Write a complete structural formula and determine the I.U.P.A.C. name for each of the carboxylic acids of molecular formula $C_4H_8O_2$.

14.36 Write the general structure of an aldehyde, a ketone, a carboxylic acid, and an ester. What similarities exist among these structures?

14.37 Write the condensed structure of each of the following carboxylic acids:
   a. 4,4-Dimethylhexanoic acid
   b. 3-Bromo-4-methylpentanoic acid
   c. 2,3-Dinitrobenzoic acid
   d. 3-Methylcyclohexanecarboxylic acid

14.38 Use I.U.P.A.C. nomenclature to write the names for each of the following carboxylic acids:

   a. benzene ring with C—OH (=O above) and NO₂ substituent

   b. benzene ring with C—OH (=O above) and CH₂CH₃ substituent

   c. cyclopentane ring with C—OH (=O above)

14.39 Provide the common and I.U.P.A.C. names for each of the following compounds:

   a. CH₃CHC—OH (OH above 2nd C, =O above last C)

   b. CH₃CHCH₂C—OH (OH above 2nd C, =O above last C)

   c. CH₃CCH₂CH₂C—OH (CH₃ above and below 2nd C, =O above last C)

   d. CH₃CH₂CCH₂C—OH (Cl above and below 3rd C, =O above last C)

14.40 Draw the structure of each of the following carboxylic acids:
   a. β-Chlorobutyric acid
   b. α,β-Dibromovaleric acid
   c. β,γ-Dihydroxybutyric acid
   d. δ-Bromo-γ-chloro-β-methylcaproic acid

14.41 Provide the I.U.P.A.C. name for each of the following aromatic carboxylic acids.

   a. benzene ring with COOH and Br

   b. benzene ring with COOH and CH₂CH₃

   c. benzene ring with COOH and OH

14.42 Provide the I.U.P.A.C. name for each of the following aromatic carboxylic acids.

   a. CH₂CH₂CH₂COOH attached to benzene ring

   b. CH₃CHCH₂COOH attached to benzene ring

   c. CH₃CHCOOH attached to benzene ring

Carboxylic Acids: Reactions

Foundations

14.43 Explain what is meant by oxidation in organic molecules and provide an example of an oxidation reaction involving an aldehyde or an alcohol.

14.44 Write a general equation showing the preparation of a carboxylic acid from an alcohol.

14.45 Write a general equation showing the dissociation of a carboxylic acid in water.

14.46 Carboxylic acids are described as weak acids. To what extent do carboxylic acids generally dissociate?

14.47 What reaction occurs when a strong base is added to a carboxylic acid?

14.48 Write a general equation showing the reaction of a strong base with a carboxylic acid.

14.49 How is a soap prepared?

14.50 How do soaps assist in the removal of oil and grease from clothing?

Applications

14.51 Write the formula of the organic product obtained through each of the following reactions.

a. $CH_3CH_2CH_2OH \xrightarrow{H_2CrO_4}$?

b. $\overset{O}{\overset{\|}{H}}CCH_2CH_2CH_3 \xrightarrow{H_2CrO_4}$?

14.52 Complete each of the following reactions by supplying the missing product(s).

a. $CH_3CH_2OH \xrightarrow{H_2CrO_4}$?

b. $\overset{O}{\overset{\|}{H}}CCH_2CH_2\overset{O}{\overset{\|}{C}}H \xrightarrow{H_2CrO_4}$?

14.53 Complete each of the following reactions by supplying the missing portion indicated by a question mark:

a. $CH_3\overset{O}{\overset{\|}{C}}-H \xrightarrow{H_2CrO_4}$?

b. $CH_3CH_2CH_2\overset{O}{\overset{\|}{C}}-OH + CH_3OH \xrightarrow{H^+, \text{heat}}$?

c. $\triangleright\!\!-\!\overset{O}{\overset{\|}{C}}-OH + ? \longrightarrow \triangleright\!\!-\!\overset{O}{\overset{\|}{C}}-OCH_3$

14.54 Complete each of the following reactions by supplying the missing part(s) indicated by the question mark(s):

a. $CH_3CH_2CH_2OH \xrightarrow{?(1)} CH_3CH_2\overset{O}{\overset{\|}{C}}-OH \underset{?(3)}{\overset{NaOH}{\rightleftharpoons}}$?(4)

$\downarrow$?(2)

$CH_3CH_2\overset{O}{\overset{\|}{C}}-O\underset{\underset{CH_3}{|}}{C}HCH_3$

b. $CH_3COOH + NaOH \longrightarrow$?

c. $CH_3CH_2CH_2CH_2CH_2COOH + NaOH \longrightarrow$?

d. ? $+ CH_3CH_2\underset{\underset{CH_3}{|}}{C}HOH \xrightarrow{H^+} CH_3\overset{O}{\overset{\|}{C}}-O\underset{\underset{CH_3}{|}}{C}HCH_2CH_3$

14.55 How might $CH_3CH_2CH_2CH_2CH_2OH$ be converted to each of the following products?

a. $CH_3CH_2CH_2CH_2CHO$

b. $CH_3CH_2CH_2CH_2COOH$

14.56 Which of the following alcohols can be oxidized to a carboxylic acid? Name the carboxylic acid produced. For those alcohols that cannot be oxidized to a carboxylic acid, name the final product.

a. Ethanol

b. 2-Propanol

c. 1-Propanol

d. 3-Pentanol

14.57 Write an equation representing the neutralization of propanoic acid with each of the following bases.

a. NaOH

b. KOH

c. $Ca(OH)_2$

14.58 Write an equation representing the neutralization of each of the following carboxylic acids with KOH.

a. 2-Hydroxypropanoic acid

b. 3-Methylbutanoic acid

c. Octanoic acid

14.59 The calcium salt of propionic acid is added to breads as a preservative that prevents mold growth. Draw the structure of the calcium salt of propionic acid. What are the common and I.U.P.A.C. names of this carboxylic acid salt?

14.60 Oxalic acid is found in rhubarb, primarily in the form of the calcium salt. Since high levels of oxalic acid are toxic, rhubarb must be boiled in water to destroy the oxalic acid before being used to make the strawberry rhubarb pie shown here. What is the I.U.P.A.C. name of oxalic acid? Write the structure of the calcium salt of oxalic acid.

Esters: Structure, Physical Properties, and Nomenclature

Foundations

14.61 Explain why esters are described as mildly polar.

14.62 Compare the boiling points of esters to those of aldehydes or ketones of similar molecular weight.

14.63 Briefly summarize the I.U.P.A.C. rules for naming esters.

14.64 How are the common names of esters derived?

Applications

14.65 Draw condensed formulas for each of the following compounds:
   a. Methyl benzoate
   b. Butyl decanoate
   c. Methyl propionate
   d. Ethyl propionate

14.66 Draw condensed formulas for each of the following compounds:
   a. Ethyl *m*-nitrobenzoate
   b. Isopropyl acetate
   c. Methyl butyrate

14.67 Use the I.U.P.A.C. Nomenclature System to name each of the following esters:

   a. $CH_3\overset{\overset{\displaystyle O}{\|}}{C}-OCH_2CH_3$

   b. $CH_3CH_2\overset{\overset{\displaystyle O}{\|}}{C}-OCH_3$

   c. $CH_3\overset{\overset{\displaystyle CH_3}{|}}{C}HCH_2\overset{\overset{\displaystyle O}{\|}}{C}-OCH_3$

   d.

14.68 Use the I.U.P.A.C. Nomenclature System to name each of the following:

   a.

   b.

   c. $\overset{\overset{\displaystyle}{}}{C}H_2CHCH_2CH_2\overset{\overset{\displaystyle O}{\|}}{C}-OCH_2CH_3$ with Br Br substituents

Esters: Reactions

Foundations

14.69 Write a general reaction showing the preparation of an ester.

14.70 Why is preparation of an ester referred to as a dehydration reaction?

14.71 Write a general reaction showing the hydrolysis of an ester using an acid catalyst.

14.72 Write a general reaction showing the base-catalyzed hydrolysis of an ester.

14.73 What is meant by a hydrolysis reaction?

14.74 Why is the salt of a carboxylic acid produced in a base-catalyzed hydrolysis of an ester?

Applications

14.75 Complete each of the following reactions by supplying the missing portion indicated with a question mark:

   a. $CH_3CH_2CH_2\overset{\overset{\displaystyle O}{\|}}{C}-OH + CH_3CH_2OH \xrightarrow{H^+,\ heat}$?

   b. $CH_3CH_2\overset{\overset{\displaystyle O}{\|}}{C}-OCH_2CH_3 + H_2O \xrightarrow{H^+,\ heat}$?

   c. $CH_3\overset{\overset{\displaystyle CH_3}{|}}{C}HCH_2CH_2\overset{\overset{\displaystyle O}{\|}}{C}-OH + ? \xrightarrow{H^+,\ heat}$

   $CH_3\overset{\overset{\displaystyle CH_3}{|}}{C}HCH_2CH_2\overset{\overset{\displaystyle O}{\|}}{C}-OCH_2CH_2CH_3$

   d. $CH_3CH_2\overset{\overset{\displaystyle Br}{|}}{C}HCH_2\overset{\overset{\displaystyle O}{\|}}{C}-OCH_2CH_3 + H_2O \xrightarrow{OH^-,\ heat}$?

14.76 Complete each of the following reactions by supplying the missing portion indicated with a question mark:

   a. $? + CH_3\overset{\overset{\displaystyle CH_3}{|}}{\underset{\underset{\displaystyle CH_3}{|}}{C}}-OH \xrightarrow{?} CH_3CH_2\overset{\overset{\displaystyle O}{\|}}{C}-O-\overset{\overset{\displaystyle CH_3}{|}}{\underset{\underset{\displaystyle CH_3}{|}}{C}}-CH_3$

   b. $CH_3CH_2CH_2CH_2COOH + CH_3CH_2CH_2CH_2OH \xrightarrow{H^+,\ heat}$?

   c. $CH_3\overset{\overset{\displaystyle CH_3}{|}}{\underset{\underset{\displaystyle CH_3}{|}}{C}}CH_2\overset{\overset{\displaystyle O}{\|}}{C}-OCH_2CH_2\overset{\overset{\displaystyle CH_3}{|}}{\underset{\underset{\displaystyle CH_3}{|}}{C}}CH_3 + H_2O \xrightarrow{H^+,\ heat}$?

   d. $CH_3CH_2\overset{\overset{\displaystyle O}{\|}}{C}-OCH_3 + H_2O \xrightarrow{OH^-,\ heat}$?

14.77 Design the synthesis of each of the following esters from organic alcohols.
   a. Isobutyl methanoate (raspberries)
   b. Pentyl butanoate (apricot)

14.78 Design the synthesis of each of the following esters from organic alcohols.
   a. Methyl butanoate (apples)
   b. Octyl ethanoate (oranges)

14.79 What is saponification? Give an example using specific molecules.

14.80 When the methyl ester of hexanoic acid is hydrolyzed in aqueous sodium hydroxide in the presence of heat, a homogeneous solution results. When the solution is acidified with dilute aqueous hydrochloric acid, a new product forms. What is the new product? Draw its structure.

14.81 The structure of salicylic acid is shown. If this acid reacts with methanol, the product is an ester, methyl salicylate. Methyl salicylate is known as oil of wintergreen and is often used as a flavoring agent. Draw the structure of the product of this reaction.

$+ CH_3OH \xrightarrow{H^+}$?

14.82 When salicylic acid reacts with acetic anhydride, one of the products is an ester, acetylsalicylic acid. Acetylsalicylic acid is the active ingredient in aspirin. Complete the following equation by drawing the structure of acetylsalicylic acid. (*Hint:* Acid anhydrides are hydrolyzed by water.)

14.83 Compound A ($C_6H_{12}O_2$) reacts with water, acid, and heat to yield compound B ($C_5H_{10}O_2$) and compound C (CH_4O). Compound B is acidic. Deduce possible structures of compounds A, B, and C.

14.84 What products are formed when methyl *o*-bromobenzoate reacts with each of the following?
   a. Aqueous acid and heat
   b. Aqueous base and heat

14.85 Write an equation for the acid-catalyzed hydrolysis of each of the following esters:
   a. Propyl propanoate
   b. Butyl methanoate
   c. Ethyl methanoate
   d. Methyl pentanoate

14.86 Write an equation for the base-catalyzed hydrolysis of each of the following esters:
   a. Pentyl methanoate
   b. Hexyl propanoate
   c. Butyl hexanoate
   d. Methyl benzoate

Acid Chlorides and Acid Anhydrides

Foundations

14.87 Describe the physical properties of acid chlorides.
14.88 Write a general equation for the formation of acid chlorides.
14.89 Describe the physical properties of acid anhydrides.
14.90 Write a general equation for the formation of acid anhydrides.

Applications

14.91 Supply the missing reagents (indicated by the question marks) necessary to complete each of the following transformations:

14.92 Supply the missing reagents (indicated by the question marks) necessary to complete each of the following transformations. All of the reactions may require more than one step to complete.

14.93 Complete each of the following reactions by supplying the missing product:

14.94 Use the I.U.P.A.C. Nomenclature System to name the products and reactants in Problem 14.93.

14.95 Write the condensed formula for each of the following compounds:
   a. Decanoic anhydride
   b. Acetic anhydride

14.96 Write the condensed formula for each of the following compounds:
   a. Valeric anhydride
   b. Benzoyl chloride

14.97 Write a condensed formula for each of the following compounds:
   a. Butanoyl chloride
   b. Hexanoyl chloride
   c. Ethanoyl chloride

14.98 Write a condensed formula for each of the following compounds:
   a. Propanoyl chloride
   b. Heptanoyl chloride
   c. Pentanoyl chloride

14.99 Write an equation representing the synthesis of methanoic anhydride.

14.100 Write an equation representing the synthesis of octanoic anhydride.

14.101 Write an equation for the reaction of each of the following acid anhydrides with ethanol.
   a. Propanoic anhydride
   b. Ethanoic anhydride
   c. Methanoic anhydride

14.102 Write an equation for the reaction of each of the following acid anhydrides with propanol. Name each of the products using the I.U.P.A.C. Nomenclature System.
   a. Butanoic anhydride
   b. Pentanoic anhydride
   c. Methanoic anhydride

Phosphoesters and Thioesters

14.103 By reacting phosphoric acid with an excess of ethanol, it is possible to obtain the mono-, di-, and triesters of phosphoric acid. Draw all three of these products.

14.104 What is meant by a phosphoanhydride bond?

14.105 We have described the molecule ATP as the body's energy storehouse. What do we mean by this designation? How does ATP actually store energy and provide it to the body as needed?

14.106 Write an equation for each of the following reactions:
   a. Ribose + phosphoric acid
   b. Methanol + phosphoric acid
   c. Adenosine diphosphate + phosphoric acid

14.107 Draw the thioester bond between the acetyl group and coenzyme A.

14.108 Explain the significance of thioester formation in the metabolic pathways involved in fatty acid and carbohydrate breakdown.

14.109 It is also possible to form esters of other inorganic acids such as sulfuric acid and nitric acid. One particularly noteworthy product is nitroglycerine, which is both highly unstable (explosive) and widely used in the treatment of the heart condition known as angina, a constricting pain in the chest usually resulting from coronary heart disease. In the latter case, its function is to alleviate the pain associated with angina. Nitroglycerine may be administered as a tablet (usually placed just beneath the tongue when needed) or as a salve or paste that can be applied to and absorbed through the skin. Nitroglycerine is the trinitroester of glycerol. Draw the structure of nitroglycerine, using the structure of glycerol.

Glycerol

14.110 Show the structure of the thioester that would be formed between coenzyme A and stearic acid.

CRITICAL THINKING PROBLEMS

1. Radioactive isotopes of an element behave chemically in exactly the same manner as the nonradioactive isotopes. As a result, they can be used as tracers to investigate the details of chemical reactions. A scientist is curious about the origin of the bridging oxygen atom in an ester molecule. She has chosen to use the radioactive isotope oxygen-18 to study the following reaction:

   Design experiments using oxygen-18 that will demonstrate whether the oxygen in the water molecule came from the —OH of the alcohol or the —OH of the carboxylic acid.

2. Triglycerides are the major lipid storage form in the human body. They are formed in an esterification reaction between glycerol (1,2,3-propanetriol) and three fatty acids (long chain carboxylic acids). Write a balanced equation for the formation of a triglyceride formed in a reaction between glycerol and three molecules of decanoic acid.

3. Chloramphenicol is a very potent, broad-spectrum antibiotic. It is reserved for life-threatening bacterial infections because it is quite toxic. It is also a very bitter tasting chemical. As a result, children had great difficulty taking the antibiotic. A clever chemist found that the taste could be improved considerably by producing the palmitate ester. Intestinal enzymes hydrolyze the ester, producing chloramphenicol, which can then be absorbed. The following structure is the palmitate ester of chloramphenicol. Draw the structure of chloramphenicol.

Chloramphenicol palmitate

4. Acetyl coenzyme A (acetyl CoA) can serve as a donor of acetyl groups in biochemical reactions. One such reaction is the formation of acetylcholine, an important neurotransmitter involved in nerve signal transmission at neuromuscular junctions. The structure of choline is shown below. Draw the structure of acetylcholine.

Choline

5. Hormones are chemical messengers that are produced in a specialized tissue of the body and travel through the bloodstream to reach receptors on cells of their target tissues. This specific binding to target tissues often stimulates a cascade of enzymatic reactions in the target cells. The work of Earl Sutherland and others led to the realization that there is a *second messenger* within the target cells. Binding of the hormone to the hormone receptor in the cell membrane triggers the enzyme adenyl cyclase to produce adenosine-3′,5′-monophosphate, which is also called *cyclic AMP*, from ATP. The reaction is summarized as follows:

$$\text{ATP} \xrightarrow{\text{Mg}^{2+},\ \text{adenyl cyclase}} \text{cyclic AMP} + \text{PP}_i + \text{H}^+$$

PP_i is the abbreviation for a pyrophosphate group, shown here:

The structure of ATP is shown here with the carbon atoms of the sugar ribose numbered according to the convention used for nucleotides:

Adenosine-5′-triphosphate

Draw the structure of adenosine-3′,5′-monophosphate.

15 Amines and Amides

Learning Goals

1 Classify amines as primary, secondary, or tertiary.

2 Describe the physical properties of amines.

3 Draw and name simple amines using systematic and common nomenclature systems.

4 Write equations representing the synthesis of amines.

5 Write equations showing the basicity and neutralization of amines.

6 Describe the structure of quaternary ammonium salts and discuss their use as antiseptics and disinfectants.

7 Discuss the biological significance of heterocyclic amines.

8 Describe the physical properties of amides.

9 Draw the structure and write the common and I.U.P.A.C. names of amides.

10 Write equations representing the preparation of amides.

11 Write equations showing the hydrolysis of amides.

12 Draw the general structure of an amino acid.

13 Draw and discuss the structure of a peptide bond.

14 Describe the function of neurotransmitters.

Outline

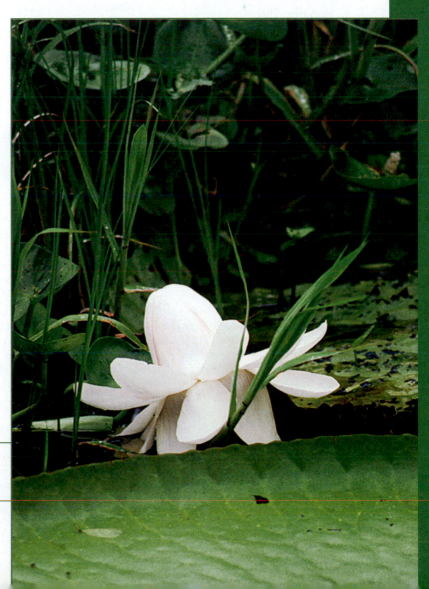

An Amazon lily. Ethnobotanists continue to search for medically active compounds from the rain forest.

Introduction

In this chapter we introduce an additional element into the structure of organic molecules. That element is nitrogen, the fourth most common atom in living systems. It is an important component of the structure of the nucleic acids, DNA and RNA, which are the molecules that carry the genetic information for living systems. It is also essential to the structure and function of proteins, molecules that carry out the majority of the work in biological systems. Some proteins serve as enzymes that catalyze the chemical reactions that allow life to exist. Other proteins, the antibodies, protect us against infection by a variety of infectious agents. Proteins are also structural components of the cell and of the body.

One class of organic molecules containing nitrogen is the amines. Amines are characterized by the presence of an amino group ($-NH_2$).

Excess nitrogen is removed from the body as urea, first synthesized by the father of organic chemistry, Friederich Wöhler (Chapter 10). Urea synthesis in the body is described in Chapter 22.

General structure of an amine

The nitrogen atom of the amino group may have one or more of its hydrogen atoms replaced by an organic group. General structures of these types of amines are shown below:

Animations

Valence Shell Electron Pair
  Repulsion Theory
VSEPR and Molecular Geometry
The Geometry of NH_3

Chemistry *Connection*

The Nicotine Patch

Smoking cigarettes is one of the most difficult habits to break—so difficult, in fact, that physicians now suspect that smoking is more than a habit: It's an addiction. The addictive drug in tobacco is *nicotine.*

Nicotine, one of the *heterocyclic amines* that we will study in this chapter (see Figure 15.4), is a highly toxic compound. In fact, it has been used as an insecticide! Small doses from cigarette smoking initially stimulate the autonomic (involuntary) nervous system. However, repeated small doses of nicotine obtained by smokers eventually depress the involuntary nervous system. As a result, the smoker *needs* another cigarette.

Some people have been able to quit smoking through behavioral modification programs, hypnosis, or sheer willpower. Others quit only after smoking has contributed to life-threatening illness, such as emphysema or a heart attack. Yet there are people who cannot quit even after a diagnosis of lung cancer.

One promising advance to help people quit smoking is the nicotine patch. The patch is applied to the smoker's skin, and

nicotine from the patch slowly diffuses through the skin and into the bloodstream. Because the body receives a constant small dose of nicotine, the smoker no longer craves a cigarette. Of course, the long-range goal is to completely cure the addiction to nicotine. This is done by decreasing doses of nicotine in the patches as the treatment period continues. Eventually, after a period of about three months, the former smoker no longer needs the patches.

There are those who criticize this therapy because nicotine is a toxic chemical. However, the benefits seem to outweigh any negative aspects. A smoker inhales not only nicotine, but also dozens of other substances that have been shown to cause cancer. When someone successfully quits smoking, the body no longer suffers the risks associated with the substances in cigarette smoke.

In this chapter we study the structure and properties of amines and their derivatives, the amides. We will see that several are important pain killers, decongestants, and antibiotics, and others are addictive drugs and carcinogens.

Amines are very common in biological systems and exhibit important physiological activity. Consider histamine. Histamine contributes to the inflammatory response that causes the symptoms of colds and allergies, including swollen mucous membranes, congestion, and excessive nasal secretions. We take antihistamines to help relieve these symptoms. Ephedrine is an antihistamine that has been extracted from the leaves of the ma-huang plant in China for over two thousand years. Today it is one of the antihistamines found in over-the-counter cold medications. This decongestant helps to shrink swollen mucous membranes and reduce nasal secretions. The structures of histamine and ephedrine are shown below.

Ephedra, from the ma-huang plant, has been sold to promote weight loss and boost energy. What antihistamines are found in ephedra?

Histamine

Ephedrine

The other group of nitrogen-containing organic compounds we will investigate in this chapter is the amides. Amides are the products of a reaction between an amine and a carboxylic acid derivative. They have the following general structure:

From the carboxylic acid

From the amine

($R = H$ or an alkyl or aryl group)

amide bond

General structure of an amide

The amino acids are the subunits from which proteins are built. They are characterized by the presence of both an amino group and a carboxyl group. When amino acids are bonded to one another to produce a protein chain, the amino group of one amino acid reacts with the carboxyl group of another amino acid. The amide bond that results is called a peptide bond.

In this chapter we will explore the chemistry of the organic molecules that contain nitrogen. In upcoming chapters we will investigate the structure and properties of the nitrogen-containing biological molecules.

The general structure of an amino acid is

$$H_2N-C-COOH$$

The amino group is highlighted in red and the carboxyl group in blue. Amino acids are the basic subunits of all proteins.

15.1 Amines

Structure and Physical Properties

Amines are organic derivatives of ammonia and, like ammonia, they are basic. In fact, amines are the most important type of organic base found in nature. We can think of them as substituted ammonia molecules in which one, two, or three of the ammonia hydrogens have been replaced by an organic group:

LEARNING GOAL

1 Classify amines as primary, secondary, or tertiary.

R substitutes for H

Ammonia

An amine

The structures drawn above and in Figure 15.1 reveal that like ammonia, amines are pyramidal. The nitrogen atom has three groups bonded to it and has a non-bonding pair of electrons.

The geometry of ammonia is described in Section 3.4.

Amines are classified according to the number of alkyl or aryl groups attached to the nitrogen. In a **primary (1°) amine,** one of the hydrogens is replaced by an organic group. In a **secondary (2°) amine,** two hydrogens are replaced. In a **tertiary (3°) amine,** three organic groups replace the hydrogens:

Ammonia

Amine

Figure 15.1

The pyramidal structure of amines. Note the similarities in structure between an amine and the ammonia molecule.

Ammonia

Methanamine
(methylamine)
(primary amine)

N-Methylmethanamine
(dimethylamine)
(secondary amine)

N, N-Dimethylmethanamine
(trimethylamine)
(tertiary amine)

Hydrogen bonding is described in Section 5.2.

The nitrogen atom is more electronegative than the hydrogen atoms in amines. As a result, the N—H bond is polar, and hydrogen bonding between amine molecules or between amine molecules and water can occur (Figure 15.2).

| EXAMPLE 15.1 | Classifying Amines as Primary, Secondary, or Tertiary |

LEARNING GOAL

1 Classify amines as primary, secondary, or tertiary.

Classify each of the following compounds as a primary, secondary, or tertiary amine.

Solution

Compare the structure of the amine with that of ammonia.

1° amine: one hydrogen replaced

2° amine: two hydrogens replaced

3° amine: three hydrogens replaced

Practice Problem 15.1

Determine whether each of the following amines is primary, secondary, or tertiary.

a. $CH_3CH_2NCH_3$ with CH_2CH_3

b. $CH_3CH_2CH_2NH_2$

c. CH_3NCH_3 with H

For Further Practice: Questions 15.29 and 15.30.

Figure 15.2
Hydrogen bonding (a) in methylamine
and (b) between methylamine and water.
Dotted lines represent hydrogen bonds.

(a)   (b)

Refer to Figure 15.2 and draw a similar figure showing the hydrogen bonding that
occurs between water and a 2° amine.

Refer to Figure 15.2 and draw hydrogen bonding between two primary amines.

The ability of primary and secondary amines to form N—H⋯N hydrogen
bonds is reflected in their boiling points (Table 15.1). Primary amines have boiling
points well above those of alkanes of similar molecular weight but considerably
lower than those of comparable alcohols. Consider the following examples:

$CH_3CH_2CH_3$	$CH_3CH_2NH_2$	CH_3CH_2OH
Propane	Ethanamine	Ethanol
M.W. = 44 g/mol	M.W. = 45 g/mol	M.W. = 46 g/mol
b.p. = –42.2°C	b.p. = 16.6°C	b.p. = 78.5°C

TABLE 15.1	**Boiling Points of Amines**		
Systematic Name	**Common Name**	**Structure**	**Boiling Point (°C)**
	Ammonia	NH_3	–33.4
Methanamine	Methylamine	CH_3NH_2	–6.3
N-Methylmethanamine	Dimethylamine	$(CH_3)_2NH$	7.4
N,N-Dimethylmethanamine	Trimethylamine	$(CH_3)_3N$	2.9
Ethanamine	Ethylamine	$CH_3CH_2NH_2$	16.6
Propanamine	Propylamine	$CH_3CH_2CH_2NH_2$	48.7
Butanamine	Butylamine	$CH_3CH_2CH_2CH_2NH_2$	77.8

Tertiary amines do not have an N—H bond. As a result they cannot form inter-molecular hydrogen bonds with other tertiary amines. Consequently, their boiling points are lower than those of primary or secondary amines of comparable molecular weight. This is seen in a comparison of the boiling points of propanamine (propylamine; M.W. = 59) and N,N-dimethylmethanamine (trimethylamine; M.W. = 59). Trimethylamine, the tertiary amine, has a boiling point of 2.9°C, whereas propylamine, the primary amine, has a boiling point of 48.7°C. Clearly the inability of trimethylamine molecules to form intermolecular hydrogen bonds results in a much lower boiling point.

$$CH_3CH_2CH_2—NH_2 \qquad CH_3CH_2—\overset{\overset{\displaystyle H}{|}}{N}—CH_3 \qquad CH_3—\overset{\overset{\displaystyle CH_3}{|}}{N}—CH_3$$

Propanamine (propylamine)
M.W. = 59 g/mol
b.p. = 48.7°C

N-Methylethanamine (ethylmethylamine)
M.W. = 59 g/mol
b.p. = 36.7°C

N,N-Dimethylmethanamine (trimethylamine)
M.W. = 59 g/mol
b.p. = 2.9°C

The intermolecular hydrogen bonds formed by primary and secondary amines are not as strong as the hydrogen bonds formed by alcohols because nitrogen is not as electronegative as oxygen. For this reason primary and secondary amines have lower boiling points than alcohols (Table 15.2).

All amines can form intermolecular hydrogen bonds with water (O—H···N). As a result, small amines (six or fewer carbons) are soluble in water. As we have noted for other families of organic molecules, water solubility decreases as the length of the hydrocarbon (hydrophobic) portion of the molecule increases.

EXAMPLE 15.2   **Predicting the Physical Properties of Amines**

Which member of each of the following pairs of molecules has the higher boiling point?

$$CH_3CH_2\overset{\overset{\displaystyle |}{\underset{\underset{\displaystyle CH_2CH_3}{|}}{N}}}{}CH_2CH_3 \qquad or \qquad CH_3CH_2CH_2CH_2CH_2CH_2NH_2$$

Solution

The molecule on the right, hexanamine, has a higher boiling point than the molecule on the left, N,N-diethylethanamine (triethylamine). Triethylamine is a tertiary amine; therefore, it has no N—H bond and cannot form intermolecular hydrogen bonds with other triethylamine molecules.

$$CH_3CH_2CH_2CH_2OH \qquad or \qquad CH_3CH_2CH_2CH_2NH_2$$

Solution

The molecule on the left, butanol, has a higher boiling point than the molecule on the right, butanamine. Nitrogen is not as electronegative as oxygen, thus the hydroxyl group is more polar than the amino group and forms stronger hydrogen bonds.

Continued—

EXAMPLE 15.2 —*Continued*

Practice Problem 15.2

Which compound in each of the following pairs would you predict to have a higher boiling point? Explain your reasoning.
a. Methanol or methylamine
b. Dimethylamine or water
c. Methylamine or ethylamine
d. Propylamine or butane

For Further Practice: Questions 15.17 and 15.18.

Nomenclature

In systematic nomenclature, primary amines are named according to the following rules:

- Determine the name of the *parent compound*, the longest continuous carbon chain containing the amine group.
- Replace the *-e* ending of the alkane chain with *-amine*. Following this pattern, the alkane becomes an alkan*amine*; for instance, ethan*e* becomes ethan*amine*.
- Number the parent chain to give the carbon bearing the amine group the lowest possible number.
- Name and number all substituents, and add them as prefixes to the "alkanamine" name.

> **LEARNING GOAL**
> **3** Draw and name simple amines using systematic and common nomenclature systems.

For instance,

$$CH_3-NH_2 \qquad CH_3CH_2CH_2-NH_2 \qquad CH_3CH_2CH_2CHCH_3$$
$$\underset{NH_2}{|}$$

Methanamine   1-Propanamine   2-Pentanamine

For secondary or tertiary amines the prefix *N*-alkyl is added to the name of the parent compound. For example,

$$CH_3-NH-CH_2CH_3 \qquad\qquad CH_3-\overset{\overset{\displaystyle CH_3}{|}}{N}-CH_3$$

N-Methylethanamine   *N,N*-Dimethylmethanamine

TABLE 15.2	Comparison of the Boiling Points of Selected Alcohols and Amines	
Name	**Molecular Weight (g/mol)**	**Boiling Point (°C)**
Methanol	32	64.5
Methanamine	31	−6.3
Ethanol	46	78.5
Ethanamine	45	16.6
Propanol	60	97.2
Propanamine	59	48.7

Several aromatic amines have special names that have also been approved for use by I.U.P.A.C. For example, the amine of benzene is given the name *aniline*. The systematic name for aniline is *benzenamine*.

Aniline or   *m*-Toluidine or   *o*-Toluidine or   *p*-Toluidine or
benzenamine   *meta*-toluidine   *ortho*-toluidine   *para*-toluidine

If additional groups are attached to the nitrogen of an aromatic amine, they are indicated with the letter *N*- followed by the name of the group.

| **EXAMPLE 15.3** | **Writing the Systematic Name for an Amine** |

Determine the systematic name for the following amine, which is used by the German cockroach as a pheromone.

$$CH_3—NH—CH_3$$

Solution

Parent compound: methane (becomes methanamine)
Additional group on N: methyl (becomes *N*-methyl)
Name: *N*-Methylmethanamine

Name the following amine.

$$CH_3CH_2CH_2—NH—CH_3$$

Solution

Parent compound: propane (becomes propanamine)
Additional group on N: methyl (becomes *N*-methyl)
Name: *N*-Methylpropanamine

Practice Problem 15.3

What is the systematic name for each of the following amines?

a. $CH_3CH_2—\underset{\underset{CH_2CH_2CH_3}{|}}{N}—CH_3$ b. $CH_3CH_2\underset{\underset{NH_2}{|}}{CH}CH_2CH_2CH_3$

c. $CH_3CH_2CH_2—\underset{\underset{CH_2CH_3}{|}}{N}—CH_3CH_2$

For Further Practice: Questions 15.21, 15.27, and 15.28.

Aniline was first isolated from the blue dye indigo, a product of the indigo plant. It is the starting material in the synthesis of hundreds of dyes. Is aniline a primary, secondary, or tertiary amine?

Common names are often used for the simple amines. The common names of the alkyl groups bonded to the amine nitrogen are followed by the ending *-amine*. Each group is listed alphabetically in one continuous word followed by the suffix *-amine*:

$$CH_3-NH_2 \qquad CH_3-NH-CH_3 \qquad CH_3-\overset{\overset{\displaystyle CH_3}{|}}{N}-CH_3$$

Methylamine   Dimethylamine   Trimethylamine

$$CH_3CH_2-NH_2 \qquad CH_3CH_2-NH-CH_3$$

Ethylamine   Ethylmethylamine

Table 15.3 compares these systems of nomenclature for a number of simple amines.

Question 15.3

Use the structure of aniline provided and draw the complete structural formula for each of the following amines.

a. *N*-Methylaniline
b. *N,N*-Dimethylaniline
c. *N*-Ethylaniline
d. *N*-Isopropylaniline

Question 15.4

Name each of the following amines using the systematic and common nomenclature systems.

a. $\overset{\overset{\displaystyle NH_2}{|}}{CH_3CHCH_2CH_3}$

b. $CH_3-\overset{\overset{\displaystyle NH_2}{|}}{\underset{\underset{\displaystyle CH_3}{|}}{C}}-CH_3$

c. $CH_3\overset{\overset{\displaystyle CH_3}{|}}{\underset{}{CH}}\overset{\overset{\displaystyle NH}{|}}{\underset{}{CHCH_2CH_3}}$

d. $CH_3\overset{\overset{\displaystyle CH_3}{|}}{\underset{\underset{\displaystyle N-CH_2CH_3}{}}{}}CHCH_3$

Question 15.5

Draw the complete structural formula for each of the following compounds.

a. 2-Propanamine
b. 3-Octanamine
c. *N*-Ethyl-2-heptanamine
d. 2-Methyl-2-pentanamine
e. 4-Chloro-5-iodo-1-nonanamine
f. *N,N*-Diethyl-1-pentanamine

TABLE 15.3	**Systematic and Common Names of Amines**	
Compound	**Systematic Name**	**Common Name**
R—NH$_2$	Alkan*amine*	Alkyl*amine*
CH$_3$—NH$_2$	Methanamine	Methylamine
CH$_3$CH$_2$—NH$_2$	Ethanamine	Ethylamine
CH$_3$CH$_2$CH$_2$—NH$_2$	1-Propanamine	Propylamine
CH$_3$—NH—CH$_3$	*N*-Methylmethanamine	Dimethylamine
CH$_3$—NH—CH$_2$CH$_3$	*N*-Methylethanamine	Ethylmethylamine
CH$_3$—$\overset{\overset{\displaystyle CH_3}{|}}{N}$—CH$_3$	*N,N*-Dimethylmethanamine	Trimethylamine

N-methylmethanamine is used by the German cockroach as a communication pheromone. What is the common name of this amine?

Question 15.6

Draw the condensed formula for each of the following compounds.
- a. Diethylmethylamine
- b. 4-Methylpentylamine
- c. N-Methylaniline
- d. Triisopropylamine
- e. Methyl-t-butylamine
- f. Ethylhexylamine

Medically Important Amines

Although amines play many different roles in our day-to-day lives, one important use is in medicine. A host of drugs derived from amines is responsible for improving the quality of life, whereas others, such as cocaine and heroin, are highly addictive.

Amphetamines, such as benzedrine and methedrine, stimulate the central nervous system. They elevate blood pressure and pulse rate and are often used to decrease fatigue. Medically, they have been used to treat depression and epilepsy. Amphetamines have also been prescribed as diet pills because they decrease the appetite. Their use is controlled by federal law because excess use of amphetamines can cause paranoia and mental illness.

1-Phenyl-2-propanamine
(Amphetamine)

Benzedrine

N-Methyl-1-phenyl-2-propanamine
(Methamphetamine)

Methedrine

Many of the medicinal amines are *analgesics* (pain relievers) or *anesthetics* (pain blockers). Novocaine and related compounds, for example, are used as local anesthetics. Demerol is a very strong pain reliever.

Novocaine

Demerol

Ephedrine, its stereoisomer pseudoephedrine, and phenylephrine (also called neosynephrine) are used as decongestants in cough syrups and nasal sprays. By shrinking of the membranes that line the nasal passages, they relieve the symptoms of congestion and stuffy nose. These compounds are very closely related to L-dopa and dopamine, which are key compounds in the function of the central nervous system.

L-Dopa, dopamine, and other key neurotransmitters are described in detail in Section 15.5.

Ephedrine

Pseudoephedrine

Phenylephrine
(neosynephrine)

Methamphetamine use and abuse are discussed in A Human Perspective: Methamphetamine found on page 524.

Recently, many states have restricted the sales of products containing ephedrine and pseudoephedrine and many drug store chains have moved these products behind the counter. The reason for these precautions is that either ephedrine or pseudoephedrine can be used as the starting material in the synthesis of methamphetamines. In response to this problem, pharmaceutical companies are replacing ephedrine and pseudoephedrine in these decongestants with phenylephrine, which cannot be used as a reactant in the synthesis of methamphetamine.

Ephedrine and pseudoephedrine are also the primary active ingredients in ephedra, a plant found in the deserts of central Asia. Ephedra is used as a stimulant in a variety of products that are sold over-the-counter as aids to boost energy, promote weight loss, and enhance athletic performance. In 2004 the Food and Drug Administration banned the use of ephedra in these over-the-counter formulations after reviewing 16,000 reports of adverse side effects including nervousness, heart irregularities, seizures, heart attacks, and 80 deaths, including that of Baltimore Orioles pitcher Steve Bechler, age 23. However, in April 2005, a Federal Judge in Texas ruled that the FDA had failed to prove that ephedra is dangerous at doses of 10 mg or lower, opening the way for the sale of ephedra-containing herbal remedies.

The *sulfa drugs,* the first chemicals used to fight bacterial infections, are synthesized from amines.

Sulfanilamide—a sulfa drug

Reactions Involving Amines
Preparation of Amines

In the laboratory, amines are prepared by the reduction of amides and nitro compounds.

LEARNING GOAL
4 Write equations representing the synthesis of amines.

Examples of amides   A nitro compound

As we will see in Section 15.3, amides are neutral nitrogen compounds that produce an amine and a carboxylic acid when hydrolyzed. Nitro compounds are prepared by the nitration of an aromatic compound.

Primary amines are readily produced by reduction of a nitro compound, as in the following reaction:

We are using the general symbol [H] to represent any reducing agent just as we used [O] to represent an oxidizing agent in previous chapters. Several different reducing agents may be used to effect the changes shown here; for example, metallic iron and acid may be used to reduce aromatic nitro compounds and LiAlH$_4$ in ether reduces amides.

In this reaction the nitro compound is nitrobenzene and the product is aniline.

Amides may also be reduced to produce primary, secondary, or tertiary amines.

$$R^1-\overset{\overset{\displaystyle O}{\|}}{C}-N\overset{\overset{\displaystyle R^2}{}}{\underset{\underset{\displaystyle R^3}{}}{}} \xrightarrow{[H]} R^1CH_2N\overset{\overset{\displaystyle R^2}{}}{\underset{\underset{\displaystyle R^3}{}}{}}$$

(R² and R³ may be a hydrogen atom or an organic group.)

Amide   Amine

If R² and R³ are hydrogen atoms, the product will be a primary amine:

$$CH_3\overset{\overset{\displaystyle O}{\|}}{C}NH_2 \xrightarrow{[H]} CH_3CH_2NH_2$$

Ethanamide   Ethanamine
                              (ethylamine)

If either R² or R³ is an organic group, the product will be a secondary amine:

$$CH_3CH_2\overset{\overset{\displaystyle O}{\|}}{C}NHCH_3 \xrightarrow{[H]} CH_3CH_2CH_2NHCH_3$$

N-Methylpropanamide   N-Methylpropanamine
                              (methylpropylamine)

If both R² and R³ are organic groups, the product will be a tertiary amine:

$$CH_3\overset{\overset{\displaystyle O}{\|}}{C}-N\overset{\overset{\displaystyle CH_3}{|}}{CH_3} \xrightarrow{[H]} CH_3CH_2\overset{\overset{\displaystyle CH_3}{|}}{N}CH_3$$

N,N-Dimethylethanamide   N,N-Dimethylethanamine

Basicity

Amines behave as weak bases, accepting H⁺, when dissolved in water. The nonbonding pair (lone pair) of electrons of the nitrogen atom can be shared with a proton (H⁺) from a water molecule, producing an **alkylammonium ion.** Hydroxide ions are also formed, so the resulting solution is basic.

Neutralization

Because amines are bases, they react with acids to form alkylammonium salts.

The reaction of methylamine with hydrochloric acid shown is typical of these reactions.

$$CH_3{-}\overset{\displaystyle H}{\underset{\displaystyle H}{N}}{:} \quad + \quad HCl \quad \longrightarrow \quad CH_3{-}\overset{\displaystyle H}{\underset{\displaystyle H}{N^+}}{-}HCl^-$$

Methylamine   Hydrochloric   Methylammonium
                 acid   chloride

Recall that the reaction of an acid and a base gives a salt (Section 8.3).

Alkylammonium salts are named by replacing the suffix -amine with ammonium. This is then followed by the name of the anion. The salts are ionic and hence are quite soluble in water.

A variety of important drugs are amines. They are usually administered as alkylammonium salts because the salts are much more soluble in aqueous solutions and in body fluids.

Question 15.7

Complete each of the following reactions by supplying the missing product(s).

a. NH_2 ⬠ $+ HBr \longrightarrow$?

b. $CH_3CH_2NHCH_3 + H_2O \longrightarrow$?
c. $CH_3NH_2 + H_2O \longrightarrow$?

Question 15.8

Complete each of the following reactions by supplying the missing product(s).

a. $CH_3NH_2 + HI \longrightarrow$?
b. $CH_3CH_2NH_2 + HBr \longrightarrow$?
c. $(CH_3CH_2)_2NH + HCl \longrightarrow$?

Alkylammonium salts can neutralize hydroxide ions. In this reaction, water is formed and the protonated amine cation is converted into an amine.

$$R{-}\overset{\displaystyle H}{\underset{\displaystyle H}{N^+}}{-}H \quad + \quad OH^- \quad \longrightarrow \quad R{-}\overset{\displaystyle H}{\underset{\displaystyle H}{N}}{:} + H{-}OH$$

Alkylammonium   Hydroxide   Amine   Water
salt   ion

Thus, by adding a strong acid to a water-insoluble amine, a water soluble alkyl-ammonium salt can be formed. The salt can just as easily be converted back to an amine by the addition of a strong base. The ability to manipulate the solubility of physiologically active amines through interconversion of the amine and its corresponding salt is extremely important in the development, manufacture, and administration of many important drugs.

Pure cocaine is an amine and a base (see Figure 15.4 and structure below). This form of cocaine, referred to as "crack" or "free base" cocaine, is generally found in the form of relatively large crystals (Figure 15.3a) that may vary in color from white to dark brown or black. As we have just seen, when an amine reacts

The local anesthetic novocaine, which is often used in dentistry and for minor surgery, is injected as an amine salt. See Medically Important Amines earlier in this section.

A Human Perspective

Methamphetamine

Methamphetamine is an addictive drug known by many names, including "speed," "crystal," "crank," "ice," and "glass." A bitter-tasting, odorless, crystalline powder, it is easily dissolved in either water or alcohol. Methamphetamine was developed early in the twentieth century and used as a decongestant in nasal and bronchial inhalers. Now it is rarely used for medical purposes.

A 2002 Health and Human Services survey revealed that twelve million Americans age twelve and older had used methamphetamine. Use of methamphetamine was once associated with white, male, blue-collar workers; but a much more diverse audience now uses the drug. Although it is still used by people in jobs such as long-distance trucking which require long hours and mental and physical alertness, it is disturbing that use of methamphetamine is becoming increasingly associated with sexual activity, teenagers attending "raves," homeless people, and runaway youths.

Methamphetamine can be smoked, taken orally, snorted, or injected, depending on the form of the drug; and it alters the mood differently depending on how it is taken. Smoking or injecting intravenously results in a "flash" or intense rush that lasts only a few minutes. This may be followed by a high that lasts several hours. Snorting and oral ingestion result in a euphoria lasting three to five minutes in the case of snorting and up to twenty minutes in the case of oral ingestion. Because the pleasurable effects are so short-lived, methamphetamine users tend to binge to try to sustain the high.

Both the intense rush and the longer-lasting euphoria are thought to result from a release of dopamine and norepinephrine into regions of the brain that control feelings of pleasure.

Once inside nerve cells (neurons), methamphetamine causes the release of dopamine and norepinephrine. At the same time, it inhibits enzymes that would normally destroy these two neurotransmitters and the excess is transported out of neurons, causing the sensations of pleasure and euphoria. The excess norepinephrine may be responsible for the increased attention and decreased fatigue associated with methamphetamine use.

Symptoms of long-term use include addiction, anxiety, violent behavior, confusion, as well as psychotic symptoms

with an acid, an alkylammonium salt is formed. When cocaine reacts with HCl, the product is cocaine hydrochloride:

"Crack" cocaine
(a base)

$+HCl \longrightarrow$

Cocaine hydrochloride
(a salt)

The salt of cocaine is a powder (Figure 15.3b) and is soluble in water. Since it is a powder, it can be snorted, and because it is water soluble, it dissolves in the fluids of the nasal mucous membranes and is absorbed into the bloodstream. This is a common form of cocaine because it is the direct product of the preparation from coca leaves (Figure 15.3c). A coca paste is made from the leaves and is mixed with HCl and water. After additional processing, the product is the salt of cocaine.

of paranoia, hallucinations, and delusions. In severe cases, paranoia causes homicidal and/or suicidal feelings. Methamphetamine also increases heart rate and blood pressure. It can cause strokes, which result from irreversible damage to blood vessels in the brain, as well as respiratory problems, irregular heartbeat, and extreme anorexia. In extreme situations, it can cause cardiovascular collapse and death.

No physical symptoms accompany withdrawal from the drug, but psychological symptoms such as depression, anxiety, aggression, and intense craving are common. Of greatest concern is the brain damage that occurs in long-term users.

Dopamine release may be the cause of the drug's long-term toxic effects. Compare the structure of the neurotransmitter dopamine (Figure 15.7) with that of methamphetamine shown below. Research in humans has shown that even three years after chronic methamphetamine use, the former user continues to have a reduced ability to transport dopamine back into nerve cells. Parkinson's disease is characterized by a decrease in the dopamine-producing neurons in the brain; so it was logical to look for similarities between methamphetamine users and those suffering from Parkinson's. In fact, the brains of methamphetamine users showed damage similar to, but not as severe as, that in Parkinson's disease. Research in animals demonstrated that up to 50% of the dopamine-producing cells in parts of the brain may be destroyed by prolonged exposure, and that serotonin-containing neurons may sustain even worse damage.

Methamphetamine use continues to rise, in part, because it is easily synthesized, or "cooked," using "recipes" that are available from many sources, including the Internet. Ephedrine, an over-the-counter decongestant drug, is the starting material.

(Pseudoephedrine, a stereoisomer of ephedrine, can also be used.) As shown in the equation below, ephedrine is simply reduced to produce methamphetamine.

$$\text{Ephedrine} \xrightarrow{[H]} \text{Methamphetamine}$$

While the synthesis involves a variety of dangerous chemicals, including anhydrous ammonia, anhydrous hydrochloric acid, sodium, and sodium hydroxide, most "meth cooks" do not have formal laboratory training. "Meth lab" fires are common, and the synthesis produces toxic wastes. The cleanup that followed the seizure of a major "meth lab" in 2003 took eight days and required fifty people. Over four million pounds of toxic soil and 133 drums of hazardous waste were removed from the site, at a cost of $226,000.

For Further Understanding

Compare the structures of methamphetamine and dopamine. Develop a hypothesis to explain why dopamine receptors also bind to and transport methamphetamine into neurons. (Hint: Receptors are proteins in cell membranes that have a pocket into which a specific molecule can fit.)

Explain why methamphetamine is soluble in alcohols and in water.

(a)

(b)

(c)

Figure 15.3

(a) Crack cocaine is a non-water-soluble base with a low melting point and a crystalline structure. (b) Powdered cocaine is a water-soluble salt of cocaine base.
(c) Cocaine is extracted from the leaves of the coca plant.

Cocaine hydrochloride salt can be converted into its base form by a process called "free basing." Although the chemistry is simple, the process is dangerous because it requires highly flammable solvents. This pure cocaine is not soluble in water. It has a relatively low melting point, however, and can be smoked. Crack, so called because of the crackling noise it makes when smoked, is absorbed into the body more quickly than the snorted powder and results in a more immediate high.

Quaternary Ammonium Salts

Quaternary ammonium salts are ammonium salts that have four organic groups bonded to the nitrogen. They have the following general structure:

$$R_4N^+X^- \quad (R = \text{any alkyl or aryl group;} \\ X^- = \text{a halide anion, most commonly } Cl^-)$$

Quaternary ammonium salts that have a very long carbon chain, sometimes called "quats," are used as disinfectants and antiseptics because they have detergent activity. Two popular quats are benzalkonium chloride (Zephiran) and cetylpyridinium chloride, found in the mouthwash Scope.

Benzalkonium chloride   Cetylpyridinium chloride

Phospholipids and biological membranes are discussed in Sections 17.3 and 17.6.

Choline is an important quaternary ammonium salt in the body. It is part of the hydrophilic "head" of the membrane phospholipid lecithin. Choline is also a precursor for the synthesis of the neurotransmitter acetylcholine.

The function of acetylcholine is described in greater detail in Section 15.5.

Choline

15.2 Heterocyclic Amines

Heterocyclic amines are cyclic compounds that have at least one nitrogen atom in the ring structure. The structures and common names of several heterocyclic amines important in nature are shown here. They are represented by their structural formulas and by abbreviated line formulas.

Imidazole   Pyridine

Pyrrole

Pyrimidine

The heterocyclic amines shown below are examples of fused ring structures. Each ring pair shares two carbon atoms in common. Thus, two fused rings share one or more common bonds as part of their ring backbones. Consider the structures of a purine, indole, and porphyrin, which are shown as structural formulas and as line diagrams.

Purine

Indole

M⁺ = metal ion

Porphyrin

The pyrimidine and purine rings are found in DNA and RNA. The porphyrin ring structure is found in hemoglobin (an oxygen-carrying blood protein), myoglobin (an oxygen-carrying protein found in muscle tissue), and chlorophyll (a photosynthetic plant pigment). The indole and pyridine rings are found in many **alkaloids,** which are naturally occurring compounds with one or more nitrogen-containing heterocyclic rings. The alkaloids include cocaine, nicotine, quinine, morphine, heroin, and LSD (Figure 15.4).

Lysergic acid diethylamide (LSD) is a hallucinogenic compound that may cause severe mental disorders. Cocaine is produced by the coca plant. In small

Coniine is produced by the poison hemlock plant. It is a neurotoxin that causes respiratory paralysis. It is the poison that was used to kill the Greek philosopher Socrates in 399 B.C. To what class of molecules does coniine belong?

The structures of purines and pyrimidines are presented in Section 20.1.

The structure of the heme group found in hemoglobin and myoglobin is presented in Section 18.9.

Figure 15.4
Structures of several heterocyclic amines with biological activity.

doses it is used as an anesthetic for the sinuses and eyes. An **anesthetic** is a drug that causes a lack of sensation in any part of the body (local anesthetic) or causes unconsciousness (general anesthetic). In higher doses, cocaine causes an intense feeling of euphoria followed by a deep depression. Cocaine is addictive because the user needs larger and larger amounts to overcome the depression. Nicotine is one of the simplest heterocyclic amines and appears to be the addictive component of cigarette smoke.

Morphine was the first alkaloid to be isolated from the sap of the opium poppy. Morphine is a strong **analgesic,** a drug that acts as a pain killer. However, it is a powerful and addictive narcotic. Codeine, also produced by the opium poppy, is a less powerful analgesic than morphine, but it is one of the most effective cough suppressants known. Heroin is produced in the laboratory by adding two acetyl groups to morphine. It was initially made in the hopes of producing a compound with the benefits of morphine but lacking the addictive qualities. However, heroin is even more addictive than morphine.

Strychnine is found in the seeds of an Asiatic tree. It is extremely toxic and was commonly used as a rat poison at one time. Quinine, isolated from the bark of South American trees, was the first effective treatment for malaria. Vitamin B_6 is one of the water-soluble vitamins required by the body.

15.3 Amides

Amides are the products formed in a reaction between a carboxylic acid derivative and ammonia or an amine. The general structure of an amide is shown here.

Ethanamide

The amide group is composed of two portions: the carbonyl group from a carboxylic acid and the amino group from ammonia or an amine. The bond between the carbonyl carbon and the nitrogen of the amine or ammonia is called the **amide bond.**

Structure and Physical Properties

Most amides are solids at room temperature. They have very high boiling points, and the simpler ones are quite soluble in water. Both of these properties are a result of strong intermolecular hydrogen bonding between the N—H bond of one amide and the C=O group of a second amide, as shown in Figure 15.5.

Unlike amines, amides are not bases (proton acceptors). The reason is that the highly electronegative oxygen atom of the carbonyl group causes a very strong attraction between the lone pair of nitrogen electrons and the carbonyl group. As a result, the unshared pair of electrons cannot "hold" a proton.

Because of the attraction of the carbonyl group for the lone pair of nitrogen electrons, the structure of the C—N bond of an amide is a *resonance hybrid.*

> **LEARNING GOAL**
> **8** Describe the physical properties of amides.

Resonance hybrids are discussed in Section 3.4.

$$ \underset{O}{\overset{R}{\underset{\|}{C}}} \overset{\overline{N}}{\underset{H\ \ H}{|}} \longleftrightarrow \underset{^-O}{\overset{R}{\underset{\|}{C}}} \overset{\overset{+}{N}}{\underset{H\ \ H}{\|}} $$

Nomenclature

The common and I.U.P.A.C. names of the amides are derived from the common and I.U.P.A.C. names of the carboxylic acids from which they were made. Remove the *-ic acid* ending of the common name or the *-oic acid* ending of the I.U.P.A.C. name of the carboxylic acid, and replace it with the ending *-amide.* Several examples

> **LEARNING GOAL**
> **9** Draw the structure and write the common and I.U.P.A.C. names of amides.

Figure 15.5
Hydrogen bonding in amides.

TABLE 15.4	I.U.P.A.C. and Common Names of Simple Amides	
Compound	**I.U.P.A.C. Name**	**Common Name**
O ‖ R—C—NH$_2$	Alkanamide (-amide replaces the -oic acid ending of the I.U.P.A.C. name of carboxylic acid)	Alkanamide (-amide replaces the -ic acid ending of the common name of carboxylic acid)
O ‖ H—C—NH$_2$	Methanamide	Formamide
O ‖ CH$_3$—C—NH$_2$	Ethanamide	Acetamide
O ‖ CH$_3$CH$_2$—C—NH$_2$	Propanamide	Propionamide
O ‖ H—C—NHCH$_3$	N-Methylmethanamide	N-Methylformamide
O ‖ CH$_3$—C—NHCH$_3$	N-Methylethanamide	N-Methylacetamide

Nomenclature of carboxylic acids is described in Section 14.1.

of the common and I.U.P.A.C. nomenclature are provided in Table 15.4 and in the following structures:

$$CH_3\overset{\displaystyle O}{\overset{\|}{C}}-NH_2 \qquad\qquad CH_3CH_2\overset{\displaystyle O}{\overset{\|}{C}}-NH_2$$

Ethan*oic acid* → Ethan*amide*   Propan*oic acid* → Propan*amide*
or   or
Acet*ic acid* → Acet*amide*   Propion*ic acid* → Propion*amide*

Substituents on the nitrogen are placed as prefixes and are indicated by *N-* followed by the name of the substituent. There are no spaces between the prefix and the amide name. For example:

$$CH_3CH_2\overset{\displaystyle O}{\overset{\|}{C}}-NH-CH_3 \qquad\qquad CH_3CH_2CH_2CH_2CH_2\overset{\displaystyle O}{\overset{\|}{C}}-NH-CH_2CH_2CH_3$$

N-Methylpropanamide   *N*-Propylhexanamide

EXAMPLE 15.4   **Naming Amides Using Common and I.U.P.A.C. Nomenclature Systems**

LEARNING GOAL

9 Draw the structure and write the common and I.U.P.A.C. names of amides.

Name the following amides using both the I.U.P.A.C. and common systems.

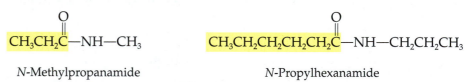

Continued—

EXAMPLE 15.4 —*Continued*

Solution

The names of amides are derived from the common and I.U.P.A.C. names of the carboxylic acids from which they were made.

	I.U.P.A.C.	**Common**
Parent carboxylic acid:	Butanoic acid (becomes butanamide)	Butyric acid (becomes butyramide)
Group on N:	propyl	propyl
Name:	N-Propylbutanamide	N-Propylbutyramide

$$CH_3CH_2\overset{\displaystyle O}{\overset{\|}{C}}-NH-CH_2CH_2CH_2CH_2CH_3$$

Solution

Parent carboxylic acid:	Propanoic acid (becomes propanamide)	Propionic acid (becomes propionamide)
Group on N:	pentyl	pentyl
Name:	N-Pentylpropanamide	N-Pentylpropionamide

Practice Problem 15.4

Provide the common and I.U.P.A.C. names for each of the following amides.

a. $CH_3CH_2CH_2CH_2\overset{\displaystyle O}{\overset{\|}{C}}-NH-CH_2CH_2CH_2CH_2CH_3$

b. $CH_3CH_2CH_2CH_2CH_2\overset{\displaystyle O}{\overset{\|}{C}}-NH-CH_2CH_2CH_2CH_3$

For Further Practice: Questions 15.51, 15.52, and 15.53.

Medically Important Amides

Barbiturates, often called "downers," are derived from amides and are used as sedatives. They are also used as anticonvulsants for epileptics and for people suffering from a variety of brain disorders that manifest themselves in neurosis, anxiety, and tension.

Barbital—a barbiturate

Phenacetin and acetaminophen are also amides. Acetaminophen is an aromatic amide that is commonly used in place of aspirin, particularly by people who are allergic to aspirin or who suffer stomach bleeding from the use of aspirin. It was first synthesized in 1893 and is the active ingredient in Tylenol and Datril. Like aspirin, acetaminophen relieves pain and reduces fever. However, unlike aspirin, it is not an anti-inflammatory drug.

A Medical Perspective

Semisynthetic Penicillins

The antibacterial properties of penicillin were discovered by Alexander Fleming in 1929. These natural penicillins produced by several species of the mold *Penicillium*, had a number of drawbacks. They were effective only against a type of bacteria referred to as Gram positive because of a staining reaction based on their cell wall structure. They were also very susceptible to destruction by bacterial enzymes called β-lactamases, and some were destroyed by stomach acid and had to be administered by injection.

To overcome these problems, chemists have produced semisynthetic penicillins by modifying the core structure. The core of penicillins is 6-aminopenicillanic acid, which consists of a thiazolidine ring fused to a β-lactam ring. In addition, there is an R group bonded via an amide bond to the core structure.

6-Aminopenicillanic acid

The β-lactam ring confers the antimicrobial properties. However, the R group determines the degree of antibacterial activity, the pharmacological properties, including the types of bacteria against which it is active, and the degree of resistance to the β-lactamases exhibited by any particular penicillin antibiotic. These are the properties that must be modified to produce penicillins that are acid resistant, effective with a broad spectrum of bacteria, and β-lactamase resistant.

Chemists simply remove the natural R group by cleaving the amide bond with an enzyme called an amidase. They then replace the R group and test the properties of the "new" anti-

biotic. Among the resulting semisynthetic penicillins are ampicillin, methicillin, and oxacillin.

Ampicillin

Methicillin

Oxacillin

For Further Understanding

Using the Internet and other resources, investigate and describe the properties of some new penicillins and the bacteria against which they are effective.

Why does changing the R group of a penicillin result in altered chemical and physiological properties?

Phenacetin was synthesized in 1887 and used as an analgesic for almost a century. Its structure and properties are similar to those of acetaminophen. However, it was banned by the U.S. Food and Drug Administration in 1983 because of the kidney damage and blood disorders that it causes.

Phenacetin   Acetaminophen

Reactions Involving Amides

Preparation of Amides

Amides are prepared from carboxylic acid derivatives, either acid chlorides or acid anhydrides. Recall that acid chlorides are made from carboxylic acids by reaction with reagents such as PCl_5.

LEARNING GOAL

10 Write equations representing the preparation of amides.

Formation of acid chlorides is described in Section 14.3.

$$\underset{\substack{\text{Carboxylic} \\ \text{acid}}}{R-\overset{\overset{\displaystyle O}{\|}}{C}-OH} \xrightarrow{PCl_5} \underset{\substack{\text{Acid} \\ \text{chloride}}}{R-\overset{\overset{\displaystyle O}{\|}}{C}-Cl} + \text{inorganic products}$$

These acid chlorides rapidly react with either ammonia or amines, as in:

$$\underset{\substack{\text{Acid} \\ \text{chloride}}}{R-\overset{\overset{\displaystyle O}{\|}}{C}-Cl} + \underset{\substack{\text{Ammonia} \\ \text{or} \\ \text{amine}}}{2NH_3} \longrightarrow \underset{\text{Amide}}{R-\overset{\overset{\displaystyle O}{\|}}{C}-NH_2} + \underset{\substack{\text{Ammonium chloride} \\ \text{or} \\ \text{alkylammonium chloride}}}{NH_4{}^+Cl^-}$$

Note that two molar equivalents of ammonia or amine are required in this reaction and that this is an acyl group transfer reaction. The **acyl group**

$$R-\overset{\overset{\displaystyle O}{\|}}{C}-$$

of the acid chloride is transferred from the Cl atom to the N atom of one of the ammonia or amine molecules. The second ammonia (or amine) reacts with the HCl formed in the transfer reaction to produce ammonium chloride or alkylammonium chloride.

The reaction between butanoyl chloride and methanamine to produce *N*-methylbutanamide is an example of an acyl group transfer reaction.

$$\underset{\substack{\text{Butanoyl} \\ \text{chloride}}}{CH_3CH_2CH_2\overset{\overset{\displaystyle O}{\|}}{C}-Cl} + \underset{\text{Methanamine}}{2CH_3NH_2} \longrightarrow$$

$$\underset{\text{N-Methylbutanamide}}{CH_3CH_2CH_2\overset{\overset{\displaystyle O}{\|}}{C}-NH-CH_3} + \underset{\substack{\text{Methylammonium} \\ \text{chloride}}}{CH_3NH_3{}^+Cl^-}$$

The reaction between an amine and an acid anhydride is also an acyl group transfer. The general equation for the synthesis of an amide in the reaction between an acid anhydride and ammonia or an amine is

$$\underset{\text{Acid anhydride}}{R-\overset{\overset{\displaystyle O}{\|}}{C}-O-\overset{\overset{\displaystyle O}{\|}}{C}-R} + \underset{\substack{\text{Ammonia} \\ \text{or} \\ \text{amine}}}{2NH_3} \longrightarrow \underset{\text{Amide}}{R-\overset{\overset{\displaystyle O}{\|}}{C}-NH_2} + \underset{\substack{\text{Carboxylic acid} \\ \text{salt}}}{R-\overset{\overset{\displaystyle O}{\|}}{C}-O^-NH_4{}^+}$$

When subjected to heat, the ammonium salt loses a water molecule to produce a second amide molecule.

A well-known commercial amide is the artificial sweetener aspartame or NutraSweet. Although the name suggests that it is a sugar, it is not a sugar at all.

Figure 15.6

The amide bond. (a) NutraSweet, the dipeptide aspartame, is a molecule composed of two amino acids joined by an amide (peptide) bond. (b) Neotame, a newly approved sweetener, is also a dipeptide. One of the amino acids has been modified so that it is safe for use by phenylketonurics.

Amino acids have both a carboxyl group and an amino group and are discussed in detail in Sections 15.4 and 18.1.

In fact, it is the methylester of a molecule composed of two amino acids, aspartic acid and phenylalanine, joined by an amide bond (Figure 15 6a).

Packages of aspartame carry the warning: "Phenylketonurics: Contains Phenylalanine." Digestion of aspartame and heating to high temperatures during cooking break both the ester bond and the amide bond, which releases the amino acid phenylalanine. People with the genetic disorder phenylketonuria (PKU) cannot metabolize this amino acid. As a result, it builds up to toxic levels that can cause mental retardation in an infant born with the condition. This no longer occurs because every child is tested for PKU at the time of birth and each is treated with a diet that limits the amount of phenylalanine to only the amount required for normal growth.

In July 2002 the Food and Drug Administration approved a new artificial sweetener that is related to aspartame. Called neotame, it has the same core structure as aspartame, but a 3,3-dimethylbutyl group has been added to the aspartic acid (Figure 15.6b). Digestion and heating still cause breakage of the ester bond, but the bulky 3,3-dimethylbutyl group blocks the breakage of the amide bond. Neotame can be used without risk by people with PKU and also retains its sweetness during cooking.

Question 15.9

What is the structure of the amine that, on reaction with the acid chlorides shown, will give each of the following products?

a. $? + CH_3\overset{\overset{\displaystyle O}{\|}}{C}-Cl \longrightarrow CH_3\overset{\overset{\displaystyle O}{\|}}{C}NHCH_3 + CH_3NH_3{}^+Cl^-$

b. $? + CH_3CH_2CH_2CH_2\underset{\underset{\displaystyle CH_2CH_3}{|}}{C}H\overset{\overset{\displaystyle O}{\|}}{C}-Cl \longrightarrow (CH_3)_2N\overset{\overset{\displaystyle O}{\|}}{C}\underset{\underset{\displaystyle CH_2CH_3}{|}}{C}HCH_2CH_2CH_2CH_3$

$+ (CH_3)_2NH_2{}^+Cl^-$

What are the structures of the acid chlorides and the amines that will react to give each of the following products?

a. *N*-Ethylhexanamide
b. *N*-Propylbutanamide

Hydrolysis of Amides

Hydrolysis of an amide results in breaking the amide bond to produce a carboxylic acid and ammonia or an amine. It is very difficult to hydrolyze the amide bond. In fact, the reaction requires heating the amide in the presence of a strong acid or base.

LEARNING GOAL
11 Write equations showing the hydrolysis of amides.

$$\underset{\substack{\text{Amide}}}{R-\overset{\overset{\displaystyle O}{\|}}{C}-NH-R^1} + \underset{\substack{\text{Strong}\\\text{acid}}}{H_3O^+} \longrightarrow \underset{\substack{\text{Carboxylic}\\\text{acid}}}{R-\overset{\overset{\displaystyle O}{\|}}{C}-OH} + \underset{\substack{\text{Alkylammonium}\\\text{ion}\\\text{or}\\\text{ammonium ion}}}{R^1-N^+H_3}$$

$$\underset{\substack{\text{Butanamide}\\\text{(butyramide)}}}{CH_3CH_2CH_2\overset{\overset{\displaystyle O}{\|}}{C}-NH_2} + H_3O^+ \longrightarrow \underset{\substack{\text{Butanoic acid}\\\text{(butyric acid)}}}{CH_3CH_2CH_2\overset{\overset{\displaystyle O}{\|}}{C}-OH} + N^+H_4$$

If a strong base is used, the products are the amine and the salt of the carboxylic acid:

$$\underset{\substack{\text{Amide}}}{R-\overset{\overset{\displaystyle O}{\|}}{C}-NH-R^1} + \underset{\substack{\text{Strong}\\\text{base}}}{NaOH} \longrightarrow \underset{\substack{\text{Carboxylic}\\\text{acid salt}}}{R-\overset{\overset{\displaystyle O}{\|}}{C}-O^-Na^+} + \underset{\substack{\text{Amine}\\\text{or}\\\text{ammonia}}}{R^1-NH_2}$$

$$\underset{\substack{\textit{N}\text{-Methylpropanamide}\\(\textit{N}\text{-methylpropionamide})}}{CH_3CH_2\overset{\overset{\displaystyle O}{\|}}{C}-NHCH_3} + NaOH \longrightarrow \underset{\substack{\text{Sodium propanoate}\\\text{(sodium propionate)}}}{CH_3CH_2\overset{\overset{\displaystyle O}{\|}}{C}-O^-Na^+} + \underset{\substack{\text{Methanamine}\\\text{(methylamine)}}}{CH_3NH_2}$$

15.4 A Preview of Amino Acids, Proteins, and Protein Synthesis

In Chapter 18 we will describe the structure of proteins, the molecules that carry out the majority of the biological processes essential to life. Proteins are polymers of amino acids. As the name suggests, amino acids have two essential functional groups, an amino group ($-NH_2$) and a carboxyl group ($-COOH$). Typically amino acids have the following general structure:

LEARNING GOAL
12 Draw the general structure of an amino acid.

LEARNING GOAL
13 Draw and discuss the structure of a peptide bond.

$$\underset{\substack{|\\R}}{\overset{\substack{H\\|}}{H_2N-C-COOH}} \qquad \text{(R may be a hydrogen atom or an organic group.)}$$

In the cell the amino group is usually protonated and the carboxyl group is usually ionized to the carboxylate anion. In the future we will represent an amino acid in the following way:

$$\overset{\displaystyle H}{\underset{\displaystyle R}{\overset{+}{H_3N}-C-COO^-}}$$

The amide bond that forms between the carboxyl group of one amino acid and the amino group of another is called the **peptide bond.**

The peptide bond is an amide bond.

$$H_3{}^+N-\overset{H}{\underset{R}{C}}-\overset{O}{C}-\overset{H}{\underset{H}{N}}-\overset{H}{\underset{R}{C}}-\overset{O}{C}-O^-$$

The joining of amino acids by amide bonds produces small *peptides* and larger *proteins*. Because protein structure and function are essential for life processes, it is fortunate indeed that the amide bonds that hold them together are not easily hydrolyzed at physiological pH and temperature.

The process of protein synthesis in the cell mimics amide formation in the laboratory; it involves acyl group transfer. There are several important differences between the chemistry in the laboratory and the chemistry in the cell. During protein synthesis, the **aminoacyl group** of the amino acid is transferred, rather than the acyl group of a carboxylic acid. In addition, the aminoacyl group is not transferred from a carboxylic acid derivative; it is transferred from a special carrier molecule called a **transfer RNA (tRNA).** When the aminoacyl group is covalently bonded to a tRNA, the resulting structure is called an *aminoacyl tRNA*:

Aminoacyl group ⟶ $H_2N-\overset{H}{\underset{R}{C}}-\overset{O}{C}-$ transfer RNA

The aminoacyl group of the aminoacyl tRNA is transferred to the amino group nitrogen to form a peptide bond. The transfer RNA is recycled by binding to another of the same kind of aminoacyl group.

More than one hundred kinds of proteins, nucleotides, and RNA molecules participate in the incredibly intricate process of protein synthesis. In Chapter 18 we will study protein structure and learn about the many functions of proteins in the life of the cell. In Chapter 20 we will study the details of protein synthesis to see how these aminoacyl transfer reactions make us the individuals that we are.

15.5 Neurotransmitters

LEARNING GOAL

14 Describe the function of neurotransmitters.

Neurotransmitters are chemicals that carry messages, or signals, from a nerve cell to a target cell, which may be another nerve cell or a muscle cell. Neurotransmitters are classified as being *excitatory*, stimulating their target cell, or *inhibitory*, decreasing activity of the target cell. One feature shared by the neurotransmitters is that they are all nitrogen-containing compounds. Some of them have rather complex structures and one, nitric oxide (NO), consists of only two atoms. Several important neurotransmitters are discussed in the following sections.

Catecholamines

All of the catecholamine neurotransmitters, including dopamine, epinephrine, and norepinephrine, are synthesized from the amino acid tyrosine (Figure 15.7). *Dopamine* is critical to good health. A deficiency in this neurotransmitter, for example, results in Parkinson's disease, a disorder characterized by tremors, monotonous speech, loss of memory and problem-solving ability, and loss of motor function. In the brain, dopamine is synthesized from L-dopa; so it would seem logical to treat Parkinson's disease with dopamine. Unfortunately, dopamine cannot cross the blood-brain barrier to enter brain cells. As a result, L-dopa, which is converted to dopamine in brain cells, is used to treat this disorder.

Just as too little dopamine causes Parkinson's disease, an excess is associated with schizophrenia. Dopamine also appears to play a role in addictive behavior. In proper amounts, it causes a pleasant, satisfied feeling. The greater the amount of dopamine, the more intense the sensation, the "high." Several drugs have been shown to increase the levels of dopamine. Among these are cocaine, heroin, amphetamines, alcohol, and nicotine. Marijuana also causes an increase in brain dopamine, raising the possibility that it, too, has the potential to produce addiction.

Both *epinephrine* (adrenaline) and *norepinephrine* are involved in the "fight or flight" response. Epinephrine stimulates the breakdown of glycogen to produce glucose, which is then metabolized to provide energy for the body. Norepinephrine is involved with the central nervous system in the stimulation of other glands and the constriction of blood vessels. All of these responses prepare the body to meet the stressful situation.

Serotonin

Serotonin is synthesized from the amino acid tryptophan (Figure 15.8). A deficiency of serotonin has been associated with depression. It is also thought to be involved in bulimia and anorexia nervosa, as well as the carbohydrate-cravings that characterize seasonal affective disorder (SAD), a depression caused by a decrease in daylight during autumn and winter.

Serotonin also affects the perception of pain, thermoregulation, and sleep. There are those who believe that a glass of warm milk will help you fall asleep. We have all noticed how sleepy we become after that big Thanksgiving turkey dinner. Both milk protein and turkey are exceptionally high in tryptophan, the precursor of serotonin!

Prozac (fluoxetine), one of the newer generation of antidepressant drugs, is one of the most widely prescribed drugs in the United States.

Prozac (fluoxetine)

Figure 15.7
The pathway for synthesis of dopamine, epinephrine, and norepinephrine.

Figure 15.8
Synthesis of serotonin from the amino acid tryptophan.

A Medical Perspective

Opiate Biosynthesis and the Mutant Poppy

Hippocrates, the Father of Medicine, left us the first record of the therapeutic use of opium (460 B.C.). Although not recorded, it is probably true that the addictive properties of opium were recognized soon thereafter!

The opium poppy *(Papaver somniferum)* is cultivated, legally and illegally, in many parts of the world. The flowers vary in color from white to deep red, but it is the seed pod that is sought after. In the seed pod is a milky fluid that contains morphine and codeine, and a small amount of an opioid called thebaine. The juice is extracted from the unripe seed pods and dried, and the opium alkaloids are extracted and purified.

In the legal pharmaceutical world, morphine and codeine are used to ease pain and spasmodic coughing. Thebaine is used as a reactant in pharmaceutical synthesis to produce a number of synthetic opioid compounds with a variety of biological effects. These include the analgesics oxycodone (brand name OxyContin), oxymorphone, and nalbuphine; naloxone, which is used to treat opioid overdosage; naltrexone, which is useful in helping people with narcotic or alcohol addictions to remain drug free; and buprenorphine, which is useful in the treatment of opiate addiction because it prevents withdrawal symptoms.

Approximately 40% of the world's legal opium poppies are grown on the Australian island state of Tasmania. This is big business, and the industry has developed an active research program to study the biochemical pathway for the synthesis of morphine and codeine. That pathway begins with the amino acid tyrosine, the same amino acid that is the initial reactant for the synthesis of dopamine and epinephrine (Section 15.5).

7 steps   6 steps   3 steps   1 step

Tyrosine → Reticuline → Thebaine → Codeine → Morphine

Through seven chemical reactions, tyrosine is converted to reticuline. Another six reactions convert reticuline to thebaine. Three chemical modifications convert thebaine to codeine, which undergoes an ester hydrolysis to produce morphine.

In the course of their studies, researchers produced a mutant strain of poppy that cannot make morphine or codeine, but does produce high levels of thebaine. "Norman," for "No Morphine," has been the most common strain of poppy grown in Tasmania since 1997. The mutation that causes Norman to produce high levels of thebaine is an alteration in one of the enzymes that catalyzes the conversion of thebaine to codeine. Since it can't be converted into codeine, large amounts of thebaine accumulate in the seed pods.

Synthetic opioids, such as naloxone and buprenorphine and the others mentioned above, have become much more important commercially than codeine and morphine. The economic value of Norman is that it produces large amounts of the

(a)

(b)

(a) Opium poppies are the source of morphine and codeine. (b) The sap of an opium poppy is white. The sap of the no-morphine mutant poppy is red.

starting material for the synthesis of these synthetic opioids, as well as the experimental synthesis of new drugs with unknown potential.

Thebaine → many steps → Naloxone

Thebaine → many steps → Buprenorphine

Consider the drug formulation Suboxone, a combination of buprenorphine and naloxone, approved for use in the United States in 2003 and produced by the British company Reckitt Benckiser. This combination of synthetic opiates calms the addict's craving for opiates and yet poses little risk of being abused.

Buprenorphine or "bupe" works by binding to the same receptors in the brain to which heroin binds; but the drug is only a partial heroin agonist, so there is no high. As a result, buprenorphine is much less addictive than drugs such a methadone, leaves patients much more clearheaded, and makes it easier for them to withdraw from the drug after a few months.

Because addicts don't get high from Suboxone, it is much less likely than methadone to be stolen and sold illegally. It has the added advantage that its effects are longer lasting than those of methadone; so addicts need only one pill every two or three days. In addition, because naloxone is an opioid antagonist, it causes instant withdrawal symptoms if an addict tries to inject Suboxone for a high. As a result, Suboxone can be given to addicts to be taken at home, rather than being dispensed only at clinics, as methadone is. With all of these features, Suboxone begins to sound like a miracle drug; but as with any addiction treatment, it will only help those who want to quit and are willing to work with counselors and support groups to resolve the underlying problems that caused the addiction in the first place.

For Further Understanding

In 1998 researchers in England reported that some individuals who had eaten poppy seed rolls or cake tested positive in an opiate drug screen. Using Internet or other resources, investigate the "poppy seed defense" and suggest guidelines for opiate testing that would protect the innocent.

OxyContin is the brand name for a formulation of oxycodone in a timed-release tablet. It is prescribed to provide up to twelve hours of relief from chronic pain. Recently, OxyContin has become a commonly abused drug and is thought to be responsible for a number of deaths. Abusers crush the time-release tablets and ingest or snort the drug, which results in a rapid and powerful high often compared to the euphoria experienced from taking heroin. Use the Internet or other resources to explain why abuse of this prescription medication has overshadowed heroin use in some areas. Consider ways to prevent such abuse.

Figure 15.9

Synthesis of histamine from the amino acid histidine.

It is a member of a class of drugs called selective serotonin reuptake inhibitors (SSRI). By inhibiting the reuptake, Prozac effectively increases the level of serotonin, relieving the symptoms of depression.

Histamine

Histamine is a neurotransmitter that is synthesized in many tissues by removing the carboxyl group from the amino acid histidine (Figure 15.9). It has many, often annoying, physiological roles. Histamine is released during the allergic response. It causes the itchy skin rash associated with poison ivy or insect bites. It also promotes the red, watery eyes and respiratory symptoms of hay fever.

Many antihistamines are available to counteract the symptoms of histamine release. These act by competing with histamine for binding to target cells. If histamine cannot bind to these target cells, the allergic response stops.

Benadryl is an antihistamine that is available as an ointment to inhibit the itchy rash response to allergens. It is also available as an oral medication to block the symptoms of systemic allergies. You need only visit the "colds and allergies" aisle of your grocery store to find dozens of medications containing antihistamines.

Histamines also stimulate secretion of stomach acid. When this response occurs frequently, the result can be chronic heartburn. The reflux of stomach acid into the esophagus can result in erosion of tissue and ulceration. The excess stomach acid may also contribute to development of stomach ulcers. The drug marketed as Tagamet (cimetidine) has proven to be an effective inhibitor of this histamine response, providing relief from chronic heartburn.

Tagamet (cimetidine)

γ-Aminobutyric Acid and Glycine

γ-Aminobutyric acid (GABA) is produced by removal of a carboxyl group from the amino acid glutamate (Figure 15.10). Both GABA and the amino acid *glycine*

Glycine

Figure 15.10

Synthesis of GABA from the amino acid glutamate.

are inhibitory neurotransmitters acting in the central nervous system. One class of tranquilizers, the benzodiazopines, relieves aggressive behavior and anxiety. These drugs have been shown to enhance the inhibitory activity of GABA, suggesting one of the roles played by this neurotransmitter.

Acetylcholine

Acetylcholine is a neurotransmitter that functions at the neuromuscular junction, carrying signals from the nerve to the muscle. It is synthesized in a reaction between the quaternary ammonium ion choline and acetyl coenzyme A (Figure 15.11). When it is released from the nerve cell, acetylcholine binds to receptors on the surface of muscle cells. This binding stimulates the muscle cell to contract. Acetylcholine is then broken down to choline and acetate ion.

$$CH_3-\overset{\overset{\displaystyle O}{\|}}{C}-O-CH_2CH_2-N^+(CH_3)_3 \longrightarrow HO-CH_2CH_2-N^+(CH_3)_3 + CH_3COO^-$$

| Acetylcholine | | Choline | Acetate |

These molecules are essentially recycled. They are taken up by the nerve cell where they are used to resynthesize acetylcholine, which is stored in the nerve cell until it is needed.

Nicotine is an agonist of acetylcholine. An agonist is a compound that binds to the receptor for another compound and causes or enhances the biological response. By binding to acetylcholine receptors, nicotine causes the sense of alertness and calm many smokers experience. Nerve cells that respond to nicotine may also signal nerve cells that produce dopamine. As noted above, the dopamine may be responsible for the addictive property of nicotine.

Inhibitors of acetylcholinesterase, the enzyme that catalyzes the breakdown of acetylcholine, are used both as poisons and as drugs. Among the most important poisons of acetylcholinesterase are a class of compounds known as organophosphates. One of these is *diisopropyl fluorophosphate* (DIFP). This molecule forms a covalently bonded intermediate with the enzyme, irreversibly inhibiting its activity.

$$(CH_3)_2CH-O-\overset{\overset{\displaystyle O}{\|}}{\underset{\underset{\displaystyle F}{|}}{P}}-O-CH(CH_3)_2$$

Diisopropyl fluorophosphate (DIFP)

Figure 15.11
Synthesis of acetylcholine.

$$HO-CH_2CH_2-N^+(CH_3)_3 + CH_3-\overset{\overset{\displaystyle O}{\|}}{C}-S-\text{Coenzyme A}$$

| Choline | Acetyl Coenzyme A |

$$CH_3-\overset{\overset{\displaystyle O}{\|}}{C}-O-CH_2CH_2-N^+(CH_3)_3 + \text{Coenzyme A}$$

Acetylcholine

Thus, it is unable to break down the acetylcholine and nerve transmission continues, resulting in muscle spasm. Death may occur as a result of laryngeal spasm. Antidotes for poisoning by organophosphates, which include many insecticides and nerve gases, have been developed. The antidotes work by reversing the effects of the inhibitor. One of these antidotes is *pyridine aldoxime methiodide* (PAM). This molecule displaces the organophosphate group from the active site of the enzyme, alleviating the effects of the poison.

Pyridine aldoxime methiodide

Succinylcholine is a competitive inhibitor of acetylcholine which is used as a muscle relaxant in surgical procedures. Competitive inhibition occurs because the two molecules have structures so similar that both can bind to the acetylcholine receptor (compare the structures below). When administered to a patient, there is more succinylcholine than acetylcholine in the synapse, and thus more succinylcholine binding to the receptor. Because it cannot stimulate muscle contraction, succinylcholine causes muscles to relax. Normal muscle contraction resumes when the drug is no longer administered.

Acetylcholine

Succinylcholine

Acetylcholine nerve transmission is discussed in further detail in A Clinical Perspective: Enzymes, Nerve Transmission, and Nerve Agents in Chapter 19.

Nitric Oxide and Glutamate

Nitric oxide (NO) is an amazing little molecule that has been shown to have many physiological functions. Among these is its ability to act as a neurotransmitter. NO is synthesized in many areas of the brain from the amino acid arginine. Research has suggested that NO works in conjunction with another neurotransmitter, the amino acid glutamate (see the structure of glutamate in Figure 15.10). Glutamate released from one nerve cell binds to receptors on its target cell. This triggers the target cell to produce NO, which then diffuses back to the original nerve cell. The NO signals the cell to release more glutamate, thus stimulating this neural pathway even further. This is a kind of positive feedback loop. It is thought that this NO-glutamate mechanism is involved in learning and the formation of memories.

Summary of Reactions

Preparation of Amines

A nitro compound $\xrightarrow{[H]}$ An aromatic primary amine

$$R^1-\underset{\underset{O}{\parallel}}{C}-N\begin{smallmatrix}R^2\\\\R^3\end{smallmatrix} \xrightarrow{[H]} R^1CH_2N\begin{smallmatrix}R^2\\\\R^3\end{smallmatrix}$$

Amide   Amine

Basicity of Amines

$$R-NH_2 + H-OH \rightleftharpoons R-\overset{\overset{H}{|}}{\underset{\underset{H}{|}}{N^+}}-H + OH^-$$

Amine   Water   Alkylammonium ion   Hydroxide ion

Neutralization of Amines

$$R-NH_2 + HCl \longrightarrow R-\overset{\overset{H}{|}}{\underset{\underset{H}{|}}{N^+}}-HCl^-$$

Amine   Acid   Alkylammonium salt

Preparation of Amides

$$R-\underset{\underset{O}{\parallel}}{C}-Cl + 2NH_3 \longrightarrow$$

Acid chloride   Ammonia or amine

$$R-\underset{\underset{O}{\parallel}}{C}-NH_2 + NH_4^+Cl^-$$

Amide   Ammonium chloride or alkylammonium chloride

$$R-\underset{\underset{O}{\parallel}}{C}-O-\underset{\underset{O}{\parallel}}{C}-R + 2NH_3 \longrightarrow$$

Acid anhydride   Ammonia or amine

$$R-\underset{\underset{O}{\parallel}}{C}-NH_2 + R-\underset{\underset{O}{\parallel}}{C}-O^-NH_4^+$$

Amide   Carboxylic acid salt

Hydrolysis of Amides

$$R-\underset{\underset{O}{\parallel}}{C}-NH-R^1 + H_3O^+ \longrightarrow R-\underset{\underset{O}{\parallel}}{C}-OH + R^1-N^+H_3$$

Amide   Strong acid   Carboxylic acid   Alkyl-ammonium ion

$$R-\underset{\underset{O}{\parallel}}{C}-NH-R^1 + NaOH \longrightarrow R-\underset{\underset{O}{\parallel}}{C}-O^-Na^+ + R^1-NH_2$$

Amide   Strong base   Carboxylic acid salt   Amine or ammonia

15.1 Amines

Amines are a family of organic compounds that contain an amino group or substituted amino group. A *primary amine* has the general formula RNH_2; a *secondary amine* has the general formula R_2NH; and a *tertiary amine* has the general formula R_3N. In the systematic nomenclature system, amines are named as *alkanamines.* In the common system they are named as *alkylamines.* Amines behave as weak bases, forming *alkylammonium ions* in water and alkylammonium salts when they react with acids. *Quaternary ammonium salts* are ammonium salts that have four organic groups bonded to the nitrogen atom.

15.2 Heterocyclic Amines

Heterocyclic amines are cyclic compounds having at least one nitrogen atom in the ring structure. *Alkaloids* are natural plant products that contain at least one heterocyclic ring. Many alkaloids have powerful biological effects.

15.3 Amides

Amides are formed in a reaction between a carboxylic acid derivative and an amine (or ammonia). The *amide bond* is the bond between the carbonyl carbon of the *acyl group* and the nitrogen of the amine. In the I.U.P.A.C. Nomenclature System, they are named by replacing the *-oic acid* ending of the carboxylic acid with the *-amide* ending. In the common system of nomenclature the *-ic acid* ending of the carboxylic acid is replaced by the *-amide* ending. Hydrolysis of an amide produces a carboxylic acid and an amine (or ammonia).

15.4 A Preview of Amino Acids, Proteins, and Protein Synthesis

Proteins are polymers of amino acids joined to one another by *amide bonds* called *peptide bonds.* During protein synthesis, the *aminoacyl group* of one amino acid is transferred from a carrier molecule called a *transfer RNA* to the amino group nitrogen of another amino acid.

15.5 Neurotransmitters

Neurotransmitters are chemicals that carry messages, or signals, from a nerve cell to a target cell, which may be another nerve cell or a muscle cell. They may be inhibitory or excitatory and all are nitrogen-containing compounds. The catecholamines include dopamine, norepinephrine, and epinephrine. Too little dopamine results in Parkinson's disease. Too much is associated with schizophrenia. Dopamine is also associated with addictive behavior. A deficiency of serotonin is associated with depression and eating

disorders. Serotonin is involved in pain perception, regulation of body temperature, and sleep. Histamine contributes to allergy symptoms. Antihistamines block histamines and provide relief from allergies. γ-Aminobutyric acid (GABA) and glycine are inhibitory neurotransmitters. It is believed that GABA is involved in control of aggressive behavior. Acetylcholine is a neurotransmitter that functions at the neuromuscular junction, carrying signals from the nerve to the muscle. Nitric oxide and glutamate function in a positive feedback loop that is thought to be involved in learning and the formation of memories.

acyl group (15.3)
alkaloid (15.2)
alkylammonium ion (15.1)
amide (15.3)
amide bond (15.3)
amine (15.1)
aminoacyl group (15.4)
analgesic (15.2)
anesthetic (15.2)

heterocyclic amine (15.2)
neurotransmitter (15.5)
peptide bond (15.4)
primary (1°) amine (15.1)
quaternary ammonium salt (15.1)
secondary (2°) amine (15.1)
tertiary (3°) amine (15.1)
transfer RNA (tRNA) (15.4)

Amines

Foundations

15.11 Compare the boiling points of amines, alkanes, and alcohols of the same molecular weight. Explain these differences in boiling points.

15.12 Describe the water solubility of amines in relation to their carbon chain length.

15.13 Describe the systematic rules for naming amines.

15.14 How are the common names of amines derived?

15.15 Describe the physiological effects of amphetamines.

15.16 Define the terms *analgesic* and *anesthetic* and list some amines that have these activities.

Applications

15.17 For each pair of compounds predict which would have greater solubility in water. Explain your reasoning.
   a. Pentane or 1-butanamine
   b. Cyclohexane or 2-pentanamine

15.18 For each pair of compounds predict which would have the higher boiling point. Explain your reasoning.
   a. Ethanamine or ethanol
   b. Butane or 1-propanamine
   c. Methanamine or water
   d. Ethylmethylamine or butane

15.19 Explain why a tertiary amine such as triethylamine has a significantly lower boiling point than its primary amine isomer, 1-hexanamine.

15.20 Draw a diagram to illustrate your answer to Problem 15.19.

15.21 Use systematic nomenclature to name each of the following amines:

a. $CH_3CH_2CHNH_2$
    |
    CH_3

b. $CH_3CH_2CH_2CHCH_2CH_3$
              |
              NH_2

c.

d. $(CH_3)_3C—NH_2$

15.22 Use systematic and common nomenclature to name each of the following amines:

a. $CH_3CH_2CH_2CH_2CH_2CH_2CH_2CH_2NH_2$

b. Cl—⟨benzene ring⟩—NH_2

c. $CH_3CHCH_2CH_3$
         |
         NH_2

d. $CH_3NCH_2CH_3$
         |
         CH_3

15.23 Draw the structure of each of the following compounds:

a. Diethylamine
b. Butylamine
c. 3-Decanamine
d. 3-Bromo-2-pentanamine
e. Triphenylamine

15.24 Draw the structure of each of the following compounds:

a. *N,N*-Dipropylaniline
b. Cyclohexanamine
c. 2-Bromocyclopentanamine
d. Tetraethylammonium iodide
e. 3-Bromobenzenamine

15.25 Draw each of the following compounds with condensed formulas:

a. 2-Pentanamine
b. 2-Bromo-1-butanamine
c. Ethylisopropylamine
d. Cyclopentanamine

15.26 Draw each of the following compounds with condensed formulas:

a. Dipentylamine
b. 3,4-Dinitroaniline
c. 4-Methyl-3-heptanamine
d. *t*-Butylpentylamine
e. 3-Methyl-3-hexanamine
f. Trimethylammonium iodide

15.27 Draw structural formulas for the eight isomeric amines that have the molecular formula $C_4H_{11}N$. Name each of the isomers using the systematic names and determine whether each isomer is a 1°, 2°, or 3° amine.

15.28 Draw all of the isomeric amines of molecular formula C_3H_9N. Name each of the isomers, using the systematic names and determine whether each isomer is a primary, secondary, or tertiary amine.

15.29 Classify each of the following amines as 1°, 2°, or 3°:

a. Cyclohexanamine
b. Dibutylamine
c. 2-Methyl-2-heptanamine
d. Tripentylamine

15.30 Classify each of the following amines as primary, secondary, or tertiary:

a. Benzenamine
b. *N*-Ethyl-2-pentanamine
c. Ethylmethylamine
d. Tripropylamine
e. *m*-Chloroaniline

15.31 Write an equation to show a reaction that would produce each of the following products:

a. ⟨benzene ring with NH_2 top and CH_3 bottom⟩

c. ⟨benzene ring with NH_2⟩

b. ⟨benzene ring with NH_2 and OH⟩

d. NH_2 / CH_2 ⟨cyclohexane ring⟩

15.32 Write an equation to show a reaction that would produce each of the following amines:

a. 1-Pentanamine
b. *N,N*-Dimethylethanamine
c. *N*-Ethylpropanamine

15.33 Complete each of the following reactions by supplying the missing reactant or product indicated by a question mark:

a. CH_3 CH_3
        | |
  $CH_3NH + ? \longrightarrow CH_3N^+H + OH^-$
                                    |
                                    H

b. CH_3 CH_3
        | |
  $CH_3CH_2N + ? \longrightarrow CH_3CH_2N^+HBr^-$
        | |
        CH_2CH_3 CH_2CH_3

c. $CH_3CH_2CH_2NH_2 + H_2O \longrightarrow ? + OH^-$

d. CH_2CH_3
        |
  $CH_3CH_2NH + HCl \longrightarrow ?$

15.34 Complete each of the following reactions by supplying the missing reactant or product indicated by a question mark:

a. $CH_3CH_2NH_2 + H_2O \longrightarrow ? + OH^-$

b. $CH_2CH_2CH_3$
                                 |
  $? + HCl \longrightarrow CH_3CH_2CH_2N^+H\ Cl^-$
                                 |
                                 H

c. CH_3
        |
  $CH_3CHNH + H_2O \longrightarrow ? + ?$
        |
        CH_3

d. $NH_3 + HBr \longrightarrow ?$

15.35 Briefly explain why the lower-molecular-weight amines (fewer than five carbons) exhibit appreciable solubility in water.

15.36 Why is the salt of an amine appreciably more soluble in water than the amine from which it was formed?

15.37 Most drugs containing amine groups are not administered as the amine but rather as the ammonium salt. Can you suggest a reason why?

15.38 Why does aspirin upset the stomach, whereas acetaminophen (Tylenol) does not?

15.39 Putrescine and cadaverine are two odoriferous amines that are produced by decaying flesh. Putrescine is 1,4-butanediamine, and cadaverine is 1,5-pentanediamine. Draw the structures of these two compounds.

15.40 How would you quickly convert an alkylammonium salt into a water-insoluble amine? Explain the rationale for your answer.

Heterocyclic Amines

15.41 Indole and pyridine rings are found in alkaloids.
　　a. Sketch each ring.
　　b. Name one compound containing each of the ring structures and indicate its use.

15.42 What is an alkaloid?

15.43 List some heterocyclic amines that are used in medicine.

15.44 Distinguish between the terms *analgesic* and *anesthetic*.

Amides

Foundations

15.45 Why do amides have very high boiling points?

15.46 Describe the water solubility of amides in relation to their carbon chain length.

15.47 Explain the I.U.P.A.C. nomenclature rules for naming amides.

15.48 How are the common names of amides derived?

15.49 Describe the physiological effects of barbiturates.

15.50 Why is acetaminophen often recommended in place of aspirin?

Applications

15.51 Use the I.U.P.A.C. and common systems of nomenclature to name the following amides:

　　a. $CH_3CH_2\overset{\overset{\displaystyle O}{\|}}{C}NH_2$

　　b. $CH_3CH_2CH_2CH_2\overset{\overset{\displaystyle O}{\|}}{C}NH_2$

　　c. $CH_3\overset{\overset{\displaystyle O}{\|}}{C}N(CH_3)_2$

15.52 Use the I.U.P.A.C. Nomenclature System to name each of the following amides:

　　a. $CH_3CH_2CHCH_2\overset{\overset{\displaystyle O}{\|}}{C}NH_2$
　　　　　　　$\underset{\displaystyle Br}{|}$

　　b. (benzene ring)$-\overset{\overset{\displaystyle O}{\|}}{C}NH_2$ with Br substituent

　　c. $CH_3\overset{\overset{\displaystyle O}{\|}}{C}HCNH_2$
　　　　　　$\underset{\displaystyle CH_3}{|}$

15.53 Draw the condensed structural formula of each of the following amides:
　　a. Ethanamide
　　b. *N*-Methylpropanamide
　　c. *N,N*-Diethylbenzamide
　　d. 3-Bromo-4-methylhexanamide
　　e. *N,N*-Dimethylacetamide

15.54 Draw the condensed formula of each of the following amides:
　　a. Acetamide
　　b. 4-Methylpentanamide
　　c. *N,N*-Dimethylpropanamide
　　d. Formamide
　　e. *N*-Ethylpropionamide

15.55 The active ingredient in many insect repellents is *N,N*-diethyl-*m*-toluamide. Draw the structure of this compound. Which carboxylic acid and amine would be released by hydrolysis of this compound?

15.56 When an acid anhydride and an amine are combined, an amide is formed. This approach may be used to synthesize acetaminophen, the active ingredient in Tylenol. Complete the following reaction to determine the structure of acetaminophen:

$$CH_3\overset{\overset{\displaystyle O}{\|}}{C}-O-\overset{\overset{\displaystyle O}{\|}}{C}CH_3 + H_2N-\text{(benzene ring)}-OH \longrightarrow ? + CH_3COOH$$

15.57 Explain why amides are neutral in the acid-base sense.

15.58 The amide bond is stabilized by resonance. Draw the contributing resonance forms of the amide bond.

15.59 Lidocaine is often used as a local anesthetic. For medicinal purposes it is often used in the form of its hydrochloride salt because the salt is water soluble. In the structure of lidocaine hydrochloride shown, locate the amide functional group.

$$NH-\overset{\overset{\displaystyle O}{\|}}{C}-CH_2-\overset{\overset{\displaystyle H}{|}}{\underset{\displaystyle CH_2CH_3}{N^+}}-CH_2CH_3 \quad Cl^-$$

Lidocaine hydrochloride

15.60 Locate the amine functional group in the structure of lidocaine. Is lidocaine a primary, secondary, or tertiary amine?

15.61 The antibiotic penicillin BT contains functional groups discussed in this chapter. In the structure of penicillin BT shown, locate and name as many functional groups as you can.

Penicillin BT

15.62 The structure of saccharin, an artificial sweetener, is shown. Circle the amide group.

Saccharin

15.63 Complete each of the following reactions by supplying the missing reactant(s) or product(s) indicated by a question mark. Provide the systematic name for all the reactants and products.

　　a. $CH_3\overset{\overset{\displaystyle O}{\|}}{C}NHCH_3 + H_3O^+ \longrightarrow ? + ?$

b. $? + H_3O^+ \longrightarrow CH_3CH_2CH_2\overset{\overset{\displaystyle O}{\|}}{C}-OH + CH_3N^+H_3$

c. $CH_3\underset{\underset{\displaystyle CH_3}{|}}{C}HCH_2\overset{\overset{\displaystyle O}{\|}}{C}NHCH_2CH_3 + ? \longrightarrow$

$CH_3\underset{\underset{\displaystyle CH_3}{|}}{C}HCH_2\overset{\overset{\displaystyle O}{\|}}{C}-OH + ?$

15.64 Complete each of the following by supplying the missing reagents. Draw the structures of each of the reactants and products.
    a. *N*-Methylpropanamide + ? ⟶ propanoic acid + ?
    b. *N,N*-Dimethylacetamide + strong acid ⟶ ? + ?
    c. Formamide + strong acid ⟶ ? + ?

15.65 Complete each of the following reactions by supplying the missing reactant(s) or product(s) indicated by a question mark.
    a. $? + 2CH_3CH_2CH_2NH_2 \longrightarrow$

$CH_3CH_2CH_2NH\overset{\overset{\displaystyle O}{\|}}{C}CH_2CH_3 +$

$CH_3CH_2\overset{\overset{\displaystyle O}{\|}}{C}-O^-\ N^+H_3-CH_2CH_2CH_3$

    b. $CH_3CH_2\overset{\overset{\displaystyle O}{\|}}{C}-Cl + 2NH_3 \longrightarrow ? + ?$

    c. $? + ? \longrightarrow CH_3CH_2CH_2\overset{\overset{\displaystyle O}{\|}}{C}NHCH_2CH_3 +$

$CH_3CH_2-NH_3{}^+Cl^-$

15.66 Write two equations for the synthesis of each of the following amides. In one equation use an acid chloride as a reactant. In the second equation use an acid anhydride.
    a. Ethanamide
    b. *N*-Propylpentanamide
    c. Propionamide

A Preview of Amino Acids, Proteins, and Protein Synthesis

Foundations

15.67 Draw the general structure of an amino acid.
15.68 What is the name of the amide bond formed between two amino acids?

Applications

15.69 The amino acid glycine has a hydrogen atom as its R group, and the amino acid alanine has a methyl group. Draw these two amino acids.
15.70 Draw a dipeptide composed of glycine and alanine. Begin by drawing glycine with its amino group on the left. Circle the amide bond.
15.71 Draw the amino acid alanine (see Problem 15.69). Place a star by the chiral carbon.
15.72 Does glycine have a chiral carbon? Explain your reasoning.
15.73 Describe acyl group transfer.
15.74 Describe the relationship between acyl group transfer and the process of protein synthesis.

Neurotransmitter

Foundations

15.75 Define the term *neurotransmitter*.
15.76 What are the two general classes of neurotransmitters? What distinguishes them from one another?

Applications

15.77 **a.** What symptoms result from a deficiency of dopamine?
    b. What is the name of this condition?
    c. What symptoms result from an excess of dopamine?
15.78 What is the starting material in the synthesis of dopamine, epinephrine, and norepinephrine?
15.79 Explain the connection between addictive behavior and dopamine.
15.80 Why is L-dopa used to treat Parkinson's disease rather than dopamine?
15.81 What is the function of epinephrine?
15.82 What is the function of norepinephrine?
15.83 What is the starting material from which serotonin is made?
15.84 What symptoms are associated with a deficiency of serotonin?
15.85 What physiological processes are affected by serotonin?
15.86 How does Prozac relieve the symptoms of depression?
15.87 What are the physiological roles of histamine?
15.88 How do antihistamines function to control the allergic response?
15.89 What type of neurotransmitters are γ-aminobutyric acid and glycine?
15.90 Explain the evidence for a relationship between γ-aminobutyric acid and aggressive behavior.
15.91 Explain the function of acetylcholine at the neuromuscular junction.
15.92 Explain why organophosphates are considered to be poisons.
15.93 How does pyridine aldoxime methiodide function as an antidote for organophosphate poisoning?
15.94 Explain the mechanism by which glutamate and NO may function to promote development of memories and learning.

CRITICAL THINKING PROBLEMS

1. Histamine is made and stored in blood cells called *mast cells*. Mast cells are involved in the allergic response. Release of histamine in response to an allergen causes dilation of capillaries. This, in turn, allows fluid to leak out of the capillary resulting in local swelling. It also causes an increase in the volume of the vascular system. If this increase is great enough, a severe drop in blood pressure may cause shock. Histamine is produced by decarboxylation (removal of the carboxylate group as CO_2) of the amino acid histidine shown below. Draw the structure of histamine.

$$H_3{}^+N-\underset{\underset{\displaystyle CH_2}{|}}{\overset{\overset{\displaystyle COO^-}{|}}{C}}-H$$

$$CH_2$$

$$\underset{\underset{\displaystyle C}{\diagdown\ \diagup}}{\overset{\overset{\displaystyle |}{C}=CH}{|}}$$

$$H^+N \qquad NH$$

$$\underset{\displaystyle H}{C}$$

2. Carnitine tablets are sold in health food stores. It is claimed that carnitine will enhance the breakdown of body fat. Carnitine is a tertiary amine found in mitochondria, cell organelles in which food molecules are completely oxidized and ATP is produced. Carnitine is involved in transporting the acyl groups of fatty acids from the cytoplasm into the mitochondria. The fatty acyl group is transferred from a fatty acyl CoA molecule and esterified to carnitine. Inside the mitochondria the reaction is reversed and the fatty acid is completely oxidized. The structure of carnitine is shown here:

$$\begin{array}{c} COO^- \\ | \\ H-C-H \\ | \\ HO-C-H \\ | \\ H-C-H \\ | \\ (CH_3)_3\overset{+}{N} \end{array}$$

Draw the acyl carnitine molecule that is formed by esterification of palmitic acid with carnitine.

3. The amino acid proline has a structure that is unusual among amino acids. Compare the general structure of an amino acid with that of proline, shown here:

$$\begin{array}{c} COO^- \\ | \\ H_2\overset{+}{N}\!-\!-\!-C-H \\ \diagup \quad\quad | \\ H_2C \quad\quad CH_2 \\ \diagdown \quad\quad \diagup \\ CH_2 \end{array}$$

What is the major difference between proline and the other amino acids? Draw the structure of a dipeptide in which the amino group of proline forms a peptide bond with the carboxyl group of alanine.

4. Bulletproof vests are made of the polymer called Kevlar. It is produced by the copolymerization of the following two monomers:

$$H_2N-\!\!\bigcirc\!\!-NH_2 \quad \text{and} \quad HO_2C-\!\!\bigcirc\!\!-CO_2H$$

Draw the structure of a portion of Kevlar polymer.

16 Carbohydrates

Learning Goals

1 Explain the difference between complex and simple carbohydrates and know the amounts of each recommended in the daily diet.

2 Apply the systems of classifying and naming monosaccharides according to the functional group and number of carbons in the chain.

3 Determine whether a molecule has a chiral center.

4 Explain stereoisomerism.

5 Identify monosaccharides as either D- or L-.

6 Draw and name the common monosaccharides using structural formulas.

7 Given the linear structure of a monosaccharide, draw the Haworth projection of its α- and β-cyclic forms and vice versa.

8 By inspection of the structure, predict whether a sugar is a reducing or a nonreducing sugar.

9 Discuss the use of the Benedict's reagent to measure the level of glucose in urine.

10 Draw and name the common disaccharides and discuss their significance in biological systems.

11 Describe the difference between galactosemia and lactose intolerance.

12 Discuss the structural, chemical, and biochemical properties of starch, glycogen, and cellulose.

Would "looking-glass milk" be nutritious?

Introduction

Emil Fischer's father, a wealthy businessman, once said that Emil was too stupid to be a businessman and had better be a student. Lucky for the field of biochemistry, Emil did just that. Originally he wanted to study physics, but his cousin Otto Fischer convinced him to study chemistry. Beginning as an organic chemist, Fischer launched a career that eventually led to groundbreaking research in biochemistry. By following his career, we get a glimpse of the entire field.

Early on, Fischer discovered the active ingredients in coffee and tea, caffeine and theobromine. Eventually he discovered their structures and synthesized them in the laboratory. In work that he carried out between 1882 and 1906, Fischer demonstrated that adenine and guanine, along with some other compounds found in plants and animals, all belonged to one family of compounds. He called these the purines. All the purines have the same core structure and differ from one another by the functional groups attached to the ring.

Purine core structure

Chemistry Connection

Chemistry Through the Looking Glass

In his children's story *Through the Looking Glass*, Lewis Carroll's heroine Alice wonders whether "looking-glass milk" would be good to drink. As we will see in this chapter, many biological molecules, such as the sugars, exist as two stereoisomers, *enantiomers*, that are mirror images of one another. Because two mirror-image forms occur, it is rather remarkable that in our bodies, and in most of the biological world, only one of the two is found. For instance, the common sugars are members of the D-family, whereas all the common amino acids that make up our proteins are members of the L-family. It is not too surprising, then, that the enzymes in our bodies that break down the sugars and proteins we eat are *stereospecific*, that is, they recognize only one mirror-image isomer. Knowing this, we can make an educated guess that "looking-glass milk" could not be digested by our enzymes and therefore would not be a good source of food for us. It is even possible that it might be toxic to us!

Pharmaceutical chemists are becoming more and more concerned with the stereochemical purity of the drugs that we take. Consider a few examples. In 1960 the drug thalidomide was commonly prescribed in Europe as a sedative. However, during that year, hundreds of women who took thalidomide during pregnancy gave birth to babies with severe birth defects. Thalidomide, it turned out, was a mixture of two enantiomers. One is a sedative; the other is a teratogen, a chemical that causes birth defects.

One of the common side effects of taking antihistamines for colds or allergies is drowsiness. Again, this is the result of the fact that antihistamines are mixtures of enantiomers. One causes drowsiness; the other is a good decongestant.

One enantiomer of the compound carvone is associated with the smell of spearmint; the other produces the aroma of caraway seeds or dill. One mirror-image form of limonene smells like lemons; the other has the aroma of oranges.

The pain reliever ibuprofen is currently sold as a mixture of enantiomers, but one is a much more effective analgesic than the other.

Taste, smell, and the biological effects of drugs in the body all depend on the stereochemical form of compounds and their interactions with cellular enzymes or receptors. As a result, chemists are actively working to devise methods of separating the isomers in pure form. Alternatively, methods of conducting stereospecific syntheses that produce only one stereoisomer are being sought. By preparing pure stereoisomers, the biological activity of a compound can be much more carefully controlled. This will lead to safer medications.

In this chapter we will begin our study of stereochemistry, the spatial arrangement of atoms in molecules, with the carbohydrates. Later, we will examine the stereochemistry of the amino acids that make up our proteins and consider the stereochemical specificity of the metabolic reactions that are essential to life.

We will study the purines in Chapter 20 where we will learn that these are two of the essential components of the genetic molecules DNA and RNA. We will see that DNA is a double helix. Each strand of the helix is made up of a backbone of alternating sugars (ribose in RNA and deoxyribose in DNA) and phosphoryl groups. The purines are one of the two types of nitrogenous bases that project into the helix. By reading the order of these nitrogenous bases, we can decipher the genetic code of an organism.

The structure of the purines and pyrimidines was first described in Section 15.2. It is discussed in much more detail in Sections 20.1 and 20.2.

In 1884 Fischer began his monumental work on sugars. In 1890 he established the stereochemical nature of all sugars, and between 1891 and 1894 he worked out the stereochemical configuration of all the known sugars and predicted all the possible stereoisomers. Stereochemistry is the study of molecules that have two mirror-image isomers. We will find, as Fischer did, that nature has "selected" only one of the two mirror-image forms for common use in biological systems. Fischer studied virtually all the sugars, but one of his greatest successes was the synthesis of glucose, fructose, and mannose, three six-carbon sugars.

The structure of simple sugars and the function in biological systems are discussed in Sections 16.2 and 16.4. Stereochemistry is discussed in Section 16.3.

Between 1899 and 1908, Fischer turned his attention to proteins. He developed methods to separate and identify individual amino acids and discovered an entirely new class, the cyclic amino acids (those with ring structures). Fischer also worked on synthesis of proteins from amino acids. He demonstrated the nature of the peptide bond and discovered how to make small peptides in the laboratory.

As we will learn in Chapter 18, amino acids are all characterized by a common core structure, having both a carboxyl group and an amino group.

$$H_3{}^+N-\underset{\underset{R}{|}}{\overset{\overset{H}{|}}{C}}-\overset{\overset{O}{||}}{C}-O^-$$

Amino acid core structure

We will study the amino acids and the structure of proteins in Chapter 18.

The peptide bond that forms between amino acids is actually an amide bond between these two groups.

See also Section 15.4, A Preview of Amino Acids, Proteins, and Protein Synthesis.

While each of the amino acids has a common core, they differ from one another in the nature of a side chain, or R group. The amino acids are classified based on the properties they acquire from these R groups.

Quite late in his career, Fischer even got around to studying the fats, a heterogenous group of substances characterized by their hydrophobic nature. In Chapter 17, we will study the incredibly diverse family of fats, or lipids.

Fischer's personal life was not as happy as his professional life. His wife died after only seven years of marriage, leaving Fischer with three sons. One son died in World War I and another committed suicide at the age of twenty-five. However, his third son, Hermann Otto Laurenz Fischer, followed in his father's footsteps, becoming a professor of biochemistry at the University of California at Berkeley.

In 1902, Fischer was awarded the Nobel Prize for his work on sugar and purine synthesis. But a glance at all of his accomplishments helps us to realize that this great man, who began his career as an organic chemist, established the field of biochemistry through his extraordinary studies of the molecules of life.

16.1 Types of Carbohydrates

We begin our study of biochemistry with the carbohydrates. Carbohydrates are produced in plants by photosynthesis (Figure 16.1). Natural carbohydrate sources such as grains and cereals, breads, sugar cane, fruits, milk, and honey are an

LEARNING GOAL

1 Explain the difference between complex and simple carbohydrates and know the amounts of each recommended in the daily diet.

important source of energy for animals. **Carbohydrates** include simple sugars having the general formula $(CH_2O)_n$, as well as long polymers of these simple sugars, for instance potato starch, and a variety of molecules of intermediate size. The simple sugar glucose, $C_6H_{12}O_6$, is the primary energy source for the brain and nervous system and can be used by many other tissues. When "burned" by cells for energy, each gram of carbohydrate releases approximately four kilocalories of energy.

A healthy diet should contain both complex carbohydrates, such as starches and cellulose, and simple sugars, such as fructose and sucrose (Figure 16.2). However, the quantity of simple sugars, especially sucrose, should be minimized because large quantities of sucrose in the diet promote obesity and tooth decay.

Complex carbohydrates are better for us than the simple sugars. Starch, found in rice, potatoes, breads, and cereals, is an excellent energy source. In addition, the complex carbohydrates, such as cellulose, provide us with an important supply of dietary fiber.

It is hard to determine exactly what percentage of the daily diet *should* consist of carbohydrates. The *actual* percentage varies widely throughout the world, from 80% in the Far East, where rice is the main component of the diet, to 40–50% in the United States. Currently, it is recommended that 45–65% of the calories in the diet should come from carbohydrates and that no more than 10% of the daily caloric intake should be sucrose.

A kilocalorie is the same as the calorie referred to in the "count-your-calories" books and on nutrition labels.

See A Human Perspective: Tooth Decay and Simple Sugars on page 554.

Question 16.1

What is the current recommendation for the amount of carbohydrates that should be included in the diet? Of the daily intake of carbohydrates, what percentage should be simple sugar?

Figure 16.1
Carbohydrates are produced by plants such as this potato in the process of photosynthesis, which uses the energy of sunlight to produce hexoses from CO_2 and H_2O.

Figure 16.2
Carbohydrates from a variety of foods are an essential component of the diet.

Distinguish between simple and complex sugars. What are some sources of complex carbohydrates?

Monosaccharides such as glucose and fructose are the simplest carbohydrates because they contain a single *(mono-)* sugar *(saccharide)* unit. **Disaccharides,** including sucrose and lactose, consist of two monosaccharide units joined through bridging oxygen atoms. Such a bond is called a **glycosidic bond. Oligosaccharides** consist of three to ten monosaccharide units joined by glycosidic bonds. The largest and most complex carbohydrates are the **polysaccharides,** which are long, often highly branched, chains of monosaccharides. Starch, glycogen, and cellulose are all examples of polysaccharides.

16.2 Monosaccharides

Monosaccharides are composed of carbon, hydrogen, and oxygen, and most are characterized by the general formula $(CH_2O)_n$, in which n is any integer from 3 to 7. As we will see, this general formula is an oversimplification because several biologically important monosaccharides are chemically modified. For instance, several blood group antigen and bacterial cell wall monosaccharides are substituted with amino groups. Many of the intermediates in carbohydrate metabolism carry phosphate groups. Deoxyribose, the monosaccharide found in DNA, has one fewer oxygen atom than the general formula above would predict.

The importance of phosphorylated sugars in metabolic reactions is discussed in Sections 14.4 and 21.3.

Monosaccharides can be named on the basis of the functional groups they contain. A monosaccharide with a ketone (carbonyl) group is a **ketose.** If an aldehyde (carbonyl) group is present, it is called an **aldose.** Because monosaccharides also contain many hydroxyl groups, they are sometimes called *polyhydroxyaldehydes* or *polyhydroxyketones.*

LEARNING GOAL

2 Apply the systems of classifying and naming monosaccharides according to the functional group and number of carbons in the chain.

Another system of nomenclature tells us the number of carbon atoms in the main skeleton. A three-carbon monosaccharide is a *triose,* a four-carbon sugar is a *tetrose,* a five-carbon sugar is a *pentose,* a six-carbon sugar is a *hexose,* and so on. Combining the two naming systems gives even more information about the structure and composition of a sugar. For example, an aldotetrose is a four-carbon sugar that is also an aldehyde.

In addition to these general names, each monosaccharide has a unique name. These names are shown in blue for the following structures. Because the monosaccharides can exist in several different isomeric forms, it is important to provide the complete name. Thus the complete names of the following structures are D-glyceraldehyde, D-glucose, and D-fructose. These names tell us that the structure represents one particular sugar and also identifies the sugar as one of two possible isomeric forms (D- or L-).

A *Human* Perspective

Tooth Decay and Simple Sugars

How many times have you heard the lecture from parents or your dentist about brushing your teeth after a sugary snack? Annoying as this lecture might be, it is based on sound scientific data that demonstrate that the cause of tooth decay is plaque and acid formed by the bacterium *Streptococcus mutans* using sucrose as its substrate.

Saliva is teeming with bacteria in concentrations up to one hundred million (10^8) per milliliter of saliva! Within minutes after you brush your teeth, sticky glycoproteins in the saliva adhere to tooth surfaces. Then millions of oral bacteria immediately bind to this surface.

Although many oral bacteria stick to the tooth surface, as the diagram shows, only *S. mutans* causes cavities. The reason for this is that this organism alone can make the enzyme *glucosyl transferase.* This enzyme acts only on the disaccharide sucrose, breaking it down into glucose and fructose. The glucose is immediately added to a growing polysaccharide called *dextran,* the glue that allows the bacteria to adhere to the tooth surface, contributing to the formation of plaque.

Now the bacteria embedded in the dextran take in the fructose and use it in the lactic acid fermentation. The lactic acid that is produced lowers the pH on the tooth surface and begins to dissolve calcium from the tooth enamel. Even though we produce about one liter of saliva each day, the acid cannot be washed away from the tooth surface because the dextran plaque is not permeable to saliva.

So what can we do to prevent tooth decay? Of course, brushing after each meal and flossing regularly reduce plaque buildup. Eating a diet rich in calcium also helps build strong tooth enamel. Foods rich in complex carbohydrates, such as fruits and vegetables, help prevent cavities in two ways. Glucosyl transferase can't use complex carbohydrates in its cavity-causing chemistry, and eating fruits and vegetables helps to mechanically remove plaque.

Perhaps the most effective way to prevent tooth decay is to avoid sucrose-containing snacks between meals. Studies have shown that eating sucrose-rich foods doesn't cause much tooth decay if followed immediately by brushing. However, even small amounts of sugar eaten between meals actively promote cavity formation.

For Further Understanding

It has been suggested that tooth decay could be prevented by a vaccine that would rid the mouth of *Streptococcus mutans.* Explain this from the point of view of the chemical reactions that are described above.

What steps could you take following a sugary snack to help prevent tooth decay, even when it is not possible to brush your teeth?

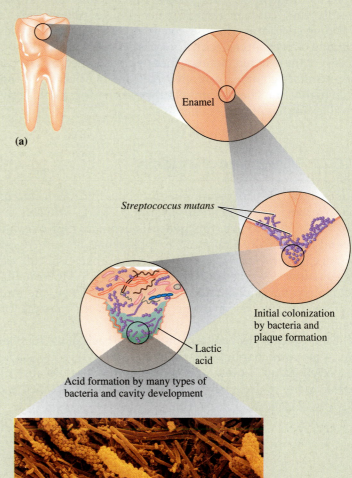

(a)

Streptococcus mutans

Enamel

Initial colonization by bacteria and plaque formation

Lactic acid

Acid formation by many types of bacteria and cavity development

(b)

(a) The complex process of tooth decay. (b) Electron micrograph of dental plaque.

$$
\begin{array}{ccc}
\text{H} & \text{H} & \text{CH}_2\text{OH} \\
| & | & | \\
\text{C}=\text{O} & \text{C}=\text{O} & \text{C}=\text{O} \\
| & | & | \\
\text{H}-\text{C}-\text{OH} & \text{H}-\text{C}-\text{OH} & \text{HO}-\text{C}-\text{H} \\
| & | & | \\
\text{CH}_2\text{OH} & \text{HO}-\text{C}-\text{H} & \text{H}-\text{C}-\text{OH} \\
 & | & | \\
 & \text{H}-\text{C}-\text{OH} & \text{H}-\text{C}-\text{OH} \\
 & | & | \\
 & \text{H}-\text{C}-\text{OH} & \text{CH}_2\text{OH} \\
 & | & \\
 & \text{CH}_2\text{OH} &
\end{array}
$$

Aldose	Aldose	Ketose
Triose	Hexose	Hexose
Aldotriose	Aldohexose	Ketohexose
D-Glyceraldehyde	D-Glucose	D-Fructose

Question *16.3*

What is the structural difference between an aldose and a ketose?

Question *16.4*

Explain the difference between:

a. A ketohexose and an aldohexose
b. A triose and a pentose

16.3 Stereoisomers and Stereochemistry

Stereoisomers

The prefixes D- and L- found in the complete name of a monosaccharide are used to identify one of two possible isomeric forms called **stereoisomers.** By definition, each member of a pair of stereoisomers must have the same molecular formula and the same bonding. How then do isomers of the D-family differ from those of the L-family? D- and L-isomers differ in the spatial arrangements of atoms in the molecule.

    Stereochemistry is the study of the different spatial arrangements of atoms. A general example of a pair of stereoisomers is shown in Figure 16.3. In this example the general molecule C-abcd is formed from the bonding of a central carbon to four different groups: a, b, c, and d. This results in two molecules rather than one. Each isomer is bonded together through the exact *same* bonding pattern, yet they are *not* identical. If they were identical, they would be superimposable one upon the other; *they are not.* They are therefore stereoisomers. These two stereoisomers have a mirror-image relationship that is analogous to the mirror-image relationship of the left and right hands (see Figure 16.3b).

    Two stereoisomers that are nonsuperimposable mirror images of one another are called a pair of **enantiomers.** Molecules that can exist in enantiomeric forms are called **chiral molecules.** The term simply means that as a result of different three-dimensional arrangements of atoms, the molecule can exist in two mirror-image forms. For any pair of nonsuperimposable mirror-image forms (enantiomers), one is always designated D- and the other L-.

    A carbon atom that has four different groups bonded to it is called a **chiral carbon** atom. Any molecule containing a chiral carbon can exist as a pair of enantiomers. Consider the simplest carbohydrate, **glyceraldehyde,** which is shown in

LEARNING GOAL
3 Determine whether a molecule has a chiral center.

LEARNING GOAL
4 Explain stereoisomerism.

LEARNING GOAL
5 Identify monosaccharides as either D- or L-.

Build models of these compounds using toothpicks and gumdrops of five different colors to prove this to yourself.

Animations
Chiral Molecules
Chiral Molecules (B)

Mirror

(a)

Nonsuperimposable mirror images:
enantiomers

Mirror

(b)

Figure 16.3
(a) A pair of enantiomers for the general molecule C-abcd. (b) Mirror-image right and left hands.

Figure 16.4. Note that the second carbon is bonded to four different groups. It is therefore a chiral carbon. As a result, we can draw two enantiomers of glyceraldehyde that are nonsuperimposable mirror images of one another. Larger biological molecules typically have more than one chiral carbon.

Rotation of Plane-Polarized Light

Stereoisomers can be distinguished from one another by their different optical properties. Each member of a pair of stereoisomers will rotate plane-polarized light in different directions.

As we learned in Chapter 2, white light is a form of electromagnetic radiation that consists of many different wavelengths (colors) vibrating in *planes* that are all perpendicular to the direction of the light beam (Figure 16.5). To measure optical properties of enantiomers, scientists use special light sources to produce *monochromatic light*, that is, light of a single wavelength. The monochromatic light is passed through a polarizing material, like a Polaroid lens, so that only waves in one plane can pass through. The light that emerges from the lens is *plane-polarized light* (Figure 16.5).

Applying these principles, scientists have developed the *polarimeter* to measure the ability of a compound to change the angle of the plane of plane-polarized light (Figure 16.5). The polarimeter allows the determination of the specific rotation of a compound, that is, the measure of its ability to rotate plane-polarized light.

Some compounds rotate light in a clockwise direction. These are said to be *dextrorotatory* and are designated by a plus sign (+) before the specific rotation value.

The polarimeter, measurement of the rotation of plane-polarized light, and the calculation of specific rotation are discussed in detail online.

Stereochemistry and
Stereoisomers Revisited

Most oxidized end

D-Glyceraldehyde

L-Glyceraldehyde

Chiral center farthest from the most oxidized end

(a)

D-Glyceraldehyde

L-Glyceraldehyde

(b)

Figure 16.4

(a) Structural formulas of D- and L-glyceraldehyde. The end of the molecule with the carbonyl group is the most oxidized end. The D- or L-configuration of a monosaccharide is determined by the orientation of the functional groups attached to the chiral carbon farthest from the oxidized end. In the D-enantiomer, the −OH is to the right. In the L-enantiomer, the −OH is to the left. (b) A three-dimensional representation of D- and L-glyceraldehyde.

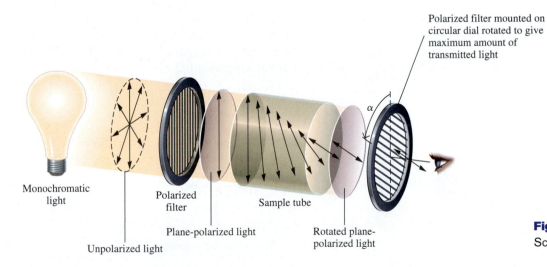

Polarized filter mounted on circular dial rotated to give maximum amount of transmitted light

Monochromatic light

Polarized filter

Sample tube

Plane-polarized light

Unpolarized light

Rotated plane-polarized light

Figure 16.5

Schematic drawing of a polarimeter.

Other substances rotate light in a counterclockwise direction. These are called *levorotatory* and are indicated by a minus sign (−) before the specific rotation value.

The Relationship Between Molecular Structure and Optical Activity

In 1848, Louis Pasteur was the first to see a relationship between the structure of a compound and the effect of that compound on plane-polarized light. In his

Stereochemistry and Stereoisomers Revisited

studies of winemaking, Pasteur noticed that salts of tartaric acid were formed as a by-product. It is a tribute to his extraordinary powers of observation that he noticed that two types of crystals were formed and that they were mirror images of one another. Using a magnifying glass and forceps, Pasteur separated the left-handed and right-handed crystals into separate piles. When he measured the optical activity of each of the mirror-image forms and of the original mixed sample, he obtained the following results:

- A solution of the original mixture of crystals was optically inactive.
- But both of the mirror-image crystals were optically active. In fact, the specific rotation produced by each was identical in magnitude but was of opposite sign.

Although Pasteur's work opened the door to understanding the relationship between structure and optical activity, it was not until 1874 that the Dutch chemist van't Hoff and the French chemist LeBel independently came up with a basis for the observed optical activity: tetrahedral carbon atoms bonded to four different atoms or groups of atoms. Thus, two enantiomers, which are identical to one another in all other chemical and physical properties, will rotate plane-polarized light to the same degree, but in opposite directions.

Fischer Projection Formulas

Emil Fischer devised a simple way to represent the structure of stereoisomers. The **Fischer Projection** is a two-dimensional drawing of a molecule that shows a chiral carbon at the intersection of two lines. The horizontal lines represent bonds projecting out of the page, and the vertical lines represent bonds that project into the page. Figure 16.6 demonstrates how to draw the Fischer Projections for the stereoisomers of bromochlorofluoromethane. In Figure 16.6a, the two isomers are represented using ball-and-stick models. The molecules are reinterpreted using the wedge-and-dash representations in Figure 16.6b. In the Fischer Projections shown in Figure 16.6c, the point at which two lines cross represents the chiral carbon. Horizontal lines replace the solid wedges indicating that the bonds are projecting toward the reader. Vertical lines replace the dashed wedges, indicating that the bonds are projecting away from the reader. For sugars, the aldehyde or ketone group, the most oxidized carbon, is always represented at the "top."

Figure 16.6

Drawing a Fischer Projection. (a) The ball-and-stick models for the stereoisomers of bromochlorofluoromethane. (b) The wedge-and-dash and (c) Fischer Projections of these molecules.

Bromochlorofluoromethane

Bromochlorofluoromethane

(a)   (b)   (c)

Drawing Fischer Projections for a Sugar

EXAMPLE 16.1

Draw the Fischer Projections for the stereoisomers of glyceraldehyde.

Solution

Review the structures of the two stereoisomers of glyceraldehyde (Figure 16.4b). The ball-and-stick models can be represented using three-dimensional wedge drawings. Remember that for sugars the most oxidized carbon (the aldehyde or ketone group) is always drawn at the top of the structure.

```
      CHO                 CHO
       |                   |
  H--C--OH           HO--C--H
       |                   |
     CH2OH              CH2OH
```

D-Glyceraldehyde   L-Glyceraldehyde

Remember that in the wedge diagram, the solid wedges represent bonds directed toward the reader. The dashed wedges represent bonds directed away from the reader and into the page. In these molecules, the center carbon is the only chiral carbon in the structure. To convert these wedge representations to a Fischer Projection, simply use a horizontal line in place of each solid wedge and use a vertical line to represent each dashed wedge. The chiral carbon is represented by the point at which the vertical and horizontal lines cross, as shown below.

```
    CHO          CHO              CHO              CHO
     |            |                |                |
 H--C--OH    H----+----OH    HO--C--H       HO----+----H
     |            |                |                |
   CH2OH        CH2OH            CH2OH            CH2OH
```

D-Glyceraldehyde   L-Glyceraldehyde

Practice Problem 16.1

Draw Fischer Projections for each of the following molecules and for their mirror images.

```
a.      CH3              b.       H              c.      CH2OH
         |                        |                       |
        C=O                      C=O                     C=O
         |                        |                       |
    H--C--OH                 H--C--OH               HO--C--H
         |                        |                       |
      CH2OH                  H--C--OH                H--C--OH
                                  |                       |
                             HO--C--H               H--C--OH
                                  |                       |
                               CH2OH                   CH2OH

d.       H               e.       CH3             f.       H
         |                        |                        |
        C=O                      C=O                      C=O
         |                        |                        |
    H--C--OH                 H--C--OH                 HO--C--H
         |                        |                        |
      CH2OH                  H--C--OH                  H--C--OH
                                  |                        |
                             H--C--OH                 HO--C--H
                                  |                        |
                               CH2OH                    CH2OH
```

For Further Practice: Questions 16.45 and 16.46.

The D- and L- System of Nomenclature

In 1891 Emil Fischer devised a nomenclature system that would allow scientists to distinguish between enantiomers. Fischer knew that the two enantiomers of glyceraldehyde rotated plane-polarized light in opposite directions, but he did not have the sophisticated tools needed to make an absolute connection between the structure and the direction of rotation of plane-polarized light. He simply decided that the (+) enantiomer would be the one with the hydroxyl group of the chiral carbon on the right. The enantiomer that rotated plane-polarized light in the (−) or levorotatory direction, he called L-glyceraldehyde (Figure 16.4).

While specific rotation is an experimental value that must be measured, the D- and L-designations of all other monosaccharides are determined by comparison of their structures with D- and L-glyceraldehyde. Sugars with more than three carbons will have more than one chiral carbon. *By convention*, it is the position of the hydroxyl group on the chiral carbon farthest from the carbonyl group (the most oxidized end of the molecule) that determines whether a monosaccharide is in the D- or L-configuration. If the —OH group is on the right, the molecule is in the D-configuration. If the —OH group is on the left, the molecule is in the L-configuration. Almost all carbohydrates in living systems are members of the D-family.

> It was not until 1952 that researchers were able to demonstrate that Fischer had guessed correctly when he proposed the structures of the (+) and (−) enantiomers of glyceraldehyde.

The structures and designations of D- and L-glyceraldehyde are defined by convention. In fact, the D- and L-terminology is generally applied only to carbohydrates and amino acids. For organic molecules, the D- and L-convention has been replaced by a new system that provides the absolute configuration of a chiral carbon. This system, called the (R) and (S) system, is described in Stereochemistry and Stereoisomers Revisited.

Stereochemistry and
Stereoisomers Revisited

$$
\begin{array}{ccc}
\text{H} & \text{H} & \text{CH}_2\text{OH} \\
| & | & | \\
\text{C}=\text{O} & \text{C}=\text{O} & \text{C}=\text{O} \\
| & | & | \\
\text{H}-\text{C}-\text{OH} & \text{H}-\text{C}-\text{OH} & \text{HO}-\text{C}-\text{H} \\
| & | & | \\
\text{CH}_2\text{OH} & \text{HO}-\text{C}-\text{H} & \text{H}-\text{C}-\text{OH} \\
& | & | \\
& \text{H}-\text{C}-\text{OH} & \text{H}-\text{C}-\text{OH} \\
& | & | \\
& \text{H}-\text{C}-\text{OH} & \text{CH}_2\text{OH} \\
& | & \\
& \text{CH}_2\text{OH} &
\end{array}
$$

D-Glyceraldehyde   D-Glucose   D-Fructose

Question 16.5

Place an asterisk beside each chiral carbon in the Fischer Projections you drew for Practice Problem 16.1 at the end of Example 16.1.

Question 16.6

In the Fischer Projections you drew for Practice Problem 16.1 at the end of Example 16.1, indicate which bonds project toward you and which project into the page.

Question 16.7

Determine the configuration (D- or L-) for each of the molecules in Practice Problem 16.1 at the end of Example 16.1.

Question 16.8

Explain the difference between the D- and L-designation and the (+) and (−) designation.

Figure 16.7
Cyclization of glucose to give α- and β-D-glucose. Note that the carbonyl carbon (C-1) becomes chiral in this process, yielding the α- and β-forms of glucose.

16.4 Biologically Important Monosaccharides

Monosaccharides, the simplest carbohydrates, have backbones of from three to seven carbons. There are many monosaccharides, but we will focus on those that are most common in biological systems. These include the five- and six-carbon sugars: glucose, fructose, galactose, ribose, and deoxyribose.

> **LEARNING GOAL**
> **6** Draw and name the common monosaccharides using structural formulas.

Glucose

Glucose is the most important sugar in the human body. It is found in numerous foods and has several common names, including dextrose, grape sugar, and blood sugar. Glucose is broken down in glycolysis and other pathways to release energy for body functions.

The concentration of glucose in the blood is critical to normal body function. As a result, it is carefully controlled by the hormones insulin and glucagon. Normal blood glucose levels are 100–120 mg/100 mL, with the highest concentrations appearing after a meal. Insulin stimulates the uptake of the excess glucose by most cells of the body, and after one to two hours, levels return to normal. If blood glucose concentrations drop too low, the individual feels lightheaded and shaky. When this happens, glucagon stimulates the liver to release glucose into the blood, reestablishing normal levels. We will take a closer look at this delicate balancing act in Section 23.6.

The molecular formula of glucose, an aldohexose, is $C_6H_{12}O_6$. The structure of glucose is shown in Figure 16.7, and the method used to draw this structure is described in Example 16.2.

Drawing the Structure of a Monosaccharide	**E X A M P L E 16.2**

Draw the structure for D-glucose.

Solution

Glucose is an aldohexose.

> **LEARNING GOAL**
> **6** Draw and name the common monosaccharides using structural formulas.

Continued—

E X A M P L E   16.2 —*Continued*

Step 1. Draw six carbons in a straight vertical line; each carbon is separated from the ones above and below it by a bond:

Step 2. The most highly oxidized carbon is, by convention, drawn as the uppermost carbon (carbon-1). In this case, carbon-1 is an aldehyde carbon:

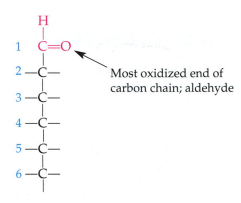

Most oxidized end of carbon chain; aldehyde

Why do diabetics need to use a blood glucose monitor like the one shown here?

Step 3. The atoms are added to the next to the last carbon atom, at the bottom of the chain, to give either the D- or L-configuration as desired. Remember, when the —OH group is to the right, you have D-glucose. When in doubt, compare your structure to D-glyceraldehyde!

D-Isomer   D-Glyceraldehyde

Compare chiral carbons farthest from the carbonyl group

Step 4. All the remaining atoms are then added to give the desired carbohydrate. For example, one would draw the following structure for D-glucose.

Continued—

EXAMPLE 16.2 —*Continued*

$$
\begin{array}{c}
H \\
| \\
C{=}O \\
| \\
H{-}C{-}OH \\
| \\
HO{-}C{-}H \\
| \\
H{-}C{-}OH \\
| \\
H{-}C{-}OH \\
| \\
CH_2OH
\end{array}
$$

D-Glucose

The positions for the hydrogen atoms and the hydroxyl groups on the remaining carbons must be learned for each sugar.

Practice Problem 16.2

Draw the structures of D-ribose and L-ribose. (Information on the structure of D-ribose is found later in this chapter.)

For Further Practice: Questions 16.9 and 16.10.

In actuality the open-chain form of glucose is present in very small concentrations in cells. It exists in cyclic form under physiological conditions because the carbonyl group at C-1 of glucose reacts with the hydroxyl group at C-5 to give a six-membered ring. In the discussion of aldehydes, we noted that the reaction between an aldehyde and an alcohol yields a **hemiacetal.** When the aldehyde portion of the glucose molecule reacts with the C-5 hydroxyl group, the product is a cyclic *intramolecular hemiacetal.* For D-glucose, two isomers can be formed in this reaction (see Figure 16.7). These isomers are called α- and β-D-glucose. The two isomers formed differ from one another in the location of the —OH attached to the hemiacetal carbon, C-1. Such isomers, differing in the arrangement of bonds around the hemiacetal carbon, are called **anomers.** In the α-anomers, the C-1 (*anomeric carbon*) hydroxyl group is below the ring, and in the β-anomers, the C-1 hydroxyl group is above the ring. Like the stereoisomers discussed previously, the α and β forms can be distinguished from one another because they rotate plane-polarized light differently.

Hemiacetal structure,

$$
\begin{array}{c}
OH \\
| \\
R^1{-}C{-}OR^2 \\
| \\
H
\end{array}
$$

and formation are described in Section 13.4.

The term *intramolecular* tells us that the reacting carbonyl and hydroxyl groups are part of the same molecule.

Question 16.9

Draw the structure of D-galactose. (Information on the structure of D-galactose is found later in this chapter.)

Question 16.10

Draw the structure of L-galactose.

In Figure 16.7 a new type of structural formula, called a **Haworth projection,** is presented. Although on first inspection it appears complicated, it is quite simple to derive a Haworth projection from a structural formula, as Example 16.3 shows.

EXAMPLE 16.3   **Drawing the Haworth Projection of a Monosaccharide from the Structural Formula**

LEARNING GOAL

7 Given the linear structure of a monosaccharide, draw the Haworth projection of its α- and β-cyclic forms and vice versa.

Draw the Haworth projections of α- and β-D-glucose.

Solution

1. Before attempting to draw a Haworth projection, look at the first steps of ring formation shown here:

Glucose   Glucose
(open chain)   (intermediates in ring formation)

Try to imagine that you are seeing the molecules shown above in three dimensions. Some of the substituent groups on the molecule will be above the ring, and some will be beneath it. The question then becomes: How do you determine which groups to place above the ring and which to place beneath the ring?

2. Look at the two-dimensional structural formula. Note the groups (drawn in blue) to the left of the carbon chain. These are placed above the ring in the Haworth projection.

α-D-Glucose   β-D-Glucose

3. Now note the groups (drawn in red) to the right of the carbon chain. These will be located beneath the carbon ring in the Haworth projection.

α-D-Glucose   β-D-Glucose

Continued—

E X A M P L E   16.3 —*Continued*

4. Thus in the Haworth projection of the cyclic form of any D-sugar the
—CH₂OH group is always "up." When the —OH group at C-1 is also
"up," *cis* to the —CH₂OH group, the sugar is β-D-glucose. When the
—OH group at C-1 is "down," *trans* to the —CH₂OH group, the sugar is
α-D-glucose.

Haworth projection
α-D-Glucose

Haworth projection
β-D-Glucose

Practice Problem 16.3

Refer to the linear structures of D-galactose and D-ribose. Draw
the Haworth projections of (a) α- and β-D-galactose and of (b) α- and
β-D-ribose. Note that D-ribose is a pentose.

For Further Practice: Questions 16.53 and 16.54.

Fructose

Fructose, also called levulose and fruit sugar, is the sweetest of all sugars. It is
found in large amounts in honey, corn syrup, and sweet fruits. The structure of
fructose is similar to that of glucose. When there is a —CH₂OH group instead of a
—CHO group at carbon-1 and a —C=O group instead of CHOH at carbon-2, the
sugar is a ketose. In this case it is D-fructose.

Cyclization of fructose produces α- and β-D-fructose:

Fructose is often called fruit sugar
because it contributes sweetness to ripe
fruits, such as these peaches. Is fructose
an aldose or a ketose?

Hemiketal structure

$$R^1-\underset{\underset{R^2}{|}}{\overset{\overset{OH}{|}}{C}}-OR^3$$

and formation are described in section 13.4.

Fructose is a ketose, or ketone sugar. Recall that the reaction between an alcohol and a ketone yields a **hemiketal.** Thus the reaction between the C-2 keto group and the C-5 hydroxyl group in the fructose molecule produces an *intramolecular hemiketal.* Fructose forms a five-membered ring structure.

Galactose

Another important hexose is **galactose.** The linear structure of D-galactose and the Haworth projections of α-D-galactose and β-D-galactose are shown here:

D-Galactose

α-D-Galactose

β-D-Galactose

Galactose is one of the components of lactose, or milk sugar. Read about galactosemia in Section 16.5 and describe the symptoms and treatment for this genetic disorder.

Galactose is found in biological systems as a component of the disaccharide lactose, or milk sugar. This is the principal sugar found in the milk of most mammals. β-D-Galactose and a modified form, β-D-*N*-acetylgalactosamine, are also components of the blood group antigens.

β-D-*N*-Acetylgalactosamine

Ribose and Deoxyribose, Five-Carbon Sugars

Ribose is a component of many biologically important molecules, including RNA and various coenzymes that are required by many of the enzymes that carry out biochemical reactions in the body. The structure of the five-carbon sugar D-ribose is shown in its open-chain form and in the α- and β-cyclic forms.

α-D-Ribose

D-Ribose

β-D-Ribose

β-D-2-Deoxyribose is one of the components of the sugar-phosphate backbone of the DNA molecule. How does this molecule differ from β-D-ribose?

DNA, the molecule that carries the genetic information of the cell, contains 2-deoxyribose. In this molecule the —OH group at C-2 has been replaced by a hydrogen, hence the designation "2-deoxy," indicating the absence of an oxygen.

β-D-2-Deoxyribose

Reducing Sugars

The aldehyde group of aldoses is readily oxidized by the Benedict's reagent. Recall that the **Benedict's reagent** is a basic buffer solution that contains Cu^{2+} ions. The Cu^{2+} ions are reduced to Cu^+ ions, which, in basic solution, precipitate as brick-red Cu_2O. The aldehyde group of the aldose is oxidized to a carboxylic acid, which undergoes an acid-base reaction to produce a carboxylate anion.

$$\text{aldose} + 2Cu^{2+}\text{ (buffer)} + 5OH^- \longrightarrow \text{carboxylate} + Cu_2O(s) + 3H_2O$$

Although ketones generally are not easily oxidized, ketoses are an exception to that rule. Because of the —OH group on the carbon next to the carbonyl group, ketoses can be converted to aldoses, under basic conditions, via an *enediol reaction*:

D-Fructose   Enediol   D-Glucose

LEARNING GOAL

8 By inspection of the structure, predict whether a sugar is a reducing or a non-reducing sugar.

LEARNING GOAL

9 Discuss the use of the Benedict's reagent to measure the level of glucose in urine.

The name of the enediol reaction is derived from the structure of the intermediate through which the ketose is converted to the aldose: It has a double bond (ene), and it has two hydroxyl groups (diol). Because of this enediol reaction, ketoses are also able to react with Benedict's reagent, which is basic. Because the metal ions in the solution are reduced, the sugars are serving as reducing agents and are called **reducing sugars.** All monosaccharides and all the common disaccharides, except sucrose, are reducing sugars.

For many years the Benedict's reagent was used to test for *glucosuria*, the presence of excess glucose in the urine. Individuals suffering from *Type I insulin-dependent diabetes mellitus* do not produce the hormone insulin, which controls the uptake of glucose from the blood. When the blood glucose level rises above 160 mg/100 mL, the kidney is unable to reabsorb the excess, and glucose is found in the urine. Although the level of blood glucose could be controlled by the injection of insulin, urine glucose levels were monitored to ensure that the amount of insulin injected was correct. The Benedict's reagent was a useful tool because the amount of Cu_2O formed, and hence the degree of color change in the reaction, is directly proportional to the amount of reducing sugar in the urine. A brick-red color indicates a very high concentration of glucose in the urine. Yellow, green, and blue-green solutions indicate decreasing amounts of glucose in the urine, and a blue solution indicates an insignificant concentration.

Use of Benedict's reagent to test urine glucose levels has largely been replaced by chemical tests that provide more accurate results. The most common technology is based on a test strip that is impregnated with the enzyme glucose oxidase and other agents that will cause a measurable color change. In one such kit, the compounds that result in color development include the enzyme peroxidase, a compound called orthotolidine, and a yellow dye. When a drop of urine is placed on the strip, the glucose oxidase catalyzes the conversion of glucose into gluconic acid and hydrogen peroxide.

See A Medical Perspective: Diabetes Mellitus and Ketone Bodies in Chapter 23.

D-Glucose   D-Gluconic acid

The enzyme peroxidase catalyzes a reaction between the hydrogen peroxide and orthotolidine. This produces a blue product. The yellow dye on the test strip simply serves to "dilute" the blue end product, thereby allowing greater accuracy of the test over a wider range of glucose concentrations. The test strip remains yellow if there is no glucose in the sample. It will vary from a pale green to a dark blue, depending on the concentration of glucose in the urine sample.

Frequently, doctors recommend that diabetics monitor their *blood* glucose levels multiple times each day because this provides a more accurate indication of how well the diabetic is controlling his or her diet. Many small, inexpensive glucose meters are available that couple the oxidation of glucose by glucose oxidase with an appropriate color change system. As with the urine test, the intensity of the color change is proportional to the amount of glucose in the blood. A photometer within the device reads the color change and displays the glucose concentration. An even newer technology uses a device that detects the electrical charge generated by the oxidation of glucose. In this case, it is the amount of electrical charge that is proportional to the glucose concentration.

Actually, glucose oxidase can only oxidize β-D-glucose. However, in the blood there is an equilibrium mixture of the α and β anomers of glucose. Fortunately, α-D-glucose is very quickly converted to β-D-glucose.

16.5 Biologically Important Disaccharides

Recall that disaccharides consist of two monosaccharides joined through an "oxygen bridge." In biological systems, monosaccharides exist in the cyclic form and, as we have seen, they are actually hemiacetals or hemiketals. Recall that when a hemiacetal reacts with an alcohol, the product is an *acetal*, and when a hemiketal reacts with an alcohol, the product is a *ketal*. In the case of disaccharides, the alcohol comes from a second monosaccharide. The acetals or ketals formed are given the general name *glycosides*, and the carbon-oxygen bonds are called *glycosidic bonds*.

Glycosidic bond formation is nonspecific; that is, it can occur between a hemiacetal or hemiketal and any of the hydroxyl groups on the second monosaccharide. However, in biological systems, we commonly see only particular disaccharides, such as maltose (Figure 16.8), lactose (see Figure 16.10), or sucrose (see Figure 16.11). These specific disaccharides are produced in cells because the reactions are catalyzed by enzymes. Each enzyme catalyzes the synthesis of one specific disaccharide, ensuring that one particular pair of hydroxyl groups on the reacting monosaccharides participates in glycosidic bond formation.

LEARNING GOAL
10 Draw and name the common disaccharides and discuss their significance in biological systems.

Acetals and ketals are described in Section 13.4:

$$
R-\underset{\underset{H}{|}}{\overset{\overset{OR}{|}}{C}}-OR \qquad R-\underset{\underset{R}{|}}{\overset{\overset{OR}{|}}{C}}-OR
$$

Acetal   Ketal

Maltose

If an α-D-glucose and a second glucose are linked, as shown in Figure 16.8, the disaccharide is **maltose,** or malt sugar. This is one of the intermediates in the hydrolysis of starch. Because the C-1 hydroxyl group of α-D-glucose is attached to C-4 of another glucose molecule, the disaccharide is linked by an α (1 → 4) glycosidic bond.

Maltose is a reducing sugar. Any disaccharide that has a hemiacetal hydroxyl group (a free —OH group at C-1) is a reducing sugar. This is because the cyclic structure can open at this position to form a free aldehyde. Disaccharides that do not contain a hemiacetal group on C-1 do not react with the Benedict's reagent and are called **nonreducing sugars.**

Lactose

Milk sugar, or **lactose,** is a disaccharide made up of one molecule of β-D-galactose and one of either α- or β-D-glucose. Galactose differs from glucose only in the configuration of the hydroxyl group at C-4 (Figure 16.9). In the cyclic form of glucose, the C-4 hydroxyl group is "down," and in galactose it is "up." In lactose the C-1

LEARNING GOAL
11 Describe the difference between galactosemia and lactose intolerance.

Figure 16.8

Glycosidic bond formed between the C-1 hydroxyl group of α-D-glucose and the C-4 hydroxyl group of β-D-glucose. The disaccharide is called β-maltose because the hydroxyl group at the reducing end of the disaccharide has the β-configuration.

α-D-Glucose   β-D-Glucose   β-Maltose

β-D-Glucose   β-D-Galactose

Figure 16.9

Comparison of the cyclic forms of glucose and galactose. Note that galactose is identical to glucose except in the position of the C-4 hydroxyl group.

A Human *Perspective*

Blood Transfusions and the Blood Group Antigens

The first blood transfusions were tried in the seventeenth century, when physicians used animal blood to replace human blood lost by hemorrhages. Unfortunately, many people died as a result of this attempted cure, and transfusions were banned in much of Europe. Transfusions from human donors were somewhat less lethal, but violent reactions often led to the death of the recipient, and by the nineteenth century, transfusions had been abandoned as a medical failure.

In 1904, Dr. Karl Landsteiner performed a series of experiments on the blood of workers in his laboratory. His results explained the mysterious transfusion fatalities, and blood transfusions were reinstated as a lifesaving clinical tool. Landsteiner took blood samples from his coworkers. He separated the blood cells from the serum, the liquid component of the blood, and mixed these samples in test tubes. When he mixed serum from one individual with blood cells of another, Landsteiner observed that, in some instances, the serum samples caused clumping, or *agglutination*, of red blood cells (RBC). (See figure below.) The agglutination reaction always indicated that the two bloods were incompatible and transfusion could lead

to life-threatening reactions. As a result of many such experiments, Landsteiner showed that there are four human blood groups, designated A, B, AB, and O.

We now know that differences among blood groups reflect differences among oligosaccharides attached to the proteins and lipids of the RBC membranes. The oligosaccharides on the RBC surface have a common core, as shown in the accompanying figure, consisting of β-D-N-acetylgalactosamine, galactose, N-acetylneuraminic acid (sialic acid), and L-fucose. It is the terminal monosaccharide of this oligosaccharide that distinguishes the cells and governs the compatibility of the blood types.

The A blood group antigen has β-D-N-acetylgalactosamine at its end, whereas the B blood group antigen has α-D-galactose. In type O blood, neither of these sugars is found on the cell surface; only the core oligosaccharide is present. Some of the oligosaccharides on type AB blood cells have a terminal β-D-N-acetylgalactosamine, whereas others have a terminal α-D-galactose.

Why does agglutination occur? The clumping reaction that occurs when incompatible bloods are mixed is an antigen-antibody reaction. Antigens are large molecules, often portions of bacteria or viruses, that stimulate the immune defenses of the body to produce protective antibodies. Antibodies bind to the foreign antigens and help to destroy them.

People with type A blood also have antibodies against type B blood (anti-B antibodies) in the blood serum. If the person with type A blood receives a transfusion of type B blood, the anti-B antibodies bind to the type B blood cells, causing clumping and destruction of those cells that can result in death. Individuals with type B blood also produce anti-A antibodies and therefore cannot receive a transfusion from a type A individual. Those with type AB blood have neither anti-A nor anti-B antibodies in their blood. (If they did, they would destroy their own red blood cells!) Type O blood has no A or B antigens on the RBC but has both anti-A and anti-B antibodies. Because of the presence of both types of antibodies, type O individuals can receive transfusions only from a person who is also type O.

Type A

Type B

Type AB

Type O

ABO blood typing kit. When antibodies bind to antigens on the cell surface, clumping occurs.

For Further Understanding

People with type AB blood are referred to as universal recipients. Explain why these people can receive blood of any of the four ABO types.

People with type O blood are referred to as universal donors. Explain why type O blood can be safely transfused into patients of any ABO blood type.

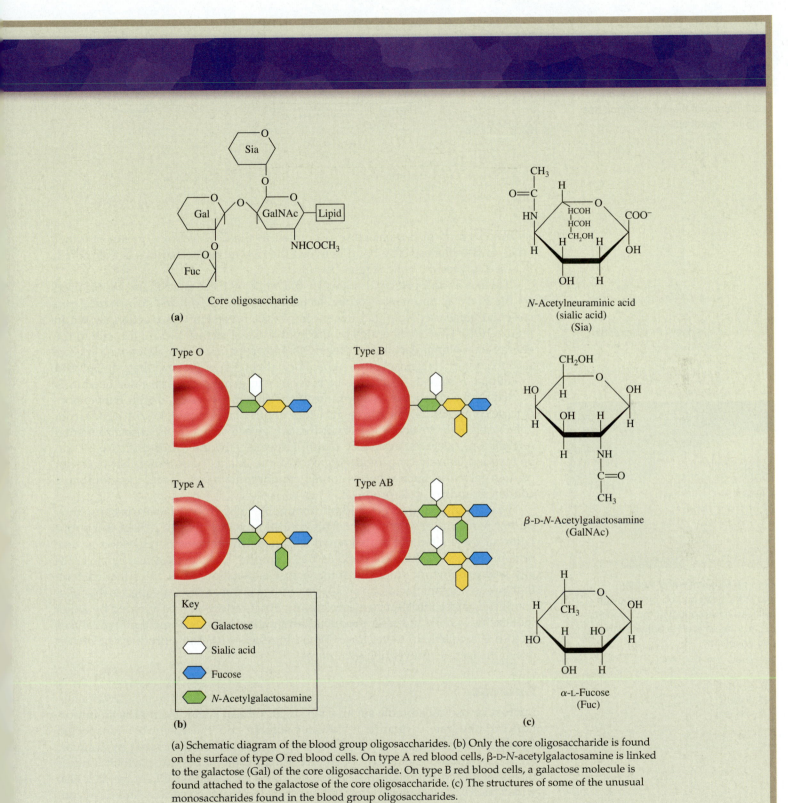

(a) Schematic diagram of the blood group oligosaccharides. (b) Only the core oligosaccharide is found on the surface of type O red blood cells. On type A red blood cells, β-D-N-acetylgalactosamine is linked to the galactose (Gal) of the core oligosaccharide. On type B red blood cells, a galactose molecule is found attached to the galactose of the core oligosaccharide. (c) The structures of some of the unusual monosaccharides found in the blood group oligosaccharides.

Figure 16.10

Glycosidic bond formed between the C-1 hydroxyl group of β-D-galactose and the C-4 hydroxyl group of β-D-glucose. The disaccharide is called β-lactose because the hydroxyl group at the reducing end of the disaccharide has the β-configuration.

β-D-Galactose   β-D-Glucose   β-Lactose

Glycolysis is discussed in Chapter 21.

hydroxyl group of β-D-galactose is bonded to the C-4 hydroxyl group of either an α- or β-D-glucose. The bond between the two monosaccharides is therefore a β(1 → 4) glycosidic bond (Figure 16.10).

Lactose is the principal sugar in the milk of most mammals. To be used by the body as an energy source, lactose must be hydrolyzed to produce glucose and galactose. Note that this is simply the reverse of the reaction shown in Figure 16.10. Glucose liberated by the hydrolysis of lactose is used directly in the energy-harvesting reactions of glycolysis. However, a series of reactions is necessary to convert galactose into a phosphorylated form of glucose that can be used in cellular metabolic reactions. In humans the genetic disease **galactosemia** is caused by the absence of one or more of the enzymes needed for this conversion. A toxic compound formed from galactose accumulates in people who suffer from galactosemia. If the condition is not treated, galactosemia leads to severe mental retardation, cataracts, and early death. However, the effects of this disease can be avoided entirely by providing galactosemic infants with a diet that does not contain galactose. Such a diet, of course, cannot contain lactose and therefore must contain no milk or milk products.

Many adults, and some children, are unable to hydrolyze lactose because they do not make the enzyme *lactase.* This condition, which affects 20% of the population of the United States, is known as **lactose intolerance.** Undigested lactose remains in the intestinal tract and causes cramping and diarrhea that can eventually lead to dehydration. Some of the lactose is metabolized by intestinal bacteria that release organic acids and CO_2 gas into the intestines, causing further discomfort. Lactose intolerance is unpleasant, but its effects can be avoided by a diet that excludes milk and milk products. Alternatively, the enzyme that hydrolyzes lactose is available in tablet form. When ingested with dairy products, it breaks down the lactose, preventing symptoms.

Both galactosemia and lactose intolerance are treated by removing milk and milk products from the diet. Explain the difference between these two conditions.

Sucrose

Sucrose is also called table sugar, cane sugar, or beet sugar. Sucrose is an important carbohydrate in plants. It is water soluble and can easily be transported through the circulatory system of the plant. It cannot be synthesized by animals. High concentrations of sucrose produce a high osmotic pressure, which inhibits the growth of microorganisms, so it is used as a preservative. Of course, it is also widely used as a sweetener. In fact, it is estimated that the average American consumes 100–125 pounds of sucrose each year. It has been suggested that sucrose in the diet is undesirable because it represents a source of empty calories; that is, it contains no vitamins or minerals. However, the only negative association that has been scientifically verified is the link between sucrose in the diet and dental caries, or cavities (see A Human Perspective: Tooth Decay and Simple Sugars on p. 554).

Sucrose is a disaccharide of α-D-glucose joined to β-D-fructose (Figure 16.11). The glycosidic linkage between α-D-glucose and β-D-fructose is quite different

Figure 16.11

Glycosidic bond formed between the C-1 hydroxyl of α-D-glucose and the C-2 hydroxyl of β-D-fructose. This bond is called an (α1 → β2) glycosidic linkage. The disaccharide formed in this reaction is sucrose.

from those that we have examined for lactose and maltose. This bond involves the anomeric carbons of *both* sugars! This bond is called an (α1 → β2) glycosidic linkage, since it involves the C-1 anomeric carbon of glucose and the C-2 anomeric carbon of fructose (noted in red in Figure 16.11). Because the (α1 → β2) glycosidic bond joins both anomeric carbons, there is no hemiacetal group. As a result, sucrose will not react with Benedict's reagent and is not a reducing sugar.

16.6 Polysaccharides

Starch

Most carbohydrates that are found in nature are large polymers of glucose. Thus a polysaccharide is a large polymer composed of many monosaccharide units (the monomers) joined in one or more chains.

Plants have the ability to use the energy of sunlight to produce monosaccharides, principally glucose, from CO_2 and H_2O. Although sucrose is the major transport form of sugar in the plant, starch (a polysaccharide) is the principal storage form in most plants. These plants store glucose in starch granules. Nearly all plant cells contain some starch granules, but in some seeds, such as corn, as much as 80% of the cell's dry weight is starch.

Starch is a heterogeneous material composed of the glucose polymers **amylose** and **amylopectin.** Amylose, which accounts for about 20% of the starch of a plant cell, is a linear polymer of α-D-glucose molecules connected by glycosidic bonds between C-1 of one glucose molecule and C-4 of a second glucose. Thus the glucose units in amylose are joined by α(1 → 4) glycosidic bonds. A single chain can contain up to four thousand glucose units. Amylose coils up into a helix that repeats every six glucose units. The structure of amylose is shown in Figure 16.12.

Amylose is degraded by two types of enzymes. They are produced in the pancreas, from which they are secreted into the small intestine, and the salivary glands, from which they are secreted into the saliva. α-*Amylase* cleaves the glycosidic bonds of amylose chains at random along the chain, producing shorter polysaccharide chains. The enzyme β-*amylase* sequentially cleaves the disaccharide maltose from the reducing end of the amylose chain. The maltose is hydrolyzed into glucose by the enzyme *maltase*. The glucose is quickly absorbed by intestinal cells and used by the cells of the body as a source of energy.

Amylopectin is a highly branched amylose in which the branches are attached to the C-6 hydroxyl groups by α(1 → 6) glycosidic bonds (Figure 16.13). The main

LEARNING GOAL

12 Discuss the structural, chemical, and biochemical properties of starch, glycogen, and cellulose.

A polymer (Section 11.5) is a large molecule made up of many small units, the monomers, held together by chemical bonds.

 Animation

Natural and Synthetic Polymers

Enzymes are proteins that serve as biological catalysts. They speed up biochemical reactions so that life processes can function. α- and β-Amylases are called α(1 → 4) glycosidases because they cleave α(1 → 4) glycosidic bonds.

Figure 16.12

Structure of amylose. (a) A linear chain of α-D-glucose joined in α(1 → 4) glycosidic linkage makes up the primary structure of amylose. (b) Owing to hydrogen bonding, the amylose chain forms a left-handed helix that contains six glucose units per turn.

(a)   (b)

chains consist of α(1 → 4) glycosidic bonds. Each branch contains 20–25 glucose units, and there are so many branches that the main chain can scarcely be distinguished.

Glycogen

Glycogen is the major glucose storage molecule in animals. The structure of glycogen is similar to that of amylopectin. The "main chain" is linked by α(1 → 4) glycosidic bonds, and it has numerous α(1 → 6) glycosidic bonds, which provide many branch points along the chain. Glycogen differs from amylopectin only by having more and shorter branches. Otherwise, the two molecules are virtually identical. The structure of glycogen is shown in Figure 16.13.

Glycogen is stored in the liver and skeletal muscle. Glycogen synthesis and degradation in the liver are carefully regulated. As we will see in Section 21.7, these two processes are intimately involved in keeping blood glucose levels constant.

Potatoes contain large amounts of starch. Describe the composition of this starch.

Cellulose

The most abundant polysaccharide, indeed the most abundant organic molecule in the world, is **cellulose,** a polymer of β-D-glucose units linked by β(1 → 4) glycosidic bonds (Figure 16.14). A molecule of cellulose typically contains about 3000 glucose units, but the largest known cellulose, produced by the alga *Valonia*, contains 26,000 glucose molecules.

Cellulose is a structural component of the plant cell wall. The unbranched structure of the cellulose polymer and the β(1 → 4) glycosidic linkages allow cellulose molecules to form long, straight chains of parallel cellulose molecules called *fibrils*. These fibrils are quite rigid and are held together tightly by hydrogen bonds; thus it is not surprising that cellulose is a cell wall structural element.

Vegetables contribute fiber to our diet. What carbohydrate provides this fiber?

In contrast to glycogen, amylose, and amylopectin, cellulose *cannot* be digested by humans. The reason is that we cannot synthesize the enzyme *cellulase*, which can hydrolyze the β(1 → 4) glycosidic linkages of the cellulose polymer. Indeed, only a few animals, such as termites, cows, and goats, are able to digest cellulose. These animals have, within their digestive tracts, microorganisms that produce the enzyme cellulase. The sugars released by this microbial digestion can then be absorbed and used by these animals. In humans, cellulose from fruits and vegetables serves as fiber in the diet.

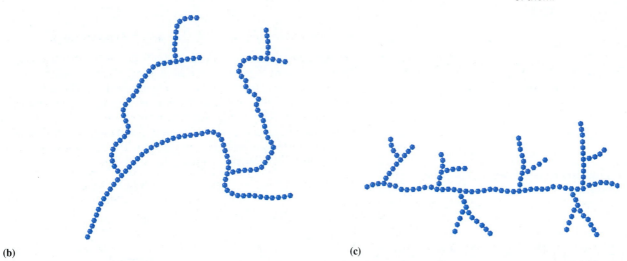

Figure 16.13
Structure of amylopectin and glycogen. (a) Both amylopectin and glycogen consist of chains of α-D-glucose molecules joined in α(1 → 4) glycosidic linkages. Branching from these chains are other chains of the same structure. Branching occurs by formation of α(1 → 6) glycosidic bonds between glucose units. (b) A representation of the branched-chain structure of amylopectin. (c) A representation of the branched-chain structure of glycogen. Glycogen differs from amylopectin only in that the branches are shorter and there are more of them.

What chemical reactions are catalyzed by α-amylase and β-amylase?

Question 16.11

What is the function of cellulose in the human diet? How does this relate to the structure of cellulose?

Question 16.12

Figure 16.14
The structure of cellulose.

A Medical Perspective

Monosaccharide Derivatives and Heteropolysaccharides of Medical Interest

Many of the carbohydrates with important functions in the human body are either derivatives of simple monosaccharides or are complex polymers of monosaccharide derivatives. One type of monosaccharide derivatives, the uronates, is formed when the terminal—CH_2OH group of a monosaccharide is oxidized to a carboxylate group. α-D-Glucuronate is a uronate of glucose:

α-D-Glucuronate

In liver cells, α-D-glucuronate is bonded to hydrophobic molecules, such as steroids, to increase their solubility in water. When bonded to the modified sugar, steroids are more readily removed from the body.

Amino sugars are a second important group of monosaccharide derivatives. In amino sugars one of the hydroxyl groups (usually on carbon-2) is replaced by an amino group. Often these are found in complex oligosaccharides that are attached to cellular proteins and lipids. The most common amino sugars, D-glucosamine and D-galactosamine, are often found in the N-acetyl form. N-acetylglucosamine is a component of bacterial cell walls and N-acetylgalactosamine is a component of the A, B, O blood group antigens (see preceding, A Human Perspective: Blood Transfusions and the Blood Group Antigens).

α-D-Glucosamine   α-D-N-Acetylglucosamine

Heteropolysaccharides are long-chain polymers that contain more than one type of monosaccharide, many of which are amino sugars. These *glycosaminoglycans* include chondroitin sulfate, hyaluronic acid, and heparin. Hyaluronic acid is abundant in the fluid of joints and in the vitreous humor of the eye. Chondroitin sulfate is an important component of cartilage; and heparin has anticoagulant function. The structures of the repeat units of these polymers are shown below.

Repeat unit of chondroitin sulfate

SUMMARY

16.1 Types of Carbohydrates

Carbohydrates are found in a wide variety of naturally occurring substances and serve as principal energy sources for the body. Dietary carbohydrates include complex carbohydrates, such as starch in potatoes, and simple carbohydrates, such as sucrose.

Carbohydrates are classified as *monosaccharides* (one sugar unit), *disaccharides* (two sugar units), *oligosaccharides* (three to ten sugar units), or *polysaccharides* (many sugar units).

16.2 Monosaccharides

Monosaccharides that have an aldehyde as their most oxidized functional group are *aldoses*, and those having a ketone group as their most oxidized functional group are *ketoses*. They may be classified as *trioses, tetroses, pentoses,* and so forth, depending on the number of carbon atoms in the carbohydrate.

16.3 Stereoisomers and Stereochemistry

Stereoisomers of monosaccharides exist because of the presence of *chiral carbon* atoms. They are classified as D- or L-depending on the arrangement of the atoms on the chiral carbon farthest from the aldehyde or ketone group. If the

Repeat unit of hyaluronic acid

Repeat unit of heparin

Two of these molecules have been studied as potential treatments for osteoarthritis, a painful, degenerative disease of the joints. The amino sugar D-glucosamine is thought to stimulate the production of collagen. Collagen is one of the main components of articular cartilage, which is the shock-absorbing cushion within the joints. With aging, some of the D-glucosamine is lost, leading to a reduced cartilage layer and to the onset and

progression of arthritis. It has been suggested that ingestion of D-glucosamine can actually "jump-start" production of cartilage and help repair eroded cartilage in arthritic joints.

It has also been suggested that chondroitin sulfate can protect existing cartilage from premature breakdown. It absorbs large amounts of water, which is thought to facilitate diffusion of nutrients into the cartilage, providing precursors for the synthesis of new cartilage. The increased fluid also acts as a shock absorber.

Studies continue on the effects that D-glucosamine and chondroitin sulfate have on degenerative joint disease. To date the studies are inconclusive because a large placebo effect is observed with sufferers of osteoarthritis. Many people in the control groups of these studies also experience relief of symptoms when they receive treatment with a placebo, such as a sugar pill.

Capsules containing D-glucosamine and chondroitin sulfate are available over the counter, and many sufferers of osteoarthritis prefer to take this nutritional supplement as an alternative to any nonsteroidal anti-inflammatory drugs (NSAID), such as ibuprofen. Although NSAIDs can reduce inflammation and pain, long-term use of NSAIDs can result in stomach ulcers, damage to auditory nerves, and kidney damage.

For Further Understanding

In Chapter 15 we learned that nonsteroidal anti-inflammatory drugs (NSAIDs), such as ibuprofen, are analgesics used to treat pain, such as that associated with osteoarthritis. Why do many people prefer to treat osteoarthritis with D-glucosamine and chondroitin sulfate rather than NSAIDs?

Explain why attaching a molecule such as α-D-glucuronate to a steroid molecule would increase its water solubility.

—OH on this carbon is to the right, the stereoisomer is of the D-family. If the —OH group is to the left, the stereoisomer is of the L-family.

Each member of a pair of stereoisomers will rotate plane-polarized light in different directions. A polarimeter is used to measure the direction of rotation of plane-polarized light. Compounds that rotate light in a clockwise direction are termed dextrorotatory and are designated by a plus sign (+). Compounds that rotate light in a counterclockwise direction are called levorotatory and are indicated by a minus sign (−).

The *Fischer Projection* is a two-dimensional drawing of a molecule that shows a chiral carbon at the intersection of two lines. Horizontal lines represent bonds projecting out

of the page and vertical lines represent bonds that project into the page. The most oxidized carbon is always represented at the "top" of the structure.

16.4 Biologically Important Monosaccharides

Important monosaccharides include *glyceraldehyde, glucose, fructose,* and *ribose.* Monosaccharides containing five or six carbon atoms can exist as five-membered or six-membered rings. Formation of a ring produces a new chiral carbon at the original carbonyl carbon, which is designated either α or β depending on the orientation of the groups. The cyclization of an aldose produces an intramolecular *hemiacetal,* and the cyclization of a ketose yields an intramolecular *hemiketal.*

Reducing sugars are oxidized by the *Benedict's reagent.* All monosaccharides and all common disaccharides, except sucrose, are reducing sugars. At one time Benedict's reagent was used to determine the concentration of glucose in urine.

16.5 Biologically Important Disaccharides

Important disaccharides include *lactose* and *sucrose.* Lactose is a disaccharide of β-D-galactose bonded β(1 → 4) with D-glucose. In *galactosemia*, defective metabolism of galactose leads to accumulation of a toxic by-product. The ill effects of galactosemia are avoided by exclusion of milk and milk products from the diet of affected infants. Sucrose is a dimer composed of α-D-glucose bonded (α1 → β2) with β-D-fructose.

16.6 Polysaccharides

Starch, the storage polysaccharide of plant cells, is composed of approximately 20% amylose and 80% amylopectin. *Amylose* is a polymer of α-D-glucose units bonded α(1 → 4). Amylose forms a helix. Amylopectin has many branches. Its main chain consists of α-D-glucose units bonded α(1 → 4). The branches are connected by α(1 → 6) glycosidic bonds.

Glycogen, the major storage polysaccharide of animal cells, resembles amylopectin, but it has more, shorter branches. The liver reserve of glycogen is used to regulate blood glucose levels.

Cellulose is a major structural molecule of plants. It is a β(1 → 4) polymer of D-glucose that can contain thousands of glucose monomers. Cellulose cannot be digested by animals because they do not produce an enzyme capable of cleaving the β(1 → 4) glycosidic linkage.

KEY TERMS

aldose (16.2)	hemiacetal (16.4)
amylopectin (16.6)	hemiketal (16.4)
amylose (16.6)	hexose (16.2)
anomer (16.4)	ketose (16.2)
Benedict's reagent (16.4)	lactose (16.5)
carbohydrate (16.1)	lactose intolerance (16.5)
cellulose (16.6)	maltose (16.5)
chiral carbon (16.3)	monosaccharide (16.1)
chiral molecule (16.3)	nonreducing sugar (16.5)
disaccharide (16.1)	oligosaccharide (16.1)
enantiomers (16.3)	pentose (16.2)
Fischer Projection (16.3)	polysaccharide (16.1)
fructose (16.4)	reducing sugar (16.4)
galactose (16.4)	ribose (16.4)
galactosemia (16.5)	saccharide (16.1)
glucose (16.4)	stereochemistry (16.3)
glyceraldehyde (16.3)	stereoisomers (16.3)
glycogen (16.6)	sucrose (16.5)
glycosidic bond (16.1)	tetrose (16.2)
Haworth projection (16.4)	triose (16.2)

Types of Carbohydrates

Foundations

16.13 What is the difference between a monosaccharide and a disaccharide?

16.14 What is a polysaccharide?

16.15 What is the general molecular formula for a simple sugar?

16.16 What biochemical process is ultimately responsible for the synthesis of sugars?

Applications

16.17 Read the labels on some of the foods in your kitchen, and see how many products you can find that list one or more carbohydrates among the ingredients in the package. Make a list of these compounds, and attempt to classify them as monosaccharides, disaccharides, or polysaccharides.

16.18 Some disaccharides are often referred to by their common names. What are the chemical names of (a) milk sugar, (b) beet sugar, and (c) cane sugar?

16.19 How many kilocalories of energy are released when 1 g of carbohydrate is "burned" or oxidized?

16.20 List some natural sources of carbohydrates.

16.21 Draw and provide the names of an aldohexose and a ketohexose.

16.22 Draw and provide the name of an aldotriose.

Monosaccharides

Foundations

16.23 Define the term *aldose.*

16.24 Define the term *ketose.*

16.25 What is a tetrose?

16.26 What is a hexose?

16.27 What is a ketopentose?

16.28 What is an aldotriose?

Applications

16.29 Identify each of the following sugars. Label each as either a hemiacetal or a hemiketal:

16.30 Draw the open-chain form of the sugars in Problem 16.29.

16.31 Draw all of the different possible aldotrioses of molecular formula $C_3H_6O_3$.

16.32 Draw all of the different possible aldotetroses of molecular formula $C_4H_8O_4$.

Stereoisomers and Stereochemistry

Foundations

16.33 Define the term *stereoisomer.*

16.34 Define the term *enantiomer.*

16.35 Define the term *chiral carbon.*

16.36 Draw an aldotetrose. Note each chiral carbon with an asterisk (*).

16.37 Explain how a polarimeter works.

16.38 What is plane-polarized light?

16.39 What is a Fischer Projection?

16.40 How would you produce a Fischer Projection beginning with a three-dimensional model of a sugar?

Applications

16.41 Is there any difference between dextrose and D-glucose?

16.42 The linear structure of D-glucose is shown in Figure 16.7. Draw its mirror image.

16.43 How are D- and L-glyceraldehyde related?

16.44 Determine whether each of the following is a D- or L-sugar:

a.
```
        O
        ‖
        CH
        |
   H—C—OH
        |
   H—C—OH
        |
     CH2OH
```

b.
```
        O
        ‖
        CH
        |
   H—C—OH
        |
   H—C—OH
        |
  HO—C—H
        |
     CH2OH
```

c.
```
        O
        ‖
        CH
        |
   H—C—OH
        |
   H—C—OH
        |
  HO—C—H
        |
     CH2OH
```

16.45 Draw a Fischer Projection formula for each of the following compounds. Indicate each of the chiral carbons with asterisks (*).

a.
```
        O
        ‖
        C—H
        |
  HO—C—H
        |
   H—C—OH
        |
  HO—C—H
        |
  HO—C—H
        |
     CH2OH
```

b.
```
        O
        ‖
        C—H
        |
   H—C—OH
        |
   H—C—OH
        |
     CH2OH
```

c.
```
        O
        ‖
        C—H
        |
  HO—C—H
        |
   H—C—OH
        |
  HO—C—H
        |
   H—C—OH
        |
  HO—C—H
        |
     CH2OH
```

16.46 Draw a Fischer Projection formula for each of the following compounds. Indicate each of the chiral carbons with asterisks (*).

a.
```
        O
        ‖
        C—H
        |
   H—C—H
        |
  HO—C—H
        |
  HO—C—H
        |
  HO—C—H
        |
     CH2OH
```

b.
```
        O
        ‖
        C—H
        |
   H—C—H
        |
   H—C—OH
        |
     CH2OH
```

c.
```
        O
        ‖
        C—H
        |
  HO—C—H
        |
  HO—C—H
        |
  HO—C—H
        |
   H—C—OH
        |
  HO—C—H
        |
     CH2OH
```

Biologically Important Monosaccharides

Foundations

16.47 Define the term *anomer*.

16.48 What is a Haworth projection?

16.49 What is a hemiacetal?

16.50 What is a hemiketal?

16.51 Explain why the cyclization of D-glucose forms a hemiacetal.

16.52 Explain why cyclization of D-fructose forms a hemiketal.

Applications

16.53 Why does cyclization of D-glucose give two isomers, α- and β-D-glucose?

16.54 Draw the structure of the open-chain form of D-fructose, and show how it cyclizes to form α- and β-D-fructose.

16.55 Which of the following would give a positive Benedict's test?
   a. Sucrose
   b. Glycogen
   c. β-Maltose
   d. α-Lactose

16.56 Why was the Benedict's reagent useful for determining the amount of glucose in the urine?

16.57 Describe what is meant by a pair of enantiomers. Draw an example of a pair of enantiomers.

16.58 What is a chiral carbon atom?

16.59 When discussing sugars, what do we mean by an intramolecular hemiacetal?

16.60 When discussing sugars, what do we mean by an intramolecular hemiketal?

Biologically Important Disaccharides

Foundations

16.61 What is a ketal?

16.62 What is an acetal?

16.63 What is a glycosidic bond?

16.64 Why are glycosidic bonds either acetals or ketals?

Applications

16.65 Maltose is a disaccharide isolated from amylose that consists of two glucose units linked α(1 → 4). Draw the structure of this molecule.

16.66 Sucrose is a disaccharide formed by linking α-D-glucose and β-D-fructose by an (α1 → β2) bond. Draw the structure of this disaccharide.

16.67 What is the major biological source of lactose?

16.68 What metabolic defect causes galactosemia?

16.69 What simple treatment prevents most of the ill effects of galactosemia?

16.70 What are the major physiological effects of galactosemia?

16.71 What is lactose intolerance?

16.72 What is the difference between lactose intolerance and galactosemia?

Polysaccharides

Foundations

16.73 What is a polymer?

16.74 What form of sugar is used as the major transport sugar in a plant?

16.75 What is the major storage form of sugar in a plant?

16.76 What is the major structural form of sugar in a plant?

Applications

16.77 What is the difference between the structure of cellulose and the structure of amylose?

16.78 How does the structure of amylose differ from that of amylopectin and glycogen?

16.79 What is the major physiological purpose of glycogen?

16.80 Where in the body do you find glycogen stored?

16.81 Where are α-amylase and β-amylase produced?

16.82 Where do α-amylase and β-amylase carry out their enzymatic functions?

CRITICAL THINKING PROBLEMS

1. The six-member glucose ring structure is not a flat ring. Like cyclohexane, it can exist in the chair conformation. Build models of the chair conformation of α- and β-D-glucose. Draw each of these structures. Which would you predict to be the more stable isomer? Explain your reasoning.

2. The following is the structure of salicin, a bitter-tasting compound found in the bark of willow trees:

Salicin

The aromatic ring portion of this structure is quite insoluble in water. How would forming a glycosidic bond between the aromatic ring and β-D-glucose alter the solubility? Explain your answer.

3. Ancient peoples used salicin to reduce fevers. Write an equation for the acid-catalyzed hydrolysis of the glycosidic bond of salicin. Compare the aromatic product with the structure of acetylsalicylic acid (aspirin). Use this information to develop a hypothesis explaining why ancient peoples used salicin to reduce fevers.

4. Chitin is a modified cellulose in which the C-2 hydroxyl group of each glucose is replaced by

$$-NHCCH_3$$

This nitrogen-containing polysaccharide makes up the shells of lobsters, crabs, and the exoskeletons of insects. Draw a portion of a chitin polymer consisting of four monomers.

5. Pectins are polysaccharides obtained from fruits and berries and used to thicken jellies and jams. Pectins are $\alpha(1 \rightarrow 4)$ linked D-galacturonic acid. D-Galacturonic acid is D-galactose in which the C-6 hydroxyl group has been oxidized to a carboxyl group. Draw a portion of a pectin polymer consisting of four monomers.

6. Peonin is a red pigment found in the petals of peony flowers. Consider the structure of peonin:

Why do you think peonin is bonded to two hexoses? What monosaccharide(s) would be produced by acid-catalyzed hydrolysis of peonin?

17 Lipids and Their Functions in Biochemical Systems

Learning Goals

1 Discuss the physical and chemical properties and biological function of each of the families of lipids.

2 Write the structures of saturated and unsaturated fatty acids.

3 Compare and contrast the structure and properties of saturated and unsaturated fatty acids.

4 Write equations representing the reactions that fatty acids undergo.

5 Describe the functions of prostaglandins.

6 Discuss the mechanism by which aspirin reduces pain.

7 Draw the structure of a phospholipid and discuss its amphipathic nature.

8 Discuss the general classes of sphingolipids and their functions.

9 Draw the structure of the steroid nucleus and discuss the functions of steroid hormones.

10 Describe the function of lipoproteins in triglyceride and cholesterol transport in the body.

11 Draw the structure of the cell membrane and discuss its functions.

Outline

Introduction

Chemistry Connection:
Lifesaving Lipids

17.1 Biological Functions of Lipids

17.2 Fatty Acids

A Human Perspective:
Mummies Made of Soap

17.3 Glycerides

17.4 Nonglyceride Lipids

A Medical Perspective:
Disorders of Sphingolipid Metabolism

A Medical Perspective:
Steroids and the Treatment of Heart Disease

17.5 Complex Lipids

17.6 The Structure of Biological Membranes

A Medical Perspective:
Liposome Delivery Systems

A Medical Perspective:
Antibiotics That Destroy Membrane Integrity

Lecithin and 1-hexadecanol are components of the pulmonary surfactant that saved this baby's life. Draw the structures of these two molecules and explain how they interact on the surface of the alveoli.

Introduction

Lipids seem to be the most controversial group of biological molecules, particularly in the fields of medicine and nutrition. We are concerned about the use of anabolic steroids by athletes. Although these hormones increase muscle mass and enhance performance, we are just beginning to understand the damage they cause to the body.

We worry about what types of dietary fat we should consume. We hear frequently about the amounts of saturated fats and cholesterol in our diets because a strong correlation has been found between these lipids and heart disease. Large quantities of dietary saturated fats may also predispose an individual to colon, esophageal, stomach, and breast cancers. As a result, we are advised to reduce our intake of cholesterol and saturated fats.

Nonetheless, lipids serve a wide variety of functions essential to living systems and are required in our diet. Standards of fat intake have not been experimentally determined. However, the most recent U.S. Dietary Guidelines recommend that dietary fat not exceed 35% of the daily caloric intake, and no more than 10% should be saturated fats. Dietary cholesterol should be no more than 300 mg/day.

Chemistry Connection

Lifesaving Lipids

In the intensive-care nursery the premature infant struggles for life. Born three and a half months early, the baby weighs only 1.6 pounds, and the lungs labor to provide enough oxygen to keep the tiny body alive. Premature infants often have respiratory difficulties because they have not yet begun to produce *pulmonary surfactant*.

Pulmonary surfactant is a combination of phospholipids and proteins that reduces surface tension in the alveoli of the lungs. (Alveoli are the small, thin-walled air sacs in the lungs.) This allows efficient gas exchange across the membranes of the alveolar cells; oxygen can more easily diffuse from the air into the tissues and carbon dioxide can easily diffuse from the tissues into the air. Without pulmonary surfactant, gas exchange in the lungs is very poor.

Pulmonary surfactant is not produced until early in the sixth month of pregnancy. Premature babies born before they have begun secretion of natural surfactant suffer from *respiratory distress syndrome (RDS)*, which is caused by the severe difficulty they have obtaining enough oxygen from the air that they breathe.

Until recently, RDS was a major cause of death among premature infants, but now a lifesaving treatment is available.

A fine aerosol of an artificial surfactant is administered directly into the trachea. The Glaxo-Wellcome Company product EXOSURF Neonatal contains the phospholipid lecithin to reduce surface tension; 1-hexadecanol, which spreads the lecithin; and a polymer called tyloxapol, which disperses both the lecithin and the 1-hexadecanol.

Artificial pulmonary surfactant therapy has dramatically reduced premature infant death caused by RDS and appears to have reduced overall mortality for all babies born weighing less than 700 g (about 1.5 pounds). Advances such as this have come about as a result of research on the makeup of body tissues and secretions in both healthy and diseased individuals. Often, such basic research provides the information needed to develop effective therapies.

In this chapter we will study the chemistry of lipids with a wide variety of structures and biological functions. Among these are the triglycerides that stock our adipose tissue, pain-producing prostaglandins, and steroids that determine our secondary sexual characteristics.

17.1 Biological Functions of Lipids

The term **lipids** actually refers to a collection of organic molecules of varying chemical composition. They are grouped together on the basis of their solubility in nonpolar solvents. Lipids are commonly subdivided into four main groups:

1. *Fatty acids* (saturated and unsaturated)
2. *Glycerides* (glycerol-containing lipids)
3. *Nonglyceride lipids* (sphingolipids, steroids, waxes)
4. *Complex lipids* (lipoproteins)

In this chapter we examine the structure, properties, chemical reactions, and biological functions of each of the lipid groups shown in Figure 17.1.

As a result of differences in their structures, lipids serve many different functions in the human body. The following brief list will give you an idea of the importance of lipids in biological processes:

- *Energy source.* Like carbohydrates, lipids are an excellent source of energy for the body. When oxidized, each gram of fat releases 9 kcal of energy, or more than twice the energy released by oxidation of a gram of carbohydrate.
- *Energy storage.* Most of the energy stored in the body is in the form of lipids (triglycerides). Stored in fat cells called *adipocytes,* these fats are a particularly rich source of energy for the body.
- *Cell membrane structural components.* Phosphoglycerides, sphingolipids, and steroids make up the basic structure of all cell membranes. These membranes control the flow of molecules into and out of cells and allow cell-to-cell communication.
- *Hormones.* The steroid hormones are critical chemical messengers that allow tissues of the body to communicate with one another. The hormonelike prostaglandins exert strong biological effects on both the cells that produce them and other cells of the body.
- *Vitamins.* The lipid-soluble vitamins, A, D, E, and K, play a major role in the regulation of several critical biological processes, including blood clotting and vision.
- *Vitamin absorption.* Dietary fat serves as a carrier of the lipid-soluble vitamins. All are transported into cells of the small intestine in association with fat molecules. Therefore a diet that is too low in fat (less than 20% of calories) can result in a deficiency of these four vitamins.
- *Protection.* Fats serve as a shock absorber, or protective layer, for the vital organs. About 4% of the total body fat is reserved for this critical function.
- *Insulation.* Fat stored beneath the skin (subcutaneous fat) serves to insulate the body from extremes of cold temperatures.

Lipid-Soluble Vitamins

Figure 17.1
The classification of lipids.

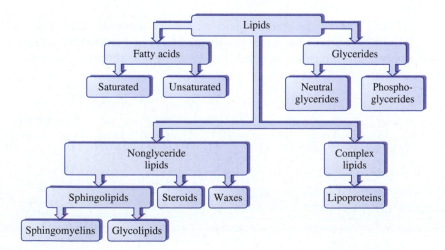

17.2 Fatty Acids

Structure and Properties

Fatty acids are long-chain monocarboxylic acids. As a consequence of their biosynthesis, fatty acids generally contain an *even number* of carbon atoms. The general formula for a **saturated fatty acid** is $CH_3(CH_2)_nCOOH$, in which n in biological systems is an even integer. If $n = 16$, the result is an 18-carbon saturated fatty acid, stearic acid, having the following structural formula:

$$
\begin{array}{c}
H-\overset{\overset{\displaystyle H}{|}}{\underset{\underset{\displaystyle H}{|}}{C}}-\overset{\overset{\displaystyle H}{|}}{\underset{\underset{\displaystyle H}{|}}{C}}-\cdots-\overset{\overset{\displaystyle H}{|}}{\underset{\underset{\displaystyle H}{|}}{C}}-\overset{\overset{\displaystyle O}{\|}}{C}-OH
\end{array}
$$

The saturated fatty acids may be thought of as derivatives of alkanes, the saturated hydrocarbons described in Chapter 10.

Note that each of the carbons in the chain is bonded to the maximum number of hydrogen atoms. To help remember the structure of a saturated fatty acid, you might think of each carbon in the chain being "saturated" with hydrogen atoms. Examples of common saturated fatty acids are given in Table 17.1. An example of an **unsaturated fatty acid** is the 18-carbon unsaturated fatty acid oleic acid, which has the following structural formula:

The unsaturated fatty acids may be thought of as derivatives of the alkenes, the unsaturated hydrocarbons discussed in Chapter 11.

In the case of unsaturated fatty acids, there is at least one carbon-to-carbon double bond. Because of the double bonds, the carbon atoms involved in these bonds are not "saturated" with hydrogen atoms. The double bonds found in almost all naturally occurring unsaturated fatty acids are in the *cis* configuration. In addition, the double bonds are not randomly located in the hydrocarbon chain. Both the placement and the geometric configuration of the double bonds are dictated by the enzymes that catalyze the biosynthesis of unsaturated fatty acids. Examples of common unsaturated fatty acids are also given in Table 17.1.

A discussion of trans-fatty acids is found in Section 11.3.

E X A M P L E 17.1   **Writing the Structural Formula of an Unsaturated Fatty Acid**

Draw the structural formula for palmitoleic acid.

Solution

The I.U.P.A.C. name of palmitoleic acid is *cis*-9-hexadecenoic acid. The name tells us that this is a 16-carbon fatty acid having a carbon-to-carbon double bond between carbons 9 and 10. The name also reveals that this is the *cis* isomer.

   16   15   14   13   12   11   10   9   8   7   6   5   4   3   2   1

Continued—

EXAMPLE 17.1 —*Continued*

Practice Problem 17.1

Draw the line formulas for (a) oleic acid and (b) linoleic acid.

For Further Practice: Questions 17.27 and 17.28.

Examination of Table 17.1 and Figure 17.2 reveals several interesting and important points about the physical properties of fatty acids.

- The melting points of saturated fatty acids increase with increasing carbon number, as is the case with alkanes. Saturated fatty acids containing ten or more carbons are solids at room temperature.
- The melting point of a saturated fatty acid is greater than that of an unsaturated fatty acid of the same chain length. The reason is that saturated fatty acid chains tend to be fully extended and to stack in a regular structure, thereby causing increased intermolecular attraction. Introduction of a *cis* double bond into the hydrocarbon chain produces a rigid 30° bend. Such "kinked" molecules cannot stack in an organized arrangement and thus have lower intermolecular attractions and lower melting points.
- As in the case for saturated fatty acids, the melting points of unsaturated fatty acids increase with increasing hydrocarbon chain length.

LEARNING GOAL

3 Compare and contrast the structure and properties of saturated and unsaturated fatty acids.

The relationship between alkane chain length and melting point is described in Section 10.2.

The relationship between alkene chain length and melting point is described in Section 11.1.

TABLE 17.1   Common Saturated and Unsaturated Fatty Acids

Common Saturated Fatty Acids

Common Name	I.U.P.A.C. Name	Melting Point (°C)	RCOOH Formula	Condensed Formula
Capric	Decanoic	32	$C_9H_{19}COOH$	$CH_3(CH_2)_8COOH$
Lauric	Dodecanoic	44	$C_{11}H_{23}COOH$	$CH_3(CH_2)_{10}COOH$
Myristic	Tetradecanoic	54	$C_{13}H_{27}COOH$	$CH_3(CH_2)_{12}COOH$
Palmitic	Hexadecanoic	63	$C_{15}H_{31}COOH$	$CH_3(CH_2)_{14}COOH$
Stearic	Octadecanoic	70	$C_{17}H_{35}COOH$	$CH_3(CH_2)_{16}COOH$
Arachidic	Eicosanoic	77	$C_{19}H_{39}COOH$	$CH_3(CH_2)_{18}COOH$

Common Unsaturated Fatty Acids

Common Name	I.U.P.A.C. Name	Melting Point (°C)	RCOOH Formula	Number of Double Bonds	Position of Double Bonds
Palmitoleic	*cis*-9-Hexadecenoic	0	$C_{15}H_{29}COOH$	1	9
Oleic	*cis*-9-Octadecenoic	16	$C_{17}H_{33}COOH$	1	9
Linoleic	*cis*,*cis*-9,12-Octadecadienoic	5	$C_{17}H_{31}COOH$	2	9, 12
Linolenic	All *cis*-9,12,15-Octadecatrienoic	−11	$C_{17}H_{29}COOH$	3	9, 12, 15
Arachidonic	All *cis*-5,8,11,14-Eicosatetraenoic	−50	$C_{19}H_{31}COOH$	4	5, 8, 11, 14

Condensed Formula

Palmitoleic	$CH_3(CH_2)_5CH{=}CH(CH_2)_7COOH$
Oleic	$CH_3(CH_2)_7CH{=}CH(CH_2)_7COOH$
Linoleic	$CH_3(CH_2)_4CH{=}CH{-}CH_2{-}CH{=}CH(CH_2)_7COOH$
Linolenic	$CH_3CH_2CH{=}CH{-}CH_2{-}CH{=}CH{-}CH_2{-}CH{=}CH(CH_2)_7COOH$
Arachidonic	$CH_3(CH_2)_4CH{=}CH{-}CH_2{-}CH{=}CH{-}CH_2{-}CH{=}CH{-}CH_2{-}CH{=}CH{-}(CH_2)_3COOH$

Figure 17.2

The melting points of fatty acids. Melting points of both saturated and unsaturated fatty acids increase as the number of carbon atoms in the chain increases. The melting points of unsaturated fatty acids are lower than those of the corresponding saturated fatty acid with the same number of carbon atoms. Also, as the number of double bonds in the chain increases, the melting points decrease.

Explain why the olive oil in the photo above is liquid at room temperature but the beef fat is solid.

TABLE **17.2**	**Similarities and Differences Between Saturated and Unsaturated Fatty Acids**	
Property	**Saturated Fatty Acid**	**Unsaturated Fatty Acid**
Chemical composition	Carbon, hydrogen, oxygen	Carbon, hydrogen, oxygen
Chemical structure	Hydrocarbon chain with a terminal carboxyl group	Hydrocarbon chain with a terminal carboxyl group
Carbon-carbon bonds within the hydrocarbon chain	Only C—C single bonds	At least one C—C double bond
Hydrocarbon chains are characteristic of what group of hydrocarbons	Alkanes	Alkenes
"Shape" of hydrocarbon chain	Linear, fully extended	Bend in carbon chain at site of C—C double bond
Physical state at room temperature	Solid	Liquid
Melting point for two fatty acids of the same hydrocarbon chain length	Higher	Lower
Relationship between melting point and chain length	Longer chain length, higher melting point	Longer chain length, higher melting point

Question *17.1*

Draw formulas for each of the following fatty acids:

a. Oleic acid   c. Linoleic acid
b. Lauric acid   d. Stearic acid

Question 17.2

What is the I.U.P.A.C. name for each of the fatty acids in Question 17.1? (*Hint*: Review the naming of carboxylic acids in Section 14.1 and Table 17.1.)

Chemical Reactions of Fatty Acids

The reactions of fatty acids are identical to those of short-chain carboxylic acids. The major reactions that they undergo include esterification, acid hydrolysis of esters, saponification, and addition at the double bond.

LEARNING GOAL
4 Write equations representing the reactions that fatty acids undergo.

Esterification

In **esterification,** fatty acids react with alcohols to form esters and water according to the following general equation:

Esterification is described in Sections 14.1 and 14.2.

$$R^1\!-\!\overset{\displaystyle O}{\overset{\|}{C}}\!-\!OH + HOR^2 \xrightarrow{H^+,\, heat} R^1\!-\!\overset{\displaystyle O}{\overset{\|}{C}}\!-\!OR^2 + H\!-\!OH$$

    Fatty acid   Alcohol   Ester   Water

Writing Equations Representing the Esterification of Fatty Acids   E X A M P L E   17.2

Write an equation representing the esterification of capric acid with propyl alcohol and write the I.U.P.A.C. names of each of the organic reactants and products.

LEARNING GOAL
4 Write equations representing the reactions that fatty acids undergo.

Solution

$$CH_3(CH_2)_8\!-\!\overset{\displaystyle O}{\overset{\|}{C}}\!-\!OH + CH_3CH_2CH_2OH \xrightarrow{H^+,\, heat}$$

    Decanoic acid   Propanol

$$CH_3(CH_2)_8\!-\!\overset{\displaystyle O}{\overset{\|}{C}}\!-\!O\!-\!CH_2CH_2CH_3 + H_2O$$

            Propyl decanoate

Practice Problem 17.2

Write the complete equation for the esterification of the following carboxylic acids and alcohols. Write the I.U.P.A.C. name for all of the organic reactants and products.
a. Lauric acid and ethyl alcohol
b. Capric acid and 1-pentanol

For Further Practice: Questions 17.31 and 17.32.

Acid Hydrolysis

Recall that hydrolysis is the reverse of esterification, producing fatty acids from esters:

Acid hydrolysis is discussed in Section 14.2.

$$R^1\!-\!\overset{\displaystyle O}{\overset{\|}{C}}\!-\!OR^2 + HO\!-\!H \xrightarrow{H^+,\, heat} R^1\!-\!\overset{\displaystyle O}{\overset{\|}{C}}\!-\!OH + R^2OH$$

    Ester   Water   Fatty acid   Alcohol

Writing Equations Representing the Acid Hydrolysis of a Fatty Acid Ester

LEARNING GOAL

4 Write equations representing the reactions that fatty acids undergo.

Write an equation representing the acid hydrolysis of methyl decanoate and write the I.U.P.A.C. names of each of the organic reactants and products.

Solution

$$CH_3(CH_2)_8{-}\overset{\overset{\displaystyle O}{\|}}{C}{-}O{-}CH_3 + H_2O \xrightarrow{\text{H}^+,\ \text{heat}} CH_3(CH_2)_8{-}\overset{\overset{\displaystyle O}{\|}}{C}{-}OH + CH_3OH$$

Methyl decanoate   Decanoic acid   Methanol

Practice Problem 17.3

Write a complete equation for the acid hydrolysis of the following esters. Write the I.U.P.A.C. name for each of the organic reactants and products.

a. Butyl propionate
b. Ethyl butyrate

For Further Practice: Questions 17.33 and 17.34.

Saponification

Saponification is described in Section 14.2.

Saponification is the base-catalyzed hydrolysis of an ester:

Ester   Base   Salt   Alcohol

The role of soaps in removal of dirt and grease is described in Section 14.2.

Examples of micelles are shown in Figures 14.4 and 23.1.

The product of this reaction, an ionized salt, is a soap. Because soaps have a long uncharged hydrocarbon tail and a negatively charged terminus (the carboxylate group), they form micelles that dissolve oil and dirt particles. Thus the dirt is emulsified and broken into small particles, and can be rinsed away.

Writing Equations Representing the Base-Catalyzed Hydrolysis of a Fatty Acid Ester

LEARNING GOAL

4 Write equations representing the reactions that fatty acids undergo.

Write an equation representing the base-catalyzed hydrolysis of ethyl dodecanoate and write the I.U.P.A.C. names of each of the organic reactants and products.

Solution

$$CH_3(CH_2)_{10}{-}\overset{\overset{\displaystyle O}{\|}}{C}{-}O{-}CH_2CH_3 + NaOH \longrightarrow$$

Ethyl dodecanoate

$$CH_3(CH_2)_{10}{-}\overset{\overset{\displaystyle O}{\|}}{C}{-}O^-Na^+ + CH_3CH_2OH$$

Sodium dodecanoate   Ethanol

Continued—

EXAMPLE 17.4 —*Continued*

Practice Problem 17.4

Write a complete equation for the following reactions. Write the
I.U.P.A.C. name for each of the organic reactants and products.
a. Butyl propionate with KOH
b. Ethyl butyrate with NaOH

For Further Practice: Questions 17.39 and 17.40.

Problems can arise when "hard" water is used for cleaning because the high
concentrations of Ca^{2+} and Mg^{2+} in such water cause fatty acid salts to precipitate.
Not only does this interfere with the emulsifying action of the soap, it also leaves
a hard scum on the surface of sinks and tubs.

$$2R-\overset{O}{\overset{\|}{C}}-O^- + Ca^{2+} \longrightarrow (R-\overset{O}{\overset{\|}{C}}-O^-)_2Ca^{2+}(s)$$

Reaction at the Double Bond (Unsaturated Fatty Acids)

Hydrogenation is an example of an addition reaction. The following is a typical
example of the addition of hydrogen to the double bonds of a fatty acid:

*Hydrogenation is described in
Section 11.5.*

$$CH_3(CH_2)_4CH=CHCH_2CH=CH(CH_2)_7COOH \xrightarrow{2H_2, Ni} CH_3(CH_2)_{16}COOH$$

Linoleic acid   Stearic acid

Writing Equations for the Hydrogenation of a Fatty Acid   **EXAMPLE 17.5**

Write an equation representing the hydrogenation of oleic acid and write
the I.U.P.A.C. names of each of the organic reactants and products.

LEARNING GOAL
4 Write equations representing the reactions
that fatty acids undergo.

Solution

$$CH_3(CH_2)_7CH=CH(CH_2)_7-\overset{O}{\overset{\|}{C}}-OH \xrightarrow{H_2, Ni} CH_3(CH_2)_{16}-\overset{O}{\overset{\|}{C}}-OH$$

cis-9-Octadecenoic acid   Octadecanoic acid

Practice Problem 17.5

Write balanced equations showing the hydrogenation of
(a) *cis*-9-hexadecenoic acid and (b) arachidonic acid.

For Further Practice: Questions 17.37 and 17.38.

Hydrogenation is used in the food industry to convert polyunsaturated vegetable oils into saturated solid fats. *Partial hydrogenation* is carried out to add hydrogen to some, but not all, double bonds in polyunsaturated oils. In this way liquid
vegetable oils are converted into solid form. Crisco is one example of a hydrogenated vegetable oil.

A Human *Perspective*

Mummies Made of Soap

In the Smithsonian Museum of Natural History in Washington, D.C., one can find a great many wonders of the natural world. None is quite as macabre as the corpse made of soap. The man in question died of yellow fever in the eighteenth century and was buried near Boston. Actually, he was buried alongside a woman, perhaps the love of his life, who has been dubbed "Soap Woman." She, too, died of yellow fever. However, the couple has been separated for quite some time, because Soap Woman has been on display at the Mutter Museum at the College of Physicians in Philadelphia since 1874.

Recently Soap Woman became a television celebrity when a CT Scan done to examine the body was filmed for "The Mummy Road Show," a presentation of the National Geographic Channel. One reason for the examination was to try to understand the conditions that caused this chemical conversion.

In particular, forensic scientists would like to understand the nature of the reaction that creates *adipocere*, the technical term for the yellowish-white, greasy, waxlike substance, which results from the saponification of fatty tissue. Some researchers hope that this information may allow determination of the postmortem interval (length of time since death). Other criminalists value the process because it helps preserve the body so well that even after long periods of time, it can be easily recognized and any wounds or injuries can be observed.

It is known that adipocere is produced when body fat is hydrolyzed (water is needed) to release fatty acids. Because the fatty acids drop the pH in the tissues, they inhibit many of the bacteria that would begin the process of decay. Certain other bacteria, particularly *Clostridium welchii*, an organism that cannot grow in the presence of oxygen, is known to speed up the formation of adipocere in moist, warm, anaerobic (oxygenless) environments. Adipocere forms first in subcutaneous tissues, including the cheeks, breasts, and buttocks. Given appropriate warmth and damp conditions, it may be seen as early as three to four weeks after death; but more commonly it is not observed until five to six months after death.

In the case of our Boston couple, one clue resides in the environment of the burial site. Apparently the groundwater running through the graves was very basic. Another clue to the mystery is that our soap couple was overweight. Certainly these two factors contributed to the saponification reactions that converted this chubby couple into blocks of soap.

For Further Understanding

Adipocere is the technical term for "soap mummification." It comes from the Latin words *adipis* or fat, as in adipose tissue, and *cera,* which means wax. Draw a triglyceride composed of the fatty acids myristic acid, stearic acid, and oleic acid. Write a balanced equation showing a possible reaction that would lead to the formation of adipocere.

Forensic scientists are studying adipocere formation as a possible source of information to determine postmortem interval (length of time since death) of bodies of murder or accident victims. Among the factors being studied are the type of soil, including pH, moisture, temperature, and presence or absence of lime. How might each of these factors influence the rate of adipocere formation and hence the determination of the postmortem interval?

Soap woman.

Hydrogenation of vegetable oils produces a mixture of *cis* and *trans* unsaturated fatty acids. The *trans* unsaturated fatty acids are thought to contribute to atherosclerosis (hardening of the arteries).

Margarine is also produced by partial hydrogenation of vegetable oils, such as corn oil or soybean oil. The extent of hydrogenation is carefully controlled so that the solid fat will be spreadable and have the consistency of butter when eaten. If too many double bonds were hydrogenated, the resulting product would have the undesirable consistency of animal fat. Artificial color is added to the product, and it may be mixed with milk to produce a butterlike appearance and flavor.

Question 17.3

Write the complete equation for the esterification of myristic acid and ethyl alcohol. Write the I.U.P.A.C. name for each of the organic reactants and products.

Question 17.4

Write the complete equation for the esterification of arachidic acid and ethyl alcohol. Write the I.U.P.A.C. name for each of the organic reactants and products.

Question 17.5

Write a complete equation for the acid hydrolysis of pentyl butyrate. Write the I.U.P.A.C. name for each of the organic reactants and products.

Question 17.6

Write a complete equation for the acid hydrolysis of butyl acetate. Write the I.U.P.A.C. name for each of the organic reactants and products.

Question 17.7

Write a complete equation for the reaction of butyl acetate and KOH. Write the I.U.P.A.C. name for each of the organic reactants and products.

Question 17.8

Write a complete equation for the reaction of methyl butyrate and NaOH. Write the I.U.P.A.C. name for each of the organic reactants and products.

Question 17.9

Write a balanced equation for the hydrogenation of linolenic acid.

Question 17.10

Write a balanced equation for the hydrogenation of 2-hexenoic acid.

Eicosanoids: Prostaglandins, Leukotrienes, and Thromboxanes

Some of the unsaturated fatty acids containing more than one double bond cannot be synthesized by the body. For many years it has been known that linolenic acid, also called α-linolenic acid to distinguish it from isomeric forms, and linoleic acid, called the **essential fatty acids,** are necessary for specific biochemical functions and must be supplied in the diet (see Table 17.1). The function of linoleic acid became clear in the 1960s when it was discovered that linoleic acid is required for the biosynthesis of **arachidonic acid,** the precursor of a class of hormonelike molecules known as **eicosanoids.** The name is derived from the Greek word *eikos,* meaning "twenty," because they are all derivatives of twenty-carbon fatty acids. The eicosanoids include three groups of structurally related compounds: the prostaglandins, the leukotrienes, and the thromboxanes.

Prostaglandins are extremely potent biological molecules with hormonelike activity. They got their name because they were originally isolated from seminal fluid produced in the prostate gland. More recently they also have been isolated from most animal tissues. Prostaglandins are unsaturated carboxylic acids consisting of a twenty-carbon skeleton that contains a five-carbon ring.

Several general classes of prostaglandins are grouped under the designations A, B, E, and F, among others. The nomenclature of prostaglandins is based on the arrangement of the carbon skeleton and the number and orientation of

A hormone is a chemical signal that is produced by a specialized tissue and is carried by the bloodstream to target tissues. Eicosanoids are referred to as hormonelike because they affect the cells that produce them, as well as other target tissues.

LEARNING GOAL
5 Describe the functions of prostaglandins.

LEARNING GOAL

6 Discuss the mechanism by which aspirin reduces pain.

Prostaglandin E_1

Prostaglandin F_1

Prostaglandin E_2

Prostaglandin F_2

Figure 17.3

The structures of four prostaglandins.

Thromboxane A_2

Leukotriene B_4

Figure 17.4

The structures of thromboxane A_2 and leukotriene B_4.

double bonds, hydroxyl groups, and ketone groups. For example, in the name PGF_2, PG stands for prostaglandin, F indicates a particular group of prostaglandins with a hydroxyl group bonded to carbon-9, and 2 indicates that there are two carbon-carbon double bonds in the compound. The examples in Figure 17.3 illustrate the general structure of prostaglandins and the current nomenclature system.

Prostaglandins are made in most tissues, and exert their biological effects on the cells that produce them and on other cells in the immediate vicinity. Because the prostaglandins and the closely related leukotrienes and thromboxanes affect so many body processes and because they often cause opposing effects in different tissues, it can be difficult to keep track of their many regulatory functions. The following is a brief summary of some of the biological processes that are thought to be regulated by the prostaglandins, leukotrienes, and thromboxanes.

1. **Blood clotting.** Blood clots form when a blood vessel is damaged, yet such clotting along the walls of undamaged vessels could result in heart attack or stroke. *Thromboxane A_2* (Figure 17.4) is produced by platelets in the blood and stimulates constriction of the blood vessels and aggregation of the platelets. Conversely, PGI_2 (prostacyclin) is produced by the cells lining the blood vessels and has precisely the opposite effect of thromboxane A_2. Prostacyclin inhibits platelet aggregation and causes dilation of blood vessels and thus prevents the untimely production of blood clots.

2. **The inflammatory response.** The inflammatory response is another of the body's protective mechanisms. When tissue is damaged by mechanical injury, burns, or invasion by microorganisms, a variety of white blood cells descend on the damaged site to try to minimize the tissue destruction. The result of this response is swelling, redness, fever, and pain. Prostaglandins are thought to promote certain aspects of the inflammatory response, especially pain and fever. Drugs such as aspirin block prostaglandin synthesis and help to relieve the symptoms. We will examine the mechanism of action of these drugs later in this section.

3. **Reproductive system.** PGE_2 stimulates smooth muscle contraction, particularly uterine contractions. An increase in the level of prostaglandins has been noted immediately before the onset of labor. PGE_2 has also been used to induce second trimester abortions. There is strong evidence that dysmenorrhea (painful menstruation) suffered by many women may be the result of an excess of two prostaglandins. Indeed, drugs, such as ibuprofen, that inhibit prostaglandin synthesis have been approved by the FDA and are found to provide relief from these symptoms.

4. **Gastrointestinal tract.** Prostaglandins have been shown to both inhibit the secretion of acid and increase the secretion of a protective mucus layer into the stomach. In this way, prostaglandins help to protect the stomach lining. Consider for a moment the possible side effect that prolonged use of a drug such as aspirin might have on the stomach—ulceration of the stomach lining. Because aspirin inhibits prostaglandin synthesis, it may actually encourage stomach ulcers by inhibiting the formation of the normal protective mucus layer, while simultaneously allowing increased secretion of stomach acid.

5. **Kidneys.** Prostaglandins produced in the kidneys cause the renal blood vessels to dilate. The greater flow of blood through the kidney results in increased water and electrolyte excretion.

6. **Respiratory tract.** Eicosanoids produced by certain white blood cells, the *leukotrienes* (see Figure 17.4), promote the constriction of the bronchi associated with asthma. Other prostaglandins promote bronchodilation.

As this brief survey suggests, the prostaglandins have numerous, often antagonistic effects. Although they do not fit the formal definition of a hormone (a substance produced in a specialized tissue and transported by the circulatory system to target tissues *elsewhere* in the body), the prostaglandins are clearly strong biological regulators with far-reaching effects.

As mentioned, prostaglandins stimulate the inflammatory response and, as a result, are partially responsible for the cascade of events that cause pain. Aspirin has long been known to alleviate such pain, and we now know that it does so by inhibiting the synthesis of prostaglandins (Figure 17.5).

The first two steps of prostaglandin synthesis (Figure 17.6), the release of arachidonic acid from the membrane and its conversion to PGH_2 by the enzyme cyclooxygenase, occur in all tissues that are able to produce prostaglandins. The conversion of PGH_2 into the other biologically active forms is tissue specific and requires the appropriate enzymes, which are found only in certain tissues.

Aspirin works by inhibiting the cyclooxygenase, which catalyzes the first step in the pathway leading from arachidonic acid to PGH_2. The acetyl group of aspirin becomes covalently bound to the enzyme, thereby inactivating it (Figure 17.5). Because the reaction catalyzed by cyclooxygenase occurs in all cells, aspirin effectively inhibits synthesis of all of the prostaglandins.

Omega-3 Fatty Acids

In 2002, the American Heart Association issued dietary guidelines that recommend that we include at least two servings of "oily" fish in our diet each week. Among the fish recommended are salmon, albacore tuna, sardines, lake trout, and

Figure 17.5

Aspirin inhibits the synthesis of prostaglandins by acetylating the enzyme cyclooxygenase. The acetylated enzyme is no longer functional.

Figure 17.6

A summary of the synthesis of several prostaglandins from arachidonic acid.

mackerel. The reason for this recommendation is that these fish contain high levels of two omega-3 fatty acids called eicosapentaenoic acid (EPA) and docosahexaenoic acid (DHA). It further recommended a third omega-3 fatty acid, α-linolenic acid, which is found in flax seed, soybeans, and canola, as well as in oil made from these plants.

The name of this group of fatty acids arises from the position of the double bond nearest the terminal *methyl group* of the molecule. In these fatty acids, it is the third carbon from the end, designated omega (ω), which is the location of the double bond.

All *cis*-9, 12, 15-Octadecatrienoic acid
(α-Linolenic acid or ALA)

All *cis*-5, 8, 11, 14, 17-Eicosapentaenoic acid
(EPA)

All cis-4, 7, 10, 13, 16, 19-Docosahexaenoic acid
(DHA)

The reason for this dietary recommendation was research that supported the idea that omega-3 fatty acids reduce the risk of cardiovascular disease by decreasing blood clot formation, blood triglyceride levels, and growth of atherosclerotic plaque. Because of these effects, arterial health improved and blood pressure decreased, as did the risk of sudden death and heart arrythmias.

In some cases, the reason for the effect can be understood. For instance, EPA is a precursor for the synthesis of prostacyclin, which inhibits clumping of platelets and thus reduces clot formation. DHA is one of the major fatty acids in the phospholipids of sperm and brain cells, as well as in the retina. It has also been shown to reduce triglyceride levels, although the mechanism is not understood.

As we learned earlier, linolenic acid is an essential fatty acid. Although it, too, seems to reduce the incidence of cardiovascular disease, it is not clear whether it acts alone or because it is the precursor of DHA and EPA.

Linoleic acid is also an essential fatty acid, required for the synthesis of arachidonic acid, the precursor for many prostaglandins. These two fatty acids are termed omega-6 fatty acids because the first double bond in the molecule is six carbon atoms from the methyl (ω) end of the molecule.

cis cis-9, 12-Octadecadienoic acid
(Linolenic acid)

All *cis*-5, 8, 11, 14-Eicosatetraenoic acid
(Arachidonic acid)

It is intriguing to note that the omega-3 fatty acids are precursors of prostaglandins that exhibit anti-inflammatory effects, and the omega-6 fatty acids are precursors to prostaglandins that have inflammatory effects. This has led researchers to suggest that the amount of omega-6 fatty acids in our diets should not exceed 4–5 times the amount of omega-3. In fact, in the United States, the diet often contains 10–30 times more omega-6 fatty acids than omega-3 fatty acids. Thus, changing the ratio of the two in the diet could increase the levels of anti-inflammatory prostaglandins and reduce the level of inflammatory prostaglandins. To encourage this dietary change, the National Institutes of Health have issued recommended daily intakes of four of these fatty acids: 650 mg/day of EPA and DHA, 2.22 g/day of α-linolenic acid, and 4.44 g/day of linoleic acid.

The American Heart Association recommends that we include two meals of fish, such as salmon, in our diets each week. However, there is concern about the amount of heavy metals and other environmental contaminants in the wild-caught fish. Go online to investigate the reasons that these contaminants accumulate in these fish.

17.3 Glycerides

Neutral Glycerides

Glycerides are lipid esters that contain the glycerol molecule and fatty acids. They may be subdivided into two classes: neutral glycerides and phosphoglycerides. Neutral glycerides are nonionic and nonpolar. Phosphoglyceride molecules have a polar region, the phosphoryl group, in addition to the nonpolar fatty acid tails. The structures of each of these types of glycerides are critical to their function.

The esterification of glycerol with a fatty acid produces a **neutral glyceride.** Esterification may occur at one, two, or all three positions, producing **mono-glycerides, diglycerides,** or **triglycerides.** You will also see these referred to as *mono-, di-,* or *triacylglycerols.*

Writing an Equation for the Synthesis of a Monoglyceride   E X A M P L E 17.6

Write a general equation for the esterification of glycerol and one fatty acid.

LEARNING GOAL

4 Write equations representing the reactions fatty acids undergo.

Solution

| Glycerol | Fatty acid | Monoglyceride | Water |

Practice Problem 17.6

Write equations for the following esterification reactions.
a. Glycerol with two molecules of stearic acid
b. Glycerol with one molecule of myristic acid

For Further Practice: Questions 17.41 and 17.42.

Although monoglycerides and diglycerides are present in nature, the most important neutral glycerides are the triglycerides, the major component of fat cells. The triglyceride consists of a glycerol backbone (shown in black) joined to three fatty acid units through ester bonds (shown in red). The formation of a tri-glyceride is shown in the following equation:

$$
\begin{array}{ccc}
\text{H} & & \text{H} \quad \text{O} \\
\text{H—C—OH} & & \text{H—C—O—C—R} \\
& \text{O} & \text{O} \\
\text{H—C—OH} + 3\text{R—C—OH} \rightleftharpoons & \text{H—C—O—C—R} + 3\text{H}_2\text{O} \\
& & \text{O} \\
\text{H—C—OH} & & \text{H—C—O—C—R} \\
\text{H} & & \text{H}
\end{array}
$$

Glycerol   Fatty acids   Triglyceride   Water

Because there are no charges (+ or −) on these molecules, they are called *neutral glycerides*. These long molecules readily stack with one another and constitute the majority of the lipids stored in the body's fat cells.

The principal function of triglycerides in biochemical systems is the storage of energy. If more energy-rich nutrients are consumed than are required for metabolic processes, much of the excess is converted to neutral glycerides and stored as triglycerides in fat cells of *adipose tissue*. When energy is needed, the triglycerides are metabolized by the body, and energy is released. For this reason, exercise, along with moderate reduction in caloric intake, is recommended for overweight individuals. Exercise, an energy-demanding process, increases the rate of metabolism of fats and results in weight loss.

Phosphoglycerides

Phospholipids are a group of lipids that are phosphate esters. The presence of the phosphoryl group results in a molecule with a polar head (the phosphoryl group) and a nonpolar tail (the alkyl chain of the fatty acid). Because the phosphoryl group ionizes in solution, a charged lipid results.

The most abundant membrane lipids are derived from glycerol-3-phosphate and are known as **phosphoglycerides.** Phosphoglycerides contain acyl groups derived from long-chain fatty acids at C-1 and C-2 of glycerol-3-phosphate. At C-3 the phosphoryl group is joined to glycerol by a phosphoester bond. The simplest phosphoglyceride contains a free phosphoryl group and is known as a **phosphatidate** (Figure 17.7). When the phosphoryl group is attached to another hydrophilic molecule, a more complex phosphoglyceride is formed. For example, *phosphatidylcholine (lecithin)* and *phosphatidylethanolamine (cephalin)* are found in the membranes of most cells (Figure 17.7).

Lecithin possesses a polar "head" and a nonpolar "tail." Thus, it is an *amphipathic* molecule. This structure is similar to that of soap and detergent molecules, discussed earlier. The ionic "head" is hydrophilic and interacts with water molecules, whereas the nonpolar "tail" is hydrophobic and interacts with nonpolar molecules. This amphipathic nature is central to the structure and function of cell membranes.

In addition to being a component of cell membranes, lecithin is the major phospholipid in pulmonary surfactant. It is also found in egg yolks and soybeans and is used as an emulsifying agent in ice cream. An **emulsifying agent** aids in the suspension of triglycerides in water. The amphipathic lecithin serves as a bridge, holding together the highly polar water molecules and the nonpolar triglycerides. Emulsification occurs because the hydrophilic head of lecithin dissolves in water and its hydrophobic tail dissolves in the triglycerides.

Cephalin is similar in general structure to lecithin; the amine group bonded to the phosphoryl group is the only difference.

Lipid metabolism is discussed in Chapter 23.

See A Human Perspective: Losing Those Unwanted Pounds of Adipose Tissue in Chapter 23.

LEARNING GOAL

7 Draw the structure of a phospholipid and discuss its amphipathic nature.

Phosphoesters are described in Section 14.4.

See the Chemistry Connection: Lifesaving Lipids, at the beginning of this chapter.

Phosphatidate

(a)

Phosphatidylcholine (lecithin)

(b)

Phosphatidylethanolamine (cephalin)

(c)

Phosphatidylserine

(d)

Figure 17.7

The structures of (a) phosphatidate and the common membrane phospholipids, (b) phosphatidylcholine (lecithin), (c) phosphatidylethanolamine (cephalin), and (d) phosphatidylserine.

Question *17.11*

Using condensed formulas, draw the mono-, di-, and triglycerides that would result from the esterification of glycerol with each of the following fatty acids.

  a. Oleic acid
  b. Capric acid

Question *17.12*

Using condensed formulas, draw the mono-, di-, and triglycerides that would result from the esterification of glycerol with each of the following fatty acids.

  a. Palmitic acid
  b. Lauric acid

17.4 Nonglyceride Lipids

Sphingolipids

LEARNING GOAL

8 Discuss the general classes of sphingolipids and their functions.

Sphingolipids are lipids that are not derived from glycerol. Like phospholipids, sphingolipids are amphipathic, having a polar head group and two nonpolar fatty acid tails, and are structural components of cellular membranes. They are derived from sphingosine, a long-chain, nitrogen-containing (amino) alcohol:

$$\text{CH}_3(\text{CH}_2)_{12}\text{CH}=\text{CH}-\overset{\overset{\displaystyle \text{OH}}{|}}{\underset{\underset{\displaystyle \text{CH}_2\text{OH}}{|}}{\underset{|}{\overset{|}{\text{C}}-\text{H}}}}$$

$$\text{H}_2\text{N}-\text{C}-\text{H}$$

Sphingosine

The sphingolipids include the sphingomyelins and the glycosphingolipids. The **sphingomyelins** are the only class of sphingolipids that are also phospholipids:

Sphingomyelin

Sphingomyelins are located throughout the body, but are particularly important structural lipid components of nerve cell membranes. They are found in abundance in the myelin sheath that surrounds and insulates cells of the central nervous system. In humans, about 25% of the lipids of the myelin sheath are sphingomyelins. Their role is essential to proper cerebral function and nerve transmission.

Glycosphingolipids, or *glycolipids,* include the cerebrosides, sulfatides, and gangliosides and are built on a ceramide backbone structure, which is a fatty acid amide derivative of sphingosine:

$$CH_3(CH_2)_{12}CH=CH-\overset{\overset{\displaystyle OH}{|}}{C}-H$$

Fatty acid — Ceramide

The *cerebrosides* are characterized by the presence of a single monosaccharide head group. Two common cerebrosides are glucocerebroside, found in the membranes of macrophages (cells that protect the body by ingesting and destroying foreign microorganisms) and galactocerebroside, found almost exclusively in the membranes of brain cells. Glucocerebroside consists of ceramide bonded to the hexose glucose; galactocerebroside consists of ceramide joined to the monosaccharide galactose.

Glucocerebroside

Galactocerebroside

Sulfatides are derivatives of galactocerebroside that contain a sulfate group. Notice that they carry a negative charge at physiological pH.

A sulfatide of galactocerebroside

Gangliosides are glycolipids that possess oligosaccharide groups, including one or more molecules of *N*-acetylneuraminic acid (sialic acid). First isolated from membranes of nerve tissue, gangliosides are found in most tissues of the body.

A ganglioside associated with Tay-Sachs disease

Steroids

LEARNING GOAL

9 Draw the structure of the steroid nucleus and discuss the functions of steroid hormones.

Lipid digestion is described in Section 23.1.

$$CH_2=\overset{\overset{\displaystyle CH_3}{|}}{C}-CH=CH_2$$

Isoprene

Steroids are a naturally occurring family of organic molecules of biochemical and medical interest. A great deal of controversy has surrounded various steroids. We worry about the amount of cholesterol in the diet and the possible health effects. We are concerned about the use of anabolic steroids by athletes wishing to build muscle mass and improve their performance. However, members of this family of molecules derived from cholesterol have many important functions in the body. The bile salts that aid in the emulsification and digestion of lipids are steroid molecules, as are the sex hormones testosterone and estrone.

The steroids are members of a large, diverse collection of lipids called the *isoprenoids*. All of these compounds are built from one or more five-carbon units called *isoprene*.

Terpene is the general term for lipids that are synthesized from isoprene units. Examples of terpenes include the steroids and bile salts, the lipid-soluble vitamins, chlorophyll, and certain plant hormones.

A Medical Perspective

Disorders of Sphingolipid Metabolism

There are a number of human genetic disorders that are caused by a deficiency in one of the enzymes responsible for the breakdown of sphingolipids. In general, the symptoms are caused by the accumulation of abnormally large amounts of these lipids within particular cells. It is interesting to note that three of these diseases, Niemann-Pick disease, Gaucher's disease, and Tay-Sachs disease are found much more frequently among Ashkenazi Jews of Northern European heritage than among other ethnic groups.

Of the four subtypes of Niemann-Pick disease, type A is the most severe. It is inherited as a recessive disorder (i.e., a defective copy of the gene must be inherited from each parent) that results in an absence of the enzyme sphingomyelinase. The absence of this enzyme causes the storage of large amounts of sphingomyelin and cholesterol in the brain, bone marrow, liver, and spleen.

Symptoms may begin when a baby is only a few months old. The parents may notice a delay in motor development and/or problems with feeding. Although the infants may develop some motor skills, they quickly begin to regress as they lose muscle strength and tone, as well as vision and hearing. The disease progresses rapidly and the children typically die within the first few years of life.

Tay-Sachs disease is a lipid storage disease caused by an absence of the enzyme hexosaminidase, which functions in ganglioside metabolism. As a result of the enzyme deficiency, the ganglioside, shown on the preceding page, accumulates in the cells of the brain causing neurological deterioration. Like Niemann-Pick disease, it is an autosomal recessive genetic trait that becomes apparent in the first few months of the life of an infant and rapidly progresses to death within a few years. Symptoms include listlessness, irritability, seizures, paralysis, loss of muscle tone and function, blindness, deafness, and delayed mental and social skills.

Gaucher's disease is an autosomal recessive genetic disorder resulting in a deficiency of the enzyme glucocerebrosidase. In the normal situation, this enzyme breaks down glucocerebroside, which is an intermediate in the synthesis and degradation of complex glycosphingolipids found in cellular membranes. In Gaucher's disease, glucocerebroside builds up in macrophages found in the liver, spleen, and bone marrow. These cells become engorged with excess lipid and displace healthy, normal cells in bone marrow. The symptoms of Gaucher's disease include severe anemia, thrombocytopenia (reduction in the number of platelets), and hepatosplenomegaly (enlargement of the spleen and liver). There can also be skeletal problems including bone deterioration and secondary fractures.

Fabry's disease is an X-linked inherited disorder caused by the deficiency of the enzyme α-galactosidase A. This disease afflicts as many as fifty thousand people worldwide. Typically, symptoms, including pain in the fingers and toes and a red rash around the waist, begin to appear when individuals reach their early twenties. A preliminary diagnosis can be confirmed by determining the concentration of the enzyme α-galactosidase A. Patients with Fabry's disease have an increased risk of kidney and heart disease, and a reduced life expectancy. Because this is an X-linked disorder, it is more common among males than females.

For Further Understanding

A defect in the enzyme sphingomyelinase is the cause of Niemann-Pick disease. Write a chemical equation to represent the reaction catalyzed by the enzyme sphingomyelinase, the cleavage of sphingomyelin to produce phosphorylcholine and ceramide. Show the structural formulas for the reactant and products of this reaction.

A defect in the enzyme glucocerebrosidase is the cause of Gaucher's disease. Write a chemical equation to represent the reaction catalyzed by the enzyme glucocerebrosidase, the cleavage of glucocerebroside to produce glucose and ceramide. Show the structural formulas for the reactant and products of this reaction.

All steroids contain the steroid nucleus (steroid carbon skeleton) as shown here:

Carbon skeleton of the steroid nucleus

Steroid nucleus

Lipid-Soluble Vitamins

A Medical Perspective

Steroids and the Treatment of Heart Disease

The foxglove plant *(Digitalis purpurea)* is an herb that produces one of the most powerful known stimulants of heart muscle. The active ingredients of the foxglove plant (digitalis) are the so-called cardiac glycosides, or *cardiotonic steroids,* which include digitoxin, digosin, and gitalin.

The structure of digitoxin, one of the cardiotonic steroids produced by the foxglove plant.

Digitalis purpurea, the foxglove plant.

These drugs are used clinically in the treatment of congestive heart failure, which results when the heart is not beating with strong, efficient strokes. When the blood is not propelled through the cardiovascular system efficiently, fluid builds up in the lungs and lower extremities (edema). The major symptoms of congestive heart failure are an enlarged heart, weakness, edema, shortness of breath, and fluid accumulation in the lungs.

This condition was originally described in 1785 by a physician, William Withering, who found a peasant woman whose folk medicine was famous as a treatment for chronic heart problems. Her potion contained a mixture of more than twenty herbs, but Dr. Withering, a botanist as well as physician, quickly discovered that foxglove was the active ingredient in the mixture. Withering used *Digitalis purpurea* successfully to treat congestive heart failure and even described some cautions in its use.

The cardiotonic steroids are extremely strong heart stimulants. A dose as low as 1 mg increases the stroke volume of the heart (volume of blood per contraction), increases the strength of the contraction, and reduces the heart rate. When the heart is pumping more efficiently because of stimulation by digitalis, the edema disappears.

Digitalis can be used to control congestive heart failure, but the dose must be carefully determined and monitored because the therapeutic dose is close to the dose that causes toxicity. The symptoms that result from high body levels of cardiotonic steroids include vomiting, blurred vision and lightheadedness, increased water loss, convulsions, and death. Only a physician can determine the initial dose and maintenance schedule for an individual to control congestive heart failure and yet avoid the toxic side effects.

For Further Understanding

Foxglove is a perennial plant, that is, it is a plant that will grow back each year for at least three years. Occasionally, foxglove first-year growth has been mistaken for comfrey, another plant with medical applications. Greeks and Romans used comfrey to treat wounds and to stop heavy bleeding, as well as for bronchial problems. Explain why the use of foxglove in place of comfrey might have fatal consequences.

Drugs such as digitalis are referred to as cardiac glycosides and as cardiotonic steroids. Explain why both these names are valid.

The steroid carbon skeleton consists of four fused rings. Each ring pair has two carbons in common. Thus two fused rings share one or more common bonds as part of their ring backbones. For example, rings A and B, B and C, and C and D are all fused in the preceding structure. Many steroids have methyl groups attached to carbons 10 and 13, as well as an alkyl, alcohol, or ketone group attached to carbon-17.

Cholesterol, a common steroid, is found in the membranes of most animal cells. It is an amphipathic molecule and is readily soluble in the hydrophobic region of membranes. It is involved in regulation of the fluidity of the membrane as a result of the nonpolar fused ring. However, the hydroxyl group is polar and functions like the polar heads of sphingolipids and phospholipids. There is a strong correlation between the concentration of cholesterol found in the blood plasma and heart disease, particularly **atherosclerosis** (hardening of the arteries). Cholesterol, in combination with other substances, contributes to a narrowing of the artery passageway. As narrowing increases, more pressure is necessary to ensure adequate blood flow, and high blood pressure *(hypertension)* develops. Hypertension is also linked to heart disease.

This breakfast is high in cholesterol and saturated fats. Why do nutritionists recommend that we limit the amount of such foods in our diets?

Cholesterol

Egg yolks contain a high concentration of cholesterol, as do many dairy products and animal fats. As a result, it has been recommended that the amounts of these products in the diet be regulated to moderate the dietary intake of cholesterol.

Bile salts are amphipathic derivatives of cholesterol that are synthesized in the liver and stored in the gallbladder. The principal bile salts in humans are cholate and chenodeoxycholate.

Bile salts are described in greater detail in Section 23.1.

Cholate

Chenodeoxycholate

Bile salts are emulsifying agents whose polar hydroxyl groups interact with water and whose hydrophobic regions bind to lipids. Following a meal, bile flows from the gallbladder to the duodenum (the uppermost region of the small intestine). Here the bile salts emulsify dietary fats into small droplets that can be more readily digested by lipases (lipid digesting enzymes) also found in the small intestine.

Steroids play a role in the reproductive cycle. In a series of chemical reactions, cholesterol is converted to the steroid *progesterone*, the most important hormone associated with pregnancy. Produced in the ovaries and in the placenta, progesterone is responsible for both the successful initiation and the successful completion of pregnancy. It prepares the lining of the uterus to accept the fertilized egg. Once the egg is attached, progesterone is involved in the development of the fetus and plays a role in the suppression of further ovulation during pregnancy.

Newer birth control pill formulations include both a progesterone and an estrogen. How do these steroids prevent pregnancy?

19-Norprogesterone

Norlutin

Progesterone

Testosterone

Estrone

Testosterone, a male sex hormone found in the testes, and *estrone*, a female sex hormone, are both produced by the chemical modification of progesterone. These hormones are involved in the development of male and female sex characteristics.

Many steroids, including progesterone, have played important roles in the development of birth control agents. 19-Norprogesterone was one of the first synthetic birth control agents. It is approximately ten times as effective as progesterone in providing birth control. However, its utility was severely limited because this compound could not be administered orally and had to be taken by injection. A related compound, norlutin (chemical name: 17-α-ethynyl-19-nortestosterone), was found to provide both the strength and the effectiveness of 19-norprogesterone and could be taken orally.

Currently "combination" oral contraceptives are prescribed most frequently. These include a progesterone and an estrogen. These newer products confer better contraceptive protection than either agent administered individually. They are also used to regulate menstruation in patients with heavy menstrual bleeding. First investigated in the late 1950s and approved by the FDA in 1961, there are at least thirty combination pills currently available. In addition, a transdermal patch for the treatment of postmenopausal osteoporosis is being investigated.

All of these compounds act by inducing a false pregnancy, which prevents ovulation. When oral contraception is discontinued, ovulation usually returns within three menstrual cycles. Although there have been problems associated with "the pill," it appears to be an effective and safe method of family planning for much of the population.

Cortisone is also important to the proper regulation of a number of biochemical processes. For example, it is involved in the metabolism of carbohydrates. Cortisone is also used in the treatment of rheumatoid arthritis, asthma, gastrointestinal disorders, many skin conditions, and a variety of other diseases. However, treatment with cortisone is not without risk. Some of the possible side effects of cortisone therapy include fluid retention, sodium retention, and potassium loss that can lead to congestive heart failure. Other side effects include muscle weakness, osteoporosis, gastrointestinal upsets including peptic ulcers, and neurological symptoms, including vertigo, headaches, and convulsions.

Animation

Mechanism of Steroid Hormone Action

Cortisone

Aldosterone is a steroid hormone produced by the adrenal cortex and secreted into the bloodstream when blood sodium ion levels are too low. Upon reaching its target tissues in the kidney, aldosterone activates a set of reactions that cause sodium ions and water to be returned to the blood. If sodium levels are elevated, aldosterone is not secreted from the adrenal cortex and the sodium ions filtered out of the blood by the kidney will be excreted.

Aldosterone

Question 17.13

Draw the structure of the steroid nucleus. Note the locations of the A, B, C, and D steroid rings.

Question 17.14

What is meant by the term *fused ring*?

Waxes

Waxes are derived from many different sources and have a variety of chemical compositions, depending on the source. Paraffin wax, for example, is composed of a mixture of solid hydrocarbons (usually straight-chain compounds). The natural waxes generally are composed of a long-chain fatty acid esterified to a long-chain alcohol. Because the long hydrocarbon tails are extremely hydrophobic, waxes are completely insoluble in water. Waxes are also solid at room temperature, owing to their high molecular weights. Two examples of waxes are myricyl palmitate, a major component of beeswax, and whale oil (spermaceti wax), from the head of the sperm whale, which is composed of cetyl palmitate.

Naturally occurring waxes have a variety of uses. Lanolin, which serves as a protective coating for hair and skin, is used in skin creams and ointments. Carnauba wax is used in automobile polish. Whale oil was once used as a fuel, in ointments, and in candles. However, synthetic waxes have replaced whale oil to a large extent, because of efforts to ban the hunting of whales.

$$CH_3(CH_2)_{14}-\overset{O}{\overset{\|}{C}}-O-(CH_2)_{29}CH_3$$

Myricyl palmitate
(beeswax)

$$CH_3(CH_2)_{14}-\overset{O}{\overset{\|}{C}}-O-(CH_2)_{15}CH_3$$

Cetyl palmitate
(whale oil)

17.5 Complex Lipids

Complex lipids are lipids that are bonded to other types of molecules. The most common and important complex lipids are plasma lipoproteins, which are responsible for the transport of other lipids in the body.

Lipids are only sparingly soluble in water, and the movement of lipids from one organ to another through the bloodstream requires a transport system that uses **plasma lipoproteins.** Lipoprotein particles consist of a core of hydrophobic lipids surrounded by amphipathic proteins, phospholipids, and cholesterol (Figure 17.8).

LEARNING GOAL

10 Describe the function of lipoproteins in triglyceride and cholesterol transport in the body.

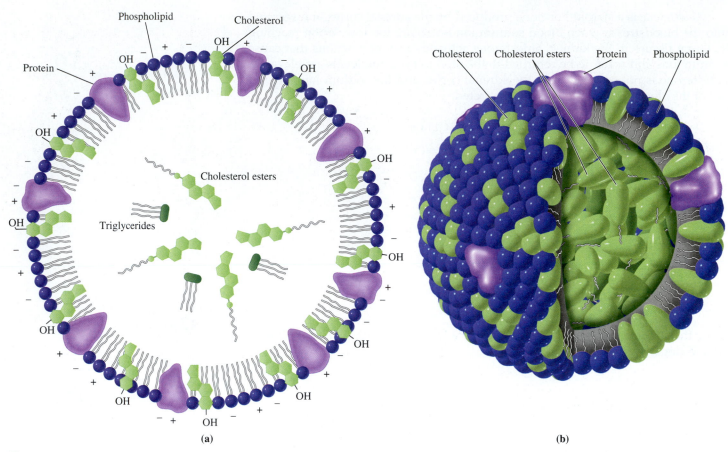

Figure 17.8

A model for the structure of a plasma lipoprotein. The various lipoproteins are composed of a shell of protein, cholesterol, and phospholipids surrounding more hydrophobic molecules such as triglycerides or cholesterol esters (cholesterol esterified to fatty acids). (a) Cross section, (b) three-dimensional view.

There are four major classes of human plasma lipoproteins:

- **Chylomicrons,** which have a density of less than 0.95 g/mL, carry dietary triglycerides from the intestine to other tissues. The remaining lipoproteins are classified by their densities.
- **Very low density lipoproteins (VLDL)** have a density of 0.95–1.019 g/mL. They bind triglycerides synthesized in the liver and carry them to adipose and other tissues for storage.
- **Low-density lipoproteins (LDL)** are characterized by a density of 1.019–1.063 g/mL. They carry cholesterol to peripheral tissues and help regulate cholesterol levels in those tissues. These are richest in cholesterol, frequently carrying 40% of the plasma cholesterol.
- **High-density lipoproteins (HDL)** have a density of 1.063–1.210 g/mL. They are bound to plasma cholesterol; however, they transport cholesterol from peripheral tissues to the liver.

A summary of the composition of each of the plasma lipoproteins is presented in Figure 17.9.

Chylomicrons are aggregates of triglycerides and protein that transport dietary triglycerides to cells throughout the body. Not all lipids in the blood are derived directly from the diet. Triglycerides and cholesterol are also synthesized in the liver and also are transported through the blood in lipoprotein packages. Triglycerides are assembled into VLDL particles that carry the energy-rich lipid molecules either to tissues requiring an energy source or to adipose tissue for

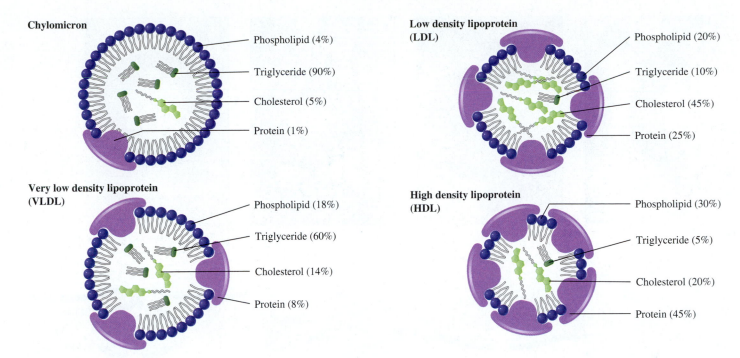

Figure 17.9

A summary of the relative amounts of cholesterol, phospholipid, protein, and triglycerides in the four classes of lipoproteins.

storage. Similarly, cholesterol is assembled into LDL particles for transport from the liver to peripheral tissues.

Entry of LDL particles into the cell is dependent on a specific recognition event and binding between the LDL particle and a protein receptor embedded within the membrane. Low-density lipoprotein receptors (LDL receptors) are found in the membranes of cells outside the liver and are responsible for the uptake of cholesterol by the cells of various tissues. LDL (lipoprotein bound to cholesterol) binds specifically to the LDL receptor, and the complex is taken into the cell by a process called *receptor-mediated endocytosis* (Figure 17.10). The membrane begins to be pulled into the cell at the site of the LDL receptor complexes. This draws the entire LDL particle into the cell. Eventually, the portion of the membrane surrounding the LDL particles pinches away from the cell membrane and forms a membrane around the LDL particles. As we will see in Section 17.6, membranes are fluid and readily flow. Thus, they can form a vesicle or endosome containing the LDL particles.

Cellular digestive organelles known as *lysosomes* fuse with the endosomes. This fusion is accomplished when the membranes of the endosome and the lysosome flow together to create one larger membrane-bound body or vesicle. Hydrolytic enzymes from the lysosome then digest the entire complex to release cholesterol into the cytoplasm of the cell. There, cholesterol inhibits its own biosynthesis and activates an enzyme that stores cholesterol in cholesterol ester droplets. High concentrations of cholesterol inside the cell also inhibit the synthesis of LDL receptors to ensure that the cell will not take up too much cholesterol. People who have a genetic defect in the gene coding for the LDL receptor do not take up as much cholesterol. As a result they accumulate LDL cholesterol in the plasma. This excess plasma cholesterol is then deposited on the artery walls, causing atherosclerosis. This disease is called *hypercholesterolemia*.

Liver lipoprotein receptors enable large amounts of cholesterol to be removed from the blood, thus ensuring low concentrations of cholesterol in the blood plasma. Other factors being equal, the person with the most lipoprotein receptors will be the least vulnerable to a high-cholesterol diet and have the least likelihood of developing atherosclerosis.

(a)

(b)

Cells have pits on the surface, which contain LDL receptors.

LDL binds to the LDL receptors in the pits.

The LDL, bound to LDL receptors, is taken into the cell by endocytosis.

Figure 17.10
Receptor-mediated endocytosis. (a) Electron micrographs of the process of receptor-mediated endocytosis. (b) Summary of the events of receptor-mediated endocytosis of LDL.

There is also evidence that high levels of HDL in the blood help reduce the incidence of atherosclerosis, perhaps because HDL carries cholesterol from the peripheral tissues back to the liver. In the liver, some of the cholesterol is used for bile synthesis and secreted into the intestines, from which it is excreted.

A final correlation has been made between diet and atherosclerosis. People whose diet is high in saturated fats tend to have high levels of cholesterol in the blood. Although the relationship between saturated fatty acids and cholesterol metabolism is unclear, it is known that a diet rich in unsaturated fats results in decreased cholesterol levels. In fact, the use of unsaturated fat in the diet results in a decrease in the level of LDL and an increase in the level of HDL. With the positive correlation between heart disease and high cholesterol levels, the current dietary recommendations include a diet that is low in fat and the substitution of unsaturated fats (vegetable oils) for saturated fats (animal fats).

Recently an inflammatory protein, the C-reactive protein (CRP), has been implicated in atherosclerosis. A test for the level of this protein in the blood is being suggested as a way to predict the risk of heart attack. A high sensitivity CRP test is now widely available.

Question 17.15

What is the mechanism of uptake of cholesterol from plasma?

Question 17.16

What is the role of lysosomes in the metabolism of plasma lipoproteins?

17.6 The Structure of Biological Membranes

LEARNING GOAL

11 Draw the structure of the cell membrane and discuss its functions.

Biological membranes are *lipid bilayers* in which the hydrophobic hydrocarbon tails are packed in the center of the bilayer and the ionic head groups are exposed on the surface to interact with water (Figure 17.11). The hydrocarbon tails of membrane phospholipids provide a thin shell of nonpolar material that prevents mixing of molecules on either side. The nonpolar tails of membrane phospholipids thus provide a barrier between the interior of the cell and its surroundings. The polar heads of lipids are exposed to water, and they are highly solvated.

The two layers of the phospholipid bilayer membrane are not identical in composition. For instance, in human red blood cells, approximately 80% of the phospholipids in the outer layer of the membrane are phosphatidylcholine and sphingomyelin; whereas phosphatidylethanolamine and phosphatidylserine make up approximately 80% of the inner layer. In addition, carbohydrate groups are found attached only to those phospholipids found on the outer layer of a membrane. Here they participate in receptor and recognition functions.

Fluid Mosaic Structure of Biological Membranes

As we have just noted, membranes are not static; they are composed of molecules in motion. The fluidity of biological membranes is determined by the proportions of saturated and unsaturated fatty acid groups in the membrane phospholipids. About half of the fatty acids that are isolated from membrane lipids from all sources are unsaturated.

Figure 17.11
(a) Representation of a phospholipid. (b) Space-filling model of a phospholipid. (c) Representation of a phospholipid bilayer membrane. (d) Line formula structure of a bilayer membrane composed of phospholipids, cholesterol, and sphingolipids.

A Medical Perspective

Liposome Delivery Systems

Liposomes were discovered by Dr. Alec Bangham in 1961. During his studies on phospholipids and blood clotting, he found that if he mixed phospholipids and water, tiny phospholipid bilayer sacs, called liposomes, would form spontaneously. Since that first observation, liposomes have been developed as efficient delivery systems for everything from antitumor and antiviral drugs, to the hair-loss therapy minoxidil!

If a drug is included in the solution during formation of liposomes, the phospholipids will form a sac around the solution. In this way the drug becomes encapsulated within the phospholipid sphere. These liposomes can be injected intravenously or applied to body surfaces. Sometimes scientists include hydrophilic molecules in the surface of the liposome. This increases the length of time that they will remain in circulation in the bloodstream. These so-called stealth liposomes are being used to carry anticancer drugs, such as doxorubicin and mitoxantrone. Liposomes are also being used as carriers for the antiviral drugs, such as AZT and ddC, that are used to treat human immunodeficiency virus infection.

A clever trick to help target the drug-carrying liposome is to include an antibody on the surface of the liposome. These antibodies are proteins designed to bind specifically to the surface of a tumor cell. Upon attaching to the surface of the tumor cell, the liposome "membrane" fuses with the cell membrane. In this way the deadly chemicals are delivered only to those cells targeted for destruction. This helps to avoid many of the unpleasant side effects of chemotherapy

treatment that occur when normal healthy cells are killed by the drug.

Another application of liposomes is in the cosmetics industry. Liposomes can be formed that encapsulate a vitamin, herbal agent, or other nutritional element. When applied to the skin, the liposomes pass easily through the outer layer of dead skin, delivering their contents to the living skin cells beneath. As with the pharmaceutical liposomes, these liposomes, sometimes called cosmeceuticals, fuse with skin cells. Thus, they directly deliver the beneficial cosmetic agent directly to the cells that can benefit the most.

Since their accidental discovery forty years ago, much has been learned about the formation of liposomes and ways to engineer them for more efficient delivery of their contents. This is another example of the marriage of serendipity, an accidental discovery, with scientific research and technological application. As the development of new types of liposomes continues, we can expect that even more ways will be found to improve the human condition.

For Further Understanding

From what you know of the structure of phospholipids, explain the molecular interactions that cause liposomes to form.

Could you use liposome technology to deliver a hydrophobic drug? Explain your answer.

(a) Cross section of a liposome, (b) three-dimensional view of a liposome, and (c) liposome fusing with cell membrane.

The unsaturated fatty acid tails of the phospholipids contribute to membrane fluidity because of the bends introduced into the hydrocarbon chain by the double bonds. Because of these "kinks," the fatty acid tails do not pack together tightly.

We also find that the percentage of unsaturated fatty acid groups in membrane lipids is inversely proportional to the temperature of the environment. Bacteria, for example, have different ratios of saturated and unsaturated fatty acids in their membrane lipids, depending on the temperatures of their surroundings. For instance, the membranes of bacteria that grow in the Arctic Ocean have high levels of unsaturated fatty acids so that their membranes remain fluid even at these frigid temperatures. Conversely, the organisms that live in the hot springs of Yellowstone National Park, with temperatures near the boiling point of water, have membranes with high levels of saturated fatty acids. This flexibility in fatty acid content enables the bacteria to maintain the same membrane fluidity over a temperature range of almost 100°C.

Generally, the body temperatures of mammals are quite constant, and the fatty acid composition of their membrane lipids is therefore usually very uniform. One interesting exception is the reindeer. Much of the year the reindeer must travel through ice and snow. Thus the hooves and lower legs must function at much colder temperatures than the rest of the body. Because of this, the percentage of unsaturation in the membranes varies along the length of the reindeer leg. We find that the proportion of unsaturated fatty acids increases closer to the hoof, permitting the membranes to function in the low temperatures of ice and snow to which the lower leg is exposed.

Thus, membranes are fluid, regardless of the environmental temperature conditions. In fact, it has been estimated that membranes have the consistency of olive oil.

Although the hydrophobic barrier created by the fluid lipid bilayer is an important feature of membranes, the proteins embedded within the lipid bilayer are equally important and are responsible for critical cellular functions. The presence of these membrane proteins was revealed by an electron microscopic technique called *freeze-fracture*. Cells are frozen to very cold temperatures and then fractured with a very fine diamond knife. Some of the cells are fractured between the two layers of the lipid bilayer. When viewed with the electron microscope, the membrane appeared to be a mosaic, studded with proteins. Because of the fluidity of membranes and the appearance of the proteins seen by electron microscopy, our concept of membrane structure is called the **fluid mosaic model** (Figure 17.12).

The bacteria growing in this hot spring in Yellowstone National Park are called thermophiles because they live at temperatures approaching the boiling point of water. What type of fatty acids do you think will be found in their membrane phospholipids?

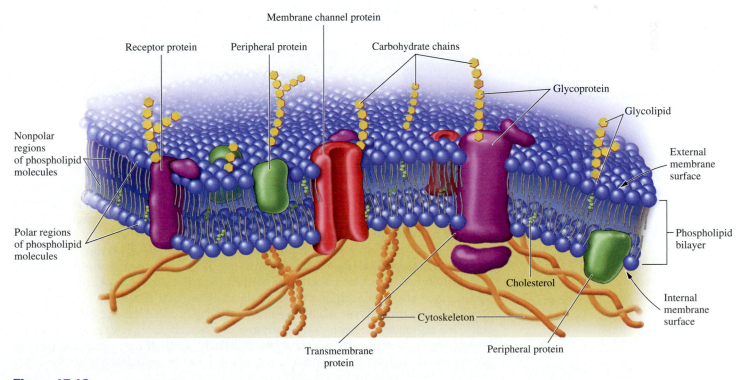

Figure 17.12
The fluid mosaic model of membrane structure.

A Medical Perspective

Antibiotics That Destroy Membrane Integrity

The "age of antibiotics" began in 1927 when Alexander Fleming discovered, quite by accident, that a product of the mold *Penicillium* can kill susceptible bacteria. We now know that penicillin inhibits bacterial growth by interfering with cell wall synthesis. Since Fleming's time, hundreds of antibiotics, which are microbial products that either kill or inhibit the growth of susceptible bacteria or fungi, have been discovered. The key to antibiotic therapy is to find a "target" in the microbe, a metabolic process or structure that the human does not have. In this way the antibiotic will selectively inhibit the disease-causing organism without harming the patient.

Many antibiotics disrupt cell membranes. The cell membrane is not an ideal target for antibiotic therapy because all cells, human and bacterial, have membranes. Therefore both types of cells are damaged. Because these antibiotics exhibit a wide range of toxic side effects when ingested, they are usually used to combat infections topically (on body surfaces). In this way, damage to the host is minimized but the inhibitory effect on the microbe is maximized.

Polymyxins are antibiotics produced by the bacterium *Bacillus polymyxa*. They are protein derivatives having one end that is hydrophobic because of an attached fatty acid. The opposite end is hydrophilic. Because of these properties, the polymyxins bind to membranes with the hydrophobic end embedded within the membrane, while the hydrophilic end remains outside the cell. As a result, the integrity of the membrane is disrupted, and leakage of cellular constituents occurs, causing cell death.

Although the polymyxins have been found to be useful in treating some urinary tract infections, pneumonias, and

Amphotericin B

Nystatin

The structures of amphotericin B and nystatin, two antifungal antibiotics.

Some of the observed proteins, called **peripheral membrane proteins,** are bound only to one of the surfaces of the membrane by interactions between ionic head groups of the membrane lipids and ionic amino acids on the surface of the peripheral protein. Other membrane proteins, called **transmembrane proteins,** are embedded within the membrane and extend completely through it, being exposed both inside and outside the cell.

Just as the phospholipid composition of the membrane is asymmetric, so too is the orientation of transmembrane proteins. Each transmembrane protein has hydrophobic regions that associate with the fatty acid tails of membrane phospholipids. Each also has a unique hydrophilic domain that is always found associated with the outer layer of the membrane and is located on the outside of the cell. This region of the protein typically has oligosaccharides covalently attached. Hence these proteins are *glycoproteins.* Similarly, each transmembrane protein has a second hydrophilic domain that is always found associated with the inner

Polymyxin
— Hydrophobic end
— Hydrophilic end

Polymyxins act like detergents, disrupting membrane integrity and killing the cell.

infections of burn patients, other antibiotics are now favored because of the toxic effects of the polymyxins on the kidney and central nervous system. Polymyxin B is still used topically and is available as an over-the-counter ointment in combination with two other antibiotics, neomycin and bacitracin.

Two other antibiotics that destroy membranes, amphotericin B and nystatin, are large ring structures that are used in treating serious systemic fungal infections. These antibiotics form complexes with ergosterol in the fungal cell membrane, and they disrupt the membrane permeability and cause leakage of cellular constituents. Neither is useful in treating bacterial infections because most bacteria have no ergosterol in their membranes. Both amphotericin B and nystatin are extremely toxic and cause symptoms that include nausea and vomiting, fever and chills, anemia, and renal failure. It is easy to understand why the use of these drugs is restricted to treatment of life-threatening fungal diseases.

For Further Understanding

Why are drugs such as amphotericin B not administered orally?

Draw the structure of amphotericin B and identify the hydrophobic and hydrophilic regions of the molecule.

layer of the membrane and projects into the cytoplasm of the cell. Typically this region of the transmembrane protein is attached to filaments of the cytoplasmic skeleton.

Membranes are dynamic structures. The mobility of proteins embedded in biological membranes was studied by labeling certain proteins in human and mouse cell membranes with red and green fluorescent dyes. The human and mouse cells were fused; in other words, special techniques were used to cause the membranes of the mouse and human cell to flow together to create a single cell. The new cell was observed through a special ultraviolet or fluorescence microscope. The red and green patches were localized within regions of their original cell membranes when the experiment began. Forty minutes later the color patches were uniformly distributed in the fused cellular membrane (Figure 17.13). This experiment suggests that we can think of the fluid mosaic membrane as an ocean filled with mobile, floating icebergs.

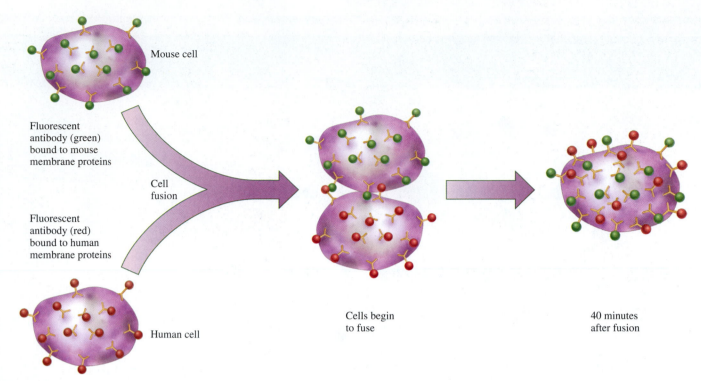

Fluorescent antibody (green) bound to mouse membrane proteins

Mouse cell

Cell fusion

Fluorescent antibody (red) bound to human membrane proteins

Human cell

Cells begin to fuse

40 minutes after fusion

Figure 17.13
Demonstration that membranes are fluid and that proteins move freely in the plane of the lipid bilayer.

SUMMARY

17.1 Biological Functions of Lipids

Lipids are organic molecules characterized by their solubility in nonpolar solvents. Lipids are subdivided into classes based on structural characteristics: *fatty acids, glycerides, nonglycerides,* and *complex lipids.* Lipids serve many functions in the body, including energy storage, protection of organs, insulation, and absorption of vitamins. Other lipids are energy sources, hormones, or vitamins. Cells store chemical energy in the form of lipids, and the cell membrane is a lipid bilayer.

17.2 Fatty Acids

Fatty acids are *saturated* and *unsaturated* carboxylic acids containing between twelve and twenty-four carbon atoms. Fatty acids with even numbers of carbon atoms occur most frequently in nature. The reactions of fatty acids are identical to those of carboxylic acids. They include esterification, production by acid hydrolysis of esters, saponification, and addition at the double bond. *Prostaglandins,* thromboxanes, and leukotrienes are derivatives of twenty-carbon fatty acids that have a variety of physiological effects.

17.3 Glycerides

Glycerides are the most abundant lipids. The triesters of glycerol (*triglycerides*) are of greatest importance. Neutral

triglycerides are important because of their ability to store energy. The ionic *phospholipids* are important components of all biological membranes.

17.4 Nonglyceride Lipids

Nonglyceride lipids consist of *sphingolipids, steroids,* and *waxes. Sphingomyelin* is a component of the myelin sheath around cells of the central nervous system. The *steroids* are important for many biochemical functions: *cholesterol* is a membrane component; testosterone, progesterone, and estrone are sex hormones; and cortisone is an anti-inflammatory steroid that is important in the regulation of many biochemical pathways.

17.5 Complex Lipids

Plasma lipoproteins are complex lipids that transport other lipids through the bloodstream. *Chylomicrons* carry dietary triglycerides from the intestine to other tissues. *Very low density lipoproteins* carry triglycerides synthesized in the liver to other tissues for storage. *Low-density lipoproteins* carry cholesterol to peripheral tissues and help regulate blood cholesterol levels. *High-density lipoproteins* transport cholesterol from peripheral tissues to the liver.

17.6 The Structure of Biological Membranes

The *fluid mosaic model* of membrane structure pictures biological membranes that are composed of lipid bilayers in which proteins are embedded. Membrane lipids contain polar head groups and nonpolar hydrocarbon tails. The hydrocarbon tails of phospholipids are derived from saturated and unsaturated long-chain fatty acids containing an even number of carbon atoms. The lipids and proteins diffuse rapidly in the lipid bilayer but seldom cross from one side to the other.

KEY TERMS

arachidonic acid (17.2)
atherosclerosis (17.4)
cholesterol (17.4)
chylomicron (17.5)
complex lipid (17.5)
diglyceride (17.3)
eicosanoid (17.2)
emulsifying agent (17.3)
essential fatty acid (17.2)
esterification (17.2)
fatty acid (17.2)
fluid mosaic model (17.6)
glyceride (17.3)
high-density lipoprotein
  (HDL) (17.5)
hydrogenation (17.2)
lipid (17.1)
low-density lipoprotein
  (LDL) (17.5)
monoglyceride (17.3)

neutral glyceride (17.3)
peripheral membrane
  protein (17.6)
phosphatidate (17.3)
phosphoglyceride (17.3)
phospholipid (17.3)
plasma lipoprotein (17.5)
prostaglandin (17.2)
saponification (17.2)
saturated fatty acid (17.2)
sphingolipid (17.4)
sphingomyelin (17.4)
steroid (17.4)
terpene (17.4)
transmembrane protein (17.6)
triglyceride (17.3)
unsaturated fatty acid (17.2)
very low density lipoprotein
  (VLDL) (17.5)
wax (17.4)

QUESTIONS AND PROBLEMS

Biological Functions of Lipids

Foundations

17.17 List the four main groups of lipids.
17.18 List the biological functions of lipids.

Applications

17.19 In terms of solubility, explain why a diet that contains no lipids can lead to a deficiency of the lipid-soluble vitamins.
17.20 Why are lipids (triglycerides) such an efficient molecule for the storage of energy in the body?

Fatty Acids

Foundations

17.21 What is the difference between a saturated and an unsaturated fatty acid?
17.22 Write the structures for a saturated and an unsaturated fatty acid.
17.23 As the length of the hydrocarbon chain of saturated fatty acids increases, what is the effect on the melting points?
17.24 As the number of carbon-carbon double bonds in fatty acids increases, what is the effect on the melting points?

Applications

17.25 Explain the relationship between fatty-acid-chain length and melting points that you described in answer to Question 17.23.
17.26 Explain the relationship that you described in answer to Question 17.24 for the effect of the number of carbon-carbon bonds in fatty acids on their melting points.
17.27 Draw the structures of each of the following fatty acids:
  a. Decanoic acid
  b. Stearic acid
17.28 Draw the structures of each of the following fatty acids:
  a. *trans*-5-Decenoic acid
  b. *cis*-5-Decenoic acid
17.29 What are the common and I.U.P.A.C. names of each of the following fatty acids?
  a. $C_{15}H_{31}COOH$
  b. $C_{11}H_{23}COOH$
17.30 What are the common and I.U.P.A.C. names of each of the following fatty acids?
  a. $CH_3(CH_2)_5CH{=}CH(CH_2)_7COOH$
  b. $CH_3(CH_2)_7CH{=}CH(CH_2)_7COOH$
17.31 Write an equation for the esterification of glycerol with three molecules of myristic acid.
17.32 Write an equation for the esterification of glycerol with three molecules of palmitic acid.
17.33 Write an equation for the acid hydrolysis of a triglyceride containing three stearic acid molecules.
17.34 Write an equation for the acid hydrolysis of a triglyceride containing three oleic acid molecules.
17.35 Write an equation for the reaction of decanoic acid with KOH.
17.36 Write an equation for the reaction of stearic acid with KOH.
17.37 Using line formulas, write an equation for the hydrogenation of all *cis*-5,8,11,14,17-eicosapentaenoic acid.
17.38 Using line formulas, write an equation for the hydrogenation of all *cis*-4,7,10,13,16,19-docosahexaenoic acid.
17.39 Write an equation for the base-catalyzed hydrolysis of a triglyceride containing a molecule of oleic acid, a molecule of lauric acid, and a molecule of palmitoleic acid.
17.40 Write an equation for the base-catalyzed hydrolysis of a triglyceride containing a molecule of capric acid, a molecule of myristic acid, and a molecule of arachidic acid.
17.41 Write an equation for the esterification of glycerol with a molecule of capric acid, a molecule of myristic acid, and a molecule of arachidic acid.
17.42 Write an equation for the esterification of glycerol with a molecule of oleic acid, a molecule of lauric acid, and a molecule of palmitoleic acid.
17.43 What is the function of the essential fatty acids?
17.44 What molecules are formed from arachidonic acid?
17.45 What is the biochemical basis for the effectiveness of aspirin in decreasing the inflammatory response?
17.46 What is the role of prostaglandins in the inflammatory response?
17.47 List four effects of prostaglandins.
17.48 What are the functions of thromboxane A_2 and leukotrienes?
17.49 What do the terms *omega-3* and *omega-6* indicate about structures of the fatty acids in those classifications?
17.50 What foods are good sources of EPA and DHA?
17.51 Summarize the health benefits associated with omega-3 fatty acids.
17.52 List some foods that are good sources of α-linolenic acid.
17.53 Explain the relationship between increased levels of omega-3 fatty acids and a decreased risk of cardiovascular disease.
17.54 Do you think that a diet higher in omega-3 fatty acids would be an effective treatment for the symptoms of arthritis? Defend your answer.
17.55 Explain the logic behind decreasing the ratio of omega-6 to omega-3 fatty acids in the diet.
17.56 What is the recommendation of the National Institutes of Health for intake of DHA, EPA, linoleic acid, and linolenic acid?

Glycerides

Foundations

17.57 Define the term *glyceride*.

17.58 Define the term *phosphatidate*.

17.59 What are emulsifying agents and what are their practical uses?

17.60 Why are triglycerides also referred to as triacylglycerols?

Applications

17.61 What do you predict the physical state would be of a triglyceride with three saturated fatty acid tails? Explain your reasoning.

17.62 What do you predict the physical state would be of a triglyceride with three unsaturated fatty acid tails? Explain your reasoning.

17.63 Draw the structure of the triglyceride molecule formed by esterification at C-1, C-2, and C-3 with hexadecanoic acid, *trans*-9-hexadecenoic acid, and *cis*-9-hexadecenoic acid, respectively.

17.64 Draw one possible structure of a triglyceride that contains the three fatty acids stearic acid, palmitic acid, and oleic acid.

17.65 Draw the structure of the phosphatidate formed between glycerol-3-phosphate that is esterified at C-1 and C-2 with capric and lauric acids, respectively.

17.66 Draw the structure of a lecithin molecule in which the fatty acyl groups are derived from stearic acid.

17.67 What are the structural differences between triglycerides (triacylglycerols) and phospholipids?

17.68 How are the structural differences between triglycerides and phospholipids reflected in their different biological functions?

Nonglyceride Lipids

Foundations

17.69 Define the term *sphingolipid*.

17.70 What are the two major types of sphingolipids?

17.71 Define the term *glycosphingolipid*.

17.72 Distinguish among the three types of glycosphingolipids, cerebrosides, sulfatides, and gangliosides.

Applications

17.73 What is the biological function of sphingomyelin?

17.74 Why are sphingomyelins amphipathic?

17.75 What is the role of cholesterol in biological membranes?

17.76 How does cholesterol contribute to atherosclerosis?

17.77 What are the biological functions of progesterone, testosterone, and estrone?

17.78 How has our understanding of the steroid sex hormones contributed to the development of oral contraceptives?

17.79 What is the medical application of cortisone?

17.80 What are the possible side effects of cortisone treatment?

17.81 A wax found in beeswax is myricyl palmitate. What fatty acid and what alcohol are used to form this compound?

17.82 A wax found in the head of sperm whales is cetyl palmitate. What fatty acid and what alcohol are used to form this compound?

17.83 What are isoprenoids?

17.84 What is a terpene?

17.85 List some important biological molecules that are terpenes.

17.86 Draw the five-carbon isoprene unit.

Complex Lipids

Foundations

17.87 What are the four major types of plasma lipoproteins?

17.88 What is the function of each of the four types of plasma lipoproteins?

Applications

17.89 There is a single, unique structure for the cholesterol molecule. What is meant by the terms *good* and *bad* cholesterol?

17.90 Distinguish among the four plasma lipoproteins in terms of their composition and their function.

17.91 What is the relationship between atherosclerosis and high blood pressure?

17.92 How is LDL taken into cells?

17.93 How does a genetic defect in the LDL receptor contribute to atherosclerosis?

17.94 What is the correlation between saturated fats in the diet and atherosclerosis?

The Structure of Biological Membranes

Foundations

17.95 What is the basic structure of a biological membrane?

17.96 Describe the fluid mosaic model of membrane structure.

17.97 Describe peripheral membrane proteins.

17.98 Describe transmembrane proteins and list some of their functions.

Applications

17.99 What is the major effect of cholesterol on the properties of biological membranes?

17.100 Why do the hydrocarbon tails of membrane phospholipids provide a barrier between the inside and outside of the cell?

17.101 What experimental observation shows that proteins diffuse within the lipid bilayers of biological membranes?

17.102 Why don't proteins turn around in biological membranes like revolving doors?

17.103 How will the properties of a biological membrane change if the fatty acid tails of the phospholipids are converted from saturated to unsaturated chains?

17.104 What is the function of unsaturation in the hydrocarbon tails of membrane lipids?

CRITICAL THINKING PROBLEMS

1. Olestra is a fat substitute that provides no calories, yet has all the properties of a naturally occurring fat. It has a creamy, tongue-pleasing consistency. Unlike other fat substitutes, olestra can withstand heating. Thus, it can be used to prepare foods such as potato chips and crackers. Olestra is a sucrose polyester and is produced by esterification of six, seven, or eight fatty acids to molecules of sucrose. Draw the structure of one such molecule having eight stearic acid acyl groups attached.

2. Liposomes can be made by vigorously mixing phospholipids (like phosphatidylcholine) in water. When the mixture is allowed to settle, spherical vesicles form that are surrounded by a phospholipid bilayer "membrane." Pharmaceutical chemists are trying to develop liposomes as a targeted drug delivery system. By adding the drug of choice to the mixture described above, liposomes form around the solution of drug. Specific proteins can be incorporated into the mixture that will end up within the phospholipid bilayers of the liposomes. These proteins are able to bind to targets on the surface of particular kinds of cells in the body. Explain why injection of liposome encapsulated pharmaceuticals might be a good drug delivery system.

3. "Cholesterol is bad and should be eliminated from the diet." Do you agree or disagree? Defend your answer.

4. Why would a phospholipid such as lecithin be a good emulsifying agent for ice cream?

5. When a plant becomes cold-adapted, the composition of the membranes changes. What changes in fatty acid and cholesterol composition would you predict? Explain your reasoning.

6. In terms of osmosis, explain why it would be preferable for a cell to store 10,000 molecules of glycogen each composed of 10^5 molecules of glucose rather than to store 10^9 individual molecules of glucose.

18 Protein Structure and Function

Learning Goals

1 List the functions of proteins.

2 Draw the general structure of an amino acid and classify amino acids based on their R groups.

3 Describe the primary structure of proteins and draw the structure of the peptide bond.

4 Draw the structures of small peptides and name them.

5 Describe the types of secondary structure of a protein.

6 Discuss the forces that maintain secondary structure.

7 Describe the structure and functions of fibrous proteins.

8 Describe the tertiary and quaternary structure of a protein.

9 List the R group interactions that maintain protein conformation.

10 List examples of proteins that require prosthetic groups and explain the way in which they function.

11 Discuss the importance of the three-dimensional structure of a protein to its function.

12 Describe the roles of hemoglobin and myoglobin.

13 Describe how extremes of pH and temperature cause denaturation of proteins.

14 Explain the difference between essential and nonessential amino acids.

Outline

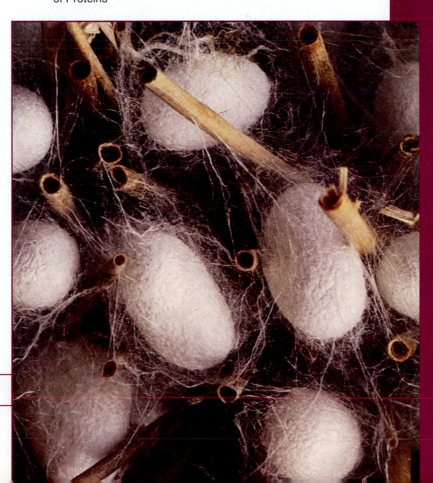

Silk fibers are harvested from the cocoons of silkworms.

Section 1-8

Introduction

In the 1800s, Johannes Mulder came up with the name **protein,** a term derived from a Greek word that means "of first importance." Indeed, proteins are a very important class of food molecules because they provide an organism not only with carbon and hydrogen, but also with nitrogen and sulfur. These latter two elements are unavailable from fats and carbohydrates, the other major classes of food molecules.

Chemistry Connection

Angiogenesis Inhibitors: Proteins That Inhibit Tumor Growth

Cancer researchers have long known that solid tumors cannot grow larger than the size of a pinhead unless they stimulate the formation of new blood vessels that provide the growing tumor with nutrients and oxygen and remove the waste products of cellular metabolism. Studies of *angiogenesis,* the formation of new blood vessels, in normal tissues have provided new weapons in the arsenal of anticancer drugs.

Angiogenesis occurs through a carefully controlled sequence of steps. Consider the process of tissue repair. One of several protein growth factors stimulates the endothelial cells that form the lining of an existing blood vessel to begin growing, dividing, and migrating into the tissue to be repaired. Threads of new endothelial cells organize themselves into hollow cylinders, or tubules. These tubules become a new network of blood vessels throughout the damaged tissue. These new blood vessels bring the needed nutrients, oxygen, and other factors to the site of damage, allowing the tissue to be repaired and healing to occur.

In addition to the growth factors that stimulate this process, there are several other proteins that inhibit the formation of new blood vessels. In fact, the normal process of angiogenesis is dependent on the appropriate balance of the stimulatory growth factors and the inhibitory proteins.

The normal events of angiogenesis are duplicated at a critical moment in the growth of a tumor. Cells of the tumor secrete one or more of the growth factors known to stimulate angiogenesis. The newly formed blood vessels provide the cells of the growing tumor with everything needed to continue growing and dividing.

Metastasis, the spreading of tumor cells to other sites in the body, also requires angiogenesis. Typically, those tumors having more blood vessels are more likely to metastasize. Clinically, treatment of these tumors has a poorer outcome.

Researchers considered all of this information known about angiogenesis and its impact on tumor formation and metastasis. They developed the hypothesis that proteins that inhibit blood vessel formation might be effective weapons against developing tumors. If this hypothesis turned out to be supported by experimental data, there would be a number of advantages to the use of angiogenesis inhibitors. Because these proteins are normally produced by the human body,

they should not have the toxic side effects caused by so many anticancer drugs. In addition, angiogenesis inhibitors can overcome the problem of cancer cell drug resistance. Most cancer cells are prone to mutations and mutant cells resistant to the anticancer drugs develop. The angiogenesis inhibitors target normal endothelial cells, which are genetically stable. As a result, drug resistance is much less likely to occur.

Endostatin is one of the anti-angiogenesis proteins. Discovered in 1997, it was found to be a protein of 20,000 g/mol, which is a fragment of the C-terminus of collagen XVIII. Experimentally, endostatin is a potent inhibitor of tumor growth. It binds to the heparin sulfate proteoglycans of the cell surface and interferes with growth factor signaling. As a result, the growth and division of endothelial cells is inhibited and new blood vessels are not formed.

Angiostatin is another anti-angiogenesis protein normally found in the human body. Discovered in 1994, it is a protein fragment of human plasminogen and has a molar mass of 50,000 g/mol. The role of angiostatin in the human body is to block the growth of diseased tissue by inhibiting the formation of blood vessels. Like endostatin, it is hoped that angiostatin will block the growth of tumors by depriving them of their blood supply.

Currently, there are about twenty angiogenesis inhibitors being tested in clinical trials involving humans. Most are in phase I or II trials, which allow scientists to determine a safe dosage and assess the severity of any side effects. Only a small number of people are involved in phase I or II trials. In phase III trials, a large number of patients are divided into two groups. One group receives standard anticancer treatment plus a placebo. The other group receives standard treatment and the new drug.

As we await the results of the clinical trials involving the proteins endostatin and angiostatin, scientists explore alternative methods to attack cancer cells. Some of these involve a class of proteins called *antibodies* that can bind specifically to cancer cells and help to inhibit or destroy them.

As we will discover, there are many different classes of proteins that carry out a variety of functions for the body. Endostatin and angiostatin serve as regulatory proteins; the antibodies serve as the body's defense system against infectious diseases. These and many other proteins are the focus of this chapter.

In addition to their dietary importance, the proteins are the most abundant macromolecules in the cell, and they carry out most of the work in a cell. Protection of the body from infection, mechanical support and strength, and catalysis of metabolic reactions—all are functions of proteins that are essential to life.

18.1 Cellular Functions of Proteins

Proteins have many biological functions, as the following short list suggests.

- **Enzymes** are biological catalysts. The majority of the enzymes that have been studied are proteins. Reactions that would take days or weeks or require extremely high temperatures without enzymes are completed in an instant. For example, the digestive enzymes *pepsin, trypsin,* and *chymotrypsin* break down proteins in our diet so that subunits can be absorbed for use by our cells.
- **Defense proteins** include **antibodies** (also called *immunoglobulins*) which are specific protein molecules produced by specialized cells of the immune system in response to foreign **antigens.** These foreign invaders include bacteria and viruses that infect the body. Each antibody has regions that precisely fit and bind to a single antigen. It helps to end the infection by binding to the antigen and helping to destroy it or remove it from the body.
- **Transport proteins** carry materials from one place to another in the body. The protein *transferrin* transports iron from the liver to the bone marrow, where it is used to synthesize the heme group for hemoglobin. The proteins *hemoglobin* and *myoglobin* are responsible for transport and storage of oxygen in higher organisms, respectively.
- **Regulatory proteins** control many aspects of cell function, including metabolism and reproduction. We can function only within a limited set of conditions. For life to exist, body temperature, the pH of the blood, and blood glucose levels must be carefully regulated. Many of the hormones that regulate body function, such as *insulin* and *glucagon,* are proteins.
- **Structural proteins** provide mechanical support to large animals and provide them with their outer coverings. Our hair and fingernails are largely composed of the protein *keratin.* Other proteins provide mechanical strength for our bones, tendons, and skin. Without such support, large, multicellular organisms like ourselves could not exist.
- **Movement proteins** are necessary for all forms of movement. Our muscles, including that most important muscle, the heart, contract and expand through the interaction of *actin* and *myosin* proteins. Sperm can swim because they have long flagella made up of proteins.
- **Nutrient proteins** serve as sources of amino acids for embryos or infants. Egg *albumin* and *casein* in milk are examples of nutrient storage proteins.

In the broadest sense, an antigen is any substance that stimulates an immune response.

18.2 The α-Amino Acids

Structure of Amino Acids

The proteins of the body are made up of some combination of twenty different subunits called **α-amino acids.** The general structure of an α-amino acid is shown in Figure 18.1. We find that nineteen of the twenty amino acids that are commonly isolated from proteins have this same general structure; they are primary amines on the α-carbon. The remaining amino acid, proline, is a secondary amine.

Notice that the α-carbon in the general structure is attached to a carboxylate group (a carboxyl group that has lost a proton, —COO⁻) and a protonated amino group (an amino group that has gained a proton, —NH₃⁺). At pH 7, conditions

Figure 18.1

General structure of an α-amino acid. All amino acids isolated from proteins, with the exception of proline, have this general structure.

Conjugate acids and bases are described in detail in Section 8.1.

Stereochemistry and Stereoisomers Revisited

Stereochemistry is discussed in Section 16.3.

required for life functions, you will not find amino acids in which the carboxylate group is protonated (—COOH) and the amino group is unprotonated (—NH$_2$). Under these conditions, the carboxyl group is in the conjugate base form (—COO$^-$), and the amino group is in its conjugate acid form (—NH$_3$$^+$). Any neutral molecule with equal numbers of positive and negative charges is called a *zwitterion*. Thus, amino acids in water exist as dipolar ions called zwitterions.

The α-carbon of each amino acid is also bonded to a hydrogen atom and a side chain, or R group. In a protein, the R groups interact with one another through a variety of weak attractive forces. These interactions participate in folding the protein chain into a precise three-dimensional shape that determines its ultimate function. They also serve to maintain that three-dimensional conformation.

Stereoisomers of Amino Acids

The α-carbon is attached to four different groups in all amino acids except glycine. The α-carbon of most α-amino acids is therefore chiral, allowing mirror-image forms, enantiomers, to exist. Glycine has two hydrogen atoms attached to the α-carbon and is the only amino acid commonly found in proteins that is not chiral.

The configuration of α-amino acids isolated from proteins is L-. This is based on comparison of amino acids with D-glyceraldehyde (Figure 18.2). In Figure 18.2a we see a comparison of D- and L-glyceraldehyde with D- and L-alanine. Notice that the most oxidized end of the molecule, in each case the carbonyl group, is drawn at the top of the molecule. In the D-isomer of glyceraldehyde, the —OH group is on the right. Similarly, in the D-isomer of alanine, the —N$^+$H$_3$ is on the right. In the L- isomers of the two compounds, the —OH and —N$^+$H$_3$ groups are on the left.

Figure 18.2

(a) Structure of D- and L-glyceraldehyde and their relationship to D- and L-alanine. (b) Ball-and-stick models of D- and L-alanine.

A Medical Perspective

Proteins in the Blood

The blood plasma of a healthy individual typically contains 60–80 g/L of protein. This protein can be separated into five classes designated α through γ. The separation is based on the overall surface charge on each of the types of protein.

The most abundant protein in the blood is albumin, making up about 55% of the blood protein. Albumin contributes to the osmotic pressure of the blood simply because it is a dissolved molecule. It also serves as a nonspecific transport molecule for important metabolites that are otherwise poorly soluble in water. Among the molecules transported through the blood by albumin are bilirubin (a waste product of the breakdown of hemoglobin), Ca^{2+}, and fatty acids (organic anions).

The α-globulins ($α_1$ and $α_2$) make up 13% of the plasma proteins. They include glycoproteins (proteins with sugar groups attached), high-density lipoproteins, haptoglobin (a transport protein for free hemoglobin), ceruloplasmin (a copper transport protein), prothrombin (a protein involved in blood clotting), and very low density lipoproteins. The most abundant is $α_1$-globulin $α_1$-antitrypsin. Although the name

leads us to believe that this protein inhibits a digestive enzyme, trypsin, the primary function of $α_1$-antitrypsin is the inactivation of an enzyme that causes damage in the lungs (see also, A Medical Perspective: $α_1$-Antitrypsin and Familial Emphysema in Chapter 19). $α_1$-Antichymotrypsin is another inhibitor found in the bloodstream. This protein, along with amyloid proteins, is found in the amyloid plaques characteristic of Alzheimer's disease (AD). As a result, it has been suggested that an overproduction of this protein may contribute to AD. In the blood, $α_1$-antichymotrypsin is also found complexed to prostate specific antigen (PSA), the protein antigen that is measured as an indicator of prostate cancer. Elevated PSA levels are observed in those with the disease. It is interesting to note that PSA is a chymotrypsin-like proteolytic enzyme.

The β-globulins represent 13% of the blood plasma proteins and include transferrin (an iron transport protein) and low-density lipoprotein. Fibrinogen, a protein involved in coagulation of blood, comprises 7% of the plasma protein. Finally, the γ-globulins, IgG, IgM, IgA, IgD, and IgE, make up the remaining 11% of the plasma proteins. The γ-globulins are synthesized by B lymphocytes, but most of the remaining plasma proteins are synthesized in the liver. In fact, a frequent hallmark of liver disease is reduced amounts of one or more of the plasma proteins.

Blood samples drawn from patients.

For Further Understanding

Develop a hypothesis to explain why albumin in the blood can serve as a nonspecific carrier for such diverse substances as bilirubin, Ca^{2+}, and fatty acids. (Hint: Consider what you know about the structures of amino acid R groups.)

Fibrinogen and prothrombin are both involved in formation of blood clots when they are converted into proteolytic enzymes. However, they are normally found in the blood in an inactive form. Develop an explanation for this observation.

By this comparison with the enantiomers of glyceraldehyde, we can define the D- and L-enantiomers of the amino acids. Figure 18.2b shows the ball-and-stick models of the D- and L-isomers of alanine.

In Chapter 16 we learned that almost all of the monosaccharides found in nature are in the D-family. Just the opposite is true of the α-amino acids. Almost all of the α-amino acids isolated from proteins in nature are members of the L-family. In other words, the orientation of the four groups around the chiral carbon of these α-amino acids resembles the orientation of the four groups around the chiral carbon of L-glyceraldehyde.

Classes of Amino Acids

Because all of the amino acids have a carboxyl group and an amino group, all differences between amino acids depend upon their side-chain R groups. The amino acids are grouped in Figure 18.3 according to the polarity of their side chains.

The side chains of some amino acids are nonpolar. They prefer contact with one another over contact with water and are said to be **hydrophobic** ("water-fearing") **amino acids.** They are generally found buried in the interior of proteins, where they can associate with one another and remain isolated from water. Nine amino acids fall into this category: alanine, valine, leucine, isoleucine, proline, glycine, methionine, phenylalanine, and tryptophan. The R group of proline is unique; it is actually bonded to the α-amino group, forming a secondary amine.

The side chains of the remaining amino acids are polar. Because they are attracted to polar water molecules, they are said to be **hydrophilic** ("water-loving") **amino acids.** The hydrophilic side chains are often found on the surfaces of proteins. The polar amino acids can be subdivided into three classes.

- *Polar, neutral amino acids* have R groups that have a high affinity for water but that are not ionic at pH 7. Serine, threonine, tyrosine, cysteine, asparagine, and glutamine fall into this category. Most of these amino acids associate with one another by hydrogen bonding; but cysteine molecules form disulfide bonds with one another, as we will discuss in Section 18.6.
- *Negatively charged amino acids* have ionized carboxyl groups in their side chains. At pH 7 these amino acids have a net charge of -1. Aspartate and glutamate are the two amino acids in this category. They are acidic amino acids because ionization of the carboxylic acid releases a proton.
- *Positively charged amino acids.* At pH 7, lysine, arginine, and histidine have a net positive charge because their side chains contain positive groups. These amino groups are basic because the side chain reacts with water, picking up a proton and releasing a hydroxide anion.

The names of the amino acids can be abbreviated by a three-letter code. These abbreviations are shown in Table 18.1.

The hydrophobic interaction between nonpolar R groups is one of the forces that helps maintain the proper three-dimensional shape of a protein.

Proline (Pro)

Hydrogen bonding (Section 5.2) is another weak interaction that helps maintain the proper three-dimensional structure of a protein. The positively and negatively charged amino acids within a protein interact with one another to form ionic bridges that also help to keep the protein chain folded in a precise way.

TABLE **18.1**	Names and Three-Letter Abbreviations of the α-Amino Acids
Amino Acid	**Three-Letter Abbreviation**
Alanine	ala
Arginine	arg
Asparagine	asn
Aspartate	asp
Cysteine	cys
Glutamic acid	glu
Glutamine	gln
Glycine	gly
Histidine	his
Isoleucine	ile
Leucine	leu
Lysine	lys
Methionine	met
Phenylalanine	phe
Proline	pro
Serine	ser
Threonine	thr
Tryptophan	trp
Tyrosine	tyr
Valine	val

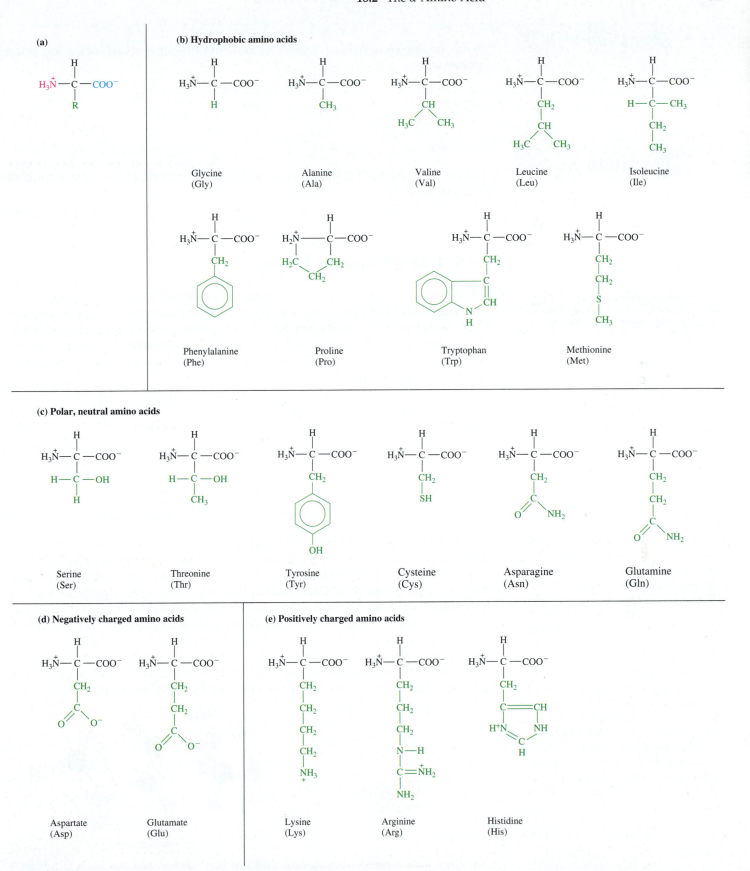

Figure 18.3

Structures of the amino acids at pH 7.0. (a) The general structure of an amino acid. Structures of (b) the hydrophobic; (c) polar, neutral; (d) negatively charged; and (e) positively charged amino acids.

Question *18.1*

Write the three-letter abbreviation and draw the structure of each of the following amino acids.

a. Glycine   d. Aspartate
b. Proline   e. Lysine
c. Threonine

Question *18.2*

Indicate whether each of the amino acids listed in Question 18.1 is polar, nonpolar, basic, or acidic.

18.3 The Peptide Bond

LEARNING GOAL

3 Describe the primary structure of proteins and draw the structure of the peptide bond.

Proteins are linear polymers of L-α-amino acids in which the carboxyl group of one amino acid is linked to the amino group of another amino acid. The **peptide bond** is an *amide bond* formed between the —COO^- group of one amino acid and the α-N^+H_3 group of another amino acid. The reaction, shown below for the amino acids glycine and alanine, is a dehydration reaction, because a water molecule is lost as the amide bond is formed.

Glycine   Alanine   Peptide bond
(amide bond)

Glycyl-alanine

To understand why the N-terminal amino acid is placed first and the C-terminal amino acid is placed last, we need to look at the process of protein synthesis. As we will see in Section 20.6, the N-terminal amino acid is the first amino acid of the protein. It forms a peptide bond involving its carboxyl group and the amino group of the second amino acid in the protein. Thus, a free amino group literally projects from the "left" end of the protein. Similarly, the C-terminal amino acid is the last amino acid added to the protein during protein synthesis. Because the peptide bond is formed between the amino group of this amino acid and the carboxyl group of the previous amino acid, a free carboxyl group projects from the "right" end of the protein chain.

The molecule formed by condensing two amino acids is called a *dipeptide.* The amino acid with a free α-N^+H_3 group is known as the amino terminal, or simply the **N-terminal amino acid,** and the amino acid with a free —COO^- group is known as the carboxyl, or **C-terminal amino acid.** Structures of proteins are conventionally written with their N-terminal amino acid on the left.

The number of amino acids in small peptides is indicated by the prefixes *di-* (two units), *tri-* (three units), *tetra-* (four units), and so forth. Peptides are named as derivatives of the C-terminal amino acid, which receives its entire name. For all other amino acids, the ending *-ine* is changed to *-yl*. Thus, the dipeptide alanyl-glycine has glycine as its C-terminal amino acid, as indicated by its full name, *glycine*:

Alanyl-glycine
(ala-gly)

Alanyl-glycine

The dipeptide formed from alanine and glycine that has alanine as its C-terminal amino acid, glycyl-alanine, is the product of the reaction shown above. These two dipeptides have the same amino acid composition, but different amino acid sequences.

The structures of small peptides can easily be drawn with practice if certain rules are followed. First note that the backbone of the peptide contains the repeating sequence

$$N—C—C—N—C—C—N—C—C$$
$$\quad\; 1 \quad 2 \qquad 1 \quad 2 \qquad 1 \quad 2$$

in which N is the α-amino group, carbon-1 is the α-carbon, and carbon-2 is the carboxyl group. Carbon-1 is always bonded to a hydrogen atom and to the R group side chain that is unique to each amino acid. Continue drawing as outlined in Example 18.1.

Writing the Structure of a Tripeptide **E X A M P L E 18.1**

Draw the structure of the tripeptide alanyl-glycyl-valine.

Solution

Step 1. Write the backbone for a tripeptide. It will contain three sets of three atoms, or nine atoms in all. Remember that the N-terminal amino acid is written to the left.

$$N—C—C \qquad N—C—C \qquad N—C—C$$
$$\text{Set 1} \qquad\;\; \text{Set 2} \qquad\;\; \text{Set 3}$$

Step 2. Add oxygens to the carboxyl carbons and hydrogens to the amino nitrogens:

```
     H       O             O             O
     |       ||            ||            ||
H — N⁺ — C — C — N — C — C — N — C — C — O⁻
     |             |             |
     H             H             H
```

Step 3. Add hydrogens to the α-carbons:

```
     H    H  O        H  O        H  O
     |    |  ||       |  ||       |  ||
H — N⁺ — C — C — N — C — C — N — C — C — O⁻
     |             |             |
     H             H             H
```

Step 4. Add the side chains. In this example (ala-gly-val) they are, from left to right, —CH₃, H, and —CH(CH₃)₂:

```
     H    H  O        H  O        H  O
     |    |  ||       |  ||       |  ||
H — N⁺ — C — C — N — C — C — N — C — C — O⁻
     |    |         |  |         |  |
     H   CH₃        H  H         H  CH
                                   /  \
                                 CH₃  CH₃
```

Practice Problem 18.1

Write the structure of each of the following peptides at pH 7.

a. Methionyl-leucyl-cysteine
b. Tyrosyl-seryl-histidine
c. Arginyl-isoleucyl-glutamine

For Further Practice: **Questions 18.37 and 18.38.**

A Human *Perspective*

The Opium Poppy and Peptides in the Brain

The seed pods of the oriental poppy contain morphine. *Morphine* is a narcotic that has a variety of effects on the body and the brain, including drowsiness, euphoria, mental confusion, and chronic constipation. Although morphine was first isolated in 1805, not until the 1850s and the advent of the hypodermic was it effectively used as a painkiller. During the American Civil War, morphine was used extensively to relieve the pain of wounds and amputations. It was at this time that the addictive properties were noticed. By the end of the Civil War, over 100,000 soldiers were addicted to morphine.

As a result of the Harrison Act (1914), morphine came under government control and was made available only by prescription. Although morphine is addictive, *heroin*, a derivative of morphine, is much more addictive and induces a greater sense of euphoria that lasts for a longer time.

Heroin

Morphine

The structures of heroin and morphine.

Why do heroin and morphine have such powerful effects on the brain? Both drugs have been found to bind to *receptors* on the surface of the cells of the brain. The function of these receptors is to bind specific chemical signals and to direct the brain cells to respond. Yet it seemed odd that the cells of our brain should have receptors for a plant chemical. This mystery was solved in 1975, when John Hughes discovered that the brain itself synthesizes small peptide hormones with a morphinelike structure. Two of these opiate peptides are called *methionine enkephalin*, or met-enkephalin, and *leucine enkephalin*, or leu-enkephalin.

These neuropeptide hormones have a variety of effects. They inhibit intestinal motility and blood flow to the gastro-intestinal tract. This explains the chronic constipation of morphine users. In addition, it is thought that these *enkephalins* play a role in pain perception, perhaps serving as a pain blockade. This is supported by the observation that they are found in higher concentrations in the bloodstream following painful stimulation. It is further suspected that they may play a role in mood and mental health. The so-called runner's high is thought to be a euphoria brought about by an excessively long or strenuous run!

Unlike morphine, the action of enkephalins is short-lived. They bind to the cellular receptor and thereby induce the cells to respond. Then they are quickly destroyed by enzymes in the brain that hydrolyze the peptide bonds of the enkephalin. Once destroyed, they are no longer able to elicit a cellular response. Morphine and heroin bind to these same receptors and induce the cells to respond. However, these drugs are not destroyed and therefore persist in the brain for long periods at concentrations high enough to continue to cause biological effects.

Many researchers are working to understand why drugs like morphine and heroin are addictive. Studies with cells in culture have suggested one mechanism for morphine tolerance and addiction. Normally, when the cell receptors bind to enkephalins, this signals the cell to decrease the production of a chemical messenger called *cyclic AMP*, or simply cAMP. (This compound is very closely related to the nucleotide

Poppies seen here growing in the wild are the source of the natural opiate drugs morphine and codeine.

Tyr-Gly-Gly-Phe-Met
Methionine enkephalin

Tyr-Gly-Gly-Phe-Leu
Leucine enkephalin

Structures of the peptide opiates leucine enkephalin and methionine enkephalin. These are the body's own opiates.

adenosine-5′-monophosphate.) The decrease in cAMP level helps to block pain and elevate one's mood. When morphine is applied to these cells they initially respond by decreasing cAMP levels. However, with chronic use of morphine the cells become desensitized; that is, they do not decrease cAMP production and thus behave as though no morphine were present. However, a greater amount of morphine will once again cause the decrease in cAMP levels. Thus addiction and the progressive need for more of the drug seem to result from biochemical reactions in the cells.

This logic can be extended to understand withdrawal symptoms. When an addict stops using the drug, he or she exhibits withdrawal symptoms that include excessive sweating, anxiety, and tremors. The cause may be that the high levels of morphine were keeping the cAMP levels low, thus reducing pain and causing euphoria. When morphine is removed completely, the cells overreact and produce huge quantities of cAMP. The result is all of the unpleasant symptoms known collectively as the *withdrawal syndrome.*

Clearly, morphine and heroin have demonstrated the potential for misuse and are a problem for society in several respects. Nonetheless, morphine remains one of the most effective painkillers known. Certainly, for people suffering from cancer, painful burns, or serious injuries, the risk of addiction is far outweighed by the benefits of relief from excruciating pain.

For Further Understanding

Compare the structures of leucine and methionine enkephalin with those of heroin and morphine. What similarities do you see that might cause them to bind to the same receptors on the surfaces of nerve cells?

What characteristics might you look for in a nonaddictive drug that could be used both to combat heroin addiction and to treat the symptoms of withdrawal?

At first it appears logical to think that a long polymer of amino acids would undergo constant change in conformation because of free rotation around the —N—C—C— single bonds of the peptide backbone. In reality, this is not the case. An explanation for this phenomenon resulted from the early X-ray diffraction studies of Linus Pauling. By interpreting the pattern formed when X-rays were diffracted by a crystal of pure protein, Pauling discovered that peptide bonds are both planar (flat) and rigid and that the N—C bonds are shorter than expected. What did all of this mean? Pauling concluded that the peptide bond has a partially double bond character because it exhibits resonance.

This means that there is free rotation around only two of the three single bonds of the peptide backbone (Figure 18.4a), which limits the number of possible conformations for any peptide. A second feature of the rigid peptide bond is that the R groups on adjacent amino acids are on opposite sides of the extended peptide chain (Figure 18.4b).

Question 18.3

Write the structure of each of the following peptides at pH 7:

a. Alanyl-phenylalanine
b. Lysyl-alanine
c. Phenylalanyl-tyrosyl-leucine

Question 18.4

Write the structure of each of the following peptides at pH 7:

a. Glycyl-valyl-serine
b. Threonyl-cysteine
c. Isoleucyl-methionyl-aspartate

18.4 The Primary Structure of Proteins

LEARNING GOAL

3 Describe the primary structure of proteins and draw the structure of the peptide bond.

The **primary structure** of a protein is the amino acid sequence of the protein chain. It results from the covalent bonding between the amino acids in the chain (peptide bonds). The primary structures of proteins are translations of information contained in genes. Each protein has a different primary structure with different amino acids in different places along the chain.

The genetic code and the process of protein synthesis are described in Sections 20.5 and 20.6.

Ultimately, it is the primary structure of a protein that will determine its biologically active form. The interactions among the R groups of the amino acids in the protein chain depend on the location of those R groups along the chain. These interactions will govern how the protein chain folds, which, in turn, dictates its final three-dimensional structure and its biological function.

Mutations and their effect on protein synthesis are discussed in Section 20.7.

Genes can change by the process of mutation during the course of evolution. A mutation in a gene can result in a change in the primary amino acid sequence of a protein. Over longer periods, more of these changes will occur. If two species of organisms diverged (became new species) very recently, the differences in the amino acid sequences of their proteins will be few. On the other hand, if they diverged millions of years ago, there will be many more differences in the amino acid sequences of their proteins. As a result, we can compare evolutionary relationships between species by comparing the primary structures of proteins present in both species.

Figure 18.4
(a) There is free rotation around only two of the three single bonds of a peptide backbone. (b) This model of an eight amino acid peptide shows that the R groups on adjacent amino acids are on opposite sides of the chain because of the rigid peptide bond.

(a)

(b)   gly   phe   gly   ala   leu   ser   gly   ala

18.5 The Secondary Structure of Proteins

The primary sequence of a protein, the chain of covalently linked amino acids, folds into regularly repeating structures that resemble designs in a tapestry. These repeating structures define the **secondary structure** of the protein. The secondary structure is the result of hydrogen bonding between the amide hydrogens and carbonyl oxygens of the peptide bonds. Many hydrogen bonds are needed to maintain the secondary structure and thereby the overall structure of the protein. Different regions of a protein chain may have different types of secondary structure. Some regions of a protein chain may have a random or nonregular structure; however, the two most common types of secondary structure are the α-helix and the β-pleated sheet because they maximize hydrogen bonding in the backbone.

LEARNING GOAL
5 Describe the types of secondary structure of a protein

LEARNING GOAL
6 Discuss the forces that maintain secondary structure.

In the photo above, some of the children have straight hair and some have curly hair. Knowing that the primary structure of a protein is dictated by the sequence of a gene, in this case a keratin gene, develop a hypothesis to explain hair curliness.

α-Helix

The most common type of secondary structure is a coiled, helical conformation known as the **α-helix** (Figure 18.5). The αhelix has several important features.

- Every amide hydrogen and carbonyl oxygen associated with the peptide backbone is involved in a hydrogen bond when the chain coils into an α-helix. These hydrogen bonds lock the α-helix into place.
- Every carbonyl oxygen is hydrogen-bonded to an amide hydrogen four amino acids away in the chain.
- The hydrogen bonds of the α-helix are parallel to the long axis of the helix (see Figure 18.5).
- The polypeptide chain in an α-helix is right-handed. It is oriented like a normal screw. If you turn a screw clockwise it goes into the wall; turned counterclockwise, it comes out of the wall.
- The repeat distance of the helix, or its pitch, is 5.4 Å, and there are 3.6 amino acids per turn of the helix.

Fibrous proteins are structural proteins arranged in fibers or sheets that have only one type of secondary structure. The **α-keratins** are fibrous proteins that form

(a)   **(b)**

(c)

Figure 18.5

The α-helix. (a) Schematic diagram showing only the helical backbone. (b) Molecular model representation. Note that all of the hydrogen bonds between C=O and N—H groups are parallel to the long axis of the helix. (c) Top view of an α-helix. The side chains of the helix point away from the long axis of the helix. The view is into the barrel of the helix.

the covering (hair, wool, nails, hooves, and fur) of most land animals. Human hair provides a typical example of the structure of the α-keratins. The proteins of hair consist almost exclusively of polypeptide chains coiled up into α-helices. A single α-helix is coiled in a bundle with two other helices to give a three-stranded superstructure called a *protofibril* that is part of an array known as a *microfibril* (Figure 18.6). These structures, which resemble "molecular pigtails," possess great mechanical strength, and they are virtually insoluble in water.

The major structural property of a coiled coil superstructure of α-helices is its great mechanical strength. This property is applied very efficiently in both the fibrous proteins of skin and those of muscle. As you can imagine, these proteins must be very strong to carry out their functions of mechanical support and muscle contraction.

β-Pleated Sheet

The second common secondary structure in proteins resembles the pleated folds of drapery and is known as **β-pleated sheet** (Figure 18.7a). All of the carbonyl oxygens and amide hydrogens in a β-pleated sheet are involved in hydrogen bonds, and the polypeptide chain is nearly completely extended. The polypeptide chains in a β-pleated sheet can have two orientations. If the N-termini are head to head, the structure is known as a *parallel* β-pleated sheet. And if the N-terminus of one chain is aligned with the C-terminus of a second chain (head to tail), the structure is known as an *antiparallel* β-pleated sheet.

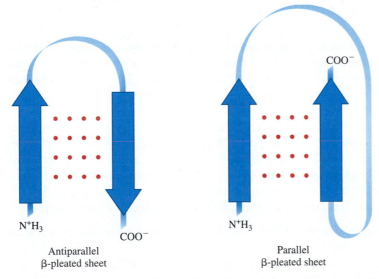

Antiparallel
β-pleated sheet

Parallel
β-pleated sheet

Some fibrous proteins are composed of β-pleated sheets. For example, the silkworm produces *silk fibroin*, a protein whose structure is an antiparallel β-pleated sheet (Figure 18.7). The polypeptide chains of a β-pleated sheet are almost completely extended, and silk does not stretch easily. Glycine accounts for nearly half of the amino acids of silk fibroin. Alanine and serine account for most of the others. The methyl groups of alanines and the hydroxymethyl groups of serines lie on opposite sides of the sheet. Thus the stacked sheets nestle comfortably, like sheets of corrugated cardboard, because the R groups are small enough to allow the stacked-sheet superstructure.

18.6 The Tertiary Structure of Proteins

Most fibrous proteins, such as silk, collagen, and the α-keratins, are almost completely insoluble in water. (Our skin would do us very little good if it dissolved in the rain.) The majority of cellular proteins, however, are soluble in the cell

Figure 18.6
Structure of the α-keratins. These proteins are assemblies of triple-helical protofibrils that are assembled in an array known as a *microfibril*. These in turn are assembled into macrofibrils. Hair is a collection of macrofibrils and hair cells.

(a)   (b)

Figure 18.7
The structure of silk fibroin is almost entirely antiparallel β-pleated sheet. (a) The molecular structure of a portion of the silk fibroin protein.
(b) A schematic representation of the antiparallel β-pleated sheet with the nestled R groups.

LEARNING GOAL
8 Describe the tertiary and quaternary structure of a protein.

LEARNING GOAL
9 List the R group interactions that maintain protein conformation.

In the next section we will see that some proteins have an additional level of structure, quaternary structure, that also influences function.

cytoplasm. Soluble proteins are usually **globular proteins.** Globular proteins have three-dimensional structures called the **tertiary structure** of the protein, which are distinct from their secondary structure. The polypeptide chain with its regions of secondary structure, α-helix and β-pleated sheet, further folds on itself to achieve the tertiary structure.

We have seen that the forces that maintain the secondary structure of a protein are hydrogen bonds between the amide hydrogen and the carbonyl oxygen of the peptide bond. What are the forces that maintain the tertiary structure of a protein? The globular tertiary structure forms spontaneously and is maintained as a result of interactions among the side chains, the R groups, of the amino acids. The structure is maintained by the following molecular interactions:

- Van der Waals forces between the R groups of nonpolar amino acids that are hydrophobic
- Hydrogen bonds between the polar R groups of the polar amino acids
- Ionic bonds (salt bridges) between the R groups of oppositely charged amino acids
- Covalent bonds between the thiol-containing amino acids. Two of the polar cysteines can be oxidized to a dimeric amino acid called *cystine* (Figure 18.8). The disulfide bond of cystine can be a cross-link between different proteins, or it can tie two segments within a protein together.

The bonds that maintain the tertiary structure of proteins are shown in Figure 18.9. The importance of these bonds becomes clear when we realize that it is the tertiary structure of the protein that defines its biological function. Most of the time, nonpolar amino acids are buried, closely packed, in the interior of a globular protein, out of contact with water. Polar and charged amino acids lie on the surfaces of globular proteins. Globular proteins are extremely compact. The tertiary structure can contain regions of α-helix and regions of β-pleated sheet. "Hinge" regions of random coil connect regions of α-helix and β-pleated sheet. Because of its cyclic structure, proline disrupts an α-helix. As a result, proline is often found in these hinge regions.

Figure 18.8
Oxidation of two cysteines to give the dimer cystine. This reaction occurs in cells and is readily reversible.

Cysteine   Cystine

$$+ 2H^+ + 2e^-$$

Figure 18.9
Summary of the weak interactions that help maintain the tertiary structure of a protein.

18.7 The Quaternary Structure of Proteins

For many proteins the functional form is not composed of a single peptide but is rather an aggregate of smaller globular peptides. For instance, the protein hemoglobin is composed of four individual globular peptide subunits: two identical α-subunits and two identical β-subunits. Only when the four peptides are bound to one another is the protein molecule functional. The association of several polypeptides to produce a functional protein defines the **quaternary structure** of a protein.

LEARNING GOAL
8 Describe the tertiary and quaternary structure of a protein.

A Human *Perspective*

Collagen, Cosmetic Procedures, and Clinical Applications

*C*ollagen is the most abundant protein in the human body, making up about one-third of the total protein content. It provides mechanical strength to bone, tendon, skin, and blood vessels. Collagen fibers in bone provide a scaffolding around which *hydroxyapatite* (a calcium phosphate polymer) crystals are arranged. Skin contains loosely woven collagen fibers that can expand in all directions. The corneas of the eyes are composed of collagen. As we consider these tissues, we realize that they have quite different properties, ranging from tensile strength (tendons) and flexibility (blood vessels) to transparency (cornea).

How could such diverse structures be composed of a single protein? The answer lies in the fact that collagen is actually a family of twenty genetically distinct, but closely related proteins. Although the differences in the amino acid sequence of these different collagen proteins allow them to carry out a variety of functions in the body, they all have a similar three-dimensional structure. Collagen is composed of three left-handed polypeptide helices that are twisted around one another to form a "superhelix" called a *triple helix*. Each of the individual peptide chains of collagen is a left-handed helix, but they are wrapped around one another in the right-handed sense.

Every third amino acid in the collagen chain is glycine. It is important to the structure because the triple-stranded helix forms as a result of interchain hydrogen bonding involving glycine. Thus, every third amino acid on one strand is in very close contact with the other two strands. Glycine has another advantage; it is the only amino acid with an R group small enough for the space allowed by the triple-stranded structure.

Collagen injections have been used in cosmetic procedures to add fullness to lips or minimize the appearance of wrinkled or sunken facial skin. This procedure is no longer very common because of the relatively high frequency of allergic reactions, including hives, redness, swelling, and flulike symptoms, to the bovine collagen used in the injections. There is also a slight risk of infection with the prion responsible for bovine spongiform encephalopathy (BSE), or "mad cow disease." Popular alternatives to collagen injections include the patient's own fat and hyaluronic acid (see also A Medical Perspective: Monosaccharide

Structure of the
collagen triple helix.

Derivatives and Heteropolysaccharides of Medical Interest in Chapter 16). Another alternative, although an expensive one, is the injection of recombinant human collagen. This reduces the incidence of allergic reactions and eliminates the risk of BSE infection.

Collagen is also used in the preparation of artificial skin for severe burn patients. The collagen is used in combination with silicones, glycosaminoglycans (such as hyaluronic acid), growth factors, and human fibroblasts, which are the most

---

The forces that hold the quaternary structure of a protein are the same as those that hold the tertiary structure. These include hydrogen bonds between polar amino acids, ionic bridges between oppositely charged amino acids, van der Waals forces between nonpolar amino acids, and disulfide bridges.

In some cases the quaternary structure of a functional protein involves binding to a nonprotein group. This additional group is called a **prosthetic group.** For common type of cell in connective tissue and which promote wound healing.

Collagen is sold over the counter as a dietary supplement to promote repair of arthritic joints, and thereby reduce the pain of arthritis and improve mobility. Often these supplements contain D-glucosamine (a monosaccharide derivative) and chondroitin sulfate (a heteropolysaccharide), which are claimed to promote joint repair, as well. Other collagen dietary supplements are sold as a means of preventing the appearance of wrinkles.

Collagen even finds its way into our diet. When partially hydrolyzed, the three polypeptide strands separate from one another and then curl up into globular random coils. The product is gelatin, found most notably in gelatin desserts such as Jello. However, gelatin is found in many other foods, as well as in dietary supplements that claim to improve fingernail and skin condition.

Two unusual, hydroxylated amino acids account for nearly one-fourth of the amino acids in collagen. These amino acids are 4-hydroxyproline and 5-hydroxylysine.

4-Hydroxyproline

5-Hydroxylysine

Structures of 4-hydroxyproline and 5-hydroxylysine, two amino acids found only in collagen.

These amino acids are an important component of the structure of collagen because they form covalent cross-linkages between adjacent molecules within the triple strand. They can also participate in interstrand hydrogen bonding to further strengthen the structure.

When collagen is synthesized, the amino acids proline and lysine are incorporated into the chain of amino acids. These are later modified by two enzymes to form 4-hydroxyproline and 5-hydroxylysine. Both of these enzymes require vitamin C to carry out these reactions. In fact, this is the major known physiological function of vitamin C. Without hydroxylation, hydrogen bonds cannot form and the triple helix is weak, resulting in fragile blood vessels

Vitamin C
(ascorbic acid)

People who are deprived of vitamin C, as were sailors on long voyages before the eighteenth century, develop *scurvy*, a disease of collagen metabolism. The symptoms of scurvy include skin lesions, fragile blood vessels, and bleeding gums. The British Navy provided the antidote to scurvy by including limes, which are rich in vitamin C, in the diets of its sailors. The epithet *limey*, a slang term for *British*, entered the English language as a result.

For Further Understanding

What feature of glycine is responsible for its importance in the hydrogen bonding that maintains the helical structure of collagen?

Collagen may have great tensile strength (as in tendons) and flexibility (as in skin and blood vessels), and may even be transparent (cornea of the eye). Propose a hypothesis to explain the biochemical differences between different forms of collagen that give rise to such different properties.

example, many of the receptor proteins on cell surfaces are **glycoproteins.** These are proteins with sugar groups covalently attached. Each of the subunits of hemoglobin is bound to an iron-containing heme group. The heme group is a large, unsaturated organic cyclic amine with an iron ion coordinated within it. As in the case of hemoglobin, the prosthetic group often determines the function of a protein. For instance, in hemoglobin it is the iron-containing heme groups that have the ability to bind reversibly to oxygen.

LEARNING GOAL

10 List examples of proteins that require prosthetic groups and explain the way in which they function.

Question 18.5

Describe the four levels of protein structure.

Question 18.6

What are the weak interactions that maintain the tertiary structure of a protein?

18.8 An Overview of Protein Structure and Function

Let's summarize the various types of protein structure and their relationship to one another (Figure 18.10).

- *Primary Structure:* The primary structure of the protein is the amino acid sequence of the protein. The primary structure results from the formation of covalent peptide bonds between amino acids. Peptide bonds are amide bonds formed between the α-carboxylate group of one amino acid and the α-amino group of another.
- *Secondary Structure:* As the protein chain grows, numerous opportunities for noncovalent interactions in the backbone of the polypeptide chain become available. These cause the chain to fold and orient itself in a variety of conformational arrangements. The secondary level of structure includes the α-helix and the β-pleated sheet, which are the result of hydrogen bonding between the amide hydrogens and carbonyl oxygens of the peptide bonds. Different portions of the chain may be involved in different types of secondary structure arrangements; some regions might be α-helix and others might be a β-pleated sheet.
- *Tertiary Structure:* When we discuss tertiary structure, we are interested in the overall folding of the entire chain. In other words, we are concerned

Figure 18.10

Summary of the four levels of protein structure, using hemoglobin as an example.

(a) Primary structure   (b) Secondary structure   (c) Tertiary structure   (d) Quaternary structure

- C
- H
- N
- O
- R groups
- Heme groups

with the further folding of the secondary structure. Are the two ends of the chain close together or far apart? What general shape is involved? Both noncovalent interactions between the R groups of the amino acids and covalent disulfide bridges play a role in determining the tertiary structure. The noncovalent interactions include hydrogen bonding, ionic bonding, and van der Waals forces.

- *Quaternary Structure:* Like tertiary structure, quaternary structure is concerned with the topological, spatial arrangements of two or more peptide chains with respect to each other. How is one chain oriented with respect to another? What is the overall shape of the final functional protein?

The quaternary structure is maintained by the same forces that are responsible for the tertiary structure. It is the tertiary and quaternary structures of the protein that ultimately define its function. Some have a fibrous structure with great mechanical strength. These make up the major structural components of the cell and the organism. Often they are also responsible for the movement of the organism. Others fold into globular shapes. Most of the transport proteins, regulatory proteins, and enzymes are globular proteins. The very precise three-dimensional structure of the transport proteins allows them to recognize a particular molecule and facilitate its entry into the cell. Similarly, it is the specific three-dimensional shape of regulatory proteins that allows them to bind to their receptors on the surfaces of the target cell. In this way they can communicate with the cell, instructing it to take some course of action. In Chapter 19 we will see that the three-dimensional structure of enzyme active sites allows them to bind to their specific reactants and speed up biochemical reactions.

As we will see with the example of sickle cell hemoglobin in the next section, an alteration of just a single amino acid within the primary structure of a protein can have far-reaching implications. When an amino acid replaces another in a peptide, there is a change in the R group at that position in the protein chain. This leads to different tertiary and perhaps quaternary structure because the nature of the noncovalent interactions is altered by changing the R group that is available for that bonding. Similarly, replacement of another amino acid with proline can disrupt important regions of secondary structure. Thus changes in the primary amino acid sequence can change the three-dimensional structure of a protein in ways that cause it to be nonfunctional. In the case of sickle cell hemoglobin, this protein malfunction can lead to death.

18.9 Myoglobin and Hemoglobin

Myoglobin and Oxygen Storage

Most of the cells of our bodies are buried in the interior of the body and cannot directly get food molecules or eliminate waste. The circulatory system solves this problem by delivering nutrients and oxygen to body cells and carrying away wastes. Our cells require a steady supply of oxygen, but oxygen is only slightly soluble in aqueous solutions. To overcome this solubility problem, we have an oxygen transport protein, **hemoglobin.** Hemoglobin is found in red blood cells and is the oxygen transport protein of higher animals. **Myoglobin** is the oxygen storage protein of skeletal muscle.

The structure of myoglobin (Mb) is shown in Figure 18.11. The **heme group** (Figure 18.12) is also an essential component of this protein. The Fe^{2+} ion in the heme group is the binding site for oxygen in both myoglobin and hemoglobin. Fortunately, myoglobin has a greater attraction for oxygen than does hemoglobin, which allows efficient transfer of oxygen from the bloodstream to the cells of the body.

9 **LEARNING GOAL**
List the R group interactions that maintain protein conformation.

11 **LEARNING GOAL**
Discuss the importance of the three-dimensional structure of a protein to its function.

The three-dimensional structure of a protein is the feature that allows it to carry out its specific biological function. However, we must always remember that it is the primary structure, the order of the R groups, that determines how the protein will fold and what the ultimate shape will be.

12 **LEARNING GOAL**
Describe the roles of hemoglobin and myoglobin.

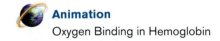
Animation
Oxygen Binding in Hemoglobin

Figure 18.11

Myoglobin. The heme group has an iron atom to which oxygen binds.

Hemoglobin and Oxygen Transport

Hemoglobin (Hb) is a tetramer composed of four polypeptide subunits: two α-subunits and two β-subunits (Figure 18.13). Because each subunit of hemoglobin contains a heme group, a hemoglobin molecule can bind four molecules of oxygen:

$$Hb \quad + \quad 4O_2 \quad \longrightarrow \quad Hb(O_2)_4$$

Deoxyhemoglobin   Oxyhemoglobin

Figure 18.12

Structure of the heme prosthetic group, which binds to myoglobin and hemoglobin.

The oxygenation of hemoglobin in the lungs and the transfer of oxygen from hemoglobin to myoglobin in the tissues are very complex processes. We begin our investigation of these events with the inhalation of a breath of air.

The oxygenation of hemoglobin in the lungs is greatly favored by differences in the oxygen partial pressure (pO_2) in the lungs and in the blood. The pO_2 in the air in the lungs is approximately 100 mm Hg; the pO_2 in oxygen-depleted blood is only about 40 mm Hg. Oxygen diffuses from the region of high pO_2 in the lungs to the region of low pO_2 in the blood. There it enters red blood cells and binds to the Fe^{2+} ions of the heme groups of deoxyhemoglobin, forming oxyhemoglobin. This binding actually helps bring more O_2 into the blood.

Oxygen Transport from Mother to Fetus

A fetus receives its oxygen from its mother by simple diffusion across the placenta. If both the fetus and the mother had the same type of hemoglobin, this transfer process would not be efficient, because the hemoglobin of the fetus and the mother would have the same affinity for oxygen. The fetus, however, has a unique type of hemoglobin, called *fetal hemoglobin*. This unique hemoglobin molecule has a greater affinity for oxygen than does the mother's hemoglobin. Oxygen is therefore efficiently transported, via the circulatory system, from the lungs of the mother to the fetus. The biosynthesis of fetal hemoglobin stops shortly after birth when the genes encoding fetal hemoglobin are switched "off" and the genes coding for adult hemoglobin are switched "on."

Hemoglobin
- ▢ α-chains
- ▢ β-chains
- ▮ Heme groups

Figure 18.13

Structure of hemoglobin. The protein contains four subunits, designated α and β. The α- and β-subunits face each other across a central cavity. Each subunit in the tetramer contains a heme group that binds oxygen.

See Chemistry Connection: Wake Up, Sleeping Gene, in Chapter 14.

Question 18.7

Why is oxygen efficiently transferred from hemoglobin in the blood to myoglobin in the muscles?

Question 18.8

How is oxygen efficiently transferred from mother to fetus?

Sickle Cell Anemia

Sickle cell anemia is a human genetic disease that first appeared in tropical west and central Africa. It afflicts about 0.4% of African Americans. These individuals produce a mutant hemoglobin known as sickle cell hemoglobin (Hb S). Sickle cell anemia receives its name from the sickled appearance of the red blood cells that form in this condition (Figure 18.14). The sickled cells are unable to pass through the small capillaries of the circulatory system, and circulation is hindered. This results in damage to many organs, especially bone and kidney, and can lead to death at an early age.

Sickle cell hemoglobin differs from normal hemoglobin by a single amino acid. In the β-chain of sickle cell hemoglobin, a valine (a hydrophobic amino acid) has replaced a glutamic acid (a negatively charged amino acid). This substitution provides a basis for the binding of hemoglobin S molecules to one another. When oxyhemoglobin S unloads its oxygen, individual deoxyhemoglobin S molecules bind to one another as long polymeric fibers. This occurs because the valine fits into a hydrophobic pocket on the surface of a second deoxyhemoglobin S molecule. The fibers generated in this way radically alter the shape of the red blood cell, resulting in the sickling effect.

Sickle cell anemia occurs in individuals who have inherited the gene for sickle cell hemoglobin from both parents. Afflicted individuals produce 90–100% defective β-chains. Individuals who inherit one normal gene and one defective gene produce both normal and altered β-chains. About 10% of African Americans carry

The genetic basis of this alteration is discussed in Chapter 20.

When hemoglobin is carrying O_2, it is called oxyhemoglobin. When it is not bound to O_2, it is called deoxyhemoglobin.

Figure 18.14
Scanning electron micrographs of normal and sickled red blood cells.

Animation
Protein Denaturation

a single copy of the defective gene, a condition known as *sickle cell trait*. Although not severely affected, they have a 50% chance of passing the gene to each of their children.

An interesting relationship exists between sickle cell trait and resistance to malaria. In some parts of Africa, up to 20% of the population has sickle cell trait. In those same parts of Africa, one of the leading causes of death is malaria. The presence of sickle cell trait is linked to an increased resistance to malaria because the malarial parasite cannot feed efficiently on sickled red blood cells. People who have sickle cell disease die young; those without sickle cell trait have a high probability of succumbing to malaria. Occupying the middle ground, people who have sickle cell trait do not suffer much from sickle cell anemia and simultaneously resist deadly malaria. Because those with sickle cell trait have a greater chance of survival and reproduction, the sickle cell hemoglobin gene is maintained in the population.

18.10 Denaturation of Proteins

We have shown that the shape of a protein is absolutely essential to its function. We have also mentioned that life can exist only within a rather narrow range of temperature and pH. How are these two concepts related? As we will see, extremes of pH or temperature have a drastic effect on protein conformation, causing the molecules to lose their characteristic three-dimensional shape. **Denaturation** occurs when the organized structures of a globular protein, the α-helix, the β-pleated sheet, and tertiary folds become completely disorganized. However, it does not alter the primary structure. Denaturation of an α-helical protein is shown in Figure 18.15.

Temperature

Consider the effect of increasing temperature on a solution of proteins—for instance, egg white. At first, increasing the temperature simply increases the rate of molecular movement, the movement of the individual molecules within the solution. Then, as the temperature continues to increase, the bonds within the proteins begin to vibrate more violently. Eventually, the weak interactions, like hydrogen bonds and hydrophobic interactions, that maintain the protein structure are disrupted. The protein molecules are denatured as they lose their characteristic three-dimensional conformation and become completely disorganized. **Coagulation** occurs as the protein molecules then unfold and become entangled. At this point they are no longer in solution; they have aggregated to become a solid (see Figure 18.15). The egg white began as a viscous solution of egg albumins; but when it was cooked, the proteins were denatured and coagulated to become solid.

Figure 18.15
The denaturation of proteins by heat.
(a) The α-helical proteins are in solution.
(b) As heat is applied, the hydrogen bonds maintaining the secondary structure are disrupted, and the protein structure becomes disorganized. The protein is denatured. (c) The denatured proteins clump together, or coagulate, and are now in an insoluble form.

α-Helical proteins   Denatured   Coagulated
in solution   proteins   proteins

(a)   (b)   (c)

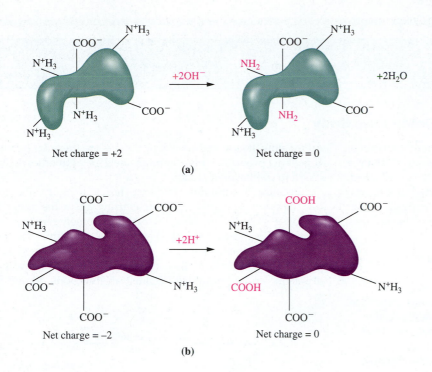

Net charge = +2

(a)

Net charge = 0

Net charge = -2

(b)

Net charge = 0

Figure 18.16
The effect of pH on proteins. (a) This protein has an overall charge of 2+. When a base is added, some of the protonated amino groups lose their protons. Now the protein is isoelectric; it has an equal number of positive and negative charges. (b) This protein has an overall charge of 2−. As acid is added, some of the carboxylate groups are protonated. The result is that the protein becomes isoelectric.

Many of the proteins of our cells, for instance, the enzymes, are in the same kind of viscous solution within the cytoplasm. To continue to function properly, they must remain in solution and maintain the correct three-dimensional configuration. If the body temperature becomes too high, or if local regions of the body are subjected to very high temperatures, as when you touch a hot cookie sheet, cellular proteins become denatured. They lose their function, and the cell or the organism dies.

pH

Because of the R groups of the amino acids, all proteins have a characteristic electric charge. Because every protein has a different amino acid composition, each will have a characteristic net electric charge on its surface. The positively and negatively charged R groups on the surface of the molecule interact with ions and water molecules, and these interactions keep the protein in solution within the cytoplasm.

The protein shown in Figure 18.16a has a net charge of 2+ because it has two extra $-N^+H_3$ groups. If we add 2 moles of base, such as NaOH, the protonated amino groups lose their protons and thus become electrically neutral. Now the net charge of the protein is zero. The pH at which a protein has an equal number of positive and negative charges, that is, a net charge of zero, is called the **isoelectric point.** The protein shown in Figure 18.16b has a net charge of 2− because of two additional carboxylate groups. When 2 moles of acid are added, the carboxylate groups become protonated. They are now electrically neutral, and the net charge on the protein is zero. As in the preceding example, the protein solution is at the isoelectric point.

When the pH of a protein solution is above the isoelectric point, all the protein molecules will have a net negative surface charge. Below the isoelectric point, they will have a net positive charge. In either case, these like-charged molecules repel one another, and this repulsion helps keep these very large molecules in solution.

At the isoelectric point the protein molecules no longer have a net surface charge. As a result they no longer strongly repel one another and are at their least

Proteins in each of the foods shown here have been denatured. What agent caused denaturation in each case?

A Medical Perspective

Immunoglobulins: Proteins That Defend the Body

A living organism is subjected to a constant barrage of bacterial, viral, parasitic, and fungal diseases. Without a defense against such perils we would soon perish. All vertebrates possess an *immune system.* In humans the immune system is composed of about 10^{12} cells, about as many as the brain or liver, which protect us from foreign invaders. This immune system has three important characteristics.

1. **It is highly specific.** The immune response to each infection is specific to, or directed against, only one disease organism or similar, related organisms.
2. **It has a memory.** Once the immune system has responded to an infection, the body is protected against reinfection by the same organism. This is the reason that we seldom suffer from the same disease more than once. Most of the diseases that we suffer recurrently, such as the common cold and flu, are actually caused by many different strains of the same virus. Each of these strains is "new" to the immune system.
3. **It can recognize "self" from "nonself."** When we are born, our immune system is already aware of all the antigens of our bodies. These it recognizes as "self" and will not attack. Every antigen that is not classified as "self" will be attacked by the immune system when it is encountered. Some individuals suffer from a defect of the immune response that allows it to attack the cells of one's own body. The result is an *autoimmune reaction* that can be fatal.

One facet of the immune response is the synthesis of *immunoglobulins,* or *antibodies,* that specifically bind a single macromolecule called an *antigen.* These antibodies are produced by specialized white blood cells called *B lymphocytes.* We are born with a variety of B lymphocytes that are capable of producing antibodies against perhaps a million different antigens. When a foreign antigen enters the body, it binds to the B lymphocyte that was preprogrammed to produce antibodies to destroy it. This stimulates the B cell to grow and divide. Then all of these new B cells produce antibodies that will bind to the disease agent and facilitate its destruction. Each B cell produces only one type of antibody with an absolute specificity for its target antigen. Many different B cells respond to each infection because the disease-causing agent is made up of many different antigens. Antibodies are made that bind to many of the antigens of the invader. This primary immune response is

rather slow. It can take a week or two before there are enough B cells to produce a high enough level of antibodies in the blood to combat an infection.

Because the immune response has a memory, the second time we encounter a disease-causing agent the antibody response is immediate. This is why it is extremely rare to suffer from mumps, measles, or chickenpox a second time. We take advantage of this property of the immune system to protect ourselves against many diseases. In the process of *vaccination* a person can be immunized against an infectious disease by

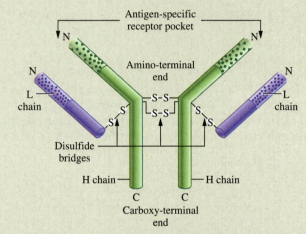

Schematic diagram of a Y-shaped immunoglobulin molecule. The binding sites for antigens are at the tips of the Y.

Schematic diagram of cross-linked immunoglobulin–antigen lattice.

soluble. Under these conditions, there is a tendency for them to clump together and precipitate out of solution. In this case, proteins may coagulate even though they are not denatured.

This is a reaction that you have probably observed in your own kitchen. When milk sits in the refrigerator for a prolonged period, the bacteria in the milk begin

(a) Sketch of immunoglobulin G showing the two heavy chains (red and blue) and the two light chains (green and yellow). (b) Space-filling model of immunoglobulin G. The color code is the same as in (a). The gray balls represent sugar groups attached to the immunoglobulin molecule.

injection of a small amount of the antigens of the virus or microorganism (the vaccine). The B lymphocytes of the body then manufacture antibodies against the antigens of the infectious agent. If the individual comes into contact with the disease-causing microorganism at some later time, the sensitized B lymphocytes "remember" the antigen and very quickly produce a large amount of specific antibody to overwhelm the microorganism or virus before it can cause overt disease.

Immunoglobulin molecules contain four peptide chains that are connected by disulfide bonds and arranged in a Y-shaped quaternary structure.

Each immunoglobulin has two identical antigen-binding sites located at the tips of the Y. Because most antigens have three or more antibody-binding sites, immunoglobulins can form large cross-linked antigen-antibody complexes that precipitate from solution.

Immunoglobulin G (IgG) is the major serum immunoglobulin. Some immunoglobulin G molecules can cross cell membranes and thus can pass between mother and fetus through the placenta, before birth. This is important because the immune system of a fetus is immature and cannot provide adequate protection from disease. Fortunately, the IgG acquired from the mother protects the fetus against most bacterial and viral infections that it might encounter before birth.

There are four additional types of antibody molecules that vary in their protein composition, but all have the same general Y shape. One of these is IgM, which is the first antibody produced in response to an infection. Secondarily, the B cell produces IgG molecules with the same antigen-binding region but a different protein composition in the rest of the molecule. IgA is the immunoglobulin responsible for protecting the body surfaces, such as the mucous membranes of the gut, the oral cavity, and the genitourinary tract. IgA is also found in mother's milk, protecting the newborn against diseases during the first few weeks of life. IgD is found in very small amounts and is thought to be involved in the regulation of antibody synthesis. The last type of immunoglobulin is IgE. For many years the function of IgE was unknown. When it was found in large quantities in the blood of people suffering from allergies, scientists realized that it is responsible for this "overblown" immunological reaction to dust particles and pollen grains.

For Further Understanding

Develop a hypothesis to explain why we may suffer from dozens of cases of the common cold caused by rhinoviruses. (Hint: Think about the structure of the proteins at the surface of the virus that serve as antigens.)

Describe the kinds of interactions that you would expect to find when an antibody binds to its cognate antigen.

to grow. They use the milk sugar, lactose, as an energy source in the process of fermentation and produce lactic acid as a by-product. As the bacteria continue to grow, the concentration of lactic acid increases. The additional acid results in the protonation of exposed carboxylate groups on the surface of the dissolved milk proteins. They become isoelectric and coagulate into a solid curd.

Lactate fermentation is discussed in Section 21.4.

See A Medical Perspective: Proteins in the Blood on page 621.

Imagine for a moment what would happen if the pH of the blood were to become too acidic or too basic. Blood is a fluid that contains water and dissolved electrolytes, a variety of cells, including the red blood cells responsible for oxygen transport, and many different proteins. These proteins include fibrinogen, which is involved in the clotting reaction; immunoglobulins, which protect us from disease; and albumins, which carry hydrophobic molecules in the blood.

When the blood pH drops too low, blood proteins become polycations. Similarly, when the blood pH rises too high, the proteins become polyanions. In either case, the proteins will unfold because of charge repulsion and loss of stabilizing ionic interactions. Under these extreme conditions, the denatured blood proteins would no longer be able to carry out their required functions. The blood cells would also die as their critical enzymes were denatured. The hemoglobin in the red blood cells would become denatured and would no longer be able to transport oxygen. Fortunately, the body has a number of mechanisms to avoid the radical changes in the blood pH that can occur as a result of metabolic or respiratory difficulties.

Organic Solvents

Polar organic solvents, such as rubbing alcohol (2-propanol), denature proteins by disrupting hydrogen bonds within the protein, in addition to forming hydrogen bonds with the solvent, water. The nonpolar regions of these solvents interfere with hydrophobic interactions in the interior of the protein molecule, thereby disrupting the conformation. Traditionally, a 70% solution of rubbing alcohol was often used as a disinfectant or antiseptic. However, recent evidence suggests that it is not an effective agent in this capacity.

Detergents

Detergents have both a hydrophobic region (the fatty acid tail) and a polar or hydrophilic region. When detergents interact with proteins, they disrupt hydrophobic interactions, causing the protein chain to unfold.

Heavy Metals

Heavy metals such as mercury (Hg^{2+}) or lead (Pb^{2+}) may form bonds with negatively charged side chain groups. This interferes with the salt bridges formed between amino acid R groups of the protein chain, resulting in loss of conformation. Heavy metals may also bind to sulfhydryl groups of a protein. This may cause a profound change in the three-dimensional structure of the protein, accompanied by loss of function.

Mechanical Stress

Stirring, whipping, or shaking can disrupt the weak interactions that maintain protein conformation. This is the reason that whipping egg whites produces a stiff meringue.

Question 18.9   How does high temperature denature proteins?

Question 18.10   How does extremely low pH cause proteins to coagulate?

18.11 Dietary Protein and Protein Digestion

Proteins, as well as carbohydrates and fats, are an energy source in the diet. As do carbohydrates and fats, proteins serve several dietary purposes. They can be oxidized to provide energy. In addition, the amino acids liberated by the hydrolysis of proteins are used directly in biosynthesis. The protein synthetic machinery of the cell can incorporate amino acids, released by the digestion of dietary protein, directly into new cellular proteins. Amino acids are also used in the biosynthesis of a large number of important molecules called the *nitrogen compounds*. This group includes some hormones, the heme groups of hemoglobin and myoglobin, and the nitrogen-containing bases found in DNA and RNA.

Digestion of dietary protein begins in the stomach. The stomach enzyme *pepsin* begins the digestion by hydrolyzing some of the peptide bonds of the protein. This breaks the protein down into smaller peptides.

Production of pepsin and other proteolytic digestive enzymes must be carefully controlled because the active enzymes would digest and destroy the cell that produces them. Thus, the stomach lining cells that make pepsin actually synthesize and secrete an inactive form called *pepsinogen.* Pepsinogen has an additional forty-two amino acids in its primary structure. These are removed in the stomach to produce active pepsin.

Protein digestion continues in the small intestine where the enzymes trypsin, chymotrypsin, elastase, and others catalyze the hydrolysis of peptide bonds at different sites in the protein. For instance, chymotrypsin cleaves peptide bonds on the carbonyl side of aromatic amino acids and trypsin cleaves peptide bonds on the carbonyl side of basic amino acids. Together these proteolytic enzymes degrade large dietary proteins into amino acids that can be absorbed by cells of the small intestine.

Amino acids can be divided into two major nutritional classes. **Essential amino acids** are those that cannot be synthesized by the body and are required in the diet. **Nonessential amino acids** are those amino acids that can be synthesized by the body and need not be included in the diet. Table 18.2 lists the essential and nonessential amino acids.

LEARNING GOAL

14 Explain the difference between essential and nonessential amino acids.

The inactive form of a proteolytic enzyme is called a proenzyme. These are discussed in Section 19.9.

The specificity of proteolytic enzymes is described in Section 19.11.

See Section 19.11 and Figure 19.12 for a more detailed picture of the action of digestive proteases.

TABLE 18.2 The Essential and Nonessential Amino Acids	
Essential Amino Acids	**Nonessential Amino Acids**
Isoleucine	Alanine
Leucine	Arginine[1]
Lysine	Asparagine
Methionine	Aspartate
Phenylalanine	Cysteine[2]
Threonine	Glutamate
Tryptophan	Glutamine
Valine	Glycine
	Histidine[1]
	Proline
	Serine
	Tyrosine[2]

[1]Histidine and arginine are essential amino acids for infants but not for healthy adults.
[2]Cysteine and tyrosine are considered to be semiessential amino acids. They are required by premature infants and adults who are ill.

Proteins are also classified as *complete* or *incomplete*. Protein derived from animal sources is generally **complete protein.** That is, it provides all of the essential and nonessential amino acids in approximately the correct amounts for biosynthesis. In contrast, protein derived from vegetable sources is generally **incomplete protein** because it lacks a sufficient amount of one or more essential amino acids. People who want to maintain a strictly vegetarian diet or for whom animal protein is often not available have the problem that no single high-protein vegetable has all of the essential amino acids to ensure a sufficient daily intake. For example, the major protein of beans contains abundant lysine and tryptophan but very little methionine, whereas corn contains considerable methionine but very little tryptophan or lysine. A mixture of corn and beans, however, satisfies both requirements. This combination, called *succotash*, was a staple of the diet of Native Americans for centuries.

Eating a few vegetarian meals each week can provide all the required amino acids and simultaneously help reduce the amount of saturated fats in the diet. Many ethnic foods apply the principle of mixing protein sources. Mexican foods such as tortillas and refried beans, Cajun dishes of spicy beans and rice, Indian cuisine of rice and lentils, and even the traditional American peanut butter sandwich are examples of ways to mix foods to provide complete protein.

Question 18.11

Why must vegetable sources of protein be mixed to provide an adequate diet?

Question 18.12

What are some common sources of dietary protein?

SUMMARY

18.1 Cellular Functions of Proteins

Proteins serve as biological catalysts (enzymes) and protective *antibodies. Transport proteins* carry materials throughout the body. Protein hormones regulate conditions in the body. Proteins also provide mechanical support and are needed for movement.

18.2 The α-Amino Acids

Proteins are made from twenty different *amino acids*, each having an α-COO^- group and an α-N^+H_3 group. They differ only in their side-chain R groups. All α-*amino acids* are chiral except glycine. Naturally occurring amino acids have the same chirality, designated L. The amino acids are grouped according to the polarity of their R groups.

18.3 The Peptide Bond

Amino acids are joined by *peptide bonds* to produce peptides and proteins. The peptide bond is an amide bond formed in the reaction between the carboxyl group of one amino acid and the amino group of another. The peptide bond is planar and relatively rigid.

18.4 The Primary Structure of Proteins

Proteins are linear polymers of amino acids. The linear sequence of amino acids defines the *primary structure* of the protein. Evolutionary relationships between species of organisms can be deduced by comparing the primary structures of their proteins.

18.5 The Secondary Structure of Proteins

The *secondary structure* of a protein is the folding of the primary sequence into an α-*helix* or β-*pleated sheet*. These structures are maintained by hydrogen bonds between the amide hydrogen and the carbonyl oxygen of the peptide bond. Usually *structural proteins*, such as the α-*keratins* and silk fibroin, are composed entirely of α-helix or β-pleated sheet.

18.6 The Tertiary Structure of Proteins

Globular proteins contain varying amounts of α-helix and β-pleated sheet folded into higher levels of structure called the *tertiary structure*. The tertiary structure of a protein is maintained by attractive forces between the R groups of amino acids. These forces include hydrophobic interactions, hydrogen bonds, ionic bridges, and disulfide bonds.

18.7 The Quaternary Structure of Proteins

Some proteins are composed of more than one peptide. They are said to have *quaternary structure*. Weak attractions between amino acid R groups hold the peptide subunits of the protein together. Some proteins require an attached, nonprotein *prosthetic group*.

18.8 An Overview of Protein Structure and Function

The primary structure of a protein dictates the way in which it folds into secondary and tertiary levels of structure. It also determines the way in which a protein may associate with other peptide subunits in the quaternary structure. An amino acid change in the primary structure may drastically affect protein folding. If the protein does not fold properly, and assume its correct three-dimensional shape, it will not be able to carry out its cellular function.

18.9 Myoglobin and Hemoglobin

Myoglobin, the oxygen storage protein of skeletal muscle, has a prosthetic group called the *heme group*. The heme group is the site of oxygen binding. *Hemoglobin* consists of four peptides. It transports oxygen from the lungs to the tissues. Myoglobin has a greater affinity for oxygen than does hemoglobin, and so oxygen is efficiently transferred from hemoglobin in the blood to myoglobin in tissues. Fetal hemoglobin has a greater affinity for oxygen than does maternal hemoglobin, and oxygen transfer occurs efficiently across the placenta from the mother to the fetus. A mutant hemoglobin is responsible for the genetic disease *sickle cell anemia*.

18.10 Denaturation of Proteins

Heat disrupts the hydrogen bonds and hydrophobic interactions that maintain protein structure. As a result, the protein unfolds and the organized structure is lost. The protein is said to be *denatured*. *Coagulation*, or clumping, occurs when the protein chains unfold and become entangled. When this occurs, proteins are no longer soluble. Changes in pH may cause proteins to become *isoelectric* (equal numbers of positive and negative charges). Isoelectric proteins coagulate because they no longer repel each other. If pH drops very low, proteins become polycations; if pH rises very high, proteins become polyanions. In either case, proteins become denatured owing to charge repulsion.

18.11 Dietary Protein and Protein Digestion

Essential amino acids must be acquired in the diet; *nonessential amino acids* can be synthesized by the body. *Complete proteins* contain all the essential and nonessential amino acids. *Incomplete proteins* are missing one or more essential amino acids. Protein digestion begins in the stomach, where proteins are degraded by the enzyme pepsin. Further digestion occurs in the small intestine by enzymes such as trypsin and chymotrypsin.

KEY TERMS

α-amino acid (18.2)
antibody (18.1)
antigen (18.1)
coagulation (18.10)
complete protein (18.11)
C-terminal amino acid (18.3)
defense proteins (18.1)
denaturation (18.10)
enzyme (18.1)
essential amino acid (18.11)
fibrous protein (18.5)
globular protein (18.6)
glycoprotein (18.7)
α-helix (18.5)
heme group (18.9)
hemoglobin (18.9)
hydrophilic amino acid (18.2)
hydrophobic amino acid (18.2)
incomplete protein (18.11)
isoelectric point (18.10)
α-keratin (18.5)
movement protein (18.1)

myoglobin (18.9)
nonessential amino
   acid (18.11)
N-terminal amino acid (18.3)
nutrient protein (18.1)
peptide bond (18.3)
β-pleated sheet (18.5)
primary structure (of a
   protein) (18.4)
prosthetic group (18.7)
protein (Intro)
quaternary structure (of a
   protein) (18.7)
regulatory protein (18.1)
secondary structure (of a
   protein) (18.5)
sickle cell anemia (18.9)
structural protein (18.1)
tertiary structure
   (of a protein) (18.6)
transport protein (18.1)

QUESTIONS AND PROBLEMS

Cellular Functions of Proteins

Foundations

18.13 Define the term *enzyme*.
18.14 Define the term *antibody*.
18.15 What is a transport protein?
18.16 What are the functions of structural proteins?

Applications

18.17 Of what significance are enzymes in the cell?
18.18 How do antibodies protect us against infection?
18.19 List two transport proteins and describe their significance to the organism.
18.20 What is the function of regulatory proteins?
18.21 Provide two examples of nutrient proteins.
18.22 Provide two examples of proteins that are required for movement.

The α-Amino Acids

Foundations

18.23 Write the basic general structure of an L-α-amino acid.
18.24 Draw the D- and L-isomers of serine. Which would you expect to find in nature?
18.25 What is a zwitterion?
18.26 Why are amino acids zwitterions at pH 7.0?

Applications

18.27 What is a chiral carbon?
18.28 Why are all of the α-amino acids except glycine chiral?
18.29 What is the importance of the R groups of the amino acids?
18.30 Describe the classification of the R groups of the amino acids, and provide an example of each class.

18.31 Write the structures of the nine amino acids that have hydrophobic side chains.

18.32 Write the structures of the aromatic amino acids. Indicate whether you would expect to find each on the surface or buried in a globular protein.

The Peptide Bond

Foundations

18.33 Define the term *peptide bond.*

18.34 What type of bond is the peptide bond? Explain why the peptide bond is rigid.

18.35 What observations led Linus Pauling and his colleagues to hypothesize that the peptide bond exists as a resonance hybrid?

18.36 Draw the resonance hybrids that represent the peptide bond.

Applications

18.37 Write the structure of each of the following peptides:
  a. His-trp-cys
  b. Gly-leu-ser
  c. Arg-ile-val

18.38 Write the structure of each of the following peptides:
  a. Ile-leu-phe
  b. His-arg-lys
  c. Asp-glu-ser

The Primary Structure of Proteins

Foundations

18.39 Define the *primary structure* of a protein.

18.40 What type of bond joins the amino acids to one another in the primary structure of a protein?

Applications

18.41 How does the primary structure of a protein determine its three-dimensional shape?

18.42 How does the primary structure of a protein ultimately determine its biological function?

18.43 Explain the relationship between the primary structure of a protein and the gene for that protein.

18.44 Explain how comparison of the primary structure of a protein from different organisms can be used to deduce evolutionary relationships between them.

The Secondary Structure of Proteins

Foundations

18.45 Define the secondary structure of a protein.

18.46 What are the two most common types of secondary structure?

Applications

18.47 What type of secondary structure is characteristic of:
  a. The α-keratins?
  b. Silk fibroin?

18.48 Describe the forces that maintain the two types of secondary structure: α-helix and β-pleated sheet.

18.49 Define fibrous proteins.

18.50 What is the relationship between the structure of fibrous proteins and their functions?

18.51 Describe a parallel β-pleated sheet.

18.52 Compare a parallel β-pleated sheet to an antiparallel β-pleated sheet.

The Tertiary Structure of Proteins

Foundations

18.53 Define the tertiary structure of a protein.

18.54 Use examples of specific amino acids to show the variety of weak interactions that maintain tertiary protein structure.

Applications

18.55 Write the structure of the amino acid produced by the oxidation of cysteine.

18.56 What is the role of cystine in maintaining protein structure?

18.57 Explain the relationship between the secondary and tertiary protein structures.

18.58 Why is the amino acid proline often found in the random coil hinge regions of the tertiary structure?

The Quaternary Structure of Proteins

Foundations

18.59 Describe the quaternary structure of proteins.

18.60 What weak interactions are responsible for maintaining quaternary protein structure?

Applications

18.61 What is a glycoprotein?

18.62 What is a prosthetic group?

An Overview of Protein Structure and Function

Applications

18.63 Why is hydrogen bonding so important to protein structure?

18.64 Explain why α-keratins that have many disulfide bonds between adjacent polypeptide chains are much less elastic and much harder than those without disulfide bonds.

18.65 How does the structure of the peptide bond make the structure of proteins relatively rigid?

18.66 The primary structure of a protein known as histone H4, which tightly binds DNA, is identical in all mammals and differs by only one amino acid between the calf and pea seedlings. What does this extraordinary conservation of primary structure imply about the importance of that one amino acid?

18.67 What does it mean to say that the structure of proteins is genetically determined?

18.68 Explain why genetic mutations that result in the replacement of one amino acid with another can lead to the formation of a protein that cannot carry out its biological function.

Myoglobin and Hemoglobin

Foundations

18.69 What is the function of hemoglobin?

18.70 What is the function of myoglobin?

18.71 Describe the structure of hemoglobin.

18.72 Describe the structure of myoglobin.

18.73 What is the function of heme in hemoglobin and myoglobin?

18.74 Write an equation representing the binding to and release of oxygen from hemoglobin.

Applications

18.75 Carbon monoxide binds tightly to the heme groups of hemoglobin and myoglobin. How does this affinity reflect the toxicity of carbon monoxide?

18.76 The blood of the horseshoe crab is blue because of the presence of a protein called hemocyanin. What is the function of hemocyanin?

18.77 Why does replacement of glutamic acid with valine alter hemoglobin and ultimately result in sickle cell anemia?

18.78 How do sickled red blood cells hinder circulation?

18.79 What is the difference between sickle cell disease and sickle cell trait?

18.80 How is it possible for sickle cell trait to confer a survival benefit on the person who possesses it?

Denaturation of Proteins

Foundations

18.81 Define the term *denaturation*.

18.82 What is the difference between denaturation and coagulation?

Applications

18.83 Why is heat an effective means of sterilization?

18.84 As you increase the temperature of an enzyme-catalyzed reaction, the rate of the reaction initially increases. It then reaches a maximum rate and finally dramatically declines. Keeping in mind that enzymes are proteins, how do you explain these changes in reaction rate?

18.85 Why is it important that blood have several buffering mechanisms to avoid radical pH changes?

18.86 Define the term *isoelectric*.

18.87 Why do proteins become polycations at extremely low pH?

18.88 Why do proteins become polyanions at very high pH?

18.89 Yogurt is produced from milk by the action of dairy bacteria. These bacteria produce lactic acid as a by-product of their metabolism. The pH decrease causes the milk proteins to coagulate. Why are food preservatives not required to inhibit the growth of bacteria in yogurt?

18.90 Wine is made from the juice of grapes by varieties of yeast. The yeast cells produce ethanol as a by-product of their fermentation. However, when the ethanol concentration reaches 12–13%, all the yeast die. Explain this observation.

Dietary Protein and Protein Digestion

Foundations

18.91 Define the term *essential amino acid*.

18.92 Define the term *nonessential amino acid*.

18.93 Define the term *complete protein*.

18.94 Define the term *incomplete protein*.

Applications

18.95 Write an equation representing the action of the proteolytic enzyme chymotrypsin. (Hint: In order to write the structure of a dipeptide that would be an appropriate reactant, you must consider what is known about where chymotrypsin cleaves a protein chain.)

18.96 Write an equation representing the action of the proteolytic enzyme trypsin. (Hint: In order to write the structure of a dipeptide that would be an appropriate reactant, you must consider what is known about where trypsin cleaves a protein chain.)

18.97 Why is it necessary to mix vegetable proteins to provide an adequate vegetarian diet?

18.98 Name some ethnic foods that apply the principle of mixing vegetable proteins to provide all of the essential amino acids.

18.99 Why must synthesis of digestive enzymes be carefully controlled?

18.100 What is the relationship between pepsin and pepsinogen?

CRITICAL THINKING PROBLEMS

1. Calculate the length of an α-helical polypeptide that is twenty amino acids long. Calculate the length of a region of antiparallel β-pleated sheet that is forty amino acids long.

2. Proteins involved in transport of molecules or ions into or out of cells are found in the membranes of all cells. They are classified as transmembrane proteins because some regions are embedded within the lipid bilayer, whereas other regions protrude into the cytoplasm or outside the cell. Review the classification of amino acids based on the properties of their R groups. What type of amino acids would you expect to find in the regions of the proteins embedded within the membrane? What type of amino acids would you expect to find on the surface of the regions in the cytoplasm or that protrude outside the cell?

3. A biochemist is trying to purify the enzyme hexokinase from a bacterium that normally grows in the Arctic Ocean at 5°C. In the next lab, a graduate student is trying to purify the same protein from a bacterium that grows in the vent of a volcano at 98°C. To maintain the structure of the protein from the Arctic bacterium, the first biochemist must carry out all her purification procedures at refrigerator temperatures. The second biochemist must perform all his experiments in a warm room incubator. In molecular terms, explain why the same kind of enzyme from organisms with different optimal temperatures for growth can have such different thermal properties.

4. The α-keratin of hair is rich in the amino acid cysteine. The location of these cysteines in the protein chain is genetically determined; as a result of the location of the cysteines in the protein, a person may have curly, wavy, or straight hair. How can the location of cysteines in α-keratin result in these different styles of hair? Propose a hypothesis to explain how a "perm" causes straight hair to become curly.

5. Calculate the number of different pentapeptides you can make in which the amino acids phenylalanine, glycine, serine, leucine, and histidine are each found. Imagine how many proteins could be made from the twenty amino acids commonly found in proteins.

19 Enzymes

Learning Goals

1 Classify enzymes according to the type of reaction catalyzed and the type of specificity.

2 Give examples of the correlation between an enzyme's common name and its function.

3 Describe the effect that enzymes have on the activation energy of a reaction.

4 Explain the effect of substrate concentration on enzyme-catalyzed reactions.

5 Discuss the role of the active site and the importance of enzyme specificity.

6 Describe the difference between the lock-and-key model and the induced fit model of enzyme-substrate complex formation.

7 Discuss the roles of cofactors and coenzymes in enzyme activity.

8 Explain how pH and temperature affect the rate of an enzyme-catalyzed reaction.

9 Describe the mechanisms used by cells to regulate enzyme activity.

10 Discuss the mechanisms by which certain chemicals inhibit enzyme activity.

11 Discuss the role of the enzyme chymotrypsin and other serine proteases.

12 Provide examples of medical uses of enzymes.

Outline

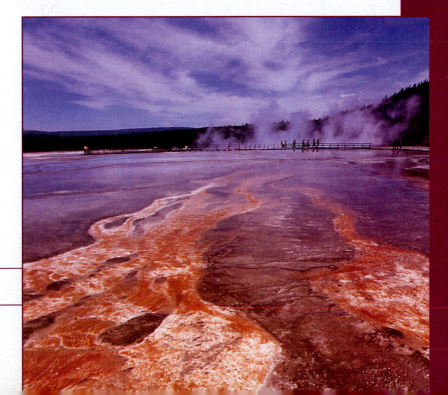

A hot-spring-fed lake at Yellowstone National Park.

Introduction

The enzymes discussed in this chapter are proteins; however, several ribonucleic acid (RNA) molecules have been demonstrated to have the ability to catalyze biological reactions. These are called *ribozymes*.

An **enzyme** is a biological molecule that serves as a catalyst for a biochemical reaction. The majority of enzymes are proteins. Without enzymes to speed up biochemical reactions, life could not exist. The life of the cell depends on the simultaneous occurrence of hundreds of chemical reactions that must take place rapidly under mild conditions. It is possible, for example, to add water to an alkene. However, this reaction is usually carried out at a temperature of 100°C in aqueous sulfuric acid. Such conditions would kill a cell. The fragile cell must carry out this same chemical reaction at body temperature (37°C) and in the absence of any strong acids or bases. How can this be accomplished? In Section 7.3 we saw that catalysts lower the energy of activation of a chemical reaction and thereby increase the rate of the reaction. This allows reactions to occur under milder conditions. The cell uses enzymes to solve the problem of chemical reactions that must occur rapidly under the mild conditions found within the cell. The enzyme facilitates a biochemical reaction, lowering the energy of activation and increasing the rate of the reaction. The efficient functioning of enzymes is essential for the life of the cell and of the organism.

The twin phenomena of high specificity and rapid reaction rates are the cornerstones of enzyme activity and the topic of this chapter. A typical cell contains thousands of different molecules, each of which is important to the chemistry of life processes. Each enzyme "recognizes" only one, or occasionally a few, of these molecules. One

Chemistry *Connection*

Super Hot Enzymes and the Origin of Life

Imagine the earth about four billion years ago: it was young then, not even a billion years old. Beginning as a red-hot molten sphere, slowly the earth's surface had cooled and become solid rock. But the interior, still extremely hot, erupted through the crust spewing hot gases and lava. Eventually these eruptions produced craggy land masses and an atmosphere composed of gases like hydrogen, carbon dioxide, ammonia, and water vapor. As the water vapor cooled, it condensed into liquid water, forming ponds and shallow seas.

At the dawn of biological life, the surface of the earth was still very hot and covered with rocky peaks and hot shallow oceans. The atmosphere was not very inviting either—filled with noxious gases and containing no molecular oxygen. Yet this is the environment where life on our planet began.

Some scientists think that they have found bacteria—living fossils—that may be very closely related to the first inhabitants of earth. These bacteria thrive at temperatures higher than the boiling point of water. Some need only H_2, CO_2, and H_2O for their metabolic processes and they quickly die in the presence of molecular oxygen.

But this lifestyle raises some uncomfortable questions. For instance, how do these bacteria survive at these extreme temperatures that would cook the life-forms with which we are more familiar? Researcher Mike Adams of the University of Georgia has found some of the answers. Adams and his

students have studied the structure of an enzyme from one of these extraordinary bacteria. He compared the structure of the super hot enzyme with that of the same enzyme purified from an organism that grows at "normal" temperatures. The overall three-dimensional structures of the two enzymes were very similar. This makes sense because they both catalyze the same reaction.

The question, then, is why is the super hot enzyme so stable at very high temperatures, while its low temperature counterpart is not. The answer lay in the tertiary structure of the enzyme. Adams observed that the three-dimensional structure of the super hot enzyme is held together by many more R group interactions than are found in the low-temperature version. These R group interactions, along with other differences, keep the protein stable and functional even at temperatures above 100°C.

In Chapter 18 we studied the structure and properties of proteins. We are now going to apply that knowledge to the study of a group of proteins that do the majority of the work for the cell. These special proteins, the enzymes, catalyze the biochemical reactions that break down food molecules to allow the cell to harvest energy. They also catalyze the biosynthetic reactions that produce the molecules required for cellular life. In this chapter we will study the properties of this extraordinary group of proteins and learn how they dramatically speed up biochemical reactions.

of the most remarkable features of enzymes is this specificity. Each can recognize and bind to a single type of substrate or reactant. The molecular size, shape, and charge distribution of both the enzyme and substrate must be compatible for this selective binding process to occur. The enzyme then transforms the substrate into the product with lightning speed. In fact, enzyme-catalyzed reactions often occur from one million to one hundred million times faster than the corresponding uncatalyzed reaction.

The enzyme catalase provides one of the most spectacular examples of the increase in reaction rates brought about by enzymes. This enzyme is required for life in an oxygen-containing environment. In this environment the process of the aerobic (oxygen-requiring) breakdown of food molecules produces hydrogen peroxide (H_2O_2). Because H_2O_2 is toxic to the cell, it must be destroyed. One molecule of catalase converts forty million molecules of hydrogen peroxide to harmless water and oxygen every second:

$$2H_2O_2 \xrightarrow[\text{(an enzyme)}]{\text{Catalase}} 2H_2O + O_2$$

Reaction occurs forty million times every second!

This is the same reaction that you witness when you pour hydrogen peroxide on a wound. The catalase released from injured cells rapidly breaks down the hydrogen peroxide. The bubbles that you see are oxygen gas released as a product of the reaction.

A wound is treated with hydrogen peroxide. Of what value is this treatment?

19.1 Nomenclature and Classification

Classification of Enzymes

Enzymes may be classified according to the type of reaction that they catalyze. The six classes are as follows.

LEARNING GOAL

1 Classify enzymes according to the type of reaction catalyzed and the type of specificity.

Oxidoreductases

Oxidoreductases are enzymes that catalyze oxidation–reduction (redox) reactions. *Lactate dehydrogenase* is an oxidoreductase that removes hydrogen from a molecule of lactate. Other subclasses of the oxidoreductases include oxidases and reductases.

Recall that redox reactions involve electron transfer from one substance to another (Section 8.5).

$$\underset{\text{Lactate}}{\begin{array}{c} COO^- \\ | \\ HO-C-H \\ | \\ CH_3 \end{array}} + NAD^+ \underset{\xrightarrow{\text{Lactate dehydrogenase}}}{\rightleftharpoons} \underset{\text{Pyruvate}}{\begin{array}{c} COO^- \\ | \\ C=O \\ | \\ CH_3 \end{array}} + NADH$$

Transferases

Transferases are enzymes that catalyze the transfer of functional groups from one molecule to another. For example, a *transaminase* catalyzes the transfer of an amino functional group, and a *kinase* catalyzes the transfer of a phosphate group. Kinases play a major role in energy-harvesting processes involving ATP. In the adrenal glands, norepinephrine is converted to epinephrine by the enzyme *phenylethanolamine-N-methyltransferase* (PNMT), a *transmethylase*.

The significance of phosphate group transfers in energy metabolism is discussed in Sections 21.1 and 21.3.

Methyl group donor + HO—⟨benzene⟩—CHCH$_2$NH$_2$ (OH) (HO) $\xrightarrow{\text{PNMT}}$ HO—⟨benzene⟩—CHCH$_2$NH—CH$_3$ (OH) (HO)

Norepinephrine   Epinephrine

Hydrolysis of esters is described in Section 14.2. The action of lipases in digestion is discussed in Section 23.1.

These kittens are adorable, but there is nothing lovely about the aroma of an untended kitty litter box. Urease, a hydrolase, catalyzes the reaction responsible for this smell. Write an equation representing the reaction catalyzed by urease and identify the offending molecule.

The reactions of the citric acid cycle are described in Section 22.4.

Hydrolases

Hydrolases catalyze hydrolysis reactions, that is, the addition of a water molecule to a bond resulting in bond breakage. These reactions are important in the digestive process. For example, *lipases* catalyze the hydrolysis of the ester bonds in triglycerides:

$$CH_2-O-\overset{\overset{O}{\|}}{C}(CH_2)_nCH_3$$
$$CH-O-\overset{\overset{O}{\|}}{C}(CH_2)_nCH_3 + 3H_2O \xrightarrow{\text{Lipase}} \begin{array}{l}CH_2OH\\CHOH\\CH_2OH\end{array} + 3CH_3(CH_2)_nCOOH$$
$$CH_2-O-\overset{\overset{O}{\|}}{C}(CH_2)_nCH_3$$

Triglyceride   Glycerol   Fatty acids

Lyases

Lyases catalyze the addition of a group to a double bond or the removal of a group to form a double bond. *Fumarase* is an example of a lyase. In the citric acid cycle, fumarase catalyzes the addition of a water molecule to the double bond of the substrate fumarate. The product is malate.

$$\begin{array}{c}COO^-\\|\\C-H\\\|\\H-C\\|\\COO^-\end{array} + H_2O \xrightarrow{\text{Fumarase}} \begin{array}{c}COO^-\\|\\HO-C-H\\|\\H-C-H\\|\\COO^-\end{array}$$

Fumarate   Malate

Citrate lyase catalyzes a far more complicated reaction in which we see the removal of a group and formation of a double bond. Specifically, citrate lyase catalyzes the removal of an acetyl group from a molecule of citrate. The products of this reaction include oxaloacetate, acetyl CoA, ADP, and an inorganic phosphate group (P_i):

$$\begin{array}{c}COO^-\\|\\CH_2\\|\\^-OOC-C-OH\\|\\CH_2\\|\\COO^-\end{array} + ATP + \text{Coenzyme A} + H_2O \xrightarrow{\text{Citrate lyase}}$$

Citrate

Recall that the squiggle (~) represents a high-energy bond.

$$\begin{array}{c}COO^-\\|\\CH_2\\|\\C=O\\|\\COO^-\end{array} + CH_3-\overset{\overset{O}{\|}}{C}\sim S-CoA + ADP + P_i$$

Oxaloacetate   Acetyl CoA

Isomerases

Isomerases rearrange the functional groups within a molecule and catalyze the conversion of one isomer into another. For example, *phosphoglycerate mutase* converts one structural isomer, 3-phosphoglycerate, into another, 2-phosphoglycerate:

This reaction is important in glycolysis, an energy-harvesting pathway described in Chapter 21.

$$\text{3-Phosphoglycerate} \quad \xrightleftharpoons[\text{Phosphoglycerate mutase}]{} \quad \text{2-Phosphoglycerate}$$

3-Phosphoglycerate   2-Phosphoglycerate

Ligases

Ligases are enzymes that catalyze a reaction in which a C—C, C—S, C—O, or C—N bond is made or broken. This is accompanied by an ATP-ADP interconversion. For example, *DNA ligase* catalyzes the joining of the hydroxyl group of a nucleotide in a DNA strand with the phosphoryl group of the adjacent nucleotide to form a phosphoester bond:

The use of DNA ligase in recombinant DNA studies is detailed in Section 20.8.

DNA strand —3′—OH + ⁻O—P(=O)(O⁻)—O—5′— DNA strand

↓ DNA ligase

DNA strand —3′—O—P(=O)(O⁻)—O—5′— DNA strand

Classifying Enzymes According to the Type of Reaction That They Catalyze · **EXAMPLE 19.1**

Classify the enzyme that catalyzes each of the following reactions, and explain your reasoning.

LEARNING GOAL
1 Classify enzymes according to the type of reaction catalyzed and the type of specificity.

Alanyl-glycine   Alanine   Glycine

Solution

The reaction occurring here involves breaking a bond, in this case a peptide bond, by adding a water molecule. The enzyme is classified as a *hydrolase*, specifically a *peptidase*.

Continued—

Glucose	Adenosine triphosphate	Glucose-6-phosphate	Adenosine diphosphate

Solution

This is the first reaction in the biochemical pathway called *glycolysis*. A phosphoryl group is transferred from a donor molecule, adenosine triphosphate, to the recipient molecule, glucose. The products are glucose-6-phosphate and adenosine diphosphate. This enzyme, called *hexokinase*, is an example of a *transferase*.

Malate	Oxaloacetate

Solution

In this reaction, the reactant malate is oxidized and the coenzyme NAD$^+$ is reduced. The enzyme that catalyzes this reaction, *malate dehydrogenase*, is an *oxidoreductase*.

Dihydroxyacetone phosphate	Glyceraldehyde-3-phosphate

Solution

Careful inspection of the structure of the reactant and the product reveals that they each have the same number of carbon, hydrogen, oxygen, and phosphorus atoms; thus, they must be structural isomers. The enzyme must be an *isomerase*. Its name is *triose phosphate isomerase*.

Continued—

EXAMPLE 19.1 —*Continued*

Practice Problem 19.1

To which class of enzymes does each of the following belong?

a. Pyruvate kinase
b. Alanine transaminase
c. Triose phosphate isomerase
d. Pyruvate dehydrogenase
e. Lactase
f. Phosphofructokinase
g. Lipase
h. Acetoacetate decarboxylase
i. Succinate dehydrogenase

For Further Practice: Questions 19.23, 19.24, and 19.25.

Nomenclature of Enzymes

The common names for some enzymes are derived from the name of the **substrate,** the reactant that binds to the enzyme and is converted into product. In many cases, the name of the enzyme is simply derived by adding the suffix -*ase* to the name of the substrate. For instance, *urease* catalyzes the hydrolysis of urea and *lactase* catalyzes the hydrolysis of the disaccharide lactose.

Names of other enzymes reflect the type of reaction that they catalyze. *Dehydrogenases* catalyze the removal of hydrogen atoms from a substrate, while *decarboxylases* catalyze the removal of carboxyl groups. *Hydrogenases* and *carboxylases* carry out the opposite reaction, adding hydrogen atoms or carboxyl groups to their substrates.

Thus, the common name of an enzyme often tells us a great deal about the function of an enzyme. Yet other enzymes have historical names that have no relationship to either the substrates or the reactions that they catalyze. A few examples include catalase, trypsin, pepsin, and chymotrypsin. In these cases, the names of the enzymes and the reactions that they catalyze must simply be memorized.

The systematic names for enzymes tell us the substrate, the type of reaction that is catalyzed, and the name of any coenzyme that is required. For instance, the systematic name of the oxidoreductase lactate dehydrogenase is lactate: NAD oxidoreductase.

LEARNING GOAL

2 Give examples of the correlation between an enzyme's common name and its function.

Coenzymes are molecules required by some enzymes to serve as donors or acceptors of electrons, hydrogen atoms, or other functional groups during a chemical reaction. Coenzymes are discussed in Section 19.7.

Question 19.1

Write an equation representing the reaction catalyzed by each of the enzymes listed in Practice Problem 19.1a through d at the end of Example 19.1. (Hint: You may need to refer to the index of this book to learn more about the substrates and the reactions that are catalyzed.)

Question 19.2

Write an equation representing the reaction catalyzed by each of the enzymes listed in Practice Problem 19.1e through i at the end of Example 19.1. (Hint: You may need to refer to the index of this book to learn more about the substrates and the reactions that are catalyzed.)

Question 19.3

What is the substrate for each of the following enzymes?

   a. Sucrase
   b. Pyruvate decarboxylase
   c. Succinate dehydrogenase

Question 19.4

What chemical reaction is mediated by each of the enzymes in Question 19.3?

19.2 The Effect of Enzymes on the Activation Energy of a Reaction

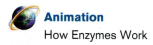

LEARNING GOAL

3 Describe the effect that enzymes have on the activation energy of a reaction.

The activation energy (Section 7.3) of a reaction is the threshold energy that must be overcome to produce a chemical reaction.

Equilibrium constants are described in Section 7.4.

Energy, rate, and equilibrium are described in Chapter 7.

Animation

How Enzymes Work

How does an enzyme speed up a chemical reaction? It changes the path by which the reaction occurs, providing a lower energy route for the conversion of the substrate into the **product,** the substance that results from the enzyme-catalyzed reaction. Thus enzymes speed up reactions by lowering the activation energy of the reaction.

Recall that every chemical reaction is characterized by an equilibrium constant. Consider, for example, the simple equilibrium

$$aA \rightleftharpoons bB$$

The equilibrium constant for this reaction, K_{eq}, is defined as

$$K_{eq} = \frac{[B]^b}{[A]^a} = \frac{[\text{product}]^b}{[\text{reactant}]^a}$$

This equilibrium constant is actually a reflection of the difference in energy between reactants and products. It is a measure of the relative stabilities of the reactants and products. No matter how the chemical reaction occurs (which path it follows), the difference in energy between the reactants and the products is always the same. **For this reason, an enzyme cannot alter the equilibrium constant for the reaction that it catalyzes.** An enzyme does, however, change the path by which the process occurs, providing a lower energy route for the conversion of the substrate into the product. An enzyme increases the rate of a chemical reaction by lowering the activation energy for the reaction (Figure 19.1). An enzyme thus increases the rate at which the reaction it catalyzes reaches equilibrium.

Figure 19.1

Diagram of the difference in energy between the reactants (*A* and *B*) and products (*C* and *D*) for a reaction. Enzymes cannot change this energy difference but act by lowering the activation energy (*E*a) for the reaction, thereby speeding up the reaction.

(a)   (b)

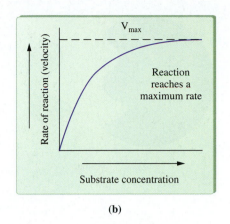

(a)   (b)

Figure 19.2
Plot of the rate or velocity, *V*, of a reaction versus the concentration of substrate, [*S*], for (a) an uncatalyzed reaction and (b) an enzyme-catalyzed reaction. For an enzyme-catalyzed reaction, the rate is at a maximum when all of the enzyme molecules are bound to the substrate. Beyond this concentration of substrate, further increases in substrate concentration have no effect on the rate of the reaction.

19.3 The Effect of Substrate Concentration on Enzyme-Catalyzed Reactions

The rates of uncatalyzed chemical reactions often double every time the substrate concentration is doubled (Figure 19.2a). Therefore as long as the substrate concentration increases, there is a direct increase in the rate of the reaction. For enzyme-catalyzed reactions, however, this is not the case. Although the rate of the reaction is initially responsive to the substrate concentration, at a certain concentration of substrate the rate of the reaction reaches a maximum value. A graph of the rate of reaction, *V*, versus the substrate concentration, [*S*], is shown in Figure 19.2b. We see that the rate of the reaction initially increases rapidly as the substrate concentration is increased but that the rate levels off at a maximum value. At its maximum rate, the active sites of all the enzyme molecules are occupied by a substrate molecule. The active site is the region of the enzyme that specifically binds the substrate and catalyzes the reaction. A new molecule of substrate cannot bind to the enzyme molecule until the substrate molecule already held in the active site is converted to product and released. Thus, it appears that the enzyme-catalyzed reaction occurs in two stages. The first, rapid step is the formation of the *enzyme-substrate complex*. The second step is slower and involves conversion of the substrate to product and the release of the product and enzyme from the resulting enzyme-product complex. It is called the *rate-limiting step* because the rate of the reaction is limited by the speed with which the substrate is converted into product and the product is released. Thus, the reaction rate is dependent on the amount of enzyme available.

LEARNING GOAL

4 Explain the effect of substrate concentration on enzyme-catalyzed reactions.

19.4 The Enzyme-Substrate Complex

The following series of reversible reactions represents the steps in an enzyme-catalyzed reaction. The first step (highlighted in blue) involves the encounter of the enzyme with its substrate and the formation of an **enzyme-substrate complex.**

LEARNING GOAL

5 Discuss the role of the active site and the importance of enzyme specificity.

$$E + S \underset{\text{Step I}}{\rightleftharpoons} ES \underset{\text{Step II}}{\rightleftharpoons} ES^* \underset{\text{Step III}}{\rightleftharpoons} EP \underset{\text{Step IV}}{\rightleftharpoons} E + P$$

| Enzyme + substrate | Enzyme– substrate complex | Transition state | Enzyme– product complex | Enzyme + product |

The part of the enzyme that binds with the substrate is called the **active site.** The characteristics of the active site that are crucial to enzyme function include the following:

Figure 19.3
(a) The lock-and-key model of enzyme-substrate binding assumes that the enzyme active site has a rigid structure that is precisely complementary in shape and charge distribution to the substrate. (b) The induced fit model of enzyme-substrate binding. As the enzyme binds to the substrate, the shape of the active site conforms precisely to the shape of the substrate. The shape of the substrate may also change.

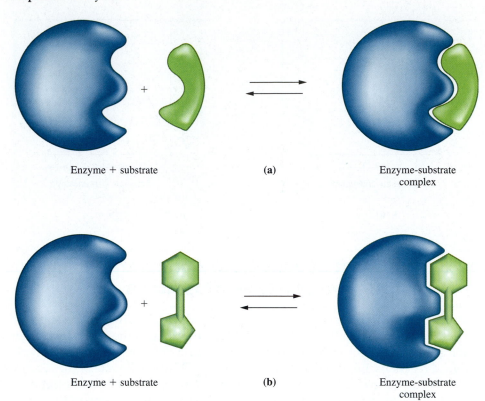

Enzyme + substrate   **(a)**   Enzyme-substrate complex

Enzyme + substrate   **(b)**   Enzyme-substrate complex

6 **LEARNING GOAL**
Describe the difference between the lock-and-key model and the induced fit model of enzyme-substrate complex formation.

The overall shape of a protein is maintained by many weak interactions. At any time a few of these weak interactions may be broken by heat energy or a local change in pH. If only a few bonds are broken, they will re-form very quickly. The overall result is that there is a brief change in the shape of the enzyme. Thus the protein or enzyme can be viewed as a flexible molecule, changing shape slightly in response to minor local changes.

- Enzyme active sites are pockets or clefts in the surface of the enzyme. The R groups in the active site that are involved in catalysis are called *catalytic groups.*
- The shape of the active site is complementary to the shape of the substrate. That is, the substrate fits neatly into the active site of the enzyme.
- An enzyme attracts and holds its substrate by weak, noncovalent interactions. The R groups involved in substrate binding, and not necessarily catalysis, make up the *binding site.*
- The conformation of the active site determines the specificity of the enzyme because only the substrate that fits into the active site will be used in a reaction.

The **lock-and-key model** of enzyme activity, shown in Figure 19.3a, was devised by Emil Fischer in 1894. At that time it was thought that the substrate simply snapped into place like a piece of a jigsaw puzzle or a key into a lock.

Today we know that proteins are flexible molecules. This led Daniel E. Koshland, Jr., to propose a more sophisticated model of the way enzymes and substrates interact. This model, proposed in 1958, is called the **induced fit model** (Figure 19.3b). In this model, the active site of the enzyme is not a rigid pocket into which the substrate fits precisely; rather, it is a flexible pocket that *approximates* the shape of the substrate. When the substrate enters the pocket, the active site "molds" itself around the substrate. This produces the perfect enzyme-substrate "fit."

Question 19.5

Compare the lock-and-key and induced fit models of enzyme-substrate binding.

Question 19.6

What is the relationship between an enzyme active site and its substrate?

19.5 Specificity of the Enzyme-Substrate Complex

For an enzyme-substrate interaction to occur, the surfaces of the enzyme and substrate must be complementary. It is this requirement for a specific fit that determines whether an enzyme will bind to a particular substrate and carry out a chemical reaction.

Enzyme specificity is the ability of an enzyme to bind only one, or a very few, substrates and thus catalyze only a single reaction. To illustrate the specificity of enzymes, consider the following reactions.

The enzyme urease catalyzes the hydrolysis of urea to carbon dioxide and ammonia as follows:

$$\underset{\text{Urea}}{H_2N-\overset{\overset{\displaystyle O}{\|}}{C}-NH_2} + H_2O \xrightarrow{\text{Urease}} CO_2 + 2NH_3$$

Methylurea, in contrast, though structurally similar to urea, is not affected by urease:

$$\underset{\text{Methylurea}}{H_2N-\overset{\overset{\displaystyle O}{\|}}{C}-NHCH_3} + H_2O \xrightarrow{\text{Urease}} \text{no reaction}$$

Not all enzymes exhibit the same degree of specificity. Four classes of enzyme specificity have been observed.

- **Absolute specificity:** An enzyme that catalyzes the reaction of only one substrate has absolute specificity. *Aminoacyl tRNA synthetases* exhibit absolute specificity. Each must attach the correct amino acid to the correct transfer RNA molecule. If the wrong amino acid is attached to the transfer RNA, it may be mistakenly added to a peptide chain, producing a nonfunctional protein.
- **Group specificity:** An enzyme that catalyzes processes involving similar molecules containing the same functional group has group specificity. *Hexokinase* is a group-specific enzyme that catalyzes the addition of a phosphoryl group to the hexose sugar glucose in the first step of glycolysis. Hexokinase can also add a phosphoryl group to several other six-carbon sugars.
- **Linkage specificity:** An enzyme that catalyzes the formation or breakage of only certain bonds in a molecule has linkage specificity. *Proteases,* such as trypsin, chymotrypsin, and elastase, are enzymes that selectively hydrolyze peptide bonds. Thus, these enzymes are linkage specific.
- **Stereochemical specificity:** An enzyme that can distinguish one enantiomer from the other has stereochemical specificity. Most of the enzymes of the human body show stereochemical specificity. Because we use only D-sugars and L-amino acids, the enzymes involved in digestion and metabolism recognize only those particular stereoisomers.

LEARNING GOAL
5 Discuss the role of the active site and the importance of enzyme specificity.

Aminoacyl tRNA synthetases are discussed in Section 20.6. Aminoacyl group transfer reactions were described in Section 15.4.

Hexokinase activity is described in Section 21.3.

Proteolytic enzymes are discussed in Section 19.11.

19.6 The Transition State and Product Formation

How does enzyme-substrate binding result in a faster chemical reaction? To answer this question, we must once again look at the steps of an enzyme-catalyzed reaction, focusing on the steps highlighted in blue:

$$\text{E + S} \underset{}{\overset{\text{Step I}}{\rightleftarrows} } \text{ES} \underset{}{\overset{\text{Step II}}{\rightleftarrows}} \text{ES*} \underset{}{\overset{\text{Step III}}{\rightleftarrows}} \text{EP} \underset{}{\overset{\text{Step IV}}{\rightleftarrows}} \text{E + P}$$

| Enzyme + substrate | Enzyme– substrate complex | Transition state | Enzyme– product complex | Enzyme + product |

After the formation of the enzyme-substrate complex in Step I, the flexible enzyme and substrate interact, changing the substrate into a configuration that is no longer energetically stable (Step II). This is the **transition state,** a state in which the substrate is in an intermediate form, having features of both the substrate and the product. This state favors conversion of the substrate into product (Step III) and release of the product (Step IV). Notice that the enzyme is completely unchanged by these events.

What kinds of transition state changes might occur in the substrate that would make a reaction proceed more rapidly?

1. The enzyme might put "stress" on a bond and thereby promote bond breakage, as in the example of sucrase. The enzyme catalyzes the hydrolysis of sucrose into glucose and fructose. The formation of the enzyme-substrate complex (Figures 19.4a and 19.4b) results in a change in the shape of the enzyme. This, in turn, may stretch or distort the bond between glucose and fructose, weakening the bond and allowing it to be broken much more easily than in the absence of the enzyme Figure 19.4c through e.

2. An enzyme may facilitate a reaction by bringing two reactants into close proximity and in the proper orientation for reaction to occur. If we look at the reaction between glucose and fructose to produce sucrose (Figure 19.5a), we see that each of the sugars has five hydroxyl groups that could undergo condensation to produce a disaccharide. By random molecular collision there is a one in twenty-five chance that the two molecules will collide in the proper orientation to produce sucrose. The probability that the two will react is actually much less than that because at body temperature, most molecular collisions will not have enough energy to overcome the energy of activation, even if the molecules are in the proper orientation. The enzyme can bring the two molecules close together in the correct alignment (Figure 19.5b), forming the transition state and greatly speeding up the reaction.

Animation

Enzyme Action and the
Hydrolysis of Sucrose

Figure 19.4
Bond breakage is facilitated by the enzyme as a result of stress on a bond. (a, b) The enzyme-substrate complex is formed. (c) In the transition state, the enzyme changes shape and thereby puts stress on the glycosidic bond holding the two monosaccharides together. This lowers the energy of activation of this reaction. (d, e) The bond is broken, and the products are released.

(a) Enzyme + substrate   (b) Enzyme-substrate complex   (c) Transition state

(d) Enzyme-product complex   (e) Enzyme + product

Glucose   Fructose   Sucrose

(a)

II. Transition state
(enzyme-substrate
complex)

+ H₂O

I. Enzyme
+ substrate

III. Enzyme-
product
complex

IV. Enzyme + product

(b)

Figure 19.5
An enzyme may lower the energy of activation required for a reaction by holding the substrates in close proximity and in the correct orientation. (a) A condensation reaction between glucose and fructose to produce sucrose. (b) The enzyme-substrate complex forms, bringing the two monosaccharides together with the correct hydroxyl groups extended toward one another.

3. The active site of an enzyme may modify the pH of the microenvironment surrounding the substrate. For instance, it may serve as a donor or an acceptor of H^+. This would cause a change in the pH in the vicinity of the substrate without disturbing the normal pH elsewhere in the cell.

Summarize three ways in which an enzyme might lower the energy of activation of a reaction.

Question 19.7

What is the transition state in an enzyme-catalyzed reaction?

Question 19.8

A Medical *Perspective*

HIV Protease Inhibitors and Pharmaceutical Drug Design

In 1981 the Centers for Disease Control in Atlanta, Georgia, recognized a new disease syndrome, acquired immune deficiency syndrome (AIDS). The syndrome is characterized by an impaired immune system, a variety of opportunistic infections and cancer, and brain damage that results in dementia. It soon became apparent that the disease was being transmitted by blood and blood products, as well as by sexual conduct.

The earliest drugs that proved effective in the treatment of HIV infections all inhibited replication of the genetic material of the virus. While these treatments were initially effective, prolonging the lives of many, it was not long before viral mutants resistant to these drugs began to appear. Clearly, a new approach was needed.

In 1989 a group of scientists revealed the three-dimensional structure of the HIV protease. This structure is shown in the accompanying figure. This enzyme is necessary for viral replication because the virus has an unusual strategy for making all of its proteins. Rather than make each protein individually, it makes large "polyproteins" that must then be cut by the HIV protease to form the final proteins required for viral replication.

Since scientists realized that this enzyme was essential for HIV replication, they decided to engineer a substance that would inhibit the enzyme by binding irreversibly to the active site, in essence plugging it up. The challenge, then, was to design a molecule that would be the plug. Researchers knew the primary structure (amino acid sequence) of the HIV protease from earlier nucleic acid sequencing studies. By 1989 they also had a very complete picture of the three-dimensional nature of the molecule, which they had obtained by X-ray crystallography.

Putting all of this information into a sophisticated computer modeling program, they could look at the protease from any angle. They could see the location of each of the R groups of each of the amino acids in the active site. This kind of information allowed the scientists to design molecules that would be complementary to the shape and charge distribution of the enzyme active site—in other words, structural analogs of the normal protease substrate. It was not long before the scientists had produced several candidates for the HIV protease inhibitor.

But, there are many tests that a drug candidate must pass before it can be introduced into the market as safe and effective. Scientists had to show that the candidate drugs would bind effectively to the HIV protease and block its function, thereby inhibiting virus replication. Properties such as the solubility, the efficiency of absorption by the body, the period of activity in the body, and the toxicity of the drug candidates all had to be determined.

By 1996 there were three protease inhibitors available to combat HIV infection. There are currently seven of these drugs on the market. In many cases development and testing of a drug candidate can take up to fifteen years. In the case of the first HIV protease inhibitors, the first three drugs were on the market in less than eight years. This is a testament both to the urgent need for HIV treatments and to the technology available to attack the problem.

The human immunodeficiency virus protease.

For Further Understanding

What particular concerns would you have, as a medical researcher, about administering a drug that is a protease inhibitor?

Often a protease inhibitor is prescribed along with an inhibitor of replication of the viral genetic material. Develop a hypothesis to explain this strategy.

19.7 Cofactors and Coenzymes

LEARNING GOAL
7 Discuss the roles of cofactors and coenzymes in enzyme activity.

In Section 18.7 we saw that some proteins require an additional nonprotein prosthetic group to function. The same is true of some enzymes. In this case, the polypeptide portion is called the **apoenzyme,** and the nonprotein group is called the **cofactor.** Together they form the active enzyme called the **holoenzyme.** Cofactors

(a)

Apoenzyme + substrate

No enzyme-substrate complex
No reaction

Cu⁺⁺
Cofactor

(b)

Functional enzyme
with active binding site

Enzyme-substrate
complex

Reaction
occurs

Figure 19.6
(a) The apoenzyme is unable to bind to its substrate. (b) When the required cofactor, in this case a copper ion, Cu^{2+}, is available, it binds to the apoenzyme. Now the active site takes on the correct configuration, the enzyme-substrate complex forms, and the reaction occurs.

may be metal ions, organic compounds, or organometallic compounds and must be bound to the enzyme to maintain the correct shape of the active site (Figure 19.6). Only when the cofactor is bound, can the enzyme bind the substrate and catalyze the reaction.

Other enzymes require the temporary binding of a **coenzyme.** Such binding is generally mediated by weak interactions like hydrogen bonds. Coenzymes are organic molecules that serve as carriers of electrons or chemical groups. In chemical reactions, they may either donate groups to the substrate or accept groups that are removed from the substrate. The example in Figure 19.7 shows a coenzyme accepting a functional group from one substrate and donating it to the second substrate in a reaction catalyzed by a transferase.

Often coenzymes contain modified vitamins as part of their structure. A **vitamin** is an organic substance that is required in the diet in only small amounts. Of the water-soluble vitamins, only vitamin C has not been associated with a coenzyme. Table 19.1 is a summary of some coenzymes and the water-soluble vitamins from which they are made.

 Water-Soluble Vitamins

TABLE 19.1	**The Water-Soluble Vitamins and their Coenzymes**	
Vitamin	**Coenzyme**	**Function**
Thiamine (B_1)	Thiamine pyrophosphate	Decarboxylation reactions
Riboflavin (B_2)	Flavin mononucleotide (FMN)	Carrier of H atoms
	Flavin adenine dinucleotide (FAD)	
Niacin (B_3)	Nicotinamide adenine dinucleotide (NAD^+)	Carrier of hydride ions
	Nicotinamide adenine dinucleotide phosphate ($NADP^+$)	
Pyridoxine (B_6)	Pyridoxal phosphate	Carriers of amino and carboxyl groups
	Pyridoxamine phosphate	
Cyanocobalamin (B_{12})	Deoxyadenosyl cobalamin	Coenzyme in amino acid metabolism
Folic acid	Tetrahydrofolic acid	Coenzyme for 1-C transfer
Pantothenic acid	Coenzyme A	Acyl group carrier
Biotin	Biocytin	Coenzyme in CO_2 fixation
Ascorbic acid	Unknown	Hydroxylation of proline and lysine in collagen

Figure 19.7

Some enzymes require a coenzyme to facilitate the reaction.

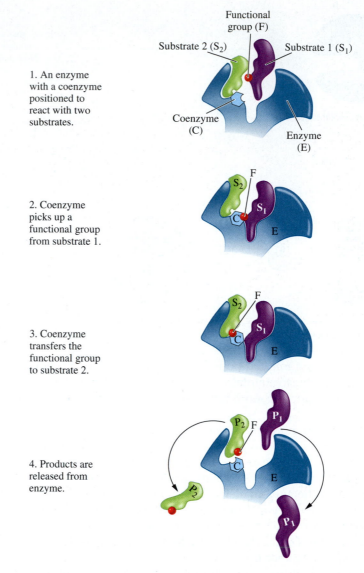

1. An enzyme with a coenzyme positioned to react with two substrates.

2. Coenzyme picks up a functional group from substrate 1.

3. Coenzyme transfers the functional group to substrate 2.

4. Products are released from enzyme.

Nicotinamide adenine dinucleotide (NAD^+), shown in Figure 19.8, is an example of a coenzyme that is of critical importance in the oxidation reactions of the cellular energy-harvesting processes. The NAD^+ molecule can accept a hydride ion, a hydrogen atom with two electrons, from the substrate of these reactions. The substrate is oxidized, and the portion of NAD^+ that is derived from the vitamin *niacin* is reduced to produce NADH. The NADH subsequently yields the hydride ion to the first acceptor in an electron transport chain. This regenerates the NAD^+ and provides electrons for the production of ATP, the chemical energy required by the cell. Also shown in Figure 19.8 is the hydride carrier $NADP^+$ and the hydrogen atom carrier FAD. Both are used in the oxidation-reduction reactions that harvest energy for the cell. Unlike NADH and $FADH_2$, NADPH serves as "reducing power" for the cell by donating hydride ions in biochemical reactions. Like NAD^+, $NADP^+$ is derived from niacin. FAD is made from the vitamin *riboflavin*.

Question *19.9*

Why does the body require the water-soluble vitamins?

Nicotinamide, derived from niacin (vitamin B$_3$)

Hydride ion (H:$^-$)

NAD$^+$
Nicotinamide adenine dinucleotide (oxidized form)

(a)

NADH
(reduced form)

NADP$^+$
Nicotinamide adenine dinucleotide phosphate (oxidized form)

(b)

FAD
Flavin adenine dinucleotide (oxidized form)

(c)

Figure 19.8

The structure of three coenzymes.
(a) The oxidized and reduced forms of nicotinamide adenine dinucleotide.
(b) The oxidized form of the closely related hydride ion carrier, nicotinamide adenine dinucleotide phosphate (NADP$^+$), which accepts hydride ions at the same position as NAD$^+$ (colored arrow). (c) The oxidized form of flavin adenine dinucleotide (FAD) accepts hydrogen atoms at the positions indicated by the colored arrows.

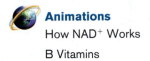

Animations

How NAD$^+$ Works

B Vitamins

Question 19.10

What are the coenzymes formed from each of the following vitamins? What are the functions of each of these coenzymes?

  a. Pantothenic acid
  b. Niacin
  c. Riboflavin

19.8 Environmental Effects

Effect of pH

LEARNING GOAL

8 Explain how pH and temperature affect the rate of an enzyme-catalyzed reaction.

Most enzymes are active only within a very narrow pH range. The cytoplasm of the cell has a pH of 7, and most enzymes function best at this pH. A plot of reaction rate versus pH for a typical enzyme is shown in Figure 19.9.

The pH at which an enzyme functions optimally is called the **pH optimum.** Making the solution more basic or more acidic sharply decreases the rate of the reaction. These pH changes alter the degree of ionization of amino acid R groups in the protein, as well as the extent to which they can hydrogen bond. This causes the enzyme to lose its biologically active configuration; it becomes *denatured.* Less drastic changes in the R groups of an enzyme active site can also destroy the ability to form the enzyme-substrate complex.

Some environments within the body must function at a pH far from 7. For instance, the pH of the stomach is approximately 2 as a result of the secretion of hydrochloric acid by cells of the stomach lining. The proteolytic digestive enzyme *pepsin* must effectively degrade proteins at this extreme pH. In the case of pepsin the enzyme has evolved an amino acid sequence that can maintain a stable tertiary structure at pH 2 and is most active in the hydrolysis of peptides that have been denatured by very low pH. Thus pepsin has a pH optimum of 2.

In a similar fashion, another proteolytic enzyme, *trypsin,* functions under the conditions of higher pH found in the intestine. Both pepsin and trypsin cleave peptide bonds by virtually identical mechanisms, yet their amino acid sequences have evolved so that they are stable and active in very different environments.

The body has used adaptation of enzymes to different environments to protect itself against one of its own destructive defense mechanisms. Within the cytoplasm of a cell are organelles called *lysosomes.* Christian de Duve, who discovered lyosomes in 1956, called them "suicide bags" because they are membrane-bound vesicles containing about fifty different kinds of hydrolases which degrade large biological molecules into small molecules that are useful for energy-harvesting reactions. For instance, some of the enzymes in the lysosomes can degrade proteins to amino acids, others hydrolyze polysaccharides into monosaccharides, still others degrade lipids and nucleic acids.

If the hydrolytic enzymes of the lysosome were accidentally released into the cytoplasm of the cell, the result would be the destruction of cellular macromolecules and death of the cell. Because of this danger, the cell invests a great deal of energy in maintaining the integrity of the lysosomal membranes. An additional protective mechanism relies on the fact that lysosomal enzymes function optimally at an acid pH (pH 4.8). Should some of these enzymes leak out of the lysosome or should a lysosome accidentally rupture, the cytoplasmic pH of 7.0–7.3 renders them inactive.

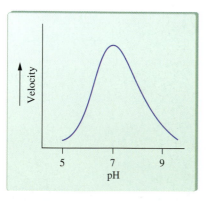

Figure 19.9

Effect of pH on the rate of an enzyme-catalyzed reaction. This enzyme functions most efficiently at pH 7. The rate of the reaction falls rapidly as the solution is made either more acidic or more basic.

Effect of Temperature

Enzymes are rapidly destroyed if the temperature of the solution rises much above 37°C, but they remain stable at much lower temperatures. This is why enzymes used for clinical assays are stored in refrigerators or freezers before use. Figure 19.10

A Medical Perspective

α_1-Antitrypsin and Familial Emphysema

Nearly two million people in the United States suffer from emphysema. Emphysema is a respiratory disease caused by destruction of the alveoli, the tiny, elastic air sacs of the lung. This damage results from the irreversible destruction of a protein called elastin, which is needed for the strength and flexibility of the walls of the alveoli. When elastin is destroyed, the small air passages in the lungs, called bronchioles, become narrower or may even collapse. This severely limits the flow of air into and out of the lung, causing respiratory distress, and in extreme conditions, death.

Some people have a genetic predisposition to emphysema. This is called familial emphysema. These individuals have a genetic defect in the gene that encodes the human plasma protein α_1-antitrypsin. As the name suggests, α_1-antitrypsin is an inhibitor of the proteolytic enzyme trypsin. But, as we have seen in this chapter, trypsin is just one member of a large family of proteolytic enzymes called the serine proteases. In the case of the α_1-antitrypsin activity in the lung, it is the inhibition of the enzyme elastase that is the critical event.

Elastase damages or destroys elastin, which in turn promotes the development of emphysema. People with normal levels of α_1-antitrypsin are protected from familial emphysema because their α_1-antitrypsin inhibits elastase and, thus, protects the elastin. The result is healthy alveoli in the lungs. However, individuals with a genetic predisposition to emphysema have very low levels of α_1-antitrypsin. This is due to a mutation that causes a single amino acid substitution in the protein chain. Because elastase in the lungs is not effectively controlled, severe lung damage characteristic of emphysema occurs.

Emphysema is also caused by cigarette smoking. Is there a link between these two forms of emphysema? The answer is yes; research has revealed that components of cigarette smoke cause the oxidation of a methionine near the amino terminus of α_1-antitrypsin. This chemical damage destroys α_1-antitrypsin activity. There are enzymes in the lung that reduce the methionine, converting it back to its original chemical form and restoring α_1-antitrypsin activity. However, it is obvious that over a long period, smoking seriously reduces the level of α_1-antitrypsin activity. The accumulated lung damage results in emphysema in many chronic smokers.

At the current time the standard treatment of emphysema is the use of inhaled oxygen. Studies have shown that intravenous infusion of α_1-antitrypsin isolated from human blood is both safe and effective. However, the level of α_1-antitrypsin in the blood must be maintained by repeated administration.

The α_1-antitrypsin gene has been cloned. In experiments with sheep it was shown that the protein remains stable when administered as an aerosol. It is still functional after it has passed through the pulmonary epithelium. This research offers hope of an effective treatment for this frightful disease.

For Further Understanding

Draw the structure of methionine and write an equation showing the reversible oxidation of this amino acid.

Develop a hypothesis to explain why an excess of elastase causes emphysema. What is the role of elastase in this disease?

---

shows the effects of temperature on enzyme-catalyzed and uncatalyzed reactions. The rate of the uncatalyzed reaction steadily increases with increasing temperature because more collisions occur with sufficient energy to overcome the energy barrier for the reaction. The rate of an enzyme-catalyzed reaction also increases with modest increases in temperature because there are increasing numbers

(a)

(b)

Figure 19.10

Effect of temperature on (a) uncatalyzed reactions and (b) enzyme-catalyzed reactions.

Animations

Protein Denaturation

Lysosomes

of collisions between the enzyme and the substrate. At the **temperature optimum,** the enzyme is functioning optimally and the rate of the reaction is maximal. Above the temperature optimum, increasing temperature begins to increase the vibrational energy of the bonds within the enzyme. Eventually, so many bonds and weak interactions are disrupted that the enzyme becomes denatured, and the reaction stops.

Because heating enzymes and other proteins destroys their three-dimensional structure, and hence their activity, a cell cannot survive very high temperatures. Thus, heat is an effective means of sterilizing medical instruments and solutions for transfusion or clinical tests. Although instruments can be sterilized by dry heat (160°C) applied for at least two hours in a dry air oven, autoclaving is a quicker, more reliable procedure. The autoclave works on the principle of the pressure cooker. Air is pumped out of the chamber, and steam under pressure is pumped into the chamber until a pressure of two atmospheres is achieved. The pressure causes the temperature of the steam, which would be 100°C at atmospheric pressure, to rise to 121°C. Within twenty minutes, all the bacteria and viruses are killed. This is the most effective means of destroying the very heat-resistant endospores that are formed by many bacteria of clinical interest. These bacteria include the genera *Bacillus* and *Clostridium,* which are responsible for such unpleasant and deadly diseases as anthrax, gas gangrene, tetanus, and botulism food poisoning.

However, not all enzymes are inactivated by heating, even to rather high temperatures. Certain bacteria live in such out-of-the-way places as coal slag heaps, which are actually burning. Others live in deep vents on the ocean floor where temperatures and pressures are extremely high. Still others grow in the hot springs of Yellowstone National Park, where they thrive at temperatures near the boiling point of water. These organisms, along with their enzymes, survive under such incredible conditions because the amino acid sequences of their proteins dictate structures that are stable at such seemingly impossible temperature extremes.

See also the Chemistry Connection: Super Hot Enzymes and the Origin of Life at the beginning of this chapter.

Question 19.11

How does a decrease in pH alter the activity of an enzyme?

Question 19.12

Heating is an effective mechanism for killing bacteria on surgical instruments. How does elevated temperature result in cellular death?

19.9 Regulation of Enzyme Activity

LEARNING GOAL

9 Describe the mechanisms used by cells to regulate enzyme activity.

Enzyme activity is often regulated by the cell. Often the reason for this is to conserve energy. If the cell runs out of chemical energy, it will die; therefore many mechanisms exist to conserve cellular energy. For instance, it is a great waste of energy to produce an enzyme if the substrate is not available. Similarly, if the product of an enzyme-catalyzed reaction is present in excess, it is a waste of energy for the enzyme to continue producing more of the unwanted product.

The simplest mechanism of enzyme regulation is to produce the enzyme only when the substrate is present. This mechanism is used by bacteria to regulate the enzymes needed to break down various sugars to yield ATP for cellular work. The bacteria have no control over their environment or over what food sources, if any, might be available. It would be an enormous waste of energy to produce all of the enzymes that are needed to break down all the possible sugars. Thus the bacteria save energy by producing the enzymes only when a specific sugar substrate is

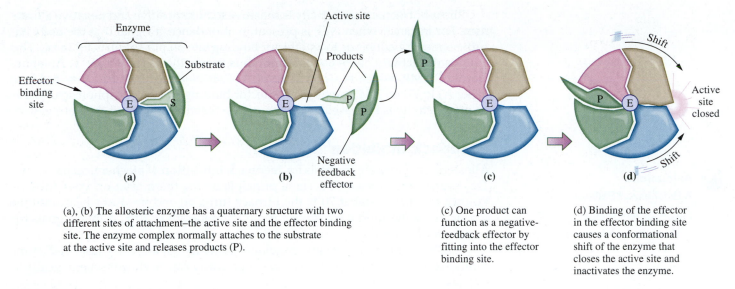

(a), (b) The allosteric enzyme has a quaternary structure with two different sites of attachment–the active site and the effector binding site. The enzyme complex normally attaches to the substrate at the active site and releases products (P).

(c) One product can function as a negative-feedback effector by fitting into the effector binding site.

(d) Binding of the effector in the effector binding site causes a conformational shift of the enzyme that closes the active site and inactivates the enzyme.

Figure 19.11
A mechanism of negative allosterism. This is an example of feedback inhibition.

available. Other mechanisms for regulating enzyme activity include use of allosteric enzymes, feedback inhibition, production of proenzymes, and protein modification. Let's take a look at these regulatory mechanisms in some detail.

Allosteric Enzymes

One type of enzyme regulation involves enzymes that have more than a single binding site. These enzymes, called **allosteric enzymes,** have active sites that can be altered by binding of small molecules called *effector molecules.* As shown in Figure 19.11, the effector binding alters the shape of the active site of the enzyme. The result can be to convert the active site to an inactive configuration, **negative allosterism,** or to convert the active site to the active configuration, **positive allosterism.** In either case, binding of the effector molecule regulates enzyme activity by determining whether it will be active or inactive.

Allosteric means "other forms."

In upcoming chapters we will study metabolic pathways. A metabolic pathway is a series of biochemical reactions that breaks down or synthesizes one or more biological molecules. One of these is glycolysis, which is the first stage of the breakdown of carbohydrates to produce ATP energy for the cell. This pathway must be responsive to the demands of the body. When more energy is required, the reactions of the pathway should occur more quickly, producing more ATP. However, if the energy demand is low, the reactions should slow down.

The third reaction in glycolysis is the transfer of a phosphoryl group from an ATP molecule to a molecule of fructose-6-phosphate. This reaction, shown here, is catalyzed by an enzyme called *phosphofructokinase:*

Fructose-6-phosphate + ATP →(Phosphofructokinase) Fructose-1,6-bisphosphate + ADP

Phosphofructokinase activity is sensitive to both positive and negative allosterism. For instance, when ATP is present in abundance, a signal that the body has sufficient energy, it binds to an effector binding site on phosphofructokinase. This inhibits the activity of the enzyme and, thus, slows the entire pathway. An abundance of AMP, which is a precursor of ATP, is evidence that the body needs to make ATP. When AMP binds to an effector binding site on phosphofructokinase, enzyme activity is increased, speeding up the reaction and the entire pathway.

Feedback Inhibition

Allosteric enzymes are the basis for **feedback inhibition** of biochemical pathways. This system functions on the same principle as the thermostat on your furnace. You set the thermostat at 70°F; the furnace turns on and produces heat until the sensor in the thermostat registers a room temperature of 70°F. It then signals the furnace to shut off.

Feedback inhibition usually regulates pathways of enzymes involved in the synthesis of a biological molecule. Such a pathway can be shown schematically as follows:

$$A \xrightarrow{E_1} B \xrightarrow{E_2} C \xrightarrow{E_3} D \xrightarrow{E_4} E \xrightarrow{E_5} F$$

In this pathway the starting material, A, is converted to B by the enzyme E_1. Enzyme E_2 immediately converts B to C, and so on until the final product, F, has been synthesized. If F is no longer needed, it is a waste of cellular energy to continue to produce it.

To avoid this waste of energy, the cell uses feedback inhibition, in which the product can shut off the entire pathway for its own synthesis. This is the result of the fact that the product, F, acts as a negative allosteric effector on one of the early enzymes of the pathway. For instance, enzyme E_1 may have an effector-binding site for F. When F is present in excess, it binds to the effector-binding site, causing the active site to close so that it cannot bind to substrate A. Thus A is not converted to B. If no B is produced, there is no substrate for enzyme E_2, and the entire pathway ceases to operate. The product, F, has turned off all the steps involved in its own synthesis, just as the heat produced by the furnace is ultimately responsible for turning off the furnace itself.

When the concentration of F drops, it will dissociate from the effector binding site. When this occurs, the enzyme is once again active. Thus, feedback inhibition is an effective metabolic on-off switch.

Proenzymes

Another means of regulating enzyme activity involves the production of the enzyme in an inactive form called a **proenzyme.** The proenzyme is converted by proteolysis (hydrolysis of the protein) to the active form when it has reached the site of its activity. On first examination it seems wasteful to add a step to the synthesis of an enzyme. But consider for a moment the very destructive nature of some of the enzymes that are necessary for life. The enzymes pepsin, trypsin, and chymotrypsin are all proteolytic enzymes of the digestive tract. They are necessary to life because they degrade dietary proteins into amino acids that are used by the cell. But what would happen to the cells that produce these enzymes if they were synthesized in active form? Those cells would be destroyed. Thus the cells of the stomach that produce pepsin actually produce an inactive proenzyme, called *pepsinogen.* Pepsinogen has an additional forty-two amino acids. In the presence of stomach acid and previously activated pepsin, the extra forty-two amino acids are cleaved off, and the proenzyme is transformed into the active enzyme. Table 19.2 lists several other proenzymes and the enzymes that convert them to active form.

Animations

A Biochemical Pathway

Feedback Inhibition of
   Biochemical Pathways

TABLE 19.2	Proenzymes of the Digestive Tract	
Proenzyme	**Activator**	**Enzyme**
Proelastase	Trypsin	Elastase
Trypsinogen	Trypsin	Trypsin
Chymotrypsinogen A	Trypsin + chymotrypsin	Chymotrypsin
Pepsinogen	Acid pH + pepsin	Pepsin
Procarboxypeptidases	Trypsin	Carboxypeptidase A, carboxypeptidase B

Protein Modification

Protein modification is another mechanism that the cell can use to turn an enzyme on or off. This is a process in which a chemical group is covalently added to or removed from the protein. This covalent modification either activates the enzyme or turns it off.

The most common type of protein modification is phosphorylation or dephosphorylation of an enzyme. Typically, the phosphoryl group is added to (or removed from) the R group of a serine, tyrosine, or threonine in the protein chain of the enzyme. Notice that these three amino acids have a free —OH in their R group, which serves as the site for the addition of the phosphoryl group.

The covalent modification of an enzyme's structure is catalyzed by other enzymes. *Protein kinases* add phosphoryl groups to a target enzyme, while *phosphatases* remove them. For some enzymes it is the phosphorylated form that is active. For instance, in adipose tissue, phosphorylation activates the enzyme triacylglycerol lipase, an enzyme that breaks triglycerides down to fatty acids and glycerol. Glycogen phosphorylase, an enzyme involved in the breakdown of glycogen, is also activated by the addition of a phosphoryl group. However, for some enzymes phosphorylation inactivates the enzyme. This is true for glycogen synthase, an enzyme involved in the synthesis of glycogen. When this enzyme is phosphorylated, it becomes inactive.

The convenient aspect of this type of regulation is the reversibility. An enzyme can quickly be turned on or off in response to environmental or physiological conditions.

19.10 Inhibition of Enzyme Activity

Many chemicals can bind to enzymes and either eliminate or drastically reduce their catalytic ability. These chemicals, called *enzyme inhibitors*, have been used for hundreds of years. When she poisoned her victims with arsenic, Lucretia Borgia was unaware that it was binding to the thiol groups of cysteine amino acids in the proteins of her victims and thus interfering with the formation of disulfide bonds needed to stabilize the tertiary structure of enzymes. However, she was well aware of the deadly toxicity of heavy metal salts like arsenic and mercury. When you take penicillin for a bacterial infection, you are taking another enzyme inhibitor. Penicillin inhibits several enzymes that are involved in the synthesis of bacterial cell walls.

Enzyme inhibitors are classified on the basis of whether the inhibition is reversible or irreversible, competitive or noncompetitive. Reversibility deals with whether the inhibitor will eventually dissociate from the enzyme, releasing it in the active form. Competition refers to whether the inhibitor is a structural analog, or look-alike, of the natural substrate. If so, the inhibitor and substrate will compete for the enzyme active site.

LEARNING GOAL
10 Discuss the mechanisms by which certain chemicals inhibit enzyme activity.

A Medical Perspective

Enzymes, Nerve Transmission, and Nerve Agents

The transmission of nerve impulses at the *neuromuscular junction* involves many steps, one of which is the activity of a critical enzyme, called *acetylcholinesterase*, which catalyzes the hydrolysis of the chemical messenger, *acetylcholine*, that initiated the nerve impulse. The need for this enzyme activity becomes clear when we consider the events that begin with a message from the nerve cell and end in the appropriate response by the muscle cell. Acetylcholine is a *neurotransmitter*, that is, a chemical messenger that transmits a message from the nerve cell to the muscle cell. Acetylcholine is stored in membrane-bound bags, called *synaptic vesicles*, in the nerve cell ending.

Acetylcholinesterase comes into play in the following way. The arrival of a nerve impulse at the end plate of the nerve axon causes an influx of Ca^{2+}. This causes the acetylcholine-containing vesicles to migrate to the nerve cell membrane that is in contact with the muscle cell. This is called the *presynaptic membrane*. The vesicles fuse with the presynaptic membrane and release the neurotransmitter. The acetylcholine then diffuses across the *nerve synapse* (the space between the nerve and muscle cells) and binds to the acetylcholine receptor protein in the *postsynaptic membrane* of the muscle cell. This receptor then opens pores in the membrane through which Na^+ and K^+ ions flow into and out of the cell, respectively. This generates the nerve impulse and causes the muscle to contract. If acetylcholine remains at the neuromuscular junction, it will continue to stimulate the muscle contraction. To stop this continued stimulation, acetylcholine is hydrolyzed, and hence, destroyed by acetylcholinesterase. When this happens, nerve stimulation ceases.

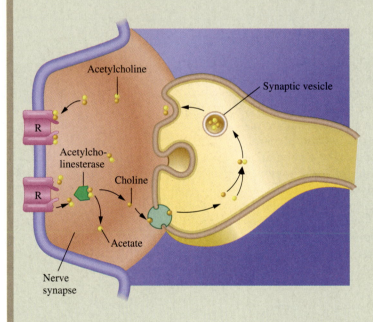

Schematic diagram of the synapse at the neuromuscular junction. The nerve impulse causes acetylcholine to be released from synaptic vesicles. Acetylcholine diffuses across the synaptic cleft and binds to a specific receptor protein (R) on the postsynaptic membrane. A channel opens that allows Na^+ ions to flow into the cell and K^+ ions to flow out of the cell. This results in muscle contraction. Any acetylcholine remaining in the synaptic cleft is destroyed by acetylcholinesterase to terminate the stimulation of the muscle cell.

$$H_3C-\overset{O}{\overset{\|}{C}}-O-CH_2CH_2-N^+-(CH_3)_3 + H_2O$$

Acetylcholine

Acetylcholinesterase

$$H_3C-\overset{O}{\overset{\|}{C}}_{O^-} + HO-CH_2CH_2-N^+-(CH_3)_3 + H^+$$

Acetate   Choline

Inhibitors of acetylcholinesterase are used both as poisons and as drugs. Among the most important inhibitors of acetylcholinesterase are a class of compounds known as *organophosphates*. One of these is the nerve agent Sarin (isopropyl-methylfluorophosphate). Sarin forms a covalently bonded

Irreversible Inhibitors

Irreversible enzyme inhibitors, such as arsenic, usually bind very tightly, sometimes even covalently, to the enzyme. This generally involves binding of the inhibitor to one of the R groups of an amino acid in the active site. Inhibitor binding may block the active site binding groups so that the enzyme-substrate complex cannot form. Alternatively, an inhibitor may interfere with the catalytic groups of the intermediate with the active site of acetylcholinesterase. Thus, it acts as an irreversible, noncompetitive inhibitor.

Sarin

HF

Sarin is covalently bonded to the serine in the active site.

Pyridine aldoxime methiodide (PAM)

Serine in the acetylcholinesterase active site

Sarin is covalently bonded to the serine in the active site.

Complex formed between sarin and PAM

Regenerated enzyme

The covalent intermediate is stable, and acetylcholinesterase is therefore inactive, no longer able to break down acetylcholine. Nerve transmission continues, resulting in muscle spasm. Death may occur as a result of laryngeal spasm. Antidotes for poisoning by organophosphates, which include many insecticides and nerve gases, have been developed. The antidotes work by reversing the effects of the inhibitor. One of these antidotes is known as *PAM*, an acronym for *pyridine aldoxime methiodide*. This molecule displaces the organophosphate group from the active site of the enzyme, alleviating the effects of the poison.

For Further Understanding

Botulinum toxin inhibits release of neurotransmitters from the presynaptic membrane. What symptoms do you predict would result from this?

Why must Na^+ and K^+ enter and exit the cell through a protein channel?

active site, thereby effectively eliminating catalysis. Irreversible inhibitors, which include snake venoms and nerve gases, generally inhibit many different enzymes.

Reversible, Competitive Inhibitors

Reversible, competitive enzyme inhibitors are often referred to as **structural analogs,** that is, they are molecules that resemble the structure and charge

See A Medical Perspective: Fooling the AIDS Virus with "Look-Alike" Nucleotides in Chapter 20.

Figure 19.12
Competitive inhibition.

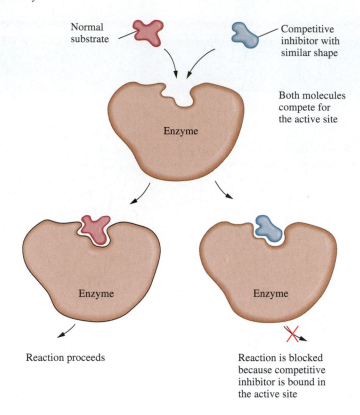

distribution of the natural substrate for a particular enzyme. Because of this resemblance, the inhibitor can occupy the enzyme active site. However, no reaction can occur, and enzyme activity is inhibited (Figure 19.12). This inhibition is competitive because the inhibitor and the substrate compete for binding to the enzyme active site. Thus, the degree of inhibition depends on their relative concentrations. If the inhibitor is in excess or binds more strongly to the active site, it will occupy the active site more frequently, and enzyme activity will be greatly decreased. On the other hand, if the natural substrate is present in excess, it will more frequently occupy the active site, and there will be little inhibition.

The sulfa drugs, the first antimicrobics to be discovered, are **competitive inhibitors** of a bacterial enzyme needed for the synthesis of the vitamin folic acid. *Folic acid* is a vitamin required for the transfer of methyl groups in the biosynthesis of methionine and the nitrogenous bases required to make DNA and RNA. Humans cannot synthesize folic acid and must obtain it from the diet. Bacteria, on the other hand, must make folic acid because they cannot take it in from the environment.

para-Aminobenzoic acid (PABA) is the substrate for an early step in folic acid synthesis. The sulfa drugs, the prototype of which was discovered in the 1930s by Gerhard Domagk, are structural analogs of PABA and thus competitive inhibitors of the enzyme that uses PABA as its normal substrate.

$$H_2N- \bigcirc -\overset{\overset{\textstyle O}{\|}}{C}-OH \qquad\qquad H_2N- \bigcirc -\overset{\overset{\textstyle O}{\|}}{\underset{\underset{\textstyle O}{\|}}{S}}-NH_2$$

　　　　p-Aminobenzoic acid　　　　　　　　　Sulfanilamide

If the correct substrate (PABA) is bound by the enzyme, the reaction occurs, and the bacterium lives. However, if the sulfa drug is present in excess over PABA, it binds more frequently to the active site of the enzyme. No folic acid will be produced, and the bacterial cell will die.

Because we obtain our folic acid from our diets, sulfa drugs do not harm us. However, bacteria are selectively killed. Luckily, we can capitalize on this property for the treatment of bacterial infections, and as a result, sulfa drugs have saved countless lives. Although bacterial infection was the major cause of death before the discovery of sulfa drugs and other antibiotics, death caused by bacterial infection is relatively rare at present.

Reversible, Noncompetitive Inhibitors

Reversible, noncompetitive enzyme inhibitors bind to R groups of amino acids or perhaps to the metal ion cofactors. Unlike the situation of irreversible inhibition, however, the binding is weak, and the enzyme activity is restored when the inhibitor dissociates from the enzyme-inhibitor complex. Binding of these inhibitors modifies the shape of the active site in much the same way that the binding of an allosteric effector does. Since this binding is nonspecific, these inhibitors inactivate a broad range of enzymes.

Why are irreversible inhibitors considered to be poisons?

Question 19.13

Explain the difference between an irreversible inhibitor and a reversible, noncompetitive inhibitor.

Question 19.14

What is a structural analog?

Question 19.15

How can structural analogs serve as enzyme inhibitors?

Question 19.16

19.11 Proteolytic Enzymes

Proteolytic enzymes break the peptide bonds that maintain the primary protein structure. *Chymotrypsin,* for example, is an enzyme that hydrolyzes dietary proteins in the small intestine. It acts specifically at peptide bonds on the carbonyl side of the peptide bond. The C-terminal amino acids of the peptides released by bond cleavage are methionine, tyrosine, tryptophan, and phenylalanine. The specificity of chymotrypsin depends upon the presence of a *hydrophobic pocket,* a cluster of hydrophobic amino acids brought together by the three-dimensional folding of the protein chain. The flat aromatic side chains of certain amino acids (tyrosine, tryptophan, phenylalanine) slide into this pocket, providing the binding specificity required for catalysis (Figure 19.13).

How can we determine which bond is cleaved by a protease such as chymotrypsin? To know which bond is cleaved, we must write out the sequence of amino acids in the region of the peptide that is being cleaved. This can be determined experimentally by amino acid sequencing techniques. Remember that the N-terminal amino acid is written to the left and the C-terminal amino acid to the right. Consider a protein having within it the sequence —Ala-Phe-Gly—. A reaction is set up in which the enzyme, chymotrypsin, is mixed with the protein substrate. After the reaction has occurred, the products are purified, and their amino acid sequences are determined. Experiments of this sort show that chymotrypsin

LEARNING GOAL
11 Discuss the role of the enzyme chymotrypsin and other serine proteases.

Catalytic domain

Hydrophobic pocket

Figure 19.13
The specificity of chymotrypsin is determined by a hydrophobic pocket that holds the aromatic side chain of the substrate. This brings the peptide bond to be cleaved into the catalytic domain of the active site.

These enzymes are called *serine proteases* because they have the amino acid serine in the catalytic region of the active site that is essential for hydrolysis of the peptide bond.

cleaves the bond between phenylalanine and glycine, which is the peptide bond on the carbonyl side of amino acids having an aromatic side chain.

$$H_3{}^+N-C-C-NH-C-C-NH-C-C-O^-$$

Peptide bond cleaved

Ala ———————— Phe ———————— Gly

The **pancreatic serine proteases** trypsin, chymotrypsin, and elastase all hydrolyze peptide bonds. These enzymes are the result of *divergent evolution* in which a single ancestral gene was first duplicated. Then each copy evolved individually. They have similar primary structures, similar tertiary structures, and virtually identical mechanisms of action. However, as a result of evolution, these enzymes all have different specificities:

- Chymotrypsin cleaves peptide bonds on the carbonyl side of aromatic amino acids and large, hydrophobic amino acids such as methionine.
- Trypsin cleaves peptide bonds on the carbonyl side of basic amino acids.
- Elastase cleaves peptide bonds on the carbonyl side of glycine and alanine.

These enzymes have different pockets for the side chains of their substrates; *different keys fit different locks*. This difference manifests itself in the substrate specificity alluded to above. For example, the binding pocket of trypsin is long, narrow, and negatively charged to accommodate lysine or arginine R groups. Yet although the binding pockets have undergone divergent evolution, the catalytic sites have remained unchanged, and the mechanism of proteolytic action is the same for all the serine proteases. In each case, the mechanism involves a serine R group.

Question 19.17

Draw the structural formulas of the following peptides and show which bond would be cleaved by chymotrypsin.

a. ala-phe-ala
b. tyr-ala-tyr

Question 19.18

Draw the structural formulas of the following peptides and show which bond would be cleaved by chymotrypsin.

a. trp-val-gly
b. phe-ala-pro

Question 19.19

Draw the structural formula of the peptide val-phe-ala-gly-leu. Which bond would be cleaved if this peptide were reacted with chymotrypsin? With elastase?

Question 19.20

Draw the structural formula of the peptide trp-val-lys-ala-ser. Show which bonds would be cleaved by trypsin, chymotrypsin, and elastase.

A Medical Perspective

Enzymes and Acute Myocardial Infarction

A patient is brought into the emergency room with acute, squeezing chest pains; shallow, irregular breathing; and pale, clammy skin. The immediate diagnosis is myocardial infarction, a heart attack. The first thoughts of the attending nurses and physicians concern the series of treatments and procedures that will save the patient's life. It is a short time later, when the patient's condition has stabilized, that the doctor begins to consider the battery of enzyme assays that will confirm the diagnosis.

Acute myocardial infarction (AMI) occurs when the blood supply to the heart muscle is blocked for an extended time. If this lack of blood supply, called *ischemia,* is prolonged, the myocardium suffers irreversible cell damage and muscle death, or infarction. When this happens, the concentration of cardiac enzymes in the blood rises dramatically as the dead cells release their contents into the bloodstream.

Three cardiac biomarkers have become the primary tools used to assess myocardial disease and suspected AMI. These are myoglobin, creatine kinase-MB (CK-MB), and cardiac troponin I. Of these three, only troponin is cardiac specific. In fact, it is so reliable that the American College of Cardiology has stated that any elevation of troponin is "abnormal and represents cardiac injury."

Myoglobin is the smallest of these three proteins and diffuses most rapidly through the vascular system. Thus, it is the first cardiac biomarker to appear, becoming elevated as early as 30 minutes after onset of chest pain. Myoglobin has another benefit in following a myocardial infarction. It is rapidly cleared from the body by the kidneys, returning to normal levels within 16 to 36 hours after a heart attack. If the physician sees this decline in myoglobin levels, followed by a subsequent rise, it is an indication that the patient has had a second myocardial infarction.

Creatine kinase-MB is one of the most important cardiac biomarkers, even though it is found primarily in muscle and brain. Levels typically rise 3 to 8 hours after chest pains begin. Within another 48 to 72 hours, the CK-MB levels return to normal.

As a result, like myoglobin, CK-MB can also be used to diagnose a second AMI.

The physician also has enzymes available to treat a heart attack patient. Most AMIs are the result of a *thrombus,* or clot, within a coronary blood vessel. The clot restricts blood flow to the heart muscle. One technique that shows promise for treatment following a coronary thrombosis, a heart attack caused by the formation of a clot, is destruction of the clot by intravenous or intracoronary injection of an enzyme called *streptokinase.* This enzyme, formerly purified from the pathogenic bacterium *Streptococcus pyogenes* but now available through recombinant DNA techniques, catalyzes the production of the proteolytic enzyme plasmin from its proenzyme, plasminogen. Plasmin can degrade a fibrin clot into subunits. This has the effect of dissolving the clot that is responsible for restricted blood flow to the heart, but there is an additional protective function as well. The subunits produced by plasmin degradation of fibrin clots are able to inhibit further clot formation by inhibiting thrombin.

Recombinant DNA technology has provided medical science with yet another, perhaps more promising, clot-dissolving enzyme. *Tissue-type plasminogen activator (TPA)* is a proteolytic enzyme that occurs naturally in the body as a part of the anticlotting mechanisms. TPA converts the proenzyme, plasminogen, into the active enzyme, plasmin. Injection of TPA within two hours of the initial chest pain can significantly improve the circulation to the heart and greatly improve the patient's chances of survival.

For Further Understanding

Why is myoglobin an effective biomarker to follow the status of the patient when using a thrombolytic agent such as TPA or streptokinase?

Aspirin has also been suggested as a treatment to enhance a patient's probability of surviving a heart attack. What mechanism can you devise to explain this suggestion?

19.12 Uses of Enzymes in Medicine

Analysis of blood serum for levels (concentrations) of certain enzymes can provide a wealth of information about a patient's medical condition. Often, such tests are used to confirm a preliminary diagnosis based on the disease symptoms or clinical picture.

For example, when a heart attack occurs, a lack of blood supplied to the heart muscle causes some of the heart muscle cells to die. These cells release their contents, including their enzymes, into the bloodstream. Simple tests can be done to

LEARNING GOAL

12 Provide examples of medical uses of enzymes.

measure the amounts of certain enzymes in the blood. Such tests, called *enzyme assays,* are very precise and specific because they are based on the specificity of the enzyme-substrate complex.

In the case of acute myocardial infarction, three cardiac biomarkers are particularly important. These are creatine kinase-MB, myoglobin, and troponin I (see also A Medical Perspective: Enzymes and Acute Myocardial Infarction). Elevated blood serum concentrations of the enzymes amylase and lipase are indications of pancreatitis, an inflammation of the pancreas. Liver diseases such as cirrhosis and hepatitis result in elevated levels of alanine aminotransferase/serum glutamate–pyruvate transaminase (ALT/SGPT) and aspartate aminotransferase/serum glutamate–oxaloacetate transaminase (AST/SGOT) in blood serum. In fact, these two enzymes also increase in concentration following heart attack, but the physician can differentiate between these two conditions by considering the relative increase in the two enzymes. If ALT/SGPT is elevated to a greater extent than AST/SGOT, it can be concluded that the problem is liver dysfunction.

Enzymes are also used as analytical reagents in the clinical laboratory owing to their specificity. They often selectively react with one substance of interest, producing a product that is easily measured. An example of this is the clinical analysis of urea in blood. The measurement of urea levels in blood is difficult because of the complexity of blood. However, if urea is converted to ammonia using the enzyme urease, the ammonia becomes an *indicator* of urea, because it is produced from urea, and it is easily measured. This test, called the *blood urea nitrogen (BUN) test,* is useful in the diagnosis of kidney malfunction and serves as one example of the utility of enzymes in clinical chemistry.

Enzyme replacement therapy can also be used in the treatment of certain diseases. One such disease, Gaucher's disease, is a genetic disorder resulting in a deficiency of the enzyme *glucocerebrosidase.* In the normal situation, this enzyme breaks down a glycolipid called *glucocerebroside,* which is an intermediate in the synthesis and degradation of complex glycosphingolipids found in cellular membranes. Glucocerebrosidase is found in the lysosomes, where it hydrolyzes glucocerebroside into glucose and ceramide.

See A Medical Perspective: Disorders of Sphingolipid Metabolism in Chapter 17.

Glucocerebroside

Glucocerebrosidase

Glucose   Ceramide

In Gaucher's disease, the enzyme is not present and glucocerebroside builds up in macrophages found in the liver, spleen, and bone marrow. These cells become engorged with excess lipid that cannot be metabolized and then displace healthy, normal cells in bone marrow. The symptoms of Gaucher's disease include severe anemia, thrombocytopenia (reduction in the number of platelets), and hepatosplenomegaly (enlargement of the spleen and liver). There can also be skeletal problems including bone deterioration and secondary fractures.

Recombinant DNA technology has been used by the Genzyme Corporation to produce the human lysosomal enzyme β-glucocerebrosidase. Given the trade name *Cerezyme,* the enzyme hydrolyzes glucocerebroside into glucose and ceramide so that the products can be metabolized normally. Patients receive Cerezyme intravenously over the course of one to two hours. The dosage and treatment schedule can be tailored to the individual. The results of testing are very encouraging. Patients experience improved red blood cell and platelet counts and reduced hepatosplenomegaly.

SUMMARY

19.1 Nomenclature and Classification

Enzymes are most frequently named by using the common system of nomenclature. The names are useful because they are often derived from the name of the substrate and/or the reaction of the substrate that is catalyzed by the enzyme. Enzymes are classified according to function. The six general classes are *oxidoreductases, transferases, hydrolases, lyases, isomerases,* and *ligases.*

19.2 The Effect of Enzymes on the Activation Energy of a Reaction

Enzymes are the biological catalysts of cells. They lower the activation energies but do not alter the equilibrium constants of the reactions they catalyze.

19.3 The Effect of Substrate Concentration on Enzyme-Catalyzed Reactions

With uncatalyzed reactions, increases in substrate concentration result in an increase in reaction rate. For enzyme-catalyzed reactions, an increase in substrate concentration initially causes an increase in reaction rate, but at a particular concentration the reaction rate reaches a maximum. At this concentration all enzyme active sites are filled with substrate.

19.4 The Enzyme-Substrate Complex

Formation of an *enzyme-substrate complex* is the first step of an enzyme-catalyzed reaction. This involves the binding of the substrate to the active site of the enzyme. The lock-and-key model of substrate binding describes the enzyme as a rigid structure into which the substrate fits precisely. The newer induced fit model describes the enzyme as a flexible molecule.

The shape of the active site approximates the shape of the substrate and then "molds" itself around the substrate.

19.5 Specificity of the Enzyme-Substrate Complex

Enzymes are also classified on the basis of their specificity. The four classifications of specificity are *absolute, group, linkage,* and *stereochemical specificity.* An enzyme with *absolute specificity* catalyzes the reaction of only a single substrate. An enzyme with group specificity catalyzes reactions involving similar substrates with the same functional group. An enzyme with *linkage specificity* catalyzes reactions involving a single kind of bond. An enzyme with *stereochemical specificity* catalyzes reactions involving only one enantiomer.

19.6 The Transition State and Product Formation

An enzyme-catalyzed reaction is mediated through an unstable *transition state.* This may involve the enzyme putting "stress" on a bond, bringing reactants into close proximity and in the correct orientation, or altering the local pH.

19.7 Cofactors and Coenzymes

Cofactors are metal ions, organic compounds, or organometallic compounds that bind to an enzyme and help maintain the correct configuration of the active site. The term *coenzyme* refers specifically to an organic group that binds transiently to the enzyme during the reaction. It accepts or donates chemical groups.

19.8 Environmental Effects

Enzymes are sensitive to pH and temperature. High temperatures or extremes of pH rapidly inactivate most enzymes by denaturing them.

19.9 Regulation of Enzyme Activity

Enzymes differ from inorganic catalysts in that they are regulated by the cell. Some of the means of enzyme regulation include allosteric regulation, *feedback inhibition*, production of inactive forms, or *proenzymes*, and *protein modification*. *Allosteric enzymes* have an effector binding site, as well as an active site. Effector binding renders the enzyme active (*positive allosterism*) or inactive (*negative allosterism*). In feedback inhibition the product of a biosynthetic pathway turns off the entire pathway via negative allosterism. In protein modification, adding or removing a covalently bound group either activates or inactivates an enzyme.

19.10 Inhibition of Enzyme Activity

Enzyme activity can be destroyed by a variety of inhibitors. *Irreversible inhibitors*, or poisons, bind tightly to enzymes and destroy their activity permanently. *Competitive inhibitors* are generally *structural analogs* of the natural substrate for the enzyme. They compete with the normal substrate for binding to the active site. When the competitive inhibitor is bound by the active site, the reaction cannot occur, and no product is produced.

19.11 Proteolytic Enzymes

Proteolytic enzymes (proteases) catalyze the hydrolysis of peptide bonds. The *pancreatic serine proteases* chymotrypsin, trypsin, and elastase have similar structures and mechanisms of action, but different substrate specificities. It is thought that they evolved from a common ancestral protease.

19.12 Uses of Enzymes in Medicine

Analysis of blood serum for unusually high levels of certain enzymes provides valuable information on a patient's condition. Such analysis is used to diagnose heart attack, liver disease, and pancreatitis. Enzymes are also used as analytical reagents, as in the blood urea nitrogen (BUN) test, and in the treatment of disease.

KEY TERMS

absolute specificity (19.5)
active site (19.4)
allosteric enzyme (19.9)
apoenzyme (19.7)
coenzyme (19.7)
cofactor (19.7)
competitive inhibitor (19.10)
enzyme (Intro)
enzyme specificity (19.5)
enzyme-substrate
  complex (19.4)
feedback inhibition (19.9)

group specificity (19.5)
holoenzyme (19.7)
hydrolase (19.1)
induced fit model (19.4)
irreversible enzyme
  inhibitor (19.10)
isomerase (19.1)
ligase (19.1)
linkage specificity (19.5)
lock-and-key model (19.4)
lyase (19.1)
negative allosterism (19.9)

oxidoreductase (19.1)
pancreatic serine
  protease (19.11)
pH optimum (19.8)
positive allosterism (19.9)
product (19.2)
proenzyme (19.9)
protein modification (19.9)
proteolytic enzyme (19.11)
reversible, competitive enzyme
  inhibitor (19.10)

reversible, noncompetitive
  enzyme inhibitor (19.10)
stereochemical
  specificity (19.5)
structural analog (19.10)
substrate (19.1)
temperature optimum (19.8)
transferase (19.1)
transition state (19.6)
vitamin (19.7)

QUESTIONS AND PROBLEMS

Nomenclature and Classification

Foundations

19.21 How are the common names of enzymes often derived?

19.22 What is the most common characteristic used to classify enzymes?

Applications

19.23 Match each of the following substrates with its corresponding enzyme:

1. Urea	a. Lipase
2. Hydrogen peroxide	b. Glucose-6-phosphatase
3. Lipid	c. Peroxidase
4. Aspartic acid	d. Sucrase
5. Glucose-6-phosphate	e. Urease
6. Sucrose	f. Aspartase

19.24 Give a systematic name for the enzyme that would act on each of the following substrates:
  a. Alanine
  b. Citrate
  c. Ampicillin
  d. Ribose
  e. Methylamine

19.25 Describe the function implied by the name of each of the following enzymes:
  a. Citrate decarboxylase
  b. Adenosine diphosphate phosphorylase
  c. Oxalate reductase
  d. Nitrite oxidase
  e. *cis-trans* Isomerase

19.26 List the six classes of enzymes based on the type of reaction catalyzed. Briefly describe the function of each class, and provide an example of each.

The Effect of Enzymes on the Activation Energy of a Reaction

Foundations

19.27 Define the term *substrate*.

19.28 Define the term *product*.

Applications

19.29 What is the activation energy of a reaction?

19.30 What is the effect of an enzyme on the activation energy of a reaction?

19.31 Write and explain the equation for the equilibrium constant of an enzyme-mediated reaction. Does the enzyme alter the K_{eq}?

19.32 If an enzyme does not alter the equilibrium constant of a reaction, how does it speed up the reaction?

The Effect of Substrate Concentration on Enzyme-Catalyzed Reactions

Foundations

19.33 What is the effect of doubling the substrate concentration on the rate of a chemical reaction?

19.34 Why doesn't the rate of an enzyme-catalyzed reaction increase indefinitely when the substrate concentration is made very large?

Applications

19.35 What is meant by the term *rate limiting step*?

19.36 How does the rate limiting step influence an enzyme-catalyzed reaction?

19.37 Draw a graph that describes the effect of increasing the concentration of the substrate on the rate of an enzyme-catalyzed reaction.

19.38 What does a graph of enzyme activity versus substrate concentration tell us about the nature of enzyme-catalyzed reactions?

The Enzyme-Substrate Complex

Foundations

19.39 Define the term *enzyme-substrate complex*.

19.40 Define the term *active site*.

19.41 What are catalytic groups of an enzyme active site?

19.42 What is the binding site of an enzyme active site?

Applications

19.43 Name three major properties of enzyme active sites.

19.44 If enzyme active sites are small, why are enzymes so large?

19.45 What is the lock-and-key model of enzyme-substrate binding?

19.46 Why is the induced fit model of enzyme-substrate binding a much more accurate model than the lock-and-key model?

Specificity of the Enzyme-Substrate Complex

Foundations

19.47 Define the term *enzyme specificity*.

19.48 What region of an enzyme is responsible for its specificity?

19.49 What is meant by the term *group specificity*?

19.50 What is meant by the term *linkage specificity*?

19.51 What is meant by the term *absolute specificity*?

19.52 What is meant by the term *stereochemical specificity*?

Applications

19.53 Provide an example of an enzyme with group specificity and explain the advantage of group specificity for that particular enzyme.

19.54 Provide an example of an enzyme with linkage specificity and explain the advantage of linkage specificity for that particular enzyme.

19.55 Provide an example of an enzyme with absolute specificity and explain the advantage of absolute specificity for that particular enzyme.

19.56 Provide an example of an enzyme with stereochemical specificity and explain the advantage of stereochemical specificity for that particular enzyme.

The Transition State and Product Formation

Foundations

19.57 Outline the four general stages in an enzyme-catalyzed reaction.

19.58 Describe the transition state.

Applications

19.59 What types of transition states might be envisioned that would decrease the energy of activation of an enzyme?

19.60 If an enzyme catalyzed a reaction by modifying the local pH, what kind of amino acid R groups would you expect to find in the active site?

Cofactors and Coenzymes

Foundations

19.61 What is the role of a cofactor in enzyme activity?

19.62 How does a coenzyme function in an enzyme-catalyzed reaction?

Applications

19.63 What is the function of NAD^+? What class of enzymes would require a coenzyme of this sort?

19.64 What is the function of FAD? What class of enzymes would require this coenzyme?

Environmental Effects

Foundations

19.65 List the factors that affect enzyme activity.

19.66 Define the optimum pH for enzyme activity.

19.67 How will each of the following changes in conditions alter the rate of an enzyme-catalyzed reaction?
  a. Decreasing the temperature from 37°C to 10°C
  b. Increasing the pH of the solution from 7 to 11
  c. Heating the enzyme from 37°C to 100°C

19.68 Why does an enzyme lose activity when the pH is drastically changed from optimum pH?

Applications

19.69 High temperature is an effective mechanism for killing bacteria on surgical instruments. How does high temperature result in cellular death?

19.70 An increase in temperature will increase the rate of a reaction if a nonenzymatic catalyst is used; however, an increase in temperature will eventually *decrease* the rate of a reaction when an enzyme catalyst is used. Explain the apparent contradiction of these two statements.

19.71 What is a lysosome?

19.72 Of what significance is it that lysosomal enzymes have a pH optimum of 4.8?

19.73 Why are enzymes that are used for clinical assays in hospitals stored in refrigerators?

19.74 Why do extremes of pH inactivate enzymes?

Regulation of Enzyme Activity

Foundations

19.75 a. Why is it important for cells to regulate the level of enzyme activity?
  b. Why must synthesis of digestive enzymes be carefully controlled?

19.76 What is an allosteric enzyme?

19.77 What is the difference between positive and negative allosterism?

19.78 a. Define feedback inhibition.
  b. Describe the role of allosteric enzymes in feedback inhibition.
  c. Is this positive or negative allosterism?

19.79 What is a proenzyme?

19.80 Three proenzymes that are involved in digestion of proteins in the stomach and intestines are pepsinogen, chymotrypsinogen, and trypsinogen. What is the advantage of producing these enzymes as inactive peptides?

Applications

19.81 The blood clotting mechanism consists of a set of proenzymes that act in a cascade that results in formation of a blood clot. Develop a hypothesis to explain the value of this mechanism for blood clotting.

19.82 What is the benefit for an enzyme such as triacylglycerol lipase to be regulated by covalent modification, in this case phosphorylation?

Inhibition of Enzyme Activity

Foundations

19.83 Define *competitive enzyme inhibition.*

19.84 How do the sulfa drugs selectively kill bacteria while causing no harm to humans?

19.85 Describe the structure of a structural analog.

19.86 How can structural analogs serve as enzyme inhibitors?

19.87 Define *irreversible enzyme inhibition.*

19.88 Why are irreversible enzyme inhibitors often called *poisons?*

Applications

19.89 Suppose that a certain drug company manufactured a compound that had nearly the same structure as a substrate for a certain enzyme but that could not be acted upon chemically by the enzyme. What type of interaction would the compound have with the enzyme?

19.90 The addition of phenylthiourea to a preparation of the enzyme polyphenoloxidase completely inhibits the activity of the enzyme.
  a. Knowing that phenylthiourea binds all copper ions, what conclusion can you draw about whether polyphenoloxidase requires a cofactor?
  b. What kind of inhibitor is phenylthiourea?

Proteolytic Enzymes

Applications

19.91 What do the similar structures of chymotrypsin, trypsin, and elastase suggest about their evolutionary relationship?

19.92 What properties are shared by chymotrypsin, trypsin, and elastase?

19.93 Draw the complete structural formula for the peptide tyr-lys-ala-phe. Show which bond would be broken when this peptide is reacted with chymotrypsin.

19.94 Repeat Question 19.93 for the peptide trp-pro-gly-tyr.

19.95 The sequence of a peptide that contains ten amino acids is as follows:

  ala-gly-val-leu-trp-lys-ser-phe-arg-pro

Which peptide bond(s) are cleaved by elastase, trypsin, and chymotrypsin?

19.96 What structural features of trypsin, chymotrypsin, and elastase account for their different specificities?

Uses of Enzymes in Medicine

19.97 List the enzymes whose levels are elevated in blood serum following a myocardial infarction.

19.98 List the enzymes whose levels are elevated as a result of hepatitis or cirrhosis of the liver.

CRITICAL THINKING PROBLEMS

1. Ethylene glycol is a poison that causes about fifty deaths a year in the United States. Treating people who have drunk ethylene glycol with massive doses of ethanol can save their lives. Suggest a reason for the effect of ethanol.

2. Generally speaking, feedback inhibition involves regulation of the first step in a pathway. Consider the following hypothetical pathway:

$$
\begin{array}{c}
C \\
\downarrow E_2 \\
A \xrightarrow{E_1} B \xrightarrow{E_4} E \xrightarrow{E_5} F \xrightarrow{E_6} G \\
\uparrow E_3 \\
D
\end{array}
$$

   Which step in this pathway do you think should be regulated? Explain your reasoning.

3. In an amplification cascade, each step greatly increases the amount of substrate available for the next step, so that a very large amount of the final product is made. Consider the following hypothetical amplification cascade:

$$
\begin{array}{c}
A_{active} \\
\downarrow \\
B_{inactive} \longrightarrow B_{active} \\
\downarrow \\
C_{inactive} \longrightarrow C_{active} \\
\downarrow \\
D_{inactive} \longrightarrow D_{active}
\end{array}
$$

   If each active enzyme in the pathway converts one hundred molecules of its substrate to active form, how many molecules of D will be produced if the pathway begins with one molecule of A?

4. L-1-(*p*-toluenesulfonyl)-amido-2-phenylethylchloromethyl ketone (TPCK, shown below) inhibits chymotrypsin, but not trypsin. Propose a hypothesis to explain this observation.

5. A graduate student is trying to make a "map" of a short peptide so that she can eventually determine the amino acid sequence. She digested the peptide with several proteases and determined the sizes of the resultant digestion products.

Enzyme	M.W. of Digestion Products
Trypsin	2000, 3000
Chymotrypsin	500, 1000, 3500
Elastase	500, 1000, 1500, 2000

Suggest experiments that would allow the student to map the order of the enzyme digestion sites along the peptide.

20 Introduction to Molecular Genetics

Learning Goals

1 Draw the general structure of DNA and RNA nucleotides.

2 Describe the structure of DNA and compare it with RNA.

3 Explain DNA replication.

4 List three classes of RNA molecules and describe their functions.

5 Explain the process of transcription.

6 List and explain the three types of post-transcriptional modifications of eukaryotic mRNA.

7 Describe the essential elements of the genetic code, and develop a "feel" for its elegance.

8 Describe the process of translation.

9 Define mutation and understand how mutations cause cancer and cell death.

10 Describe the tools used in the study of DNA and in genetic engineering.

11 Describe the process of polymerase chain reaction and discuss potential uses of the process.

12 Discuss strategies for genome analysis and DNA sequencing.

These twin girls are identical. Explain why this is so.

Introduction

Look around at the students in your chemistry class. They all share many traits: upright stance, a head with two eyes, a nose, and a mouth facing forward, one ear on each side of the head, and so on. You would have no difficulty listing the similarities that define you and your classmates as Homo sapiens.

As you look more closely at the individuals you begin to notice many differences. Eye color, hair color, skin color, the shape of the nose, height, body build: all these traits, and many more, show amazing variety from one person to the next. Even within one family, in which the similarities may be more pronounced, each individual has a unique appearance. In fact, only identical twins look exactly alike—well, most of the time.

The molecule responsible for all these similarities and differences is deoxyribonucleic acid (DNA). Tightly wound up in structures called chromosomes in the nucleus of the cell, DNA carries the genetic code to produce the thousands of different proteins that make us who we are. These proteins include enzymes that are responsible for production of the pigment melanin. The more melanin we are genetically programmed to make, the darker our hair, eyes, and skin will be. Others are structural proteins. The gene for α-keratin that makes up hair determines whether our hair will be wavy, straight, or curly. Thousands of genes carry the genetic information for thousands of proteins that dictate our form and, some believe, our behavior.

Chemistry Connection

Molecular Genetics and Detection of Human Genetic Disease

It is estimated that 3–5% of the human population suffers from a serious genetic defect. That's 200 million people! But what if genetic disease could be detected and "cured"? Two new technologies, *gene therapy* and *preimplantation diagnosis,* may help us realize this dream.

For a couple with a history of a genetic disease in the family, pregnancy is a time of anxiety. Through *genetic counseling* these couples can learn the probability that their child has the disease. For about 200 genetic diseases the uncertainty can be eliminated. *Amniocentesis* (removal of 10–20 mL of fluid from the sac around the fetus) and *chorionic villus sampling* (removal of cells from a fetal membrane) are two procedures that are used to obtain fetal cells for genetic testing. Fetal cells are cultured and tested by enzyme assays and DNA tests to look for genetic diseases. If a genetic disease is diagnosed, the parents must make a difficult decision: to abort the fetus or to carry the child to term and deal with the effects of the genetic disease.

The power of modern molecular genetics is obvious in our ability to find a "bad" gene from just a few cells. But scientists have developed an even more impressive way to test for genetic disease before the embryo implants into the uterine lining. This technique, called *preimplantation diagnosis,* involves fertilizing a human egg and allowing the resulting zygote to divide in a sterile petri dish. When the zygote consists of 8–16 cells, *one* cell is removed for genetic testing. Only genetically normal embryos are implanted in the mother. Thus, the genetic diseases that we can detect could be eliminated from the population by preimplantation diagnosis because only a zygote with "good" genes is used.

Gene therapy is a second way in which genetic diseases may one day be eliminated. Foreign genes, including growth hormone, have been introduced into fertilized mouse eggs and the zygotes implanted in female mice. The baby mice born with the foreign growth hormone gene were about three times larger than their normal littermates! One day, this kind of technology may be used to introduce normal genes into human fertilized eggs carrying a defective gene, thereby replacing the defective gene with a normal one.

In this chapter we will examine the molecules that carry and express our genetic information, DNA and RNA. Only by understanding the structure and function of these molecules have we been able to develop the amazing array of genetic tools that currently exists. We hope that as we continue to learn more about human genetics, we will be able to detect and one day correct most of the known genetic diseases.

Genetic traits are passed from one generation to the next. When a sperm fertilizes an ovum, a zygote is created from a single set of maternal chromosomes and a single set of paternal chromosomes. As this fertilized egg divides, each daughter cell will receive one copy of each of these chromosomes. The genes on these chromosomes will direct fetal development from that fertilized cell to a newborn with all the characteristics we recognize as human.

In this chapter we will explore the structure of DNA and the molecular events that translate the genetic information of a gene into the structure of a protein.

20.1 The Structure of the Nucleotide

Even before the philosopher Aristotle observed that "like begets like," humans were curious about the way in which family likenesses are passed from one generation to the next. In the 1860s Gregor Mendel combined astute observations, careful experimental design, and mathematical analysis to explain inheritance. Presented at a meeting in 1865 and published in 1866, Mendel's brilliant work was largely ignored by a scientific community that simply could not understand it.

At about the same time (1869), Friedrich Miescher discovered a substance in the nuclei of white blood cells recovered from pus. Chemical analysis indicated that, in addition to carbon, hydrogen, and oxygen, nuclein contained 14% nitrogen and 3% phosphorus. As microscopes improved in the last decades of the nineteenth century, biologists were able to peer into the nuclei of cells. They observed structures, later called chromosomes, which seemed to play a critical role in the process of cell division. Interestingly, egg and sperm cells were observed to have only half the chromosomes of the cells that produced them.

Chemical analysis of chromosomes indicated that they were composed of both protein and nuclein. But which of these molecules represented the genetic material? Most were convinced that the answer to this question was protein. The reasoning was that the genetic material must have a structure that would allow it to encode the enormous variation seen in the biological world. Both nuclein and protein were known to be polymers. However, proteins were polymers of twenty different subunits, the amino acids. Based on the results of Phoebus Levene, working with Emil Fischer and Albrecht Kossel, nuclein was composed of only four subunits. It appeared to lack the complexity required of a molecule responsible for the great diversity seen among plants, animals, and microbes.

In 1950, the genetic information was demonstrated to be nuclein, now called deoxyribonucleic acid, or DNA. In 1953, just over fifty years ago, James Watson and Francis Crick published a paper describing the structure of the DNA molecule.

Nucleotide Structure

From the work of Watson and Crick, as well as that of Miescher, Levene, and many others, we now know that **deoxyribonucleic acid (DNA)** and **ribonucleic acid (RNA)** are long polymers of **nucleotides.** Every nucleotide is composed of a nitrogenous base, a five-carbon sugar, and at least one phosphoryl group.

Nitrogenous bases are heterocyclic amines; that is, they are cyclic compounds with at least one nitrogen atom in the ring structure. There are two types of nitrogenous bases: purines and pyrimidines. **Purines,** which include adenine and guanine, consist of a six-member ring fused to a five-member ring. **Pyrimidines,** which include thymine, cytosine, and uracil, consist of a single six-member ring. Structures of these molecules are shown in Figure 20.1.

LEARNING GOAL

1 Draw the general structure of DNA and RNA nucleotides.

Phosphate group	Sugars	Bases			
		Purines (double ring)		**Pyrimidines** (single ring)	

2'-Deoxyribose (in DNA)   Adenine (A)   Thymine (T) (in DNA)   Uracil (U) (in RNA)

Ribose (in RNA)   Guanine (G)   Cytosine (C)

Figure 20.1

The components of nucleic acids include phosphate groups, the five-carbon sugars ribose and deoxyribose, and purine and pyrimidine nitrogenous bases. The ring positions of the sugars are designated with primes (') to distinguish them from the ring positions of the bases.

The five-carbon sugar in RNA is ribose, and the sugar in DNA is 2'-deoxyribose. The only difference between these two sugars is found at the 2'-carbon. Ribose has a hydroxyl group attached to this carbon, while deoxyribose has a hydrogen atom.

Each nucleotide consists of either ribose or deoxyribose, one of the five nitrogenous bases, and one or more phosphoryl groups (Figure 20.2). A nucleotide with the sugar ribose is a **ribonucleotide,** and one having the sugar 2'-deoxyribose is a **deoxyribonucleotide.** Because there are two cyclic molecules in a nucleotide, the sugar, and the base, we need an easy way to describe the ring atoms of each. For this reason the ring atoms of the sugar are designated with a prime to distinguish them from the atoms of the base (Figures 20.1 and 20.2).

The covalent bond between the sugar and the phosphoryl group is a phospho-ester bond formed by a condensation reaction between the 5'-OH of the sugar and an —OH of the phosphoryl group. The bond between the base and the sugar is a β-N-glycosidic linkage that joins the 1'-carbon of the sugar and a nitrogen atom of the base (N-9 of purines and N-1 of pyrimidines).

Question 20.1

Referring to the structures in Figures 20.1 and 20.2, draw the structures for nucleotides consisting of the following units.

a. Ribose, adenine, two phosphoryl groups
b. 2'-Deoxyribose, guanine, three phosphoryl groups

Question 20.2

Referring to the structures in Figures 20.1 and 20.2, draw the structures for nucleotides consisting of the following units.

a. 2'-Deoxyribose, thymine, one phosphoryl group
b. Ribose, cytosine, three phosphoryl groups
c. Ribose, uracil, one phosphoryl group

(a)

Phosphate | 2′ Deoxyribose

Deoxyribonucleotide

Phosphate | Ribose

Ribonucleotide

Figure 20.2

(a) The general structures of a deoxyribonucleotide and a ribonucleotide. (b) A specific example of a ribonucleotide, adenosine triphosphate.

(b)

Adenosine triphosphate

Adenosine diphosphate

Adenosine monophosphate

Adenosine

Adenine

Phosphoester bond

Phosphate groups

Ribose

20.2 The Structure of DNA and RNA

A single strand of DNA is a polymer of nucleotides bonded to one another by 3′–5′ phosphodiester bonds. The backbone of the polymer is called the *sugar-phosphate backbone* because it is composed of alternating units of the five-carbon sugar 2′-deoxyribose and phosphoryl groups in phosphodiester linkage. A nitrogenous base is bonded to each sugar by an *N*-glycosidic linkage (Figure 20.3).

DNA Structure: The Double Helix

James Watson and Francis Crick were the first to describe the three-dimensional structure of DNA in 1953. They deduced the structure by building models based on the experimental results of others. Irwin Chargaff observed that the amount of adenine in any DNA molecule is equal to the amount of thymine. Similarly, he found that the amounts of cytosine and guanine are also equal. The X-ray diffraction studies of Rosalind Franklin and Maurice Wilkens revealed several repeat distances that characterize the structure of DNA: 0.34 nm, 3.4 nm, and 2 nm. (Look at the structure of DNA in Figure 20.4 to see the significance of these measurements.)

With this information, Watson and Crick concluded that DNA is a **double helix** of two strands of DNA wound around one another. The structure of the double helix is often compared to a spiral staircase. The sugar-phosphate backbones of the two strands of DNA spiral around the outside of the helix like the handrails on a spiral staircase. The nitrogenous bases extend into the center at right angles

LEARNING GOAL

2 Describe the structure of DNA and compare it with RNA.

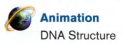

Animation

DNA Structure

Figure 20.3

The covalent, primary structure of DNA.

Backbone   Bases

Thymine (T)

Phosphodiester linkage

Adenine (A)

Cytosine (C)

Guanine (G)

Single nucleotide

Phosphate

Sugar (deoxyribose)

Base-pairing explains Chargaff's observation that the amount of adenine always equals the amount of thymine and the amount of guanine always equals the amount of cytosine for any DNA sample.

to the axis of the helix. You can imagine the nitrogenous bases forming the steps of the staircase. The structure of this elegant molecule is shown in Figure 20.4.

One noncovalent attraction that helps maintain the double helix structure is hydrogen bonding between the nitrogenous bases in the center of the helix. Adenine forms two hydrogen bonds with thymine, and cytosine forms three hydrogen bonds with guanine (Figure 20.4). These are called **base pairs.** The two strands of DNA are **complementary strands** because the sequence of bases on one automatically determines the sequence of bases on the other. When there is an adenine on one strand, there will always be a thymine in the same location on the opposite strand.

The diameter of the double helix is 2.0 nm. This is dictated by the dimensions of the purine-pyrimidine base pairs. The helix completes one turn every ten base pairs. One complete turn is 3.4 nm. Thus, each base pair advances the helix by 0.34 nm.

One last important feature of the DNA double helix is that the two strands are **antiparallel strands,** as this example shows:

Key Features

- Two strands of DNA form a right-handed double helix.

- The bases in opposite strands hydrogen bond according to the AT/GC rule.

- The 2 strands are antiparallel with regard to their 5′ to 3′ directionality.

- There are ~10.0 nucleotides in each strand per complete 360° turn of the helix.

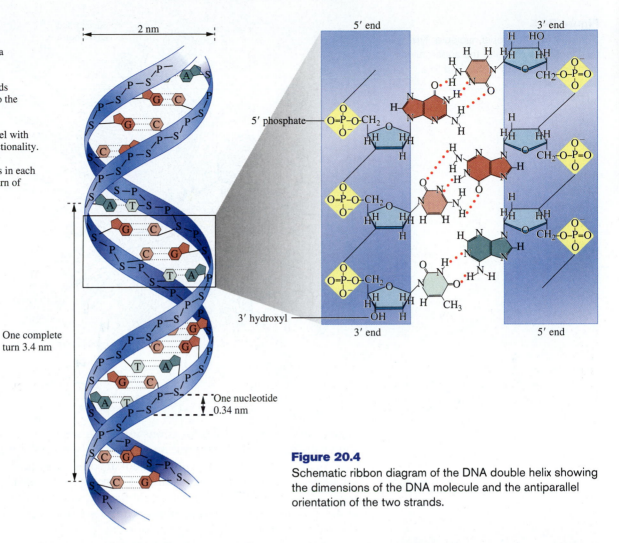

Figure 20.4

Schematic ribbon diagram of the DNA double helix showing the dimensions of the DNA molecule and the antiparallel orientation of the two strands.

In other words, the two strands of the helix run in opposite directions (see Figure 20.4). Only when the two strands are antiparallel can the base pairs form the hydrogen bonds that hold the two strands together.

Chromosomes

Chromosomes are pieces of DNA that carry the genetic instructions, or genes, of an organism. Organisms such as the prokaryotes have only a single chromosome and its structure is relatively simple. Others, the eukaryotes, have many chromosomes, each of which has many different levels of structure. The complete set of genetic information in all the chromosomes of an organism is called the **genome.**

Prokaryotes are organisms with a simple cellular structure in which there is no true nucleus surrounded by a nuclear membrane and there are no true membrane-bound organelles. This group includes all of the bacteria. In these organisms the chromosome is a circular DNA molecule that is supercoiled, which means that the helix is coiled on itself. The supercoiled DNA molecule is attached to a complex of proteins at roughly forty sites along its length, forming a series of loops. This structure, called the nucleoid, can be seen in Figure 20.5.

Eukaryotes are organisms that have cells containing a true nucleus enclosed by a nuclear membrane. They also have a variety of membrane-bound organelles that

This karyotype shows the 23 pairs of chromosomes of humans. What type of genetic disorders could be identified by karyotyping?

Figure 20.5
Structure of a bacterial nucleoid. The nucleoid is made up of the supercoiled, circular chromosome attached to a protein core.

Supercoiled
DNA loops

Protein core

segregate different cellular functions into different compartments. As an example, the reactions of aerobic respiration are located within the mitochondria.

All animals, plants, and fungi are eukaryotes. The number and size of the chromosomes of eukaryotes vary from one species to the next. For instance, humans have 23 pairs of chromosomes, while the Adder's Tongue fern has 631 pairs of chromosomes. But the chromosome structure is the same for all those organisms that have been studied.

Eukaryotic chromosomes are very complex structures (Figure 20.6). The first level of structure is the **nucleosome,** which consists of a strand of DNA wrapped around a small disk made up of histone proteins. At this level the DNA looks like beads along a string. The string of beads then coils into a larger structure called the *30 nm fiber.* These, in turn, are further coiled into a *200 nm fiber.* Other proteins are probably involved in the organization of the 200 nm fiber. The full complexities of the eukaryotic chromosome are not yet understood, but there are probably many such levels of coiled structures.

Some human genetic disorders are characterized by unusual chromosome numbers. For instance, Down syndrome is characterized by an extra copy of chromosome 21. The presence of this additional chromosome causes the traits associated with Down syndrome, including varying degrees of mental retardation, a flattened face, and short stature. The presence of an additional chromosome 18 causes Edward syndrome and an extra chromosome 13 causes Patau syndrome. Both of these are extremely rare and result in extreme mental and physical defects and early death. The presence of extra copies of the sex chromosomes, X or Y, is not lethal. Males with two X chromosomes and one Y suffer from Klinefelter syndrome and show sexual immaturity and breast development. Males with an extra Y chromosome are unusually tall, as are women with three X chromosomes. A woman with only a single X chromosome experiences Turner syndrome, including short stature, a webbed neck, and sexual immaturity. All other abnormalities in chromosome number are thought to be lethal to the fetus. In fact, it is thought that 50% of all miscarriages are the result of abnormal chromosome numbers.

Chromosome

Chromatin

1 μm
Chromosome diameter

Condensed fiber
(30 nm diameter)

Histone protein scaffold

Nucleosome (11 nm diameter)

DNA (2 nm diameter)

Figure 20.6
The eukaryotic chromosome has many levels of structure.

RNA Structure

The sugar-phosphate backbone of RNA consists of ribonucleotides, also linked by 3'–5' phosphodiester bonds. These phosphodiester bonds are identical to those found in DNA. However, RNA molecules differ from DNA molecules in three basic properties.

- RNA molecules are usually *single-stranded*.
- The sugar-phosphate backbone of RNA consists of *ribonucleotides* linked by 3'–5' phosphodiester bonds. Thus the sugar *ribose* is found in place of 2'-deoxyribose.
- The nitrogenous base *uracil* (U) replaces thymine (T).

A Medical Perspective

Fooling the AIDS Virus with "Look-Alike" Nucleotides

The virus that is responsible for the acquired immune deficiency syndrome (AIDS) is called the *human immunodeficiency virus*, or *HIV*. HIV is a member of a family of viruses called *retroviruses*, all of which have single-stranded RNA as their genetic material. The RNA is copied by a viral enzyme called *reverse transcriptase* into a double-stranded DNA molecule. This process is the opposite of the central dogma, which states that the flow of genetic information is from DNA to RNA. But these viruses reverse that flow, RNA to DNA. For this reason, these viruses are called retroviruses, which literally means "backward viruses." The process of producing a DNA copy of the RNA is called *reverse transcription*.

Because our genetic information is DNA and it is expressed by the classical DNA $\longrightarrow$ RNA $\longrightarrow$ protein pathway, our cells have no need for a reverse transcriptase enzyme. Thus, the HIV reverse transcriptase is a good target for antiviral chemotherapy because inhibition of reverse transcription should kill the virus but have no effect on the human host. Many drugs have been tested for the ability to selectively inhibit HIV reverse transcription. Among these is the DNA chain terminator 3'-azido-2', 3'-dideoxythymidine, commonly called *AZT* or *zidovudine*.

How does AZT work? It is one of many drugs that looks like one of the normal nucleosides. These are called *nucleoside analogs*. A nucleoside is just a nucleotide without any phosphate groups attached. The analog is phosphorylated by the cell and then tricks a polymerase, in this case viral reverse transcriptase, into incorporating it into the growing DNA chain in place of the normal phosphorylated nucleoside. AZT is a nucleoside analog that looks like the nucleoside thymidine except that in the 3' position of the deoxyribose sugar there is an azido group ($—N_3$) rather than the 3'-OH group. Compare the structures of thymidine and AZT shown in the accompanying figure. The 3'-OH group is necessary for further DNA polymerization because it

is there that the phosphoester linkage must be made between the growing DNA strand and the next nucleotide. If an azido group or some other group is present at the 3' position, the nucleotide analog can be incorporated into the growing DNA strand, but further chain elongation is blocked, as shown in the following figure. If the viral RNA cannot be reverse transcribed into the DNA form, the virus will not be able to replicate and can be considered to be dead.

Comparison of the structures of the normal nucleoside, 2'-deoxythymidine, and the nucleoside analog, 3'-azido-2', 3'-dideoxythymidine.

AZT is particularly effective because the HIV reverse transcriptase actually prefers it over the normal nucleotide, thymidine. Nonetheless, AZT is not a cure. At best it prolongs the life of a person with AIDS for a year or two. Eventually, however, AZT has a negative effect on the body. The cells of our bone marrow are constantly dividing to produce new blood cells: red blood cells to carry oxygen to the tissues, white blood cells of the immune system, and platelets for blood clotting. For cells to divide, they must replicate their DNA. The DNA polymerases of these dividing cells also accidentally incorporate AZT into the growing DNA chains with the result that cells of

Although RNA molecules are single-stranded, base pairing between uracil and adenine and between guanine and cytosine can still occur. We will show the importance of this property as we examine the way in which RNA molecules are involved in the expression of the genetic information in DNA.

20.3 DNA Replication

LEARNING GOAL

3 Explain DNA replication.

DNA must be replicated before a cell divides so that each daughter cell inherits a copy of each gene. A cell that is missing a critical gene will die, just as an individual with a genetic disease, a defect in an important gene, may die early in life. Thus it is essential that the process of DNA replication produces an absolutely accurate copy of the original genetic information. If mistakes are made in critical genes, the result may be lethal mutations.

Direction of movement

Reverse transcriptase

Viral RNA

Newly synthesized viral DNA

5′

3′
N₃

Next nucleotide can't be added because of the 3′–N₃

The mechanism by which AZT inhibits HIV reverse transcriptase. Incorporation of AZT into the growing HIV DNA strand in place of thymidine results in DNA chain termination; the azido group on the 3′ carbon of the sugar cannot react to produce the phosphoester linkage required to add the next nucleotide.

the bone marrow begin to die. This can result in anemia and even further depression of the immune response.

Another problem that has arisen with prolonged use of AZT is that AZT-resistant mutants of the virus appear. It is well known that HIV is a virus that mutates rapidly. Some of these mutant forms of the virus have an altered reverse transcriptase that will no longer use AZT. When these mutants appear, AZT is no longer useful in treating the infection.

Fortunately, research with other nucleoside analogs, alternative types of antiviral treatments, and combinations of drugs has provided a more effective means of treating HIV infection.

For Further Understanding

AZT is a very effective competitive inhibitor of reverse transcriptase. Review what you have learned about competitive inhibition and structural analogs in Chapter 19 and develop a hypothesis to explain why AZT is so effective.

As noted in this perspective, resistance to AZT is a common problem. This has led to the simultaneous use of two or more drugs in the treatment of HIV AIDS, for instance, AZT and a protease inhibitor. Explain why multiple drug therapy reduces the problem of viral drug resistance.

The structure of the DNA molecule suggested the mechanism for its accurate replication. Since adenine can base pair only with thymine and cytosine with guanine, Watson and Crick first suggested that an enzyme could "read" the nitrogenous bases on one strand of a DNA molecule and add complementary bases to a strand of DNA being synthesized. The product of this mechanism would be a new DNA molecule in which one strand is the original, or parent, strand and the second strand is a newly synthesized, or daughter, strand. This mode of DNA replication is called **semiconservative replication** (Figure 20.7).

Experimental evidence for this mechanism of DNA replication was provided by an experiment designed by Matthew Meselson and Franklin Stahl in 1958. *Escherichia coli* cells were grown in a medium in which $^{15}NH_4^+$ was the sole nitrogen source. ^{15}N is a nonradioactive, heavy isotope of nitrogen. Thus, growing the cells in this medium resulted in all of the cellular DNA containing this heavy isotope.

Electron micrograph of HIV.

Isotopes are atoms of the same element having the same number of protons but different numbers of neutrons and, therefore, different mass numbers.

Figure 20.7

In semiconservative DNA replication, each parent strand serves as a template for the synthesis of a new daughter strand.

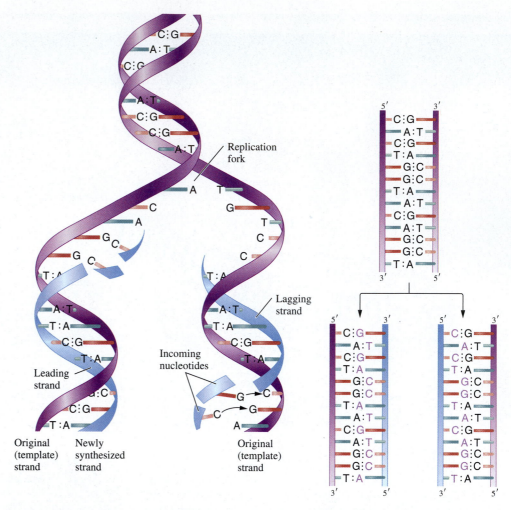

(a) The mechanism of DNA replication

(b) The products of replication

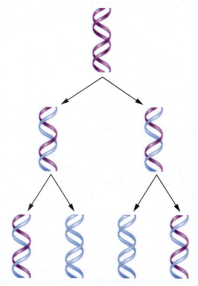

Figure 20.8

Representation of the Meselson and Stahl experiment. The DNA from cells grown in medium containing $^{15}NH_4^+$ is shown in purple. After a single generation in medium containing $^{14}NH_4^+$, the daughter DNA molecules have one ^{15}N-labeled parent strand and one ^{14}N-labeled daughter strand (blue). After a second generation in $^{14}NH_4^+$ containing medium, there are equal numbers of ^{14}N/^{15}N DNA molecules and ^{14}N/^{14}N DNA molecules.

 Animation

Meselson and Stahl Experiment

 Animations

How Nucleotides Are Added in
    DNA Replication
DNA Replication Fork
Bidirectional DNA Replication
DNA Replication (E. coli)

The cells containing only $^{15}NH_4^+$ were then added to a medium containing only the abundant isotope of nitrogen, $^{14}NH_4^+$, and were allowed to grow for one cycle of cell division. When the daughter DNA molecules were isolated and analyzed, it was found that each was made up of one strand of "heavy" DNA, the parental strand, and one strand of "light" DNA, the new daughter strand. After a second round of cell division, half of the isolated DNA contained no ^{15}N and half contained a 50/50 mixture of ^{14}N and ^{15}N-labeled DNA (Figure 20.8). This demonstrated conclusively that each parental strand of the DNA molecule serves as the template for the synthesis of a daughter strand and that each newly synthesized DNA molecule is composed of one parental strand and one newly synthesized daughter strand.

Bacterial DNA Replication

The bacterial chromosome is a circular molecule of DNA made up of about three million nucleotides. DNA replication begins at a unique sequence on the circular chromosome known as the **replication origin** (Figure 20.9). Replication occurs bidirectionally at the rate of about five hundred new nucleotides every second! The point at which the new deoxyribonucleotide is added to the growing daughter strand is called the **replication fork** (Figure 20.9). It is here that the DNA has been opened to allow binding of the various proteins and enzymes responsible for DNA replication. Since DNA synthesis occurs bidirectionally, there are two replication

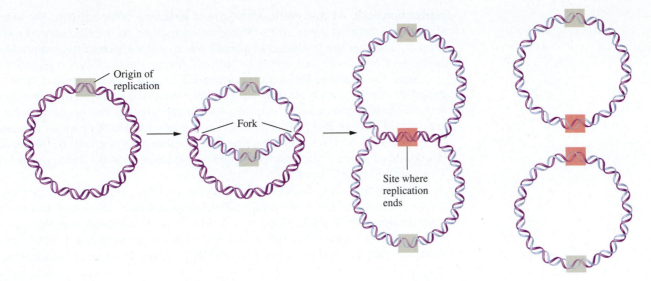

Figure 20.9
Bacterial chromosome replication.

Functions of key proteins involved with DNA replication

- **DNA helicase** breaks the hydrogen bonds between the DNA strands.

- **Topoisomerase** alleviates positive supercoiling.

- **Single-strand binding proteins** keep the parental strands apart.

- **Primase** synthesizes an RNA primer.

- **DNA polymerase III** synthesizes a daughter strand of DNA.

- **DNA polymerase I** excises the RNA primers and fills in with DNA.

- **DNA ligase** covalently links the DNA fragments together.

Figure 20.10
Because the two strands of DNA are antiparallel and DNA polymerase can only catalyze 5′ ⟶ 3′ replication, only one of the two DNA strands (top strand) can be read continuously to produce a daughter strand. The other must be synthesized in segments that are extended away from the direction of movement of the replication fork (bottom strand). These discontinuous segments are later covalently joined together by DNA ligase.

forks moving in opposite directions. Replication is complete when the replication forks collide approximately half way around the circular chromosome.

The first step in DNA replication (Figure 20.10) is the separation of the strands of DNA. The protein *helicase* does this by breaking the hydrogen bonds between the base pairs. This, in turn, causes supercoiling of the molecule. This stress is relieved by the enzyme *topoisomerase*, which travels along the DNA ahead of the replication fork. At this point, *single-strand binding protein* binds to the separated strands, preventing them from coming back together. In the next step, the enzyme *primase* catalyzes the synthesis of a small piece of RNA (ten to twelve nucleotides) called an *RNA primer* that serves to "prime" the process of DNA replication.

Now the enzyme **DNA polymerase III** "reads" each parental strand, also called the *template,* and catalyzes the polymerization of a complementary daughter strand. Deoxyribonucleotide triphosphate molecules are the precursors for DNA replication (Figure 20.11). In this reaction, a pyrophosphate group is released as a phosphoester bond is formed between the 5'-phosphoryl group of the nucleotide being added to the chain and the 3'-OH of the nucleotide on the daughter strand. This is called 5' to 3' synthesis.

One complicating factor in the process of DNA replication is the fact that the two strands of DNA are antiparallel to one another. DNA polymerase III can only catalyze DNA chain elongation in the 5' to 3' direction, yet the replication fork proceeds in one direction, while both strands are replicated simultaneously. Another complication is the need for an RNA primer to serve as the starting point

Figure 20.11

The reaction catalyzed by DNA polymerase.

Leading strand

3′

5′

5′

3′

Lagging strand

DNA
polymerase III

Helicase

Topoisomerase

3′

5′

Primase

RNA
primer

RNA
primer

ssDNA-binding
proteins

Figure 20.12

Model of the complex events occurring at the replication fork. In this representation, the replication fork is moving to the right. On the leading (top) strand, DNA polymerase III synthesizes DNA in the 5′ to 3′ direction continuously. Thus, replication proceeds in the same direction as the movement of the replication fork. On the lagging strand, DNA polymerase III also synthesizes DNA in the 5′ to 3′ direction. However, since the DNA strands are antiparallel, DNA polymerase III must read this strand in short segments (discontinuously) and in the opposite direction of the movement of the replication fork.

for DNA replication. As a result of these two obstacles, there are different mechanisms for replication of the two strands. One strand, called the **leading strand,** is replicated continuously. The opposite strand, called the **lagging strand,** is replicated discontinuously.

The two mechanisms are shown in Figures 20.10 and 20.12. For the leading strand, a single RNA primer is produced at the replication origin and DNA polymerase III continuously catalyzes the addition of nucleotides in the 5′ to 3′ direction, beginning with addition of the first nucleotide to the RNA primer.

On the lagging strand, many RNA primers are produced as the replication fork proceeds along the molecule. DNA polymerase III catalyzes DNA chain elongation from each of these primers. When the new strand "bumps" into a previous one, synthesis stops at that site. Meanwhile, at the replication fork, a new primer is being synthesized by primase. The final steps of synthesis on the lagging strand involve removal of the primers, repair of the gaps, and sealing of the fragments into an intact strand of DNA. The enzyme DNA polymerase I catalyzes the removal of the RNA primer and its replacement with DNA nucleotides. In the final step of the process, the enzyme DNA ligase catalyzes the formation of a phosphoester bond between the two adjacent fragments. It is little wonder that this is referred to as lagging strand replication! A more accurate model of the replication fork is shown in Figure 20.12.

Because it is critical to produce an accurate copy of the parental DNA, it is very important to avoid errors in the replication process. In addition to catalyzing the replication of new DNA, DNA polymerase III is able to proofread the newly synthesized strand. If the wrong nucleotide has been added to the growing DNA strand, it is removed and replaced with the correct one. In this way, a faithful copy of the parental DNA is ensured.

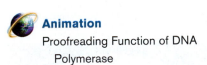 **Animation**

Proofreading Function of DNA
Polymerase

Eukaryotic DNA Replication

DNA replication in eukaryotes is more complex. The human genome consists of approximately three billion nucleotide pairs. Just one chromosome may be nearly one hundred times longer than a bacterial chromosome. To accomplish this huge job, DNA replication begins at many replication origins and proceeds bidirectionally along each chromosome.

Animation

Simple Gene Expression

20.4 Information Flow in Biological Systems

The **central dogma** of molecular biology states that in cells the flow of genetic information contained in DNA is a one-way street that leads from DNA to RNA to protein. The process by which a single strand of DNA serves as a template for the synthesis of an RNA molecule is called **transcription.** The word *transcription* is derived from the Latin word *transcribere* and simply means "to make a copy." Thus, in this process, part of the information in the DNA is copied into a strand of RNA. The process by which the message is converted into protein is called **translation.** Unlike transcription the process of translation involves converting the information from one language to another. In this case the genetic information in the linear sequence of nucleotides is being translated into a protein, a linear sequence of amino acids. The expression of the information contained in DNA is fundamental to the growth, development, and maintenance of all organisms.

Classes of RNA Molecules

Three classes of RNA molecules are produced by transcription: messenger RNA, transfer RNA, and ribosomal RNA.

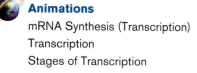

LEARNING GOAL

4 List three classes of RNA molecules and describe their functions.

1. **Messenger RNA (mRNA)** carries the genetic information for a protein from DNA to the ribosomes. It is a complementary RNA copy of a gene on the DNA.
2. **Ribosomal RNA (rRNA)** is a structural and functional component of the ribosomes, which are "platforms" on which protein synthesis occurs. There are three types of rRNA molecules in bacterial ribosomes and four in the ribosomes of eukaryotes.
3. **Transfer RNA (tRNA)** translates the genetic code of the mRNA into the primary sequence of amino acids in the protein. In addition to the primary structure, tRNA molecules have a cloverleaf-shaped secondary structure resulting from base pair hydrogen bonding (A—U and G—C) and a roughly L-shaped tertiary structure (Figure 20.13). The sequence CCA is found at the 3' end of the tRNA. The 3'—OH group of the terminal nucleotide, adenosine, can be covalently attached to an amino acid. Three nucleotides at the base of the cloverleaf structure form the **anticodon.** As we will discuss in more detail in Section 20.6, this triplet of bases forms hydrogen bonds to a **codon** (complementary sequence of bases) on a messenger RNA (mRNA) molecule on the surface of a ribosome during protein synthesis. This hydrogen bonding of codon and anticodon brings the correct amino acid to the site of protein synthesis at the appropriate location in the growing peptide chain.

Animations

mRNA Synthesis (Transcription)
Transcription
Stages of Transcription

Transcription

Transcription, shown in Figure 20.14, is catalyzed by the enzyme **RNA polymerase.** The process occurs in three stages. The first, called *initiation,* involves binding of RNA polymerase to a specific nucleotide sequence, the **promoter,** at the beginning of a gene. This interaction of RNA polymerase with specific promoter DNA sequences allows RNA polymerase to recognize the start point for transcription. It also determines which DNA strand will be transcribed. Unlike DNA replication, transcription produces a complementary copy of only one of the two strands of DNA. As it binds to the DNA, RNA polymerase separates the two strands of DNA so that it can "read" the base sequence of the DNA.

The second stage, chain elongation, begins as the RNA polymerase "reads" the DNA template strand and catalyzes the polymerization of a complementary RNA copy. With each step, RNA polymerase transfers a complementary ribonucleotide to the end of the growing RNA chain and catalyzes the formation of a 3'–5' phosphodiester bond between the 5' phosphoryl group of the incoming ribonucleotide and the 3' hydroxyl group of the last ribonucleotide of the growing RNA chain.

LEARNING GOAL

5 Explain the process of transcription.

Before the invention of the printing press, monks in monasteries transcribed copies of scripture to preserve the texts. Explain how this process is similar to the process of transcription in the cell.

Figure 20.13
Structure of tRNA. The primary structure of a tRNA is the linear sequence of ribonucleotides. Here we see the hydrogen-bonded secondary structure of a tRNA showing the three loops and the amino acid accepting end.

Initiation

Elongation/synthesis of the RNA transcript

Termination

Figure 20.14
The stages of transcription.

The final stage of transcription is termination. The RNA polymerase finds a termination sequence at the end of the gene and releases the newly formed RNA molecule.

Question 20.3

What is the function of RNA polymerase in the process of transcription?

Question 20.4

What is the function of the promoter sequence in the process of transcription?

Post-transcriptional Processing of RNA

LEARNING GOAL

6 List and explain the three types of post-transcriptional modifications of eukaryotic mRNA.

In bacteria, which are prokaryotes, termination releases a mature mRNA for translation. In fact, because prokaryotes have no nuclear membrane separating the DNA from the cytoplasm, translation begins long before the mRNA is completed. In eukaryotes, transcription produces a **primary transcript** that must undergo extensive **post-transcriptional modification** before it is exported out of the nucleus for translation in the cytoplasm.

Eukaryotic primary transcripts undergo three post-transcriptional modifications. These are the addition of a 5' cap structure and a 3' poly(A) tail, and RNA splicing.

In the first modification, a **cap structure** is enzymatically added to the 5' end of the primary transcript. The cap structure (Figure 20.15) consists of 7-methyl-guanosine attached to the 5' end of the RNA by a 5'–5' triphosphate bridge. The first two nucleotides of the mRNA are also methylated. The cap structure is required for efficient translation of the final mature mRNA.

The second modification is the enzymatic addition of a **poly(A)** tail to the 3' end of the transcript. *Poly(A) polymerase* uses ATP and catalyzes the stepwise polymerization of one hundred to two hundred adenosine nucleotides on the

Figure 20.15
The 5'-methylated cap structure of eukaryotic mRNA.

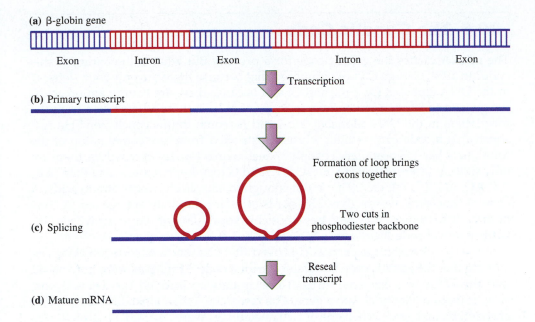

(a) β-globin gene

Exon   Intron   Exon   Intron   Exon

Transcription

(b) Primary transcript

Formation of loop brings
exons together

Two cuts in
phosphodiester backbone

(c) Splicing

Reseal
transcript

(d) Mature mRNA

Figure 20.16
Schematic diagram of mRNA splicing.
(a) The β-globin gene contains protein
coding exons, as well as noncoding
sequences called introns. (b) The
primary transcript of the DNA carries
both the introns and the exons.
(c) The introns are looped out, the
phosphodiester backbone of the mRNA
is cut twice, and the pieces are tied
together. (d) The final mature mRNA
now carries only the coding sequences
(exons) of the gene.

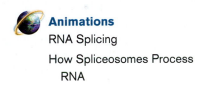

Animations

RNA Splicing

How Spliceosomes Process
RNA

3′ end of the RNA. The poly(A) tail protects the 3′ end of the mRNA from enzymatic degradation and thus prolongs the lifetime of the mRNA.

The third modification, **RNA splicing,** involves the removal of portions of the primary transcript that are not protein coding. Bacterial genes are continuous; all the nucleotide sequences of the gene are found in the mRNA. However, study of the gene structure of eukaryotes revealed a fascinating difference. Eukaryotic genes are discontinuous; there are *extra* DNA sequences within these genes that do not encode any amino acid sequences for the protein. These sequences are called *intervening sequences* or **introns.** The primary transcript contains both the introns and the protein coding sequences, called **exons.** The presence of introns in the mRNA would make it impossible for the process of translation to synthesize the correct protein. Therefore they must be removed, which is done by the process of RNA splicing.

As you can imagine, RNA splicing must be very precise. If too much, or too little, RNA is removed, the mRNA will not carry the correct code for the protein. Thus, there are "signals" in the DNA to mark the boundaries of the introns. The sequence GpU is always found at the intron's 5′ boundary and the sequence ApG is found at the 3′ boundary.

Recognition of the splice boundaries and stabilization of the splicing complex requires the assistance of particles called *spliceosomes.* Spliceosomes are composed of a variety of *small nuclear ribonucleoproteins* (snRNPs, read "snurps"). Each snRNP consists of a small RNA and associated proteins. The RNA components of different snRNPs are complementary to different sequences involved in splicing. By hydrogen bonding to a splice boundary or intron sequences the snRNPs recognize and bring together the sequences involved in the splicing reactions.

One of the first eukaryotic genes shown to contain introns was the gene for the β subunit of adult hemoglobin (Figure 20.16). On the DNA, the gene for β-hemoglobin is 1200 nucleotides long, but only 438 nucleotides carry the genetic information for protein. The remaining sequences are found in two introns of 116 and 646 nucleotides that are removed by splicing before translation. It is interesting that the larger intron is longer than the final β-hemoglobin mRNA! In the genes that have been studied, introns have been found to range in size from 50 to 20,000 nucleotides in length, and there may be many throughout a gene. Thus a typical human gene might be 10–30 times longer than the final mRNA.

20.5 The Genetic Code

The mRNA carries the genetic code for a protein. But what is the nature of this code? In 1954, George Gamow proposed that because there are only four "letters" in the DNA alphabet (A, T, G, and C) and because there are twenty amino acids, the genetic code must contain words made of at least three letters taken from the four letters in the DNA alphabet. How did he come to this conclusion? He reasoned that a code of two-letter words constructed from any combination of the four letters has a "vocabulary" of only sixteen words (4^2). In other words, there are only sixteen different ways to put A, T, C, and G together two bases at a time (AA, AT, AC, AG, TT, TA, etc.). That is not enough to encode all twenty amino acids. A code of four-letter words gives 256 words (4^4), far more than are needed. A code of three-letter words, however, has a possible vocabulary of sixty-four words (4^3), sufficient to encode the twenty amino acids but not too excessive.

A series of elegant experiments proved that Gamow was correct by demonstrating that the genetic code is, indeed, a triplet code. Mutations were introduced into the DNA of a bacterial virus. These mutations inserted (or deleted) one, two, or three nucleotides into a gene. The researchers then looked for the protein encoded by that gene. When one or two nucleotides were inserted, no protein was produced. However, when a third base was inserted, the sense of the mRNA was restored, and the protein was made. You can imagine this experiment by using a sentence composed of only three-letter words. For instance,

THE CAT RAN OUT

What happens to the "sense" of the sentence if we insert one letter?

THE FCA TRA NOU T

The reading frame of the sentence has been altered, and the sentence is now nonsense. Can we now restore the sense of the sentence by inserting a second letter?

THE FAC ATR ANO UT

No, we have not restored the sense of the sentence. Once again, we have altered the reading frame, but because our code has only three-letter words, the sentence is still nonsense. If we now insert a third letter, it should restore the correct reading frame:

THE FAT CAT RAN OUT

Indeed, by inserting three new letters we have restored the sense of the message by restoring the reading frame. This is exactly the way in which the message of the mRNA is interpreted. Each group of three nucleotides in the sequence of the mRNA is called a *codon,* and each codes for a single amino acid. If the sequence is interrupted or changed, it can change the amino acid composition of the protein that is produced or even result in the production of no protein at all.

As we noted, a three-letter genetic code contains sixty-four words, called *codons,* but there are only twenty amino acids. Thus, there are forty-four more codons than are required to specify all of the amino acids found in proteins. Three of the codons—UAA, UAG, and UGA—specify termination signals for the process of translation. But this still leaves us with forty-one additional codons. What is the function of the "extra" code words? Francis Crick (recall Watson and Crick and the double helix) proposed that the genetic code is a **degenerate code.** The term *degenerate* is used to indicate that different triplet codons may serve as code words for the same amino acid.

The complete genetic code is shown in Figure 20.17. We can make several observations about the genetic code. First, methionine and tryptophan are the

FIRST BASE	SECOND BASE				THIRD BASE
	U	C	A	G	
U	UUU Phenylalanine	UCU Serine	UAU Tyrosine	UGU Cysteine	U
U	UUC Phenylalanine	UCC Serine	UAC Tyrosine	UGC Cysteine	C
U	UUA Leucine	UCA Serine	UAA STOP	UGA STOP	A
U	UUG Leucine	UCG Serine	UAG STOP	UGG Tryptophan	G
C	CUU Leucine	CCU Proline	CAU Histidine	CGU Arginine	U
C	CUC Leucine	CCC Proline	CAC Histidine	CGC Arginine	C
C	CUA Leucine	CCA Proline	CAA Glutamine	CGA Arginine	A
C	CUG Leucine	CCG Proline	CAG Glutamine	CGG Arginine	G
A	AUU Isoleucine	ACU Threonine	AAU Asparagine	AGU Serine	U
A	AUC Isoleucine	ACC Threonine	AAC Asparagine	AGC Serine	C
A	AUA Isoleucine	ACA Threonine	AAA Lysine	AGA Arginine	A
A	AUG (START) Methionine	ACG Threonine	AAG Lysine	AGG Arginine	G
G	GUU Valine	GCU Alanine	GAU Aspartic acid	GGU Glycine	U
G	GUC Valine	GCC Alanine	GAC Aspartic acid	GGC Glycine	C
G	GUA Valine	GCA Alanine	GAA Glutamic acid	GGA Glycine	A
G	GUG Valine	GCG Alanine	GAG Glutamic acid	GGG Glycine	G

Cys
ACG tRNA

Asn
UUG tRNA

Figure 20.17
The genetic code. The table shows the possible codons found in mRNA. To read the universal biological language from this chart, find the first base in the column on the left, the second base from the row across the top, and the third base from the column to the right. This will direct you to one of the sixty-four squares in the matrix. Within that square you will find the codon and the amino acid that it specifies. In the cell this message is decoded by tRNA molecules like those shown to the right of the table.

only amino acids that have a single codon. All others have at least two codons, and serine and leucine have six codons each. The genetic code is also somewhat mutation-resistant. For those amino acids that have multiple codons, the first two bases are often identical and thus identify the amino acid, and only the third position is variable. Mutations—changes in the nucleotide sequence—in the third position therefore often have no effect on the amino acid that is incorporated into a protein.

Why is the genetic code said to be degenerate?

Question 20.5

Why is the genetic code said to be mutation-resistant?

Question 20.6

20.6 Protein Synthesis

LEARNING GOAL
Describe the process of translation.

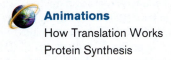

Animations
How Translation Works
Protein Synthesis

The process of protein synthesis is called *translation*. It involves translating the genetic information from the sequence of nucleotides into the sequence of amino acids in the primary structure of a protein. Figure 20.18 shows the relationship through which the nucleotide sequence of a DNA molecule is transcribed into a complementary sequence of ribonucleotides, the mRNA molecule. Each mRNA has a short untranslated region followed by the sequences that carry the information for the order of the amino acids in the protein that will be produced in the process of translation. That genetic information is the sequence of codons along the mRNA. The decoding process is carried out by tRNA molecules.

Translation is carried out on **ribosomes,** which are complexes of ribosomal RNA (rRNA) and proteins. Each ribosome is made up of two subunits: a small and a large ribosomal subunit (Figure 20.19a). In eukaryotic cells, the small ribosomal subunit contains one rRNA molecule and thirty-three different ribosomal proteins, and the large subunit contains three rRNA molecules and about forty-nine different proteins.

Protein synthesis involves the simultaneous action of many ribosomes on a single mRNA molecule. These complexes of many ribosomes along a single mRNA are known as *polyribosomes* or **polysomes** (Figure 20.19b). Each ribosome is synthesizing one copy of the protein molecule encoded by the mRNA. Thus, many copies of a protein are simultaneously produced.

The Role of Transfer RNA

The codons of mRNA must be read if the genetic message is to be translated into protein. The molecule that decodes the information in the mRNA molecule into the primary structure of a protein is transfer RNA (tRNA). To decode the genetic message into the primary sequence of a protein, the tRNA must faithfully perform two functions.

First, the tRNA must covalently bind one, and only one, specific amino acid. There is at least one transfer RNA for each amino acid. All tRNA molecules have the sequence CCA at their 3′ ends. This is the site where the amino acid will be covalently

Animation
Aminoacyl tRNA Synthetase

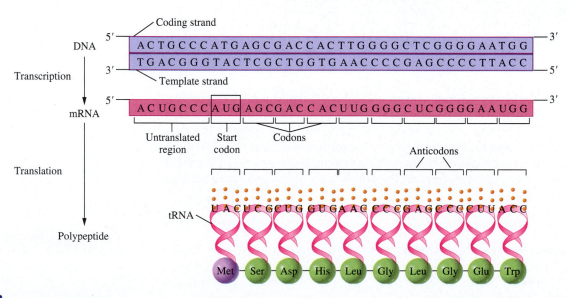

Figure 20.18
Messenger RNA (mRNA) is an RNA copy of one strand of a gene in the DNA. Each codon on the mRNA that specifies a particular amino acid is recognized by the complementary anticodon on a transfer RNA (tRNA).

attached to the tRNA molecule. Each tRNA is specifically recognized by the active site of an enzyme called an **aminoacyl tRNA synthetase.** This enzyme also recognizes the correct amino acid and covalently links the amino acid to the 3′ end of the tRNA molecule (Figure 20.20). The resulting structure is called an **aminoacyl tRNA.** The covalently bound amino acid will be transferred from the tRNA to a growing polypeptide chain during protein synthesis.

Second, the tRNA must be able to recognize the appropriate codon on the mRNA that calls for that amino acid. This is mediated through a sequence of three bases called the *anticodon,* which is located at the bottom of the tRNA cloverleaf (refer to Figure 20.13). The anticodon sequence for each tRNA is complementary to the codon on the mRNA that specifies a particular amino acid. As you can see in Figure 20.18, the anticodon-codon complementary hydrogen bonding will bring the correct amino acid to the site of protein synthesis.

How are codons related to anticodons?

Question 20.7

If the sequence of a codon on the mRNA is 5′-AUG-3′, what will the sequence of the anticodon be? Remember that the hydrogen bonding rules require antiparallel strands. It is easiest to write the anticodon first 3′ → 5′ and then reverse it to the 5′ → 3′ order.

Question 20.8

The Process of Translation

Initiation

The first stage of protein synthesis is *initiation.* Proteins called **initiation factors** assist in the formation of a translation complex composed of an mRNA molecule, the small and large ribosomal subunits, and the initiator tRNA. This initiator tRNA recognizes the codon AUG and carries the amino acid methionine.

The ribosome has two sites for binding tRNA molecules. The first site, called the **peptidyl tRNA binding site (P-site),** holds the peptidyl tRNA, the growing

LEARNING GOAL

8 Describe the process of translation.

Animation

Translation Initiation

Figure 20.20
Aminoacyl tRNA synthetase binds the amino acid in one region of the active site and the appropriate tRNA in another. The acylation reaction occurs and the aminoacyl tRNA is released.

Aminoacyl tRNA synthetase

Amino acid

ATP

An amino acid and ATP bind to the enzyme. AMP is covalently bound to the amino acid and pyrophosphate is released.

Pyrophosphate

The correct tRNA binds to the enzyme. The amino acid becomes covalently attached to the tRNA. AMP is released.

tRNA

AMP

The "charged" tRNA is released.

peptide bound to a tRNA molecule. The second site, called the **aminoacyl tRNA binding site (A-site),** holds the aminoacyl tRNA carrying the next amino acid to be added to the peptide chain. Each of the tRNA molecules is hydrogen bonded to the mRNA molecule by codon-anticodon complementarity. The entire complex is further stabilized by the fact that the mRNA is also bound to the ribosome. Figure 20.21a shows the series of events that result in the formation of the initiation complex. The initiator methionyl tRNA occupies the P-site in this complex.

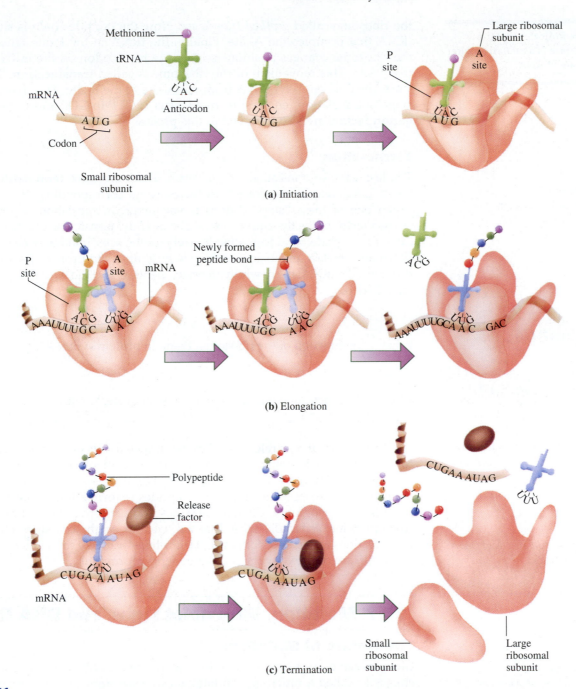

(a) Initiation

(b) Elongation

(c) Termination

Figure 20.21

(a) Formation of an initiation complex sets protein synthesis in motion. The mRNA and proteins called initiation factors bind to the small ribosomal subunit. Next, a charged methionyl tRNA molecule binds, and finally, the initiation factors are released, and the large subunit binds. (b) The elongation phase of protein synthesis involves addition of new amino acids to the C-terminus of the growing peptide. An aminoacyl tRNA molecule binds at the empty A-site, and the peptide bond is formed. The uncharged tRNA molecule is released, and the peptidyl tRNA is shifted to the P-site as the ribosome moves along the mRNA. (c) Termination of protein synthesis occurs when a release factor binds the stop codon on mRNA. This leads to the hydrolysis of the ester bond linking the peptide to the peptidyl tRNA molecule in the P-site. The ribosome then dissociates into its two subunits, releasing the mRNA and the newly synthesized peptide.

Chain Elongation

The second stage of translation is *chain elongation*. This occurs in three steps that are repeated until protein synthesis is complete. We enter the action after a tetra-peptide has already been assembled, and a peptidyl tRNA occupies the P-site (Figure 20.21b).

The first event is binding of an aminoacyl-tRNA molecule to the empty A-site. Next, peptide bond formation occurs. This is catalyzed by an enzyme on

Animation

Translation Elongation

Animation

Translation Termination

the ribosome called *peptidyl transferase.* Now the peptide chain is shifted to the tRNA that occupies the A-site. Finally, the tRNA in the P-site falls away, and the ribosome changes positions so that the next codon on the mRNA occupies the A-site. This movement of the ribosome is called **translocation.** The process shifts the new peptidyl tRNA from the A-site to the P-site. The chain elongation stage of translation requires the hydrolysis of GTP to GDP and P_i. Several **elongation factors** are also involved in this process.

Termination

The last stage of translation is *termination.* There are three **termination codons**—UAA, UAG, and UGA—for which there are no corresponding tRNA molecules. When one of these "stop" codons is encountered, translation is terminated. A **release factor** binds the empty A-site. The peptidyl transferase that had previously catalyzed peptide bond formation hydrolyzes the ester bond between the peptidyl tRNA and the last amino acid of the newly synthesized protein (Figure 20.21c). At this point the tRNA, the newly synthesized peptide, and the two ribosomal subunits are released.

Question *20.9*

What is the function of the ribosomal P-site in protein synthesis?

Question *20.10*

What is the function of the ribosomal A-site in protein synthesis?

Post-translational proteolytic cleavage of digestive enzymes is discussed in Section 19.11.

The peptide that is released following translation is not necessarily in its final functional form. In some cases the peptide is proteolytically cleaved before it becomes functional. Synthesis of digestive enzymes uses this strategy. Sometimes the protein must associate with other peptides to form a functional protein, as in the case of hemoglobin. Cellular enzymes add carbohydrate or lipid groups to some proteins, especially those that will end up on the cell surface. These final modifications are specific for particular proteins and, like the sequence of the protein itself, are directed by the cellular genetic information.

The quaternary structure of hemoglobin is described in Section 18.9.

20.7 Mutation, Ultraviolet Light, and DNA Repair

The Nature of Mutations

9 **LEARNING GOAL**
Define mutation and understand how mutations cause cancer and cell death.

Changes can occur in the nucleotide sequence of a DNA molecule. Such a genetic change is called a **mutation.** Mutations can arise from mistakes made by DNA polymerase during DNA replication. They also result from the action of chemicals, called **mutagens,** that damage the DNA.

Mutations are classified by the kind of change that occurs in the DNA. The substitution of a single nucleotide for another is called a **point mutation:**

ATGGACTTC:	normal DNA sequence
ATGCACTTC:	point mutation

Sometimes a single nucleotide or even large sections of DNA are lost. These are called **deletion mutations:**

ATGGACTTC:	normal DNA sequence
ATGTTC:	deletion mutation

Occasionally, one or more nucleotides are added to a DNA sequence. These are called **insertion mutations:**

Animations
Addition and Deletion Mutations
Mutation by Base Substitution

| ATGGACTTC: | normal DNA sequence |
| ATGCTCGACTTC: | insertion mutation |

The Results of Mutations

Some mutations are **silent mutations;** that is, they cause no change in the protein. Often, however, a mutation has a negative effect on the health of the organism. The effect of a mutation depends on how it alters the genetic code for a protein. Consider the two codons for glutamic acid: GAA and GAG. A point mutation that alters the third nucleotide of GAA to GAG will still result in the incorporation of glutamic acid at the correct position in the protein. Similarly, a GAG to GAA mutation will also be silent.

Many mutations are not silent. There are approximately four thousand human genetic disorders that result from such mutations. These occur because the mutation in the DNA changes the codon and results in incorporation of the wrong amino acid into the protein. This causes the protein to be nonfunctional or to function improperly.

Consider the human genetic disease sickle cell anemia. In the normal β-chain of hemoglobin, the sixth amino acid is glutamic acid. In the β-chain of sickle cell hemoglobin, the sixth amino acid is valine. How did this amino acid substitution arise? The answer lies in examination of the codons for glutamic acid and valine:

| Glutamic acid: | GAA or GAG |
| Valine: | GUG, GUC, GUA, or GUU |

A point mutation of A → U in the second nucleotide changes some codons for glutamic acid into codons for valine:

GAA	⟶	GUA
GAG	⟶	GUG
Glutamic acid codon		Valine codon

This mutation in a single codon leads to the change in amino acid sequence at position 6 in the β-chain of human hemoglobin from glutamic acid to valine. The result of this seemingly minor change is sickle cell anemia in individuals who inherit two copies of the mutant gene.

Question 20.11

The sequence of a gene on the mRNA is normally AUGCCCGACUUU. A point mutation in the gene results in the mRNA sequence AUGCGCGACUUU. What are the amino acid sequences of the normal and mutant proteins? Would you expect this to be a silent mutation?

Question 20.12

The sequence of a gene on the mRNA is normally AUGCCCGACUUU. A point mutation in the gene results in the mRNA sequence AUGCCGGACUUU. What are the amino acid sequences of the normal and mutant proteins? Would you expect this to be a silent mutation?

A Medical Perspective

The Ames Test for Carcinogens

Each day we come into contact with a variety of chemicals, including insecticides, food additives, hair dyes, automobile emissions, and cigarette smoke. Some of these chemicals have the potential to cause cancer. How do we determine whether these agents are harmful? More particularly, how do we determine whether they cause cancer?

If we consider the example of cigarette smoke, we see that it can be years, even centuries, before a relationship is seen between a chemical and cancer. Europeans and Americans have been smoking since Sir Walter Raleigh introduced tobacco into England in the seventeenth century. However, it was not until three centuries later that physicians and scientists demonstrated the link between smoking and lung cancer. Obviously, this epidemiological approach takes too long, and too many people die. Alternatively, we can test chemicals by treating laboratory animals, such as mice, and observing them for various kinds of cancer. However, this, too, can take years, is expensive, and requires the sacrifice of many laboratory animals. How, then, can chemicals be tested for carcinogenicity (the ability to cause cancer) quickly and inexpensively? In the 1970s it was recognized that most carcinogens are also mutagens. That is, they cause cancer by causing mutations in the DNA, and the mutations cause the cells of the body to lose growth control. Bruce Ames, a biochemist and bacterial geneticist, developed a test using mutants of the bacterium *Salmonella typhimurium* that can demonstrate in 48–72 hours whether a chemical is a mutagen and thus a suspected carcinogen.

Ames chose several mutants of *S. typhimurium* that cannot grow unless the amino acid histidine is added to the growth medium. The Ames test involves subjecting these bacteria to a chemical and determining whether the chemical causes reversion of the mutation. In other words, the researcher is looking for a mutation that reverses the original mutation. When a reversion occurs, the bacteria will be able to grow in the absence of histidine.

The details of the Ames test are shown in the accompanying figure. Both an experimental and a control test are done. The control test contains no carcinogen and will show the number of spontaneous revertants that occur in the culture. If there are many colonies on the surface of the experimental plate and only a few colonies on the negative control plate, it can be concluded that the chemical tested is a mutagen. It is therefore possible that the chemical is also a carcinogen.

The Ames test has greatly accelerated our ability to test new compounds for mutagenic and possibly carcinogenic effects. However, once the Ames test identifies a mutagenic compound, testing in animals must be done to show conclusively that the compound also causes cancer.

For Further Understanding

A researcher carried out the Ames test in which an experimental sample was exposed to a suspected mutagen and a control sample was not. A sample from each tube was grown on a medium containing no histidine. On the experimental plate, he observed forty-three colonies and on the control plate, he observed thirty-one colonies. He concluded that the substance is a mutagen. When he reported his data and conclusion to his supervisor, she told him that his conclusions were not valid. How can the researcher modify his experimental procedure to obtain better data?

Suppose that the mutation in a strain of *S. typhimurium* produces the codon UUA instead of UUC. What is the amino acid change caused by this mutation? What base substitutions could correct the mutant codon so that it once again calls for the correct amino acid? What base substitutions would not correct the mutant codon?

Mutagens and Carcinogens

Any chemical that causes a change in the DNA sequence is called a *mutagen.* Often, mutagens are also **carcinogens,** cancer-causing chemicals. Most cancers result from mutations in a single normal cell. These mutations result in the loss of normal growth control, causing the abnormal cell to proliferate. If that growth is not controlled or destroyed, it will result in the death of the individual. We are exposed to many carcinogens in the course of our lives. Sometimes we are exposed to a carcinogen by accident, but in some cases it is by choice. There are about three thousand chemical components in cigarette smoke, and several are potent mutagens. As a result, people who smoke have a much greater chance of lung cancer than those who don't.

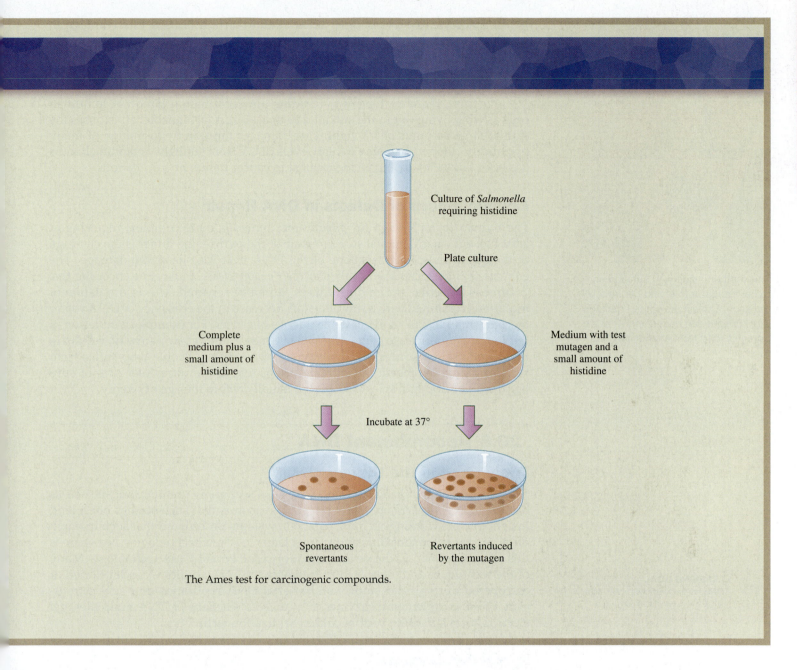

The Ames test for carcinogenic compounds.

Ultraviolet Light Damage and DNA Repair

Ultraviolet (UV) light is another agent that causes damage to DNA. Absorption of UV light by DNA causes adjacent pyrimidine bases to become covalently linked. The product is called a **pyrimidine dimer.** As a result of pyrimidine dimer formation, there is no hydrogen bonding between these pyrimidine molecules and the complementary bases on the other DNA strand. This stretch of DNA cannot be replicated or transcribed!

Bacteria such as *Escherichia coli* have four different mechanisms to repair ultraviolet light damage. However, even a repair process can make a mistake. Mutations occur when the UV damage repair system makes an error and causes a change in the nucleotide sequence of the DNA.

 Animation

Thymine Dimer Formation
and Repair

Pyrimidine dimers were originally called thymine dimers because thymine is more commonly involved in these reactions than cytosine.

This mother is applying sun screen to her daughter to shield her from UV radiation. Explain the kind of damage that UV light can cause and what the potential long-term effects may be.

Animation

Restriction Endonucleases

In medicine, the pyrimidine dimerization reaction is used to advantage in hospitals where germicidal (UV) light is used to kill bacteria in the air and on environmental surfaces, such as in a vacant operating room. This cell death is caused by pyrimidine dimer formation on a massive scale. The repair systems of the bacteria are overwhelmed, and the cells die.

Of course, the same type of pyrimidine dimer formation can occur in human cells as well. Lying out in the sun all day to acquire a fashionable tan exposes the skin to large amounts of UV light. This damages the skin by formation of many pyrimidine dimers. Exposure to high levels of UV from sunlight or tanning booths has been linked to a rising incidence of skin cancer in human populations.

Consequences of Defects in DNA Repair

The human repair system for pyrimidine dimers is quite complex, requiring at least five enzymes. The first step in repair of the pyrimidine dimer is the cleavage of the sugar-phosphate backbone of the DNA near the site of the damage. The enzyme that performs this is called a *repair endonuclease*. If the gene encoding this enzyme is defective, pyrimidine dimers cannot be repaired. The accumulation of mutations combined with a simultaneous decrease in the efficiency of DNA repair mechanisms leads to an increased incidence of cancer. For example, a mutation in the repair endonuclease gene, or in other genes in the repair pathway, results in the genetic skin disorder called *xeroderma pigmentosum*. People who suffer from xeroderma pigmentosum are extremely sensitive to the ultraviolet rays of sunlight and develop multiple skin cancers, usually before the age of twenty.

20.8 Recombinant DNA

Tools Used in the Study of DNA

Scientists are often asked why they study such seemingly unimportant subjects as bacterial DNA replication. One very good reason is that such studies often lend insight into the workings of human genetic systems. A second is that such research often produces the tools that allow great leaps into new technologies. Nowhere is this more true than in the development of recombinant DNA technology. Many of the techniques and tools used in recombinant DNA studies were developed or discovered during basic studies on bacterial DNA replication and gene expression. These include many enzymes that catalyze reactions of DNA molecules, gel electrophoresis, cloning vectors, and hybridization techniques.

Restriction Enzymes

Restriction enzymes, often called restriction endonucleases, are bacterial enzymes that "cut" the sugar-phosphate backbone of DNA molecules at specific nucleotide sequences. The first of these enzymes to be purified and studied was called EcoR1. The name is derived from the genus and species name of the bacteria from which it was isolated, in this case *Escherichia coli,* or *E. coli.* The following is the specific nucleotide sequence recognized by EcoR1:

When EcoR1 cuts the DNA at this site, it does so in a staggered fashion. Specifically, it cuts between the G and the first A on both strands. Cutting produces two DNA fragments with the following structure:

TABLE 20.1	Common Restriction Enzymes and Their Recognition Sequences
Restriction Enzyme	**Recognition Sequence**
BamHI	5'-G/GATCC-3' 3'-CCTAG/G-5'
HindIII	5'-A/AGCTT-3' 3'-TTCGA/A-5'
SalI	5'-G/TCGAC-3' 3'-CAGCT/G-5'
BglII	5'-A/GATCT-3' 3'-TCTAG/A-5'
PstI	5'-CTGCA/G-3' 3'-G/ACGTC-5'

These staggered termini are called *sticky ends* because they can reassociate with one another by hydrogen bonding. This is a property of the DNA fragments generated by restriction enzymes that is very important to gene cloning.

Examples of other restriction enzymes and their specific recognition sequences are listed in Table 20.1. The sites on the sugar-phosphate backbone that are cut by the enzymes are indicated by slashes.

These enzymes are used to digest large DNA molecules into smaller fragments of specific size. Because a restriction enzyme always cuts at the same site, DNA from a particular individual generates a reproducible set of DNA fragments. This is convenient for the study or cloning of DNA from any source.

Agarose Gel Electrophoresis

One means of studying the DNA fragments produced by restriction enzyme digestion is agarose gel electrophoresis. The digested DNA sample is placed in a sample well in the gel, and an electrical current is applied. The negative charge of the phosphoryl groups in the sugar-phosphate backbone causes the DNA fragment to move through the gel away from the negative electrode (cathode) and toward the positive electrode (anode). The smaller DNA fragments move more rapidly than the larger ones, and as a result the DNA fragments end up distributed throughout the gel according to their size. The sizes of each fragment can be determined by comparison with the migration pattern of DNA fragments of known size.

Animation
Electrophoresis

Hybridization

Agarose gel electrophoresis allows the determination of the size of a DNA fragment. However, in recombinant DNA research it is also important to identify what gene is carried by a particular DNA fragment.

Hybridization is a technique used to identify the presence of a gene on a particular DNA fragment. This technique is based on the fact that complementary DNA sequences will hydrogen bond, or hybridize, to one another. In fact, even RNA can be used in hybridization studies. RNA can hybridize to DNA molecules or to other RNA molecules.

One technique, called Southern blotting, involves hybridization of DNA fragments from an agarose gel (Figure 20.22a). DNA digested by a restriction enzyme is run on an agarose gel. Next the DNA fragments are transferred by blotting onto a special membrane filter. Figure 20.22b shows an apparatus that uses an electric field to transfer DNA from a gel onto a filter. In the next step, the DNA molecules on the filter are "melted" into single DNA strands so that they are ready

Animation
Southern Blot

A sample of chromosomal DNA is digested into small fragments with a restriction enzyme.

The fragments are separated by gel electrophoresis, and then denatured.

Gel

As shown in part b, the DNA bands are transferred to a nitrocellulose filter.

Nitrocellulose filter (DNA bands would not be visible at this stage.)

The filter is placed in a solution containing a radiolabeled probe. Excess probe is washed away, and the filter is exposed to X-ray film.

Lid

Cathode plate

Blotting paper

Gel

Nitrocellulose filter

Blotting paper

Anode plate

Base

(a) The steps in Southern blotting

(b) The transfer step in a Southern blotting experiment

Figure 20.22

Southern blot hybridization. (a) DNA is digested and the fragments are separated by gel electrophoresis. The DNA is transferred from the gel onto a filter and "melted" into single strands. A radioactive probe is applied and the filter exposed to X-ray film. (b) An electroblotting apparatus for the transfer of DNA from a gel onto a filter.

for hybridization. The filter is then bathed in a solution containing a radioactive DNA or RNA molecule. This probe will hybridize to any DNA fragments on the filter that are complementary to it. X-ray film is used to detect any bands where the radioactive probe hybridized, thus locating the gene of interest.

DNA Cloning Vectors

DNA cloning experiments combine these technologies with a few additional tricks to isolate single copies of a gene and then produce billions of copies. To produce multiple copies of a gene, it may be joined to a **cloning vector.** A cloning vector is a piece of DNA having its own replication origin so that it can be replicated inside a host cell. Often the bacterium *E. coli* serves as the host cell in which the vector carrying the cloned DNA is replicated in abundance.

There are two major kinds of cloning vectors. The first are bacterial virus or phage vectors. These are bacterial viruses that have been genetically altered to allow the addition of cloned DNA fragments. These viruses have all the genes required to replicate one hundred to two hundred copies of the virus (and cloned fragment) per infected cell.

The second commonly used vector is a plasmid vector. Plasmids are extra pieces of circular DNA found in most kinds of bacteria. The plasmids that are used as cloning vectors often contain antibiotic resistance genes that are useful in the selection of cells containing a plasmid.

Each plasmid has its own replication origin to allow efficient DNA replication in the bacterial host cell. Most plasmid vectors also have a *selectable marker,* often a gene for resistance to an antibiotic. Finally, plasmid vectors have a gene that has several restriction enzyme sites useful for cloning. The valuable feature of this gene is that it is inactivated when a DNA fragment has been cloned into it. Thus, cells containing a plasmid carrying a cloned DNA fragment can be recognized by their ability to grow in the presence of antibiotic and by loss of function of the gene into which the cloned DNA has been inserted.

Antibiotic resistance causes countless problems in the treatment of bacterial infections.

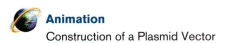

Animation
Construction of a Plasmid Vector

Genetic Engineering

Now that we have assembled most of the tools needed for a cloning experiment, we must decide which gene to clone. The example that we will use is the cloning of the β-globin genes for normal and sickle cell hemoglobin. DNA from an individual with normal hemoglobin is digested with a restriction enzyme. This is the target DNA. The vector DNA must be digested with the same enzyme (Figure 20.23). In our example, the restriction enzyme cuts within the *lacZ* gene, which codes for the enzyme β-galactosidase.

The digested vector and target DNA are mixed together under conditions that encourage the sticky ends of the target and vector DNA to hybridize with one another. The sticky ends are then covalently linked by the enzyme DNA ligase. This enzyme catalyzes the formation of phosphoester bonds between the two pieces of DNA.

Animation
Steps in Cloning a Gene

Now the recombinant DNA molecules are introduced into bacterial cells by a process called transformation. Next, the cells of the transformation mixture are plated on a solid nutrient agar medium containing the antibiotic ampicillin and the β-galactosidase substrate X-gal (5-bromo-4-chloro-3-indolyl-β-D-galactoside). Only those cells containing the antibiotic resistance gene will survive and grow into bacterial colonies. Cells with an intact *lacZ* gene will produce the enzyme β-galactosidase. The enzyme will hydrolyze X-gal to produce a blue product that will cause the colonies to appear blue. If the *lacZ* gene has been inactivated by insertion of a cloned gene, no β-galactosidase will be produced and the colonies will be white.

Now hybridization can be used to detect the clones that carry the β-globin gene. A replica of the experimental plate is made by transferring some cells from each colony onto a membrane filter. These cells are gently broken open so that the

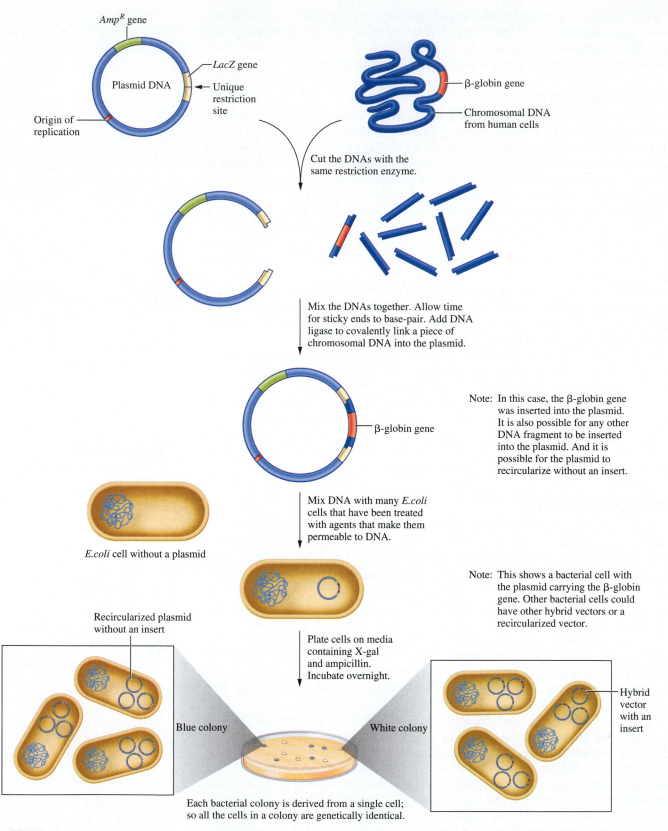

Figure 20.23
Cloning of eukaryotic DNA into a plasmid cloning vector.

TABLE 20.2	A Brief List of Medically Important Proteins Produced by Genetic Engineering

Protein	Medical Condition Treated
Insulin	Insulin-dependent diabetes
Human growth hormone	Pituitary dwarfism
Factor VIII	Type A hemophilia
Factor IX	Type B hemophilia
Tissue plasminogen factor	Stroke, myocardial infarction
Streptokinase	Myocardial infarction
Interferon	Cancer, some virus infections
Interleukin-2	Cancer
Tumor necrosis factor	Cancer
Atrial natriuretic factor	Hypertension
Erythropoietin	Anemia
Thymosin α-1	Stimulate immune system
Hepatitis B virus (HBV) vaccine	Prevent HBV viral hepatitis
Influenza vaccine	Prevent influenza infection

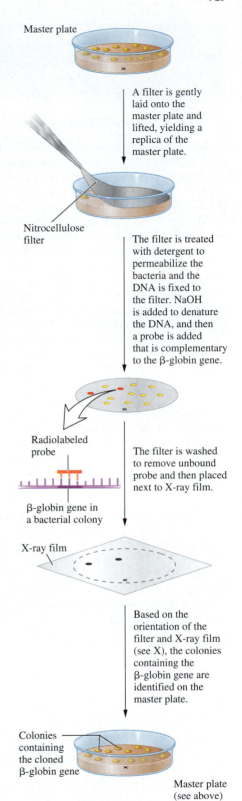

Master plate

A filter is gently laid onto the master plate and lifted, yielding a replica of the master plate.

Nitrocellulose filter

The filter is treated with detergent to permeabilize the bacteria and the DNA is fixed to the filter. NaOH is added to denature the DNA, and then a probe is added that is complementary to the β-globin gene.

Radiolabeled probe

β-globin gene in a bacterial colony

The filter is washed to remove unbound probe and then placed next to X-ray film.

X-ray film

Based on the orientation of the filter and X-ray film (see X), the colonies containing the β-globin gene are identified on the master plate.

Colonies containing the cloned β-globin gene

Master plate (see above)

Figure 20.24

Colony blot hybridization for detection of cells carrying a plasmid clone of the β-chain gene of hemoglobin.

released DNA becomes attached to the membrane. When hybridization is carried out on these filters, the radioactive probe will hybridize only to the complementary sequences of the β-globin gene. When the membrane filter is exposed to X-ray film, a "spot" will appear on the developed film only at the site of a colony carrying the desired clone. By going back to the original plate, we can select cells from that colony and grow them for further study (Figure 20.24).

The same procedure can be used to clone the β-chain gene of sickle cell hemoglobin. Then the two can be studied and compared to determine the nature of the genetic defect.

This simple example makes it appear that all gene cloning is very easy and straightforward. This has proved to be far from the truth. Genetic engineers have had to overcome many obstacles to clone eukaryotic genes of particular medical interest. One of the first obstacles encountered was the presence of introns within eukaryotic genes. Bacteria that are used for cloning lack the enzymatic machinery to splice out introns. Molecular biologists found that a DNA copy of a eukaryotic mRNA could be made by using the enzyme reverse transcriptase from a family of viruses called retroviruses (see A Medical Perspective: Fooling the AIDS Virus with "Look-Alike" Nucleotides, Section 20.2). Such a DNA copy of the mRNA carries all the protein-coding sequences of a gene but none of the intron sequences. Thus bacteria are able to transcribe and translate the cloned DNA and produce valuable products for use in medicine and other applications.

This is only one of the many technical problems that have been overcome by the amazing developments in recombinant DNA technology. A brief but impressive list of medically important products of genetic engineering is presented in Table 20.2.

20.9 Polymerase Chain Reaction

A bacterium originally isolated from a hot spring in Yellowstone National Park provides the key to a powerful molecular tool for the study of DNA. Polymerase chain reaction (PCR) allows scientists to produce unlimited amounts of any gene of interest and the bacterium *Thermus aquaticus* produces a heat-stable DNA polymerase (Taq polymerase) that allows the process to work.

A Human *Perspective*

DNA Fingerprinting

Four U.S. Army helicopters swept over the field of illicit coca plants (*Erythroxylum* spp.) growing in a mountainous region of northern Colombia. When the soldiers were certain that the fields were unguarded, a fifth helicopter landed. From it emerged Dr. Jim Saunders, currently director of the Molecular Biology, Biochemistry, and Bioinformatics Program at Towson University, and a team of researchers from the Agricultural Research Service of the U.S. Department of Agriculture (ARS-USDA). Quickly the scientists gathered leaves from mature plants, as well as from seedlings growing in a coca nursery, and returned to the helicopter with their valuable samples. With a final sweep over the field, the Army helicopters sprayed herbicides to kill the coca plants.

From 1997 to 2001, this scene was repeated in regions of Colombia known to have the highest coca production. The reason for these collections was to study the genetic diversity of the coca plants being grown for the illegal production of cocaine. The tool selected for this study was DNA fingerprinting.

DNA fingerprinting was developed in the 1980s by Alec Jeffries of the University of Leicester in England. The idea grew out of basic molecular genetic studies of the human genome. Scientists observed that some DNA sequences varied greatly from one person to the next. Such hypervariable regions are made up of variable numbers of repeats of short DNA sequences. They are located at many sites on different chromosomes. Each person has a different number of repeats and when his or her DNA is digested with restriction enzymes, a unique set of DNA fragments is generated. Jeffries invented DNA fingerprinting by developing a set of DNA probes that detect these variable number tandem repeats (VNTRs) when used in hybridization with Southern blots.

Although several variations of DNA fingerprinting exist, the basic technique is quite simple. DNA is digested with restriction enzymes, producing a set of DNA fragments. These are separated by electrophoresis through an agarose gel. The DNA fragments are then transferred to membrane filters and hybridized with the radioactive probe DNA. The bands that hybridize the radioactive probe are visualized by exposing the membrane to X-ray film and developing a "picture" of the gel. The result is what Jeffries calls a *DNA fingerprint*, a set of twenty-five to sixty DNA bands that are unique to an individual.

Coca nursery next to a mature field in Colombia.

The human genome consists of approximately three billion base pairs of DNA. But suppose you are interested in studying only one gene, perhaps the gene responsible for muscular dystrophy or cystic fibrosis. It's like looking for a needle in a haystack. Using PCR, a scientist can make millions of copies of the gene of interest, while ignoring the thousands of other genes on human chromosomes.

The secret to this specificity is the synthesis of a DNA primer, a short piece of single-stranded DNA that will specifically hybridize to the beginning of a particular gene. DNA polymerases require a primer for initiation of DNA synthesis because they act by adding new nucleotides to the 3'—OH of the last nucleotide of the primer.

To perform PCR, a small amount of DNA is mixed with Taq polymerase, the primer, and the four DNA nucleotide triphosphates. The mixture is then placed in an instrument called a *thermocycler*. The temperature in the thermocycler is raised to 94–96°C for several minutes to separate the two strands of DNA. The

LEARNING GOAL

11 Describe the process of polymerase chain reaction and discuss potential uses of the process.

An example of a DNA fingerprint used in a criminal case. The DNA sample designated *V* is that of the victim and the sample designated *D* is that of the defendant. The samples labeled *jeans* and *shirt* were taken from the clothing of the defendant. The DNA bands from the defendant's clothing clearly match the DNA bands of the victim, providing evidence of the guilt of the defendant.

DNA fingerprinting is now routinely used for paternity testing, testing for certain genetic disorders, and identification of the dead in cases where no other identification is available. DNA fingerprints are used as evidence in criminal cases involving rape and murder. In such cases, the evidence may be little more than a hair with an intact follicle on the clothing of the victim.

Less widely known is the use of DNA fingerprinting to study genetic diversity in natural populations of plants and animals. The greater the genetic diversity, the healthier the population is likely to be. Populations with low genetic diversity face a far higher probability of extinction under adverse conditions. Customs officials have used DNA fingerprinting to determine whether confiscated elephant tusks were taken illegally from an endangered population of elephants or were obtained from a legally harvested population.

In the case of the coca plants, ARS wanted to know whether the drug cartels were developing improved strains that might be hardier or more pest resistant or that have a higher concentration of cocaine. Their conclusions, which you can read in *Phytochemistry* (*64: 187–197, 2003*), were that the drug cartels have introduced significant genetic modification into coca plants in Colombia in the last two decades. In addition, some of these new variants, those producing the highest levels of cocaine, have been transplanted to other regions of the country. All of this indicates that the cocaine agribusiness is thriving.

For Further Understanding

As this sampling of applications suggests, DNA fingerprinting has become an invaluable tool in law enforcement, medicine, and basic research. What other applications of this technology can you think of?

Do some research on the development of DNA fingerprinting as a research and forensics tool. What is the probability that two individuals will have the same DNA fingerprint? How are these probabilities determined?

temperature is then dropped to 50–56°C to allow the primers to hybridize to the target DNA. Finally, the temperature is raised to 72°C to allow Taq polymerase to act, reading the template DNA strand and polymerizing a daughter strand extended from the primer. At the end of this step, the amount of the gene has doubled (Figure 20.25).

Now the three steps are repeated. With each cycle the amount of the gene is doubled. Theoretically after thirty cycles, you have one billion times more DNA than you started with!

PCR can be used in genetic screening to detect the gene responsible for muscular dystrophy. It can also be used to diagnose disease. For instance, it can be used to amplify small amounts of HIV in the blood. It can also be used by forensic scientists to amplify DNA from a single hair follicle or a tiny drop of blood at a crime scene.

Animations

PCR Reactions

Polymerase Chain Reaction

DNA Fingerprinting

Discuss strategies for genome analysis and DNA sequencing.

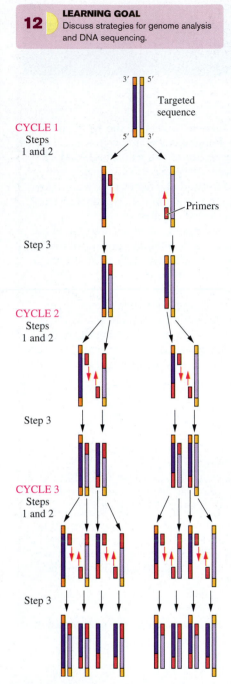

CYCLE 1
Steps
1 and 2

Step 3

CYCLE 2
Steps
1 and 2

Step 3

CYCLE 3
Steps
1 and 2

Step 3

Figure 20.25
Polymerase chain reaction.

Animation
Sanger Sequencing

20.10 The Human Genome Project

In 1990 the Department of Energy and the National Institutes of Health began the Human Genome Project (HGP), a multinational project that would extend into the next millennium. The goals of the HGP were to identify all of the genes in human DNA and to sequence the entire three billion nucleotide pairs of the genome. In order to accomplish these goals, enormous computer databases had to be developed to store the information and computer software had to be designed to analyze it.

Initially, the HGP planned to complete the work by the year 2005. However, as a result of technological advances made by those in the project, a working draft of the human genome was published in February 2001 and the successful completion of the project was announced on April 14, 2003.

Genetic Strategies for Genome Analysis

The strategy for HGP was rather straightforward. In order to determine the DNA sequence of the human genome, genomic libraries had to be produced. A *genomic library* is a set of clones representing the entire genome. The DNA sequences of each of these clones could then be determined. Of course, once the sequence of each of these clones is determined, there is no way to know how they are arranged along the chromosomes.

A second technique, called *chromosome walking,* provides both DNA sequence information, as well as a method for identifying the DNA sequences next to it on the chromosome. This method requires clones that are overlapping. To accomplish this, libraries of clones are made using many different restriction enzymes. The DNA sequence of a fragment is determined. Then that information is used to develop a probe for any clones in the library that are overlapping. Each time a DNA fragment is sequenced, the information is used to identify overlapping clones. This process continues, allowing scientists to walk along the chromosome in two directions until the entire sequence is cloned, mapped, and sequenced.

DNA Sequencing

The method of DNA sequencing that is used is based on a technique developed by Frederick Sanger. A cloned piece of DNA is separated into its two strands. Each of these will serve as a template strand to carry out DNA replication in test tubes. A primer strand is also needed. This is a short piece of DNA that will hybridize to the template strand. The primer is the starting point for addition of new nucleotides during DNA synthesis.

The DNA is then placed in four test tubes with all of the enzymes and nucleotides required for DNA synthesis. In addition, each tube contains an unusual nucleotide, called a dideoxynucleotide. These nucleotides differ from the standard nucleotides by having a hydrogen atom at the 3' position of the deoxyribose, rather than a hydroxyl group. When a dideoxynucleotide is incorporated into a growing DNA chain, it acts as a chain terminator. Because it does not have a 3'-hydroxyl group, no phosphoester bond can be formed with another nucleotide and no further polymerization can occur.

Each of the four tubes containing the DNA, enzymes, and an excess of the nucleotides required for replication will also have a small amount of one of the four dideoxynucleotides. In the tube that receives dideoxyadenosine triphosphate (ddA), for example, DNA synthesis will begin. As replication proceeds, either the standard nucleotide or ddA will be incorporated into the growing strand. Since the standard nucleotide is present in excess, the dideoxynucleotide will be incorporated infrequently and randomly. This produces a family of DNA fragments that terminate at the location of one of the deoxyadenosines in the molecule.

A Medical Perspective

A Genetic Approach to Familial Emphysema

Familial emphysema is a human genetic disease resulting from the inability to produce the protein α_1-antitrypsin. See also A Medical Perspective: α_1-Antitrypsin and Familial Emphysema, in Chapter 19. In individuals who have inherited one or two copies of the α_1-antitrypsin gene, this serum protein protects the lungs from the enzyme elastase. Normally, elastase fights bacteria and helps in the destruction and removal of dead lung tissue. However, the enzyme can also cause lung damage. By inhibiting elastase, α_1-antitrypsin prevents lung damage. Individuals who have inherited two defective α_1-antitrypsin genes do not produce this protein and suffer from familial, or A1AD, emphysema. In the absence of α_1-antitrypsin, the elastase and other proteases cause the severe lung damage characteristic of emphysema.

A1AD is the second most common genetic disorder in Caucasians. It is estimated that there are 100,000 sufferers in the United States and that one in five Americans carries the gene. The disorder, discovered in 1963, is often misdiagnosed as asthma or chronic obstructive pulmonary disease. In fact, it is estimated that fewer than 5% of the sufferers are diagnosed with A1AD.

The α_1-antitrypsin gene has been cloned. Early experiments with sheep showed that the protein remains stable when administered as an aerosol and remains functional after it has passed through the pulmonary epithelium. This research offers hope of an effective treatment for this disease.

The current treatment involves weekly IV injections of α_1-antitrypsin. The supply of the protein, purified from human plasma that has been demonstrated to be virus free, is rather limited. Thus, the injections are expensive. In addition, they are painful. These two factors cause some sufferers to refuse the treatment.

Recently Dr. Terry Flotte and his colleagues at the University of Florida have taken a new approach. They have cloned the gene for α_1-antitrypsin into the DNA of adeno-associated virus. This virus is an ideal vector for human gene replacement therapy because it replicates only in cells that are not dividing and it does not stimulate a strong immune or inflammatory response. The researchers injected the virus carrying the cloned α_1-antitrypsin gene into the muscle tissue of mice, then tested for the level of α_1-antitrypsin in the blood. The results were very promising. Effective levels of α_1-antitrypsin were produced in the muscle cells of the mice and secreted into the bloodstream. Furthermore, the level of α_1-antitrypsin remained at therapeutic levels for more than four months. The research team is planning tests with larger animals and eventually will confirm their results in human trials.

For Further Understanding

Of the three treatments described in this perspective, which do you think has the highest probability of success in the long term? Defend your answer.

Why is it impractical to "replace" the defective gene in an adult suffering from a genetic disease such as A1AD?

The same reaction is done with each of the dideoxynucleotides. The DNA fragments are then separated by gel electrophoresis on a DNA sequencing gel. The four reactions are placed in four wells, side by side, on the gel. Following electrophoresis, the DNA sequence can be read directly from the gel, as shown in Figure 20.26.

When chain termination DNA sequencing was first done, radioactive isotopes were used to label the DNA strands. However, new technology has resulted in automated systems that employ dideoxynucleotides that are labeled with fluorescent dyes, a different color for each dideoxynucleotide. Because each reaction (A, G, C, and T) will be a different color, all the reactions can be done in a single reaction mixture and the products separated on a single lane of a sequencing gel. A computer then "reads" the gel by distinguishing the color of each DNA band. The sequence information is directly stored into a databank for later analysis.

There is currently a vast amount of DNA information available on the Internet. The complete genomes of many bacteria have been reported, as well as the sequence information generated by the Human Genome Project. Because we know the genetic code, we can predict the amino acid sequence of proteins encoded by the genes. Researchers can also compare the sequences of normal genes with those of people suffering from genetic disorders. The enormity of the DNA information

Figure 20.26

DNA sequencing by chain termination requires a template DNA strand and a radioactive primer. These are placed into each of four reaction mixtures that contain DNA polymerase, the four DNA nucleotides (dATP, dCTP, dGTP, and TTP), as well as one of the four dideoxynucleotides. Following the reaction, the products are separated on a DNA sequencing gel. The sequence is read from an autoradiograph of the gel.

DNA sequence of original strand

available, as well as the many types of analysis that need to be carried out, have given rise to an entirely new branch of science. The field of **bioinformatics** is a marriage of computer information sciences and DNA technology that is helping to devise methods for understanding, analyzing, and applying the DNA sequence information that we are gathering.

SUMMARY

20.1 The Structure of the Nucleotide

DNA and RNA are polymers of nucleotides, which are composed of a five-carbon sugar (ribose in RNA and 2'-deoxyribose in DNA), a nitrogenous base, and one, two, or three phosphoryl groups. There are two kinds of nitrogenous bases, the *purines* (adenine and guanine) and the *pyrimidines* (cytosine, thymine, and uracil). *Deoxyribonucleotides* are the subunits of DNA. *Ribonucleotides* are the subunits of RNA.

20.2 The Structure of DNA and RNA

Nucleotides are joined by 3'–5' phosphodiester bonds in both DNA and RNA. DNA is a *double helix*, two strands of DNA wound around one another. The sugar-phosphate backbone is on the outside of the helix, and complementary pairs of bases extend into the center of the helix. The *base*

pairs are held together by hydrogen bonds. Adenine base pairs with thymine, and cytosine base pairs with guanine. The two strands of DNA in the helix are antiparallel to one another. RNA is single stranded.

20.3 DNA Replication

DNA replication involves synthesis of a faithful copy of the DNA molecule. It is semiconservative; each daughter molecule consists of one parental strand and one newly synthesized strand. *DNA polymerase III* "reads" each parental strand and synthesizes the complementary daughter strand according to the rules of base pairing.

20.4 Information Flow in Biological Systems

The *central dogma* states that the flow of biological information in cells is DNA → RNA → protein. There are three classes of RNA: *messenger RNA, transfer RNA,* and *ribosomal RNA. Transcription* is the process by which RNA molecules are synthesized. *RNA polymerase* catalyzes the synthesis of RNA. Transcription occurs in three stages: initiation, elongation, and termination. Eukaryotic genes contain *introns*, sequences that do not encode protein. These are removed from the primary transcript by the process of *RNA splicing*. The final mRNA contains only the protein coding sequences or *exons*. This final mRNA also has an added 5′ cap structure and 3′ poly(A) tail.

20.5 The Genetic Code

The genetic code is a triplet code. Each code word is called a *codon* and consists of three nucleotides. There are sixty-four codons in the genetic code. Of these, three are *termination codons* (UAA, UAG, and UGA), and the remaining sixty-one specify an amino acid. Most amino acids have several codons. As a result, the genetic code is said to be *degenerate.*

20.6 Protein Synthesis

The process of protein synthesis is called *translation*. The genetic code words on the mRNA are decoded by tRNA. Each tRNA has an *anticodon* that is complementary to a codon on the mRNA. In addition the tRNA is covalently linked to its correct amino acid. Thus hydrogen bonding between codon and anticodon brings the correct amino acid to the site of protein synthesis. Translation also occurs in three stages called initiation, chain elongation, and termination.

20.7 Mutation, Ultraviolet Light, and DNA Repair

Any change in a DNA sequence is a *mutation*. Mutations are classified according to the type of DNA alteration, including *point mutations, deletion mutations,* and *insertion mutations*. Ultraviolet light (UV) causes formation of *pyrimidine dimers*. Mistakes can be made during pyrimidine dimer repair, causing UV-induced mutations. Germicidal (UV) lamps are used to kill bacteria on environmental surfaces. UV damage to skin can result in skin cancer.

20.8 Recombinant DNA

Several tools are required for genetic engineering, including *restriction enzymes, agarose gel electrophoresis, hybridization,* and *cloning vectors*. Cloning a DNA fragment involves digestion of the target and vector DNA with a restriction enzyme. DNA ligase joins the target and vector DNA covalently, and the recombinant DNA molecules are introduced into bacterial cells by transformation. The desired clone is located by using antibiotic selection and hybridization. Many eukaryotic genes have been cloned for the purpose of producing medically important proteins.

20.9 Polymerase Chain Reaction

Using a heat-stable DNA polymerase produced by the bacterium *Thermus aquaticus* and specific DNA primers, polymerase chain reaction allows the amplification of DNA sequences that are present in small quantities. This technique is useful in genetic screening, diagnosis of viral or bacterial disease, and forensic science.

20.10 The Human Genome Project

The Human Genome Project has identified and mapped the genes of the human genome and determined the complete DNA sequence of each of the chromosomes. To do this, genomic libraries were generated and the DNA sequences of the clones were determined. To map the sequences along each chromosome, chromosome walking was used. DNA sequencing involves reactions in which DNA polymerase copies specific DNA sequences. Nucleotide analogues that cause chain termination (dideoxynucleotides) are incorporated randomly into the growing DNA chain. This generates a family of DNA fragments that differ in size by one nucleotide. DNA sequencing gels separate these fragments and provide DNA sequence data.

KEY TERMS

aminoacyl tRNA (20.6)
aminoacyl tRNA binding site
  of ribosome (A-site) (20.6)
aminoacyl tRNA
  synthetase (20.6)
anticodon (20.4)
antiparallel strands (20.2)
base pairs (20.2)
bioinformatics (20.10)
cap structure (20.4)
carcinogen (20.7)
central dogma (20.4)
chromosome (20.2)
cloning vector (20.8)
codon (20.4)
complementary strands (20.2)
degenerate code (20.5)
deletion mutation (20.7)
deoxyribonucleic acid
  (DNA) (20.1)
deoxyribonucleotide (20.1)
DNA polymerase III (20.3)
double helix (20.2)
elongation factor (20.6)
eukaryote (20.2)
exon (20.4)
genome (20.2)
hybridization (20.8)
initiation factor (20.6)
insertion mutation (20.7)
intron (20.4)
lagging strand (20.3)
leading strand (20.3)
messenger RNA
  (mRNA) (20.4)

mutagen (20.7)
mutation (20.7)
nucleosome (20.2)
nucleotide (20.1)
peptidyl tRNA binding site of
  ribosome (P-site) (20.6)
point mutation (20.7)
poly(A) tail (20.4)
polysome (20.6)
post-transcriptional
  modification (20.4)
primary transcript (20.4)
prokaryote (20.2)
promoter (20.4)
purine (20.1)
pyrimidine (20.1)
pyrimidine dimer (20.7)
release factor (20.6)
replication fork (20.3)
replication origin (20.3)
restriction enzyme (20.8)
ribonucleic acid (RNA) (20.1)
ribonucleotide (20.1)
ribosomal RNA (rRNA) (20.4)
ribosome (20.6)
RNA polymerase (20.4)
RNA splicing (20.4)
semiconservative
  replication (20.3)
silent mutation (20.7)
termination codon (20.6)
transcription (20.4)
transfer RNA (tRNA) (20.4)
translation (20.4)
translocation (20.6)

QUESTIONS AND PROBLEMS

The Structure of the Nucleotide

Foundations

20.13 What is a heterocyclic amine?

20.14 What components of nucleic acids are heterocyclic amines?

Applications

20.15 Draw the structure of the purine ring, and indicate the nitrogen that is bonded to sugars in nucleotides.

20.16 a. Draw the ring structure of the pyrimidines.
   b. In a nucleotide, which nitrogen atom of pyrimidine rings is bonded to the sugar?

20.17 ATP is the universal energy currency of the cell. What components make up the ATP nucleotide?

20.18 One of the energy-harvesting steps of the citric acid cycle results in the production of GTP. What is the structure of the GTP nucleotide?

The Structure of DNA and RNA

Foundations

20.19 The two strands of a DNA molecule are antiparallel. What is meant by this description?

20.20 List three differences between DNA and RNA.

Applications

20.21 What is the significance of the following repeat distances in the structure of the DNA molecule: 0.34 nm, 3.4 nm, and 2 nm?

20.22 Except for the functional groups attached to the rings, the nitrogenous bases are largely flat, hydrophobic molecules. Explain why the arrangement of the purines and pyrimidines found in DNA molecules is very stable.

20.23 How many hydrogen bonds link the adenine-thymine base pair?

20.24 How many hydrogen bonds link the guanine-cytosine base pair?

20.25 Write the structure that results when deoxycytosine-5′-monophosphate is linked by a 3′ ⟶ 5′ phosphodiester bond to thymidine-5′-monophosphate.

20.26 Write the structure that results when adenosine-5′-monophosphate is linked by a 3′ ⟶ 5′ phosphodiester bond to uridine-5′-monophosphate.

20.27 Describe the structure of the prokaryotic chromosome.

20.28 Describe the structure of the eukaryotic chromosome.

DNA Replication

Foundations

20.29 What is meant by semiconservative DNA replication?

20.30 Draw a diagram illustrating semiconservative DNA replication.

Applications

20.31 What are the two primary functions of DNA polymerase III?

20.32 a. Why is DNA polymerase said to be template-directed?
   b. Why is DNA replication a self-correcting process?

20.33 If a DNA strand had the nucleotide sequence

   5′-ATGCGGCTAGAATATTCCA-3′

what would the sequence of the complementary daughter strand be?

20.34 If the sequence of a double-stranded DNA is

   5′- G A A T T C C T T A A G G A T C G A T C -3′
   | | | | | | | | | | | | | | | | | | | | | |
   3′- C T T A A G G A A T T C C T A G C T A G -5′

what would the sequence of the two daughter DNA molecules be after DNA replication? Indicate which strands are newly synthesized and which are parental.

20.35 What is the replication origin of a DNA molecule?

20.36 What is occurring at the replication fork?

20.37 What is the function of the enzyme helicase?

20.38 What is the function of the enzyme primase?

20.39 What role does the RNA primer play in DNA replication?

20.40 Explain the differences between leading strand and lagging strand replication.

Information Flow in Biological Systems

Foundations

20.41 What is the central dogma of molecular biology?

20.42 What are the roles of DNA, RNA, and protein in information flow in biological systems?

20.43 On what molecule is the anticodon found?

20.44 On what molecule is the codon found?

Applications

20.45 If a gene had the nucleotide sequence

   5′-TACCTAGCTCTGGTCATTAAGGCAGTA-3′

what would the sequence of the mRNA be?

20.46 If a mRNA had the nucleotide sequence

   5′-AUGCCCUUUCAUUACCCGGUA-3′

what was the sequence of the DNA strand that was transcribed?

20.47 What is meant by the term *RNA splicing?*

20.48 The following is the unspliced transcript of a eukaryotic gene:

exon 1   intron A   exon 2   intron B   exon 3   intron C   exon 4
────────■■■■■■──────────■■■■■■──────────■■■■■■────────

What would the structure of the final mature mRNA look like, and which of the above sequences would be found in the mature mRNA?

20.49 List the three classes of RNA molecules.

20.50 What is the function of each of the classes of RNA molecules?

20.51 What is the function of the spliceosome?

20.52 What are snRNPs? How do they facilitate RNA splicing?

20.53 What is a poly(A) tail?

20.54 What is the purpose of the poly(A) tail on eukaryotic mRNA?

20.55 What is the cap structure?

20.56 What is the function of the cap structure on eukaryotic mRNA?

The Genetic Code

Foundations

20.57 How many codons constitute the genetic code?

20.58 What is meant by a triplet code?

20.59 What is meant by the reading frame of a gene?

20.60 What happens to the reading frame of a gene if a nucleotide is deleted?

Applications

20.61 Which two amino acids are encoded by only one codon?

20.62 Which amino acids are encoded by six codons?

20.63 An essential gene has the codon 5'-UUU-3' in a critical position. If this codon is mutated to the sequence 5'-UUA-3', what is the expected consequence for the cell?

20.64 An essential gene has the codon 5'-UUA-3' in a critical position. If this codon is mutated to the sequence 5'-UUG-3', what is the expected consequence for the cell?

Protein Synthesis

Foundations

20.65 What is the function of ribosomes?

20.66 What are the two tRNA binding sites on the ribosome?

Applications

20.67 Explain the relationship between the sequence of nucleotides of a gene in the DNA and the sequence of amino acids in the protein encoded by that gene.

20.68 Explain how a change in the sequence of nucleotides of a gene, a mutation, may alter the sequence of amino acids in the protein encoded by that gene.

20.69 Briefly describe the three stages of translation: initiation, elongation, and termination.

20.70 What peptide sequence would be formed from the mRNA 5'-AUGUGUAGUGACCAACCGAUUUCACUGUGA-3'?

The following diagram shows the reaction that produces an aminoacyl tRNA, in this case methionyl tRNA. Use the following diagram to answer Questions 20.71 and 20.72.

20.71 By what type of bond is an amino acid linked to a tRNA molecule in an aminoacyl tRNA molecule?

20.72 Draw the structure of an alanine residue bound to the 3' position of adenine at the 3' end of alanyl tRNA.

The amino acyl linkage is formed between the 3'—OH of the tRNA and the carboxylate group of the amino acid, methionine

Mutation, Ultraviolet Light, and DNA Repair

Foundations

20.73 Define the term *point mutation*.

20.74 What are deletion and insertion mutations?

Applications

20.75 Why are some mutations silent?

20.76 Which is more likely to be a silent mutation, a point mutation or a deletion mutation? Explain your reasoning.

20.77 What damage does UV light cause in DNA, and how does this lead to mutations?

20.78 Explain why UV lights are effective germicides on environmental surfaces.

20.79 What is a carcinogen? Why are carcinogens also mutagens?

20.80 a. What causes the genetic disease xeroderma pigmentosum?
  b. Why are people who suffer from xeroderma pigmentosum prone to cancer?

Recombinant DNA

Foundations

20.81 What is a restriction enzyme?

20.82 Of what value are restriction enzymes in recombinant DNA research?

20.83 What is a selectable marker?

20.84 What is a cloning vector?

Applications

20.85 Name three products of recombinant DNA that are of value in the field of medicine.

20.86 a. What is the ultimate goal of genetic engineering?
  b. What ethical issues does this goal raise?

Polymerase Chain Reaction

20.87 After ten cycles of polymerase chain reaction, how many copies of target DNA would you have for each original molecule in the mixture?

20.88 List several practical applications of polymerase chain reaction.

The Human Genome Project

Foundations

20.89 What were the major goals of the Human Genome Project?

20.90 What are the potential benefits of the information gained in the Human Genome Project?

20.91 What is a genome library?

20.92 What is meant by the term *chromosome walking*?

20.93 What is a dideoxynucleotide?

20.94 How does a dideoxynucleotide cause chain termination in DNA replication?

Applications

20.95 A researcher has determined the sequence of the following pieces of DNA. Using this sequence information, map the location of these pieces relative to one another.
  a. 5' AGCTCCTGATTTCATACAGTTTCTACTACCTACTA 3'
  b. 5' AGACATTCTATCTACCTAGACTATGTTCAGAA 3'
  c. 5' TTCAGAACTCATTCAGACCTACTACTATACCTTGG GAGCTCCT 3'
  d. 5' ACCTACTAGACTATACTACTACTAAGGGGACTATT CCAGACTT 3'

20.96 Draw a DNA sequencing gel that would represent the sequence shown below. Be sure to label which lanes of the gel represent each of the four dideoxynucleotides in the chain termination reaction mixture.

5' GACTATCCTAG 3'

CRITICAL THINKING PROBLEMS

1. It has been suggested that the triplet genetic code evolved from a two-nucleotide code. Perhaps there were fewer amino acids in the ancient proteins. Examine the genetic code in Figure 20.17. What features of the code support this hypothesis?

2. The strands of DNA can be separated by heating the DNA sample. The input heat energy breaks the hydrogen bonds between base pairs, allowing the strands to separate from one another. Suppose that you are given two DNA samples. One has a G + C content of 70% and the other has a G + C content of 45%. Which of these samples will require a higher temperature to separate the strands? Explain your answer.

3. A mutation produces a tRNA with a new anticodon. Originally the anticodon was 5′-CCA-3′; the mutant anticodon is 5′-UCA-3′. What effect will this mutant tRNA have on cellular translation?

4. You have just cloned an EcoR1 fragment that is 1650 base pairs (bp) and contains the gene for the hormone leptin. Your first job is to prepare a restriction enzyme map of the recombinant plasmid. You know that you have cloned into a plasmid vector that is 805 bp and that has only one EcoR1 site (the one into which you cloned). There are no other restriction enzyme sites in the plasmid. The following table shows the restriction enzymes used and the DNA fragment sizes that result. Draw a map of the circular recombinant plasmid and a representation of the gel from which the fragment sizes were obtained.

Restriction Enzymes	DNA Fragment Sizes (bp)
EcoR1	805, 1650
EcoR1 + BamHI	450, 805, 1200
EcoR1 + SalI	200, 805, 1450
BamHI + SalI	200, 250, 805, 1200

5. A scientist is interested in cloning the gene for blood clotting factor VIII into bacteria so that large amounts of the protein can be produced to treat hemophiliacs. Knowing that bacterial cells cannot carry out RNA splicing, she clones a complementary DNA copy of the factor VIII mRNA and introduces this into bacteria. However, there is no transcription of the cloned factor VIII gene. How could the scientist engineer the gene so that the bacterial cell RNA polymerase will transcribe it?

21 Carbohydrate Metabolism

Learning Goals

1 Discuss the importance of ATP in cellular energy transfer processes.

2 Describe the three stages of catabolism of dietary proteins, carbohydrates, and lipids.

3 Discuss glycolysis in terms of its two major segments.

4 Looking at an equation representing any of the chemical reactions that occur in glycolysis, describe the kind of reaction that is occurring and the significance of that reaction to the pathway.

5 Describe the mechanism of regulation of the rate of glycolysis. Discuss particular examples of that regulation.

6 Discuss the practical and metabolic roles of fermentation reactions.

7 List several products of the pentose phosphate pathway that are required for biosynthesis.

8 Compare glycolysis and gluconeogenesis.

9 Summarize the regulation of blood glucose levels by glycogenesis and glycogenolysis.

Outline

Introduction

Chemistry Connection:
The Man Who Got Tipsy from Eating Pasta

21.1 ATP: The Cellular Energy Currency

21.2 Overview of Catabolic Processes

21.3 Glycolysis

A Medical Perspective:
Genetic Disorders of Glycolysis

21.4 Fermentations

A Human Perspective:
Fermentations: The Good, the Bad, and the Ugly

21.5 The Pentose Phosphate Pathway

21.6 Gluconeogenesis: The Synthesis of Glucose

21.7 Glycogen Synthesis and Degradation

A Medical Perspective:
Diagnosing Diabetes

A Human Perspective:
Glycogen Storage Diseases

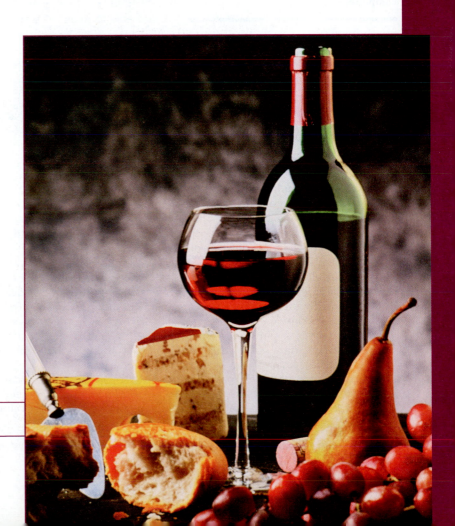

Some familiar fermentation products.

Introduction

Just as we need energy to run, jump, and think, the cell needs a ready supply of cellular energy for the many functions that support these activities. Cells need energy for active transport, to move molecules between the environment and the cell. Energy is also needed for biosynthesis of small metabolic molecules and production of macromolecules from these intermediates. Finally, energy is required for mechanical work, including muscle contraction and motility of sperm cells. Table 21.1 lists some examples of each of these energy-requiring processes.

We need a supply of energy-rich food molecules that can be degraded, or oxidized, to provide this needed cellular energy. Our diet includes three major sources of energy: carbohydrates, fats, and proteins. Each of these types of large biological molecules must be broken down into its basic subunits—simple sugars, fatty acids and glycerol, and amino acids—before they can be taken into the cell and used to produce cellular energy. Of these classes of food molecules, carbohydrates are the most readily used. The pathway for the first stages of carbohydrate breakdown is called glycolysis. We find the same pathway in organisms as different as the simple bacterium and humans.

In this chapter we are going to examine the steps of this ancient energy-harvesting pathway. We will see that it is responsible for the capture of some of the bond energy of carbohydrates and the storage of that energy in the molecular form of adenosine triphosphate (ATP). Glycolysis actually releases and stores very little (2.2%) of the potential energy of glucose, but the pathway also serves as a source of biosynthetic building blocks. It also modifies the carbohydrates in such a way that other pathways are able to release as much as 40% of the potential energy.

Recall that the potential energy of a compound is the bond energy of that compound.

Chemistry Connection

The Man Who Got Tipsy from Eating Pasta

Imagine becoming drunk after eating a plate of spaghetti or a bag of potato chips. That is exactly what happened to Charles Swaart while he was stationed in Tokyo after World War II. Suddenly, he would be completely drunk—without having swallowed a drop of alcohol.

During the next two decades, Swaart continued to have unexplainable bouts of drunkenness and horrible hangovers. The problem was so serious that his liver was being destroyed. But in 1964, Swaart heard of a man in Japan who suffered from the same mysterious—and embarrassing—symptoms. After twenty-five years, physicians diagnosed the problem. There was a mutant strain of the yeast *Candida albicans* living in the gastrointestinal tract of the Japanese man. These yeast cells were using carbohydrates from the man's diet to make ethanol. The metabolic pathways used by these yeast cells were glycolysis and alcohol fermentation, two of the pathways that we will study in this chapter.

Swaart took advantage of the therapy used in Japan. He had to try several antibiotics over the years. But finally, in 1975, all of the mutant yeast cells in his intestine were killed, and his life returned to normal.

Why was it so difficult for physicians to solve this medical mystery? Nonfermenting *Candida albicans* is a regular inhabitant of the human gut. It took some very clever scientific detective work to find this mutant ethanol-producing strain. The scientists even have a hypothesis about where the mutant yeast came from. They think that the radiation released in one of the atomic bomb blasts at Nagasaki or Hiroshima may have caused the mutation.

In Chapter 20 we learned about the kinds of DNA damage that cause mutations. In this chapter we begin our study of the chemical reactions used by all organisms to provide energy for cellular work.

TABLE 21.1	The Types of Cellular Work That Require Energy

Biosynthesis: Synthesis of Metabolic Intermediates and Macromolecules

Synthesis of glucose from CO_2 and H_2O in the process of photosynthesis in plants
Synthesis of amino acids
Synthesis of nucleotides
Synthesis of lipids
Protein synthesis from amino acids
Synthesis of nucleic acids
Synthesis of organelles and membranes

Active Transport: Movement of Ions and Molecules

Transport of H^+ to maintain constant pH
Transport of food molecules into the cell
Transport of K^+ and Na^+ into and out of nerve cells for transmission of nerve impulses
Secretion of HCl from parietal cells into the stomach
Transport of waste from the blood into the urine in the kidneys
Transport of amino acids and most hexose sugars into the blood from the intestine
Accumulation of calcium ions in the mitochondria

Motility

Contraction and flexion of muscle cells
Separation of chromosomes during cell division
Ability of sperm to swim via flagella
Movement of foreign substances out of the respiratory tract by cilia on the epithelial lining of the trachea
Translocation of eggs into the fallopian tubes by cilia in the female reproductive tract

21.1 ATP: The Cellular Energy Currency

The degradation of fuel molecules, called **catabolism,** provides the energy for cellular energy-requiring functions, including **anabolism,** or biosynthesis. Actually, the energy of a food source can be released in one of two ways: as heat or, more important to the cell, as chemical bond energy. We can envision two alternative modes of aerobic degradation of the simple sugar glucose. Imagine that we simply set the glucose afire. This would result in its complete oxidation to CO_2 and H_2O and would release 686 kcal/mol of glucose. Yet in terms of a cell, what would be accomplished? Nothing. All of the potential energy of the bonds of glucose is lost as heat and light.

The cell uses a different strategy. With a series of enzymes, biochemical pathways in the cell carry out a step-by-step oxidation of glucose. Small amounts of energy are released at several points in the pathway and that energy is harvested and saved in the bonds of a molecule that has been called the *universal energy currency.* This molecule is **adenosine triphosphate (ATP).**

ATP serves as a "go-between" molecule that couples the *exergonic* (energy releasing) reactions of catabolism and the endergonic (energy requiring) reactions of anabolism. To understand how this molecule harvests the energy and releases it for energy-requiring reactions, we must take a look at the structure of this amazing molecule (Figure 21.1). ATP is a **nucleotide,** which means that it is a molecule composed of a nitrogenous base; a five-carbon sugar; and one, two, or three phosphoryl groups.

LEARNING GOAL
1 Discuss the importance of ATP in cellular energy transfer processes.

Nitrogenous bases are heterocyclic amines. Their structure and functions are discussed in Section 15.2. The five-carbon sugars are discussed in Section 16.4. More information on the structure of nucleotides is found in Section 20.1.

Figure 21.1

The structure of the universal energy currency, ATP.

Nature's high energy bonds, including phosphoanhydride and phosphoester bonds, are discussed in Section 14.4.

See Sections 14.4 and 20.1 for further information on the structure of ATP and hydrolysis of the phosphoanhydride bonds.

In ATP, a phosphoester bond joins the first phosphoryl group to the five-carbon sugar ribose. The next two phosphoryl groups are joined to one another by phosphoanhydride bonds (Figure 21.1). Recall that the phosphoanhydride bond is a *high-energy bond*. When it is broken or hydrolyzed, a large amount of energy is released. When the phosphoanhydride bond of ATP is broken, the energy that is released can be used for cellular work. These high-energy bonds are indicated as squiggles ($\sim$) in Figure 21.1.

Hydrolysis of ATP yields adenosine diphosphate (ADP), an inorganic phosphate group (P_i), and energy (Figure 21.2). The energy released by this hydrolysis of ATP is then used to drive biological processes, for instance, the phosphorylation of glucose or fructose.

An example of the way in which the energy of ATP is used can be seen in the first step of glycolysis, the anaerobic degradation of glucose to harvest chemical energy. The first step involves the transfer of a phosphoryl group, $-PO_3^{2-}$, from ATP to the C-6 hydroxyl group of glucose (Figure 21.2). This reaction is catalyzed by the enzyme hexokinase.

This reaction can be dissected to reveal the role of ATP as a source of energy. Although this is a coupled reaction, we can think of it as a two-step process.

Figure 21.2

The hydrolysis of the phosphoanhydride bond of ATP releases inorganic phosphate and energy. In this coupled reaction catalyzed by an enzyme, the phosphoryl group and some of the released energy are transferred to β-D-glucose.

The first step is the hydrolysis of ATP to ADP and phosphate, abbreviated P_i. This is an exergonic reaction that *releases* about 7 kcal/mol of energy:

$$ATP + H_2O \longrightarrow ADP + P_i + 7\,kcal/mol$$

The second step, the synthesis of glucose-6-phosphate from glucose and phosphate, is an endergonic reaction that *requires* 3.0 kcal/mol:

$$3.0\,kcal/mol + glucose + P_i \longrightarrow glucose\text{-}6\text{-}phosphate + H_2O$$

These two chemical reactions can then be added to give the equation showing the way in which ATP hydrolysis is *coupled* to the phosphorylation of glucose:

$$ATP + H_2O \longrightarrow ADP + P_i + 7\,kcal/mol$$
$$3.0\,kcal/mol + glucose + P_i \longrightarrow glucose\text{-}6\text{-}phosphate + H_2O$$
$$\overline{\text{Net: } ATP + glucose \longrightarrow glucose\text{-}6\text{-}phosphate + ADP + 4\,kcal/mol}$$

Because the hydrolysis of ATP releases more energy than is required to synthesize glucose-6-phosphate from glucose and phosphate, there is an overall energy release in this process and the reaction proceeds spontaneously to the right. The product, glucose-6-phosphate, has more energy than the reactant, glucose, because it now carries some of the energy from the original phosphoanhydride bond of ATP.

The primary function of all catabolic pathways is to harvest the chemical energy of fuel molecules and to store that energy by the production of ATP. This continuous production of ATP is what provides the stored potential energy that is used to power most cellular functions.

Question 21.1

Why is ATP called the universal energy currency?

Question 21.2

List five biological activities that require ATP.

21.2 Overview of Catabolic Processes

Although carbohydrates, fats, and proteins can all be degraded to release energy, carbohydrates are the most readily used energy source. We will begin by examining the oxidation of the hexose glucose. In Chapters 22 and 23 we will see how the pathways of glucose oxidation are also used for the degradation of fats and proteins.

Any catabolic process must begin with a supply of nutrients. When we eat a meal, we are eating quantities of carbohydrates, fats, and proteins. From this point the catabolic processes can be broken down into a series of stages. The three stages of catabolism are summarized in Figure 21.3.

LEARNING GOAL

2 ▸ Describe the three stages of catabolism of dietary proteins, carbohydrates, and lipids.

Stage I: Hydrolysis of Dietary Macromolecules into Small Subunits

The purpose of the first stage of catabolism is to degrade large food molecules into their component subunits. These subunits—simple sugars, amino acids, fatty acids, and glycerol—are then taken into the cells of the body for use as an energy source. The process of digestion is summarized in Figure 21.4.

Polysaccharides are hydrolyzed to monosaccharides. This process begins in the mouth, where the enzyme amylase begins the hydrolysis of starch. Digestion continues in the small intestine, where pancreatic amylase further hydrolyzes the

Figure 21.3

The three stages of the conversion of food into cellular energy in the form of ATP.

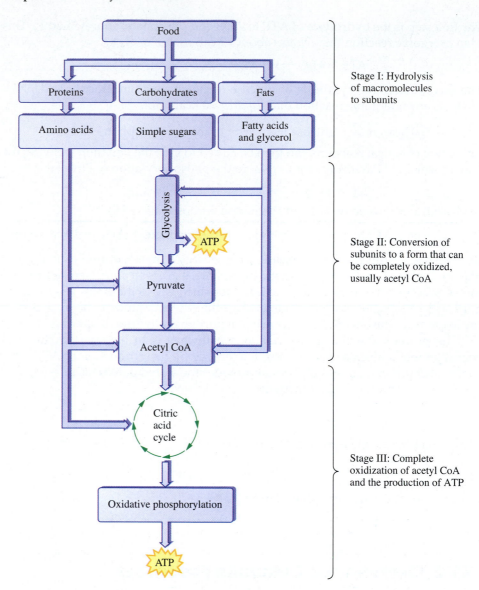

Stage I: Hydrolysis of macromolecules to subunits

Stage II: Conversion of subunits to a form that can be completely oxidized, usually acetyl CoA

Stage III: Complete oxidization of acetyl CoA and the production of ATP

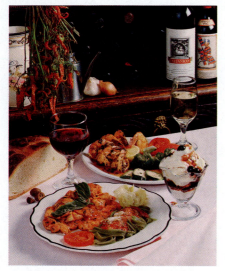

Describe the path of the carbohydrates, lipids, and proteins in this meal from digestion through the biochemical energy-harvesting reactions.

In the laboratory, a strong acid or base and high temperatures are required for hydrolysis of amide bonds (Section 15.3). However, this reaction proceeds quickly under physiological conditions when catalyzed by enzymes (Section 19.11).

The digestion and transport of fats are considered in greater detail in Chapter 23.

starch into maltose (a disaccharide of glucose). Maltase catalyzes the hydrolysis of maltose, producing two glucose molecules. Similarly, sucrose is hydrolyzed to glucose and fructose by the enzyme sucrase, and lactose (milk sugar) is degraded into the monosaccharides glucose and galactose by the enzyme lactase in the small intestine. The monosaccharides are taken up by the epithelial cells of the intestine in an energy-requiring process called *active transport.*

The digestion of proteins begins in the stomach, where the low pH denatures the proteins so that they are more easily hydrolyzed by the enzyme pepsin. They are further degraded in the small intestine by trypsin, chymotrypsin, elastase, and other proteases. The products of protein digestion—amino acids and short oligopeptides—are taken up by the cells lining the intestine. This uptake also involves an active transport mechanism.

Digestion of fats does not begin until the food reaches the small intestine, even though there are lipases in both the saliva and stomach fluid. Fats arrive in the duodenum, the first portion of the small intestine, in the form of large fat globules. Bile salts produced by the liver break these up into an emulsion of tiny fat droplets. Because the small droplets have a greater surface area, the lipids are now more accessible to the action of pancreatic lipase. This enzyme hydrolyzes the fats into fatty acids and glycerol, which are taken up by intestinal cells by a transport

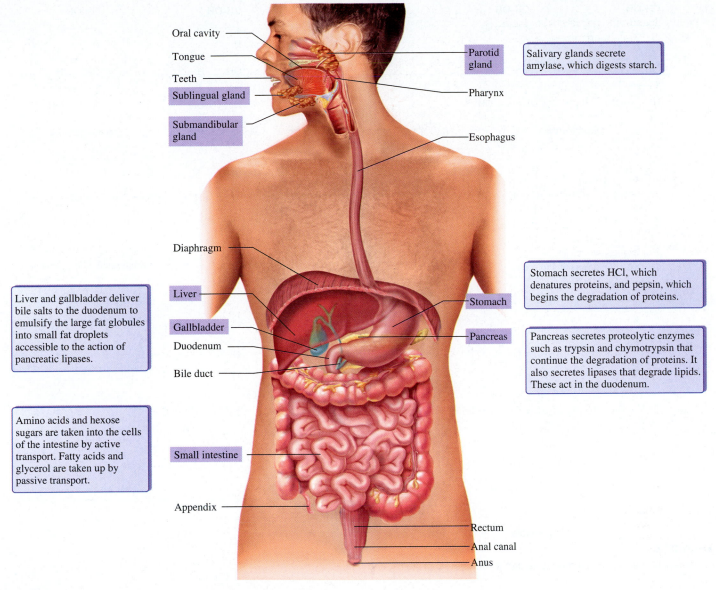

Oral cavity

Tongue

Teeth

Sublingual gland

Submandibular gland

Parotid gland

Salivary glands secrete amylase, which digests starch.

Pharynx

Esophagus

Diaphragm

Liver

Liver and gallbladder deliver bile salts to the duodenum to emulsify the large fat globules into small fat droplets accessible to the action of pancreatic lipases.

Gallbladder

Duodenum

Bile duct

Stomach

Stomach secretes HCl, which denatures proteins, and pepsin, which begins the degradation of proteins.

Pancreas

Pancreas secretes proteolytic enzymes such as trypsin and chymotrypsin that continue the degradation of proteins. It also secretes lipases that degrade lipids. These act in the duodenum.

Amino acids and hexose sugars are taken into the cells of the intestine by active transport. Fatty acids and glycerol are taken up by passive transport.

Small intestine

Appendix

Rectum

Anal canal

Anus

Figure 21.4
An overview of the digestive processes that hydrolyze carbohydrates, proteins, and fats.

process that does not require energy. This process is called *passive transport*. A summary of these hydrolysis reactions is shown in Figure 21.5.

Stage II: Conversion of Monomers into a Form That Can Be Completely Oxidized

The monosaccharides, amino acids, fatty acids, and glycerol must now be assimilated into the pathways of energy metabolism. The two major pathways are glycolysis and the citric acid cycle (see Figure 21.3). Sugars usually enter the glycolysis pathway in the form of glucose or fructose. They are eventually converted to acetyl CoA, which is a form that can be completely oxidized in the citric acid cycle. Amino groups are removed from amino acids, and the remaining carbon skeletons enter the catabolic processes at many steps of the citric acid cycle. Fatty acids are converted to acetyl CoA and enter the citric acid cycle in that form. Glycerol, produced by the hydrolysis of fats, is converted to glyceraldehyde-3-phosphate, one of the intermediates of glycolysis, and enters energy metabolism at that level.

The citric acid cycle is considered in detail in Section 22.4.

Disaccharide   +   Water   ⟶   Monosaccharide   +   Monosaccharide

Peptide
(portion of protein molecule)   +   Water   ⟶   Amino acid   +   Amino acid

Triglyceride   +   Water   ⟶   Fatty acids   +   Glycerol

Figure 21.5
A summary of the hydrolysis reactions of carbohydrates, proteins, and fats.

Stage III: The Complete Oxidation of Nutrients and the Production of ATP

Oxidative phosphorylation is described in Section 22.6.

Acetyl CoA carries acetyl groups, two-carbon remnants of the nutrients, to the citric acid cycle. Acetyl CoA enters the cycle, and electrons and hydrogen atoms are harvested during the complete oxidation of the acetyl group to CO_2. Coenzyme A is released (recycled) to carry additional acetyl groups to the pathway. The electrons and hydrogen atoms that are harvested are used in the process of oxidative phosphorylation to produce ATP.

Question *21.3*   Briefly describe the three stages of catabolism.

Question *21.4*   Discuss the digestion of dietary carbohydrates, lipids, and proteins.

21.3 Glycolysis

An Overview

Glycolysis, also known as the Embden-Meyerhof Pathway, is a pathway for carbohydrate catabolism that begins with the substrate D-glucose. The very fact that all organisms can use glucose as an energy source for glycolysis suggests that glycolysis was the first successful energy-harvesting pathway that evolved on the earth. The pathway evolved at a time when the earth's atmosphere was *anaerobic;* no free oxygen was available. As a result, glycolysis requires no oxygen; it is an anaerobic process. Further, it must have evolved in very simple, single-celled organisms, much like bacteria. These organisms did not have complex organelles in the cytoplasm to carry out specific cellular functions. Thus glycolysis was a process carried out by enzymes that were free in the cytoplasm. To this day, glycolysis remains an anaerobic process carried out by cytoplasmic enzymes, even in cells as complex as our own.

The ten steps of glycolysis, catalyzed by ten enzymes, are outlined in Figure 21.6. The first reactions of glycolysis involve an energy investment. ATP molecules are hydrolyzed, energy is released, and phosphoryl groups are added to the hexose sugars. In the remaining steps of glycolysis, energy is harvested to produce a net gain of ATP.

The three major products of glycolysis are seen in Figure 21.6. These are chemical energy in the form of ATP, chemical energy in the form of NADH, and two three-carbon pyruvate molecules. Each of these products is considered below:

- **Chemical energy as ATP.** Four ATP molecules are formed by the process of **substrate-level phosphorylation.** This means that a high-energy phosphoryl group from one of the substrates in glycolysis is transferred to ADP to form ATP. The two substrates involved in these transfer reactions are 1,3-bisphosphoglycerate and phosphoenolpyruvate (see Figure 21.6, steps 7 and 10). Although four ATP molecules are produced during glycolysis, the *net* gain is only two ATP molecules because two ATP molecules are used early in glycolysis (Figure 21.6, steps 1 and 3). The two ATP molecules produced represent only 2.2% of the potential energy of the glucose molecule. Thus glycolysis is not a very efficient energy-harvesting process.

- **Chemical energy in the form of reduced NAD^+, NADH. Nicotinamide adenine dinucleotide (NAD^+)** is a coenzyme derived from the vitamin niacin. The reduced form of NAD^+, NADH, carries hydride anions, hydrogen atoms with two electrons ($H:^-$), removed during the oxidation of glyceraldehyde-3-phosphate (see Figure 21.6, step 6). Under aerobic conditions, the electrons and hydrogen atom are transported from the cytoplasm into the mitochondria. Here they enter an electron transport system for the generation of ATP by **oxidative phosphorylation.** Under anaerobic conditions, NADH is used as a source of electrons in fermentation reactions.

- **Two pyruvate molecules.** At the end of glycolysis the six-carbon glucose molecule has been converted into two three-carbon pyruvate molecules. The fate of the pyruvate also depends on whether the reactions are occurring in the presence or absence of oxygen. Under aerobic conditions it is used to produce acetyl CoA destined for the citric acid cycle and complete oxidation. Under anaerobic conditions it is used as an electron acceptor in fermentation reactions.

In any event these last two products must be used in some way so that glycolysis can continue to function and produce ATP. There are two reasons for this. First, if pyruvate were allowed to build up, it would cause glycolysis to stop,

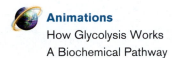

Animations
How Glycolysis Works
A Biochemical Pathway

The structure of NAD^+ and the way it functions as a hydride anion carrier are shown in Figure 19.8 and described in Section 19.7.

Animation
How NAD^+ Works

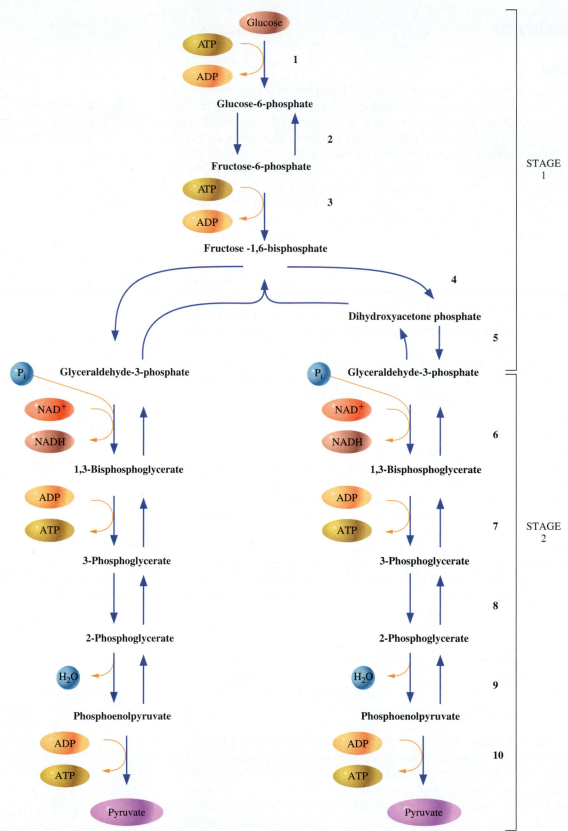

Figure 21.6
A summary of the reactions of glycolysis. These reactions occur in the cell cytoplasm.

thereby stopping the production of ATP. Thus pyruvate must be used in some kind of follow-up reaction, aerobic or anaerobic. Second, in step 6, glyceraldehyde-3-phosphate is oxidized and NAD^+ is reduced (accepts the hydride anion). The cell has only a small supply of NAD^+. If all the NAD^+ is reduced, none will be available for this reaction, and glycolysis will stop. Therefore NADH must be reoxidized so that glycolysis can continue to produce ATP for the cell.

Reactions of Glycolysis

Glycolysis can be divided into two major segments. The first is the investment of ATP energy. Without this investment, glucose would not have enough energy for glycolysis to continue, and there would be no ATP produced. This segment includes the first five reactions of the pathway. The second major segment involves the remaining reactions of the pathway (steps 6–10), those that result in a net energy yield.

> **3** **LEARNING GOAL**
> Discuss glycolysis in terms of its two major segments.

> **4** **LEARNING GOAL**
> Looking at an equation representing any of the chemical reactions that occur in glycolysis, describe the kind of reaction that is occurring and the significance of that reaction to the pathway.

Reaction 1

The substrate, glucose, is phosphorylated by the enzyme *hexokinase* in a coupled phosphorylation reaction. The source of the phosphoryl group is ATP. At first this reaction seems contrary to the overall purpose of catabolism, the *production* of ATP. The expenditure of ATP in these early reactions must be thought of as an "investment." The cell actually goes into energy "debt" in these early reactions, but this is absolutely necessary to get the pathway started.

The enzyme name can tell us a lot about the reaction (see Section 19.1). The suffix *-kinase* tells us that the enzyme is a transferase that will transfer a phosphoryl group, in this case from an ATP molecule to the substrate. The prefix *hexo-* gives us a hint that the substrate is a six-carbon sugar. Hexokinase predominantly phosphorylates the six-carbon sugar glucose.

Glucose + ATP → (Hexokinase) → Glucose-6-phosphate + ADP + H⁺

Reaction 2

The glucose-6-phosphate formed in the first reaction is rearranged to produce the structural isomer fructose-6-phosphate. The enzyme *phosphoglucose isomerase* catalyzes this isomerization. The result is that the C-1 carbon of the six-carbon sugar is exposed; it is no longer part of the ring structure. Examination of the open-chain structures reveals that this isomerization converts an aldose into a ketose.

The enzyme name, phosphoglucose isomerase, provides clues to the reaction that is being catalyzed (Section 19.1). *Isomerase* tells us that the enzyme will catalyze the interconversion of one isomer into another. *Phosphoglucose* suggests that the substrate is a phosphorylated form of glucose.

Glucose-6-phosphate ⇌ (Phosphoglucose isomerase) → Fructose-6-phosphate

A Medical Perspective

Genetic Disorders of Glycolysis

Imagine always having difficulty with physical exercise. Imagine the coach telling you to get tough and run that lap again, since you were last! Imagine being accused of being lazy because you didn't carry your share of the camping gear. Imagine all that and not knowing why it is that you can't keep up with your friends or the others in your physical education class. This has been the fate of thousands of people who suffer from a metabolic myopathy—a muscle (*myo-*) disorder (*-pathy*) caused by an inability to extract the energy from the food that they eat.

The onset of fatigue during exercise is called *exercise intolerance*. It is one of the major symptoms of a metabolic myopathy. But simple fatigue is just the mildest of the symptoms. Overexertion may cause episodes of muscle breakdown (rhabdomyolysis) in which the muscle cells, unable to provide enough ATP energy for themselves, begin to die. The muscle breakdown causes greatly elevated blood levels of creatine kinase. Creatine kinase is an abundant enzyme in muscle that is critical in energy metabolism (see A Human Perspective: Exercise and Energy Metabolism, in Chapter 22). When muscle cells die, this enzyme is released into the bloodstream. Another symptom is myoglobinuria (myoglobin in the urine). Recall that myoglobin is the oxygen storage protein in muscle. When muscles die, myoglobin ends up in the urine, turning it the color of cola soft drinks. Myoglobinuria may even cause kidney damage. Accompanying these clinical symptoms, many people describe intense muscle pain. They describe it as a

cramp, but it is not a cramp, since the muscle is not able to contract because of the lack of energy. Rather, the pain is caused by cell death and tissue damage that result from an inability to produce enough ATP.

There are three glycolytic enzyme deficiencies that lead to metabolic myopathy. The first is phosphofructokinase deficiency or Tarui's disease. Although this is not a sex-linked disorder, the great majority of sufferers are males (nine males to one female). The disorder is most frequently found in U.S. Ashkenazi Jews and Italian families. Onset of symptoms typically occurs between the ages of twenty and forty, although some severe cases have been reported in infants and young children. Patients experiencing the late-onset form of Tarui's

Imagine not having the energy to keep up with the others in your aerobics class. Exercise intolerance is caused by a deficiency of one of the enzymes of glycolysis.

This is an enediol reaction. It occurs through exactly the same steps as the conversion of fructose to glucose that we discussed in Section 16.4.

Glucose-6-phosphate
(an aldose)

Phosphoglucose isomerase

Fructose-6-phosphate
(a ketose)

The suffix *-kinase* in the name of the enzyme tells us that this is a coupled reaction: ATP is hydrolyzed and a phosphoryl group is transferred to another molecule. The prefix *phosphofructo-* tells us the other molecule is a phosphorylated form of fructose.

Reaction 3

A second energy "investment" is catalyzed by the enzyme *phosphofructokinase*. The phosphoanhydride bond in ATP is hydrolyzed, and a phosphoester linkage disease typically experienced exercise intolerance when they were younger. Vigorous exercise results in myoglobinuria and severe muscle pain. Meals high in carbohydrates worsen the exercise intolerance. Early-onset disease is often associated with respiratory failure, cardiomyopathy (heart muscle disease), seizures, and cortical blindness.

Phosphoglycerate kinase deficiency is a sex-linked genetic disorder (located on the X chromosome). As a result, far more males than females suffer from this disease. There are many clinical features associated with this deficiency, although only rarely are they all found in the same patient. These symptoms range from mental retardation and seizures to a slowly progressive myopathy.

Phosphoglycerate mutase deficiency has been mapped on chromosome 7. The disorder is found predominantly in U.S. African American, Italian, and Japanese families. The clinical features include exercise intolerance, muscle pain, and myoglobinuria following more intense exercise.

Since each of these disorders is caused by the lack of an enzyme, scientists are trying to design a way to replace the lost activity. Oral medication will not work because enzymes are proteins. They would simply be digested, like any other dietary protein. Enzyme replacement therapy is one approach that is being studied. This would involve periodic injections of the enzyme into the bloodstream, a treatment just like the injection of insulin by diabetics. Enzyme replacement therapy would require a large supply of the enzyme. Following the

model of insulin, the gene for the enzyme could be cloned into bacteria. The bacteria would then produce the protein, which would be purified for use by humans.

Another strategy is to introduce the gene for the missing enzyme into the patient's cells. This is called gene therapy. This method would also require that the gene for the enzyme be cloned. The DNA would then have to be introduced into the body using a safe procedure that would promote entry into target cells. There are still many obstacles to overcome before this type of treatment will be a reality for sufferers of metabolic myopathy.

A number of physicians are using a commonsense approach to the management of these disorders. Logic tells us that if a person cannot harvest energy from carbohydrates in the diet, perhaps a diet high in protein and lipids might be beneficial. As with any condition of this sort, it is important to consult a physician who understands the metabolic disorder and who will design and supervise a customized diet.

For Further Understanding

Write equations showing the reactions catalyzed by each of the enzymes discussed in this perspective and explain how the absence of each will impair ATP production.

Applying what you learned in Chapter 20, describe an experiment to clone the gene for phosphoglycerate kinase.

between the phosphoryl group and the C-1 hydroxyl group of fructose-6-phosphate is formed. The product is fructose-1,6-bisphosphate.

Fructose-6-phosphate + ATP $\xrightarrow{\text{Phospho-fructokinase}}$ Fructose-1,6-bisphosphate + ADP + H$^+$

Reaction 4

Fructose-1,6-bisphosphate is split into two three-carbon intermediates in a reaction catalyzed by the enzyme *aldolase*. The products are glyceraldehyde-3-phosphate (G3P) and dihydroxyacetone phosphate.

In aldol condensation, aldehydes and ketones react to form a larger molecule (Section 13.4). This reaction is a reverse aldol condensation. The large ketone sugar fructose-1,6-bisphosphate is broken down into dihydroxyacetone phosphate (a ketone) and glyceraldehyde-3-phosphate (an aldehyde).

The double reaction arrows tell us that this is a reversible reaction. The reverse reaction is an aldol condensation (Section 13.4) that we will study in the pathway for glucose synthesis called gluconeogenesis (Section 21.6).

Fructose-1,6-bisphosphate $\xrightarrow{\text{Aldolase}}$ Dihydroxyacetone phosphate + Glyceraldehyde-3-phosphate

Reaction 5

Because G3P is the only substrate that can be used by the next enzyme in the pathway, the dihydroxyacetone phosphate is rearranged to become a second molecule of G3P. The enzyme that mediates this isomerization is *triose phosphate isomerase*.

The enzyme name hints that two isomers of a phosphorylated three-carbon sugar are going to be interconverted (Section 19.1). The ketone dihydroxyacetone phosphate and its isomeric aldehyde, phosphoglyceraldehyde-3-phosphate are interconverted through an enediol intermediate.

Dihydroxyacetone phosphate $\xrightarrow{\text{Triose phosphate isomerase}}$ Glyceraldehyde-3-phosphate

Reaction 6

In this reaction the aldehyde glyceraldehyde-3-phosphate is oxidized to a carboxylic acid in a reaction catalyzed by *glyceraldehyde-3-phosphate dehydrogenase*. This is the first step in glycolysis that harvests energy, and it involves the reduction of the coenzyme nicotinamide adenine dinucleotide (NAD^+). This reaction occurs in two steps. First, NAD^+ is reduced to NADH as the aldehyde group of glyceraldehyde-3-phosphate is oxidized to a carboxyl group. Second, an inorganic phosphate group is transferred to the carboxyl group to give 1,3-bisphosphoglycerate. Notice that the new bond is denoted with a squiggle ($\sim$), indicating that this is a high-energy bond. This, and all remaining reactions of glycolysis, occur twice for each glucose because each glucose has been converted into two molecules of glyceraldehyde-3-phosphate.

The name glyceraldehyde-3-phosphate dehydrogenase tells us that the substrate glyceraldehyde-3-phosphate is going to be oxidized. In this reaction, we see that the aldehyde group has been oxidized to a carboxylate group (Section 13.4).

Actually the intermediate of the oxidation reaction is a high-energy thioester formed between the enzyme and the substrate (Section 14.4). When this bond is hydrolyzed, enough energy is released to allow the formation of a bond between an oxygen atom of an inorganic phosphate group and the substrate.

Glyceraldehyde-3-phosphate $+ NAD^+ + P_i \xrightarrow{\text{Glyceraldehyde-3-phosphate dehydrogenase}}$ 1,3-Bisphosphoglycerate $+ NADH$

Reaction 7

In this reaction, energy is harvested in the form of *ATP*. The enzyme *phosphoglycerate kinase* catalyzes the transfer of the phosphoryl group of 1,3-bisphosphoglycerate to ADP. This is the first substrate-level phosphorylation of glycolysis, and it produces ATP and 3-phosphoglycerate. It is a coupled reaction in which the high-energy bond is hydrolyzed and the energy released is used to drive the synthesis of ATP.

1,3-Bisphosphoglycerate   3-Phosphoglycerate

Once again, the enzyme name reveals a great deal about the reaction. The suffix -*kinase* tells us that a phosphoryl group will be transferred. In this case, a phosphoester bond in the substrate 1,3-bisphosphoglycerate is hydrolyzed and ADP is phosphorylated. Note that this is a reversible reaction.

Reaction 8

3-Phosphoglycerate is isomerized to produce 2-phosphoglycerate in a reaction catalyzed by the enzyme *phosphoglycerate mutase.* The phosphoryl group attached to the third carbon of 3-phosphoglycerate is transferred to the second carbon.

3-Phosphoglycerate   2-Phosphoglycerate

The suffix -*mutase* indicates another type of isomerase. Notice that the chemical formulas of the substrate and reactant are the same. The only difference is in the location of the phosphoryl group.

Reaction 9

In this step the enzyme *enolase* catalyzes the dehydration (removal of a water molecule) of 2-phosphoglycerate. The energy-rich product is phosphoenolpyruvate, the highest energy phosphorylated compound in metabolism.

2-Phosphoglycerate   Phosphoenolpyruvate

In Section 13.4 we learned that aldehydes and ketones exist in an equilibrium mixture of two tautomers called the keto and enol forms. The dehydration of 2-phosphoglycerate produces the molecule phosphoenolpyruvate, which is in the enol form. In this case, the enol is extremely unstable. Because of this instability, the phosphoester bond in the product is a high-energy bond; in other words, a great deal of energy is released when this bond is broken.

Reaction 10

Here we see the final substrate-level phosphorylation in the pathway, which is catalyzed by *pyruvate kinase.* Phosphoenolpyruvate serves as a donor of the phosphoryl group that is transferred to ADP to produce ATP. This is another coupled reaction in which hydrolysis of the phosphoester bond in phosphoenolpyruvate provides energy for the formation of the phosphoanhydride bond of ATP. The final product of glycolysis is pyruvate.

Phosphoenolpyruvate   Pyruvate

The enzyme name indicates that a phosphoryl group will be transferred (kinase) and that the product will be pyruvate. Pyruvate is a keto tautomer and is much more stable than the enol (Section 13.4).

It should be noted that reactions 6 through 10 occur twice per glucose molecule, because the starting six-carbon sugar is split into two three-carbon molecules. Thus in reaction 6, two NADH molecules are generated, and a total of four ATP molecules are made (steps 7 and 10). The net ATP gain from this pathway is, however, only two ATP molecules because there was an energy investment of two ATP molecules in the early steps of the pathway. This investment was paid back by the two ATP molecules produced by substrate-level phosphorylation in step 7. The actual energy yield is produced by substrate-level phosphorylation in reaction 10.

Question 21.5

What is substrate-level phosphorylation?

Question 21.6

What are the major products of glycolysis?

Question 21.7

Describe an overview of the reactions of glycolysis.

Question 21.8

How do the names of the first three enzymes of the glycolytic pathway relate to the reactions they catalyze?

Regulation of Glycolysis

5 **LEARNING GOAL**
Describe the mechanism of regulation of the rate of glycolysis. Discuss particular examples of that regulation.

There are additional mechanisms that regulate the rate of glycolysis, but we will focus on those that involve principles studied previously (Section 19.9).

Energy-harvesting pathways, such as glycolysis, are responsive to the energy needs of the cell. Reactions of the pathway speed up when there is a demand for ATP. They slow down when there is abundant ATP to meet the energy requirements of the cell.

One of the major mechanisms for the control of the rate of glycolysis is the use of *allosteric enzymes.* In addition to the active site, which binds the substrate, allosteric enzymes have an effector binding site, which binds a chemical signal that alters the rate at which the enzyme catalyzes the reaction. Effector binding may increase (positive allosterism) or decrease the rate of reaction (negative allosterism).

The chemical signals, or effectors, that indicate the energy needs of the cell include molecules such as ATP. When the ATP concentration is high, the cell must have sufficient energy. Similarly, ADP and AMP, which are precursors of ATP, are indicators that the cell is in need of ATP. In fact, all of these molecules are allosteric effectors that alter the rate of irreversible reactions catalyzed by enzymes in the glycolytic pathway.

The enzyme hexokinase, which catalyzes the phosphorylation of glucose, is allosterically inhibited by the product of the reaction it catalyzes, glucose-6-phosphate. A buildup of this product indicates that the reactions of glycolysis are not proceeding at a rapid rate, presumably because the cell has enough energy.

Phosphofructokinase, the enzyme that catalyzes the third reaction in glycolysis, is a key regulatory enzyme in the pathway. ATP is an allosteric inhibitor of phosphofructokinase, whereas AMP and ADP are allosteric activators. Another allosteric inhibitor of phosphofructokinase is citrate. As we will see in the next chapter, citrate is the first intermediate in the citric acid cycle, a pathway that results in the complete oxidation of the pyruvate. A high concentration of citrate signals that sufficient substrate is entering the citric acid cycle. The inhibition of phosphofructokinase by citrate is an example of *feedback inhibition*: the product, citrate, allosterically inhibits the activity of an enzyme early in the pathway.

The last enzyme in glycolysis, pyruvate kinase, is also subject to allosteric regulation. In this case, fructose-1,6-bisphosphate, the product of the reaction catalyzed by phosphofructokinase, is the allosteric activator. Thus, activation of phosphofructokinase results in the activation of pyruvate kinase. This is an example of *feedforward activation* because the product of an earlier reaction causes activation of an enzyme later in the pathway.

21.4 Fermentations

LEARNING GOAL

6 Discuss the practical and metabolic roles of fermentation reactions.

In the overview of glycolysis, we noted that the pyruvate produced must be used up in some way so that the pathway will continue to produce ATP. Similarly, the NADH produced by glycolysis in step 6 (see Figure 21.6) must be reoxidized at a later time, or glycolysis will grind to a halt as the available NAD$^+$ is used up. If the cell is functioning under aerobic conditions, NADH will be reoxidized, and pyruvate will be completely oxidized by aerobic respiration. Under anaerobic conditions, however, different types of fermentation reactions accomplish these purposes. **Fermentations** are catabolic reactions that occur with no net oxidation. Pyruvate or an organic compound produced from pyruvate is reduced as NADH is oxidized. We will examine two types of fermentation pathways in detail: lactate fermentation and alcohol fermentation.

Aerobic respiration is discussed in Chapter 22.

Lactate Fermentation

Lactate fermentation is familiar to anyone who has performed strenuous exercise. If you exercise so hard that your lungs and circulatory system can't deliver enough oxygen to the working muscles, your aerobic (oxygen-requiring) energy-harvesting pathways are not able to supply enough ATP to your muscles. But the muscles still demand energy. Under these anaerobic conditions, lactate fermentation begins. In this reaction, the enzyme *lactate dehydrogenase* reduces pyruvate to lactate. NADH is the reducing agent for this process (Figure 21.7). As pyruvate is reduced, NADH is oxidized, and NAD$^+$ is again available, permitting glycolysis to continue.

The lactate produced in the working muscle passes into the blood. Eventually, if strenuous exercise is continued, the concentration of lactate becomes so high that this fermentation can no longer continue. Glycolysis, and thus ATP production, stops. The muscle, deprived of energy, can no longer function. This point of exhaustion is called the **anaerobic threshold.**

Of course, most of us do not exercise to this point. When exercise is finished, the body begins the process of reclaiming all of the potential energy that was lost in the form of lactate. The liver takes up the lactate from the blood and converts it back to pyruvate. Now that a sufficient supply of oxygen is available, the pyruvate can be completely oxidized in the much more efficient aerobic energy-harvesting reactions to replenish the store of ATP. Alternatively, the pyruvate may be converted to glucose and used to restore the supply of liver and muscle glycogen. This exchange of metabolites between the muscles and liver is called the *Cori Cycle.*

A variety of bacteria are able to carry out lactate fermentation under anaerobic conditions. This is of great importance in the dairy industry, because these

When you exercise beyond the ability of your heart and lungs to provide oxygen to muscle, the lactate fermentation kicks in. Write an equation representing the reaction catalyzed by lactate dehydrogenase and explain how this reaction enables muscle to continue working.

The Cori Cycle is described in Section 21.6 and shown in Figure 21.13.

Figure 21.7
The final reaction of lactate fermentation.

A Human *Perspective*

Fermentations: The Good, the Bad, and the Ugly

In this chapter we have seen that fermentation is an anaerobic, cytoplasmic process that allows continued ATP generation by glycolysis. ATP production can continue because the pyruvate produced by the pathway is utilized in the fermentation and because NAD$^+$ is regenerated.

The stable end products of alcohol fermentation are CO_2 and ethanol. These have been used by humankind in a variety of ways, including the production of alcoholic beverages, bread making, and alternative fuel sources.

If alcohol fermentation is carried out by using fruit juices in a vented vat, the CO_2 will escape, and the result will be a still wine (not bubbly). But conditions must remain anaerobic; otherwise, fermentation will stop, and aerobic energy-harvesting reactions will ruin the wine. Fortunately for vintners (wine makers), when a vat is fermenting actively, enough CO_2 is produced to create a layer that keeps the oxygen-containing air away from the fermenting juice, thus maintaining an anaerobic atmosphere.

Now suppose we want to make a sparkling wine, such as champagne. To do this, we simply have to trap the CO_2 produced. In this case the fermentation proceeds in a sealed bottle, a very strong bottle. Both the fermentation products, CO_2 and ethanol, accumulate. Under pressure within the sealed bottle the CO_2 remains in solution. When the top is "popped," the pressure is released, and the CO_2 comes out of solution in the form of bubbles.

In either case the fermentation continues until the alcohol concentration reaches 12–13%. At that point the yeast "stews in its own juices"! That is, 12–13% ethanol kills the yeast cells that produce it. This points out a last generalization about fermentations. The stable fermentation end product, whether it is lactate or ethanol, eventually accumulates to a concentration

that is toxic to the organism. Muscle fatigue is the early effect of lactate buildup in the working muscle. In the same way, continued accumulation of the fermentation product can lead to concentrations that are fatal if there is no means of getting rid of the toxic product or of getting away from it. For single-celled organisms the result is generally death. Our bodies have evolved in such a way that lactate buildup contributes to muscle fatigue that causes the exerciser to stop the exercise. Then the lactate is removed from the blood and converted to glucose by the process of gluconeogenesis.

The production of bread, wine, and cheese depends on fermentation processes.

As we saw in A Human Perspective: Tooth Decay and Simple Sugars (Chapter 16), the lactate produced by oral bacteria is responsible for the gradual removal of calcium from tooth enamel and the resulting dental cavities.

These applications and other fermentations are described in A Human Perspective: Fermentations: The Good, the Bad, and the Ugly.

organisms are used to produce yogurt and some cheeses. The tangy flavor of yogurt is contributed by the lactate produced by these bacteria. Unfortunately, similar organisms also cause milk to spoil.

Alcohol Fermentation

Alcohol fermentation has been appreciated, if not understood, since the dawn of civilization. The fermentation process itself was discovered by Louis Pasteur during his studies of the chemistry of winemaking and "diseases of wines." Under anaerobic conditions, yeast are able to ferment the sugars produced by fruit and grains. The sugars are broken down to pyruvate by glycolysis. This is followed by the two reactions of alcohol fermentation. First, *pyruvate decarboxylase* removes CO_2 from the pyruvate producing acetaldehyde (Figure 21.8). Second, *alcohol dehydrogenase* catalyzes the reduction of acetaldehyde to ethanol but, more important, reoxidizes NADH in the process. The regeneration of NAD$^+$ allows glycolysis to continue, just as in the case of lactate fermentation.

Another application of alcohol fermentation is the use of yeast in bread making. When we mix the water, sugar, and dried yeast, the yeast cells begin to grow and carry out the process of fermentation. This mixture is then added to the flour, milk, shortening, and salt, and the dough is placed in a warm place to rise. The yeast continues to grow and ferment the sugar, producing CO_2 that causes the bread to rise. Of course, when we bake the bread, the yeast cells are killed, and the ethanol evaporates, but we are left with a light and airy loaf of bread.

Today, alcohol produced by fermentation is being considered as an alternative fuel to replace the use of some fossil fuels. Geneticists and bioengineers are trying to develop strains of yeast that can survive higher alcohol concentrations and thus convert more of the sugar of corn and other grains into alcohol.

Bacteria perform a variety of other fermentations. The propionibacteria produce propionic acid and CO_2. The acid gives Swiss cheese its characteristic flavor, and the CO_2 gas produces the characteristic holes in the cheese. Other bacteria, the clostridia, perform a fermentation that is responsible in part for the horrible symptoms of gas gangrene. When these bacteria are inadvertently introduced into deep tissues by a puncture wound, they find a nice anaerobic environment in which to grow. In fact, these organisms are *obligate anaerobes*; that is, they are killed by even a small amount of oxygen. As they grow, they perform a fermentation called the *butyric acid, butanol, acetone fermentation*. This results in the formation of CO_2, the gas associated with gas gangrene. The CO_2 infiltrates the local tissues and helps to maintain an anaerobic environment because oxygen from the local blood supply cannot enter the area of the wound. Now able to grow well, these bacteria produce a variety of toxins and enzymes that cause extensive tissue death and necrosis. In addition, the fermentation produces acetic acid, ethanol, acetone, isopropanol, butanol, and butyric acid (which is responsible, along with the necrosis, for the characteristic foul smell of gas gangrene). Certainly, the presence of these organic chemicals in the wound enhances tissue death.

Gas gangrene is very difficult to treat. Because the bacteria establish an anaerobic region of cell death and cut off the local circulation, systemic antibiotics do not infiltrate the wound and kill the bacteria. Even our immune response is stymied. Treatment usually involves surgical removal of the necrotic tissue accompanied by antibiotic therapy. In some cases a hyperbaric oxygen chamber is employed. The infected extremity is placed in an environment with a very high partial pressure of oxygen. The oxygen forced into the tissues is poisonous to the bacteria, and they die.

These are but a few examples of the fermentations that have an effect on humans. Regardless of the specific chemical reactions, all fermentations share the following traits:

- They use pyruvate produced in glycolysis.
- They reoxidize the NADH produced in glycolysis.
- They are self-limiting because the accumulated stable fermentation end product eventually kills the cell that produces it.

For Further Understanding

Write condensed structural formulas for each of the fermentation products made by clostridia in gas gangrene. Identify the functional groups and provide the I.U.P.A.C. name for each.

Explain the importance of utilizing pyruvate and reoxidizing NADH to the ability of a cell to continue producing ATP.

Figure 21.8
The final two reactions of alcohol fermentation.

The two products of alcohol fermentation, then, are ethanol and CO_2. We take advantage of this fermentation in the production of wines and other alcoholic beverages and in the process of breadmaking.

Question 21.9

How is the alcohol fermentation in yeast similar to lactate production in skeletal muscle?

Question 21.10

Why must pyruvate be used and NADH be reoxidized so that glycolysis can continue?

21.5 The Pentose Phosphate Pathway

The **pentose phosphate pathway** is an alternative pathway for glucose oxidation. It provides the cell with energy in the form of reducing power for biosynthesis. Specifically, NADPH is produced in the oxidative stage of this pathway. NADPH is the reducing agent required for many biosynthetic pathways.

The details of the pentose phosphate pathway will not be covered in this text. But an overview of the key reactions will allow us to understand the importance of the pathway (Figure 21.9). We can consider the pathway in three stages. The first is the oxidative stage, which can be summarized as

glucose-6-phosphate + $2NADP^+$ + $H_2O \longrightarrow$

ribulose-5-phosphate + $2NADPH$ + CO_2

These reactions provide the NADPH required for biosynthesis.

The second stage involves isomerization reactions that convert ribulose-5-phosphate into ribose-5-phosphate or xylulose-5-phosphate. The pathway's name reflects the production of these phosphorylated five-carbon sugars (pentose phosphates).

The third stage is a complex series of reactions involving C—C bond breakage and formation. The result of these reactions is the formation of two molecules of fructose-6-phosphate and one molecule of glyceraldehyde-3-phosphate from three molecules of pentose phosphate.

In addition to providing reducing power (NADPH), the pentose phosphate pathway provides sugar phosphates that are required for biosynthesis. For instance, ribose-5-phosphate is used for the synthesis of nucleotides such as ATP. The four-carbon sugar phosphate, erythrose-4-phosphate, produced in the third

Figure 21.9

Summary of the major stages of the pentose phosphate pathway.

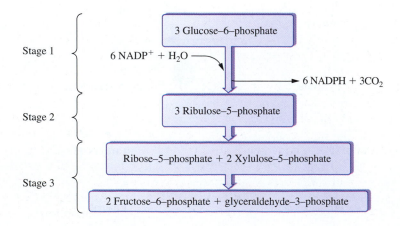

stage of the pentose phosphate pathway is a precursor of the amino acids phenyl-alanine, tyrosine, and tryptophan.

The pentose phosphate pathway is most active in tissues involved in choles-terol and fatty acid biosynthesis. These two processes require abundant NADPH. Thus the liver, which is the site of cholesterol synthesis and a major site for fatty acid biosynthesis, and adipose (fat) tissue, where active fatty acid synthesis also occurs, have very high levels of pentose phosphate pathway enzymes.

The pathway for fatty acid biosynthesis is discussed in Section 23.4.

21.6 Gluconeogenesis: The Synthesis of Glucose

Under normal conditions, we have enough glucose to satisfy our needs. However, under some conditions the body must make glucose. This is necessary following strenuous exercise to replenish the liver and muscle supplies of glycogen. It also occurs during starvation so that the body can maintain adequate blood glucose levels to supply the brain cells and red blood cells. Under normal conditions these two tissues use only glucose for energy.

LEARNING GOAL

8 Compare glycolysis and gluconeogenesis.

Glucose is produced by the process of **gluconeogenesis,** the production of glucose from noncarbohydrate starting materials (Figure 21.10). Gluconeogenesis, an anabolic pathway, occurs primarily in the liver. Lactate, all the amino acids except leucine and lysine, and glycerol from fats can all be used to make glucose. However, the amino acids and glycerol are generally used only under starvation conditions.

Under extreme conditions of starvation the brain eventually switches to the use of ketone bodies. Ketone bodies are produced, under certain circumstances, from the breakdown of lipids (Section 23.3).

At first glance, gluconeogenesis appears to be simply the reverse of gly-colysis (compare Figures 21.10 and 21.6), because the intermediates of the two pathways are identical. But this is not the case, because steps 1, 3, and 10 of gly-colysis are irreversible, and therefore the reverse reactions must be carried out by other enzymes. In step 1 of glycolysis, hexokinase catalyzes the phosphorylation of glucose. In gluconeogenesis the dephosphorylation of glucose-6-phosphate is carried out by the enzyme *glucose-6-phosphatase,* which is found in the liver but not in muscle. Similarly, reaction 3, the phosphorylation of fructose-6-phosphate catalyzed by phosphofructokinase, is irreversible. That step is bypassed in gluco-neogenesis by using the enzyme *fructose-1,6-bisphosphatase.* Finally, the phos-phorylation of ADP catalyzed by pyruvate kinase, step 10 of glycolysis, cannot be reversed. The conversion of pyruvate to phosphoenolpyruvate actually involves two enzymes and some unusual reactions. First, the enzyme *pyruvate carboxylase* adds CO_2 to pyruvate. The product is the four-carbon compound oxaloacetate. Then *phosphoenolpyruvate carboxykinase* removes the CO_2 and adds a phosphoryl group. The donor of the phosphoryl group in this unusual reaction is **guanosine triphosphate (GTP).** This is a nucleotide like ATP, except that the nitrogenous base is guanine.

This last pair of reactions is complicated by the fact that pyruvate carboxyl-ase is found in the mitochondria, whereas phosphoenolpyruvate carboxykinase is found in the cytoplasm. As we will see in Chapters 22 and 23, mitochondria are organelles in which the final oxidation of food molecules occurs and large amounts of ATP are produced. A complicated shuttle system transports the oxalo-acetate produced in the mitochondria through the two mitochondrial membranes and into the cytoplasm. There, phosphoenolpyruvate carboxykinase catalyzes its conversion to phosphoenolpyruvate.

If glycolysis and gluconeogenesis were not regulated in some fashion, the two pathways would occur simultaneously, with the disastrous effect that nothing would get done. Three convenient sites for this regulation are the three bypass reactions. Step 3 of glycolysis is catalyzed by the enzyme phosphofructokinase. This enzyme is stimulated by high concentrations of AMP, ADP, and inorganic phosphate, signals that the cell needs energy. When the enzyme is active, glycoly-sis proceeds. On the other hand, when ATP is plentiful, phosphofructokinase is

Figure 21.10

Comparison of the reactions of glycolysis and gluconeogenesis.

inhibited, and fructose-1,6-bisphosphatase is stimulated. The net result is that in times of energy excess (high concentrations of ATP), gluconeogenesis will occur.

As we have seen, the conversion of lactate into glucose is important in mammals. As the muscles work, they produce lactate, which is converted back to glucose in the liver. The glucose is transported into the blood and from there back to the muscle. In the muscle it can be catabolized to produce ATP, or it can be used to replenish the muscle stores of glycogen. This cyclic process between the liver and skeletal muscles is called the **Cori Cycle** and is shown in Figure 21.11. Through this cycle, gluconeogenesis produces enough glucose to restore the depleted muscle glycogen reservoir within forty-eight hours.

Question *21.11*   What are the major differences between gluconeogenesis and glycolysis?

Question *21.12*   What do the three irreversible reactions of glycolysis have in common?

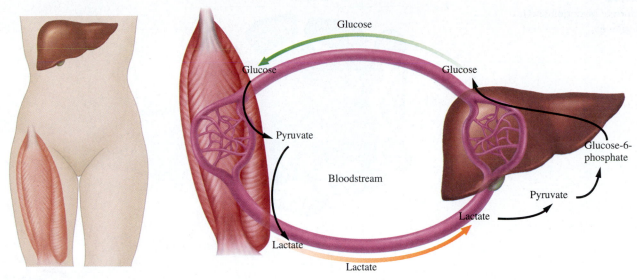

Figure 21.11
The Cori Cycle.

21.7 Glycogen Synthesis and Degradation

Glucose is the sole source of energy of mammalian red blood cells and the major source of energy for the brain. Neither red blood cells nor the brain can store glucose; thus a constant supply must be available as blood glucose. This is provided by dietary glucose and by the production of glucose either by gluconeogenesis or by **glycogenolysis,** the degradation of glycogen. Glycogen is a long-branched-chain polymer of glucose. Stored in the liver and skeletal muscles, it is the principal storage form of glucose.

The total amount of glucose in the blood of a 70-kg (approximately 150-lb) adult is about 20 g, but the brain alone consumes 5–6 g of glucose per hour. Breakdown of glycogen in the liver mobilizes the glucose when hormonal signals register a need for increased levels of blood glucose. Skeletal muscle also contains substantial stores of glycogen, which provide energy for rapid muscle contraction. However, this glycogen is not able to contribute to blood glucose because muscle cells do not have the enzyme glucose-6-phosphatase. Because glucose cannot be formed from the glucose-6-phosphate, it cannot be released into the bloodstream.

The Structure of Glycogen

Glycogen is a highly branched glucose polymer in which the "main chain" is linked by α $(1 \rightarrow 4)$ glycosidic bonds. The polymer also has numerous α $(1 \rightarrow 6)$ glycosidic bonds, which provide many branch points along the chain. This structure is shown schematically in Figure 21.12. **Glycogen granules** with a diameter of 10–40 nm are found in the cytoplasm of liver and muscle cells. These granules exist in complexes with the enzymes that are responsible for glycogen synthesis and degradation. The structure of such a granule is also shown in Figure 21.12.

Glycogenolysis: Glycogen Degradation

Two hormones control glycogenolysis, the degradation of glycogen. These are **glucagon,** a peptide hormone synthesized in the pancreas, and *epinephrine*, produced in the adrenal glands. Glucagon is released from the pancreas in response to low

Marathon runners often carbo-load in the days before a race. The goal is to build stores of muscle glycogen. Carbo-loading involves reduced exercise the week before the race, along with a diet that is as high as 70% carbohydrate. Explain how this helps build the runner's endurance.

Figure 21.12

The structure of glycogen and a glycogen granule.

blood glucose, and epinephrine is released from the adrenal glands in response to a threat or a stress. Both situations require an increase in blood glucose, and both hormones function by altering the activity of two enzymes, glycogen phosphorylase and glycogen synthase. *Glycogen phosphorylase* is involved in glycogen degradation and is activated; *glycogen synthase* is involved in glycogen synthesis and is inactivated. The steps in glycogen degradation are summarized as follows.

Step 1. The enzyme glycogen phosphorylase catalyzes *phosphorolysis* of a glucose at one end of a glycogen polymer (Figure 21.13). The reaction involves the displacement of a glucose unit of glycogen by a phosphate

Glucose-1-phosphate

Glycogen phosphorylase

Glycogen

Glycogen

General reaction:

$$\text{Glycogen (glucose)}_x + n\ HPO_4^{2-} \xrightarrow[\text{phosphorylase}]{\text{Glycogen}} \text{Glycogen (glucose)}_{x-n} + n\ \text{glucose-1-phosphate}$$

Figure 21.13
The action of glycogen phosphorylase in glycogenolysis.

group. As a result of phosphorolysis, glucose-1-phosphate is produced without using ATP as the phosphoryl group donor.

Step 2. Glycogen contains many branches bound to the α (1 → 4) backbone by α (1 → 6) glycosidic bonds. These branches must be removed to allow the complete degradation of glycogen. The extensive action of glycogen phosphorylase produces a smaller polysaccharide with a single glucose bound by an α (1 → 6) glycosidic bond to the main chain. The enzyme *α (1 → 6) glycosidase,* also called the *debranching enzyme,* hydrolyzes the α (1 → 6) glycosidic bond at a branch point and frees one molecule of glucose (Figure 21.14). This molecule of glucose can be phosphorylated and utilized in glycolysis, or it may be released into the bloodstream for use elsewhere. Hydrolysis of the branch bond liberates another stretch of α (1 → 4)-linked glucose for the action of glycogen phosphorylase.

Step 3. Glucose-1-phosphate is converted to glucose-6-phosphate by *phosphoglucomutase* (Figure 21.15). Glucose originally stored in glycogen enters glycolysis through the action of phosphoglucomutase. Alternatively, in the liver and kidneys it may be dephosphorylated for transport into the bloodstream.

Figure 21.14

The action of α (1 → 6) glycosidase (debranching enzyme) in glycogen degradation.

Figure 21.15

The action of phosphoglucomutase in glycogen degradation.

Question *21.13*

Explain the role of glycogen phosphorylase in glycogenolysis.

Question *21.14*

How does the action of glycogen phosphorylase and phosphoglucomutase result in an energy savings for the cell if the product, glucose-6-phosphate, is used directly in glycolysis?

Glycogenesis: Glycogen Synthesis

The hormone **insulin,** produced by the pancreas in response to high blood glucose levels, stimulates the synthesis of glycogen, **glycogenesis.** Insulin is perhaps one of the most influential hormones in the body because it directly alters the metabolism and uptake of glucose in all but a few cells.

When blood glucose rises, as after a meal, the beta cells of the pancreas secrete insulin. It immediately accelerates the uptake of glucose by all the cells of the body except the brain and certain blood cells. In these cells the uptake of glucose is insulin-independent. The increased uptake of glucose is especially marked in the liver, heart, skeletal muscle, and adipose tissue.

In the liver, insulin promotes glycogen synthesis and storage by inhibiting glycogen phosphorylase, thus inhibiting glycogen degradation. It also stimulates glycogen synthase and glucokinase, two enzymes that are involved in glycogen synthesis.

Although glycogenesis and glycogenolysis share some reactions in common, the two pathways are not simply the reverse of one another. Glycogenesis involves some very unusual reactions, which we will now examine in detail.

The first reaction of glycogen synthesis in the liver traps glucose within the cell by phosphorylating it. In this reaction, catalyzed by the enzyme *glucokinase*, ATP serves as a phosphoryl donor, and glucose-6-phosphate is formed:

Glucose   Glucose-6-phosphate

The second reaction of glycogenesis is the reverse of one of the reactions of glycogenolysis. The glucose-6-phosphate formed in the first step is isomerized to glucose-1-phosphate. The enzyme that catalyzes this step is phosphoglucomutase:

Glucose-6-phosphate   Glucose-1-phosphate

The glucose-1-phosphate must now be activated before it can be added to the growing glycogen chain. The high-energy compound that accomplishes this is the nucleotide **uridine triphosphate (UTP).** In this reaction, mediated by the enzyme *pyrophosphorylase*, the C-1 phosphoryl group of glucose is linked to the α-phosphoryl group of UTP to produce UDP-glucose:

Glucose-1-phosphate   +   UTP   UDP-glucose   +   Pyrophosphate

A Medical Perspective

Diagnosing Diabetes

When diagnosing diabetes, doctors take many factors and symptoms into consideration. However, there are two primary tests that are performed to determine whether an individual is properly regulating blood glucose levels. First and foremost is the fasting blood glucose test. A person who has fasted since midnight should have a blood glucose level between 70 and 110 mg/dL in the morning. If the level is 140 mg/dL on at least two occasions, a diagnosis of diabetes is generally made.

The second commonly used test is the glucose tolerance test. For this test the subject must fast for at least ten hours. A beginning blood sample is drawn to determine the fasting blood glucose level. This will serve as the background level for the test. The subject ingests 50–100 g of glucose (40 g/m² body surface), and the blood glucose level is measured at thirty minutes, and at one, two, and three hours after ingesting the glucose.

A graph is made of the blood glucose levels over time. For a person who does not have diabetes, the curve will show a peak of blood glucose at approximately one hour. There will be a reduction in the level, and perhaps a slight hypoglycemia (low blood glucose level) over the next hour. Thereafter, the blood glucose level stabilizes at normal levels.

An individual is said to have impaired glucose tolerance if the blood glucose level remains between 140 and 200 mg/dL two hours after ingestion of the glucose solution. This suggests that there is a risk of the individual developing diabetes and is reason to prescribe periodic testing to allow early intervention.

If the blood glucose level remains at or above 200 mg/dL after two hours, a tentative diagnosis of diabetes is made. However, this result warrants further testing on subsequent days to rule out transient problems, such as the effect of medications on blood glucose levels.

It was recently suggested that the upper blood glucose level of 200 mg/dL should be lowered to 180 mg/dL as the standard to diagnose impaired glucose tolerance and diabetes. This would allow earlier detection and intervention. Considering the grave nature of long-term diabetic complications, it is thought to be very beneficial to begin treatment at an early stage to maintain constant blood glucose levels. For more information on diabetes, see A Medical Perspective: Diabetes Mellitus and Ketone Bodies, in Chapter 23.

For Further Understanding

Draw a graph representing blood glucose levels for a normal glucose tolerance test.

Draw a similar graph for an individual who would be diagnosed as diabetic.

This is accompanied by the release of a pyrophosphate group (PP_i). The structure of UDP-glucose is seen in Figure 21.16.

The UDP-glucose can now be used to extend glycogen chains. The enzyme glycogen synthase breaks the phosphoester linkage of UDP-glucose and forms an

Figure 21.16

The structure of UDP-glucose.

α (1 → 4) glycosidic bond between the glucose and the growing glycogen chain. UDP is released in the process.

UDP-glucose

Glycogen primer
(*n* residues)

Glycogen synthase

Glycogen
(*n* + 1 residues)

UDP

Finally, we must introduce the α (1 → 6) glycosidic linkages to form the branches. The branches are quite important to proper glycogen utilization. As Figure 21.17 shows, the *branching enzyme* removes a section of the linear α (1 → 4) linked glycogen and reattaches it in α (1 → 6) glycosidic linkage elsewhere in the chain.

Describe the way in which glucokinase traps glucose inside liver cells.

Question *21.15*

Describe the reaction catalyzed by the branching enzyme.

Question *21.16*

Compatibility of Glycogenesis and Glycogenolysis

As was the case with glycolysis and gluconeogenesis, it would be futile for the cell to carry out glycogen synthesis and degradation simultaneously. The results achieved by the action of one pathway would be undone by the other. This problem is avoided by a series of hormonal controls that activate the enzymes of one pathway while inactivating the enzymes of the other pathway.

When the blood glucose level is too high, a condition known as **hyperglycemia,** insulin stimulates the uptake of glucose via a transport mechanism. It further stimulates the trapping of the glucose by the elevated activity of glucokinase. Finally, it activates glycogen synthase, the last enzyme in the synthesis of glycogen chains. To further accelerate storage, insulin *inhibits* the first enzyme in glycogen degradation, glycogen phosphorylase. The net effect, seen in Figure 21.18, is that glucose is removed from the bloodstream and converted into glycogen in the liver. When the glycogen stores are filled, excess glucose is converted to fat and stored in adipose tissue.

Glucagon is produced in response to low blood glucose levels, a condition known as **hypoglycemia,** and has an effect opposite to that of insulin. It stimulates glycogen phosphorylase, which catalyzes the first stage of glycogen degradation. This accelerates glycogenolysis and release of glucose into the bloodstream. The effect is further enhanced because glucagon inhibits glycogen synthase. The opposing effects of insulin and glucagon are summarized in Figure 21.18.

This elegant system of hormonal control ensures that the reactions involved in glycogen degradation and synthesis do not compete with one another. In this way they provide glucose when the blood level is too low, and they cause the storage of glucose in times of excess.

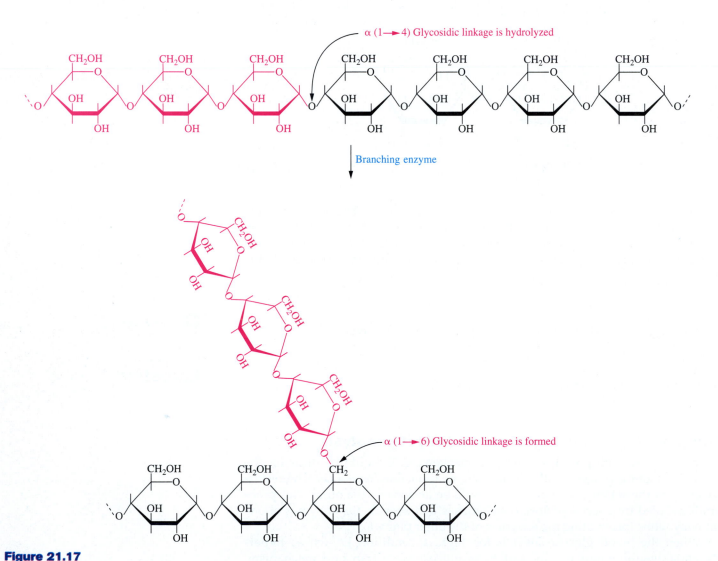

Figure 21.17

The action of the branching enzyme in glycogen synthesis.

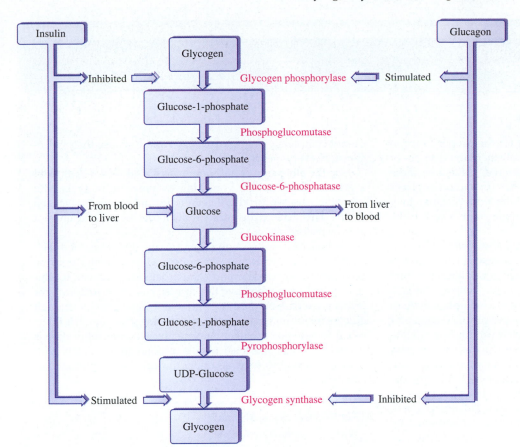

Figure 21.18
The opposing effects of the hormones insulin and glucagon on glycogen metabolism.

Explain how glucagon affects the synthesis and degradation of glycogen.

Question *21.17*

How does insulin affect the storage and degradation of glycogen?

Question *21.18*

A Human *Perspective*

Glycogen Storage Diseases

Glycogen metabolism is important for the proper function of many aspects of cellular metabolism. Many diseases of glycogen metabolism have been discovered. Generally, these are diseases that result in the excessive accumulation of glycogen in the liver, muscle, and tubules of the kidneys. Often they are caused by defects in one of the enzymes involved in the degradation of glycogen.

One example is an inherited defect of glycogen metabolism known as *von Gierke's disease.* This disease results from a defective gene for glucose-6-phosphatase, which catalyzes the final step of gluconeogenesis and glycogenolysis. People who lack glucose-6-phosphatase cannot convert glucose-6-phosphate to glucose. As we have seen, the liver is the primary source of blood glucose, and much of this glucose is produced by gluconeogenesis. Glucose-6-phosphate, unlike glucose, cannot cross the cell membrane, and the liver of a person suffering from von Gierke's disease cannot provide him or her with glucose. The blood sugar level falls precipitously low between meals. In addition, the lack of glucose-6-phosphatase also affects glycogen metabolism. Because glucose-6-phosphatase is absent, the supply of glucose-6-phosphate in the liver is large. This glucose-6-phosphate can also be converted to glycogen. A person suffering from von Gierke's disease has a massively enlarged liver as a result of enormously increased stores of glycogen.

Defects in other enzymes of glycogen metabolism also exist. *Cori's disease* is caused by a genetic defect in the debranching enzyme. As a result, individuals who have this disease

cannot completely degrade glycogen and thus use their glycogen stores very inefficiently.

On the other side of the coin, *Andersen's disease* results from a genetic defect in the branching enzyme. Individuals who have this disease produce very long, unbranched glycogen chains. This genetic disorder results in decreased efficiency of glycogen storage.

A final example of a glycogen storage disease is *McArdle's disease.* In this syndrome, the muscle cells lack the enzyme glycogen phosphorylase and cannot degrade glycogen to glucose. Individuals who have this disease have little tolerance for physical exercise because their muscles cannot provide enough glucose for the necessary energy-harvesting processes. It is interesting to note that the liver enzyme glycogen phosphorylase is perfectly normal, and these people respond appropriately with a rise of blood glucose levels under the influence of glucagon or epinephrine.

For Further Understanding

Write equations showing the reactions catalyzed by the enzymes that are defective in each of the genetic disorders described in this perspective.

There are different forms of glycogen phosphorylase, one found in the liver and the other in skeletal muscle. Discuss the differences you would expect between a defect in the muscle enzyme and a defect in the liver enzyme.

SUMMARY

21.1 ATP: The Cellular Energy Currency

Adenosine triphosphate, ATP, is a *nucleotide* composed of adenine, the sugar ribose, and a triphosphate group. The energy released by the hydrolysis of the phosphoanhydride bond between the second and third phosphoryl groups provides the energy for most cellular work.

21.2 Overview of Catabolic Processes

The body needs a supply of ATP to carry out life processes. To provide this ATP, we consume a variety of energy-rich food molecules: carbohydrates, lipids, and

proteins. In the digestive tract these large molecules are degraded into smaller molecules (monosaccharides, glycerol, fatty acids, and amino acids) that are absorbed by our cells. These molecules are further broken down to generate ATP.

21.3 Glycolysis

Glycolysis is the pathway for the *catabolism* of glucose that leads to pyruvate. It is an anaerobic process carried out by enzymes in the cytoplasm of the cell. The net harvest of ATP during glycolysis is two molecules of ATP per molecule of glucose. Two molecules of NADH are also produced. The rate of glycolysis responds to the energy demands of the cell. The regulation of glycolysis occurs through the allosteric enzymes hexokinase, phosphofructokinase, and pyruvate kinase.

21.4 Fermentations

Under anaerobic conditions the NADH produced by glycolysis is used to reduce pyruvate to lactate in skeletal muscle (lactate *fermentation*) or to convert acetaldehyde to ethanol in yeast (alcohol fermentation).

21.5 The Pentose Phosphate Pathway

The *pentose phosphate pathway* is an alternative pathway for glucose degradation that is particularly abundant in the liver and adipose tissue. It provides the cell with a source of NADPH to serve as a reducing agent for biosynthetic reactions. It also provides ribose-5-phosphate for nucleotide synthesis and erythrose-4-phosphate for biosynthesis of the amino acids tryptophan, tyrosine, and phenylalanine.

21.6 Gluconeogenesis: The Synthesis of Glucose

Gluconeogenesis is the pathway for glucose synthesis from noncarbohydrate starting materials. It occurs in mammalian liver. Glucose can be made from lactate, all the amino acids except lysine and leucine, and glycerol. Gluconeogenesis is not simply the reversal of glycolysis. Three steps in glycolysis in which ATP is produced or consumed are bypassed by different enzymes in gluconeogenesis. All other enzymes in gluconeogenesis are shared with glycolysis.

21.7 Glycogen Synthesis and Degradation

Glycogenesis is the pathway for the synthesis of glycogen, and *glycogenolysis* is the pathway for the degradation of glycogen. The concentration of blood glucose is controlled by the liver. A high blood glucose level causes secretion of *insulin*. This hormone stimulates glycogenesis and inhibits glycogenolysis. When blood glucose levels are too low, the hormone *glucagon* stimulates gluconeogenesis and glycogen degradation in the liver.

KEY TERMS

adenosine triphosphate (ATP) (21.1)
anabolism (21.1)
anaerobic threshold (21.4)
catabolism (21.1)
Cori Cycle (21.6)
fermentation (21.4)
glucagon (21.7)
gluconeogenesis (21.6)

glycogen (21.7)
glycogenesis (21.7)
glycogen granule (21.7)
glycogenolysis (21.7)
glycolysis (21.3)
guanosine triphosphate (GTP) (21.6)
hyperglycemia (21.7)
hypoglycemia (21.7)

insulin (21.7)
nicotinamide adenine dinucleotide (NAD^+) (21.3)
nucleotide (21.1)
oxidative phosphorylation (21.3)

pentose phosphate pathway (21.5)
substrate-level phosphorylation (21.3)
uridine triphosphate (UTP) (21.7)

QUESTIONS AND PROBLEMS

ATP: The Cellular Energy Currency

Foundations

21.19 What molecule is primarily responsible for conserving the energy released in catabolism?

21.20 Describe the structure of ATP.

Applications

21.21 Write a reaction showing the hydrolysis of the terminal phosphoanhydride bond of ATP.

21.22 What is meant by the term *high-energy bond*?

21.23 What is meant by a coupled reaction?

21.24 Compare and contrast anabolism and catabolism in terms of their roles in metabolism and their relationship to ATP.

Overview of Catabolic Processes

Foundations

21.25 What is the most readily used energy source in the diet?

21.26 What is a hydrolysis reaction?

Applications

21.27 Write an equation showing the hydrolysis of maltose.

21.28 Write an equation showing the hydrolysis of sucrose.

21.29 Write an equation showing the hydrolysis of lactose.

21.30 How are monosaccharides transported into a cell?

21.31 Write an equation showing the hydrolysis of a triglyceride consisting of glycerol, oleic acid, linoleic acid, and stearic acid.

21.32 How are fatty acids taken up into the cell?

21.33 Write an equation showing the hydrolysis of the dipeptide alanyl-leucine.

21.34 How are amino acids transported into the cell?

Glycolysis

Foundations

21.35 Define glycolysis and describe its role in cellular metabolism.

21.36 What are the end products of glycolysis?

21.37 Why does glycolysis require a supply of NAD^+ to function?

21.38 Why must the NADH produced in glycolysis be reoxidized to NAD^+?

21.39 What is the net energy yield of ATP in glycolysis?

21.40 How many molecules of ATP are produced by substrate-level phosphorylation during glycolysis?

21.41 Explain how muscle is able to carry out rapid contraction for prolonged periods even though its supply of ATP is sufficient only for a fraction of a second of rapid contraction.

21.42 Where in the muscle cell does glycolysis occur?

21.43 Write the balanced chemical equation for glycolysis.

21.44 Write a chemical equation for the transfer of a phosphoryl group from ATP to fructose-6-phosphate.

21.45 Which glycolysis reactions are catalyzed by each of the following enzymes?
  a. Hexokinase
  b. Pyruvate kinase
  c. Phosphoglycerate mutase
  d. Glyceraldehyde-3-phosphate dehydrogenase

21.46 Fill in the blanks:
  a. _____ molecules of ATP are produced per molecule of glucose that is converted to pyruvate.
  b. Two molecules of ATP are consumed in the conversion of _____ to fructose-1,6-bisphosphate.
  c. NAD^+ is _____ to NADH in the first energy-releasing step of glycolysis.
  d. The second substrate-level phosphorylation in glycolysis is phosphoryl group transfer from phosphoenolpyruvate to _____.

Applications

21.47 Examine the following pair of reactions and use them to answer Questions 21.47–21.50. What type of enzyme would catalyze each of these reactions?

21.48 To which family of organic molecules do a and d belong? To which family of organic molecules do b and c belong?

21.49 What is the name of the type of intermediate formed in each of these reactions?

21.50 Draw the intermediate that would be formed in each of these reactions.

21.51 When an enzyme has the term *kinase* in the name, what type of reaction do you expect it to catalyze?

21.52 What features do the reactions catalyzed by hexokinase and phosphofructokinase share in common?

21.53 What is the role of NAD^+ in a biochemical oxidation reaction?

21.54 Write the equation for the reaction catalyzed by glyceraldehyde-3-phosphate dehydrogenase. Highlight the chemical changes that show this to be an oxidation reaction.

21.55 The enzyme that catalyzes step 9 of glycolysis is called enolase. What is the significance of that name?

21.56 Draw the enol tautomer of pyruvate.

21.57 What is the importance of the regulation of glycolysis?

21.58 Explain the role of allosteric enzymes in control of glycolysis.

21.59 What molecules serve as allosteric effectors of phosphofructokinase?

21.60 What molecule serves as an allosteric inhibitor of hexokinase?

21.61 Explain the role of citrate in the feedback inhibition of glycolysis.

21.62 Explain the feedforward activation mechanism that results in the activation of pyruvate kinase.

Fermentations

Foundations

21.63 Write a balanced chemical equation for the conversion of acetaldehyde to ethanol.

21.64 Write a balanced chemical equation for the conversion of pyruvate to lactate.

Applications

21.65 After running a 100-m dash, a sprinter had a high concentration of muscle lactate. What process is responsible for production of lactate?

21.66 If the muscle of an organism had no lactate dehydrogenase, could anaerobic glycolysis occur in those muscle cells? Explain your answer.

21.67 What food products are the result of lactate fermentation?

21.68 Explain the value of alcohol fermentation in bread making.

21.69 What enzyme catalyzes the reduction of pyruvate to lactate?

21.70 What enzymes catalyze the conversion of pyruvate to ethanol and carbon dioxide?

21.71 A child was brought to the doctor's office suffering from a strange set of symptoms. When the child exercised hard, she became giddy and behaved as though drunk. What do you think is the metabolic basis of these symptoms?

21.72 A family started a batch of wine by adding yeast to grape juice and placing the mixture in a sealed bottle. Two weeks later, the bottle exploded. What metabolic reactions—and specifically, what product of those reactions—caused the bottle to explode?

The Pentose Phosphate Pathway

21.73 Describe the three stages of the pentose phosphate pathway.

21.74 Write an equation to summarize the pentose phosphate pathway.

21.75 Of what value are the ribose-5-phosphate and erythrose-4-phosphate that are produced in the pentose phosphate pathway?

21.76 Of what value is the NADPH that is produced in the pentose phosphate pathway?

Gluconeogenesis

21.77 Define gluconeogenesis and describe its role in metabolism.

21.78 What is the role of guanosine triphosphate in gluconeogenesis?

21.79 What organ is primarily responsible for gluconeogenesis?

21.80 What is the physiological function of gluconeogenesis?

21.81 Lactate can be converted to glucose by gluconeogenesis. To what metabolic intermediate must lactate be converted so that it can be a substrate for the enzymes of gluconeogenesis?

21.82 L-Alanine can be converted to pyruvate. Can L-alanine also be converted to glucose? Explain your answer.

21.83 Explain why gluconeogenesis is not simply the reversal of glycolysis.

21.84 In step 10 of glycolysis, phosphoenolpyruvate is converted to pyruvate, and ATP is produced by substrate-level phosphorylation. How is this reaction bypassed in gluconeogenesis?

21.85 Which steps in the glycolysis pathway are irreversible?

21.86 What enzymatic reactions of gluconeogenesis bypass the irreversible steps of glycolysis?

Glycogen Synthesis and Degradation

Foundations

21.87 What organs are primarily responsible for maintaining the proper blood glucose level?

21.88 Why must the blood glucose level be carefully regulated?

21.89 What does the term *hypoglycemia* mean?

21.90 What does the term *hyperglycemia* mean?

Applications

21.91 a. What enzymes involved in glycogen metabolism are stimulated by insulin?
   b. What effect does this have on glycogen metabolism?
   c. What effect does this have on blood glucose levels?

21.92 a. What enzyme is stimulated by glucagon?
   b. What effect does this have on glycogen metabolism?
   c. What effect does this have on blood glucose levels?

21.93 Explain how a defect in glycogen metabolism can cause hypoglycemia.

21.94 What defects of glycogen metabolism would lead to a large increase in the concentration of liver glycogen?

CRITICAL THINKING PROBLEMS

1. An enzyme that hydrolyzes ATP (an ATPase) bound to the plasma membrane of certain tumor cells has an abnormally high activity. How will this activity affect the rate of glycolysis?

2. Explain why no net oxidation occurs during anaerobic glycolysis followed by lactate fermentation.

3. A certain person was found to have a defect in glycogen metabolism. The liver of this person could (a) make glucose-6-phosphate from lactate and (b) synthesize glucose-6-phosphate from glycogen but (c) could not synthesize glycogen from glucose-6-phosphate. What enzyme is defective?

4. A scientist added phosphate labeled with radioactive phosphorus (^{32}P) to a bacterial culture growing anaerobically (without O_2). She then purified all the compounds produced during glycolysis. Look carefully at the steps of the pathway. Predict which of the intermediates of the pathway would be the first one to contain radioactive phosphate. On which carbon of this compound would you expect to find the radioactive phosphate?

5. A two-month-old baby was brought to the hospital suffering from seizures. He deteriorated progressively over time, showing psychomotor retardation. Blood tests revealed a high concentration of lactate and pyruvate. Although blood levels of alanine were high, they did not stimulate gluconeogenesis. The doctor measured the activity of pyruvate carboxylase in the baby and found it to be only 1% of the normal level. What reaction is catalyzed by pyruvate carboxylase? How could this deficiency cause the baby's symptoms and test results?

22 Aerobic Respiration and Energy Production

Learning Goals

1 Name the regions of the mitochondria and the function of each region.

2 Describe the reaction that results in the conversion of pyruvate to acetyl CoA, describing the location of the reaction and the components of the pyruvate dehydrogenase complex.

3 Summarize the reactions of aerobic respiration.

4 Looking at an equation representing any of the chemical reactions that occur in the citric acid cycle, describe the kind of reaction that is occurring and the significance of that reaction to the pathway.

5 Explain the mechanisms for the control of the citric acid cycle.

6 Describe the process of oxidative phosphorylation.

7 Describe the conversion of amino acids to molecules that can enter the citric acid cycle.

8 Explain the importance of the urea cycle and describe its essential steps.

9 Discuss the cause and effect of hyperammonemia.

10 Summarize the role of the citric acid cycle in catabolism and anabolism.

Outline

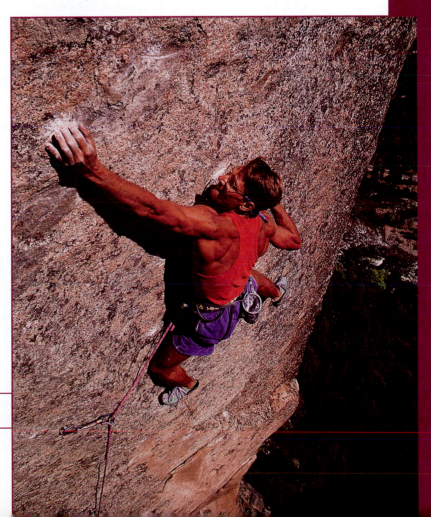

Rock climbing demands a great deal of energy.

Introduction

> An organelle is a compartment within the cytoplasm that has a specialized function.

As we have seen, the anaerobic glycolysis pathway begins the breakdown of glucose and produces a small amount of ATP and NADH. But it is aerobic catabolic pathways that complete the oxidation of glucose to CO_2 and H_2O and provide most of the ATP needed by the body. In fact, this process, called aerobic respiration, produces thirty-six ATP molecules using the energy harvested from each glucose molecule that enters glycolysis. These reactions occur in metabolic pathways located in mitochondria, the cellular "power plants." Mitochondria are a type of membrane-enclosed cell organelle.

Here, in the mitochondria, the final oxidations of carbohydrates, lipids, and proteins occur. Here, also, the electrons that are harvested in these oxidation reactions are used to make ATP. In these remarkably efficient reactions, nearly 40% of the potential energy of glucose is stored as ATP.

Chemistry Connection

Mitochondria from Mom

In this chapter we will be studying the amazing, intricate set of reactions that allow us to completely degrade fuel molecules such as sugars and amino acids. These oxygen-requiring reactions occur in cellular organelles called *mitochondria*.

We are used to thinking of the organelles as a collection of membrane-bound structures that are synthesized under the direction of the genetic information in the nucleus of the cell. Not so with the mitochondria. These organelles have their own genetic information and are able to make some of their own proteins. They grow and multiply in a way very similar to the simple bacteria. This, along with other information on the structure and activities of mitochondria, has led researchers to conclude that the mitochondria are actually the descendants of bacteria captured by eukaryotic cells millions of years ago.

Recent studies of the mitochondrial genetic information (DNA) have revealed fascinating new information. For instance, although each of us inherited half our genetic information from our mothers and half from our fathers, each of us inherited all of our mitochondria from our mothers. The reason for this is that when the sperm fertilizes the egg, only the sperm nucleus enters the cell.

The observation that all of our mitochondria are inherited from our mothers led Dr. A. Wilson to study the mitochondrial DNA of thousands of women around the world. He thought that by looking for similarities and differences in the mitochondrial DNA he would be able to identify a "Mitochondrial Eve"—the mother of all humanity. He didn't really think that he could identify a single woman who would have lived tens of thousands of years ago. But he hoped to determine the location

of the first population of human women to help answer questions about the origin of humankind. Although the idea was a good one, the study had a number of experimental flaws. Currently, a hot debate is going on among hundreds of scientists about the Mitochondrial Eve. This controversy should encourage better experiments and analysis to help us identify our origins and to better understand the workings of the mitochondria.

Like the mitochondria themselves, some genetic diseases of energy metabolism are maternally inherited. One such disease, Leber's hereditary optic neuropathy (LHON), causes blindness and heart problems. People with LHON have a reduced ability to make ATP. As a result, sensitive tissues that demand a great deal of energy eventually die. LHON sufferers eventually lose their sight because the optic nerve dies from lack of energy.

Researchers have identified and cloned a mutant mitochondrial gene that is responsible for LHON. The defect is a mutant form of *NADH dehydrogenase*. NADH dehydrogenase is a huge, complex enzyme that accepts electrons from NADH and sends them on through an electron transport system. Passage of electrons through the electron transport system allows the synthesis of ATP. If NADH dehydrogenase is defective, passage of electrons through the electron transport system is less efficient, and less ATP is made. In LHON sufferers, the result is eventual blindness.

In this chapter and the next, we will study some of the important biochemical reactions that occur in the mitochondria. A better understanding of the function of healthy mitochondria will eventually allow us to help those suffering from LHON and other mitochondrial genetic diseases.

22.1 The Mitochondria

Mitochondria are football-shaped organelles that are roughly the size of a bacterial cell. They are surrounded by an **outer mitochondrial membrane** and an **inner mitochondrial membrane** (Figure 22.1). The space between the two membranes is the **intermembrane space,** and the space inside of the inner membrane is the **matrix space.** The enzymes of the citric acid cycle, of the β-oxidation pathway for the breakdown of fatty acids, and for the degradation of amino acids are all found in the mitochondrial matrix space.

LEARNING GOAL

1 ▷ Name the regions of the mitochondria and the function of each region.

Structure and Function

The outer mitochondrial membrane has many small pores through which small molecules (less than 10,000 g/mol) can pass. Thus, the small molecules to be oxidized for the production of ATP can easily enter the mitochondrial intermembrane space.

The inner membrane is highly folded to create a large surface area. The folded membranes are known as **cristae.** The inner mitochondrial membrane is almost completely impermeable to most substances. For this reason it has many transport proteins to bring particular fuel molecules into the matrix space. Also embedded within the inner mitochondrial membrane are the protein electron carriers of the *electron transport system* and *ATP synthase.* ATP synthase is a large complex of many proteins that catalyzes the synthesis of ATP.

Origin of the Mitochondria

Not only are mitochondria roughly the size of bacteria, they have several other features that have led researchers to suspect that they may once have been free-living bacteria that were "captured" by eukaryotic cells. They have their own genetic information (DNA). They also make their own ribosomes that are very similar to those of bacteria. These ribosomes allow the mitochondria to synthesize some of their own proteins. Finally, mitochondria are actually self-replicating; they grow in size and divide to produce new mitochondria. All of these characteristics suggest that the mitochondria that produce the majority of the ATP for our cells evolved from bacteria "captured" perhaps as long as 1.5×10^9 years ago.

As we saw in Chapter 20, ribosomes are complexes of protein and RNA that serve as small platforms for protein synthesis.

Figure 22.1
Structure of the mitochondrion.
(a) Electron micrograph of mitochondria.
(b) Schematic drawing of the mitochondrion.

(a)

(b)

A Human *Perspective*

Exercise and Energy Metabolism

The Olympic sprinters get set in the blocks. The gun goes off, and roughly ten seconds later the 100-m dash is over. Elsewhere, the marathoners line up. They will run 26 miles and 385 yards in a little over two hours. Both sports involve running, but they utilize very different sources of energy.

Let's look at the sprinter first. The immediate source of energy for the sprinter is stored ATP. But the quantity of stored ATP is very small, only about three ounces. This allows the sprinter to run as fast as he or she can for about three seconds. Obviously, another source of stored energy must be tapped, and that energy store is *creatine phosphate:*

$$^-O-\overset{\overset{\displaystyle O}{\|}}{\underset{\underset{\displaystyle O^-}{|}}{P}}-\overset{\overset{\displaystyle H}{|}}{N}-\overset{\overset{\displaystyle NH}{\|}}{C}-\overset{}{\underset{\underset{\displaystyle CH_3}{|}}{N}}-CH_2-C\overset{\nearrow O^-}{\underset{\searrow O}{}}$$

The structure of creatine phosphate.

Creatine phosphate, stored in the muscle, donates its high-energy phosphate to ADP to produce new supplies of ATP.

This will keep our runner in motion for another five or six seconds before the store of creatine phosphate is also depleted. This is almost enough energy to finish the 100-m dash, but in reality, all the runners are slowing down, owing to energy depletion, and the winner is the sprinter who is slowing down the least!

Consider a longer race, the 400-m or the 800-m. These runners run at maximum capacity for much longer. When they have depleted their ATP and creatine phosphate stores, they must synthesize more ATP. Of course, the cells have been making ATP all the time, but now the demand for energy is much greater. To supply this increased demand, the anaerobic energy-generating reactions (glycolysis and lactate fermentation, Chapter 21) and aerobic processes (citric acid cycle

and oxidative phosphorylation) begin to function much more rapidly. Often, however, these athletes are running so strenuously that they cannot provide enough oxygen to the exercising muscle to allow oxidative phosphorylation to function efficiently. When this happens, the muscles must rely on glycolysis and lactate fermentation to provide *most* of the energy requirement. The chemical by-product of these anaerobic processes, lactate, builds up in the muscle and diffuses into the bloodstream. However, the concentration of lactate inevitably builds up in the working muscle and causes muscle fatigue and, eventually, muscle failure. Thus, exercise that depends primarily on anaerobic ATP production cannot continue for very long.

The marathoner presents us with a different scenario. This runner will deplete his or her stores of ATP and creatine phosphate as quickly as a short-distance runner. The anaerobic glycolytic pathway will begin to degrade glucose provided by the blood at a more rapid rate, as will the citric acid cycle and oxidative phosphorylation. The major difference in ATP production between the long-distance runner and the short- or middle-distance runner is that the muscles of the long-distance runner derive almost all the energy through aerobic pathways. These individuals continue to run long distances at a pace that allows them to supply virtually all the oxygen needed by the exercising muscle. In fact, only aerobic pathways can provide a constant supply of ATP for exercise that goes on for hours. Theoretically, under such conditions our runner could run indefinitely, utilizing first his or her stored glycogen and eventually stored lipids. Of course, in reality, other factors such as dehydration and fatigue place limits on the athlete's ability to continue.

From this we can conclude that long-distance runners must have a great capacity to produce ATP aerobically, in the mitochondria, whereas short- and middle-distance runners need a great capacity to produce energy anaerobically, in the

$$^-O-\overset{\overset{\displaystyle O}{\|}}{\underset{\underset{\displaystyle O^-}{|}}{P}}-\overset{\overset{\displaystyle H}{|}}{N}-\overset{\overset{\displaystyle NH}{\|}}{C}-\overset{}{\underset{\underset{\displaystyle CH_3}{|}}{N}}-CH_2-C\overset{\nearrow O^-}{\underset{\searrow O}{}} + ADP \xrightarrow{\text{Creatine kinase}} H_2N-\overset{\overset{\displaystyle NH}{\|}}{C}-N-CH_2-C\overset{\nearrow O^-}{\underset{\searrow O}{}} + ATP$$

Phosphoryl group transfer from creatine phosphate to ADP is catalyzed by the enzyme creatine kinase.

Question *22.1*	What is the function of the mitochondria?
Question *22.2*	How do the mitochondria differ from the other components of eukaryotic cells?

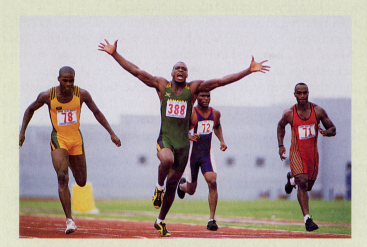

The sprinter relies on fast-twitch muscle fibers.

The marathon runner largely uses slow-twitch muscle fibers.

cytoplasm of the muscle cells. It is interesting to note that the muscles of these runners reflect these diverse needs.

When one examines muscle tissue that has been surgically removed, one finds two predominant types of muscle fibers. *Fast-twitch muscle fibers* are large, relatively plump, pale cells. They have only a few mitochondria but contain a large reserve of glycogen and high concentrations of the enzymes that are needed for glycolysis and lactate fermentation. These muscle fibers fatigue rather quickly because fermentation is inefficient, quickly depleting the cell's glycogen store and causing the accumulation of lactate.

Slow-twitch muscle fiber cells are about half the diameter of fast-twitch muscle cells and are red. The red color is a result of the high concentrations of myoglobin in these cells. Recall that myoglobin stores oxygen for the cell (Section 18.9) and facilitates rapid diffusion of oxygen throughout the cell. In addition, slow-twitch muscle fiber cells are packed with mitochondria. With this abundance of oxygen and mitochondria these cells have the capacity for extended ATP production via aerobic pathways—ideal for endurance sports like marathon racing.

It is not surprising, then, that researchers have found that the muscles of sprinters have many more fast-twitch muscle fibers and those of endurance athletes have many more slow-twitch muscle fibers. One question that many researchers are trying to answer is whether the type of muscle fibers an individual has is a function of genetic makeup or training. Is a marathon runner born to be a long-distance runner, or are his or her abilities due to the type of training the runner undergoes? There is no doubt that the training regimen for an endurance runner does indeed increase the number of slow-twitch muscle fibers and that of a sprinter increases the number of fast-twitch muscle fibers. But there is intriguing new evidence to suggest that the muscles of endurance athletes have a greater proportion of slow-twitch muscle fibers before they ever begin training. It appears that some of us truly were born to run.

For Further Understanding

It has been said that the winner of the 100-m race is the one who is slowing down the least. Explain this observation in terms of energy-harvesting pathways.

Design an experiment to safely test whether the type of muscle fibers a runner has are the result of training or genetic makeup.

Draw a schematic diagram of a mitochondrion, and label the parts of this organelle.

Question *22.3*

Describe the evidence that suggests that mitochondria evolved from free-living bacteria.

Question *22.4*

22.2 Conversion of Pyruvate to Acetyl CoA

LEARNING GOAL

2 Describe the reaction that results in the conversion of pyruvate to acetyl CoA, describing the location of the reaction and the components of the pyruvate dehydrogenase complex.

Coenzyme A is described in Sections 12.9 and 14.4.

Water-Soluble Vitamins

Thioester bonds are discussed in Section 14.4.

Animation
B Vitamins

As we saw in Chapter 21, under *anaerobic* conditions, glucose is broken down into two pyruvate molecules that are then converted to a stable fermentation product. This limited degradation of glucose releases very little of the potential energy of glucose. Under *aerobic* conditions the cells can use oxygen and completely oxidize glucose to CO_2 in a metabolic pathway called the *citric acid cycle*.

This pathway is often referred to as the *Krebs cycle* in honor of Sir Hans Krebs who worked out the steps of this cyclic pathway from his own experimental data and that of other researchers. It is also called the *tricarboxylic acid (TCA) cycle* because several of the early intermediates in the pathway have three carboxyl groups.

Once pyruvate enters the mitochondria, it must be converted to a two-carbon acetyl group. This acetyl group must be "activated" to enter the reactions of the citric acid cycle. Activation occurs when the acetyl group is bonded to the thiol group of coenzyme A. **Coenzyme A** is a large thiol derived from ATP and the vitamin pantothenic acid (Figure 22.2). It is an acceptor of acetyl groups (in red in Figure 22.2), which are bonded to it through a high-energy thioester bond. The acetyl coenzyme A (**acetyl CoA**) formed is the "activated" form of the acetyl group.

Figure 22.3 shows us the reaction that converts pyruvate to acetyl CoA. First, pyruvate is decarboxylated, which means that it loses a carboxyl group that is released as CO_2. Next it is oxidized, and the hydride anion that is removed is accepted by NAD^+. Finally, the remaining acetyl group, $CH_3CO—$, is linked to coenzyme A by a thioester bond. This very complex reaction is carried out by three enzymes and five coenzymes that are organized together in a single bundle called the **pyruvate dehydrogenase complex** (see Figure 22.3). This organization allows the substrate to be passed from one enzyme to the next as each chemical reaction occurs. A schematic representation of this "disassembly line" is shown in Figure 22.3b.

This single reaction requires four coenzymes made from four different vitamins, in addition to the coenzyme lipoamide. These are thiamine pyrophosphate, derived from thiamine (Vitamin B_1); FAD, derived from riboflavin (Vitamin B_2); NAD^+, derived from niacin; and coenzyme A, derived from pantothenic acid. Obviously, a deficiency in any of these vitamins would seriously reduce the amount of acetyl CoA that our cells could produce. This, in turn, would limit the amount of ATP that the body could make and would contribute to vitamin-deficiency diseases. Fortunately, a well-balanced diet provides an adequate supply of these and other vitamins.

Acetyl coenzyme A
(acetyl CoA)

Figure 22.2
The structure of acetyl CoA. The bond between the acetyl group and coenzyme A is a high-energy thioester bond.

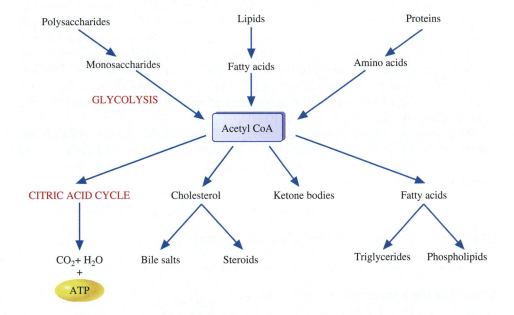

(a)

Pyruvate   Coenzyme A   Acetyl coenzyme A

(b)

Pyruvate

Pyruvate dehydrogenase complex

Acetyl CoA

CO_2

Coenzyme A

$H:^-$

NAD^+ → NADH

Figure 22.3

The decarboxylation and oxidation of pyruvate to produce acetyl CoA. (a) The overall reaction in which CO_2 and an $H:^-$ are removed from pyruvate and the remaining acetyl group is attached to coenzyme A. This requires the concerted action of three enzymes and five coenzymes. (b) The pyruvate dehydrogenase complex that carries out this reaction is actually a cluster of enzymes and coenzymes. The substrate is passed from one enzyme to the next as the reaction occurs.

Polysaccharides   Lipids   Proteins

Monosaccharides   Fatty acids   Amino acids

GLYCOLYSIS

Acetyl CoA

CITRIC ACID CYCLE   Cholesterol   Ketone bodies   Fatty acids

$CO_2 + H_2O$ + ATP   Bile salts   Steroids   Triglycerides   Phospholipids

Figure 22.4

The central role of acetyl CoA in cellular metabolism.

In Figure 22.4, we see that acetyl CoA is a central character in cellular metabolism. It is produced by the degradation of glucose, fatty acids, and some amino acids. The major function of acetyl CoA in energy-harvesting pathways is to carry the acetyl group to the citric acid cycle, in which it will be used to produce large amounts of ATP. In addition to these catabolic duties, the acetyl group of acetyl CoA can also be used for *anabolic* or biosynthetic reactions to produce cholesterol and fatty acids. It is through this intermediate, acetyl CoA, that all the energy sources (fats, proteins, and carbohydrates) are interconvertible.

Question 22.5

What vitamins are required for acetyl CoA production from pyruvate?

Question 22.6

What is the major role of coenzyme A in catabolic reactions?

22.3 An Overview of Aerobic Respiration

LEARNING GOAL

3 Summarize the reactions of aerobic respiration.

Aerobic respiration is the oxygen-requiring breakdown of food molecules and production of ATP. The different steps of aerobic respiration occur in different compartments of the mitochondria.

The enzymes for the citric acid cycle are found in the mitochondrial matrix space. The first enzyme catalyzes a reaction that joins the acetyl group of acetyl CoA (two carbons) to a four-carbon molecule (oxaloacetate) to produce citrate (six carbons). The remaining enzymes catalyze a series of rearrangements, decarboxylations (removal of CO_2), and oxidation–reduction reactions. The eventual products of this cyclic pathway are two CO_2 molecules and oxaloacetate—the molecule we began with.

At several steps in the citric acid cycle, a substrate is oxidized. In three of these steps, a pair of electrons is transferred from the substrate to NAD^+, producing NADH (three NADH molecules per turn of the cycle). At another step a pair of electrons is transferred from a substrate to FAD, producing $FADH_2$ (one $FADH_2$ molecule per turn of the cycle).

The electrons are passed from NADH or $FADH_2$, through an electron transport system located in the inner mitochondrial membrane, and finally to the terminal electron acceptor, molecular oxygen (O_2). The transfer of electrons through the electron transport system causes protons (H^+) to be pumped from the mitochondrial matrix into the intermembrane compartment. The result is a high-energy H^+ reservoir.

In the final step, the energy of the H^+ reservoir is used to make ATP. This last step is carried out by the enzyme complex ATP synthase. As protons flow back into the mitochondrial matrix through a pore in the ATP synthase complex, the enzyme catalyzes the synthesis of ATP.

This long, involved process is called *oxidative phosphorylation*, because the energy of electrons from the *oxidation* of substrates in the citric acid cycle is used to *phosphorylate* ADP and produce ATP. The details of each of these steps will be examined in upcoming sections.

Remember (Section 19.7) that it is really the hydride anion with its pair of electrons ($H:^-$) that is transferred to NAD^+ to produce NADH. Similarly, a pair of hydrogen atoms (and thus two electrons) are transferred to FAD to produce $FADH_2$.

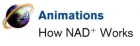
Animations

How NAD^+ Works

A Biochemical Pathway

Question 22.7

What is meant by the term *oxidative phosphorylation*?

Question 22.8

What does the term *aerobic respiration* mean?

22.4 The Citric Acid Cycle (The Krebs Cycle)

LEARNING GOAL

4 Looking at an equation representing any of the chemical reactions that occur in the citric acid cycle, describe the kind of reaction that is occurring and the significance of that reaction to the pathway.

Reactions of the Citric Acid Cycle

The **citric acid cycle** is the final stage of the breakdown of carbohydrates, fats, and amino acids released from dietary proteins (Figure 22.5). To understand this important cycle, let's follow the fate of the acetyl group of an acetyl CoA as it

Figure 22.5
The reactions of the citric acid cycle.

Animation
How the Krebs Cycle Works

Aldol condensation reactions are reactions between aldehydes and ketones to form larger molecules (Section 13.4).

passes through the citric acid cycle. The numbered steps listed below correspond to the steps in the citric acid cycle that are summarized in Figure 22.5.

Reaction 1.   This is a condensation reaction between the acetyl group of acetyl CoA and oxaloacetate. Actually, this is another biological example of an aldol condensation reaction. It is catalyzed by the enzyme *citrate synthase*. The product that is formed is citrate:

$$\text{Oxaloacetate} + \text{Acetyl CoA} + H_2O \xrightarrow{\text{Citrate synthase}} \text{Citrate} + HS-CoA + H^+$$

Oxaloacetate   Acetyl CoA   Citrate   Coenzyme A

Notice that the conversion of citrate to cis-aconitate is a biological example of the dehydration of an alcohol to produce an alkene (Section 12.5). The conversion of cis-aconitate to isocitrate is a biochemical example of the hydration of an alkene to produce an alcohol (Sections 11.5 and 12.5).

Reaction 2.   The enzyme *aconitase* catalyzes the dehydration of citrate, producing *cis*-aconitate. The same enzyme, aconitase, then catalyzes addition of a water molecule to the *cis*-aconitate, converting it to isocitrate. The net effect of these two steps is the isomerization of citrate to isocitrate:

$$\text{Citrate} \xrightarrow{\text{Aconitase}} \textit{cis-}\text{Aconitate} + H_2O \xrightarrow{\text{Aconitase}} \text{Isocitrate}$$

Citrate   *cis*-Aconitate   Isocitrate

Reaction 3.   The first oxidative step of the citric acid cycle is catalyzed by *isocitrate dehydrogenase*. It is a complex reaction in which three things happen:

The oxidation of a secondary alcohol produces a ketone (Sections 12.5 and 13.4).

   a. the hydroxyl group of isocitrate is oxidized to a ketone,
   b. carbon dioxide is released, and

The structure of NAD$^+$ and its reduction to NADH are shown in Figure 19.8.

   c. NAD$^+$ is reduced to NADH.

The product of this oxidative decarboxylation reaction is α-ketoglutarate:

Remember, in organic (and thus biochemical) reactions, oxidation can be recognized as a gain of oxygen or loss of hydrogen (Section 12.6).

$$\text{Isocitrate} + NAD^+ \xrightarrow{\text{Isocitrate dehydrogenase}} \text{α-Ketoglutarate} + CO_2 + \text{NADH}$$

Isocitrate   α-Ketoglutarate

The pyruvate dehydrogenase complex was described in Section 22.2 and shown in Figure 22.3.

Reaction 4.   Coenzyme A enters the picture again as the *α-ketoglutarate dehydrogenase* complex carries out a complex series of reactions similar to those catalyzed by the pyruvate dehydrogenase complex. The same coenzymes are required and, once again, three chemical events occur:

a. α-ketoglutarate loses a carboxylate group as CO_2,

b. it is oxidized and NAD^+ is reduced to NADH, and

c. coenzyme A combines with the product, succinate, to form succinyl CoA. The bond formed between succinate and coenzyme A is a high-energy thioester bond.

$$
\begin{array}{c}
COO^- \\
| \\
CH_2 \\
| \\
CH_2 \\
| \\
C=O \\
| \\
COO^-
\end{array}
+ NAD^+ + \text{Coenzyme A}
\xrightarrow{\substack{\alpha\text{-Ketoglutarate} \\ \text{dehydrogenase} \\ \text{complex}}}
\begin{array}{c}
COO^- \\
| \\
CH_2 \\
| \\
CH_2 \\
| \\
C{\sim}S{-}CoA \\
\| \\
O
\end{array}
+ CO_2 + \text{NADH}
$$

α-Ketoglutarate   Succinyl CoA

Reaction 5. Succinyl CoA is converted to succinate in this step, which once more is chemically very involved. The enzyme *succinyl CoA synthase* catalyzes a coupled reaction in which the high-energy thioester bond of succinyl CoA is hydrolyzed and an inorganic phosphate group is added to GDP to make GTP:

$$
\begin{array}{c}
COO^- \\
| \\
CH_2 \\
| \\
CH_2 \\
| \\
C{\sim}S{-}CoA \\
\| \\
O
\end{array}
+ GDP + P_i
\xrightarrow{\substack{\text{Succinyl CoA} \\ \text{synthase}}}
\begin{array}{c}
COO^- \\
| \\
CH_2 \\
| \\
CH_2 \\
| \\
COO^-
\end{array}
+ \text{GTP} + \text{Coenzyme A}
$$

Succinyl CoA   Succinate

Another enzyme, *dinucleotide diphosphokinase*, then catalyzes the transfer of a phosphoryl group from GTP to ADP to make ATP:

$$
\text{GTP} + ADP \xrightarrow{\substack{\text{Dinucleotide} \\ \text{diphosphokinase}}} GDP + \text{ATP}
$$

Reaction 6. *Succinate dehydrogenase* then catalyzes the oxidation of succinate to fumarate in the next step. The oxidizing agent, *flavin adenine dinucleotide (FAD)*, is reduced in this step:

The structure of FAD was shown in Figure 19.8.

$$
\begin{array}{c}
COO^- \\
| \\
CH_2 \\
| \\
CH_2 \\
| \\
COO^-
\end{array}
+ FAD
\xrightarrow{\substack{\text{Succinate} \\ \text{dehydrogenase}}}
\begin{array}{c}
COO^- \\
| \\
C{-}H \\
\| \\
H{-}C \\
| \\
COO^-
\end{array}
+ \text{FADH}_2
$$

Succinate   Fumarate

We studied hydrogenation of alkenes to produce alkanes in Section 11.5. This is simply the reverse.

Reaction 7. Addition of H_2O to the double bond of fumarate gives malate. The enzyme *fumarase* catalyzes this reaction:

This reaction is a biological example of the hydration of an alkene to produce an alcohol (Sections 11.5 and 12.5).

Fumarate   Malate

Reaction 8. In the final step of the citric acid cycle, *malate dehydrogenase* catalyzes the reduction of NAD^+ to NADH and the oxidation of malate to oxaloacetate. Because the citric acid cycle "began" with the addition of an acetyl group to oxaloacetate, we have come full circle.

This reaction is a biochemical example of the oxidation of a secondary alcohol to a ketone, which we studied in Sections 12.5 and 14.4.

Malate   Oxaloacetate

22.5 Control of the Citric Acid Cycle

LEARNING GOAL

5 Explain the mechanisms for the control of the citric acid cycle.

Allosteric enzymes bind to *effectors*, such as ATP or ADP, that alter the shape of the enzyme active site, either stimulating the rate of the reaction (positive allosterism) or inhibiting the reaction (negative allosterism). For more detail, see Section 19.9.

Just like glycolysis, the citric acid cycle is responsive to the energy needs of the cell. The pathway speeds up when there is a greater demand for ATP, and it slows down when ATP energy is in excess. In the last chapter we saw that several of the enzymes that catalyze the reactions of glycolysis are *allosteric enzymes*. Similarly, four enzymes or enzyme complexes involved in the complete oxidation of pyruvate are allosteric enzymes. Because the control of the pathway must be precise, there are several enzymatic steps that are regulated. These are summarized in Figure 22.6 and below:

1. *Conversion of pyruvate to acetyl CoA.* The pyruvate dehydrogenase complex is inhibited by high concentrations of ATP, acetyl CoA, and NADH. Of course, the presence of these compounds in abundance signals that the cell has an adequate supply of energy, and thus energy metabolism is slowed.
2. *Synthesis of citrate from oxaloacetate and acetyl CoA.* The enzyme citrate synthase is an allosteric enzyme. In this case, the negative effector is ATP. Again, this is logical because an excess of ATP indicates that the cell has an abundance of energy.
3. *Oxidation and decarboxylation of isocitrate to α-ketoglutarate.* Isocitrate dehydrogenase is also an allosteric enzyme; however, the enzyme is controlled by the positive allosteric effector, ADP. ADP is a signal that the levels of ATP must be low, and therefore the rate of the citric acid cycle should be increased. Interestingly, isocitrate dehydrogenase is also *inhibited* by high levels of NADH and ATP.
4. *Conversion of α-ketoglutarate to succinyl CoA.* The α-ketoglutarate dehydrogenase complex is inhibited by high levels of the products of the reactions that it catalyzes, namely, NADH and succinyl CoA. It is further inhibited by high concentrations of ATP.

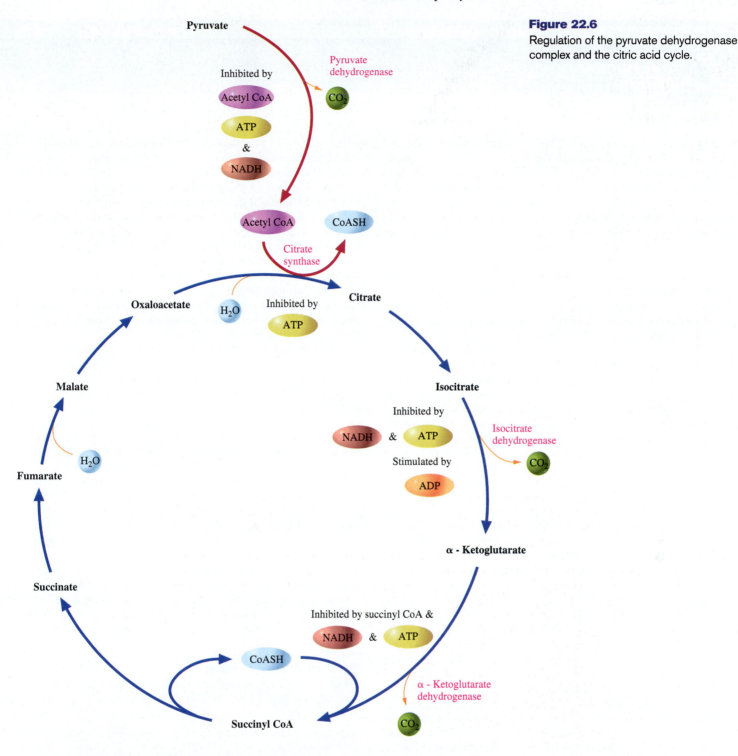

22.6 Oxidative Phosphorylation

In Section 22.3 we noted that the electrons carried by NADH can be used to produce three ATP molecules, and those carried by FADH$_2$ can be used to produce two ATP molecules. We turn now to the process by which the energy of electrons carried by these coenzymes is converted to ATP energy. It is a series of reactions called **oxidative phosphorylation,** which couples the oxidation of NADH and FADH$_2$ to the phosphorylation of ADP to generate ATP.

LEARNING GOAL
6 Describe the process of oxidative phosphorylation.

A Human Perspective

Brown Fat: The Fat That Makes You Thin?

Humans have two types of fat, or adipose, tissue. *White fat* is distributed throughout the body and is composed of aggregations of cells having membranous vacuoles containing stored triglycerides. The size and number of these storage vacuoles determine whether a person is overweight or not. The other type of fat is *brown fat*. Brown fat is a specialized tissue for heat production, called *nonshivering thermogenesis*. As the name suggests, this is a means of generating heat in the absence of the shivering response. The cells of brown fat look nothing like those of white fat. They do contain small fat vacuoles; however, the distinguishing feature of brown fat is the huge number of mitochondria within the cytoplasm. In addition, brown fat tissue contains a great many blood vessels. These provide oxygen for the thermogenic metabolic reactions.

Brown fat is most pronounced in newborns, cold-adapted mammals, and hibernators. One major difficulty faced by a newborn is temperature regulation. The baby leaves an environment in which he or she was bathed in fluid of a constant 37°C, body temperature. Suddenly, the child is thrust into a world that is much colder and in which he or she must generate his or her own warmth internally. By having a good reserve of active brown fat to generate that heat, the newborn is protected against cold shock at the time of birth. However, this thermogenesis literally burns up most of the brown fat tissue, and adults typically have so little brown fat that it can be found only by using a special technique called *thermography*, which detects temperature differences throughout a body. However, in some individuals, brown fat is very highly developed. For instance, the Korean diving women who spend 6–7 hours every day diving for pearls in cold water have a massive amount of brown fat to warm them by nonshivering thermogenesis. Thus, development of brown fat is a mechanism of cold adaptation.

When it was noticed that such cold-adapted individuals were seldom overweight, a correlation was made between the amount of brown fat in the body and the tendency to become overweight. Studies done with rats suggest that, to some degree, fatness is genetically determined. In other words, you are as lean as your genes allow you to be. In these studies, cold-adapted and non-cold-adapted rats were fed cafeteria food—as much as they wanted—and their weight gain was monitored. In every case the cold-adapted rats, with their greater quantity of brown fat, gained significantly less weight than their non-cold-adapted counterparts, despite the fact that they ate as

much as the non-cold-adapted rats. This and other studies led researchers to conclude that brown fat burns excess fat in a highly caloric diet.

How does brown fat generate heat and burn excess calories? For the answer we must turn to the mitochondrion. In addition to the ATP synthase and the electron transport system proteins that are found in all mitochondria, there is a protein in the inner mitochondrial membrane of brown fat tissue called *thermogenin*. This protein has a channel in the center through which the protons (H^+) of the intermembrane space could pass back into the mitochondrial matrix. Under normal conditions, this channel is plugged by a GDP molecule so that it remains closed and the proton gradient can continue to drive ATP synthesis by oxidative phosphorylation.

When brown fat is "turned on," by cold exposure or in response to certain hormones, there is an immediate increase in the rate of glycolysis and β-oxidation of the stored fat (Chapter 23). These reactions produce acetyl CoA, which then fuels the citric acid cycle. The citric acid cycle, of course, produces NADH and $FADH_2$, which carry electrons to the electron transport system. Finally, the electron transport system pumps protons into the intermembrane space. Under usual conditions, the energy of the proton gradient would be used to synthesize ATP. However, when brown fat is stimulated, the GDP that had plugged the pore in thermogenin is lost. Now protons pass freely back into the matrix space, and the proton gradient is dissipated. The energy of the gradient, no longer useful for generating ATP, is released as *heat*, the heat that warms and protects newborns and cold-adapted individuals.

Brown fat is just one of the body's many systems for maintaining a constant internal environment regardless of the conditions in the external environment. Such mechanisms, called *homeostatic mechanisms*, are absolutely essential to allow the body to adapt to and survive in an ever-changing environment.

For Further Understanding

Hibernators eat a great deal in preparation for their long winter nap. Explain their lifestyle in terms of energy requirements and heat production.

Explain why cold-adapted mammals can eat a high caloric diet and remain thin.

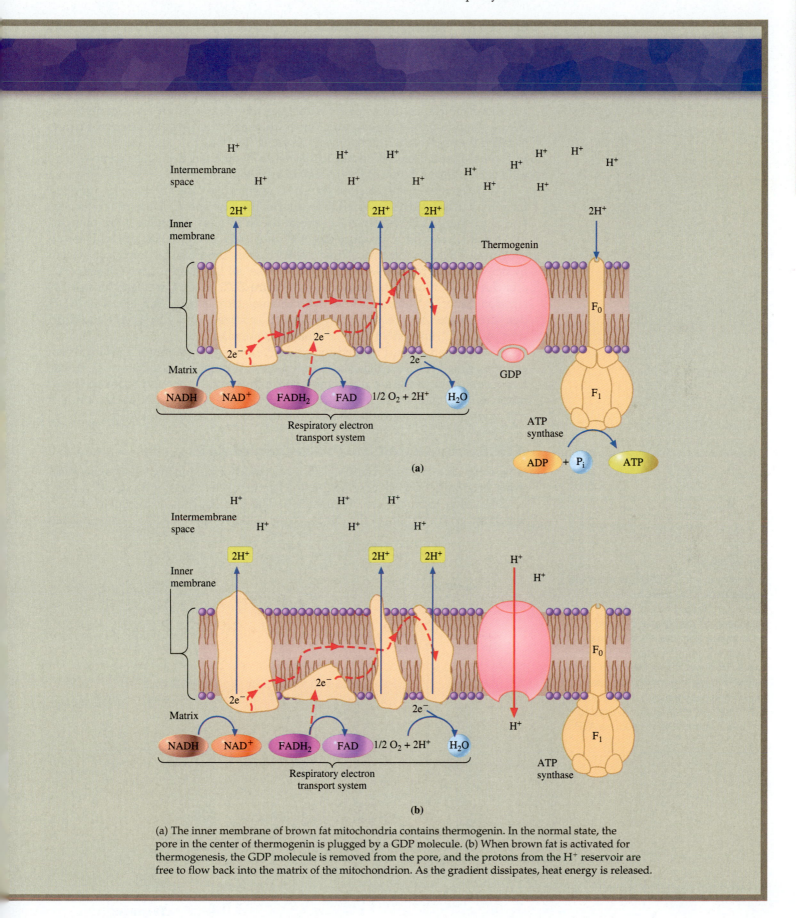

(a) The inner membrane of brown fat mitochondria contains thermogenin. In the normal state, the pore in the center of thermogenin is plugged by a GDP molecule. (b) When brown fat is activated for thermogenesis, the GDP molecule is removed from the pore, and the protons from the H^+ reservoir are free to flow back into the matrix of the mitochondrion. As the gradient dissipates, heat energy is released.

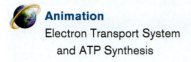

Animation

Electron Transport System
and ATP Synthesis

Electron Transport Systems and the Hydrogen Ion Gradient

Before we try to understand the mechanism of oxidative phosphorylation, let's first look at the molecules that carry out this complex process. Embedded within the mitochondrial inner membrane are **electron transport systems.** These are made up of a series of electron carriers, including coenzymes and cytochromes. All these molecules are located within the membrane in an arrangement that allows them to pass electrons from one to the next. This array of electron carriers is called the *respiratory electron transport system* (Figure 22.7). As you would expect in such sequential oxidation-reduction reactions, the electrons lose some energy with each transfer. Some of this energy is used to make ATP.

At three sites in the electron transport system, protons (H^+) can be pumped from the mitochondrial matrix to the intermembrane space. These H^+ contribute to a high-energy H^+ reservoir. At each of the three sites, enough H^+ are pumped into the H^+ reservoir to produce one ATP molecule. The first site is NADH dehydrogenase. Because electrons from NADH enter the electron transport system by being transferred to NADH dehydrogenase, all three sites actively pump H^+, and three ATP molecules are made (see Figure 22.7). $FADH_2$ is a less "powerful" electron donor. It transfers its electrons to an electron carrier that follows NADH dehydrogenase. As a result, when $FADH_2$ is oxidized, only the second and third sites pump H^+, and only two ATP molecules are made.

The last component needed for oxidative phosphorylation is a multiprotein complex called **ATP synthase,** also called the F_0F_1 **complex** (see Figure 22.7). The F_0 portion of the molecule is a channel through which H^+ pass. It spans the inner mitochondrial membrane, as shown in Figure 22.7. The F_1 part of the molecule is an enzyme that catalyzes the phosphorylation of ADP to produce ATP.

ATP Synthase and the Production of ATP

How does all this complicated machinery actually function? NADH carries electrons, originally from glucose, to the first carrier of the electron transport system, NADH dehydrogenase (see Figure 22.7). There, NADH is oxidized to NAD^+, which returns to the site of the citric acid cycle to be reduced again. As the dashed

Figure 22.7

Electrons flow from NADH to molecular oxygen through a series of electron carriers embedded in the inner mitochondrial membrane. Protons are pumped from the mitochondrial matrix space into the intermembrane space. This results in a hydrogen ion reservoir in the intermembrane space. As protons pass through the channel in ATP synthase, their energy is used to phosphorylate ADP and produce ATP.

red line shows, the pair of electrons is passed to the next electron carrier, and H^+ are pumped to the intermembrane compartment. The electrons are passed sequentially through the electron transport system, and at two additional sites, H^+ from the matrix are pumped into the intermembrane compartment. With each transfer the electrons lose some of their potential energy. It is this energy that is used to transport H^+ across the inner mitochondrial membrane and into the H^+ reservoir. As mentioned above, $FADH_2$ donates its electrons to a carrier of lower energy and fewer H^+ are pumped into the reservoir.

Finally, the electrons arrive at the last carrier. They now have too little energy to accomplish any more work, but they *must* be donated to some final electron acceptor so that the electron transport system can continue to function. In aerobic organisms the **terminal electron acceptor** is molecular oxygen, O_2, and the product is water.

As the electron transport system continues to function, a high concentration of protons builds up in the intermembrane space. This creates an H^+ gradient across the inner mitochondrial membrane. Such a gradient is an enormous energy source, like water stored behind a dam. The mitochondria make use of the potential energy of the gradient to synthesize ATP energy.

ATP synthase harvests the energy of this gradient by making ATP. H^+ pass through the F_0 channel back into the matrix. This causes F_1 to become an active enzyme that catalyzes the phosphorylation of ADP to produce ATP. In this way the energy of the H^+ reservoir is harvested to make ATP.

The importance of keeping the electron transport system functioning becomes obvious when we consider what occurs in cyanide poisoning. Cyanide binds to the heme group iron of cytochrome oxidase, instantly stopping electron transfers and causing death within minutes!

Write a balanced chemical equation for the reduction of NAD^+.

Question *22.9*

Write a balanced chemical equation for the reduction of FAD.

Question *22.10*

Summary of the Energy Yield

One turn of the citric acid cycle results in the production of two CO_2 molecules, three NADH molecules, one $FADH_2$ molecule, and one ATP molecule. Oxidative phosphorylation yields three ATP molecules per NADH molecule and two ATP molecules per $FADH_2$ molecule. The only exception to these energy yields is the NADH produced in the cytoplasm during glycolysis. Oxidative phosphorylation yields only two ATP molecules per cytoplasmic NADH molecule. The reason for this is that energy must be expended to shuttle electrons from NADH in the cytoplasm to $FADH_2$ in the mitochondrion.

Knowing this information and keeping in mind that two turns of the citric acid cycle are required, we can sum up the total energy yield from the complete oxidation of one glucose molecule.

 Energy Yield from Aerobic Respiration: Some Alternatives

In some tissues of the body there is a more efficient shuttle system that results in the production of three ATP per cytoplasmic NADH. This system is described online.

Determining the Yield of ATP from Aerobic Respiration	**E X A M P L E 22.1**

Calculate the number of ATP produced by the complete oxidation of one molecule of glucose.

Solution

Glycolysis:

Substrate-level phosphorylation	2 ATP
2 NADH × 2 ATP/cytoplasmic NADH	4 ATP

Continued—

Conversion of 2 pyruvate molecules to 2 acetyl CoA molecules:

2 NADH × 3 ATP/NADH	6 ATP

Citric acid cycle (two turns):

2 GTP × 1 ATP/GTP	2 ATP
6 NADH × 3 ATP/NADH	18 ATP
2 FADH$_2$ × 2 ATP/FADH$_2$	4 ATP
	36 ATP

This represents an energy harvest of about 40% of the potential energy of glucose.

Practice Problem 22.1

Calculate the number of ATP produced by the complete oxidation of pyruvate.

For Further Practice: Questions 22.41, 22.42, and 22.44.

Aerobic metabolism is very much more efficient than anaerobic metabolism. The abundant energy harvested by aerobic metabolism has had enormous consequences for the biological world. Much of the energy released by the oxidation of fuels is not lost as heat but conserved in the form of ATP. Organisms that possess abundant energy have evolved into multicellular organisms and developed specialized functions. As a consequence of their energy requirements, all multicellular organisms are aerobic.

22.7 The Degradation of Amino Acids

LEARNING GOAL

7 Describe the conversion of amino acids to molecules that can enter the citric acid cycle.

Carbohydrates are not our only source of energy. As we saw in Chapter 21, dietary protein is digested to amino acids that can also be used as an energy source, although this is not their major metabolic function. Most of the amino acids used for energy come from the diet. In fact, it is only under starvation conditions, when stored glycogen has been depleted, that the body begins to burn its own protein, for instance from muscle, as a fuel.

The fate of the mixture of amino acids provided by digestion of protein depends upon a balance between the need for amino acids for biosynthesis and the need for cellular energy. Only those amino acids that are not needed for protein synthesis are eventually converted into citric acid cycle intermediates and used as fuel.

The degradation of amino acids occurs primarily in the liver and takes place in two stages. The first stage is the removal of the α-amino group, and the second is the degradation of the carbon skeleton. In land mammals the amino group generally ends up in urea, which is excreted in the urine. The carbon skeletons can be converted into a variety of compounds, including citric acid cycle intermediates, pyruvate, acetyl CoA, or acetoacetyl CoA. The degradation of the carbon skeletons is summarized in Figure 22.8. Deamination reactions and the fate of the carbon skeletons of amino acids are the focus of this section.

Removal of α-Amino Groups: Transamination

The first stage of amino acid degradation, the removal of the α-amino group, is usually accomplished by a **transamination** reaction. **Transaminases** catalyze the transfer of the α-amino group from an α-amino acid to an α-keto acid:

$$H-\underset{\underset{R^1}{|}}{\overset{\overset{+}{NH_3}}{\overset{|}{C}}}-COO^- + \underset{\underset{R^2}{|}}{\overset{\overset{O}{\|}}{C}}-COO^- \underset{\xleftarrow{\hspace{1cm}}}{\overset{\text{Transaminase}}{\xrightarrow{\hspace{1cm}}}} \underset{\underset{R^1}{|}}{\overset{\overset{O}{\|}}{C}}-COO^- + H-\underset{\underset{R^2}{|}}{\overset{\overset{+}{NH_3}}{\overset{|}{C}}}-COO^-$$

| Donor amino acid | Acceptor keto acid | | α-Keto acid of amino acid | New amino acid |

The α-amino group of a great many amino acids is transferred to α-ketoglutarate to produce the amino acid glutamate and a new keto acid. This glutamate family of transaminases is especially important because the α-keto acid corresponding to glutamate is α-ketoglutarate, a citric acid cycle intermediate. The glutamate transaminases thus provide a direct link between amino acid degradation and the citric acid cycle.

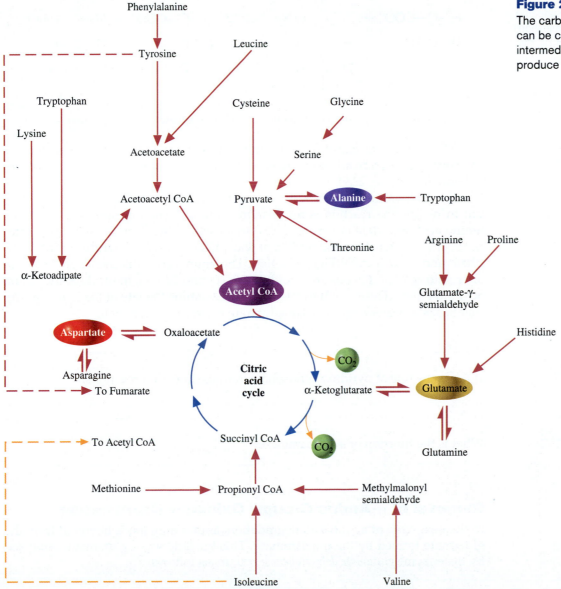

Figure 22.8

The carbon skeletons of amino acids can be converted to citric acid cycle intermediates and completely oxidized to produce ATP energy.

CH₂OH structure — Pyridoxine (vitamin B₆)

Figure 22.9
The structure of pyridoxal phosphate, the coenzyme required for all transamination reactions, and pyridoxine, vitamin B₆, the vitamin from which it is derived.

Pyridoxal phosphate

For more information on these vitamins and the coenzymes that are made from them, look online.

Water-Soluble Vitamins

Aspartate transaminase catalyzes the transfer of the α-amino group of aspartate to α-ketoglutarate, producing oxaloacetate and glutamate:

$$\text{Aspartate} + \alpha\text{-Ketoglutarate} \rightleftharpoons \text{Oxaloacetate} + \text{Glutamate}$$

Another important transaminase in mammalian tissues is *alanine transaminase*, which catalyzes the transfer of the α-amino group of alanine to α-ketoglutarate and produces pyruvate and glutamate:

$$\text{Alanine} + \alpha\text{-Ketoglutarate} \rightleftharpoons \text{Pyruvate} + \text{Glutamate}$$

All of the more than fifty transaminases that have been discovered require the coenzyme **pyridoxal phosphate.** This coenzyme is derived from vitamin B₆ (pyridoxine, Figure 22.9).

The transamination reactions shown above appear to be a simple transfer, but in reality, the reaction is much more complex. The transaminase binds the amino acid (aspartate in Figure 22.10a) in its active site. Then the α-amino group of aspartate is transferred to pyridoxal phosphate, producing pyridoxamine phosphate and oxaloacetate (Figure 22.10b). The amino group is then transferred to an α-keto acid, in this case, α-ketoglutarate (Figure 22.10c), to produce the amino acid glutamate (Figure 22.10d). Next we will examine the fate of the amino group that has been transferred to α-ketoglutarate to produce glutamate.

Question 22.11   What is the role of pyridoxal phosphate in transamination reactions?

Question 22.12   What is the function of a transaminase?

Removal of α-Amino Groups: Oxidative Deamination

In the next stage of amino acid degradation, ammonium ion is liberated from the glutamate formed by the transaminase. This breakdown of glutamate, catalyzed by the enzyme *glutamate dehydrogenase*, occurs as follows:

Glutamate + NAD$^+$ + H$_2$O ⇌ NH$_4^+$ + α-Ketoglutarate + NADH

This is an example of an **oxidative deamination,** an oxidation-reduction process in which NAD$^+$ is reduced to NADH and the amino acid is deaminated (the amino group is removed). A summary of the deamination reactions described is shown in Figure 22.11.

Figure 22.10
The mechanism of transamination.

Figure 22.11
Summary of the deamination of an
α-amino acid and the fate of the
ammonium ion (NH_4^+).

The Fate of Amino Acid Carbon Skeletons

The carbon skeletons produced by these and other deamination reactions enter glycolysis or the citric acid cycle at many steps. For instance, we have seen that transamination converts aspartate to oxaloacetate and alanine to pyruvate. The positions at which the carbon skeletons of various amino acids enter the energy-harvesting pathways are summarized in Figure 22.8.

22.8 The Urea Cycle

LEARNING GOAL

8 Explain the importance of the urea cycle and describe its essential steps.

Oxidative deamination produces large amounts of ammonium ion. Because ammonium ions are extremely toxic, they must be removed from the body, regardless of the energy expenditure required. In humans, they are detoxified in the liver by converting the ammonium ions into urea. This pathway, called the **urea cycle,** is the method by which toxic ammonium ions are kept out of the blood. The excess ammonium ions incorporated in urea are excreted in the urine (Figure 22.12).

Reactions of the Urea Cycle

The five reactions of the urea cycle are shown in Figure 22.12, and details of the reactions are summarized as follows.

Step 1. The first step of the cycle is a reaction in which CO_2 and NH_4^+ form carbamoyl phosphate. This reaction, which also requires ATP and H_2O, occurs in the mitochondria and is catalyzed by the enzyme *carbamoyl phosphate synthase.*

$$CO_2 + NH_4^+ + 2ATP + H_2O \longrightarrow H_2N-\overset{\overset{\displaystyle O}{\|}}{C}-O-\overset{\overset{\displaystyle O}{\|}}{\underset{\underset{\displaystyle O^-}{|}}{P}}-O^- + 2ADP + P_i + 3H^+$$

Carbamoyl phosphate

The urea cycle involves several unusual amino acids that are not found in polypeptides.

Step 2. The carbamoyl phosphate now condenses with the amino acid ornithine to produce the amino acid citrulline. This reaction also occurs in the mitochondria and is catalyzed by the enzyme *ornithine transcarbamoylase.*

Ornithine   Carbamoyl phosphate   Citrulline

Figure 22.12

The urea cycle converts ammonium ions into urea, which is less toxic. The intracellular locations of the reactions are indicated. Citrulline, formed in the reaction between ornithine and carbamoyl phosphate, is transported out of the mitochondrion and into the cytoplasm. Ornithine, a substrate for the formation of citrulline, is transported from the cytoplasm into the mitochondrion.

The abbreviation PP$_i$ represents the pyrophosphate group, which consists of two phosphate groups joined by a phosphoanhydride bond:

$$
\begin{array}{ccc}
 & O & & O \\
 & \parallel & & \parallel \\
^-O & -P & -O- & P-O^- \\
 & | & & | \\
 & O^- & & O^-
\end{array}
$$

Step 3. Citrulline is transported into the cytoplasm and now condenses with aspartate to produce argininosuccinate. This reaction, which requires energy released by the hydrolysis of ATP, is catalyzed by the enzyme *argininosuccinate synthase*.

Citrulline   Aspartate   →(ATP → AMP + PP$_i$)   Argininosuccinate

Step 4. Now the argininosuccinate is cleaved to produce the amino acid arginine and the citric acid cycle intermediate fumarate. This reaction is catalyzed by the enzyme *argininosuccinate lyase*.

Argininosuccinate   →   Arginine   +   Fumarate

Step 5. Finally, arginine is hydrolyzed to generate urea, the product of the reaction to be excreted, and ornithine, the original reactant in the cycle. *Arginase* is the enzyme that catalyzes this reaction.

Arginine   Water   →   Urea   +   Ornithine

(+ H$_2$O)   (H$_2$N—C(=O)—NH$_2$)

Note that one of the amino groups in urea is derived from the ammonium ion and the second is derived from the amino acid aspartate.

There are genetically transmitted diseases that result from a deficiency of one of the enzymes of the urea cycle. The importance of the urea cycle is apparent

LEARNING GOAL

9 Discuss the cause and effect of hyperammonemia.

A Medical Perspective

Pyruvate Carboxylase Deficiency

Pyruvate carboxylase is the enzyme that converts pyruvate to oxaloacetate.

$$Pyruvate + CO_2 + ATP + H_2O \longrightarrow$$
$$Oxaloacetate + ADP + 2H^+$$

This reaction is important because it provides oxaloacetate for the citric acid cycle when the supplies have run low because of the demands of biosynthesis. It is also the enzyme that catalyzes the first step in gluconeogenesis, the pathway that provides the body with needed glucose in times of starvation or periods of exercise that deplete glycogen stores. But somehow these descriptions don't fill us with a sense of the importance of this enzyme and its jobs. It is not until we investigate a case study of a child born with pyruvate carboxylase deficiency that we see the full impact of this enzyme.

Pyruvate carboxylase deficiency is found in about 1 in 250,000 births; however, there is an increased incidence in native North American Indians who speak the Algonquin dialect and in the French. There are two types of genetic disorders that have been described. In the neonatal form of the disease, there is a complete absence of the enzyme. Symptoms are apparent at birth and the child is born with brain abnormalities. In the infantile form, the patient develops symptoms early in infancy. Again, it is neurological symptoms that draw attention to the condition. The infants do not develop mental or psychomotor skills. They may develop seizures and/or respiratory depression. In both cases, it is the brain that suffers the greatest damage. In fact, this is the case in most of the disorders that reduce energy metabolism because the brain has such high energy requirements.

Biochemically, patients exhibit quite a variety of symptoms. They show acidosis (low blood pH) due to accumulations of lactate and extremely high pyruvate concentrations in the blood. Blood levels of alanine are also high and large doses of alanine do not stimulate gluconeogenesis. Furthermore, a patient's cells accumulate lipid.

We can understand each of these symptoms by considering the pathways affected by the absence of this single enzyme.

Lactic acidosis results from the fact that the body must rely on glycolysis and lactate fermentation for most of its energy needs. Alanine levels are high because it isn't being transaminated to pyruvate efficiently, because pyruvate levels are so high. In addition, alanine can't be converted to glucose by gluconeogenesis. Although the excess alanine is taken up by the liver and converted to pyruvate, the pyruvate can't be converted to glucose. Lipids accumulate because a great deal of pyruvate is converted to acetyl CoA. However, the acetyl CoA is not used to produce citrate as a result of the absence of oxaloacetate. So, the acetyl CoA is thus used to synthesize fatty acids, which are stored as triglycerides.

Dietary intervention has been tried. One such regimen is to supplement with aspartic acid and glutamic acid. The theory behind this treatment is as complex as the many symptoms of the disorder. Both amino acids can be aminated (amino groups added) in non-nervous tissue. This produces asparagine and glutamine, both of which are able to cross the blood-brain barrier. Glutamine is deaminated to glutamate, which is then transaminated to α-ketoglutarate, indirectly replenishing oxaloacetate. Asparagine can be deaminated to aspartate, which can be converted to oxaloacetate. This serves as a second supply of oxaloacetate. To date, these attempts at dietary intervention have not proved successful. Perhaps in time research will provide the tools for enzyme replacement therapy or gene therapy that could alleviate the symptoms.

For Further Understanding

Write an equation showing the reaction catalyzed by pyruvate carboxylase using structural formulas for pyruvate and oxaloacetate.

Supplementing the diet with asparagine and glutamine was tried as a treatment for pyruvate carboxylase deficiency. Write equations showing the reactions that convert these amino acids into citric acid cycle intermediates.

when we consider the terrible symptoms suffered by afflicted individuals. A deficiency of urea cycle enzymes causes an elevation of the concentration of NH_4^+, a condition known as **hyperammonemia.** If there is a complete deficiency of one of the enzymes of the urea cycle, the result is death in early infancy. If there is a partial deficiency of one of the enzymes of the urea cycle, the result may be retardation, convulsions, and vomiting. In these milder forms of hyperammonemia, a low-protein diet leads to a lower concentration of NH_4^+ in blood and less severe clinical symptoms.

What is the purpose of the urea cycle?

Where do the reactions of the urea cycle occur?

22.9 Overview of Anabolism: The Citric Acid Cycle as a Source of Biosynthetic Intermediates

LEARNING GOAL

10 ▶ Summarize the role of the citric acid cycle in catabolism and anabolism.

So far, we have talked about the citric acid cycle only as an energy-harvesting mechanism. We have seen that dietary carbohydrates and amino acids enter the pathway at various stages and are oxidized to generate NADH and $FADH_2$, which, by means of oxidative phosphorylation, are used to make ATP.

However, the role of the citric acid cycle in cellular metabolism involves more than just **catabolism.** It plays a key role in **anabolism,** or biosynthesis, as well. Figure 22.13 shows the central role of glycolysis and the citric acid cycle as energy-harvesting reactions, as well as their role as a source of biosynthetic precursors.

As you may already suspect from the fact that amino acids can be converted into citric acid cycle intermediates, these same citric acid cycle intermediates can also be used as starting materials for the synthesis of amino acids. Oxaloacetate provides the carbon skeleton for the one-step synthesis of the amino acid aspartate by the transamination reaction:

$$\text{oxaloacetate} + \text{glutamate} \rightleftharpoons \text{aspartate} + \alpha\text{-ketoglutarate}$$

Aside from providing aspartate for protein synthesis, this reaction provides aspartate for the urea cycle.

Asparagine is made from aspartate by the amination reaction

$$\text{aspartate} + NH_4^+ + \text{ATP} \longrightarrow \text{asparagine} + \text{AMP} + PP_i + H^+$$

α-Ketoglutarate serves as the starting carbon chain for the family of amino acids including glutamate, glutamine, proline, and arginine. Glutamate is especially important because it serves as the donor of the α-amino group of almost all other amino acids. It is synthesized from NH_4^+ and α-ketoglutarate in a reaction mediated by glutamate dehydrogenase. This is the reverse of the reaction shown in Figure 22.11 and previously described. In this case the coenzyme that serves as the reducing agent is NADPH.

$$NH_4^+ + \alpha\text{-ketoglutarate} + \text{NADPH} \rightleftharpoons \text{L-glutamate} + NADP^+ + H_2O$$

Glutamine, proline, and arginine are synthesized from glutamate.

Examination of Figure 22.13 reveals that serine, glycine, and cysteine are synthesized from 3-phosphoglycerate; alanine is synthesized from pyruvate; and tyrosine is produced from phosphoenolpyruvate and the four-carbon sugar erythrose-4-phosphate, which, in turn, is synthesized from glucose-6-phosphate in the pentose phosphate pathway. In addition to the amino acid precursors, glycolysis and the citric acid cycle also provide precursors for lipids and the nitrogenous bases that are required to make DNA, the molecule that carries the genetic information. They also generate precursors for heme, the prosthetic group that is required for hemoglobin, myoglobin, and the cytochromes.

Clearly, the reactions of glycolysis and the citric acid cycle are central to both anabolic and catabolic cellular activities. Metabolic pathways that function in both anabolism and catabolism are called **amphibolic pathways.** Consider for a moment the difficulties that the dual nature of these pathways could present to

The nine amino acids not shown in Figure 22.13 (histidine, isoleucine, leucine, lysine, methionine, phenylalanine, threonine, tryptophan, and valine) are called the essential amino acids (Section 18.11) because they cannot be synthesized by humans. Arginine is an essential amino acid for infants and adults under physical stress.

Actually, in humans tyrosine is made from the essential amino acid phenylalanine.

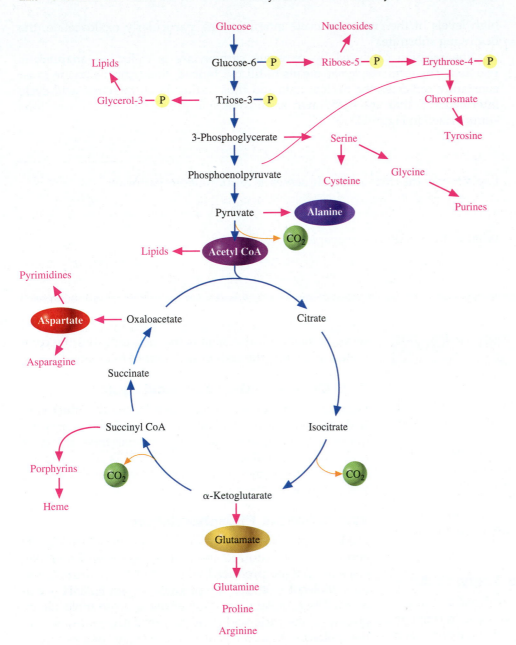

Figure 22.13
Glycolysis, the pentose phosphate pathway, and the citric acid cycle also provide a variety of precursors for the biosynthesis of amino acids, nitrogenous bases, and porphyrins.

the cell. When the cell is actively growing, there is a great demand for biosynthetic precursors to build new cell structures. A close look at Figure 22.13 shows us that periods of active cell growth and biosynthesis may deplete the supply of citric acid cycle intermediates. The problem is, the processes of growth and biosynthesis also require a great deal of ATP!

The solution to this problem is to have an alternative pathway for oxaloacetate synthesis that can produce enough oxaloacetate to supply the anabolic and catabolic requirements of the cell. Although bacteria and plants have several mechanisms, the only way that mammalian cells can produce more oxaloacetate is by the carboxylation of pyruvate, a reaction that is also important in gluconeogenesis. This reaction is

$$\text{pyruvate} + CO_2 + ATP \longrightarrow \text{oxaloacetate} + ADP + P_i$$

The enzyme that catalyzes this reaction is *pyruvate carboxylase.* It is a conjugated protein having as its covalently linked prosthetic group the vitamin *biotin.* This enzyme is "turned on" by high levels of acetyl CoA, a signal that the cell requires

Carboxylation of pyruvate during gluconeogenesis is discussed in Section 21.6.

See also A Medical Perspective: Pyruvate Carboxylase Deficiency earlier in this chapter.

high levels of the citric acid cycle intermediates, particularly oxaloacetate, the beginning substrate.

The reaction catalyzed by pyruvate carboxylase is called an **anaplerotic reaction.** The term *anaplerotic* means to fill up. Indeed, this critical enzyme must constantly replenish the oxaloacetate and thus indirectly all the citric acid cycle intermediates that are withdrawn as biosynthetic precursors for the reactions summarized in Figure 22.13.

Question 22.15   Explain how the citric acid cycle serves as an amphibolic pathway.

Question 22.16   What is the function of an anaplerotic reaction?

SUMMARY

22.1 The Mitochondria

The *mitochondria* are aerobic cell organelles that are responsible for most of the ATP production in eukaryotic cells. They are enclosed by a double membrane. The outer membrane permits low-molecular-weight molecules to pass through. The *inner mitochondrial membrane,* by contrast, is almost completely impermeable to most molecules. The inner mitochondrial membrane is the site where *oxidative phosphorylation* occurs. The enzymes of the *citric acid cycle,* of amino acid catabolism, and of fatty acid oxidation are located in the *matrix space* of the mitochondrion.

22.2 Conversion of Pyruvate to Acetyl CoA

Under aerobic conditions, pyruvate is oxidized by the pyruvate dehydrogenase complex. Acetyl CoA, formed in this reaction, is a central molecule in both catabolism and anabolism.

22.3 An Overview of Aerobic Respiration

Aerobic respiration is the oxygen-requiring degradation of food molecules and production of ATP. *Oxidative phosphorylation* is the process that uses the high-energy electrons harvested by oxidation of substrates of the citric acid cycle to produce ATP.

22.4 The Citric Acid Cycle (The Krebs Cycle)

The *citric acid cycle* is the final pathway for the degradation of carbohydrates, amino acids, and fatty acids. The citric acid cycle occurs in the matrix of the mitochondria. It is a cyclic

series of biochemical reactions that accomplishes the complete oxidation of the carbon skeletons of food molecules.

22.5 Control of the Citric Acid Cycle

Because the rate of ATP production by the cell must vary with the amount of available oxygen and the energy requirements of the body at any particular time, the citric acid cycle is regulated at several steps. This allows the cell to generate more energy when needed, as for exercise, and less energy when the body is at rest.

22.6 Oxidative Phosphorylation

Oxidative phosphorylation is the process by which NADH and $FADH_2$ are oxidized and ATP is produced. Two molecules of ATP are produced when $FADH_2$ is oxidized, and three molecules of ATP are produced when NADH is oxidized. The complete oxidation of one glucose molecule by glycolysis, the citric acid cycle, and oxidative phosphorylation yields thirty-six molecules of ATP versus two molecules of ATP for anaerobic degradation of glucose by glycolysis and fermentation.

22.7 The Degradation of Amino Acids

Amino acids are oxidized in the mitochondria. The first step of amino acid catabolism is deamination, the removal of the amino group. The carbon skeletons of amino acids are converted into molecules that can enter the citric acid cycle.

22.8 The Urea Cycle

In the *urea cycle* the toxic ammonium ions released by deamination of amino acids are incorporated in urea, which is excreted in the urine.

22.9 Overview of Anabolism: The Citric Acid Cycle as a Source of Biosynthetic Intermediates

In addition to its role in *catabolism,* the citric acid cycle also plays an important role in cellular *anabolism,* or biosynthetic reactions. Many of the citric acid cycle intermediates are precursors for the synthesis of amino acids and macromolecules required by the cell. A pathway that functions in both catabolic and anabolic reactions is called an *amphibolic pathway.*

KEY TERMS

acetyl CoA (22.2)
aerobic respiration (22.3)
amphibolic pathway (22.9)
anabolism (22.9)
anaplerotic reaction (22.9)
ATP synthase (22.6)
catabolism (22.9)
citric acid cycle (22.4)
coenzyme A (22.2)
cristae (22.1)
electron transport
   system (22.6)
F_0F_1 complex (22.6)
hyperammonemia (22.8)
inner mitochondrial
   membrane (22.1)

intermembrane space (22.1)
matrix space (22.1)
mitochondria (22.1)
outer mitochondrial
   membrane (22.1)
oxidative deamination (22.7)
oxidative phosphorylation
   (22.6)
pyridoxal phosphate (22.7)
pyruvate dehydrogenase
   complex (22.2)
terminal electron
   acceptor (22.6)
transaminase (22.7)
transamination (22.7)
urea cycle (22.8)

QUESTIONS AND PROBLEMS

The Mitochondria

Foundations

22.17 Define the term *mitochondrion.*
22.18 Define the term *cristae.*

Applications

22.19 What is the function of the intermembrane compartment of the mitochondria?
22.20 What biochemical processes occur in the matrix space of the mitochondria?
22.21 In what important way do the inner and outer mitochondrial membranes differ?
22.22 What kinds of proteins are found in the inner mitochondrial membrane?

Conversion of Pyruvate to Acetyl CoA

Foundations

22.23 What is coenzyme A?
22.24 What is the role of coenzyme A in the reaction catalyzed by pyruvate dehydrogenase?
22.25 In the reaction catalyzed by pyruvate dehydrogenase, pyruvate is decarboxylated. What is meant by the term *decarboxylation?*
22.26 In the reaction catalyzed by pyruvate dehydrogenase, pyruvate is also oxidized. What substance is reduced when pyruvate is oxidized? What is the product of that reduction reaction?

Applications

22.27 Under what metabolic conditions is pyruvate converted to acetyl CoA?
22.28 Write a chemical equation for the production of acetyl CoA from pyruvate. Under what conditions does this reaction occur?
22.29 How could a deficiency of riboflavin, thiamine, niacin, or pantothenic acid reduce the amount of ATP the body can produce?
22.30 In what form are the vitamins riboflavin, thiamine, niacin, and pantothenic acid needed by the pyruvate dehydrogenase complex?

The Citric Acid Cycle

Foundations

22.31 The reaction catalyzed by citrate synthase is an aldol condensation. Define the term *aldol condensation.*
22.32 The pair of reactions catalyzed by aconitase results in the conversion of isocitrate to its isomer citrate. What are isomers?
22.33 How is oxidation of an organic molecule often recognized?
22.34 The hydroxyl group of isocitrate is oxidized to a ketone, α-ketoglutarate, in the third reaction of the citric acid cycle. Write this reaction and circle the chemical change that reveals that this is an oxidation reaction.
22.35 The reaction catalyzed by succinate dehydrogenase is a dehydrogenation reaction. What is meant by the term *dehydrogenation reaction?*
22.36 The reaction catalyzed by fumarase is an example of the hydration of an alkene to produce an alcohol. Write the equation for this reaction. What is meant by the term *hydration reaction?*
22.37 Label each of the following statements as true or false:
   a. Both glycolysis and the citric acid cycle are aerobic processes.
   b. Both glycolysis and the citric acid cycle are anaerobic processes.
   c. Glycolysis occurs in the cytoplasm, and the citric acid cycle occurs in the mitochondria.
   d. The inner membrane of the mitochondrion is virtually impermeable to most substances.
22.38 Fill in the blanks:
   a. The proteins of the electron transport system are found in the _____, the enzymes of the citric acid cycle are found in the _____, and the hydrogen ion reservoir is found in the _____ of the mitochondria.
   b. The infoldings of the inner mitochondrial membrane are called _____.
   c. Energy released by oxidation in the citric acid cycle is conserved in the form of phosphoanhydride bonds in _____.
   d. The purpose of the citric acid cycle is the _____ of the acetyl group.
22.39 a. To what metabolic intermediate is the acetyl group of acetyl CoA transferred in the citric acid cycle?
   b. What is the product of this reaction?
22.40 To what final products is the acetyl group of acetyl CoA converted during oxidation in the citric acid cycle?
22.41 How many ions of NAD^+ are reduced to molecules of NADH during one turn of the citric acid cycle?
22.42 How many molecules of FAD are converted to $FADH_2$ during one turn of the citric acid cycle?
22.43 What is the net yield of ATP for anaerobic glycolysis?
22.44 How many molecules of ATP are produced by the complete degradation of glucose via glycolysis, the citric acid cycle, and oxidative phosphorylation?

22.45 What is the function of acetyl CoA in the citric acid cycle?

22.46 What is the function of oxaloacetate in the citric acid cycle?

22.47 GTP is formed in one step of the citric acid cycle. How is this GTP converted into ATP?

22.48 What is the chemical meaning of the term *decarboxylation?*

Applications

22.49 The first reaction in the citric acid cycle is an aldol condensation. Write the equation for this reaction and explain its significance.

22.50 A bacterial culture is given ^{14}C-labeled pyruvate as its sole source of carbon and energy. The following is the structure of the radiolabeled pyruvate.

$$\overset{\displaystyle O \quad\ O}{\underset{\displaystyle}{*CH_3-\overset{\displaystyle \|}{C}-\overset{\displaystyle \|}{C}-O^-}}$$

Follow the fate of the radioactive carbon through the reactions of the citric acid cycle.

22.51 The enzyme aconitase catalyzes the isomerization of citrate into isocitrate. Discuss the two reactions catalyzed by aconitase in terms of the chemistry of alcohols and alkenes.

22.52 A bacterial culture is given ^{14}C-labeled pyruvate as its sole source of carbon and energy. The following is the structure of the radiolabeled pyruvate.

$$\overset{\displaystyle O \quad\ O}{\underset{\displaystyle}{CH_3-\overset{\displaystyle \|}{C}*-\overset{\displaystyle \|}{C}-O^-}}$$

Follow the fate of the radioactive carbon through the reactions of the citric acid cycle.

22.53 In the oxidation of malate to oxaloacetate, what is the structural evidence that an oxidation reaction has occurred? What functional groups are involved?

22.54 In the oxidation of succinate to fumarate, what is the structural evidence that an oxidation reaction has occurred? What functional groups are involved?

22.55 To what class of enzymes does dinucleotide diphosphokinase belong? Explain your answer.

22.56 To what class of enzymes does succinate dehydrogenase belong? Explain your answer.

Control of the Citric Acid Cycle

Foundations

22.57 Define the term *allosteric enzyme.*

22.58 Define the term *effector.*

22.59 Define *negative allosterism.*

22.60 Define *positive allosterism.*

Applications

22.61 What four allosteric enzymes or enzyme complexes are responsible for the regulation of the citric acid cycle?

22.62 Which of the four allosteric enzymes or enzyme complexes in the citric acid cycle are under negative allosteric control? Which are under positive allosteric control?

22.63 What is the importance of the regulation of the citric acid cycle?

22.64 Explain the role of allosteric enzymes in control of the citric acid cycle.

22.65 What molecule serves as a signal to increase the rate of the reactions of the citric acid cycle?

22.66 What molecules serve as signals to decrease the rate of the reactions of the citric acid cycle?

Oxidative Phosphorylation

Foundations

22.67 Define the term *electron transport system.*

22.68 What is the terminal electron acceptor in aerobic respiration?

Applications

22.69 How many molecules of ATP are produced when one molecule of NADH is oxidized by oxidative phosphorylation?

22.70 How many molecules of ATP are produced when one molecule of $FADH_2$ is oxidized by oxidative phosphorylation?

22.71 What is the source of energy for the synthesis of ATP in mitochondria?

22.72 What is the name of the enzyme that catalyzes ATP synthesis in mitochondria?

22.73 What is the function of the electron transport systems of the mitochondria?

22.74 What is the cellular location of the electron transport systems?

22.75 **a.** Compare the number of molecules of ATP produced by glycolysis to the number of ATP molecules produced by oxidation of glucose by aerobic respiration.
b. Which pathway produces more ATP? Explain.

22.76 At which steps in the citric acid cycle do oxidation–reduction reactions occur?

The Degradation of Amino Acids

Foundations

22.77 What chemical transformation is carried out by transaminases?

22.78 Write a chemical equation for the transfer of an amino group from alanine to α-ketoglutarate, catalyzed by a transaminase.

22.79 Why is the glutamate family of transaminases so important?

22.80 What biochemical reaction is catalyzed by glutamate dehydrogenase?

Applications

22.81 Into which citric acid cycle intermediate is each of the following amino acids converted?
a. Alanine
b. Glutamate
c. Aspartate
d. Phenylalanine
e. Threonine
f. Arginine

22.82 What is the net ATP yield for degradation of each of the amino acids listed in Problem 22.81?

22.83 Write a balanced equation for the synthesis of glutamate that is mediated by the enzyme glutamate dehydrogenase.

22.84 Write a balanced equation for the transamination of aspartate.

The Urea Cycle

22.85 What metabolic condition is produced if the urea cycle does not function properly?

22.86 What is hyperammonemia? How are mild forms of this disease treated?

22.87 The structure of urea is

$$\overset{\displaystyle O}{\underset{\displaystyle}{NH_2-\overset{\displaystyle \|}{C}-NH_2}}$$

a. What substances are the sources of each of the amino groups in the urea molecule?
b. What substance is the source of the carbonyl group?

22.88 What is the energy source used for the urea cycle?

**Overview of Anabolism: The Citric Acid Cycle
as a Source of Biosynthetic Intermediates**

Foundations
22.89 Define the term *anabolism.*
22.90 Define the term *catabolism.*

Applications
22.91 From which citric acid cycle intermediate is the amino acid glutamate synthesized?
22.92 What amino acids are synthesized from α-ketoglutarate?
22.93 What is the role of the citric acid cycle in biosynthesis?
22.94 How are citric acid cycle intermediates replenished when they are in demand for biosynthesis?
22.95 What is meant by the term *essential amino acid?*
22.96 What are the nine essential amino acids?
22.97 Write a balanced equation for the reaction catalyzed by pyruvate carboxylase.
22.98 How does the reaction described in Problem 22.97 allow the citric acid cycle to fulfill its roles in both catabolism and anabolism?

CRITICAL THINKING PROBLEMS

1. A one-month-old baby boy was brought to the hospital showing severely delayed development and cerebral atrophy. Blood tests showed high levels of lactate and pyruvate. By three months of age, very high levels of succinate and fumarate were found in the urine. Fumarase activity was absent in the liver and muscle tissue. The baby died at five months of age. This was the first reported case of fumarase deficiency and the defect was recognized too late for effective therapy to be administered. What reaction is catalyzed by fumarase? How would a deficiency of this mitochondrial enzyme account for the baby's symptoms and test results?

2. A certain bacterium can grow with ethanol as its only source of energy and carbon. Propose a pathway to describe how ethanol can enter a pathway that would allow ATP production and synthesis of precursors for biosynthesis.

3. Fluoroacetate has been used as a rat poison and can be fatal when eaten by humans. Patients with fluoroacetate poisoning accumulate citrate and fluorocitrate within the cells. What enzyme is inhibited by fluoroacetate? Explain your reasoning.

4. The pyruvate dehydrogenase complex is activated by removal of a phosphoryl group from pyruvate dehydrogenase. This reaction is catalyzed by the enzyme pyruvate dehydrogenase phosphate phosphatase. A baby is born with a defect in this enzyme. What effects would this defect have on the rate of each of the following pathways: aerobic respiration, glycolysis, lactate fermentation? Explain your reasoning.

5. Pyruvate dehydrogenase phosphate phosphatase is stimulated by Ca^{2+}. In muscles, the Ca^{2+} concentration increases dramatically during muscle contraction. How would the elevated Ca^{2+} concentration affect the rate of glycolysis and the citric acid cycle?

6. Liver contains high levels of nucleic acids. When excess nucleic acids are degraded, ribose-5-phosphate is one of the degradation products that accumulate in the cell. Can this substance be used as a source of energy? What pathway would be used?

7. In birds, arginine is an essential amino acid. Can birds produce urea as a means of removing ammonium ions from the blood? Explain your reasoning.

23 Fatty Acid Metabolism

Learning Goals

1. Summarize the digestion and storage of lipids.

2. Describe the degradation of fatty acids by β-oxidation.

3. Explain the role of acetyl CoA in fatty acid metabolism.

4. Understand the role of ketone body production in β-oxidation.

5. Compare β-oxidation of fatty acids and fatty acid biosynthesis.

6. Describe the regulation of lipid and carbohydrate metabolism in relation to the liver, adipose tissue, muscle tissue, and the brain.

7. Summarize the antagonistic effects of glucagon and insulin.

A tasty source of energy.

Introduction

The metabolism of fatty acids and lipids revolves around the fate of acetyl CoA. We saw in Chapter 22 that, under aerobic conditions, pyruvate is converted to acetyl CoA, which feeds into the citric acid cycle. Fatty acids are also degraded to acetyl CoA and oxidized by the citric acid cycle, as are certain amino acids. Moreover, acetyl CoA is itself the starting material for the biosynthesis of fatty acids, cholesterol, and steroid hormones. Acetyl CoA is thus a key intermediary in lipid metabolism.

Chemistry Connection

Obesity: A Genetic Disorder?

Approximately a third of all Americans are obese; that is, they are more than 20% overweight. One million are morbidly obese; they carry so much extra weight that it threatens their health. Many obese people simply eat too much and exercise too little, but others actually gain weight even though they eat fewer calories than people of normal weight. This observation led many researchers to the hypothesis that obesity in some people is a genetic disorder.

This hypothesis was supported by the 1950 discovery of an obesity mutation in mice. Selective breeding produced a strain of genetically obese mice from the original mutant mouse. The hypothesis was further strengthened by the results of experiments performed in the 1970s by Douglas Coleman. Coleman connected the circulatory systems of a genetically obese mouse and a normal mouse. The obese mouse started eating less and lost weight. Coleman concluded that there was a substance in the blood of normal mice that signals the brain to decrease the appetite. Obese mice, he hypothesized, can't produce this "satiety factor," and thus they continue to eat and gain weight.

In 1987, Jeffrey Friedman assembled a team of researchers to map and then clone the obesity gene that was responsible for appetite control. In 1994, after seven years of intense effort, the scientists achieved their goal, but they still had to demonstrate that the protein encoded by the cloned obesity gene did, indeed, have a metabolic effect. The gene was modified to be compatible with the genetic system of bacteria so that they could be used to manufacture the protein. When the engineered gene was then introduced into bacteria, they produced an abundance of the protein product. The protein was then purified in preparation for animal testing.

The researchers calculated that a normal mouse has about 12.5 mg of the protein in its blood. They injected that amount into each of ten mice that were so fat they couldn't squeeze into the feeding tunnels used for normal mice. The day after the first injection, graduate student Jeff Halaas observed that the mice had eaten less food. Injections were given daily, and each day the obese mice ate less. After two weeks of treatment, each of the ten mice had lost about 30% of its weight. In addition, the mice had become more active and their metabolisms had speeded up.

When normal mice underwent similar treatment, their body fat fell from 12.2% to 0.67%, which meant that these mice had no extra fat tissue. The 0.67% of their body weight represented by fat was accounted for by the membranes that surround each of the cells of their bodies! Because of the dramatic results, Friedman and his colleagues called the protein leptin, from the Greek word *leptos,* meaning slender.

The leptin protein is a hormone that functions as a signal in a metabolic thermostat. Fat cells produce leptin and secrete it into the bloodstream. As a result, the leptin concentration in a normal person is proportional to the amount of body fat. The blood concentration of the hormone is monitored by the hypothalamus, a region known to control appetite and set metabolic rates. When the concentration reaches a certain level, it triggers the hypothalamus to suppress the appetite. If no leptin or only small amounts of are produced, the hypothalamus "thinks" that the individual has too little body fat or is starving. Under these circumstances it does not send a signal to suppress hunger and the individual continues to eat.

The human leptin gene also has been cloned and shown to correct genetic obesity in mice. Unfortunately, the dramatic results achieved with mice were *not* observed with humans. Why? It seems that nearly all of the obese volunteers already produced an abundance of leptin. In fact, fewer than ten people have been found, to date, who do not produce leptin. In fact, many obese people have very high levels of leptin in the blood. It appears that, as with type 2 diabetes and insulin, these people are no longer sensitive to the leptin produced by their fat cells.

Ghrelin is another hormone that influences appetite. Produced in the stomach, ghrelin stimulates appetite. As you would predict, the level of ghrelin is high before a meal and decreases following a meal. Many obese people have high levels of ghrelin and therefore experience constant hunger.

Obestatin is another hormone that has been discovered to influence body weight. This hormone decreases appetite. Interestingly, both ghrelin and obestatin are encoded in a single gene. When the protein is produced, it is cleaved into the two peptide hormones.

Clearly lipid metabolism in animals is a complex process and is not yet fully understood. In this chapter we will study other aspects of lipid metabolism: the pathways for fatty acid degradation and biosynthesis and the processes by which dietary lipids are digested and excess lipids are stored.

23.1 Lipid Metabolism in Animals

Digestion and Absorption of Dietary Triglycerides

Triglycerides are highly hydrophobic ("water fearing"). Because of this they must be processed before they can be digested, absorbed, and metabolized. Because processing of dietary lipids occurs in the small intestine, the water soluble **lipases,** enzymes that hydrolyze triglycerides, that are found in the stomach and in the saliva are not very effective. In fact, most dietary fat arrives in the duodenum, the first part of the small intestine, in the form of fat globules. These fat globules stimulate the secretion of bile from the gallbladder. **Bile** is composed of micelles of lecithin, cholesterol, protein, bile salts, inorganic ions, and bile pigments. **Micelles** (Figure 23.1) are aggregations of molecules having a polar region and a nonpolar region. The nonpolar ends of bile salts tend to bunch together when placed in water. The hydrophilic ("water loving") regions of these molecules interact with water. Bile salts are made in the liver and stored in the gallbladder, awaiting the stimulus to be secreted into the duodenum. The major bile salts in humans are cholate and chenodeoxycholate (Figure 23.2).

LEARNING GOAL

1 Summarize the digestion and storage of lipids.

See Sections 14.1 and 17.2 for a discussion of micelles.

Figure 23.1

The structure of a micelle formed from the phospholipid lecithin. The straight lines represent the long hydrophobic fatty acid tails, and the spheres represent the hydrophilic heads of the phospholipid.

Figure 23.2

Structures of the most common bile acids in human bile: cholate and chenodeoxycholate.

Cholate

Chenodeoxycholate

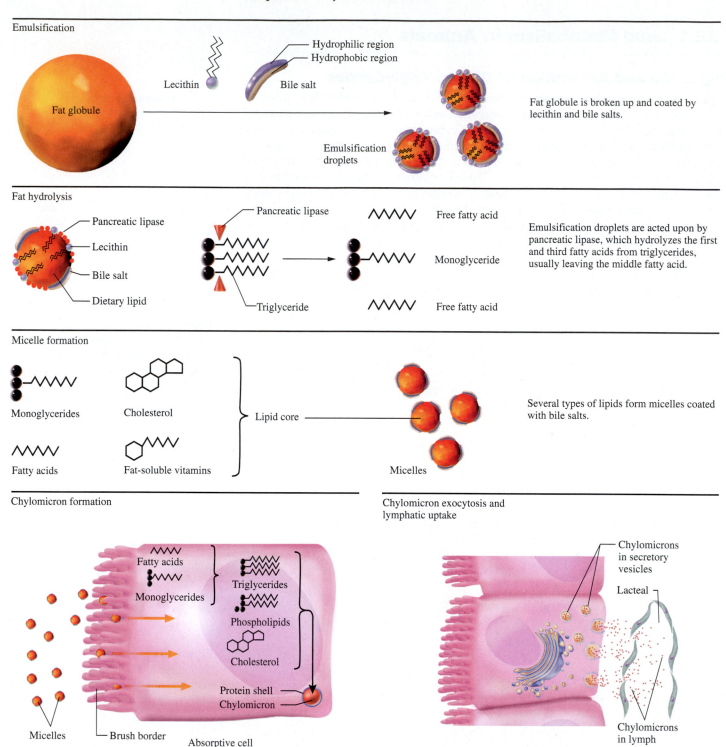

Emulsification

Hydrophilic region
Hydrophobic region

Lecithin   Bile salt

Fat globule

Emulsification droplets

Fat globule is broken up and coated by lecithin and bile salts.

Fat hydrolysis

Pancreatic lipase

Pancreatic lipase   Free fatty acid

Lecithin

Bile salt   Monoglyceride

Dietary lipid   Triglyceride   Free fatty acid

Emulsification droplets are acted upon by pancreatic lipase, which hydrolyzes the first and third fatty acids from triglycerides, usually leaving the middle fatty acid.

Micelle formation

Monoglycerides   Cholesterol

Lipid core

Fatty acids   Fat-soluble vitamins

Micelles

Several types of lipids form micelles coated with bile salts.

Chylomicron formation   **Chylomicron exocytosis and lymphatic uptake**

Fatty acids

Triglycerides

Monoglycerides

Phospholipids

Cholesterol

Protein shell
Chylomicron

Micelles   Brush border
Absorptive cell

Chylomicrons in secretory vesicles

Lacteal

Chylomicrons in lymph

Intestinal cells absorb lipids from micelles, resynthesize triglycerides, and package triglycerides, cholesterol, and phospholipids into protein-coated chylomicrons.

Golgi complex packages chylomicrons into secretory vesicles; chylomicrons are released from basal cell membrane by exocytosis and enter the lacteal (lymphatic capillary).

Figure 23.3

Stages of lipid digestion in the intestinal tract.

Cholesterol is almost completely insoluble in water, but the conversion of cholesterol to bile salts creates *detergents* whose polar heads make them soluble in the aqueous phase and whose hydrophobic tails bind triglycerides. After a meal is eaten, bile flows through the common bile duct into the duodenum, where bile

Figure 23.4
The action of pancreatic lipase in the hydrolysis of dietary lipids.

salts emulsify the fat globules into tiny droplets. This increases the surface area of the lipid molecules, allowing them to be more easily hydrolyzed by lipases (Figure 23.3).

Much of the lipid in these droplets is in the form of **triglycerides,** or triacylglycerols, which are fatty acid esters of glycerol. A protein called **colipase** binds to the surface of the lipid droplets and helps pancreatic lipases to stick to the surface and hydrolyze the ester bonds between the glycerol and fatty acids of the triglycerides (Figure 23.4). In this process, two of the three fatty acids are liberated, and the monoglycerides and free fatty acids produced mix freely with the micelles of bile. These micelles are readily absorbed through the membranes of the intestinal epithelial cells (Figure 23.3).

Triglycerides are described in Section 17.3.

Surprisingly, the monoglycerides and fatty acids are then reassembled into triglycerides that are combined with protein to produce the class of plasma lipoproteins called **chylomicrons** (Figure 23.3). These collections of lipid and protein are secreted into small lymphatic vessels and eventually arrive in the bloodstream. In the bloodstream the triglycerides are once again hydrolyzed to produce glycerol and free fatty acids that are then absorbed by the cells. If the body needs energy, these molecules are degraded to produce ATP. If the body does not need energy, these energy-rich molecules are stored.

Plasma lipoproteins are described in Section 17.5.

Lipid Storage

Fatty acids are stored in the form of triglycerides. Most of the body's triglyceride molecules are stored as fat droplets in the cytoplasm of **adipocytes** (fat cells) that make up **adipose tissue.** Each adipocyte contains a large fat droplet that accounts for nearly the entire volume of the cell. Other cells, such as those of cardiac muscle, contain a few small fat droplets. In these cells the fat droplets are surrounded by mitochondria. When the cells need energy, triglycerides are hydrolyzed to release fatty acids that are transported into the matrix space of the mitochondria. There the fatty acids are completely oxidized, and ATP is produced.

The fatty acids provided by the hydrolysis of triglycerides are a very rich energy source for the body. The complete oxidation of fatty acids releases much more energy than the oxidation of a comparable amount of glycogen.

How do bile salts aid in the digestion of dietary lipids?

Question *23.1*

Why must dietary lipids be processed before enzymatic digestion can be effective?

Question *23.2*

Figure 23.5

The last carbon of the chain is called the ω-carbon (omega-carbon), so the attached phenyl group is an ω-phenyl group. (a) Oxidation of ω-phenyl-labeled fatty acids occurs two carbons at a time. Fatty acids having an even number of carbon atoms are degraded to phenyl acetate and "acetate." (b) Oxidation of ω-phenyl-labeled fatty acids that contain an odd number of carbon atoms yields benzoate and "acetate."

ω-Phenyl-labeled fatty acid with an even number of carbon atoms

Phenyl acetate   Acetate

(a)

ω-Phenyl-labeled fatty acid having an odd number of carbon atoms

Benzoate   Acetate

(b)

LEARNING GOAL

2 Describe the degradation of fatty acids by β-oxidation.

LEARNING GOAL

3 Explain the role of acetyl CoA in fatty acid metabolism.

This pathway is called β-oxidation because it involves the stepwise oxidation of the β-carbon of the fatty acid.

Review Section 22.6 for the ATP yields that result from oxidation of FADH$_2$ and NADH.

23.2 Fatty Acid Degradation

An Overview of Fatty Acid Degradation

Early in the twentieth century, a very clever experiment was done to determine how fatty acids are degraded. Recall from Chapter 9 that radioactive elements can be attached to biological molecules and followed through the body. A German biochemist, Franz Knoop, devised a similar kind of labeling experiment long before radioactive tracers were available. Knoop fed dogs fatty acids in which the usual terminal methyl group had a phenyl group attached to it. Such molecules are called ω-labeled (omega-labeled) fatty acids (Figure 23.5). When he isolated the metabolized fatty acids from the urine of the dogs, he found that phenyl acetate was formed when the fatty acid had an even number of carbon atoms in the chain. But benzoate was formed when the fatty acid had an odd number of carbon atoms. Knoop interpreted these data to mean that the degradation of fatty acids occurs by the removal of two-carbon acetate groups from the carboxyl end of the fatty acid. We now know that the two-carbon fragments produced by the degradation of fatty acids are not acetate, but acetyl CoA. The pathway for the breakdown of fatty acids into acetyl CoA is called **β-oxidation.**

The β-oxidation cycle (steps 2–5, Figure 23.6) consists of a set of four reactions whose overall form is similar to the last four reactions of the citric acid cycle. Each trip through the sequence of reactions releases acetyl CoA and returns a fatty acyl CoA molecule that has two fewer carbons. One molecule of FADH$_2$, equivalent to two ATP molecules, and one molecule of NADH, equivalent to three ATP molecules, are produced for each cycle of β-oxidation.

EXAMPLE 23.1   **Predicting the Products of β-Oxidation of a Fatty Acid**

LEARNING GOAL

2 Describe the degradation of fatty acids by β-oxidation.

What products would be produced by the β-oxidation of 10-phenyldecanoic acid?

Solution

This ten-carbon fatty acid would be broken down into four acetyl CoA molecules and one phenyl acetate molecule. Because four cycles through β-oxidation are required to break down a ten-carbon fatty acid, four NADH molecules and four FADH$_2$ molecules would also be produced.

Continued—

E X A M P L E 23.1 —*Continued*

Practice Problem 23.1

What products would be formed by β-oxidation of each of the following fatty acids?
a. 9-Phenylnonanoic acid
b. 8-Phenyloctanoic acid
c. 7-Phenylheptanoic acid
d. 12-Phenyldodecanoic acid

Further Practice: Questions 23.41 and 23.42.

Figure 23.6
The reactions in β-oxidation of fatty acids.

A Human *Perspective*

Losing Those Unwanted Pounds of Adipose Tissue

Weight, or overweight, is a topic of great concern to the American populace. A glance through almost any popular magazine quickly informs us that by today's standards, "beautiful" is synonymous with "thin." The models in all these magazines are extremely thin, and there are literally dozens of ads for weight-loss programs. Americans spend millions of dollars each year trying to attain this slim ideal of the fashion models.

Studies have revealed that this slim ideal is often below a desirable, healthy body weight. In fact, the suggested weight for a 6-foot tall male between 18 and 39 years of age is 179 pounds. For a 5'6'' female in the same age range, the desired weight is 142 pounds. For a 5'1'' female, 126 pounds is recommended. Just as being too thin can cause health problems, so too can obesity.

What is obesity, and does it have disadvantages beyond aesthetics? An individual is considered to be obese if his or her body weight is more than 20% above the ideal weight for his or her height. The accompanying table lists desirable body weights, according to sex, age, height, and body frame.

Overweight carries with it a wide range of physical problems, including elevated blood cholesterol levels; high blood pressure; increased incidence of diabetes, cancer, and heart disease; and increased probability of early death. It often causes psychological problems as well, such as guilt and low self-esteem.

Many factors may contribute to obesity. These include genetic factors, a sedentary lifestyle, and a preference for high-calorie, high-fat foods. However, the real concern is how to lose weight. How can we lose weight wisely and safely and keep the weight off for the rest of our lives? Unfortunately, the answer is *not* the answer that most people want to hear. The prevalence and financial success of the quick-weight-loss programs suggest that the majority of people want a program that is rapid and effortless. Unfortunately, most programs that promise dramatic weight reduction with little effort are usually ineffective or, worse, unsafe. The truth is that weight loss and management are best obtained by a program involving three elements.

	Men*					Women**		
Height					**Height**			
Feet	Inches	Small Frame	Medium Frame	Large Frame	Feet	Inches	Small Frame	Medium Frame	Large Frame
5	2	128–134	131–141	138–150	4	10	102–111	109–121	118–131
5	3	130–136	133–143	140–153	4	11	103–113	111–123	120–134
5	4	132–138	135–145	142–156	5	0	104–115	113–126	122–137
5	5	134–140	137–148	144–160	5	1	106–118	115–129	125–140
5	6	136–142	139–151	146–164	5	2	108–121	118–132	128–143
5	7	138–145	142–154	149–168	5	3	111–124	121–135	131–147
5	8	140–148	145–157	152–172	5	4	114–127	124–138	134–151
5	9	142–151	148–160	155–176	5	5	117–130	127–141	137–155
5	10	144–154	151–163	158–180	5	6	120–133	130–144	140–159
5	11	146–157	154–166	161–184	5	7	123–136	133–147	143–163
6	0	149–160	157–170	164–188	5	8	126–139	136–150	146–167
6	1	152–164	160–174	168–192	5	9	129–142	139–153	149–170
6	2	155–168	164–178	172–197	5	10	132–145	142–156	152–173
6	3	158–172	167–182	176–202	5	11	135–148	145–159	155–176
6	4	162–176	171–187	181–207	6	0	138–151	148–162	158–179

*Weights at ages 25–59 based on lowest mortality. Weight in pounds according to frame (in indoor clothing weighing 5 lb, shoes with 1'' heels).
**Weights at ages 25–59 based on lowest mortality. Weight in pounds according to frame (in indoor clothing weighing 3 lb, shoes with 1'' heels).
Reprinted with permission of the Metropolitan Life Insurance Companies *Statistical Bulletin.*

Reduced caloric intake and exercise are the keys to permanent weight loss.

1. *Reduced caloric intake.* A pound of body fat is equivalent to 3500 Calories (kilocalories). So if you want to lose 2 pounds each week, a reasonable goal, you must reduce your caloric intake by 1000 Calories per day. Remember that diets recommending fewer than 1200 Calories per day are difficult to maintain because they are not very satisfying and may be unsafe because they don't provide all the required vitamins and minerals. The best way to decrease Calories is to reduce fat and increase complex carbohydrates in the diet.

2. *Exercise.* Increase energy expenditures by 200–400 Calories each day. You may choose walking, running, or mowing the lawn; the type of activity doesn't matter, as long as you get moving. Exercise has additional benefits. It increases cardiovascular fitness, provides a psychological lift, and may increase the base rate at which you burn calories after exercise is finished.

3. *Behavior modification.* For some people, overweight is as much a psychological problem as it is a physical problem, and half the battle is learning to recognize the triggers that cause overeating. Several principles of behavior modification have been found to be very helpful.

  a. Keep a diary. Record the amount of foods eaten and the circumstances—for instance, a meal at the kitchen table or a bag of chips in the car on the way home.

  b. Identify your eating triggers. Do you eat when you feel stress, boredom, fatigue, joy?

  c. Develop a plan for avoiding or coping with your trigger situations or emotions. You might exercise when you feel that stress-at-the-end-of-the-day trigger or carry a bag of carrot sticks for the midmorning-boredom trigger.

  d. Set realistic goals, and reward yourself when you reach them. The reward should not be food related.

Traditionally, there has been no "quick fix" for safe, effective weight control. A commitment has to be made to modify existing diet and exercise habits. Most important, those habits have to be avoided forever and replaced by new, healthier behaviors and attitudes.

Frustrated by attempts to modify their diet and exercise, growing numbers of people are turning to bariatric surgery, such as the gastric bypass and the reversible laparoscopic stomach banding (lap-band) surgery. One study suggests that the gastric bypass surgery works not only because it reduces the size of the stomach, but because it also reduces the amount of ghrelin produced. Ghrelin is a hormone produced in the stomach that stimulates appetite.

Another interesting approach to the problem of weight loss has been the development of an antiobesity vaccine. A group of researchers at the Scripps Research Institute has developed a vaccine that slowed weight gain in rats and reduced the amount of stored body fat. The vaccine works by stimulating the body to produce antibodies against ghrelin. These bind the hormone, preventing it from reaching its target in the brain. While this research holds out hope for a safe obesity treatment, it should be noted that the rats in the study were given low-fat, relatively tasteless chow and were rather lean to begin with. It is not clear whether the vaccine would be effective in humans eating a "Western" diet high in calories and fat.

For Further Understanding

In terms of the energy-harvesting reactions we have studied in Chapters 22 and 23, explain how reduced caloric intake and increase in activity level contribute to weight loss.

If you increased your energy expenditure by 200 Calories per day and did not change your eating habits, how long would it take you to lose 10 pounds?

The Reactions of β-Oxidation

The enzymes that catalyze the β-oxidation of fatty acids are located in the matrix space of the mitochondria. Special transport mechanisms are required to bring fatty acid molecules into the mitochondrial matrix. Once inside, the fatty acids are degraded by the reactions of β-oxidation. As we will see, these reactions interact with oxidative phosphorylation and the citric acid cycle to produce ATP.

Reaction 1. The first step is an *activation* reaction that results in the production of a fatty acyl CoA molecule. A thioester bond is formed between coenzyme A and the fatty acid:

$$CH_3-(CH_2)_n-CH_2-CH_2-\overset{\overset{O}{\|}}{C}-OH \quad \xrightarrow[\text{Coenzyme A}]{\text{ATP} \quad \text{AMP} + PP_i}$$

Fatty acid

$$CH_3-(CH_2)_n-CH_2-CH_2-\overset{\overset{O}{\|}}{C}\sim S-CoA \qquad \text{thioester bond}$$

Fatty acyl CoA

This reaction requires energy in the form of ATP, which is cleaved to AMP and pyrophosphate. This involves hydrolysis of two phosphoanhydride bonds. Here again we see the need to invest a small amount of energy so that a much greater amount of energy can be harvested later in the pathway. Coenzyme A is also required for this step. The product, a fatty acyl CoA, has a *high-energy* thioester bond between the fatty acid and coenzyme A. *Acyl-CoA ligase,* which catalyzes this reaction, is located in the outer membrane of the mitochondria. The mechanism that brings the fatty acyl CoA into the mitochondrial matrix involves a carrier molecule called *carnitine.* The first step, catalyzed by the enzyme *carnitine acyltransferase I,* is the transfer of the fatty acyl group to carnitine, producing acylcarnitine and coenzyme A. Next a carrier protein located in the mitochondrial inner membrane transfers the acylcarnitine into the mitochondrial matrix. There *carnitine acyltransferase II* catalyzes the regeneration of fatty acyl CoA, which now becomes involved in the remaining reactions of β-oxidation.

Acyl group transfer reactions are described in Section 14.4.

Reaction 2. The next reaction is an *oxidation* reaction that removes a pair of hydrogen atoms from the fatty acid. These are used to reduce FAD to produce FADH₂. This *dehydrogenation* reaction is catalyzed by the enzyme *acyl-CoA dehydrogenase* and results in the formation of a carbon-carbon double bond:

$$CH_3-(CH_2)_n-CH_2-CH_2-\overset{\overset{O}{\|}}{C}\sim S-CoA \quad \xrightarrow{\text{FAD} \quad \text{FADH}_2}$$

$$CH_3-(CH_2)_n-\overset{\overset{H}{|}}{C}=\overset{\overset{}{}}{C}-\overset{\overset{O}{\|}}{C}\sim S-CoA$$

Oxidative phosphorylation yields two ATP molecules for each molecule of $FADH_2$ produced by this oxidation–reduction reaction.

Reaction 3. The third reaction involves the *hydration* of the double bond produced in reaction 2. As a result the β-carbon is hydroxylated. This reaction is catalyzed by the enzyme *enoyl-CoA hydrase.*

$$CH_3-(CH_2)_n-\overset{\overset{\displaystyle H}{|}}{C}=\overset{\overset{\displaystyle O}{\|}}{\underset{\underset{\displaystyle H}{|}}{C}}-C\sim S-CoA \quad \xrightarrow{\quad H_2O \quad}$$

$$CH_3-(CH_2)_n-\overset{\overset{\displaystyle OH}{|}}{\underset{\underset{\displaystyle H}{|}}{C}}-CH_2-\overset{\overset{\displaystyle O}{\|}}{C}\sim S-CoA$$

Reaction 4. In this *oxidation* reaction the hydroxyl group of the β-carbon is now dehydrogenated. NAD^+ is reduced to form NADH that is subsequently used to produce three ATP molecules by oxidative phosphorylation. L-β-*Hydroxyacyl-CoA dehydrogenase* catalyzes this reaction.

$$CH_3-(CH_2)_n-\overset{\overset{\displaystyle OH}{|}}{\underset{\underset{\displaystyle H}{|}}{C}}-CH_2-\overset{\overset{\displaystyle O}{\|}}{C}\sim S-CoA \quad \xrightarrow{\quad NAD^+ \quad NADH \quad}$$

$$CH_3-(CH_2)_n-\overset{\overset{\displaystyle O}{\|}}{C}-CH_2-\overset{\overset{\displaystyle O}{\|}}{C}\sim S-CoA$$

Reaction 5. The final reaction, catalyzed by the enzyme *thiolase,* is the cleavage that releases acetyl CoA. This is accomplished by *thiolysis,* attack of a molecule of coenzyme A on the β-carbon. The result is the release of acetyl CoA and a fatty acyl CoA that is two carbons shorter than the beginning fatty acid:

$$CH_3-(CH_2)_n-\overset{\overset{\displaystyle O}{\|}}{C}-CH_2-\overset{\overset{\displaystyle O}{\|}}{C}\sim S-CoA \quad \xrightarrow{\quad CoA \quad}$$

$$CH_3-(CH_2)_{n-2}-CH_2-CH_2-\overset{\overset{\displaystyle O}{\|}}{C}\sim S-CoA$$
$$+$$
$$\overset{\overset{\displaystyle O}{\|}}{C}-CH_3$$
$$\underset{S-CoA}{\,}$$

The shortened fatty acyl CoA is further oxidized by cycling through reactions 2–5 until the fatty acid carbon chain is completely degraded to acetyl CoA. The acetyl CoA produced by β-oxidation of fatty acids then enters the reactions of the citric acid cycle. Of course, this eventually results in the production of 12 ATP molecules per molecule of acetyl CoA released during β-oxidation.

As an example of the energy yield from β-oxidation, the balance sheet for ATP production when the sixteen-carbon-fatty acid palmitic acid is degraded by β-oxidation is summarized in Figure 23.7. Complete oxidation of palmitate results in production of 129 molecules of ATP, *three and one half times more energy than results from the complete oxidation of an equivalent amount of glucose.*

Figure 23.7

Complete oxidation of palmitic acid yields 129 molecules of ATP. Note that the activation step is considered to be an expenditure of two high-energy phosphoanhydride bonds because ATP is hydrolyzed to AMP + PP$_i$.

EXAMPLE 23.2

Calculating the Amount of ATP Produced in Complete Oxidation of a Fatty Acid

2 LEARNING GOAL
Describe the degradation of fatty acids by β-oxidation.

How many molecules of ATP are produced in the complete oxidation of stearic acid, an eighteen-carbon saturated fatty acid?

Solution

Step 1 (activation)	− 2 ATP

Steps 2–5:

8 FADH$_2$ × 2 ATP/FADH$_2$	16 ATP
8 NADH × 3 ATP/NADH	24 ATP

9 acetyl CoA (to citric acid cycle):

9 × 1 GTP × 1 ATP/GTP	9 ATP
9 × 3 NADH × 3 ATP/NADH	81 ATP
9 × 1 FADH$_2$ × 2 ATP/FADH$_2$	18 ATP
	146 ATP

Continued—

E X A M P L E 23.2 —*Continued*

Practice Problem 23.2

Write out the sequence of steps for β-oxidation of butyryl CoA. What is the energy yield from the complete degradation of butyryl CoA via β-oxidation, the citric acid cycle, and oxidative phosphorylation?

For Further Practice: Questions 23.43 and 23.44.

23.3 Ketone Bodies

For the acetyl CoA produced by the β-oxidation of fatty acids to efficiently enter the citric acid cycle, there must be an adequate supply of oxaloacetate. If glycolysis and β-oxidation are occurring at the same rate, there will be a steady supply of pyruvate (from glycolysis) that can be converted to oxaloacetate. But what happens if the supply of oxaloacetate is too low to allow all of the acetyl CoA to enter the citric acid cycle? Under these conditions, acetyl CoA is converted to the so-called **ketone bodies:** β-hydroxybutyrate, acetone, and acetoacetate (Figure 23.8).

LEARNING GOAL
4 Understand the role of ketone body production in β-oxidation.

See Section 22.9 for a review of the reactions that provide oxaloacetate.

Ketosis

Ketosis, abnormally high levels of blood ketone bodies, is a situation that arises under some pathological conditions, such as starvation, a diet that is extremely low in carbohydrates (as with the high-protein diets), or uncontrolled **diabetes mellitus.** The carbohydrate intake of a diabetic is normal, but the carbohydrates cannot get into the cell to be used as fuel. Thus diabetes amounts to starvation in the midst of plenty. In diabetes the very high concentration of ketone acids in the blood leads to **ketoacidosis.** The ketone acids are relatively strong acids and therefore readily dissociate to release H^+. Under these conditions the blood pH becomes acidic, which can lead to death.

Diabetes mellitus is a disease characterized by the appearance of glucose in the urine as a result of high blood glucose levels. The disease is usually caused by the inability to produce the hormone insulin.

Ketogenesis

The pathway for the production of ketone bodies (Figure 23.9) begins with a "reversal" of the last step of β-oxidation. When oxaloacetate levels are low, the enzyme that normally carries out the last reaction of β-oxidation now catalyzes the fusion of two acetyl CoA molecules to produce acetoacetyl CoA:

$$2CH_3-\overset{\overset{O}{\|}}{C}\!\sim\!S-CoA \;\rightleftharpoons\; CH_3-\overset{\overset{O}{\|}}{C}-CH_2-\overset{\overset{O}{\|}}{C}\!\sim\!S-CoA$$

Acetyl CoA   CoA   Acetoacetyl CoA

$$CH_3-\underset{\underset{H}{|}}{\overset{\overset{OH}{|}}{C}}-CH_2-C\overset{\diagup O^-}{\diagdown_O} \qquad CH_3-\overset{\overset{O}{\|}}{C}-CH_3 \qquad CH_3-\overset{\overset{O}{\|}}{C}-CH_2-C\overset{\diagup O^-}{\diagdown_O}$$

β-Hydroxybutyrate   Acetone   Acetoacetate

Figure 23.8
Structures of ketone bodies.

Figure 23.9
Summary of the reactions involved in ketogenesis.

Acetoacetyl CoA can react with a third acetyl CoA molecule to yield β-hydroxy-β-methylglutaryl CoA (HMG-CoA):

If HMG-CoA were formed in the cytoplasm, it would serve as a precursor for cholesterol biosynthesis. But ketogenesis, like β-oxidation, occurs in the mitochondrial matrix, and here HMG-CoA is cleaved to yield acetoacetate and acetyl CoA:

$$^-OOC-CH_2-\underset{\underset{CH_3}{|}}{\overset{\overset{OH}{|}}{C}}-CH_2-\overset{\overset{O}{\|}}{C}\sim S-CoA \longrightarrow {}^-OOC-CH_2-\underset{\underset{CH_3}{|}}{\overset{\overset{O}{\|}}{C}} + CH_3-\overset{\overset{O}{\|}}{C}\sim S-CoA$$

HMG-CoA   Acetoacetate   Acetyl CoA

In very small amounts, acetoacetate spontaneously loses carbon dioxide to give acetone. This is the reaction that causes the "acetone breath" that is often associated with uncontrolled diabetes mellitus.

$$^-OOC-CH_2-\underset{\underset{CH_3}{|}}{\overset{\overset{O}{\|}}{C}} + H^+ \xrightarrow{} CH_3-\overset{\overset{O}{\|}}{C}-CH_3$$
$$\searrow CO_2$$

Acetoacetate   Acetone

More frequently, it undergoes NADH-dependent reduction to produce β-hydroxybutyrate:

$$^-OOC-CH_2-\underset{\underset{CH_3}{|}}{\overset{\overset{O}{\|}}{C}} \quad\xrightarrow[\text{NADH} \quad \text{NAD}^+]{}\quad {}^-OOC-CH_2-\underset{\underset{H}{|}}{\overset{\overset{OH}{|}}{C}}-CH_3$$

Acetoacetate   β-Hydroxybutyrate

Acetoacetate and β-hydroxybutyrate are produced primarily in the liver. These metabolites diffuse into the blood and are circulated to other tissues, where they may be reconverted to acetyl CoA and used to produce ATP. In fact, the heart muscle derives most of its metabolic energy from the oxidation of ketone bodies, not from the oxidation of glucose. Other tissues that are best adapted to the use of glucose will increasingly rely on ketone bodies for energy when glucose becomes unavailable or limited. This is particularly true of the brain.

What conditions lead to excess production of ketone bodies?

Question *23.3*

What is the cause of the characteristic "acetone breath" that is associated with uncontrolled diabetes mellitus?

Question *23.4*

23.4 Fatty Acid Synthesis

All organisms possess the ability to synthesize fatty acids. In humans the excess acetyl CoA produced by carbohydrate degradation is used to make fatty acids that are then stored as triglycerides.

LEARNING GOAL

5 Compare β-oxidation of fatty acids and fatty acid biosynthesis.

A Comparison of Fatty Acid Synthesis and Degradation

On first examination, fatty acid synthesis appears to be simply the reverse of β-oxidation. Specifically, the fatty acid chain is constructed by the sequential addition of two-carbon acetyl groups (Figure 23.10). Although the chemistry of fatty acid synthesis and breakdown are similar, there are several major differences between β-oxidation and fatty acid biosynthesis. These are summarized as follows.

Figure 23.10
Summary of fatty acid synthesis. Malonyl ACP is produced in two reactions: carboxylation of acetyl CoA to produce malonyl CoA and transfer of the malonyl acyl group from malonyl CoA to ACP.

- **Intracellular location.** The enzymes responsible for fatty acid biosynthesis are located in the cytoplasm of the cell, whereas those responsible for the degradation of fatty acids are in the mitochondria.
- **Acyl group carriers.** The activated intermediates of fatty acid biosynthesis are bound to a carrier molecule called the **acyl carrier protein (ACP)** (Figure 23.11). In β-oxidation the acyl group carrier was coenzyme A. However, there are important similarities between these two carriers. Both contain the **phosphopantetheine** group, which is made from the vitamin pantothenic acid. In both cases the fatty acyl group is bound by a thioester bond to the phosphopantetheine group.
- **Enzymes involved.** Fatty acid biosynthesis is carried out by a multienzyme complex known as *fatty acid synthase.* The enzymes responsible for fatty acid degradation are not physically associated in such complexes.
- **Electron carriers.** NADH and $FADH_2$ are produced by fatty acid oxidation, whereas NADPH is the reducing agent for fatty acid biosynthesis. As a general rule, *NADH is produced by catabolic reactions, and NADPH is the reducing agent of biosynthetic reactions.* These two coenzymes differ only by the presence of a phosphate group bound to the ribose ring of NADPH (Figure 23.12). The enzymes that use these coenzymes, however, are easily able to distinguish them on this basis.

Phosphopantetheine prosthetic group of ACP

Phosphopantetheine group of coenzyme A

Question 23.5

List the four major differences between β-oxidation and fatty acid biosynthesis that reveal that the two processes are not just the reverse of one another.

Question 23.6

What chemical group is part of coenzyme A and acyl carrier protein and allows both molecules to form thioester bonds to fatty acids?

LEARNING GOAL

6 Describe the regulation of lipid and carbohydrate metabolism in relation to the liver, adipose tissue, muscle tissue, and the brain.

23.5 The Regulation of Lipid and Carbohydrate Metabolism

The metabolism of fatty acids and carbohydrates occurs to a different extent in different organs. As we will see in this section, the regulation of these two related aspects of metabolism is of great physiological importance.

The Liver

The liver provides a steady supply of glucose for muscle and brain and plays a major role in the regulation of blood glucose concentration. This regulation is under hormonal control. Recall that the hormone insulin causes blood glucose to be taken up by the liver and stored as glycogen (*glycogenesis*). In this way the liver reduces the blood glucose levels when they are too high.

The hormone glucagon, on the other hand, stimulates the breakdown of glycogen and the release of glucose into the bloodstream. Lactate produced by muscles under anaerobic conditions is also taken up by liver cells and is converted to glucose by gluconeogenesis. Both glycogen degradation (*glycogenolysis*) and *gluconeogenesis* are pathways that produce glucose for export to other organs when energy is needed (Figure 23.13).

The liver also plays a central role in lipid metabolism. When excess fuel is available, the liver synthesizes fatty acids. These are used to produce triglycerides that are transported from the liver to adipose tissues by very low density lipoprotein (VLDL) complexes. In fact, VLDL complexes provide adipose tissue with its major source of fatty acids. This transport is particularly active when more calories are eaten than are burned! During fasting or starvation conditions, however, the liver converts fatty acids to acetoacetate and other ketone bodies. The liver cannot use these ketone bodies because it lacks an enzyme for the conversion of acetoacetate to acetyl CoA. Therefore the ketone bodies produced by the liver are exported to other organs where they are oxidized to make ATP.

Figure 23.12
Structure of NADPH. The phosphate group shown in red is the structural feature that distinguishes NADPH from NADH.

A Medical Perspective

Diabetes Mellitus and Ketone Bodies

More than one person, found unconscious on the streets of some metropolis, has been carted to jail only to die of complications arising from uncontrolled diabetes mellitus. Others are fortunate enough to arrive in hospital emergency rooms. A quick test for diabetes mellitus–induced coma is the odor of acetone on the breath of the afflicted person. Acetone is one of several metabolites produced by diabetics that are known collectively as *ketone bodies.*

The term *diabetes* was used by the ancient Greeks to designate diseases in which excess urine is produced. Two thousand years later, in the eighteenth century, the urine of certain individuals was found to contain sugar, and the name *diabetes mellitus* (Latin: *mellitus,* sweetened with honey) was given to this disease. People suffering from diabetes mellitus waste away as they excrete large amounts of sugar-containing urine.

The cause of insulin-dependent diabetes mellitus is an inadequate production of insulin by the body. Insulin is secreted in response to high blood glucose levels. It binds to the membrane receptor protein on its target cells. Binding increases the rate of transport of glucose across the membrane and stimulates glycogen synthesis, lipid biosynthesis, and protein synthesis. As a result, the blood glucose level is reduced. Clearly, the inability to produce sufficient insulin seriously impairs the body's ability to regulate metabolism.

Individuals suffering from diabetes mellitus do not produce enough insulin to properly regulate blood glucose levels. This generally results from the destruction of the β-cells of the islets of Langerhans. One theory to explain the mysterious disappearance of these cells is that a virus infection stimulates the immune system to produce antibodies that cause the destruction of the β-cells.

In the absence of insulin the uptake of glucose into the tissues is not stimulated, and a great deal of glucose is eliminated in the urine. Without insulin, then, adipose cells are unable to take up the glucose required to synthesize triglycerides. As a result, the rate of fat hydrolysis is much greater than the rate of fat resynthesis, and large quantities of free fatty acids are liberated into the bloodstream. Because glucose is not being efficiently taken into cells, carbohydrate metabolism slows, and there is an increase in the rate of lipid catabolism. In the liver this lipid catabolism results in the production of ketone bodies: acetone, acetoacetate, and β-hydroxybutyrate.

A similar situation can develop from improper eating, fasting, or dieting—any situation in which the body is not provided with sufficient energy in the form of carbohydrates. These ketone bodies cannot all be oxidized by the citric acid cycle, which is limited by the supply of oxaloacetate. The acetone concentration in blood rises to levels so high that acetone can be detected in the breath of untreated diabetics. The elevated concentration of ketones in the blood can overwhelm the buffering capacity of the blood, resulting in ketoacidosis. Ketones, too, will be excreted through the kidney. In fact, the presence of excess ketones in the urine can raise the osmotic concentration of the urine so that it behaves as an "osmotic diuretic," causing the excretion of enormous amounts of water. As a result, the patient may become severely dehydrated. In extreme cases the combination of dehydration and ketoacidosis may lead to coma and death.

It has been observed that diabetics also have a higher than normal level of glucagon in the blood. As we have seen, glucagon stimulates lipid catabolism and ketogenesis. It may be that the symptoms previously described result from both the deficiency of insulin and the elevated glucagon levels. The absence of insulin may cause the elevated blood glucose and fatty acid levels, whereas the glucagon, by stimulating ketogenesis, may be responsible for the ketoacidosis and dehydration.

There is no cure for diabetes. However, when the problem is the result of the inability to produce active insulin, blood glucose levels can be controlled moderately well by the injection of human insulin produced from the cloned insulin gene. Unfortunately, one or even a few injections of insulin each day cannot mimic the precise control of blood glucose accomplished by the pancreas.

As a result, diabetics suffer progressive tissue degeneration that leads to early death. One primary cause of this degeneration is atherosclerosis, the deposition of plaque on the walls of blood vessels. This causes a high frequency of strokes, heart attack, and gangrene of the feet and lower extremities, often necessitating amputation. Kidney failure causes the death of about 20% of diabetics under forty years of age, and diabetic retinopathy (various kinds of damage to the retina of the eye) ranks fourth among the leading causes of blindness in the United States. Nerves are also damaged, resulting in neuropathies that can cause pain or numbness, particularly of the feet.

The metabolic events that occur in uncontrolled diabetes and that can lead to coma and death.

There is no doubt that insulin injections prolong the life of diabetics, but only the presence of a fully functioning pancreas can allow a diabetic to live a life free of the complications noted here. At present, pancreas transplants do not have a good track record. Only about 50% of the transplants are functioning after one year. It is hoped that improved transplantation techniques will be developed so that diabetics can live a normal life span, free of debilitating disease.

For Further Understanding

The Atkins' low carbohydrate diet recommends that dieters test their urine for the presence of ketone bodies as an indicator that the diet is working. In terms of lipid and carbohydrate metabolism, explain why ketone bodies are being produced and why this is an indication that the diet is working.

An excess of ketone bodies in the blood causes ketoacidosis. Consider the chemical structure of the ketone bodies and explain why they are acids.

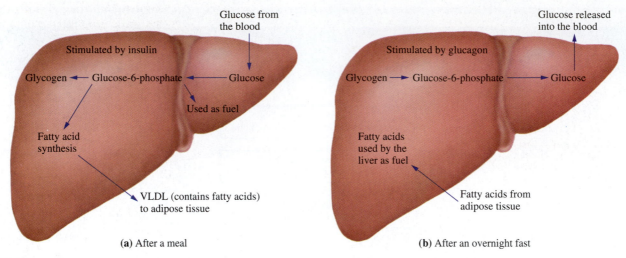

Figure 23.13
The liver controls the concentration of blood glucose.

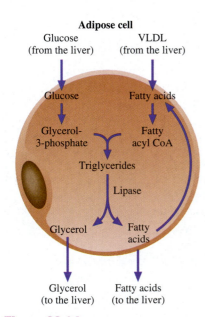

Figure 23.14
Synthesis and degradation of triglycerides in adipose tissue.

Gluconeogenesis is described in Section 21.6.

Adipose Tissue

Adipose tissue is the major storage depot of fatty acids. Triglycerides produced by the liver are transported through the bloodstream as components of VLDL complexes. The triglycerides are hydrolyzed by the same lipases that act on chylomicrons, and the fatty acids are absorbed by adipose tissue. The synthesis of triglycerides in adipose tissue requires glycerol-3-phosphate. However, adipose tissue is unable to make glycerol-3-phosphate and depends on glycolysis for its supply of this molecule. Thus adipose cells must have a ready source of glucose to synthesize and store triglycerides.

Triglycerides are constantly being hydrolyzed and resynthesized in the cells of adipose tissue. Lipases that are under hormonal control determine the rate of hydrolysis of triglycerides into fatty acids and glycerol. If glucose is in limited supply, there will not be sufficient glycerol-3-phosphate for the resynthesis of triglycerides, and the fatty acids and glycerol are exported to the liver for further processing (Figure 23.14).

Muscle Tissue

The energy demand of resting muscle is generally supplied by the β-oxidation of fatty acids. The heart muscle actually prefers ketone bodies over glucose.

Working muscle, however, obtains energy by degradation of its own supply of glycogen. Glycogen degradation produces glucose-6-phosphate, which is directly funneled into glycolysis. If the muscle is working so hard that it doesn't get enough oxygen, it produces large amounts of lactate. This fermentation end product, as well as alanine (from catabolism of proteins and transamination of pyruvate), is exported to the liver. Here they are converted to glucose by gluconeogenesis (Figure 23.15).

The Brain

Under normal conditions the brain uses glucose as its sole source of metabolic energy. When the body is in the resting state, about 60% of the free glucose of the body is used by the brain. Starvation depletes glycogen stores, and the amount of glucose available to the brain drops sharply. The ketone bodies acetoacetate and β-hydroxybutyrate are then used by the brain as an alternative energy source. Fatty acids are transported in the blood in complexes with proteins and cannot cross the blood-brain barrier to be used by brain cells as

Figure 23.15
Metabolic relationships between liver and muscle.

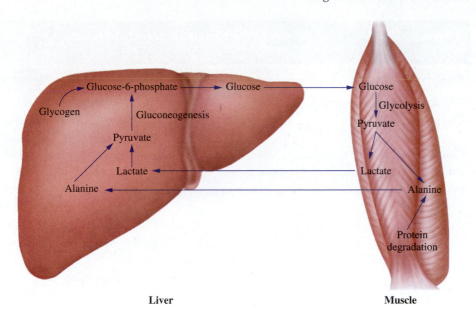

Liver   Muscle

an energy source. But ketone bodies, which have a free carboxyl group, are soluble in blood and can enter the brain.

How does the liver regulate blood glucose levels?

Question *23.7*

Why is regulation of blood glucose levels important to the efficient function of the brain?

Question *23.8*

23.6 The Effects of Insulin and Glucagon on Cellular Metabolism

The hormone **insulin** is produced by the β-cells of the islets of Langerhans in the pancreas. It is secreted from these cells in response to an increase in the blood glucose level. Insulin lowers the concentration of blood glucose by causing a number of changes in metabolism (Table 23.1).

The simplest way to lower blood glucose levels is to stimulate storage of glucose, both as glycogen and as triglycerides. *Insulin therefore activates biosynthetic processes and inhibits catabolic processes.*

Insulin acts only on those cells, known as *target cells,* that possess a specific insulin receptor protein in their plasma membranes. The major target cells for insulin are liver, adipose, and muscle cells.

The blood glucose level is normally about 10 mM. However, a substantial meal increases the concentration of blood glucose considerably and stimulates insulin secretion. Subsequent binding of insulin to the plasma membrane insulin receptor increases the rate of transport of glucose across the membrane and into cells.

Insulin exerts a variety of effects on all aspects of cellular metabolism:

- **Carbohydrate metabolism.** Insulin stimulates glycogen synthesis. At the same time it inhibits glycogenolysis and gluconeogenesis. The overall result of these activities is the storage of excess glucose.

LEARNING GOAL

7 Summarize the antagonistic effects of glucagon and insulin.

The effect of insulin on glycogen metabolism is described in Section 21.7.

TABLE 23.1	Comparison of the Metabolic Effects of Insulin and Glucagon		
Actions		**Insulin**	**Glucagon**
Cellular glucose transport		Increased	No effect
Glycogen synthesis		Increased	Decreased
Glycogenolysis in liver		Decreased	Increased
Gluconeogenesis		Decreased	Increased
Amino acid uptake and protein synthesis		Increased	No effect
Inhibition of amino acid release and protein degradation		Decreased	No effect
Lipogenesis		Increased	No effect
Lipolysis		Decreased	Increased
Ketogenesis		Decreased	Increased

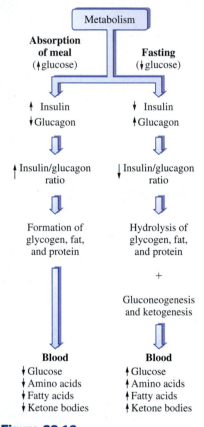

Figure 23.16

A summary of the antagonistic effects of insulin and glucagon.

- **Protein metabolism.** Insulin stimulates transport and uptake of amino acids, as well as the incorporation of amino acids into proteins.
- **Lipid metabolism.** Insulin stimulates uptake of glucose by adipose cells, as well as the synthesis and storage of triglycerides. As we have seen, storage of lipids requires a source of glucose, and insulin helps the process by increasing the available glucose. At the same time, insulin inhibits the breakdown of stored triglycerides.

As you may have already guessed, insulin is only part of the overall regulation of cellular metabolism in the body. A second hormone, **glucagon,** is secreted by the α-cells of the islets of Langerhans in response to decreased blood glucose levels. The effects of glucagon, generally the opposite of the effects of insulin, are summarized in Table 23.1. Although it has no direct effect on glucose uptake, glucagon inhibits glycogen synthesis and stimulates glycogenolysis and gluconeogenesis. It also stimulates the breakdown of fats and ketogenesis.

The antagonistic effects of these two hormones, seen in Figure 23.16, are critical for the maintenance of adequate blood glucose levels. During fasting, low blood glucose levels stimulate production of glucagon, which increases blood glucose by stimulating the breakdown of glycogen and the production of glucose by gluconeogenesis. This ensures a ready supply of glucose for the tissues, especially the brain. On the other hand, when blood glucose levels are too high, insulin is secreted. It stimulates the removal of the excess glucose by enhancing uptake and inducing pathways for storage.

Question 23.9

Summarize the effects of the hormone insulin on carbohydrate, lipid, and amino acid metabolism.

Question 23.10

Summarize the effects of the hormone glucagon on carbohydrate and lipid metabolism.

SUMMARY

23.1 Lipid Metabolism in Animals

Dietary lipids *(triglycerides)* are emulsified into tiny fat droplets in the intestine by the action of *bile* salts. Pancreatic *lipase* catalyzes the hydrolysis of triglycerides into monoglycerides and fatty acids. These are absorbed by intestinal epithelial cells, reassembled into triglycerides, and combined with protein to form *chylomicrons*. Chylomicrons are transported to the cells of the body through the bloodstream. Fatty acids are stored as triglycerides (triacylglycerols) in fat droplets in the cytoplasm of *adipocytes*.

23.2 Fatty Acid Degradation

Fatty acids are degraded to acetyl CoA in the mitochondria by the β-*oxidation* pathway, which involves five steps: (1) the production of a fatty acyl CoA molecule, (2) oxidation of the fatty acid by an FAD-dependent dehydrogenase, (3) hydration, (4) oxidation by an NAD$^+$-dependent dehydrogenase, and (5) cleavage of the chain with release of acetyl CoA and a fatty acyl CoA that is two carbons shorter than the beginning fatty acid. The last four reactions are repeated until the fatty acid is completely degraded to acetyl CoA.

23.3 Ketone Bodies

Under some conditions, fatty acid degradation occurs more rapidly than glycolysis. As a result, a large amount of acetyl CoA is produced from fatty acids, but little oxaloacetate is generated from pyruvate. When oxaloacetate levels are too low, the excess acetyl CoA is converted to the *ketone bodies* acetone, acetoacetate, and β-hydroxybutyrate.

23.4 Fatty Acid Synthesis

Fatty acid biosynthesis occurs by the sequential addition of acetyl groups and, on first inspection, appears to be a simple reversal of the β-oxidation pathway. Although the biochemical reactions are similar, fatty acid synthesis differs from β-oxidation in the following ways: It occurs in the cytoplasm, utilizes *acyl carrier protein* and NADPH, and is carried out by a multienzyme complex, fatty acid synthase.

23.5 The Regulation of Lipid and Carbohydrate Metabolism

Lipid and carbohydrate metabolism occur to different extents in different organs. The liver regulates the flow of metabolites to brain, muscle, and adipose tissue and ultimately controls the concentration of blood glucose. *Adipose tissue* is the major storage depot for fatty acids. Triglycerides are constantly hydrolyzed and resynthesized in adipose tissue. Muscle oxidizes glucose, fatty acids, and ketone bodies.

The brain uses glucose as a fuel except in prolonged fasting or starvation, when it will use ketone bodies as an energy source.

23.6 The Effects of Insulin and Glucagon on Cellular Metabolism

Insulin stimulates biosynthetic processes and inhibits catabolism in liver, muscle, and adipose tissue. Insulin is synthesized in the β-cells of the pancreas and is secreted when the blood glucose levels become too high. The insulin receptor protein binds to the insulin. This binding mediates a variety of responses in target tissues, including the storage of glucose and lipids. *Glucagon* is secreted when blood glucose levels are too low. It has the opposite effects on metabolism, including the breakdown of lipids and glycogen.

KEY TERMS

acyl carrier protein (ACP) (23.4)
adipocyte (23.1)
adipose tissue (23.1)
bile (23.1)
chylomicron (23.1)
colipase (23.1)
diabetes mellitus (23.3)
glucagon (23.6)

insulin (23.6)
ketoacidosis (23.3)
ketone bodies (23.3)
ketosis (23.3)
lipase (23.1)
micelle (23.1)
β-oxidation (23.2)
phosphopantetheine (23.4)
triglyceride (23.1)

QUESTIONS AND PROBLEMS

Lipid Metabolism in Animals

Foundations

23.11 Describe the composition of bile.
23.12 Why are bile salts referred to as detergents?
23.13 Define the term *micelle*.
23.14 In Figure 23.1, a micelle composed of the phospholipid lecithin is shown. Why is lecithin a good molecule for the formation of micelles?
23.15 Define the term *triglyceride*.
23.16 Draw the structure of a triglyceride composed of glycerol, palmitoleic acid, linolenic acid, and oleic acid.
23.17 Review the information on chylomicrons in Chapter 17. Describe the composition of chylomicrons?
23.18 What is an adipocyte?

Applications

23.19 What is the major storage form of fatty acids?
23.20 What tissue is the major storage depot for lipids?
23.21 What is the outstanding structural feature of an adipocyte?
23.22 What is the major metabolic function of adipose tissue?
23.23 What is the general reaction catalyzed by lipases?

23.24 Why are the lipases that are found in saliva and in the stomach not very effective at digesting triglycerides?

23.25 List three major biological molecules for which acetyl CoA is a precursor.

23.26 Why are triglycerides more efficient energy-storage molecules than glycogen?

23.27 What are chylomicrons, and what is their function?

23.28 **a.** What are very low density lipoproteins?
b. Compare the function of VLDLs with that of chylomicrons.

23.29 What is the function of the bile salts in the digestion of dietary lipids?

23.30 What is the function of colipase in the digestion of dietary lipids?

23.31 Describe the stages of lipid digestion.

23.32 Describe the transport of lipids digested in the lumen of the intestines to the cells of the body.

Fatty Acid Degradation

Foundations

23.33 What is the energy source for the activation of a fatty acid in preparation for β-oxidation?

23.34 Which bond in fatty acyl CoA is a high energy bond?

23.35 What is carnitine?

23.36 Explain the mechanism by which a fatty acyl group is brought into the mitochondrial matrix.

23.37 Explain why the reaction catalyzed by acyl-CoA dehydrogenase is an example of an oxidation reaction.

23.38 What is the reactant that is oxidized in the reaction catalyzed by acyl-CoA dehydrogenase? What is the reactant that is reduced in this reaction?

23.39 What is the product of the hydration of an alkene?

23.40 Which reaction in β-oxidation is a hydration reaction? What is the name of the enzyme that catalyzes this reaction? Write an equation representing this reaction.

Applications

23.41 What products are formed when the ω-phenyl-labeled carboxylic acid 14-phenyltetradecanoic acid is degraded by β-oxidation?

23.42 What products are formed when the ω-phenyl-labeled carboxylic acid 5-phenylpentanoic acid is degraded by β-oxidation?

23.43 Calculate the number of ATP molecules produced by complete β-oxidation of the fourteen-carbon saturated fatty acid tetradecanoic acid (common name: myristic acid).

23.44 **a.** Write the sequence of steps that would be followed for one round of β-oxidation of hexanoic acid.
b. Calculate the number of ATP molecules produced by complete β-oxidation of hexanoic acid.

23.45 How many molecules of ATP are produced for each molecule of $FADH_2$ that is generated by β-oxidation?

23.46 How many molecules of ATP are produced for each molecule of NADH generated by β-oxidation?

23.47 What is the fate of the acetyl CoA produced by β-oxidation?

23.48 How many ATP molecules are produced from each acetyl CoA molecule generated in β-oxidation that enters the citric acid cycle?

Ketone Bodies

Foundations

23.49 What are ketone bodies?

23.50 What are the chemical properties of ketone bodies?

23.51 What is ketosis?

23.52 Define *ketoacidosis*.

23.53 In what part of the cell does ketogenesis occur? Be specific.

23.54 What would be the fate of HMG-CoA produced in ketogenesis if it were produced in the cell cytoplasm?

Applications

23.55 Draw the structures of acetoacetate and β-hydroxybutyrate.

23.56 Describe the relationship between the formation of ketone bodies and β-oxidation.

23.57 Why do uncontrolled diabetics produce large amounts of ketone bodies?

23.58 How does the presence of ketone bodies in the blood lead to ketoacidosis?

23.59 When does the heart use ketone bodies?

23.60 When does the brain use ketone bodies?

Fatty Acid Synthesis

Foundations

23.61 Where in the cell does fatty acid biosynthesis occur?

23.62 What is the acyl group carrier in fatty acid biosynthesis?

23.63 What enzyme is involved in fatty acid biosynthesis?

23.64 What is the reducing agent for fatty acid biosynthesis?

Applications

23.65 **a.** What is the role of the phosphopantetheine group in fatty acid biosynthesis?
b. From what molecule is phosphopantetheine made?

23.66 What molecules involved in fatty acid degradation and fatty acid biosynthesis contain the phosphopantetheine group?

23.67 How does the structure of fatty acid synthase differ from that of the enzymes that carry out β-oxidation?

23.68 In what cellular compartments do fatty acid biosynthesis and β-oxidation occur?

The Regulation of Lipid and Carbohydrate Metabolism

Foundations

23.69 Define the term *glycogenesis*.

23.70 Define the term *glycogenolysis*.

23.71 Define the term *gluconeogenesis*.

23.72 How are the fatty acids synthesized in the liver transported to adipose tissue?

Applications

23.73 Which pathway provides the majority of the ATP for *resting* muscle?

23.74 Why is the liver unable to utilize ketone bodies as an energy source?

23.75 What is the major metabolic function of the liver?

23.76 What is the fate of lactate produced in skeletal muscle during rapid contraction?

23.77 What are the major fuels of the heart, brain, and liver?

23.78 Why can't the brain use fatty acids as fuel?

23.79 Briefly describe triglyceride metabolism in an adipocyte.

23.80 What is the source of the glycerol molecule that is used in the synthesis of triglycerides?

The Effects of Insulin and Glucagon on Cellular Metabolism

Foundations

23.81 In general, what is the effect of insulin on catabolic and anabolic or biosynthetic processes?

23.82 What is the trigger that causes insulin to be secreted into the bloodstream?

23.83 What is meant by the term *target cell*?

23.84 What are the primary target cells of insulin?

23.85 What is the trigger that causes glucagon to be secreted into the bloodstream?

23.86 What are the primary target cells of glucagon?

Applications

23.87 Where is insulin produced?

23.88 Where is glucagon produced?

23.89 How does insulin affect carbohydrate metabolism?

23.90 How does glucagon affect carbohydrate metabolism?

23.91 How does insulin affect lipid metabolism?

23.92 How does glucagon affect lipid metabolism?

23.93 Why is it said that diabetes mellitus amounts to starvation in the midst of plenty?

23.94 What is the role of the insulin receptor in controlling blood glucose levels?

CRITICAL THINKING PROBLEMS

1. Suppose that fatty acids were degraded by sequential oxidation of the α-carbon. What product(s) would Knoop have obtained with fatty acids with even numbers of carbon atoms? What product(s) would he have obtained with fatty acids with odd numbers of carbon atoms?

2. Oil-eating bacteria can oxidize long-chain alkanes. In the first step of the pathway, the enzyme monooxygenase catalyzes a reaction that converts the long-chain alkane into a primary alcohol. Data from research studies indicate that three more reactions are required to allow the primary alcohol to enter the β-oxidation pathway. Propose a pathway that would convert the long-chain alcohol into a product that could enter the β-oxidation pathway.

3. A young woman sought the advice of her physician because she was 30 pounds overweight. The excess weight was in the form of triglycerides carried in adipose tissue. Yet when the woman described her diet, it became obvious that she actually ate very moderate amounts of fatty foods. Most of her caloric intake was in the form of carbohydrates. This included candy, cake, beer, and soft drinks. Explain how the excess calories consumed in the form of carbohydrates ended up being stored as triglycerides in adipose tissue.

4. Olestra is a fat substitute that provides no calories, yet has a creamy, tongue-pleasing consistency. Because it can withstand heating, it can be used to prepare foods such as potato chips and crackers. Recently the Food and Drug Administration approved olestra for use in prepared foods. Olestra is a sucrose polyester produced by esterification of six, seven, or eight fatty acids to molecules of sucrose. Develop a hypothesis to explain why olestra is not a source of dietary calories.

5. Carnitine is a tertiary amine found in mitochondria that is involved in transporting the acyl groups of fatty acids from the cytoplasm into the mitochondria. The fatty acyl group is transferred from a fatty acyl CoA molecule and esterified to carnitine. Inside the mitochondria, the reaction is reversed and the fatty acid enters the β-oxidation pathway.

   A seventeen-year-old male went to a university medical center complaining of fatigue and poor exercise tolerance. Muscle biopsies revealed droplets of triglycerides in his muscle cells. Biochemical analysis showed that he had only one-fifth of the normal amount of carnitine in his muscle cells.

   What effect will carnitine deficiency have on β-oxidation? What effect will carnitine deficiency have on glucose metabolism?

6. Acetyl CoA carboxylase catalyzes the formation of malonyl CoA from acetyl CoA and the bicarbonate anion, a reaction that requires the hydrolysis of ATP. Write a balanced equation showing this reaction.

   The reaction catalyzed by acetyl CoA carboxylase is the ratelimiting step in fatty acid biosynthesis. The malonyl group is transferred from coenzyme A to acyl carrier protein; similarly, the acetyl group is transferred from coenzyme A to acyl carrier protein. This provides the two beginning substrates of fatty acid biosynthesis shown in Figure 23.10.

   Consider the following case study. A baby boy was brought to the emergency room with severe respiratory distress. Examination revealed muscle pathology, poor growth, and severe brain damage. A liver biopsy revealed that the child didn't make acetyl CoA carboxylase. What metabolic pathway is defective in this child? How is this defect related to the respiratory distress suffered by the baby?

Glossary

A

absolute specificity (19.5) the property of an enzyme that allows it to bind and catalyze the reaction of only one substrate

accuracy (1.3) the nearness of an experimental value to the true value

acetal (13.4) the family of organic compounds formed via the reaction of two molecules of alcohol with an aldehyde in the presence of an acid catalyst; acetals have the following general structure:

$$\begin{array}{c} OR^2 \\ | \\ R^1\!-\!C\!-\!OR^3 \\ | \\ H \end{array}$$

acetyl coenzyme A (acetyl CoA) (14.4, 22.2) a molecule composed of coenzyme A and an acetyl group; the intermediate that provides acetyl groups for complete oxidation by aerobic respiration

acid (8.1) a substance that behaves as a proton donor

acid anhydride (14.3) the product formed by the combination of an acid chloride and a carboxylate ion; structurally they are two carboxylic acids with a water molecule removed:

$$\begin{array}{c} O \quad\quad O \\ \| \quad\quad \| \\ (Ar)\ R\!-\!C\!-\!O\!-\!C\!-\!R\ (Ar) \end{array}$$

acid-base reaction (4.3) reaction that involves the transfer of a hydrogen ion (H^+) from one reactant to another

acid chloride (14.3) member of the family of organic compounds with the general formula

$$\begin{array}{c} O \\ \| \\ (Ar)\ R\!-\!C\!-\!Cl \end{array}$$

activated complex (7.3) the arrangement of atoms at the top of the potential energy barrier as a reaction proceeds

activation energy (7.3) the threshold energy that must be overcome to produce a chemical reaction

active site (19.4) the cleft in the surface of an enzyme that is the site of substrate binding

acyl carrier protein (ACP) (23.4) the protein that forms a thioester linkage with fatty acids during fatty acid synthesis

acyl group (14: Intro, 15.3) the functional group found in carboxylic acid derivatives that contains the carbonyl group attached to one alkyl or aryl group:

$$\begin{array}{c} O \\ \| \\ (Ar)\ R\!-\!C\!- \end{array}$$

addition polymer (11.5) polymers prepared by the sequential addition of a monomer

addition reaction (11.5, 13.4) a reaction in which two molecules add together to form a new molecule; often involves the addition of one molecule to a double or triple bond in an unsaturated molecule; e.g., the addition of alcohol to an aldehyde or ketone to form a hemiacetal or hemiketal

adenosine triphosphate (ATP) (14.4, 21.1) a nucleotide composed of the purine adenine, the sugar ribose, and three phosphoryl groups; the primary energy storage and transport molecule used by the cells in cellular metabolism

adipocyte (23.1) a fat cell

adipose tissue (23.1) fatty tissue that stores most of the body lipids

aerobic respiration (22.3) the oxygen-requiring degradation of food molecules and production of ATP

alcohol (12.1) an organic compound that contains a hydroxyl group (—OH) attached to an alkyl group

aldehyde (13.1) a class of organic molecules characterized by a carbonyl group; the carbonyl carbon is bonded to a hydrogen atom and to another hydrogen or an alkyl or aryl group. Aldehydes have the following general structure:

$$\begin{array}{cc} O & O \\ \| & \| \\ (Ar)\!-\!C\!-\!H & R\!-\!C\!-\!H \end{array}$$

aldol condensation (13.4) a reaction in which aldehydes or ketones react to form a larger molecule

aldose (16.2) a sugar that contains an aldehyde (carbonyl) group

aliphatic hydrocarbon (10.1) any member of the alkanes, alkenes, and alkynes or the substituted alkanes, alkenes, and alkynes

alkali metal (2.4) an element within Group IA (1) of the periodic table

alkaline earth metal (2.4) an element within Group IIA (2) of the periodic table

alkaloid (15.2) a class of naturally occurring compounds that contain one or more nitrogen heterocyclic rings; many of the alkaloids have medicinal and other physiological effects

alkane (10.2) a hydrocarbon that contains only carbon and hydrogen and is bonded together through carbon-hydrogen and carbon-carbon single bonds; a saturated hydrocarbon with the general molecular formula C_nH_{2n+2}

alkene (11.1) a hydrocarbon that contains one or more carbon-carbon double bonds; an unsaturated hydrocarbon with the general formula C_nH_{2n}

alkyl group (10.2) a hydrocarbon group that results from the removal of one hydrogen from the original hydrocarbon (e.g., methyl, —CH_3; ethyl, —CH_2CH_3)

alkyl halide (10.5) a substituted hydrocarbon with the general structure R—X, in which R— represents any alkyl group and X = a halogen (F—, Cl—, Br—, or I—)

alkylammonium ion (15.1) the ion formed when the lone pair of electrons of the nitrogen atom of an amine is shared with a proton (H^+) from a water molecule

alkyne (11.1) a hydrocarbon that contains one or more carbon-carbon triple bonds; an unsaturated hydrocarbon with the general formula C_nH_{2n-2}

allosteric enzyme (19.9) an enzyme that has an effector binding site and an active site; effector binding changes the shape of the active site, rendering it either active or inactive

alpha particle (9.1) a particle consisting of two protons and two neutrons; the alpha particle is identical to a helium nucleus

amide bond (15.3) the bond between the carbonyl carbon of a carboxylic acid and the amino nitrogen of an amine

amides (15.3) the family of organic compounds formed by the reaction between a carboxylic acid derivative and an amine and characterized by the amide group

amines (15.1)   the family of organic molecules with the general formula RNH_2, R_2NH, or R_3N (R— can represent either an alkyl or aryl group); they may be viewed as substituted ammonia molecules in which one or more of the ammonia hydrogens has been substituted by a more complex organic group

α-amino acid (18.2)   the subunits of proteins composed of an α-carbon bonded to a carboxylate group, a protonated amino group, a hydrogen atom, and a variable R group

aminoacyl group (15.4)   the functional group that is characteristic of an amino acid; the aminoacyl group has the following general structure:

$$H_3\overset{+}{N}-\overset{\overset{\displaystyle H}{|}}{\underset{\underset{\displaystyle R}{|}}{C}}-\overset{\overset{\displaystyle O}{\|}}{C}-$$

aminoacyl tRNA (20.6)   the transfer RNA covalently linked to the correct amino acid

aminoacyl tRNA binding site of ribosome (A-site) (20.6)   a pocket on the surface of a ribosome that holds the aminoacyl tRNA during translation

aminoacyl tRNA synthetase (20.6)   an enzyme that recognizes one tRNA and covalently links the appropriate amino acid to it

amorphous solid (5.3)   a solid with no organized, regular structure

amphibolic pathway (22.9)   a metabolic pathway that functions in both anabolism and catabolism

amphiprotic (8.1)   a substance that can behave either as a Brønsted acid or a Brønsted base

amylopectin (16.6)   a highly branched form of amylose; the branches are attached to the C-6 hydroxyl by $\alpha(1 \rightarrow 6)$ glycosidic linkage; a component of starch

amylose (16.6)   a linear polymer of α-D-glucose molecules bonded in $\alpha(1 \rightarrow 4)$ glycosidic linkage that is a component of starch; a polysaccharide storage form

anabolism (21.1, 22.9)   all of the cellular energy-requiring biosynthetic pathways

anaerobic threshold (21.4)   the point at which the level of lactate in the exercising muscle inhibits glycolysis and the muscle, deprived of energy, ceases to function

analgesic (15.2)   any drug that acts as a painkiller, e.g., aspirin, acetaminophen

anaplerotic reaction (22.9)   a reaction that replenishes a substrate needed for a biochemical pathway

anesthetic (15.2)   a drug that causes a lack of sensation in part of the body (local anesthetic) or causes unconsciousness (general anesthetic)

angular structure (3.4)   a planar molecule with bond angles other than 180°

anion (2.1)   a negatively charged atom or group of atoms

anode (8.5)   the positively charged electrode in an electrical cell

anomers (16.4)   isomers of cyclic monosaccharides that differ from one another in the arrangement of bonds around the hemiacetal carbon

antibodies (18.1)   immunoglobulins; specific glycoproteins produced by cells of the immune system in response to invasion by infectious agents

anticodon (20.4)   a sequence of three ribonucleotides on a tRNA that are complementary to a codon on the mRNA; codon-anticodon binding results in delivery of the correct amino acid to the site of protein synthesis

antigen (18.1)   any substance that is able to stimulate the immune system; generally a protein or large carbohydrate

antiparallel strands (20.2)   a term describing the polarities of the two strands of the DNA double helix; on one strand the sugar-phosphate backbone advances in the $5' \rightarrow 3'$ direction; on the opposite, complementary strand the sugar-phosphate backbone advances in the $3' \rightarrow 5'$ direction

apoenzyme (19.7)   the protein portion of an enzyme that requires a cofactor to function in catalysis

aqueous solution (6.1)   any solution in which the solvent is water

arachidonic acid (17.2)   a fatty acid derived from linoleic acid; the precursor of the prostaglandins

aromatic hydrocarbon (10.1, 11.6)   an organic compound that contains the benzene ring or a derivative of the benzene ring

Arrhenius theory (8.1)   a theory that describes an acid as a substance that dissociates to produce H^+ and a base as a substance that dissociates to produce OH^-

artificial radioactivity (9.5)   radiation that results from the conversion of a stable nucleus to another, unstable nucleus

atherosclerosis (17.4)   deposition of excess plasma cholesterol and other lipids and proteins on the walls of arteries, resulting in decreased artery diameter and increased blood pressure

atom (2.1)   the smallest unit of an element that retains the properties of that element

atomic mass (2.1)   the mass of an atom expressed in atomic mass units

atomic mass unit (4.1) $1/12$ of the mass of a ^{12}C atom, equivalent to 1.661×10^{-24} g

atomic number (2.1)   the number of protons in the nucleus of an atom; it is a characteristic identifier of an element

atomic orbital (2.3, 2.5)   a specific region of space where an electron may be found

ATP synthase (22.6)   a multiprotein complex within the inner mitochondrial membrane that uses the energy of the proton (H^+) gradient to produce ATP

autoionization (8.1)   also known as *self-ionization*, the reaction of a substance, such as water, with itself to produce a positive and a negative ion

Avogadro's law (5.1)   a law that states that the volume is directly proportional to the number of moles of gas particles, assuming that the pressure and temperature are constant

Avogadro's number (4.1) 6.022×10^{23} particles of matter contained in 1 mol of a substance

axial atom (10.4)   an atom that lies above or below a cycloalkane ring

B

background radiation (9.6)   the radiation that emanates from natural sources

barometer (5.1)   a device for measuring pressure

base (8.1)   a substance that behaves as a proton acceptor

base pair (20.2)   a hydrogen-bonded pair of bases within the DNA double helix; the standard base pairs always involve a purine and a pyrimidine; in particular, adenine always base pairs with thymine and cytosine with guanine

Benedict's reagent (16.4)   a buffered solution of Cu^{2+} ions that can be used to test for reducing sugars or to distinguish between aldehydes and ketones

Benedict's test (13.4)   a test used to determine the presence of reducing sugars or to distinguish between aldehydes and ketones; it requires a buffered solution of Cu^{2+} ions that are reduced to Cu^+, which precipitates as brick-red Cu_2O

beta particle (9.1)   an electron formed in the nucleus by the conversion of a neutron into a proton

bile (23.1)   micelles of lecithin, cholesterol, bile salts, protein, inorganic ions, and bile pigments that aid in lipid digestion by emulsifying fat droplets

binding energy (9.3)   the energy required to break down the nucleus into its component parts

bioinformatics (20.10)   an interdisciplinary field that uses computer information sciences and DNA technology to devise methods for understanding, analyzing, and applying DNA sequence information

boat conformation (10.4)   a form of a six-member cycloalkane that resembles a rowboat. It is less stable than the chair conformation because the hydrogen atoms are not perfectly staggered

boiling point (3.3)   the temperature at which the vapor pressure of a liquid is equal to the atmospheric pressure

bond energy (3.4) the amount of energy necessary to break a chemical bond

Boyle's law (5.1) a law stating that the volume of a gas varies inversely with the pressure exerted if the temperature and number of moles of gas are constant

breeder reactor (9.4) a nuclear reactor that produces its own fuel in the process of providing electrical energy

Brønsted-Lowry theory (8.1) a theory that describes an acid as a proton donor and a base as a proton acceptor

buffer capacity (8.4) a measure of the ability of a solution to resist large changes in pH when a strong acid or strong base is added

buffer solution (8.4) a solution containing a weak acid or base and its salt (the conjugate base or acid) that is resistant to large changes in pH upon addition of strong acids or bases

buret (8.3) a device calibrated to deliver accurately known volumes of liquid, as in a titration

C

C-terminal amino acid (18.3) the amino acid in a peptide that has a free α-CO_2^- group; the last amino acid in a peptide

calorimetry (7.2) the measurement of heat energy changes during a chemical reaction

cap structure (20.4) a 7-methylguanosine unit covalently bonded to the 5′ end of a mRNA by a 5′–5′ triphosphate bridge

carbinol carbon (12.4) that carbon in an alcohol to which the hydroxyl group is attached

carbohydrate (16.1) generally sugars and polymers of sugars; the primary source of energy for the cell

carbonyl group (13: Intro) the functional group that contains a carbon-oxygen double bond: —C=O; the functional group found in aldehydes and ketones

carboxyl group (14.1) the —COOH functional group; the functional group found in carboxylic acids

carboxylic acid (14.1) a member of the family of organic compounds that contain the —COOH functional group

carboxylic acid derivative (14.2) any of several families of organic compounds, including the esters and amides, that are derived from carboxylic acids and have the general formula

Z = —OR or OAr for the esters, and
Z = —NH$_2$ for the amides

carcinogen (20.7) any chemical or physical agent that causes mutations in the DNA that lead to uncontrolled cell growth or cancer

catabolism (21.1, 22.9) the degradation of fuel molecules and production of ATP for cellular functions

catalyst (7.3) any substance that increases the rate of a chemical reaction (by lowering the activation energy of the reaction) and that is not destroyed in the course of the reaction

cathode (8.5) the negatively charged electrode in an electrical cell

cathode rays (2.2) a stream of electrons that is given off by the cathode (negative electrode) in a cathode ray tube

cation (2.1) a positively charged atom or group of atoms

cellulose (16.6) a polymer of β-D-glucose linked by β(1 → 4) glycosidic bonds

central dogma (20.4) a statement of the directional transfer of the genetic information in cells: DNA → RNA → Protein

chain reaction (9.4) the process in a fission reactor that involves neutron production and causes subsequent reactions accompanied by the production of more neutrons in a continuing process

chair conformation (10.4) the most energetically favorable conformation for a six-member cycloalkane; so-called for its resemblance to a lawn chair

Charles's law (5.1) a law stating that the volume of a gas is directly proportional to the temperature of the gas, assuming that the pressure and number of moles of the gas are constant

chemical bond (3.1) the attractive force holding two atomic nuclei together in a chemical compound

chemical equation (4.3) a record of chemical change, showing the conversion of reactants to products

chemical formula (4.2) the representation of a compound or ion in which elemental symbols represent types of atoms and subscripts show the relative numbers of atoms

chemical property (1.2) characteristics of a substance that relate to the substance's participation in a chemical reaction

chemical reaction (1.2) a process in which atoms are rearranged to produce new combinations

chemistry (1.1) the study of matter and the changes that matter undergoes

chiral carbon (16.3) a carbon atom bonded to four different atoms or groups of atoms

chiral molecule (16.3) molecule capable of existing in mirror-image forms

cholesterol (17.4) a twenty-seven-carbon steroid ring structure that serves as the precursor of the steroid hormones

chromosome (20.2) a piece of DNA that carries the genetic instructions, or genes, of an organism

chylomicron (17.5, 23.1) a plasma lipoprotein (aggregate of protein and triglycerides) that carries triglycerides from the intestine to all body tissues via the bloodstream

cis-trans isomers (10.3) isomers that differ from one another in the placement of substituents on a double bond or ring

citric acid cycle (22.4) a cyclic biochemical pathway that is the final stage of degradation of carbohydrates, fats, and amino acids. It results in the complete oxidation of acetyl groups derived from these dietary fuels

cloning vector (20.8) a DNA molecule that can carry a cloned DNA fragment into a cell and that has a replication origin that allows the DNA to be replicated abundantly within the host cell

coagulation (18.10) the process by which proteins in solution are denatured and aggregate with one another to produce a solid

codon (20.4) a group of three ribonucleotides on the mRNA that specifies the addition of a specific amino acid onto the growing peptide chain

coenzyme (19.7) an organic group required by some enzymes; it generally serves as a donor or acceptor of electrons or a functional group in a reaction

coenzyme A (22.2) a molecule derived from ATP and the vitamin pantothenic acid; coenzyme A functions in the transfer of acetyl groups in lipid and carbohydrate metabolism

cofactor (19.7) an inorganic group, usually a metal ion, that must be bound to an apoenzyme to maintain the correct configuration of the active site

colipase (23.1) a protein that aids in lipid digestion by binding to the surface of lipid droplets and facilitating binding of pancreatic lipase

colligative property (6.4) property of a solution that is dependent only on the concentration of solute particles

colloidal suspension (6.1) a heterogeneous mixture of solute particles in a solvent; distribution of solute particles is not uniform because of the size of the particles

combination reaction (4.3) a reaction in which two substances join to form another substance

combined gas law (5.1) an equation that describes the behavior of a gas when volume, pressure, and temperature may change simultaneously

combustion (10.5) the oxidation of hydrocarbons by burning in the presence of air to produce carbon dioxide and water

competitive inhibitor (19.10) a structural analog; a molecule that has a structure very similar to the natural substrate of an enzyme, competes with the natural substrate for binding to the enzyme active site, and inhibits the reaction

complementary strands (20.2) the opposite strands of the double helix are hydrogen-bonded to one another such that adenine and thymine or guanine and cytosine are always paired

complete protein (18.11)   a protein source that contains all the essential and nonessential amino acids

complex lipid (17.5)   a lipid bonded to other types of molecules

compound (1.2)   a substance that is characterized by constant composition and that can be chemically broken down into elements

concentration (1.5, 6.2)   a measure of the quantity of a substance contained in a specified volume of solution

concentration gradient (6.4)   region where concentration decreases over distance

condensation (5.2)   the conversion of a gas to a liquid

condensation polymer (14.2)   a polymer, which is a large molecule formed by combination of many small molecules (monomers) that results from joining of monomers in a reaction that forms a small molecule, such as water or an alcohol

condensed formula (10.2)   a structural formula showing all of the atoms in a molecule and placing them in a sequential arrangement that details which atoms are bonded to each other; the bonds themselves are not shown

conformations, conformers (10.4)   discrete, distinct isomeric structures that may be converted, one to the other, by rotation about the bonds in the molecule

conjugate acid (8.1)   substance that has one more proton than the base from which it is derived

conjugate acid-base pair (8.1)   two species related to each other through the gain or loss of a proton

conjugate base (8.1)   substance that has one less proton than the acid from which it is derived

constitutional isomers (10.2)   two molecules having the same molecular formulas, but different chemical structures

Cori Cycle (21.6)   a metabolic pathway in which the lactate produced by working muscle is taken up by cells in the liver and converted back to glucose by gluconeogenesis

corrosion (8.5)   the unwanted oxidation of a metal

covalent bond (3.1)   a pair of electrons shared between two atoms

covalent solid (5.3)   a collection of atoms held together by covalent bonds

cristae (22.1)   the folds of the inner membrane of the mitochondria

crystal lattice (3.2)   a unit of a solid characterized by a regular arrangement of components

crystalline solid (5.3)   a solid having a regular repeating atomic structure

curie (9.7)   the quantity of radioactive material that produces 3.7×10^{10} nuclear disintegrations per second

cycloalkane (10.3)   a cyclic alkane; a saturated hydrocarbon that has the general formula C_nH_{2n}

D

Dalton's law (5.1)   also called the law of partial pressures; states that the total pressure exerted by a gas mixture is the sum of the partial pressures of the component gases

data (1.1)   a group of facts resulting from an experiment

decomposition reaction (4.3)   the breakdown of a substance into two or more substances

defense proteins (18.1)   proteins that defend the body against infectious diseases. Antibodies are defense proteins

degenerate code (20.5)   a term used to describe the fact that several triplet codons may be used to specify a single amino acid in the genetic code

dehydration (of alcohols) (12.5)   a reaction that involves the loss of a water molecule, in this case the loss of water from an alcohol and the simultaneous formation of an alkene

deletion mutation (20.7)   a mutation that results in the loss of one or more nucleotides from a DNA sequence

denaturation (18.10)   the process by which the organized structure of a protein is disrupted, resulting in a completely disorganized, nonfunctional form of the protein

density (1.5)   mass per unit volume of a substance

deoxyribonucleic acid (DNA) (20.1)   the nucleic acid molecule that carries all of the genetic information of an organism; the DNA molecule is a double helix composed of two strands, each of which is composed of phosphate groups, deoxyribose, and the nitrogenous bases thymine, cytosine, adenine, and guanine

deoxyribonucleotide (20.1)   a nucleoside phosphate or nucleotide composed of a nitrogenous base in β-N-glycosidic linkage to the 1' carbon of the sugar 2'-deoxyribose and with one, two, or three phosphoryl groups esterified at the hydroxyl of the 5' carbon

diabetes mellitus (23.3)   a disease caused by the production of insufficient levels of insulin and characterized by the appearance of very high levels of glucose in the blood and urine

dialysis (6.6)   the removal of waste material via transport across a membrane

diffusion (6.4)   net movement of solute or solvent molecules from a region of high concentration to a region of low concentration

diglyceride (17.3)   the product of esterification of glycerol at two positions

dipole-dipole interactions (5.2)   attractive forces between polar molecules

disaccharide (16.1)   a sugar composed of two monosaccharides joined through an oxygen atom bridge

dissociation (3.3)   production of positive and negative ions when an ionic compound dissolves in water

disulfide (12.9)   an organic compound that contains a disulfide group (—S—S—)

DNA polymerase III (20.3)   the enzyme that catalyzes the polymerization of daughter DNA strands using the parental strand as a template

double bond (3.4)   a bond in which two pairs of electrons are shared by two atoms

double helix (20.2)   the spiral staircase-like structure of the DNA molecule characterized by two sugar-phosphate backbones wound around the outside and nitrogenous bases extending into the center

double-replacement reaction (4.3)   a chemical change in which cations and anions "exchange partners"

dynamic equilibrium (7.4)   the state that exists when the rate of change in the concentration of products and reactants is equal, resulting in no net concentration change

E

eicosanoid (17.2)   any of the derivatives of twenty-carbon fatty acids, including the prostaglandins, leukotrienes, and thromboxanes

electrolysis (8.5)   an electrochemical process that uses electrical energy to cause nonspontaneous oxidation-reduction reactions to occur

electrolyte (3.3, 6.1)   a material that dissolves in water to produce a solution that conducts an electrical current

electrolytic solution (3.3)   a solution composed of an electrolytic solute dissolved in water

electromagnetic radiation (2.3)   energy that is propagated as waves at the speed of light

electromagnetic spectrum (2.3)   the complete range of electromagnetic waves

electron (2.1)   a negatively charged particle outside of the nucleus of an atom

electron affinity (2.7)   the energy released when an electron is added to an isolated atom

electron configuration (2.5)   the arrangement of electrons around a nucleus of an atom, ion, or a collection of nuclei of a molecule

electron density (2.3)   the probability of finding the electron in a particular location

electron transport system (22.6)   the series of electron transport proteins embedded in the inner mitochondrial membrane that accept high-energy electrons from NADH and FADH$_2$ and transfer them in stepwise fashion to molecular oxygen (O$_2$)

electronegativity (3.1) a measure of the tendency of an atom in a molecule to attract shared electrons

element (1.2) a substance that cannot be decomposed into simpler substances by chemical or physical means

elimination reaction (12.5) a reaction in which a molecule loses atoms or ions from its structure

elongation factor (20.6) proteins that facilitate the elongation phase of translation

emulsifying agent (17.3) a bipolar molecule that aids in the suspension of fats in water

enantiomers (16.3) stereoisomers that are nonsuperimposable mirror images of one another

endothermic reaction (7.1) a chemical or physical change in which energy is absorbed

energy (1.1) the capacity to do work

energy level (2.3) one of numerous atomic regions where electrons may be found

enthalpy (7.1) a term that represents heat energy

entropy (7.1) a measure of randomness or disorder

enzyme (18.1, 19: Intro) a protein that serves as a biological catalyst

enzyme specificity (19.5) the ability of an enzyme to bind to only one, or a very few, substrates and thus catalyze only a single reaction

enzyme-substrate complex (19.4) a molecular aggregate formed when the substrate binds to the active site of the enzyme

equatorial atom (10.4) an atom that lies in the plane of a cycloalkane ring

equilibrium constant (7.4) number equal to the ratio of the equilibrium concentrations of products to the equilibrium concentrations of reactants, each raised to the power corresponding to its coefficient in the balanced equation

equilibrium reaction (7.4) a reaction that is reversible and the rates of the forward and reverse reactions are equal

equivalence point (8.3) the situation in which reactants have been mixed in the molar ratio corresponding to the balanced equation

equivalent (6.3) the number of grams of an ion corresponding to Avogadro's number of electrical charges

error (1.3) the difference between the true value and the experimental value for data or results

essential amino acid (18.11) an amino acid that cannot be synthesized by the body and must therefore be supplied by the diet

essential fatty acids (17.2) the fatty acids linolenic and linoleic acids that must be supplied in the diet because they cannot be synthesized by the body

ester (14.2) a carboxylic acid derivative formed by the reaction of a carboxylic acid and an alcohol. Esters have the following general formula:

esterification (17.2) the formation of an ester in the reaction of a carboxylic acid and an alcohol

ether (12.8) an organic compound that contains two alkyl and/or aryl groups attached to an oxygen atom; R—O—R, Ar—O—R, and Ar—O—Ar

eukaryote (20.2) an organism having cells containing a true nucleus enclosed by a nuclear membrane and having a variety of membrane-bound organelles that segregate different cellular functions into different compartments

evaporation (5.2) the conversion of a liquid to a gas below the boiling point of the liquid

exon (20.4) protein-coding sequences of a gene found on the final mature mRNA

exothermic reaction (7.1) a chemical or physical change that releases energy

extensive property (1.2) a property of a substance that depends on the quantity of the substance

F

F$_0$F$_1$ complex (22.6) an alternative term for the ATP synthase, the multiprotein complex in the inner mitochondrial membrane that uses the energy of the proton gradient to produce ATP

fatty acid (14.1, 17.2) any member of the family of continuous-chain carboxylic acids that generally contain four to twenty carbon atoms; the most concentrated source of energy used by the cell

feedback inhibition (19.9) the process whereby excess product of a biosynthetic pathway turns off the entire pathway for its own synthesis

fermentation (12.3, 21.4) anaerobic (in the absence of oxygen) catabolic reactions that occur with no net oxidation. Pyruvate or an organic compound produced from pyruvate is reduced as NADH is oxidized

fibrous protein (18.5) a protein composed of peptides arranged in long sheets or fibers

Fischer Projection (16.3) a two-dimensional drawing of a molecule, which shows a chiral carbon at the intersection of two lines and horizontal lines representing bonds projecting out of the page and vertical lines representing bonds that project into the page

fission (9.4) the splitting of heavy nuclei into lighter nuclei accompanied by the release of large quantities of energy

fluid mosaic model (17.6) the model of membrane structure that describes the fluid

nature of the lipid bilayer and the presence of numerous proteins embedded within the membrane

formula (3.2) the representation of the fundamental compound unit using chemical symbols and numerical subscripts

formula unit (4.2) the smallest collection of atoms from which the formula of a compound can be established

formula weight (4.2) the mass of a formula unit of a compound relative to a standard (carbon-12)

free energy (7.1) the combined contribution of entropy and enthalpy for a chemical reaction

fructose (16.4) a ketohexose that is also called levulose and fruit sugar; the sweetest of all sugars, abundant in honey and fruits

fuel value (7.2) the amount of energy derived from a given mass of material

functional group (10.1) an atom (or group of atoms and their bonds) that imparts specific chemical and physical properties to a molecule

fusion (9.4) the joining of light nuclei to form heavier nuclei, accompanied by the release of large amounts of energy

G

galactose (16.4) an aldohexose that is a component of lactose (milk sugar)

galactosemia (16.5) a human genetic disease caused by the inability to convert galactose to a phosphorylated form of glucose (glucose-1-phosphate) that can be used in cellular metabolic reactions

gamma ray (9.1) a high-energy emission from nuclear processes, traveling at the speed of light; the high-energy region of the electromagnetic spectrum

gaseous state (1.2) a physical state of matter characterized by a lack of fixed shape or volume and ease of compressibility

genome (20.2) the complete set of genetic information in all the chromosomes of an organism

geometric isomer (10.3, 11.3) an isomer that differs from another isomer in the placement of substituents on a double bond or a ring

globular protein (18.6) a protein composed of polypeptide chains that are tightly folded into a compact spherical shape

glucagon (21.7, 23.6) a peptide hormone synthesized by the α-cells of the islets of Langerhans in the pancreas and secreted in response to low blood glucose levels; glucagon promotes glycogenolysis and gluconeogenesis and thereby increases the concentration of blood glucose

gluconeogenesis (21.6) the synthesis of glucose from noncarbohydrate precursors

glucose (16.4) an aldohexose, the most abundant monosaccharide; it is a component of many disaccharides, such as lactose and sucrose, and of polysaccharides, such as cellulose, starch, and glycogen

glyceraldehyde (16.3)   an aldotriose that is the simplest carbohydrate; phosphorylated forms of glyceraldehyde are important intermediates in cellular metabolic reactions

glyceride (17.3)   a lipid that contains glycerol

glycogen (16.6, 21.7)   a long, branched polymer of glucose stored in liver and muscles of animals; it consists of a linear backbone of α-D-glucose in α(1 → 4) linkage, with numerous short branches attached to the C-6 hydroxyl group by α(1 → 6) linkage

glycogenesis (21.7)   the metabolic pathway that results in the addition of glucose to growing glycogen polymers when blood glucose levels are high

glycogen granule (21.7)   a core of glycogen surrounded by enzymes responsible for glycogen synthesis and degradation

glycogenolysis (21.7)   the biochemical pathway that results in the removal of glucose molecules from glycogen polymers when blood glucose levels are low

glycolysis (21.3)   the enzymatic pathway that converts a glucose molecule into two molecules of pyruvate; this anaerobic process generates a net energy yield of two molecules of ATP and two molecules of NADH

glycoprotein (18.7)   a protein bonded to sugar groups

glycosidic bond (16.1)   the bond between the hydroxyl group of the C-1 carbon of one sugar and a hydroxyl group of another sugar

group (2.4)   any one of eighteen vertical columns of elements; often referred to as a family

group specificity (19.5)   an enzyme that catalyzes reactions involving similar substrate molecules having the same functional groups

guanosine triphosphate (GTP) (21.6)   a nucleotide composed of the purine guanosine, the sugar ribose, and three phosphoryl groups

H

half-life ($t_{1/2}$) (9.3)   the length of time required for one-half of the initial mass of an isotope to decay to products

halogen (2.4)   an element found in Group VIIA (17) of the periodic table

halogenation (10.5, 11.5)   a reaction in which one of the C—H bonds of a hydrocarbon is replaced with a C—X bond (X = Br or Cl generally)

Haworth projection (16.4)   a means of representing the orientation of substituent groups around a cyclic sugar molecule

α-helix (18.5)   a right-handed coiled secondary structure maintained by hydrogen bonds

between the amide hydrogen of one amino acid and the carbonyl oxygen of an amino acid four residues away

heme group (18.9)   the chemical group found in hemoglobin and myoglobin that is responsible for the ability to carry oxygen

hemiacetal (13.4, 16.4)   the family of organic compounds formed via the reaction of one molecule of alcohol with an aldehyde in the presence of an acid catalyst; hemiacetals have the following general structure:

$$\begin{array}{c} OH \\ | \\ R^1\!\!-\!\!C\!\!-\!\!OR^2 \\ | \\ H \end{array}$$

hemiketal (13.4, 16.4)   the family of organic compounds formed via the reaction of one molecule of alcohol with a ketone in the presence of an acid catalyst; hemiketals have the following general structure:

$$\begin{array}{c} OH \\ | \\ R^1\!\!-\!\!C\!\!-\!\!OR^3 \\ | \\ R^2 \end{array}$$

hemoglobin (18.9)   the major protein component of red blood cells; the function of this red, iron-containing protein is transport of oxygen

Henderson-Hasselbalch equation (8.4)   an equation for calculating the pH of a buffer system:

$$pH = pKa + \log\frac{[\text{conjugate base}]}{[\text{weak acid}]}$$

Henry's law (6.1)   a law stating that the number of moles of a gas dissolved in a liquid at a given temperature is proportional to the partial pressure of the gas

heterocyclic amine (15.2)   a heterocyclic compound that contains nitrogen in at least one position in the ring skeleton

heterocyclic aromatic compound (11.7)   cyclic aromatic compound having at least one atom other than carbon in the structure of the aromatic ring

heterogeneous mixture (1.2)   a mixture of two or more substances characterized by nonuniform composition

hexose (16.2)   a six-carbon monosaccharide

high-density lipoprotein (HDL) (17.5)   a plasma lipoprotein that transports cholesterol from peripheral tissue to the liver

holoenzyme (19.7)   an active enzyme consisting of an apoenzyme bound to a cofactor

homogeneous mixture (1.2)   a mixture of two or more substances characterized by uniform composition

hybridization (20.8)   a technique for identifying DNA or RNA sequences that is based on specific hydrogen bonding between a radioactive probe and complementary DNA or RNA sequences

hydrate (4.2)   any substance that has water molecules incorporated in its structure

hydration (11.5, 12.5)   a reaction in which water is added to a molecule, e.g., the addition of water to an alkene to form an alcohol

hydrocarbon (10.1)   a compound composed solely of the elements carbon and hydrogen

hydrogen bonding (5.2)   the attractive force between a hydrogen atom covalently bonded to a small, highly electronegative atom and another atom containing an unshared pair of electrons

hydrogenation (11.5, 13.4, 17.2)   a reaction in which hydrogen (H_2) is added to a double or a triple bond

hydrohalogenation (11.5)   the addition of a hydrohalogen (HCl, HBr, or HI) to an unsaturated bond

hydrolase (19.1)   an enzyme that catalyzes hydrolysis reactions

hydrolysis (14.2)   a chemical change that involves the reaction of a molecule with water; the process by which molecules are broken into their constituents by addition of water

hydronium ion (8.1)   a protonated water molecule, H_3O^+

hydrophilic amino acid (18.1)   "water loving"; a polar or ionic amino acid that has a high affinity for water

hydrophobic amino acid (18.2)   "water fearing"; a nonpolar amino acid that prefers contact with other nonpolar molecules over contact with water

hydroxyl group (12.1)   the —OH functional group that is characteristic of alcohols

hyperammonemia (22.8)   a genetic defect in one of the enzymes of the urea cycle that results in toxic or even fatal elevation of the concentration of ammonium ions in the body

hyperglycemia (21.7)   blood glucose levels that are higher than normal

hypertonic solution (6.4)   the more concentrated solution of two separated by a semipermeable membrane

hypoglycemia (21.7)   blood glucose levels that are lower than normal

hypothesis (1.1)   an attempt to explain observations in a commonsense way

hypotonic solution (6.4)   the more dilute solution of two separated by a semipermeable membrane

I

ideal gas (5.1)   a gas in which the particles do not interact and the volume of the individual gas particles is assumed to be negligible

ideal gas law (5.1)   a law stating that for an ideal gas the product of pressure and volume is proportional to the product of the number of moles of the gas and its temperature; the proportionality constant for an ideal gas is symbolized R

incomplete protein (18.11) a protein source that does not contain all the essential and nonessential amino acids

indicator (8.3) a solute that shows some condition of a solution (such as acidity or basicity) by its color

induced fit model (19.4) the theory of enzyme-substrate binding that assumes that the enzyme is a flexible molecule and that both the substrate and the enzyme change their shapes to accommodate one another as the enzyme-substrate complex forms

initiation factors (20.6) proteins that are required for formation of the translation initiation complex, which is composed of the large and small ribosomal subunits, the mRNA, and the initiator tRNA, methionyl tRNA

inner mitochondrial membrane (22.1) the highly folded, impermeable membrane within the mitochondrion that is the location of the electron transport system and ATP synthase

insertion mutation (20.7) a mutation that results in the addition of one or more nucleotides to a DNA sequence

insulin (21.7, 23.6) a hormone released from the pancreas in response to high blood glucose levels; insulin stimulates glycogenesis, fat storage, and cellular uptake and storage of glucose from the blood

intensive property (1.2) a property of a substance that is independent of the quantity of the substance

intermembrane space (22.1) the region between the outer and inner mitochondrial membranes, which is the location of the proton (H^+) reservoir that drives ATP synthesis

intermolecular force (3.5) any attractive force that occurs between molecules

intramolecular force (3.5) any attractive force that occurs within molecules

intron (20.4) a noncoding sequence within a eukaryotic gene that must be removed from the primary transcript to produce a functional mRNA

ion (2.1) an electrically charged particle formed by the gain or loss of electrons

ionic bonding (3.1) an electrostatic attractive force between ions resulting from electron transfer

ionic solid (5.3) a solid composed of positive and negative ions in a regular three-dimensional crystalline arrangement

ionization energy (2.7) the energy needed to remove an electron from an atom in the gas phase

ionizing radiation (9.1) radiation that is sufficiently high in energy to cause ion formation upon impact with an atom

ion pair (3.1) the simplest formula unit for an ionic compound

ion product for water (8.1) the product of the hydronium and hydroxide ion concentrations in pure water at a specified temperature; at 25°C, it has a value of 1.0×10^{-14}

irreversible enzyme inhibitor (19.10) a chemical that binds strongly to the R groups of an amino acid in the active site and eliminates enzyme activity

isoelectric point (18.10) a situation in which a protein has an equal number of positive and negative charges and therefore has an overall net charge of zero

isoelectronic (2.6) atoms, ions, and molecules containing the same number of electrons

isomerase (19.1) an enzyme that catalyzes the conversion of one isomer to another

isomers (3.4) molecules having the same molecular formula but different chemical structures

isotonic solution (6.4) a solution that has the same solute concentration as another solution with which it is being compared; a solution that has the same osmotic pressure as a solution existing within a cell

isotope (2.1) atom of the same element that differs in mass because it contains different numbers of neutrons

I.U.P.A.C. Nomenclature System (10.2) the International Union of Pure and Applied Chemistry (I.U.P.A.C.) standard, universal system for the nomenclature of organic compounds

K

α-keratin (18.5) a member of the family of fibrous proteins that form the covering of most land animals; major components of fur, skin, beaks, and nails

ketal (13.4) the family of organic compounds formed via the reaction of two molecules of alcohol with a ketone in the presence of an acid catalyst; ketals have the following general structure:

$$
\begin{array}{c}
OR^3 \\
| \\
R^1\!-\!C\!-\!OR^4 \\
| \\
R^2
\end{array}
$$

ketoacidosis (23.3) a drop in the pH of the blood caused by elevated levels of ketone bodies

ketone (13.1) a family of organic molecules characterized by a carbonyl group; the carbonyl carbon is bonded to two alkyl groups, two aryl groups, or one alkyl and one aryl group; ketones have the following general structures:

ketone bodies (23.3) acetone, acetoacetone, and β-hydroxybutyrate produced from fatty acids in the liver via acetyl CoA

ketose (16.2) a sugar that contains a ketone (carbonyl) group

ketosis (23.3) an abnormal rise in the level of ketone bodies in the blood

kinetic energy (1.5) the energy resulting from motion of an object [kinetic energy = 1/2(mass)(velocity)²]

kinetic-molecular theory (5.1) the fundamental model of particle behavior in the gas phase

kinetics (7.3) the study of rates of chemical reactions

L

lactose (16.5) a disaccharide composed of β-D-galactose and either α-or β-D-glucose in β(1 → 4) glycosidic linkage; milk sugar

lactose intolerance (16.5) the inability to produce the digestive enzyme lactase, which degrades lactose to galactose and glucose

lagging strand (20.3) in DNA replication, the strand that is synthesized discontinuously from numerous RNA primers

law (1.1) a summary of a large quantity of information

law of conservation of mass (4.3) a law stating that, in chemical change, matter cannot be created or destroyed

leading strand (20.3) in DNA replication, the strand that is synthesized continuously from a single RNA primer

LeChatelier's principle (7.4) a law stating that when a system at equilibrium is disturbed, the equilibrium shifts in the direction that minimizes the disturbance

lethal dose (LD$_{50}$) (9.7) the quantity of toxic material (such as radiation) that causes the death of 50% of a population of an organism

Lewis symbol (3.1) representation of an atom or ion using the atomic symbol (for the nucleus and core electrons) and dots to represent valence electrons

ligase (19.1) an enzyme that catalyzes the joining of two molecules

linear structure (3.4) the structure of a molecule in which the bond angles about the central atom(s) is (are) 180°

line formula (10.2) the simplest representation of a molecule in which it is assumed that there is a carbon atom at any location where two or more lines intersect, there is a carbon at the end of any line, and each carbon is bonded to the correct number of hydrogen atoms

linkage specificity (19.5) the property of an enzyme that allows it to catalyze reactions involving only one kind of bond in the substrate molecule

lipase (23.1) an enzyme that hydrolyzes the ester linkage between glycerol and the fatty acids of triglycerides

lipid (17.1) a member of the group of biological molecules of varying composition that are classified together on the basis of their solubility in nonpolar solvents

liquid state (1.2)   a physical state of matter characterized by a fixed volume and the absence of a fixed shape

lock-and-key model (19.4)   the theory of enzyme-substrate binding that depicts enzymes as inflexible molecules; the substrate fits into the rigid active site in the same way a key fits into a lock

London forces (5.2)   weak attractive forces between molecules that result from short-lived dipoles that occur because of the continuous movement of electrons in the molecules

lone pair (3.4)   an electron pair that is not involved in bonding

low-density lipoprotein (LDL) (17.5)   a plasma lipoprotein that carries cholesterol to peripheral tissues and helps to regulate cholesterol levels in those tissues

lyase (19.1)   an enzyme that catalyzes a reaction involving double bonds

M

maltose (16.5)   a disaccharide composed of α-D-glucose and a second glucose molecule in α(1 → 4) glycosidic linkage

Markovnikov's rule (11.5)   the rule stating that a hydrogen atom, adding to a carbon-carbon double bond, will add to the carbon having the larger number of hydrogens attached to it

mass (1.5)   a quantity of matter

mass number (2.1)   the sum of the number of protons and neutrons in an atom

matrix space (22.1)   the region of the mitochondrion within the inner membrane; the location of the enzymes that carry out the reactions of the citric acid cycle and β-oxidation of fatty acids

matter (1.1)   the material component of the universe

melting point (3.3, 5.3)   the temperature at which a solid converts to a liquid

messenger RNA (20.4)   an RNA species produced by transcription and that specifies the amino acid sequence for a protein

metal (2.4)   an element located on the left side of the periodic table (left of the "staircase" boundary)

metallic bond (5.3)   a bond that results from the orbital overlap of metal atoms

metallic solid (5.3)   a solid composed of metal atoms held together by metallic bonds

metalloid (2.4)   an element along the "staircase" boundary between metals and nonmetals; metalloids exhibit both metallic and nonmetallic properties

metastable isotope (9.2)   an isotope that will give up some energy to produce a more stable form of the same isotope

micelle (23.1)   an aggregation of molecules having nonpolar and polar regions; the nonpolar regions of the molecules aggregate, leaving the polar regions facing the surrounding water

mitochondria (22.1)   the cellular "power plants" in which the reactions of the citric acid cycle, the electron transport system, and ATP synthase function to produce ATP

mixture (1.2)   a material composed of two or more substances

molality (6.4)   the number of moles of solute per kilogram of solvent

molar mass (4.1)   the mass in grams of 1 mol of a substance

molar volume (5.1)   the volume occupied by 1 mol of a substance

molarity (6.3)   the number of moles of solute per liter of solution

mole (4.1)   the amount of substance containing Avogadro's number of particles

molecular formula (10.2)   a formula that provides the atoms and number of each type of atom in a molecule but gives no information regarding the bonding pattern involved in the structure of the molecule

molecular solid (5.3)   a solid in which the molecules are held together by dipole-dipole and London forces (van der Waals forces)

molecule (3.2)   a unit in which the atoms of two or more elements are held together by chemical bonds

monatomic ion (3.2)   an ion formed by electron gain or loss from a single atom

monoglyceride (17.3)   the product of the esterification of glycerol at one position

monomer (11.5)   the individual molecules from which a polymer is formed

monosaccharide (16.1)   the simplest type of carbohydrate consisting of a single saccharide unit

movement protein (18.1)   a protein involved in any aspect of movement in an organism, for instance actin and myosin in muscle tissue and flagellin that composes bacterial flagella

mutagen (20.7)   any chemical or physical agent that causes changes in the nucleotide sequence of a gene

mutation (20.7)   any change in the nucleotide sequence of a gene

myoglobin (18.9)   the oxygen storage protein found in muscle

N

N-terminal amino acid (18.3)   the amino acid in a peptide that has a free α-N^+H_3 group; the first amino acid of a peptide

natural radioactivity (9.5)   the spontaneous decay of a nucleus to produce high-energy particles or rays

negative allosterism (19.9)   effector binding inactivates the active site of an allosteric enzyme

neurotransmitter (15.5)   a chemical that carries a message, or signal, from a nerve cell to a target cell

neutral glyceride (17.3)   the product of the esterification of glycerol at one, two, or three positions

neutralization (8.3)   the reaction between an acid and a base

neutron (2.1)   an uncharged particle, with the same mass as the proton, in the nucleus of an atom

nicotinamide adenine dinucleotide (NAD^+) (21.3)   a molecule synthesized from the vitamin niacin and the nucleotide ATP and that serves as a carrier of hydride anions; a coenzyme that is an oxidizing agent used in a variety of metabolic processes

noble gas (2.4)   elements in Group VIIIA (18) of the periodic table

nomenclature (3.2)   a system for naming chemical compounds

nonelectrolyte (3.3, 6.1)   a substance that, when dissolved in water, produces a solution that does not conduct an electrical current

nonessential amino acid (18.11)   any amino acid that can be synthesized by the body

nonmetal (2.4)   an element located on the right side of the periodic table (right of the "staircase" boundary)

nonreducing sugar (16.5)   a sugar that cannot be oxidized by Benedict's or Tollens' reagent

normal boiling point (5.2)   the temperature at which a substance will boil at 1 atm of pressure

nuclear equation (9.2)   a balanced equation accounting for the products and reactants in a nuclear reaction

nuclear imaging (9.5)   the generation of images of components of the body (organs, tissues) using techniques based on the measurement of radiation

nuclear medicine (9.5)   a field of medicine that uses radioisotopes for diagnostic and therapeutic purposes

nuclear reactor (9.5)   a device for conversion of nuclear energy into electrical energy

nucleosome (20.2)   the first level of chromosome structure consisting of a strand of DNA wrapped around a small disk of histone proteins

nucleotide (20.1, 21.1)   a molecule composed of a nitrogenous base, a five-carbon sugar, and one, two, or three phosphoryl groups

nucleus (2.1)   the small, dense center of positive charge in the atom

nuclide (9.1) any atom characterized by an atomic number and a mass number

nutrient protein (18.1) a protein that serves as a source of amino acids for embryos or infants

nutritional Calorie (7.2) equivalent to one kilocalorie (1000 calories); also known as a large Calorie

O

octet rule (2.6) a rule predicting that atoms form the most stable molecules or ions when they are surrounded by eight electrons in their highest occupied energy level

oligosaccharide (16.1) an intermediate-sized carbohydrate composed of from three to ten monosaccharides

order of the reaction (7.3) the exponent of each concentration term in the rate equation

osmolarity (6.4) molarity of particles in solution; this value is used for osmotic pressure calculations

osmosis (6.4) net flow of a solvent across a semipermeable membrane in response to a concentration gradient

osmotic pressure (6.4) the net force with which water enters a solution through a semipermeable membrane; alternatively, the pressure required to stop net transfer of solvent across a semipermeable membrane

outer mitochondrial membrane (22.1) the membrane that surrounds the mitochondrion and separates it from the contents of the cytoplasm; it is highly permeable to small "food" molecules

β-oxidation (23.2) the biochemical pathway that results in the oxidation of fatty acids and the production of acetyl CoA

oxidation (8.5, 12.6, 13.4, 14.1) a loss of electrons; in organic compounds it may be recognized as a loss of hydrogen atoms or the gain of oxygen

oxidation-reduction reaction (4.3) also called redox reaction, a reaction involving the transfer of one or more electrons from one reactant to another

oxidative deamination (22.7) an oxidation-reduction reaction in which NAD^+ is reduced and the amino acid is deaminated

oxidative phosphorylation (21.3, 22.6) production of ATP using the energy of electrons harvested during biological oxidation-reduction reactions

oxidizing agent (8.5) a substance that oxidizes, or removes electrons from, another substance; the oxidizing agent is reduced in the process

oxidoreductase (19.1) an enzyme that catalyzes an oxidation-reduction reaction

P

pancreatic serine proteases (19.11) a family of proteolytic enzymes, including trypsin, chymotrypsin, and elastase, that arose by divergent evolution

parent compound or parent chain (10.2) in the I.U.P.A.C. Nomenclature System the parent compound is the longest carbon-carbon chain containing the principal functional group in the molecule that is being named

partial pressure (5.1) the pressure exerted by one component of a gas mixture

particle accelerator (9.5) a device for production of high-energy nuclear particles based on the interaction of charged particles with magnetic and electrical fields

pentose (16.2) a five-carbon monosaccharide

pentose phosphate pathway (21.5) an alternative pathway for glucose degradation that provides the cell with reducing power in the form of NADPH

peptide bond (15.4, 18.3) the amide bond between two amino acids in a peptide chain

peptidyl tRNA binding site of ribosome (P-site) (20.6) a pocket on the surface of the ribosome that holds the tRNA bound to the growing peptide chain

percent yield (4.5) the ratio of the actual and theoretical yields of a chemical reaction multiplied by 100%

period (2.4) any one of seven horizontal rows of elements in the periodic table

periodic law (2.4) a law stating that properties of elements are periodic functions of their atomic numbers (Note that Mendeleev's original statement was based on atomic masses.)

peripheral membrane protein (17.6) a protein bound to either the inner or the outer surface of a membrane

phenol (12.7) an organic compound that contains a hydroxyl group (—OH) attached to a benzene ring

phenyl group (11.6) a benzene ring that has had a hydrogen atom removed, C_6H_5—

pH optimum (19.8) the pH at which an enzyme catalyzes the reaction at maximum efficiency

phosphatidate (17.3) a molecule of glycerol with fatty acids esterified to C-1 and C-2 of glycerol and a free phosphoryl group esterified at C-3

phosphoanhydride (14.4) the bond formed when two phosphate groups react with one another and a water molecule is lost

phosphoester (14.4) the product of the reaction between phosphoric acid and an alcohol

phosphoglyceride (17.3) a molecule with fatty acids esterified at the C-1 and C-2 positions of glycerol and a phosphoryl group esterified at the C-3 position

phospholipid (17.3) a lipid containing a phosphoryl group

phosphopantetheine (23.4) the portion of coenzyme A and the acyl carrier protein that is derived from the vitamin pantothenic acid

pH scale (8.2) a numerical representation of acidity or basicity of a solution; $pH = -\log[H_3O^+]$

physical change (1.2) a change in the form of a substance but not in its chemical composition; no chemical bonds are broken in a physical change

physical property (1.2) a characteristic of a substance that can be observed without the substance undergoing change (examples include color, density, melting and boiling points)

plasma lipoprotein (17.5) a complex composed of lipid and protein that is responsible for the transport of lipids throughout the body

β-pleated sheet (18.5) a common secondary structure of a peptide chain that resembles the pleats of an Oriental fan

point mutation (20.7) the substitution of one nucleotide pair for another within a gene

polar covalent bonding (3.4) a covalent bond in which the electrons are not equally shared

polar covalent molecule (3.4) a molecule that has a permanent electric dipole moment resulting from an unsymmetrical electron distribution; a dipolar molecule

poly(A) tail (20.4) a tract of 100–200 adenosine monophosphate units covalently attached to the 3' end of eukaryotic messenger RNA molecules

polyatomic ion (3.2) an ion containing a number of atoms

polymer (11.5) a very large molecule formed by the combination of many small molecules (called monomers) (e.g., polyamides, nylons)

polyprotic substance (8.3) a substance that can accept or donate more than one proton per molecule

polysaccharide (16.1) a large, complex carbohydrate composed of long chains of monosaccharides

polysome (20.6) complexes of many ribosomes all simultaneously translating a single mRNA

positive allosterism (19.9) effector binding activates the active site of an allosteric enzyme

positron (9.2) particle that has the same mass as an electron but opposite (+) charge

post-transcriptional modification (20.4) alterations of the primary transcripts produced in eukaryotic cells; these include addition of a poly(A) tail to the 3' end of the mRNA, addition of the cap structure to the 5' end of the mRNA, and RNA splicing

potential energy (1.5)   stored energy or energy caused by position or composition

precipitate (6.1)   an insoluble substance formed and separated from a solution

precision (1.3)   the degree of agreement among replicate measurements of the same quantity

pressure (5.1)   a force per unit area

primary (1°) alcohol (12.4)   an alcohol with the general formula RCH_2OH

primary (1°) amine (15.1)   an amine with the general formula RNH_2

primary (1°) carbon (10.2)   a carbon atom that is bonded to only one other carbon atom

primary structure (of a protein) (18.4)   the linear sequence of amino acids in a protein chain determined by the genetic information of the gene for each protein

primary transcript (20.4)   the RNA product of transcription in eukaryotic cells, before post-transcriptional modifications are carried out

product (4.3, 19.2)   the chemical species that results from a chemical reaction and that appears on the right side of a chemical equation

proenzyme (19.9)   the inactive form of a proteolytic enzyme

prokaryote (20.2)   an organism with simple cellular structure in which there is no true nucleus enclosed by a nuclear membrane and there are no true membrane-bound organelles in the cytoplasm

promoter (20.4)   the sequence of nucleotides immediately before a gene that is recognized by the RNA polymerase and signals the start point and direction of transcription

properties (1.2)   characteristics of matter

prostaglandins (17.2)   a family of hormonelike substances derived from the twenty-carbon fatty acid, arachidonic acid; produced by many cells of the body, they regulate many body functions

prosthetic group (18.7)   the nonprotein portion of a protein that is essential to the biological activity of the protein; often a complex organic compound

protein (18: Intro)   a macromolecule whose primary structure is a linear sequence of α-amino acids and whose final structure results from folding of the chain into a specific three-dimensional structure; proteins serve as catalysts, structural components, and nutritional elements for the cell

protein modification (19.9)   a means of enzyme regulation in which a chemical group is covalently added to or removed from a protein. The chemical modification either turns the enzyme on or turns it off

proteolytic enzyme (19.11)   an enzyme that hydrolyzes the peptide bonds between amino acids in a protein chain

proton (2.1)   a positively charged particle in the nucleus of an atom

pure substance (1.2)   a substance with constant composition

purine (20.1)   a family of nitrogenous bases (heterocyclic amines) that are components of DNA and RNA and consist of a six-sided ring fused to a five-sided ring; the common purines in nucleic acids are adenine and guanine

pyridoxal phosphate (22.7)   a coenzyme derived from vitamin B_6 that is required for all transamination reactions

pyrimidine (20.1)   a family of nitrogenous bases (heterocyclic amines) that are components of nucleic acids and consist of a single six-sided ring; the common pyrimidines of DNA are cytosine and thymine; the common pyrimidines of RNA are cytosine and uracil

pyrimidine dimer (20.7)   UV-light induced covalent bonding of two adjacent pyrimidine bases in a strand of DNA

pyruvate dehydrogenase complex (22.2)   a complex of all the enzymes and coenzymes required for the synthesis of CO_2 and acetyl CoA from pyruvate

Q

quantization (2.3)   a characteristic that energy can occur only in discrete units called quanta

quaternary ammonium salt (15.1)   an amine salt with the general formula $R_4N^+A^-$ (in which R— can be an alkyl or aryl group or a hydrogen atom and A^- can be any anion)

quaternary (4°) carbon (10.2)   a carbon atom that is bonded to four other carbon atoms

quaternary structure (of a protein) (18.7)   aggregation of more than one folded peptide chain to yield a functional protein

R

rad (9.7)   abbreviation for *radiation absorbed dose*, the absorption of 2.4×10^{-3} calories of energy per kilogram of absorbing tissue

radioactivity (9.1)   the process by which atoms emit high-energy particles or rays; the spontaneous decomposition of a nucleus to produce a different nucleus

radiocarbon dating (9.3)   the estimation of the age of objects through measurement of isotopic ratios of carbon

Raoult's law (6.4)   a law stating that the vapor pressure of a component is equal to its mole fraction times the vapor pressure of the pure component

rate constant (7.3)   the proportionality constant that relates the rate of a reaction and the concentration of reactants

rate equation (7.3)   expresses the rate of a reaction in terms of reactant concentration and a rate constant

rate of chemical reaction (7.3)   the change in concentration of a reactant or product per unit time

reactant (4.3)   starting material for a chemical reaction, appearing on the left side of a chemical equation

reducing agent (8.5)   a substance that reduces, or donates electrons to, another substance; the reducing agent is itself oxidized in the process

reducing sugar (16.4)   a sugar that can be oxidized by Benedict's or Tollens' reagents; includes all monosaccharides and most disaccharides

reduction (8.5, 12.6)   the gain of electrons; in organic compounds it may be recognized by a gain of hydrogen or loss of oxygen

regulatory proteins (18.1)   proteins that control cell functions such as metabolism and reproduction

release factor (20.6)   a protein that binds to the termination codon in the empty A-site of the ribosome and causes the peptidyl transferase to hydrolyze the bond between the peptide and the peptidyl tRNA

rem (9.7)   abbreviation for *roentgen equivalent for man,* the product of rad and RBE

replication fork (20.3)   the point at which new nucleotides are added to the growing daughter DNA strand

replication origin (20.3)   the region of a DNA molecule where DNA replication always begins

representative element (2.4)   member of the groups of the periodic table designated as A

resonance (3.4)   a condition that occurs when more than one valid Lewis structure can be written for a particular molecule

resonance form (3.4)   one of a number of valid Lewis structures for a particular molecule

resonance hybrid (3.4)   a description of the bonding in a molecule resulting from a superimposition of all valid Lewis structures (resonance forms)

restriction enzyme (20.8)   a bacterial enzyme that recognizes specific nucleotide sequences on a DNA molecule and cuts the sugar-phosphate backbone of the DNA at or near that site

result (1.1)   the outcome of a designed experiment, often determined from individual bits of data

reversible, competitive enzyme inhibitor (19.10)   a chemical that resembles the structure and charge distribution of the natural substrate and competes with it for the active site of an enzyme

reversible, noncompetitive enzyme inhibitor (19.10)   a chemical that binds weakly to an amino acid R group of an enzyme and inhibits activity; when the inhibitor dissociates, the enzyme is restored to its active form

reversible reaction (7.4)   a reaction that will proceed in either direction, reactants to products or products to reactants

ribonucleic acid (RNA) (20.1)   single-stranded nucleic acid molecules that are composed of phosphoryl groups, ribose, and the nitrogenous bases uracil, cytosine, adenine, and guanine

ribonucleotide (20.1) a ribonucleoside phosphate or nucleotide composed of a nitrogenous base in β-N-glycosidic linkage to the 1′ carbon of the sugar ribose and with one, two, or three phosphoryl groups esterified at the hydroxyl of the 5′ carbon of the ribose

ribose (16.4) a five-carbon monosaccharide that is a component of RNA and many coenzymes

ribosomal RNA (rRNA) (20.4) the RNA species that are structural and functional components of the small and large ribosomal subunits

ribosome (20.6) an organelle composed of a large and a small subunit, each of which is made up of ribosomal RNA and proteins; the platform on which translation occurs and that carries the enzymatic activity that forms peptide bonds

RNA polymerase (20.4) the enzyme that catalyzes the synthesis of RNA molecules using DNA as the template

RNA splicing (20.4) removal of portions of the primary transcript that do not encode protein sequences

röentgen (9.7) the dose of radiation producing 2.1×10^9 ions in 1 cm³ of air at 0°C and 1 atm of pressure

S

saccharide (16.1) a sugar molecule

saponification (14.2, 17.2) a reaction in which a soap is produced; more generally, the hydrolysis of an ester by an aqueous base

saturated fatty acid (17.2) a long-chain monocarboxylic acid in which each carbon of the chain is bonded to the maximum number of hydrogen atoms

saturated hydrocarbon (10.1) an alkane; a hydrocarbon that contains only carbon and hydrogen bonded together through carbon-hydrogen and carbon-carbon single bonds

saturated solution (6.1) one in which undissolved solute is in equilibrium with the solution

scientific method (1.1) the process of studying our surroundings that is based on experimentation

scientific notation (1.3) a system used to represent numbers as powers of ten

secondary (2°) alcohol (12.4) an alcohol with the general formula R_2CHOH

secondary (2°) amine (15.1) an amine with the general formula R_2NH

secondary (2°) carbon (10.2) a carbon atom that is bonded to two other carbon atoms

secondary structure (of a protein) (18.5) folding of the primary structure of a protein into an α-helix or a β-pleated sheet; folding is maintained by hydrogen bonds between the amide hydrogen and the carbonyl oxygen of the peptide bond

selectively permeable membrane (6.4) a membrane that restricts diffusion of some ions and molecules (based on size and charge) across the membrane

semiconservative DNA replication (20.3) DNA polymerase "reads" each parental strand of DNA and produces a complementary daughter strand; thus, all newly synthesized DNA molecules consist of one parental and one daughter strand

semipermeable membrane (6.4) a membrane permeable to the solvent but not the solute; a material that allows the transport of certain substances from one side of the membrane to the other

shielding (9.6) material used to provide protection from radiation

sickle cell anemia (18.9) a human genetic disease resulting from inheriting mutant hemoglobin genes from both parents

significant figures (1.3) all digits in a number known with certainty and the first uncertain digit

silent mutation (20.7) a mutation that changes the sequence of the DNA but does not alter the amino acid sequence of the protein encoded by the DNA

single bond (3.4) a bond in which one pair of electrons is shared by two atoms

single-replacement reaction (4.3) also called substitution reaction, one in which one atom in a molecule is displaced by another

soap (14.2) any of a variety of the alkali metal salts of fatty acids

solid state (1.2) a physical state of matter characterized by its rigidity and fixed volume and shape

solubility (3.5, 6.1) the amount of a substance that will dissolve in a given volume of solvent at a specified temperature

solute (6.1) a component of a solution that is present in lesser quantity than the solvent

solution (6.1) a homogeneous (uniform) mixture of two or more substances

solvent (6.1) the solution component that is present in the largest quantity

specific gravity (1.5) the ratio of the density of a substance to the density of water at 4°C or any specified temperature

specific heat (7.2) the quantity of heat (calories) required to raise the temperature of 1 g of a substance one degree Celsius

spectroscopy (2.3) the measurement of intensity and energy of electromagnetic radiation

speed of light (2.3) 2.99×10^8 m/s in a vacuum

sphingolipid (17.4) a phospholipid that is derived from the amino alcohol sphingosine rather than from glycerol

sphingomyelin (17.4) a sphingolipid found in abundance in the myelin sheath that surrounds and insulates cells of the central nervous system

standard solution (8.3) a solution whose concentration is accurately known

standard temperature and pressure (STP) (5.1) defined as 273 K and 1 atm

stereochemical specificity (19.5) the property of an enzyme that allows it to catalyze reactions involving only one enantiomer of the substrate

stereochemistry (16.3) the study of the spatial arrangement of atoms in a molecule

stereoisomers (10.3, 16.3) a pair of molecules having the same structural formula and bonding pattern but differing in the arrangement of the atoms in space

steroid (17.4) a lipid derived from cholesterol and composed of one five-sided ring and three six-sided rings; the steroids include sex hormones and anti-inflammatory compounds

structural analog (19.10) a chemical having a structure and charge distribution very similar to those of a natural enzyme substrate

structural formula (10.2) a formula showing all of the atoms in a molecule and exhibiting all bonds as lines

structural isomers (10.2) molecules having the same molecular formula but different chemical structures

structural protein (18.1) a protein that provides mechanical support for large plants and animals

sublevel (2.5) a set of equal-energy orbitals within a principal energy level

substituted hydrocarbon (10.1) a hydrocarbon in which one or more hydrogen atoms is replaced by another atom or group of atoms

substitution reaction (10.5, 11.6) a reaction that results in the replacement of one group for another

substrate (19.1) the reactant in a chemical reaction that binds to an enzyme active site and is converted to product

substrate-level phosphorylation (21.3) the production of ATP by the transfer of a phosphoryl group from the substrate of a reaction to ADP

sucrose (16.5) a disaccharide composed of α-D-glucose and β-D-fructose in ($\alpha 1 \longrightarrow \beta 2$) glycosidic linkage; table sugar

supersaturated solution (6.1) a solution that is more concentrated than a saturated solution (Note that such a solution is not at equilibrium.)

surface tension (5.2) a measure of the strength of the attractive forces at the surface of a liquid

surfactant (5.2) a substance that decreases the surface tension of a liquid

surroundings (7.1) the universe outside of the system

suspension (6.1) a heterogeneous mixture of particles; the suspended particles are larger than those found in a colloidal suspension

system (7.1) the process under study

T

temperature (1.5) a measure of the relative "hotness" or "coldness" of an object

temperature optimum (19.8) the temperature at which an enzyme functions optimally and the rate of reaction is maximal

terminal electron acceptor (22.6) the final electron acceptor in an electron transport system that removes the low-energy electrons from the system; in aerobic organisms the terminal electron acceptor is molecular oxygen

termination codon (20.6) a triplet of ribonucleotides with no corresponding anticodon on a tRNA; as a result, translation will end, because there is no amino acid to transfer to the peptide chain

terpene (17.4) the general term for lipids that are synthesized from isoprene units; the terpenes include steroids, bile salts, lipid-soluble vitamins, and chlorophyll

tertiary (3°) alcohol (12.4) an alcohol with the general formula R_3COH

tertiary (3°) amine (15.1) an amine with the general formula R_3N

tertiary (3°) carbon (10.2) a carbon atom that is bonded to three other carbon atoms

tertiary structure (of a protein) (18.6) the globular, three-dimensional structure of a protein that results from folding the regions of secondary structure; this folding occurs spontaneously as a result of interactions of the side chains or R groups of the amino acids

tetrahedral structure (3.4) a molecule consisting of four groups attached to a central atom that occupy the four corners of an imagined regular tetrahedron

tetrose (16.2) a four-carbon monosaccharide

theoretical yield (4.5) the maximum amount of product that can be produced from a given amount of reactant

theory (1.1) a hypothesis supported by extensive testing that explains and predicts facts

thermodynamics (7.1) the branch of science that deals with the relationship between energies of systems, work, and heat

thioester (14.4) the product of a reaction between a thiol and a carboxylic acid

thiol (12.9) an organic compound that contains a thiol group (—SH)

titration (8.3) the process of adding a solution from a buret to a sample until a reaction is complete, at which time the volume is accurately measured and the concentration of the sample is calculated

Tollens' test (13.4) a test reagent (silver nitrate in ammonium hydroxide) used to distinguish aldehydes and ketones; also called the Tollens' silver mirror test

tracer (9.5) a radioisotope that is rapidly and selectively transmitted to the part of the body for which diagnosis is desired

transaminase (22.7) an enzyme that catalyzes the transfer of an amino group from one molecule to another

transamination (22.7) a reaction in which an amino group is transferred from one molecule to another

transcription (20.4) the synthesis of RNA from a DNA template

transferase (19.1) an enzyme that catalyzes the transfer of a functional group from one molecule to another

transfer RNA (tRNA) (15.4, 20.4) small RNAs that bind to a specific amino acid at the 3′ end and mediate its addition at the appropriate site in a growing peptide chain; accomplished by recognition of the correct codon on the mRNA by the complementary anticodon on the tRNA

transition element (2.4) any element located between Groups IIA (2) and IIIA (13) in the long periods of the periodic table

transition state (19.6) the unstable intermediate in catalysis in which the enzyme has altered the form of the substrate so that it now shares properties of both the substrate and the product

translation (20.4) the synthesis of a protein from the genetic code carried on the mRNA

translocation (20.6) movement of the ribosome along the mRNA during translation

transmembrane protein (17.6) a protein that is embedded within a membrane and crosses the lipid bilayer, protruding from the membrane both inside and outside the cell

transport protein (18.1) a protein that transports materials across the cell membrane or throughout the body

triglyceride (17.3, 23.1) triacylglycerol; a molecule composed of glycerol esterified to three fatty acids

trigonal pyramidal molecule (3.4) a nonplanar structure involving three groups bonded to a central atom in which each group is equidistant from the central atom

triose (16.2) a three-carbon monosaccharide

triple bond (3.4) a bond in which three pairs of electrons are shared by two atoms

U

uncertainty (1.3) the degree of doubt in a single measurement

unit (1.4) a determinate quantity (of length, time, etc.) that has been adopted as a standard of measurement

unsaturated fatty acid (17.2) a long-chain monocarboxylic acid having at least one carbon-to-carbon double bond

unsaturated hydrocarbon (10.1, 11: Intro) a hydrocarbon containing at least one multiple (double or triple) bond

urea cycle (22.8) a cyclic series of reactions that detoxifies ammonium ions by incorporating them into urea, which is excreted from the body

uridine triphosphate (UTP) (21.7) a nucleotide composed of the pyrimidine uracil, the sugar ribose, and three phosphoryl groups and that serves as a carrier of glucose-1-phosphate in glycogenesis

V

valence electron (2.5) electron in the outermost shell (principal quantum level) of an atom

valence shell electron pair repulsion theory (VSEPR) (3.4) a model that predicts molecular geometry using the premise that electron pairs will arrange themselves as far apart as possible, to minimize electron repulsion

van der Waals forces (5.2) a general term for intermolecular forces that include dipole-dipole and London forces

vapor pressure of a liquid (5.2) the pressure exerted by the vapor at the surface of a liquid at equilibrium

very low density lipoprotein (VLDL) (17.5) a plasma lipoprotein that binds triglycerides synthesized by the liver and carries them to adipose tissue for storage

viscosity (5.2) a measure of the resistance to flow of a substance at constant temperature

vitamin (19.7) an organic substance that is required in the diet in small amounts; water-soluble vitamins are used in the synthesis of coenzymes required for the function of cellular enzymes; lipid-soluble vitamins are involved in calcium metabolism, vision, and blood clotting

voltaic cell (8.5) an electrochemical cell that converts chemical energy into electrical energy

W

wax (17.4) a collection of lipids that are generally considered to be esters of long-chain alcohols

weight (1.5) the force exerted on an object by gravity

weight/volume percent [% (W/V)] (6.2) the concentration of a solution expressed as a ratio of grams of solute to milliliters of solution multiplied by 100%

weight/weight percent [% (W/W)] (6.2) the concentration of a solution expressed as a ratio of mass of solute to mass of solution multiplied by 100%

Z

Zaitsev's rule (12.5) states that in an elimination reaction, the alkene with the greatest number of alkyl groups on the double-bonded carbon (the more highly substituted alkene) is the major product of the reaction

Answers to Odd-Numbered Problems

Chapter 1

1.1 a. Physical property
b. Chemical property
c. Physical property

1.3 a. Three
b. Three
c. Four
d. Two
e. Three

1.5 a. 2.4×10^{-3}
b. 1.80×10^{-2}
c. 2.24×10^{2}

1.7 a. 8.09
b. 5.9
c. 20.19

1.9 a. 51
b. 8.0×10^{1}
c. 1.6×10^{2}

1.11 a. Chemistry is the study of matter and the changes that matter undergoes.
b. Matter is the material component of the universe.
c. Energy is the ability to do work.

1.13 a. Potential energy is stored energy, or energy due to position or composition.
b. Kinetic energy is the energy resulting from motion of an object.
c. Data are a group of facts resulting from an experiment.

1.15 a. Gram (or kilogram)
b. Liter
c. Meter

1.17 Mass is an independent quantity while weight is dependent on gravity.

1.19 Density is mass per volume. Specific gravity is the ratio of the density of a substance to the density of water at 4°C.

1.21 The scientific method is an organized way of doing science.

1.23 A theory.

1.25 Freeze the contents of two beakers, one containing pure water, and the other containing salt dissolved in water. Slowly warm each container and measure the temperature of each as they convert from the solid to the liquid state. This temperature is the melting point, hence, the freezing point. Note that the melting and freezing point of a solution are the same temperature.

1.27 A physical property is a characteristic of a substance that can be observed without the substance undergoing a change in chemical composition.

1.29 Chemical properties of matter include flammability and toxicity.

1.31 A pure substance has constant composition with only a single substance whereas a mixture is composed of two or more substances.

1.33 Mixtures are composed of two or more substances. A homogeneous mixture has uniform composition while a heterogeneous mixture has non-uniform composition.

1.35 A gas is made up of particles that are widely separated. A gas will expand to fill any container and it has no definite shape or volume.

1.37 a. Chemical reaction
b. Physical change
c. Physical change

1.39 a. Physical property
b. Chemical property

1.41 a. Pure substance
b. Pure substance
c. Mixture

1.43 a. Homogeneous
b. Homogeneous
c. Homogeneous

1.45 a. Extensive property
b. Extensive property
c. Intensive property

1.47 An element is a pure substance that cannot be changed into a simpler form of matter by any chemical reaction. An atom is the smallest unit of an element that retains the properties of that element.

1.49 a. Iron, oxygen, and carbon are just a few of the more than 100 possible elements.
b. Sodium chloride, water, sucrose, ethyl alcohol

1.51 a. 3
b. 3
c. 3
d. 4
e. 4
f. 3

1.53 a. 3.87×10^{-3}
b. 5.20×10^{-2}
c. 2.62×10^{-3}
d. 2.43×10^{1}
e. 2.40×10^{2}
f. 2.41×10^{0}

1.55 a. Precision is a measure of the agreement of replicate results.
b. Accuracy is the degree of agreement between the true value and measured value.

1.57 a. 1.5×10^{4}
b. 2.41×10^{-1}
c. 5.99
d. 1139.42
e. 7.21×10^{3}

1.59 a. 1.23×10^{1}
b. 5.69×10^{-2}
c. -1.527×10^{3}
d. 7.89×10^{-7}

e. 9.2×10^7

f. 5.280×10^{-3}

g. 1.279×10^0

h. -5.3177×10^2

1.61 **a.** 3,240

b. 0.000150

c. 0.4579

d. $-683,000$

e. -0.0821

f. 299,790,000

g. 1.50

h. 602,200,000,000,000,000,000,000

1.63 **a.** 32 oz

b. 1.0×10^{-3} t

c. 9.1×10^2 g

d. 9.1×10^5 mg

e. 9.1×10^1 da

1.65 **a.** 6.6×10^{-3} lb

b. 1.1×10^{-1} oz

c. 3.0×10^{-3} kg

d. 3.0×10^2 cg

e. 3.0×10^3 mg

1.67 **a.** 10.0°C

b. 283.2 K

1.69 **a.** 293.2 K

b. 68.0°F

1.71 13.4 m²

1.73 4 L

1.75 101°F

1.77 5 cm is shorter than 5 in.

1.79 5.0 μg is smaller that 5.0 mg.

1.81 6.00 g/mL

1.83 6.20×10^2 L

1.85 1.08×10^3 g

1.87 teak

1.89 0.789

1.91 Lead has the lowest density and platinum has the greatest density.

1.93 12.6 mL

Chapter 2

2.1 Electron density is the probability that an electron will be found in a particular region of an atomic orbital.

2.3 **a.** Zr (zirconium)

b. 22.99

c. Cr (chromium)

d. At (astatine)

2.5 **a.** Helium, atomic number = 2, mass = 4.00 amu

b. Fluorine, atomic number = 9, mass = 19.00 amu

c. Manganese, atomic number = 25, mass = 54.94 amu

2.7 **a.** [Ne] $3s^2, 3p^4$

b. [Ar] $4s^2$

2.9 **a.** Ca^{2+} and Ar are isoelectronic

b. Sr^{2+} and Kr are isoelectronic

c. S^{2-} and Ar are isoelectronic

d. Mg^{2+} and Ne are isoelectronic

e. P^{3-} and Ar are isoelectronic

2.11 **a.** (Smallest) F, N, Be (largest)

b. (Lowest) Be, N, F (highest)

c. (Lowest) Be, N, F, (highest)

2.13 **a.** 8 protons, 8 electrons, 8 neutrons

b. 16 neutrons

2.15

Particle	Mass	Charge
a. electron	5.4×10^{-4} amu	-1
b. proton	1.00 amu	$+1$
c. neutron	1.00 amu	0

2.17 **a.** An ion is a charged atom or group of atoms formed by the loss or gain of electrons.

b. A loss of electrons by a neutral species results in a cation.

c. A gain of electrons by a neutral species results in an anion.

2.19 From the periodic table, all isotopes of Rn have 86 protons. Isotopes differ in the number of neutrons.

2.21 **a.** 34

b. 46

2.23 **a.** ^1_1H

b. $^{14}_6C$

2.25

Atomic Symbol	# Protons	# Neutrons	# Electrons	Charge
a. $^{23}_{11}Na$	11	12	11	0
b. $^{32}_{16}S^{2-}$	16	16	18	2−
c. $^{16}_8O$	8	8	8	0
d. $^{24}_{12}Mg^{2+}$	12	12	10	2+
e. $^{39}_{19}K^+$	19	20	18	1+

2.27 **a.** Neutrons

b. Protons

c. Protons, neutrons

d. Ion

e. Nucleus, negative

2.29 63.55 amu

2.31 • All matter consists of tiny particles called atoms.

• Atoms cannot be created, divided, destroyed, or converted to any other type of atom.

• All atoms of a particular element have identical properties.

• Atoms of different elements have different properties.

• Atoms combine in simple whole-number ratios.

• Chemical change involves joining, separating, or rearranging atoms.

2.33 **a.** Chadwick—demonstrated the existence of the neutron in 1932.

b. Goldstein—identified positive charge in the atom.

2.35 **a.** Dalton—developed the Law of Multiple Proportions; determined the relative atomic weights of the elements known at that time; developed the first scientific atomic theory.

b. Crookes—developed the cathode ray tube and discovered "cathode rays;" characterized electron properties.

2.37 Our understanding of the nucleus is based on the gold foil experiment performed by Geiger and interpreted by Rutherford. In this experiment, Geiger bombarded a piece of gold foil with alpha particles, and observed that some alpha particles passed straight through the foil, others were deflected and some simply bounced back. This led Rutherford to propose that the atom consisted of a small, dense nucleus (alpha particles bounced back), surrounded by a cloud of electrons (some alpha particles were deflected). The size of the nucleus is small when compared to the volume of the atom (alpha particles were able to pass through the foil).

2.39 Crookes used the cathode ray tube. He observed particles emitted by the cathode and traveling toward the anode. This ray was deflected by an electric field. Thomson measured the curvature of the ray influenced by the electric and magnetic fields. This measurement provided the mass to charge ratio of the negative particle. Thomson also gave the particle the name, electron.

2.41 A cathode ray is the negatively charged particle formed in a cathode ray tube.

2.43
Radiowave ↑
Microwave |
Infrared | Increasing
Visible | Wavelength
Ultraviolet |
X-ray |
Gamma ray |

2.45 Infrared radiation has greater energy than microwave radiation.

2.47 Spectroscopy is the measurement of intensity and energy of electromagnetic radiation.

2.49 According to Bohr, Planck, and others, electrons exist only in certain allowed regions, quantum levels, outside of the nucleus.

2.51 • Electrons are found in orbits at discrete distances from the nucleus.
• The orbits are quantized—they are of discrete energies.
• Electrons can only be found in these orbits, never in between (they are able to jump instantaneously from orbit to orbit).
• Electrons can undergo transitions—if an electron absorbs energy, it will jump to a higher orbit; when the electron falls back to a lower orbit, it will release energy.

2.53 Bohr's atomic model was the first to successfully account for electronic properties of atoms, specifically, the interaction of atoms and light (spectroscopy).

2.55 a. Sodium
b. Potassium
c. Magnesium

2.57 Group IA (or 1) is known collectively as the alkali metals and consists of lithium, sodium, potassium, rubidium, cesium, and francium.

2.59 Group VIIA (or 17) is known collectively as the halogens and consists of fluorine, chlorine, bromine, iodine, and astatine.

2.61 a. True
b. True

2.63 a. Na, Ni, Al
b. Na, Al
c. Na, Ni, Al
d. Ar

2.65

2.67

Atom	Total electrons	Valence electrons	Principal energy level number
a. H	1	1	1
b. Na	11	1	3
c. B	5	3	2
d. F	9	7	2
e. Ne	10	0 (or 8)	2
f. He	2	0 (or 2)	1

2.69 A principal energy level is designated $n = 1, 2, 3$, and so forth. It is similar to Bohr's orbits in concept. A sublevel is a part of a principal energy level and is designated s, p, d, and f.

2.71 The s orbital represents the probability of finding an electron in a region of space surrounding the nucleus.

2.73 Three p orbitals (p_x p_y p_z) can exist in a given principal energy level.

2.75 A $3p$ orbital is a higher energy orbital than a $2p$ orbital because it is a part of a higher energy principal energy level.

2.77 $2\,e^-$ for $n = 1$
$8\,e^-$ for $n = 2$
$18\,e^-$ for $n = 3$

2.79 a. $3p$ orbital
b. $3s$ orbital
c. $3d$ orbital
d. $4s$ orbital
e. $3d$ orbital
f. $3p$ orbital

2.81 a. Not possible
b. Possible
c. Not possible
d. Not possible

2.83 a. Li^+
b. O^{2-}
c. Ca^{2+}
d. Br^-
e. S^{2-}
f. Al^{3+}

2.85 a. Isoelectronic
b. Isoelectronic

2.87 a. Na^+
b. S^{2-}
c. Cl^-

2.89 a. $1s^2, 2s^2, 2p^6, 3s^2, 3p^6$
b. $1s^2, 2s^2, 2p^6$

2.91 a. (Smallest) F, O, N (Largest)
b. (Smallest) Li, K, Cs (Largest)
c. (Smallest) Cl, Br, I (Largest)

2.93 a. (Smallest) O, N, F (Largest)
b. (Smallest) Cs, K, Li (Largest)
c. (Smallest) I, Br, Cl (Largest)

2.95 A positive ion is always smaller than its parent atom because the positive charge of the nucleus is shared among fewer electrons in the ion. As a result, each electron is pulled closer to the nucleus and the volume of the ion decreases.

2.97 The fluoride ion has a completed octet of electrons and an electron configuration resembling its nearest noble gas.

2.99 Cl^- is larger because it has a smaller nuclear (positive) charge.

Chapter 3

3.1 **a.** Potassium cyanide
   b. Magnesium sulfide
   c. Magnesium acetate

3.3 **a.** The bonded nuclei are closer together when a double bond exists, in comparison to a single bond.
   b. The bond strength increases as the bond order increases. Therefore, a double bond is stronger than a single bond.

3.5 **a.** $H:\overset{..}{\underset{H}{P}}:H$ $\overset{\overset{..}{P}}{H\overset{|}{\underset{H}{\wedge}}H}$

   b. $H:\overset{..}{\underset{H}{Si}}:H$ $\overset{\overset{H}{|}}{H\overset{Si}{\underset{H}{\wedge}}H}$

3.7 **a.** Oxygen is more electronegative than sulfur; the bond is polar. The electrons are pulled toward the oxygen atom.
   b. Nitrogen is more electronegative than carbon; the bond is polar. The electrons are pulled toward the nitrogen atom.
   c. There is no electronegativity difference between two identical atoms; the bond is nonpolar.
   d. Chlorine is more electronegative than iodine; the bond is polar. The electrons are pulled toward the chlorine atom.

3.9 **a.** Nonpolar
   b. Polar
   c. Polar
   d. Nonpolar

3.11 **a.** H_2O
   b. CO
   c. NH_3
   d. ICl

3.13 **a.** Ionic
   b. Covalent
   c. Covalent
   d. Covalent

3.15 **a.** Covalent
   b. Covalent
   c. Covalent
   d. Ionic

3.17 **a.** $Li\cdot + :\overset{..}{\underset{..}{Br}}\cdot \longrightarrow Li^+ + :\overset{..}{\underset{..}{Br}}:^-$

   b. $\cdot Mg\cdot + 2:\overset{..}{\underset{..}{Cl}}\cdot \longrightarrow Mg^{2+} + 2:\overset{..}{\underset{..}{Cl}}:^-$

3.19 **a.** $:\overset{}{\underset{.}{S}}\cdot + 2H\cdot \longrightarrow \overset{..}{\underset{\underset{H}{..}}{S}}:H$

   b. $\cdot\overset{..}{P}\cdot + 3H\cdot \longrightarrow H:\overset{}{\underset{H}{P}}:H$

3.21 He has two valence electrons (electron configuration $1s^2$) and a complete $N = 1$ level. It has a stable electron configuration, with no tendency to gain or lose electrons, and satisfies the octet rule ($2\ e^-$ for period 1). Hence, it is nonreactive.
   $He:$

3.23 **a.** Sodium ion
   b. Copper(I) ion (or cuprous ion)
   c. Magnesium ion
   d. Iron(II) ion (or ferrous ion)
   e. Iron(III) ion (or ferric ion)

3.25 **a.** Sulfide ion
   b. Chloride ion
   c. Carbonate ion

3.27 **a.** Magnesium chloride
   b. Aluminum chloride

3.29 **a.** Nitrogen dioxide
   b. Sulfur trioxide

3.31 **a.** Sulfur dioxide
   b. Sulfur trioxide

3.33 **a.** K^+
   b. Br^-

3.35 **a.** SO_4^{2-}
   b. NO_3^-

3.37 **a.** $NaCl$
   b. $MgBr_2$

3.39 **a.** $AgCN$
   b. NH_4Cl

3.41 **a.** CuO
   b. Fe_2O_3

3.43 **a.** Al_2O_3
   b. Li_2S

3.45 **a.** SiO_2
   b. SO_2

3.47 **a.** $NaNO_3$
   b. $Mg(NO_3)_2$

3.49 **a.** NH_4I
   b. $(NH_4)_2SO_4$

3.51 Ionic solid state compounds exist in regular, repeating, three-dimensional structures; the crystal lattice. The crystal lattice is made up of positive and negative ions. Solid state covalent compounds are made up of molecules which may be arranged in a regular crystalline pattern or in an irregular (amorphous) structure.

3.53 The boiling points of ionic solids are generally much higher than those of covalent solids.

3.55 KCl would be expected to exist as a solid at room temperature; it is an ionic compound, and ionic compounds are characterized by high melting points.

3.57 Water will have a higher boiling point. Water is a polar molecule with strong intermolecular attractive forces, whereas carbon tetrachloride is a nonpolar molecule with weak intermolecular attractive forces. More energy, hence, a higher temperature is required to overcome the attractive forces among the water molecules.

3.59 **a.** $H\cdot$
   b. $He:$
   c. $\cdot\overset{.}{\underset{.}{C}}\cdot$
   d. $\cdot\overset{..}{N}\cdot$

3.61 **a.** Li^+
   b. Mg^{2+}
   c. $:\overset{..}{\underset{..}{Cl}}:^-$
   d. $:\overset{..}{\underset{..}{P}}:^{3-}$

3.63 **a.** $:\overset{..}{\underset{..}{Cl}}:\overset{..}{N}:\overset{..}{\underset{..}{Cl}}:$
        $:\overset{..}{\underset{..}{Cl}}:$

   b. $H:\overset{}{\underset{\underset{H}{H}}{C}}:\overset{..}{\underset{..}{O}}:H$

   c. $:\overset{..}{\underset{.}{S}}::C::\overset{..}{\underset{.}{S}}:$

3.65 a. $:\!\ddot{C}l\!:\!\ddot{N}\!:\!\ddot{C}l\!:$
$:\!\ddot{C}l\!:$

Pyramidal,
polar
water soluble

b. H:$\ddot{C}$:$\ddot{O}$:H
    H

Tetrahedral around C,
angular around O
polar
water soluble

c. $:\!\ddot{S}\!:\!:\!C\!:\!:\!\ddot{S}\!:$
Linear
nonpolar
not water soluble

3.67 $[:\!N\!:\!:\!:\!O\!:]^+$

3.69 $[:\!\ddot{O}\!:\!H]^-$

3.71 Resonance can occur when more than one valid Lewis structure can be written for a molecule. Each individual structure which can be drawn is a resonance form. The true nature of the structure for the molecule is the resonance hybrid, which consists of the "average" of the resonance forms.

3.73 $\left[\begin{array}{c} H\!:\!\ddot{O}\!: \\ H\!:\!\ddot{C}\!:\!\ddot{C}\!:\!\ddot{O}\!: \\ H \end{array}\right]^- \longleftrightarrow \left[\begin{array}{c} H\!:\!\ddot{O}\!: \\ H\!:\!\ddot{C}\!:\!C\!:\!:\!\ddot{O} \\ H \end{array}\right]^-$

3.75 H:$\ddot{C}$:$\ddot{C}$:$\ddot{O}$:H
   H H

3.77 H:$\ddot{C}$:$\overset{\ddot{O}}{C}$:$\ddot{C}$:H
   H   H

3.79 a. Polar covalent
b. Polar covalent
c. Ionic
d. Ionic
e. Ionic

3.81 a. $[:\!C\!\equiv\!N\!:]^-$
b. $[:\!Si\!\equiv\!P\!:]^-$

c), d), and e) are ionic compounds.

3.83 $:\!\ddot{C}l\!:\!Be\!:\!\ddot{C}l\!:$

3.85 A molecule containing no polar bonds *must* be nonpolar. A molecule containing polar bonds may or may not itself be polar. It depends upon the number and arrangement of the bonds.

3.87 Polar compounds have strong intermolecular attractive forces. Higher temperatures are needed to overcome these forces and convert the solid to a liquid; hence, we predict higher melting points for polar compounds when compared to nonpolar compounds.

3.89 Yes

Chapter 4

4.1 a. DR
b. SR
c. DR
d. D

4.3 Examples of other packaging units include a *ream* of paper (500 sheets of paper), a six-pack of soft drinks, a case of canned goods (24 cans), to name a few.
4.5 a. 28.09 g.
b. 107.9 g.
4.7 39.95 g.
4.9 6.0×10^{19} carbon atoms
4.11 1.7×10^{-22} mol AS
4.13 40.36 g Ne
4.15 4.00 g He/mol He
4.17 a. 5.00 mol He
b. 1.7 mol Na
c. 4.2×10^{-2} mol Cl_2
4.19 1.62×10^3 g Ag
4.21 8.37×10^{22} Ag atoms
4.23 A molecule is a single unit comprised of atoms joined by covalent bonds. An ion-pair is composed of positive and negatively charged ions joined by electrostatic attraction, the ionic bond. The ion pairs, unlike the molecule, do not form single units; the electrostatic charge is directed to other ions in a crystal lattice, as well.
4.25 a. 58.44 g/mol
b. 142.04 g/mol
c. 357.49 g/mol
4.27 32.00 g/mol O_2
4.29 249.70 grams
4.31 a. 0.257 mol NaCl
b. 0.106 mol Na_2SO_4
4.33 a. 18.02 g H_2O
b. 116.9 g NaCl
4.35 a. 40.0 g He
b. 2.02×10^2 g H
4.37 a. 2.43 g Mg
b. 10.0 g $CaCO_3$
4.39 a. 4.00 g NaOH
b. 9.81 g H_2SO_4
4.41 a. 0.420 mol KBr
b. 0.415 mol $MgSO_4$
4.43 a. 6.57×10^{-1} mol CS_2
b. 2.14×10^{-1} mol $Al_2(CO_3)_3$
4.45 The ultimate basis for a correct chemical equation is the law of conservation of mass. No mass may be gained or lost in a chemical reaction, and the chemical equation must reflect this fact.
4.47 The subscript tells us the number of atoms or ions contained in one unit of the compound.
4.49 Heat is necessary for the reaction to occur.
4.51 Yes, PbI_2
4.53 If we change the subscript we change the identity of the compound.
4.55 A reactant is the starting material for a chemical reaction.
4.57 A product is the chemical species that results from a chemical reaction.
4.59 a. $2C_2H_6(g) + 7O_2(g) \rightarrow 4CO_2(g) + 6H_2O(g)$
b. $6K_2O(s) + P_4O_{10}(s) \rightarrow 4K_3PO_4(s)$
c. $MgBr_2(aq) + H_2SO_4(aq) \rightarrow 2HBr(g) + MgSO_4(aq)$
4.61 a. $Ca(s) + F_2(g) \rightarrow CaF_2(s)$
b. $2Mg(s) + O_2(g) \rightarrow 2MgO(s)$
c. $3H_2(g) + N_2(g) \rightarrow 2NH_3(g)$
4.63 a. $2C_4H_{10}(g) + 13O_2(g) \rightarrow 10H_2O(g) + 8CO_2(g)$
b. $Au_2S_3(s) + 3H_2(g) \rightarrow 2Au(s) + 3H_2S(g)$
c. $Al(OH)_3(s) + 3HCl(aq) \rightarrow AlCl_3(aq) + 3H_2O(l)$
d. $(NH_4)_2Cr_2O_7(s) \rightarrow Cr_2O_3(s) + N_2(g) + 4H_2O(g)$
e. $C_2H_5OH(l) + 3O_2(g) \rightarrow 2CO_2(g) + 3H_2O(g)$

4.65 a. $N_2(g) + 3H_2(g) \rightarrow 2NH_3(g)$
   b. $HCl(aq) + NaOH(aq) \rightarrow NaCl(aq) + H_2O(l)$
4.67 a. $C_6H_{12}O_6(s) + 6O_2(g) \rightarrow 6H_2O(l) + 6CO_2(g)$
   b. $Na_2CO_3(s) \xrightarrow{\Delta} Na_2O(s) + CO_2(g)$
4.69 Conservation of mass
4.71 Changing subscripts changes the identity of the compound. Changing coefficients changes the relative number of moles of compound.
4.73 27.7 g B_2H_6
4.75 104 g $CrCl_3$
4.77 a. $N_2(g) + 3H_2(g) \rightarrow 2NH_3(g)$
   b. Three moles of H_2 will react with one mole of N_2
   c. One mole of N_2 will produce two moles of the product NH_3
   d. 1.50 mol H_2
   e. 17.0 g NH_3
4.79 a. 149.21 g/mol
   b. 1.20×10^{24} O atoms
   c. 32.00 g O
   d. 10.7 g O
4.81 7.39 g O_2
4.83 6.77×10^4 g CO_2
4.85 70.6 g $C_{10}H_{22}$
4.87 9.13×10^2 g N_2
4.89 92.6%
4.91 6.85×10^2 g N_2

Chapter 5

5.1 a. 0.954 atm
   b. 0.382 atm
   c. 0.730 atm
5.3 2.91 atm
5.5 Evaporation is the conversion of a liquid to a gas at a temperature lower than the boiling point of the liquid. Condensation is the conversion of a gas to a liquid at a temperature lower than the boiling point of the liquid.
5.7 $CO_2 < CH_3Cl < CH_3OH$; Only CH_3OH exhibits London Dispersion Forces, dipole-dipole interactions, and hydrogen bonding. Hence, CH_3OH has the strongest intermolecular forces and therefore, the highest boiling point.
5.9 In all cases, gas particles are much further apart than similar particles in the liquid or solid state. In most cases, particles in the liquid state are, on average, farther apart than those in the solid state. Water is the exception; liquid water's molecules are closer together than they are in the solid state.
5.11 Pressure is a force/unit area. Gas particles are in continuous, random motion. Collisions with the walls of the container result in a force (mass × acceleration) on the walls of the container. The sum of these collisional forces constitutes the pressure exerted by the gas.
5.13 Gases are easily compressed simply because there is a great deal of space between particles; they can be pushed closer together (compressed) because the space is available.
5.15 Gas particles are in continuous, random motion. They are free (minimal attractive forces between particles) to roam, up to the boundary of their container.
5.17 Gases exhibit more ideal behavior at low pressures. At low pressures, gas particles are more widely separated and therefore the attractive forces between particles are less. The ideal gas model assumes negligible attractive forces between gas particles.
5.19 The kinetic molecular theory states that the average kinetic energy of the gas particles increases as the temperature

increases. Kinetic energy is proportional to (velocity)^2. Therefore, as the temperature increases the gas particle velocity increases and the rate of mixing increases as well.
5.21 The volume of the balloon is directly proportional to the pressure the gas exerts on the inside surface of the balloon. As the balloon cools, its pressure drops, and the balloon contracts. (Recall that the speed, hence the force exerted by the molecules, decreases as the temperature decreases.)
5.23 Boyle's law states that the volume of a gas varies inversely with the gas pressure if the temperature and the number of moles of gas are held constant.
5.25 Volume will decrease according to Boyle's law. Volume is inversely proportional to the pressure exerted on the gas.
5.27 1 atm
5.29 5 L-atm
5.31 5.23 atm
5.33 Charles's law states that the volume of a gas varies directly with the absolute temperature if pressure and number of moles of gas are constant.
5.35 The Kelvin scale is the only scale that is directly proportional to molecular motion, and it is the motion that determines the physical properties of gases.
5.37 No. The volume is proportional to the temperature in K, not Celsius.
5.39 0.96 L
5.41 1.51 L
5.43 • Volume and temperature are *directly* proportional; increasing T *increases* V.
   • Volume and pressure are *inversely* proportional; decreasing P *increases* V.
   Therefore, both variables work together to *increase* the volume.
5.45 $V_f = \dfrac{P_i V_i T_f}{P_f T_i}$
5.47 1.82×10^{-2} L
5.49 Avogadro's law states that equal volumes of any ideal gas contain the same number of moles if measured at constant temperature and pressure.
5.51 6.00 L
5.53 No. One mole of an ideal gas will occupy exactly 22.4 L; however, there is no completely ideal gas and careful measurement will show a different volume.
5.55 Standard temperature is 273K.
5.57 0.80 mol
5.59 22.4 L
5.61 0.276 mol
5.63 5.94×10^{-2} L
5.65 22.4 L
5.67 9.08×10^3 L
5.69 172°C
5.71 Dalton's law states that the total pressure of a mixture of gases is the sum of the partial pressures of the component gases.
5.73 0.74 atm
5.75 Intermolecular forces in liquids are considerably stronger than intermolecular forces in gases. Particles are, on average, much closer together in liquids and the strength of attraction is inversely proportional to the distance of separation.
5.77 The vapor pressure of a liquid increases as the temperature of the liquid increases.
5.79 Viscosity is the resistance to flow caused by intermolecular attractive forces. Complex molecules may become entangled and not slide smoothly across one another.

5.81 All molecules exhibit London forces.

5.83 Only methanol exhibits hydrogen bonding. Methanol has an oxygen atom bonded to a hydrogen atom, a necessary condition for hydrogen bonding.

5.85 Solids are essentially incompressible because the average distance of separation among particles in the solid state is small. There is literally no space for the particles to crowd closer together.

5.87 **a.** High melting temperature, brittle
b. High melting temperature, hard

5.89 Beryllium

5.91 Mercury

Chapter 6

6.1 A chemical analysis must be performed in order to determine the identity of all components, a qualitative analysis. If only one component is found, it is a pure substance; two or more components indicates a true solution.

6.3 After the container of soft drink is opened, CO_2 diffuses into the surrounding atmosphere; consequently the partial pressure of CO_2 over the soft drink decreases and the equilibrium

$$CO_2 \text{ (g)} \rightleftharpoons CO_2 \text{ (aq)}$$

shifts to the left, lowering the concentration of CO_2 in the soft drink.

6.5 0.125 mol HCl

6.7 Pure water

6.9 **a.** 2.00% NaCl
b. 6.60% $C_6H_{12}O_6$

6.11 **a.** 5.00% ethanol
b. 10.0% ethanol

6.13 **a.** 21.0% NaCl
b. 3.75% NaCl

6.15 **a.** 2.25 g NaCl
b. 3.13 g $NaC_2H_3O_2$

6.17 19.5% KNO_3

6.19 1.00 g sugar

6.21 0.04% (w/w) solution is more concentrated.

6.23 2.0×10^{-3} ppt

6.25 In calculation of chemical quantities, the coefficients of the balanced chemical equation represent relative number of moles, not relative mass.

6.27 Laboratories' managers often purchase concentrated solutions for practical reasons, such as economy and conservation of storage space.

6.29 0.50 M

6.31 **a.** 0.342 M NaCl
b. 0.367 M $C_6H_{12}O_6$

6.33 **a.** 1.46 g NaCl
b. 9.00 g $C_6H_{12}O_6$

6.35 0.266 L

6.37 5.00×10^{-2} L

6.39 20.0 M

6.41 0.900 M

6.43 158 g glucose

6.45 0.0500 M

6.47 0.10 eq/L

6.49 A colligative property is a solution property that depends on the concentration of solute particles rather than the identity of the particles.

6.51 Salt is an ionic substance that dissociates in water to produce positive and negative ions. These ions (or particles) lower the freezing point of water. If the concentration of salt particles is large, the freezing point may be depressed below the surrounding temperature, and the ice would melt.

6.53 Chemical properties depend on the identity of the substance, whereas colligative properties depend on concentration, not identity.

6.55 Raoult's law states that when a solute is added to a solvent, the vapor pressure of the solvent decreases in proportion to the concentration of the solute.

6.57 One mole of $CaCl_2$ produces three moles of particles in solution whereas one mole of NaCl produces two moles of particles in solution. Therefore, a one molar $CaCl_2$ solution contains a greater number of particles than a one molar NaCl solution and will produce a greater freezing-point depression.

6.59 Sucrose

6.61 Sucrose

6.63 24 atm

6.65 A → B

6.67 No net flow

6.69 Hypertonic

6.71 Hypotonic

6.73 Water is often termed the "universal solvent" because it is a polar molecule and will dissolve, at least to some extent, most ionic and polar covalent compounds. The majority of our body mass is water and this water is an important part of the nutrient transport system due to its solvent properties. This is true in other animals and plants as well. Because of its ability to hydrogen bond, water has a high boiling point and a low vapor pressure. Also, water is abundant and easily purified.

6.75

6.77 1.0×10^{-3} osmol

6.79 The shelf life is a function of the stability of the ammonia-water solution. The ammonia can react with the water to convert to the extremely soluble and stable ammonium ion. Also, ammonia and water are polar molecules. Polar interactions, particularly hydrogen bonding, are strong and contribute to the long-term solution stability.

6.81

6.83 Polar; like dissolves like (H_2O is polar)

6.85 The number of particles in solution is dependent on the degree of dissociation.

6.87 In dialysis, sodium ions move from a region of high concentration to a region of low concentration. If we wish to remove (transport) sodium ions from the blood, they can move to a region of lower concentration, the dialysis solution.

6.89 Elevated concentrations of sodium ion in the blood may cause confusion, stupor, or coma.

6.91 Elevated concentrations of sodium ion in the blood may occur whenever large amounts of water are lost. Diarrhea, diabetes, and certain high-protein diets are particularly problematic.

6.93 4.0×10^{-2} M

Chapter 7

7.1 **a.** Exothermic
b. Exothermic

7.3 He(g)

7.5 $\Delta G = (+) - T(-)$
    ΔG must always be positive.

7.7 2.7×10^3 J

7.9   Heat energy produced by the friction of striking the match provides the activation energy necessary for this combustion process.

7.11 If the enzyme catalyzed a process needed to sustain life, the substance interfering with that enzyme would be classified as a poison.

7.13 At rush hour, approximately the same number of passengers enter and exit the train at any given stop. Throughout the trip, the number of passengers on the train may be essentially unchanged, but the identity of the individual passengers is continually changing.

7.15 Measure the concentrations of products and reactants at a series of times until no further concentration change is observed.

7.17 Product formation

7.19 joule

7.21 An exothermic reaction is one in which energy is released during chemical change.

7.23 A fuel must release heat in the combustion (oxidation) process.

7.25 The temperature of the water (or solution) is measured in a calorimeter. If the reaction being studied is exothermic, released energy heats the water and the temperature increases. In an endothermic reaction, heat flows from the water to the reaction and the water temperature decreases.

7.27 Double-walled containers, used in calorimeters, provide a small airspace between the part of the calorimeter (inside wall) containing the sample solution and the outside wall, contacting the surroundings. This makes heat transfer more difficult.

7.29 Free energy is the combined contribution of entropy and enthalpy for a chemical reaction.

7.31 The first law of thermodynamics, the law of conservation of energy, states that the energy of the universe is constant.

7.33 Enthalpy is a measure of heat energy.

7.35 1.20×10^3 cal

7.37 5.02×10^3 J

7.39 **a.** Entropy increases.
    b. Entropy increases.

7.41 $\Delta G = (-) - T(+)$
    ΔG must always be negative and the process is always spontaneous.

7.43 Isopropyl alcohol quickly evaporates (liquid $\rightarrow$ gas) after being applied to the skin. Conversion of a liquid to a gas requires heat energy. The heat energy is supplied by the skin. When this heat is lost, the skin temperature drops.

7.45 Decomposition of leaves and twigs to produce soil.

7.47 The activated complex is the arrangement of reactants in an unstable transition state as a chemical reaction proceeds. The activated complex must form in order to convert reactants to products.

7.49 The rate of a reaction is the change in concentration of a reactant or product per unit time. The rate constant is the proportionality constant that relates rate and concentration. The order is the exponent of each concentration term in the rate equation.

7.51 Increase

7.53 A catalyst increases the rate of a reaction without itself undergoing change.

7.55

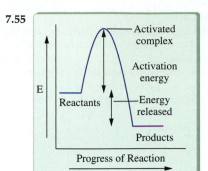

Non-catalyzed reaction
High activation energy

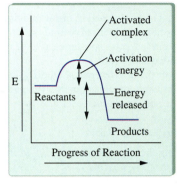

Catalyzed reaction
Lower activation energy

7.57 Enzymes are biological catalysts. The enzyme lysozyme catalyzes a process that results in the destruction of the cell walls of many harmful bacteria. This helps to prevent disease in organisms. The breakdown of foods to produce material for construction and repair of body tissue, as well as energy, is catalyzed by a variety of enzymes. For example, amylase begins the hydrolysis of starch in the mouth.

7.59 An increase in concentration of reactants means that there are more molecules in a certain volume. The probability of collision is enhanced because they travel a shorter distance before meeting another molecule. The rate is proportional to the number of collisions per unit time.

7.61 Rate = k $[N_2O_4]$

7.63 A catalyst speeds up a chemical reaction by facilitating the formation of the activated complex, thus lowering the activation energy, the energy barrier for the reaction.

7.65 Products

7.67 Ice and water at 0°C

7.69 Le Chaltelier's principle states that when a system at equilibrium is disturbed, the equilibrium shifts in the direction that minimizes the disturbance.

7.71 A dynamic equilibrium has fixed concentrations of all reactants and products—these concentrations do not change with time. However, the process is dynamic because products and reactants are continuously being formed and consumed. The concentrations do not change because the rates of production and consumption are equal.

7.73 $K_{eq} = \dfrac{[NO_2]^2}{[N_2O_4]}$

7.75 A physical equilibrium describes physical change; examples include the equilibrium between ice and water, or the equilibrium vapor pressure of a liquid. A chemical equilibrium describes chemical change; examples include the reactions shown in questions 7.87 and 7.88.

7.77 $K_{eq} = \dfrac{[NH_3]^2}{[N_2][H_2]^3}$

7.79 $K_{eq} = \dfrac{[H_2S]^2}{[H_2]^2 [S_2]}$

7.81 7.7×10^{-5} M

7.83 **a.** Equilibrium shifts to the left.
  b. No change
  c. No change

7.85 **a.** False
  b. False

7.87 **a.** PCl_3 increases
  b. PCl_3 decreases
  c. PCl_3 decreases
  d. PCl_3 decreases
  e. PCl_3 remains the same

7.89 Decrease

7.91 $K_{eq} = \dfrac{[CO][H_2]}{[H_2O]}$

7.93 False

7.95 Removing the cap allows CO_2 to escape into the atmosphere. This corresponds to the removal of product (CO_2):
$$CO_2 \ (l) \rightleftharpoons CO_2 \ (g)$$
The equilibrium shifts to the right, dissolved CO_2 is lost, and the beverage goes "flat."

Chapter 8

8.1 **a.** $HF(aq) + H_2O \ (l) \rightleftharpoons H_3O^+ \ (aq) + F^- \ (aq)$
  b. $NH_3(aq) + H_2O \ (l) \rightleftharpoons NH_4^+ \ (aq) + OH^- \ (aq)$

8.3 **a.** HF and F^-; H_2O and H_3O^+
  b. NH_3 and NH_4^+; H_2O and OH^-

8.5 1.0×10^{-11} M

8.7 $CO_2 + H_2O \rightleftharpoons H_2CO_3 \rightleftharpoons H_3O^+ + HCO_3^-$
An increase in the partial pressure of CO_2 is a stress on the left side of the equilibrium. The equilibrium will shift to the right in an effort to decrease the concentration of CO_2. This will cause the molar concentration of H_2CO_3 to increase.

8.9 In Question 8.7, the equilibrium shifts to the right. Therefore the molar concentration of H_3O^+ should increase.
  In Question 8.8, the equilibrium shifts to the left. Therefore the molar concentration of H_3O^+ should decrease.

8.11 4.74

8.13 4.87

8.15 $Ca \rightarrow Ca^{2+} + 2 \ e^-$ (oxidation ½ reaction)
$S + 2 \ e^- \rightarrow S^{2-}$ (reduction ½ reaction)
$Ca + S \rightarrow CaS$ (complete reaction)

8.17 **a.** An Arrhenius acid is a substance that dissociates, producing hydrogen ions.
  b. A Brønsted-Lowry acid is a substance that behaves as a proton donor.

8.19 The Brønsted-Lowry theory provides a broader view of acid-base theory than does the Arrhenius theory. Brønsted-Lowry emphasizes the role of the solvent in the dissociation process.

8.21 **a.** $HNO_2 \ (aq) + H_2O \ (l) \rightleftharpoons H_3O^+ \ (aq) + NO_2^- \ (aq)$
  b. $HCN \ (aq) + H_2O \ (l) \rightleftharpoons H_3O^+ \ (aq) + CN^- \ (aq)$

8.23 **a.** HNO_3
  b. CN^-
  c. HNO_3

8.25 **a.** Weak
  b. Weak
  c. Weak

8.27 **a.** CN^- and HCN; NH_3 and NH_4^+
  b. CO_3^{2-} and HCO_3^-; Cl^- and HCl

8.29 Concentration refers to the quantity of acid or base contained in a specified volume of solvent. Strength refers to the degree of dissociation of the acid or base.

8.31 **a.** Brønsted acid
  b. Brønsted base
  c. Both

8.33 **a.** Brønsted acid
  b. Both
  c. Brønsted base

8.35 HCN

8.37 I^-

8.39 **a.** 1.0×10^{-7} M
  b. 1.0×10^{-11} M

8.41 **a.** Neutral
  b. Basic

8.43 **a.** 2.00
  b. 4.00

8.45 **a.** $[H_3O^+] = 1.0 \times 10^{-1}$ M
  b. $[H_3O^+] = 1.0 \times 10^{-5}$ M

8.47 11.00

8.49 **a.** $[H_3O^+] = 5.0 \times 10^{-2}$ M
    $[OH^-] = 2.0 \times 10^{-13}$ M
  b. $[H_3O^+] = 2.0 \times 10^{-10}$ M
    $[OH^-] = 5.0 \times 10^{-5}$ M

8.51 A neutralization reaction is one in which an acid and a base react to produce water and a salt (a "neutral" solution).

8.53 **a.** $[H_3O^+] = 1.0 \times 10^{-6}$ M
    $[OH^-] = 1.0 \times 10^{-8}$ M
  b. $[H_3O^+] = 6.3 \times 10^{-6}$ M
    $[OH^-] = 1.6 \times 10^{-9}$ M
  c. $[H_3O^+] = 1.6 \times 10^{-8}$ M
    $[OH^-] = 6.3 \times 10^{-7}$ M

8.55 **a.** 1×10^2
  b. 1×10^4
  c. 1×10^{10}

8.57 **a.** $[H_3O^+] = 1.0 \times 10^{-5}$
  b. $[H_3O^+] = 1.0 \times 10^{-12}$
  c. $[H_3O^+] = 3.2 \times 10^{-6}$

8.59 **a.** pH = 6.00
  b. pH = 8.00
  c. pH = 3.25

8.61 pH = 3.12

8.63 pH = 10.74

8.65 $HNO_3 \ (aq) + NaOH \ (aq) \rightarrow H_2O \ (l) + NaNO_3 \ (aq)$

8.67 $H^+ \ (aq) + OH^- \ (aq) \rightarrow H_2O \ (l)$ or
$H_3O^+ \ (aq) + OH^- \ (aq) \rightarrow 2H_2O \ (l)$

8.69 Color change tells us that the proper volume of titrant has been added.

8.71 0.1800 M

8.73 13.33 mL

8.75 **a.** NH_3 and NH_4Cl can form a buffer solution.
  b. HNO_3 and KNO_3 cannot form a buffer solution.

8.77 **a.** A buffer solution contains components (a weak acid and its salt or a weak base and its salt) that enable the solution to resist large changes in pH when acids or bases are added.
  b. Acidosis is a medical condition characterized by higher-than-normal levels of CO_2 in the blood and lower-than-normal blood pH.

8.79 **a.** Addition of strong acid is equivalent to adding H_3O^+. This is a stress on the right side of the equilibrium and the equilibrium will shift to the left. Consequently the $[CH_3COOH]$ increases.
  b. Water, in this case, is a solvent and does not appear in the equilibrium expression. Hence, it does not alter the position of the equilibrium.

8.81 $[H_3O^+] = 2.32 \times 10^{-7}$ M

8.83 The very fact that an acid has a K_a value means that it is a weak acid. The smaller the value of K_a, the weaker the acid.

8.85 pH = 4.74

8.87 $11.2 = \dfrac{[HCO_3^-]}{[H_2CO_3]}$

8.89 The species oxidized *loses* electrons.

8.91 During an oxidation-reduction reaction the species *oxidized* is the reducing agent.

8.93 Cl_2 +   2KI $\rightarrow 2KCl + I_2$
    substance reduced   substance oxidized
     oxidizing agent   reducing agent

8.95 $2KI \rightarrow 2K^+ + 2e^- + I_2$ (oxidation $\frac{1}{2}$ reaction)

$Cl_2 + 2e^- \rightarrow 2Cl^-$ (reduction $\frac{1}{2}$ reaction)

8.97 An oxidation-reduction reaction must take place to produce electron flow in a voltaic cell.

8.99 Storage battery.

Chapter 9

9.1 X-ray, ultraviolet, visible, infrared, microwave, and radiowave.

9.3 1/4 of the radioisotope remains after 2 half-lives.

9.5 Isotopes with short half-lives release their radiation rapidly. There is much more radiation per unit time observed with short half-life substances; hence, the signal is stronger and the sensitivity of the procedure is enhanced.

9.7 The rem takes into account the relative biological effect of the radiation in addition to the quantity of radiation. This provides a more meaningful estimate of potential radiation damage to human tissue.

9.9 Natural radioactivity is the spontaneous decay of a nucleus to produce high-energy particles or rays.

9.11 Two protons and two neutrons

9.13 An electron with a -1 charge.

9.15 A positron has a positive charge and a beta particle has a negative charge.

9.17 • charge, $\alpha = +2$, $\beta = -1$
• mass, $\alpha = 4$ amu, $\beta = 0.000549$ amu
• velocity, $\alpha = 10\%$ of the speed of light, $\beta = 90\%$ of the speed of light

9.19 Chemical reactions involve joining, separating and rearranging atoms; valence electrons are critically involved. Nuclear reactions involve only changes in nuclear composition.

9.21 ^4_2He

9.23 $^{235}_{92}U$

9.25 $^1_1H \; 1 - 1 = 0$ neutrons
$^2_1H \; 2 - 1 = 1$ neutron
$^3_1H \; 3 - 1 = 2$ neutrons

9.27 $^{15}_7N$

9.29 Alpha and beta particles are matter; gamma radiation is pure energy. Alpha particles are large and relatively slow moving. They are the least energetic and least penetrating. Gamma radiation moves at the speed of light, is highly energetic, and is most penetrating.

9.31 A helium atom has two electrons; an α particle has no electrons.

9.33 $^{60}_{27}Co \rightarrow ^{60}_{28}Ni + ^{\;0}_{-1}\beta + \gamma$

9.35 $^{23}_{11}Na + ^2_1H \rightarrow ^{24}_{11}Na + ^1_1H$

9.37 $^{24}_{10}Ne \rightarrow \beta + ^{24}_{11}Na$

9.39 $^{140}_{55}Cs \rightarrow ^{140}_{56}Ba + ^{\;0}_{-1}e$

9.41 $^{209}_{83}Bi + ^{54}_{24}Cr \rightarrow ^{262}_{107}Bh + ^1_0n$

9.43 $^{27}_{12}Mg \rightarrow ^{\;0}_{-1}e + ^{27}_{13}Al$

9.45 $^{12}_7N \rightarrow ^{12}_6C + ^0_1e$

9.47 Natural radioactivity is a spontaneous process; artificial radioactivity is nonspontaneous and results from a nuclear reaction that produces an unstable nucleus.

9.49 • Nuclei for light atoms tend to be most stable if their neutron/proton ratio is close to 1.
• Nuclei with more than 84 protons tend to be unstable.
• Isotopes with a "magic number" of protons or neutrons (2, 8, 20, 50, 82, or 126 protons or neutrons) tend to be stable.
• Isotopes with even numbers of protons or neutrons tend to be more stable.

9.51 $^{20}_8O$; Oxygen-20 has $20 - 8 = 12$ neutrons, an n/p of 12/8, or 1.5. The n/p is probably too high for stability even though it does have a "magic number" of protons and an even number of protons and neutrons.

9.53 $^{48}_{24}Cr$; Chromium-48 has $48 - 24 = 24$ neutrons, an n/p of 24/24, or 1.0. It also has an even number of protons and neutrons. It would probably be stable.

9.55 0.40 mg of iodine-131 remains

9.57 13 mg of iron-59 remains

9.59 201 hours

9.61 Radiocarbon dating is a process used to determine the age of objects. The ratio of the masses of the stable isotope, carbon-12, and unstable isotope, carbon-14, is measured. Using this value and the half-life of carbon-14, the age of the coffin may be calculated.

9.63 Fission

9.65 **a.** The fission process involves the breaking down of large, unstable nuclei into smaller, more stable nuclei. This process releases some of the binding energy in the form of heat and/or light.
b. The heat generated during the fission process could be used to generate steam, which is then used to drive a turbine to create electricity.

9.67 $^3_1H + ^1_1H \rightarrow ^4_2He + $ energy

9.69 A "breeder" reactor creates the fuel which can be used by a conventional fission reactor during its fission process.

9.71 The reaction in a fission reactor that involves neutron production and causes subsequent reactions accompanied by the production of more neutrons in a continuing process.

9.73 High operating temperatures

9.75 Radiation therapy provides sufficient energy to destroy molecules critical to the reproduction of cancer cells.

9.77 Background radiation, radiation from natural sources, is emitted by the sun as cosmic radiation, and from naturally radioactive isotopes found throughout our environment.

9.79 $^{108}_{47}Ag + ^4_2He \rightarrow ^{112}_{49}In$

9.81 **a.** Technetium-99 m is used to study the heart (cardiac output, size, and shape), kidney (follow-up procedure for kidney transplant), and liver and spleen (size, shape, presence of tumors).
b. Xenon-133 is used to locate regions of reduced ventilation and presence of tumors in the lung.

9.83 Level decreases

9.85 Positive effect

9.87 Positive effect

9.89 Yes

9.91 Relative biological effect is a measure of the damage to biological tissue caused by different forms of radiation.

9.93 **a.** The curie is the amount of radioactive material needed to produce 3.7×10^{10} atomic disintegrations per second.

    b. The roentgen is the amount of radioactive material needed to produce 2×10^9 ion-pairs when passing through 1 cc of air at 0°C.

9.95 A film badge detects gamma radiation by darkening photographic film in proportion to the amount of radiation exposure over time. Badges are periodically collected and evaluated for their level of exposure. This mirrors the level of exposure of the personnel wearing the badges.

Chapter 10

10.1 The student could test the solubility of the substance in water and in an organic solvent, such as hexane. Solubility in hexane would suggest an organic substance; whereas solubility in water would indicate an inorganic compound. The student could also determine the melting and boiling points of the substance. If the melting and boiling points are very high, an inorganic substance would be suspected.

10.3 **a.**

    b.

    c.

10.5 Three of the six axial hydrogen atoms of cyclohexane lie above the ring. The remaining three hydrogen atoms lie below the ring.

10.7 **a.** The monobromination of propane will produce two products, as shown in the following two equations:

$$CH_3CH_2CH_3 + Br_2 \xrightarrow{\text{Light or heat}} CH_3CH_2CH_2Br + HBr$$

$$CH_3CH_2CH_3 + Br_2 \xrightarrow{\text{Light or heat}} CH_3CHBrCH_3 + HBr$$

    b. The monochlorination of butane will produce two products, as shown in the following two equations:

$$CH_3CH_2CH_2CH_3 + Cl_2 \xrightarrow{\text{Light or heat}} CH_3CH_2CH_2CH_2Cl + HCl$$

$$CH_3CH_2CH_2CH_3 + Cl_2 \xrightarrow{\text{Light or heat}} CH_3CH_2CHClCH_3 + HCl$$

    c. The monochlorination of cyclobutane:

    d. The monobromination of pentane will produce three products as shown in the following equations:

$$CH_3CH_2CH_2CH_2CH_3 + Br_2 \xrightarrow{\text{Light or heat}} CH_3CH_2CH_2CH_2CH_2Br + HBr$$

$$CH_3CH_2CH_2CH_2CH_3 + Br_2 \xrightarrow{\text{Light or heat}} CH_3CH_2CH_2CHBrCH_3 + HBr$$

$$CH_3CH_2CH_2CH_2CH_3 + Br_2 \xrightarrow{\text{Light or heat}} CH_3CH_2CHBrCH_2CH_3 + HBr$$

10.9 The number of organic compounds is nearly limitless because carbon forms stable covalent bonds with other carbon atoms in a variety of different patterns. In addition, carbon can form stable bonds with other elements and functional groups, producing many families of organic compounds, including alcohols, aldehydes, ketones, esters, ethers, amines, and amides. Finally, carbon can form double or triple bonds with other carbon atoms to produce organic molecules with different properties.

10.11 The allotropes of carbon include graphite, diamond, and buckminsterfullerene.

10.13 Because ionic substances often form three-dimensional crystals made up of many positive and negative ions, they generally have much higher melting and boiling points than covalent compounds.

10.15 **a.** $LiCl > H_2O > CH_4$

    b. $NaCl > C_3H_8 > C_2H_6$

10.17 **a.** LiCl would be a solid; H_2O would be a liquid; and CH_4 would be a gas.

    b. NaCl would be a solid; both C_3H_8 and C_2H_6 would be gases.

10.19 **a.** Water-soluble inorganic compounds

    b. Inorganic compounds

    c. Organic compounds

    d. Inorganic compounds

    e. Organic compounds

10.21 **a.**

    b.

10.23 **a.** $CH_3CH_2CHCH_3$ with CH_3 branch

    b. $CH_3CH_2CHCH_2CH_2CHCH_3$ with two CH_3 branches

10.25 **a.**

    b.

    c.

10.27 **a.**

    b.

    c.

10.29 Structure b is not possible because there are five bonds to carbon-2. Structure d is not possible because there are five bonds to carbon-3. Structure e is not possible because there are five bonds to carbon-2 and carbon-3 and only 3 bonds to carbon-4. Structure f is not possible because there are five bonds to carbon-3.

10.31 a.

b.

10.33

a. **b.**

10.35 An alcohol

An aldehyde

A ketone

A carboxylic acid

An amine

10.37 a. C_nH_{2n+2}
b. C_nH_{2n-2}
c. C_nH_{2n}
d. C_nH_{2n}
e. C_nH_{2n-2}

10.39 Alkanes have only carbon-to-carbon and carbon-to-hydrogen single bonds, as in the molecule ethane:

Alkenes have at least one carbon-to-carbon double bond, as in the molecule ethene:

Alkynes have at least one carbon-to-carbon triple bond, as in the molecule ethyne:

$$H-C\equiv C-H$$

10.41 a. A carboxylic acid:

$$CH_3CH_2-\overset{\overset{\displaystyle O}{\|}}{C}-OH$$

b. An amine: $CH_3CH_2CH_2-NH_2$
c. An alcohol: $CH_3CH_2CH_2-OH$
d. An ether: $CH_3CH_2-O-CH_2CH_3$

10.43

Carboxyl group

Ester

Aspirin
Acetylsalicylic acid

10.45 Hydrocarbons are nonpolar molecules, and hence are not soluble in water.

10.47 a. heptane > hexane > butane > ethane
b. $CH_3CH_2CH_2CH_2CH_2CH_2CH_2CH_2CH_3$ > $CH_3CH_2CH_2CH_2CH_3$ > $CH_3CH_2CH_3$

10.49 a. Heptane and hexane would be liquid at room temperature; butane and ethane would be gases.
b. $CH_3CH_2CH_2CH_2CH_2CH_2CH_2CH_2CH_3$ and $CH_3CH_2CH_2CH_2CH_3$ would be liquids at room temperature; $CH_3CH_2CH_3$ would be a gas.

10.51 Nonane: $CH_3CH_2CH_2CH_2CH_2CH_2CH_2CH_2CH_3$
Pentane: $CH_3CH_2CH_2CH_2CH_3$
Propane: $CH_3CH_2CH_3$

10.53 a.
$$\underset{\underset{\displaystyle |}{Br}}{CH_3CHCH_2CH_3}$$

b.
$$CH_3-\overset{\overset{\displaystyle Cl}{|}}{\underset{\underset{\displaystyle CH_3}{|}}{C}}-CH_3$$

c.
$$CH_3-\overset{\overset{\displaystyle CH_3}{|}}{\underset{\underset{\displaystyle CH_3}{|}}{C}}-CH_2CH_2CH_2CH_3$$

10.55 **a.** 2,2-Dibromobutane:

```
    H  Br  H   H
    |   |  |   |
H—C—C—C—C—H
    |   |  |   |
    H  Br  H   H
```

b. 2-Iododecane:

```
    H  I  H  H  H  H  H  H  H  H
    |  |  |  |  |  |  |  |  |  |
H—C—C—C—C—C—C—C—C—C—C—H
    |  |  |  |  |  |  |  |  |  |
    H  H  H  H  H  H  H  H  H  H
```

c. 1,2-Dichloropentane:

```
     H  Cl  H  H  H
     |   |  |  |  |
Cl—C—C—C—C—C—H
     |   |  |  |  |
     H   H  H  H  H
```

d. 1-Bromo-2-methylpentane:

```
        H
        |
      H—C—H
        |
    H   |   H  H  H
    |   |   |  |  |
H—C—C—C—C—C—H
    |   |   |  |  |
    Br  H   H  H  H
```

10.57 **a.** 3-Methylpentane
　　　　b. 2,5-Dimethylhexane
　　　　c. 1-Bromoheptane
　　　　d. 1-Chloro-3-methylbutane

10.59 **a.** 2-Chloropropane
　　　　b. 2-Iodobutane
　　　　c. 2,2-Dibromopropane
　　　　d. 1-Chloro-2-methylpropane
　　　　e. 2-Iodo-2-methylpropane

10.61 **a.** The straight chain isomers of molecular formula C_4H_9Br:

```
    H  H  H  H              H  H  Br  H
    |  |  |  |              |  |   |  |
H—C—C—C—C—Br        H—C—C—C—C—H
    |  |  |  |              |  |   |  |
    H  H  H  H              H  H   H  H
```

b. The straight chain isomers of molecular formula $C_4H_8Br_2$:

```
    H  H  H  Br             H  H  Br  H
    |  |  |  |              |  |   |  |
H—C—C—C—C—Br        H—C—C—C—C—Br
    |  |  |  |              |  |   |  |
    H  H  H  H              H  H   H  H

    H  Br  H  H            H  H  H  H
    |   |  |  |            |  |  |  |
H—C—C—C—C—Br       Br—C—C—C—C—Br
    |   |  |  |            |  |  |  |
    H   H  H  H            H  H  H  H

    H  H  Br  H            H  Br  Br  H
    |  |   |  |            |   |   |  |
H—C—C—C—C—H         H—C—C—C—C—H
    |  |   |  |            |   |   |  |
    H  H  Br  H            H   H   H  H
```

10.63 **a.** 2-Chlorohexane
　　　　b. 1,4-Dibromobutane
　　　　c. 3-Chloropentane
　　　　d. 2-Methylheptane

10.65 **a.** The first pair of molecules are constitutional isomers: hexane and 2-methylpentane.
　　　　b. The second pair of molecules are identical. Both are heptane.

10.67 **a.** Incorrect: 3-Methylhexane
　　　　b. Incorrect: 2-Methylbutane
　　　　c. Incorrect: 3-Methylheptane
　　　　d. Correct

10.69 **a.**

The name given in the problem is correct.

b.

The correct name is 3-methylhexane.

c. I—$CH_2CH_2CH_2CH_2CH_2$—I　The name given in the problem is correct.

d.

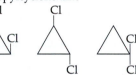

The correct name is 4-ethylnonane.

e.

The name given in the problem is correct.

10.71 Cycloalkanes are a family of molecules having carbon-to-carbon bonds in a ring structure.

10.73 The general formula for a cycloalkane is C_nH_{2n}.

10.75 **a.** Chlorocyclopropane
　　　　b. *cis*-1, 2-Dichlorocyclopropane
　　　　c. *trans*-1, 2-Dichlorocyclopropane
　　　　d. Bromocyclobutane

10.77 **a.** 1-Bromo-2-methylcyclobutane:

b. Iodocyclopropane:

10.79 **a.** *trans*-1-Bromo-2-ethylcyclobutane
　　　　b. *trans*-1, 2-Dimethylcyclopropane
　　　　c. Propylcyclohexane

10.81

10.83 **a.** Incorrect—1, 2-Dibromocyclobutane
　　　　b. Incorrect—1, 2-Diethylcyclobutane
　　　　c. Correct
　　　　d. Incorrect—1, 2, 3-Trichlorocyclohexane

10.85 **a.**

b.

c.

d.

10.87 **a.** *cis*-1, 2-Dibromocyclopentane
　　　　b. *trans*-1, 3-Dibromocyclopentane
　　　　c. *cis*-1, 2-Dimethylcyclohexane
　　　　d. *cis*-1, 2-Dimethylcyclopropane

10.89 Conformational isomers are distinct isomeric structures that may be converted into one another by rotation about the bonds in the molecule.

10.91 In the chair conformation the hydrogen atoms, and thus the electron pairs of the C— H bonds, are farther from one another. As a result, there is less electron repulsion and the structure is more stable (more energetically favored). In the boat conformation, the electron pairs are more crowded. This causes greater electron repulsion, producing a less stable, less energetically favored conformation.

10.93 Because conformations are freely and rapidly interconverted, they cannot be separated from one another.

10.95 One conformation is more stable than the other because the electron pairs of the carbon-hydrogen bonds are farther from one another.

10.97 Combustion is the oxidation of hydrocarbons by burning in the presence of air to produce carbon dioxide and water.

10.99 a. $C_3H_8 + 5O_2 \rightarrow 4H_2O + 3CO_2$
  b. $C_7H_{16} + 11O_2 \rightarrow 8H_2O + 7CO_2$
  c. $C_9H_{20} + 14O_2 \rightarrow 10H_2O + 9CO_2$
  d. $2C_{10}H_{22} + 31O_2 \rightarrow 22H_2O + 20CO_2$

10.101 a. $8CO_2 + 10H_2O$

b.

Br—C—CH₃ + CH₃CHCH₂Br + 2 HBr

c. Cl_2 + light

10.103 The following molecules are all isomers of C_6H_{14}.

CH₃CH₂CH₂CH₂CH₂CH₃   CH₃CHCH₂CH₂CH₃
                                          |
                                        CH₃

        Hexane   2-Methylpentane

CH₃CH₂CHCH₂CH₃   CH₃CHCHCH₃
         | |
        CH₃   CH₃

    3-Methylpentane   2,3-Dimethylbutane

CH₃CCH₂CH₃
 |
CH₃
 |
CH₃

    2,2-Dimethylbutane

a. 2, 3-Dimethylbutane produces only two monobrominated derivatives: 1-bromo-2, 3-dimethylbutane and 2-bromo-2, 3-dimethylbutane.

b. Hexane produces three monobrominated products: 1-bromohexane, 2-bromohexane, and 3-bromohexane. 2, 2-Dimethylbutane also produces three monobrominated products: 1-bromo-2, 2-dimethylbutane, 2-bromo-3, 3-dimethylbutane, and 1-bromo-3, 3-dimethylbutane.

c. 3-Methylpentane produces four monobrominated products: 1-bromo-3-methylpentane, 2-bromo-3-methylpentane, 3-bromo-3-methylpentane, and 1-bromo-2-ethylbutane.

10.105 The hydrocarbon is cyclooctane, having a molecular formula of C_8H_{16}.

$+ 12O_2 \longrightarrow 8CO_2 + 8H_2O$

Chapter 11

11.1 a.

Br—C—C—C≡C—C—C—H

b.

H—C—C≡C—C—H

c. Cl—C≡C—Cl

d.

H—C≡C—C—C—C—C—C—C—I

11.3 a.

cis-3-Hexene   *trans*-3-Hexene

b.

trans-2,3-Dibromo-2-butene   *cis*-2,3-Dibromo-2-butene

11.5 Molecule c can exist as *cis*- and *trans*-isomers because there are two different groups on each of the carbon atoms attached by the double bond.

11.7 a.   b.   c.

11.9 The hydrogenation of the *cis* and *trans* isomers of 2-pentene would produce the same product, pentane.

11.11 a.

$H_3C—C≡C—CH_3 + 2 H_2 \xrightarrow{Ni}$ H—C—C—C—C—H

        2-Butyne   Butane

b.

$H_3C—C≡C—CH_2CH_3 + 2 H_2 \xrightarrow{Ni}$ H—C—C—C—C—C—H

        2-Pentyne   Pentane

11.13 a. $CH_3CH=CH_2 + Br_2 \longrightarrow$ H—C—C—C—H

b. $CH_3CH=CHCH_3 + Br_2 \longrightarrow$ H—C—C—C—C—H

11.15 a. $CH_3C\equiv CCH_3 + 2Cl_2 \longrightarrow$

H—C—C—C—C—H (with H, Cl, Cl, H on top and H, Cl, Cl, H on bottom)

b.

$CH_3C\equiv CCH_2CH_3 + 2Cl_2 \longrightarrow$

H—C—C—C—C—C—H (with H, Cl, Cl, H, H on top and H, Cl, Cl, H, H on bottom)

11.17 a. $CH_3CH=CHCH_3 + H_2O \xrightarrow{H^+}$

$CH_3CHCH_2CH_3$ (only product)
      |
     OH

b. $CH_2=CHCH_2CH_2CHCH_3 + H_2O \xrightarrow{H^+}$
                              |
                             CH_3

$CH_3CHCH_2CH_2CHCH_3$ (major product)
      | |
     OH CH_3

$CH_2=CHCH_2CH_2CHCH_3 + H_2O \xrightarrow{H^+}$
                          |
                         CH_3

$CH_2CH_2CH_2CH_2CHCH_3$ (minor product)
  | |
 OH CH_3

c. $CH_3CH_2CH_2CH=CHCH_2CH_3 + H_2O \xrightarrow{H^+}$

$CH_3CH_2CH_2CHCH_2CH_2CH_3$
                |
               OH

$CH_3CH_2CH_2CH=CHCH_2CH_3 + H_2O \xrightarrow{H^+}$

$CH_3CH_2CH_2CH_2CHCH_2CH_3$
                    |
                   OH

These products will be formed in approximately equal amounts.

d. $CH_3CHClCH=CHCHClCH_3 + H_2O \xrightarrow{H^+}$

$CH_3CHClCHCH_2CHClCH_3$ (only product)
          |
         OH

11.19 a. $H_3CC\equiv CH + H_2O \xrightarrow{H^+}$

H—C—C=C—H (with H, H on top; H, OH bottom)

$\longrightarrow$ H—C—C—C—H (with H, H, H top; H, H, O)

Or

$H_3CC\equiv CH + H_2O \xrightarrow{H^+}$ H—C—C=C—H (H, OH top; H, H)

$\longrightarrow$ H—C—C—C—H (H, O, H top; H, H)

b.

$H_3CC\equiv CCH_2CH_3 + H_2O \xrightarrow{H^+}$

H—C—C=C—C—C—H (H, H, H, H on top; H, OH, H, H)

$\longrightarrow$ H—C—C—C—C—C—H (H, H, H, H top; H, H, O, H, H)

Or

$H_3CC\equiv CCH_2CH_3 + H_2O \xrightarrow{H^+}$

H—C—C=C—C—C—H (H, OH, H, H top; H, H, H, H)

$\longrightarrow$ H—C—C—C—C—C—H (H, O, H, H, H top; H, H, H, H)

11.21 a.

b.

c.

d.

e.

f.

11.23 The longer the carbon chain of an alkene, the higher the boiling point.

11.25 The general formula for an alkane is C_nH_{2n+2}.
The general formula for an alkene is C_nH_{2n}.
The general formula for an alkyne is C_nH_{2n-2}.

11.27 Ethene is a planar molecule. All of the bond angles are 120°.

11.29 In alkanes, such as ethane, the four bonds around each carbon atom have tetrahedral geometry. The bond angles are 109.5°. In alkenes, such as ethene, each carbon is bonded by two single bonds and one double bond. The molecule is planar and each bond angle is approximately 120°.

11.31 Ethyne is a linear molecule. All of the bond angles are 180°.

11.33 In alkanes, such as ethane, the four bonds around each carbon atom have tetrahedral geometry. The bond angles are 109.5°. In alkenes, such as ethene, each carbon is bonded by two single bonds and one double bond. The molecule is planar and each bond angle is approximately 120°. In alkynes, such as ethyne, each carbon is bonded by one single bond and one triple bond. The molecule is linear and the bond angles are 180°.

11.35 a. 2-Pentyne > Propyne > Ethyne
b. 3-Decene > 2-Butene > Ethene

11.37 Identify the longest carbon chain containing the carbon-to-carbon double or triple bond. Replace the –ane suffix of the alkane name with –ene for an alkene or -yne for an alkyne.

Number the chain to give the lowest number to the first of the two carbons involved in the double or triple bond. Determine the name and carbon number of each substituent group and place that information as a prefix in front of the name of the parent compound.

11.39 Geometric isomers of alkenes differ from one another in the placement of substituents attached to each of the carbon atoms of the double bond. Of the pair of geometric isomers, the *cis*- isomer, is the one in which identical groups are on the same side of the double bond.

11.41 **a.**

$$CH_3 \quad\quad CH_2CH_2CH_3$$
$$C{=}C$$
$$CH_3 \quad\quad H$$

b.

$$CH_3CH_2 \quad\quad H$$
$$C{=}C$$
$$H \quad\quad CH_2CH_2CH_3$$

c.

$$Cl$$
$$CH_2 \quad\quad CH_2CH_3$$
$$C{=}C$$
$$H \quad\quad H$$

d.

$$CH_3$$
$$CH_3CCl \quad\quad CH_2CH_2CH_3$$
$$C{=}C$$
$$H \quad\quad H$$

e.

$$CH_3$$
$$CH_3CH \quad\quad H$$
$$C{=}C$$
$$H \quad\quad CH{-}CHCH_2CH_3$$
$$\quad\quad\quad Br \quad CH_3$$

11.43 **a.** 3-Methyl-1-pentene
b. 7-Bromo-1-heptene
c. 5-Bromo-3-heptene
d. 1-*t*-Butyl-4-methylcyclohexene

11.45 **a.** 1,3,5-Trifluoropentane: $CH_2{-}CH_2{-}CH{-}CH_2{-}CH_2$
(F on each: F, F, F)

b. *cis*-2-Octene:

$$H \quad\quad H$$
$$C{=}C$$
$$H_3C \quad\quad CH_2CH_2CH_2CH_2CH_3$$

c. Dipropylacetylene: $CH_3CH_2CH_2{-}C{\equiv}C{-}CH_2CH_2CH_3$

11.47 **a.** 2,3-Dibromobutane could not exist as *cis* and *trans* isomers.
b.

$$CH_3 \quad\quad CH_2CH_2CH_2CH_3$$
$$C{=}C$$
$$H \quad\quad H$$
cis-2-Heptene

$$CH_3 \quad\quad H$$
$$C{=}C$$
$$H \quad\quad CH_2CH_2CH_2CH_3$$
trans-2-Heptene

c.

$$CH_3 \quad\quad CH_3$$
$$C{=}C$$
$$Br \quad\quad Br$$
cis-2,3-Dibromo-2-butene

$$CH_3 \quad\quad Br$$
$$C{=}C$$
$$Br \quad\quad CH_3$$
trans-2,3-Dibromo-2-butene

d. Propene cannot exist as *cis* and *trans* isomers.

11.49 Alkenes b and c would not exhibit *cis-trans* isomerism.
11.51 Alkenes b and d can exist as both *cis*- and *trans*- isomers.
11.53 **a.** 1,5-Nonadiene
b. 1,4,7-Nonatriene
c. 2,5-Octadiene
d. 4-Methyl-2,5-heptadiene

11.55

$$R \quad\quad R$$
$$C{=}C \quad + H_2 \quad \xrightarrow[\text{heat or pressure}]{\text{Pt, Pd, or Ni}} \quad R{-}\overset{H}{\underset{R}{C}}{-}\overset{R}{\underset{H}{C}}{-}R$$
$$R \quad\quad R$$

11.57

$$R \quad\quad R$$
$$C{=}C \quad + X_2 \quad \longrightarrow \quad R{-}\overset{X}{\underset{R}{C}}{-}\overset{R}{\underset{X}{C}}{-}R$$
$$R \quad\quad R$$

11.59

$$R \quad\quad R$$
$$C{=}C \quad + H_2O \quad \xrightarrow{H^+} \quad R{-}\overset{H}{\underset{R}{C}}{-}\overset{R}{\underset{OH}{C}}{-}R$$
$$R \quad\quad R$$

11.61 The primary difference between complete hydrogenation of an alkene and an alkyne is that 2 moles of H_2 are required for the complete hydrogenation of an alkyne.

11.63 Addition of bromine (Br_2) to an alkene results in a color change from red to colorless. If equimolar quantities of Br_2 are added to hexene, the reaction mixture will change from red to colorless. This color change will not occur if cyclohexane is used.

11.65 **a.** H_2 **d.** $19O_2 \rightarrow 12CO_2 + 14H_2O$
b. H_2O **e.** Cl_2
c. HBr   **f.** (cyclopentene structure)

11.67 **a.**

$$H_3C{-}C{\equiv}C{-}CH_3 + 2H_2 \quad \xrightarrow[\text{heat or pressure}]{\text{Pt, Pd, or Ni}} \quad H_3C{-}\overset{H}{\underset{H}{C}}{-}\overset{H}{\underset{H}{C}}{-}CH_3$$
2-Butyne

b.

$$CH_3CH_2{-}C{\equiv}C{-}CH_3 + 2X_2 \quad \longrightarrow \quad CH_3CH_2{-}\overset{X}{\underset{X}{C}}{-}\overset{X}{\underset{X}{C}}{-}CH_3$$
2-Pentyne

11.69 **a.** Reactant—*cis*-2-butene; Only product—butane
b. Reactant—1-butene; Major product—2-butanol
c. Reactant—2-butene; Only product—2, 3-dichlorobutane
d. Reactant—1-pentene; Major product—2-bromopentane

11.71 $CH_2{=}CHCH_2CH_2CH_3$, $CH_3CH{=}CHCH_2CH_3$,

$$CH_3\overset{}{\underset{CH_3}{C}}{=}CHCH_3, \quad CH_2{=}\overset{}{\underset{CH_3}{C}}CH_2CH_3, \quad CH_2{=}CHCHCH_3 \; (CH_3)$$

11.73 **a.**

$$Br$$
$$CH_3CHCH_2CH_3$$

b.

$$CH_3CH_2{-}\overset{I}{\underset{CH_3}{C}}{-}CH_2CH_2CH_3 + CH_3CHCHCH_2CH_2CH_3 \;(I\; CH_3)$$
(major product)   (minor product)

c. Cl (cyclopentane structure)

11.75 A polymer is a macromolecule composed of repeating structural units called *monomers*.

11.77

$$n \; \overset{F}{\underset{F}{C}}{=}\overset{F}{\underset{F}{C}} \quad \longrightarrow \quad \left[\overset{F}{\underset{F}{C}}{-}\overset{F}{\underset{F}{C}}\right]_n$$
Tetrafluoroethene   Teflon

11.79 a.

$$CH_3C{=}CCH_2CH_3 + H_2O \xrightarrow{H^+} CH_3CHCH_2CH_2CH_3$$

(with H substituents on the double bond carbons; OH product)

2-Pentene

$$CH_3C{=}CCH_2CH_3 + H_2O \xrightarrow{H^+} CH_3CH_2CHCH_2CH_3$$

(OH product)

These products will be formed in approximately equal amounts.

b.

$$CH_2C{=}C{-}H + H_2O \xrightarrow{H^+} CH_2CHCH_3 \text{ (major product)}$$

Br (on CH$_2$); Br OH on products

3-Bromo-1-propene

$$CH_2C{=}C{-}H + H_2O \xrightarrow{H^+} CH_2CH_2CHOH \text{ (minor product)}$$

Br

c.

3,4-Dimethylcyclohexene

$$+ H_2O \xrightarrow{H^+}$$

(cyclohexane ring with OH and CH$_3$ groups)

$$+ H_2O \xrightarrow{H^+}$$

(cyclohexane ring with HO and CH$_3$ groups)

These products will be formed in approximately equal amounts.

11.81 a.

$$CH_2{=}CHCH_2CHCH_3 + H_2O \xrightarrow{H^+} CH_2CH_2CH_2CHCH_3$$

with CH$_3$ and OH

(This is the minor product of this reaction.)

b.

$$CH_3C{=}CCH_2CH_2CH_3 + HBr \longrightarrow CH_3CH_2CHCH_2CH_2CH_3$$

Br

OR

$$CH_3CH_2C{=}CCH_2CH_3 + HBr \longrightarrow CH_3CH_2CHCH_2CH_2CH_3$$

Br

c.

$$+ HBr \longrightarrow$$

(cyclohexene with CH$_3$ to cyclohexane with Br and CH$_3$)

d.

$$-CH_2CH_3 + H_2O \xrightarrow{H^+}$$

(cyclopentene with CH$_2$CH$_3$ to cyclopentane with CH$_2$CH$_3$ and OH)

11.83 a.

$$CH_2{=}CHCH_2CH{=}CHCH_3 + 2H_2 \xrightarrow[\text{heat}]{Pt} CH_3(CH_2)_4CH_3$$

1,4-Hexadiene   Hexane

b.

$$CH_3CH{=}CHCH{=}CHCH{=}CHCH_3 + 3H_2$$

2,4,6-Octatriene $\xrightarrow[\text{heat}]{Ni} CH_3(CH_2)_6CH_3$

Octane

c.

$$+ 2H_2 \xrightarrow[\text{Pressure}]{Pd}$$

1,3-Cyclohexadiene   Cyclohexane

d.

$$+ 3H_2 \xrightarrow[\text{heat}]{Ni}$$

1,3,5-Cyclooctatriene   Cyclooctane

11.85 The term aromatic hydrocarbon was first used as a term to describe the pleasant-smelling resins of tropical trees.

11.87 Resonance hybrids are molecules for which more than one valid Lewis structure can be written.

11.89 a. (benzene ring with CH$_3$, Br, Br)   **b.** (benzene ring with CH$_2$CH$_3$, CH$_2$CH$_3$, CH$_2$CH$_3$)

c. CH$_3$CHCH$_3$ (benzene with isopropyl)   **d.** (benzene ring with CH$_3$, Br, Cl)

11.91 a. OH (benzene with OH and CH$_3$)   **b.** CH$_2$CH$_2$CH$_3$ (benzene with propyl)

c. NO$_2$ (benzene with NO$_2$, O$_2$N, NO$_2$)   **d.** CH$_3$ (benzene with CH$_3$, Cl)

11.93 Kekulé proposed that single and double carbon-carbon bonds alternate around the benzene ring. To explain why benzene does not react like other unsaturated compounds, he proposed that the double and single bonds shift positions rapidly.

11.95 An addition reaction involves addition of a molecule to a double or triple bond in an unsaturated molecule. In a substitution reaction, one chemical group replaces another.

11.97

$$+ Cl_2 \xrightarrow{FeCl_3} \text{(chlorobenzene)} + HCl$$

11.99

Pyrimidine

11.101

Purine

Chapter 12

12.1 a. 2-Methyl-1-propanol

$$CH_3CHCH_2OH$$
$$|$$
$$CH_3$$

b. 2-Chlorocyclopentanol

c. 2,4-Dimethylcyclohexanol

d. 2,3-Dichloro-3-hexanol

$$OH$$
$$|$$
$$CH_3CHCCH_2CH_2CH_3$$
$$|\ |$$
$$Cl\ Cl$$

12.3 a. The reactant is propene. The major product is 2-propanol and the minor product is 1-propanol. The major product is a secondary alcohol (2-propanol) and the minor product is a primary alcohol (1-propanol). Propene cannot exist as both *cis-* and *trans-* isomers.

b. The reactant is ethene. The product is ethanol. Ethanol is a primary alcohol. Ethene cannot exist as both *cis-* and *trans-* isomers.

c. The reactant is 3-hexene. The product is 3-hexanol, which is a secondary alcohol. The reactant 3-hexene can be found as both *cis-* and *trans-* isomers.

12.5 a. Ethanol is a primary alcohol.
b. 2-Propanol is a secondary alcohol.
c. 4-Methyl-3-hexanol is a secondary alcohol.
d. 2-Methyl-2-pronanol is a tertiary alcohol.

12.7 a. The reactant is 2-butanol and the product is butanone.
b. The reactant is 2-pentanol and the product is 2-pentanone.

12.9 Simple phenols are somewhat soluble in water because they have the polar hydroxyl group.

12.11 Ethers have much lower boiling points than alcohols because ether molecules cannot hydrogen bond to one another.

12.13 The longer the hydrocarbon tail of an alcohol becomes, the less water soluble it will be.

12.15 a < d < c < b

12.17 a. CH_3CH_2OH
b. $CH_3CH_2CH_2CH_2OH$
c. CH_3CHCH_3
$$|$$
$$OH$$

12.19 The I.U.P.A.C. rules for the nomenclature of alcohols require you to name the parent compound, that is the longest continuous carbon chain bonded to the —OH group. Replace the *–e* ending of the parent alkane with *–ol* of the alcohol. Number the parent chain so that the carbon bearing the hydroxyl group has the lowest possible number. Name and number all other substituents. If there is more than one hydroxyl group, the *–ol* ending will be modified to reflect the number. If there are two — OH groups, the suffix *–diol* is used; if it has three — OH groups, the suffix *–triol* is used, etc.

12.21 a. 1-Heptanol
b. 2-Propanol
c. 2, 2-Dimethylpropanol

12.23 a. 3-Hexanol:

b. 1,2,3-Pentanetriol:

c. 2-Methyl-2-pentanol:

12.25 a. Cyclopentanol
b. Cyclooctanol
c. 3-Methylcyclohexanol

12.27 a. Methyl alcohol
b. Ethyl alcohol
c. Ethylene glycol
d. Propyl alcohol

12.29 a. 4-Methyl-2-hexanol

$$CH_3$$
$$|$$
$$CH_3CHCH_2CHCH_2CH_3$$
$$|$$
$$OH$$

b. Isobutyl alcohol

$$CH_3$$
$$|$$
$$CH_3CHCH_2OH$$

c. 1,5-Pentanediol

$$CH_2CH_2CH_2CH_2CH_2$$
$$|\qquad\qquad\qquad |$$
$$OH\qquad\qquad\quad OH$$

d. 2-Nonanol

$$CH_3CHCH_2CH_2CH_2CH_2CH_2CH_3$$
$$|$$
$$OH$$

e. 1,3,5-Cyclohexanetriol

12.31 Denatured alcohol is 100% ethanol to which benzene or methanol is added. The additive makes the ethanol unfit to drink and prevents illegal use of pure ethanol.

12.33 Fermentation is the anaerobic degradation of sugar that involves no net oxidation. The alcohol fermentation, carried out by yeast, produces ethanol and carbon dioxide.

12.35 When the ethanol concentration in a fermentation reaches 12–13%, the yeast producing the ethanol are killed by it. To produce a liquor of higher alcohol concentration, the product of the original fermentation must be distilled.

12.37 The carbinol carbon is the one to which the hydroxyl group is bonded.

12.39 a. Primary
b. Secondary
c. Tertiary
d. Tertiary
e. Tertiary

12.41 a. Primary alcohol
b. Secondary alcohol
c. Primary alcohol
d. Primary alcohol
e. Secondary alcohol

12.43 a. 2-Nonanol is a secondary alcohol:
$CH_3CHOHCH_2CH_2CH_2CH_2CH_2CH_3$
2. 2-Heptanol is a secondary alcohol:
$CH_3CHOHCH_2CH_2CH_2CH_2CH_3$
c. 2-Undecanol is a secondary alcohol:
$CH_3CHOHCH_2CH_2CH_2CH_2CH_2CH_2CH_2CH_2CH_3$

12.45

12.47

12.49

12.51 a. 2-Pentanol (major product), 1-pentanol (minor product)
b. 2-Pentanol and 3-pentanol
c. 3-Methyl-2-butanol (major product), 3-methyl-1-butanol (minor product)
d. 3, 3-Dimethyl-2-butanol (major product), 3, 3-dimethyl-1-butanol (minor product)

12.53 a.

These products will be formed in approximately equal amounts.

b.

c. $CH_2{=}CHCH_2CH_2CH_2CH_2CH_2CH_3 + H_2O \xrightarrow{H^+}$
1-Octene

2-Octanol (major product)
or
1-Octanol (minor product)

d.

1-Methylcyclohexene

1-Methylcyclohexanol (major product)   2-Methylcyclohexanol (minor product)

12.55 a. 2-Butanone
b. N.R.
c. Cyclohexanone
d. N.R.

12.57 a. 3-Pentanone
b. Propanal (Upon further oxidation, propanoic acid would be formed.)
c. 4-Methyl-2-pentanone
d. N.R.
e. 3-Phenylpropanal (Upon further oxidation, 3-phenylpropanoic acid will be formed.)

12.59 $CH_3CH_2OH \xrightarrow{liver\ enzymes} CH_3{-}\overset{O}{\overset{\|}{C}}{-}H$
Ethanol   Ethanal

The product, ethanal, is responsible for the symptoms of a hangover.

12.61 The reaction in which a water molecule is added to 1-butene is a hydration reaction.

$CH_3CH_2CH{=}CH_2 + H_2O \xrightarrow{H^+} CH_3CH_2CHCH_3$ (OH)
1-Butene   2-Butanol

12.63

$CH_3CH{=}CH_2 \xrightarrow{H_2O,\ H^+} CH_3{-}CH{-}CH_3 \xrightarrow{[O]} CH_3{-}C{-}CH_3$
Propene (propylene)   2-Propanol (isopropanol)   Propanone (acetone)

12.65 a.

$$CH_3CH_2CH_2CH_2\overset{\displaystyle O}{\overset{\|}{C}}\text{-}H + H_2 \xrightarrow{\text{Catalyst}} CH_3CH_2CH_2CH_2CH_2OH$$

   Pentanal   1-Pentanol

b.

$$CH_3CH_2\overset{\displaystyle O}{\overset{\|}{C}}CH_2CH_3 + H_2 \xrightarrow{\text{Catalyst}} CH_3CH_2CHOHCH_2CH_3$$

   3-Pentanone   3-Pentanol

12.67 Oxidation is a loss of electrons, whereas reduction is a gain of electrons.

12.69

$$CH_3CH_2CH_3 < CH_3CH_2CH_2OH < CH_3CH_2\overset{\displaystyle O}{\overset{\|}{C}}\text{—}H < CH_3CH_2\overset{\displaystyle O}{\overset{\|}{C}}\text{—}OH$$

12.71 Phenols are compounds with an —OH attached to a benzene ring.

12.73   Picric acid: 2,4,6,-Trinitrotoluene:

   Picric acid is water-soluble because of the polar hydroxyl group that can form hydrogen bonds with water.

12.75 Hexachlorophene, hexylresorcinol, and *o*-phenylphenol are phenol compounds used as antiseptics or disinfectants.

12.77 Ethers have much lower boiling points than alcohols of similar molecular weight, but higher boiling points than alkanes of similar molecular weight. The boiling points are higher than alkanes because the R-O-R bond is polar. However, there is no —OH group, so ether molecules cannot hydrogen bond to one another. This is the reason that the boiling points are lower than alcohols of similar molecular weight.

12.79 Alcohols of molecular formula $C_4H_{10}O$

$$CH_3CH_2CH_2CH_2OH, \qquad CH_3\overset{\displaystyle OH}{\overset{|}{CH}}CH_2CH_3,$$

$$CH_3\underset{\underset{\displaystyle CH_3}{|}}{CH}CH_2OH, \qquad CH_3\text{—}\overset{\displaystyle OH}{\underset{\underset{\displaystyle CH_3}{|}}{\overset{|}{C}}}\text{—}CH_3$$

   Ethers of molecular formula $C_4H_{10}O$
$$CH_3\text{—}O\text{—}CH_2CH_2CH_3 \qquad CH_3CH_2\text{—}O\text{—}CH_2CH_3$$
$$CH_3\text{—}O\text{—}\underset{\underset{\displaystyle CH_3}{|}}{CH}CH_3$$

12.81 Penthrane: 2, 2-Dichloro-1, 1-difluoro-1-methoxyethane
   Enthrane: 2-Chloro-1-(difluoromethoxy)-1, 1, 2-trifluoroethane

12.83 a. $CH_3CH_2\text{—}O\text{—}CH_2CH_3 + H_2O$
b. $CH_3CH_2\text{—}O\text{—}CH_2CH_3 + CH_3\text{—}O\text{—}CH_3 + CH_3\text{—}O\text{—}CH_2CH_3 + H_2O$
c. $CH_3\text{—}O\text{—}CH_3 + CH_3\text{—}O\text{—}\underset{\underset{\displaystyle CH_3}{|}}{CH}CH_3 +$

$$CH_3\underset{\underset{\displaystyle CH_3}{|}}{CH}\text{—}O\text{—}\underset{\underset{\displaystyle CH_3}{|}}{CH}CH_3 + H_2O$$

d.

12.85 a. 2-Ethoxypentane
b. 2-Methoxybutane
c. 1-Ethoxybutane
d. Methoxycyclopentane

12.87 a. Methyl propyl ether:
$$CH_3\text{—}O\text{—}CH_2CH_2CH_3$$

b. Methyl octyl ether:
$$CH_3\text{—}O\text{—}CH_2CH_2CH_2CH_2CH_2CH_2CH_2CH_3$$

c. Diisopropyl ether:
$$CH_3\text{—}\underset{\underset{\displaystyle CH_3}{|}}{CH}\text{—}O\text{—}\underset{\underset{\displaystyle CH_3}{|}}{CH}\text{—}CH_3$$

d. Ethyl pentyl ether:
$$CH_3CH_2\text{—}O\text{—}CH_2CH_2CH_2CH_2CH_3$$

12.89 Thiols contain the sulfhydryl group (—SH). The sulfhydryl group is similar to the hydroxyl group (—OH) of alcohols, except that a sulfur atom replaces the oxygen atom.

12.91 Cystine:

$$H_3\overset{+}{N}\text{—}\underset{\underset{\displaystyle CH_2}{|}}{\overset{\overset{\displaystyle H}{|}}{C}}\text{—}COO^-$$
$$\underset{\underset{\underset{\underset{\underset{\displaystyle H}{|}}{H_3\overset{+}{N}\text{—}C\text{—}COO^-}}{|}}{CH_2}}{\underset{\underset{\displaystyle S}{|}}{S}}$$

12.93 a. 1-Propanethiol
b. 2-Butanethiol
c. 2-Methyl-2-butanethiol
d. 1, 4-Cyclohexanedithiol

Chapter 13

13.1 a.
$$CH_3\text{—}\overset{\displaystyle O}{\overset{\|}{C}}\text{—}CH_3$$

b.
$$CH_3\overset{\displaystyle OH}{\overset{|}{CH}}CH_2CH_2CH_3$$

13.3 a.
$$CH_3CH_2\overset{\displaystyle O}{\overset{\|}{C}}\text{—}OH$$

b.
$$CH_3\overset{\displaystyle O}{\overset{\|}{C}}\text{—}OH$$

13.5 a. 2,3-Dichloropentanal

Cl O
| ||
CH₃CH₂CHCHC—H
|
Cl

b. 2-Bromobutanal

O
||
CH₃CH₂CHC—H
|
Br

c. 4-Methylhexanal

CH₃ O
| ||
CH₃CH₂CHCH₂CH₂C—H

d. Butanal

O
||
CH₃CH₂CH₂C—H

e. 2,4-Dimethylpentanal

CH₃ O
| ||
CH₃CHCH₂CHC—H
|
CH₃

13.7 a. 3-Iodobutanone
b. 4-Methyl-2-octanone
c. 3-Methylbutanone
d. 2-Methyl-3-pentanone
e. 2-Fluoro-3-pentanone

13.9

O
||
CH₃—C—H

13.11 a. Reduction
b. Reduction
c. Reduction
d. Oxidation
e. Reduction

13.13 a. Hemiacetal
b. Ketal
c. Acetal
d. Hemiketal

13.15

O
||
2CH₃CH₂C—H

OH⁻
——→

OH O
| ||
CH₃CH₂CHCHC—H
|
CH₃

Propanal   3-Hydroxy-2-methylpentanal

13.17 As the carbon chain length increases, the compounds become less polar and more hydrocarbonlike. As a result, their solubility in water decreases.

13.19 A good solvent should dissolve a wide range of compounds. Simple ketones are considered to be universal solvents because they have both a polar carbonyl group and nonpolar side chains. As a result, they dissolve organic compounds and are also miscible in water.

13.21

O
/ \
H   H
⋮   ⋮
O   O
|| ||
CH₃—C—H   H—C—CH₃

13.23 Alcohols have higher boiling points than aldehydes or ketones of comparable molecular weights because alcohol molecules can form intermolecular hydrogen bonds with one another. Aldehydes and ketones cannot form intermolecular hydrogen bonds.

13.25 To name an aldehyde using the I.U.P.A.C. Nomenclature System, identify and name the longest carbon chain containing the carbonyl group. Replace the final *-e* of the alkane name with *-al*. Number and name all substituents as usual. Remember that the carbonyl carbon is always carbon-1 and does not need to be numbered in the name of the compound.

13.27 The common names of aldehydes are derived from the same Latin roots as the corresponding carboxylic acids. For instance, methanal is formaldehyde; ethanal is acetaldehyde; propanal is propionaldehyde, *etc.*

Substituted aldehydes are named as derivatives of the straight-chain parent compound. Greek letters are used to indicate the position of substituents. The carbon nearest the carbonyl group is the α-carbon, the next is the β-carbon, and so on.

13.29 a.

O
||
H—C—H

b.

H H H H H H H O
| | | | | | | ||
H—C—C—C—C—C—C—C—C—H
| | | | | | |
Br Br H H H H H

13.31 a. 3-Chloro-2-pentanone

H H H O H
| | | || |
H—C—C—C—C—C—H
| | | |
H H Cl H

b. Benzaldehyde

13.33 a. Butanone
b. 2-Ethylhexanal

13.35 a. 3-Nitrobenzaldehyde
b. 3,4-Dihydroxycyclopentanone

13.37 7-Hydroxy-3, 7-dimethyloctanal

13.39 a. 4,6-Dimethyl-3-heptanone
b. 3,3-Dimethylcyclopentanone

13.41 a. Acetone
b. Ethyl methyl ketone
c. Acetaldehyde
d. Propionaldehyde
e. Methyl isopropyl ketone

13.43 a.

O
||
CH₃CHCH₂C—H
|
OH

b.

O
||
CH₃CH₂CH₂CHC—H
|
CH₃

c.

O
||
CH₃CH₂CHCH₂CH₂C—H
|
Br

d.

O
||
CH₃CH₂CHCH₂C—H
|
I

e.

CH₃ O
| ||
CH₃CH₂CH₂CH₂CHCHC—H
|
OH

13.45 Acetone is a good solvent because it can dissolve a wide range of compounds. It has both a polar carbonyl group and nonpolar side chains. As a result, it dissolves organic compounds and is also miscible in water.

13.47 The liver

13.49 In organic molecules, oxidation may be recognized as a gain of oxygen or a loss of hydrogen. An aldehyde may be oxidized to form a carboxylic acid as in the following example in which ethanal is oxidized to produce ethanoic acid.

$$H_3C-\overset{\overset{\displaystyle O}{\|}}{C}-H \xrightarrow{[O]} H_3C-\overset{\overset{\displaystyle O}{\|}}{C}-OH$$

Ethanal   Ethanoic acid

13.51 Addition reactions of aldehydes or ketones are those in which a second molecule is added to the double bond of the carbonyl group. An example is the addition of the alcohol ethanol to the aldehyde ethanal.

$$H_3C-\overset{\overset{\displaystyle O}{\|}}{C}-H + CH_3CH_2OH \xrightarrow{H^+} H_3C-\overset{\overset{\displaystyle OH}{|}}{\underset{\underset{\displaystyle H}{|}}{C}}-OCH_2CH_3$$

Ethanal   Ethanol   Hemiacetal

13.53 The following equation represents the oxidation of an aldehyde. The product is a carboxylic acid.

$$R-\overset{\overset{\displaystyle O}{\|}}{C}-H \xrightarrow{[O]} R-\overset{\overset{\displaystyle O}{\|}}{C}-OH$$

Aldehyde   Carboxylic acid

13.55 Ketones cannot be oxidized further.

$$R-\overset{\overset{\displaystyle O}{\|}}{C}-R \xrightarrow{[O]} \text{No reaction}$$

Ketone

13.57 The following general equation represents the addition of two alcohol molecules to an aldehyde:

$$R-\overset{\overset{\displaystyle O}{\|}}{C}-H + R'OH \underset{}{\overset{H^+}{\rightleftharpoons}} R-\overset{\overset{\displaystyle OH}{|}}{\underset{\underset{\displaystyle H}{|}}{C}}-OR' + R''OH \underset{}{\overset{H^+}{\rightleftharpoons}} R-\overset{\overset{\displaystyle OR''}{|}}{\underset{\underset{\displaystyle H}{|}}{C}}-OR'$$

Aldehyde   Hemiacetal   Acetal

13.59 The following general equation represents the addition of two alcohol molecules to a ketone:

$$R-\overset{\overset{\displaystyle O}{\|}}{C}-R + R'OH \underset{}{\overset{H^+}{\rightleftharpoons}} R-\overset{\overset{\displaystyle OH}{|}}{\underset{\underset{\displaystyle R}{|}}{C}}-OR' + R''OH \underset{}{\overset{H^+}{\rightleftharpoons}} R-\overset{\overset{\displaystyle OR''}{|}}{\underset{\underset{\displaystyle R}{|}}{C}}-OR'$$

Ketone   Hemiketal   Ketal

13.61 a.

$$CH_3CH_2\overset{\overset{\displaystyle OH}{|}}{C}HCH_3 \xrightarrow{[O]} CH_3CH_2\overset{\overset{\displaystyle O}{\|}}{C}CH_3$$

2-Butanol   Butanone

b.

$$CH_3\overset{\overset{\displaystyle CH_3}{|}}{C}HCH_2OH \xrightarrow{[O]} CH_3\overset{\overset{\displaystyle CH_3}{|}}{C}H-\overset{\overset{\displaystyle O}{\|}}{C}-H$$

2-Methyl-1-propanol   Methylpropanal

Note that methylpropanal can be further oxidized to methylpropanoic acid.

c.

Cyclopentanol $\xrightarrow{[O]}$ Cyclopentanone

13.63 $R-CH_2OH \xrightarrow{[O]} R-\overset{\overset{\displaystyle O}{\|}}{C}-H \xrightarrow{[O]} R-\overset{\overset{\displaystyle O}{\|}}{C}-OH$

Primary alcohol   Aldehyde   Carboxylic acid

13.65 a. Reduction reaction

$$CH_3-\overset{\overset{\displaystyle O}{\|}}{C}-H \longrightarrow CH_3CH_2OH$$

Ethanal   Ethanol

b. Reduction reaction

Cyclohexanone   Cyclohexanol

c. Oxidation reaction

$$CH_3\overset{\overset{\displaystyle OH}{|}}{C}HCH_3 \longrightarrow CH_3-\overset{\overset{\displaystyle O}{\|}}{C}-CH_3$$

2-Propanol   Propanone

13.67 a. The following equation represents the hydrogenation of 3-methylbutanone. The product is 3-methyl-2-butanol.

$$CH_3\overset{\overset{\displaystyle O}{\|}}{\underset{\underset{\displaystyle CH_3}{|}}{C}}HCCH_3 + H_2 \xrightarrow{Pt} CH_3\overset{\overset{\displaystyle OH}{|}}{\underset{\underset{\displaystyle CH_3}{|}}{C}}HCHCH_3$$

3-Methylbutanone   3-Methyl-2-butanol

b. The following equation represents the hydrogenation of 2-pentanone. The product is 2-pentanol.

$$CH_3CH_2CH_2\overset{\overset{\displaystyle O}{\|}}{C}CH_3 + H_2 \xrightarrow{Pt} CH_3CH_2CH_2\overset{\overset{\displaystyle OH}{|}}{C}HCH_3$$

2-Pentanone   2-Pentanol

c. The following equation represents the hydrogenation of 1-chloropropanone. The product is 1-chloro-2-propanol.

$$CH_2\overset{\overset{\displaystyle O}{\|}}{\underset{\underset{\displaystyle Cl}{|}}{C}}CH_3 + H_2 \xrightarrow{Pt} CH_2\overset{\overset{\displaystyle OH}{|}}{\underset{\underset{\displaystyle Cl}{|}}{C}}HCH_3$$

1-Chloropropanone   1-Chloro-2-propanol

13.69 a. The following equation represents the hydrogenation of butanol. The product is 1-butanol.

$$CH_3CH_2CH_2\overset{\overset{\displaystyle O}{\|}}{C}H + H_2 \xrightarrow{Pt} CH_3CH_2CH_2CH_2OH$$

Butanal   1-Butanol

b. The following equation represents the hydrogenation of 3-methylpentanal. The product is 3-methyl-1-pentanol.

$$CH_3CH_2\overset{}{\underset{\underset{\displaystyle CH_3}{|}}{C}}HCH_2\overset{\overset{\displaystyle O}{\|}}{C}H + H_2 \xrightarrow{Pt} CH_3CH_2\overset{}{\underset{\underset{\displaystyle CH_3}{|}}{C}}HCH_2CH_2OH$$

3-Methylpentanal   3-Methyl-1-pentanol

c. The following equation represents the hydrogenation of 2-methylpropanal. The product is 2-methyl-1-propanol.

$$CH_3\overset{}{\underset{\underset{\displaystyle CH_3}{|}}{C}}H\overset{\overset{\displaystyle O}{\|}}{C}H + H_2 \xrightarrow{Pt} CH_3\overset{}{\underset{\underset{\displaystyle CH_3}{|}}{C}}HCH_2OH$$

13.71 Only (c) 3-methylbutanal and (f) acetaldehyde would give a positive Tollens' test.

13.73 a.

$$CH_3-\overset{\overset{\displaystyle O}{\|}}{C}-CH_3 + CH_3CH_2OH \xrightarrow{H^+} CH_3-\overset{\overset{\displaystyle OH}{|}}{\underset{\underset{\displaystyle OCH_2CH_3}{|}}{C}}-CH_3$$

b.

$$CH_3-\overset{\overset{\displaystyle O}{\|}}{C}-H + CH_3CH_2OH \xrightarrow{H^+} CH_3-\overset{\overset{\displaystyle OH}{|}}{\underset{\underset{\displaystyle OCH_2CH_3}{|}}{C}}-H$$

13.75 Hemiacetal

13.77 Acetal

13.79 a.

$$CH_3-\overset{\overset{\displaystyle O}{\|}}{C}-CH_3 + 2\,CH_3OH \xrightarrow{H^+} CH_3-\overset{\overset{\displaystyle OCH_3}{|}}{\underset{\underset{\displaystyle OCH_3}{|}}{C}}-CH_3 + H_2O$$

b.

$$CH_3-\overset{\overset{\displaystyle O}{\|}}{C}-H + 2\,CH_3OH \xrightarrow{H^+} CH_3-\overset{\overset{\displaystyle OCH_3}{|}}{\underset{\underset{\displaystyle OCH_3}{|}}{C}}-H + H_2O$$

13.81 a.

$$H-\overset{\overset{\displaystyle O}{\|}}{C}-OH$$

b.

$$CH_3-\overset{\overset{\displaystyle O}{\|}}{C}-OH$$

13.83 a. Methanal
b. Propanal

13.85 a. False
b. True
c. False
d. False

13.87

$$2\,CH_3-\overset{\overset{\displaystyle O}{\|}}{C}-H \xrightarrow{OH^-} CH_3\overset{\underset{\underset{\displaystyle OH}{|}}{}}{C}HCH_2\overset{\overset{\displaystyle O}{\|}}{C}-H$$

Ethanal   3-Hydroxybutanal

13.89

$$CH_3-\overset{\overset{\displaystyle O}{\|}}{C}-CH_3$$

Keto form of Propanone

Enol form of Propanone (H₂C=C(OH)CH₃ structure)

13.91 a.

$$CH_3CH_2CH_2-\overset{\overset{\displaystyle OH}{|}}{\underset{\underset{\displaystyle OCH_2CH_3}{|}}{C}}-CH_3$$

b. (phenyl)–C(OH)(OCH₂CH₃)–CH₃

c. cyclopentane with OH and OCH₂CH₃

13.93 (1) $2\,CH_3CH_2OH$ (2) $KMnO_4/OH^-$ (3) $CH_3CH=CH_2$

Chapter 14

14.1 a. Ketone
b. Ketone
c. Alkane

14.3 The carboxyl group consists of two very polar groups, the carbonyl group and the hydroxyl group. Thus, carboxylic acids are very polar, in addition to which, they can hydrogen bond to one another. Aldehydes are polar, as a result of the carbonyl group, but cannot hydrogen bond to one another. As a result, carboxylic acids have higher boiling points than aldehydes of the same carbon chain length.

14.5 a. 3-Methylcyclohexanecarboxylic acid
b. 2-Ethylcyclopentanecarboxylic acid

14.7 a. $CH_3COOH + CH_3CH_2CH_2OH$
Ethanoic acid   1-Propanol
b. $CH_3CH_2CH_2CH_2CH_2COO^-K^+ + CH_3CH_2CH_2OH$
Potassium hexanoate   1-Propanol
c. $CH_3CH_2CH_2CH_2COO^-Na^+ + CH_3OH$
Sodium pentanoate   Methanol
d. $CH_3CH_2CH_2CH_2CH_2COOH + CH_3CHCH_2CH_2CH_3$ (with OH)
Hexanoic acid   2-Pentanol

14.9 a.

$$CH_3CHC-OH \xrightarrow{SOCl_2} CH_3CHC-Cl + \text{inorganic products}$$ (each with O double bond and CH₃ branch)

2-Methylpropanoic acid   2-Methylpropanoyl chloride

b.

$$CH_3CH_2CH_2CH_2CH_2C-OH \xrightarrow{SOCl_2}$$

Hexanoic acid

$$CH_3CH_2CH_2CH_2CH_2C-Cl + \text{inorganic products}$$

Hexanoyl chloride

14.11 a.

$$H-\overset{\overset{\displaystyle O}{\|}}{C}-OH \xrightarrow{PCl_3} H-\overset{\overset{\displaystyle O}{\|}}{C}-Cl + \text{inorganic products}$$

Formic acid   Formyl chloride

b.

$$CH_3CH_2-\overset{\overset{\displaystyle O}{\|}}{C}-OH \xrightarrow{PCl_3} CH_3CH_2-\overset{\overset{\displaystyle O}{\|}}{C}-Cl + \text{inorganic products}$$

Propionic acid   Propionyl chloride

14.13 a.

$$CH_3CHCH_2C-Cl \xrightarrow{CH_3CHCH_2C-O^- / CH_3 \text{ (3-Methylbutanoate ion)}}$$

3-Methylbutanoyl chloride

$$CH_3CHCH_2C-O-CCH_2CHCH_3 + Cl^-$$

3-Methylbutanoic anhydride

b.

$$H-\overset{\overset{\displaystyle O}{\|}}{C}-Cl \xrightarrow{CH_3C-O^- \text{ (Ethanoate ion)}} H-C-O-C-CH_3 + Cl^-$$

Methanoyl chloride   Ethanoic methanoic anhydride

14.15 Aldehydes are polar, as a result of the carbonyl group, but cannot hydrogen bond to one another. Alcohols are polar and can hydrogen bond as a result of the polar hydroxyl group. The carboxyl group of the carboxylic acids consists of both of these groups: the carbonyl group and the hydroxyl group. Thus, carboxylic acids are more polar than either aldehydes or alcohols, in addition to which, they can hydrogen bond to one another. As a result, carboxylic acids have higher boiling points than aldehydes or alcohols of comparable molar mass.

14.17 **a.** 3-Hexanone
b. 3-Hexanone
c. Hexane

14.19

Propanoic acid > 2-Butanol >

Butanal > 2-Methylbutane

14.21 **a.** Heptanoic acid
b. 1-Propanol
c. Pentanoic acid
d. Butanoic acid

14.23 The smaller carboxylic acids are water-soluble. They have sharp, sour tastes and unpleasant aromas.

14.25 Citric acid is found naturally in citrus fruits. It is added to foods to give them a tart flavor or to act as a food preservative and antioxidant. Adipic acid imparts a tart flavor to soft drinks and is a preservative.

14.27 Determine the name of the parent compound, that is the longest carbon chain containing the carboxyl group. Change the -e ending of the alkane name to -oic acid. Number the chain so that the carboxyl carbon is carbon-1. Name and number substituents in the usual way.

14.29 The I.U.P.A.C. name for adipic acid is hexanedioic acid.

14.31 **a.**

b.

c.

14.33 **a.** I.U.P.A.C. name: Methanoic acid
    Common name: Formic acid
b. I.U.P.A.C. name: 3-Methylbutanoic acid
    Common name: β-Methylbutyric acid
c. I.U.P.A.C. name: Cyclopentanecarboxylic acid
    Common name: Cyclovalericcarboxylic acid

14.35

Butanoic acid   Methylpropanoic acid

14.37 **a.**

$CH_3CH_2CCH_2CH_2COOH$ (with CH_3 groups)

b.

$CH_3CHCHCH_2COOH$ (with CH_3 and Br)

c.

COOH with O_2N and NO_2

d.

COOH with H_3C

14.39 **a.** I.U.P.A.C. name: 2-Hydroxypropanoic acid
    Common name: α-Hydroxypropionic acid
b. I.U.P.A.C. name: 3-Hydroxybutanoic acid
    Common name: β-Hydroxybutyric acid
c. I.U.P.A.C. name: 4, 4-Dimethylpentanoic acid
    Common name: γ, γ-Dimethylvaleric acid
d. I.U.P.A.C. name: 3, 3-Dichloropentanoic acid
    Common name: β, β-Dichlorovaleric acid

14.41 **a.** 3-Bromobenzoic acid (or *meta*-bromobenzoic acid or *m*-bromobenzoic acid)
b. 2-Ethylbenzoic acid (or *ortho*-ethylbenzoic acid or *o*-bromobenzoic acid)
c. 4-Hydroxybenzoic acid (or *para*-hydroxybenzoic acid or *p*-hydroxybenzoic acid)

14.43 In organic molecules, oxidation may be recognized as a gain of oxygen or a loss of hydrogen. An aldehyde may be oxidized to form a carboxylic acid as in the following example in which ethanal is oxidized to produce ethanoic acid.

$$H_3C-\overset{O}{\overset{\|}{C}}-H \xrightarrow{[O]} H_3C-\overset{O}{\overset{\|}{C}}-OH$$

Ethanal   Ethanoic acid

14.45 The following general equation represents the dissociation of a carboxylic acid.

$$R-\overset{O}{\overset{\|}{C}}-OH \rightleftarrows R-\overset{O}{\overset{\|}{C}}-O^- + H^+$$

14.47 When a strong base is added to a carboxylic acid, neutralization occurs.

14.49 Soaps are made from water, a strong base, and natural fats or oils.

14.51 **a.**

$$CH_3CH_2-\overset{O}{\overset{\|}{C}}-H \longrightarrow CH_3CH_2-\overset{O}{\overset{\|}{C}}-OH$$

Propanal would be the first oxidation product. However, it would quickly be oxidized further to propanoic acid.

b.

$$HO-\overset{O}{\overset{\|}{C}}-CH_2CH_2CH_2CH_3$$

14.53 a. CH_3COOH

b.

$$CH_3CH_2CH_2-\overset{\overset{O}{\|}}{C}-O-CH_3 + H_2O$$

c. CH_3OH

14.55 a. The oxidation of 1-pentanol yields pentanal.
 b. Continued oxidation of pentanal yields pentanoic acid.

14.57 a. The following equation represents the neutralization of propanoic acid with NaOH.

$$CH_3CH_2COOH + NaOH \rightarrow CH_3CH_2COO^-Na^+ + H_2O$$

 b. The following equation represents the neutralization of propanoic acid with KOH.

$$CH_3CH_2COOH + KOH \rightarrow CH_3CH_2COO^-K^+ + H_2O$$

 c. The following equation represents the neutralization of propanoic acid with $Ca(OH)_2$.

$$2CH_3CH_2COOH + Ca(OH)_2 \rightarrow [CH_3CH_2COO^-]_2Ca^{+2} + 2H_2O$$

14.59 The structure of the calcium salt of propionic acid is $[CH_3CH_2COO^-]_2Ca^{+2}$. The common name of this salt is calcium propionate and the I.U.P.A.C. name is calcium propanoate.

14.61 Esters are mildly polar as a result of the polar carbonyl group within the structure.

14.63 Esters are formed in the reaction of a carboxylic acid with an alcohol. The name is derived by using the alkyl or aryl portion of the alcohol I.U.P.A.C. name as the first name. The -*ic* acid ending of the I.U.P.A.C. name of the carboxylic acid is replaced with -*ate* and follows the name of the aryl or alkyl group.

14.65 a.

$$\overset{\overset{O}{\|}}{C}-OCH_3$$

(attached to benzene ring)

 b.

$$CH_3CH_2CH_2CH_2CH_2CH_2CH_2CH_2CH_2-\overset{\overset{O}{\|}}{C}-O-CH_2CH_2CH_2CH_3$$

 c.

$$CH_3CH_2-\overset{\overset{O}{\|}}{C}-O-CH_3$$

 d.

$$CH_3CH_2-\overset{\overset{O}{\|}}{C}-O-CH_2CH_3$$

14.67 a. Ethyl ethanoate
 b. Methyl propanoate
 c. Methyl-3-methylbutanoate
 d. Cyclopentyl benzoate

14.69 The following equation shows the general reaction for the preparation of an ester:

$$\underset{\substack{\text{Carboxylic} \\ \text{acid}}}{R-\overset{\overset{O}{\|}}{C}-OH} + \underset{\text{Alcohol}}{R-OH} \overset{H^+,\ heat}{\rightleftharpoons} \underset{\text{Ester}}{R-\overset{\overset{O}{\|}}{C}-OR} + \underset{\text{Water}}{H_2O}$$

14.71 The following equation shows the general reaction for the acid-catalyzed hydrolysis of an ester:

$$\underset{\text{Ester}}{R-\overset{\overset{O}{\|}}{C}-OR} + \underset{\text{Water}}{H_2O} \overset{H^+,\ heat}{\rightleftharpoons} \underset{\substack{\text{Carboxylic} \\ \text{acid}}}{R-\overset{\overset{O}{\|}}{C}-OH} + \underset{\text{Alcohol}}{R-OH}$$

14.73 A hydrolysis reaction is the cleavage of any bond by the addition of a water molecule.

14.75 a.

$$CH_3CH_2CH_2-\overset{\overset{O}{\|}}{C}-O-CH_2CH_3$$

 b.

$$CH_3CH_2-\overset{\overset{O}{\|}}{C}-OH + CH_3CH_2OH$$

 c. $CH_3CH_2CH_2OH$

 d.

$$CH_3CH_2\overset{\overset{Br}{|}}{C}HCH_2-\overset{\overset{O}{\|}}{C}-O^- + CH_3CH_2OH$$

14.77 a. Isobutyl methanoate is made from isobutyl alcohol (I.U.P.A.C. name 2-methyl-1-propanol) and methanoic acid.

$$\underset{\substack{\text{Isobutyl alcohol}}}{CH_3\overset{\overset{CH_3}{|}}{C}HCH_2OH} + \underset{\substack{\text{Methanoic acid}}}{HCOOH} \rightarrow \underset{\substack{\text{Isobutyl methanoate}}}{HCO\overset{\overset{O}{\|}}{}CH_2\overset{\overset{CH_3}{|}}{C}HCH_3}$$

Isobutyl alcohol is an allowed starting material, but methanoic acid is not. However, it can easily be produced by the oxidation of its corresponding alcohol, methanol:

$$\underset{\text{Methanol}}{CH_3OH} \overset{[O]}{\rightarrow} \underset{\text{Methanal}}{HCHO} \overset{[O]}{\rightarrow} \underset{\text{Methanoic acid}}{HCOOH}$$

 b. Pentyl butanoate is made from pentanol and butanoic acid.

$$\underset{\text{Pentanol}}{CH_3(CH_2)_3CH_2OH} + \underset{\text{Butanoic acid}}{CH_3CH_2CH_2COOH}$$

$$\rightarrow \underset{\text{Pentyl butanoate}}{CH_3CH_2CH_2\overset{\overset{O}{\|}}{C}OCH_2(CH_2)_3CH_3}$$

Pentanol is an allowed starting material but butanoic acid is not. However, it can easily be produced by the oxidation of its corresponding alcohol, 1-butanol:

$$\underset{\text{Butanol}}{CH_3CH_2CH_2CH_2OH} \overset{[O]}{\rightarrow} \underset{\text{Butanal}}{CH_3CH_2CH_2CHO} \overset{[O]}{\rightarrow}$$

$$\rightarrow \underset{\text{Butanoic acid}}{CH_3CH_2CH_2COOH}$$

14.79 Saponification is a reaction in which a soap is produced. More generally, it is the hydrolysis of an ester in the presence of a base. The following reaction shows the base-catalyzed hydrolysis of an ester:

$$CH_3(CH_2)_{14}-\overset{\overset{O}{\|}}{C}-O-CH_3 + NaOH \longrightarrow$$

$$CH_3(CH_2)_{14}-\overset{\overset{O}{\|}}{C}-O^-Na^+ + CH_3OH$$

14.81

$$\underset{\text{Salicylic acid}}{\overset{\overset{O}{\|}}{C}-OH}\ + CH_3OH \overset{H^+}{\longrightarrow}$$

$$\underset{\text{Methyl salicylate}}{\overset{\overset{O}{\|}}{C}-OCH_3}\ + H_2O$$

14.83 Compound A is

$$CH_3CH_2CH_2CH_2-\overset{\overset{\displaystyle O}{\|}}{C}-O-CH_3$$

Compound B is

$$CH_3CH_2CH_2CH_2-\overset{\overset{\displaystyle O}{\|}}{C}-OH$$

Compound C is CH_3OH

14.85 a.

$$CH_3CH_2-\overset{\overset{\displaystyle O}{\|}}{C}-OCH_2CH_2CH_3 \underset{}{\overset{H^+, heat}{\rightleftharpoons}} CH_3CH_2-\overset{\overset{\displaystyle O}{\|}}{C}-OH$$

Propyl propanoate   Propanoic acid

$$+ \ CH_3CH_2CH_2OH$$

1-Propanol

b.

$$H-\overset{\overset{\displaystyle O}{\|}}{C}-OCH_2CH_2CH_3 \underset{}{\overset{H^+, heat}{\rightleftharpoons}} H-\overset{\overset{\displaystyle O}{\|}}{C}-OH$$

Butyl methanoate   Methanoic acid

$$+ \ CH_3CH_2CH_2CH_2OH$$

1-Butanol

c.

$$H-\overset{\overset{\displaystyle O}{\|}}{C}-OCH_2CH_3 \underset{}{\overset{H^+, heat}{\rightleftharpoons}} H-\overset{\overset{\displaystyle O}{\|}}{C}-OH$$

Ethyl methanoate   Methanoic acid

$$+ \ CH_3CH_2OH$$

Ethanol

d.

$$CH_3CH_2CH_2CH_2-\overset{\overset{\displaystyle O}{\|}}{C}-OCH_3 \underset{}{\overset{H^+, heat}{\rightleftharpoons}}$$

Methyl pentanoate

$$CH_3CH_2CH_2CH_2-\overset{\overset{\displaystyle O}{\|}}{C}-OH \ + \ CH_3OH$$

Pentanoic acid   Methanol

14.87 Acid chlorides are noxious, irritating chemicals. They are slightly polar and have boiling points similar to comparable aldehydes or ketones. They cannot be dissolved in water because they react violently with it.

14.89 Acid anhydrides have much lower boiling points than carboxylic acids of comparable molecular weight. They are also less soluble in water, and often react with it.

14.91 a. PCl_3, PCl_5, or $SOCl_2$

b.

$$CH_3-\overset{\overset{\displaystyle O}{\|}}{C}-O^-$$

c.

$$\text{cyclohexyl}-\overset{\overset{\displaystyle O}{\|}}{C}-O^-$$

14.93 a.

$$\text{phenyl}-\overset{\overset{\displaystyle O}{\|}}{C}-OH \ + \ HCl$$

b.

$$2 \ CH_3-\overset{\overset{\displaystyle O}{\|}}{C}-OH$$

14.95 a.

$$CH_3(CH_2)_8-\overset{\overset{\displaystyle O}{\|}}{C}-O-\overset{\overset{\displaystyle O}{\|}}{C}-(CH_2)_8CH_3$$

b.

$$CH_3-\overset{\overset{\displaystyle O}{\|}}{C}-O-\overset{\overset{\displaystyle O}{\|}}{C}-CH_3$$

14.97 a. Butanoyl chloride

$$CH_3CH_2CH_2\overset{\overset{\displaystyle O}{\|}}{C}-Cl$$

b. Hexanoyl chloride

$$CH_3(CH_2)_4\overset{\overset{\displaystyle O}{\|}}{C}-Cl$$

c. Ethanoyl chloride

$$CH_3\overset{\overset{\displaystyle O}{\|}}{C}-Cl$$

14.99 The following equation represents the synthesis of methanoic anhydride:

$$HC\overset{\overset{\displaystyle O}{\|}}{O^-} + HC\overset{\overset{\displaystyle O}{\|}}{-}Cl \rightarrow H\overset{\overset{\displaystyle O}{\|}}{C}-O-\overset{\overset{\displaystyle O}{\|}}{C}H$$

Methanoate   Methanoic   Methanoic
anion   chloride   anhydride

14.101 a.

$$CH_3CH_2OH + CH_3CH_2-\overset{\overset{\displaystyle O}{\|}}{C}-O-\overset{\overset{\displaystyle O}{\|}}{C}-CH_2CH_3 \longrightarrow$$

$$CH_3CH_2-\overset{\overset{\displaystyle O}{\|}}{C}-OCH_2CH_3$$

$$+$$

$$CH_3CH_2-\overset{\overset{\displaystyle O}{\|}}{C}-OH$$

b.

$$CH_3CH_2OH + CH_3-\overset{\overset{\displaystyle O}{\|}}{C}-O-\overset{\overset{\displaystyle O}{\|}}{C}-CH_3 \longrightarrow$$

$$CH_3-\overset{\overset{\displaystyle O}{\|}}{C}-OCH_2CH_3 + CH_3-\overset{\overset{\displaystyle O}{\|}}{C}-OH$$

c.

$$CH_3CH_2OH + H-\overset{\overset{\displaystyle O}{\|}}{C}-O-\overset{\overset{\displaystyle O}{\|}}{C}-H \longrightarrow$$

$$H-\overset{\overset{\displaystyle O}{\|}}{C}-OCH_2CH_3 + H-\overset{\overset{\displaystyle O}{\|}}{C}-OH$$

14.103 a. Monoester:

$$HO-\overset{\overset{\displaystyle O}{\|}}{\underset{\underset{\displaystyle OH}{|}}{P}}-OCH_2CH_3$$

b. Diester:

$$HO-\overset{\overset{\displaystyle O}{\|}}{\underset{\underset{\displaystyle OCH_2CH_3}{|}}{P}}-OCH_2CH_3$$

c. Triester:

$$CH_3CH_2-O-\overset{\overset{\displaystyle O}{\|}}{\underset{\underset{\displaystyle OCH_2CH_3}{|}}{P}}-OCH_2CH_3$$

14.105 ATP is the molecule used to store the energy released in metabolic reactions. The energy is stored in the phosphoanhydride bonds between two phosphoryl groups. The energy is released when the bond is hydrolyzed. A portion of the energy can be transferred to another molecule if the phosphoryl group is transferred from ATP to the other molecule.

14.107

$$CH_3-\overset{\overset{\displaystyle O}{\|}}{C}{\sim}S-COENZYME \ A$$

The squiggle denotes a high energy bond.

14.109

$$\begin{array}{c} \overset{\displaystyle H}{\underset{\displaystyle |}{|}} \\ H-\underset{\displaystyle |}{\overset{\displaystyle |}{C}}-O-NO_2 \\ H-\underset{\displaystyle |}{\overset{\displaystyle |}{C}}-O-NO_2 \\ H-\underset{\displaystyle |}{\overset{\displaystyle |}{C}}-O-NO_2 \\ \overset{\displaystyle |}{H} \end{array}$$

Chapter 15

15.1

```
            O··········H
        H       H        O
         \     /          \
   CH₃—N      N—H          H
        /      \
       H   CH₃H₃C   CH₃
```

15.3 a.

```
        H
   ◯—N—CH₃
```

c.

```
        H
   ◯—N—CH₂CH₃
```

b.

```
        CH₃
   ◯—N—CH₃
```

d.

```
       H  CH₃
   ◯—N—CHCH₃
```

15.5 a.

```
       H H H
       | | |
   H—C—C—C—H
       | | |
       H N H
         |
       H   H
```

b.

```
         H   H
          \ /
   H H  N  H H H H H
   | |  |  | | | | |
   H—C—C—C—C—C—C—C—C—H
   | |  |  | | | | |
   H H  H  H H H H H
```

c.

```
   H H H H H H H
   | | | | | | |
   H—C—C—C—C—C—C—C—H
   | | | | | | |
   H |  H H H H H
     N—H
     |
   H—C—H
     |
   H—C—H
     |
     H
```

d.

```
       H   H
        \ /
   H H  N  H H H
   | |  |  | | |
   H—C—C—C—C—C—H
   | |  |  | | |
   H H  H  H H H
     |
   H—C—H
     |
     H
```

e.

```
   H   H
    \ /
     N  H H H H H H H H
     |  | | | | | | | |
   H—C—C—C—C—C—C—C—C—C—H
     |  | | | | | | | |
     H  H H Cl I H H H H
```

f.

```
   H H    H H
   | |    | |
   H—C—C—N—C—C—H
   | |  |  | |
   H H  |  H H
        H H H H
        | | | |
     H—C—C—C—C—C—H
        | | | | |
        H H H H H
```

15.7 a.

```
   ⬠—NH₃⁺ Br⁻
```

b.

```
              H
              |
   CH₃CH₂—N⁺—CH₃ + OH⁻
              |
              H
```

c. CH₃—N⁺H₃ + OH⁻

15.9 a. CH₃—NH₂

b.
```
       CH₃
       |
   CH₃—NH
```

15.11 The nitrogen atom is more polar than the hydrogen atom in amines; thus, the N—H bond is polar and hydrogen bonding can occur between primary or secondary amine molecules. Thus, amines have a higher boiling point than alkanes, which are nonpolar. Because nitrogen is not as electronegative as oxygen, the N—H bond is not as polar as the O—H. As a result, intermolecular hydrogen bonds between primary and secondary amine molecules are not as strong as the hydrogen bonds between alcohol molecules. Thus, alcohols have a higher boiling point.

15.13 In systematic nomenclature, primary amines are named by determining the name of the parent compound, the longest continuous carbon chain containing the amine group. The -e ending of the alkane chain is replaced with -amine. Thus, an alkane becomes an alkanamine. The parent chain is then numbered to give the carbon bearing the amine group the lowest possible number. Finally, all substituents are named and numbered and added as prefixes to the "alkanamine" name.

15.15 Amphetamines elevate blood pressure and pulse rate. They also decrease the appetite.

15.17 a. 1-Butanamine would be more soluble in water because it has a polar amine group that can form hydrogen bonds with water molecules.

b. 2-Pentanamine would be more soluble in water because it has a polar amine group that can form hydrogen bonds with water molecules.

15.19 Triethylamine molecules cannot form hydrogen bonds with one another, but 1-hexanamine molecules are able to do so.

15.21 a. 2-Butanamine

b. 3-Hexanamine

c. Cyclopentanamine

d. 2-Methyl-2-propanamine

15.23 a. CH₃CH₂—NH—CH₂CH₃

b. CH₃CH₂CH₂CH₂NH₂

c.
```
   CH₃CH₂CHCH₂CH₂CH₂CH₂CH₂CH₃
         |
         NH₂
```

d.
```
        Br
        |
   CH₃CHCHCH₂CH₃
       |
       NH₂
```

e.

```
   ◯—N—◯
      |
      ◯
```

15.25 a.
```
   CH₃CHCH₂CH₂CH₃
       |
       NH₂
```

b.
```
            Br
            |
   CH₃CH₂CHCH₂NH₂
```

c.
```
   CH₃CH₂—NH—CHCH₃
              |
              CH₃
```

d.
```
   ⬠—NH₂
```

15.27 $CH_3CH_2CH_2CH_2NH_2$ $CH_3CH_2CHCH_3$
                                                              |
                                                             NH_2

1-Butanamine   2-Butanamine
(Primary amine)   (Primary amine)

                                                    CH_3
                                                     |
$CH_3CHCH_2NH_2$ $CH_3—C—CH_3$
          | |
        CH_3 NH_2

2-Methyl-1-propanamine   2-Methyl-2-propanamine
(Primary amine)   (Primary amine)

         CH_3
          |
$CH_3CH_2—N—CH_3$ $CH_3CH_2—NH—CH_2CH_3$

N,N-Dimethylethanamine   N-Ethylethanamine
(Tertiary amine)   (Secondary amine)

CH_3CHCH_3
      |
    $NH—CH_3$ $CH_3CH_2CH_2—NH—CH_3$

N-Methyl-2-propanamine   N-Methyl-1-propanamine
(Secondary amine)   (Secondary amine)

15.29 **a.** Primary
         b. Secondary
         c. Primary
         d. Tertiary

15.31 **a.**

b.

c.

d.

15.33 **a.** H_2O
         b. HBr
         c. $CH_3CH_2CH_2—N^+H_3$
         d. $CH_3CH_2—N^+H_2Cl^-$
                            |
                         CH_2CH_3

15.35 Lower molecular weight amines are soluble in water because the N—H bond is polar and can form hydrogen bonds with water molecules.

15.37 Drugs containing amine groups are generally administered as ammonium salts because the salt is more soluble in water and, hence, in body fluids.

15.39 Putrescine (1,4-Butanediamine):

$CH_2CH_2CH_2CH_2$
  | |
NH_2 NH_2

Cadaverine (1, 5-Pentanediamine):

$CH_2CH_2CH_2CH_2CH_2$
  | |
NH_2 NH_2

15.41 **a.**

Pyridine   Indole

b. The indole ring is found in lysergic acid diethylamide, which is a hallucinogenic drug. The pyridine ring is found in vitamin B_6, an essential water-soluble vitamin.

15.43 Morphine, codeine, quinine, and vitamin B_6

15.45 Amides have very high boiling points because the amide group consists of two very polar functional groups, the carbonyl group and the amino group. Strong intermolecular hydrogen bonding between the N—H bond of one amide and the C=O group of a second amide results in very high boiling points.

15.47 The I.U.P.A.C. names of amides are derived from the I.U.P.A.C. names of the carboxylic acids from which they are derived. The *-oic acid* ending of the carboxylic acid is replaced with the *-amide* ending.

15.49 Barbiturates are often called "downers" because they act as sedatives. They are sometimes used as anticonvulsants for epileptics and people suffering from other disorders that manifest as neurosis, anxiety, or tension.

15.51 **a.** I.U.P.A.C. name: Propanamide
            Common name: Propionamide
         b. I.U.P.A.C. name: Pentanamide
            Common name: Valeramide
         c. I.U.P.A.C. name: N,N-Dimethylethanamide
            Common name: N,N-Dimethylacetamide

15.53 **a.**
                        O
                        ‖
            $CH_3—C—NH_2$
         b.
                             O
                             ‖
            $CH_3CH_2—C—NH—CH_3$
         c.

         d.
                        CH_3 O
                          | ‖
            $CH_3CH_2CHCHCH_2—C—NH_2$
                                |
                              Br
         e.
                     O
                     ‖
            $CH_3—C—N—CH_3$
                          |
                        CH_3

15.55 *N, N*-Diethyl-*m*-toluamide:

Hydrolysis of this compound would release the carboxylic acid *m*-toluic acid and the amine *N*-ethylethanamine (diethylamine).

15.57 Amides are not proton acceptors (bases) because the highly electronegative carbonyl oxygen has a strong attraction for the nitrogen lone pair of electrons. As a result they cannot "hold" a proton.

15.59 Amide group

Lidocaine hydrochloride

15.61

Penicillin BT

15.63 a.

N-Methylethanamide

$$CH_3COOH + CH_3N^+H_3$$

Ethanoic acid   Methanamine

b.

N-Methylbutanamide

$$CH_3CH_2CH_2COOH + CH_3N^+H_3$$

Butanoic acid   Methanamine

c.

N-Ethyl-3-methylbutanamide

$$CH_3CHCH_2COOH + CH_3CH_2N^+H_3$$
     |
    CH_3

3-Methylbutanoic acid   Ethanamine

15.65 a.

b.

c.

15.67

15.69 Glycine: Alanine:

15.71

15.73 In an acyl group transfer reaction, the acyl group of an acid chloride is transferred from the Cl of the acid chloride to the N of an amine or ammonia. The product is an amide.

15.75 A chemical that carries messages or signals from a nerve to a target cell

15.77 a. Tremors, monotonous speech, loss of memory and problem-solving ability, and loss of motor function
  b. Parkinson's disease
  c. Schizophrenia, intense satiety sensations

15.79 In proper amounts, dopamine causes a pleasant, satisfied feeling. This feeling becomes intense as the amount of dopamine increases. Several drugs, including cocaine, heroin, amphetamines, alcohol, and nicotine increase the levels of dopamine. It is thought that the intense satiety response this brings about may contribute to addiction to these substances.

15.81 Epinephrine is a component of the flight or fight response. It stimulates glycogen breakdown to provide the body with glucose to supply the needed energy for this stress response.

15.83 The amino acid tryptophan

15.85 Perception of pain, thermoregulation, and sleep

15.87 Promotes the itchy skin rash associated with poison ivy and insect bites; the respiratory symptoms characteristic of hay fever; secretion of stomach acid

15.89 Inhibitory neurotransmitters

15.91 When acetylcholine is released from a nerve cell, it binds to receptors on the surface of muscle cells. This binding stimulates the muscle cell to contract. To stop the contraction, the acetylcholine is then broken down to choline and acetate ion. This is catalyzed by the enzyme acetylcholinesterase.

15.93 Organophosphates inactivate acetylcholinesterase by binding covalently to it. Since acetylcholine is not broken down, nerve transmission continues, resulting in muscle spasm. Pyridine aldoxime methiodide (PAM) is an antidote to organophosphate poisoning because it displaces the organophosphate, thereby allowing acetycholinesterase to function.

Chapter 16

16.1 It is currently recommended that 45–55% of the calories in the diet should be carbohydrates. Of that amount, no more than 10% should be simple sugars.

16.3 An aldose is a sugar with an aldehyde functional group. A ketose is a sugar with a ketone functional group.

16.5

a.

b.

c.

```
      CH2OH              CH2OH
      |                  |
      C=O                C=O
      |                  |
  HO—*—H             H—*—OH
      |                  |
   H—*—OH            HO—*—H
      |                  |
   H—*—OH            HO—*—H
      |                  |
      CH2OH              CH2OH
```

d.

```
      CHO                CHO
      |                  |
   H—*—OH            HO—*—H
      |                  |
      CH2OH              CH2OH
```

e.

```
      CH3                CH3
      |                  |
      C=O                C=O
      |                  |
   H—*—OH            HO—*—H
      |                  |
   H—*—OH            HO—*—H
      |                  |
   H—*—OH            HO—*—H
      |                  |
      CH2OH              CH2OH
```

f.

```
      CHO                CHO
      |                  |
  HO—*—H             H—*—OH
      |                  |
   H—*—OH            HO—*—H
      |                  |
  HO—*—H             H—*—OH
      |                  |
      CH2OH              CH2OH
```

16.7   **a.** D-   **b.** L-   **c.** D-   **d.** D-   **e.** D-   **f.** L-

16.9

```
      CHO
      |
   H——OH
      |
  HO——H
      |
  HO——H
      |
   H——OH
      |
      CH2OH
```
D-Galactose

16.11 α-Amylase and β-amylase are digestive enzymes that break down the starch amylose. α-Amylase cleaves glycosidic bonds of the amylose chain at random, producing shorter polysaccharide chains. β-Amylase sequentially cleaves maltose (a disaccharide of glucose) from the reducing end of the polysaccharide chain.

16.13 A monosaccharide is the simplest sugar and consists of a single saccharide unit. A disaccharide is made up of two monosaccharides joined covalently by a glycosidic bond.

16.15 The molecular formula for a simple sugar is $(CH_2O)_n$. Typically n is an integer from 3 to 7.

16.17 Mashed potato flakes, rice, and corn starch contain amylose and amylopectin, both of which are polysaccharides. A candy bar contains sucrose, a disaccharide. Orange juice contains fructose, a monosaccharide. It may also contain sucrose if the label indicates that sugar has been added.

16.19 Four

16.21

```
      O                  CH2OH
      ||                 |
      C—H                C=O
      |                  |
   H—C—OH            HO—C—H
      |                  |
  HO—C—H             H—C—OH
      |                  |
  HO—C—H             H—C—OH
      |                  |
   H—C—OH                CH2OH
      |
      CH2OH
```
D-Galactose   D-Fructose
(An aldohexose)   (A ketohexose)

16.23 An *aldose* is a sugar that contains an aldehyde (carbonyl) group.

16.25 A tetrose is a sugar with a four-carbon backbone.

16.27 A ketopentose is a sugar with a five-carbon backbone and containing a ketone (carbonyl) group.

16.29   **a.** β-D-Glucose is a hemiacetal.
       b. β-D-Fructose is a hemiketal.
       c. α-D-Galactose is a hemiacetal.

16.31

```
      O                  O
      ||                 ||
      C—H                C—H
      |                  |
   H—C—OH            HO—C—H
      |                  |
      CH2OH              CH2OH
```
D-Glyceraldehyde   L-Glyceraldehyde

16.33 Stereoisomers are a pair of molecules that have the same structural formula and bonding pattern but that differ in the arrangement of the atoms in space.

16.35 A chiral carbon is one that is bonded to four different chemical groups.

16.37 A polarimeter converts monochromatic light into monochromatic plane-polarized light. This plane-polarized light is passed through a sample and into an analyzer. If the sample is optically active, it will rotate the plane of the light. The degree and angle of rotation are measured by the analyzer.

16.39 A Fischer Projection is a two-dimensional drawing of a molecule that shows a chiral carbon at the intersection of two lines. Horizontal lines at the intersection represent bonds projecting out of the page and vertical lines represent bonds that project into the page.

16.41 Dextrose is a common name used for D-glucose.

16.43 D- and L-Glyceraldehyde are a pair of enantiomers, that is, they are nonsuperimposable mirror images of one another.

16.45   **a.**

```
      O
      ||
      C—H
      |
  HO—*—H
      |
   H—*—OH
      |
  HO—*—H
      |
  HO—*—H
      |
      CH2OH
```

b.

```
      O
      ||
      C—H
      |
   H—*—OH
      |
   H—*—OH
      |
      CH2OH
```

c.

```
      O
      ||
      C—H
      |
  HO—*—H
      |
   H—*—OH
      |
  HO—*—H
      |
   H—*—OH
      |
  HO—*—H
      |
      CH2OH
```

16.47 Anomers are isomers that differ in the arrangement of bonds around the hemiacetal carbon.

16.49 A hemiacetal is a member of the family of organic compounds formed in the reaction of one molecule of alcohol with an aldehyde. They have the following general structure:

```
      OH
      |
   R——OR
      |
      H
```

16.51 The reaction between an aldehyde and an alcohol yields a hemiacetal. Thus, when the aldehyde portion of a glucose molecule reacts with the C-5 hydroxyl group, the product is an intramolecular hemiacetal.

16.53 When the carbonyl group at C-1 of D-glucose reacts with the C-5 hydroxyl group, a new chiral carbon is created (C-1). In the α-isomer of the cyclic sugar, the C-1 hydroxyl group is below the ring; and in the β-isomer, the C-1 hydroxyl group is above the ring.

16.55 β-Maltose and α-lactose would give positive Benedict's tests. Glycogen would give only a weak reaction because there are fewer reducing ends for a given mass of the carbohydrate.

16.57 Enantiomers are stereoisomers that are nonsuperimposable mirror images of one another. For instance:

D-Glyceraldehyde   L-Glyceraldehyde

16.59 An aldehyde sugar forms an intramolecular hemiacetal when the carbonyl group of the monosaccharide reacts with a hydroxyl group on one of the other carbon atoms.

16.61 A ketal is the product formed in the reaction of two molecules of alcohol with a ketone. They have the following general structure:

16.63 A glycosidic bond is the bond formed between the hydroxyl group of the C-1 carbon of one sugar and a hydroxyl group of another sugar.

16.65

β-Maltose

16.67 Milk

16.69 Eliminating milk and milk products from the diet

16.71 Lactose intolerance is the inability to produce the enzyme lactase that hydrolyzes the milk sugar lactose into its component monosaccharides, glucose and galactose.

16.73 A polymer is a very large molecule formed by the combination of many small molecules, called monomers.

16.75 Starch

16.77 The glucose units of amylose are joined by α $(1 \rightarrow 4)$ glycosidic bonds and those of cellulose are bonded together by β $(1 \rightarrow 4)$ glycosidic bonds.

16.79 Glycogen serves as a storage molecule for glucose.

16.81 The salivary glands and the pancreas

Chapter 17

17.1 a. $CH_3(CH_2)_7CH=CH(CH_2)_7COOH$
b. $CH_3(CH_2)_{10}COOH$
c. $CH_3(CH_2)_4CH=CH-CH_2-CH=CH(CH_2)_7COOH$
d. $CH_3(CH_2)_{16}COOH$

17.3

$$CH_3(CH_2)_{12}\overset{O}{\underset{}{C}}-OH + CH_3CH_2OH \xrightarrow[heat]{H^+} CH_3(CH_2)_{12}\overset{O}{\underset{}{C}}-O-CH_2CH_3$$
Tetradecanoic acid   Ethanol   Ethyl tetradecanoate

17.5

$$CH_3CH_2CH_2\overset{O}{\underset{}{C}}-O-CH_2(CH_2)_3CH_3 \xrightarrow[heat]{H^+}$$
Pentyl butanoate

$$CH_3CH_2CH_2\overset{O}{\underset{}{C}}-OH + CH_3(CH_2)_3CH_2OH$$
Butanoic acid   Pentanol

17.7

$$CH_3\overset{O}{\underset{}{C}}-O-CH_2(CH_2)_2CH_3 + KOH \rightarrow$$
Butyl ethanoate

$$CH_3\overset{O}{\underset{}{C}}-O^-K^+ + CH_3(CH_2)_2CH_2OH$$
Potassium   Butanol
ethanoate

17.9

$$CH_3CH_2CH=CHCH_2CH=CHCH_2CH=CH(CH_2)_7COOH + 3H_2$$
All *cis*-9, 12, 15-Octadecatrienoic acid

$$\downarrow Ni$$

$$CH_3(CH_2)_{16}COOH$$
Octadecanoic acid

17.11 a.

b.

17.13

Steroid nucleus

17.15 Receptor-mediated endocytosis

17.17 Fatty acids, glycerides, nonglyceride lipids, and complex lipids

17.19 Lipid-soluble vitamins are transported into cells of the small intestine in association with dietary fat molecules. Thus, a diet low in fat reduces the amount of vitamins A, D, E, and K that enters the body.

17.21 A saturated fatty acid is one in which the hydrocarbon tail has only carbon-to-carbon single bonds. An unsaturated fatty acid has at least one carbon-to-carbon double bond.

17.23 The melting points increase.

17.25 The melting points of fatty acids increase as the length of the hydrocarbon chains increase. This is because the intermolecular attractive forces, including van der Waals forces, increase as the length of the hydrocarbon chain increases.

17.27 **a.** Decanoic acid $CH_3(CH_2)_8COOH$
        b. Stearic acid $CH_3(CH_2)_{16}COOH$

17.29 **a.** I.U.P.A.C. name: Hexadecanoic acid
              Common name: Palmitic acid
        b. I.U.P.A.C. name: Dodecanoic acid
              Common name: Lauric acid

17.31 Esterification of glycerol with three molecules of myristic acid:

17.33 Acid hydrolysis of a triglyceride containing three stearic acid molecules (tristearoyl glycerate):

17.35 Reaction of decanoic acid with KOH:

17.37 This line drawing of EPA shows the bends or "kinks" introduced into the molecule by the double bonds.

All *cis*-5,8,11,14,17-Eicosapentaenoic acid (EPA)

$+ 5H_2$ | Ni

Eicosanoic acid

17.39

17.41

17.43 The essential fatty acid linoleic acid is required for the synthesis of arachidonic acid, a precursor for the synthesis of the prostaglandins, a group of hormonelike molecules.

17.45 Aspirin effectively decreases the inflammatory response by inhibiting the synthesis of all prostaglandins. Aspirin works by inhibiting cyclooxygenase, the first enzyme in prostaglandin biosynthesis. This inhibition results from the transfer of an acetyl group from aspirin to the enzyme. Because cyclooxygenase is found in all cells, synthesis of all prostaglandins is inhibited.

17.47 Smooth muscle contraction, enhancement of fever and swelling associated with the inflammatory response, bronchial dilation, inhibition of secretion of acid into the stomach.

17.49 The name of these fatty acids arises from the position of the double bond nearest the terminal *methyl group* of the molecule. The terminal methyl group is designated omega (ω). In ω-3 fatty acids the double bond nearest the ω methyl group is three carbons along the chain. In ω-6 fatty acids, the nearest double bond is six carbons from the end.

17.51 Omega-3 fatty acids reduce the risk of cardiovascular disease by decreasing blood clot formation, blood triglyceride levels, and growth of atherosclerotic plaque.

17.53 The decrease in blood clot formation, along with the reduced blood triglyceride levels and decreased atherosclerotic plaque result in improved arterial health. This, in turn, results in lower blood pressure and a decreased risk of sudden death and heart arrythmias.

17.55 Omega-3 fatty acids are precursors of prostaglandins that exhibit anti-inflammatory effects. On the other hand, omega-6 fatty acids are precursors to prostaglandins that have inflammatory effects. To reduce the inflammatory response contribution to cardiovascular disease, it is logical to increase the amount of omega-3 fatty acids in the diet and to decrease the amount of omega-6 fatty acids.

17.57 A glyceride is a lipid ester that contains the glycerol molecule and from 1 to 3 fatty acids.

17.59 An emulsifying agent is a molecule that aids in the suspension of triglycerides in water. They are amphipathic molecules, such as lecithin, that serve as bridges holding together the highly polar water molecules and the nonpolar triglycerides.

17.61 A triglyceride with three saturated fatty acid tails would be a solid at room temperature. The long, straight fatty acid tails would stack with one another because of strong intermolecular and intramolecular attractions.

17.63

$$CH_3(CH_2)_{14}-\overset{\overset{\displaystyle O}{\|}}{C}-O-\underset{1}{CH_2}$$

$$CH_2(CH_2)_6-\overset{\overset{\displaystyle O}{\|}}{C}-O-\underset{2}{CH}$$

$$CH_2(CH_2)_6-\overset{\overset{\displaystyle O}{\|}}{C}-O-\underset{3}{CH_2}$$

17.65

$$CH_3CH_2CH_2CH_2CH_2CH_2CH_2CH_2CH_2-\overset{\overset{\displaystyle O}{\|}}{C}-O-\underset{1}{CH_2}$$

$$CH_3CH_2CH_2CH_2CH_2CH_2CH_2CH_2CH_2CH_2-\overset{\overset{\displaystyle O}{\|}}{C}-O-\underset{2}{CH}$$

$$\underset{3}{CH_2}-O-\overset{\overset{\displaystyle O}{\|}}{\underset{\underset{\displaystyle O^-}{|}}{P}}-O^-$$

17.67 Triglycerides consist of three fatty acids esterified to the three hydroxyl groups of glycerol. In phospholipids there are only two fatty acids esterified to glycerol. A phosphoryl group is esterified (phosphoester linkage) to the third hydroxyl group.

17.69 A sphingolipid is a lipid that is not derived from glycerol, but rather from sphingosine, a long-chain, nitrogen-containing (amino) alcohol. Like phospholipids, sphingolipids are amphipathic.

17.71 A glycosphingolipid or glycolipid is a lipid that is built on a ceramide backbone structure. Ceramide is a fatty acid derivative of sphingosine.

17.73 Sphingomyelins are important structural lipid components of nerve cell membranes. They are found in the myelin sheath that surrounds and insulates cells of the central nervous system.

17.75 Cholesterol is readily soluble in the hydrophobic region of biological membranes. It is involved in regulating the fluidity of the membrane.

17.77 Progesterone is the most important hormone associated with pregnancy. Testosterone is needed for development of male secondary sexual characteristics. Estrone is required for proper development of female secondary sexual characteristics.

17.79 Cortisone is used to treat rheumatoid arthritis, asthma, gastrointestinal disorders, and many skin conditions.

17.81 Myricyl palmitate (beeswax) is made up of the fatty acid palmitic acid and the alcohol myricyl alcohol— $CH_3(CH_2)_{28}CH_2OH$.

17.83 Isoprenoids are a large, diverse collection of lipids that are synthesized from the isoprene unit:

$$CH_2{=}\overset{\overset{\displaystyle CH_3}{|}}{C}{-}CH{=}CH_2$$

17.85 Steroids and bile salts, lipid-soluble vitamins, certain plant hormones, and chlorophyll

17.87 Chylomicrons, high-density lipoproteins, low-density lipoproteins, and very low density lipoproteins

17.89 The terms "good" and "bad" cholesterol refer to two classes of lipoprotein complexes. The high-density lipoproproteins, or HDL, are considered to be "good" cholesterol because a correlation has been made between elevated levels of HDL and a reduced incidence of atherosclerosis. Low-density lipoproteins, or LDL, are considered to be "bad" cholesterol because evidence suggests that a high level of LDL is associated with increased risk of atherosclerosis.

17.91 Atherosclerosis results when cholesterol and other substances coat the arteries causing a narrowing of the passageways. As the passageways become narrower, greater pressure is required to provide adequate blood flow. This results in higher blood pressure (hypertension).

17.93 If the LDL receptor is defective, it cannot function to remove cholesterol-bearing LDL particles from the blood. The excess cholesterol, along with other substances, will accumulate along the walls of the arteries, causing atherosclerosis.

17.95 The basic structure of a biological membrane is a bilayer of phospholipid molecules arranged so that the hydrophobic hydrocarbon tails are packed in the center and the hydrophilic head groups are exposed on the inner and outer surfaces.

17.97 A peripheral membrane protein is bound to only one surface of the membrane, either inside or outside the cell.

17.99 Cholesterol is freely soluble in the hydrophobic layer of a biological membrane. It moderates the fluidity of the membrane by disrupting the stacking of the fatty acid tails of membrane phospholipids.

17.101 L. Frye and M. Edidin carried out studies in which specific membrane proteins on human and mouse cells were labeled with red and green fluorescent dyes, respectively. The human and mouse cells were fused into single-celled hybrids and were observed using a microscope with an ultraviolet light source. The ultraviolet light caused the dyes to fluoresce. Initially the dyes were localized in regions of the membrane representing the original human or mouse cell. Within an hour, the proteins were evenly distributed throughout the membrane of the fused cell.

17.103 If the fatty acyl tails of membrane phospholipids are converted from saturated to unsaturated, the fluidity of the membrane will increase.

Chapter 18

18.1 a. Glycine (gly):

b. Proline (pro):

c. Threonine (thr):

d. Aspartate (asp):

e. Lysine (lys):

18.3 a. Alanyl-phenylalanine:

b. Lysyl-alanine:

c. Phenylalanyl-tyrosyl-leucine:

18.5 The primary structure of a protein is the amino acid sequence of the protein chain. Regular, repeating folding of the peptide chain caused by hydrogen bonding between the amide nitrogens and carbonyl oxygens of the peptide bond is the secondary structure of a protein. The two most common types of secondary structure are the α-helix and the β-pleated sheet. Tertiary structure is the further folding of the regions of α-helix and β-pleated sheet into a compact, spherical structure. Formation and maintenance of the tertiary structure results from weak attractions between amino acid R groups. The binding of two or more peptides to produce a functional protein defines the quaternary structure.

18.7 Oxygen is efficiently transferred from hemoglobin to myoglobin in the muscle because myoglobin has a greater affinity for oxygen.

18.9 High temperature disrupts the hydrogen bonds and other weak interactions that maintain protein structure.

18.11 Vegetables vary in amino acid composition. No single vegetable can provide all of the amino acid requirements of the body. By eating a variety of different vegetables, all the amino acid requirements of the human body can be met.

18.13 An enzyme is a protein that serves as a biological catalyst, speeding up biological reactions.

18.15 A transport protein is a protein that transports materials across the cell membrane or throughout the body.

18.17 Enzymes speed up reactions that might take days or weeks to occur on their own. They also catalyze reactions that might require very high temperatures or harsh conditions if carried out in the laboratory. In the body, these reactions occur quickly under physiological conditions.

18.19 Transferrin is a transport protein that carries iron from the liver to the bone marrow, where it is used to produce the heme group for hemoglobin and myoglobin. Hemoglobin transports oxygen in the blood.

18.21 Egg albumin is a nutrient protein that serves as a source of protein for the developing chick. Casein is the nutrient storage protein in milk, providing protein, a source of amino acids, for mammals.

18.23 The general structure of an L-α-amino acid:

$$H_3{}^+N-\underset{\underset{R}{|}}{\overset{\overset{\displaystyle COO^-}{|}}{C}}-H$$

18.25 A zwitterion is a neutral molecule with equal numbers of positive and negative charges. Under physiological conditions, amino acids are zwitterions.

18.27 A chiral carbon is one that has four different atoms or groups of atoms attached to it.

18.29 Interactions between the R groups of the amino acids in a polypeptide chain are important for the formation and maintenance of the tertiary and quaternary structures of proteins.

18.31

Glycine   Alanine   Valine

Leucine   Isoleucine   Phenylalanine

Proline   Tryptophan   Methionine

18.33 A peptide bond is an amide bond between two amino acids in a peptide chain.

18.35 Linus Pauling and his colleagues carried out X-ray diffraction studies of protein. Interpretation of the the pattern formed when X-rays were diffracted by a crystal of pure protein led Pauling to conclude that peptide bonds are both planar (flat) and rigid and that the N—C bonds are shorter that expected. In other words, they deduced that the peptide bond has a partially double bond character because it exhibits resonance. There is no free rotation about the amide bond because the carbonyl group of the amide bond

has a strong attraction for the amide nitrogen lone pair of electrons. This can best be described using a resonance model:

The partially double bonded character of the resonance structure restricts free rotation.

18.37 **a.** His-trp-cys:

b. Gly-leu-ser:

c. Arg-ile-val:

18.39 The primary structure of a protein is the sequence of amino acids bonded to one another by peptide bonds.

18.41 The primary structure of a protein determines its three dimensional shape because the location of R groups along the protein chain is determined by the primary structure. The interactions among the R groups, based on their location in the chain, will govern how the protein folds. This, in turn, dictates its three-dimensional structure and biological function.

18.43 The genetic information in the DNA dictates the order in which amino acids will be added to the protein chain. The order of the amino acids is the primary structure of the protein.

18.45 The secondary structure of a protein is the folding of the primary structure into an α-helix or β-pleated sheet.

18.47 **a.** α-Helix
b. β-Pleated sheet

18.49 A fibrous protein is one that is composed of peptides arranged in long sheets or fibers.

18.51 A parallel β-pleated sheet is one in which the hydrogen bonded peptide chains have their amino-termini aligned head-to-head.

18.53 The tertiary structure of a protein is the globular, three-dimensional structure of a protein that results from folding the regions of secondary structure.

18.55

$$H_3{}^+N-\overset{\overset{\textstyle H}{|}}{C}-COO^-$$
$$|$$
$$CH_2$$
$$|$$
$$S$$
$$|$$
$$S$$
$$|$$
$$CH_2$$
$$|$$
$$H_3{}^+N-\underset{\underset{\textstyle H}{|}}{C}-COO^-$$

18.57 The tertiary structure is a level of folding of a protein chain that has already undergone secondary folding. The regions of α-helix and β-pleated sheet are folded into a globular structure.

18.59 Quaternary protein structure is the aggregation of two or more folded peptide chains to produce a functional protein.

18.61 A glycoprotein is a protein with covalently attached sugars.

18.63 Hydrogen bonding maintains the secondary structure of a protein and contributes to the stability of the tertiary and quaternary levels of structure.

18.65 The peptide bond exhibits resonance, which results in a partially double bonded character. This causes the rigidity of the peptide bond.

18.67 The code for the primary structure of a protein is carried in the genetic information (DNA).

18.69 The function of hemoglobin is to carry oxygen from the lungs to oxygen-demanding tissues throughout the body. Hemoglobin is found in red blood cells.

18.71 Hemoglobin is a protein composed of four subunits—two α-globin and two β-globin subunits. Each subunit holds a heme group, which in turn carries an Fe^{2+} ion.

18.73 The function of the heme group in hemoglobin and myoglobin is to bind to molecular oxygen.

18.75 Because carbon monoxide binds tightly to the heme groups of hemoglobin, it is not easily removed or replaced by oxygen. As a result, the effects of oxygen deprivation (suffocation) occur.

18.77 When sickle cell hemoglobin (HbS) is deoxygenated, the amino acid valine fits into a hydrophobic pocket on the surface of another HbS molecule. Many such sickle cell hemoglobin molecules polymerize into long rods that cause the red blood cell to sickle. In normal hemoglobin, glutamic acid is found in the place of the valine. This negatively charged amino acid will not "fit" into the hydrophobic pocket.

18.79 When individuals have one copy of the sickle cell gene and one copy of the normal gene, they are said to carry the *sickle cell trait*. These individuals will not suffer serious side effects, but may pass the trait to their offspring. Individuals with two copies of the sickle cell globin gene

exhibit all the symptoms of the disease and are said to have *sickle cell anemia.*

18.81 *Denaturation* is the process by which the organized structure of a protein is disrupted, resulting in a completely disorganized, nonfunctional form of the protein.

18.83 Heat is an effective means of sterilization because it destroys the proteins of microbial life-forms, including fungi, bacteria, and viruses.

18.85 Even relatively small fluctuations in blood pH can be life threatening. It is likely that these small changes would alter the normal charges on the proteins and modify their interactions. These changes can render a protein incapable of carrying out its functions.

18.87 Proteins become polycations at low pH because the additional protons will protonate the carboxylate groups. As these negative charges are neutralized, the charge on the proteins will be contributed only by the protonated amino groups ($-N^+H_3$).

18.89 The low pH of the yogurt denatures the proteins of microbial contaminants, inhibiting their growth.

18.91 An essential amino acid is one that must be provided in the diet because it cannot be synthesized in the body.

18.93 A complete protein is one that contains all of the essential and nonessential amino acids.

18.95 Chymotrypsin catalyzes the hydrolysis of peptide bonds on the carbonyl side of aromatic amino acids.

Phenylalanyl-alanine   Phenylalanine

Alanine

18.97 In a vegetarian diet, vegetables are the only source of dietary protein. Because individual vegetable sources do not provide all the needed amino acids, vegetables must be mixed to provide all the essential and nonessential amino acids in the amounts required for biosynthesis.

18.99 Synthesis of digestive enzymes must be carefully controlled because the active enzyme would digest, and thus destroy, the cell that produces it.

Chapter 19

19.1 a. Pyruvate kinase catalyzes the transfer of a phosphoryl group from phosphoenolpyruvate to adenosine diphosphate.

Phosphoenolpyruvate   Pyruvate
                          kinase

Pyruvate

b. Alanine transaminase catalyzes the transfer of an amino group from alanine to α-ketoglutarate, producing pyruvate and glutamate.

$$H_3{}^+N-CH-\overset{\overset{O}{\|}}{C}-O^- \qquad H_3C-\overset{\overset{O}{\|}}{C}-\overset{\overset{O}{\|}}{C}-O^-$$
$$\underset{CH_3}{|}$$

Alanine   **Alanine transaminase** →   Pyruvate

+   +

$$O=C-\overset{H}{\underset{|}{C}}-\overset{H}{\underset{|}{C}}-\overset{\overset{O}{\|}}{C}-O^- \qquad H_3{}^+N-\overset{H}{\underset{|}{C}}-\overset{H}{\underset{|}{C}}-\overset{H}{\underset{|}{C}}-\overset{\overset{O}{\|}}{C}-O^-$$

α-Ketoglutarate   Glutamate

c. Triose phosphate isomerase catalyzes the isomerization of the ketone dihydroxyacetone phosphate to the aldehyde glyceraldehyde-3-phosphate.

$$O\sim\text{\textcircled{P}} \quad H$$
$$H_2C-C-C-OH \xrightarrow{\text{Triose phosphate isomerase}} H_2C-C-C-H$$

Dihydroxyacetone phosphate   Glyceraldehyde-3-phosphate

d. Pyruvate dehydrogenase catalyzes the oxidation and decarboxylation of pyruvate, producing acetyl coenzyme A and CO_2.

$$H_3C-\overset{\overset{O}{\|}}{C}-\overset{\overset{O}{\|}}{C}-O^- \quad + \text{ H-S-CoA} \xrightarrow{\text{Pyruvate dehydrogenase}}$$

Pyruvate   Coenzyme A

$$H_3C-\overset{\overset{O}{\|}}{C}\sim\text{S-CoA} \ + \ CO_2$$

Acetyl coenzyme A

19.3 a. Sucrose   **c.** Succinate
b. Pyruvate

19.5 The induced fit model assumes that the enzyme is flexible. Both the enzyme and the substrate are able to change shape to form the enzyme-substrate complex. The lock-and-key model assumes that the enzyme is inflexible (the lock) and the substrate (the key) fits into a specific rigid site (the active site) on the enzyme to form the enzyme-substrate complex.

19.7 An enzyme might distort a bond, thereby catalyzing bond breakage. An enzyme could bring two reactants into close proximity and in the proper orientation for the reaction to occur. Finally, an enzyme could alter the pH of the microenvironment of the active site, thereby serving as a transient donor or acceptor of H^+.

19.9 Water-soluble vitamins are required by the body for the synthesis of coenzymes that are required for the function of a variety of enzymes.

19.11 A decrease in pH will change the degree of ionization of the R groups within a peptide chain. This disturbs the weak interactions that maintain the structure of an enzyme, which

may denature the enzyme. Less drastic alterations in the charge of R groups in the active site of the enzyme can inhibit enzyme-substrate binding or destroy the catalytic ability of the active site.

19.13 Irreversible inhibitors bind very tightly, sometimes even covalently, to an R group in enzyme active sites. They generally inhibit many different enzymes. The loss of enzyme activity impairs normal cellular metabolism, resulting in death of the cell or the individual.

19.15 A structural analog is a molecule that has a structure and charge distribution very similar to that of the natural substrate of an enzyme. Generally they are able to bind to the enzyme active site. This inhibits enzyme activity because the normal substrate must compete with the structural analog to form an enzyme-substrate complex.

19.17 a.

Bond cleaved by chymotrypsin

$$H_3N^+-\overset{H}{\underset{CH_3}{C}}-\overset{O}{C}-N-\overset{H}{\underset{CH_2}{C}}-\overset{O}{C}-N-\overset{H}{\underset{CH_3}{C}}-COO^-$$

ala-phe-ala

b.

Bond cleaved by chymotrypsin

$$H_3N^+-\overset{H}{\underset{CH_2}{C}}-\overset{O}{C}-N-\overset{H}{\underset{CH_3}{C}}-\overset{O}{C}-N-\overset{H}{\underset{CH_2}{C}}-COO^-$$

tyr-ala-tyr

19.19   Chymotrypsin   Elastase   Elastase

$$H_3{}^+N-C-C-N-C-C-N-C-C-N-C-C-N-C-COO^-$$

19.21 The common name of an enzyme is often derived from the name of the substrate and/or the type of reaction that it catalyzes.

19.23 1. Urease
2. Peroxidase
3. Lipase
4. Aspartase
5. Glucose-6-phosphatase
6. Sucrase

19.25 **a.** Citrate decarboxylase catalyzes the cleavage of a carboxyl group from citrate.
   b. Adenosine diphosphate phosphorylase catalyzes the addition of a phosphate group to ADP.
   c. Oxalate reductase catalyzes the reduction of oxalate.
   d. Nitrite oxidase catalyzes the oxidation of nitrite.
   e. *cis-trans* Isomerase catalyzes interconversion of *cis* and *trans* isomers.

19.27 A substrate is the reactant in an enzyme-catalyzed reaction that binds to the active site of the enzyme and is converted into product.

19.29 The activation energy of a reaction is the energy required for the reaction to occur.

19.31 The equilibrium constant for a chemical reaction is a reflection of the difference in energy of the reactants and products.
   Consider the following reaction:
$$aA + bB \rightarrow cC + dD$$
   The equilibrium constant for this reaction is:
$$K_{eq} = [D]^d[C]^c/[A]^a[B]^b = [\text{products}]/[\text{reactants}]$$
   Because the difference in energy between reactants and products is the same regardless of what path the reaction takes, an enzyme does not alter the equilibrium constant of a reaction.

19.33 The rate of an uncatalyzed chemical reaction typically doubles every time the substrate concentration is doubled.

19.35 The rate-limiting step is that step in an enzyme-catalyzed reaction that is the slowest, and hence limits the speed with which the substrate can be converted into product.

19.37

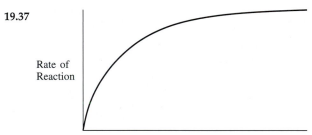

19.39 The enzyme-substrate complex is the molecular aggregate formed when the substrate binds to the active site of an enzyme.

19.41 The catalytic groups of an enzyme active site are those functional groups that are involved in carrying out catalysis.

19.43 Enzyme active sites are pockets in the surface of an enzyme that include R groups involved in binding and R groups involved in catalysis. The shape of the active site is complementary to the shape of the substrate. Thus, the conformation of the active site determines the specificity of the enzyme. Enzyme-substrate binding involves weak, noncovalent interactions.

19.45 The lock-and-key model of enzyme-substrate binding was proposed by Emil Fischer in 1894. He thought that the active site was a rigid region of the enzyme into which the substrate fit perfectly. Thus, the model purports that the substrate simply snaps into place within the active site, like two pieces of a jigsaw puzzle fitting together.

19.47 Enzyme specificity is the ability of an enzyme to bind to only one, or a very few, substrates and thus catalyze only a single reaction.

19.49 Group specificity means that an enzyme catalyzes reactions involving similar molecules having the same functional group.

19.51 Absolute specificity means that an enzyme catalyzes the reaction of only one substrate.

19.53 Hexokinase has group specificity. The advantage is that the cell does not need to encode many enzymes to carry out the phosphorylation of six-carbon sugars. Hexokinase can carry out many of these reactions.

19.55 Methionyl tRNA synthetase has absolute specificity. This is the enzyme that attaches the amino acid methionine to the transfer RNA (tRNA) that will carry the amino acid to the site of protein synthesis. If the wrong amino acid were attached to the tRNA, it could be incorporated into the protein, destroying its correct three-dimensional structure and biological function.

19.57 The first step of an enzyme-catalyzed reaction is the formation of the enzyme-substrate complex. In the second step, the transition state is formed. This is the state in which the substrate assumes a form intermediate between the original substrate and the product. In step 3 the substrate is converted to product and the enzyme-product complex is formed. Step 4 involves the release of the product and regeneration of the enzyme in its original form.

19.59 In a reaction involving bond breaking, the enzyme might distort a bond, producing a transition state in which the bond is stressed. An enzyme could bring two reactants into close proximity and in the proper orientation for the reaction to occur, producing a transition state in which the proximity of the reactants facilitates bond formation. Finally, an enzyme could alter the pH of the microenvironment of the active site, thereby serving as a transient donor or acceptor of H^+.

19.61 A cofactor helps maintain the shape of the active site of an enzyme.

19.63 $NAD^+/NADH$ serves an acceptor/donor of hydride anions in biochemical reactions. $NAD^+/NADH$ serves as a coenzyme for oxidoreductases.

19.65 Changes in pH or temperature affect the activity of enzymes, as can changes in the concentration of substrate and the concentrations of certain ions.

19.67 Each of the following answers assumes that the enzyme was purified from an organism with optimal conditions for life near 37°C, pH 7.
   a. Decreasing the temperature from 37°C to 10°C will cause the rate of an enzyme-catalyzed reaction to decrease because the frequency of collisions between enzyme and substrate will decrease as the rate of molecular movement decreases.
   b. Increasing the pH from 7 to 11 will generally cause a decrease in the rate of an enzyme-catalyzed reaction. In fact, most enzymes would be denatured by a pH of 11 and enzyme activity would cease.
   c. Heating an enzyme from 37°C to 100°C will destroy enzyme activity because the enzyme would be denatured by the extreme heat.

19.69 High temperature denatures bacterial enzymes and structural proteins. Because the life of the cell is dependent on the function of these proteins, the cell dies.

19.71 A lysosome is a membrane-bound vesicle in the cytoplasm of cells that contains approximately fifty types of hydrolytic enzymes.

19.73 Enzymes used for clinical assays in hospitals are typically stored at refrigerator temperatures to ensure that they are not denatured by heat. In this way they retain their activity for long periods.

19.75 **a.** Cells regulate the level of enzyme activity to conserve energy. It is a waste of cellular energy to produce an enzyme if its substrate is not present or if its product is in excess.

b. Production of proteolytic digestive enzymes must be carefully controlled because the active enzyme could destroy the cell that produces it. Thus, they are produced in an inactive form in the cell and are only activated at the site where they carry out digestion.

19.77 In positive allosterism, binding of the effector molecule turns the enzyme on. In negative allosterism, binding of the effector molecule turns the enzyme off.

19.79 A proenzyme is the inactive form of an enzyme that is converted to the active form at the site of its activity.

19.81 Blood clotting is a critical protective mechanism in the body, preventing excessive loss of blood following an injury. However, it can be a dangerous mechanism if it is triggered inappropriately. The resulting clot could cause a heart attack or stroke. By having a cascade of proteolytic reactions leading to the final formation of the clot, there are many steps at which the process can be regulated. This ensures that it will only be activated under the appropriate conditions.

19.83 *Competitive enzyme inhibition* occurs when a structural analog of the normal substrate occupies the enzyme active site so that the reaction cannot occur. The structural analog and the normal substrate compete for the active site. Thus, the rate of the reaction will depend on the relative concentrations of the two molecules.

19.85 A structural analog has a shape and charge distribution that are very similar to those of the normal substrate for an enzyme.

19.87 Irreversible inhibitors bind tightly to and block the active site of an enzyme and eliminate catalysis at the site.

19.89 The compound would be a competitive inhibitor of the enzyme.

19.91 The structural similarities among chymotrypsin, trypsin, and elastase suggest that these enzymes evolved from a single ancestral gene that was duplicated. Each copy then evolved independently.

19.93

tyr-lys-ala-phe

19.95 Elastase will cleave the peptide bonds on the carbonyl side of alanine and glycine. Trypsin will cleave the peptide bonds on the carbonyl side of lysine and arginine. Chymotrypsin will cleave the peptide bonds on the carbonyl side of tryptophan and phenylalanine.

19.97 Creatine kinase-MB, and aspartate aminotransferase (AST/SGOT)

Chapter 20

20.1 **a.** Adenosine diphosphate:

b. Deoxyguanosine triphosphate:

20.3 The RNA polymerase recognizes the promoter site for a gene, separates the strands of DNA, and catalyzes the polymerization of an RNA strand complementary to the DNA strand that carries the genetic code for a protein. It recognizes a termination site at the end of the gene and releases the RNA molecule.

20.5 The genetic code is said to be degenerate because several different triplet codons may serve as code words for a single amino acid.

20.7 The nitrogenous bases of the codons are complementary to those of the anticodons. As a result they are able to hydrogen bond to one another according to the base pairing rules.

20.9 The ribosomal P-site holds the peptidyl tRNA during protein synthesis. The peptidyl tRNA is the tRNA carrying the growing peptide chain. The only exception to this is during initiation of translation when the P-site holds the initiator tRNA.

20.11 The normal mRNA sequence, AUG-CCC-GAC-UUU, would encode the peptide sequence methionine-proline-aspartate-phenylalanine. The mutant mRNA sequence, AUG-CGC-GAC-UUU, would encode the mutant peptide sequence methionine-arginine-aspartate-phenylalanine. This would not be a silent mutation because a hydrophobic amino acid (proline) has been replaced by a positively charged amino acid (arginine).

20.13 A heterocyclic amine is a compound that contains nitrogen in at least one position of the ring skeleton.

20.15 It is the N-9 of the purine that forms the *N*-glycosidic bond with C-1 of the five-carbon sugar. The general structure of the purine ring is shown below:

20.17 The ATP nucleotide is composed of the five-carbon sugar ribose, the purine adenine, and a triphosphate group.

20.19 The two strands of DNA in the double helix are said to be *antiparallel* because they run in opposite directions. One strand progresses in the $5' \rightarrow 3'$ direction, and the opposite strand progresses in the $3' \rightarrow 5'$ direction.

20.21 The DNA double helix is 2 nm in width. The nitrogenous bases are stacked at a distance of 0.34 nm from one another. One complete turn of the helix is 3.4 nm, or 10 base pairs.

20.23 Two

20.25

20.27 The prokaryotic chromosome is a circular DNA molecule that is supercoiled, that is, the helix is coiled on itself.

20.29 The term *semiconservative DNA replication* refers to the fact that each parental DNA strand serves as the template for the synthesis of a daughter strand. As a result, each of the daughter DNA molecules is made up of one strand of the original parental DNA and one strand of newly synthesized DNA.

20.31 The two primary functions of DNA polymerase III are to read a template DNA strand and catalyze the polymerization of a new daughter strand, and to proofread the newly synthesized strand and correct any errors by removing the incorrectly inserted nucleotide and adding the proper one.

20.33 3'-TACGCCGATCTTATAAGGT-5'

20.35 The *replication origin* of a DNA molecule is the unique sequence on the DNA molecule where DNA replication begins.

20.37 The enzyme helicase separates the strands of DNA at the origin of DNA replication so that the proteins involved in replication can interact with the nitrogenous base pairs.

20.39 The RNA primer "primes" DNA replication by providing a 3'−OH which can be used by DNA polymerase III for the addition of the next nucleotide in the growing DNA chain.

20.41 DNA → RNA → Protein

20.43 Anticodons are found on transfer RNA molecules.

20.45 3'-AUGGAUCGAGACCAGUAAUUCCGUCAU-5'.

20.47 *RNA splicing* is the process by which the noncoding sequences (introns) of the primary transcript of a eukaryotic mRNA are removed and the protein coding sequences (exons) are spliced together.

20.49 Messenger RNA, transfer RNA, and ribosomal RNA

20.51 Spliceosomes are small ribonucleoprotein complexes that carry out RNA splicing.

20.53 The *poly(A) tail* is a stretch of 100–200 adenosine nucleotides polymerized onto the 3' end of a mRNA by the enzyme poly(A) polymerase.

20.55 The *cap structure* is made up of the nucleotide 7-methylguanosine attached to the 5' end of a mRNA by a 5'-5' triphosphate bridge. Generally the first two nucleotides of the mRNA are also methylated.

20.57 Sixty-four

20.59 The reading frame of a gene is the sequential set of triplet codons that carries the genetic code for the primary structure of a protein.

20.61 Methionine and tryptophan

20.63 The codon 5'-UUU-3' encodes the amino acid phenylalanine. The mutant codon 5'-UUA-3' encodes the amino acid leucine. Both leucine and phenylalanine are hydrophobic amino acids, however, leucine has a smaller R group. It is possible that the smaller R group would disrupt the structure of the protein.

20.65 The ribosomes serve as a platform on which protein synthesis can occur. They also carry the enzymatic activity that forms peptide bonds.

20.67 The sequence of DNA nucleotides in a gene is transcribed to produce a complementary sequence of RNA nucleotides in a messenger RNA (mRNA). In the process of translation the sequence of the mRNA is read sequentially in words of three nucleotides (codons) to produce a protein. Each codon calls for the addition of a particular amino acid to the growing peptide chain. Through these processes, the sequence of nucleotides in a gene determines the sequence of amino acids in the primary structure of a protein.

20.69 In the initiation of translation, initiation factors, methionyl tRNA (the initiator tRNA), the mRNA, and the small and large ribosomal subunits form the initiation complex. During the elongation stage of translation, an aminoacyl tRNA binds to the A-site of the ribosome. Peptidyl transferase catalyzes the formation of a peptide bond and the peptide chain is transferred to the tRNA in the A-site. Translocation shifts the peptidyl tRNA from the A-site into the P-site, leaving the A-site available for the next aminoacyl tRNA. In the termination stage of translation, a termination codon is encountered. A release factor binds to the empty A-site and peptidyl transferase catalyzes the hydrolysis of the bond between the peptidyl tRNA and the completed peptide chain.

20.71 An ester bond

20.73 A point mutation is the substitution of one nucleotide pair for another in a gene.

20.75 Some mutations are silent because the change in the nucleotide sequence does not alter the amino acid sequence of the protein. This can happen because there are many amino acids encoded by multiple codons.

20.77 UV light causes the formation of pyrimidine dimers, the covalent bonding of two adjacent pyrimidine bases. Mutations occur when the UV damage repair system makes an error during the repair process. This causes a change in the nucleotide sequence of the DNA.

20.79 **a.** A *carcinogen* is a compound that causes cancer. Cancers are caused by mutations in the genes responsible for controlling cell division.

     b. Carcinogens cause DNA damage that results in changes in the nucleotide sequence of the gene. Thus, carcinogens are also mutagens.

20.81 A *restriction enzyme* is a bacterial enzyme that "cuts" the sugar–phosphate backbone of DNA molecules at a specific nucleotide sequence.

20.83 A selectable marker is a genetic trait that can be used to detect the presence of a plasmid in a bacterium. Many plasmids have antibiotic resistance genes as selectable markers. Bacteria containing the plasmid will be able to grow in the presence of the antibiotic; those without the plasmid will be killed.

20.85 Human insulin, interferon, human growth hormone, and human blood clotting factor VIII

20.87 1024 copies

20.89 The goals of the Human Genome Project were to identify and map all of the genes of the human genome and to determine the DNA sequences of the complete three billion nucleotide pairs.

20.91 A genome library is a set of clones that represents all of the DNA sequences in the genome of an organism.

20.93 A dideoxynucleotide is one that has hydrogen atoms rather than hydroxyl groups bonded to both the 2′ and 3′ carbons of the five-carbon sugar.

20.95 Sequences that these DNA sequences have in common are highlighted in bold.

a. 5′ **AGCTCCT**GATTTCATACAGTTTCTACT **ACCTACTA** 3′

b. 5′ AGACATTCTATCTACCTAGACTATG**TTCAGAA** 3′

c. 5′**TTCAGAA**CTCATTCAGACCTACTACTATACCTTGGG **AGCTCCT** 3′

d. 5′**ACCTACTA**GACTATACTACTACTAAGGGGACTATT CCAGACTT 3′

The 5′ end of sequence (a) is identical to the 3′ end of sequence (c). The 3′ end of sequence (a) is identical to the 5′ end of sequence (d). The 3′ end of sequence (b) is identical to the 5′ end of sequence (c). From 5′ to 3′, the sequences would form the following map:

```
    b
 _____
         c
     _____
            a
       _____
               d
          _____
```

Chapter 21

21.1 ATP is called the universal energy currency because it is the major molecule used by all organisms to store energy.

21.3 The first stage of catabolism is the digestion (hydrolysis) of dietary macromolecules in the stomach and intestine.

In the second stage of catabolism, monosaccharides, amino acids, fatty acids, and glycerol are converted by metabolic reactions into molecules that can be completely oxidized.

In the third stage of catabolism, the two-carbon acetyl group of acetyl CoA is completely oxidized by the reactions of the citric acid cycle. The energy of the electrons harvested in these oxidation reactions is used to make ATP.

21.5 Substrate level phosphorylation is one way the cell can make ATP. In this reaction, a high-energy phosphoryl group of a substrate in the reaction is transferred to ADP to produce ATP.

21.7 Glycolysis is a pathway involving ten reactions. In reactions 1–3, energy is invested in the beginning substrate, glucose. This is done by transferring high-energy phosphoryl groups from ATP to the intermediates in the pathway. The product

is fructose-1,6-bisphosphate. In the energy-harvesting reactions of glycolysis, fructose-1,6-bisphosphate is split into two three-carbon molecules that begin a series of rearrangement, oxidation-reduction, and substrate-level phosphorylation reactions that produce four ATP, two NADH, and two pyruvate molecules. Because of the investment of two ATP in the early steps of glycolysis, the net yield of ATP is two.

21.9 Both the alcohol and lactate fermentations are anaerobic reactions that use the pyruvate and re-oxidize the NADH produced in glycolysis.

21.11 Gluconeogenesis (synthesis of glucose from noncarbohydrate sources) appears to be the reverse of glycolysis (the first stage of carbohydrate degradation) because the intermediates in the two pathways are the same. However, reactions 1, 3, and 10 of glycolysis are not reversible reactions. Thus, the reverse reactions must be carried out by different enzymes.

21.13 The enzyme glycogen phosphorylase catalyzes the phosphorolysis of a glucose unit at one end of a glycogen molecule. The reaction involves the displacement of the glucose by a phosphate group. The products are glucose-1-phosphate and a glycogen molecule that is one glucose unit shorter.

21.15 Glucokinase traps glucose within the liver cell by phosphorylating it. Because the product, glucose-6-phosphate, is charged, it cannot be exported from the cell.

21.17 Glucagon indirectly stimulates glycogen phosphorylase, the first enzyme of glycogenolysis. This speeds up glycogen degradation. Glucagon also inhibits glycogen synthase, the first enzyme in glycogenesis. This inhibits glycogen synthesis.

21.19 ATP

21.21

Adenosine triphosphate

Adenosine diphosphate

Inorganic phosphate group

21.23 A coupled reaction is one that can be thought of as a two-step process. In a coupled reaction, two reactions occur simultaneously. Frequently one of the reactions releases the energy that drives the second, energy-requiring, reaction.

21.25 Carbohydrates

21.27 The following equation represents the hydrolysis of maltose:

β-Maltose

β-D-Glucose

21.29 The following equation represents the hydrolysis of lactose:

β-D-Glucose

β-D-Galactose

21.31 The hydrolysis of a triglyceride containing oleic acid, stearic acid, and linoleic acid is represented in the following equations:

Glycerol

+

Oleic acid

+

Stearic acid

+

Linoleic acid

21.33 The hydrolysis of the dipeptide alanyl leucine is represented in the following equation:

Alanyl leucine

Alanine

+

Leucine

21.35 Glycolysis is the enzymatic pathway that converts a glucose molecule into two molecules of pyruvate. The pathway generates a net energy yield of two ATP and two NADH. Glycolysis is the first stage of carbohydrate catabolism.

21.37 Glycolysis requires NAD^+ for reaction 6 in which glyceraldehyde-3-phosphate dehydrogenase catalyzes the oxidation of glyceraldehyde-3-phosphate. NAD^+ is reduced.

21.39 Two ATP per glucose

21.41 Although muscle cells have enough ATP stored for only a few seconds of activity, glycolysis speeds up dramatically when there is a demand for more energy. If the cells have a sufficient supply of oxygen, aerobic respiration (the citric acid cycle and oxidative phosphorylation) will contribute large amounts of ATP. If oxygen is limited, the lactate fermentation will speed up. This will use up the pyruvate and re-oxidize the NADH produced by glycolysis and allow continued synthesis of ATP for muscle contraction.

21.43 $C_6H_{12}O_6 + 2ADP + 2P_i + 2NAD^+ \rightarrow$
Glucose

$$2C_3H_3O_3 + 2ATP + 2NADH + 2H_2O$$
Pyruvate

21.45 **a.** Hexokinase catalyzes the phosphorylation of glucose.
b. Pyruvate kinase catalyzes the transfer of a phosphoryl group from phosphoenolpyruvate to ADP.
c. Phosphoglycerate mutase catalyzes the isomerization reaction that converts 3-phosphoglycerate to 2-phosphoglycerate.
d. Glyceraldehyde-3-phosphate dehydrogenase catalyzes the oxidation and phosphorylation of glyceraldehyde-3-phosphate and the reduction of NAD^+ to NADH.

21.47 Isomerase

21.49 Enediol

21.51 A kinase transfers a phosphoryl group from one molecule to another.

21.53 NAD^+ is reduced, accepting a hydride anion.

21.55 Enolase catalyzes a reaction that produces an enol; in this particular reaction, it is phosphoenolpyruvate.

21.57 To optimize efficiency and minimize waste, it is important that energy-harvesting pathways, such as glycolysis, respond to the energy demands of the cell. If energy in the form of ATP is abundant, there is no need for the pathway to continue at a rapid rate. When this is the case, allosteric enzymes that catalyze the reactions of the pathway are inhibited by binding to their negative effectors. Similarly, when there is a great demand for ATP, the pathway speeds up as a result of the action of allosteric enzymes binding to positive effectors.

21.59 ATP and citrate are allosteric inhibitors of phosphofructokinase, whereas AMP and ADP are allosteric activators.

21.61 Citrate, which is the first intermediate in the citric acid cycle, is an allosteric inhibitor of phosphofructokinase. The citric acid cycle is a pathway that results in the complete oxidation of the pyruvate produced by glycolysis. A high concentration of citrate signals that sufficient substrate is entering the citric acid cycle. The inhibition of phosphofructokinase by citrate is an example of *feedback inhibition*: the product, citrate, allosterically inhibits the activity of an enzyme early in the pathway.

21.63

CH$_3$—C—H (Acetaldehyde) → CH$_3$CH$_2$OH (Ethanol), with NADH → NAD$^+$ over the arrow

21.65 The lactate fermentation

21.67 Yogurt and some cheeses

21.69 Lactate dehydrogenase

21.71 This child must have the enzymes to carry out the alcohol fermentation. When the child exercised hard, there was not enough oxygen in the cells to maintain aerobic respiration. As a result, glycolysis and the alcohol fermentation were responsible for the majority of the ATP production by the child. The accumulation of alcohol (ethanol) in the child caused the symptoms of drunkenness.

21.73 The first stage of the pentose phosphate pathway is an oxidative stage in which glucose-6-phosphate is converted to ribulose-5-phosphate. Two NADPH molecules and one CO_2 molecule are also produced in these reactions. The second stage of the pentose phosphate pathway involves isomerization reactions that convert ribulose-5-phosphate into other five-carbon sugars, ribose-5-phosphate and xylulose-5-phosphate. The third stage of the pathway involves a complex series of rearrangement reactions that results in the production of two fructose-6-phosphate and one glyceraldehyde-3-phosphate molecules from three molecules of pentose phosphate.

21.75 The ribose-5-phosphate is used for the biosynthesis of nucleotides. The erythrose-4-phosphate is used for the biosynthesis of aromatic amino acids.

21.77 Gluconeogenesis is production of glucose from noncarbohydrate starting materials. This pathway can provide glucose when starvation or strenuous exercise leads to a depletion of glucose from the body.

21.79 The liver

21.81 Lactate is first converted to pyruvate.

21.83 Because steps 1, 3, and 10 of glycolysis are irreversible, gluconeogenesis is not simply the reverse of glycolysis. The reverse reactions must be carried out by different enzymes.

21.85 Steps 1, 3, and 10 of glycolysis are irreversible. Step 1 is the transfer of a phosphoryl group from ATP to carbon-6 of glucose and is catalyzed by hexokinase. Step 3 is the transfer of a phosphoryl group from ATP to carbon-1 of fructose-6-phosphate and is catalyzed by phosphofructokinase. Step 10 is the substrate-level phosphorylation in which a phosphoryl group is transferred from phosphoenolpyruvate to ADP and is catalyzed by pyruvate kinase.

21.87 The liver and pancreas

21.89 *Hypoglycemia* is the condition in which blood glucose levels are too low.

21.91 **a.** Insulin stimulates glycogen synthase, the first enzyme in glycogen synthesis. It also stimulates uptake of glucose from the bloodstream into cells and phosphorylation of glucose by the enzyme glucokinase.
b. This traps glucose within liver cells and increases the storage of glucose in the form of glycogen.
c. These processes decrease blood glucose levels.

21.93 Any defect in the enzymes required to degrade glycogen or export glucose from liver cells will result in a reduced ability of the liver to provide glucose at times when blood glucose levels are low. This will cause hypoglycemia.

Chapter 22

22.1 Mitochondria are the organelles responsible for aerobic respiration.

22.3

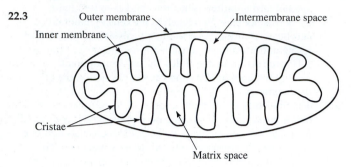

Outer membrane, Intermembrane space, Inner membrane, Cristae, Matrix space

22.5 Pyruvate is converted to acetyl CoA by the pyruvate dehydrogenase complex. This huge enzyme complex requires four coenzymes, each of which is made from a different vitamin. The four coenzymes are thiamine pyrophosphate (made from thiamine), FAD (made from riboflavin), NAD^+ (made from niacin), and coenzyme A (made from the vitamin pantothenic acid). The coenzyme lipoamide is also involved in this reaction.

22.7 *Oxidative phosphorylation* is the process by which the energy of electrons harvested from oxidation of a fuel molecule is used to phosphorylate ADP to produce ATP.

22.9 $NAD^+ + H:^- \rightarrow NADH$

22.11 During transamination reactions, the α-amino group is transferred to the coenzyme pyridoxal phosphate. In the last part of the reaction, the α-amino group is transferred from pyridoxal phosphate to an α-keto acid.

22.13 The purpose of the urea cycle is to convert toxic ammonium ions to urea, which is excreted in the urine of land animals.

22.15 An amphibolic pathway is a metabolic pathway that functions both in anabolism and catabolism. The citric acid cycle is amphibolic because it has a catabolic function—it completely oxidizes the acetyl group carried by acetyl CoA to provide electrons for ATP synthesis. Because citric acid cycle intermediates are precursors for the biosynthesis of many other molecules, it also serves a function in anabolism.

22.17 The mitochondrion is an organelle that serves as the cellular power plant. The reactions of the citric acid cycle, the electron transport system, and ATP synthase function together within the mitochondrion to harvest ATP energy for the cell.

22.19 The intermembrane compartment is the location of the high-energy proton (H^+) reservoir produced by the electron transport system. The energy of this H^+ reservoir is used to make ATP.

22.21 The outer mitochondrial membrane is freely permeable to substances of molar mass less than 10,000 g/mol. The inner mitochondrial membrane is highly impermeable. Embedded within the inner mitochondrial membrane are the electron carriers of the electron transport system, and ATP synthase, the multisubunit enzyme that makes ATP.

22.23 Coenzyme A is a molecule derived from ATP and the vitamin pantothenic acid. It functions in the transfer of acetyl groups in lipid and carbohydrate metabolism.

22.25 Decarboxylation is a chemical reaction in which a carboxyl group is removed from a molecule.

22.27 Under aerobic conditions pyruvate is converted to acetyl CoA.

22.29 The coenzymes NAD^+, FAD, thiamine pyrophosphate, and coenzyme A are required by the pyruvate dehydrogenase complex for the conversion of pyruvate to acetyl CoA. These coenzymes are synthesized from the vitamins niacin, riboflavin, thiamine, and pantothenic acid, respectively. If the vitamins are not available, the coenzymes will not be available and pyruvate cannot be converted to acetyl CoA. Because the complete oxidation of the acetyl group of acetyl CoA produces the vast majority of the ATP for the body, ATP production would be severely inhibited by a deficiency of any of these vitamins.

22.31 An aldol condensation is a reaction in which aldehydes or ketones react to form larger molecules.

22.33 Oxidation in an organic molecule is often recognized as the loss of hydrogen atoms or a gain of oxygen atoms.

22.35 A dehydrogenation reaction is an oxidation reaction in which protons and electrons are removed from a molecule.

22.37 a. False   c. True
         b. False   d. True

22.39 a. The acetyl group of acetyl CoA is transferred to oxaloacetate.
         b. The product is citrate.

22.41 Three

22.43 Two ATP per glucose

22.45 The function of acetyl CoA in the citric acid cycle is to bring the two-carbon remnant (acetyl group) of pyruvate from glycolysis and transfer it to oxaloacetate. In this way the acetyl group enters the citric acid cycle for the final stages of oxidation.

22.47 The high-energy phosphoryl group of the GTP is transferred to ADP to produce ATP. This reaction is catalyzed by the enzyme dinucleotide diphosphokinase.

22.49

$$\begin{array}{c} COO^- \\ | \\ C=O \\ | \\ CH_2 \\ | \\ COO^- \end{array} \quad + \quad H_3C-\overset{\overset{\displaystyle O}{\|}}{C}-S{\sim}CoA + H_2O \quad \longrightarrow$$

Oxaloacetate   Acetyl CoA

$$\begin{array}{c} COO^- \\ | \\ CH_2 \\ | \\ HO-C-COO^- \\ | \\ CH_2 \\ | \\ COO^- \end{array} \quad + \quad HS-CoA + H^+$$

Citrate   Coenzyme A

The importance of this reaction is that it brings the acetyl group, the two-carbon remnants of the glucose molecule, into the citric acid cycle to be completely oxidized. Through these reactions, and subsequent oxidative phosphorylation, the majority of the cellular ATP energy is provided.

22.51 The conversion of citrate to *cis*-aconitate is an example of the dehydration of an alcohol to produce an alkene (double bond). The conversion of *cis*-aconitate to isocitrate is an example of the hydration of an alkene, that is, the addition of water to the double bond, to produce an alcohol (—OH).

22.53 This reaction is an example of the oxidation of a secondary alcohol to a ketone. The two functional groups are the hydroxyl group of the alcohol and the carbonyl group of the ketone.

22.55 It is a kinase because it transfers a phosphoryl group from one molecule to another. Kinases are a specific type of transferase.

22.57 An allosteric enzyme is one that has an effector binding site and an active site. Effector binding can change the shape of the active site, causing it to be active or inactive.

22.59 Negative allosterism is a means of enzyme regulation in which effector binding inactivates the active site of the allosteric enzyme.

22.61 The citric acid cycle is regulated by the following four enzymes or enzyme complexes: pyruvate dehydrogenase complex, citrate sythase, isocitrate dehydrogenase, and the α-ketoglutarate dehydrogenase complex.

22.63 Energy-harvesting pathways, such as the citric acid cycle, must be responsive to the energy needs of the cell. If the energy requirements are high, as during exercise, the reactions must speed up. If energy demands are low and ATP is in excess, the reactions of the pathway slow down.

22.65 ADP

22.67 The electron transport system is series of electron transport proteins embedded in the inner mitochondrial membrane that accept high-energy electrons from NADH and $FADH_2$ and transfer them in stepwise fashion to molecular oxygen (O_2).

22.69 Three ATP

22.71 The oxidation of a variety of fuel molecules, including carbohydrates, the carbon skeletons of amino acids, and fatty acids provides the electrons. The energy of these electrons is used to produce an H^+ reservoir. The energy of this proton reservoir is used for ATP synthesis.

22.73 The electron transport system passes electrons harvested during oxidation of fuel molecules to molecular oxygen. At three sites protons are pumped from the mitochondrial matrix into the intermembrane compartment. Thus, the electron transport system builds the high-energy H^+ reservoir that provides energy for ATP synthesis.

22.75 **a.** Two ATP per glucose (net yield) are produced in glycolysis, whereas the complete oxidation of glucose in aerobic respiration (glycolysis, the citric acid cycle, and oxidative phosphorylation) results in the production of thirty-six ATP per glucose.
b. Thus, aerobic respiration harvests nearly 40% of the potential energy of glucose, and anaerobic glycolysis harvests only about 2% of the potential energy of glucose.

22.77 Transaminases transfer amino groups from amino acids to ketoacids.

22.79 The glutamate family of transaminases is very important because the ketoacid corresponding to glutamate is α-ketoglutarate, one of the citric acid cycle intermediates. This provides a link between the citric acid cycle and amino acid metabolism. These transaminases provide amino groups for amino acid synthesis and collect amino groups during catabolism of amino acids.

22.81 **a.** Pyruvate
b. α-Ketoglutarate
c. Oxaloacetate
d. Acetyl CoA
e. Succinate
f. α-Ketoglutarate

22.83

$$\underset{\text{α-Ketoglutarate}}{\begin{array}{c}O\\\parallel\\C-COO^-\\\vert\\CH_2\\\vert\\CH_2\\\vert\\COO^-\end{array}} + NADPH + N^+H_4 \longrightarrow \underset{\text{Glutamate}}{\begin{array}{c}N^+H_3\\\vert\\H-C-COO^-\\\vert\\CH_2\\\vert\\CH_2\\\vert\\COO^-\end{array}} + NADP^+ + H_2O$$

Ammonia

22.85 Hyperammonemia

22.87 **a.** The source of one amino group of urea is the ammonium ion and the source of the other is the α-amino group of the amino acid aspartate.
b. The carbonyl group of urea is derived from CO_2.

22.89 Anabolism is a term used to describe all of the cellular energy-requiring biosynthetic pathways.

22.91 α-Ketoglutarate

22.93 Citric acid cycle intermediates are the starting materials for the biosynthesis of many biological molecules.

22.95 An essential amino acid is one that cannot be synthesized by the body and must be provided in the diet.

22.97

$$\underset{\text{Pyruvate}}{\begin{array}{c}O\\\parallel\\C-COO^-\\\vert\\CH_3\end{array}} + CO_2 + ATP \longrightarrow \underset{\text{Oxaloacetate}}{\begin{array}{c}O\\\parallel\\C-COO^-\\\vert\\CH_2\\\vert\\COO^-\end{array}} + ADP + P_i$$

Chapter 23

23.1 Because dietary lipids are hydrophobic, they arrive in the small intestine as large fat globules. The bile salts emulsify these fat globules into tiny fat droplets. This greatly increases the surface area of the lipids, allowing them to be more accessible to pancreatic lipases and thus more easily digested.

23.3 Starvation, a diet low in carbohydrates, and diabetes mellitus are conditions that lead to the production of ketone bodies.

23.5 **(1)** Fatty acid biosynthesis occurs in the cytoplasm whereas β-oxidation occurs in the mitochondria.
(2) The acyl group carrier in fatty acid biosynthesis is acyl carrier protein while the acyl group carrier in β-oxidation is coenzyme A.

(3) The seven enzymes of fatty acid biosynthesis are associated as a multienzyme complex called *fatty acid synthase*. The enzymes involved in β-oxidation are not physically associated with one another.

(4) NADPH is the reducing agent used in fatty acid biosynthesis. NADH and FADH$_2$ are produced by β-oxidation.

23.7 The liver regulates blood glucose levels under the control of the hormones insulin and glucagon. When blood glucose levels are too high, insulin stimulates the uptake of glucose by liver cells and the storage of the glucose in glycogen polymers. When blood glucose levels are too low, the hormone glucagon stimulates the breakdown of glycogen and release of glucose into the bloodstream. Glucagon also stimulates the liver to produce glucose for export into the bloodstream by the process of gluconeogenesis.

23.9 Insulin stimulates uptake of glucose and amino acids by cells, glycogen and protein synthesis, and storage of lipids. It inhibits glycogenolysis, gluconeogenesis, breakdown of stored triglycerides, and ketogenesis.

23.11 Bile consists of micelles of lecithin, cholesterol, bile salts, proteins, inorganic ions, and bile pigments that aid in lipid digestion by emulsifying fat droplets.

23.13 A micelle is an aggregation of molecules having nonpolar and polar regions; the nonpolar regions of the molecules aggregate, leaving the polar regions facing the surrounding water.

23.15 A triglyceride is a molecule composed of glycerol esterified to three fatty acids.

23.17 A chylomicron is a plasma lipoprotein that carries triglycerides from the intestine to all body tissues via the bloodstream. That function is reflected in the composition of the chylomicron, which is approximately 85% triglycerides, 9% phospholipids, 3% cholesterol esters, 2% protein, and 1% cholesterol.

23.19 Triglycerides

23.21 The large fat globule that takes up nearly the entire cytoplasm

23.23 Lipases catalyze the hydrolysis of the ester bonds of triglycerides.

23.25 Acetyl CoA is the precursor for fatty acids, several amino acids, cholesterol, and other steroids.

23.27 Chylomicrons are plasma lipoproteins (aggregates of protein and triglycerides) that carry dietary triglycerides from the intestine to all tissues via the bloodstream.

23.29 Bile salts serve as detergents. Fat globules stimulate their release from the gallbladder. The bile salts then emulsify the lipids, increasing their surface area and making them more accessible to digestive enzymes (pancreatic lipases).

23.31 When dietary lipids in the form of fat globules reach the duodenum, they are emulsified by bile salts. The triglycerides in the resulting tiny fat droplets are hydrolyzed into monoglycerides and fatty acids by the action of pancreatic lipases, assisted by colipase. The monoglycerides and fatty acids are absorbed by cells lining the intestine.

23.33 The energy source for the activation of a fatty acid entering β-oxidation is the hydrolysis of ATP into AMP and PP$_i$ (pyrophosphate group), an energy expense of two high-energy phosphoester bonds.

23.35 Carnitine is a carrier molecule that brings fatty acyl groups into the mitochondrial matrix.

23.37 The following equation represents the reaction catalyzed by acyl-CoA dehydrogenase. Notice that the reaction involves the loss of two hydrogen atoms. Thus, this is an oxidation reaction.

23.39 An alcohol is the production of the hydration of an alkene.

23.41 Six acetyl CoA, one phenyl acetate, six NADH, and six FADH$_2$

23.43 112 ATP

23.45 Two ATP

23.47 The acetyl CoA produced by β-oxidation will enter the citric acid cycle.

23.49 Ketone bodies include the compounds acetone, acetoacetone, and β-hydroxybutyrate, which are produced from fatty acids in the liver via acetyl CoA.

23.51 Ketosis is an abnormal rise in the level of ketone bodies in the blood.

23.53 Matrix of the mitochondrion.

23.55

Acetoacetate   β-Hydroxybutyrate

23.57 In those suffering from uncontrolled diabetes, the glucose in the blood cannot get into the cells of the body. The excess glucose is excreted in the urine. Body cells degrade fatty acids because glucose is not available. β-oxidation of fatty acids yields enormous quantities of acetyl CoA, so much acetyl CoA, in fact, that it cannot all enter the citric acid cycle because there is not enough oxaloacetate available. Excess acetyl CoA is used for ketogenesis.

23.59 Ketone bodies are the preferred energy source of the heart.

23.61 Cytoplasm

23.63 Fatty acid synthase

23.65 The phosphopantetheine group allows formation of a high-energy thioester bond with a fatty acid. It is derived from the vitamin pantothenic acid.

23.67 Fatty acid synthase is a huge multienzyme complex consisting of the seven enzymes involved in fatty acid synthesis. It is found in the cell cytoplasm. The enzymes involved in β-oxidation are not physically associated with one another. They are free in the mitochondrial matrix space.

23.69 Glycogenesis is the synthesis of the polymer glycogen from glucose monomers.

23.71 Gluconeogenesis is the synthesis of glucose from noncarbohydrate precursors.

23.73 β-oxidation of fatty acids

23.75 The major metabolic function of the liver is to regulate blood glucose levels.

23.77 Ketone bodies are the major fuel for the heart. Glucose is the major energy source of the brain, and the liver obtains most of its energy from the oxidation of amino acid carbon skeletons.

23.79 Fatty acids are absorbed from the bloodstream by adipocytes. Using glycerol-3-phosphate, produced as a by-product of glycolysis, triglycerides are synthesized. Triglycerides are constantly being hydrolyzed and resynthesized in adipocytes. The rates of hydrolysis and synthesis are determined by lipases that are under hormonal control.

23.81 In general, insulin stimulates anabolic processes, including glycogen synthesis, uptake of amino acids and protein synthesis, and triglyceride synthesis. At the same time, catabolic processes such as glycogenolysis are inhibited.

23.83 A target cell is one that has a receptor for a particular hormone.

23.85 Decreased blood glucose levels trigger the secretion of glucagon into the bloodstream.

23.87 Insulin is produced in the β-cells of the islets of Langerhans in the pancreas.

23.89 Insulin stimulates the uptake of glucose from the blood into cells. It enhances glucose storage by stimulating glycogenesis and inhibiting glycogen degradation and gluconeogenesis.

23.91 Insulin stimulates synthesis and storage of triglycerides.

23.93 Untreated diabetes mellitus is starvation in the midst of plenty because blood glucose levels are very high. However, in the absence of insulin, blood glucose can't be taken up into cells. The excess glucose is excreted into the urine while the cells of the body are starved for energy.

Credits

Design Elements
Microscope: © Vol. 85/PhotoDisc/Getty Images; stethoscope: © Vol. OS48/PhotoDisc/Getty Images; eye: © Vol. OS2/PhotoDisc/Getty Images; pill bottle: © Vol. 168 Corbis RF; pills: © Vol. 168 Corbis.

Chapter 1
Opener: © PhotoDisc/Getty Images; p. 3: © Digital Vision/PunchStock RF; p. 4: © Corbis Royalty Free; p. 5: © Creatas/PunchStock RF; p. 6: © Laguna Designs/SPL/Photo Researchers; 1.2a: © Geoff Tompkinson/SPL/Photo Researchers, Inc.; 1.2b: © T.J. Florian/Rainbow; 1.2c: © David Parker/Segate Microelectronics/SPL/Photo Researchers, Inc.; 1.2d: Courtesy APHIS, PPG, Otis Methods Development Center, USDA; 1.3a, c: © The McGraw-Hill Companies, Inc./Louis Rosenstock, photographer; 1.3b: © The McGraw-Hill Companies, Inc./Jeff Topping, photographer; 1.4(left, right): © The McGraw-Hill Companies, Inc./Ken Karp, photographer; 1.6c: © Doug Martin/Photo Researchers, Inc.; p. 21: © DV545/Getty Images RF; 1.8a-c, 1.10a-d: © The McGraw-Hill Companies, Inc./Louis Rosenstock, photographer; p. 30: © Tony Freeman/PhotoEdit; 1.12: © The McGraw-Hill Companies, Inc./Louis Rosenstock, photographer; p. 35: © Vol. 12/Getty Images RF.

Chapter 2
Opener: © The McGraw-Hill Companies, Inc./Louis Rosenstock, photographer; 2.1: © IBM Corporation, Almaden Research Center; 2.6: © Yoav Levy/Phototake; p. 12: © Earth Satellite Corp. /SPL/Photo Researchers, Inc.; p. 53: © Dan McCoy/Rainbow; p. 55: © Vol. 2/PhotoDisc/Getty Images; p. 59: © The McGraw-Hill Companies, Inc./Stephen Frisch, Photographer; p. 61(Peanuts): © C Squared Studios/Getty Images RF; Medical Perspective p. 61(Almonds, sesame, sunflower): © Vol. 83/Corbis; p. 61(Oyster): © PhotoDisc/Getty Images; p. 61(Crab): © Ingram Publishing/Alamy RF.

Chapter 3
Opener: © PhotoLink/Getty Images RF; 3.2c: © The McGraw Hill Companies, Inc./Trent Stephens, photographer; p. 94(top): © PhotoDisc/Getty Images; p. 94(bottom): © BananaStock/Jupiter Images RF; 3.14: © Charles D. Winter/Photo Researchers, Inc.

Chapter 4
Opener: © SPL/Photo Researchers, Inc.; 4.1: © The McGraw-Hill Companies, Inc./Louis Rosenstock, photographer; p. 138: © The McGraw-Hill Companies/Stephen Frisch, Photographer; 4.3, p. 137: © The McGraw-Hill Companies, Inc./Louis Rosenstock, photographer; p. 144: © Stockbyte/Getty RF; p. 153: © Vol 72/PhotoDisc/Getty Images.

Chapter 5
Opener © Corbis/Royalty Free; p. 160: © Hulton-Deutsch/Collection/Corbis; 5.2a, b: © The McGraw-Hill Companies, Inc./Louis Rosenstock, photographer; 5.5: © Peter Stef Lambert/Stone/Getty Images; p. 176 © BrandX/Picture Quest RF.

Chapter 6
Opener: © C. Paxton & J. Farrow / Photo Researchers, Inc.; 6.1: © Kip Peticolas/Fundamental Photographs; p. 191: © J.W. Mowbray/Photo Researchers, Inc.; 6.2: © The McGraw-Hill Companies, Inc./Louis Rosenstock, photographer; Fig. 6.4a-c: © David M. Phillips/Visuals Unlimited; p. 206(left) © RMF/Visuals Unlimited; p. 206(right): Courtesy of Rita Colwell, National Science Foundation; 6.5a,b, p. 209: © The McGraw-Hill Companies, Inc./Louis Rosenstock, photographer; p. 210: © AJPhoto/Photo Researchers, Inc.

Chapter 7
Opener: Courtesy Honda; 7.9: © The McGraw-Hill Companies, Inc./Ken Karp, photographer; p. 232(top, bottom): © The McGraw-Hill Companies, Inc./Louis Rosenstock, photographer; p. 233: © Design Pics/Punchstock RF; 7.15, 7.16: © The McGraw-Hill Companies, Inc./Ken Karp, photographer.

Chapter 8
Opener: © Comstock/Jupiter Images RF; 8.1: © Dr. E.R. Degginger/Color Pic Inc.; p. 259: © Digital Vision/Getty Images RF; 8.3a(left): © The McGraw-Hill Companies, Inc./Louis Rosenstock, photographer; 8.3b(right): © Richard Menga/Fundamental Photographs; 8.6a, b: © The McGraw-Hill Companies, Inc./Stephen Frisch, photographer; p. 269, 270a,b, 271: © The McGraw-Hill Companies, Inc./Louis Rosenstock, photographer; p. 276: © The McGraw Hill Companies, Inc./Auburn University Photographic Services; p. 279: © PhotoDisc/Getty Images; p. 279a: © The McGraw-Hill Companies, Inc./Louis Rosenstock, photographer; p. 279b: © Bonnie Kamin/PhotoEdit; p. 279c: © Tony Freeman/PhotoEdit; p. 280: © AAA/Photo Phototake; 8.7(left, right): © The McGraw-Hill Companies, Inc./Stephen Frisch, photographer.

Chapter 9
Opener: © Kathy McLaughlin/The Image Works, Inc.; 9.3: © Gianni Tortoli/Photo Researchers, Inc.; p. 302: NASA; 9.5: © U.S. DOE/Science Source/Photo Researchers, Inc.; p. 305: © Stockbyte/Punchstock; 9.7: © Blair Seitz/Photo Researchers, Inc.; p. 308(top) © The McGraw-Hill Companies, Inc./Louis Rosenstock, photographer; p. 308(bottom): © SIU/Biomed/Custom Medical Stock Photo; 9.8b: Courtesy Bristol-Myers Squibb Medical Imaging; p. 311: © Tony Freeman/PhotoEdit; 9.9: © U.S. DOE/Mark Marten/Photo Researchers, Inc.; 9.10: © The McGraw-Hill Companies, Inc./Louis Rosenstock, photographer; 9.11: © Scott Camazine/Photo Researchers, Inc.

Chapter 10
Opener: © Digital Vision/RF; p. 321: © Vol. 0S22/PhotoDisc/Getty Images; p. 322: © Fotosearch RF; p. 323: Courtesy Ocean Drilling Program/NOAA; p. 333: © Digital Vision/PunchStock; p. 335: © Vol. 59/PhotoDisc/Getty Images; p. 343(top): © Corbis RF; p. 343(bottom) © OS23/Getty Images; p. 344: © Corbis RF; p. 345: © Corbis RF; p. 348: © Vol. 27/PhotoDisc/Getty Images.

Chapter 11
Opener: © David Frazier/Corbis RF; p. 358(top): © Vol. 12/PhotoDisc/Getty Images; p. 358(bottom): © Vol. 27/PhotoDisc/Getty Images; p. 360(top): © Vol. 19/PhotoDisc/Getty Images; p. 360(bottom): © Banana Stock/Punchstock RF; p. 364: © Vol. 77/PhotoDisc/Getty Images; p. 365: © Buddy Mays/Corbis; 11.1(Geranium): © Larry Lefever/Grant Heilman Photograph, Inc.; 11.1(Oranges): © Michelle Garrett/Corbis; 11.1(Bayberry): © Walter H. Hodge/Peter Arnold, Inc.; 11.1(Lily): © Hal Horwitz/Corbis; p. 375: © PhotoDisc/Getty Images; 11.5: © The McGraw-Hill Companies, Inc./Ken Karp, photographer; p. 381: © Superstock/Alamy; p. 383: © Vol. 67 PhotoDisc/Getty Images; p. 386: © Alan Detrick/Photo Researchers, Inc.

Chapter 12

Opener: © Vol. 30/PhotoDisc/Getty Images; p. 402: © Corbis Royalty Free; p. 405 © Digital Vision RF; 12.3: © Vol. 12/PhotoDisc/Getty Images; p. 408: © Vol. 18/PhotoDisc/Getty Images; 12.4(top): © Corbis RF; p. 409(bottom) © Comstock/Punchstock RF; p. 420: © Vol. 88/PhotoDisc/Getty Images; 12.8: © Corbis RF; p. 425: © Vol. OS22/PhotoDisc/Getty Images; p. 430: © Getty RF; p. 431: © Dr. Morley Read/SPL/Photo Researchers, Inc.

Chapter 13

Opener: © Layer, W./Peter Arnold, Inc.; p. 440: © Vol. 2/PhotoDisc/Getty Images; p. 442: © Vol. 12/PhotoDisc/Getty Images; p. 445(Almonds): © B. Borrell Cassals/FLPA/Corbis; p. 445(Vanilla): © Eisenhut & Mayer/Getty Images; p. 445(Cinnamon): © Rita Maas/Getty Images; p. 445(Lemon grass): © Corbis RF; p. 445(Blue berries): © Charles Krebs/Corbis; p. 445(Mushroom): © James Noble/Corbis; p. 447: © Vol. 18/PhotoDisc/Getty Images; 13.4a-d: © Richard Menga/Fundamental Photographs; 13.5: © Rob and Ann Simpson/Visuals Unlimited; p. 450: © Corbis RF; p. 453: © Hanson Carroll/Peter Arnold, Inc.; p. 463: © Getty RF.

Chapter 14

Opener: © Ullstein-Superclic/Peter Arnold, Inc.; p. 469: © Alamy RF; p. 471, 476 © Corbis RF; p. 477 © Paul W. Johnson/Biological Photo Service; p. 486(Pineapple): © Vol. 19/PhotoDisc/Getty Images; p. 486(Raspberries): © Vol. OS49/PhotoDisc/Getty Images; p. 486(Bananas): © Vol. 30/PhotoDisc/Getty Images; p. 487(Oranges, Apples): © Corbis RF; p. 487(Apricots): © Vol. 121/Corbis; p. 487(Strawberries): © Vol. 83/ Corbis; p. 500 © Vol. 81/Corbis; p. 501(Bee): © Vol. 101/

Corbis; p. 501(Looper): © Norm Thomas/Photo Researchers, Inc.; p. 501(Borer): © William Weber/Visuals Unlimited; p. 504: © Brand X/Getty RF; p. 505 © Getty RF; p. 506(top): © Vol. 30/Getty RF; p. 506(bottom): © SS04/GettyRF.

Chapter 15

Opener: © Vol. 6/PhotoDisc/Getty Images; p. 513: © Vol. 59/PhotoDisc/Getty Images; p. 518: © Duncan Smith/SPL/Photo Researchers Inc.; p. 519: © Creatas/Punchstock RF; p. 524: © The McGraw Hill Companies, Inc./Gary He, photographer; 15.3a,b: © Vol. 94/Corbis; 15.3c: © Gregory G. Dimijian/Photo Researchers, Inc.; p. 527: © Steven P. Lynch; p. 538a: © Vol. 9/PhotoDisc/Getty Images; p. 538b: © Phil Larkin, CSIRO Plant Industry.

Chapter 16

Opener: © The McGraw Hill Higher Education/Louis Rosenstock, photographer; 16.1: USDA; 16.2: © Vol. 67/PhotoDisc/Getty Images; p. 554b: © Stanley Flegler/Visuals Unlimited; 16.3b: © Corbis RF; p. 562: © Corbis RF; p. 565: USDA; p. 566: © The McGraw-Hill Companies, Inc./Jill Braaten, photographer; p. 567: © Getty RF; p. 570: © Jean Claude Revy-ISM/Phototake; p. 572: © Pixtal/Superstock; p. 574(top): © Getty Royalty Free; 574(bottom) © Corbis Royalty Free.

Chapter 17

Opener: © Corbis Royalty Free; p. 586(top): © Getty RF; p. 586(bottom): © Vol. 20/PhotoDisc/Getty Images; p. 590: © Roadsideamerica.com, Kirby, Smith & Wilkins; p. 595: © Vol. 20/PhotoDisc/Getty Images; p. 602: © Hans Pfletschinger/Peter Arnold, Inc.; p. 603(top): © Vol. 12/PhotoDisc/Getty Images; p. 603(bottom): © Getty RF 17.10a: © James Dennis/Phototake; 611: © Corbis RF.

Chapter 18

Opener: © Fritz Polking/Peter Arnold, Inc.; p. 621: © Vol. 18/PhotoDisc/Getty Images; p. 626: © Vol. 270/Corbis; p. 630: © Vol. 258/PhotoDisc/Getty Images; 18.14: © Meckes/Ottawa/Photo Researchers, Inc.; p. 641(top): © Getty Royalty Free; p. 641(middle): © Vol. 20/PhotoDisc/Getty Images; p. 641(bottom): © The McGraw-Hill Companies, Inc./Bob Coyle, photographer.

Chapter 19

Opener: © J. Schmidt/National Parks Service; p. 653: © Gregg Otto/Visuals Unlimited; p. 654: © Comstock/Alamy RF; p. 663: © Merck/Phil Degginger/Color-Pic, Inc.

Chapter 20

Opener: © Corbis RF; p. 691: © Vol. 72/PhotoDisc/Getty Images; p. 695: © Vol. 29/PhotoDisc/Getty Images; p. 700: Library of Congress; p. 714: © EP77/PhotoDisc/Getty Images; p. 720: © Dr. Charles S. Helling/USDA; p. 721: Courtesy of Orchid Cellmark, Germantown, Maryland.

Chapter 21

Opener: © PhotoLink/Getty Images RF; p. 734: © Corbis RF; p. 740: © Jupiter Images RF; p. 745: © SS07/Getty RF; p. 746: © The McGraw Hill Companies, Inc/Louis Rosenstock, photographer; p. 751: © Photolink/Getty RF.

Chapter 22

Opener: © Chris Falkenstein/PhotoDisc/Getty Images; 22.1: © CNRI/Phototake; p. 769(left): © DV287/Getty Images; p. 769(right): © PhotoLink/Getty RF.

Chapter 23

Opener: Courtesy Katherine Denniston; p. 805: © Vol. 11/PhotoDisc/Getty Images.

Index

PRINCIPAL FUNCTIONAL GROUPS IN ORGANIC COMPOUNDS

Type of Compound	Structural Formula	Condensed Formula	Chapter Reference	Example Structural Formula	Example IUPAC Name	Example Common Name
Alcohol	R—O—H	ROH	12	CH_3CH_2—O—H	Ethanol	Ethyl alcohol
Aldehyde	R—C(=O)—H	RCHO	12	CH_3C(=O)—H	Ethanal	Acetaldehyde
Amide	R—C(=O)—N(H)—H	$RCONH_2$	13	CH_3C(=O)—N(H)—H	Ethanamide	Acetamide
Amine	R—N(H)—H	RNH_2	13	CH_3CH_2N(H)—H	Ethanamine	Ethyl amine
Carboxylic acid	R—C(=O)—O—H	RCOOH	13	CH_3C(=O)—O—H	Ethanoic acid	Acetic acid
Ester	R—C(=O)—O—R'	RCOOR'	13	CH_3C(=O)—OCH_3	Methyl ethanoate	Methyl acetate
Ether	R—O—R'	ROR'	12	CH_3OCH_3	Methoxymethane	Dimethyl ether
Halide	—Cl (or —Br, —F, —I)	RCl	10	CH_3CH_2Cl	Chloroethane	Ethyl chloride
Ketone	R—C(=O)—R'	RCOR'	12	CH_3C(=O)CH_3	Propanone	Acetone